GEOMETRY

For Enjoyment and Challenge

REVISED EDITION

GEOMETRY

For Enjoyment and Challenge

REVISED EDITION

Richard Rhoad
New Trier High School
Winnetka, Illinois

George Milauskas
Barrington High School
Barrington, Illinois

Robert Whipple
New Trier High School
Winnetka, Illinois

McDougal, Littell & Company
Evanston, Illinois
New York Sacramento

Consultants

William Duncker — Chairperson, Mathematics Department,
Buffalo Grove High School, Buffalo Grove, Illinois

Dean L. Gould — Chairperson, Mathematics Department,
Maine Township High School West, Des Plaines, Illinois

Charles L. Hamberg — Mathematics Department,
Adlai E. Stevenson High School, Prairie View, Illinois

Charles R. Leonard — Director of Instruction, Mathematics,
Homewood-Flossmoor Community High School,
Flossmoor, Illinois

Gerhard Robien — Chairperson, Mathematics Department,
Lake Forest High School East, Lake Forest, Illinois

Jacob D. Turban — Mathematics Department,
Elk Grove High School, Elk Grove Village, Illinois

Ray Venn — Chairperson, Mathematics Department
Evanston Township High School, Evanston, Illinois

ISBN: 0–88343–916–6

Box 1667, Evanston, Illinois 60204

Printed in the United States of America

85 86 87 / 9 8 7 6 5 4

CONTENTS

Chapter 10 CIRCLES 409

Chapter 11 AREA 481

Chapter 12 VOLUMES 547

InterSection: Shape-Preserving Changes 575

Chapter 13 COORDINATE GEOMETRY 579

Chapter 1

INTRODUCTION TO GEOMETRY

Why Study Geometry?

Here are some reasons:

Reason 1 **GEOMETRY IS USEFUL.** In order to construct buildings, build roads, design and use instruments, plan new physical inventions, or operate planes and rockets, you must have a working knowledge of geometric principles. Engineers, architects, painters, carpenters, plumbers, teachers, electricians, machinists, home builders, dressmakers, skilled tradespeople, and many others rely on geometry.

Reason 2 **GEOMETRY IS CHALLENGING.** Why do you think so many people enjoy riddles and crossword puzzles? Why do so many young people like to rebuild cars? How do you feel when someone says, "I dare you"? All of us like challenges of one type or another, and geometry will present you with more kinds of challenges than any other math course you may take. Has math been difficult for you? You will find that you really can learn geometry, and you will even be excited about its challenges. That is its special appeal. There will be hard work, to be sure, but you may be like others before you—you may enjoy math for the first time in your life. The chances are good!

On the other hand, has math been easy for you? This year you can expect new excitement in a variety of challenges, some of them tough enough for anyone. You will apply new, powerful rules to familiar objects, discovering relationships you never thought of before.

Here's your first challenge.
How many squares are in the figure?
(The answer is at the bottom of this page.)

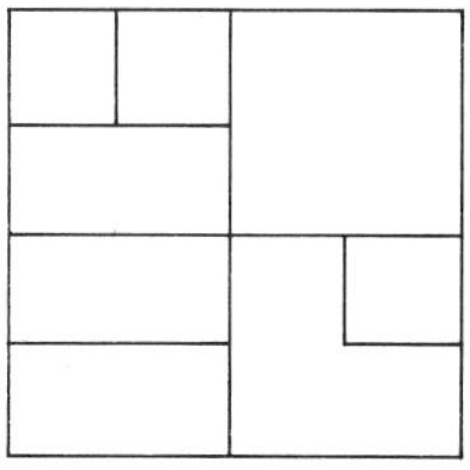

Reason 3 **GEOMETRY WILL HELP YOU LEARN WHAT PROOF MEANS.** All of us take many things for granted. As we become educated people, we would hope to rely more on reason and proof and less on superstition, prejudice, and guesswork. Many of you will learn for the first time how to prove a statement formally. All of you will learn the fundamentals of proof in a logical structure of geometry. You will also acquire some bases for value judgments in other courses.

Reason 4 **GEOMETRY WILL OFFER YOU A PRACTICAL SETTING IN WHICH TO USE YOUR ARITHMETIC AND ALGEBRA.** Here is a geometry problem that is largely algebra.

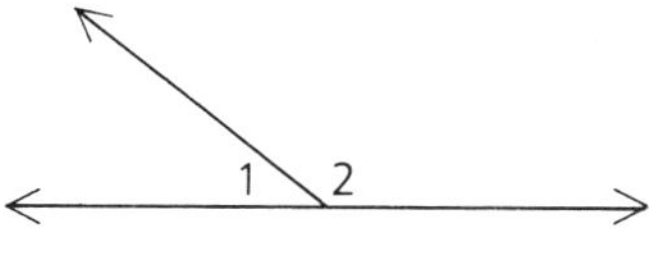

If angle 2 is five times as large as angle 1, can you discover the size of the angles?

Common sense and algebra could lead us to:

$$x + 5x = 180$$

and we could solve the problem.

Challenge: Where did the x, 5x, and 180 come from?

Why study geometry?

You have four reasons. Can you think of any others?

There are 9 squares in the figure .

Section 1.1 Getting Started

In order to begin your study of geometry, you need to become acquainted with:

A POINTS

B LINES

C LINE SEGMENTS

D RAYS

E ANGLES

F TRIANGLES

1.1 (A) Points

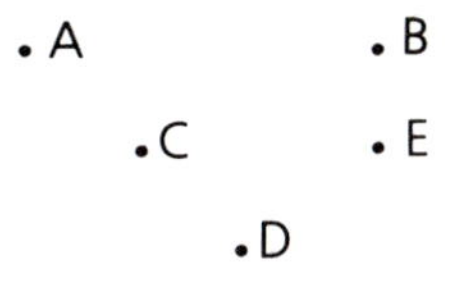

In the figure to the left, five points are represented by five dots. The names of the points are A, B, C, D and E. (We use only a capital letter to name a point.)

1.1 (B) Lines

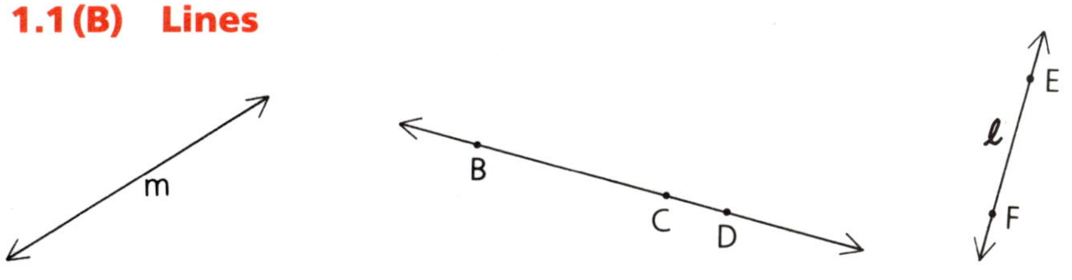

In the figure above, you see three lines represented. Lines are made up of points and are considered to be **STRAIGHT**. The arrows on each end indicate that the line extends infinitely far in both directions.

ALL LINES ARE STRAIGHT AND EXTEND INFINITELY FAR IN BOTH DIRECTIONS.

The line on the left is called line m.
The line in the middle can be called by any of the following names:

$\overleftrightarrow{BD}$ $\overleftrightarrow{BC}$ $\overleftrightarrow{CD}$ $\overleftrightarrow{CB}$ $\overleftrightarrow{DB}$ $\overleftrightarrow{DC}$

Thus, you can also name a line by using any **TWO** points on the line.
The line on the right can be called by any of three names:

line ℓ $\overleftrightarrow{EF}$ $\overleftrightarrow{FE}$

1.1 (C) Line Segments

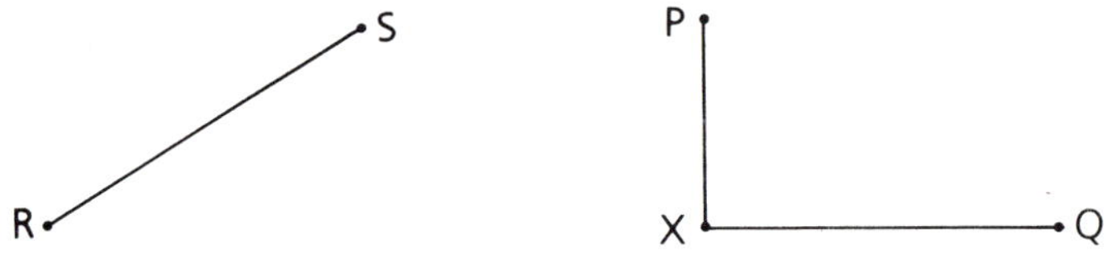

In the figure above, there are three line segments, or simply segments, represented. Segments are also made up of points and are considered to be **STRAIGHT.** A segment has a definite beginning and ending.

The segment on the left is called by either of two names: $\overline{RS}$, or $\overline{SR}$. To name a segment, you use its two end points.

On the right there are two segments. The vertical (up and down) segment is called by either of two names: $\overline{PX}$, or $\overline{XP}$.

The horizontal (crosswise) segment can also be named in either of two ways. Can you name these in two ways?

1.1 (D) Rays

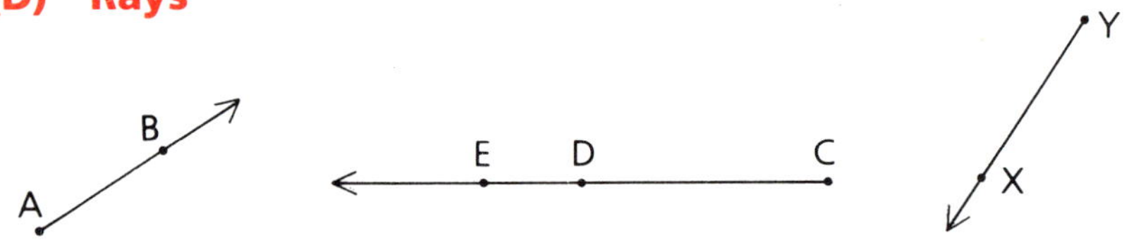

In the figure above, you see three rays represented. Rays are also made up of points and are considered to be **STRAIGHT.** A ray begins at a point called its **ENDPOINT** and then extends infinitely far in *one* direction.

The ray on the left is called $\overrightarrow{AB}$.

The endpoint must be named first to make it clear where the ray begins.

The ray in the middle can be given either of two names: $\overrightarrow{CD}$, or $\overrightarrow{CE}$.

Again, the endpoint must be named first.

The ray on the right can be named in only one way. Do you know what the name is?

1.1 (E) Angles

D **An ANGLE is made up of two rays with a COMMON ENDPOINT. This common endpoint is called the VERTEX of the angle. The rays are called SIDES of the angle.**

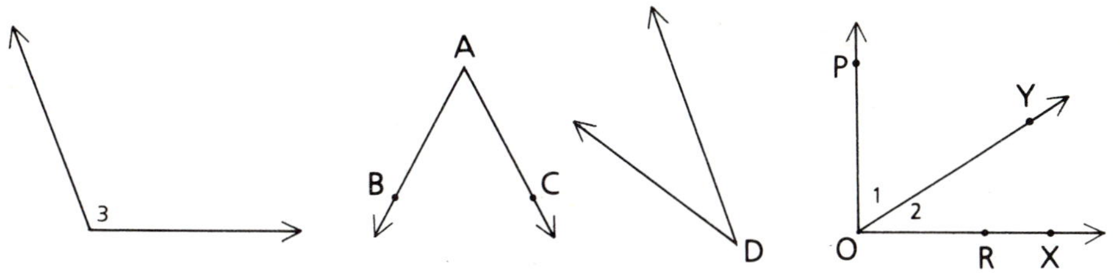

The angle on the left is called $\angle 3$. The 3 placed inside the angle near the vertex names it.

The angle second to the left can be called any of three names:

$$\angle BAC \qquad \angle CAB \qquad \angle A$$

Notice, when we use three letters, the vertex must be named in the middle.

In the third diagram, the angle is called $\angle D$.

In the fourth diagram, there are three angles. Can you tell which angle is $\angle O$? Rather than to specify how to tell, the rule is *never to name an angle in that way when confusion could result*.

$\angle 1$ can also be called $\angle POY$ or $\angle YOP$.

$\angle 2$ can also be called $\angle YOR$, $\angle YOX$, $\angle ROY$, or $\angle XOY$.

There is another angle in the fourth diagram, which can be named $\angle POR$. See if you can find three other names for this angle.

1.1 (F) Triangles

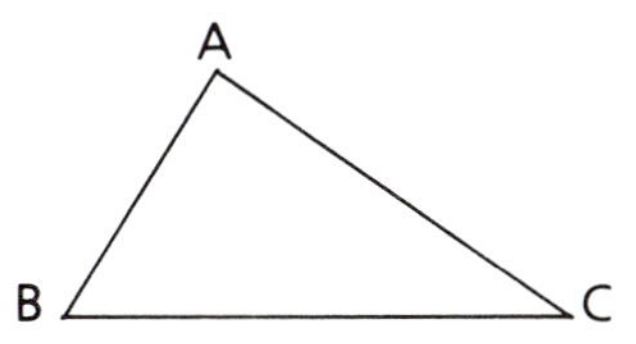

We shall call the above figure triangle ABC ($\triangle ABC$). A triangle has three segments as sides. You may wonder whether or not we may refer to $\angle B$ above since we have no arrows on the diagram. The answer is yes. We often, as a matter of fact, **IMAGINE** that the arrows are there and talk about rays, lines, and angles in the diagram. So a triangle ($\triangle$) not only has three sides, but has three angles as well. Can you name the angle at the top of the diagram in three ways?

Section 1.1 Sample Problems

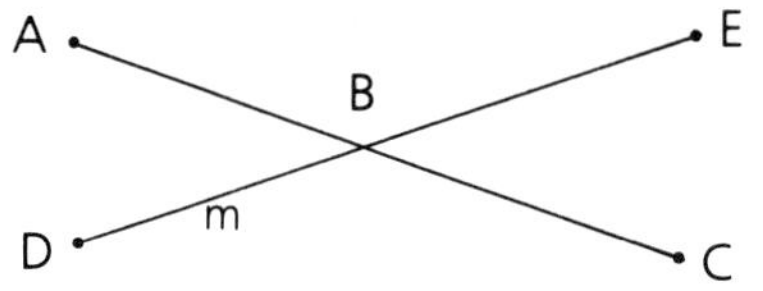

1 **a** How many lines are shown? Ans. 2

(Imagine that there are arrows on the diagram.)

b Name these lines.

Ans. line m, $\overleftrightarrow{DB}$, $\overleftrightarrow{DE}$, $\overleftrightarrow{BD}$, $\overleftrightarrow{BE}$, $\overleftrightarrow{EB}$, or $\overleftrightarrow{ED}$

$\overleftrightarrow{AB}$, $\overleftrightarrow{AC}$, $\overleftrightarrow{BA}$, $\overleftrightarrow{BC}$, $\overleftrightarrow{CA}$, or $\overleftrightarrow{CB}$

c Where do $\overleftrightarrow{AC}$ and $\overleftrightarrow{DE}$ intersect? Ans. $\{B\}$

d Where does $\overrightarrow{AC}$ intersect $\overleftrightarrow{BC}$? (Or $\overrightarrow{AC} \cap \overleftrightarrow{BC} =$ _?_)

Ans. $\overrightarrow{AC}$. Remember sets? If P and Q are sets of points,
$P \cap Q = \{$all points in P *and* in Q$\}$.

e What is the union of $\overrightarrow{BA}$ and $\overrightarrow{BD}$? (Or $\overrightarrow{BA} \cup \overrightarrow{BD} =$ _?_)

Ans. $\angle ABD$.
$P \cup Q = \{$all points in P *or* in Q or in both.$\}$

2 **a** Name the ray shown with endpoint at A and going in the direction of C. Ans. $\overrightarrow{AB}$ or $\overrightarrow{AC}$

b Name the segment joining A with B. Ans. $\overline{AB}$ or $\overline{BA}$

3 Draw a diagram in which the intersection of $\overrightarrow{AB}$ with $\overrightarrow{CA}$ is $\overline{AC}$ (or $\overrightarrow{AB} \cap \overrightarrow{CA} = \overline{AC}$).

Ans. 

4 Draw a diagram in which $\triangle ABC \cap \overrightarrow{DE} = \{F\}$.

Ans.

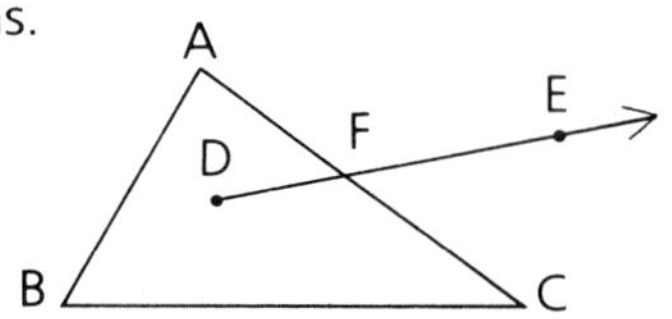

There are other correct answers, and a lot of wrong ones.

Section 1.1 Problem Set A

1 In three different ways, name the line shown.

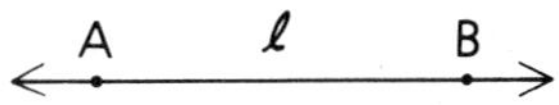

2 In four different ways, name the angle shown.

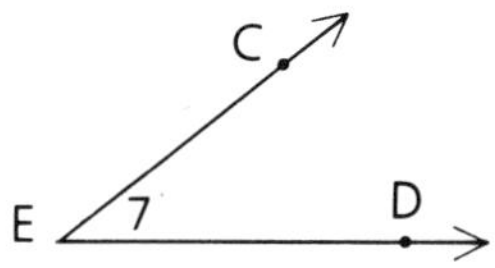

3 Can the ray shown be called $\overrightarrow{XY}$?

4 Name the sides of $\triangle RST$.

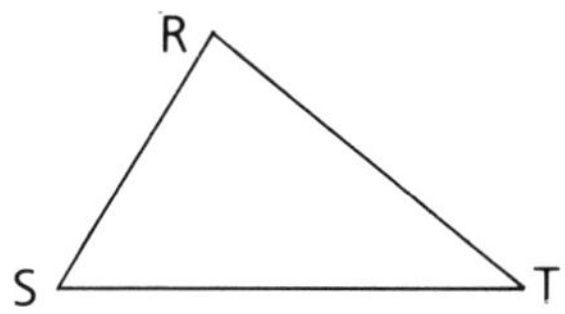

5 a Name the intersection of $\overline{AB}$ and $\overline{BC}$, or $\overline{AB} \cap \overline{BC}$.

b Name the union of $\overrightarrow{EC}$ and $\overrightarrow{EA}$, or $\overrightarrow{EC} \cup \overrightarrow{EA}$.

c $\overleftrightarrow{AC} \cap \overleftrightarrow{DB} = ?$

d $\overline{DC} \cap \overline{AB} = ?$

e $\overrightarrow{AC} \cap \overrightarrow{EC} = ?$

f $\overrightarrow{BA} \cup \overrightarrow{BC} = ?$

g $\overline{EC} \cup \overline{CB} \cup \overline{BE} = ?$

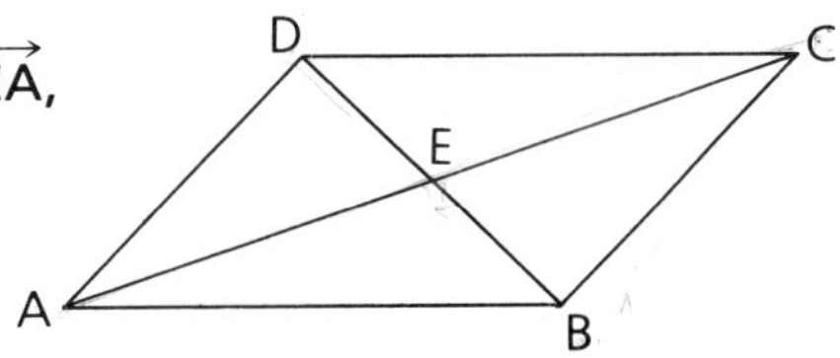

6 a Name $\angle OPR$ in all other ways.

b Where is the vertex of $\angle TOS$?

c How many angles have their vertex at R?

d Name $\angle TSP$ in all other ways.

e How many triangles are there in the figure?

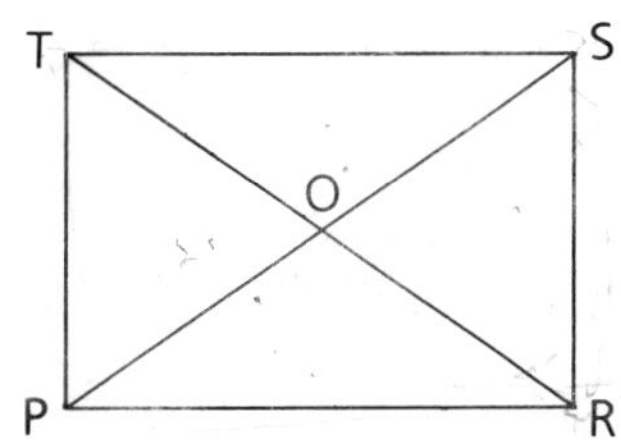

7 a A line is made up of ___?___.

b An angle is the union of two ___?___ with a common ___?___.

8 Draw $\triangle ABC$ with D exactly in the middle of $\overline{BC}$. (Later in this chapter, we shall call D the midpoint of $\overline{BC}$.)

Section 1.1 Problem Set B

9 a In $\triangle HJK$, $\overline{HJ}$ is twice as long as $\overline{JK}$ and exactly as long as $\overline{HK}$. If the length of $\overline{HJ}$ is 15, find the perimeter (distance around) of $\triangle HJK$.

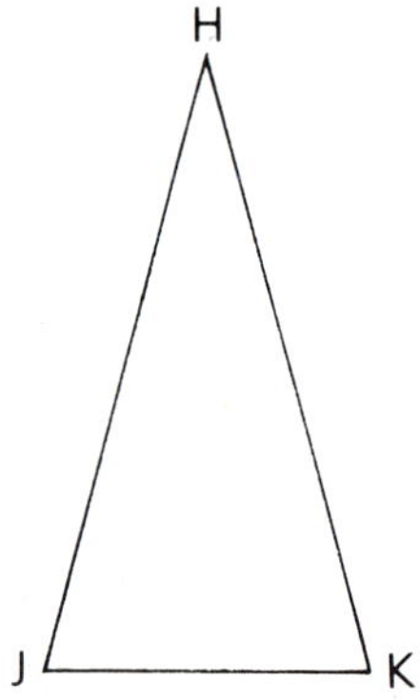

b If the sides compare with each other as in part a, and the perimeter of $\triangle HJK$ is 50, how long is $\overline{HK}$?

c If the length of $\overline{HJ}$ is 4x, and the length of $\overline{HK}$ is 3x, and the length of $\overline{JK}$ is 2x, and the perimeter of $\triangle HJK$ is 63, find the length of $\overline{HJ}$.

10 $\overline{AB} \cap \overline{CD} = \overline{CB}$. Draw a diagram for which this statement is true.

Section 1.1 Problem Set C

11 Draw a diagram in which the intersection of $\angle AEF$ and $\angle DPC$ is $\overrightarrow{ED}$.

12 a What percentage of the triangles in the diagram at the right have $\overline{CT}$ as a side?

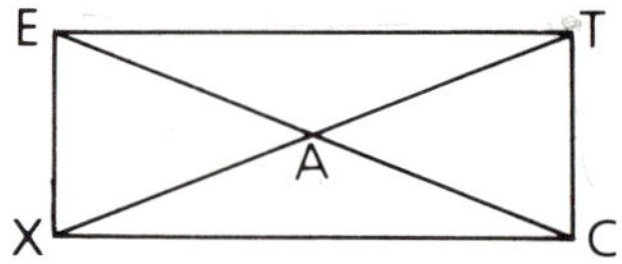

b What percentage have $\overline{AC}$ as a side?

Section 1.2 Measurement of Segments and Angles

In this section we shall learn about:

A MEASURING SEGMENTS
B MEASURING ANGLES
C CLASSIFYING ANGLES BY SIZE
D PARTS OF A DEGREE
E CONGRUENT ANGLES AND SEGMENTS

1.2 (A) Measuring Segments

Segments are measured by using such instruments as a ruler or meterstick. We may use any convenient length as a unit of measure. Some of the units that have been used in the past or at present are inches, feet, yards, millimeters, centimeters, and meters. The measure of $\overline{AB}$ is written as AB.

1.2 (B) Measuring Angles

Angles are commonly measured by a **PROTRACTOR**, shown in the figure measuring a 117° angle.

This year we shall measure angles ($\angle$s) in degrees (°). In later courses you may use other units such as radians or grads.

The measure, or **SIZE**, of an angle is the **AMOUNT OF TURNING** you would do if you were at the vertex, looked along one side, and then turned to look along the other side. An engineer's transit works that way.

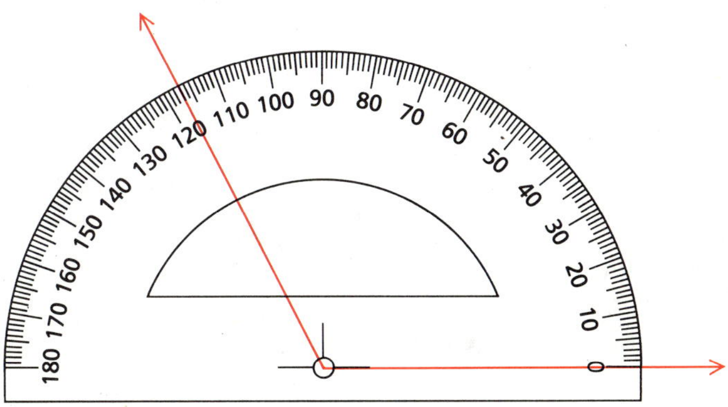

In some math courses there are negative angles, zero angles, and angles larger than 180°. This year we shall limit ourselves to angles of size more than 0° and less than or equal to 180°.

If you turn all the way around (to face your starting direction) you will have turned 360°. You can use that fact to estimate the size of an angle.

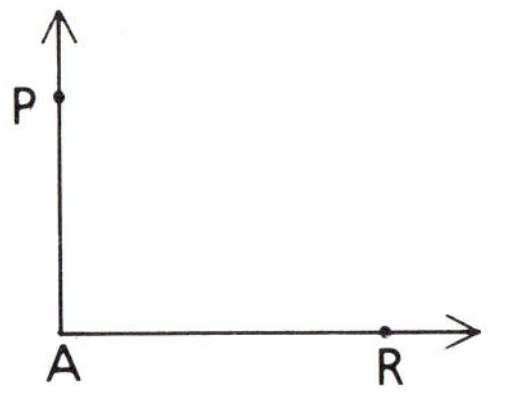

Side $\overrightarrow{AP}$ appears to have been turned $\frac{1}{4}$ of the way around from $\overrightarrow{AR}$, so you could guess that $\angle A$ is approximately a 90° angle.

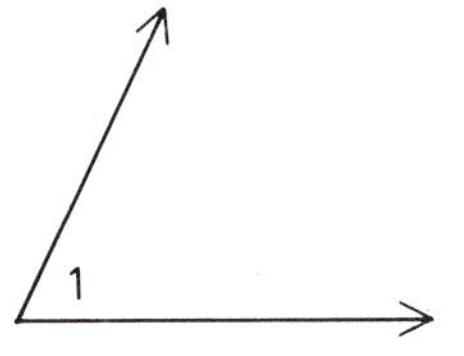

Angle 1 required less than a quarter turn, and a good guess would be 60°.

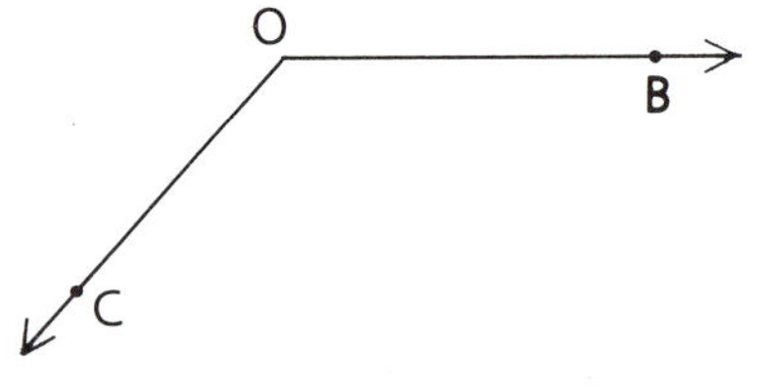

Angle BOC required more than a quarter turn and could be estimated at 130°.

Your teacher may want you to use a protractor to get more accurate measurements of angles. If so, the teacher will give you instruction in class. A good way to get practice in the correct use of a protractor is to draw a few triangles of different shapes. Measure the three angles of each triangle. The sum of your three measures should be 180° in each case. (We shall prove this statement in Chapter 7.)

1.2 (C) Classifying Angles by Size

As shown below, we classify angles according to their measures into four categories:

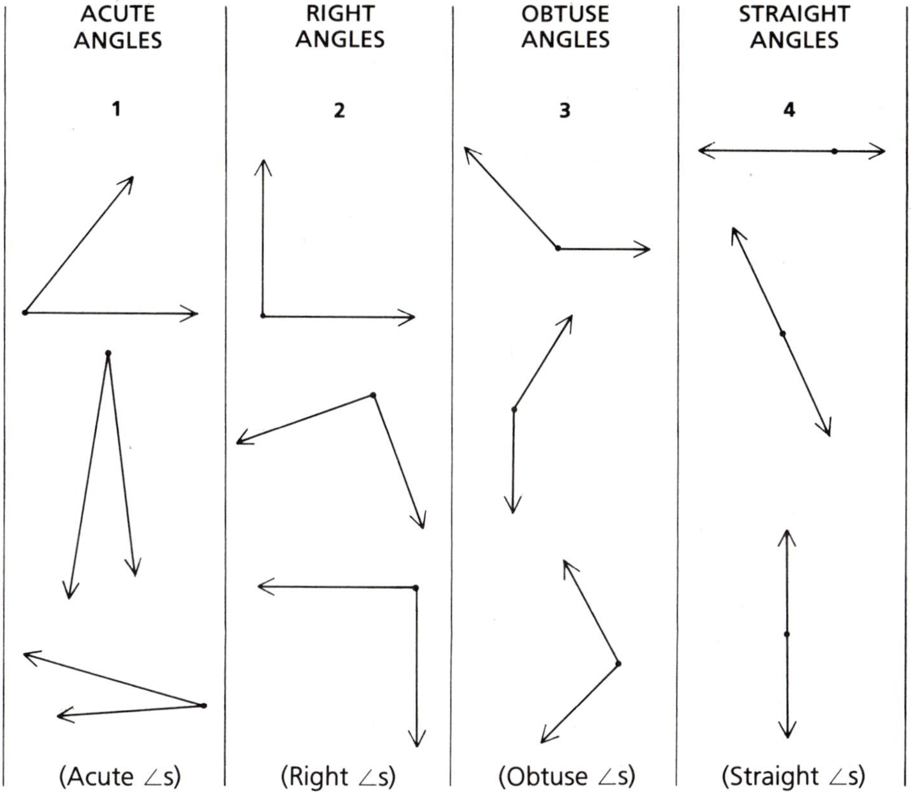

D **An ACUTE ANGLE (diagram 1) is an angle greater than 0° and less than 90°.**

A RIGHT ANGLE (diagram 2) is an angle of exactly 90°.

An OBTUSE ANGLE (diagram 3) is an angle greater than 90° and less than 180°.

A STRAIGHT ANGLE (diagram 4) is an angle of exactly 180°. As you can see, a straight angle is an angle that forms a straight line.

1.2 (D) Parts of a Degree

All of us know that time is divided into hours, minutes, and seconds. Similarly, each degree (°) of an angle is divided into 60 minutes (′) and each minute of an angle is divided into 60 seconds (″).

To summarize, $60' = 1^\circ$ (60 minutes equals 1 degree)
$60'' = 1'$ (60 seconds equals 1 minute)

Thus, $87\frac{1}{2}^\circ = 87^\circ 30'$
$60.4^\circ = 60^\circ 24'$
$90^\circ = 89^\circ 60'$ (since $60' = 1^\circ$)
$180^\circ = 179^\circ 59' 60''$ (since $60'' = 1'$ and $60' = 1^\circ$)

Study the following examples closely.

■ 1 Change $41\frac{2}{5}^\circ$ to degrees and minutes.

Since there are $60'$ in 1°, then

$\frac{2}{5}^\circ$ is $\frac{2}{5}(60)$ minutes or $24'$.

Hence, $41\frac{2}{5}^\circ = 41^\circ 24'$

■ 2 ∠ABC is a right angle.

$\angle ABD = 67^\circ 21' 37''$.

Find : ∠DBC

Subtract $67^\circ 21' 37''$ from 90° as follows:

$$\begin{array}{r} 89^\circ 59' 60'' \\ -67^\circ 21' 37'' \\ \hline 22^\circ 38' 23'' \end{array} \quad \text{(Note: } 90^\circ = 89^\circ\ 59' 60''\text{)}$$

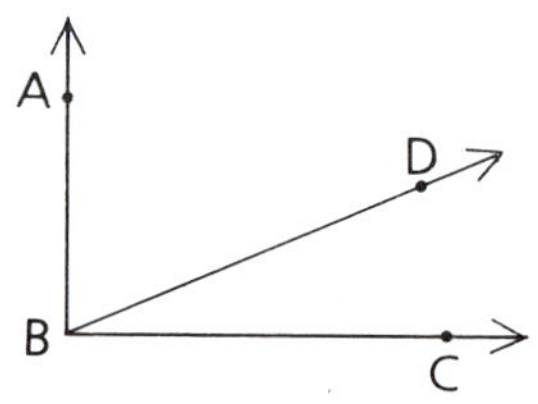

■ 3 Change $60^\circ 45'$ to degrees.

We must change $45'$ to a fractional part of a degree. Since $60' = 1^\circ$, 45 is placed over 60, and the fraction is reduced.

$\frac{45}{60} = \frac{3}{4}$ So $60^\circ 45' = 60\frac{3}{4}^\circ$

1.2 (E) Congruent Angles and Segments

D **Congruent (≅) angles are angles that have the same measure.**

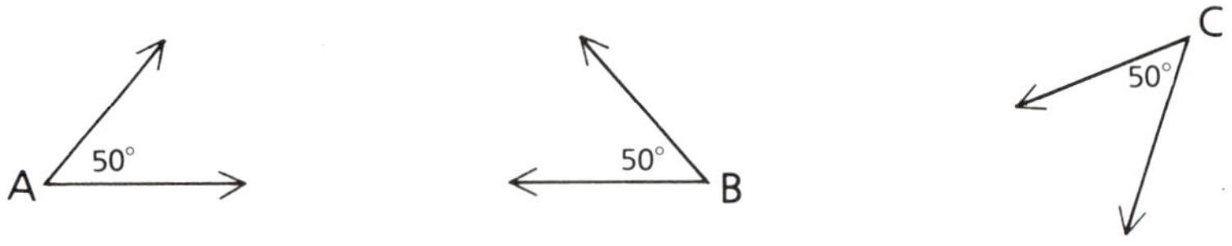

In the diagram above, ∠s A, B, and C are CONGRUENT.
We write ∠A ≅ ∠B ≅ ∠C.

In the same way:

D **Congruent (≅) segments are segments that have the same length.**

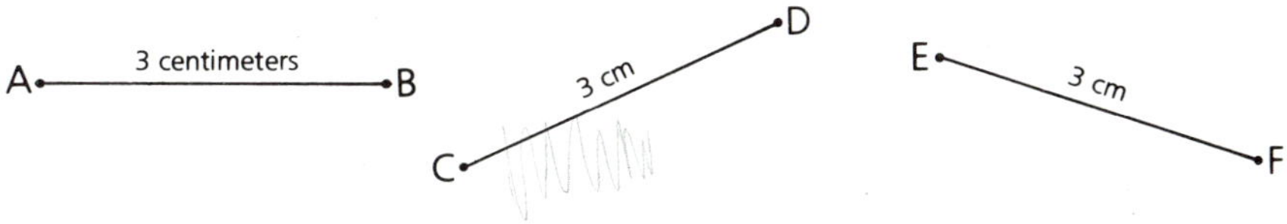

In the diagram above, segments $\overline{AB}$, $\overline{CD}$ and $\overline{EF}$ are congruent.
We write $\overline{AB} \cong \overline{CD} \cong \overline{EF}$.

Often, we use identical "tick marks" to indicate congruent angles and segments. In the diagram below, the identical tick marks indicate there are four pairs of congruent parts. Can you name them? (The answers are at the bottom of the page.)

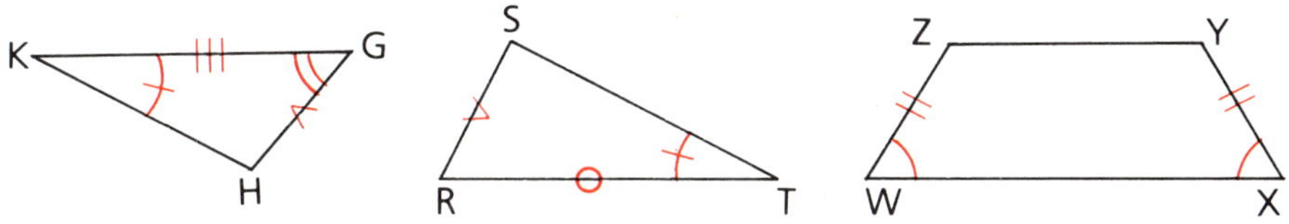

Section 1.2 Sample Problems

1

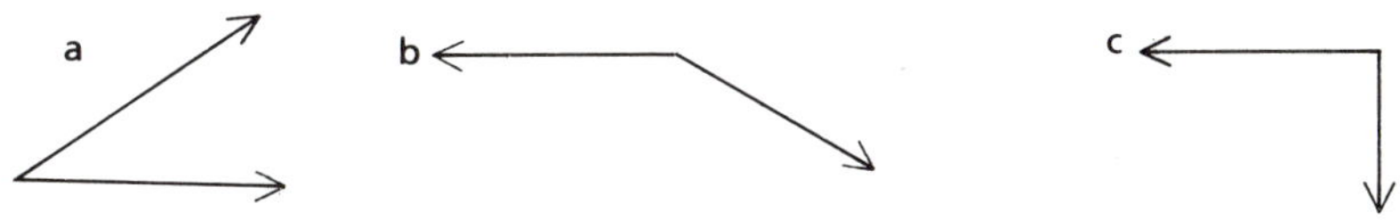

Classify each of the above angles as acute, right, or obtuse. Then estimate the number of degrees in each angle.

Ans. **a** acute, 40° **b** obtuse, 150° **c** right, 90°

2

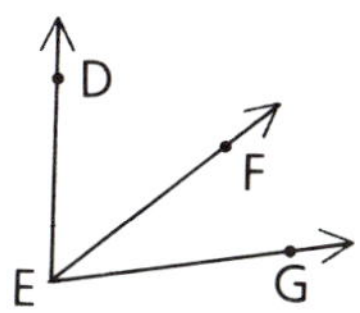

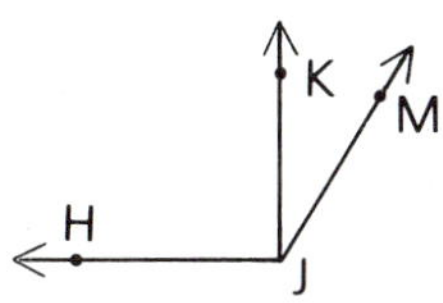

∠DEG = 80° ∠DEF = 50° ∠HJM = 120° ∠HJK = 90°

Draw a conclusion about ∠FEG and ∠KJM.

Solution: **∠FEG = 30° and ∠KJM = 30°, so ∠FEG ≅ ∠KJM**

Ans. $\overline{GH} \cong \overline{SR}$, ∠K ≅ ∠T, $\overline{WZ} \cong \overline{XY}$, ∠W ≅ ∠X

3 Given: $\angle ABC$ is a right angle

$\angle 1 = (3x+4)^\circ$

$\angle 2 = (x+6)^\circ$

Find: $m\angle 1$ (measure of $\angle 1$)

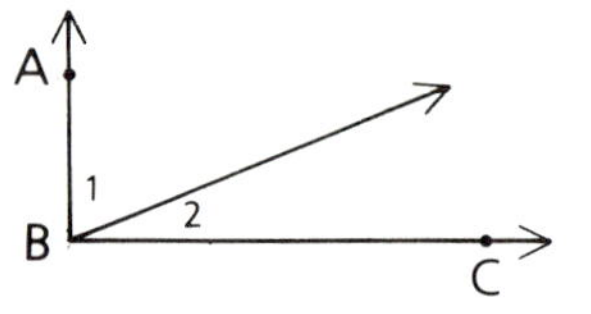

Solution: Since $\angle ABC$ is a right angle, $\angle 1 + \angle 2 = 90^\circ$. Therefore, we have the following equation:

$$\begin{aligned}(3x + 4) + (x + 6) &= 90\\ 4x + 10 &= 90\\ 4x &= 80\\ x &= 20\end{aligned}$$

Although we have solved for "x", this is not the answer to the problem. We must find the measure of $\angle 1$ by substituting 20 for "x" in $\angle 1 = (3x + 4)^\circ$

$$\begin{aligned}3x + 4 &= 3(20) + 4\\ &= 60 + 4\\ &= 60 + 4\\ m\angle 1 &= 64\end{aligned}$$

4 Find the measure of the angle made by the hands of a clock at:

a 4:00

Solution: Since there are 12 intervals on the clock to account for 360°, each interval accounts for 30°. From 12 to 4 there are 4 intervals, or 4(30°), or 120°.

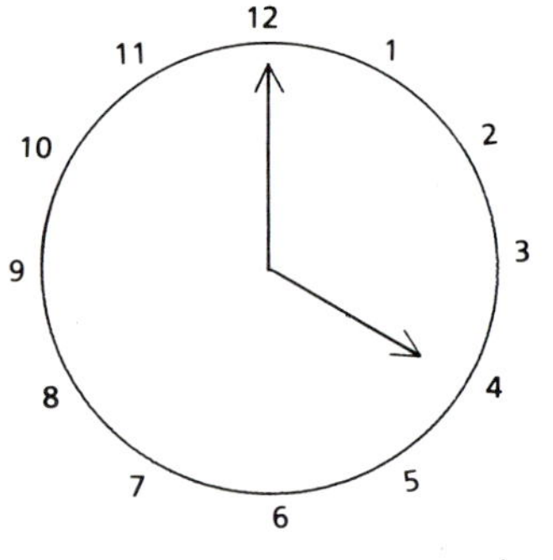

b 5:15

Solution: You might expect the answer to be 60°, but remember that the hour hand is on 5 only when the minute hand is on 12. Thus at 5:15 the hour hand is $\frac{1}{4}$ of the way from 5 to 6.

Since $\frac{1}{4}(30^\circ) = 7\frac{1}{2}^\circ$, there is an additional $7\frac{1}{2}^\circ$ to add on to the 60°.

$60 + 7\frac{1}{2} = 67\frac{1}{2}^\circ$

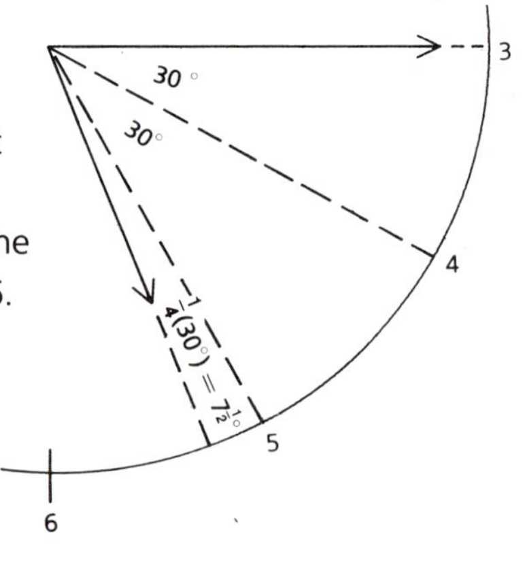

Section 1.2 Problem Set A

1 Change $61\frac{2}{3}^\circ$ to degrees and minutes.

2 Change 71.7° to degrees and minutes.

3 Change $84^\circ 50'$ to degrees.

4 Which two angles below *appear* to be congruent?

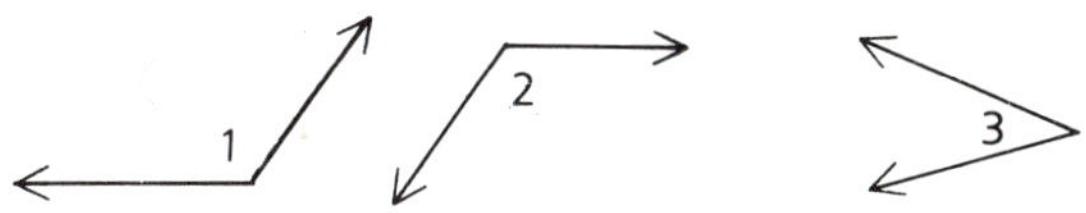

5 Which angle *appears* larger, $\angle RXS$ or $\angle XRS$?

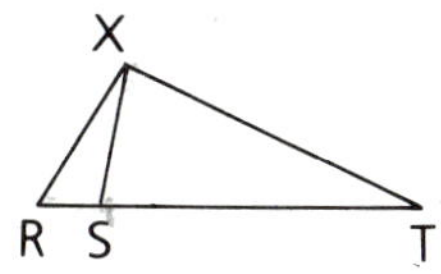

6 Given: There is a right angle at each corner of PRST. (Later in the course you will learn that PRST is a rectangle.)

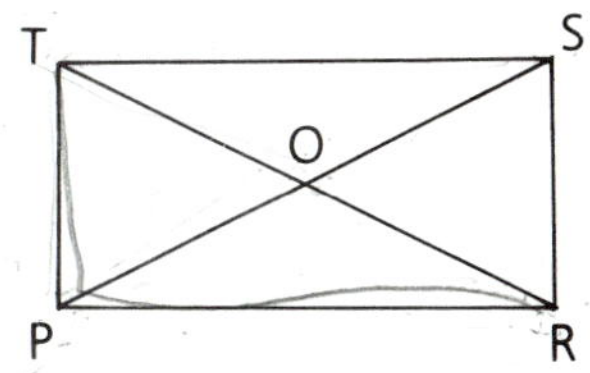

a If $\angle TPO = 60^\circ$, how large is $\angle RPO$?

b If $\angle PTO = 70^\circ$, how large is $\angle STO$?

c If $\angle TOP = 50^\circ$, how large is $\angle POR$?

d Classify $\angle TOS$ according to size.

7 **a** Which angle *appears* to have the same measure as $\angle 1$?

b Which angle *appears* larger, $\angle 2$ or $\angle 3$?

c Does $\angle 3$ *appear* to have the same size as $\angle 4$ or $\angle 5$?

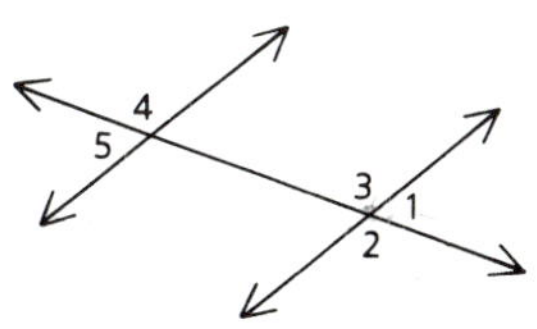

8 If $\angle CBD \cong \angle DBE$, find $m\angle A$.

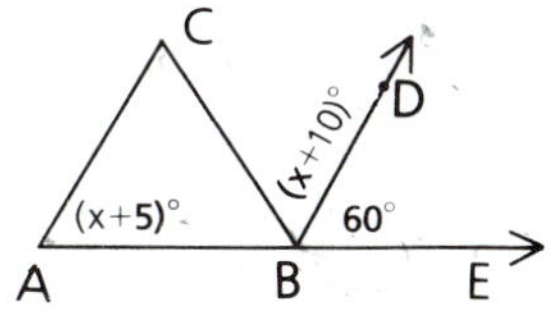

9 Find the measure of the angle made by the hands of a clock at:

a 3:00 **b** 4:30 **c** 7:20 **d** 1:45

Section 1.2 Problem Set B

10 a How many triangles (△s) are in the diagram?

b How many angles (∠s) in the figure appear to be right?

c How many angles in the figure appear to be acute?

d How many angles in the figure appear to be obtuse?

e Name the straight angles in the figure.

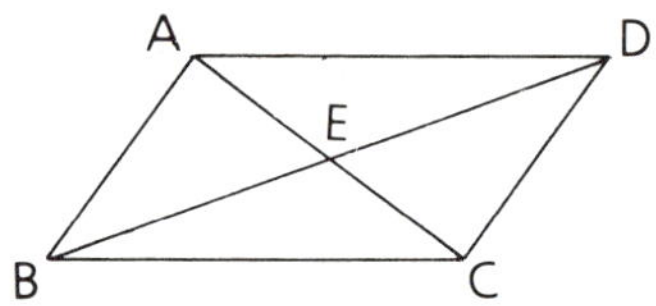

11 The perimeter (distance around) of ABCD is 66, and $\overline{DC}$ is twice as long as $\overline{CB}$. How long is $\overline{AB}$?

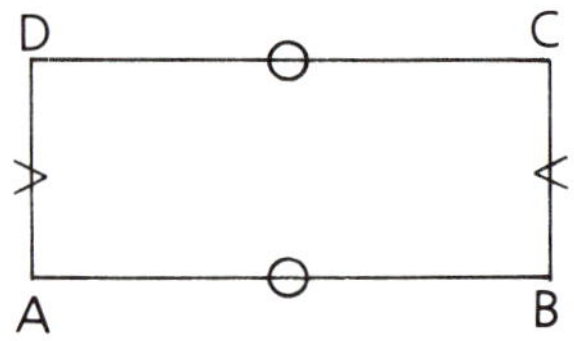

12 $\overline{XS} \cong \overline{YT}$, $\overline{YS} \cong \overline{XT}$

$XT = 2r + 5$

$XS = 3m + 7$

$YS = 3\frac{1}{2}r + 2$

$YT = 4.2m + 5$

Solve for r and m.

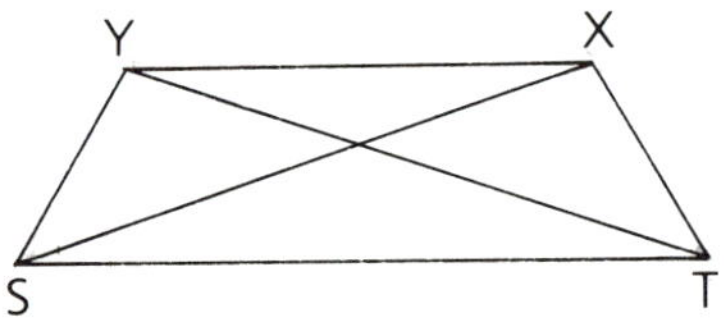

13 $\angle 1 \cong \angle 2$

$m\angle 1 = x + 14$

$m\angle 2 = y - 3$

Solve for y in terms of x.

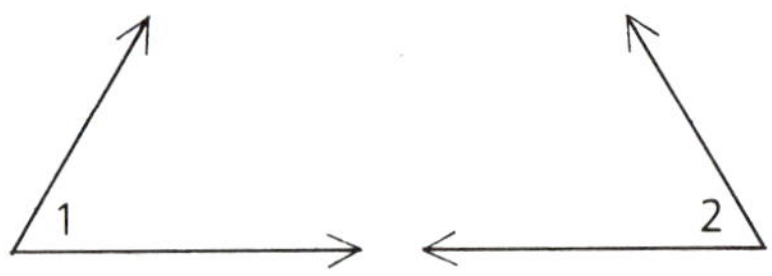

14 If ∠POA is a right angle and if ∠POC is three times as large as ∠COA, find m∠POC.

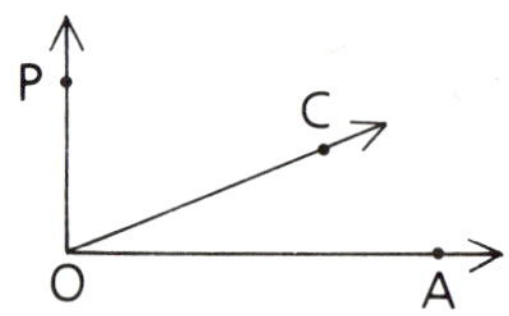

15 Rod drew an angle, measured it with his protractor, and said it was 60°. Mary measured the angle with her protractor and said it was really $60\frac{1}{2}^\circ$. Who was right?

16 Bonnie held a piece of paper behind her. She told Jerry a segment on the paper was about 1 unit long. However, she said that if you blinked your eyes three times, you could see the same segment really had a measure of about 2.5. Jerry laughed, so Bonnie bet him a quarter. Jerry had to pay up. Why?

Section 1.2 Problem Set C

17 $\angle ABC$ and $\angle CBD$ have the same measure.
If $\angle ABC = (\frac{3x}{2} + 2)^\circ$ and
$\angle CBD = (2x - 29\frac{1}{4})^\circ$,
is $\angle ABD$ a straight angle?

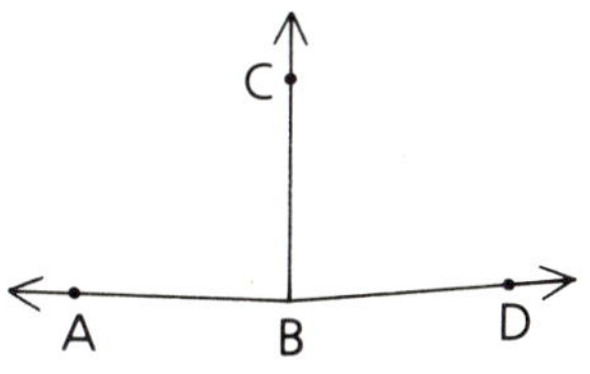

18 $\angle TRS$ is a straight angle
$\angle TRX$ is a right angle
$m\angle TRS = 2x + 5y$
$m\angle XRS = 3x + 3y$
Solve for x and y.

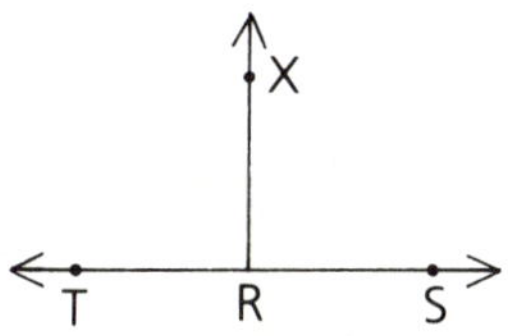

19 Maxie and Minnie were taking a stroll in the Arizona desert, when a spaceship from Mars landed. A Martian walked up to Maxie and Minnie and pointed to figure 1: "XLr8r, XLr8r, XLr8r plus YBcaws, YBcaws," she said. As Maxie and Minnie looked at each other, the Martian pointed to figure 2 and said: "YBcaws plus XLr8r, XLr8r, XLr8r." Any ideas as to what an XLr8r might be?

Figure 1
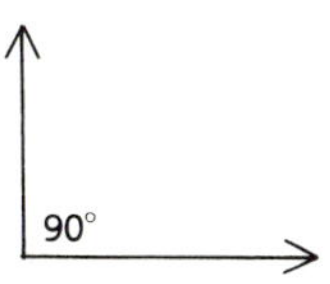

Figure 2
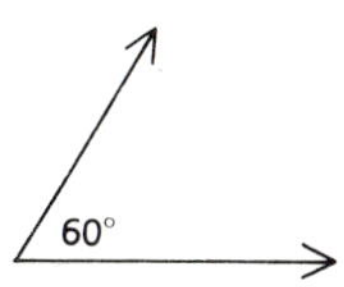

20 Change $13\frac{31}{90}^\circ$ to degrees, minutes, and seconds.

21 Change $21^\circ 30' 10''$ to degrees.

Section 1.3 Collinearity, Betweenness, and Assumptions

In this section we shall discuss:

A COLLINEARITY

B BETWEENNESS OF POINTS

C ASSUMPTIONS FROM DIAGRAMS

1.3 (A) Collinearity

D **Points that lie on the same line are called COLLINEAR. Points that do not lie on the same line are called NON-COLLINEAR.**

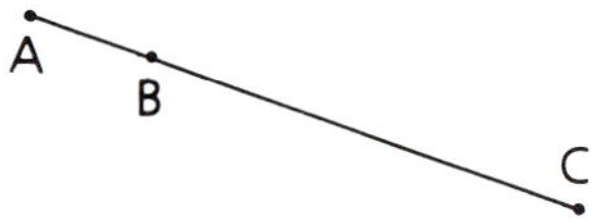

COLLINEAR POINTS

NON-COLLINEAR POINTS

In the diagram, R, S, and T are *collinear* points.

P, O, and X are *collinear*.

M, O, X, and Y are *non-collinear*.

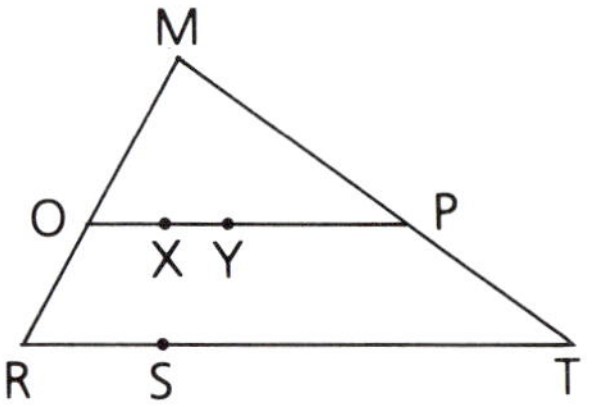

1.3 (B) Betweenness of Points

In order to say that one point is **BETWEEN** two other points, all points must be **COLLINEAR.**

T is between A and R

We do *not* say O is between X and Y

1.3 (C) Assumptions from Diagrams

You may wonder what you should and should not assume when you look at a diagram. In this book:

In General*

YOU SHOULD ASSUME	YOU SHOULD NOT ASSUME
Straight lines and angles	Right angles
Collinearity of points	Congruent segments
Betweenness of points	Congruent angles
Relative location of points	Relative size of segments and angles

The following example will help your understanding.

■ Given: Diagram as shown
Question: What should you assume?

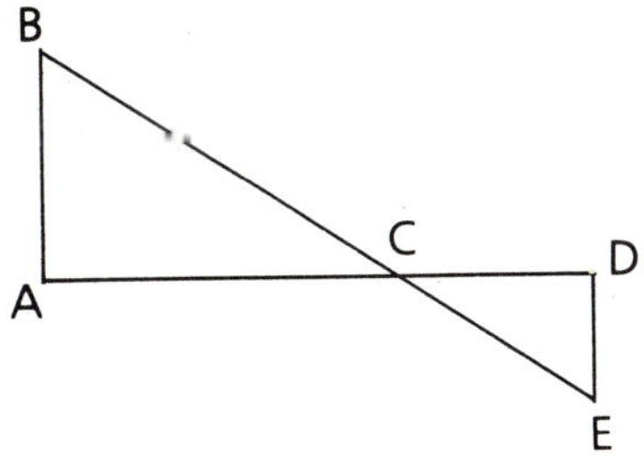

Some possible answers are given below :

DO ASSUME	DO NOT ASSUME
$\overleftrightarrow{ACD}$ and $\overleftrightarrow{BCE}$ are straight lines	$\angle$BAC is a right $\angle$
$\angle$BCE is a straight angle	$\overline{CD} \cong \overline{DE}$
C, D, E are non-collinear	$\angle B \cong \angle E$
C is between B and E	$\angle$CDE is an obtuse angle
E is to the right of A	$\overline{BC}$ is longer than $\overline{CE}$

Reread and study this section carefully, for it is important that you know these agreements about assumptions from a diagram.

* There are occasional exceptions. See Section 1.2 Problem Set C, Problem 17.

Section 1.3 Sample Problems

1 (Answer Yes or No). Is X between P and R?

a 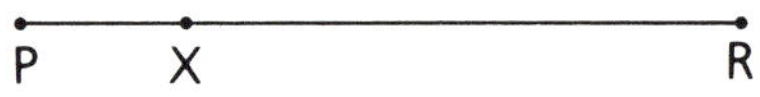

a Ans. Yes

b 

b Ans. No

c 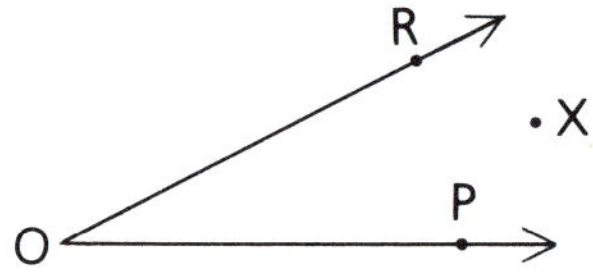

c Ans. No

2 Draw a diagram in which A, B, and C are collinear, A, D, and E are also collinear, and B, C, and D are noncollinear.

Ans. 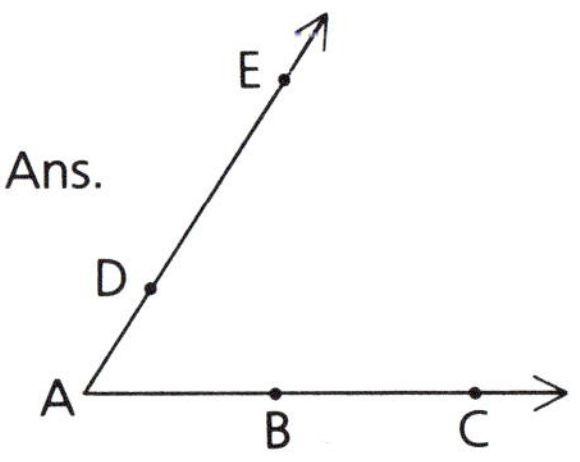

3 **a** In the diagram, may you assume S, T, and V are collinear?

b May you assume $\angle S = 90^\circ$?

a Ans. Yes **b** Ans. No

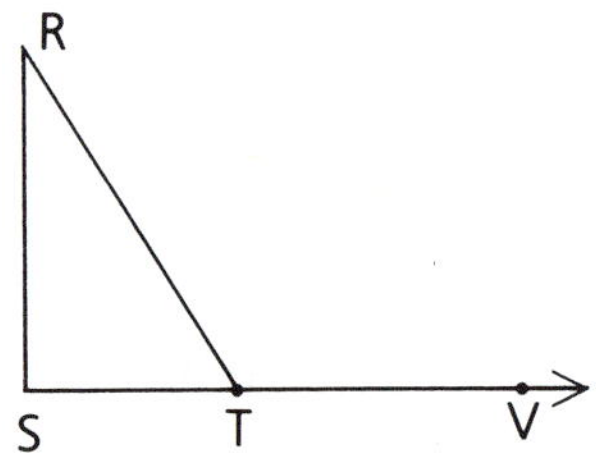

Section 1.3 Problem Set A

1 Find: $m\angle ABC$

(Find the measure of $\angle ABC$.)

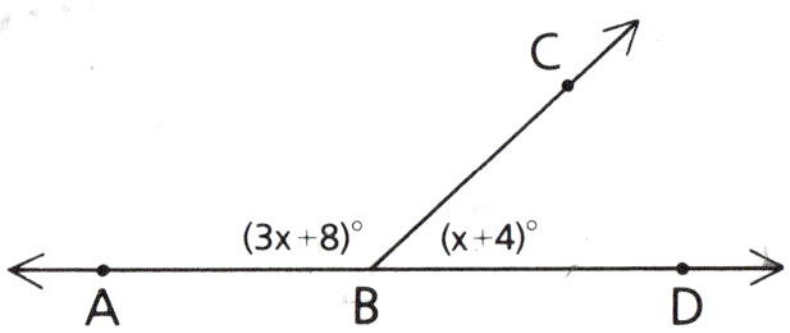

2 Draw a diagram of four points, no three of which are collinear.

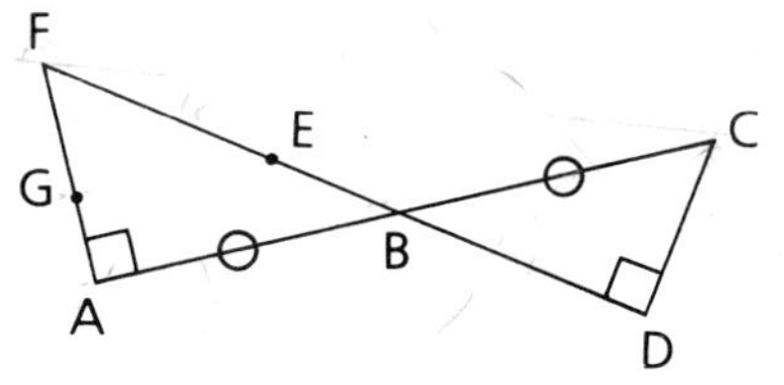

3 a Name all points collinear with E and F.

b Are G, E, and D collinear? Are F and C collinear?

c What other point is collinear with A and F?

d Which two segments do the tick marks indicate are congruent?

e Is $\angle A \cong \angle D$?

f Is $\angle F \cong \angle ABF$?

g Name the angle with vertex at C in all possible ways.

h Where do $\overleftrightarrow{AC}$ and $\overleftrightarrow{FE}$ intersect?

i $\overline{AG} \cap \overline{GF} =$ __?__

j $\overline{AG} \cup \overline{GF} =$ __?__

k B lies on a ray whose endpoint is E. Name this ray in all possible ways.

l Name all points in the diagram between F and D.

4 a May we assume that $\angle$s E, F, G, and H are right angles?

b May we assume that E, F, and G are non-collinear?

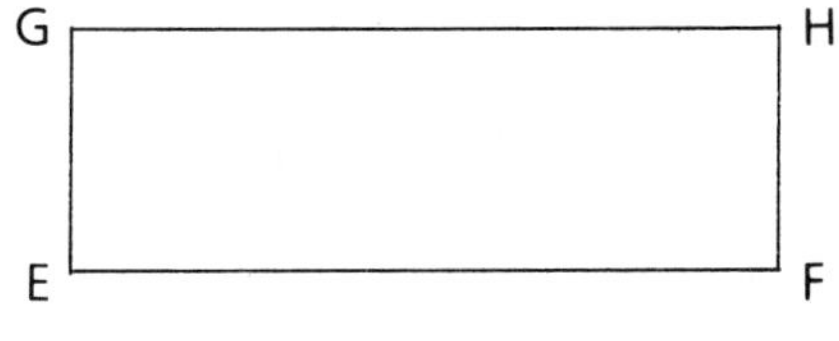

5 $\angle ABD = 3x^\circ$

$\angle DBC = x^\circ$

Find: $m\angle ABD$.

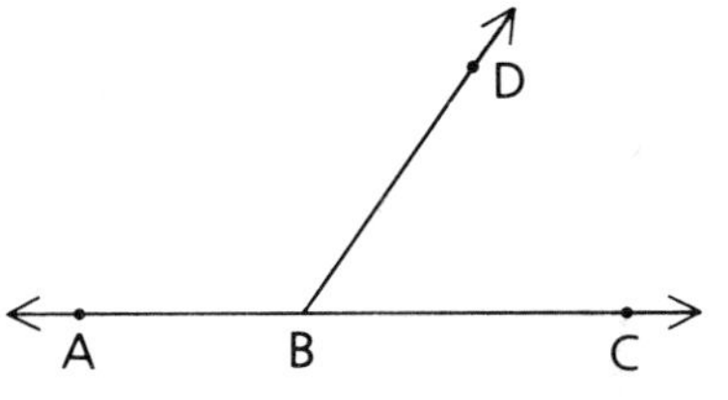

Section 1.3 Problem Set B

6 Given: 4 collinear points A, K, O, and Y; K is between O and A; the length of $\overline{AO}$ added to the length of $\overline{AY}$ is equal to the length of $\overline{OY}$ (OA + AY = OY); and A is to the right of O. Show a correct diagram.

7 F is between A and E, F is also between R and S, and A, E, R, and S are non-collinear. Show a correct diagram.

8 AB = 16, BC = 8, AC = 24. Which point is between the other two?

Section 1.3 Problem Set C

9 Given: $m\angle 1 = (2x + 40)$
$m\angle 2 = (2y + 40)$
$m\angle 3 = (x + 2y)$
Find: $m\angle 1$, $m\angle 2$, and $m\angle 3$

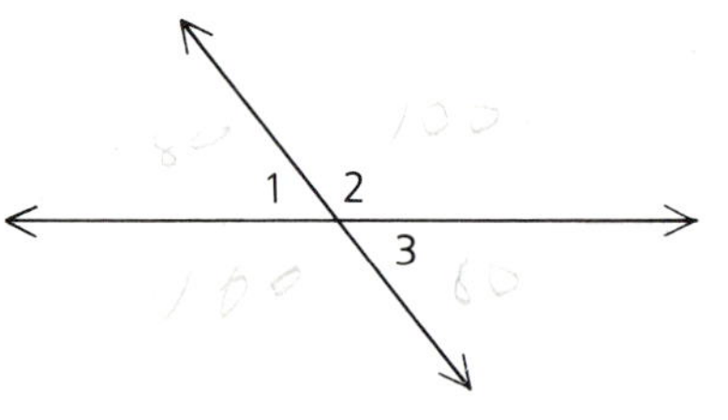

10 When Brock Clock was asked what time it was, he said, "Well, the minute hand is pointing directly at one of the twelve numbers on the clock, the hour hand is pointing toward a spot whose nearest number is at least five more than the number the minute hand is pointing toward, the angle formed by the hands is acute, the sun is shining in the east, and it is *not* five minutes past the hour." Wow! What time is it?

11 To the nearest second, when is the next time after 12:00 that the hour hand and minute hand of a clock are together?

Section 1.4 Beginning Proofs

Much of the enjoyment and challenge in geometry lies in "proving things." In this section we give examples of the **TWO COLUMN PROOF**. The **TWO COLUMN PROOF** is the major type of proof we shall use this year.

We shall also introduce our first **THEOREMS**.

D **A THEOREM is a mathematical statement that can be proved.**

We shall also introduce in this section a procedure we shall follow numerous times in this textbook:

1 We present a theorem or theorems.
2 We prove the theorem(s). (Note: Although all theorems presented can be proved, we shall omit the proofs of certain theorems.)
3 We use these theorems to help prove **SAMPLE PROBLEMS**.
4 The student is then given the challenge of using these theorems to prove homework problems. Theorems will save you much time if you *learn* them and then *use* them.

We now present our first two **THEOREMS**.

Theorem 1

T **If two angles are right angles, then they are congruent.**

Given: $\angle A$ is a right $\angle$
$\angle B$ is a right $\angle$

Prove: $\angle A \cong \angle B$

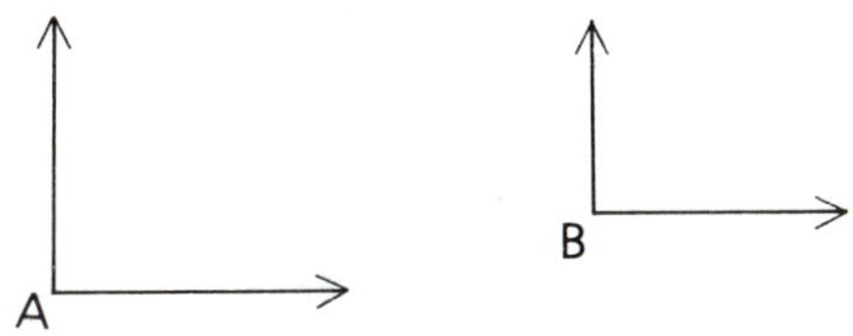

Proof of Theorem 1:

Statements	Reasons
1 $\angle A$ is a right angle	1 Given
2 $m\angle A = 90$	2 If an angle is a right angle, then its measure is equal to 90
3 $\angle B$ is a right angle	3 Given
4 $m\angle B = 90$	4 Same as 2
5 $\angle A \cong \angle B$	5 If two angles have the same measure, then they are congruent (See steps 2 and 4)

Theorem 2

T **If two angles are straight angles, then they are congruent.**

Given: ∠ABC is a straight angle
∠DEF is a straight angle

Prove: ∠ABC ≅ ∠DEF

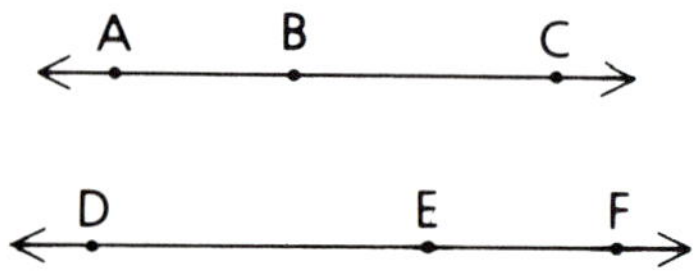

Proof of Theorem 2:

Statements	Reasons
1 ∠ABC is a straight angle	1 Given
2 m∠ABC = 180	2 If an angle is a straight angle, then its measure is equal to 180
3 ∠DEF is a straight angle	3 Given
4 m∠DEF = 180	4 Same as 2
5 ∠ABC ≅ ∠DEF	5 If two angles have the same measure, then they are congruent (See steps 2 and 4)

Now that we have presented and proved two **THEOREMS**, we are ready to *use* them to help prove some **SAMPLE PROBLEMS.**

We will use the **THEOREMS** themselves as reasons in our proofs.

You should also use the **THEOREMS** as reasons in your homework problems.

Remember, the purpose of a **THEOREM** is to shorten your work. So, when doing homework problems, do *not* use the proof of a **THEOREM** as a guide. Use the Sample Problems as a guide.

Section 1.4 Sample Problems

1 **Given:** **∠A is a right angle**
∠C is a right angle

Conclusion: **∠A ≅ ∠C**

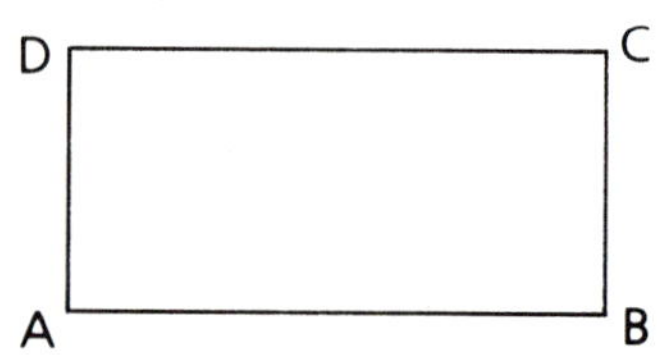

Statements	Reasons
1 ∠A is a right angle	1 Given
2 ∠C is a right angle	2 Given
3 ∠A ≅ ∠C	3 If two angles are right angles, then they are congruent

You will recognize the reason for statement 3 as Theorem 1. While it may seem easier to you merely to write "Theorem 1," *do not do so*. Eventually, such a shortcut would make it harder for you to learn the concepts of geometry.

2 **Given: Diagram as shown**

Conclusion: ∠EFG ≅ ∠HFJ

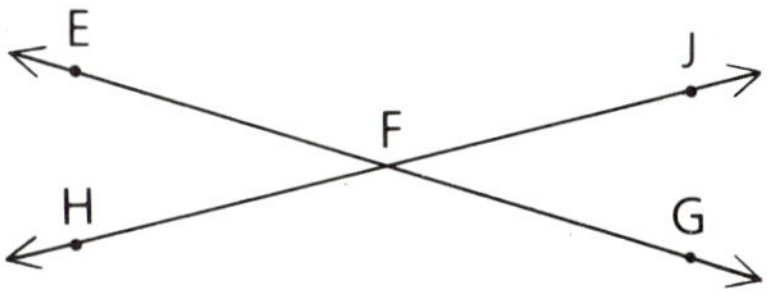

Statements	Reasons
1 Diagram as shown	1 Given
2 ∠EFG is a straight angle	2 Assumed from diagram
3 ∠HFJ is a straight angle	3 Assumed from diagram
4 ∠EFG ≅ ∠HFJ	4 If two angles are straight angles, then they are congruent

3 **Given: ∠RST = 50°**

∠TSV = 40°

∠X is a right angle

Prove: ∠RSV ≅ ∠X

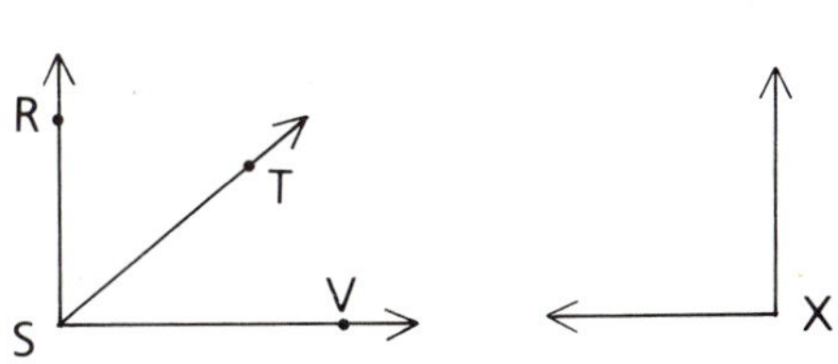

Statements	Reasons
1 ∠RST = 50°	1 Given
2 ∠TSV = 40°	2 Given
3 ∠RSV = 90°	3 Addition (50° + 40° = 90°)
4 ∠RSV is a right angle	4 If an angle is a 90° angle, it is a right angle
5 ∠X is a right angle	5 Given
6 ∠RSV ≅ ∠X	6 If two angles are right angles, then they are congruent

Section 1.4 Problem Set A

For **1, 2** draw each figure and state the information given and to be proved. Then complete each proof by filling in the missing reasons.

1 Given: $\angle 1$ is a right $\angle$
$\angle 2$ is a right $\angle$
Prove: $\angle 1 \cong \angle 2$

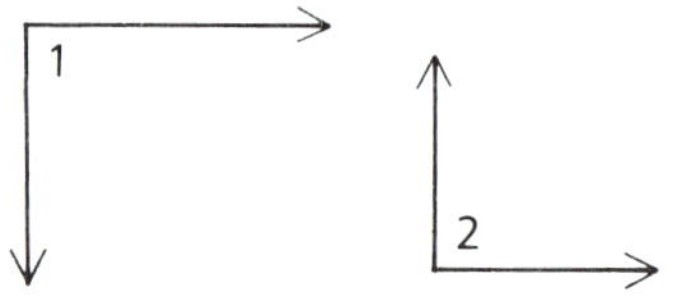

Statements	Reasons
1 $\angle 1$ is a right angle	1 ______
2 $\angle 2$ is a right angle	2 ______
3 $\angle 1 \cong \angle 2$	3 ______

2 Given: Diagram as shown
Prove: $\angle AGD \cong \angle EGB$

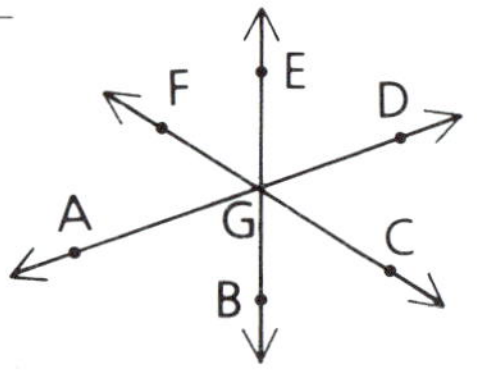

Statements	Reasons
1 Diagram as shown	1 ______
2 $\angle AGD$ is a straight angle	2 ______
3 $\angle EGB$ is a straight angle	3 ______
4 $\angle AGD \cong \angle EGB$	4 ______

For **3–8** use the two column form of proof.

3 Given: $\angle A$ is a right angle
$\angle B$ is a right angle
Prove: $\angle A \cong \angle B$

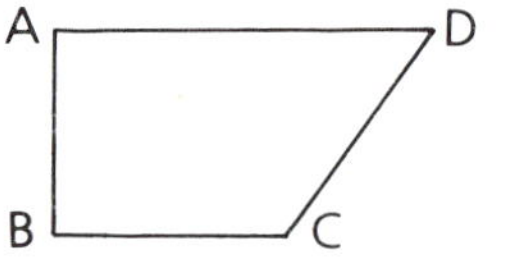

4 Given: $\angle CDE = 110°$
$\angle FGH = 110°$
Conclusion: $\angle CDE \cong \angle FGH$

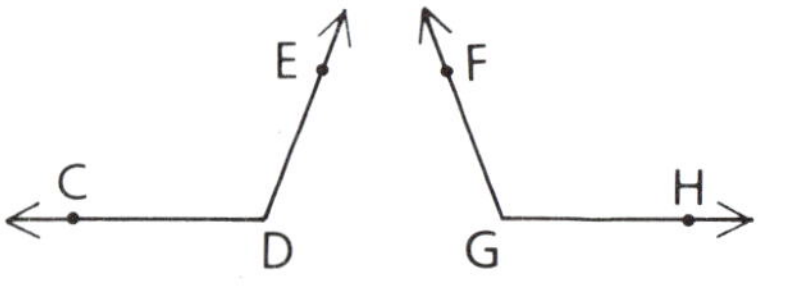

5 Given: $JK = 2\frac{1}{2}$ centimeters
$NO = 2\frac{1}{2}$ centimeters
Conclusion: $\overline{JK} \cong \overline{NO}$

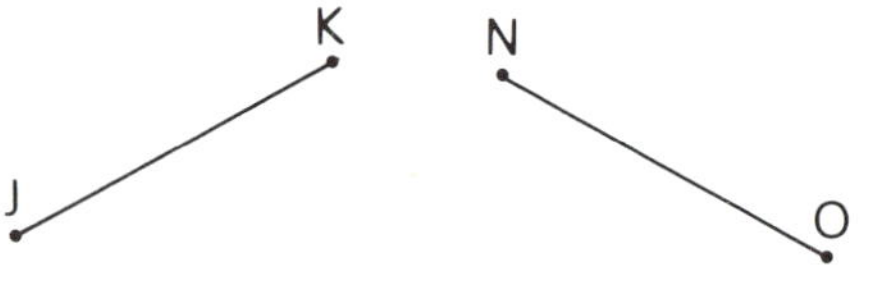

6 Given: Diagram as shown

Prove: $\angle APR \cong \angle SPB$

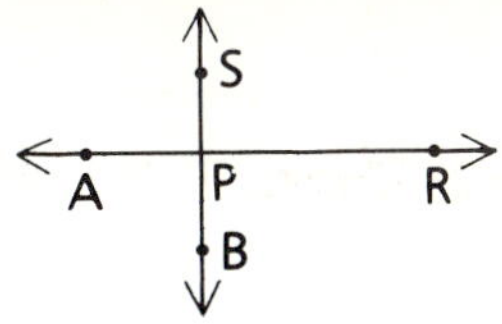

7 Given: $\angle TVA = 71°$

Conclusion: $\angle TVA$ is acute

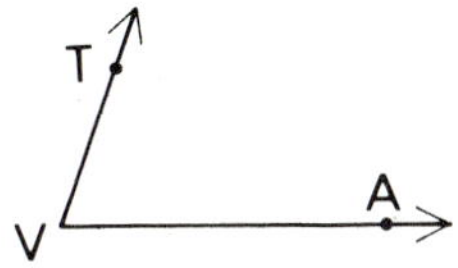

8 Given: $\angle 1 = 20°$

$\angle 2 = 40°$

$\angle 3 = 30°$

Prove: $\angle XYZ$ is a right angle

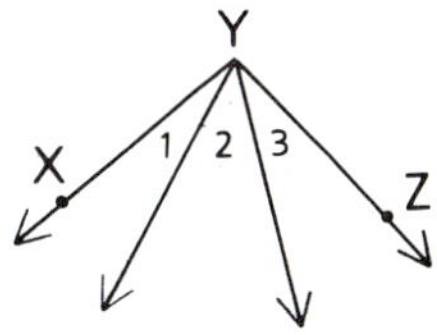

Problem Set 1.4 B

9 Given: $\angle ABD = 10°$

$\angle ABC = 100°$

$\angle EFY = 70°20'$

$\angle XFY = 19°40'$

Prove: $\angle DBC \cong \angle XFE$

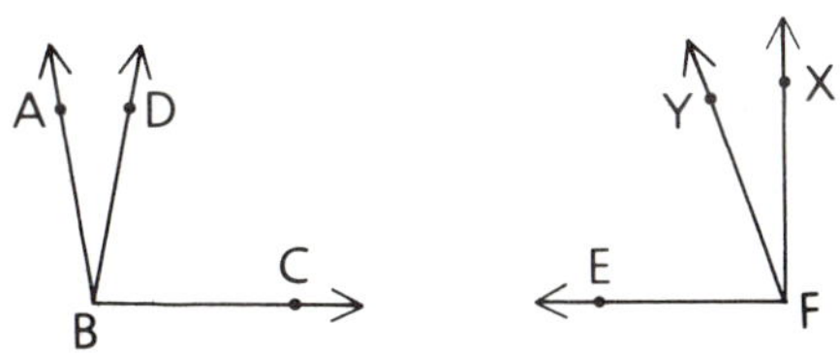

10 Given: $\angle GKH = 38°$

$\angle KHJ = 38°$

$\angle HKJ = 51°$

$\angle GHK = 51°$

Conclusion: $\angle GHJ \cong \angle GKJ$

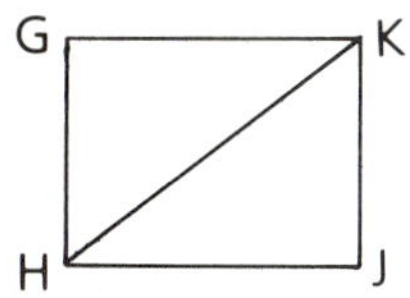

Problem Set 1.4 C

11 Given: $\angle 1 = (x + 7)°$

$\angle 2 = (2x - 3)°$

$\angle ABC = (x^2)°$

$\angle D = (5x - 4)°$

Show that $\angle ABC \cong \angle D$

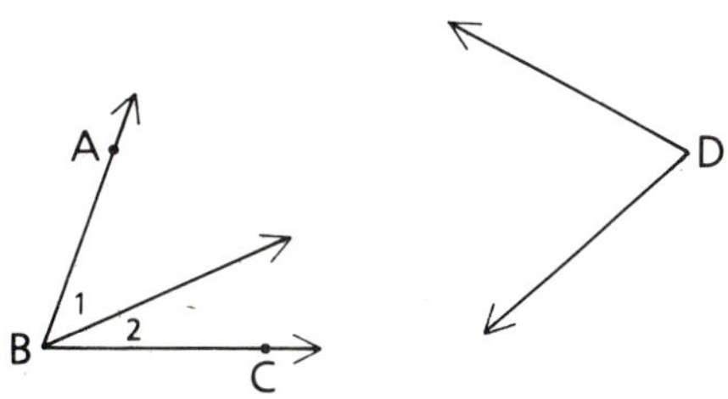

Section 1.5 Division of Segments and Angles

In this section we present:

A MIDPOINT AND BISECTORS OF A SEGMENT
B TRISECTION POINTS AND TRISECTORS OF A SEGMENT
C BISECTING AN ANGLE
D TRISECTING AN ANGLE

1.5(A) Midpoint and Bisectors of a Segment

D **A point (segment, ray, or line) that divides a segment into two congruent segments BISECTS the segment. The point is called the MIDPOINT of a segment.**

X is *not* a midpoint

Y is *not* a midpoint

Only *segments* have midpoints. It will *not* make any sense to say that rays or lines have a midpoint. Do you understand why?

How many midpoints does $\overline{PQ}$ have?

How many bisectors does $\overline{PQ}$ have?

Study the following examples:

■ 1 Given: $\overline{XY}$ bisects $\overline{AC}$ at B

Conclusions that can be drawn:
B is the midpoint of $\overline{AC}$
$\overline{AB} \cong \overline{BC}$

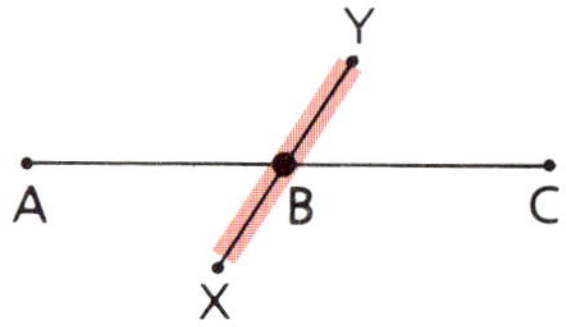

■ 2 Given: D is the midpoint of $\overline{FE}$

Conclusions that can be drawn:
$\overline{FD} \cong \overline{DE}$
Point D bisects $\overline{FE}$
$\overrightarrow{DG}$ bisects $\overline{FE}$

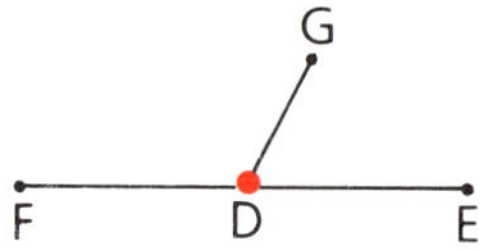

■ 3 Given: $\overline{OK} \cong \overline{KP}$

Conclusions that can be drawn:
K is the midpoint of $\overline{OP}$
$\overleftrightarrow{JM}$ is a bisector of $\overline{OP}$
Point K bisects $\overline{OP}$

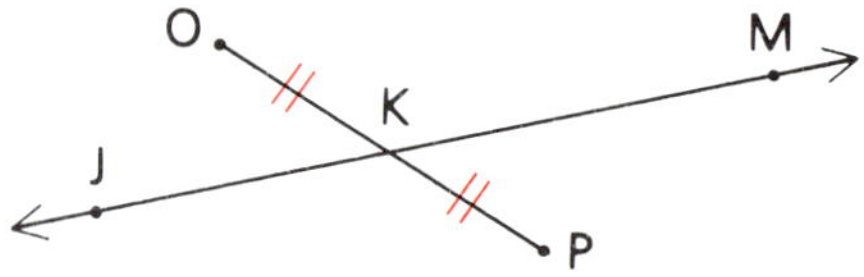

1.5 (B) Trisection Points and Trisecting a Segment

D **Two points (segments, rays, or lines) that divide a segment into three congruent segments TRISECT the segment. The two points are called the TRISECTION POINTS of a segment.**

Again, only segments have trisection points. Rays and lines do *not* have trisection points.

Below are examples of trisection.

■ 1 Given: $\overline{AR} \cong \overline{RS} \cong \overline{SC}$

Conclusions that can be drawn:

R and S are trisection points of $\overline{AC}$

$\overline{AC}$ has been trisected

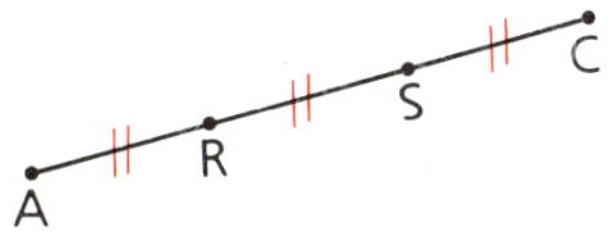

■ 2 Given: E and F are trisection points of $\overline{DG}$

Conclusions that can be drawn:

$\overline{DE} \cong \overline{EF} \cong \overline{FG}$

$\overline{HE}$ and $\overline{HF}$ are trisectors of $\overline{DG}$

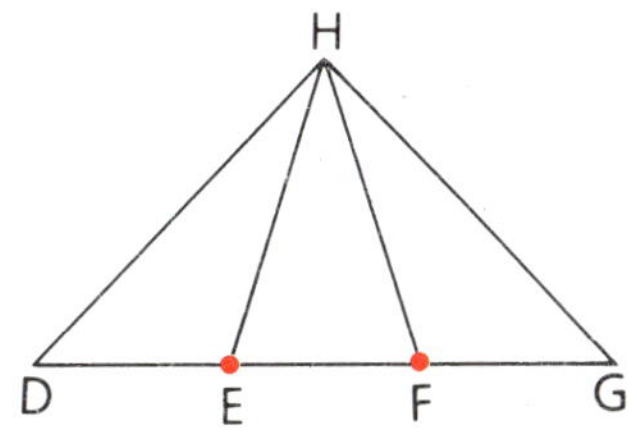

1.5 (C) Bisecting an Angle

D **A ray that divides an angle into two congruent angles BISECTS the angle. The dividing ray is called the BISECTOR of the angle.**

Below are some examples of the bisection of angles.

■ 1

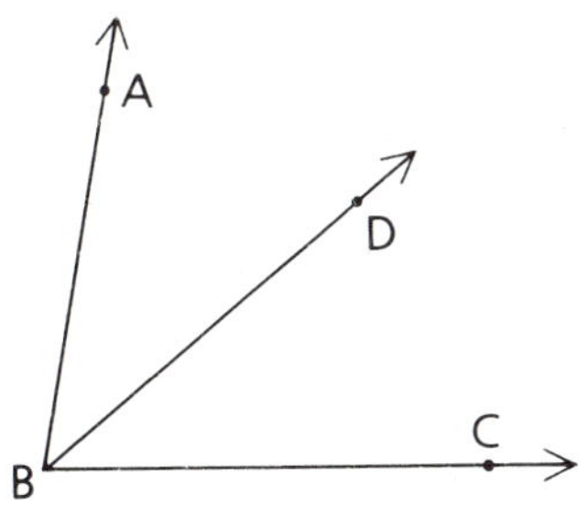

Given: $\angle ABD \cong \angle DBC$

Conclusion: $\overrightarrow{BD}$ (not $\overrightarrow{DB}$) is the bisector of $\angle ABC$

■ 2

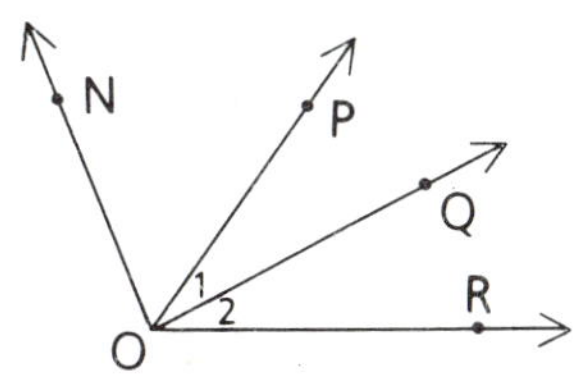

Given: $\angle NOP \cong \angle POR$

$\overrightarrow{OQ}$ bisects $\angle POR$

Conclusions: $\overrightarrow{OP}$ (not $\overrightarrow{PO}$) is the bisector of $\angle NOR$

$\angle 1 \cong \angle 2$

1.5 (D) Trisecting an Angle

D **Two rays that divide an angle into three congruent angles TRISECT the angle. The two dividing rays are called TRISECTORS of the angle.**

Below are some examples of the trisection of angles.

■ 1

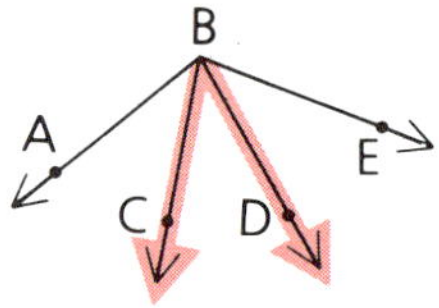

Given: $\angle ABC \cong \angle CBD \cong \angle DBE$

Prove: $\overrightarrow{BC}$ and $\overrightarrow{BD}$ trisect $\angle ABE$

■ 2

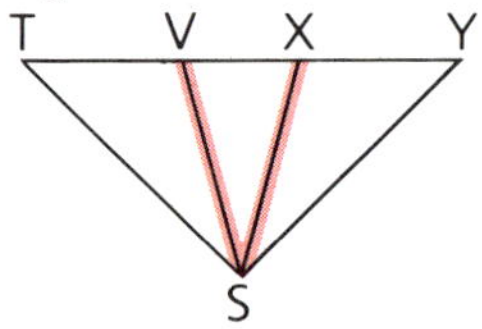

Given: $\overrightarrow{SV}$ and $\overrightarrow{SX}$ are trisectors of $\angle TSY$

Prove: $\angle TSV \cong \angle VSX \cong \angle XSY$

Section 1.5 Sample Problems

1 **Is S the midpoint of $\overline{RT}$?**

Ans. No, the points are not collinear.

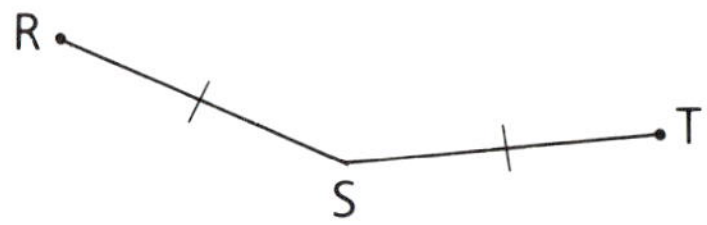

2 **If $\overrightarrow{BD}$ bisects $\angle ABC$, does $\overrightarrow{DB}$ bisect $\angle ADC$?**

Ans. No. We would need more information.

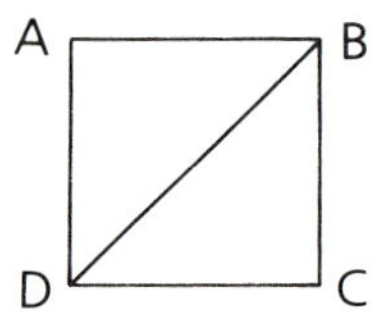

3 **If B and C trisect $\overline{AD}$, do $\overrightarrow{EB}$ and $\overrightarrow{EC}$ trisect $\angle AED$?**

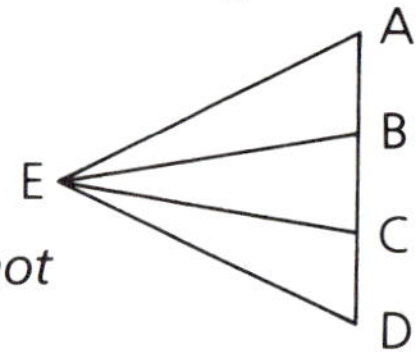

Ans. *NO!* It is true that $\overline{AB} \cong \overline{BC} \cong \overline{CD}$, but that does *not* mean an *angle* has been trisected.
Later we can show $\overrightarrow{EB}$ and $\overrightarrow{EC}$ do *not* trisect $\angle AED$.

4 **Given: $\overrightarrow{PS}$ bisects $\angle RPO$**

Prove: $\angle RPS \cong \angle OPS$

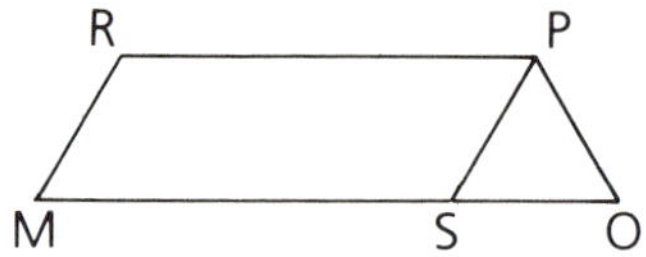

Statements	Reasons
1 $\overrightarrow{PS}$ bisects $\angle RPO$	1 Given
2 $\angle RPS \cong \angle OPS$	2 If a ray bisects an angle, then it divides the angle into two congruent angles

5 Given: $\overleftrightarrow{CM}$ bisects $\overline{AB}$
(Note: In Chapter 3 we shall call $\overline{CM}$ a median of the triangle)

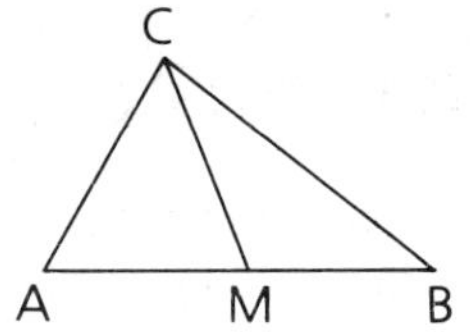

Conclusion: $\overline{AM} \cong \overline{MB}$

Statements	Reasons
1 $\overleftrightarrow{CM}$ bisects $\overline{AB}$	1 Given
2 $\overline{AM} \cong \overline{MB}$	2 If a line bisects a segment, it divides the segment into two congruent segments

6 Given: DH = HF (This means the segments have equal length)
Prove: H is the midpoint of $\overline{DF}$

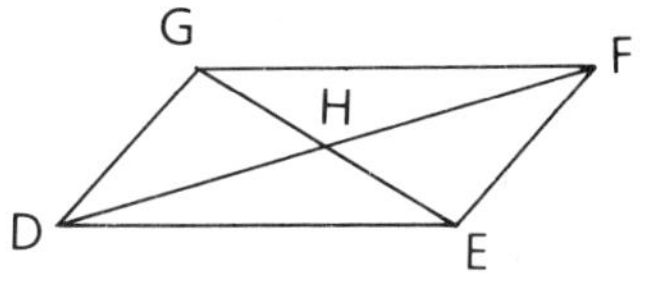

Statements	Reasons
1 DH = HF	1 Given
2 H is the midpoint of $\overline{DF}$	2 If a point divides a segment into two segments of equal length, it is called the midpoint

7 From left to right $\overline{EH}$ is divided in the ratio of 5:3:2 by F and G. If EH = 30, find FG and name the midpoint of $\overline{EH}$.

Solution: From the ratios given, EF = 5x, FG = 3x, GH = 2x

$$5x + 3x + 2x = 30$$
$$10x = 30$$
$$x = 3, \text{ and } FG = 9$$

Since EF = 15 and FH = 15, F is the midpoint of $\overline{EH}$.

8 Given: $\overrightarrow{KO}$ bisects $\angle JKM$
$\angle JKM = 41°37'$

Find: $m\angle OKM$

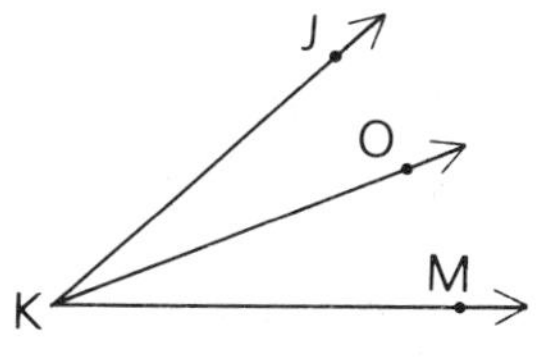

Solution: $\frac{1}{2}(41°37') = 20\frac{1}{2}° \; 18\frac{1}{2}'$
or $20° \; 48\frac{1}{2}'$ (since $\frac{1}{2}° = 30'$)
or $20° \; 48' \; 30''$ (since $\frac{1}{2}' = 30''$)

Section 1.5 Problem Set A

1 Name the congruent segments:

a

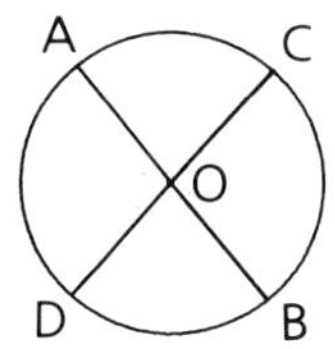

O is the midpoint of $\overline{CD}$

b

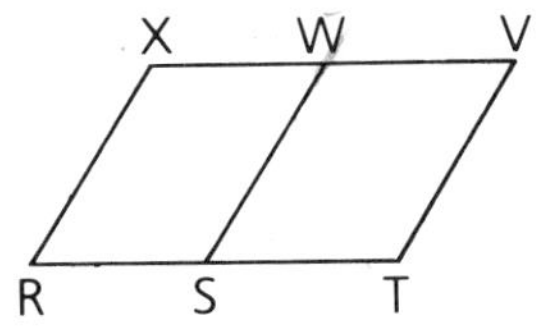

$\overline{SW}$ bisects $\overline{XV}$

2 Name the congruent angles:

a

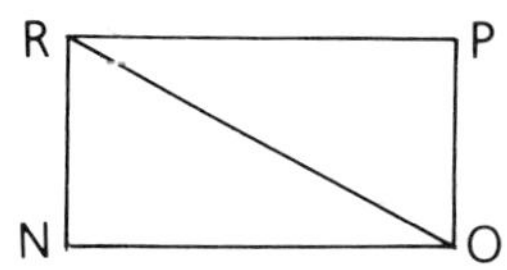

$\overrightarrow{RO}$ bisects $\angle NRP$

b

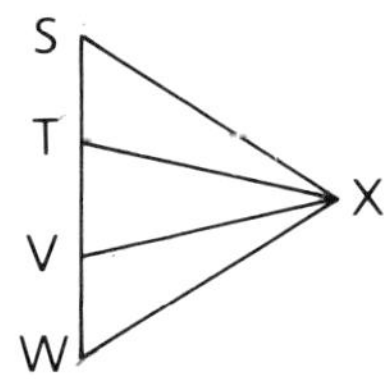

$\overrightarrow{XT}$ and $\overrightarrow{XV}$ trisect $\angle SXW$

3 Name the midpoint:

a

$\overline{MA} \cong \overline{AP}$

b

4 Name the angle bisector:

a

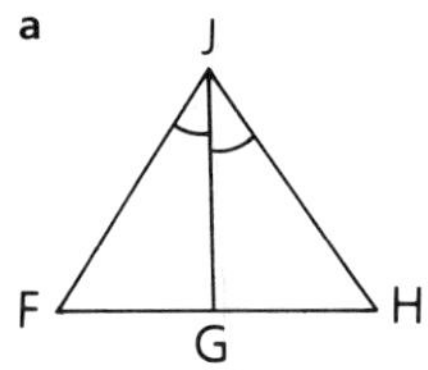

b

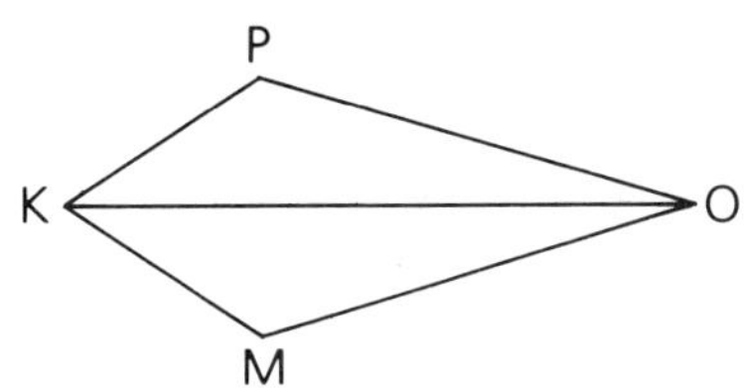

$m\angle POK = m\angle MOK$

5 Find $\angle XTZ$, if $\angle XTY =$

a 60°

b $48^\circ 50'$

c $36\frac{1}{2}^\circ$

d $29\frac{1}{4}^\circ$

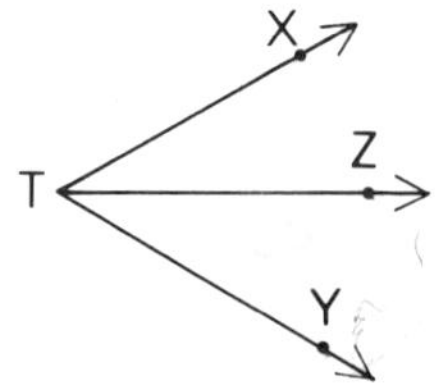

$\overrightarrow{TZ}$ bisects $\angle XTY$

6 B and C are trisection points of $\overline{AD}$

AD = 12

- a Find AB
- b Find AC
- c If AB = x + 3, solve for x
- d If AB = x + 3 and AE = 3x + 6, find AE
- e C is the midpoint of what segment?
- f Do $\overrightarrow{EB}$ and $\overrightarrow{EC}$ trisect $\angle AED$?

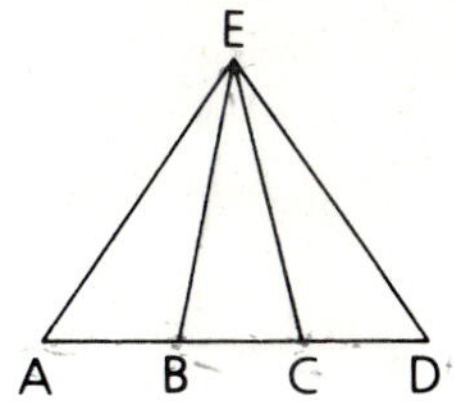

7 $m\angle FGJ = 3x - 5$

$m\angle JGH = x + 27$

$\overrightarrow{GJ}$ bisects $\angle FGH$

Find: $m\angle FGJ$

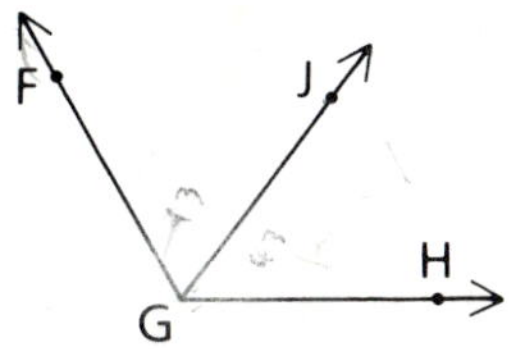

8 $\angle ABC = 90^\circ$

$\angle 1 = (2x + 10)^\circ$

$\angle 2 = (x + 20)^\circ$

$\angle 3 = (3x)^\circ$

Has $\angle ABC$ been trisected?

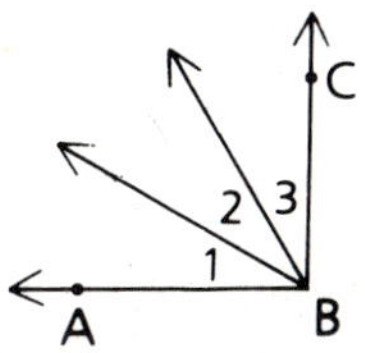

In **9, 10** the final reason for each problem is stated incorrectly. Correct the wording of the FINAL REASON for each problem.

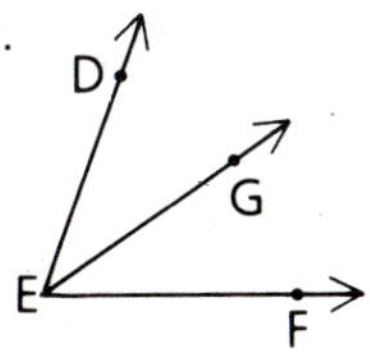

9 Given: $\angle DEG \cong \angle FEG$

Prove: $\overrightarrow{EG}$ bisects $\angle DEF$

Statements	Reasons
1 $\angle DEG \cong \angle FEG$	1 Given
2 $\overrightarrow{EG}$ bisects $\angle DEF$	2 *Correct This Reason:* If a ray divides an angle into two angles, the ray bisects the angle

10 Given: $\overline{KJ} \cong \overline{HJ}$

Prove: J is the midpoint of $\overline{HK}$

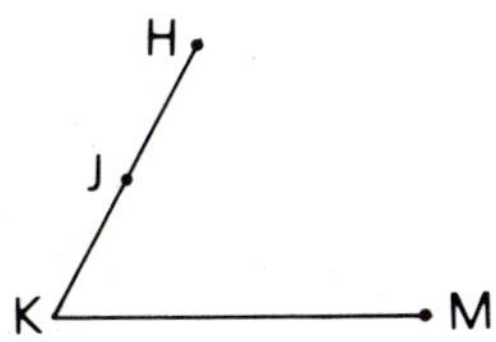

Statements	Reasons
1 $\overline{KJ} \cong \overline{HJ}$	1 Given
2 J is the midpoint of $\overline{HK}$	2 *Correct This Reason:* If a point is the midpoint of a segment, it divides the segment into two congruent segments

For **11–16** do each proof in the TWO COLUMN FORM.

11 Given: $\overrightarrow{WS}$ bisects $\angle RWP$

Prove: $\angle RWS \cong \angle PWS$

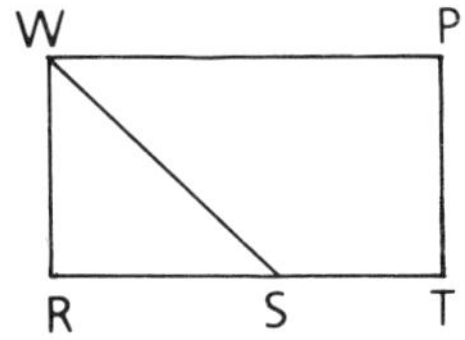

12 Given: $\overline{XY} \cong \overline{YZ}$

Prove: Y is the midpoint of $\overline{XZ}$

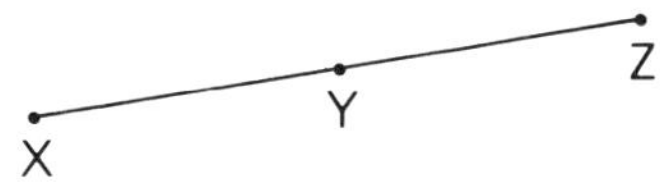

13 Given: $\angle AEB \cong \angle BEC \cong \angle CED$

Conclusion: $\overrightarrow{EB}$ and $\overrightarrow{EC}$ trisect $\angle AED$

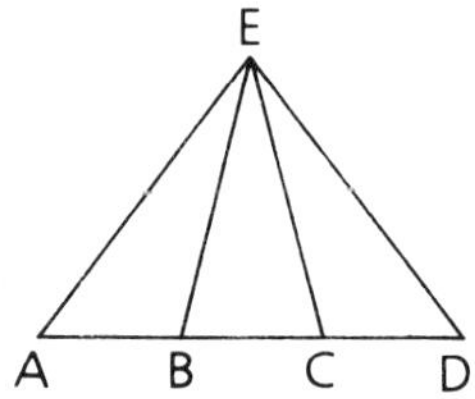

14 Given: $\angle 1 \cong \angle 2$

Conclusion: $\overrightarrow{HK}$ bisects $\angle FHJ$

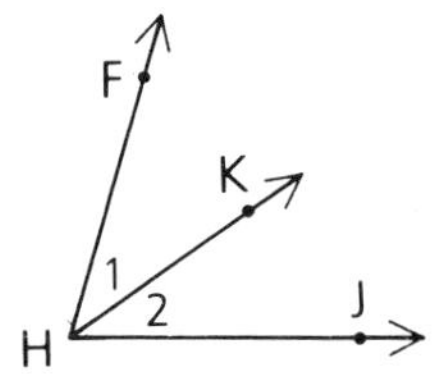

15 Given: $\angle TXW$ is a right angle

$\angle TYV$ is a right angle

Prove: $\angle TXW \cong \angle TYV$

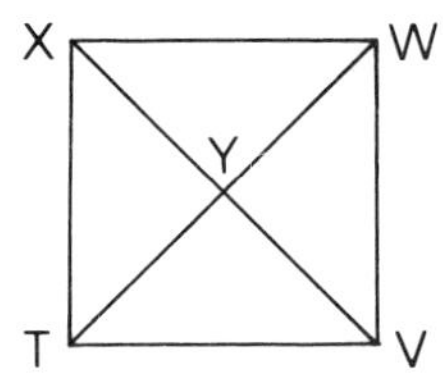

16 Given: B is the midpoint of $\overline{AC}$

Prove: $\overline{AB} \cong \overline{BC}$

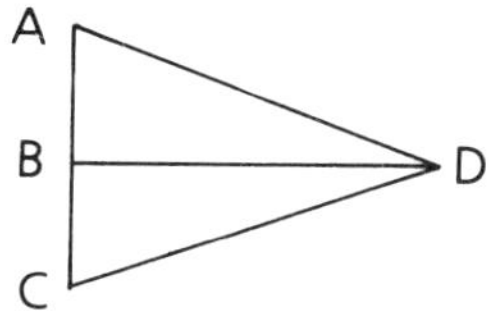

17 Rays $\overrightarrow{AB}$ and $\overrightarrow{AC}$ each bisect $\angle DAE$. Draw a sketch to show how this could happen.

Section 1.5 Problem Set B

18 $\overrightarrow{OG}$ and $\overrightarrow{OH}$ divide straight angle FOJ into three angles whose measures have respective ratios of 4:3:2

Find: $m\angle FOG$

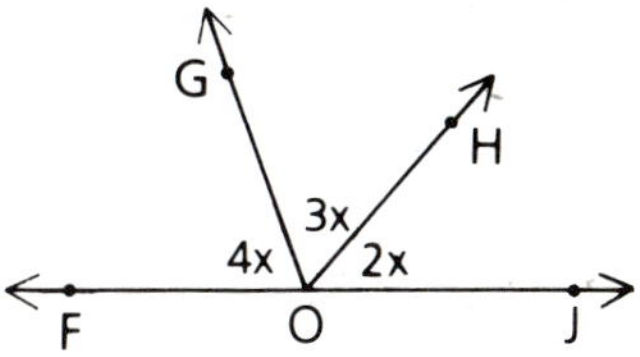

19 $\overleftrightarrow{TP}$ bisects $\overline{VS}$ and $\overline{MR}$

$\overline{VM} \cong \overline{SR}$

$MP = 9, \; VT = 6$

Perimeter of MRSV = 62

Find: VM

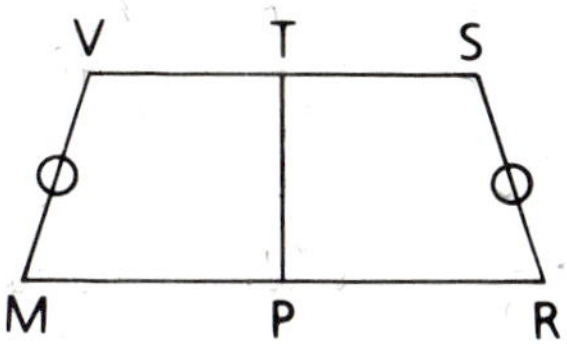

20 $OM = x + 8$

$MP = 2x - 6$

$OP = 44$

Is M the midpoint of $\overline{OP}$?

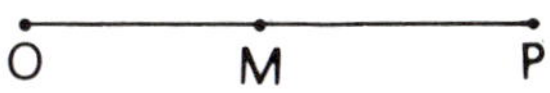

Section 1.5 Problem Set C

21 $\overrightarrow{OP}$ and $\overrightarrow{OR}$ trisect $\angle NOS$

$m\angle NOP = 3x - 4y$

$m\angle POR = x - y$

$m\angle ROS = y - 10$

Find: $m\angle ROS$

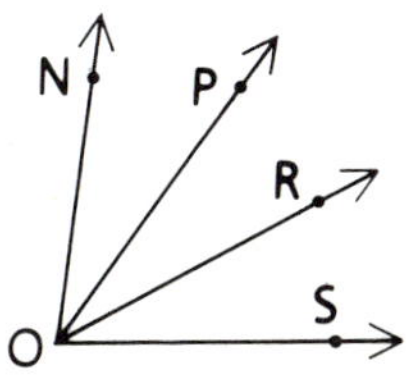

22 $\angle BAC = 120°$. D, E, and F are in the interior of $\angle BAC$ as shown. $\overrightarrow{AD}$ bisects $\angle BAF$. $\overrightarrow{AE}$ bisects $\angle CAF$.

Find: $m\angle DAE$

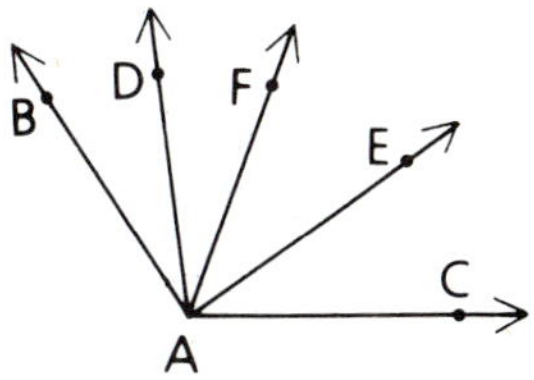

23 The measures of two angles have the ratio of 5:2. The larger of the angles is 30 more than half the difference of the angles. Find the measure of each angle.

Section 1.6 Paragraph Proofs

Although most of the proofs we shall encounter this year will use the **TWO COLUMN FORM**, we now introduce **PARAGRAPH PROOFS**. They are important because the proofs in journals, more advanced courses, and other subjects are in that form.

The **SAMPLE PROBLEMS** that follow demonstrate how to use **PARAGRAPH PROOFS**, including how to show that in some cases a problem cannot be proved true or can be proved false.

Section 1.6 Sample Problems

For **1–4** use a **PARAGRAPH PROOF** for each problem.

1 **Given:** $\angle O = 67\frac{1}{2}^\circ$
$\angle P = 67^\circ 30'$
Prove: $\angle O \cong \angle P$

Proof: Since there are 60 minutes in 1 degree, $67^\circ 30'$ equals $67\frac{1}{2}^\circ$. Since $\angle O$ and $\angle P$ have the same measure, they are congruent.

2 **Prove:** $\angle DBC \cong \angle E$

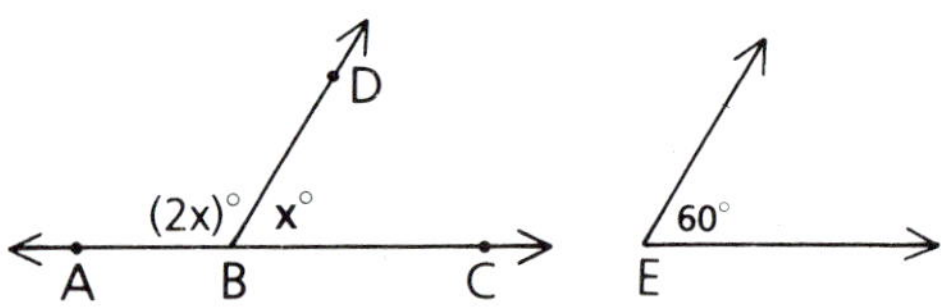

Proof: $\angle ABC$ is a straight angle from the diagram.

From algebra,
$$\begin{aligned} 2x + x &= 180 \\ 3x &= 180 \\ x &= 60 \end{aligned}$$

Since $\angle DBC$ and $\angle E$ each contain 60°, they are congruent.

3 **Given:** $\angle 1$ is acute
$\angle 2$ is acute
Conclusion: $\angle 1 \cong \angle 2$

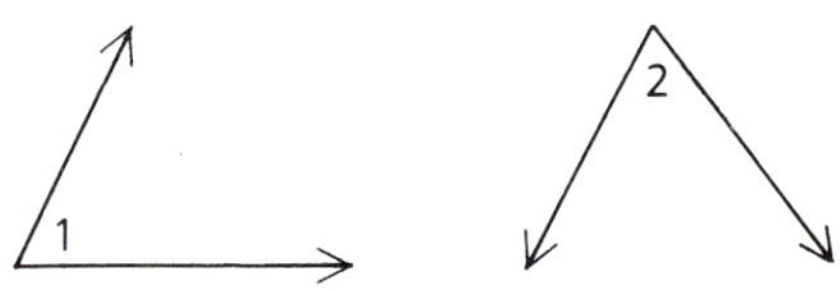

Discussion: This problem *cannot be proved,* since two acute angles need not be congruent.

4 Given: $\angle D = 90^\circ$
$\angle E$ is obtuse

Prove: $\angle D \cong \angle E$

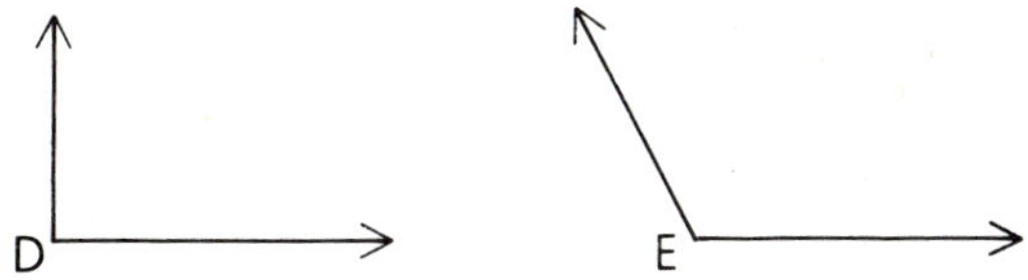

Discussion: This problem *can be proved false.* Since $\angle E$ is obtuse, its measure is more than 90. Since $\angle D$ and $\angle E$ have different measures, they are not congruent ($\angle D \ncong \angle E$).

Section 1.6 Problem Set A

For **1–6** use a PARAGRAPH PROOF for each problem.

1 Given: $\angle V = 119\frac{2}{3}^\circ$
$\angle S = 119^\circ 40'$

Conclusion: $\angle V \cong \angle S$

2 Given: Diagram as shown

Prove: $\angle FEH \cong \angle JKM$

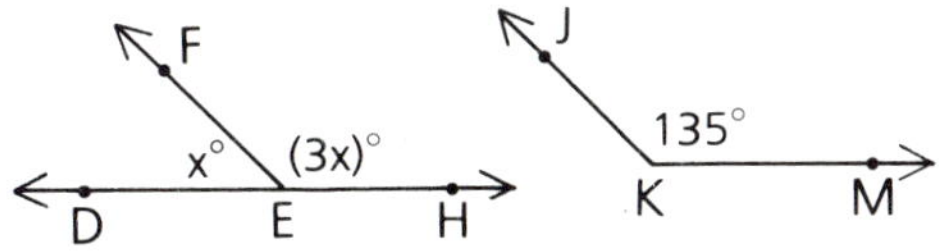

3 Given: Diagram as shown, $\angle OPT = 90^\circ$

Prove: The measure of $\angle VAY$ is twice that of $\angle RPT$

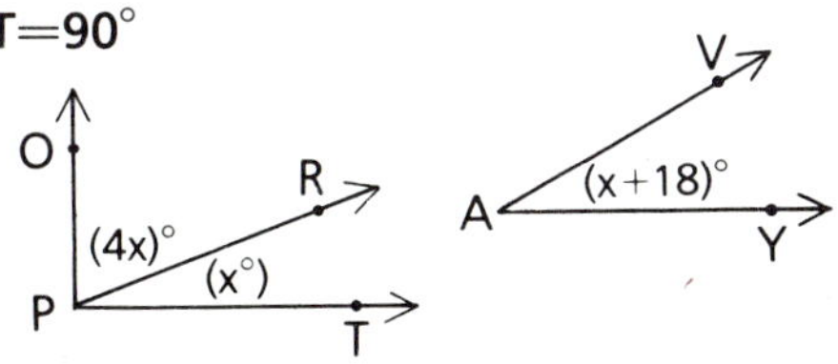

4 Given: $AB = x + 4$
$BC = 2x$
$AC = 16$

Conclusion: $\overline{AB} \cong \overline{BC}$

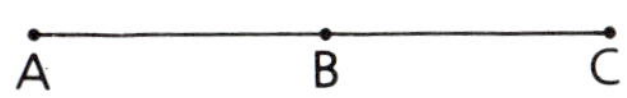

5 Given: $\angle D$ is obtuse
$\angle C$ contains more than 90° ($\angle C > 90^\circ$)

Conclusion: $\angle D \cong \angle C$

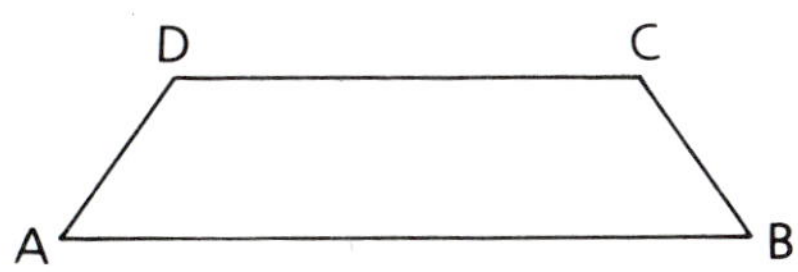

6 Given: $\angle 1$ is obtuse, $\angle 2$ is acute
Prove: $\angle 1 \cong \angle 2$

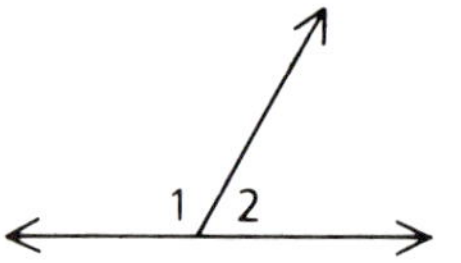

Section 1.6 Problem Set B

For **7, 8** use PARAGRAPH PROOFS.

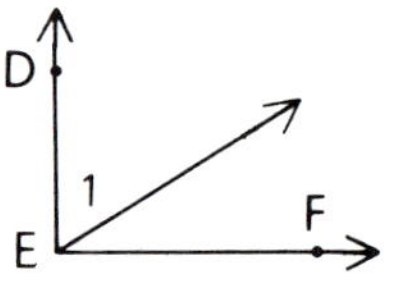

7 Given: $\angle DEF$ is a right angle
Prove: $\angle 1$ is acute

8 Prove the following statement: If an obtuse angle is bisected, each of the two resulting angles is acute.

Section 1.6 Problem Set C

For **9, 10** use PARAGRAPH PROOFS.

9 Given: Diagram as shown
$\overrightarrow{AC}$ bisects $\angle BAD$
$\overrightarrow{AE}$ bisects $\angle DAF$
Prove: $\angle CAE$ is a right angle

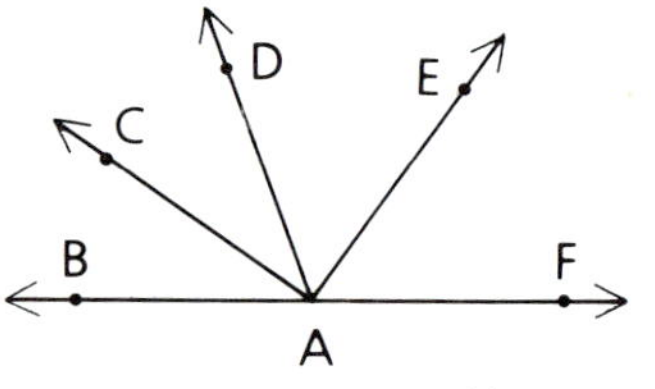

10 Given: $m\angle J + m\angle H + m\angle JKH = 180$
Diagram as shown
Prove: a $m\angle 1 = m\angle J + m\angle H$
b $m\angle 1 > m\angle J$

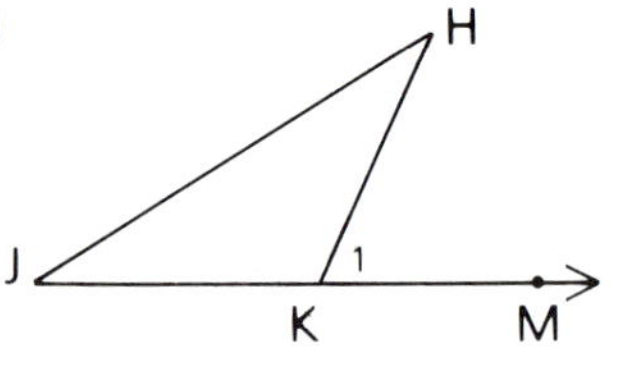

Section 1.6 Problem Set D

11 Use a PARAGRAPH PROOF.
Given: $m\angle A + m\angle ABC + m\angle ACB = 180$
$m\angle D + m\angle DBC + m\angle DCB = 180$
$\overrightarrow{BD}$ bisects $\angle ABC$, $\overrightarrow{CD}$ bisects $\angle ACB$
Prove: $m\angle D = 90 + \frac{1}{2}m\angle A$

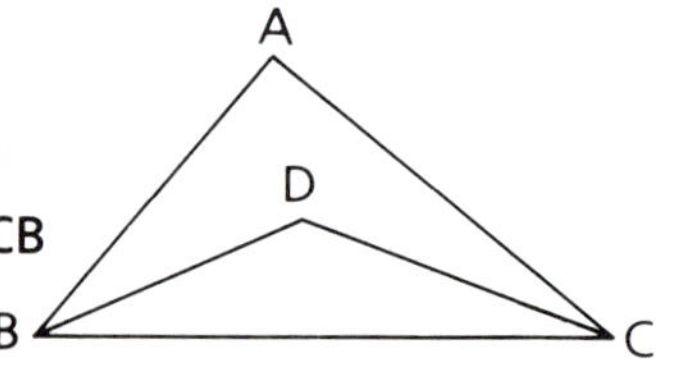

Section 1.7 Deductive Structure

You have just spent a few days doing **TWO COLUMN PROOFS** and **PARAGRAPH PROOFS.** We feel it important that you learn how to prove a few things so that you can appreciate some theory of proofs.

A **DEDUCTIVE STRUCTURE** is a system of thought in which conclusions are justified on the basis of previously assumed or proved statements.

Every **DEDUCTIVE STRUCTURE** contains four **INGREDIENTS**:

1 UNDEFINED TERMS
2 ASSUMPTIONS, or POSTULATES
3 DEFINITIONS
4 THEOREMS and other CONCLUSIONS

Undefined Terms

Examples of **UNDEFINED TERMS** we have already studied are *points* and *lines.* While we did *not* give a definition for a point or a line, we and your teacher did "describe" them in the hope that everyone would then have the same idea of what they were.

Assumptions, or Postulates

A **POSTULATE** is an **ASSUMPTION.** As yet, we have not formalized any **POSTULATES** even though we have already made use of many from algebra. **POSTULATES** will be marked in the left margin with a

P

Definitions

A **DEFINITION** states the meaning of a concept. We have already had many **DEFINITIONS,** such as the definitions of acute angles, right angles, obtuse angles, and straight angles.

One very important characteristic of a **DEFINITION** is that a **DEFINITION** is **REVERSIBLE.** For example, the definition of the midpoint of a segment can be used in either of two ways:

1 If a point is the midpoint of a line segment, then the point divides the segment into two congruent segments.

2 If a point divides a line segment into two congruent segments, then the point is the midpoint of the line segment.

In some problems the definition of midpoint must be stated as in 1 above. In other problems the definition must be **REVERSED,** as in 2 above.

Definitions are marked in the left margin with a

D

Theorems

A **THEOREM** is a mathematical statement that can be *proved.* The presentation of each theorem will be headed by its number, for example,

Theorem 78

Also, in the left margin the statement of the theorem will be marked,

T

You will prove some theorems and other relationships such as homework problems. As you work, remember that you must prove conclusions by using reasons *previously* assumed or proved. Thus, you *cannot* use **THEOREM 1** in order to *prove* **THEOREM 1.**

THEOREMS and **POSTULATES** are *not always* reversible. For example, "If two angles are right angles, then they are congruent" is *true.* The reverse statement, "If two angles are congruent, then they are right angles" is *false*.

In summary:

DEFINITIONS are always reversible.

THEOREMS and **POSTULATES** are *not* always reversible.

If you are to be successful in writing proofs, you must *memorize* postulates, definitions, and theorems. There is no easier way.

A complete mastery of the use of **DEDUCTIVE STRUCTURE** is not possible in a short time. However, we do wish to point out the most common error that students make—using a *reverse* statement at the *wrong* time.

For example, suppose you know,

"If it's opening day in baseball, then Baggy Bat is at Wrigley Field."

Now suppose you also know that Baggy is at Wrigley Field. Can you deductively conclude that it's opening day in baseball? The answer is NO. You would be making the error of assuming the *reverse* statement was true at the *wrong* time. The reverse statement, "If Baggy Bat is at Wrigley Field, then it's opening day in baseball" is NOT necessarily true.

Remember, only definitions are always reversible. Theorems and postulates are *not* always reversible.

The major purpose of this section is to acquaint you with some terminology. In Chapter 2 and Chapter 3 we think you will grow to appreciate and understand this section even more.

The homework problems in this section are quite different, and we think you will enjoy them.

Section 1.7 Problem Set A

1 Which of the following is (are) always reversible?

- **a** definitions
- **b** theorems
- **c** postulates

2 (Answer Yes or No)

- **a** Do we prove theorems?
- **b** Do we prove definitions?

3 Tell whether each of the following statements is a theorem or a definition.

- **a** If two angles are right angles, then they are congruent.
- **b** If a ray bisects an angle, then it divides the angle into two congruent angles.

4 Write the reverse of the following definition.

If a point is the midpoint of a segment, then it divides the segment into two congruent segments.

For **5, 6** comment on the reasoning.

5 The school colors are orange and black, so I'll wear my orange skirt to the game and everyone will notice me.

6 I've flipped this silver dollar 5 times and it's come up "heads" each time. Thus, the odds are more than 50-50 that it will come up "tails" next time.

Section 1.7 Problem Set B

For **7–11** study each of the arguments and state whether or not the conclusion is deducible. If it is not, comment on the error in the reasoning.

7 If a student at Niles High has Room 303 for home room, the student is a Freshman. Joe Jacobs is a student at Niles High and has Room 303 for home room. Therefore, Joe Jacobs is a Freshman.

8 If the three angles of a triangle are acute, then the triangle is acute. In triangle ABC, angle A and angle B are acute. Therefore, triangle ABC is acute.

9 All school buses stop at railroad crossings. This vehicle stopped at the Santa Fe railroad crossing. Therefore, this vehicle is a school bus.

10 All cloudy days are depressing. Therefore, since I was depressed on Thursday, it was cloudy then.

11 If two angles of a triangle are congruent, then the sides opposite them are congruent. In $\triangle ABC$, $\angle A \cong \angle B$. Therefore, in $\triangle ABC$, $\overline{BC} \cong \overline{AC}$.

Section 1.7 Problem Set C

12 Given the following five statements:

1 Spoof is the set of all purrs.
2 Spoof contains at least two distinct purrs.
3 Every lilt is a set of purrs and contains at least two distinct purrs.
4 If A and B are any two distinct purrs, there is one and only one lilt that contains them.
5 No lilt contains all the purrs.

a Show each of the following:

i There is at least one lilt.
ii There are at least three purrs.
iii There are at least three lilts.

b If lilt "girt" contains purr "pil" and purr "til," and lilt "mirt" contains purr "pil" and purr "til" then lilt "girt" is the same as lilt "mirt" except in one case. What is this case?

13 The Bronx Zoo has a green lizard, a red crocodile, and a purple monkey. They are the only animals of their kind in existence. One violently windy Saturday, their name tags blew off, and their scrapbook was torn to shreds. Inasmuch as they were to appear on television at 7:30 Sunday morning, the night watchman had to replace their name tags. He managed to piece together the following information from the mangled scrapbook:

1 Wendy could not get along with the lizard.
2 Katie playfully took a bite out of the monkey's ear one month ago.
3 Wendy never cast a red reflection in the mirror.
4 Jody had the personality of a crocodile, but she wasn't one.

Match up the animals with their names.

Section 1.8 Probability Problems

A knowledge of probability is obviously important to an insurance company, to a card player or a backgammon expert, or to an operator of a gambling casino. Moreover, setting up and solving probability problems demands precision and requires organized, ordered thinking needed by secretaries, accountants, doctors, filing clerks, computer programmers, and geometry students.

While probability problems will *NOT* be a major part of this textbook, we shall occasionally place some probability problems in the Problem Sets.

We now introduce TWO BASIC STEPS FOR PROBABILITY PROBLEMS.

TWO BASIC STEPS FOR PROBABILITY PROBLEMS

1 **Determine ALL POSSIBILITIES in a logical manner. Count them.**

2 **Determine the number of these possibilities that are "favorable." We shall call these WINNERS.**

Probability is calculated by the following formula.

$$\textbf{PROBABILITY} = \frac{\textbf{Number of WINNERS}}{\textbf{Number of POSSIBILITIES}}$$

Section 1.8 Sample Problems

1 **If one of the four points is picked at random, what is the probability that the point lies on the angle?**

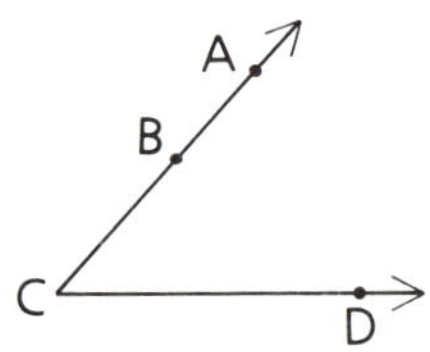

Solution: **We follow the TWO BASIC STEPS by listing ALL POSSIBILITIES and circling the WINNERS.**

Ⓐ Ⓑ Ⓒ Ⓓ

$$\frac{\text{WINNERS}}{\text{ALL POSSIBILITIES}} = \frac{4}{4} = 1$$

2 If two of the four points are selected at random, what is the probability that *both* will lie on $\overrightarrow{CA}$?

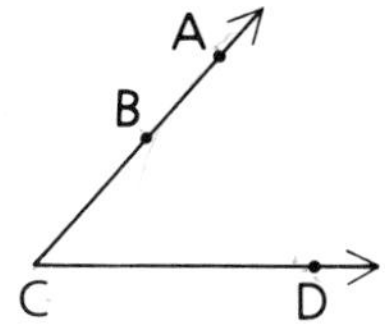

Solution: We follow the TWO BASIC STEPS by listing ALL POSSIBILITIES and circling the WINNERS.

Notice how we have attempted to list the possibilities in an orderly manner.

AB	**BC**	CD
AC	BD	
AD		

$$\frac{\text{WINNERS}}{\text{ALL POSSIBILITIES}} = \frac{3}{6} = \frac{1}{2}$$

3 If three of the four points are selected at random, what is the probability that the *ordered* selection will correctly name the angle shown?

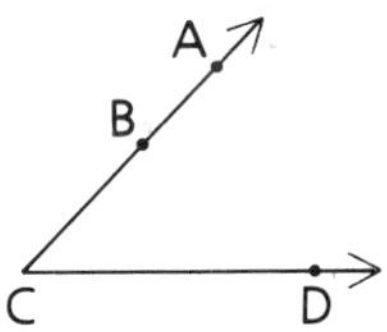

Solution: We follow the TWO BASIC STEPS by listing ALL POSSIBILITIES and circling the WINNERS. This problem is harder than either of the first two examples because the *order* of the points is important. Notice how we have listed the possibilities in an orderly manner.

ABC	BAC	CAB	DAB
ABD	BAD	CAD	DAC
ACB	BCA	CBA	DBA
ACD	**BCD**	CBD	DBC
ADB	BDA	CDA	**DCA**
ADC	BDC	CDB	**DCB**

$$\frac{\text{WINNERS}}{\text{ALL POSSIBILITIES}} = \frac{4}{24} = \frac{1}{6}$$

Section 1.8 Problem Set A

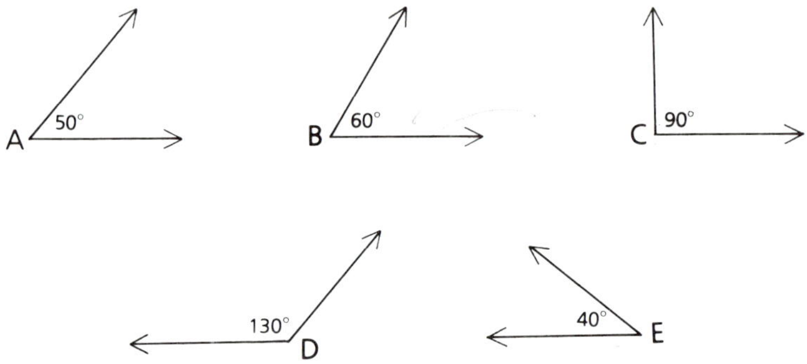

1 If one of the five angles is selected at random, what is the probability that the angle is acute?

2 If one of the five angles is selected at random, what is the probability that the angle is right?

3 If one of the five angles is selected at random, what is the probability that the angle is obtuse?

4 If one of the five angles is selected at random, what is the probability that the angle is straight?

Section 1.8 Problem Set B

For **5–8** use the five angles given in Problem Set A.

5 If two of the five angles are selected at random, what is the probability that they will *both* be acute?

6 If two of the five angles are selected at random, what is the probability that one of them is obtuse?

7 If two of the five angles are selected at random, what is the probability that one is right and the other is obtuse?

8 An angle is drawn at random from the five angles and then replaced. A second drawing is then made at random. (You might draw the same angle again.) What is the probability that *both* angles were acute?

Section 1.8 Problem Set C

9 Using the five angles given in Problem Set A, if two angles are selected at random, what is the probability that neither angle is acute?

10 Given: the 4 points to the right and their labels A, B, C, and D such that A and two of the other points are always collinear while the four points are always non-collinear. In how many ways that are different can the diagram be labeled?

11 Given points A, B, C, D, and E as shown.

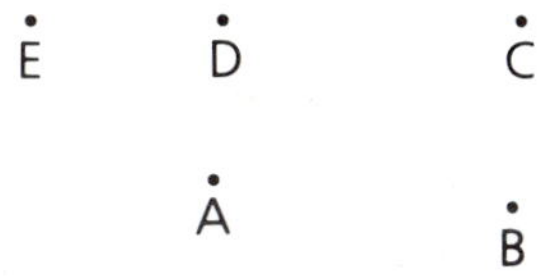

a If two of these points are selected at random, what is the probability that they will be collinear?

b If three of these points are selected at random, what is the probability that they will be collinear?

c If four of these points are selected at random, what is the probability that they will be collinear?

Chapter 1 Study Guide

At the end of each chapter, there is a CHAPTER STUDY GUIDE that will alert you to the essential concepts of the chapter. The guide is written as a list of objectives, or goals. As you study and review this chapter, be sure that you are able to:

1 Describe undefined terms such as: points, lines, and betweenness of points. (pp. 3, 18)

2 Describe each of the following terms:

- **a** segments (p.4)
- **b** rays (p. 4)
- **c** triangles (p.5)
- **d** perimeter (p. 8)
- **e** measures of segments and angles (p. 9)
- **f** tick marks (p. 13)

3 Describe the following terms from algebra:

- **a** intersection (p. 6)
- **b** union (p. 6)
- **c** empty set (p. 7)

4 Write complete definitions of the following terms:

- **a** angle (p. 4)
- **b** acute angle (p. 11)
- **c** right angle (p. 11)
- **d** obtuse angle (p. 11)
- **e** straight angle (p. 11)
- **f** congruent segments and angles (p. 12)
- **g** collinear and non-collinear points (p. 18)
- **h** midpoints and trisection points (p. 28)
- **i** bisection of a segment or angle (p. 28)
- **j** trisection of a segment or angle (p. 29)

5 State the two theorems presented in this chapter. (pp. 23, 24)

6 Explain whether definitions are always reversible. (p. 40) Are theorems and postulates always reversible?

7 Solve the following types of problems:

- **a** linear algebraic equations (p. 8)
- **b** paragraph proofs (p. 36)
- **c** clock problems (p. 14)
- **d** two column proofs (p. 23)
- **e** computation with degrees, minutes, and seconds (pp. 11, 12)

8 Describe what you can, and cannot, assume from a diagram. (p. 19)

9 Describe the two steps that are used to solve probability problems. (p. 44)

Chapter 1 Review Problems A

1 a Name the line containing A,E,F,C in all possible ways.

b Name the sides of ∠ABC.

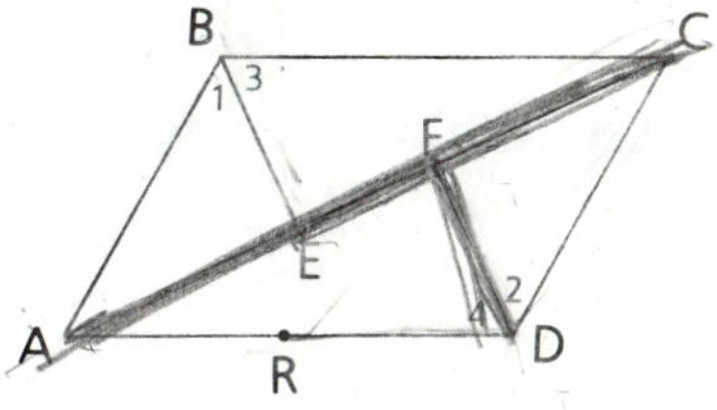

c R is between what two points?

d What side do ∠2 and ∠4 have in common?

e Name the horizontal ray with its endpoint at C.

f Estimate the size of:

1 ∠BAD

2 ∠2

3 ∠ABC

g What angle of △AFD can be named by using a number?

h Are angles FCD and DCE different angles?

i Which angle in the problem is ∠B?

j $\overline{AE} \cap \overline{BE} =$ __?__

k $\overrightarrow{EC} \cap \overrightarrow{FA} =$ __?__

l $\overrightarrow{BA} \cup \overrightarrow{BE} =$ __?__

m $\overleftrightarrow{AC} \cap \overleftrightarrow{DR} =$ __?__

n $\angle AFD \cap \overline{CE} =$ __?__

2 Estimate the following angles as acute, right, obtuse, or straight.

a ∠H

b ∠G

c ∠GFE

d ∠DEF

e ∠HDF

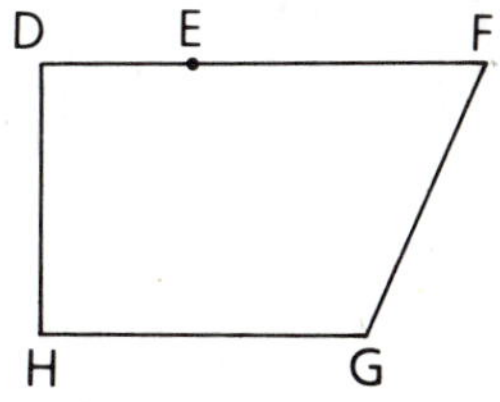

f Which angle may we assume has the size it appears to have?

3 $43°15'17'' + 25°49'18'' =$ ___?___

4 $90° - 39°17'' =$ ___?___

5 a Which two segments must be congruent?

b Which two angles must be congruent?

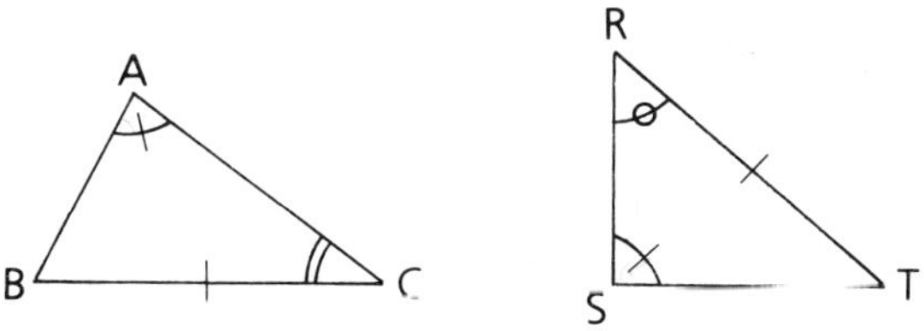

6 a If $\angle EFG$ is obtuse and $\angle HJK$ is right, is $\angle 1 \cong \angle 2$?

b If $\angle EFG \cong \angle HJK$, is $\angle 1 \cong \angle 2$?

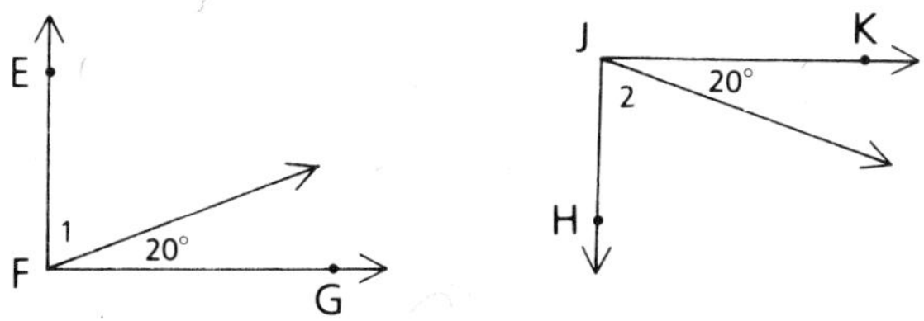

7 If $\angle A \cong \angle B$, find $m\angle A$.

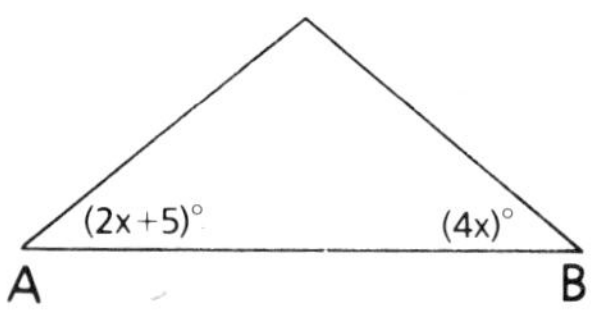

8 $m\angle 1 = x°$
$m\angle 2 = (3x)°$
$m\angle 3 = (2x)°$
Find the measure of each angle.

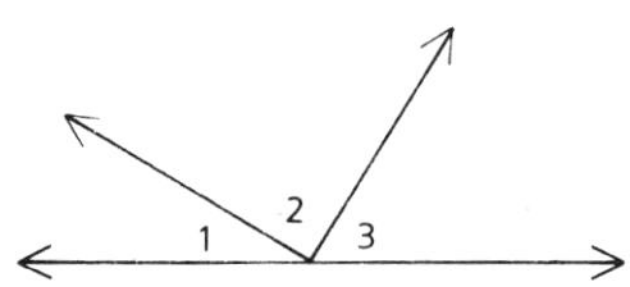

9 Is it possible for $\angle NOR$ and $\angle POS$ both to be right angles?

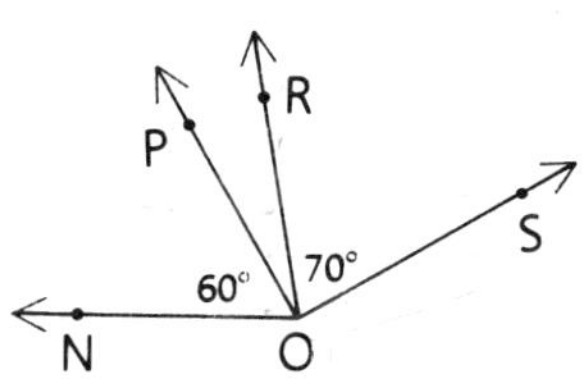

For **10, 11** draw each figure and write the "given" and "prove" information. Then complete each proof by filling in the missing reasons.

10 **Given:** $\angle ABC = 130°$
$\angle ABD = 60°$

Prove: $\angle DBC$ **is acute**

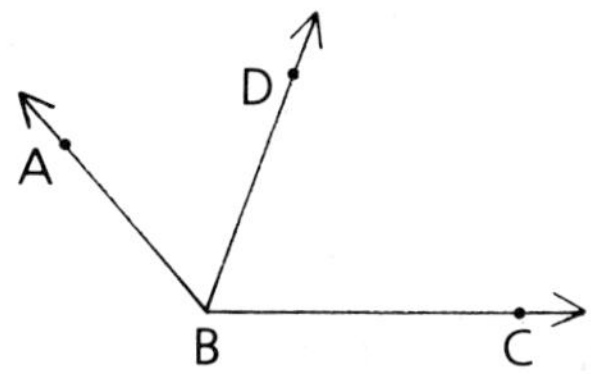

Statements	Reasons
1 $\angle ABC = 130°$	1 ______
2 $\angle ABD = 60°$	2 ______
3 $\angle DBC = 70°$	3 ______
4 $\angle DBC$ is acute	4 ______

11 **Given: Diagram as shown**

Prove: $\angle EFG \cong \angle HJK$

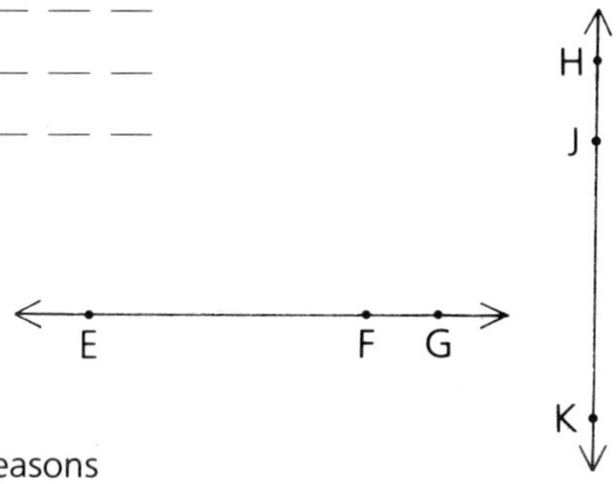

Statements	Reasons
1 Diagram as shown	1 ______
2 $\angle EFG$ is a straight angle	2 ______
3 $\angle HJK$ is a straight angle	3 ______
4 $\angle EFG \cong \angle HJK$	4 ______

For **12–14** prove each problem in two column form.

12 **Given:** $\angle X$ **is a right angle**
$\angle Y$ **is a right angle**

Prove: $\angle X \cong \angle Y$

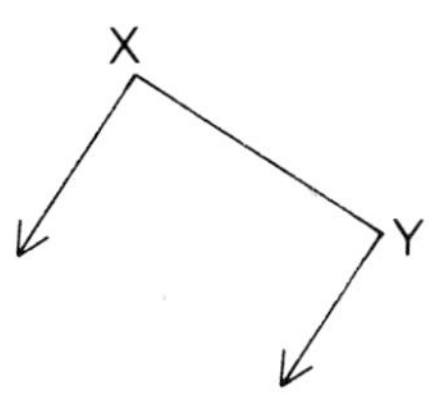

13 **Given:** $\overline{AB} \cong \overline{BC}$

Prove: B is the midpoint of $\overline{AC}$

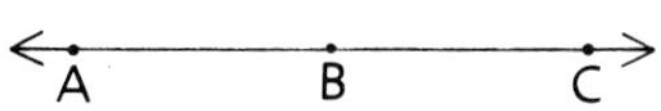

14 **Given:** $\overrightarrow{DF}$ **and** $\overrightarrow{DG}$ **trisect** $\angle EDH$

Conclusion: $\angle EDF \cong \angle FDG \cong \angle GDH$

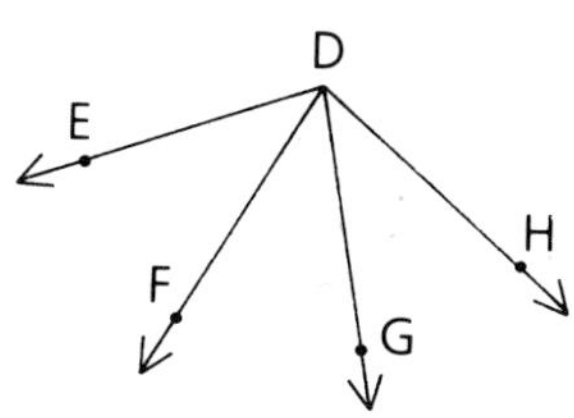

15 Given: $\overrightarrow{TW}$ bisects $\angle VTX$

Prove: $\angle VTW \cong \angle XTW$

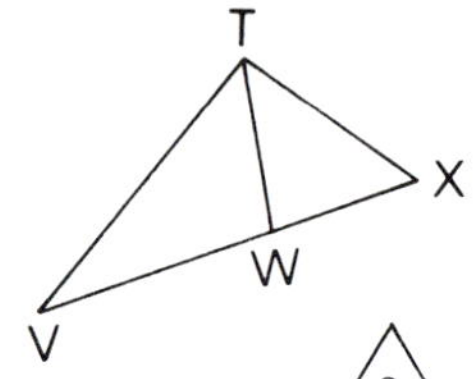

16 Prove in paragraph form:

Given: $\angle 1 = 61.6^\circ$

$\angle 2 = 61\frac{3}{5}^\circ$

Prove: $\angle 1 \cong \angle 2$

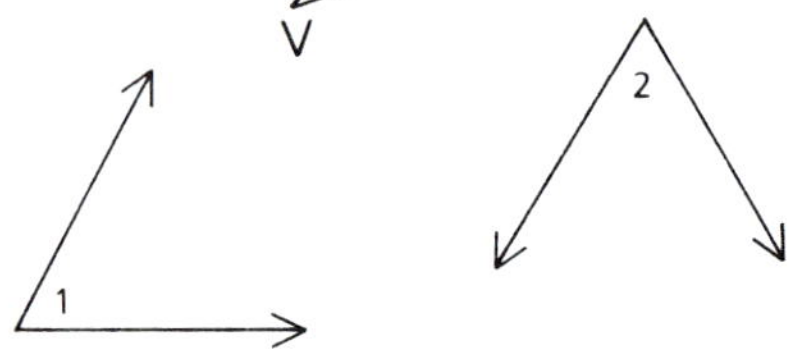

17 Prove in paragraph form:

Given: $\overline{AB}$ is x units long

$\overline{CD}$ is (x + 3) units long

Prove: $\overline{AB} \cong \overline{CD}$

18 $\overrightarrow{OR}$ and $\overrightarrow{OS}$ trisect $\angle TOP$

$\angle TOP = 40.2^\circ$

Find: m $\angle POR$

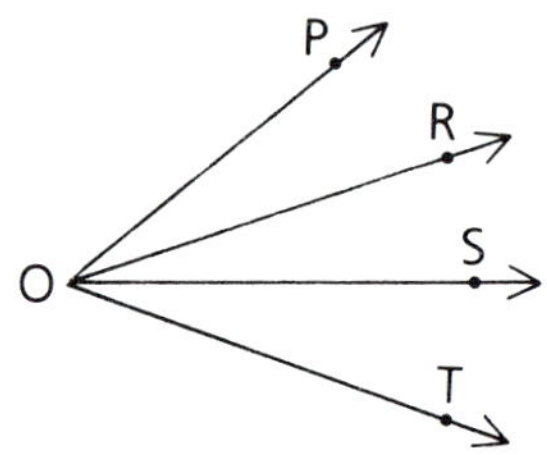

19 How large is the angle made by the hands of a clock at:

a 1:00 b 11:20

Chapter 1 Review Problems B

20 WY = 25

The ratio of WX to XY is 3:2,

that is, $\frac{WX}{XY} = \frac{3}{2}$

Find WX

21 Angle A is 6 more than twice as much as angle B. If their sum is 42°, find the measure of angle A.

22 The perimeter of PRST is 10 more than 5RS.

If PR = 26, find RS.

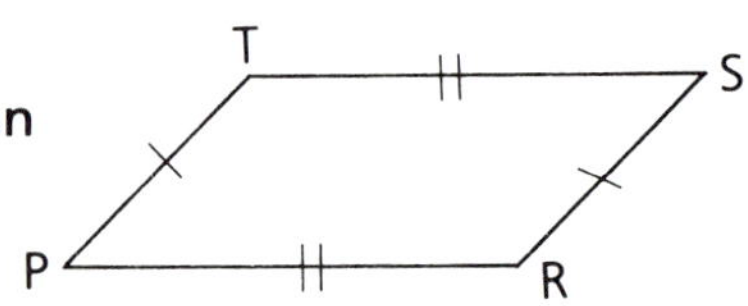

23 $\angle DEG = (x + 3y)^\circ$
$\angle GEF = (2x + y)^\circ$
$\angle DEF$ is a right $\angle$

a Solve for y in terms of x.

b If $\angle DEG \cong \angle GEF$, solve for x and y.

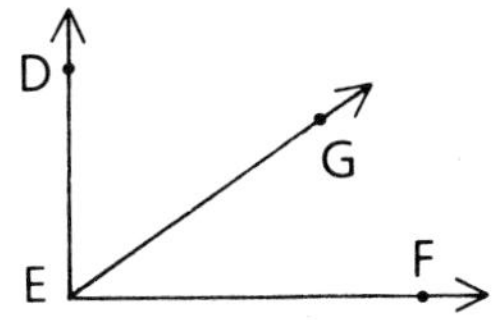

24 Prove in two column form:
Given: $\angle ABC$ is a right angle
$\angle DBC = 20^\circ$
$\angle FEG = 40^\circ$
$\angle GEH = 30^\circ$
Prove: $\angle ABD \cong \angle FEH$

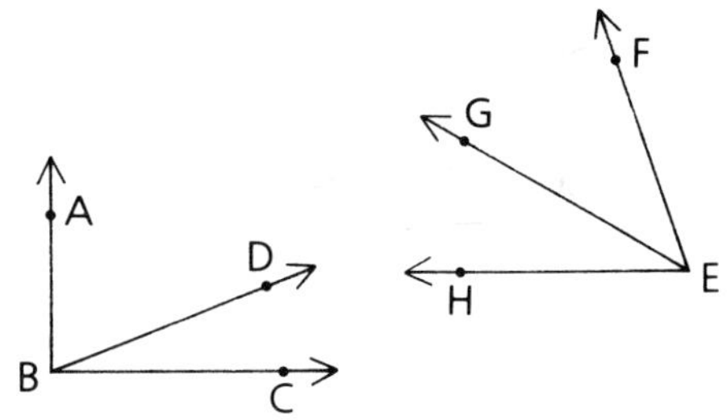

25 Prove in paragraph form:
Given: $\angle OMK = 50^\circ$
$\angle OKM = (2x)^\circ$
$\angle OKJ = (5x + 5)^\circ$
Conclusion: $\angle OKJ \cong \angle OMN$

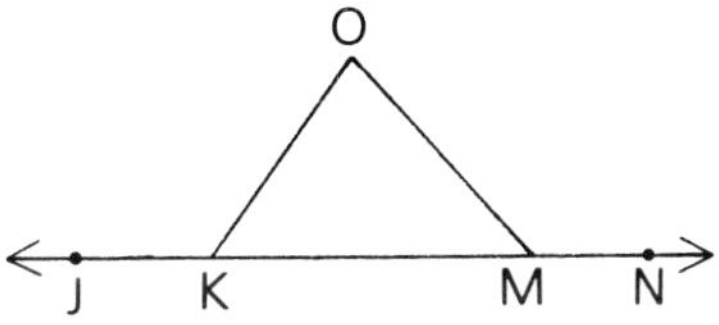

26 Find $m\angle 1$

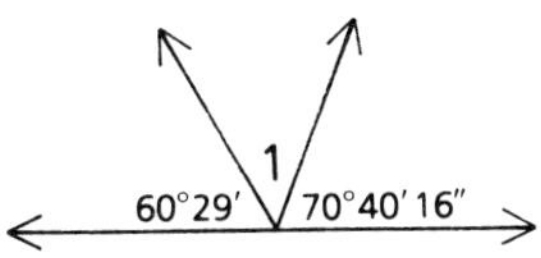

Chapter 1 Review Problems C

27 The length of $\overline{RS}$ measured in centimeters is 198 more than $\overline{RS}$ measured in meters. Find the length of $\overline{RS}$ in meters.

28 Given: $\angle ABC$ is a right angle
$\angle DBE$ is a right angle

Prove: $\angle ABD \cong \angle CBE$
(Use the paragraph style)

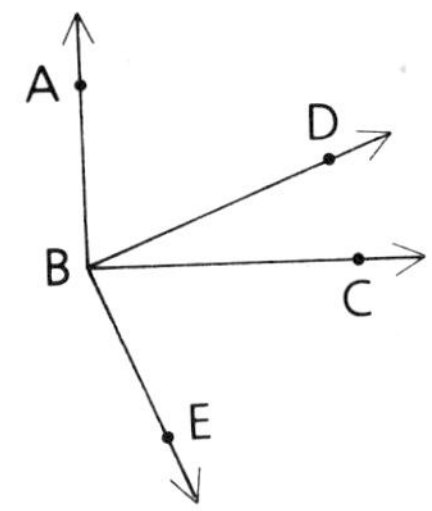

29 Draw a diagram in which $\overleftrightarrow{AB}$ and $\overleftrightarrow{CD}$ intersect at E, but in which $\angle AEC$ does not appear to be congruent to $\angle DEB$.

30 Given: a green and gray silk bag that contains 4 acute angles, 3 right angles, 2 obtuse angles, and 1 straight angle

a if an angle is drawn at random, what is the probability that the angle is a right angle?

b if two angles are drawn at random, what is the probability that both are acute?

c if two angles are drawn at random, what is the probability that one is obtuse and the other is right?

d if one angle is drawn at random, then put back in the bag, and a second drawing made, what is the probability that both angles drawn are acute?

e if three angles are drawn at random, and Bart Mackeriv offers to bet you a thousand dollars that at least one of them is obtuse, should you accept the bet?

31 At 3:00 the hands of a clock form an angle of 90°. To the nearest second, at what time will the hands of the clock next form a 90° angle?

Chapter 1 Review Problems D

32 Given 6 points on a sheet of paper such that any four of them are non-collinear.

a What is the maximum number of lines determined?

b What is the minimum number of lines determined?

33 $\angle A = (x^2 + 3y)^\circ$
$\angle D = (2y + 5)^\circ$
$\angle DBA = (2x + 3y)^\circ$
$\angle DBC = (6y - 2x)^\circ$

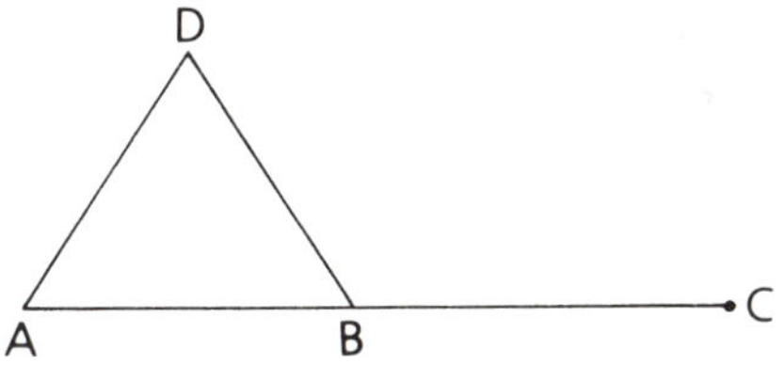

If one assumes the three angles of a triangle total 180°, find:

$m\angle A$ $m\angle D$ $m\angle DBA$ $m\angle DBC$

34 To the nearest second, at what time after 2:00 will the hands of a clock form an angle $2\frac{1}{2}$ times as large as their angle at 2:00?

The Royal Road: Euclid and Geometry

The history of geometry is as old as the history of building, drawing, and surveying. One of the first people who organized the concepts of geometry was Euclid, a Greek mathematician who lived around 300 B.C. He arranged all the known properties of geometry as a logical progression of axioms, postulates, and theorems (which Euclid called "propositions"). His work consisted of 13 books, called *The Elements*. Parts of *The Elements* were still being used as a basic textbook in the twentieth century!

Many of Euclid's proofs were quite difficult to follow. There is a story that Ptolemy I, king of Alexandria, asked Euclid if there was a secret to understanding geometry, a single key to learn all the definitions and proofs. Euclid's answer was: "There is no royal road to geometry!"

There is still no easy way to learn the terminology and skills of geometry. However, episodes from the history of geometry, and examples of different ways to study geometry, can be thought of as geometry's royal road!

Chapter 2

EARLY CONCEPTS AND PROOFS

A Look Back and a Look Ahead

If you feel somewhat confused at this time, you need not feel bad nor discouraged. Some confusion is inevitable at the start of geometry. There are new terms and ideas to learn, the art of proof may be new to many of you, and of course, you have forgotten some of the algebra needed to solve some of the geometry problems. Then too, memorizing isn't everyone's idea of amusement.

Be patient! Read carefully, study the sample problems closely, and the confusion will begin to go away. Remember, also, to see your teacher for help as you need it.

In Chapter 1, you studied many new ideas and problems. You also proved problems by means of TWO COLUMN PROOFS and PARAGRAPH PROOFS. We introduced you to DEDUCTIVE STRUCTURE, and you saw that with order and organization, PROBABILITY PROBLEMS can be interesting and their solutions found.

In Chapter 2, you will increase your knowledge and use of the **TWO COLUMN PROOF.** Again, there will be some **DEFINITIONS** and some **THEOREMS** to memorize and use. Toward the end of Chapter 2, the proofs will be a little longer and a little harder. Although you may not think so now, proofs only become enjoyable as they become more challenging.

Section 2.1 Perpendicularity

D **Lines, rays, or segments that intersect at right angles are PERPENDICULAR (⊥).**

Below are some examples of perpendicularity.

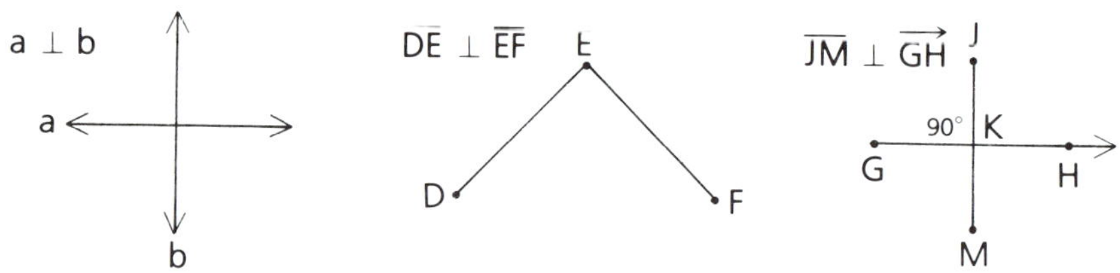

Perpendicularity, right angles, and 90° all go together.

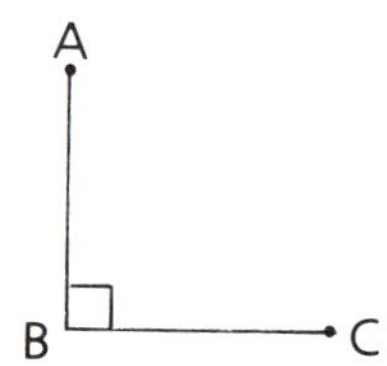

In the figure at the right,
the mark inside the angle (⌐) indicates
that $\angle B$ is a right angle.
It is also true that $\overline{AB} \perp \overline{BC}$ and $\angle B = 90°$.

DO NOT ASSUME PERPENDICULARITY FROM A DIAGRAM!

In $\triangle DEF$ it appears that $\overline{DE} \perp \overline{EF}$,
but we may *not* assume so.

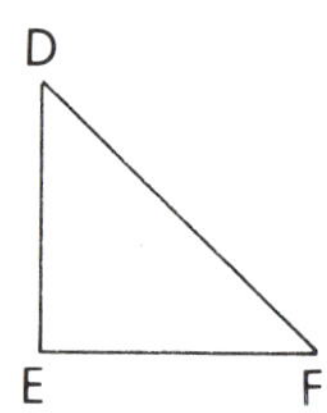

Section 2.1 Sample Problems

1 Given: $\overline{AB} \perp \overline{BC}$
$\overline{DC} \perp \overline{BC}$

Conclusion: $\angle B \cong \angle C$

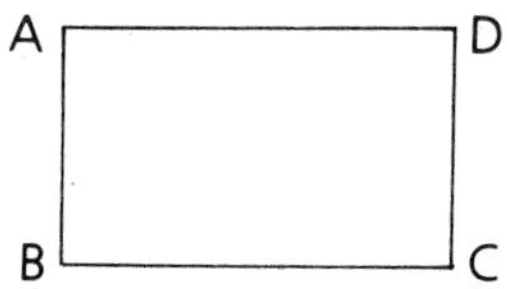

Statements	Reasons
1 $\overline{AB} \perp \overline{BC}$	1 Given
2 $\angle B$ is a right angle	2 If two segments are $\perp$, they form a right angle
3 $\overline{DC} \perp \overline{BC}$	3 Given
4 $\angle C$ is a right angle	4 Same as 2
5 $\angle B \cong \angle C$	5 If angles are right angles, they are $\cong$

2 Given: $\overleftrightarrow{EH} \perp \overleftrightarrow{HG}$

Name all the right angles.

Ans. $\angle EHG$ (Not $\angle EFH$ nor $\angle HFG$)

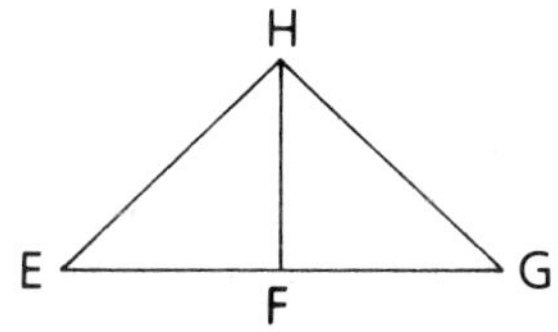

3 Given: $\overrightarrow{KJ} \perp \overrightarrow{KM}$
$\angle JKO$ is 4 times as large as $\angle MKO$

Find: $m\angle JKO$

Solution: Let $m\angle MKO = x$

$$m\angle JKO = 4x$$
$$4x + x = 90$$
$$5x = 90$$
$$x = 18$$

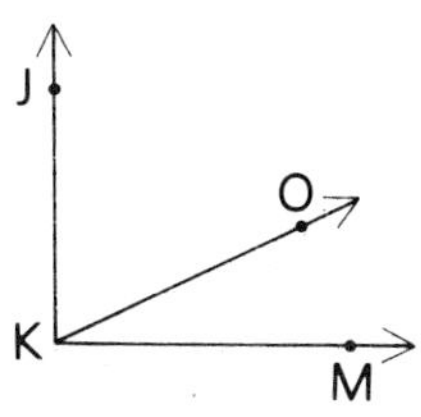

Substituting 18 for x, we get $m\angle JKO = 72$

4 Given: $a \perp b$
$c \not\perp d$ (c is not $\perp$ to d)

Conclusion: $\angle 1 \cong \angle 2$

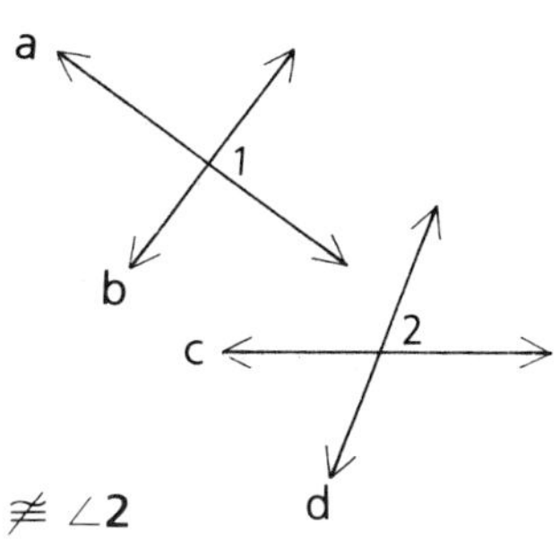

Discussion: This conclusion is FALSE.
Since $a \perp b$, $\angle 1 = 90°$. Since $c \not\perp d$, $\angle 2 \neq 90°$.
Since $\angle 1$ and $\angle 2$ have different measures, $\angle 1 \not\cong \angle 2$

Section 2.1 Problem Set A

1 Name all the angles in the figures to the right that *appear* to be right angles.

a

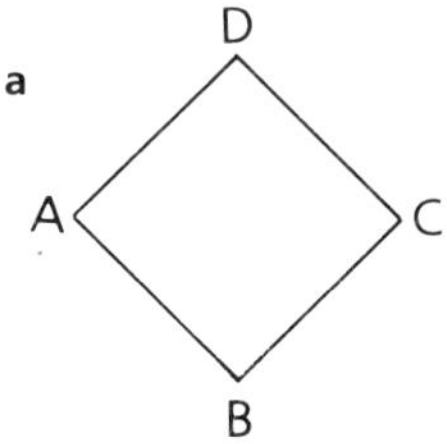

b

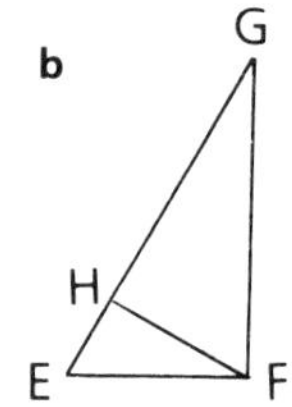

2 In each of the following, name the angles that can be proved to be right angles.

a

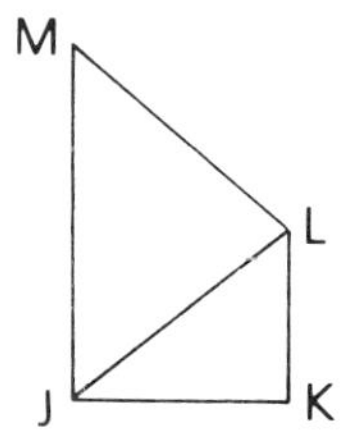

Given: $\overline{JM} \perp \overline{JK}$

b

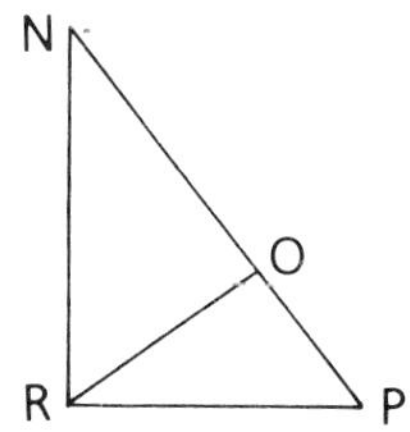

Given: $\overrightarrow{RO} \perp \overrightarrow{PN}$

c

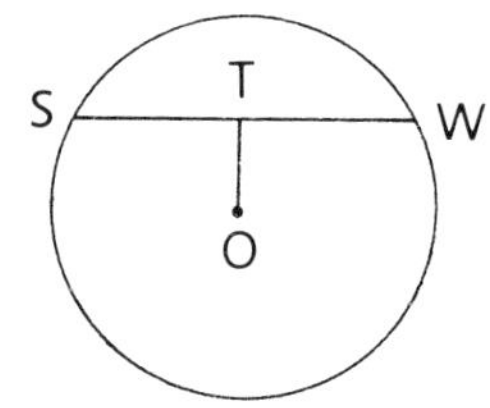

Given: $\overline{OT} \perp \overline{SW}$

3 In each of the following, find the measure of $\angle 1$.

a

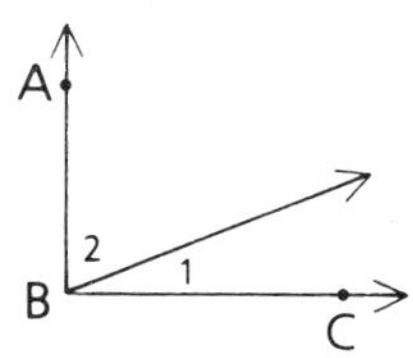

Given: $\overline{AB} \perp \overline{BC}$

$\angle 2 = 70^\circ$

b

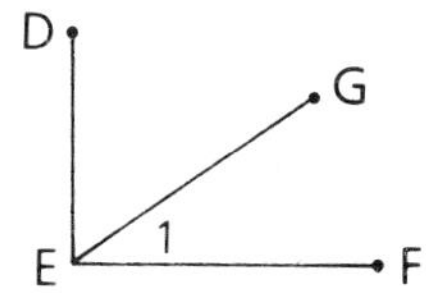

Given: $\overleftrightarrow{DE} \perp \overleftrightarrow{EF}$

$\overrightarrow{EG}$ bisects $\angle DEF$

c Given: $\overline{TR} \perp \overleftrightarrow{PS}$

$\angle PRV = 68^\circ 17' 34''$

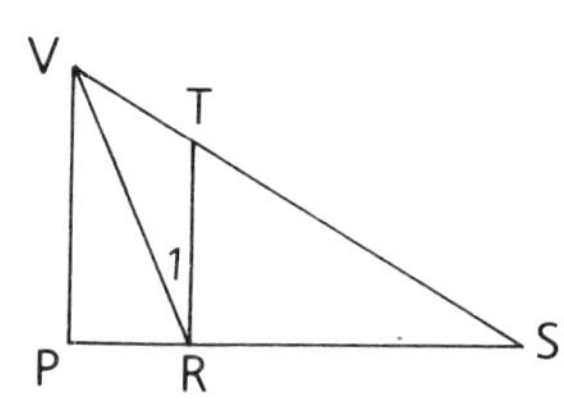

4 $\angle 1$ is five times as large as $\angle 2$

Find: $m\angle 1$

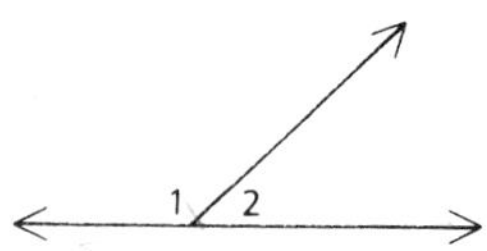

5 Given: $a \perp b$

Prove: $\angle 1 \cong \angle 2$

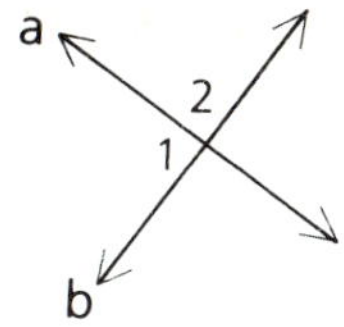

6 Given: $\angle ACB = 90^\circ$

$\overline{AD} \perp \overline{BD}$

Prove: $\angle C \cong \angle D$

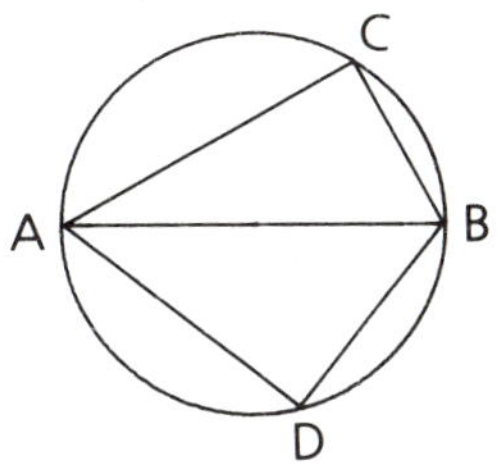

7 Given: $\overline{EF} \perp \overline{FG}$

Conclusion: $\angle F \cong \angle H$

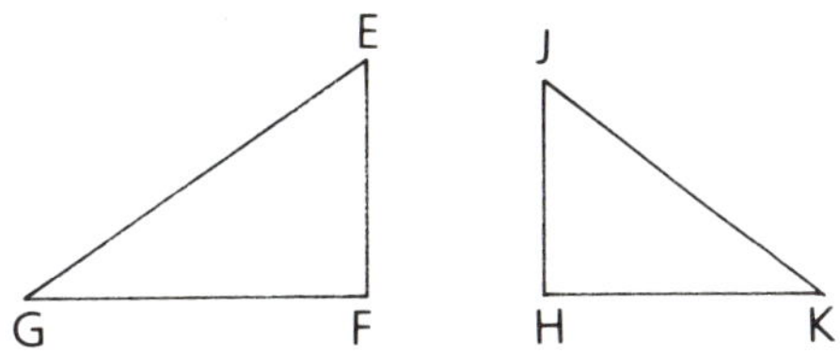

What is wrong with the following paragraph proof?

Proof: Since $\overline{EF} \perp \overline{FG}$, $\angle F$ is a right angle. Similarly, $\angle H$ is a right angle. Since $\angle F$ and $\angle H$ are both right angles, they are congruent.

8 Given: $\angle MOR = (3x + 7)^\circ$

$\angle ROP = (4x - 1)^\circ$

$\overline{MO} \perp \overline{OP}$

Which angle is larger, $\angle MOR$ or $\angle ROP$?

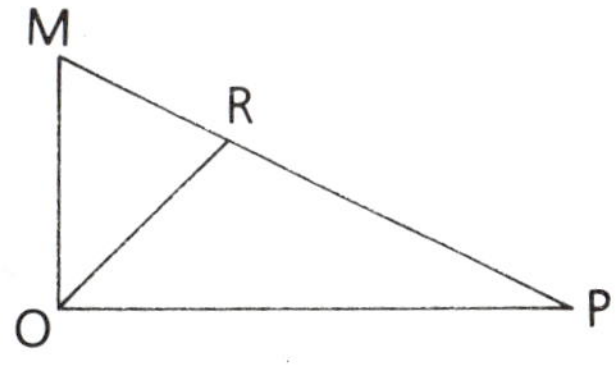

Section 2.1 Problem Set B

9 $\overleftrightarrow{AB} \perp \overleftrightarrow{BC}$

Angles 1, 2, 3 are in the ratio of 1:2:3 respectively. Find the measure of each angle.

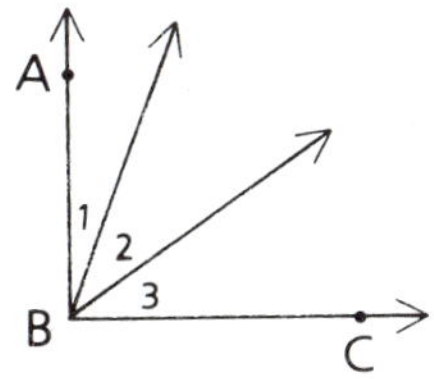

10 Line DE is perpendicular to line EF. The resulting angle is trisected, then one of the new angles is bisected, and then one of those angles is trisected. How large is the smallest angle?

11 Use a paragraph proof:

Given: $\angle HGJ = 37°20'$

$\angle KGJ = 52°40'$

$\overline{KJ} \perp \overline{HJ}$

Conclusion: $\angle HGK \cong \angle HJK$

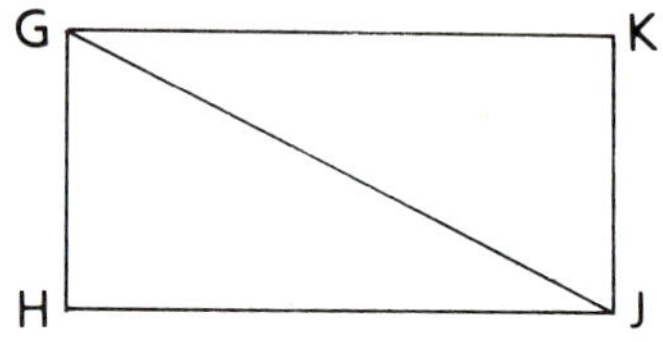

12 When two lines intersect and are not perpendicular, they are called OBLIQUE LINES. Write a paragraph proof of:

Given: $\angle AEC = (2x + 5)°$

$\angle AED = (3x - 35)°$

Prove: $\overleftrightarrow{AB}$ and $\overleftrightarrow{CD}$ are oblique

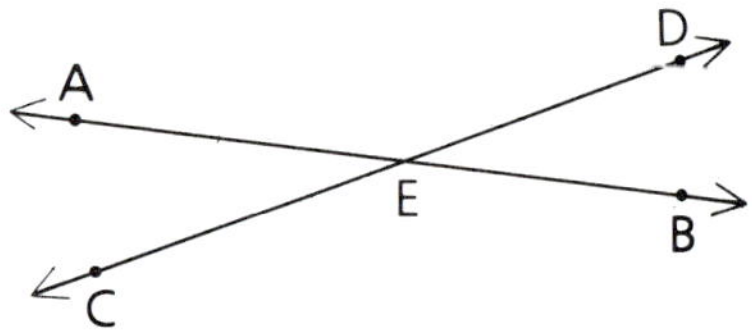

Section 2.1 Problem Set C

13 $\overleftrightarrow{AB} \perp \overline{AC}$ and $\overrightarrow{AD} \perp \overline{AC}$. Explain how this could happen in either of two different ways.

14 Given: $\overline{AB} \perp \overline{BC}$

$\angle ABO = (2x + y)°$

$\angle OBC = (6x + 8)°$

$\angle AOB = (23y + 90)°$

$\angle BOC = (4x + 4)°$

Find: $m\angle ABO$

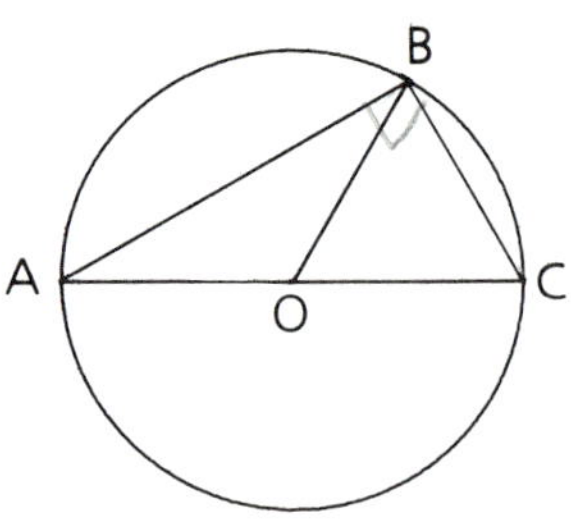

15 Find the measure of the angle formed by the hands of a clock at 5:55 AM.

Section 2.2 Complementary and Supplementary Angles

D **COMPLEMENTARY angles are two angles whose sum is 90° (or a right angle). Each of the two angles is the COMPLEMENT of the other.**

Below are examples of pairs of complementary angles:

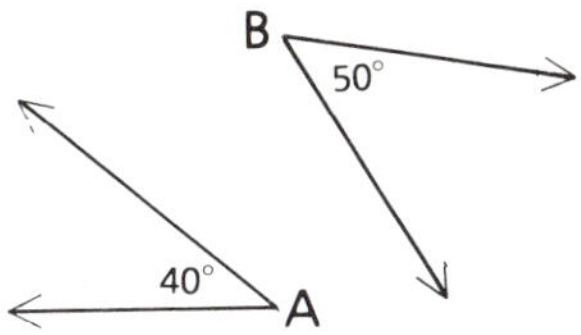

∠A and ∠B are complementary

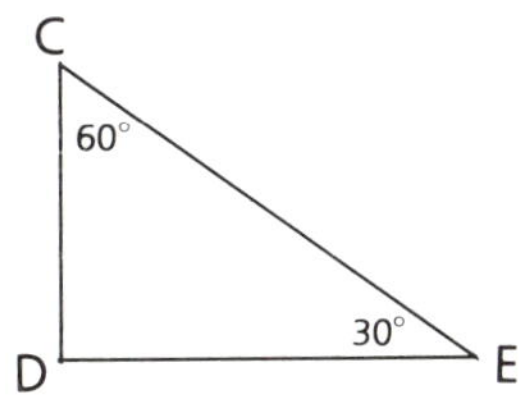

∠C is comp. to ∠E

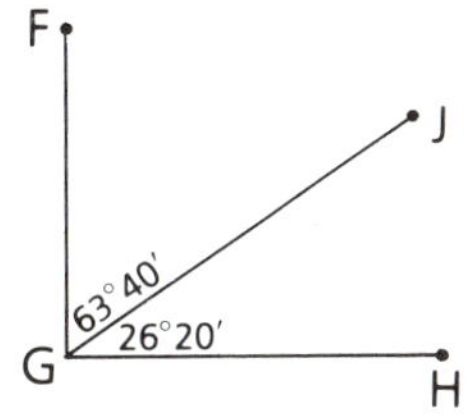

∠FGJ is the comp. of ∠JGH

In the first diagram, ∠A is the complement of ∠B. Also, ∠B is the complement of ∠A. In the third diagram, you can see that the two complementary angles together form a right angle, as sometimes will happen.

D **SUPPLEMENTARY angles are two angles whose sum is 180° (or a straight angle). Each of the two angles is the SUPPLEMENT of the other.**

Below are examples of pairs of supplementary angles.

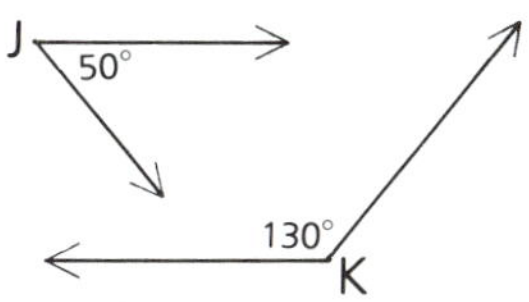

∠J and ∠K are supplementary

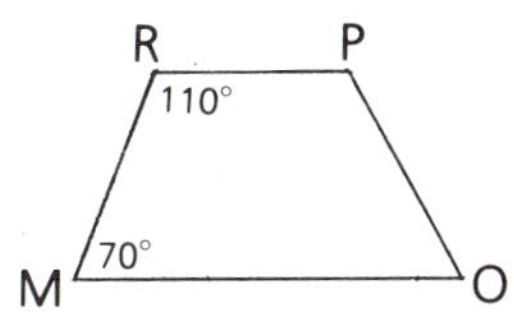

∠M is supp. to ∠R

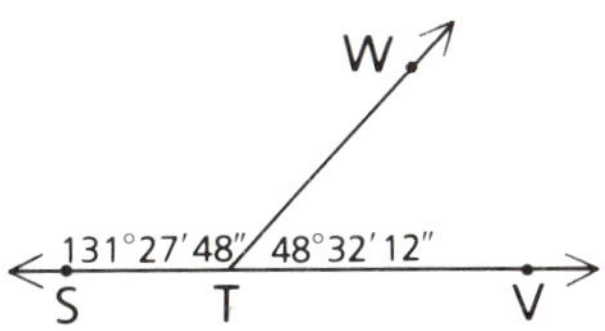

∠STW is the supp. of ∠WTV

In the first diagram, ∠J is the supplement of ∠K, and vice versa. In the middle diagram, which angle is the supplement of ∠M?

Sometimes, two supplementary angles will together form a straight angle. See if you can verify that ∠STW + ∠WTV = 180°.

Section 2.2 Sample Problems

1 **Given:** $\angle TVK$ **is a right** $\angle$

Prove: $\angle 1$ **is comp. to** $\angle 2$

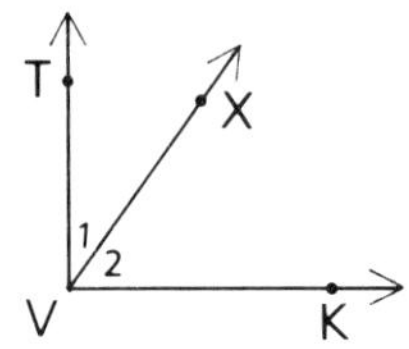

Statements	Reasons
1 $\angle TVK$ is a right $\angle$	1 Given
2 $\angle 1$ is comp. to $\angle 2$	2 If the sum of two angles is a right angle, they are comp.

2 **Given:** **Diagram as shown**

Conclusion: $\angle 1$ **is supp. to** $\angle 2$

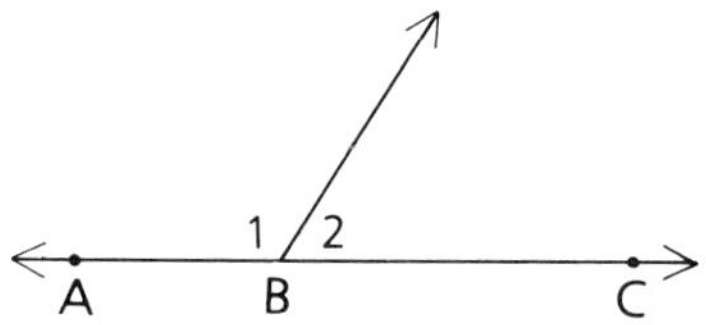

Statements	Reasons
1 Diagram as shown	1 Given
2 $\angle ABC$ is a straight angle	2 Assumed from diagram
3 $\angle 1$ is supp. to $\angle 2$	3 If the sum of two angles forms a straight angle, they are supp.

3 **One of two complementary angles is three more than twice the other. Find the measure of each.**

Solution: Let $x =$ measure of smaller angle

$2x + 3 =$ measure of larger angle

$x + 2x + 3 = 90$ (The sum of two comp. $\angle$s is 90°)

$3x + 3 = 90$

$3x = 87$

$x = 29$

$2x + 3 = 61$

The measures of the two angles are 29 and 61.

4 The measure of the supplement of an angle is 60 degrees less than 3 times the measure of the complement of the angle. Find the measure of the complement.

Solution: Let x = measure of the angle

$90 - x$ = measure of the complement (Do you know why?)

$180 - x$ = measure of the supplement (Do you know why?)

Note: This is a KEY SAMPLE PROBLEM. The definitions at the start of the solution (x, $90 - x$, $180 - x$) are used in many problems throughout the book.

$$\begin{aligned} 180 - x &= 3(90 - x) - 60 \\ 180 - x &= 270 - 3x - 60 \\ 180 - x &= 210 - 3x \\ 2x &= 30 \\ x &= 15 \\ 90 - x &= 75 \end{aligned}$$

The measure of the complement is 75.

Section 2.2 Problem Set A

1 Which two angles are complementary?

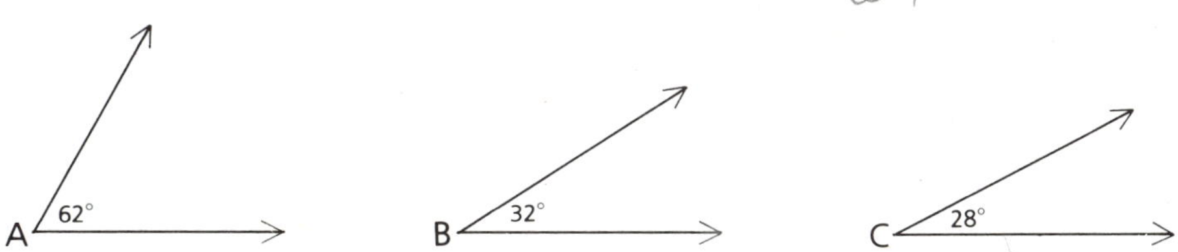

2 What is the supplement of a 70° angle?

3 $\angle 1$ is complementary to $\angle 3$. If $\angle 3 = y°$, how large is $\angle 1$?

4 Find the complement of a $61°21'13''$ angle.

5 One of two complementary angles is twice the other. Find the measures of the angles.

6 Draw the figure and state the *given* and *prove* information. Then complete the proof by filling in the missing statements.

Given: $\angle 1$ is comp. to $\angle 2$

Prove: $\overleftrightarrow{AB} \perp \overleftrightarrow{BC}$

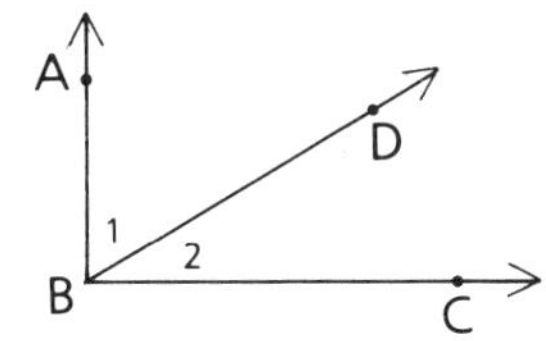

Statements	Reasons
1 __________	1 Given
2 __________	2 If a ray divides an $\angle$ into two comp. angles, then the original $\angle$ is a right $\angle$
3 __________	3 If two lines intersect to form a right $\angle$, the two lines are $\perp$

7 Given: $\overleftrightarrow{CD} \perp \overleftrightarrow{DE}$

Prove: $\angle CDF$ is comp. to $\angle FDE$

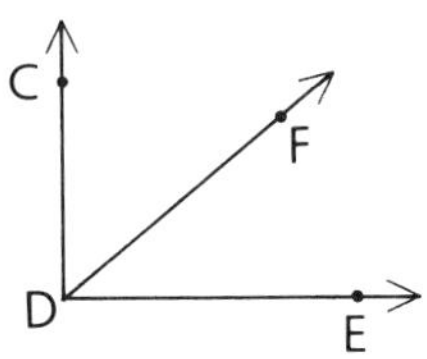

8 Given: Diagram as shown

Prove: $\angle GHK$ is supp. to $\angle KHJ$

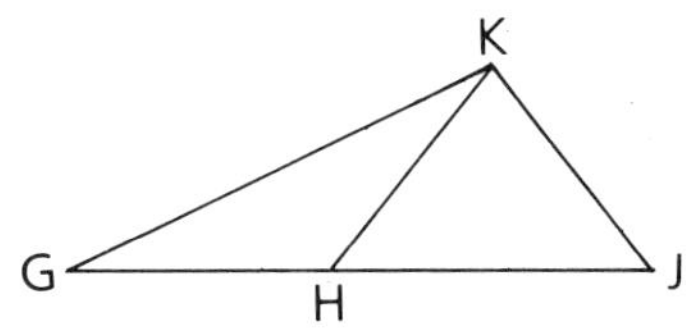

9 Given: $\angle MRO$ is comp. to $\angle PRO$

Prove: $\angle MRP$ is a right angle

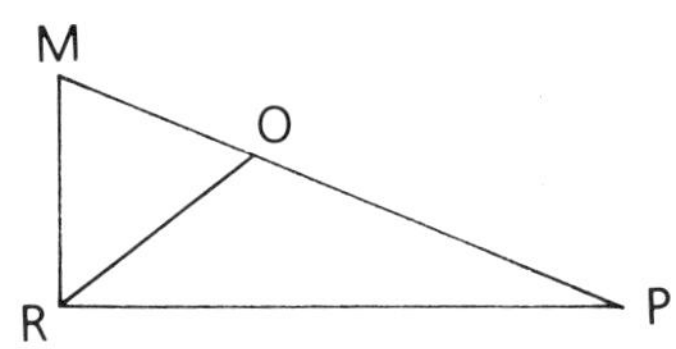

10 Find the measure of $\angle XVS$

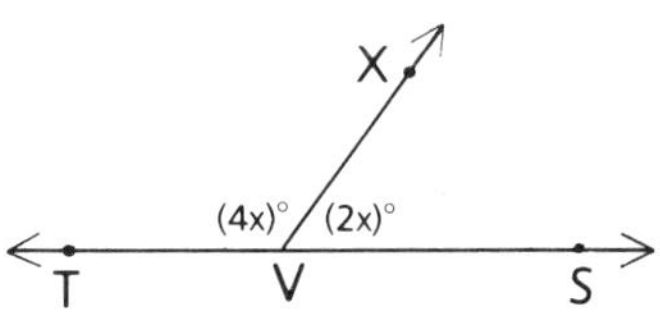

11 One of two supplementary angles is 70° more than the second. Find the measure of the larger angle.

Section 2.2 Problem Set B

12 Two supplementary angles are in the ratio of 11:7. Find the measure of each.

13 The larger of two supplementary angles exceeds 7 times the smaller by 4°. Find the measure of the larger angle.

14 Write a paragraph proof to show that ∠ABF is complementary to ∠EBD.

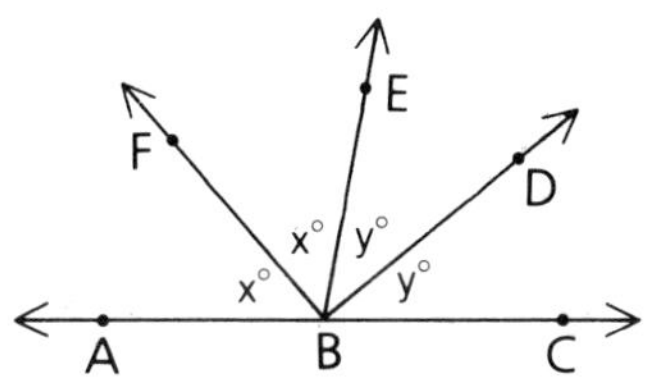

15 One of two complementary angles added to one-half of the other yields 72°. Find the measure of one-half of the larger.

16 Given: $\overline{XY} \perp \overline{YW}$
$\overline{AB} \perp \overline{BC}$

Find: m ∠DBC

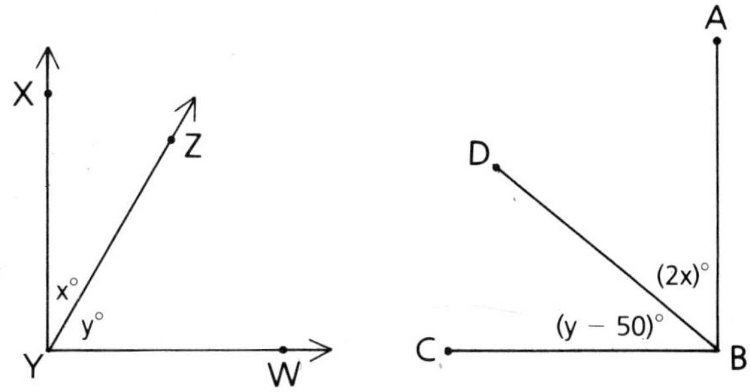

17 The supplement of an angle is 4 times the complement of the angle. Find the measure of the complement.

18 Five times the complement of an angle less twice the supplement is 40°. Find the measure of the supplement.

19 The measure of the supplement of an angle is 30° less than 5 times the measure of the complement. Find the measure of two-fifths of the complement.

20 Arnex has a 30°, a 60°, a 150°, a 45°, and a 135° angle in his pocket. He takes out two of the five angles.

Find the probability that—

a the angles are supplementary.

b the angles are complementary.

Section 2.2 Problem Set C

21 The supplement of an angle is 60 degrees less than twice the supplement of the complement of the angle. Find the measure of the complement.

22 Debbie has drawn distinct rays $\overrightarrow{BA}$, $\overrightarrow{BC}$, $\overrightarrow{BD}$, $\overrightarrow{BE}$, and $\overrightarrow{BF}$ on a piece of paper such that $\angle ABC$ is a straight angle.

a What is the minimum number of pairs of complementary angles that she could have drawn?

b What is the maximum number of pairs of complementary angles that she could have drawn?

c What is the minimum number of pairs of supplementary angles that she could have drawn?

d What is the maximum number of pairs of supplementary angles that she could have drawn?

Section 2.3 Drawing Conclusions

There wouldn't be much progress in this world if all we did was justify conclusions that someone else had already drawn. In a similar manner, you won't make much progress as a student this year if all you can do is justify conclusions the textbook has already stated. Although the following procedure may not work every time, it will be helpful to you in DRAWING CONCLUSIONS.

> PROCEDURE FOR DRAWING CONCLUSIONS
>
> **1 Memorize *theorems, definitions,* and *postulates.***
> **2 Look for KEY words (or symbols) in the given information.**
> **3 Think of all the *theorems, definitions,* and *postulates* we have had that use those KEYS.**
> **4 Decide which *theorem, definition,* or *postulate* is needed to draw a conclusion.**
> **5 Draw a conclusion and give a reason to justify the conclusion. Be certain that you have *not* used the *reverse* of the correct reason.**

■ Example:

Given: $\overrightarrow{AB}$ bisects $\angle CAD$

Conclusion: __?__

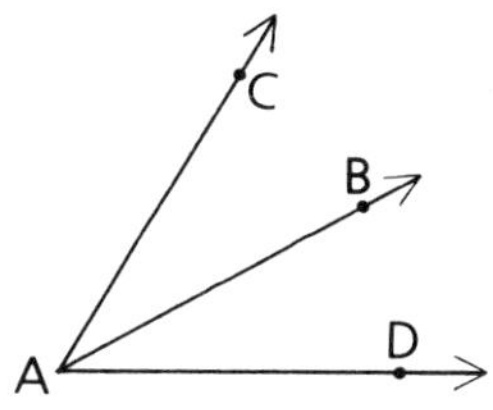

Thinking Process:

The KEY word is *bisects*.
The KEY symbols are ($\longrightarrow$) and ($\angle$).
Only the definition of bisector of an angle uses those KEYS.
The proper conclusion is that $\angle CAB \cong \angle DAB$.

Statements	Reasons
1 $\overrightarrow{AB}$ bisects $\angle CAD$	1 Given
2 $\angle CAB \cong \angle DAB$	2 If a ray bisects an $\angle$, then it divides the $\angle$ into two $\cong$ angles

Notice! The IF part of the reason matches GIVEN information and the THEN part matches the CONCLUSION being justified. *Be sure not to reverse that order.*

Section 2.3 Sample Problems

In **1–3** write a two column proof for each in which you supply your own correct conclusion and reason.

1 **Given:** $\angle$**A is a right angle**
$\angle$**B is a right angle**

Conclusion: ___?___

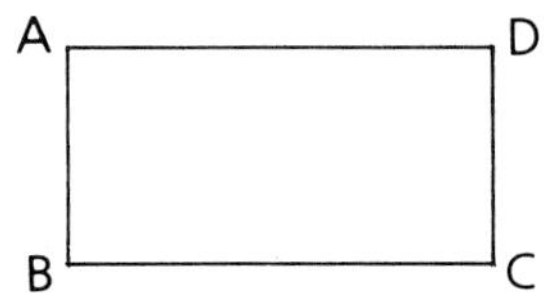

Statements	Reasons
1 $\angle$A is a right angle	1 Given
2 $\angle$B is a right angle	2 Given
3 $\angle A \cong \angle B$	3 If two $\angle$s are right $\angle$s, then they are $\cong$

2 **Given:** **E is the midpoint of** $\overline{SG}$

Conclusion: ___?___

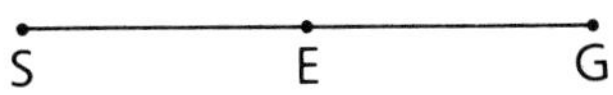

Statements	Reasons
1 E is the midpoint of $\overline{SG}$	1 Given
2 $\overline{SE} \cong \overline{EG}$	2 If a point is the midpoint of a segment, the point divides the segment into two $\cong$ segments

3 **Given:** $\angle$**PRS is a right angle**

Conclusion: ___?___

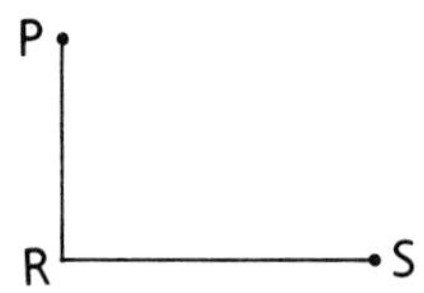

Statements	Reasons
1 $\angle$PRS is a right $\angle$	1 Given
2 $\overleftrightarrow{PR} \perp \overleftrightarrow{RS}$	2 If two lines intersect to form a right $\angle$, they are $\perp$

In sample problem 3, we could have drawn a different conclusion. Do you know what the other conclusion is?

Section 2.3 Problem Set A

In **1–7** write a two column proof for each in which you supply your own correct conclusion and reason.

1 Given: $\overleftrightarrow{AB} \perp \overleftrightarrow{BC}$

Conclusion: ___?___

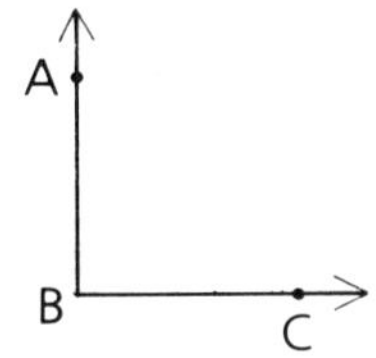

2 Given: $\angle DEF$ is comp. to $\angle HEF$

Conclusion: ___?___

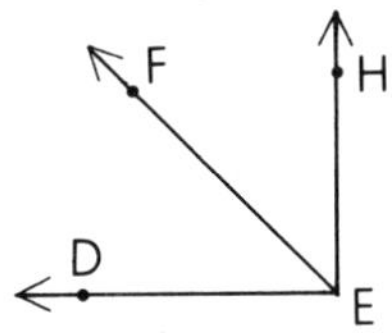

3 Given: $\angle WXZ \cong \angle YXZ$

Conclusion: ___?___

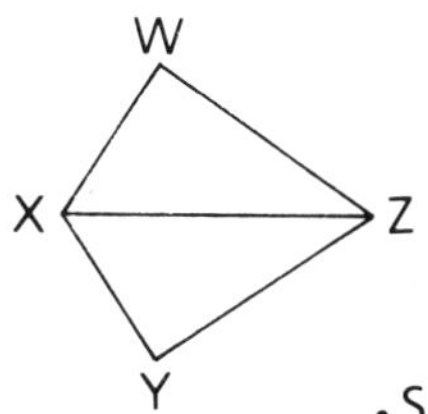

4 Given: $\overrightarrow{QS}$ and $\overrightarrow{QT}$ trisect $\angle PQR$

Conclusion: ___?___

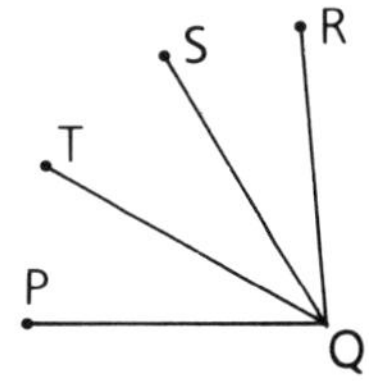

5 Given: E is the midpoint of $\overline{AC}$

Conclusion: ___?___

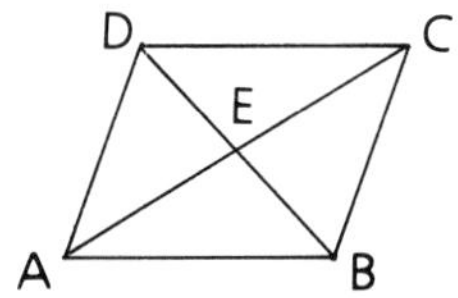

6 Given: A and R trisect $\overline{CD}$

Conclusion: ___?___

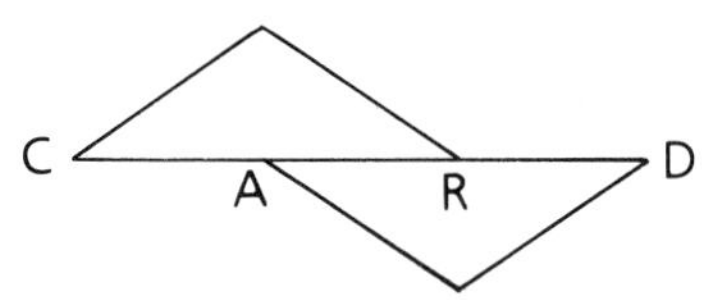

7 Given: Diagram as shown

Conclusion: ___?___

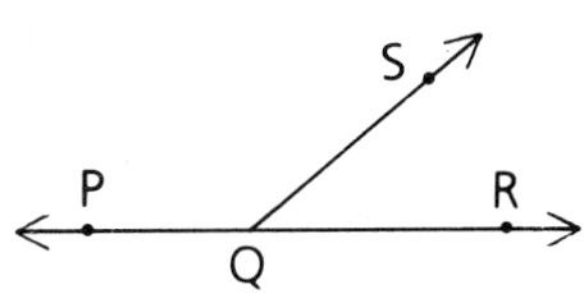

Section 2.3 Problem Set B

In **8–12** draw at least two conclusions for each and give reasons for them in two column proof form.

8 Given: $\overleftrightarrow{WZ}$ bisects $\overline{VY}$

Conclusions: ___?___

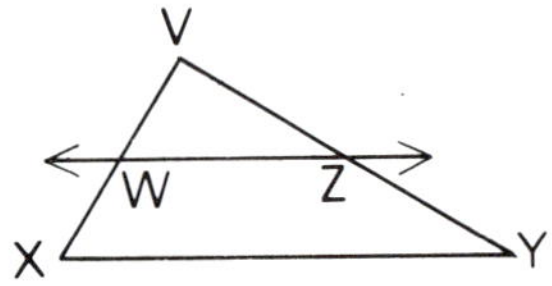

9 Given: $\overline{PA} \perp \overline{AR}$

Conclusions: ___?___

10 Given: $\overleftrightarrow{CG}$ bisects $\overline{BD}$

Conclusions: ___?___

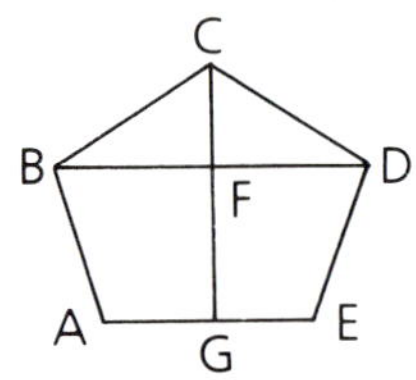

11 Given: $\angle AEN \cong \angle GEN \cong \angle GEL$

Conclusions: ___?___

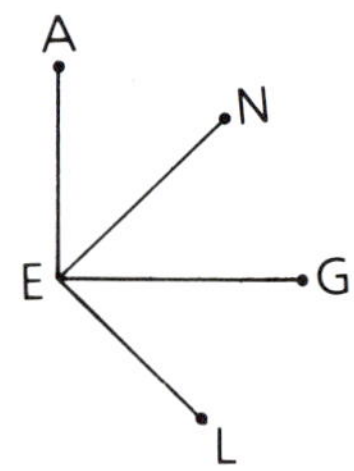

12 Given: $m\angle PQS = 90$

Conclusions: ___?___

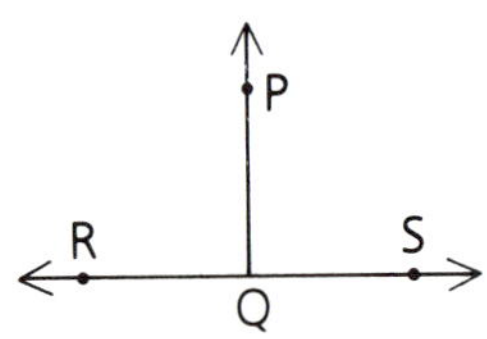

Section 2.3 Problem Set C

13 Given: Two intersecting lines as shown

Conclusions: (Find as many as you can.)

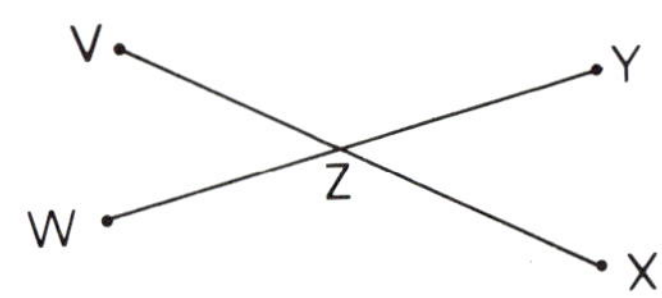

14 The right angle of a right triangle is bisected. Draw a diagram and set up the given information. Then discuss all possible conclusions.

Section 2.4 Theorems about Supplementary and Complementary Angles

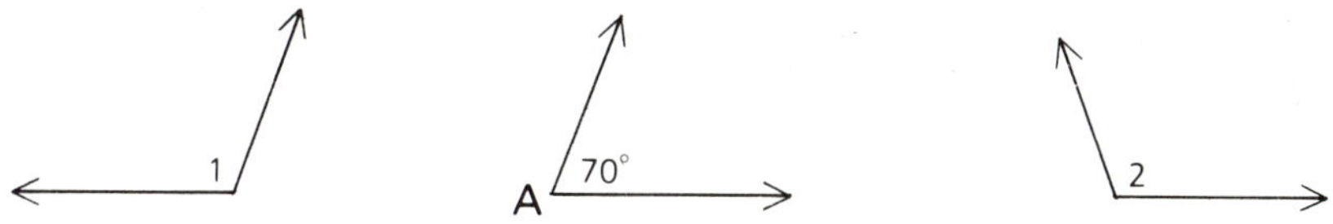

In the diagrams above, $\angle 1$ is supplementary to $\angle A$, and $\angle 2$ is also supplementary to $\angle A$. How large is $\angle 1$? Now calculate $\angle 2$. How does $\angle 1$ compare with $\angle 2$?

The above illustrates (but does not prove) the following theorem:

Theorem 3

T **If angles are supplementary to the same angle, they are congruent.**

Given: $\angle 3$ is supp. to $\angle 4$
$\angle 5$ is supp. to $\angle 4$

Prove: $\angle 3 \cong \angle 5$

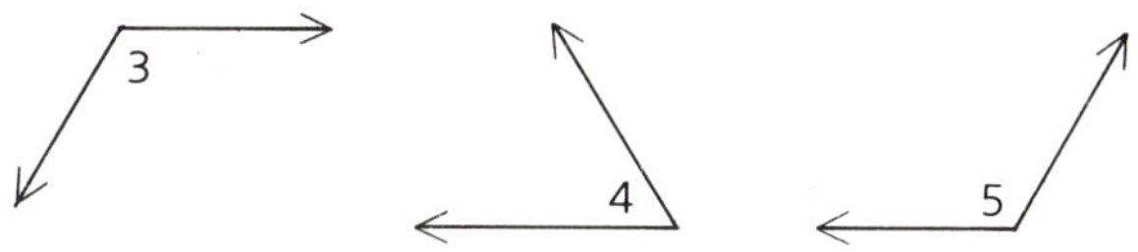

Proof of Theorem 3: $\angle 3$ is supp. to $\angle 4$, so $m\angle 3 + m\angle 4 = 180$

Thus, $m\angle 3 = 180 - m\angle 4$

$\angle 5$ is supp. to $\angle 4$, so $m\angle 5 + m\angle 4 = 180$

Thus, $m\angle 5 = 180 - m\angle 4$

Since $\angle 3$ and $\angle 5$ have the same measure, $\angle 3 \cong \angle 5$.

A companion to Theorem 3 follows.

Theorem 4

T **If angles are supplementary to congruent angles, they are congruent.**

Given: $\angle F$ is supp. to $\angle G$
$\angle H$ is supp. to $\angle J$
$\angle G \cong \angle J$

Conclusion: $\angle F \cong \angle H$

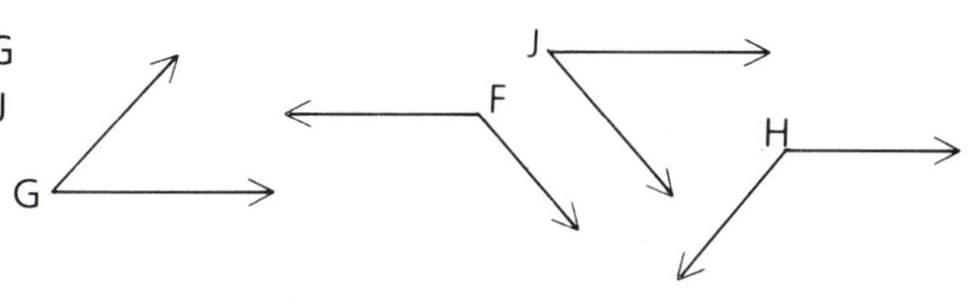

The proof of Theorem 4 is similar to that of Theorem 3.

Two similar theorems hold for complementary angles.

Theorem 5

T **If angles are complementary to the same angle, they are congruent.**

Theorem 6

T **If angles are complementary to congruent angles, they are congruent.**

Section 2.4 Sample Problems

1 Given: $\angle 1$ is supp. to $\angle 2$
$\angle 3$ is supp. to $\angle 4$
$\angle 1 \cong \angle 4$

Conclusion: $\angle 2 \cong \angle 3$

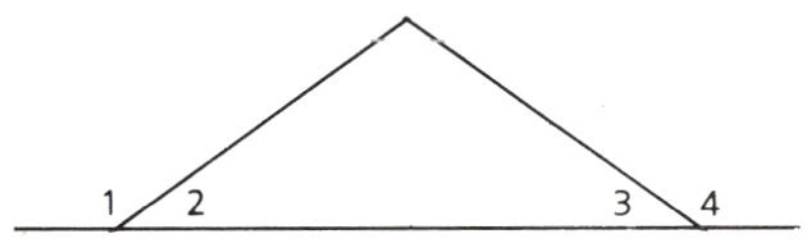

Statements	Reasons
1 $\angle 1$ is supp. to $\angle 2$	1 Given
2 $\angle 3$ is supp. to $\angle 4$	2 Given
3 $\angle 1 \cong \angle 4$	3 Given
4 $\angle 2 \cong \angle 3$	4 If angles are supplementary to $\cong$ angles, they are $\cong$ Short form: Supplements of $\cong$ angles are $\cong$

2 Given: $\angle A$ is comp. to $\angle C$
$\angle DBC$ is comp. to $\angle C$

Conclusion: __?__

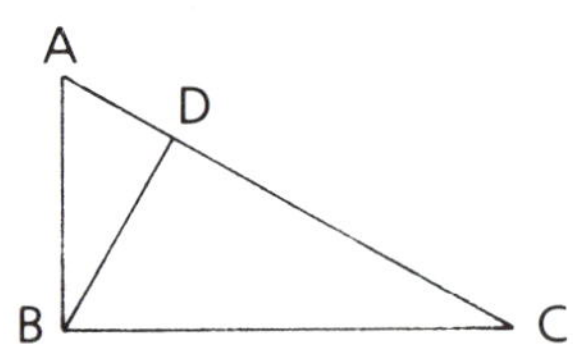

Statements	Reasons
1 $\angle A$ is comp. to $\angle C$	1 Given
2 $\angle DBC$ is comp. to $\angle C$	2 Given
3 $\angle A \cong \angle DBC$	3 If angles are complementary to the same $\angle$, they are $\cong$ Or: Complements of the same $\angle$ are $\cong$

3 **Given: Diagram as shown**

Prove: ∠HFE ≅ ∠GFJ

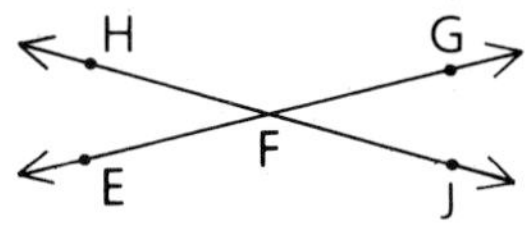

Statements	Reasons
1 Diagram as shown	1 Given
2 ∠EFG is a straight ∠	2 Assumed from diagram
3 ∠HFE is supp. to ∠HFG	3 If two angles form a straight angle, they are supplementary
4 ∠HFJ is a straight ∠	4 Same as 2
5 ∠HFG is supp. to ∠GFJ	5 Same as 3
6 ∠HFE ≅ ∠GFJ	6 If angles are supplementary to the same ∠, they are ≅ Or: Supplements of the same ∠ are ≅

4 **Given:** $\overline{KM} \perp \overline{MO}$

$\overline{PO} \perp \overline{MO}$

∠KMR ≅ ∠POR

Prove: ∠ROM ≅ ∠RMO

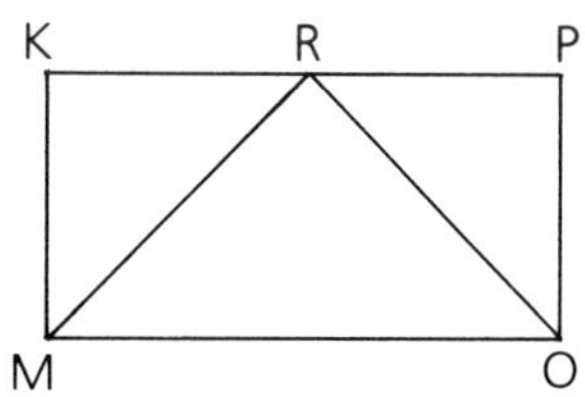

Statements	Reasons
1 $\overline{KM} \perp \overline{MO}$	1 Given
2 ∠KMO is a right ∠	2 If segments are ⊥, they form right angles
3 ∠KMR is comp. to ∠RMO	3 If two angles form a right angle, they are complementary
4 In a similar manner, ∠POR is comp. to ∠ROM	4 Reasons 1–3
5 ∠KMR ≅ ∠POR	5 Given
6 ∠ROM ≅ ∠RMO	6 If angles are complementary to congruent angles, they are ≅ Or: Complements of ≅ angles are ≅

Section 2.4 Problem Set A

1 $\angle 2$ is comp. to $\angle 3$
$\angle 4 = 131°$

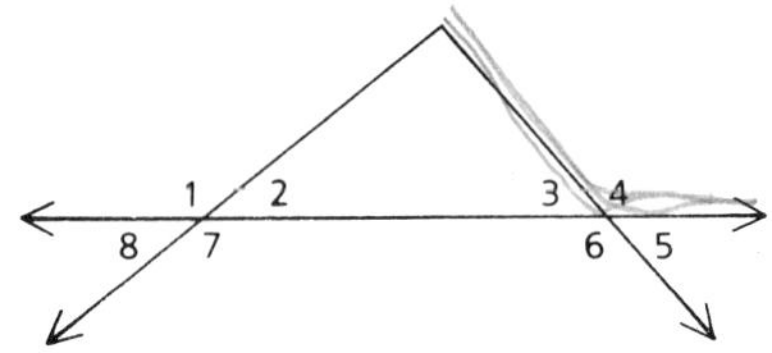

Find the measure of each of these angles:

a $\angle 3$ **c** $\angle 5$ **e** $\angle 1$ **g** $\angle 7$

b $\angle 6$ **d** $\angle 2$ **f** $\angle 8$

2 Given: $\angle 1$ is supp. $\angle 3$
$\angle 2$ is supp. $\angle 3$
Prove: $\angle 1 \cong \angle 2$

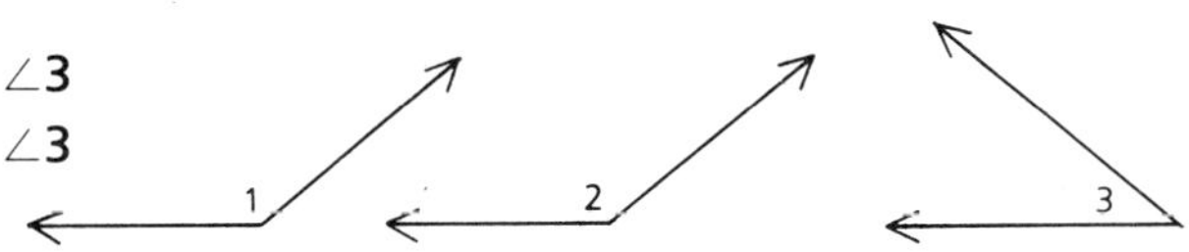

3 Given: $\angle 4$ is comp. $\angle 6$
$\angle 5$ is comp. $\angle 6$
Prove: $\angle 4 \cong \angle 5$

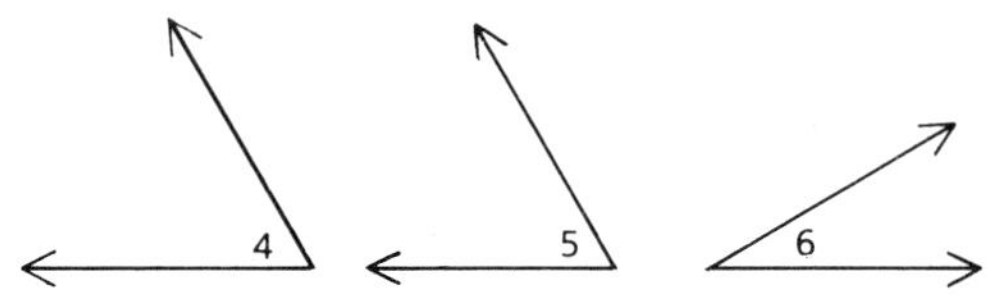

4 One of two supplementary angles is 4 times the other. Find the larger angle.

5 One of two complementary angles is 20 more than the other. Find the measure of each.

6 Given: $\angle 4$ is supp. $\angle 6$
$\angle 5$ is supp. $\angle 7$
$\angle 4 \cong \angle 5$
Conclusion: ___?___

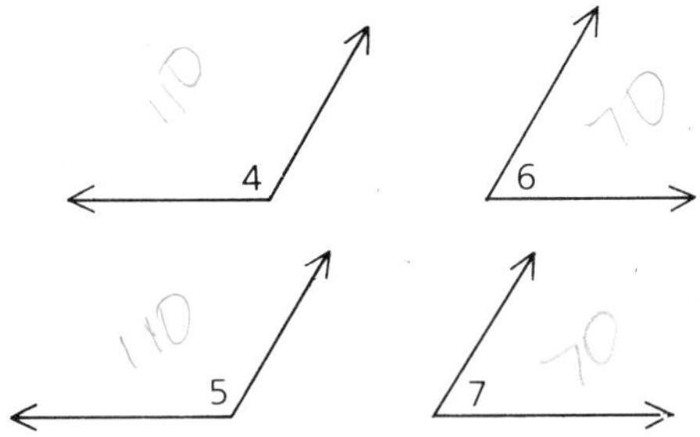

7 Given: $\angle FKJ$ is a right $\angle$
$\angle HJK$ is a right $\angle$
$\angle GKJ \cong \angle GJK$
Conclusion: $\angle FKG \cong \angle HJG$

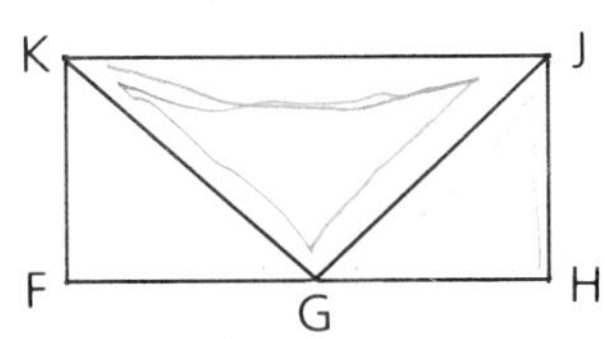

8 Given: Diagram as shown
$\angle 6 \cong \angle 7$

Prove: $\angle 5 \cong \angle 8$

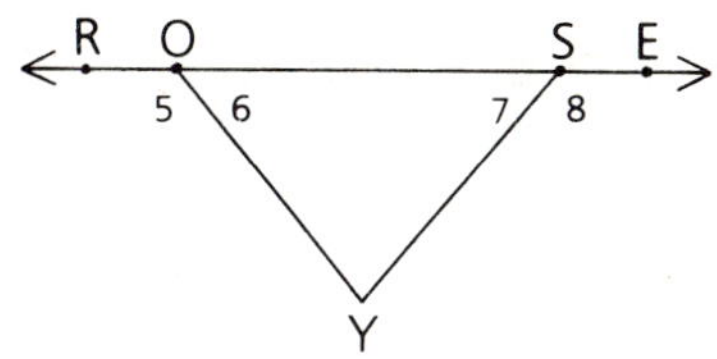

9 Given: $\overrightarrow{SV}$ bisects $\angle RST$

Conclusion: $\angle RSV \cong \angle TSV$

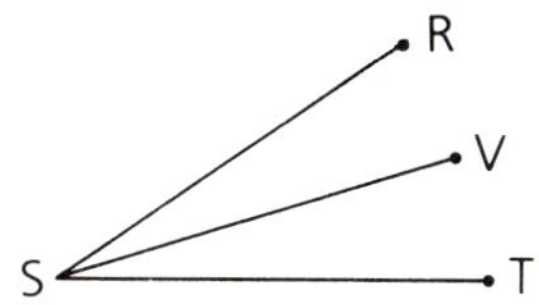

Section 2.4 Problem Set B

10 Given: $\overleftrightarrow{OA} \perp \overleftrightarrow{OC}$
$\overleftrightarrow{OB} \perp \overleftrightarrow{OD}$

Prove: $\angle 1 \cong \angle 3$

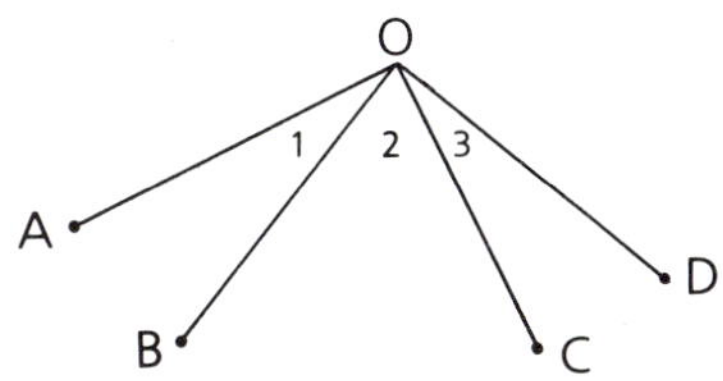

11 Given: $\angle F$ is comp. to $\angle FGJ$
$\angle H$ is comp. to $\angle HGJ$
$\overrightarrow{GJ}$ bisects $\angle FGH$

Conclusion: $\angle F \cong \angle H$

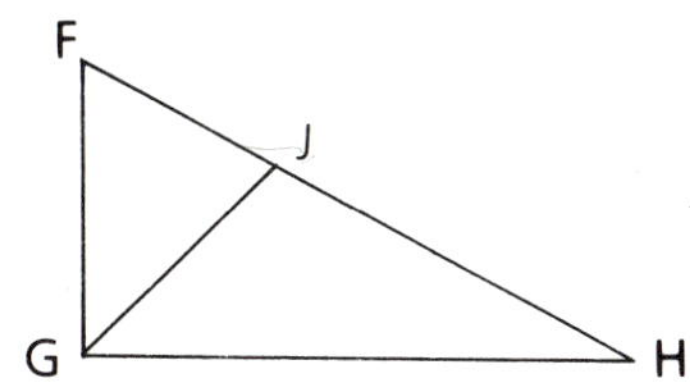

12 The measure of the supp. of an $\angle$ exceeds 3 times the measure of the comp. of the $\angle$ by 10°. Find the measure of the comp.

13 Given: $\angle 1$ is comp. to $\angle 4$
$\angle 2$ is comp. to $\angle 3$
$\overrightarrow{RT}$ bisects $\angle SRV$

Prove: $\overrightarrow{TR}$ bisects $\angle STV$

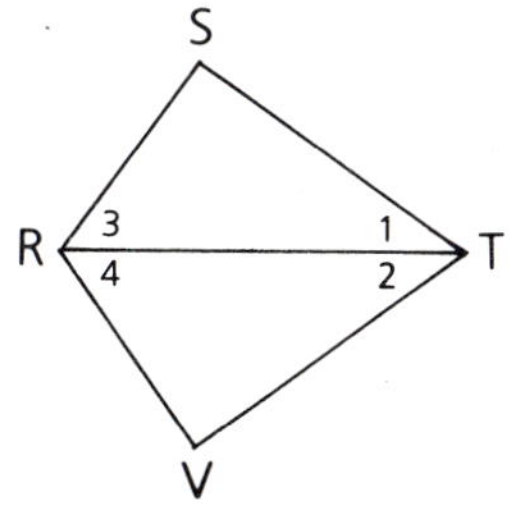

14 If three times the supp. of an $\angle$ is subtracted from seven times the comp. of the $\angle$, the answer is the same as that obtained from trisecting a right $\angle$. Find the supplement.

15 Given: $\angle WXZ \cong \angle VXY$

Conclusion: $\angle 1 \cong \angle 3$

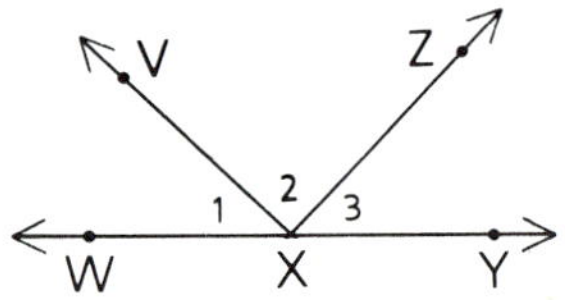

16 Given: $\angle PQR$ supp. $\angle QRS$, $\angle QRS$ supp. $\angle TWX$

$\angle PQR = (5x - 48)^\circ$, $\angle TWX = (2x + 30)^\circ$

Find: m $\angle QRS$

Section 2.4 Problem Set C

17 Given: $\angle E$ is comp. to $\angle ABE$

$\angle D$ is comp. to $\angle DBC$

Prove: $\angle E \cong \angle D$

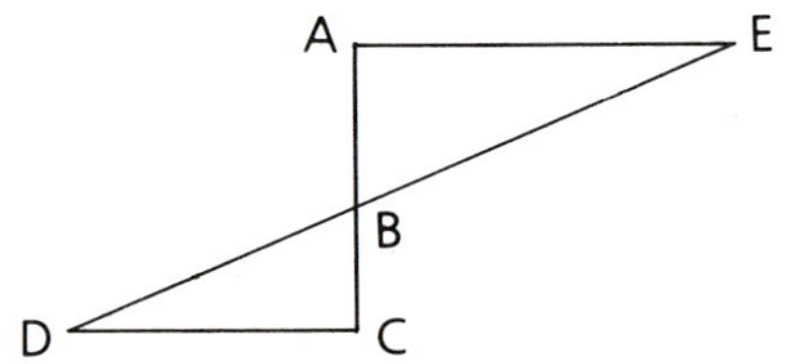

18 $\angle 1 = (x^2 + 3y)^\circ$

$\angle 2 = (20y + 3)^\circ$

$\angle 3 = (3y + 4x)^\circ$

Find: $m\angle 1$

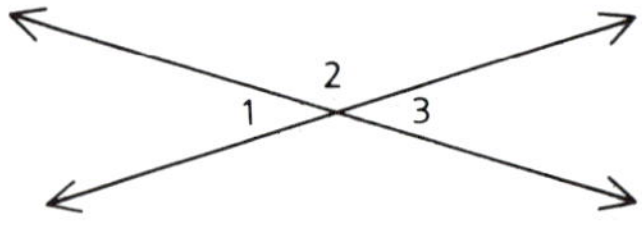

19 The ratio of an angle to its supplement is 3:7. Find the ratio of the angle to its complement.

Section 2.5 Addition and Subtraction Properties

2.5(A) Addition Properties

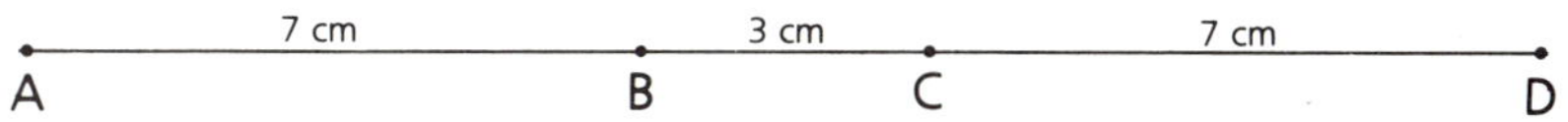

In the above diagram, AB = CD. Do you think AC = BD? Suppose that BC were 12 cm. Would AC = BD?

Your answers should be *yes*. Did you use the *addition property of equality* (AB + *BC* = CD + *BC*)?

Theorem 7

T **If a segment is added to two congruent segments, the sums are congruent. (ADDITION PROPERTY)**

Given: $\overline{PQ} \cong \overline{RS}$

Conclusion: $\overline{PR} \cong \overline{QS}$

Proof: $\overline{PQ} \cong \overline{RS}$, so by definition of congruent segments, PQ = RS.

Now, the Addition Property of Equality says that we may add QR to both sides, so PQ + QR = RS + QR. Substituting, we get PR = QS, or $\overline{PR} \cong \overline{QS}$ by definition of congruent segments (reversed).

Would a similar relationship hold for angles? Do you believe that ∠EFH would always be congruent to ∠JFG?

The next theorem confirms a "yes." Its proof is like that of Theorem 7.

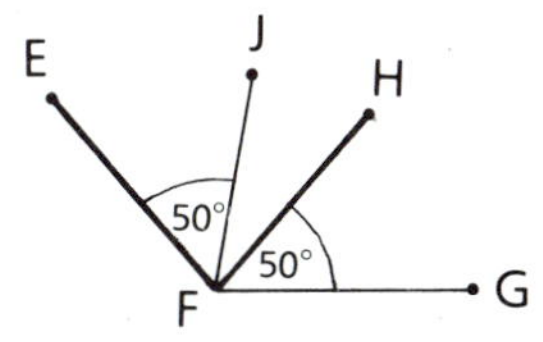

Theorem 8

T **If an angle is added to two congruent angles, the sums are congruent. (ADDITION PROPERTY)**

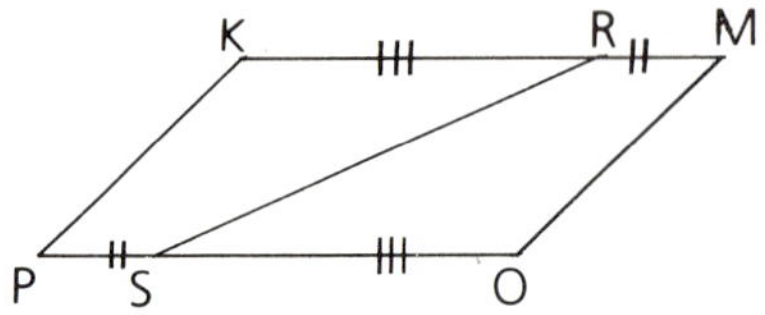

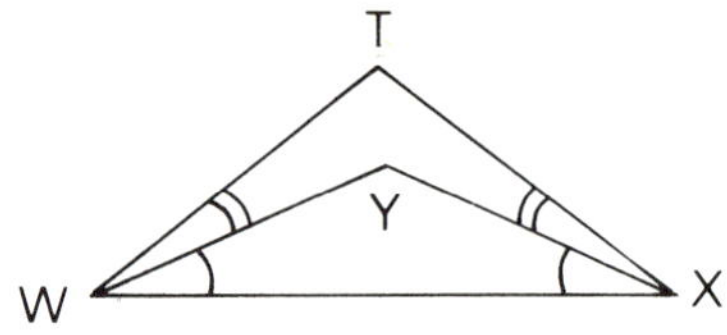

In the figures above, identical tick marks indicate congruent parts.

Do you think $\overline{KM}$ would always be congruent to $\overline{PO}$?

In the diagram on the right, do you think ∠TWX is always congruent to ∠TXW? The answer is yes.

Again, the proof of Theorem 7 is a model for proofs of the next two theorems.

Theorem 9

T **If congruent segments are added to congruent segments, the results are congruent. (ADDITION PROPERTY)**

Theorem 10

T **If congruent angles are added to congruent angles, the results are congruent. (ADDITION PROPERTY)**

2.5(B) Subtraction Properties

We now have four types of addition properties. Because subtraction is addition of the opposite we can expect four types of subtraction properties.

If AC = BD, does AB = CD?

Let AC = 12 and BC = 3

How long is $\overline{BD}$?

Is AB = CD?

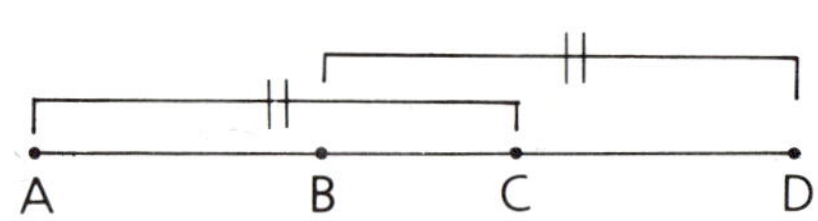

If ∠EFH ≅ ∠GFJ, is ∠EFG ≅ ∠HFJ?

Let m∠EFH = 50 and m∠GFH = 10

How large is ∠GFJ?

Is ∠EFG ≅ ∠HFJ?

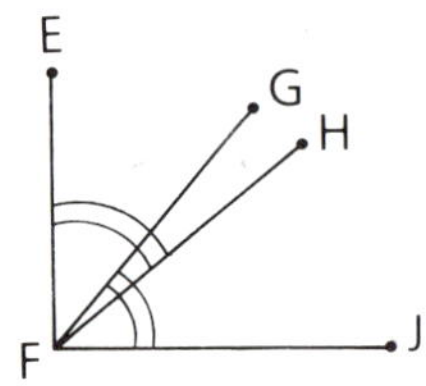

If KO = KP and NO = RP, does KN = KR?

Try this on your own and see what you think.

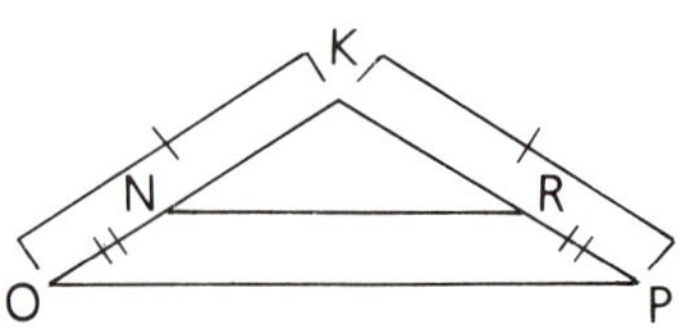

If $\angle STE \cong \angle WET$ and $\angle STW \cong \angle WES$, is $\angle WTE \cong \angle SET$?

Try this on your own also.

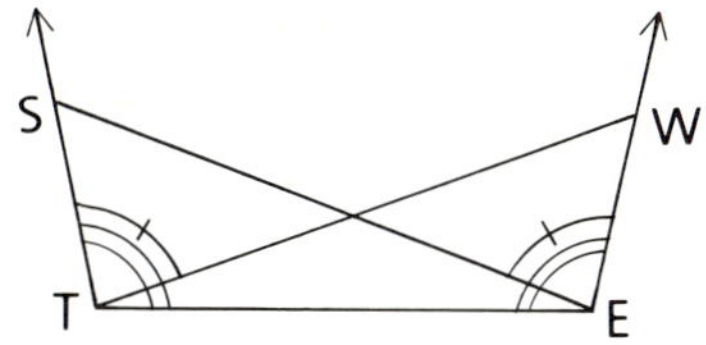

Your results should agree with the next two theorems.

Theorems 11 and 12

T **If a segment (or angle) is subtracted from congruent segments (or angles), the differences are congruent. (SUBTRACTION PROPERTY)**

T **If congruent segments (or angles) are subtracted from congruent segments (or angles), the differences are congruent. (SUBTRACTION PROPERTY)**

Section 2.5 Sample Problems

1 Given: $\overline{AB} \cong \overline{FE}$
$\overline{BC} \cong \overline{ED}$

Prove: $\overline{AC} \cong \overline{FD}$

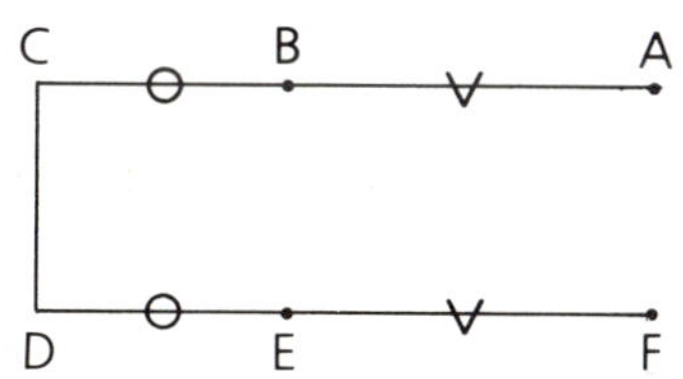

Statements	Reasons
1 $\overline{AB} \cong \overline{FE}$	1 Given
2 $\overline{BC} \cong \overline{ED}$	2 Given
3 $\overline{AC} \cong \overline{FD}$	3 If $\cong$ segments are added to $\cong$ segments, the sums are $\cong$ (ADDITION PROPERTY)

2 Given: $\overline{GJ} \cong \overline{HK}$

Conclusion: $\overline{GH} \cong \overline{JK}$

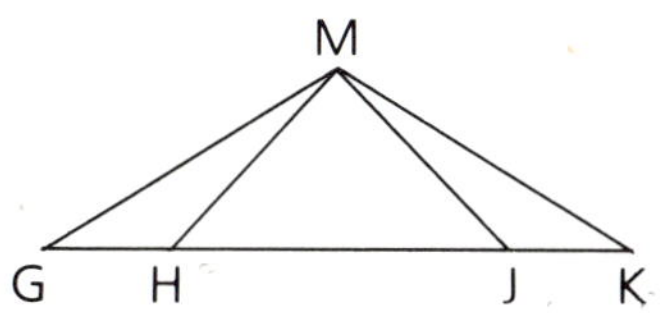

Statements	Reasons
1 $\overline{GJ} \cong \overline{HK}$	1 Given
2 $\overline{GH} \cong \overline{JK}$	2 If a segment ($\overline{HJ}$) is subtracted from $\cong$ segments, the results are $\cong$ (SUBTRACTION PROPERTY)

3 **Given:** ∠NOP ≅ ∠NPO
∠ROP ≅ ∠RPO
Prove: ∠NOR ≅ ∠NPR

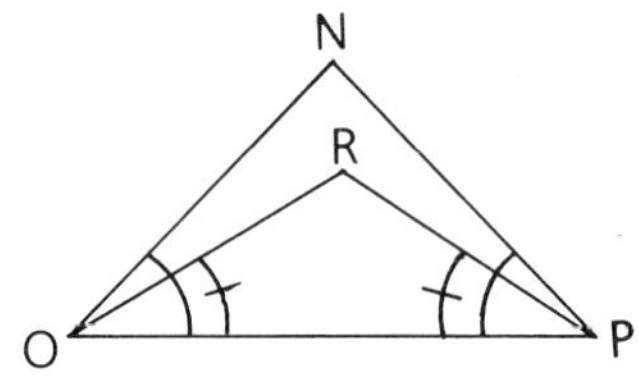

Statements	Reasons
1 ∠NOP ≅ ∠NPO	1 Given
2 ∠ROP ≅ ∠RPO	2 Given
3 ∠NOR ≅ ∠NPR	3 If ≅ angles are subtracted from ≅ angles, the results are ≅ (SUBTRACTION PROPERTY)

4 **Given:** $\overline{AB} \cong \overline{CD}$
Conclusion: ___?___

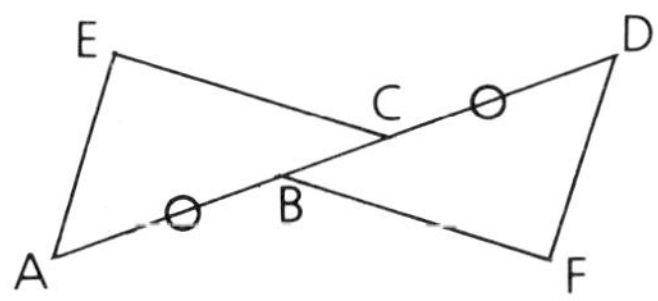

Statements	Reasons
1 $\overline{AB} \cong \overline{CD}$	1 Given
2 $\overline{AC} \cong \overline{BD}$	2 If a segment ($\overline{BC}$) is added to ≅ segments, the results are ≅ (ADDITION PROPERTY)

5 **Given:** ∠HEF is supp. to ∠EHG
∠GFE is supp. to ∠FGH
∠EHF ≅ ∠FGE
∠GHF ≅ ∠HGE
Conclusion: ∠ HEF ≅ ∠GFE

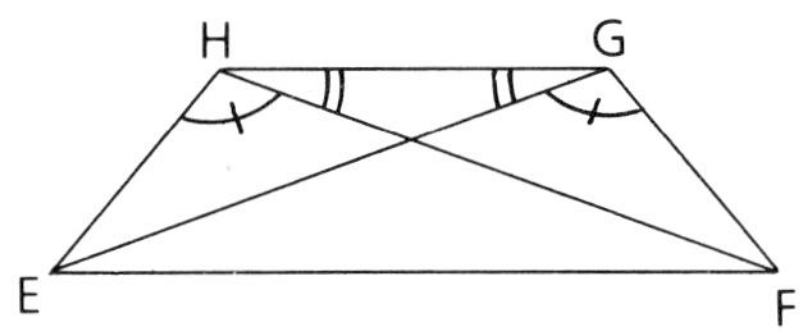

Statements	Reasons
1 ∠HEF is supp. to ∠EHG	1 Given
2 ∠GFE is supp. to ∠FGH	2 Given
3 ∠EHF ≅ ∠FGE	3 Given
4 ∠GHF ≅ ∠HGE	4 Given
5 ∠EHG ≅ ∠FGH	5 If ≅ angles are added to ≅ angles, the sums are ≅ (ADDITION PROPERTY)
6 ∠HEF ≅ ∠GFE	6 If angles are supplementary to ≅ angles, they are ≅ Alternate wording: Supplements of ≅ ∠s are ≅

The Royal Road: Archimedes

Archimedes was a Greek mathematician and inventor who lived in Sicily around 250 B.C. One of his mathematical achievements was to show that the value of π was between $3\frac{10}{71}$ and $3\frac{10}{70}$. We still use his value $3\frac{1}{7}$ as a useful approximation for π.

Use a calculator with a π key for these questions:

1 Which is larger, $3\frac{10}{71}$ or $3\frac{10}{70}$? What is the difference?

2 Which value is closer to π? By how much?

3 Suppose the cross-section of a wire is one square millimeter. How many wires would there be in a huge cable whose radius is one meter? (Find three values from the area formula $A = \pi r^2$, using $3\frac{10}{71}$, $3\frac{1}{7}$, and the built-in calculator value for π.)

4 How does the number of wires vary, based on the different values for π?

Section 2.5 Problem Set A

1 Name the angles or segments that are $\cong$ by the addition property.

a

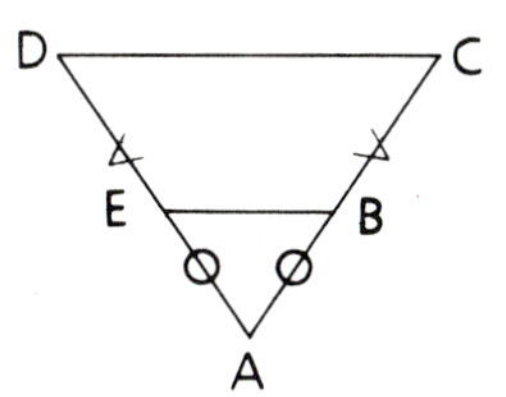

b

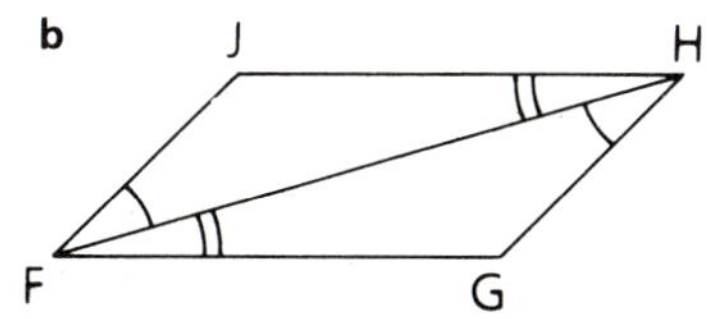

2 Name the angles or segments that are $\cong$ by the subtraction property.

a

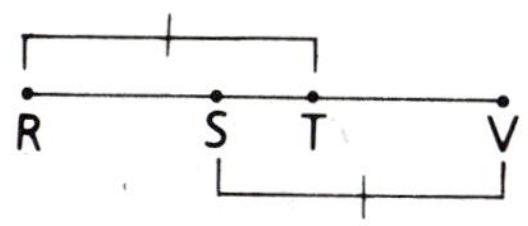

b

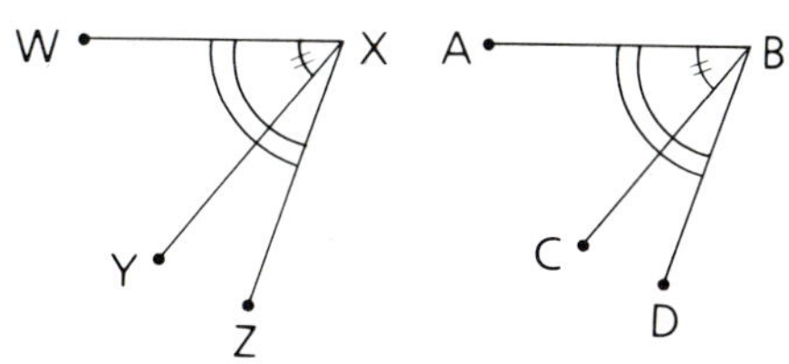

3 Given: $\overline{PQ} \cong \overline{SR}$
$\overline{QN} \cong \overline{RN}$

Conclusion: $\overline{PN} \cong \overline{SN}$

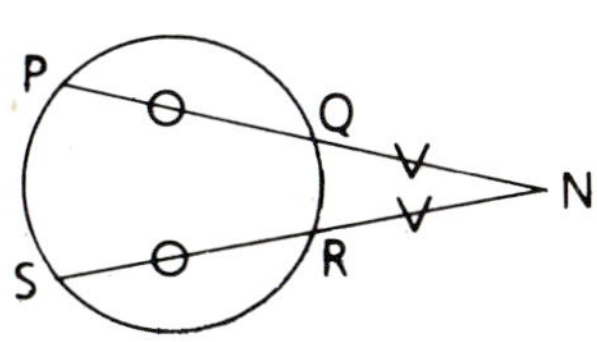

4 Given: $\angle TEV \cong \angle XEW$

Prove: $\angle TEW \cong \angle XEV$

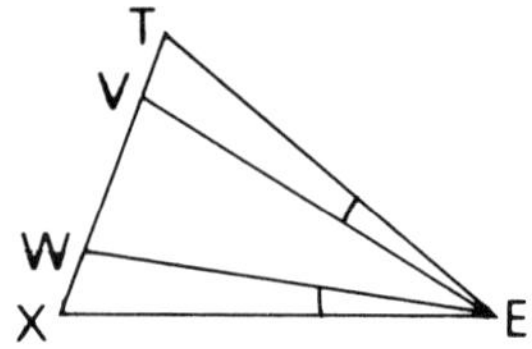

5 Given: $\overline{AC} \cong \overline{DF}$

$\overline{BC} \cong \overline{EF}$

Prove: $\overline{AB} \cong \overline{DE}$

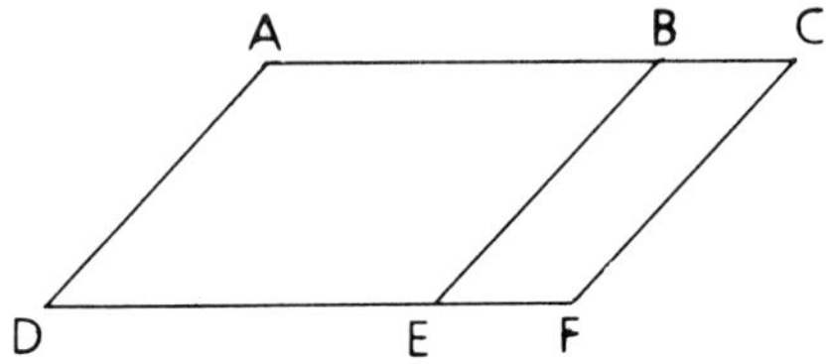

6 $\overline{GH} \cong \overline{JK}$, $GH = x + 10$

$HJ = 8$, $JK = 2x - 4$

Find GJ

7 Given: $\angle PNO \cong \angle PON$

$\angle 1 \cong \angle 2$

Conclusion: ___?___

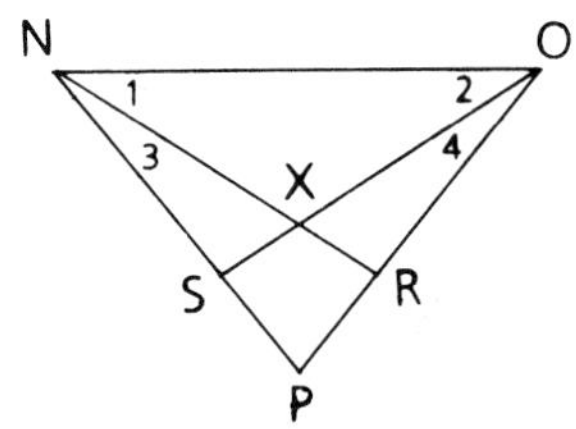

8 Given: $\angle T$ is comp. to $\angle W$

$\angle X$ is comp. to $\angle Z$

$\angle Z \cong \angle W$

Prove: ___?___

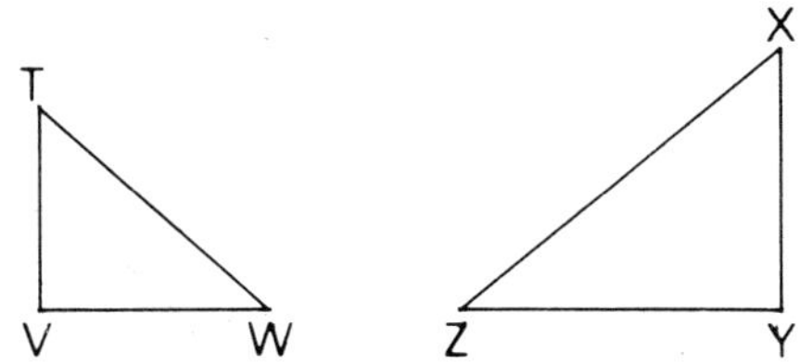

9 Given: $\overline{QR} \cong \overline{ST}$, $QS = 5x + 17$

$RT = 10 - 2x$, $RS = 3$

Find: QS, QT

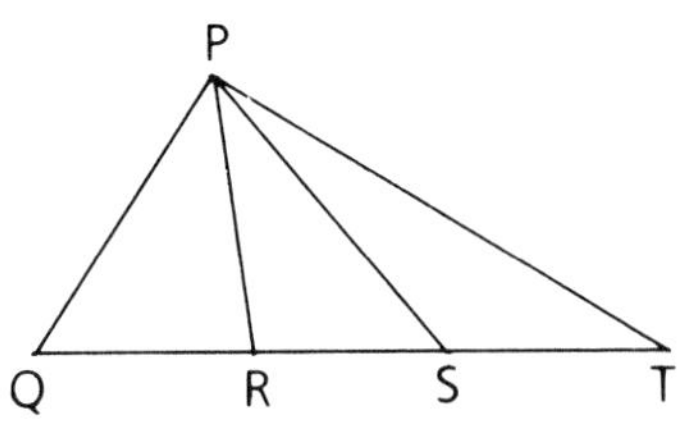

10 Given: $\angle BAD$ is a right $\angle$

$\overline{CA} \perp \overline{AE}$

Prove: $\angle BAC \cong \angle EAD$

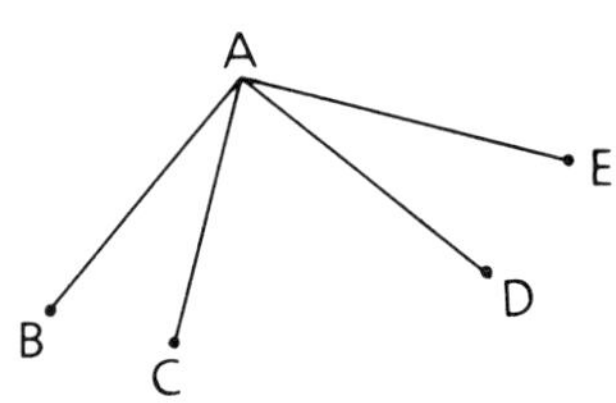

Section 2.5 Problem Set B

11 Given: $\angle BAD \cong \angle FAD$
$\overrightarrow{AD}$ bisects $\angle CAE$
Conclusion: $\angle BAC \cong \angle FAE$

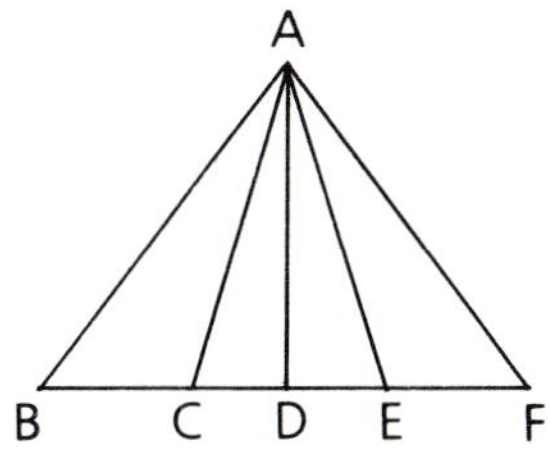

12 Given: J and K are trisection points of $\overline{HM}$
$\overline{GH} \cong \overline{MO}$
Conclusion: $\overline{GJ} \cong \overline{KO}$

13 Given: $\angle NPR$ is a right $\angle$
$\overline{WE} \perp \overline{ET}$
$\angle SPR \cong \angle XET$
Prove: $\angle NPS \cong \angle WEX$

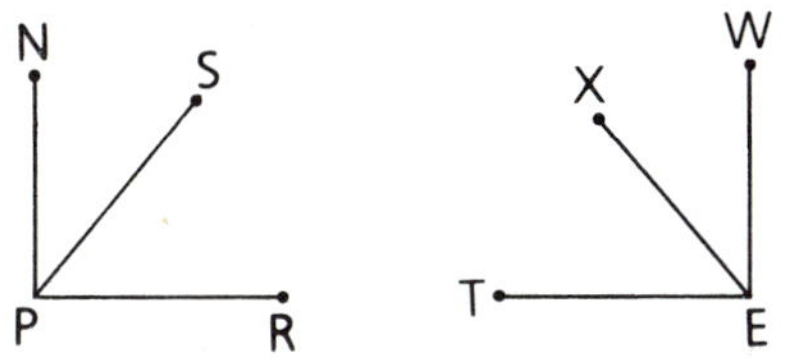

14 $\angle A$ is complementary to $\angle B$
$\angle C$ is complementary to $\angle B$
$\angle A = (3x + y)^\circ$
$\angle B = (x + 4y + 2)^\circ$
$\angle C = (3y - 3)^\circ$
Find $m\angle B$

Section 2.5 Problem Set C

15 Given: $\overrightarrow{BF}$ bisects $\angle DBE$

a Does $\overrightarrow{BF}$ bisect $\angle CBA$?

b What did you discover about $\angle ABC$ and $\overrightarrow{BF}$?

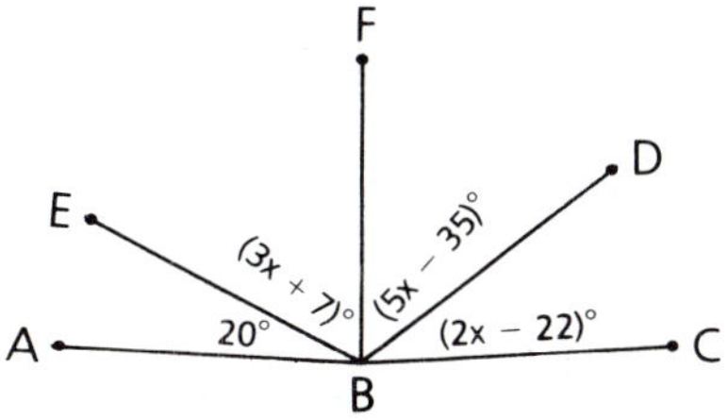

Section 2.6 Multiplication and Division Properties

In the figure above, B, C, F, G are trisection points. If AB = EF = 3, what can we say about $\overline{AD}$ and $\overline{EH}$?

If $\overline{AB} \cong \overline{EF}$, is $\overline{AD} \cong \overline{EH}$?

In the figure at the right, the rays $\overrightarrow{KO}$ and $\overrightarrow{PS}$ are angle bisectors.

If m∠JKO = m∠NPS = 25, what is true about ∠JKM and ∠NPR?

If ∠JKO ≅ ∠NPS, is ∠JKM ≅ ∠NPR?

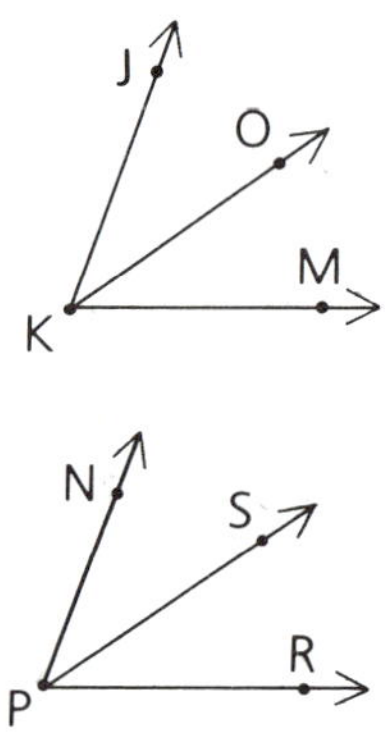

The examples above illustrate a property that has a proof similar to the proof of THEOREM 7.

Theorem 13

T **If segments (or angles) are congruent, their like multiples are congruent. (MULTIPLICATION PROPERTY)**

Also, because division is multiplication by the reciprocal of the denominator, it is easy to prove the next theorem.

Theorem 14

T **If segments (or angles) are congruent, their like divisions are congruent. (DIVISION PROPERTY)**

Section 2.6 Sample Problems

1 **Given:** $\overline{MP} \cong \overline{NS}$
O is the midpoint of $\overline{MP}$
R is the midpoint of $\overline{NS}$

Prove: $\overline{MO} \cong \overline{NR}$

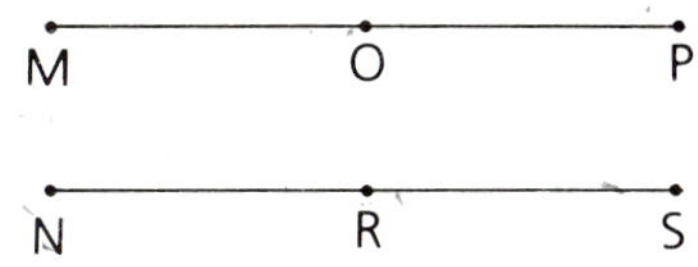

Statements	Reasons
1 $\overline{MP} \cong \overline{NS}$	1 Given
2 O is the midpoint of $\overline{MP}$	2 Given
3 R is the midpoint of $\overline{NS}$	3 Given
4 $\overline{MO} \cong \overline{NR}$	4 If segments are $\cong$, their like divisions (or halves) are $\cong$ (DIVISION PROPERTY)

2 **Given:** $\angle TRY \cong \angle ABE$
$\overrightarrow{RW}$ and $\overrightarrow{RX}$ trisect $\angle TRY$
$\overrightarrow{BC}$ and $\overrightarrow{BD}$ trisect $\angle ABE$

Conclusion: $\angle TRW \cong \angle CBD$

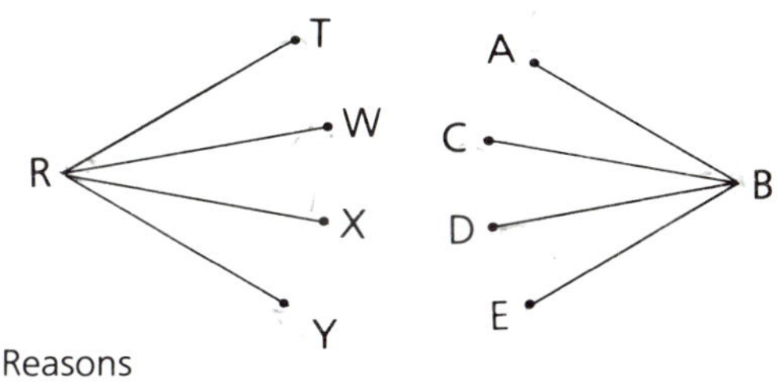

Statements	Reasons
1 $\angle TRY \cong \angle ABE$	1 Given
2 $\overrightarrow{RW}$ and $\overrightarrow{RX}$ trisect $\angle TRY$	2 Given
3 $\overrightarrow{BC}$ and $\overrightarrow{BD}$ trisect $\angle ABE$	3 Given
4 $\angle TRW \cong \angle CBD$	4 If $\angle$s are $\cong$, their like divisions (or thirds) are $\cong$ (DIVISION PROPERTY)

3 **Given:** $\overline{MK} \cong \overline{FG}$
$\overline{KG}$ bisects $\overline{MJ}$ and $\overline{FH}$

Prove: $\overline{MJ} \cong \overline{FH}$

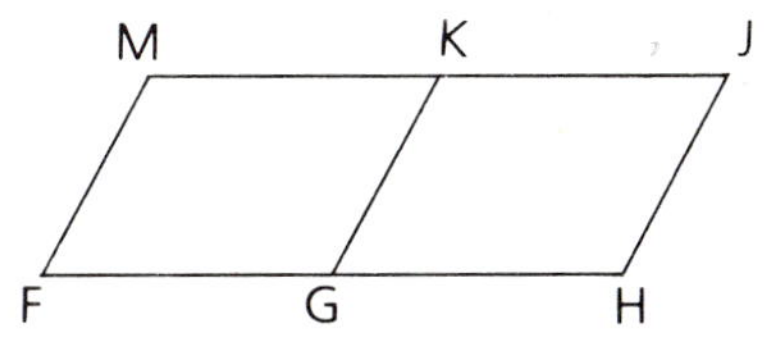

Statements	Reasons
1 $\overline{MK} \cong \overline{FG}$	1 Given
2 $\overline{KG}$ bisects $\overline{MJ}$ and $\overline{FH}$	2 Given
3 $\overline{MJ} \cong \overline{FH}$	3 If segments are $\cong$, like multiples (doubles) are $\cong$ (MULTIPLICATION PROPERTY)

4 Given: $\angle NOP \cong \angle RPO$
$\overrightarrow{PT}$ bisects $\angle RPO$
$\overrightarrow{OS}$ bisects $\angle NOP$
$\angle NSO$ is comp. to $\angle 1$
$\angle RTP$ is comp. to $\angle 3$

Prove: $\angle NSO \cong \angle RTP$

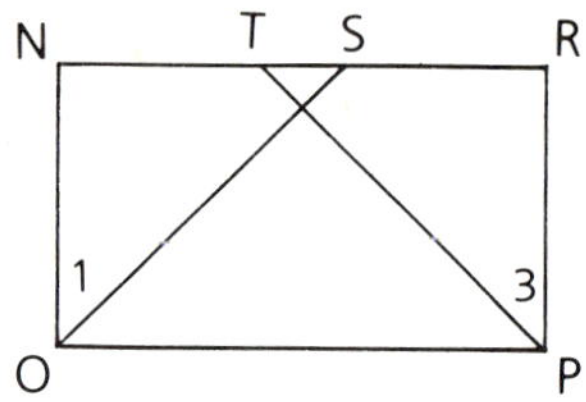

Statements	Reasons
1 $\angle NOP \cong \angle RPO$	1 Given
2 $\overrightarrow{PT}$ bisects $\angle RPO$	2 Given
3 $\overrightarrow{OS}$ bisects $\angle NOP$	3 Given
4 $\angle 1 \cong \angle 3$	4 Halves of $\cong$ angles are $\cong$ (An alternate form of DIVISION PROPERTY)
5 $\angle NSO$ is comp. to $\angle 1$	5 Given
6 $\angle RTP$ is comp. to $\angle 3$	6 Given
7 $\angle NSO \cong \angle RTP$	7 Complements of $\cong$ $\angle$s are $\cong$

Section 2.6 Problem Set A

1 Solve for x

a $\angle HGJ \cong \angle ONP$
$\overrightarrow{GJ}$ and $\overrightarrow{NP}$ are $\angle$ bisectors
$\angle HGK = 50°$
$\angle ONR = (2x + 10)°$

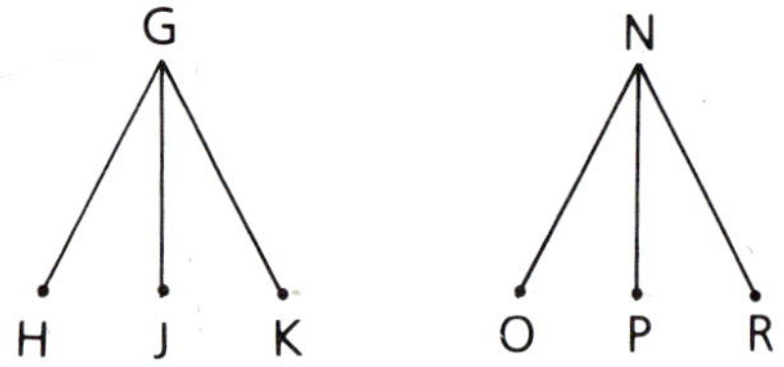

b $\overline{SW} \cong \overline{SZ}$
$\overleftrightarrow{TX}$ and $\overleftrightarrow{VY}$ trisect $\overline{SW}$ and $\overline{SZ}$
$ST = 12$
$YZ = x - 4$

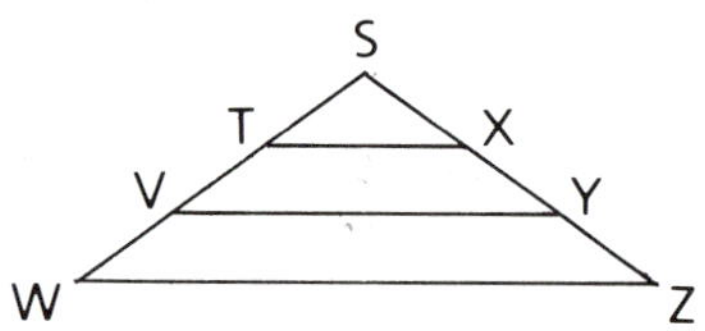

2 Given: $\overline{DF} \cong \overline{GJ}$
E is the midpoint of $\overline{DF}$
H is the midpoint of $\overline{GJ}$

Prove: $\overline{DE} \cong \overline{GH}$

3 Given: $\angle KMR \cong \angle VTW$

$\overrightarrow{MR}$ and $\overrightarrow{MP}$ trisect $\angle KMO$

$\overrightarrow{TX}$ and $\overrightarrow{TW}$ trisect $\angle STV$

Prove: $\angle KMO \cong \angle STV$

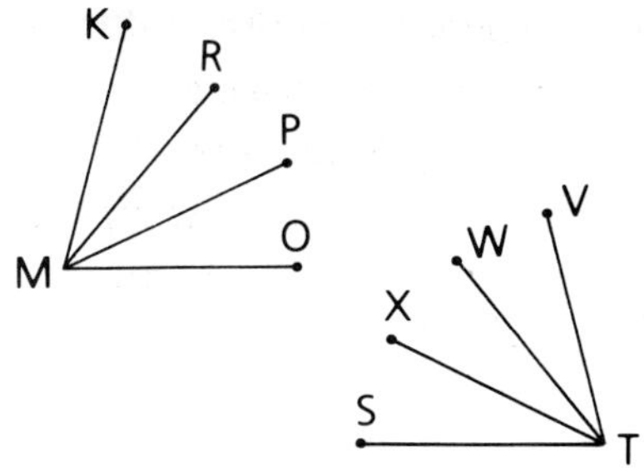

4 Given: $\angle AFE \cong \angle DEF$

$\overrightarrow{FC}$ bisects $\angle AFE$

$\overrightarrow{EB}$ bisects $\angle DEF$

Conclusion: $\angle 1 \cong \angle 2$

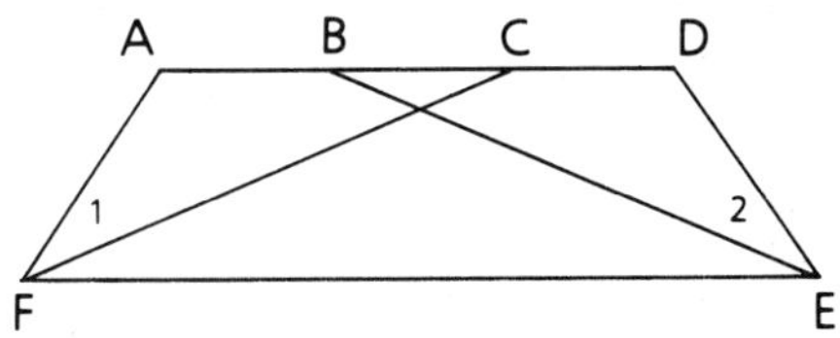

5 Given: $\overline{JK} \cong \overline{MK}$

$\overleftrightarrow{OP}$ bisects $\overline{JK}$ and $\overline{MK}$

Prove: $\overline{JO} \cong \overline{PK}$

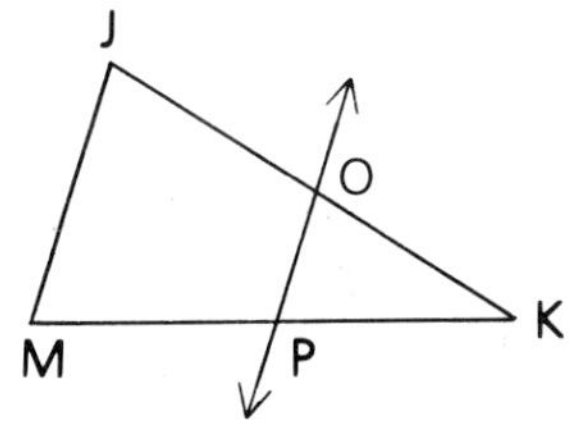

6 Given: M is the midpoint of $\overline{GH}$

Conclusion: $\overline{GM} \cong \overline{MH}$

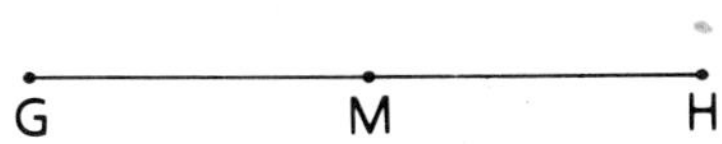

7 Given: $\angle TNR \cong \angle TRN$

$\angle NRS \cong \angle RNS$

Conclusion: ___?___

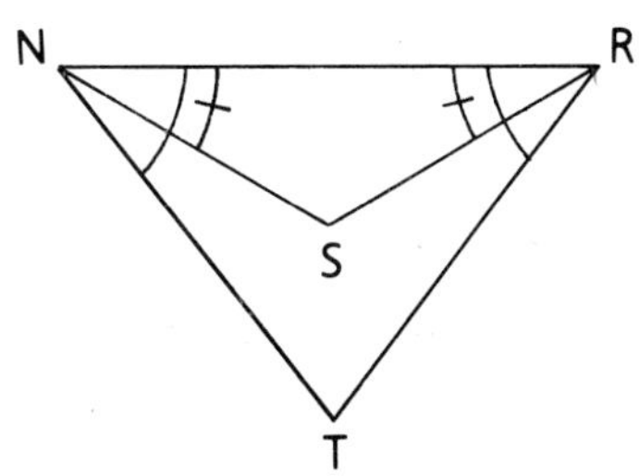

8 Copy the figure and the *given* and *prove* information. Then complete the proof by filling in the missing reasons.

Given: $\overline{VW} \cong \overline{AB}$, $\overline{WX} \cong \overline{BC}$
X is the midpt. of $\overline{VZ}$
C is the midpt. of $\overline{AD}$

Prove: $\overline{VZ} \cong \overline{AD}$

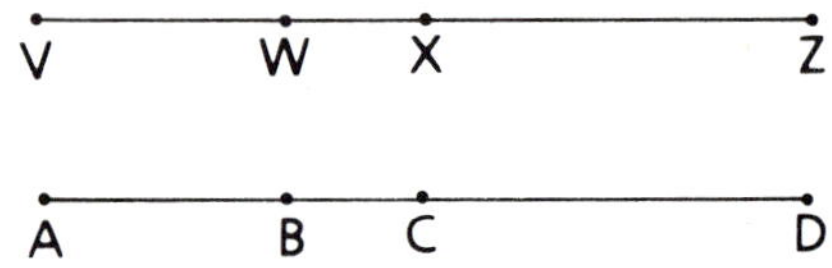

Statements	Reasons
1 $\overline{VW} \cong \overline{AB}$	1 ______
2 $\overline{WX} \cong \overline{BC}$	2 ______
3 $\overline{VX} \cong \overline{AC}$	3 ______
4 X is midpt. of $\overline{VZ}$	4 ______
5 C is midpt. of $\overline{AD}$	5 ______
6 $\overline{VZ} \cong \overline{AD}$	6 ______

Section 2.6 Problem Set B

9 Given: $\overline{SZ} \cong \overline{ST}$
$\overline{XY} \cong \overline{VW}$
Y is the midpoint of $\overline{ZX}$
V is the midpoint of $\overline{TW}$

Prove: $\overline{SX} \cong \overline{SW}$

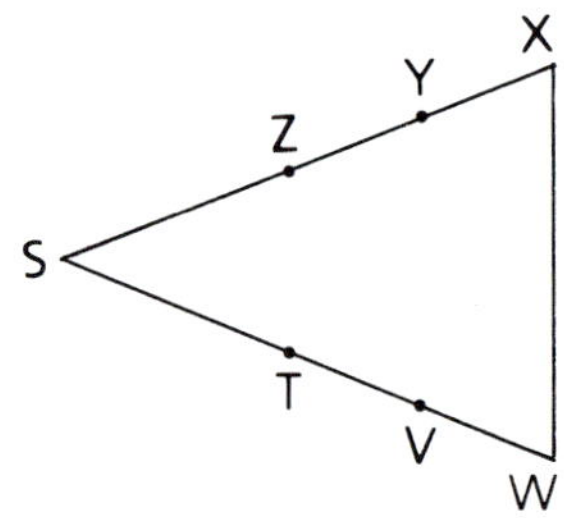

10 Given: $\overrightarrow{PR}$ bisects $\angle QPS$
$\overrightarrow{KO}$ bisects $\angle JKM$
$\angle 1$ is supp. to $\angle JKM$
$\angle 1$ is supp. to $\angle QPS$

Conclusion: $\angle 2 \cong \angle 3$

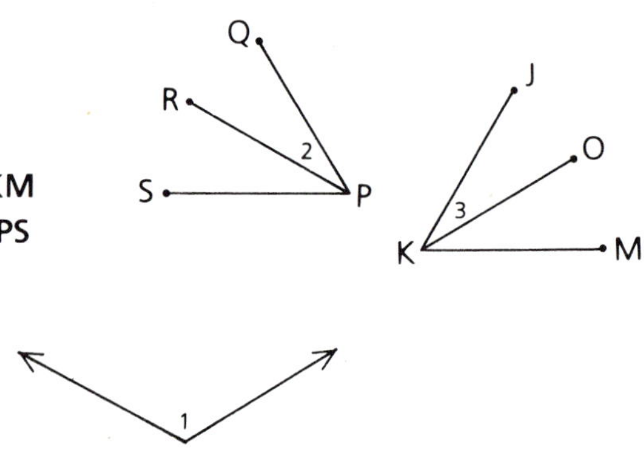

11 Given: $\angle 1 \cong \angle 2$
$\overrightarrow{BG}$ bisects $\angle ABF$
$\overrightarrow{CE}$ bisects $\angle FCD$
Prove: $\angle 3 \cong \angle 4$

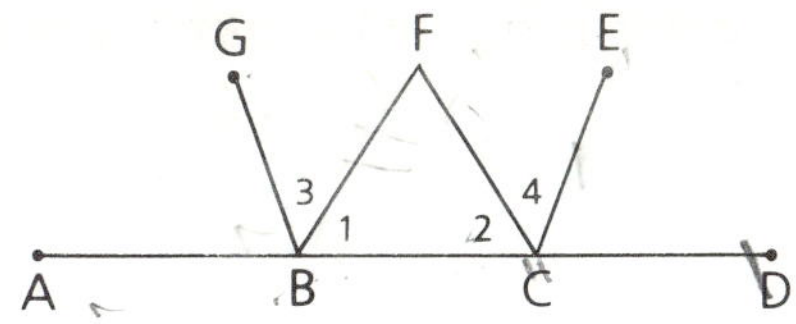

12 Given: $\angle HJK$ is a right $\angle$
$\overline{MK} \perp \overline{KJ}$
$\angle PJO \cong \angle PKO$
$\overrightarrow{JO}$ bisects $\angle PJK$
$\overrightarrow{KO}$ bisects $\angle PKJ$
Conclusion: $\angle 1 \cong \angle 2$

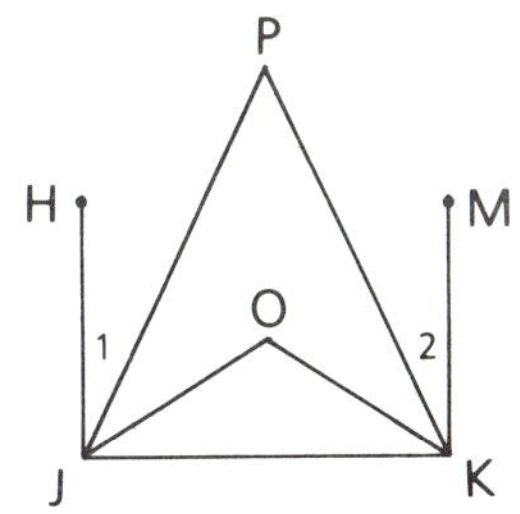

13 If four times the supplement of an angle is added to eight times the complement, the result is 3 straight angles. Find the measure of the angle that is supplementary to the complement.

Section 2.6 Problem Set C

14 Given: $\angle 1 \cong \angle 2$
$\overrightarrow{PE}$ bis $\angle APN$
$\overrightarrow{NE}$ bis $\angle ANP$
Prove: $\angle XPE \cong \angle ENY$

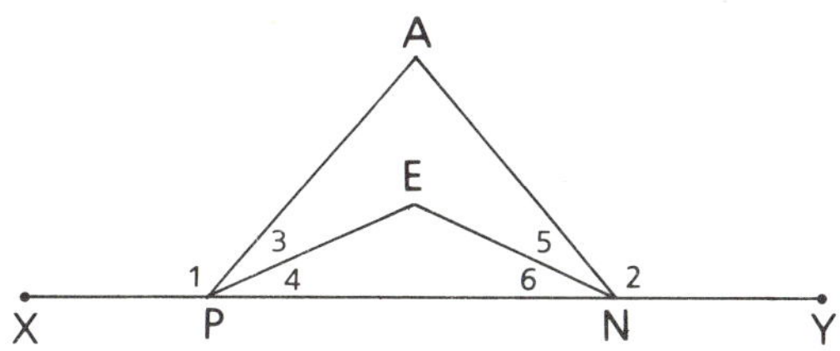

15 $\angle T$ is comp. to $\angle YXT$,
$\angle YRS \cong \angle YXT$,
$m\angle T = x + y$,
$m\angle YXT = 4y + 2$,
$m\angle YRS = 2x - 6$,
and $\angle PRS = 100°$.

Does $\overrightarrow{RX}$ bisect $\angle PRS$?

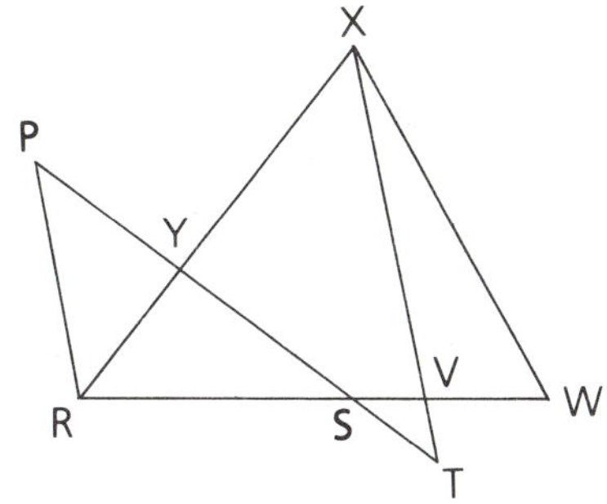

Section 2.7 Transitive and Substitution Properties

2.7 (A) Transitive Properties

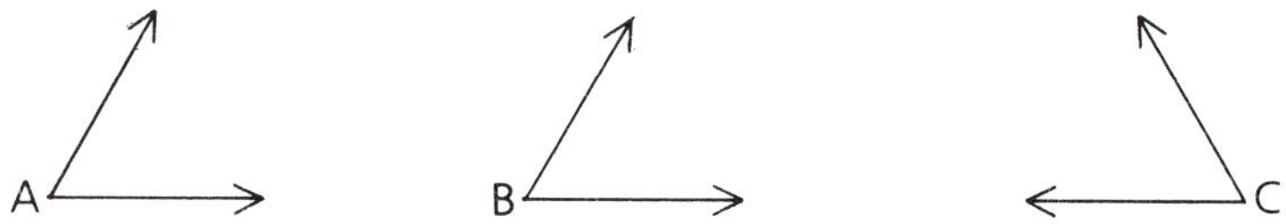

Suppose that $\angle A \cong \angle B$ and $\angle A \cong \angle C$. Do you think $\angle B \cong \angle C$?

The transitive property of Algebra can be used to prove this general rule:

Theorem 15

T **If angles (or segments) are congruent to the same angle (or segment), the angles (or segments) are congruent. (TRANSITIVE PROPERTY)**

THEOREM 15 can be used twice to prove the next theorem.

Theorem 16

T **If angles (or segments) are congruent to congruent angles (or segments), the angles (or segments) are congruent. (TRANSITIVE PROPERTY)**

2.7 (B) Substitution Property

In algebra last year and also in problems we have worked this year, we have solved for a variable such as x and then SUBSTITUTED that value. For example:

- If $\angle A \cong \angle B$, find $m\angle A$.

  Solution: $2x - 4 = x + 10$

  $x = 14$

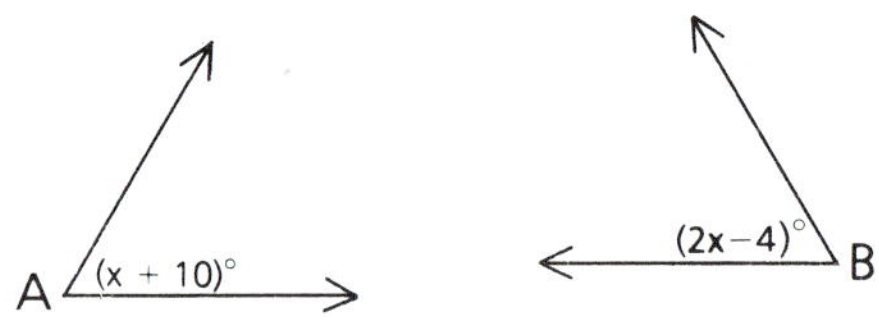

SUBSTITUTE $x = 14$ into $m\angle A = x + 10$. Thus, $m\angle A = 14 + 10 = 24$

Another example of the SUBSTITUTION property is this one:

- If $\angle 1$ is comp. to $\angle 2$ and $\angle 2 \cong \angle 3$, then $\angle 1$ is comp. to $\angle 3$ by SUBSTITUTION.

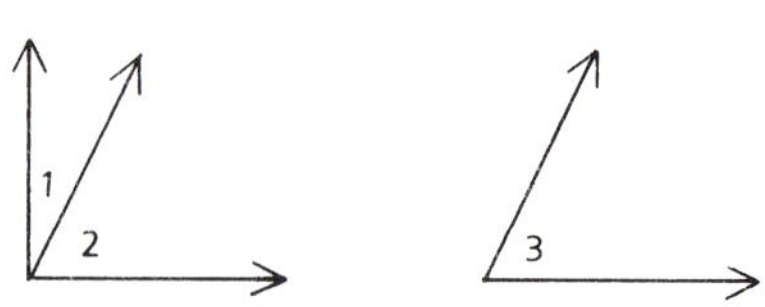

Section 2.7 Sample Problems

1 **Given:** $\overline{FG} \cong \overline{KJ}$

$\overline{GH} \cong \overline{KJ}$

Prove: $\overleftrightarrow{KG}$ bisects $\overline{FH}$

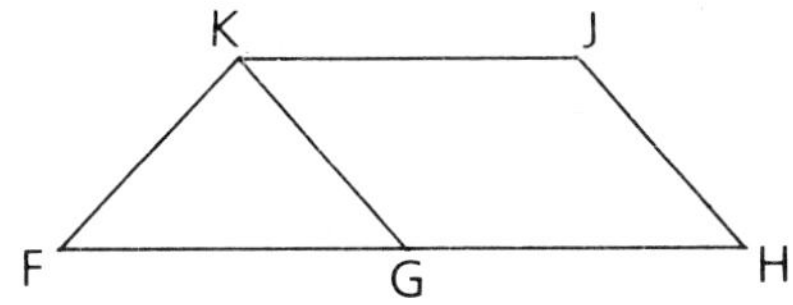

Statements	Reasons
1 $\overline{FG} \cong \overline{KJ}$	1 Given
2 $\overline{GH} \cong \overline{KJ}$	2 Given
3 $\overline{FG} \cong \overline{GH}$	3 If segments are each $\cong$ to the same segment, they are $\cong$ (TRANSITIVE PROPERTY)
4 $\overleftrightarrow{KG}$ bisects $\overline{FH}$	4 If a line divides a segment into two $\cong$ segments, it bisects the segment

2 **Given:** $\angle 1 + \angle 2 = 90^\circ$

$\angle 1 \cong \angle 3$

Prove: $\angle 3 + \angle 2 = 90^\circ$

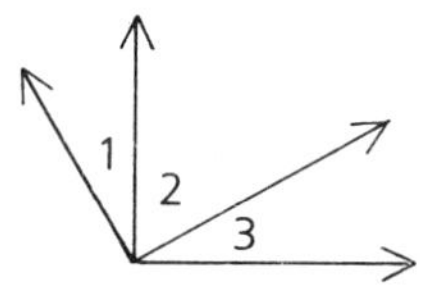

Statements	Reasons
1 $\angle 1 + \angle 2 = 90^\circ$	1 Given
2 $\angle 1 \cong \angle 3$	2 Given
3 $\angle 3 + \angle 2 = 90^\circ$	3 Substitution (step 2 into step 1)

3 $\angle P \cong \angle R$, and $\angle Q \cong \angle R$

Express $m\angle Q$ **in terms of x and a**

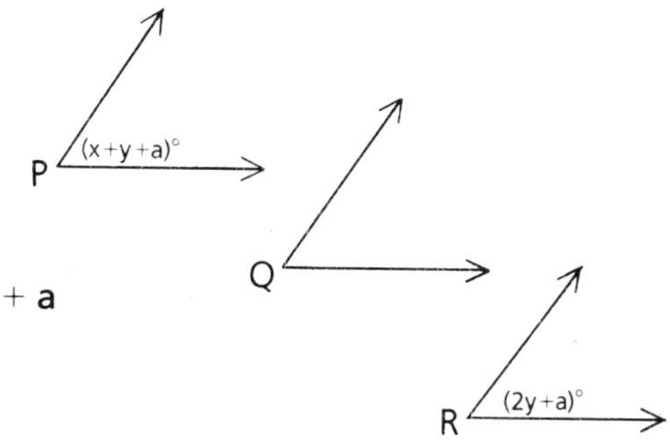

Solution:
$$
\begin{aligned}
2y + a &= x + y + a \\
2y &= x + y \\
y &= x
\end{aligned}
$$

$$m\angle P = x + y + a = x + x + a$$

$$m\angle Q = 2x + a$$

Section 2.7 Problem Set A

1 Given: $\angle X \cong \angle Y$
$\angle X \cong \angle Z$
Conclusion: $\angle Y \cong \angle Z$

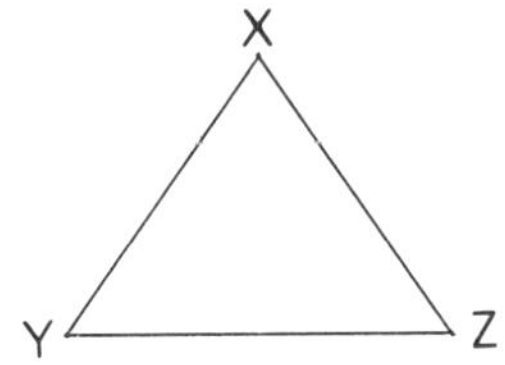

2 Given: $\angle 1 \cong \angle 2$
$\angle 2 \cong \angle 3$
Conclusion: $\angle 1 \cong \angle 3$

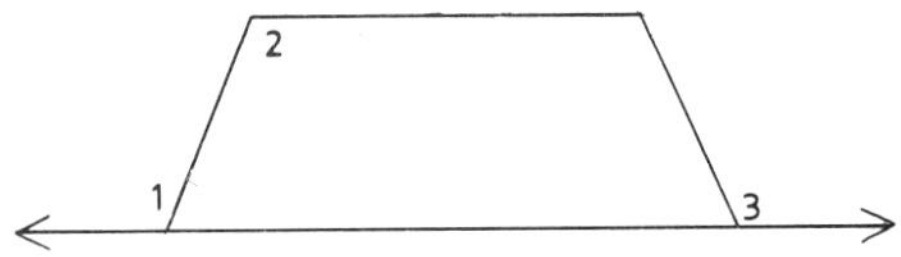

3 Given: $\angle 1 \cong \angle 3$
$\angle 2 \cong \angle 3$
$\angle 2 \cong \angle 4$
Prove: $\angle 1 \cong \angle 4$

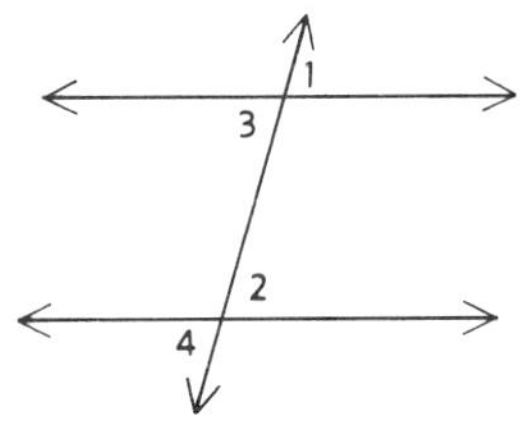

4 Given: $BC + BE = AD$
$BE = EF$
Prove: $BC + EF = AD$

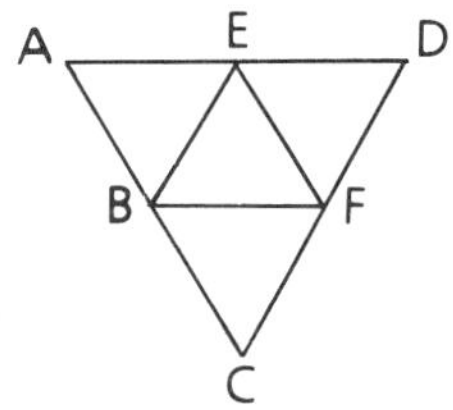

5 Given: O is the midpoint of $\overline{NP}$
R is the midpoint of $\overline{SP}$
$\overline{NP} \cong \overline{SP}$
Conclusion: $\overline{SR} \cong \overline{NO}$

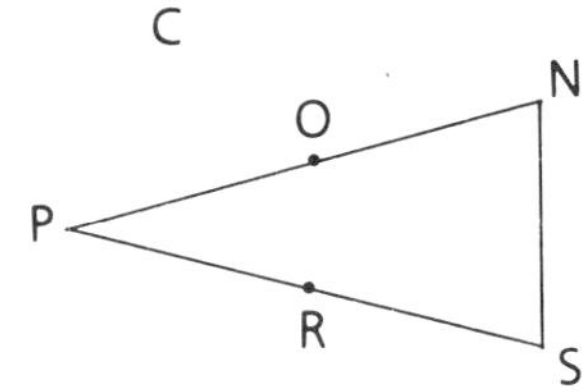

6 Given: $\overline{GJ} \cong \overline{HK}$
Conclusion: $\overline{GH} \cong \overline{JK}$

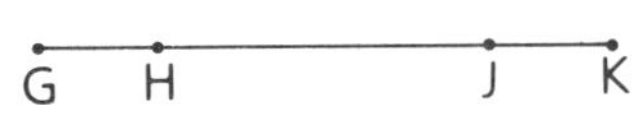

7 Given: $\angle OMP \cong \angle RPM$
$\overrightarrow{MP}$ bisects $\angle OMR$
$\overrightarrow{PM}$ bisects $\angle OPR$
Prove: $\angle OMR \cong \angle OPR$

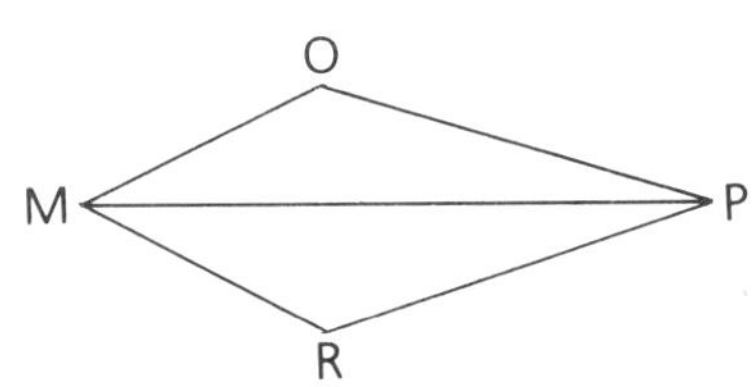

8 The complement of an angle is 24 degrees more than twice the measure of the angle. Find the measure of the complement.

9 $\angle W \cong \angle STV$
$\overrightarrow{TV}$ bisects $\angle STW$
$\angle W = (2x - 5)^\circ$
$\angle VTW = (x + 15)^\circ$
Find: $m\angle STW$

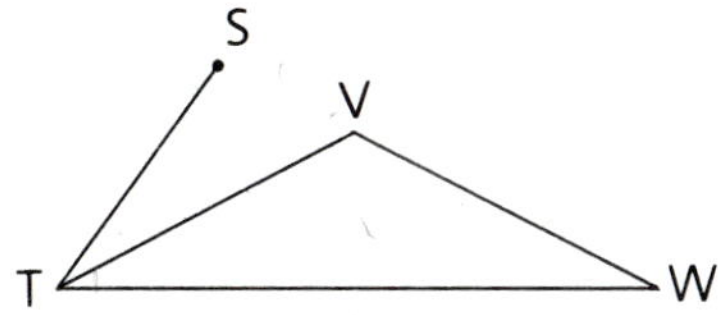

Section 2.7 Problem Set B

10 Given: $\overline{VW} \cong \overline{RS}$
$\overline{XY} \cong \overline{RS}$
Prove: $\overline{VX} \cong \overline{WY}$

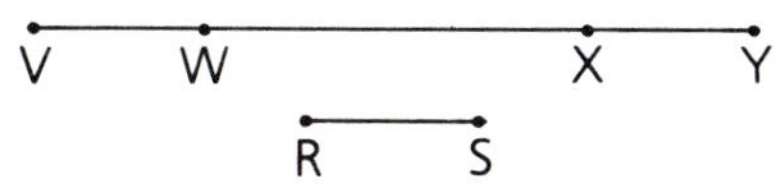

11 Given: $\angle 1 \cong \angle 2$
Conclusion: $\angle 1$ is supp. to $\angle 3$

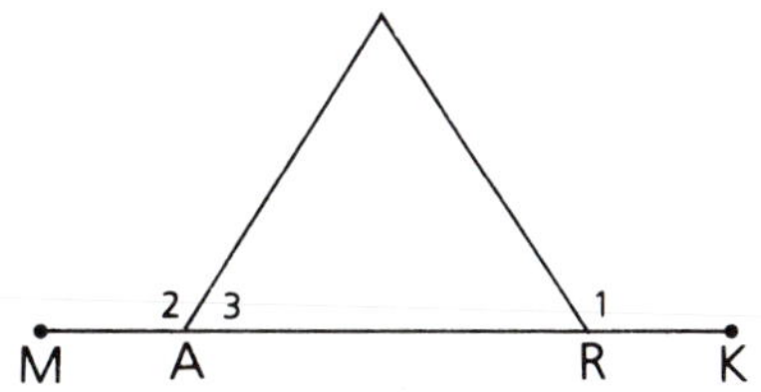

12 Given: $\angle A$ is comp. to $\angle ADB$
$\angle C$ is comp. to $\angle CDB$
$\overrightarrow{DB}$ bisects $\angle ADC$
Conclusion: $\angle A \cong \angle C$

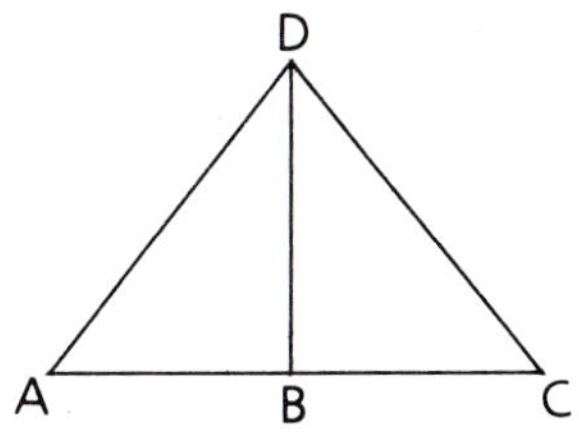

13 Given: $\angle A$ is a right angle
$\angle B$ is a right angle
$\angle B \cong \angle D$
Prove: $\angle A \cong \angle D$

14 Find the measures of each of the following angles in terms of x and y.

a $\angle HFK$

b $\angle EFK$

c $\angle HFG$

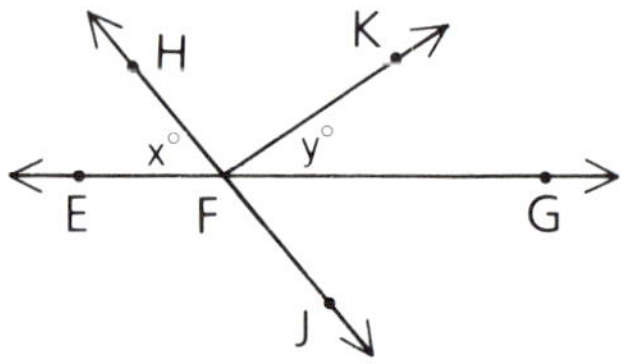

15 When one-half of the supplement of an angle is added to the complement of the angle, the result is 120°. Find the measure of the complement.

Section 2.7 Problem Set C

16 Prove that it is impossible for a third of the supplement of an angle to be added to the complement of that angle and obtain 150°.

17 Given: $\overline{AB} \perp \overline{BC}$

$\angle 1 \not\cong \angle 3$

Prove: $\angle DBE$ is a right angle

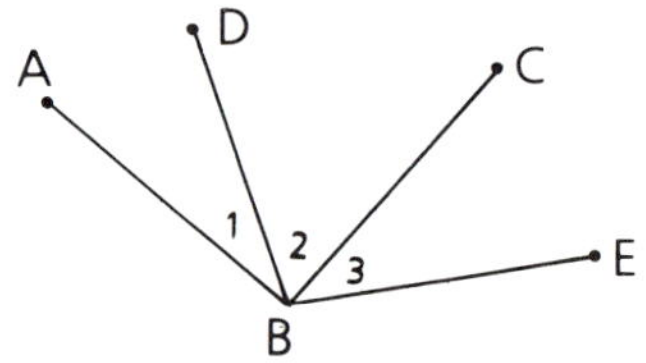

Section 2.7 Problem Set D

18 Find the set of measures of angles for which the following set of conditions is true:

Each angle must have a complement, and three-fourths of the supplement of that angle must have a complement.

Section 2.8 Vertical Angles

2.8 (A) Opposite Rays

D **Two collinear rays that have a common endpoint and extend in different directions are called OPPOSITE RAYS.**

Examples of OPPOSITE RAYS:

$\overrightarrow{AB}$ and $\overrightarrow{AC}$ are opposite.

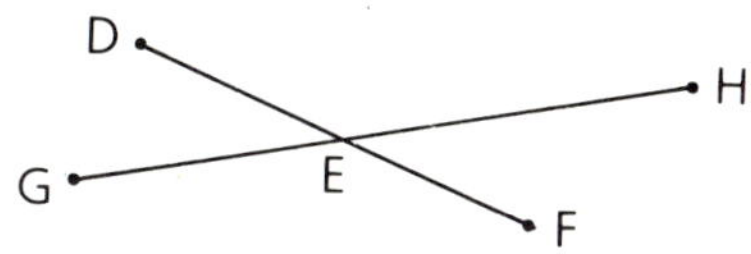

$\overrightarrow{ED}$ and $\overrightarrow{EF}$ are opposite.

$\overrightarrow{EG}$ and $\overrightarrow{EH}$ are opposite.

Examples of rays that are *not* opposite rays are shown below.

$\overrightarrow{JK}$ and $\overrightarrow{MO}$ are not parts of the same line.
$\overrightarrow{PT}$ and $\overrightarrow{RS}$ are not opposite because they do not have a common endpoint.

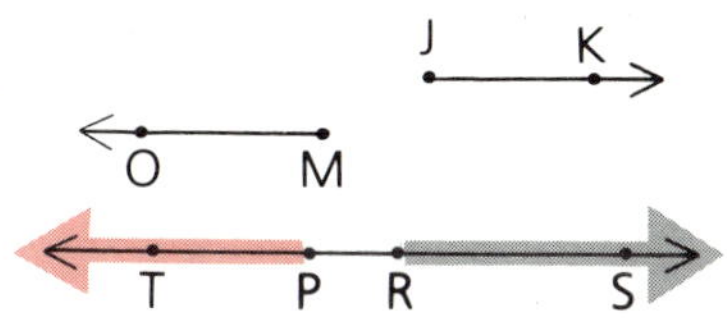

2.8 (B) Vertical Angles

D **VERTICAL ANGLES are two angles in which the sides of one are opposite rays to the sides of the other.**

Whenever two lines intersect, the VERTICAL ANGLES are the angles across from one another. In the diagram:

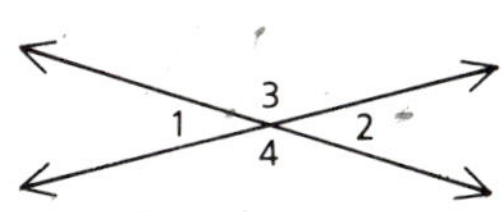

$\angle 1$ and $\angle 2$ are vertical angles.

$\angle 3$ and $\angle 4$ are vertical angles.

Are $\angle 3$ and $\angle 2$ vertical angles? How do vertical angles compare in size?

Theorem 17

T **Vertical angles are congruent.**

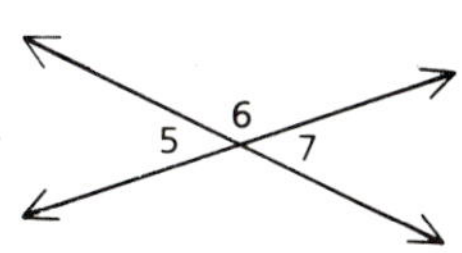

Given: Diagram as shown

Prove: $\angle 5 \cong \angle 7$

We proved Theorem 17 in Section 2.4, Sample Problem 3.

Section 2.8 Sample Problems

1 Given: $\angle 2 \cong \angle 3$

Prove: $\angle 1 \cong \angle 3$

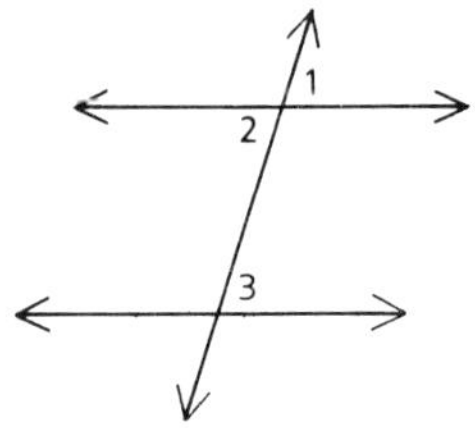

Statements	Reasons
1 $\angle 2 \cong \angle 3$	1 Given
2 $\angle 1 \cong \angle 2$	2 Vertical angles are congruent
3 $\angle 1 \cong \angle 3$	3 If angles are $\cong$ to the same $\angle$, they are $\cong$ (TRANSITIVE PROPERTY)

2 $m\angle 4 = 2x + 5$

$m\angle 5 = x + 30$

Find: $m\angle 4$

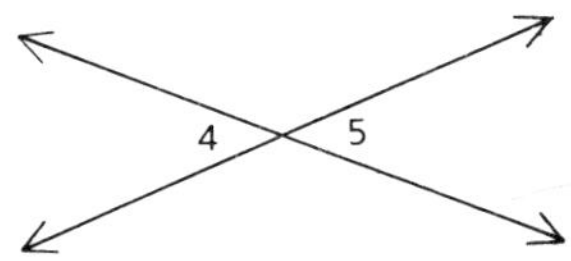

Solution:
$$\begin{aligned} 2x + 5 &= x + 30 \\ x &= 25 \\ 2x + 5 &= 55, \text{ so } m\angle 4 = 55 \end{aligned}$$

3 Given: $\angle O$ is comp. to $\angle 2$

$\angle J$ is comp. to $\angle 1$

Conclusion: $\angle O \cong \angle J$

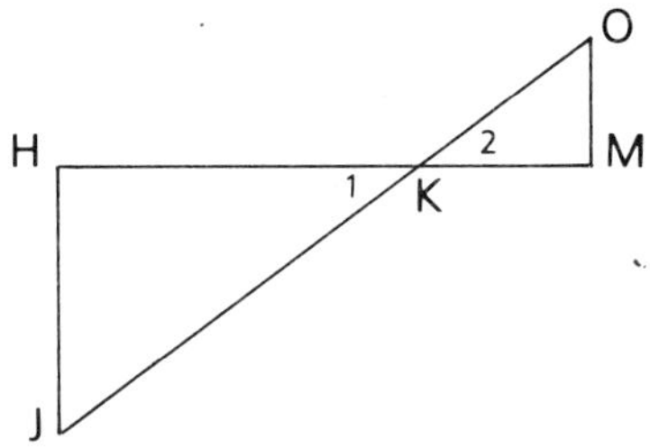

Statements	Reasons
1 $\angle O$ is comp. to $\angle 2$	1 Given
2 $\angle J$ is comp. to $\angle 1$	2 Given
3 $\angle 1 \cong \angle 2$	3 Vertical angles are congruent
4 $\angle O \cong \angle J$	4 If angles are complementary to $\cong$ angles, they are $\cong$ Or: Complements of $\cong$ angles are $\cong$

Section 2.8 Problem Set A

1 a Name 3 pairs of opposite rays

b Name 2 pairs of vertical angles

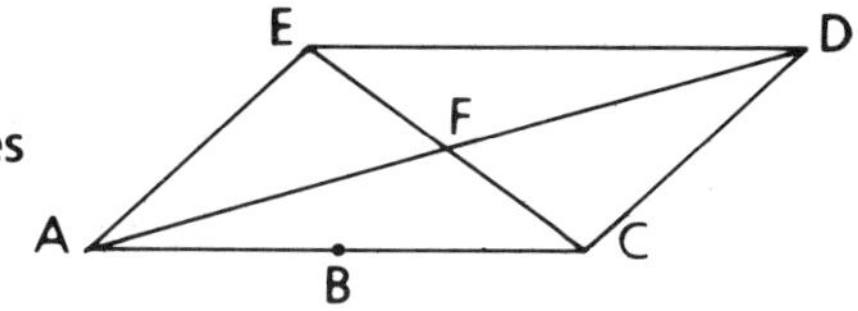

2 $\angle 1 = 60°32'$

Find: $\angle 2$

$\angle 3$

$\angle 4$

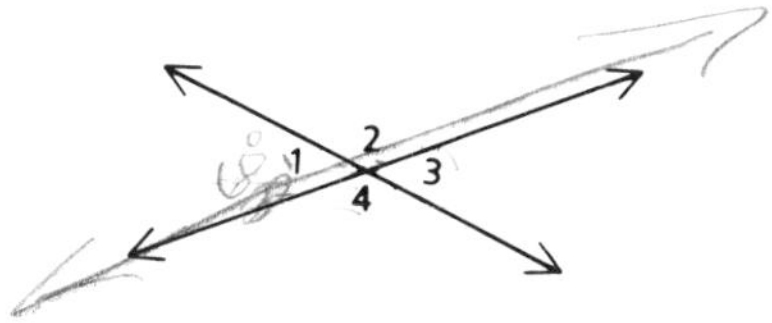

3 $\angle 5 = (2x + 7)°$

$\angle 6 = (x + 25)°$

Find: $m\angle 5$

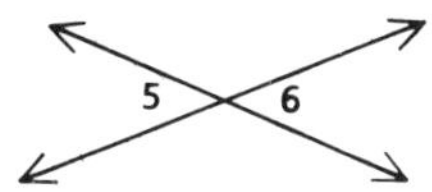

4 Given: $\angle A \cong \angle ACB$

Prove: $\angle A \cong \angle DCE$

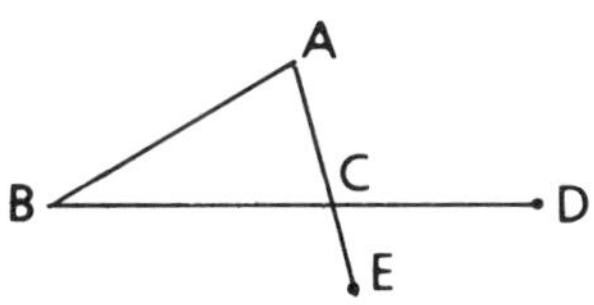

5 Given: $\angle 1 \cong \angle 4$

Conclusion: $\angle 2 \cong \angle 3$

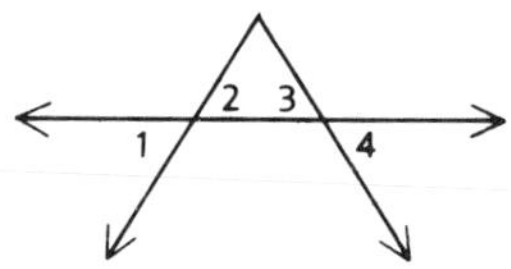

6 Given: $\overline{FH} \cong \overline{GJ}$

Prove: $\overline{FG} \cong \overline{HJ}$

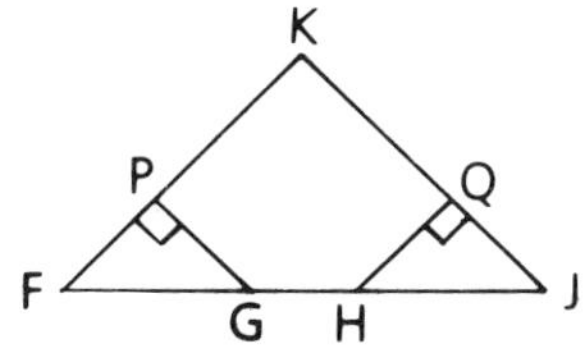

7 Is this possible?

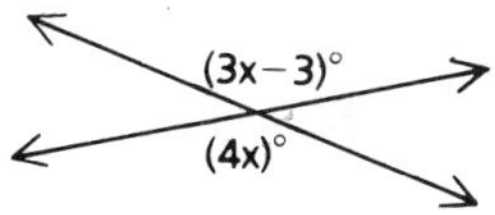

8 Given: $\angle 4 \cong \angle 6$

Prove: $\angle 5 \cong \angle 6$

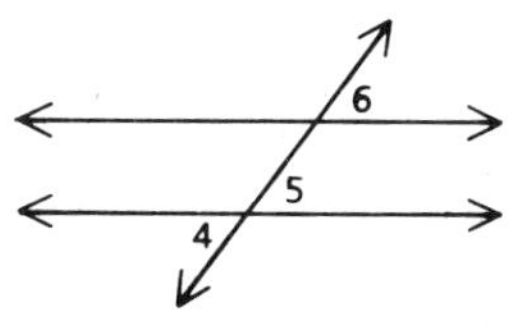

Section 2.8 Problem Set B

9 Given: $\angle 1 \cong \angle 3$

Prove: $\angle 2$ is supp. to $\angle 3$

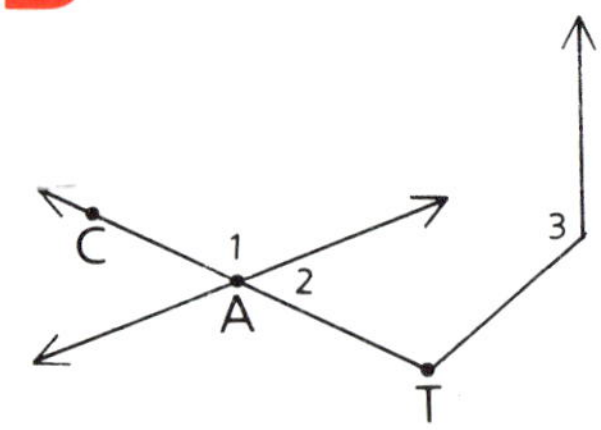

10 Given: $\angle V \cong \angle YRX$

$\angle Y \cong \angle TRV$

Prove: $\angle V \cong \angle Y$

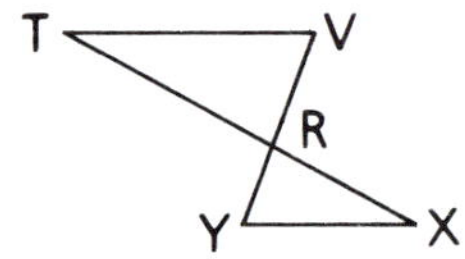

11 Given: $\overleftrightarrow{GD}$ bisects $\angle CBE$

Conclusion: $\angle 1 \cong \angle 2$

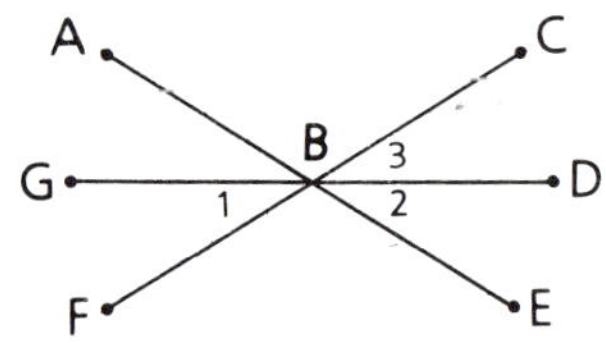

12 Angles 4, 5, and 6 have the respective ratios of 2:5:3. Find the measure of each angle.

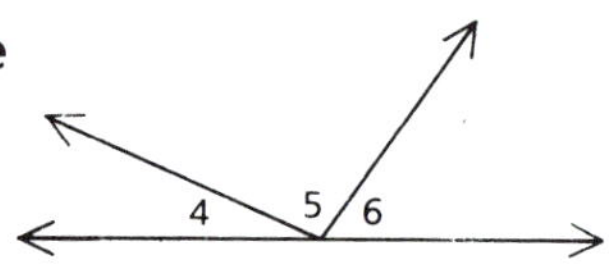

13 If a pair of vertical angles are supp., what can you conclude about the angles?

Section 2.8 Problem Set C

14 Given: $\angle 2$ is comp. to $\angle 6$

$\angle 1$ is comp. to $\angle 3$

$\angle 3$ is comp. to $\angle 4$

Prove: $\angle ABC \cong \angle EDF$

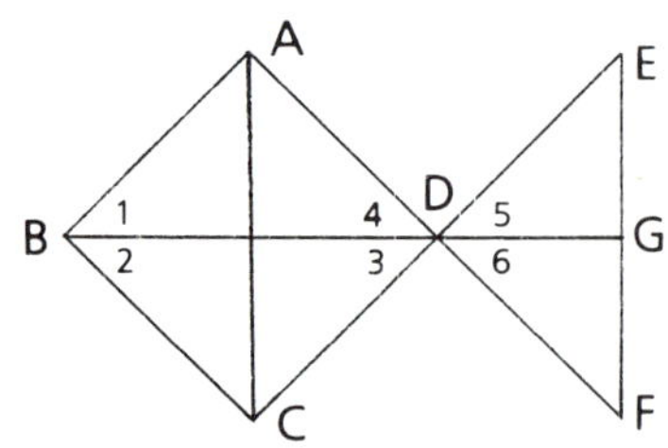

15 Find: $m\angle 1$

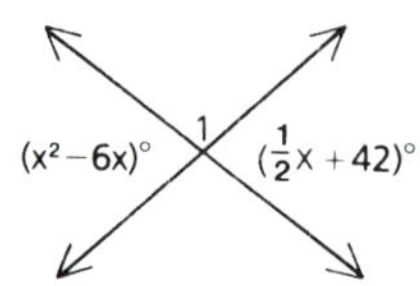

Chapter 2 Study Guide

As you study and review this chapter, be sure that you are able to:

1 Describe each of the following definitions and theorems by:

(i) writing the full statement;
(ii) writing the alternate form of the statement (if there is one);
(iii) writing the reverse (if it is a definition);
(iv) drawing a sketch (as shown for **a** and **d**).

a def. of ⊥ lines (p. 58)

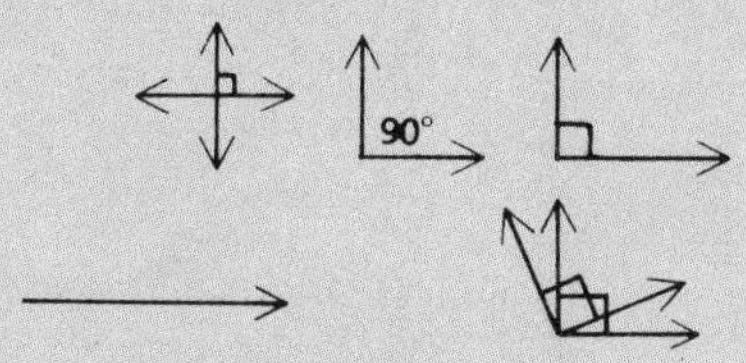

b def. of oblique lines (p. 62)
c def. of comp. ∠s (p. 63)
d a theorem about comp. ∠s (p. 74)
e another theorem about comp. ∠s (p. 74)
f def. of supp. ∠s (p. 63)
g a theorem about supp. ∠s (p. 73)
h another theorem about supp. ∠s (p. 73)
i a theorem about vertical ∠s (p. 97)

2 Follow the five part procedure for drawing conclusions. (p. 69)

3 State the theorems needed to draw the stated conclusions.

a Given: ∠1 ≅ ∠4 and ∠3 ≅ ∠4
Conclusion: ∠1 ≅ ∠3 (p. 92)

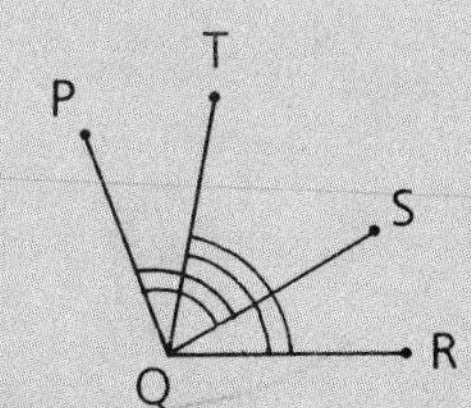

b Given: Marked diagram
Conclusion: ∠PQT ≅ SQR (p. 81)

c Given: Marked diagram
∠ABC ≅ ∠ADC
Conclusion: ∠1 ≅ ∠2 (p. 86)

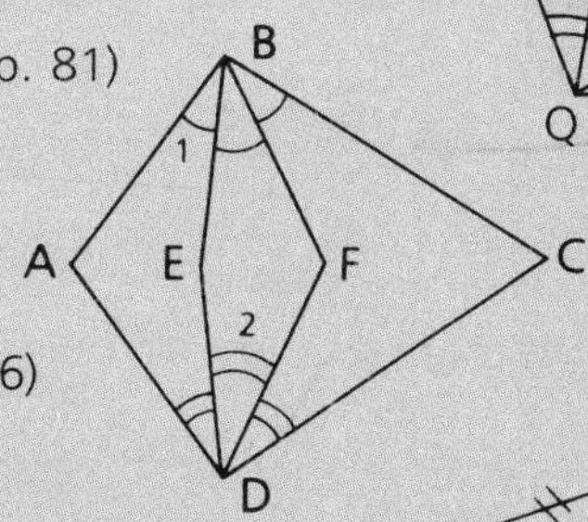

d Given: Marked diagram
Conclusion: $\overline{EL} \cong \overline{EG}$ (p. 80)

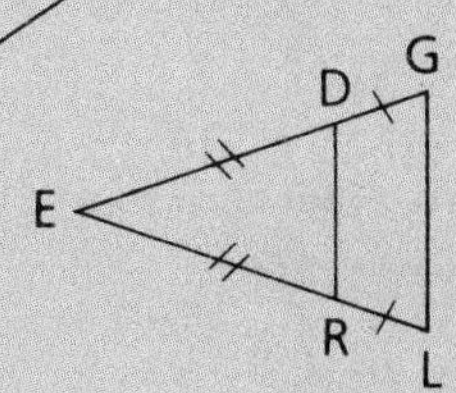

4 Show how you decide when to use the following properties.

a Addition (pp. 79, 80)
b Subtraction (p. 81)
c Multiplication (p. 86)
d Division (p. 86)
e Transitive (p. 92)

Chapter 2 Review Problems **A**

1 Given: $\overline{JK} \perp \overline{KM}$

Prove: $\angle JKO$ is comp. to $\angle OKM$

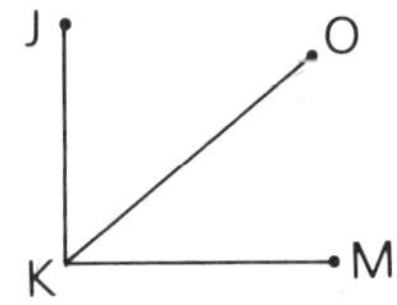

2 Given: $\overline{PV} \cong \overline{PR}$

$\overline{VT} \cong \overline{RS}$

Conclusion: $\overline{PT} \cong \overline{PS}$

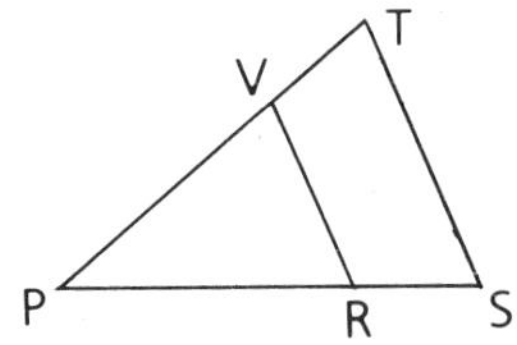

3 Given: $\angle WXT \cong \angle YXZ$

Prove: $\angle WXZ \cong \angle TXY$

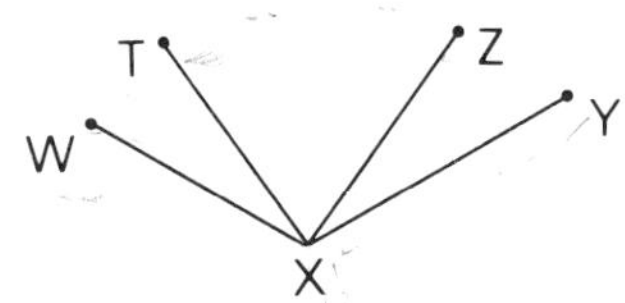

4 Given: $\overline{FG} \cong \overline{JH}$

N is the midpt. of $\overline{FG}$

O is the midpt. of $\overline{JH}$

Prove: $\overline{NG} \cong \overline{OH}$

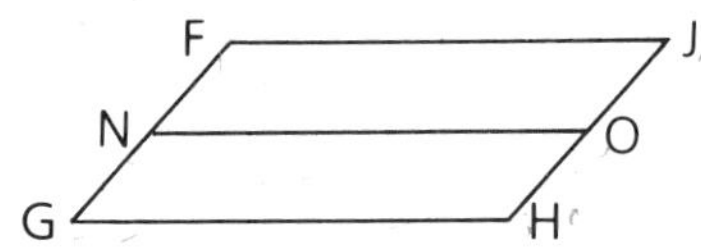

5 Given: $\overline{RV}$ and $\overline{SW}$ trisect $\overline{PT}$ and $\overline{PX}$

$\overline{ST} \cong \overline{WX}$

Conclusion: $\overline{PT} \cong \overline{PX}$

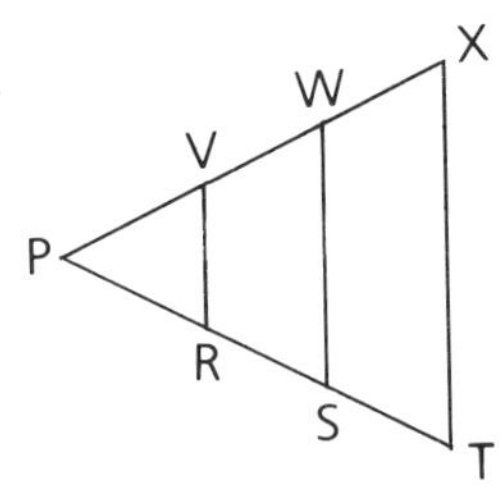

6 Given: Diagram as shown

$\angle 1 \cong \angle 4$

Prove: $\angle 2 \cong \angle 3$

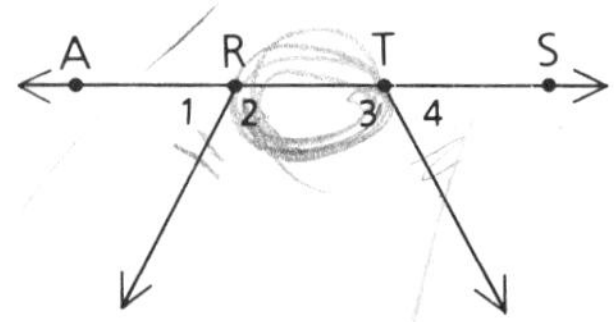

7 Given: $\angle A$ is supp. to $\angle D$

$\angle A \cong \angle C$

Prove: $\angle C$ is supp. to $\angle D$

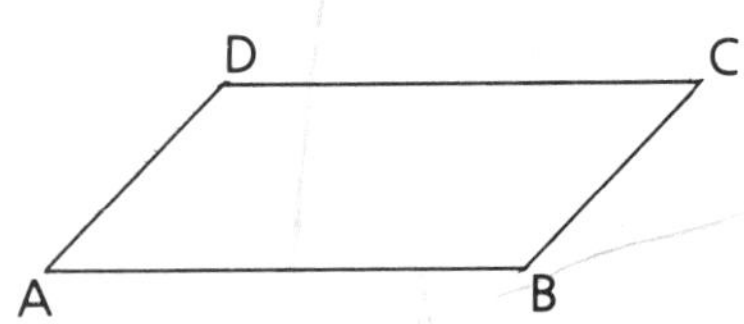

8 Given: $\angle 1 \cong \angle 3$

Conclusion: $\angle 1 \cong \angle 2$

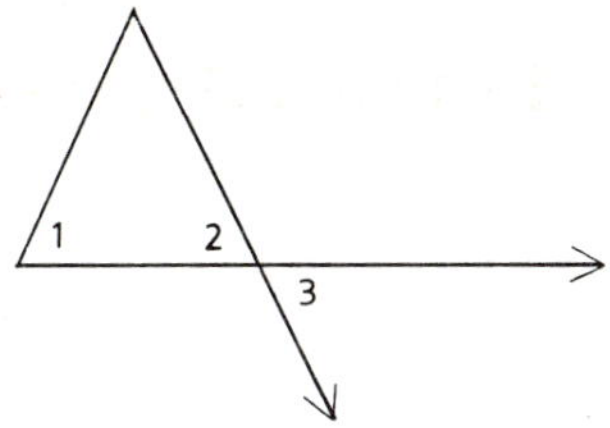

9 Given: $\angle EGF \cong \angle EFG$

$\angle EGH \cong \angle EFJ$

Conclusion: $\angle HGF \cong \angle JFG$

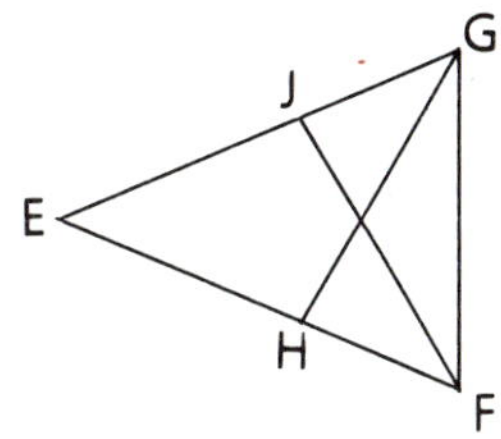

10 Given: $\angle ABD$ is a right $\angle$

$\angle CBE$ is a right $\angle$

Conclusion: $\angle ABE \cong \angle CBD$

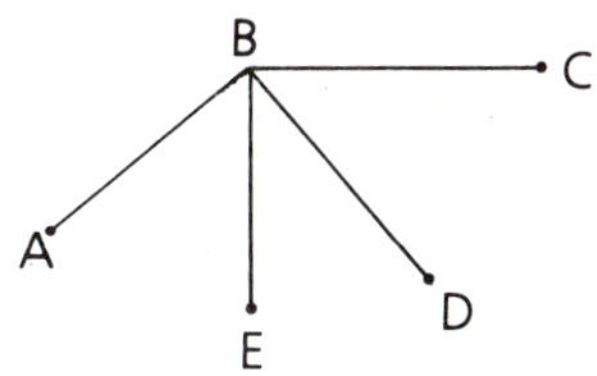

11 Point E divides $\overline{DF}$ into segments having the ratio (from left to right) of 5:2. If DF = 21 cm, find EF.

12 One of two complementary angles has a measure that is six more than twice the other. Find the measure of the larger.

13 The measure of the supplement of an angle is 5 times that of its complement. Find the measure of the complement.

14 Two non-perpendicular intersecting lines are called __?__

15 Given: DA = 12

A is the midpoint of $\overline{DE}$

I and N are trisection points of $\overline{DE}$

Find: m $\overline{AN}$

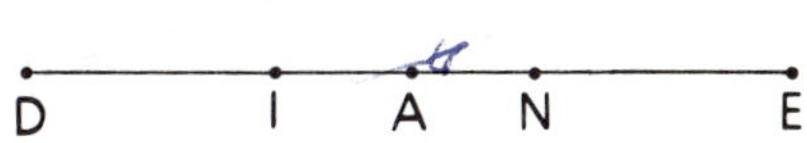

16 Find the supplement and complement of each angle:

a 83° **b** $42^\circ 15' 38''$ **c** 97°

Chapter 2 Review Problems B

17 Given: $\angle 1$ is comp. to $\angle 3$
$\angle 4$ is comp. to $\angle 2$

Conclusion: $\angle 1 \cong \angle 4$

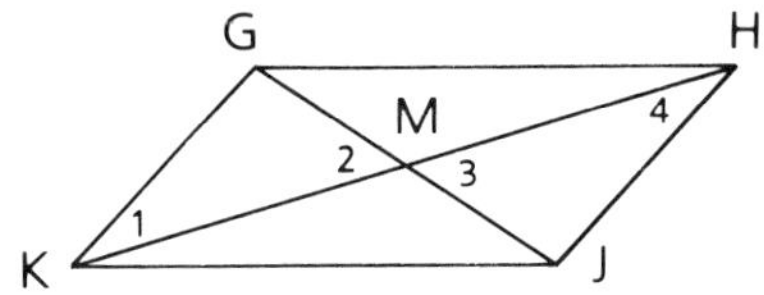

18 Given: O is the midpoint of $\overline{NP}$
$\overline{RN} \cong \overline{PO}$

Conclusion: $\overline{RN} \cong \overline{NO}$

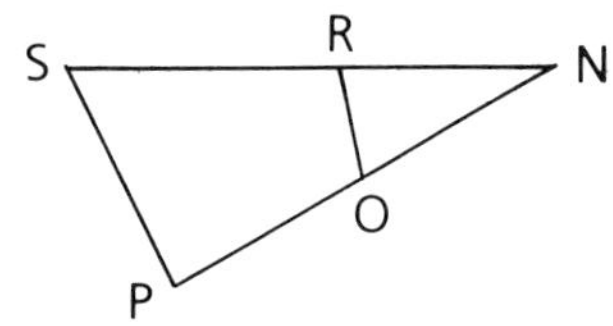

19 Given: $\overrightarrow{VY}$ bisects $\angle TVZ$
$\overrightarrow{ZY}$ bisects $\angle TZV$
$\angle TVZ \cong \angle TZV$

Conclusion: $\angle 3 \cong \angle 1$

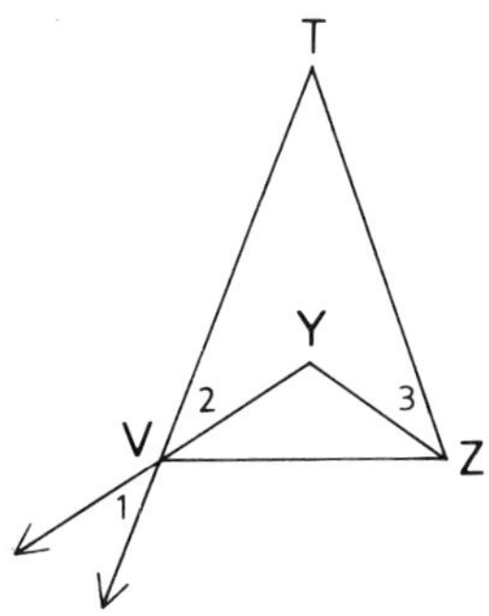

20 Given: $\overrightarrow{BC}$ bisects $\angle DBE$

Prove: $\angle ABD \cong \angle ABE$

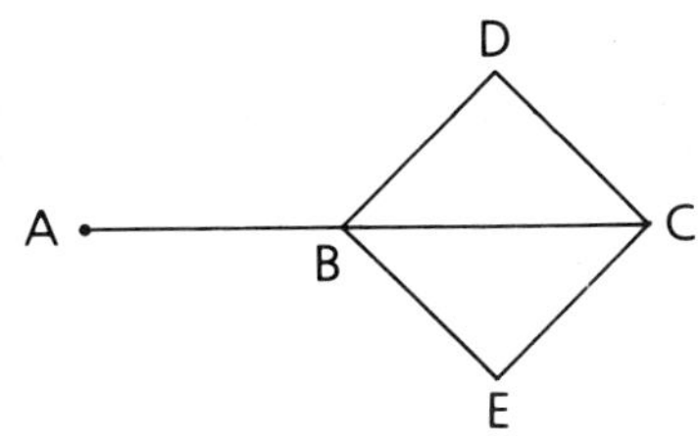

21 Given: $\angle F \cong \angle 1$
$\angle J \cong \angle 2$
$\overline{FK} \perp \overline{KH}$
$\overline{GK} \perp \overline{KJ}$

Prove: $\angle F \cong \angle J$

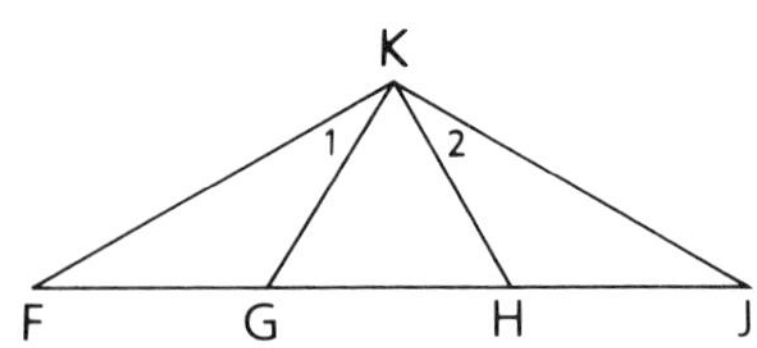

22 Given: $\angle NOP \cong \angle SRP$
$\angle NOP$ is comp. to $\angle POR$
$\angle SRP$ is comp. to $\angle PRO$

Prove: $\angle POR \cong \angle PRO$

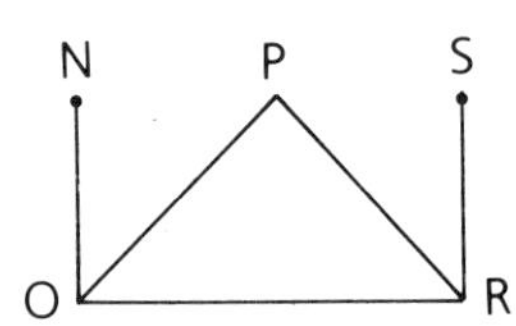

23 Solve for x and y

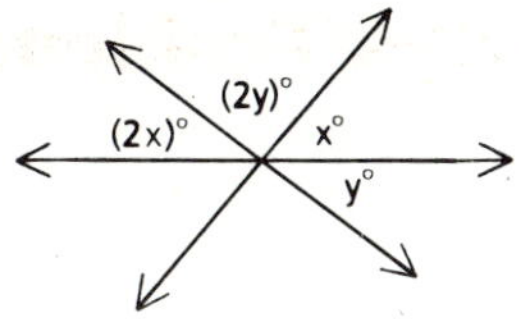

24 $\overline{VS} \cong \overline{VR}$
$\overline{WT}$ bisects $\overline{VS}$ and $\overline{VR}$
Find the perimeter of $\triangle VRS$

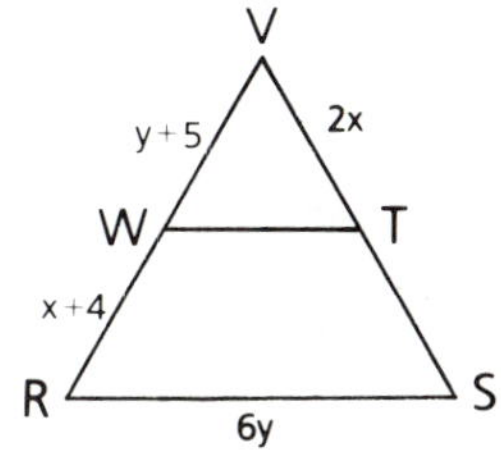

25 The measure of the supplement of an angle exceeds twice the measure of the complement of the angle by 20. Find the measure of half of the complement.

26 Solve for y in terms of x

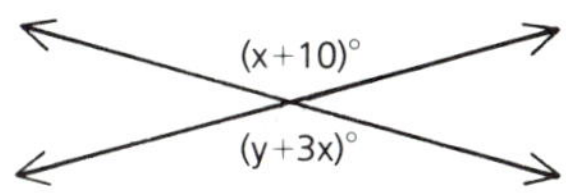

Chapter 2 Review Problems C

27 Tippy Van Winkle is awakened from a deep sleep by the cuckoo of a clock which sounds every half hour. Before Tippy can look at the clock, his brother Bippy enters the room and offers to bet $10 that the hands of the clock form an acute angle. Assuming that the hands have not moved since the cuckoo, how much should Tippy put up against Bippy's $10 so that it is an even bet?

28 Use a paragraph proof:

Given: $\angle ABD$ is supp. to $\angle EDB$
$\overrightarrow{BC}$ bisects $\angle ABD$
$\overrightarrow{DC}$ bisects $\angle BDE$

Prove: $\angle CBD$ is comp. to $\angle BDC$

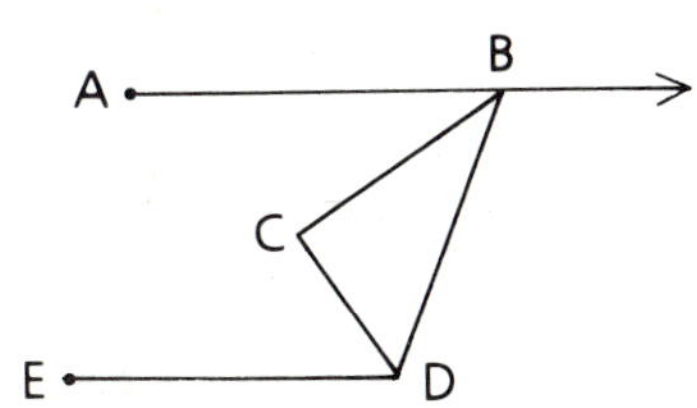

Chapter 3

CONGRUENT TRIANGLES

The Enjoyment of Congruent Triangles

In Chapters 1 and 2, you were introduced to numerous properties of segments and angles. You learned some of the art of proof and the use of algebra and arithmetic in geometry. Although you may still be uneasy about proofs, you will find your confidence improving as you work daily with triangles in this chapter. Also, since congruent triangles serve as the springboard to the properties of other figures, this chapter may well be the first of many profitable and enjoyable chapters that lie ahead.

Short Forms

Proofs of congruent triangles will become unnecessarily long to write unless we shorten the language of some of the reasons. Accordingly, from now on, the theorems and postulates of Chapter 2 that have short forms will be quoted only in short form. The short forms were presented in capital letters and in parentheses.

You may want to review the following properties from Chapter 2.

ADDITION PROPERTY	MULTIPLICATION PROPERTY	TRANSITIVE PROPERTY
DIVISION PROPERTY	SUBTRACTION PROPERTY	SUBSTITUTION

We shall also need the **REFLEXIVE PROPERTY**.

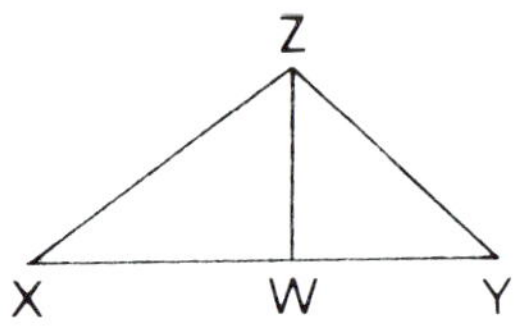

To see what it means, look at the figure.
Which is longer, $\overline{ZW}$ of $\triangle XZW$ or $\overline{ZW}$ of $\triangle YZW$?

Which is larger, $\angle X$ of $\triangle XZW$ or $\angle X$ of $\triangle XYZ$?

Obviously, $\overline{ZW} \cong \overline{ZW}$ and $\angle X \cong \angle X$, regardless of which triangle they are in.

P **Any segment (or angle) is congruent to itself. (REFLEXIVE PROPERTY)**

In the following two problems, see if you can justify each conclusion with one of the seven properties named above:

Given: M and N are midpoints.
$\overline{DC} \cong \overline{AB}$, and $\overline{AB} \cong \overline{DB}$
$\angle 1 \cong \angle 4$, and $\angle 2 \cong \angle 3$

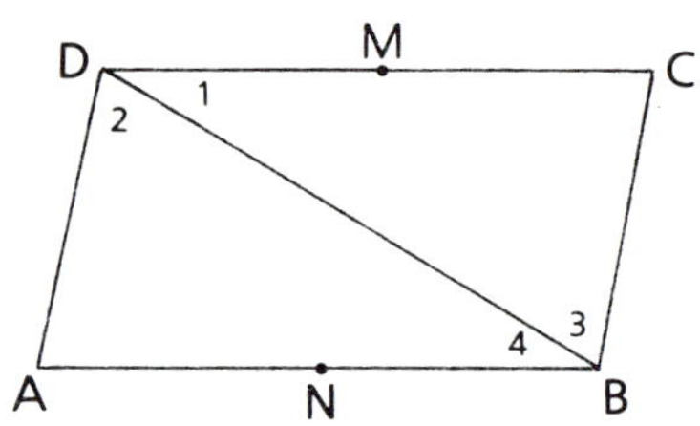

Conclusions:
- a $\angle ADC \cong \angle ABC$
- b $\overline{CM} \cong \overline{AN}$
- c $\overline{BD} \cong \overline{DB}$
- d $\overline{DC} \cong \overline{DB}$

Given: $\overrightarrow{FP}$ and $\overrightarrow{GP}$ are angle bisectors
$\angle 5 \cong \angle 7$, $\angle 5$ is an acute angle
$\overline{PF} \cong \overline{PG}$, and $\overline{QG} \cong \overline{FR}$

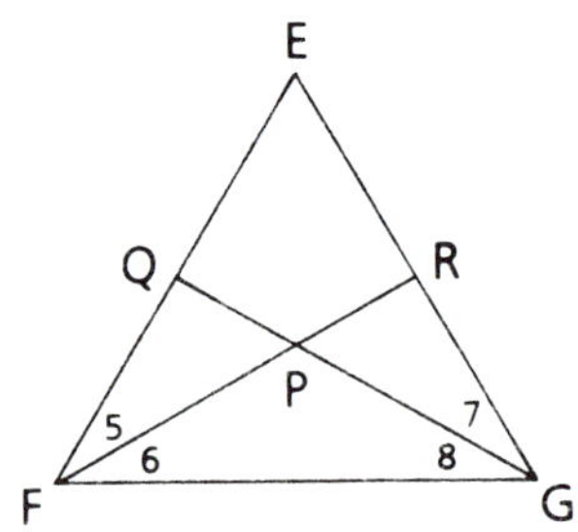

Conclusions:
- e $\angle QFG \cong \angle RGF$
- f $\overline{QP} \cong \overline{PR}$
- g $\angle 7$ is an acute angle
- h $\angle FER \cong \angle GEQ$

Answers:
- a ADDITION PROPERTY
- b DIVISION PROPERTY
- c REFLEXIVE PROPERTY
- d TRANSITIVE PROPERTY
- e MULTIPLICATION PROPERTY
- f SUBTRACTION PROPERTY
- g SUBSTITUTION
- h REFLEXIVE PROPERTY

Section 3.1 Three Ways To Prove That Triangles Are Congruent

In this section we discuss:

A THE MEANING OF CONGRUENT TRIANGLES
B INCLUDED ANGLES AND INCLUDED SIDES
C SSS
D SAS
E ASA

3.1 (A) The Meaning of Congruent Triangles

If we give you two congruent triangles, $\triangle ABC \cong \triangle DEF$, then we will mean ALL of the following:

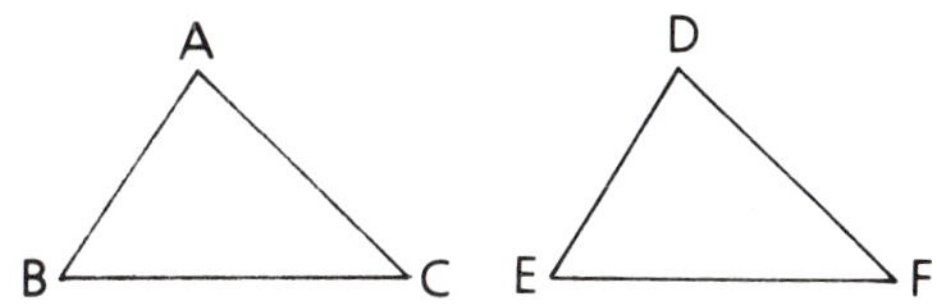

$\triangle ABC$ has the same size and shape as $\triangle DEF$

Either triangle will fit exactly on the other so that A could become the same point as D, B the same point as E, and C the same point as F

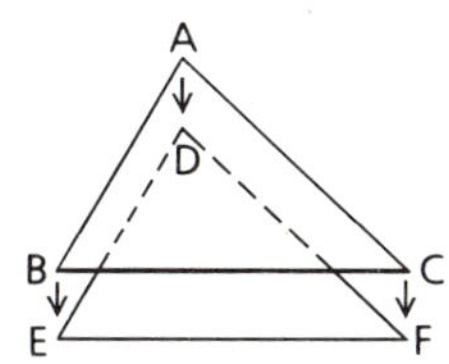

$\angle A \cong \angle D$, $\angle B \cong \angle E$, $\angle C \cong \angle F$

$\overline{AB} \cong \overline{DE}$, $\overline{BC} \cong \overline{EF}$, $\overline{CA} \cong \overline{FD}$

D **CONGRUENT TRIANGLES $\Longleftrightarrow$ *all* pairs of *corresponding* parts are congruent.**

An arrow symbol ($\Longrightarrow$) means *implies* (if . . . , then . . .).
The double arrow symbol ($\Longleftrightarrow$) means the statement is reversible.

Would $\triangle ABC \cong \triangle EFD$ be correct? No! Corresponding letters must *match* in the correspondence.

CORRECT: $\triangle ABC \cong \triangle DEF$ INCORRECT: $\triangle ABC \cong \triangle EDF$

$\triangle ABC \cong \triangle EDF$ is incorrect, because $\triangle ABC$ cannot be placed on $\triangle EDF$ so that A falls on E, B on D, and C on F. Would $\triangle EDF \cong \triangle BAC$ be correct?

3.1 (B) Included Angles and Included Sides

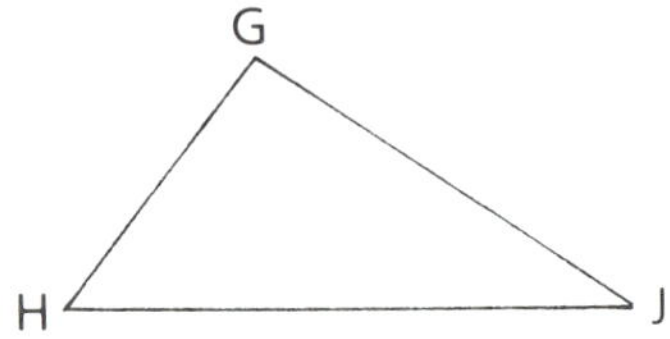

In the above figure, $\angle H$ is the included angle for $\overline{GH}$ and $\overline{HJ}$, and $\overline{HJ}$ and $\overline{JG}$ "include" $\angle J$. Can you name the sides that include $\angle G$?

Using similar language for a side of a triangle, we say that $\overline{GH}$ is included by $\angle G$ and $\angle H$. We also say that $\angle G$ and $\angle J$ include $\overline{GJ}$. Can you name the two angles that include $\overline{HJ}$?

We must be careful to name the triangle in which we are operating and then talk about sides and angles of that triangle. For example:

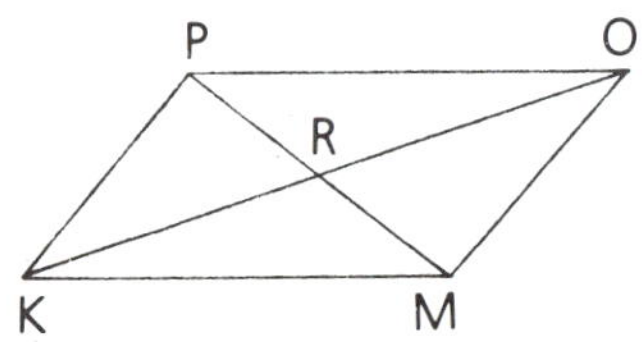

In $\triangle RKM$, $\angle OKM$ is included by $\overline{RK}$ and $\overline{KM}$.
In $\triangle PMK$, $\overline{PM}$ is included by $\angle KPM$ and $\angle KMP$.
In $\triangle KOM$, can you name the sides that include $\angle MOR$?

3.1 (C) SSS

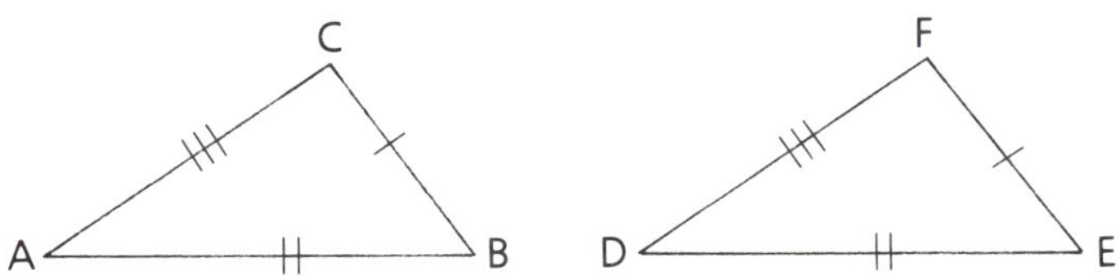

The tick marks on $\triangle ABC$ and $\triangle DEF$ show sufficient conditions to know that $\triangle ABC \cong \triangle DEF$. The set of conditions is called SSS, where each S stands for a pair of congruent corresponding sides, such as $\overline{AC} \cong \overline{DF}$. The full wording is:

P **If there exists a correspondence between the vertices of two triangles such that three sides of one triangle are congruent respectively to the corresponding sides of the second triangle, the two triangles are congruent. (SSS)**

The SSS relationship can be proved by methods that are not part of this course; we shall assume it and use its short form, SSS.

In the figure, is $\triangle GHJ \cong \triangle GKJ$ by SSS? Let us see. From the tick marks we have SS, and that is not enough. But wait! Since $\overline{GJ} \cong \overline{GJ}$ by the reflexive property, we actually do have SSS.

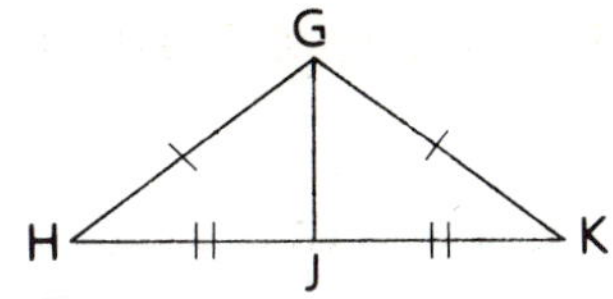

Notice, $\overline{GJ}$ is a side of each of the triangles, $\triangle GHJ$ and $\triangle GKJ$. The fact that $\overline{GJ}$ is a side of each triangle is important, for someone could have said that $\overline{HK} \cong \overline{HK}$ by the reflexive property. Although it is true that $\overline{HK} \cong \overline{HK}$ by the reflexive property, that would not have helped us prove that $\triangle GHJ \cong \triangle GKJ$. Do you see why?

3.1(D) SAS

P **The SAS postulate.**

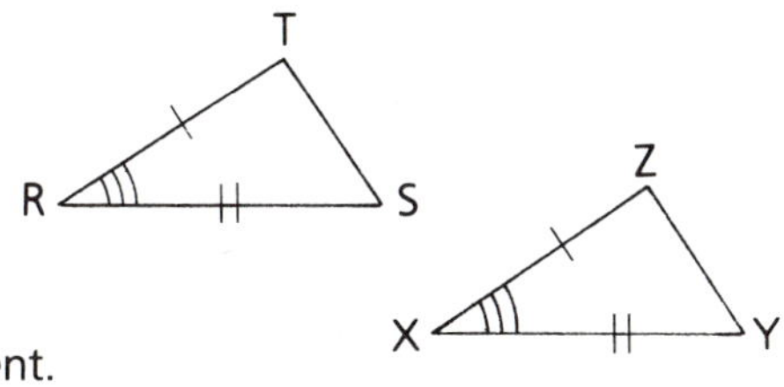

As with SSS, it can be proved that SAS describes sufficient conditions to conclude that two triangles are congruent. The A in the middle of SAS means that the corresponding *Angle* in each triangle is *included* by the two Sides. We make SAS a postulate and use only the short form, SAS.

3.1(E) ASA

P **The ASA postulate.**

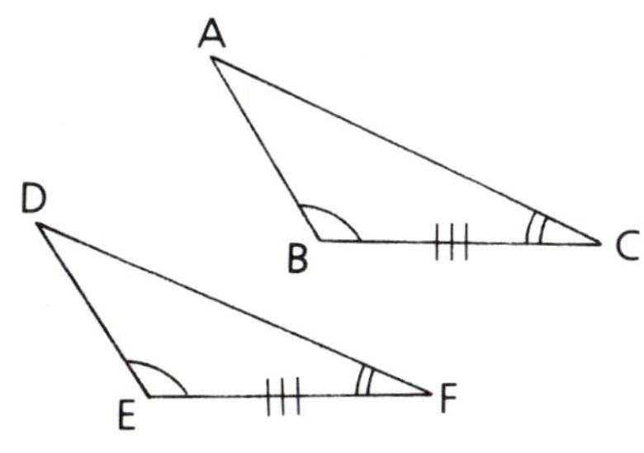

Again, as with SSS, it can be proved that ASA describes sufficient conditions to claim that two triangles are congruent. The S in the *middle* of ASA matches the arrangement of marked parts in each triangle; the *Side* in each triangle is the *included side.* We postulate ASA. Its full wording is much like that of SSS, so we use only its short form.

If you are curious, you are certainly wondering if SSS, SAS, and ASA are the only shortcuts for proving that triangles are congruent. Not quite. However, SSS, SAS, and ASA are enough to get us started on proofs that triangles are congruent.

Study the sample problems carefully before you attempt the Problem Set. Notice that we call SSS, SAS, and ASA methods of proof. Any definition, postulate, or theorem will be called a method if it is a key reason in proofs.

Section 3.1 Sample Problems

In **1–4** you are given the congruent angles or sides shown by the tick marks. Name the additional congruent sides or angles needed to prove that the triangles are congruent by the specified method.

	Method	Answers
1	**a** SSS	**a** $\overline{AC} \cong \overline{DF}$
	b SAS	**b** $\angle B \cong \angle E$

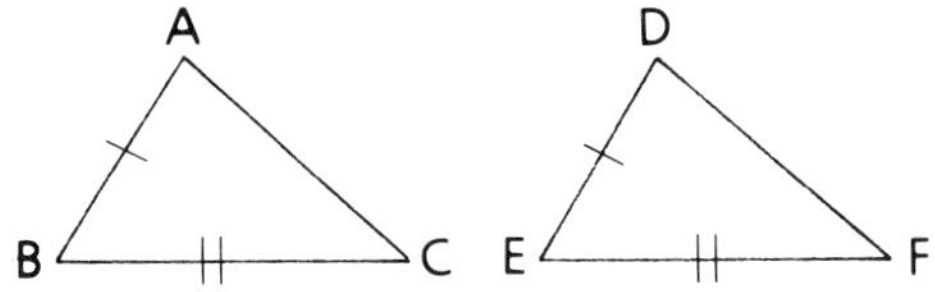

2	**a** SAS	**a** $\overline{GJ} \cong \overline{OM}$
	b ASA	**b** $\angle H \cong \angle K$

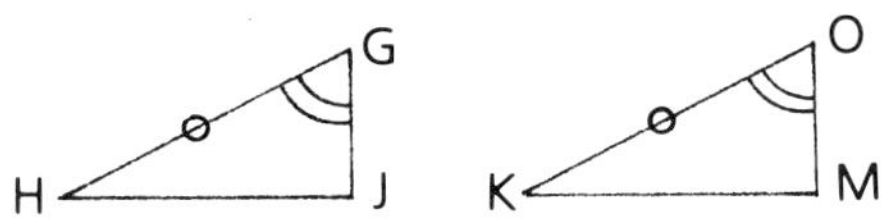

3 Prove: $\triangle PWT \cong \triangle SVR$

a SAS		**a** $\overline{TW} \cong \overline{RV}$
b ASA		**b** $\angle TPW \cong \angle RSV$

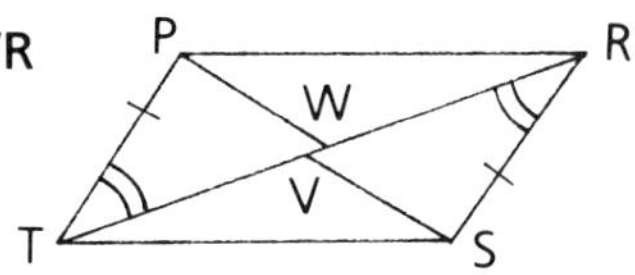

4 Prove: $\triangle AEC \cong \triangle DEB$

a SSS		**a** $\overline{AC} \cong \overline{BD}$
b SAS		**b** $\angle AEC \cong \angle DEB$

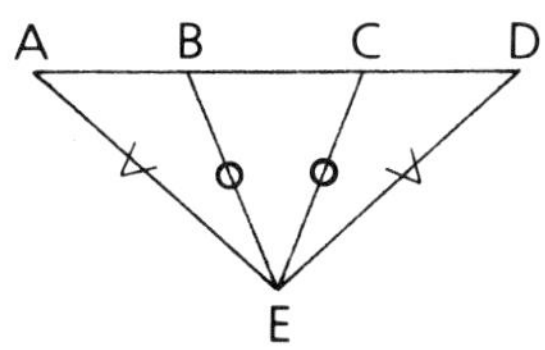

5 Given: $\overline{AD} \cong \overline{CD}$

B is midpoint of $\overline{AC}$

Conclusion: $\triangle ABD \cong \triangle CBD$

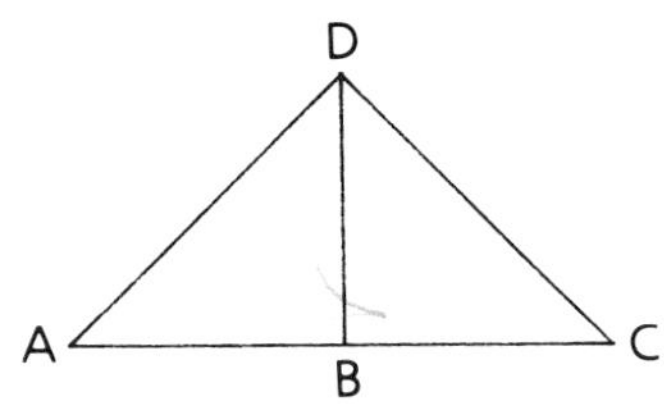

Statements	Reasons
1 $\overline{AD} \cong \overline{CD}$	1 Given
2 B is midpoint of $\overline{AC}$	2 Given
3 $\overline{AB} \cong \overline{CB}$	3 If a point is the midpoint of a segment, it divides the segment into two $\cong$ segments
4 $\overline{BD} \cong \overline{BD}$	4 Reflexive Property
5 $\triangle ABD \cong \triangle CBD$	5 SSS (1, 3, 4)*

* Note: In parentheses we placed the numbers of the statements in which we obtained each of the parts needed.

6 Using the tick marks for each pair of triangles, name the method (SSS, SAS, ASA), if any, that will prove the triangles are congruent.

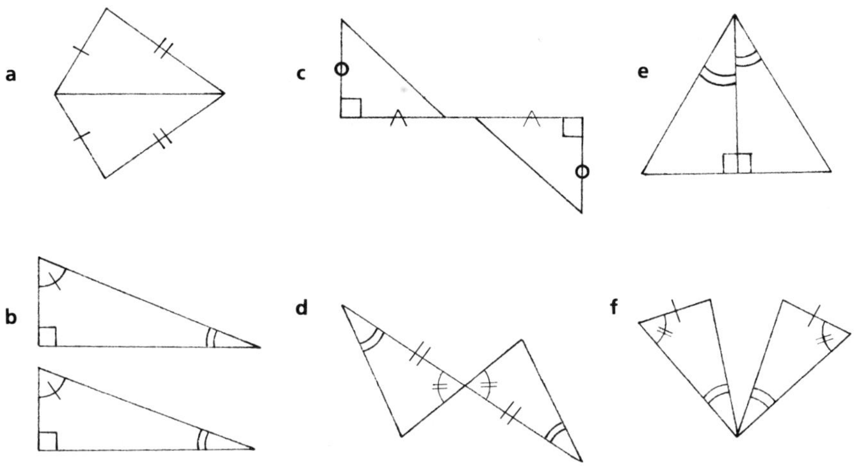

Answers; **a** SSS **c** SAS **e** ASA
b None **d** ASA **f** None

7 Given: $\angle 3 \cong \angle 6$
$\overline{KR} \cong \overline{PR}$
$\angle KRO \cong \angle PRM$

Prove: $\triangle KRM \cong \triangle PRO$

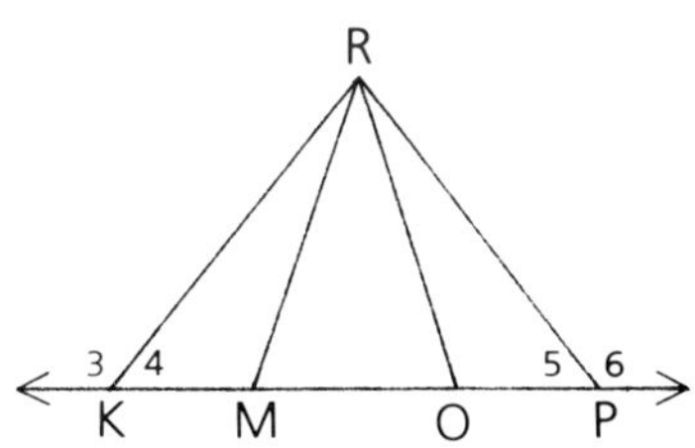

Statements	Reasons
1 $\angle 3 \cong \angle 6$	1 Given
2 $\angle 3$ is supp. to $\angle 4$	2 If two angles form a straight$\angle$ (assumed from diagram), they are supplementary *
3 $\angle 5$ is supp. to $\angle 6$	3 Same as 2
4 $\angle 4 \cong \angle 5$	4 Angles each supp. to $\cong$ $\angle$s are $\cong$
5 $\overline{KR} \cong \overline{PR}$	5 Given
6 $\angle KRO \cong \angle PRM$	6 Given
7 $\angle KRM \cong \angle PRO$	7 Subtraction Property
8 $\triangle KRM \cong \triangle PRO$	8 ASA (4, 5, 7)

* The assumption of straight angles and the fact that two angles are supplementary may now be combined in one step.

Section 3.1 Problem Set A

1 Using the ≅ sides or angles shown by the tick marks, indicate the additional information needed to support the specified method of proving that the △s are ≅.

		Triangles	Method	Needed information
a	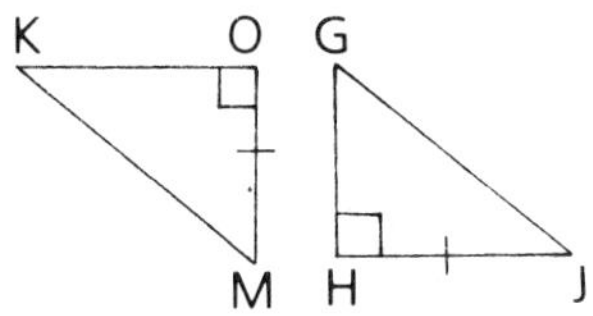	△HGJ ≅ △OKM	SAS	________
			ASA	________
b	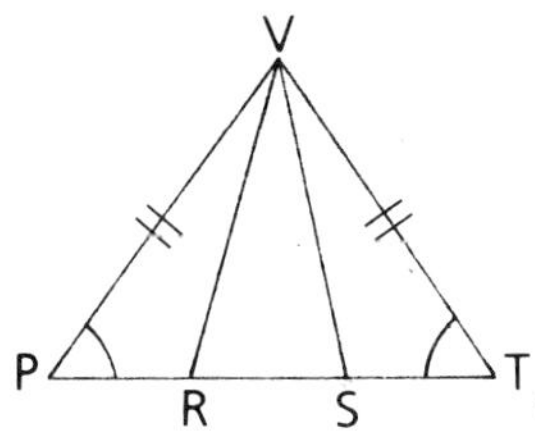	△PSV ≅ △TRV	SAS	________
			ASA	________
c	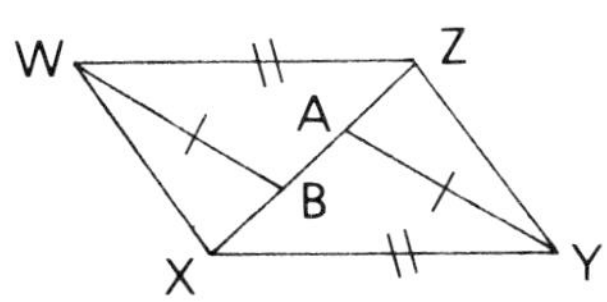	△WBZ ≅ △YAX	SSS	________
			SAS	________
d	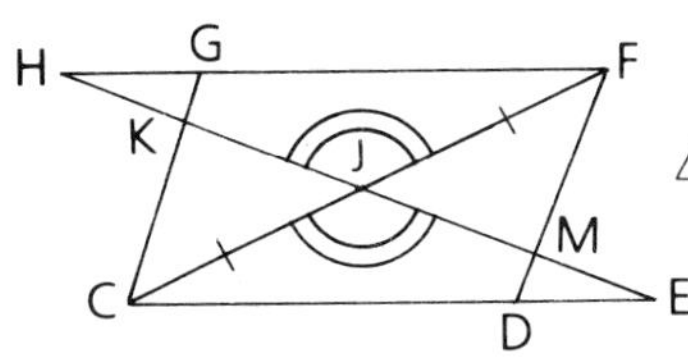	△HJF ≅ △EJC	SAS	________
			ASA	________

2 Using the tick marks for each pair of △s, name the method (SSS, SAS, ASA), if any, that will prove the △s are ≅.

a

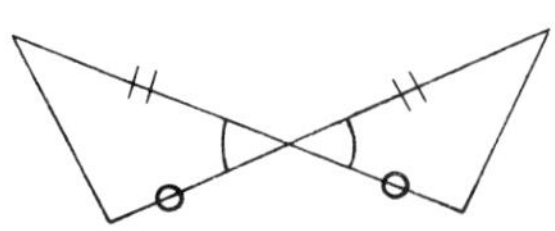

b

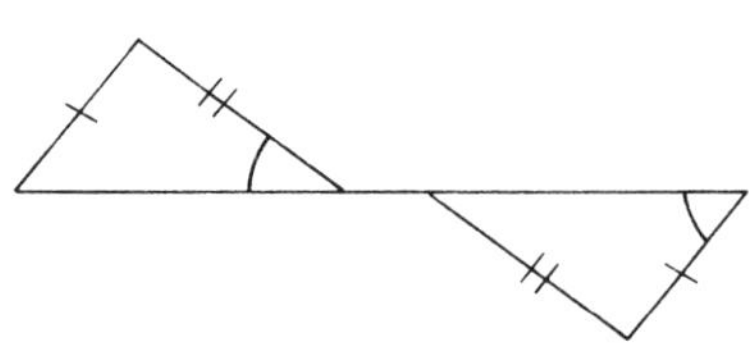

c

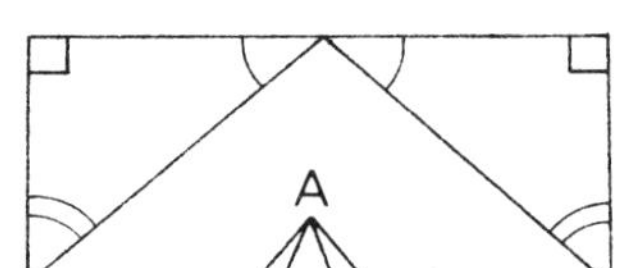

d

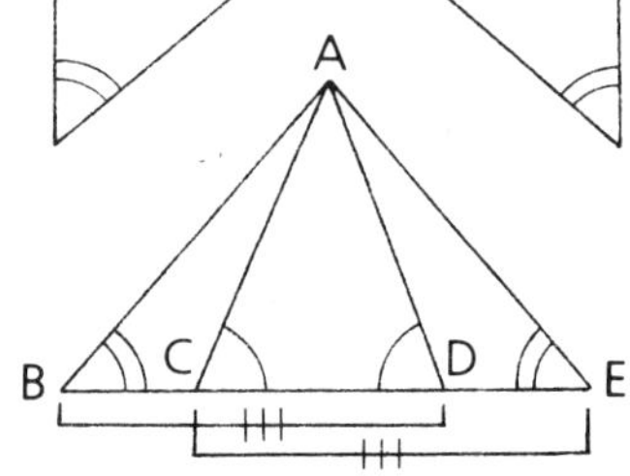

△ABD and △AEC

3 Given: $\overline{AB} \cong \overline{CB}$

$\angle ABD \cong \angle CBD$

Prove: $\triangle ABD \cong \triangle CBD$

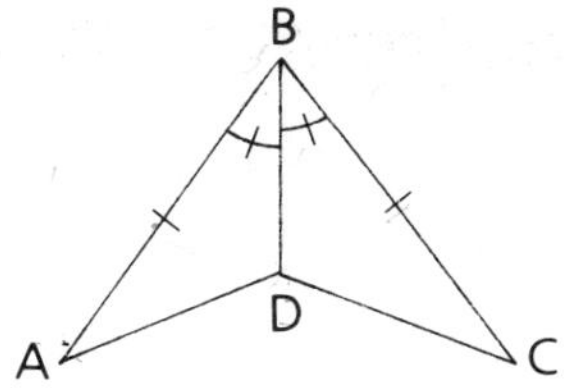

4 Given: $\angle 1 \cong \angle 2$

$\overline{EF} \cong \overline{HF}$

Prove: $\triangle EFJ \cong \triangle HFG$

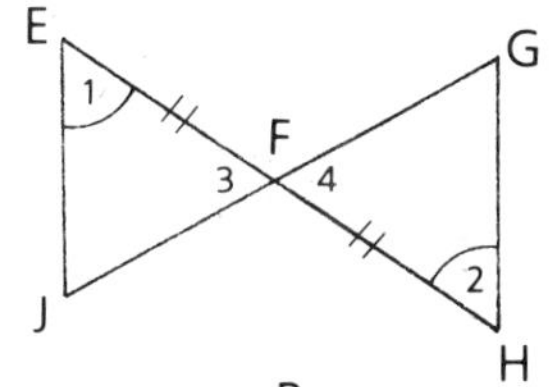

5 Given: $\overline{RO} \perp \overline{MP}$

$\overline{MO} \cong \overline{OP}$

Prove: $\triangle MRO \cong \triangle PRO$

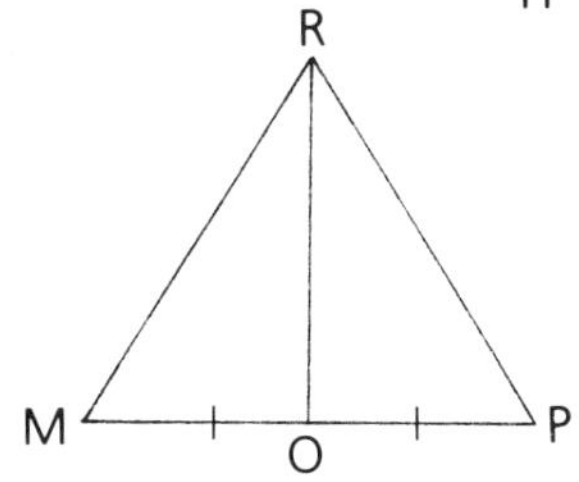

6 Given: $\overleftrightarrow{SV}$ bisects $\angle TVB$ and $\angle TSB$

Prove: $\triangle TSV \cong \triangle BSV$

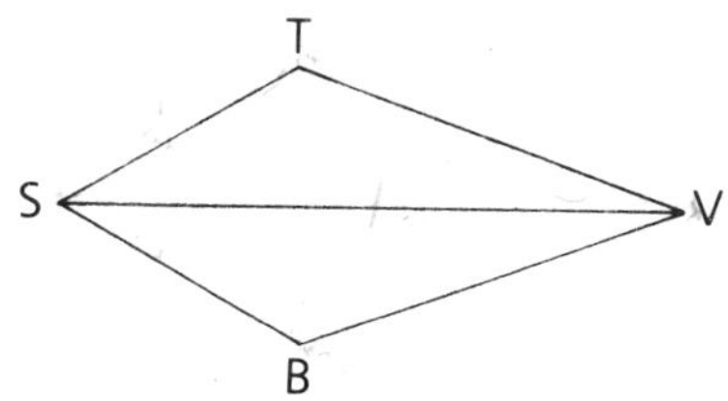

7 Given: $\overline{TV} \cong \overline{XW}$

$\overline{VA} \cong \overline{WA}$

$\overline{TA} \cong \overline{XA}$

Prove: $\triangle TVA \cong \triangle XWA$

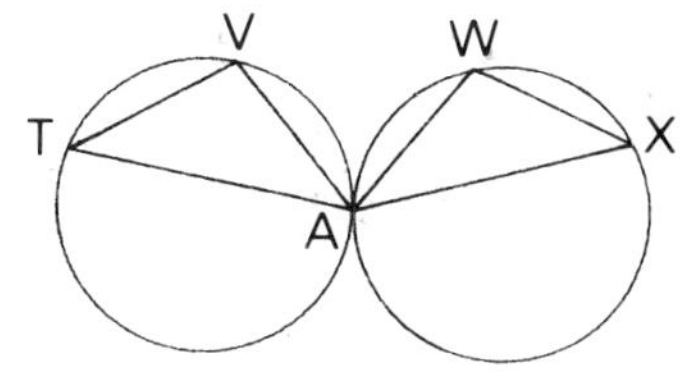

8 Given: $\overline{BC} \cong \overline{FE}$

$\overline{DC} \cong \overline{DE}$

$\angle 5 \cong \angle 6$

Prove: $\triangle BDG \cong \triangle FDG$

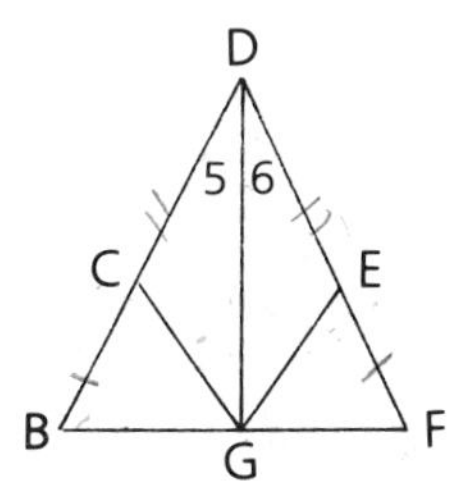

9 Given: $\angle N$ is comp. to $\angle NPO$
$\angle S$ is comp. to $\angle SPR$
$\angle NPO \cong \angle SPR$
$\overline{NP} \cong \overline{SP}$
Conclusion: $\triangle NOP \cong \triangle SRP$

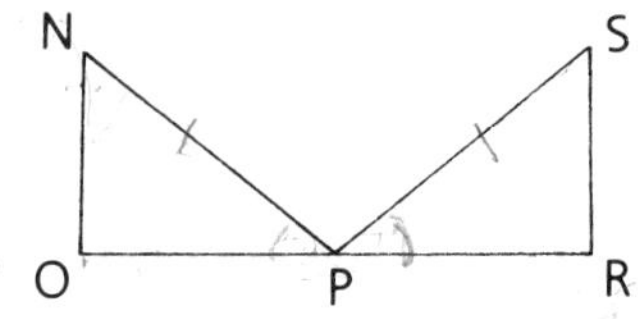

10 Given: O is the midpt. of $\overline{AY}$
O is the midpt. of $\overline{ZX}$
Conclusion: $\triangle ZOA \cong \triangle XOY$

11 Given: $\overline{EO} \cong \overline{KM}$
$\overline{FO} \cong \overline{JM}$
$\overline{EG} \cong \overline{KH}$
F is the midpt. of $\overline{EG}$
J is the midpt. of $\overline{KH}$
Conclusion: $\triangle EFO \cong \triangle KJM$

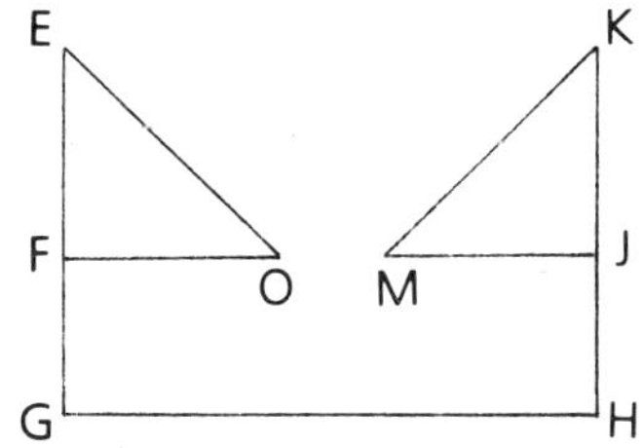

12 Given: $\angle 1 \cong \angle 4$
$\overline{PR} \cong \overline{TS}$
$\overline{NP} \cong \overline{NT}$
Prove: $\triangle NPR \cong \triangle NTS$

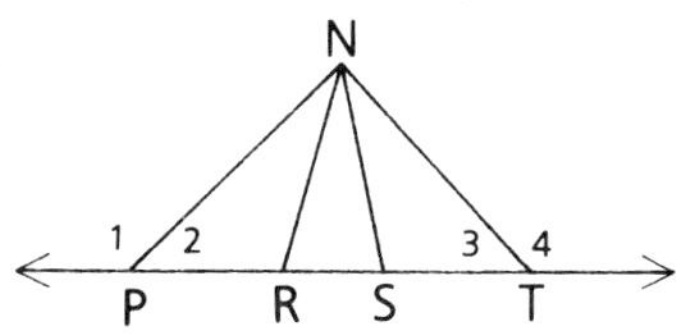

13 Given: $\overline{GH} \cong \overline{KJ}$
$\overline{HM} \cong \overline{JO}$
$\overline{GO} \cong \overline{KM}$
Prove: $\triangle GOJ \cong \triangle KMH$

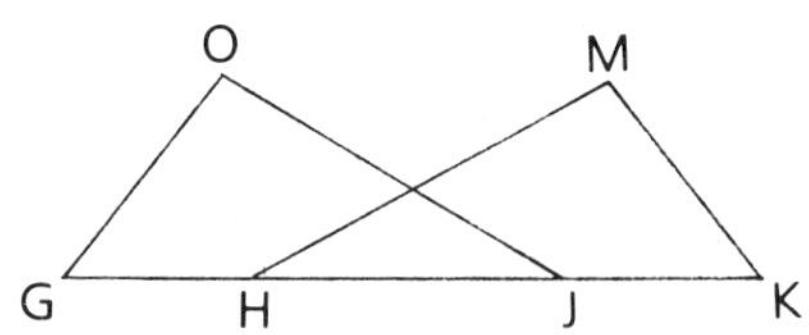

14 Given: $\angle R \cong \angle N$
$\overline{RP} \cong \overline{NT}$
$\overline{RT} \cong \overline{NP}$
$\overline{TS} \cong \overline{OP}$
Conclusion: $\triangle NOT \cong \triangle RSP$

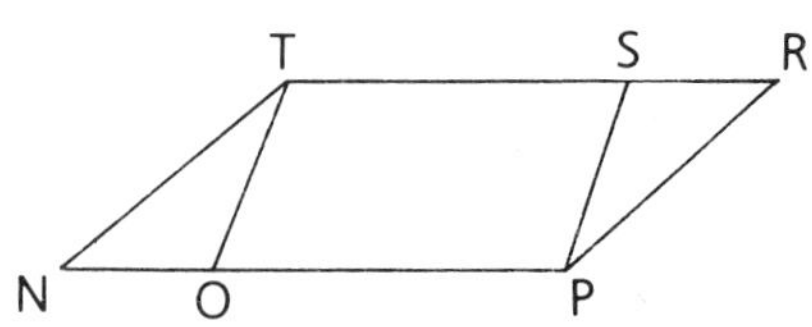

Section 3.1 Problem Set B

15 Given: $\angle 1 \cong \angle 6$

$\overline{BC} \cong \overline{EC}$

Conclusion: $\triangle ABC \cong \triangle DEC$

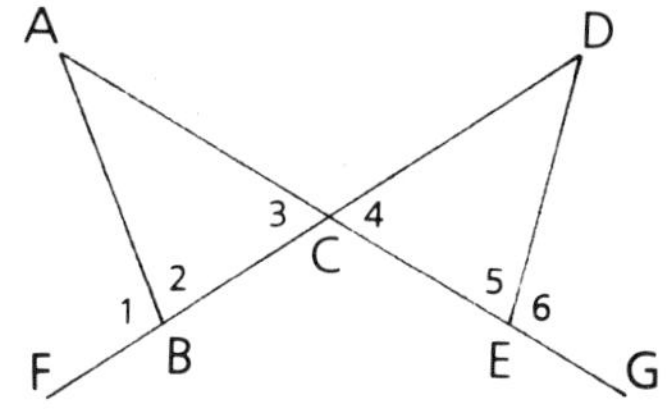

16 Given: $\overline{FH} \cong \overline{FK}$

$\angle H \cong \angle K$

G is the midpt. of $\overline{FH}$

M is the midpt. of $\overline{FK}$

J is midpt. of $\overline{HK}$

Conclusion: $\triangle GHJ \cong \triangle MKJ$

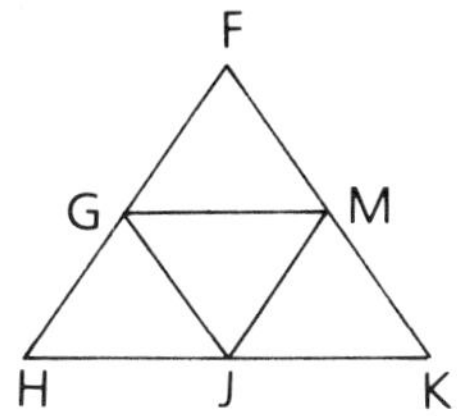

17 Given: $\overline{PR} \cong \overline{NT}$

$\overline{NO} \cong \overline{SR}$

O is $\frac{1}{3}$ of the way from N to P

S is $\frac{1}{3}$ of the way from R to T

Prove: $\triangle NRT \cong \triangle RNP$

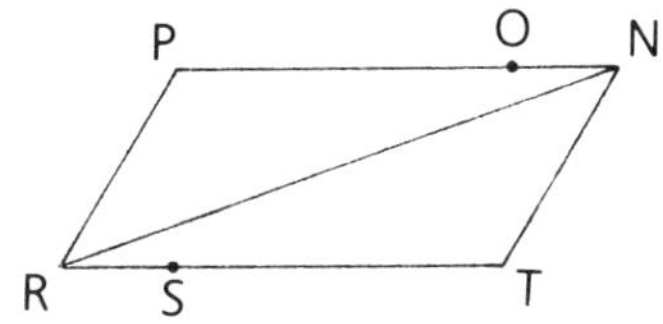

18 Given: S and T trisect $\overline{RV}$

$\angle R \cong \angle V$

$\angle BST \cong \angle BTS$

Conclusion: $\triangle BRS \cong \triangle BVT$

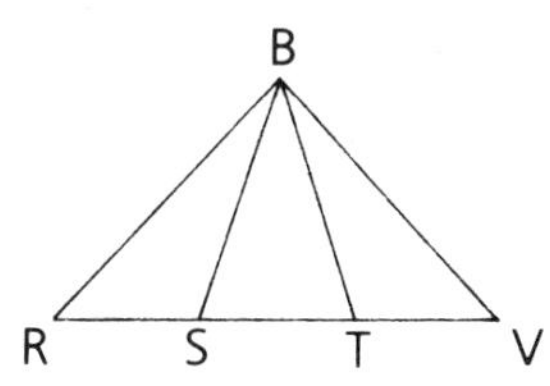

19 Given: $\overline{RS} \cong \overline{RT}$

Conclusion: $\triangle RST \cong \triangle RTS$

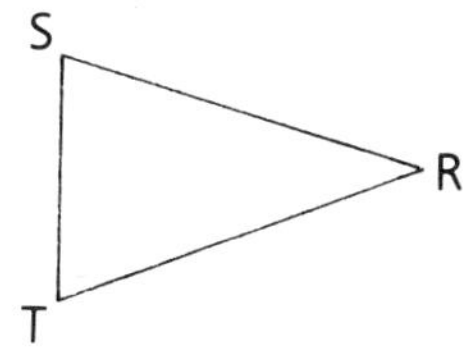

20 Given: $\overrightarrow{PY}$ bisects $\angle VPZ$

$\angle VPY = (2x + 7)^\circ$

$\angle ZPY = (3x - 9)^\circ$

$PZ = \frac{1}{2}x + 5$

$PV = x - 3$

Prove: $\triangle VPY \cong \triangle ZPY$

(Use a paragraph proof.)

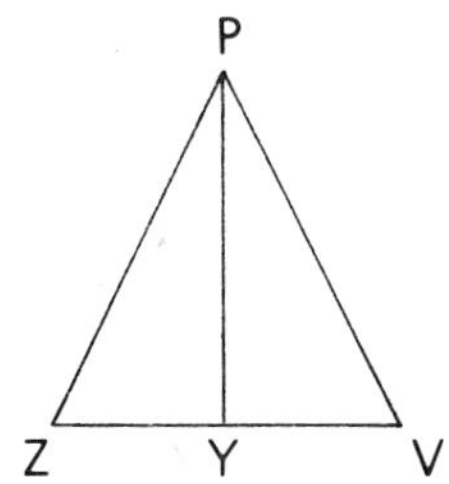

21 Given: $\angle 3 \cong \angle 1$, $\angle 4 \cong \angle 2$
$\angle DAC \cong \angle 3$, $\angle BAC \cong \angle 1$
$\overline{AD} \cong \overline{AB}$

Prove: $\triangle CAD \cong \triangle CAB$

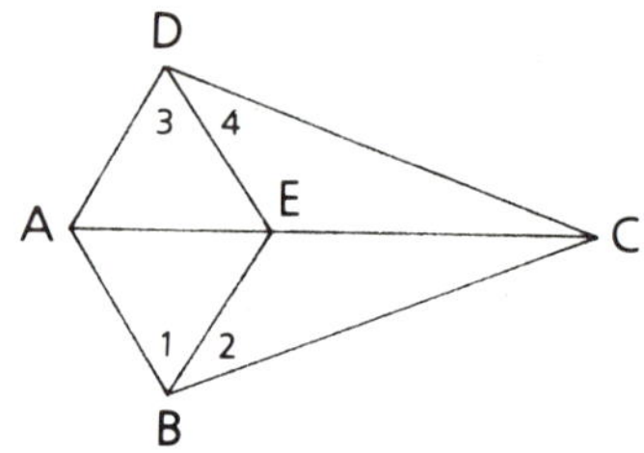

Section 3.1 Problem Set C

22 Given: $\overline{AB} \cong \overline{AE}$
$\overrightarrow{AE}$ and $\overrightarrow{AC}$ trisect $\angle BAD$
$\overline{AB} \perp \overline{BC}$
$\overline{AE} \perp \overline{DE}$

Conclusion: $\triangle ABC \cong \triangle AED$

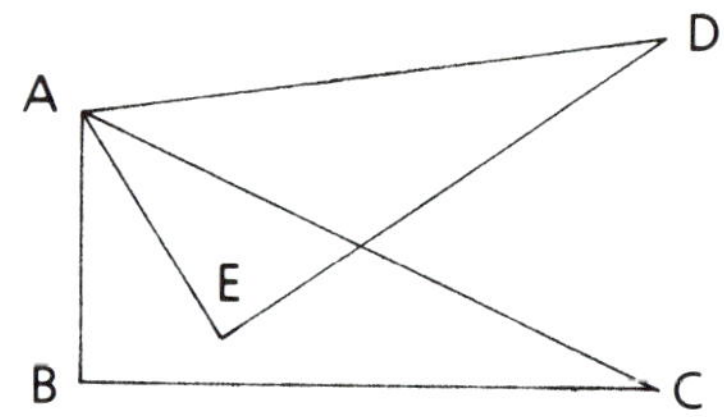

23 Given: $\overline{JH} \cong \overline{FG}$
K and M are midpoints
$\angle HKF \cong \angle FMH$
$\angle KJG \cong \angle MGJ$
$\angle JGH \cong \angle FJG$

Conclusion: $\triangle FJK \cong \triangle HGM$

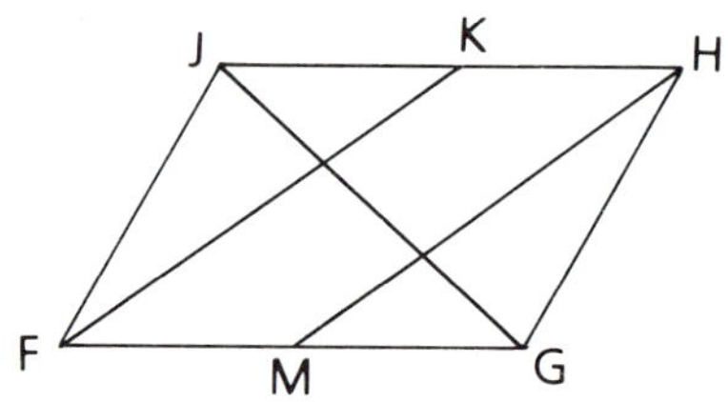

24 Given: $\overline{AR} \cong \overline{TR}$, $\overline{AV} \cong \overline{TS}$
S is the midpt. of $\overline{TR}$
$\angle PAT \cong \angle TRP$
$\angle T \cong \angle P$
$\angle RAT$ is comp. to $\angle T$
$\angle ARP$ is comp. to $\angle P$

Prove: $\triangle AOV \cong \triangle RSV$

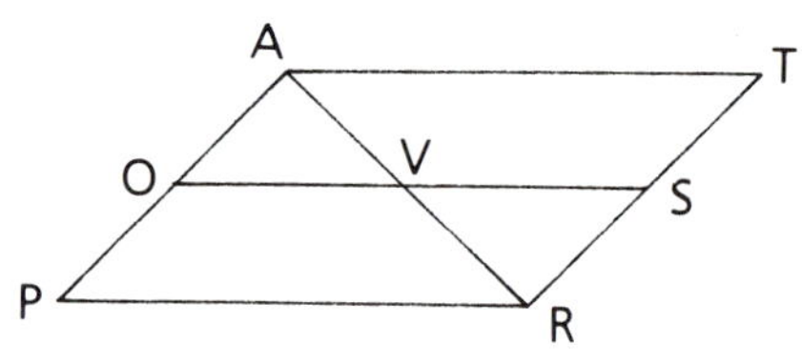

25 Given: 2 sides of $\triangle BEG$ are $\cong$
G is the midpt. of $\overline{BC}$
$\overline{EB} \not\cong \overline{GC}$
E is the midpt. of $\overline{AB}$
$\overline{AE} \not\cong \overline{EG}$
$\overline{GC} \cong \overline{GF}$

Prove: $\triangle BEG \cong \triangle CFG$ (Use a paragraph proof)

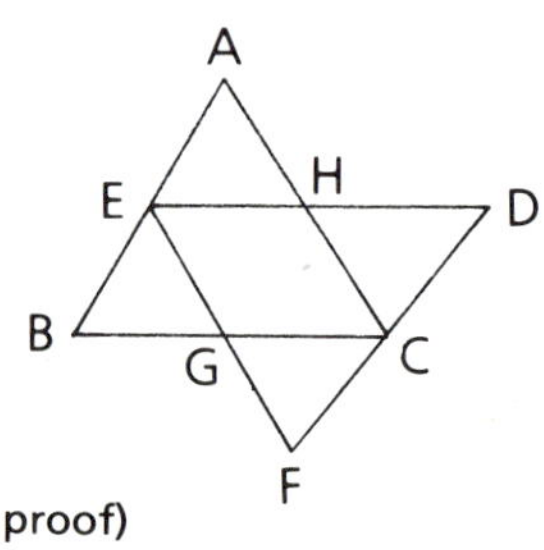

Section 3.2 CPCTC

In this section we shall discuss:

A CPCTC

B CIRCLES

3.2 (A) CPCTC

In the figure, suppose $\triangle ABC \cong \triangle DEF$. Would it then be true that $\angle B \cong \angle E$? If you refer back to Section 3.1(A), you will find that we have already answered yes to this question in the definition of congruent triangles.

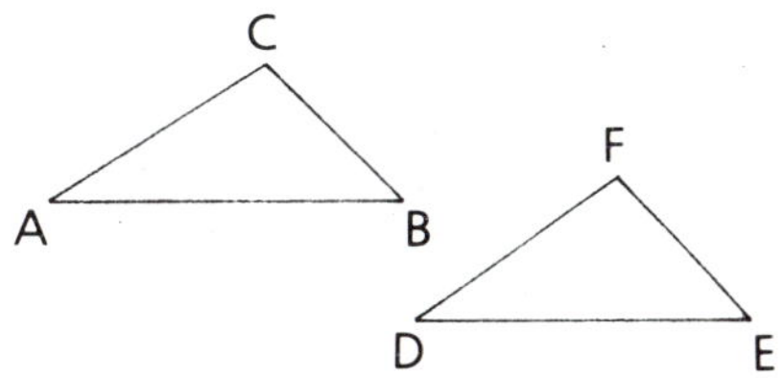

In the portions of the book that follow, we shall often draw such a conclusion *after* knowing that some triangles are congruent. We shall use CPCTC as the reason. CPCTC is short for "Corresponding Parts of Congruent Triangles are Congruent." By corresponding *parts,* we shall mean only the matching angles or sides of the respective triangles.

3.2 (B) Circles

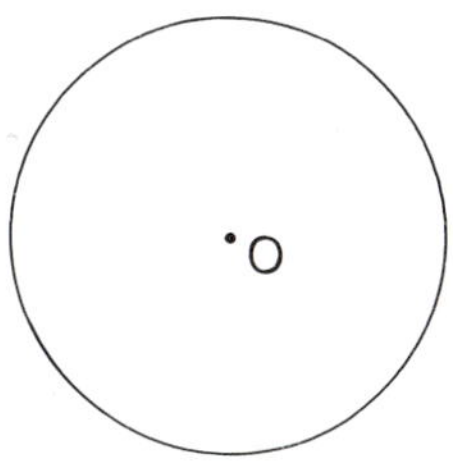

In the circle to the right, point O is the center. By definition, every point of the circle is the same distance from the center. The center, however, is not an element of the circle; the circle consists only of the "rim." A circle is named by its center: this circle is called circle O (or $\odot$O).

In circle P ($\odot$P), A, B, and C lie on the circle. $\overline{PA}$ is called a **RADIUS**, and $\overline{PA}$, $\overline{PB}$, and $\overline{PC}$ are called **RADII.**

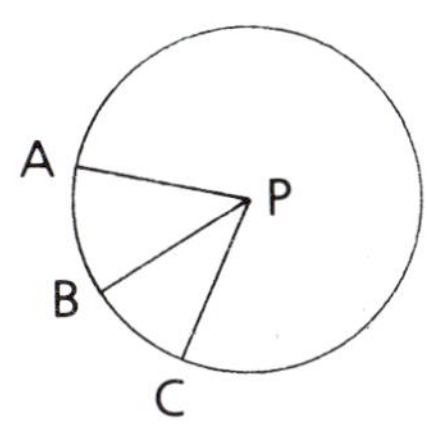

Theorem 18

T **All radii of a circle are congruent.**

Section 3.2 Sample Problems

1 **Given:** ⊙ P

Conclusion: $\overline{AB} \cong \overline{CD}$

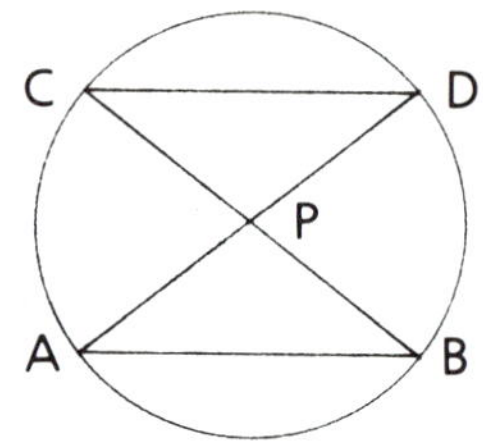

Statements	Reasons
1 ⊙ P	1 Given
2 $\overline{PA} \cong \overline{PB} \cong \overline{PC} \cong \overline{PD}$	2 All radii of a circle are ≅
3 ∠CPD ≅ ∠APB	3 Vertical angles are congruent
4 △CPD ≅ △APB	4 SAS (2, 3, 2)
5 $\overline{AB} \cong \overline{CD}$	5 CPCTC (Corresponding parts of congruent triangles are congruent)

2 **Given:** ⊙ O

∠T is comp. to ∠MOT

∠S is comp. to ∠POS

Prove: $\overline{MO} \cong \overline{PO}$

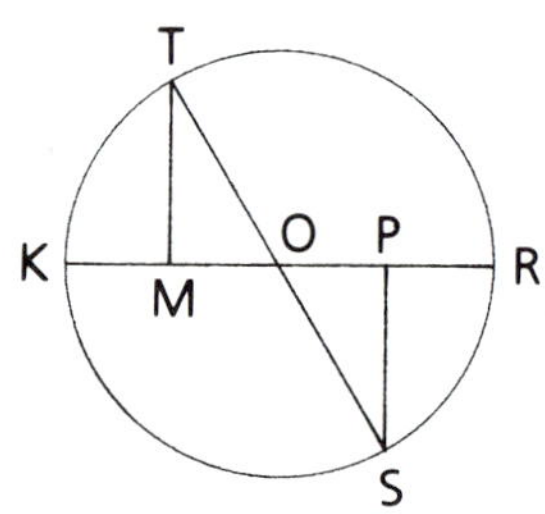

Statements	Reasons
1 ⊙ O	1 Given
2 $\overline{OT} \cong \overline{OS}$	2 All radii of a circle are ≅
3 ∠T is comp. to ∠MOT	3 Given
4 ∠S is comp. to ∠POS	4 Given
5 ∠MOT ≅ ∠POS	5 Vertical angles are congruent
6 ∠T ≅ ∠S	6 Complements of ≅ ∠s are ≅
7 △MOT ≅ △POS (watch the correspondence)	7 ASA (5, 2, 6)
8 $\overline{MO} \cong \overline{PO}$	8 CPCTC

Section 3.2 Problem Set A

1 Given: $\overline{AB} \cong \overline{DE}$
$\overline{BC} \cong \overline{EF}$
$\overline{AC} \cong \overline{DF}$
Prove: $\angle A \cong \angle D$

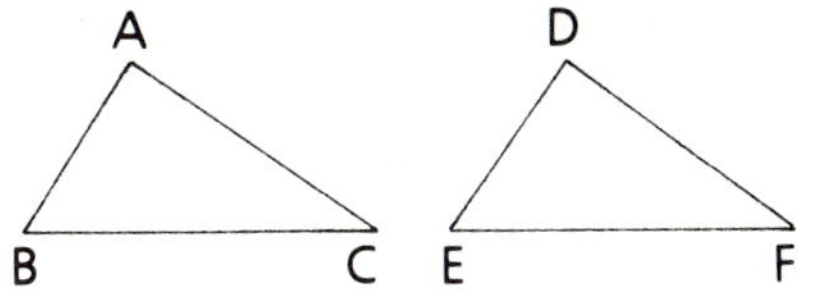

2 Given: $\angle HGJ \cong \angle KJG$
$\angle KGJ \cong \angle HJG$
Conclusion: $\overline{HG} \cong \overline{KJ}$

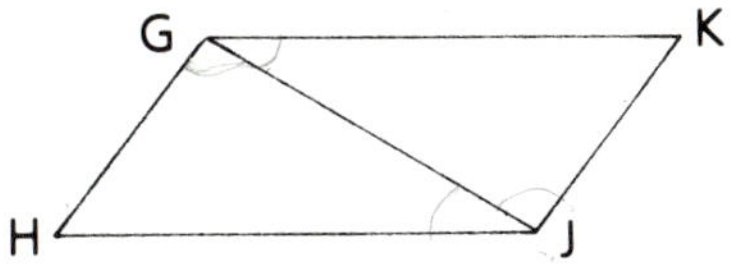

3 Given: $\odot$ O
$\overline{RO} \perp \overline{MP}$
Prove: $\overline{MR} \cong \overline{PR}$

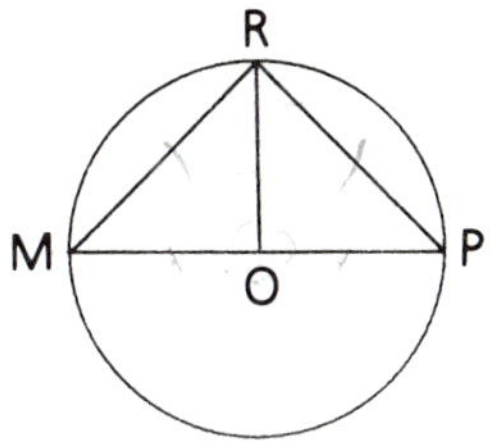

4 Given: T and R trisect $\overline{SW}$
$\overline{XS} \cong \overline{XW}$
$\angle S \cong \angle W$
Prove: $\overline{XT} \cong \overline{XR}$

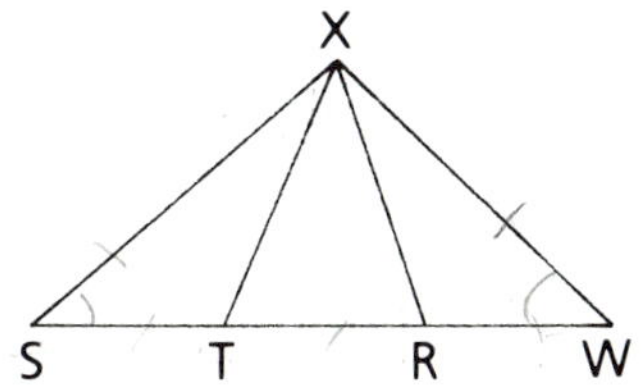

5 Given: $\angle B \cong \angle Y$
C is the midpt. of $\overline{BY}$
Conclusion: $\overline{AB} \cong \overline{YZ}$

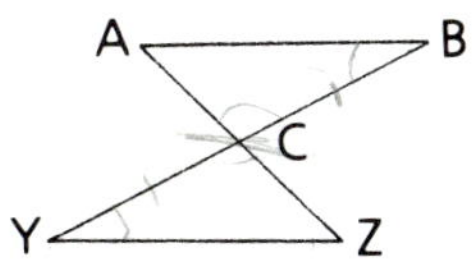

6 Given: $\odot$ O
$\overline{CD} \cong \overline{DE}$
Prove: $\angle COD \cong \angle DOE$

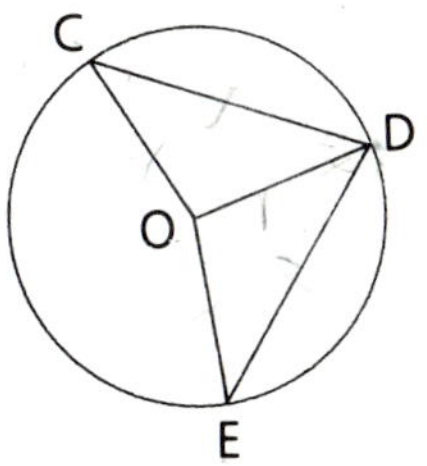

7 Given: $\overleftrightarrow{FH}$ bisects $\angle GFJ$ and $\angle GHJ$

Conclusion: $\overline{FG} \cong \overline{FJ}$

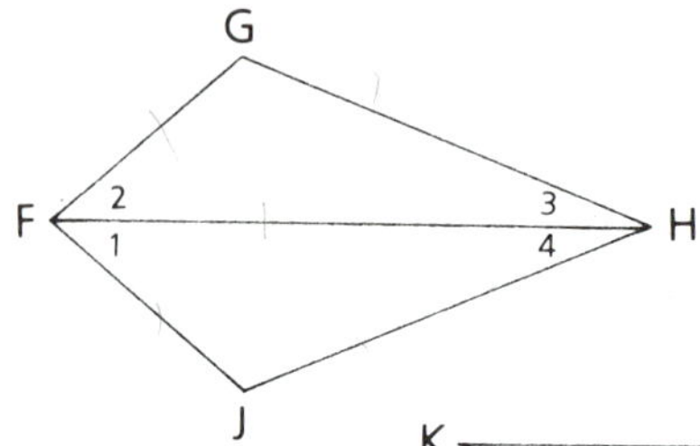

8 Given: $\angle M \cong \angle R$

$\angle RPS \cong \angle MOK$

$\overline{MP} \cong \overline{RO}$

Conclusion: $\overline{KM} \cong \overline{RS}$

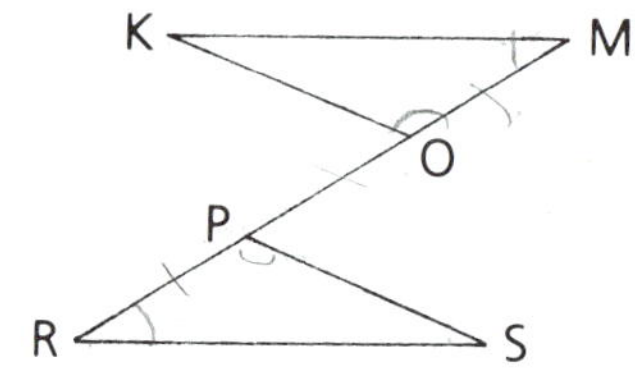

9 $\triangle ABC \cong \triangle DEF$

$\angle A = 90°$, $\angle B = 50°$, $\angle C = 40°$

$m\angle E = x + 30$, $m\angle F = \frac{y}{2} - 10$

Solve for x and y

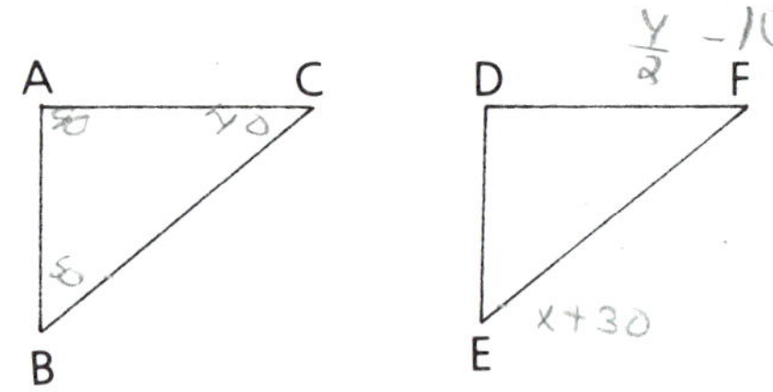

Section 3.2 Problem Set B

10 Given: $\overline{GO} \cong \overline{JK}$

$\overline{GJ} \cong \overline{OK}$

$\angle G \cong \angle K$

H is the midpt. of $\overline{GJ}$

M is the midpt. of $\overline{OK}$

$OK = 27$

$m\angle GOH = x + 24$, $m\angle GHO = 2y - 7$

$m\angle JMK = 3y - 23$, $m\angle MJK = 4x - 105$

Find: $m\angle GOH$, $m\angle GHO$, and GH

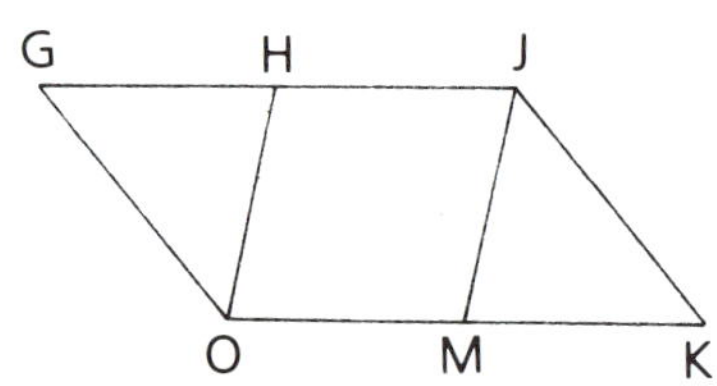

11 Given: $\angle A \cong \angle E$

$\overline{AB} \cong \overline{BE}$

$\overline{FB} \perp \overline{AE}$

$\angle 2 \cong \angle 3$

Prove: $\overline{CB} \cong \overline{DB}$

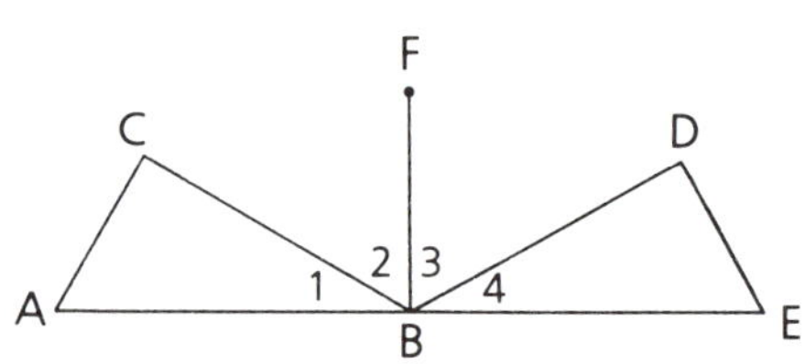

12 Given: $\angle 5 \cong \angle 6$
$\angle JHG \cong \angle O$
$\overline{GH} \cong \overline{MO}$
Conclusion: $\angle J \cong \angle P$

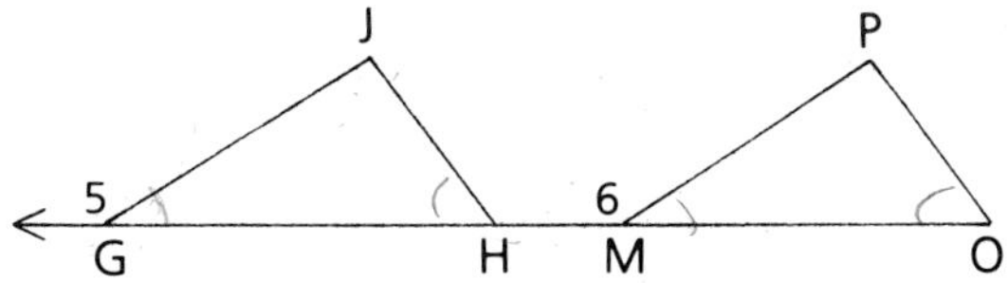

13 Given: $\angle RST \cong \angle RVT$
$\angle RVS \cong \angle TSV$
Conclusion: $\overline{RS} \cong \overline{VT}$

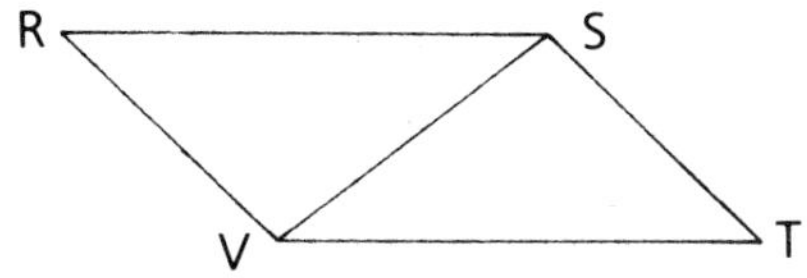

14 Given: $\angle 7 \cong \angle 8$
$\overline{ZY} \cong \overline{WX}$
Prove: $\angle W \cong \angle Y$

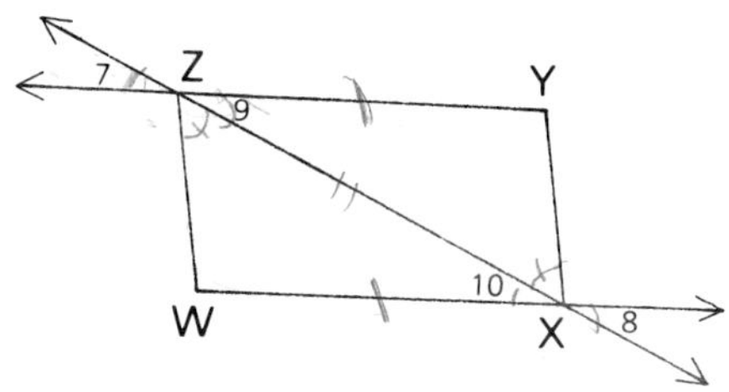

15 Given: $\angle AEC \cong \angle DEB$
$\overline{BE} \cong \overline{CE}$
$\angle ABE \cong \angle DCE$
Prove: $\overline{AB} \cong \overline{CD}$

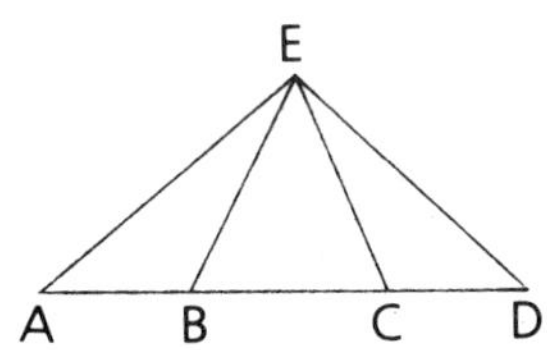

16 Given: $\overline{KG} \cong \overline{GJ}$
$\angle 2 \cong \angle 4$
$\angle 1$ is comp. to $\angle 2$
$\angle 3$ is comp. to $\angle 4$
$\angle FGJ \cong \angle HGK$
Conclusion: $\overline{FG} \cong \overline{HG}$

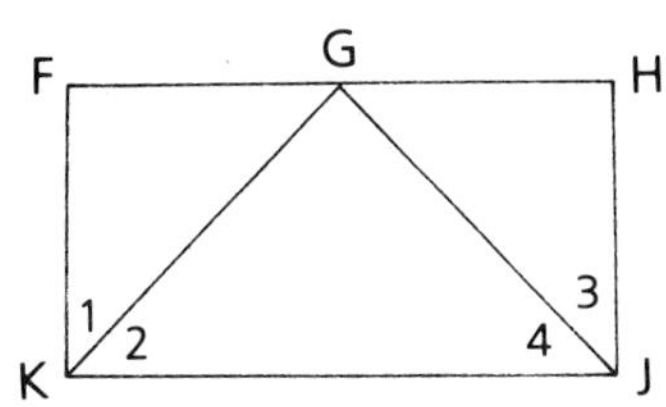

17 Given: $\overrightarrow{SO}$ bisects $\angle MSP$
T is $\frac{1}{3}$ of the way from M to S
R is $\frac{1}{3}$ of the way from P to S
$\overline{MT} \cong \overline{RP}$
Prove: $\angle M \cong \angle P$

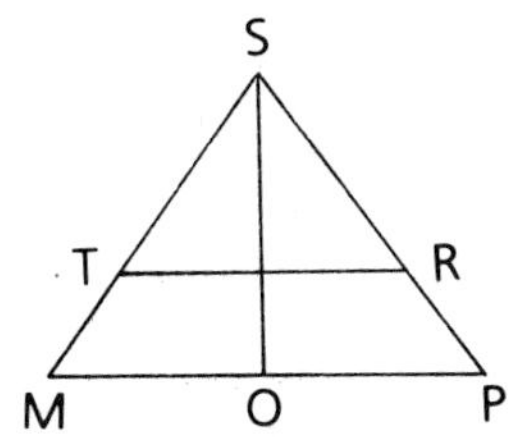

18 Given: $\overline{ZW} \cong \overline{YX}$, $\overline{CY} \cong \overline{WD}$
$\overline{ZC} \cong \overline{DX}$, $\overline{ZB} \cong \overline{AX}$
A and B are midpoints
Conclusion: $\angle W \cong \angle Y$

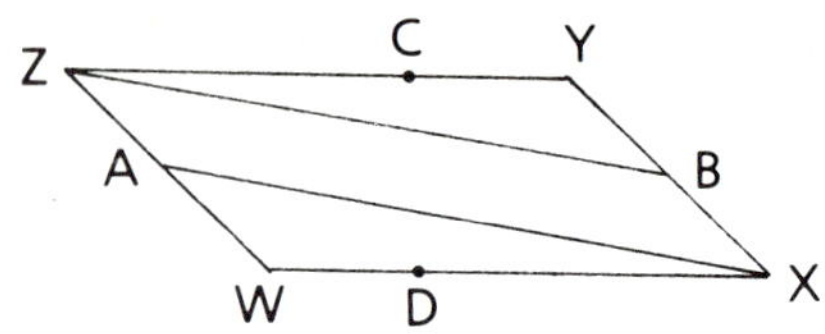

Section 3.2 Problem Set C

19 Given: $\overline{AE} \cong \overline{FC}$
$\overline{FB} \cong \overline{DE}$
$\angle CFB \cong \angle AED$
Prove: $\angle 1 \cong \angle 2$

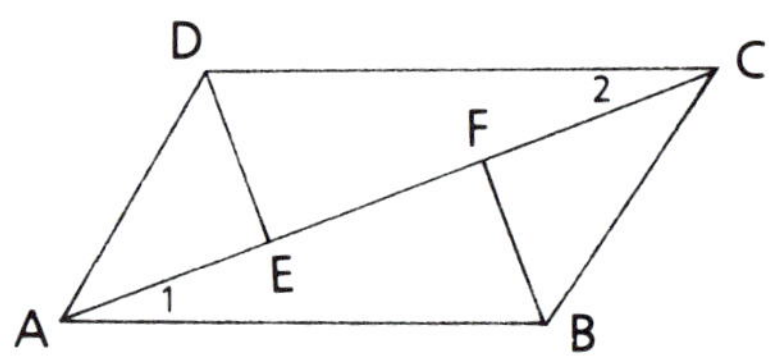

20 If $\overline{GJ}$ and $\overline{KH}$ bisect each other, then $\angle MHO$ is larger than $\angle K$.
Write a paragraph proof.

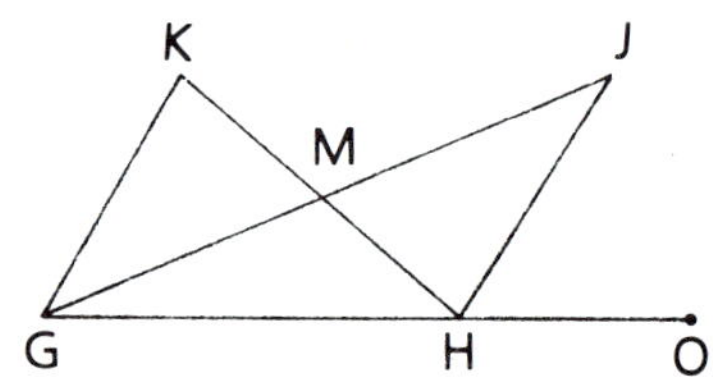

21 Given: S and T trisect $\overline{RV}$
$\overline{PR} \perp \overline{RV}$
$\overline{KV} \perp \overline{RV}$
$\angle 3 \cong \angle 4$
$\angle PTV \cong \angle KSR$
Prove: $\overline{RX} \cong \overline{VX}$

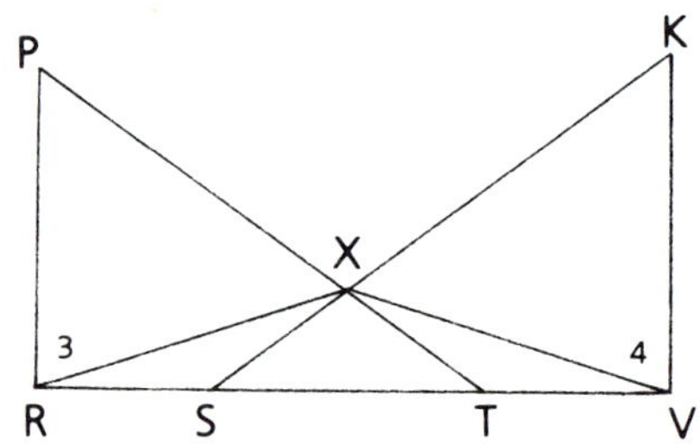

22 $\triangle ABC \cong \triangle DEF$
$m\angle A = 4x - 3y$
$m\angle C = \frac{1}{2}x + y$
$m\angle F = 2x - 2y$
$m\angle D = 3y + 30$
$AC = .2x + \frac{3}{5}y$
Find: $m\angle A$, $m\angle C$, DF

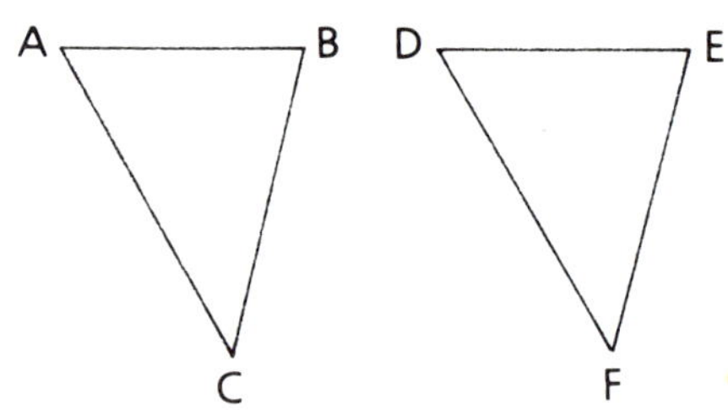

Section 3.3 Beyond CPCTC

In this section we shall discuss:

A MEDIANS OF TRIANGLES
B ALTITUDES OF TRIANGLES
C AUXILIARY LINES
D PROOFS BEYOND CPCTC

3.3 (A) Medians of Triangles

Three **MEDIANS** are shown:

$\overline{AD}$ is a median of $\triangle ABC$
$\overline{EH}$ is a median of $\triangle EFG$
$\overline{FJ}$ is a median of $\triangle EFG$

Every triangle has three medians.

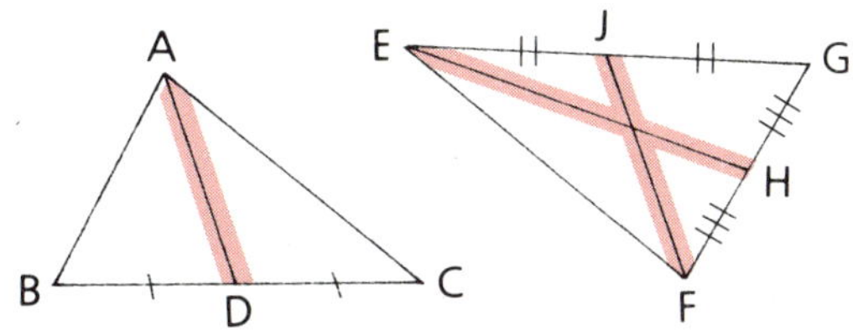

D **A MEDIAN of a triangle is a line segment drawn from any vertex of the triangle to the midpoint of the opposite side. (Alternate forms: A MEDIAN of a triangle divides one side into two congruent segments; the MEDIAN bisects that side.)**

3.3 (B) Altitudes of Triangles

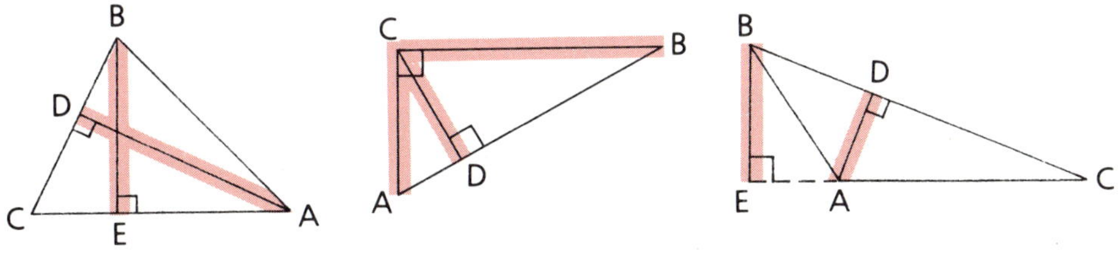

In the figure on the left, $\overline{AD}$ and $\overline{BE}$ are **ALTITUDES** of $\triangle ABC$.

In the middle figure, $\overline{AC}$ and $\overline{BC}$ and $\overline{CD}$ are **ALTITUDES** of $\triangle ABC$. You will notice that two of the altitudes are sides of the original triangle in this case.

In the figure on the right, $\overline{AD}$ and $\overline{BE}$ are **ALTITUDES** of $\triangle ABC$. Notice that altitude $\overline{BE}$ falls outside the triangle. Where does the third altitude lie?

Every triangle has three altitudes.
Could an altitude of a $\triangle$ be a median as well?

D **An ALTITUDE of a triangle is a line segment drawn from any vertex of the triangle perpendicular to the opposite side, extended, if necessary. (Alternate form: An ALTITUDE of a triangle forms right angles (90°) with one of the sides.)**

3.3 (C) Auxiliary Lines

Given: $\overline{AB} \cong \overline{AC}$

$\overline{BD} \cong \overline{CD}$

Conclusion: $\angle ABD \cong \angle ACD$

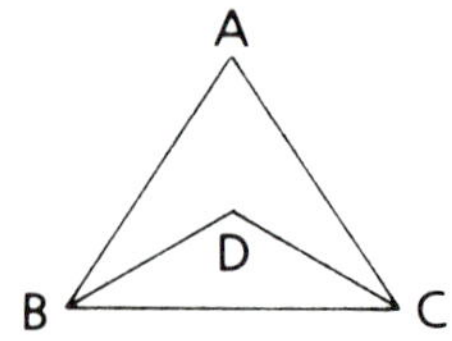

The above proof would be easy if a line segment were drawn from A to D.

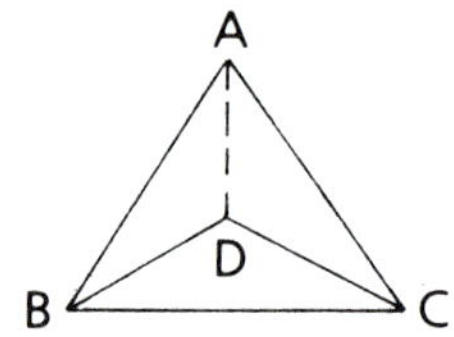

We could then proceed to prove $\triangle ABD \cong \triangle ACD$ by SSS, and $\angle ABD \cong \angle ACD$ by CPCTC.

As the year progresses, we will often need lines, rays, or segments that do not appear in the original figure. Such lines are called **AUXILIARY LINES.** Most auxiliary lines merely connect two points already in the diagram, although you will see other types of auxiliary lines later in the course. Each auxiliary line in a proof needs a reason that such a line can be drawn.

It is a postulate that *one and only one line, ray, or segment can be drawn through any two distinct points.* The word *determine* represents the idea that for two given points there **IS** a line through them, and there is **NO MORE THAN ONE** such line.

P **Two points determine one line (ray, or segment).**

3.3 (D) Proofs beyond CPCTC

Given: $\overline{AD} \cong \overline{CD}$

$\angle ADB \cong \angle CDB$

Prove: $\overline{DB}$ is the median to $\overline{AC}$

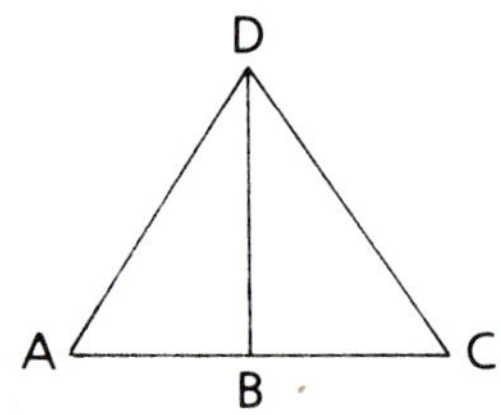

In the problem above, we can prove $\triangle ABD \cong \triangle CBD$ by SAS. Do you see how? Then $\overline{AB} \cong \overline{CB}$ by CPCTC. Now we shall go one step beyond CPCTC. Since $\overline{AB} \cong \overline{CB}$, we may call $\overline{DB}$ a median of $\triangle ACD$, and the problem is proved.

Many types of proofs go beyond CPCTC. By using CPCTC first we can prove there are altitudes, angle bisectors, midpoints, etc. You will see some examples in the Sample Problems to follow.

A fascinating type of problem occurs when we prove one pair of triangles are congruent and then use CPCTC to help us prove another pair of triangles are congruent. Such problems are called **DETOUR PROBLEMS** and are studied in detail in Chapter 4.

Section 3.3 Sample Problems

1 **Given:** $\overline{AC} \cong \overline{BC}$
$\overline{AD} \cong \overline{BD}$
Prove: $\overrightarrow{CD}$ bisects $\angle ACB$

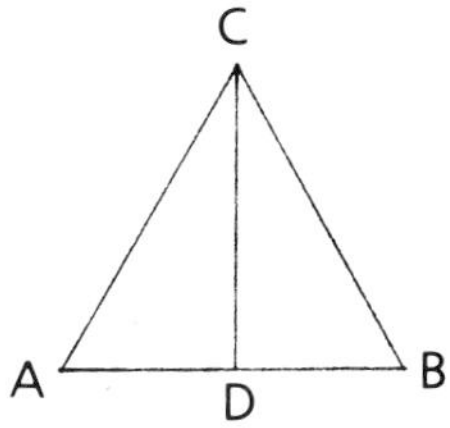

Statements	Reasons
1 $\overline{AC} \cong \overline{BC}$	1 Given
2 $\overline{AD} \cong \overline{BD}$	2 Given
3 $\overline{CD} \cong \overline{CD}$	3 Reflexive Property
4 $\triangle ACD \cong \triangle BCD$	4 SSS (1, 2, 3)
5 $\angle ACD \cong \angle BCD$	5 CPCTC
6 $\overrightarrow{CD}$ bisects $\angle ACB$	6 If a ray divides an angle into two $\cong$ angles, the ray bisects the angle

2 **Given:** $\overline{CD}$ and $\overline{BE}$ are altitudes of $\triangle ABC$
$\overline{AD} \cong \overline{AE}$
Prove: $\overline{DB} \cong \overline{EC}$

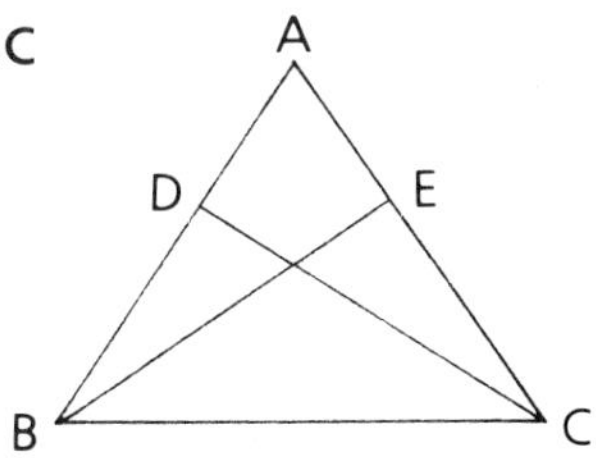

Statements	Reasons
1 $\overline{CD}$ and $\overline{BE}$ are altitudes of $\triangle ABC$	1 Given
2 $\angle ADC$ is a right $\angle$	2 An altitude of a $\triangle$ forms right $\angle$s with the side to which it is drawn
3 $\angle AEB$ is a right $\angle$	3 Same as 2
4 $\angle ADC \cong \angle AEB$	4 If $\angle$s are right $\angle$s, they are $\cong$
5 $\angle A \cong \angle A$	5 Reflexive
6 $\overline{AD} \cong \overline{AE}$	6 Given
7 $\triangle ADC \cong \triangle AEB$	7 ASA (4, 6, 5)
8 $\overline{AB} \cong \overline{AC}$	8 CPCTC
9 $\overline{DB} \cong \overline{EC}$	9 Subtraction Property (6 from 8)

3 **Given:** G is the midpt. of $\overline{FH}$

$\overline{EF} \cong \overline{EH}$

Prove: $\angle 1 \cong \angle 2$

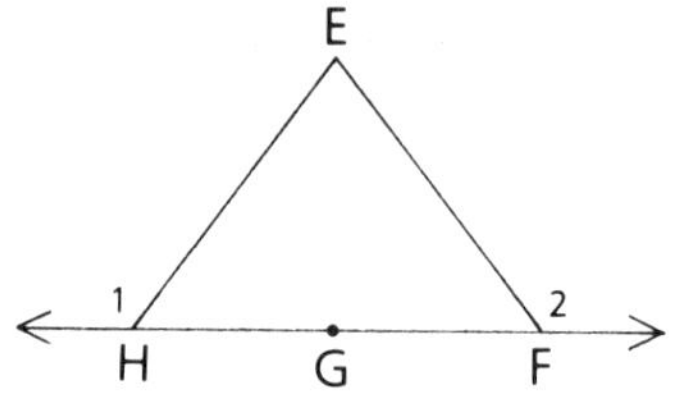

Statements	Reasons
1 G is the midpt. of $\overline{FH}$	1 Given
2 $\overline{FG} \cong \overline{HG}$	2 If a point is the midpt. of a segment, it divides the segment into two $\cong$ segments
3 $\overline{EF} \cong \overline{EH}$	3 Given
4 Draw $\overline{EG}$	4 Two points determine a segment
5 $\overline{EG} \cong \overline{EG}$	5 Reflexive Property
6 $\triangle EFG \cong \triangle EHG$	6 SSS (2, 3, 5)
7 $\angle EFG \cong \angle EHG$	7 CPCTC
8 $\angle 2$ is supp. to $\angle EFG$	8 If two angles form a straight angle, they are supplementary
9 $\angle 1$ is supp. to $\angle EHG$	9 Same as 8
10 $\angle 1 \cong \angle 2$	10 Supplements of $\cong$ $\angle$s are $\cong$

4 **Given:** $\angle T \cong \angle Y$

$\angle SVZ \cong \angle SXZ$

$\overline{TV} \cong \overline{YX}$

Conclusion: $\overline{SZ}$ is the median to $\overline{TY}$

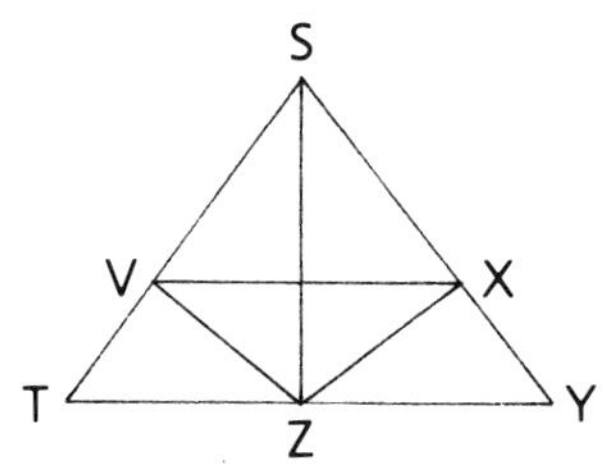

Statements	Reasons
1 $\angle T \cong \angle Y$	1 Given
2 $\angle SVZ \cong \angle SXZ$	2 Given
3 $\angle SVZ$ is supp. to $\angle TVZ$	3 If two angles form a straight angle, they are supplementary
4 $\angle SXZ$ is supp. to $\angle YXZ$	4 Same as 3
5 $\angle TVZ \cong \angle YXZ$	5 Supplements of $\cong$ $\angle$s are $\cong$
6 $\overline{TV} \cong \overline{YX}$	6 Given
7 $\triangle TVZ \cong \triangle YXZ$	7 ASA (1, 6, 5)
8 $\overline{TZ} \cong \overline{YZ}$	8 CPCTC
9 $\overline{SZ}$ is the median to $\overline{TY}$	9 If a segment from a vertex of a $\triangle$ divides the opposite side into two $\cong$ segments, it is the median

The Royal Road: The Bridge of Donkeys

In Euclid's *Elements,* Proposition 5 of Book I proves that the base angles of an isosceles triangle are congruent. Euclid's proof came to be called *the bridge of donkeys;* if you could not follow the proof, you could not pass over the "bridge" to further study of geometry.

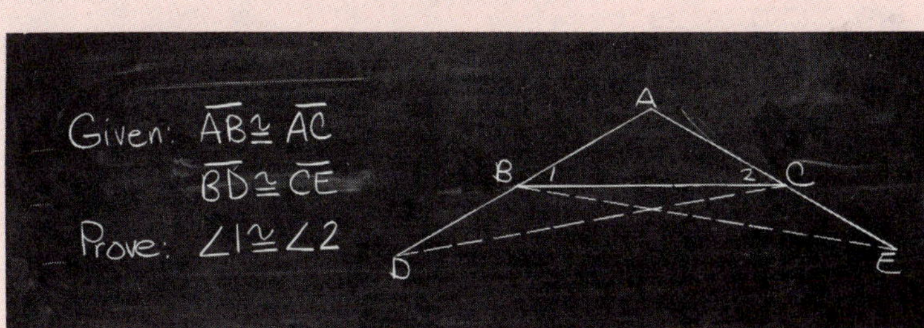

First show AD = AE, and so △ADC = △AEB. Then show △BCD ≅ △CBE by SSS. Finally, show that ∠1 and ∠2 are supplements of congruent angles. Then you will be across the bridge!

Section 3.3 Problem Set A

1 In each of the following problems, identify $\overline{AD}$ as a median, altitude, neither, or both according to what can be proved:

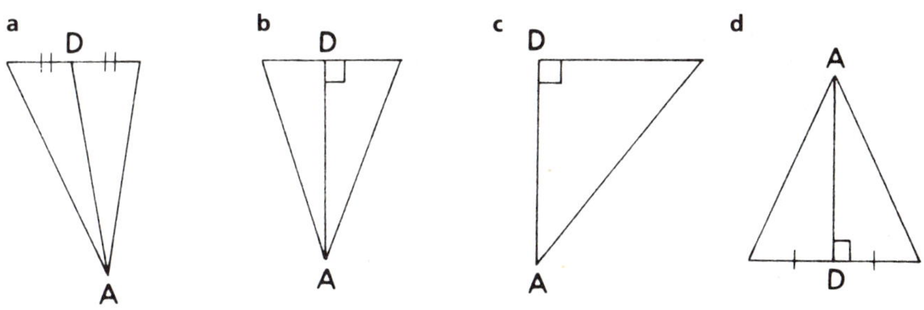

2 Given: $\overline{HJ} \cong \overline{KJ}$

$\angle MJH \cong \angle MJK$

Prove: $\overrightarrow{MJ}$ bisects $\angle HMK$

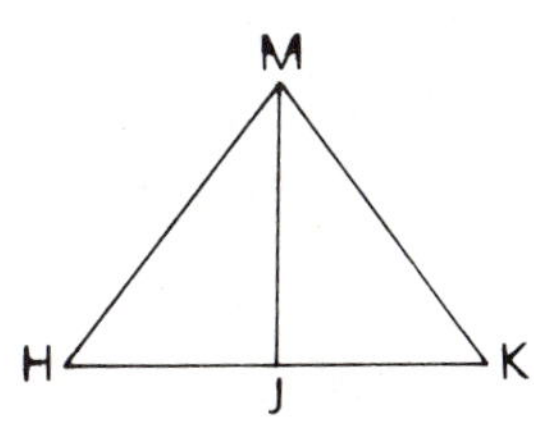

3 Given: $\overline{NR} \cong \overline{PR}$

$\overrightarrow{RO}$ bisects $\angle NRP$

Prove: $\overrightarrow{OR}$ bisects $\angle NOP$

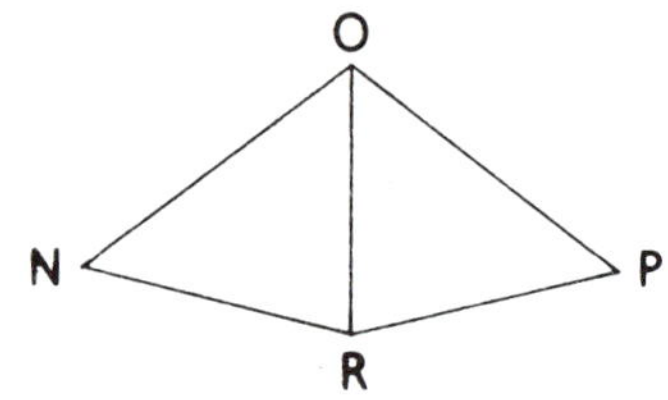

4 Given: $\angle CFD \cong \angle EFD$

$\overline{FD}$ is an altitude

Prove: $\overline{FD}$ is a median

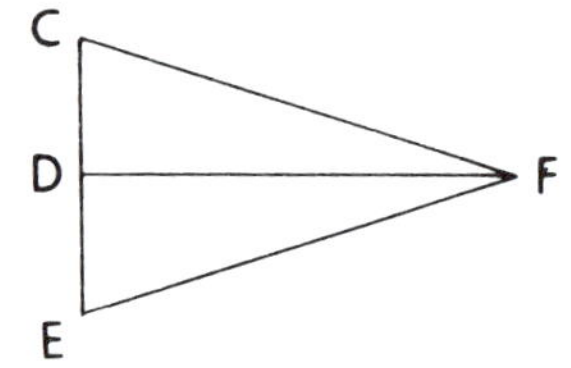

5 Given: $\odot$ O

$\overline{GJ} \cong \overline{HJ}$

Prove: $\angle G \cong \angle H$

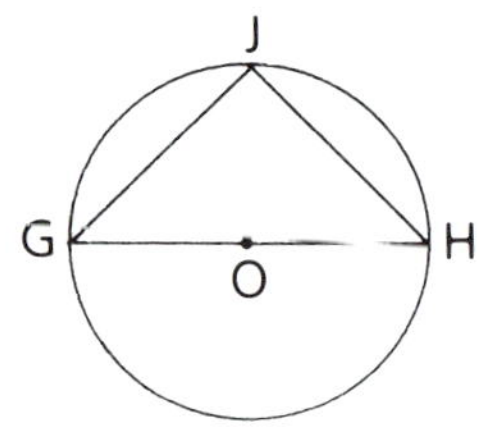

6 Given: $\overline{KP}$ is a median

$\overline{MK} \cong \overline{RK}$

Conclusion: $\angle 3 \cong \angle 4$

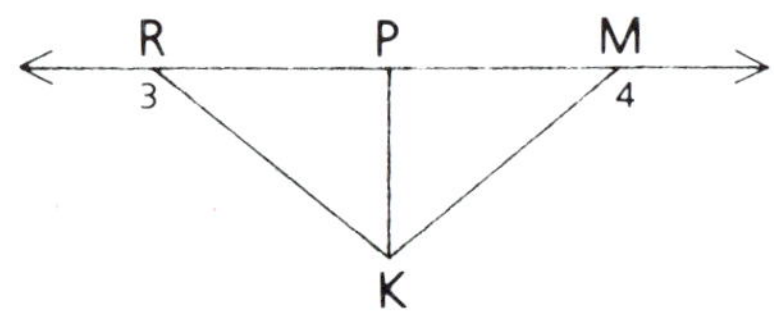

7 Given: $\overline{TW}$ is a median

$ST = x + 10$

$SW = 2x + 3$

$WV = 3x - 1$

Find: SW, WV, and ST

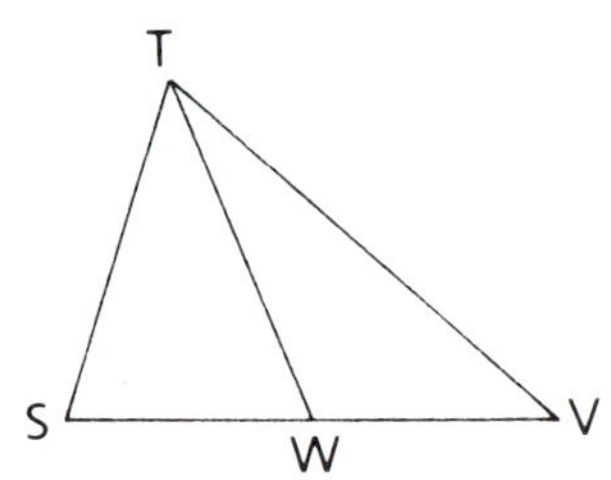

Section 3.3 Problem Set B

8 Given: $\angle AEB \cong \angle DEC$

$\overline{AE} \cong \overline{DE}$

$\angle A \cong \angle D$

Conclusion: $\overline{AC} \cong \overline{BD}$

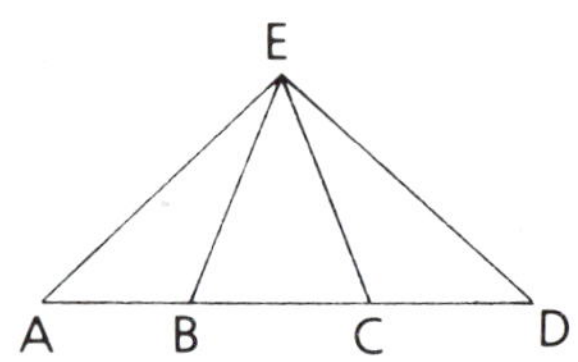

9 Given: $\odot$ O

$\angle NOG \cong \angle POG$

Conclusion: $\overrightarrow{RO}$ bisects $\angle NRP$

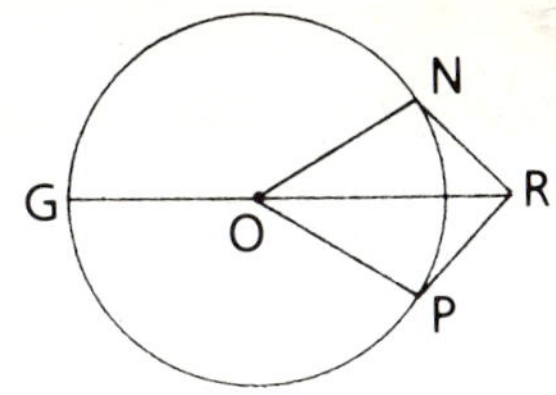

10 Given: $\overline{AZ} \cong \overline{ZB}$

Z is the midpt. of $\overline{XY}$

$\angle AZX \cong \angle BZY$

$\overline{XW} \cong \overline{YW}$

Prove: $\overline{AW} \cong \overline{BW}$

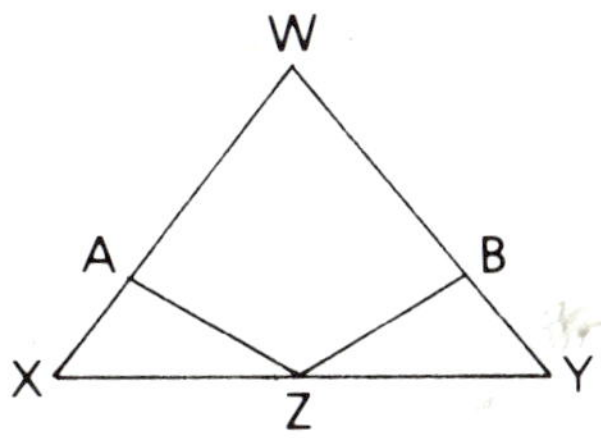

11 Given: $\overrightarrow{DF}$ bisects $\angle CDE$

$\overrightarrow{EF}$ bisects $\angle CED$

G is the midpt. of $\overline{DE}$

$\overline{DF} \cong \overline{EF}$

Prove: $\angle CDE \cong \angle CED$

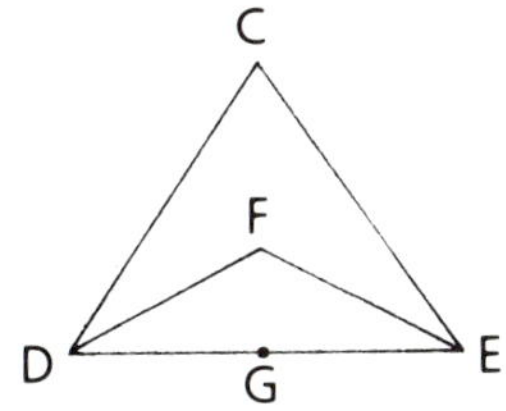

Section 3.3 Problem Set C

12 Given: $\overline{AC}$ is the altitude to $\overline{BD}$

$\overline{AC}$ is a median

$\angle BAC$ is comp. to $\angle D$

Conclusion: $\angle DAC$ is comp. to $\angle B$

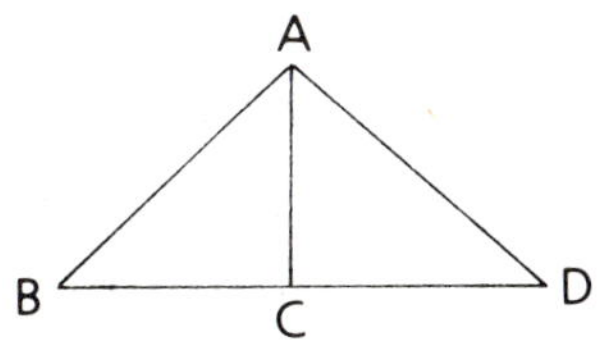

13 Given: $\overline{EJ} \cong \overline{FH}$

$\overline{EK} \cong \overline{GH}$

$\angle JKG \cong \angle KGF$

Prove: $\angle JKF \cong \angle FGJ$

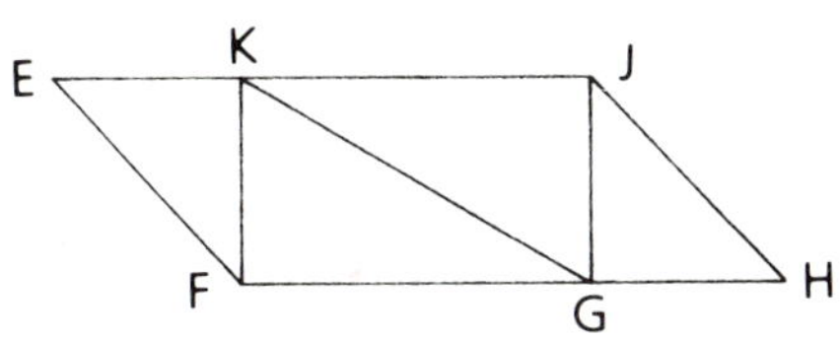

14 $\odot$ O and $\odot$ P

Perimeter of $\triangle AOP = 80$

$OC + DP = 16$

$\overline{CD}$ is 2 units longer than $\overline{OC}$

Find: $OB + BP$

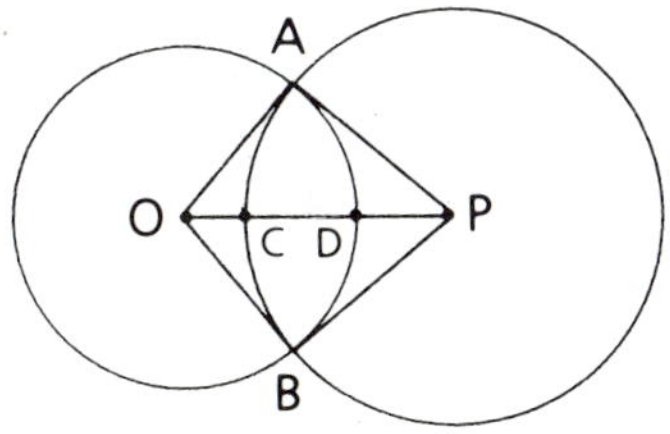

Section 3.4 Overlapping Triangles

Given: $\overline{DB} \cong \overline{AC}$
$\overline{AD} \cong \overline{BC}$

Conclusion: $\angle D \cong \angle C$

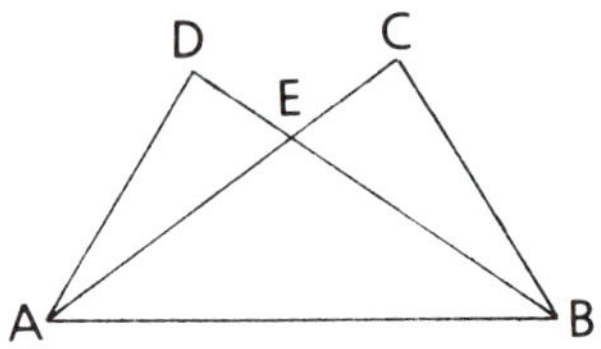

Consider the above problem. At first glance you would probably think of proving that $\triangle ADE \cong \triangle BCE$, thus getting $\angle D \cong \angle C$ by CPCTC.

Soon you would realize that there is not enough information to prove that $\triangle ADE \cong \triangle BCE$. There must be another way.

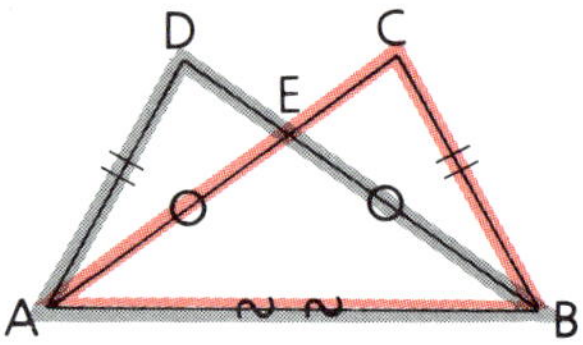

In this case the problem could be done by finding two other triangles to which $\angle D$ and $\angle C$ belong. You would notice the overlapping triangles and soon see that $\triangle ABD \cong \triangle BAC$ by SSS, and $\angle D \cong \angle C$ by CPCTC.

At first, overlapping triangles may be difficult, and you may want to outline triangles in color as in the sample problems. Just be willing to draw figures more than once or twice to find the triangles that serve best.

All the problems of the next set use overlapping triangles. Thereafter the triangles of interest may, or may not, overlap.

Section 3.4 Sample Problems

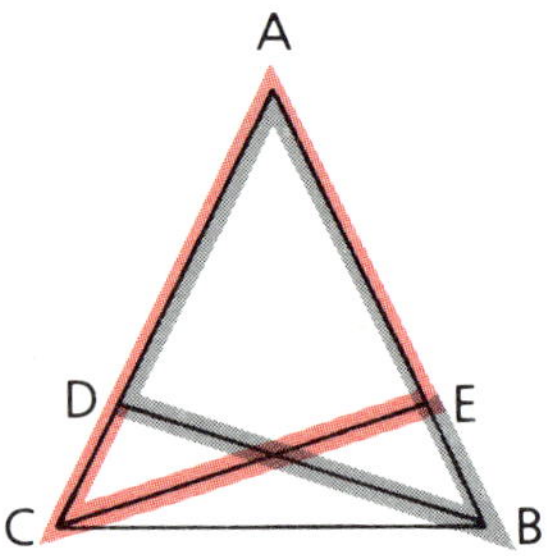

1 **Given:** $\overline{AC} \cong \overline{AB}$
$\overline{AE} \cong \overline{AD}$

Conclusion: $\overline{CE} \cong \overline{BD}$

Statements	Reasons
1 $\overline{AC} \cong \overline{AB}$	1 Given
2 $\overline{AE} \cong \overline{AD}$	2 Given
3 $\angle A \cong \angle A$	3 Reflexive Property
4 $\triangle ADB \cong \triangle AEC$	4 SAS (1, 3, 2)
5 $\overline{CE} \cong \overline{BD}$	5 CPCTC

2 Given: $\overline{FH} \cong \overline{MJ}$
G is the midpt. of $\overline{FH}$
K is the midpt. of $\overline{MJ}$
$\angle GHJ \cong \angle KJH$

Prove: $\overline{GJ} \cong \overline{HK}$

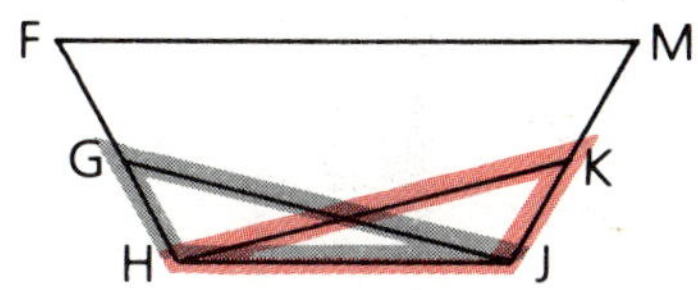

Statements	Reasons
1 $\overline{FH} \cong \overline{MJ}$	1 Given
2 G is the midpt. of $\overline{FH}$	2 Given
3 K is the midpt. of $\overline{MJ}$	3 Given
4 $\overline{GH} \cong \overline{KJ}$	4 Division Property
5 $\angle GHJ \cong \angle KJH$	5 Given
6 $\overline{HJ} \cong \overline{HJ}$	6 Reflexive Property
7 $\triangle GHJ \cong \triangle KJH$	7 SAS (4, 5, 6)
8 $\overline{GJ} \cong \overline{HK}$	8 CPCTC

Section 3.4 Problem Set A

1 Given: $\overline{AB} \cong \overline{DC}$
$\overline{AC} \cong \overline{DB}$

Prove: $\triangle ABC \cong \triangle DCB$

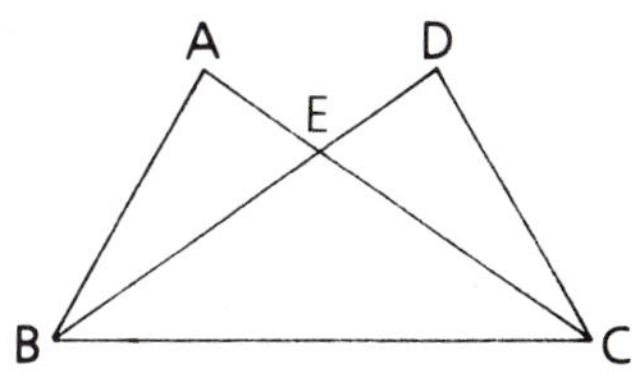

2 Given: $\angle FGH$ is a right $\angle$
$\angle JHG$ is a right $\angle$
$\overline{FG} \cong \overline{JH}$

Prove: $\triangle FGH \cong \triangle JHG$

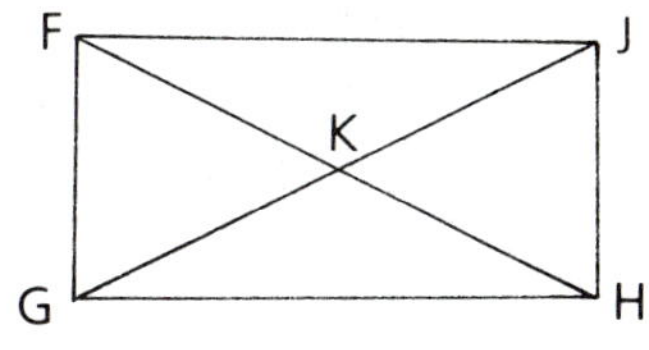

3 Given: $\overline{PM} \cong \overline{RM}$
$\angle SPM \cong \angle ORM$

Prove: $\triangle PSM \cong \triangle ROM$

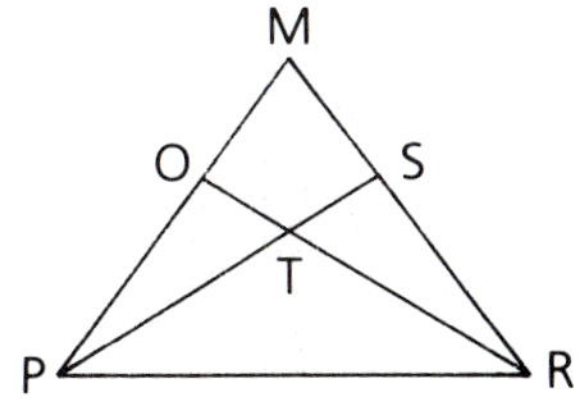

4 Given: $\angle 1 \cong \angle 3$

$\angle 2 \cong \angle 4$

Conclusion: $\overline{BC} \cong \overline{ED}$

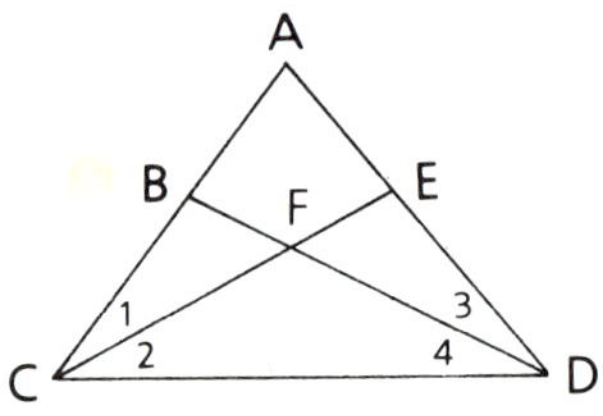

5 Given: $\overline{JH} \cong \overline{KH}$

$\overline{HG} \cong \overline{HM}$

$\angle 5 \cong \angle 6$

Conclusion: $\triangle JHG \cong \triangle KHM$

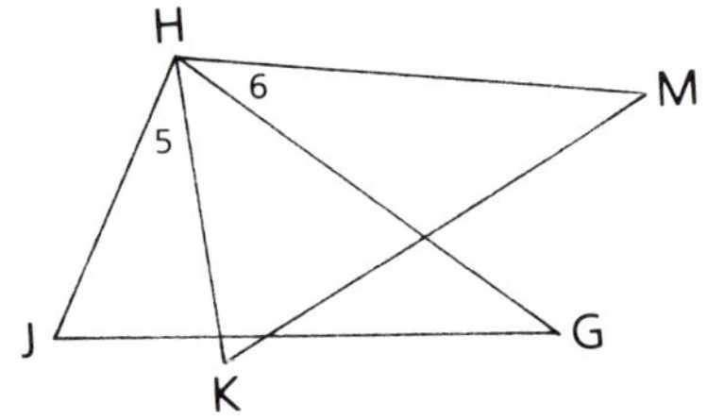

Section 3.4 Problem Set B

6 Given: $\angle 1$ is comp. to $\angle 2$

$\angle 3$ is comp. to $\angle 4$

$\angle 1 \cong \angle 3$

Conclusion: $\overline{AB} \cong \overline{CD}$

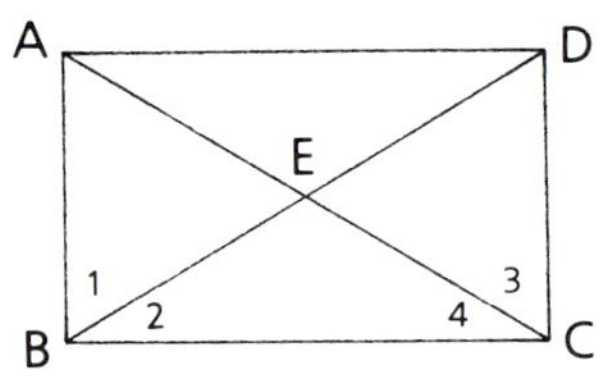

7 Given: Figure NOPRS is equilateral (all sides are congruent)

$\angle OPR \cong \angle PRS$

$\overline{PT} \cong \overline{TR}$

Prove: $\overline{OT} \cong \overline{ST}$

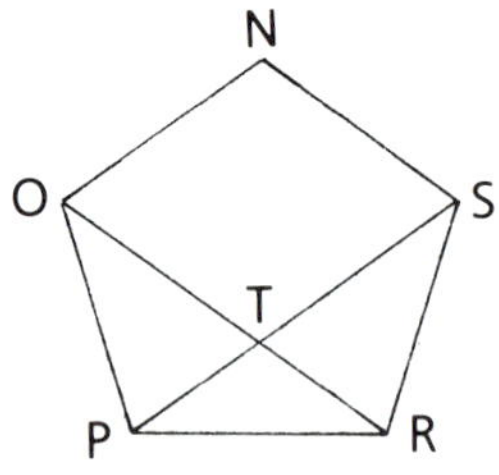

8 Given: $\overline{YW}$ bisects $\overline{AX}$

$\angle A \cong \angle X$

$\angle 5 \cong \angle 6$

Conclusion: $\overline{ZW} \cong \overline{YW}$

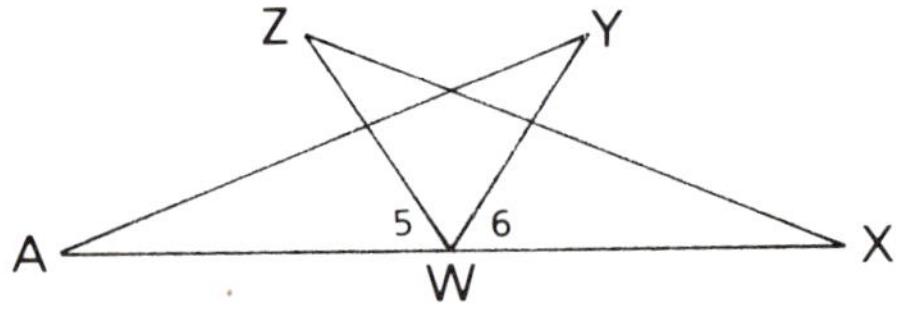

9 Given: B is the midpt. of $\overline{AC}$
E is the midpt. of $\overline{AD}$
$\angle 7 \cong \angle 8$
$\angle ECD \cong \angle BDC$
Prove: $\overline{AC} \cong \overline{AD}$

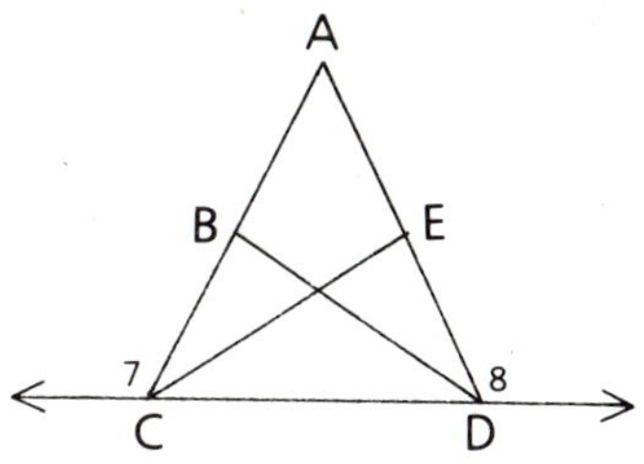

10 Given: $\angle 9 \cong \angle 10$
$\angle GFH \cong \angle HJG$
Conclusion: $\overline{FG} \cong \overline{JH}$

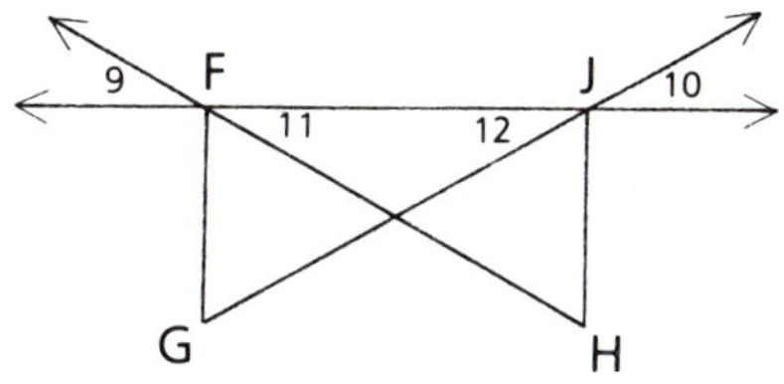

Section 3.4 Problem Set C

11 Given: $\overline{HO} \cong \overline{MO}$
$\overline{JO} \cong \overline{KO}$
$\overline{HJ}$ is an altitude of $\triangle HJK$
$\overline{MK}$ is an altitude of $\triangle MKJ$
Prove: $\angle 1 \cong \angle 2$

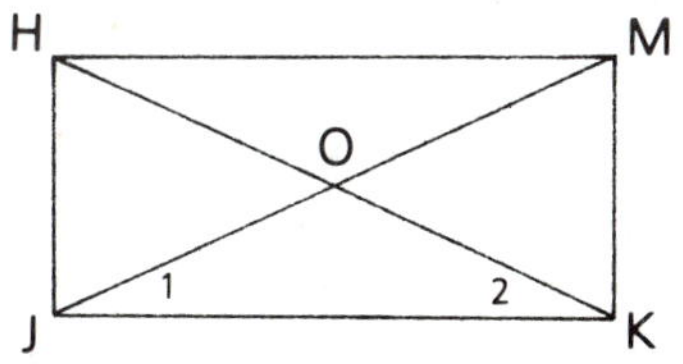

12 Given: $\overline{NR} \cong \overline{NV}$
P and Q are midpoints
$\angle R \cong \angle V$
$\overline{PX} \cong \overline{QX}$
Prove: $\triangle XST$ is isosceles
(at least two sides are $\cong$)

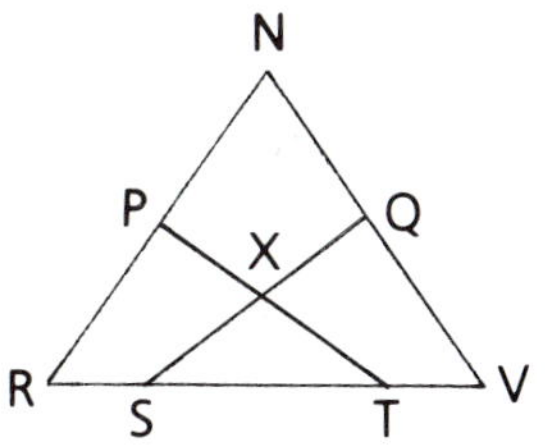

13 Given: $\overline{YD} \cong \overline{ZD}$
$\overline{BD} \cong \overline{CD}$
E is the midpt. of $\overline{YZ}$
Conclusion: $\angle BYZ \cong \angle CZY$

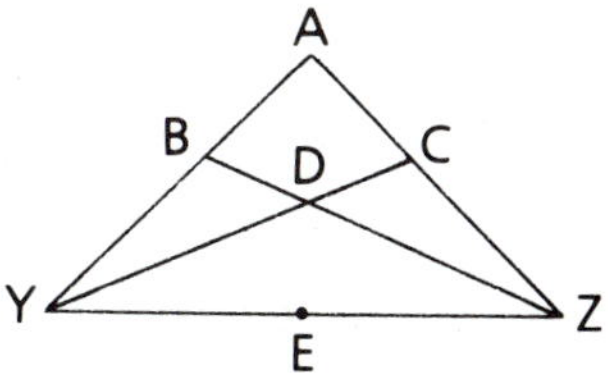

Section 3.5 Types of Triangles

In this section we present the meaning of:

1 SCALENE TRIANGLE
2 ISOSCELES TRIANGLE
3 EQUILATERAL TRIANGLE
4 EQUIANGULAR TRIANGLE
5 ACUTE TRIANGLE
6 RIGHT TRIANGLE
7 OBTUSE TRIANGLE
8 PARTS OF ISOSCELES TRIANGLES
9 PARTS OF RIGHT TRIANGLES

D **A SCALENE TRIANGLE is a triangle in which no two sides are congruent.**

Scalene Triangle

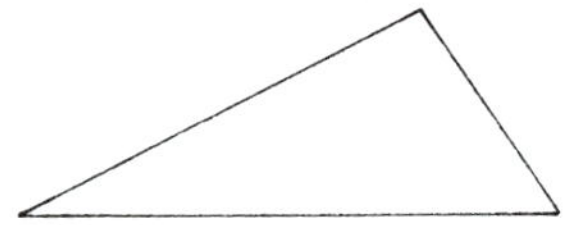

D **An ISOSCELES TRIANGLE is a triangle in which at least two sides are congruent.**

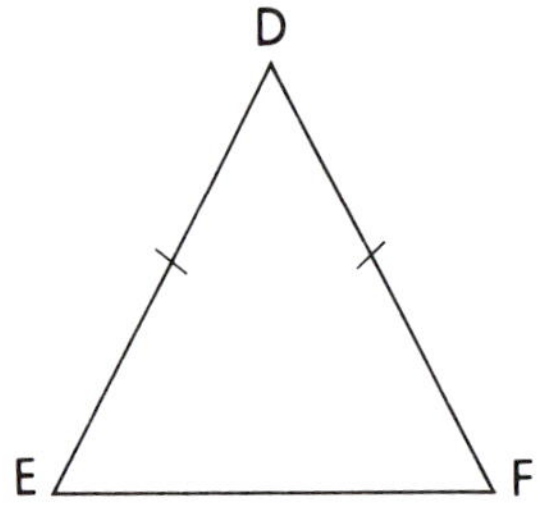

Isosceles Triangles

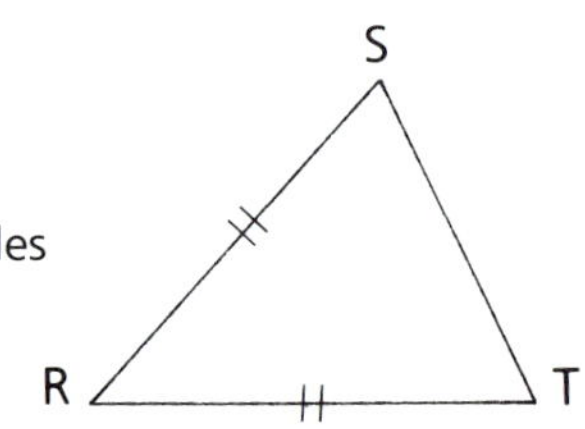

In $\triangle DEF$ above, $\overline{DE} \cong \overline{DF}$. $\overline{DE}$ and $\overline{DF}$ are called LEGS of the isosceles triangle, $\overline{EF}$ is called the BASE, $\angle E$ and $\angle F$ are called BASE ANGLES, and $\angle D$ is called the VERTEX ANGLE. Can you name these parts in $\triangle RST$?

D **An EQUILATERAL TRIANGLE is a triangle in which all sides are congruent.**

Equilateral Triangle

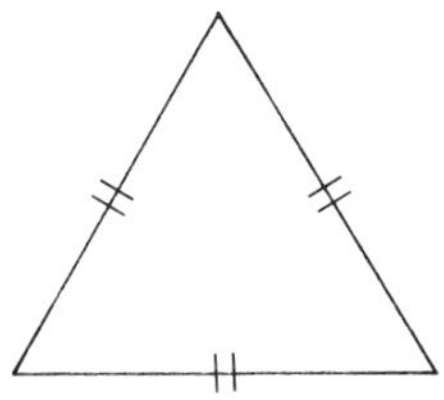

The word "equilateral" is also applied to *any* figure in which all sides are congruent.

D **An EQUIANGULAR TRIANGLE is a triangle in which all angles are congruent.**

Equiangular Triangle

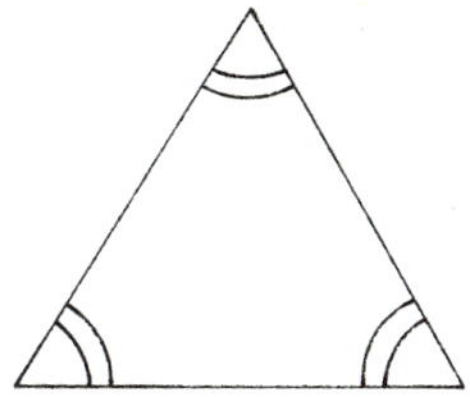

The word "equiangular" is also applied to *any* figure in which all angles are congruent.

From the diagrams you may wonder if there is any real difference between an equilateral and an equiangular triangle. You will learn the answer in Section 3.6, where you will also consider the question for equilateral and equiangular figures of any number of sides.

D **An ACUTE TRIANGLE is a triangle in which all angles are acute.**

Acute Triangle

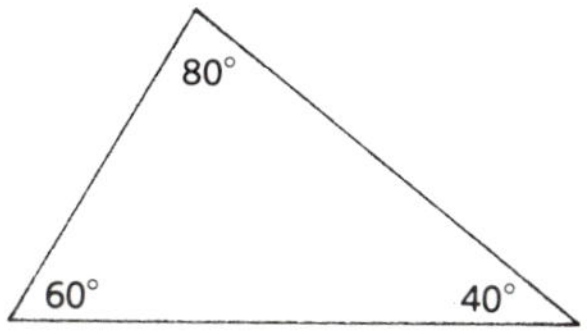

D **A RIGHT TRIANGLE is a triangle in which one of the angles is a right angle. The side opposite the right angle is called the HYPOTENUSE. The sides that form the right angle are called LEGS.**

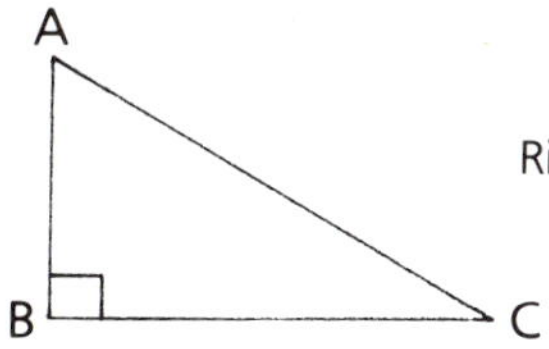

Right Triangles

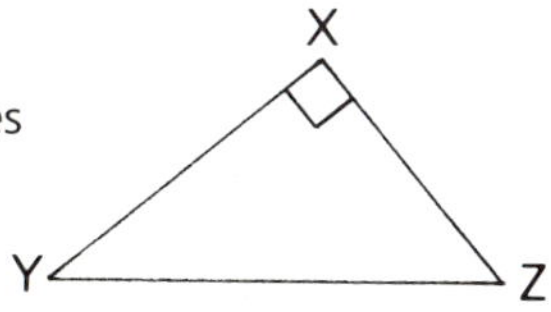

In $\triangle ABC$ above, $\overline{AB}$ and $\overline{BC}$ are called LEGS, and $\overline{AC}$ is called the **HYPOTENUSE**. Can you name those parts in $\triangle XYZ$?

D **An OBTUSE TRIANGLE is a triangle in which one of the angles is an obtuse angle.**

Obtuse Triangle

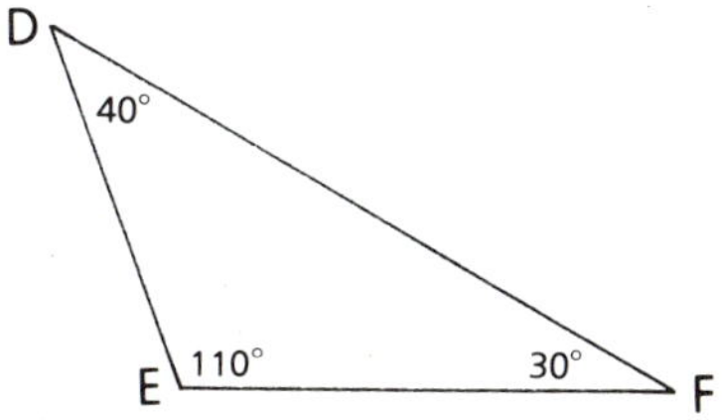

Section 3.5 Sample Problems

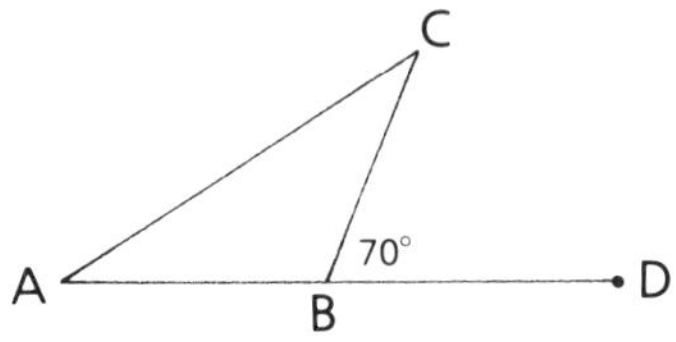

1 Given: $\angle CBD = 70°$

Prove: $\triangle ABC$ is obtuse

Proof: Since $\angle CBD = 70°$ and $\angle ABD$ is a straight $\angle$, $\angle ABC = 110°$. Since $\triangle ABC$ contains an obtuse angle, it is an obtuse triangle.

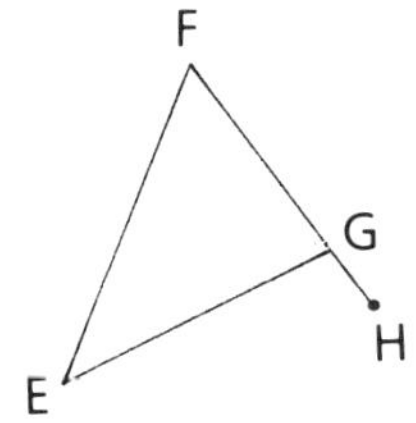

2 Given: $EG = FH$

$EF > EG$

Prove: $\triangle EFG$ is scalene

Proof: Since $EG = FH$ and $\overline{FH}$ is clearly longer than $\overline{FG}$, that means $\overline{EG}$ is also longer than $\overline{FG}$. $EF > EG$ is given, so $\overline{EF}$ is also longer than $\overline{FG}$. Since no two sides of $\triangle EFG$ are congruent, the triangle is scalene.

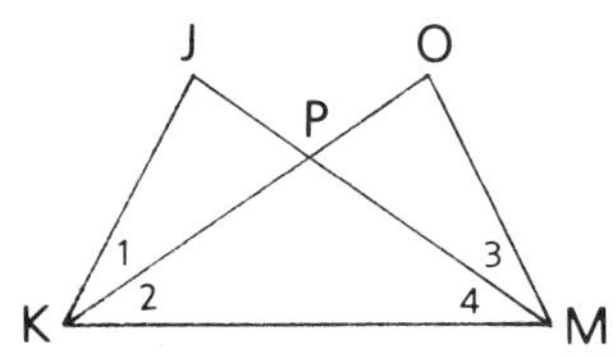

3 Given: $\angle 1 \cong \angle 3$

$\angle 2 \cong \angle 4$

$\overline{JP} \cong \overline{PO}$

Prove: $\triangle KPM$ is isosceles

Statements	Reasons
1 $\angle 1 \cong \angle 3$	1 Given
2 $\angle 2 \cong \angle 4$	2 Given
3 $\angle JKM \cong \angle OMK$	3 Addition Property
4 $\overline{KM} \cong \overline{KM}$	4 Reflexive Property
5 $\triangle JKM \cong \triangle OMK$	5 ASA (2, 4, 3)
6 $\overline{JM} \cong \overline{KO}$	6 CPCTC
7 $\overline{JP} \cong \overline{PO}$	7 Given
8 $\overline{KP} \cong \overline{MP}$	8 Subtraction Property
9 $\triangle KPM$ is isosceles	9 If a triangle has at least two sides congruent, it is isosceles

Section 3.5 Problem Set A

1 Classify each of the triangles as scalene, isosceles, or equilateral.

a

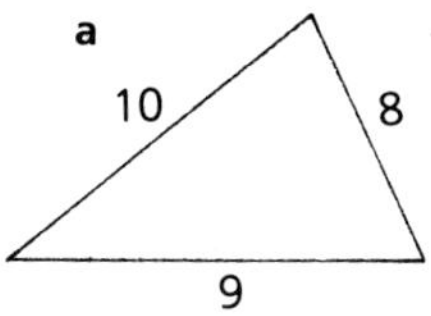

b

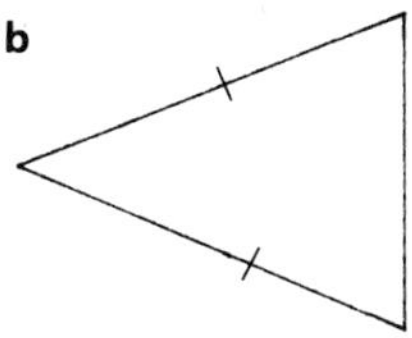

c

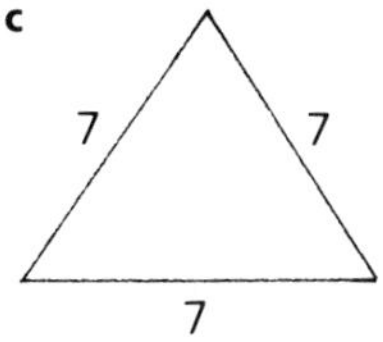

d

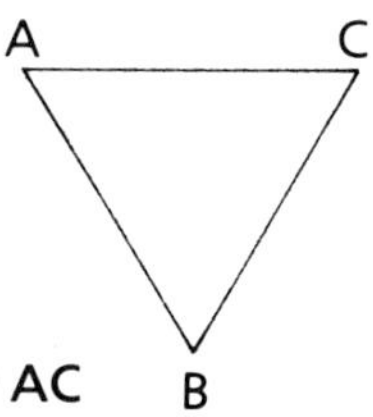

$AB > AC$
$AC > BC$

e

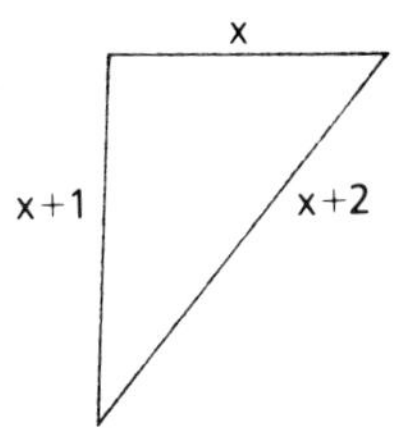

f

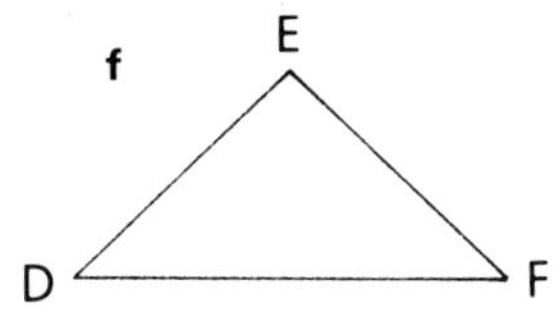

$DE < DF$
$EF = DE$

2 Classify each of the triangles as acute, right, or obtuse.

a

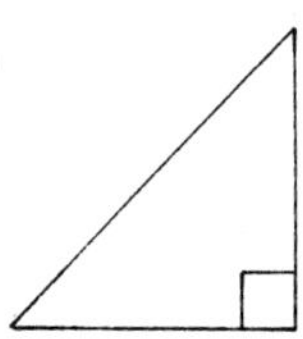

b

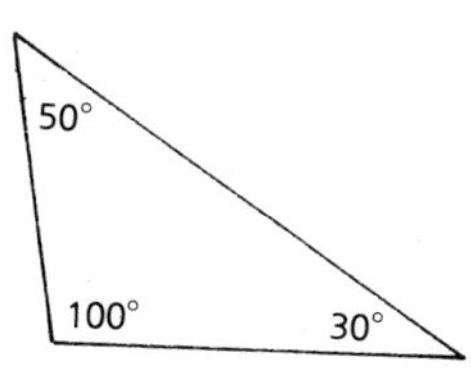

c

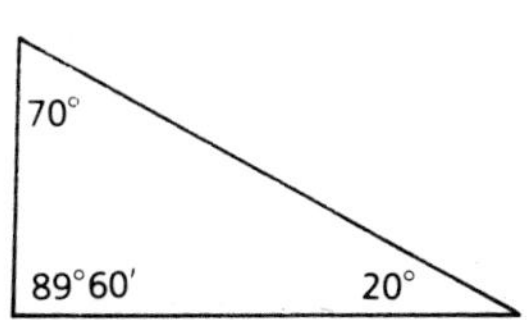

d

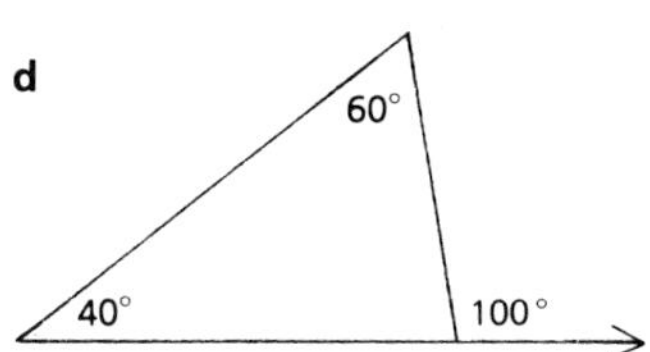

e $\overline{GH} \perp \overline{HJ}$

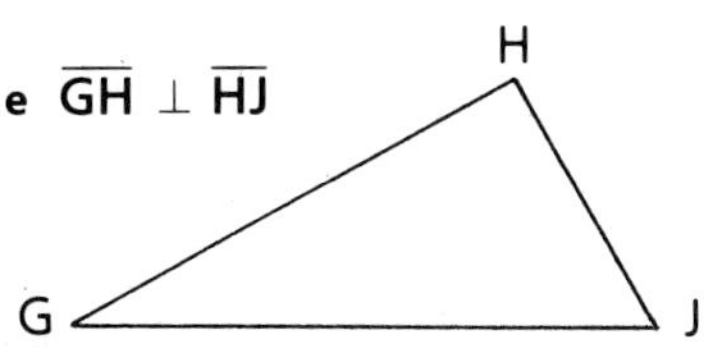

f $\frac{1}{2}m\angle K = 30°$
$\frac{1}{3}m\angle M = 20°$
$\frac{1}{4}m\angle O = 15°$

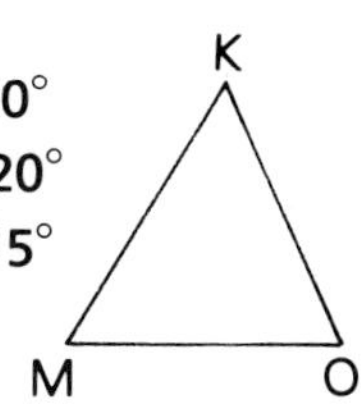

3 Using the figure as marked, write a paragraph proof that $\triangle ABC$ is acute.

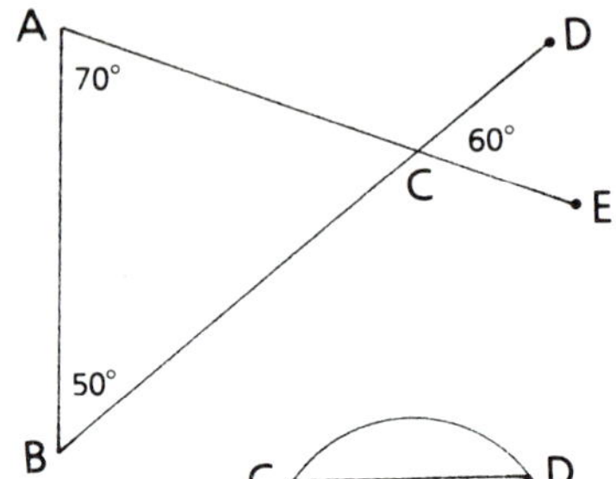

4 Given: $\odot$ O

Prove: $\triangle COD$ is isosceles

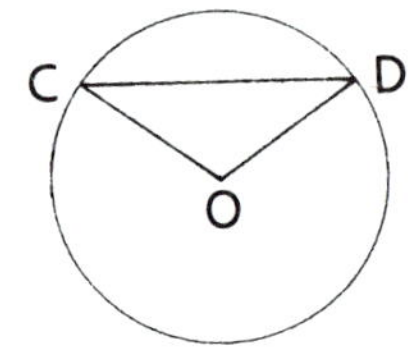

5 If the perimeter of $\triangle EFG$ is 32, is $\triangle EFG$ scalene, isosceles, or equilateral?

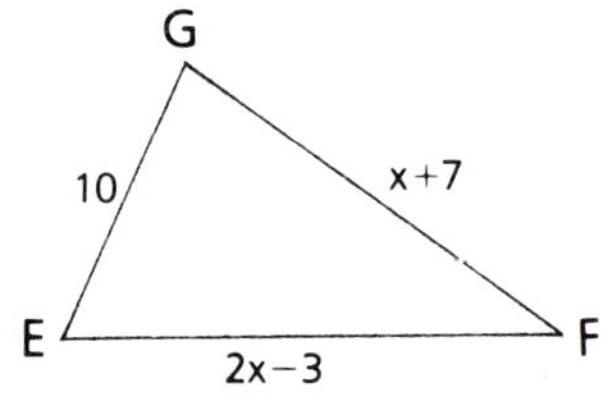

6 If $\triangle HJK$ is equilateral, solve for x and y.

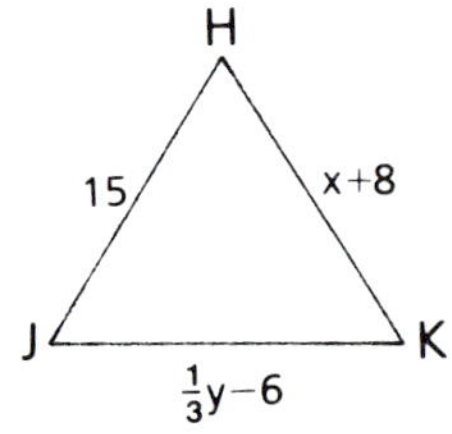

Section 3.5 Problem Set B

7 Given: $\overline{AD}$ and $\overline{CD}$ are legs of isosceles $\triangle ACD$
B is the midpt. of $\overline{AC}$

Prove: $\angle A \cong \angle C$

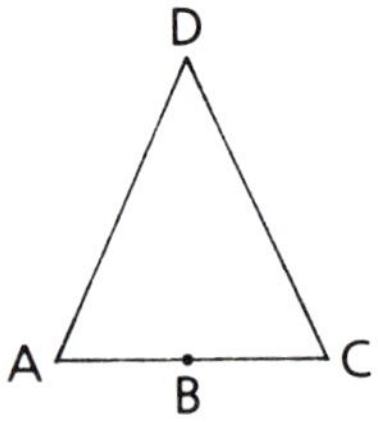

8 Given: $\overline{BI} \cong \overline{RD}$, $\overline{RI} \cong \overline{BD}$
$\angle 3$ is comp. to $\angle 2$

Prove: $\triangle RIB$ is a right $\triangle$

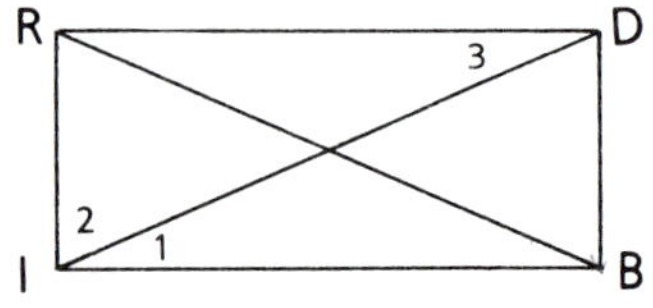

9 Given: $\overline{JF} \cong \overline{JG}$
F and G trisect $\overline{EH}$
$\angle EFJ \cong \angle HGJ$
Conclusion: $\triangle EHJ$ is isosceles

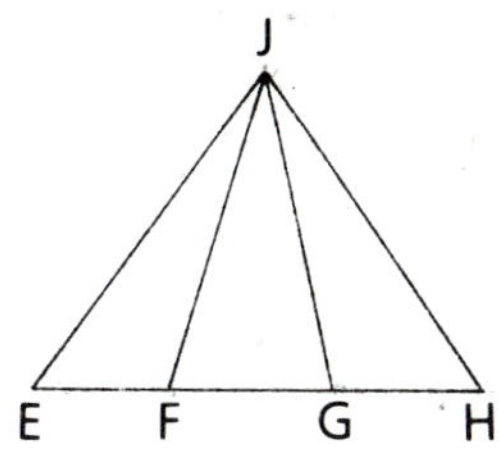

10 $RS = x + 7$
$RT = 3x + 5$
$ST = 9 - x$
If $\triangle RST$ is isosceles, is it also equilateral?

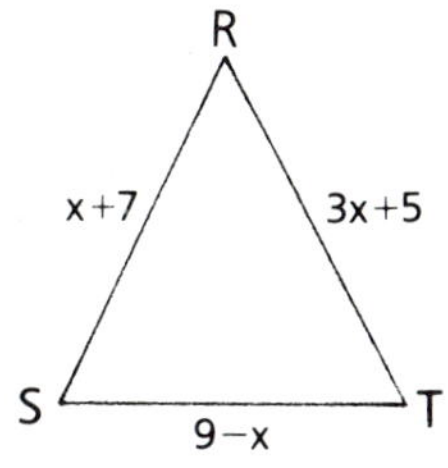

11 If $\triangle VSY$ is isosceles and the perimeter of $\triangle VSY$ is less than 45, which side of $\triangle VSY$ is the base?

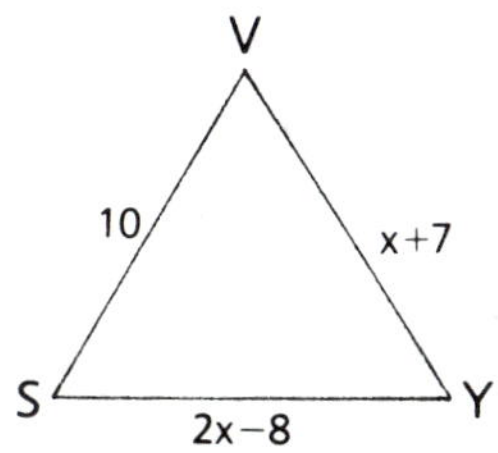

12 $AB = x + 3$
$AC = 3x + 2$
$BC = 2x + 3$
Perimeter of $\triangle ABC = 20$
Show that $\triangle ABC$ is scalene

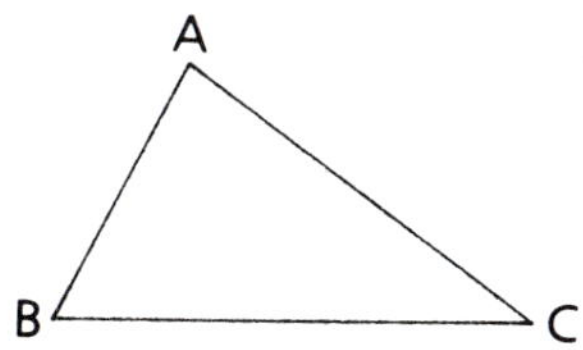

Section 3.5 Problem Set C

13 Given: $\overline{AB}$ and $\overline{AC}$ are the legs of isosceles $\triangle ABC$
$m\angle 1 = (5x)$
$m\angle 3 = (2x + 12)$
Find: $m\angle 2$

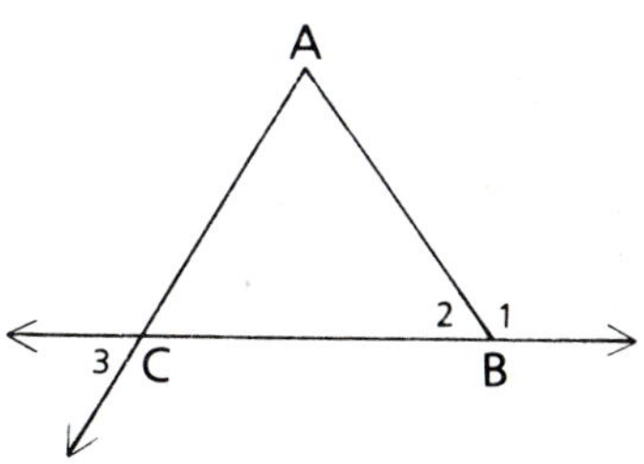

Section 3.6 Isosceles Triangle Theorems

THEOREM 19

T **If two sides of a triangle are congruent, the base angles opposite them are congruent. (if  then )**

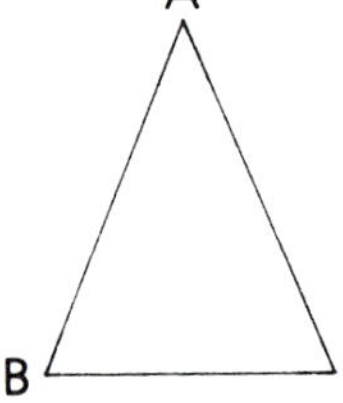

Given: $\overline{AB} \cong \overline{AC}$

Prove: $\angle B \cong \angle C$

Proof of Theorem 19:

Statements	Reasons
1 $\overline{AB} \cong \overline{AC}$	1 Given
2 $\overline{BC} \cong \overline{BC}$	2 Reflexive Property
3 $\triangle ABC \cong \triangle ACB$	3 SSS (1, 2, 1)
4 $\angle B \cong \angle C$	4 CPCTC

We have become accustomed to proving that one triangle is congruent to another triangle. As you can see, we proved the above theorem by proving that one triangle is congruent to itself (its mirror image).

We shall use the same type of proof for Theorem 20.

THEOREM 20

T **If the base angles of a triangle are congruent, the legs opposite them are congruent. (if  then )**

Given: $\angle D \cong \angle E$

Conclusion: $\overline{DF} \cong \overline{EF}$

Proof of Theorem 20:

Statements	Reasons
1 $\angle D \cong \angle E$	1 Given
2 $\overline{DE} \cong \overline{DE}$	2 Reflexive Property
3 $\triangle DEF \cong \triangle EDF$	3 ASA (1, 2, 1)
4 $\overline{DF} \cong \overline{EF}$	4 CPCTC

Because of Theorem 20, we know now that a triangle is isosceles if it has either *two or more sides* or *two or more angles* congruent. Thus we have:

TWO WAYS TO PROVE THAT A TRIANGLE IS ISOSCELES

D **1 If at least two sides of a triangle are congruent, the triangle is isosceles.**

T **2 If at least two angles of a triangle are congruent, the triangle is isosceles.**

Let us now consider a question we raised in Section 3.5 about equilateral and equiangular triangles.

Given: $\overline{GH} \cong \overline{HJ} \cong \overline{GJ}$

Is $\angle H \cong \angle J \cong \angle G$?

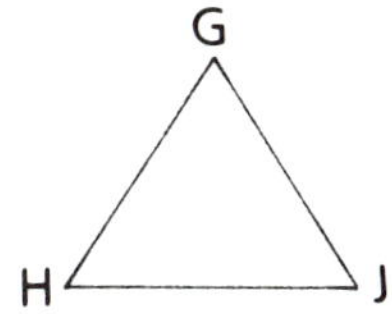

If $\overline{GH} \cong \overline{HJ}$, which two angles must be congruent? If $\overline{HJ} \cong \overline{GJ}$, which two angles must be congruent? Does that mean $\triangle GHJ$ is equiangular? Could you also prove that an equiangular triangle is equilateral?

Because of what can be proved above, we shall use *equilateral triangle* and *equiangular triangle* interchangeably for the rest of the text.

Unfortunately, we *cannot* use the words equilateral and equiangular interchangeably for *all* figures. For example, figure ABCD is equilateral, but not equiangular. Figure EFGH is equiangular, but not equilateral.

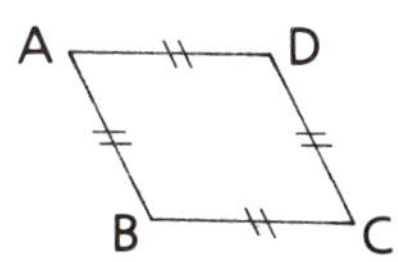

Section 3.6 Sample Problems

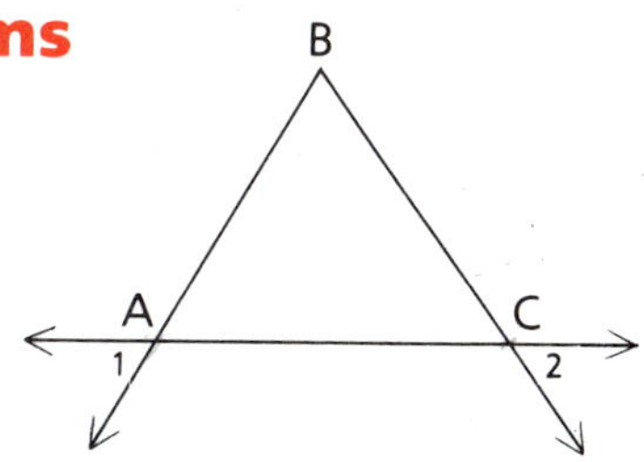

1 **Given:** $\overline{AB} \cong \overline{BC}$

Prove: $\angle 1 \cong \angle 2$

Statements	Reasons
1 $\overline{AB} \cong \overline{BC}$	1 Given
2 $\angle BAC \cong \angle BCA$	2 If △ then △
3 $\angle 1 \cong \angle BAC$	3 Vertical angles are congruent
4 $\angle 2 \cong \angle BCA$	4 Same as 3
5 $\angle 1 \cong \angle 2$	5 Transitive Property

2 **Given:** $\angle E \cong \angle H$

$\overline{EF} \cong \overline{GH}$

Conclusion: $\overline{DF} \cong \overline{DG}$

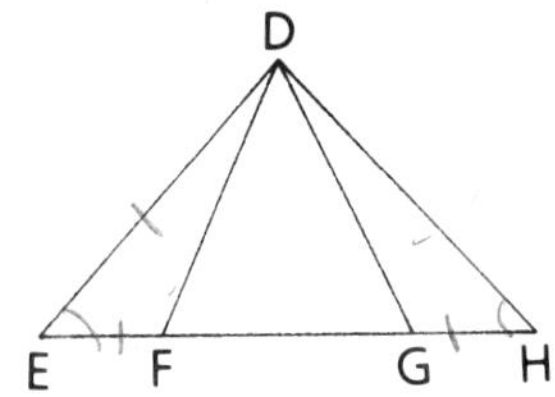

Statements	Reasons
1 $\angle E \cong \angle H$	1 Given
2 $\overline{DE} \cong \overline{DH}$	2 If △ then △
3 $\overline{EF} \cong \overline{GH}$	3 Given
4 $\triangle DEF \cong \triangle DHG$	4 SAS (2, 1, 3)
5 $\overline{DF} \cong \overline{DG}$	5 CPCTC

3 **Prove: The angle bisector of the vertex angle of an isosceles triangle is also the median to the base.**

Note: In such problems *you* must set up the proof, supplying your own diagram.

Given: $\triangle JOM$ **is isosceles with**

$\angle JOM$ **the vertex angle**

$\overrightarrow{OK}$ **bisects** $\angle JOM$

Conclusion: $\overline{OK}$ **is the median to the base**

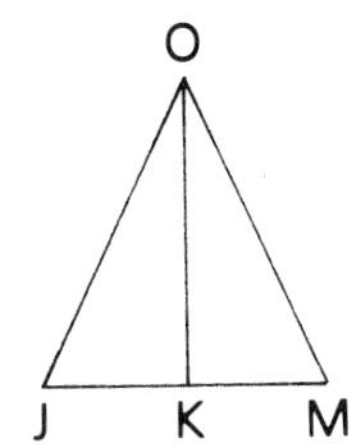

Statements	Reasons
1 $\triangle JOM$ is isosceles with $\angle JOM$ the vertex angle	1 Given
2 $\overline{OJ} \cong \overline{OM}$	2 The legs of an isosceles △ are ≅
3 $\overrightarrow{OK}$ bisects $\angle JOM$	3 Given
4 $\angle JOK \cong \angle MOK$	4 If a ray bisects an ∠, it divides the ∠ into two ≅ angles
5 $\overline{OK} \cong \overline{OK}$	5 Reflexive Property
6 $\triangle JOK \cong \triangle MOK$	6 SAS (2, 4, 5)
7 $\overline{JK} \cong \overline{MK}$	7 CPCTC
8 $\overline{OK}$ is the median to the base	8 If a segment from a vertex of a △ divides the opposite side into two ≅ segments, it is the median

4 Given: $\angle 3 \cong \angle 4$
$\overline{BX} \cong \overline{AY}$
$\overline{BW} \cong \overline{AZ}$

Conclusion: $\triangle WTZ$ is isosceles

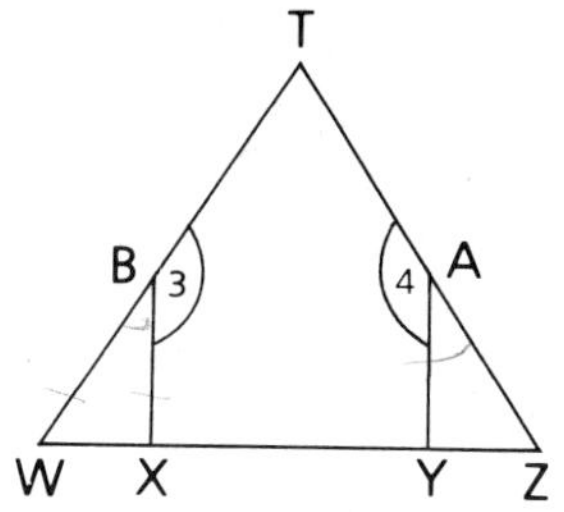

Statements	Reasons
1 $\angle 3 \cong \angle 4$	1 Given
2 $\angle 3$ is supp. to $\angle WBX$	2 If two angles form a straight angle, they are supplementary
3 $\angle 4$ is supp. to $\angle YAZ$	3 Same as 2
4 $\angle WBX \cong \angle YAZ$	4 Angles each supp. to $\cong \angle$s are $\cong$
5 $\overline{BX} \cong \overline{AY}$	5 Given
6 $\overline{BW} \cong \overline{AZ}$	6 Given
7 $\triangle BWX \cong \triangle AZY$	7 SAS (5, 4, 6)
8 $\angle W \cong \angle Z$	8 CPCTC
9 $\triangle WTZ$ is isosceles	9 If at least two angles of a $\triangle$ are $\cong$, the $\triangle$ is isosceles

5 $\triangle JKM$ is isosceles with $\overline{KM}$ the base
$\angle K = (3x + 10)^\circ$
$\angle M = (2\frac{1}{2}x + 18)^\circ$
Find: $m\angle M$

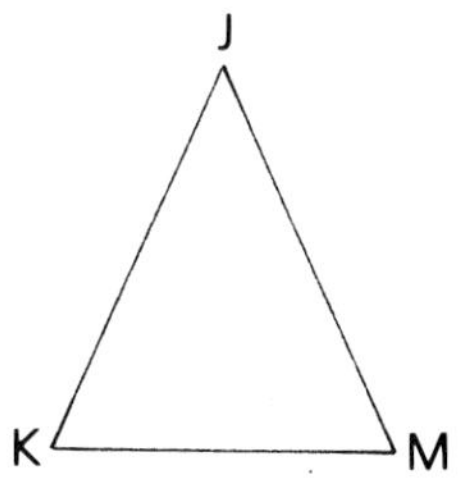

Solution: Since $\triangle JKM$ is isosceles with $\overline{KM}$ the base, $\angle K \cong \angle M$.

$$3x + 10 = 2\tfrac{1}{2}x + 18$$
$$\frac{1}{2}x + 10 = 18$$
$$\frac{1}{2}x = 8$$
$$x = 16$$

Substituting for x, $\angle M = 58^\circ$

Section 3.6 Problem Set A

1 Given: $\overline{AB} \cong \overline{AC}$

Conclusion: $\angle 1 \cong \angle 2$

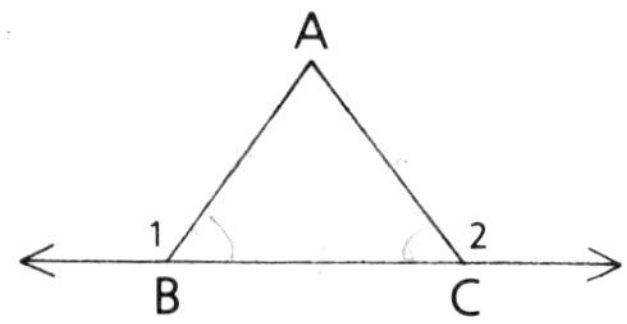

2 Given: $\angle KRM \cong \angle PRO$

$\overline{KR} \cong \overline{PR}$

Prove: $\overline{RM} \cong \overline{RO}$

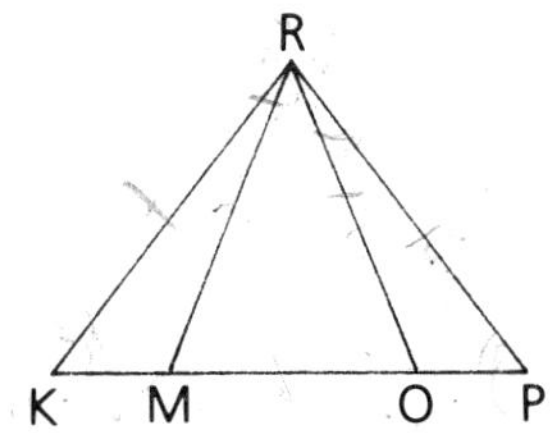

3 Given: $\overline{SX} \cong \overline{TY}$

$\overline{WX} \cong \overline{YZ}$

$\overline{SW} \cong \overline{TZ}$

Prove: $\overline{RW} \cong \overline{RZ}$

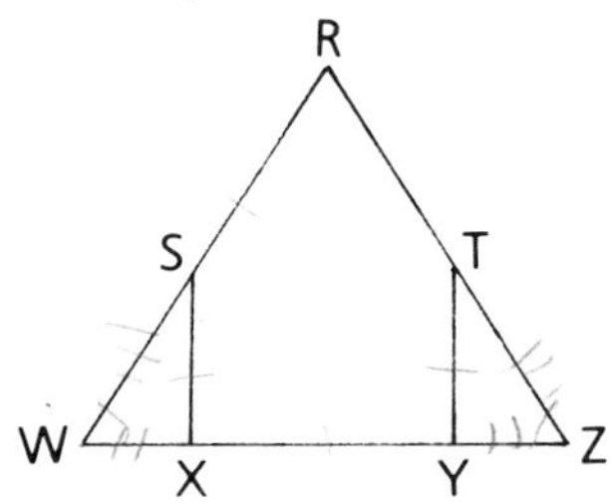

4 Given: $\angle 3 \cong \angle 6$

$\angle 3$ is comp. to $\angle 4$

$\angle 6$ is comp. to $\angle 5$

Prove: $\triangle EBC$ is isosceles

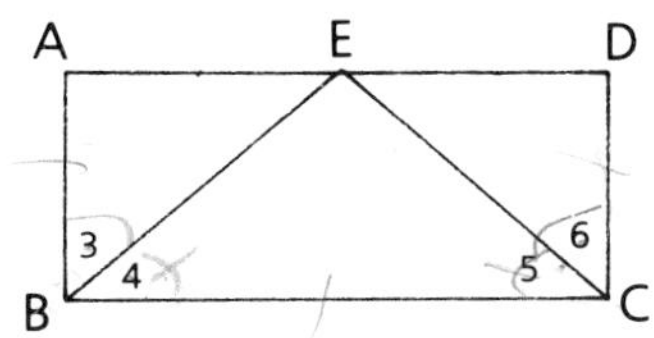

5 Given: $\overline{FH} \cong \overline{GJ}$

$\triangle FKJ$ is isosceles with $\overline{FK} \cong \overline{JK}$

Prove: $\triangle FKH \cong \triangle JKG$

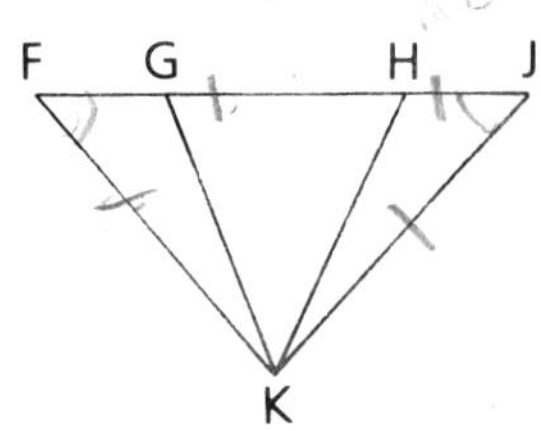

6 Given: $\angle 5 \cong \angle 6$

$\overline{JG}$ is the altitude to $\overline{FH}$

Prove: $\triangle FJH$ is isosceles

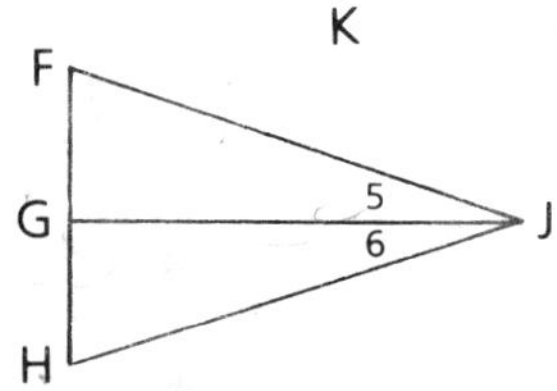

7 Given: $\overline{OP} \cong \overline{RS}$

$\overline{KO} \cong \overline{KS}$

M is the midpt. of $\overline{OK}$

T is the midpt. of $\overline{KS}$

Prove: $\overline{MP} \cong \overline{TR}$

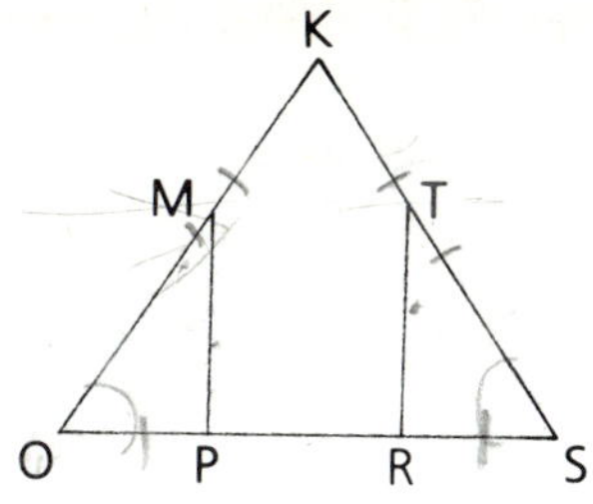

8 Given: ⊙ O

$\overline{OX} \cong \overline{XW}$

Prove: $\triangle XOW$ is equilateral

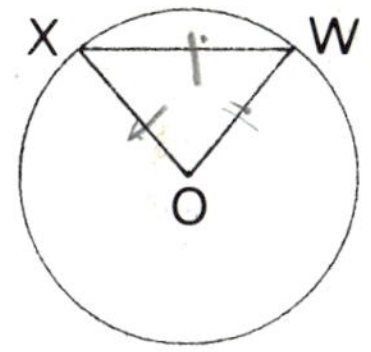

9 $\overline{AC} \perp \overline{BC}$

$\angle C = (3x)^\circ$

$BC = x + 20$

$AC = 2x - 20$

Is $\triangle ABC$ isosceles?

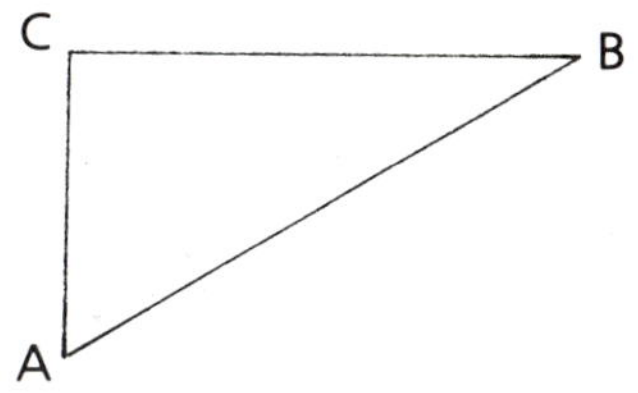

10 $\overline{GH} \cong \overline{GJ}$

$\angle 1 = (3x)^\circ$

$\angle 2 = x^\circ$

Find: $m\angle J$

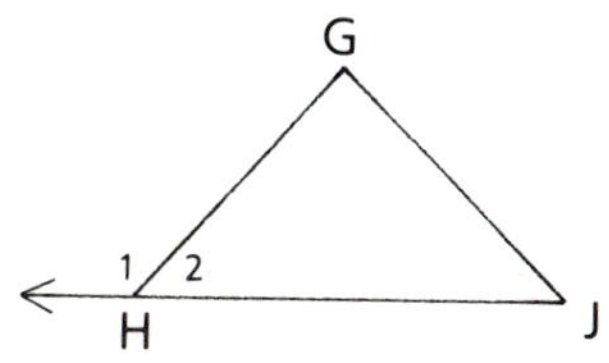

Section 3.6 Problem Set B

11 Given: $\overline{BE} \cong \overline{BD}$

$\overline{BE} \perp \overline{AE}$

$\angle BDC = 90^\circ$

Prove: $\angle AED \cong \angle CDE$

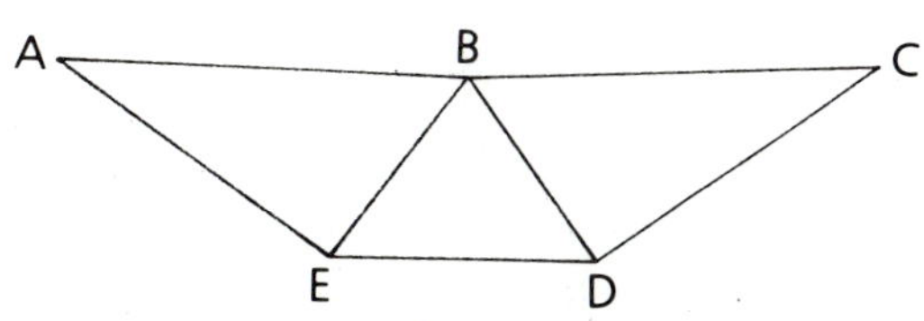

12 Prove: The median to the base of an isosceles triangle also bisects the vertex angle.

13 Given: $\overline{HK} \cong \overline{JM}$
$\overline{GJ} \cong \overline{JK}$
$\overline{OK} \cong \overline{JK}$
$\overline{GJ}$ and $\overline{OK}$ are $\perp$ to $\overline{HM}$

Prove: $\triangle FHM$ is isosceles

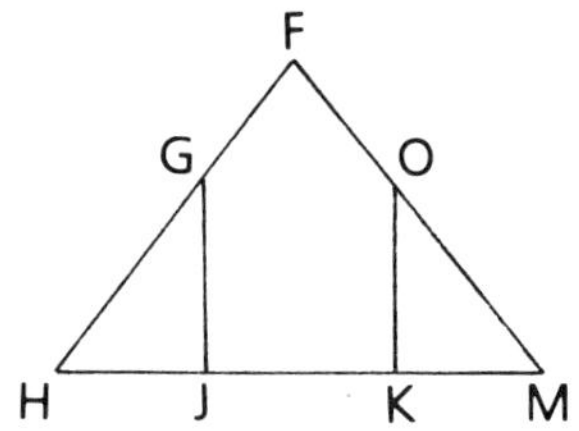

14 Given: $\overline{PR} \cong \overline{ST}$
$\overline{NP} \cong \overline{VT}$
$\angle P \cong \angle T$

Prove: $\triangle WRS$ is isosceles

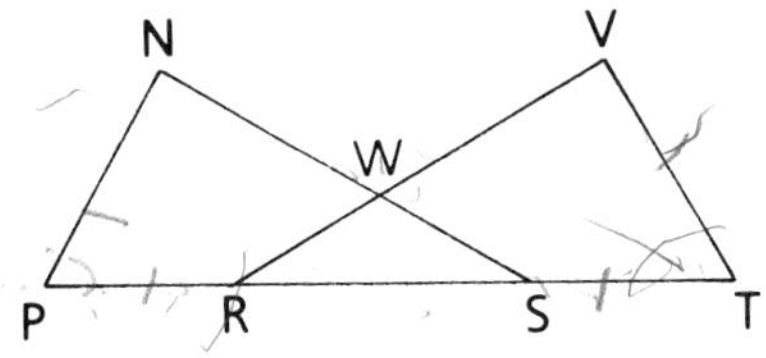

15 Given: $\overline{YZ}$ is the base of an isosceles triangle
$\angle 2 \cong \angle Z$
$\angle 1 \cong \angle Y$

Prove: $\overrightarrow{XA}$ bisects $\angle BXZ$

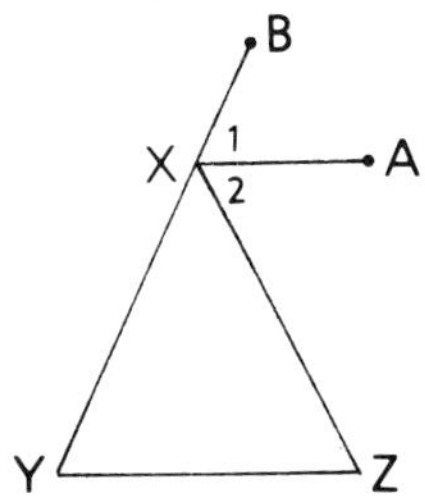

16 Given: $\overline{CE} \cong \overline{CF}$
$\angle F \cong \angle 3$
$\angle E$ is supp. to $\angle 5$

Prove: $\triangle CDG$ is isosceles

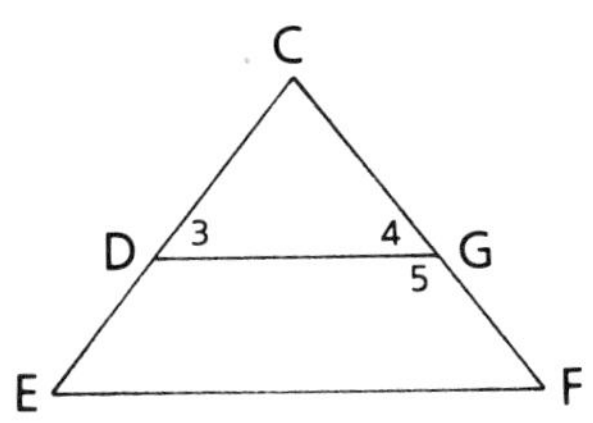

17 Given: $\overline{HJ} \cong \overline{MK}$
$\angle HJK \cong \angle MKJ$

Conclusion: $\triangle JOK$ is isosceles

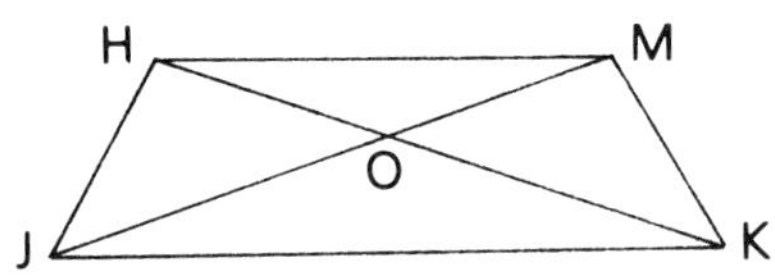

18 Given: $\triangle PSV$ is equilateral
R is the midpt. of $\overline{SP}$
T is the midpt. of $\overline{SV}$
$RT = \frac{1}{2}SV$

Prove: $\triangle RST$ is equiangular

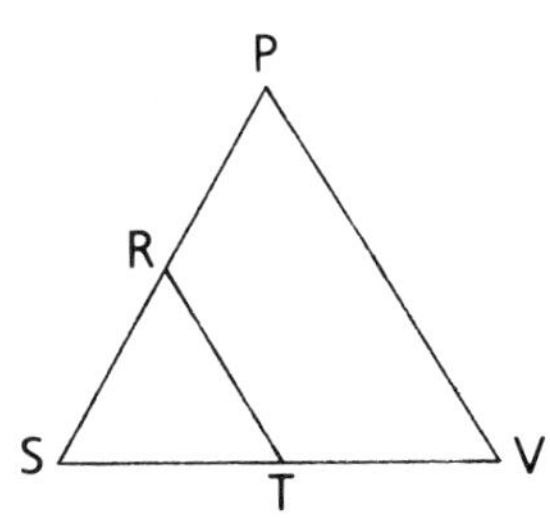

19 $\angle A$ is the vertex of an isosceles $\triangle$
$\angle B$ has twice as many degrees as $\overline{BC}$ has centimeters
$\angle C$ has three times as many degrees as $\overline{AB}$ has centimeters
$m\angle B = x + 6$
$m\angle C = 2x - 54$
Find the perimeter of $\triangle ABC$

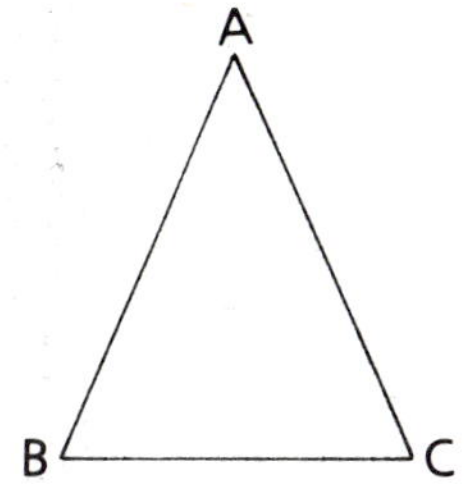

Section 3.6 Problem Set C

20 Given: $\overline{AB}$ is the base of isosceles triangle EAB
$\angle 1$ is comp. to $\angle 2$
$\angle 4$ is comp. to $\angle 3$
Prove: $\angle 5 \cong \angle 6$

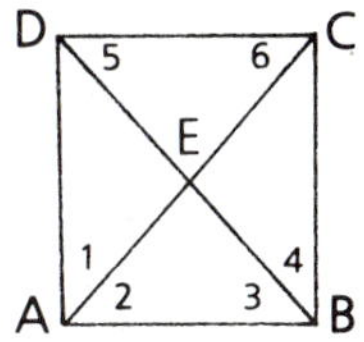

21 Given: $\overline{FG} \cong \overline{JH}$
$\angle FGH \cong \angle JHG$
Conclusion: $\triangle FKJ$ is isosceles

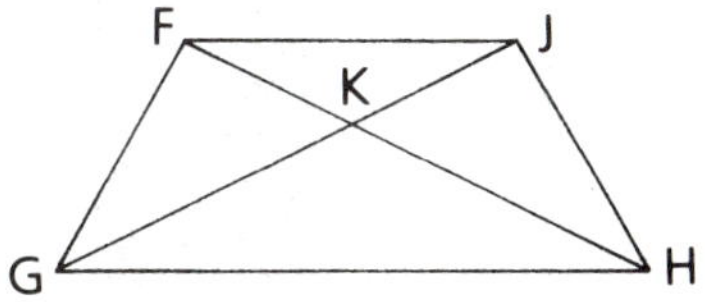

22 Given: $\odot O$
$\odot P$
$\overleftrightarrow{AB}$ bisects $\angle$s OAP and OBP
Prove: Figure AOBP is equilateral

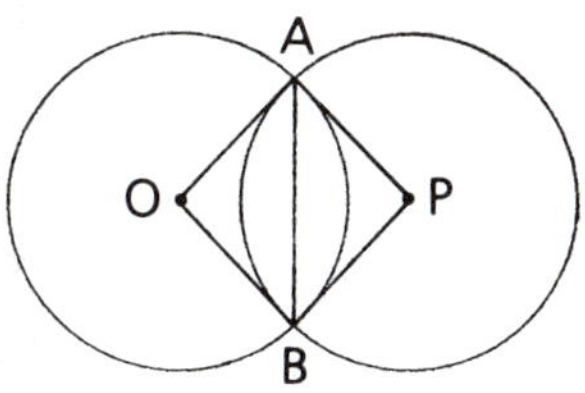

23 Given: Figure XSTOW is equilateral and equiangular
Prove: $\triangle YTO$ is isosceles

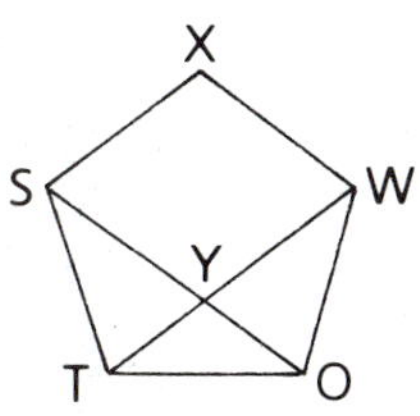

24 Given: $\triangle FED$ is equilateral
$\overline{GE} \perp \overline{DE}$
$m\angle FEG = x + y$
$m\angle D = 3x - 6$
$m\angle F = 6y + 12$
Solve for: x, y, and $\angle F$

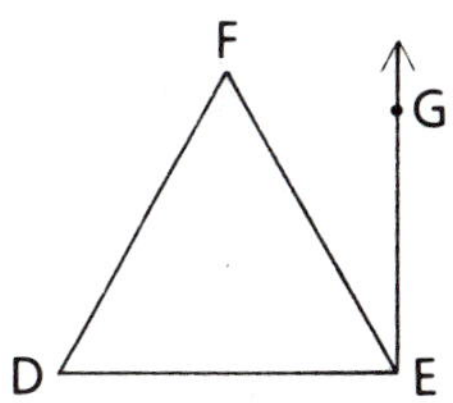

Section 3.7 HL

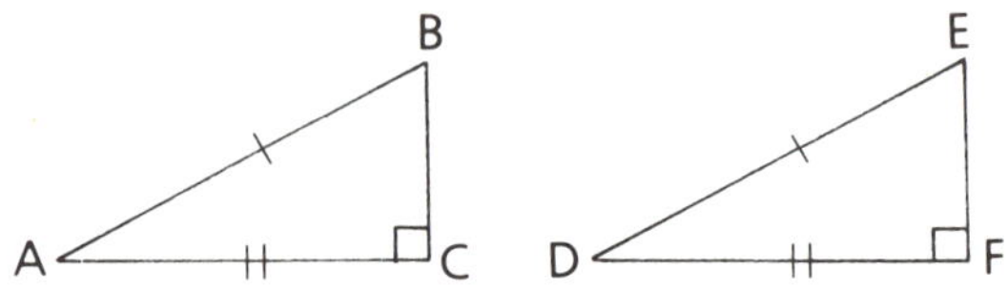

In the figure above, $\triangle ABC \cong \triangle DEF$ by a new postulate called HL. We could prove that this new method is true, but since the technique involved is not our present concern, we shall assume:

P **If there exists a correspondence between the vertices of the two RIGHT triangles such that the *Hypotenuse* and a *Leg* of one RIGHT triangle are congruent to the corresponding parts of a second RIGHT triangle, the RIGHT triangles are congruent. (HL)**

It is important to note that HL *has meaning only for* RIGHT *triangles.* Therefore, in writing proofs we must establish that the triangles are right triangles. We do this by inserting steps stating that each triangle contains a right angle. Naturally, any triangle containing a right angle is a right triangle. Did you notice that, again, you must meet three conditions to conclude that two triangles are congruent?

Section 3.7 Sample Problems

1 Given: $\overline{OF}$ is an altitude
$\odot O$
Conclusion: $\overline{EF} \cong \overline{FG}$

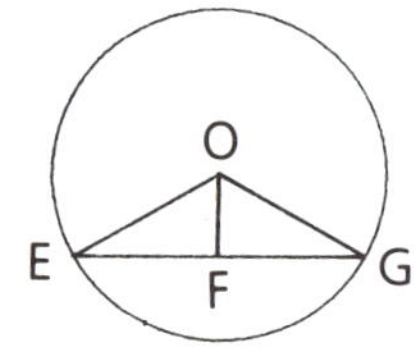

Statements	Reasons
1 $\overline{OF}$ is an altitude	1 Given
2 $\angle EFO$ and $\angle GFO$ are right $\angle$s	2 An altitude of a $\triangle$ forms right $\angle$s with the side to which it is drawn
3 $\overline{OF} \cong \overline{OF}$	3 Reflexive Property
4 $\odot O$	4 Given
5 $\overline{OE} \cong \overline{OG}$	5 All radii of a circle are $\cong$
6 $\triangle OEF \cong \triangle OGF$	6 HL (2, 5, 3)
7 $\overline{EF} \cong \overline{FG}$	7 CPCTC

2 **Given:** $\overline{BC} \perp \overline{AC}$
$\overline{BD} \perp \overline{AD}$
$\overline{AC} \cong \overline{AD}$

Prove: $\overrightarrow{AB}$ **bisects** $\angle CAD$

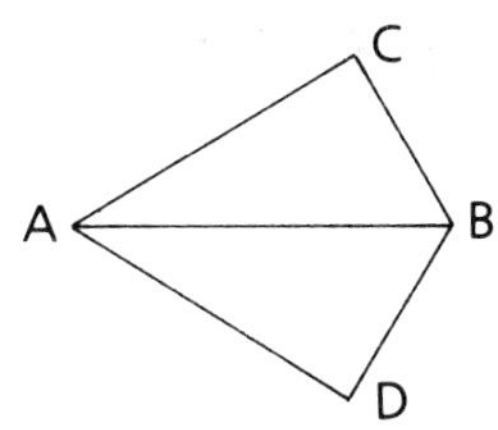

Statements	Reasons
1 $\overline{BC} \perp \overline{AC}$	1 Given
2 $\angle ACB$ is a right $\angle$	2 If two segments are $\perp$, they form right $\angle$s
3 $\overline{BD} \perp \overline{AD}$	3 Given
4 $\angle BDA$ is a right $\angle$	4 Same as 2
5 $\overline{AC} \cong \overline{AD}$	5 Given
6 $\overline{AB} \cong \overline{AB}$	6 Reflexive Property
7 $\triangle ACB \cong \triangle ADB$	7 HL (2, 4, 6, 5)
8 $\angle CAB \cong \angle DAB$	8 CPCTC
9 $\overrightarrow{AB}$ bisects $\angle CAD$	9 A ray that divides an $\angle$ into two $\cong$ angles bisects the $\angle$

3 **Prove: Corresponding angle bisectors of $\cong$ triangles are $\cong$.**

Note: Offhand, it looks like a two step proof using CPCTC. Not so! *Corresponding Parts* refers only to sides and angles of the $\cong$ $\triangle$s.

Again, we must make up our own figure and set up the proof.

Given: $\triangle KPR \cong \triangle SAW$
$\overrightarrow{RM}$ **bisects** $\angle KRP$
$\overrightarrow{WT}$ **bisects** $\angle SWA$

Prove: $\overline{RM} \cong \overline{WT}$

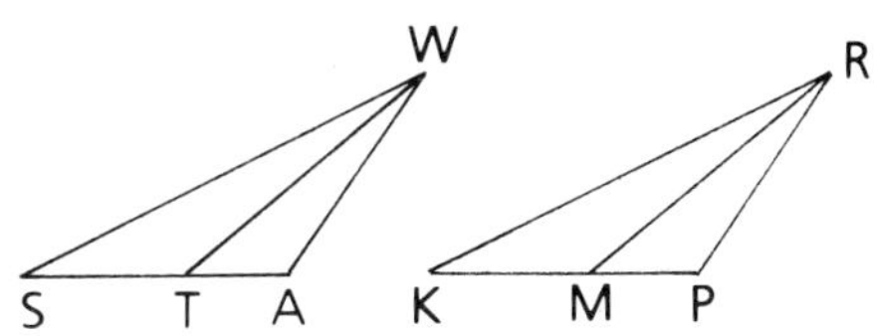

Statements	Reasons
1 $\triangle KPR \cong \triangle SAW$	1 Given
2 $\overline{KR} \cong \overline{SW}$	2 CPCTC
3 $\angle K \cong \angle S$	3 CPCTC
4 $\angle KRP \cong \angle SWA$	4 CPCTC
5 $\overrightarrow{RM}$ bisects $\angle KRP$	5 Given
6 $\overrightarrow{WT}$ bisects $\angle SWA$	6 Given
7 $\angle KRM \cong \angle SWT$	7 Division Property
8 $\triangle KRM \cong \triangle SWT$	8 ASA (3, 2, 7)
9 $\overline{RM} \cong \overline{WT}$	9 CPCTC

Section 3.7 Problem Set **A**

1 Given: $\overline{GJ}$ is the altitude to $\overline{HK}$

$\overline{HG} \cong \overline{KG}$

Prove: $\triangle HGJ \cong \triangle KGJ$

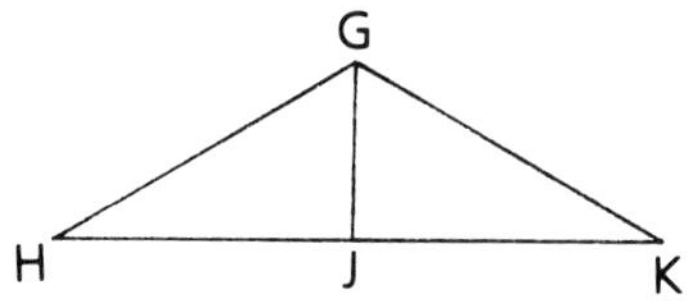

2 Given: $\overline{MO} \perp \overline{OP}$

$\overline{RP} \perp \overline{OP}$

$\overline{MP} \cong \overline{RO}$

Prove: $\triangle MOP \cong \triangle RPO$

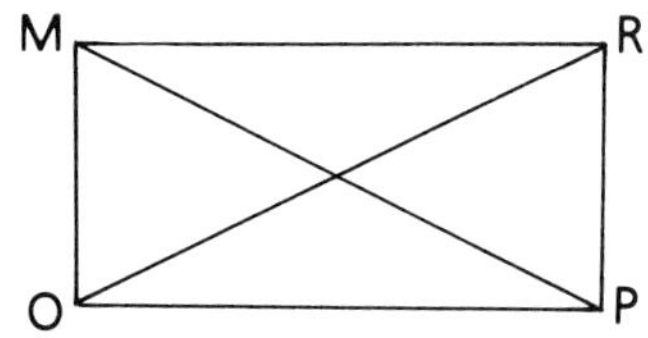

3 Given: $\odot O$

$\overline{YO} \perp \overline{YX}$

$\overline{ZO} \perp \overline{ZX}$

Conclusion: $\overline{YX} \cong \overline{ZX}$

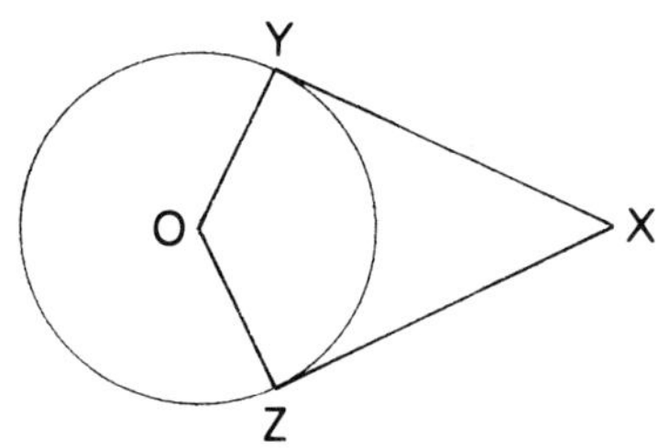

4 Given: $\overline{AE} \cong \overline{CF}$

$\overline{AB} \cong \overline{CD}$

$\angle BFA$ is a right angle

$\angle DEC$ is a right angle

Prove: $\angle CDE \cong \angle ABF$

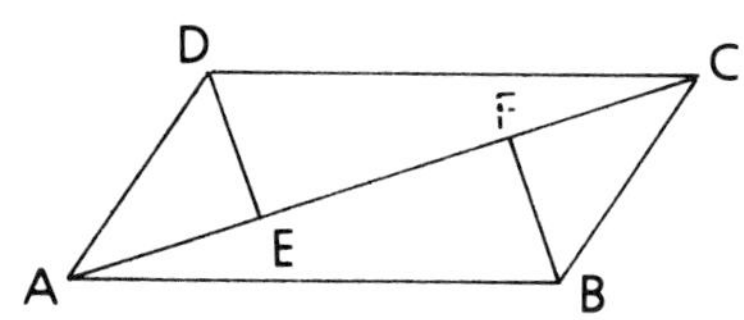

5 Set up and prove: The altitude to the base of an isosceles triangle divides the triangle into two congruent triangles.

6 Given: $\overline{GH} \cong \overline{GK}$

$\overline{GJ}$ is an altitude

Prove: $\overrightarrow{GJ}$ bisects $\angle HGK$

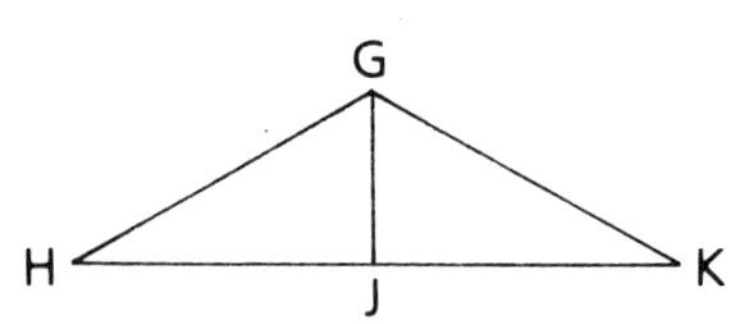

Section 3.7 Problem Set B

7 Prove: The altitude to the base of an equilateral triangle is the median to the base.

8 Given: $\overline{BD} \perp \overline{CF}$

$\overline{GE} \perp \overline{CF}$

$\overline{CE} \cong \overline{DF}$

$\overline{BC} \cong \overline{GF}$

Prove: $\triangle ACF$ is isosceles

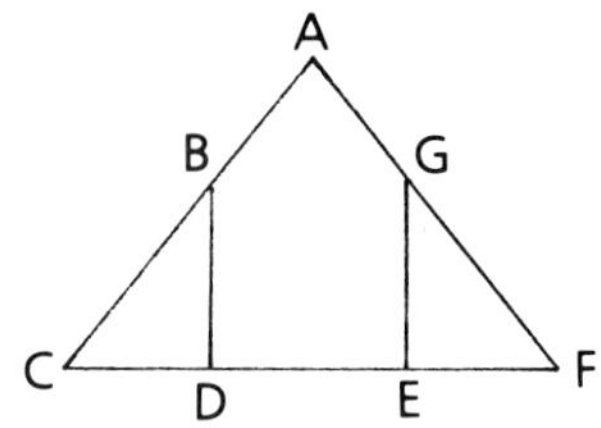

9 Given: $\overline{RK} \perp \overline{HR}$

$\overline{JO} \perp \overline{PM}$

$\overline{PH} \cong \overline{PM}$

$\overline{PR} \cong \overline{PO}$

Conclusion: $\overline{RK} \cong \overline{JO}$

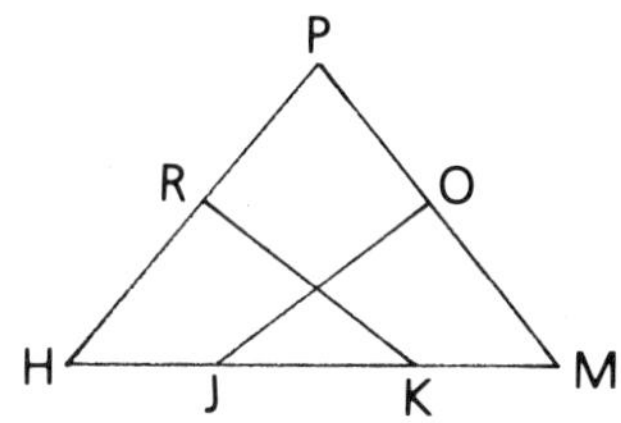

10 Given: $\odot$ P

$\overline{ST} \cong \overline{VT}$

Prove: $\angle PST \cong \angle PVT$

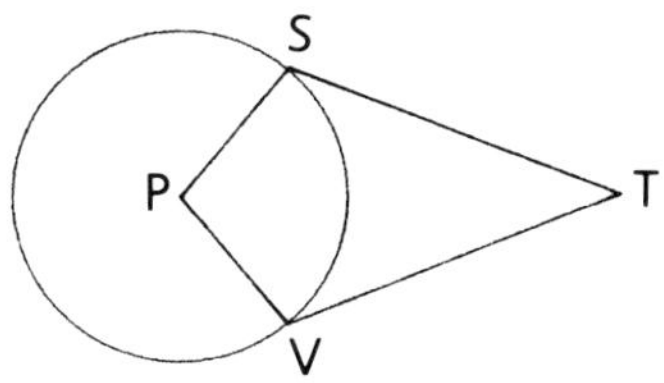

11 Prove: Corresponding medians of $\cong$ triangles are $\cong$.

12 Given: $\overline{CD} \cong \overline{EF}$

$\overline{JF} \perp \overline{JD}$

$\overline{CH} \perp \overline{HE}$

$\overline{CH} \cong \overline{JF}$

Prove: $\overline{JD} \cong \overline{HE}$

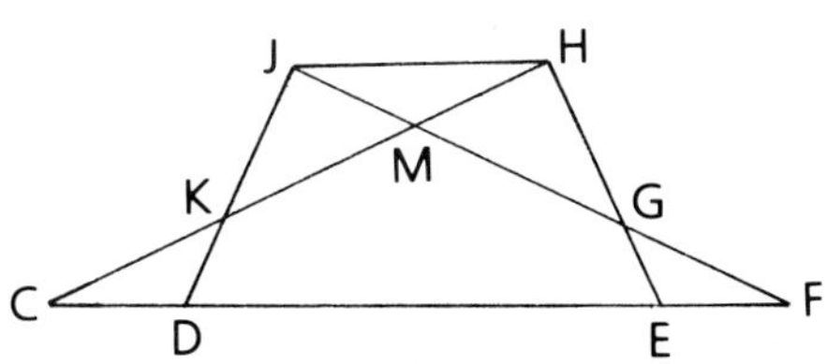

Section 3.7 Problem Set C

13 Given: $\overline{BE} \perp \overline{AD}$, $\overline{AC} \perp \overline{BD}$
$\overline{AC} \cong \overline{BE}$, $\overline{DE} \cong \overline{EC}$
Prove: $\triangle DEC$ is equilateral

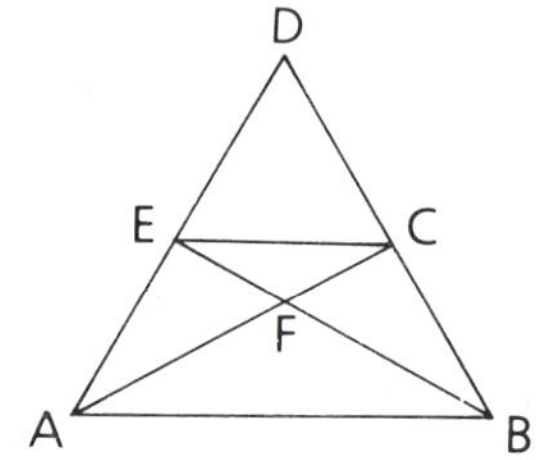

14 Given: K is midpt. of $\overline{HM}$
$\overline{GH} \cong \overline{NM}$, $\overline{HJ} \cong \overline{MP}$
$\overline{HJ} \perp \overline{GK}$, $\overline{MP} \perp \overline{KN}$
Conclusion: $\overline{GK} \cong \overline{NK}$

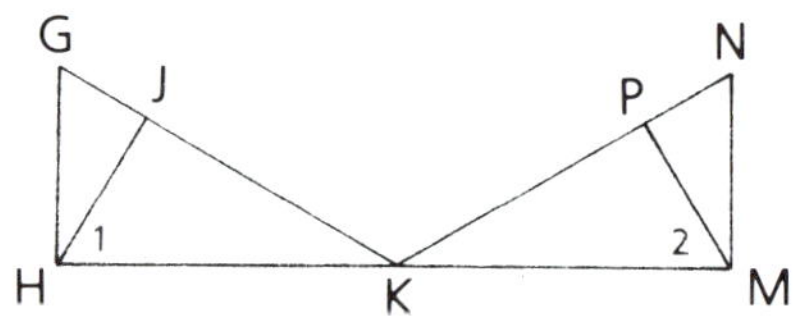

15 Given: $\angle R$ and $\angle W$ are right $\angle$s
$\overline{RX} \cong \overline{WX}$
S is $\frac{3}{7}$ of the way from R to T
V is $\frac{4}{7}$ of the way from T to W
Prove: $\overline{ST} \cong \overline{TV}$

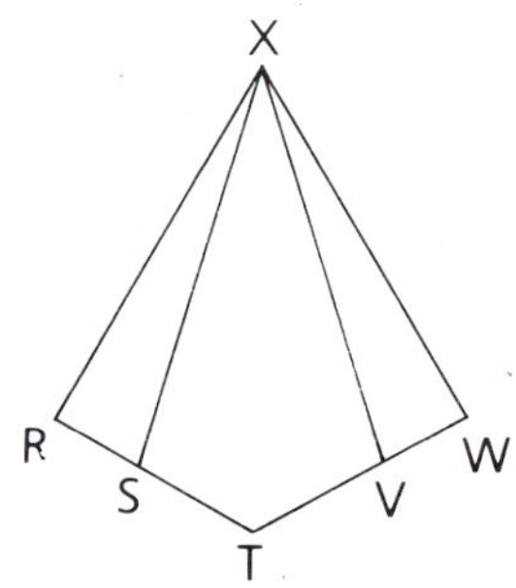

Section 3.7 Problem Set D

16 Given: $\overline{DT} \cong \overline{RK}$
$\overline{DE} \cong \overline{KV}$
$\angle ERV$ and $\angle VTE$ are right $\angle$s
Prove: **a** $\overline{ER} \cong \overline{TV}$
b $\overline{DR} \cong \overline{TK}$

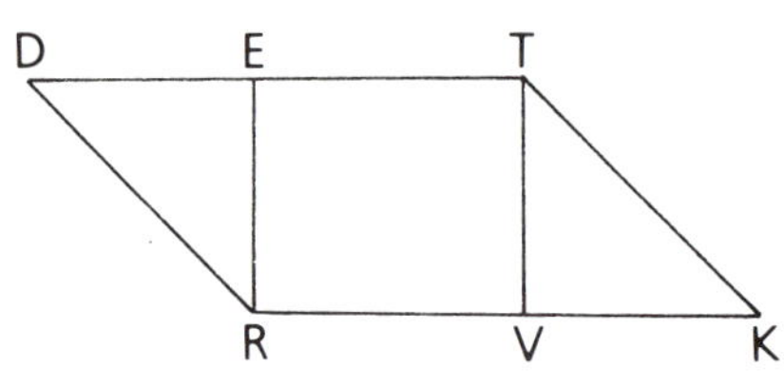

Chapter 3 Study Guide

As you study and review this chapter, be sure that you are able to:

1 Draw a figure to illustrate the use of each property:

- **a** Addition Property (p. 108)
- **b** Subtraction Property (p. 108)
- **c** Multiplication Property (p. 108)
- **d** Division Property (p. 108)
- **e** Transitive Property (p. 108)
- **f** Substitution (p. 108)
- **g** Reflexive Property (p. 108)

2 Explain how to decide when to use each of the following ways to prove that two triangles are congruent.

- **a** SSS (p. 110)
- **b** SAS (p. 111)
- **c** ASA (p. 111)
- **d** HL (right triangles only) (p. 150)

3 State what *CPCTC* means and describe when it is used. (p. 115)

4 Describe by drawings and words the use of this vocabulary:

- **a** correspondence (p. 109)
- **b** included angles and included sides (p. 110)
- **c** auxiliary lines and segments (p. 126)
- **d** overlapping triangles (p. 132)

5 Describe a circle, a radius, and the center of a circle. (p. 119)

6 State what the word *radii* means. (p. 119)

7 Draw and describe the following triangles and their parts:

- **a** scalene △ (p. 136)
- **b** isosceles △ (p. 136): its legs, vertex angle, base, base angles
- **c** equilateral and equiangular △s (p. 136)
- **d** acute △ (p. 137)
- **e** right △ (p. 137): its legs and hypotenuse
- **f** obtuse △ (p. 137)

8 Draw a scalene triangle and label a *median* and an *altitude*. (p. 125)

9 Describe the meaning of:

- **a** If , then  (p. 142)
- **b** If , then  (p. 142)

Chapter 3 Review Problems A

1 For each of the following, write:

A if the statement is always true
S if the statement is sometimes true
N if the statement is never true

a Two triangles are congruent if two sides and an angle of one are congruent to the corresponding parts of the other.

b If two sides of a right triangle are $\cong$ to the corresponding parts of another right triangle, the triangles are $\cong$.

c The three altitudes of a triangle all fall outside the $\triangle$.

d A median of a triangle does *not* contain the midpoint of the side to which it is drawn.

e A right triangle is congruent to an obtuse triangle.

2 Given: $\overline{AB} \perp \overline{BC}$
$\overline{DC} \perp \overline{BC}$
$\angle 1 \cong \angle 2$

Conclusion: $\overline{AC} \cong \overline{DB}$

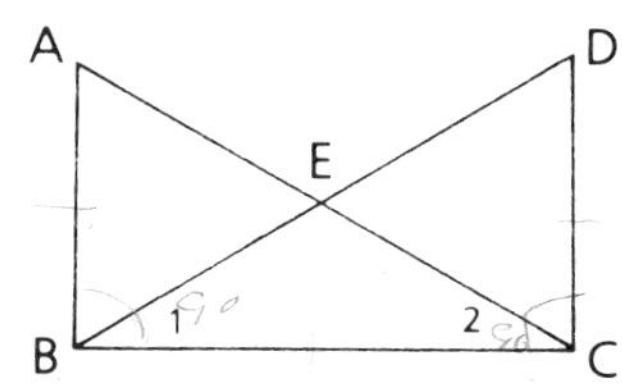

3 Given: $\odot$ O
$\overline{OG} \perp \overline{FH}$

Conclusion: $\overline{FG} \cong \overline{GH}$

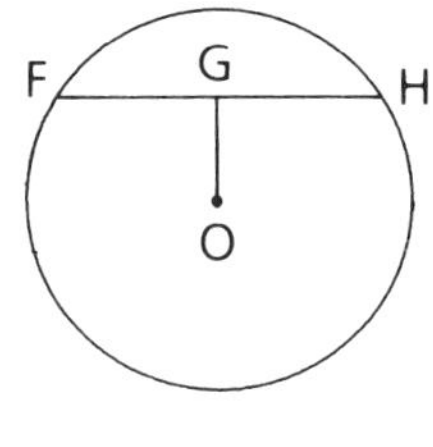

4 Given: $\overline{PK}$ and $\overleftrightarrow{JM}$ bisect each other at R

Prove: $\overline{PJ} \cong \overline{MK}$

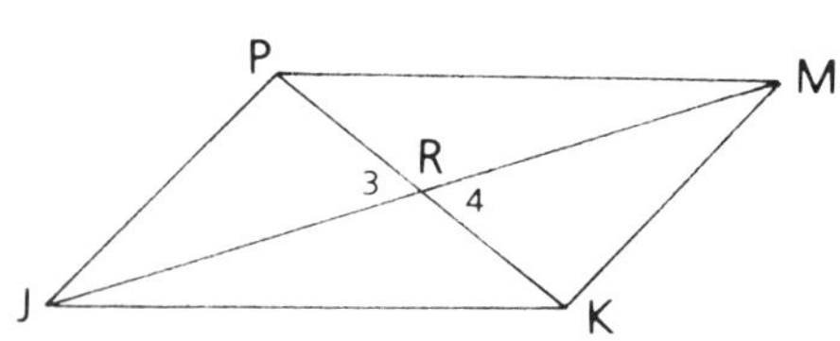

5 Given: $\overline{DG} \cong \overline{JF}$
$\overline{DE} \cong \overline{JH}$
$\overline{EG} \cong \overline{HF}$

Prove: $\triangle HCE$ is isosceles

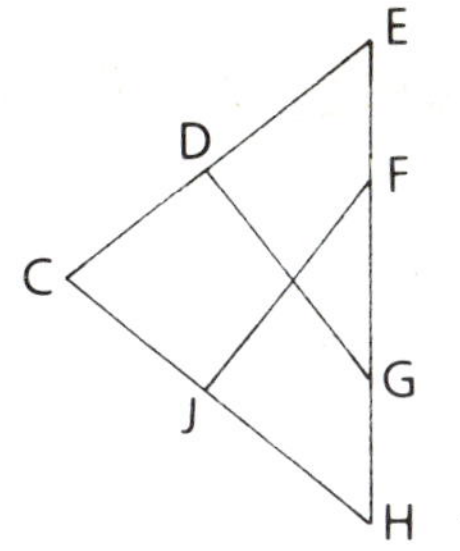

6 Given: $\odot P$
$\overrightarrow{PR}$ bisects $\angle KPM$

Conclusion: $\overline{PR}$ is a median

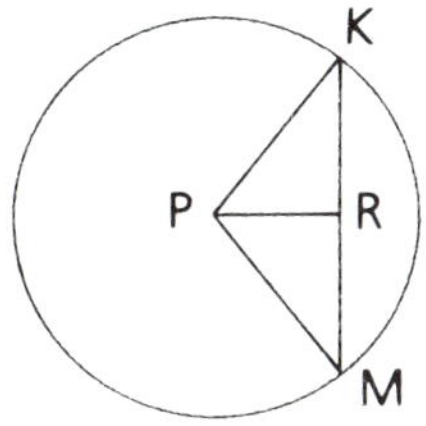

7 $\triangle HGF$ is equilateral

a $\angle F = (x + 32)^\circ$
$\angle H = (2x + 4)^\circ$

Solve for x and find $m\angle G$

b Perimeter of $\triangle HGF = 6y + 24$
$HG = 3y - 7$

Find the perimeter of $\triangle HGF$

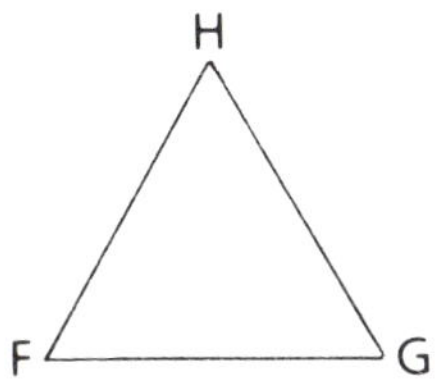

Chapter 3 Review Problems **B**

8 Kate and Jaclyn wished to find the distance from N on one side of a lake to P on the other side. So they put stakes at N, P, and T. Then they extended $\overline{PT}$ to S making sure $\overline{PT} \cong \overline{TS}$. They followed a similar process in extending $\overline{NT}$ to R. They then measured $\overline{SR}$ and found it to be 70 meters. They concluded that NP was 70 meters.

Prove that they were correct.

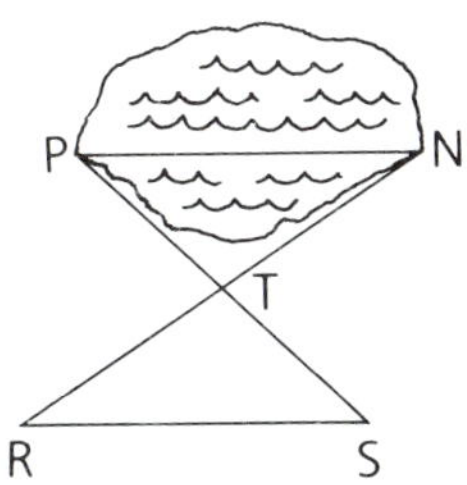

9 Given: $\overline{AD} \cong \overline{BC}$
$\angle DAB \cong \angle CBA$
Prove: $\triangle ABE$ is isosceles

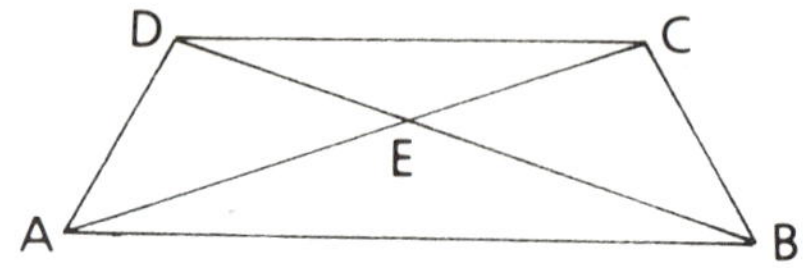

10 Given: $\overline{FJ}$ is the base of an isosceles $\triangle$
$\overline{FG} \cong \overline{JH}$
O is the midpt. of $\overline{MF}$
K is the midpt. of $\overline{MJ}$
Conclusion: $\overline{OH} \cong \overline{KG}$

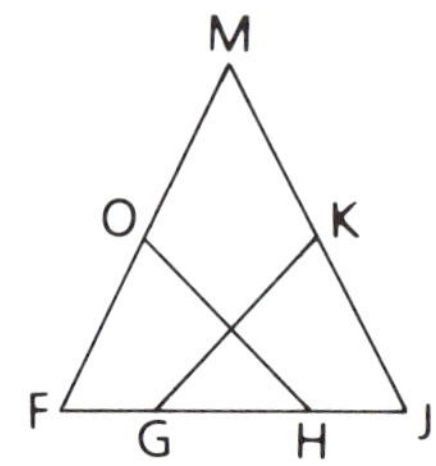

11 Given: $\overline{VX} \cong \overline{VZ}$
Y is the midpt. of $\overline{XZ}$
Prove: $\overline{WY} \cong \overline{YA}$

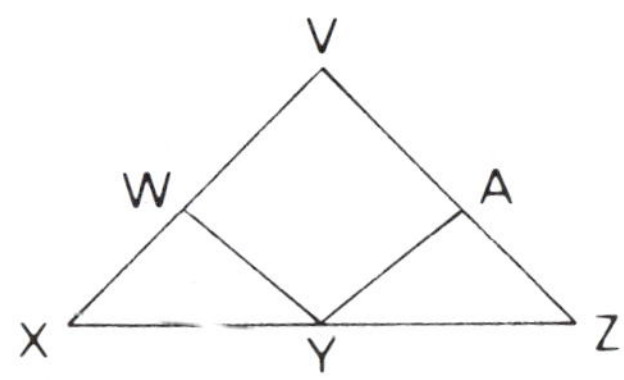

12 $\triangle RST \cong \triangle DFE$ (Draw your own diagram for this.)

$\angle R = 50°$ $\angle T = 40°$ $\angle E = (y + 10)°$
$\angle S = 90°$ $\angle D = (x + 20)°$ $\angle F = (z - 30)°$

Solve for x, y, z

Chapter 3 Review Problems C

13 Given: $\triangle FJH$ is isosceles with base $\overline{JH}$
K and G are midpoints
$FK = 2x + 3$
$GH = 5x - 9$
$JH = 4x$
Find the perimeter of $\triangle FHJ$

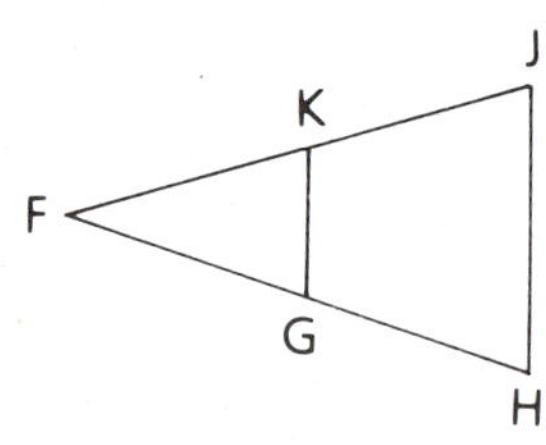

14 Given: $\overline{AC} \cong \overline{BC}$
$\angle 1 \cong \angle 3$
Prove: $\triangle DFE$ is isosceles

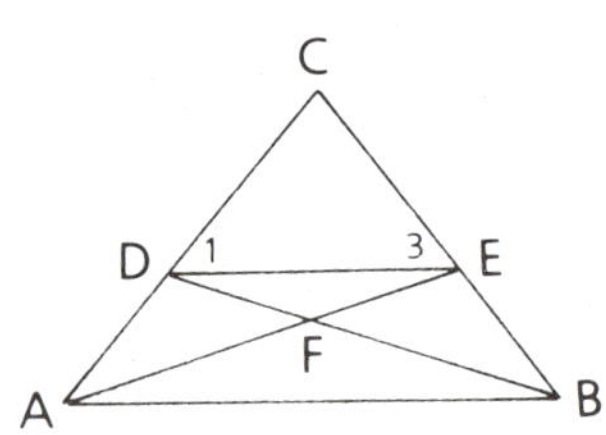

15 Given: $\angle RHG \cong \angle KHJ$
$\overline{OH}$ is median to $\overline{GJ}$
$\overline{OH}$ is altitude to $\overline{GJ}$

Conclusion: $\angle P \cong \angle M$

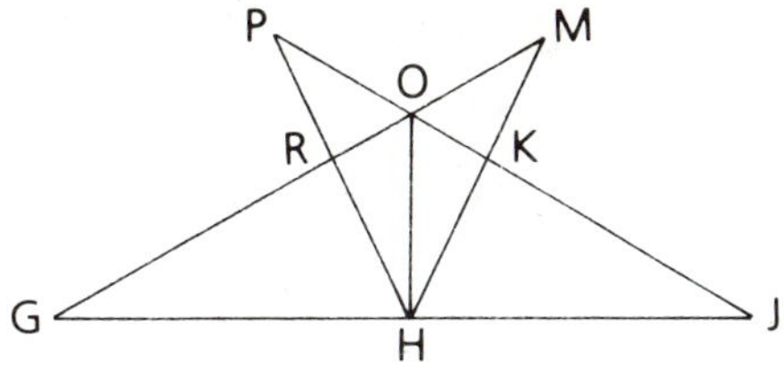

16 Given: $\overline{ZY}$ bisects $\overline{SX}$
$\overline{ST} \cong \overline{WX}$
$\overline{SX}$ bisects $\overline{ZY}$

Prove: $\angle SZW \cong \angle XYT$

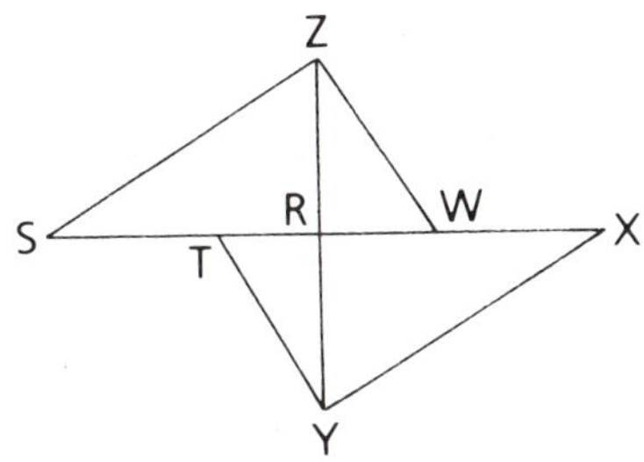

Chapters 1–3 Cumulative Review Problems A

1
a Name four collinear points.
b Name three non-collinear points.
c Name a pair of opposite rays.
d Estimate the size of $\angle ABE$.
e Does $\triangle FED$ appear to be acute, right, or obtuse?
f Which segment appears to be horizontal?
g Do $\overline{ED}$ and $\overline{BD}$ appear to be perpendicular or oblique?

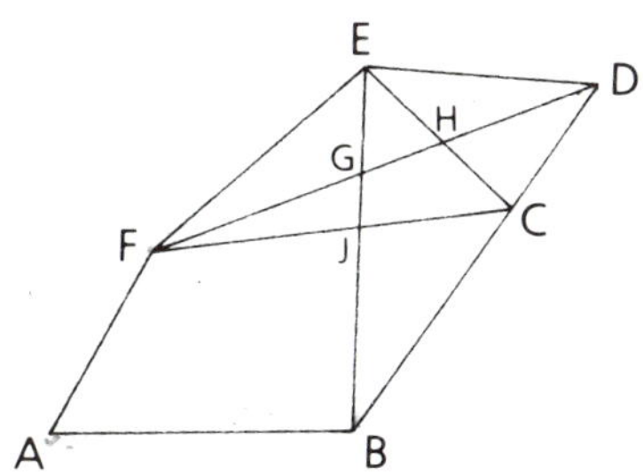

h $\overline{BC} \cap \overline{CD} =$ _____.
i $\overline{BG} \cap \overline{EJ} =$ _____.
j $\overrightarrow{AF} \cup \overrightarrow{AB} =$ _____.
k $\overleftrightarrow{BC} \cap \overleftrightarrow{ED} =$ _____.
l $\overline{BC} \cap \overline{ED} =$ _____.
m $\overrightarrow{FD} \cap \overrightarrow{HG} =$ _____.
n $\overline{GH} \cup \overline{EH} \cup \overline{GE} =$ _____.

2 What name is given to a triangle with no two sides $\cong$?

3 Three-fifths of a degree is equivalent to how many minutes?

4 Find the complement of $43^\circ 17' 51''$.

5 How large is the angle formed by the hands of a clock at 11:20?

6 One of two supplementary angles is 8 degrees more than the other. Find the measure of the larger.

7 Given: $\angle A$ is comp. to $\angle BCA$
$\angle D$ is comp. to $\angle DBC$
$\angle D \cong \angle BCA$

Prove: $\angle A \cong \angle DBC$

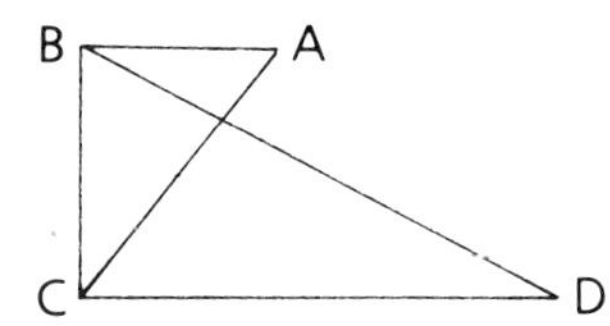

8 Given: $\overline{PS} \perp \overline{SR}$
$\angle QRP$ is comp. to $\angle PRS$

Prove: $\angle S \cong \angle QRS$

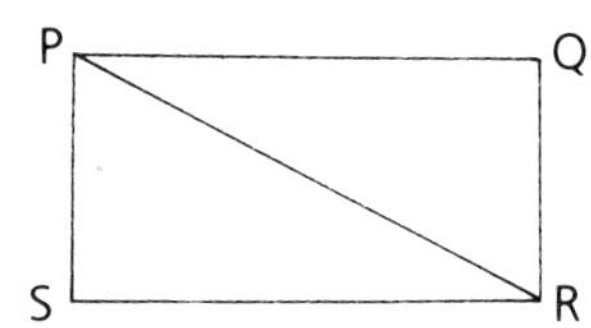

9 Given: $\angle 1 \cong \angle 2$
$\angle 1 \cong \angle 3$

Conclusion: $\overrightarrow{FH}$ bisects $\angle EFG$

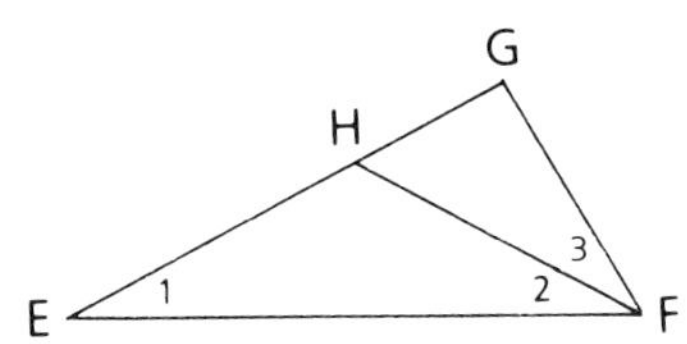

10 Given: $\odot$ Q
$\overline{RP} \cong \overline{PS}$

Conclusion: $\overline{PQ}$ bisects $\angle RPS$

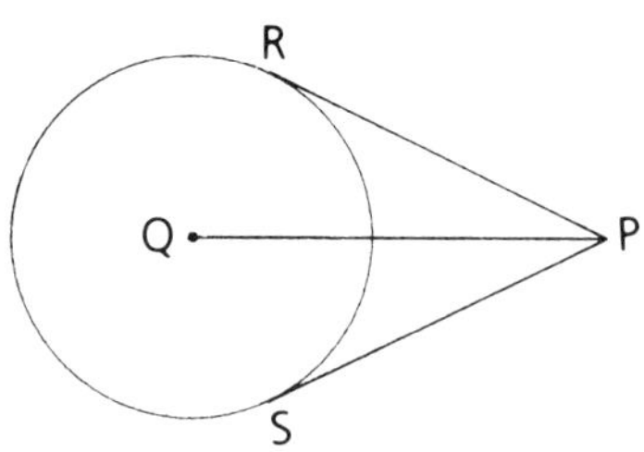

11 Given: Diagram as shown
$\angle 6$ is supp. to $\angle 7$

Conclusion: $\angle 6 \cong \angle 8$

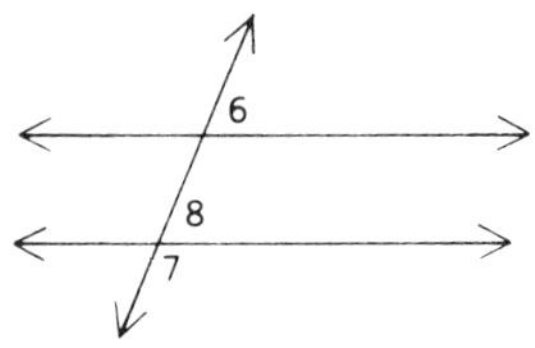

Chapters 1–3 Cumulative Review Problems B

12 Given: $\angle T \cong \angle W$
$\angle TSW \cong \angle XSV$
$\overline{ST} \cong \overline{SW}$

Conclusion: $\overline{SX} \cong \overline{SV}$

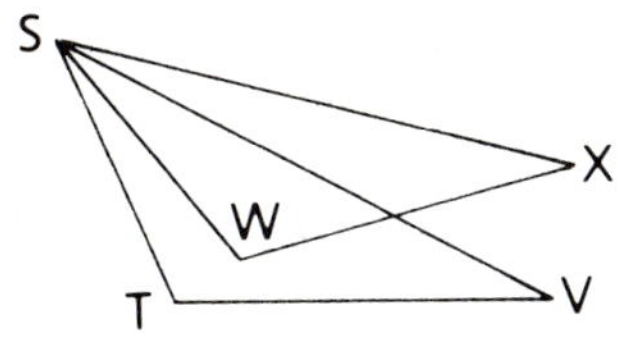

13 Given: $\overline{CE} \cong \overline{DF}$, and $\overline{BD} \cong \overline{GE}$
$\overline{BD} \perp \overline{CF}$
$\overline{GE} \perp \overline{CF}$

Conclusion: $\triangle ACF$ is isosceles

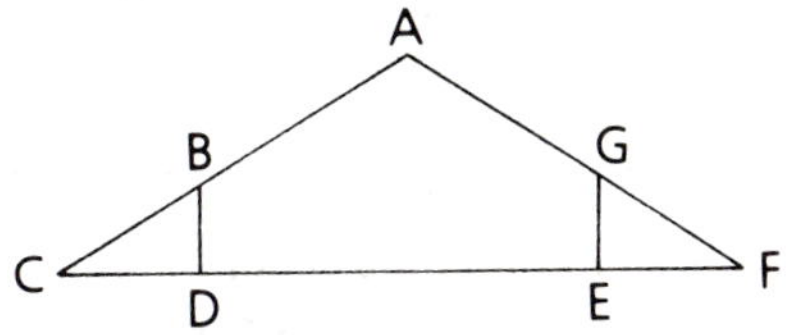

14 Given: $\triangle ZWX$ is isosceles with base $\overline{WX}$
$\overrightarrow{WR}$ bisects $\angle XWZ$
$\overrightarrow{XR}$ bisects $\angle ZXW$

Prove: $\angle XWR \cong \angle RXW$

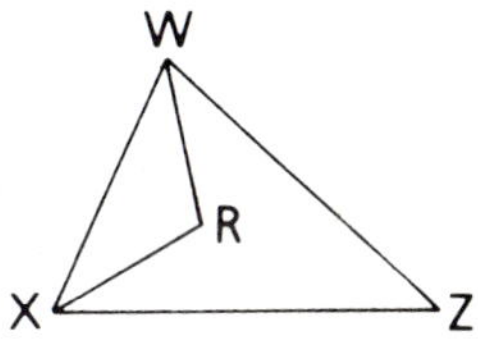

15 If angles 1, 2, and 3 are in the ratio of 6:5:4, find their measures.

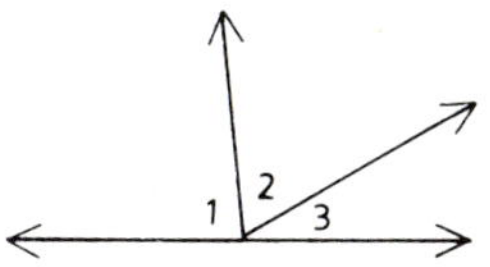

16 $AB = 2r + 7$
$CD = 3r - 1$
$BC = 6$
C is the midpoint of $\overline{AD}$

Find AC

17 Prove: The segments drawn from the midpoint of the base of an isosceles triangle to the midpoints of the legs are $\cong$.

18 The measure of the supplement of an angle exceeds twice the measure of the complement of the angle by 40°. Find the measure of half of the complement.

19 Two segments have the ratio of 5:3, and the longer exceeds the shorter by 14 meters. Find the length of the longer.

20 Given: $\angle AEC \cong \angle BED$
$\overline{AE} \cong \overline{ED}$

Conclusion: $\overline{AB} \cong \overline{CD}$

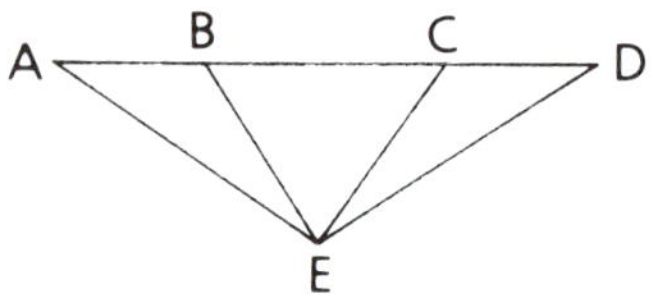

21 Given: $\angle 1 \cong \angle 5$
$\angle 2 \cong \angle 6$

Conclusion: $\triangle ABC$ is isosceles

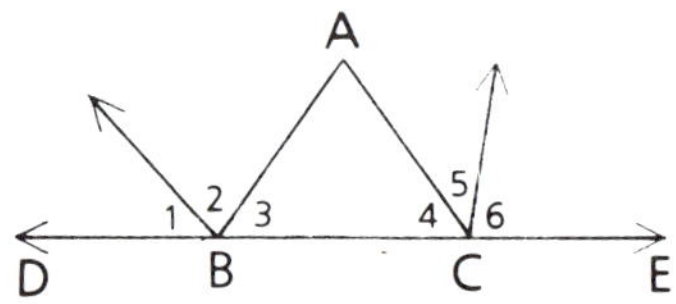

Chapters 1–3 Cumulative Review Problems

C

22 Given: $\overline{OD} \cong \overline{OE}$ in $\odot$ O
$\angle DOB \cong \angle EOA$

Conclusion: $\overline{CD} \cong \overline{CE}$

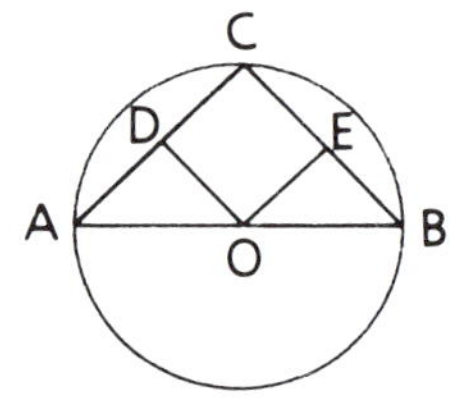

23 Given: $\angle NOT \cong \angle POV$
O is a midpoint
$\angle N \cong \angle P$

Prove: $\overline{ST} \cong \overline{RV}$

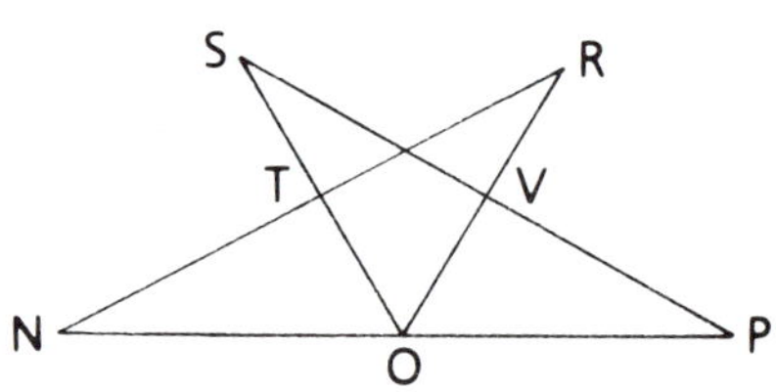

24 $\angle YXW \cong \angle ZWX$, $\overline{ZW} \cong \overline{YX}$
$AW = 3y + 7$, $AY = 2x + y$
$AX = \frac{1}{2}x + 3$, $ZX = 7$
Find WA and AY

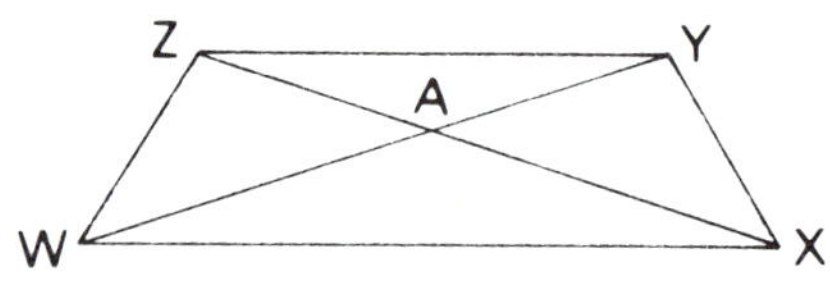

25 If two of the ten segments in problem 33 are selected at random, what is the probability that the two segments are collinear?

InterSection: Shipshape

A place for everything, and everything in its place. These ideas are especially important for people who work for moving companies and people who design building interiors.

Geometry Works: For Packing and Moving

Moving companies usually sell boxes for packing. This particular box, made from a single sheet of corrugated cardboard with one glued seam, can be stored "flat" to save space.

Cut out a model of the box, 32 cm by 72 cm. Can you follow the instructions?

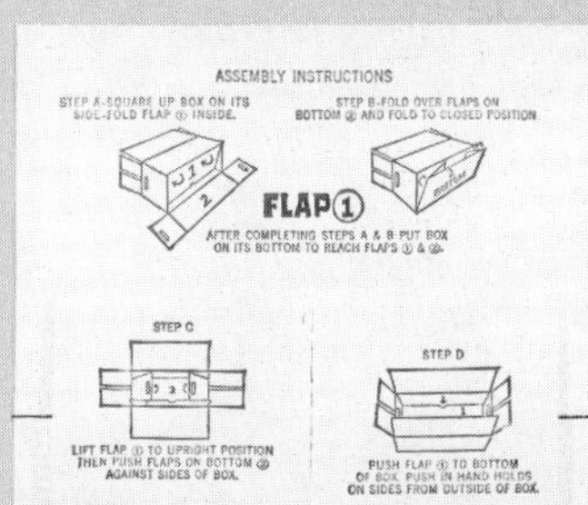

The Novack family wanted to give their old couch to their daughter for her first apartment. Before they moved the couch, they wanted to be certain it would pass through the 26-inch-wide door of the new living room.

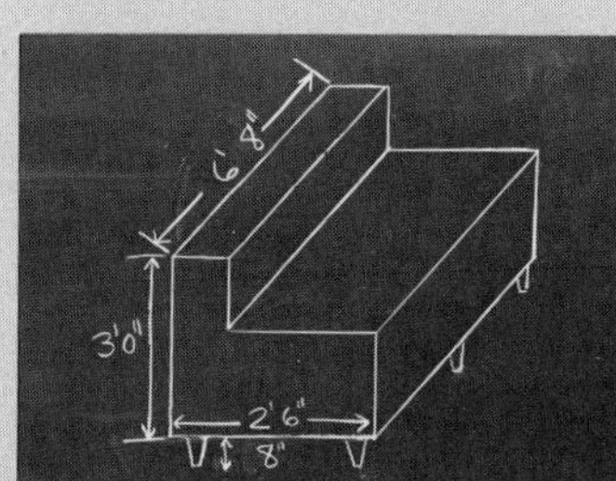

They drew an accurate scale model of the end of the couch and the door opening. Would the couch fit? Would they have to remove the back legs of the couch? the front legs, too?

Geometry Works: For Builders

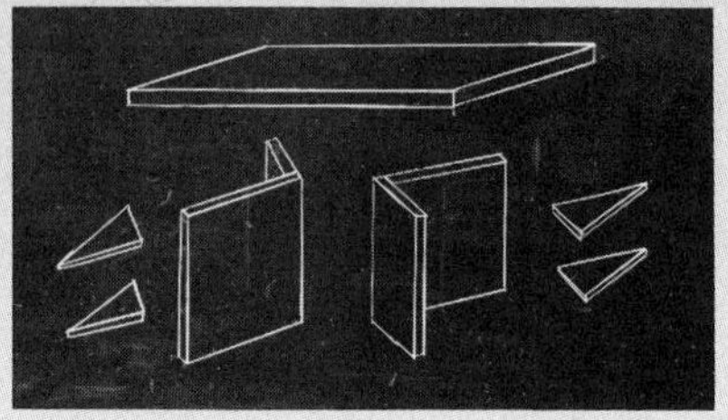

Martina and her mother had just moved to a new apartment. Martina decided to build a desk for her room. She used ¾-inch birch plywood for all the pieces.

When she cut all the pieces, she used glue and metal L-brackets to join the pieces together. She covered the top and edges with green felt and painted the shelves and legs.

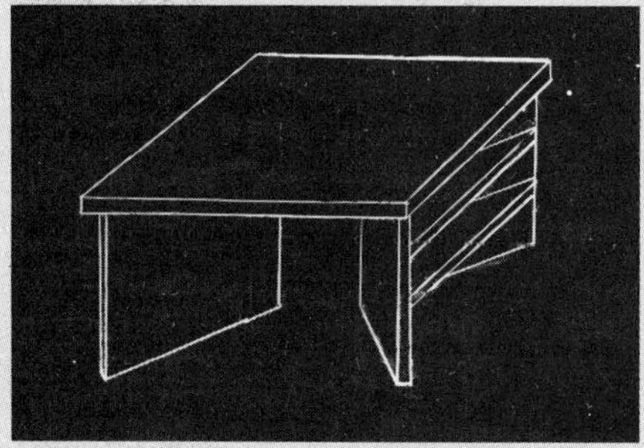

Martina was able to cut all the pieces from a single 4-foot by 8-foot sheet of plywood. Can you show how she might have done it? What could you design and build with the left over pieces of wood?

The Student Council voted to construct a ramp to the commons area for students in wheelchairs. The ramp had to take the place of two steps, each six inches high.

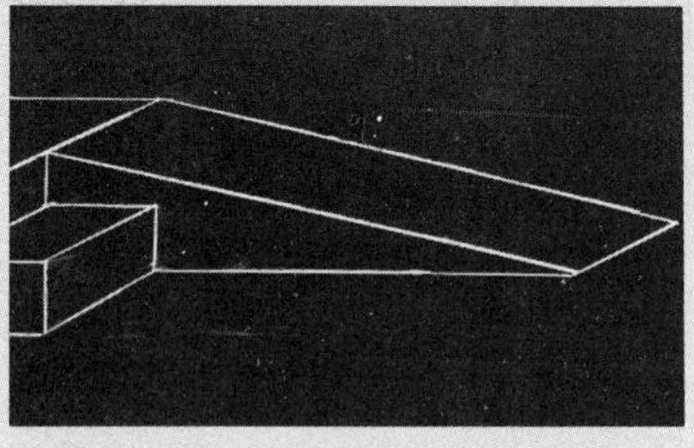

The council suggested that the school build a ramp twelve feet long and four feet wide.

Draw a model of the proposed ramp. What is the measure of the angle determined by the ramp and the floor? Do you know if a student in a wheelchair could go up and down this ramp without assistance?

Projects

1 You have just moved into a new house. Your new room is small, just large enough for a bed and a desk, but your closet is very large.

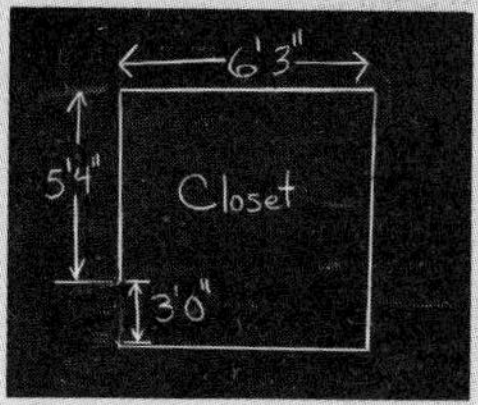

Design your closet, including a bureau, shelves, and bars for hangers. Do you need mostly shelves or mostly clothes bars? Will two bars on a wall, for "short" clothes like shirts, be better than one bar for trousers or dresses? How many linear feet of shelves and clothes bars would you need?

2 The kitchen in your new house needs to be remodeled. The realtor has suggested two plans for the kitchen:

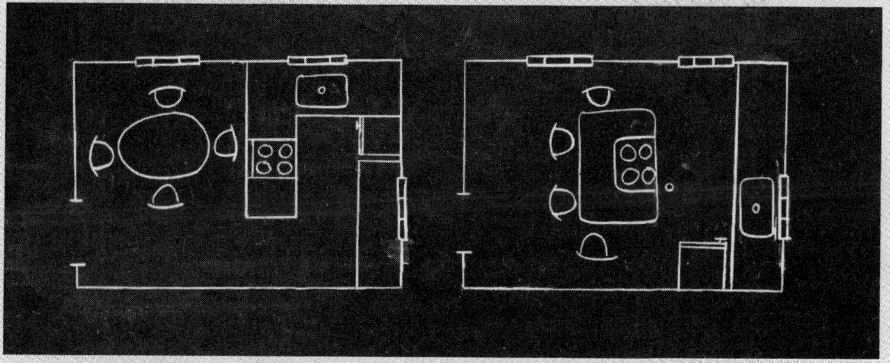

How would you design the kitchen? How many "work areas" do you want? Do you like an "island" design? An "eat-in" counter?

3 This program inputs the length, width, and height of packing boxes and prints out the volume of the box. Expand the program to print the area of the bottom of the box.

Density is weight (mass) divided by volume. Modify the program to accept weight as an input and calculate both the density of the box and the weight per unit area on the bottom of the box.

```
10 INPUT L,W,H
20 PRINT "VOLUME IS:", L*W*H
```

Chapter 4

PERPENDICULARITY

Section 4.1 Detour Problems

Analyze carefully the following problem:

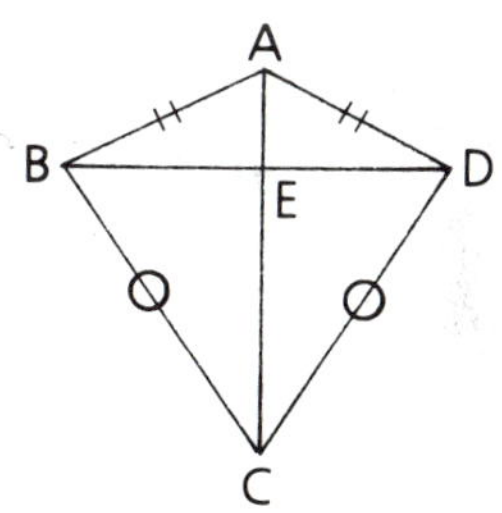

Given: $\overline{AB} \cong \overline{AD}$
$\overline{BC} \cong \overline{CD}$

Prove: $\triangle ABE \cong \triangle ADE$

Notice that among the given parts, only $\overline{AB} \cong \overline{AD}$ seems to be usable. There does not seem to be enough information to prove that $\triangle ABE \cong \triangle ADE$. We must therefore prove something else first, taking a little **DETOUR** to pick up the congruent parts we need. Thus we obtain the proof below.

	Statements	Reasons
	1 $\overline{AB} \cong \overline{AD}$	1 Given
DETOUR	2 $\overline{BC} \cong \overline{CD}$	2 Given
DETOUR	3 $\overline{AC} \cong \overline{AC}$	3 Reflexive Property
DETOUR	4 $\triangle ABC \cong \triangle ADC$	4 SSS (1, 2, 3)
	5 $\angle BAE \cong \angle DAE$	5 CPCTC
	6 $\overline{AE} \cong \overline{AE}$	6 Reflexive Property
	7 $\triangle ABE \cong \triangle ADE$	7 SAS (1, 5, 6)

The procedure to use in **DETOUR** problems is always the same:

1 Determine which pair of triangles you must prove to be congruent. (In the preceding problem it is the final conclusion, $\triangle ABE \cong \triangle ADE$.)

2 Attempt to prove that these triangles are congruent. If you cannot do so for lack of enough given information, take a **DETOUR**.

3 Observe which congruent parts are needed. (Remember that you have many ways to prove that triangles are congruent; consider them all.)

4 Find another pair of triangles that:
   a you *can* readily prove are congruent, and
   b contain a pair of missing parts needed in the main proof (item 3).

5 Prove that the pair of triangles in item 4 are congruent.

6 Use CPCTC and complete the proof as planned in item 1.

Section 4.1 Sample Problems

1 **Given:** $\overleftrightarrow{PQ}$ **bisects** $\overline{YZ}$
Q is midpt. of $\overline{WX}$
$\angle Y \cong \angle Z$, $\overline{WZ} \cong \overline{XY}$
Conclusion: $\angle WQP \cong \angle XQP$

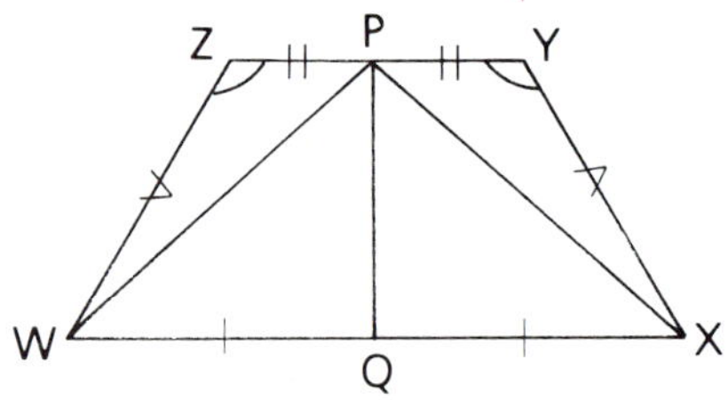

Analysis:

a To reach the conclusions we must prove that $\triangle WQP \cong \triangle XQP$.

b There is not enough information (by the tick marks).

c We must **DETOUR** through another pair of triangles. Can you see which pair of triangles we should use? Check your choice against the following proof.

Statements	Reasons
1 $\overleftrightarrow{PQ}$ bisects $\overline{YZ}$	1 Given
2 $\overline{ZP} \cong \overline{PY}$	2 If a line bisects a segment, then it divides the segment into two ≅ segments
3 $\angle Z \cong \angle Y$	3 Given
4 $\overline{WZ} \cong \overline{XY}$	4 Given
5 $\triangle ZWP \cong \triangle YXP$	5 SAS (2, 3, 4)
6 $\overline{WP} \cong \overline{PX}$	6 CPCTC
7 Q is the midpt. $\overline{WX}$	7 Given
8 $\overline{WQ} \cong \overline{QX}$	8 The midpoint of a segment divides the segment into two ≅ segments
9 $\overline{PQ} \cong \overline{PQ}$	9 Reflexive Property
10 $\triangle WQP \cong \triangle XQP$	10 SSS (6, 8, 9)
11 $\angle WQP \cong \angle XQP$	11 CPCTC

DETOUR (steps 1–5)

2 **Given:** $\overline{AB} \cong \overline{DC}$, $\overline{AD} \cong \overline{BC}$
$\overrightarrow{AE}$ **bisects** $\angle BAD$
$\overrightarrow{CF}$ **bisects** $\angle BCD$

Conclusion: $\overline{ED} \cong \overline{BF}$

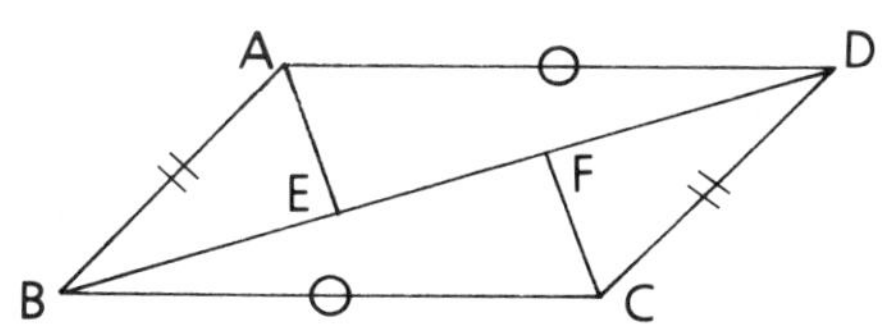

Statements	Reasons
1 $\overline{AB} \cong \overline{DC}$	1 Given
2 $\overline{AD} \cong \overline{BC}$	2 Given
3 $\overline{BD} \cong \overline{BD}$	3 Reflexive Property
4 $\triangle ABD \cong \triangle CDB$	4 SSS (1, 2, 3)
5 $\angle ADB \cong \angle CBD$	5 CPCTC
6 $\angle DAB \cong \angle BCD$	6 CPCTC
7 $\overrightarrow{AE}$ bisects $\angle BAD$	7 Given
8 $\overrightarrow{CF}$ bisects $\angle BCD$	8 Given
9 $\angle DAE \cong \angle BCF$	9 Division Property
10 $\triangle AED \cong \triangle CFB$	10 ASA (5, 2, 9)
11 $\overline{ED} \cong \overline{BF}$	11 CPCTC

DETOUR (steps 1–4)

Section 4.1 Problem Set A

1 Copy the entire problem and fill in the correct statements and reasons.

Given: $\overline{WX} \cong \overline{YZ}$, $\overline{XP} \cong \overline{QZ}$
P and Q trisect $\overline{WY}$

Conclusion: $\overline{WZ} \cong \overline{XY}$

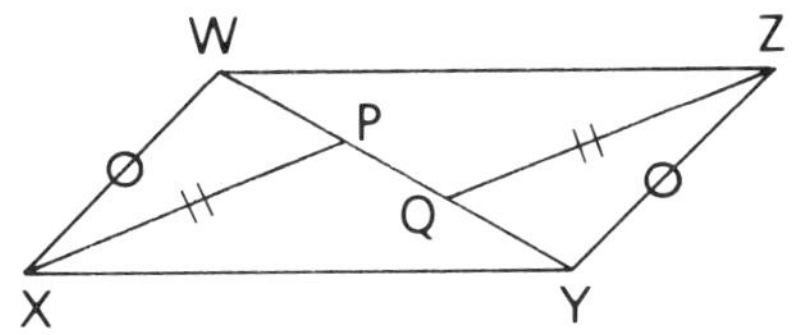

Statements	Reasons
1 $\overline{WX} \cong \overline{YZ}$	1 Given
2 $\overline{XP} \cong \overline{QZ}$	2 Given
3 P, Q trisect $\overline{WY}$	3 Given
4 ______	4 ______
5 $\triangle WXP \cong \triangle YZQ$	5 ______
6 ______	6 CPCTC
7 $\overline{WY} \cong \overline{WY}$	7 ______
8 ______	8 ______
9 $\overline{WZ} \cong \overline{XY}$	9 CPCTC

2 Given: $\overline{WX} \cong \overline{WZ}$, and $\overline{XY} \cong \overline{ZY}$

Prove: $\triangle XAY \cong \triangle ZAY$

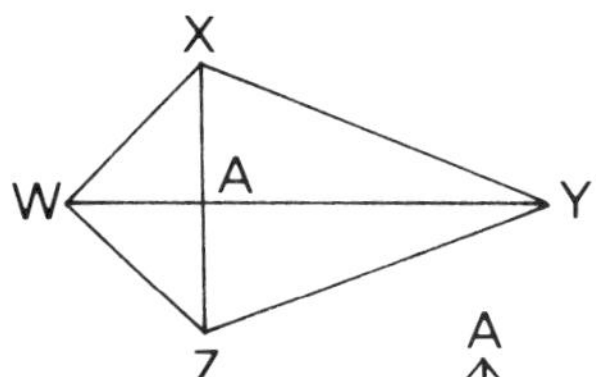

3 Given: $\overline{AC}$ is the $\perp$ bisector of $\overline{BD}$
(This is two statements.)

Prove: $\triangle ABD$ is isosceles

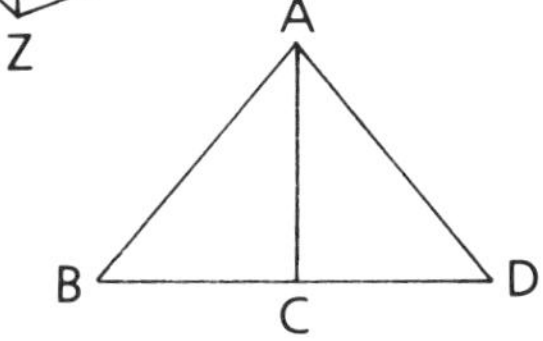

4 Given: A is *equidistant* from B and D
(That means AB = AD.)
$\overrightarrow{AC}$ bisects $\angle BAD$

Prove: $\overline{AC}$ bisects $\overline{BD}$

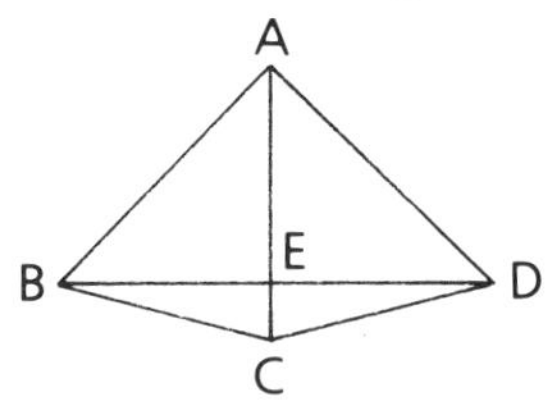

5 Given: $\overline{MN} \cong \overline{NS}$
$\overline{MP} \cong \overline{PS}$

Prove: $\angle MQP \cong \angle SQP$

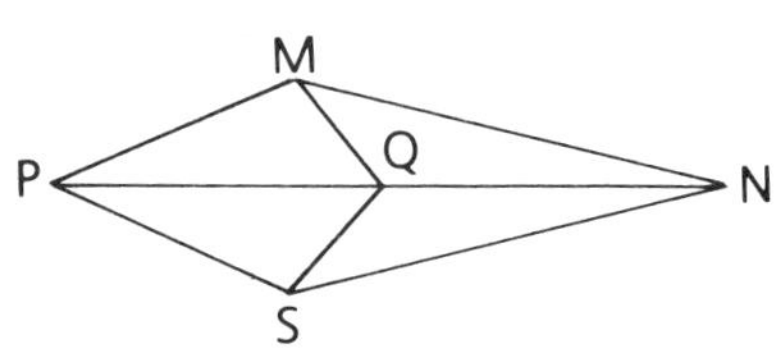

Section 4.1 Problem Set B

6 Given: $\triangle ABC$ is isosceles, base $\overline{BC}$
$\overline{AD} \perp \overline{BC}$
Prove: $\triangle BEC$ is isosceles

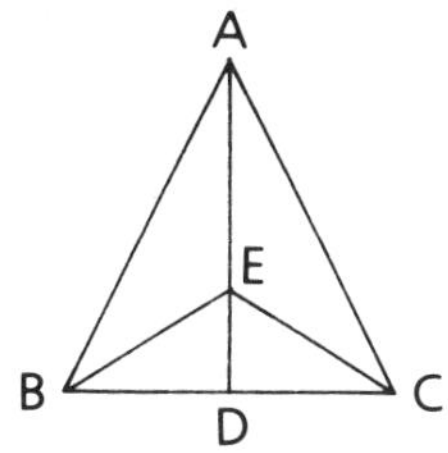

7 Given: $\odot$ O, $\overline{WX} \cong \overline{WY}$
Prove: $\overleftrightarrow{WZ}$ bisects $\overline{XY}$

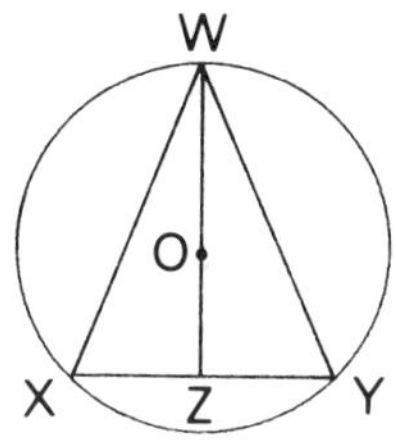

8 Given: $\overline{AD} \cong \overline{BC}$, $\overline{AF} \cong \overline{EC}$
$\overleftrightarrow{BD} \perp \overleftrightarrow{AF}$ and $\overleftrightarrow{EC}$
Conclusion: $\overline{AB} \cong \overline{DC}$

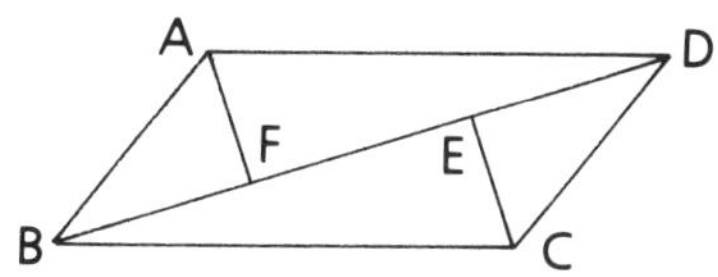

9 Given: $\overline{PR} \cong \overline{PU}$
$\overline{QR} \cong \overline{QU}$
$\overline{RS} \cong \overline{UT}$
Conclusion: $\angle 1 \cong \angle 2$

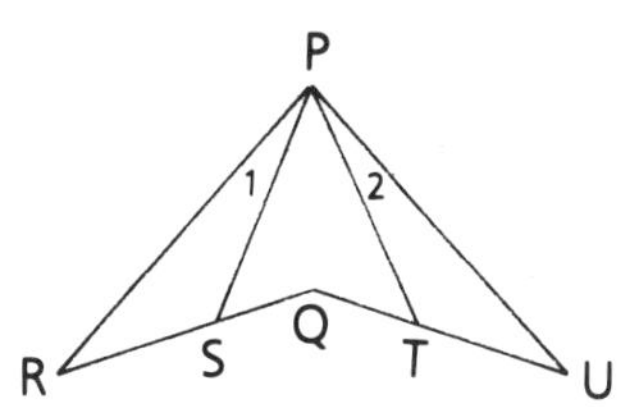

10 Given: $\overline{AB} \cong \overline{AC}$
$\overline{AD} \cong \overline{AE}$
Prove: $\triangle FBC$ is isosceles

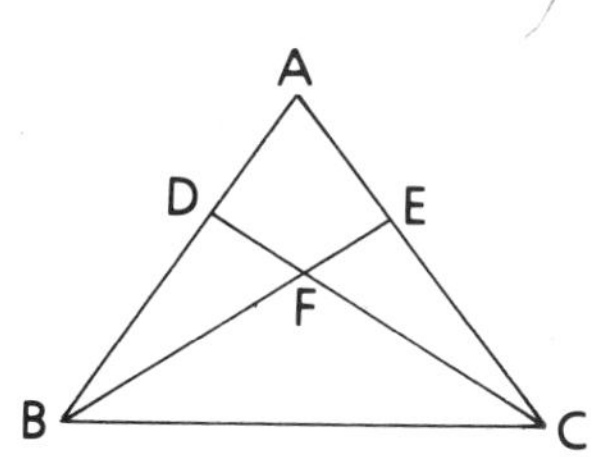

11 Given: T is the midpt. of $\overline{MN}$
$\angle PMT$, $\angle QNT$ are right $\angle$s
$\overline{MR} \cong \overline{SN}$, $\angle 1 \cong \angle 2$
Conclusion: $\angle P \cong \angle Q$

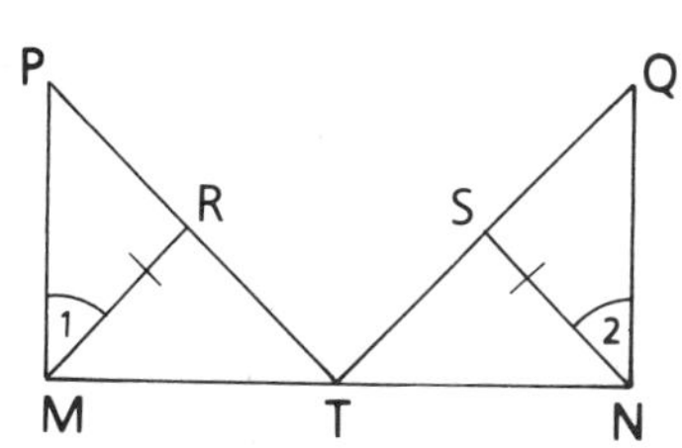

12 Given: $\odot O$, $\angle B \cong \angle C$

Prove: $\overline{AO}$ bisects $\overline{BC}$

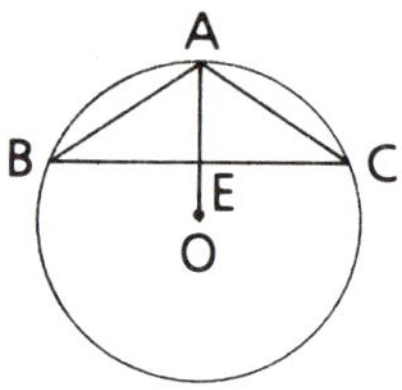

Section 4.1 Problem Set C

13 Given: $\overline{AB} \cong \overline{AC}$

$\overrightarrow{BD}$ bisects $\angle ABE$

$\overrightarrow{CD}$ bisects $\angle ACE$

Conclusion: $\overline{AE}$ bisects $\overline{BC}$

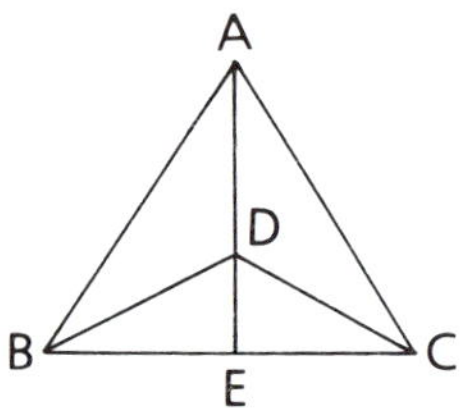

14 Given: $\overline{PT} \cong \overline{PU}$

$\overline{PR} \cong \overline{PS}$

Prove: $\overrightarrow{PQ}$ bisects $\angle RPS$

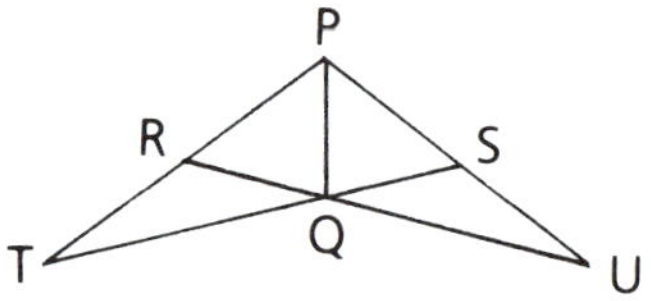

15 Given: $\overline{AD} \cong \overline{DB}$

$\overline{AE} \cong \overline{BC}$

$\overline{CD} \cong \overline{ED}$

Prove: $\triangle AFB$ is isosceles

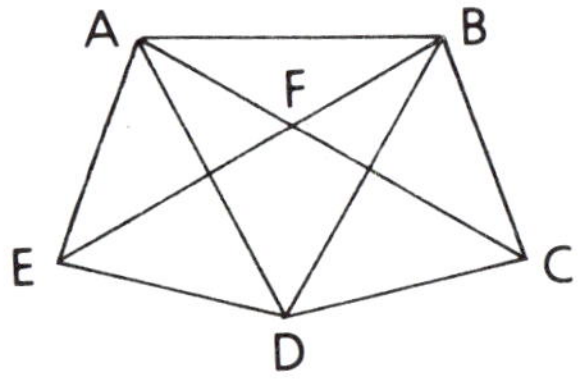

Section 4.1 Problem Set D

16 Given: $\triangle LTM$ is isosceles with $\overline{LT} \cong \overline{TM}$

$\overline{TQ}$ is a median

Conclusion: $\overline{PS} \cong \overline{SR}$

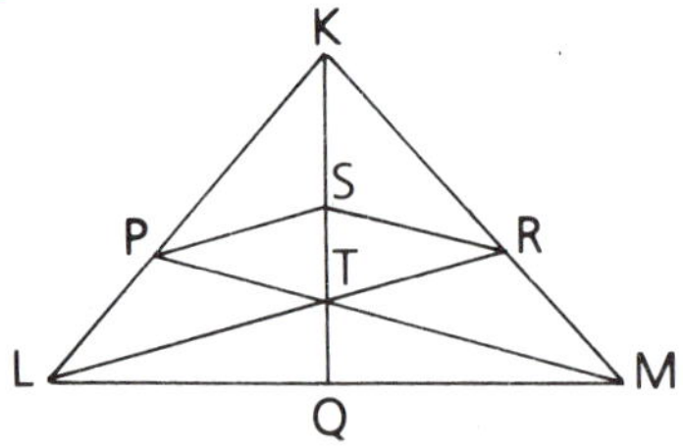

17 Given: $\overline{AD} \cong \overline{EB}$

$\overline{BD} \cong \overline{AE}$

$\overline{AJ} \cong \overline{FB}$

Conclusion: $\overline{JG} \cong \overline{FH}$

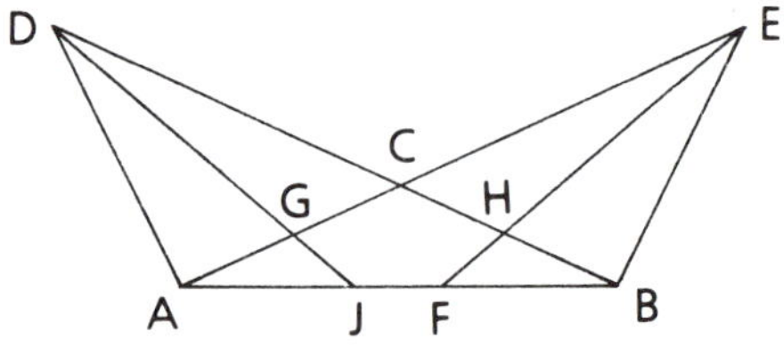

Section 4.2 The Case of the Missing Diagram

Many geometry problems are stated without a diagram. Also, at times you may be unsure of precisely what is *given,* and what should be *proved.*

■ Example 1: If two altitudes of a triangle are congruent, then the triangle is isosceles.

This example is stated in "If . . . , then. . . ." form. Such a sentence is a *conditional statement.* The **given** *conditions* are *mainly* in the **if clause,** and the *conclusion* (what we are to *prove*) is in the **then clause.**

The diagram we draw should represent the given information, but otherwise should be *as general as possible.* For instance, in the set-up below for Example 1, we have *not* drawn the altitudes as bisecting the sides, because bisections were *not* given. To do so would *overdetermine* the problem.

A Set-up of Example 1:

Given: $\overline{BD}$ and $\overline{CE}$ are altitudes to $\overline{AC}$ and $\overline{AD}$ of $\triangle ACD$

$\overline{BD} \cong \overline{CE}$

Prove: $\triangle ACD$ is isosceles

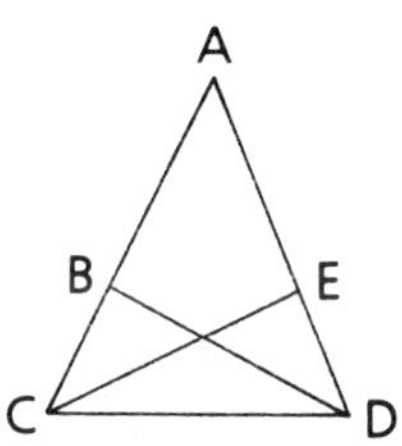

Sometimes the word *then* is left out of a conditional sentence, or the word order is reversed, or both. The *if clause* ALWAYS follows the word *if* and always contains given conditions. Occasionally, *some* of the given conditions appear in the *then clause,* as shown in the next example.

■ Example 2: The medians of a triangle are congruent if the triangle is equilateral.

First, we are *given* an equilateral triangle, so we draw one. Second, we are to prove something about the medians, so the medians are also *given.* We draw them. We letter our diagram any way we wish and write our *given* and *conclusion* statements in terms of our diagram.

A Set-up of Example 2:

Given: $\triangle XYZ$ is equilateral

$\overline{PZ}$, $\overline{RY}$, and $\overline{QX}$ are medians

Prove: $\overline{PZ} \cong \overline{RY} \cong \overline{QX}$

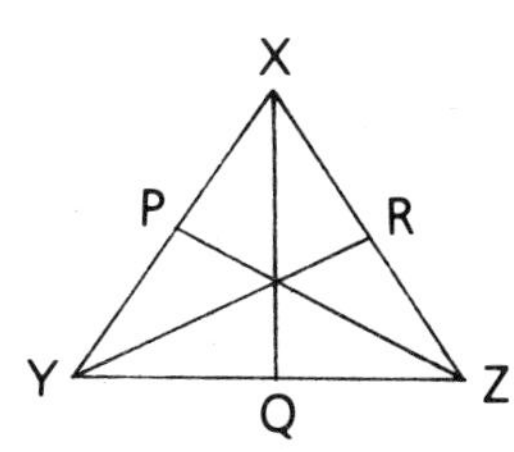

■ Example 3: The altitude to the base of an isosceles triangle bisects the vertex angle.

This sentence is a conditional statement with *if* and *then* left out. The main clue is the order from left to right, beginning with the given information.

First, we are *given* the altitude to the base. Second, we are *given* the isosceles triangle. We must *prove* that the altitude bisects the vertex angle of the isosceles triangle.

A Set-up of Example 3:

Given: $\triangle PQR$ is isosceles with base $\overline{QR}$
$\overline{PM}$ is an altitude

Prove: $\overline{PM}$ bisects $\angle QPR$

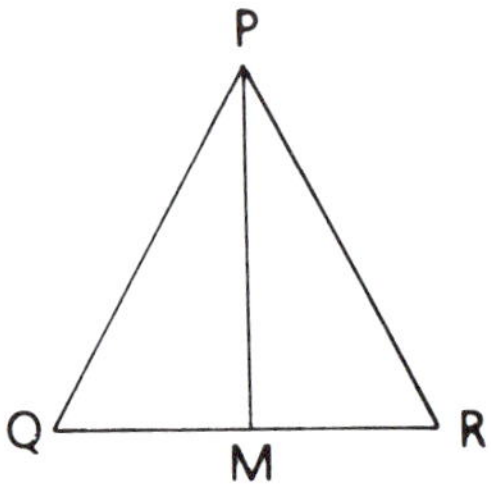

Why was it necessary to specify in the given information that $\overline{QR}$ is the base of $\triangle PQR$? Why was it *not* necessary to specify that $\angle QPR$ is the vertex angle?

■ Example 4: If two angles of one triangle are congruent to two angles of a second triangle, then the remaining pair of angles are congruent.

A Set-up of Example 4:

Given: $\angle A \cong \angle D$
$\angle B \cong \angle E$

Prove: $\angle C \cong \angle F$

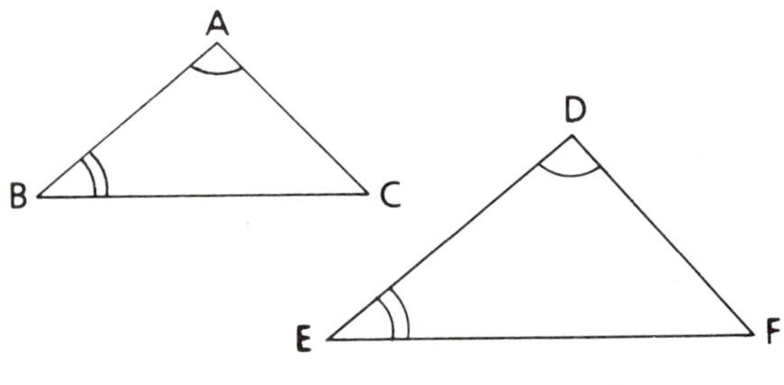

Notice that the triangles were drawn as *scalene* figures since nothing was given to indicate otherwise. Also, the triangles were drawn with *different* sizes; the angles could be congruent without the triangles being congruent.

Section 4.2 Problem Set A

In **1–4** draw your own diagram and write the "given" and "conclusion" in terms of your diagram. *Do not* write the proof. In **5–7** also supply a proof.

1 **Given: An isosceles triangle with the median to the base**
Prove: The median is the perpendicular bisector of the base.

2 Given: A four-sided polygon with all four sides congruent (This figure is called a RHOMBUS.)

Conclusion: The lines joining opposite vertices are $\perp$.

3 Given: Segments drawn perpendicular to the sides of an angle from a point on the bisector of the angle

Conclude: These two segments are congruent.

4 The bisector of the vertex angle of an isosceles triangle is perpendicular to the base.

5 The altitude to a side of a scalene triangle forms two congruent angles with that side of the triangle.

6 The median to the base of an isosceles triangle divides the triangle into two congruent triangles.

7 If the base of an isosceles triangle is extended in both directions, then the exterior angles formed are congruent.

The Royal Road: Computer Graphics

The graphics capabilities of many microcomputer systems can be used to display most geometric figures. The following program will display a scalene triangle and then an isosceles triangle.

```
98   REM : SCALENE THEN ISOSCELES
99   HGR
100  HPLOT 50,20 TO 20,120 TO 210,30 TO 50,20
101  FOR PAUSE = 1 TO 1000: NEXT PAUSE
120  HGR
200  HPLOT 120,20 TO 20,100 TO 220,100 TO 120,20
```

1 Add to the program to print "scalene" or "isosceles" under the appropriate triangle.
2 Extend the program to display and label other types of triangles: equilateral, right, obtuse, and acute.
3 Modify the program from part 2 to allow a user to request and keep any two of the triangles.

Section 4.2 Problem Set B

In **8–12** set up and complete each proof:

8 If the median to a side of a triangle is also an altitude to that side, then the triangle is isosceles.

9 The line segments joining the vertex angle of an isosceles triangle to the trisection points of the base are congruent.

10 If the line joining a pair of opposite vertices of a four-sided polygon bisects both angles, then the remaining two angles are congruent.

11 If two triangles are congruent, then a pair of corresponding medians are congruent.

12 The angle bisectors of the base angles of an isosceles triangle meet to form another isosceles triangle with the base.

Section 4.2 Problem Set C

In **13–15** set up and complete each proof:

13 If each pair of opposite sides of a four-sided figure are congruent, then the segments joining opposite vertices bisect each other.

14 If a point on the base of an isosceles triangle is equidistant from the midpoints of the legs, then that point is the midpoint of the base.

15 If a point in the interior of an angle (between the sides) is equidistant from the sides of the angle, then the ray joining the vertex of the angle to this point bisects the angle. (Note: The distance from a point to a line is always taken as the length of the perpendicular segment from the point to the line.)

Section 4.3 Adjacent Angles

In this section we discuss:

A ADJACENT ANGLES

B A THEOREM ABOUT PERPENDICULAR LINES

4.3 (A) Adjacent Angles

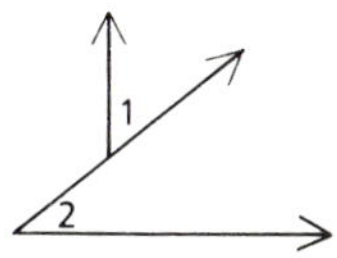

∠1 and ∠2 are *nonadjacent* angles

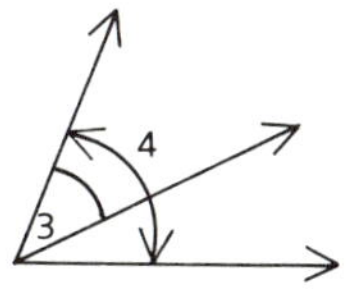

∠3 and ∠4 are *nonadjacent* angles

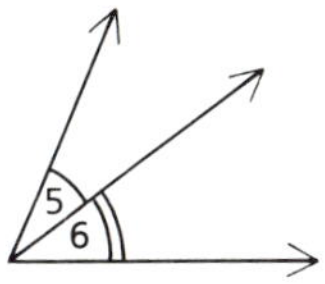

∠5 and ∠6 are *adjacent* angles

In English, the word *adjacent* means bordering or neighboring.
In geometry, *adjacent angles* are bordering angles in a special way.

D **If a ray divides an angle into two angles (*not necessarily congruent*), the two newly formed angles are called ADJACENT ANGLES.**

4.3 (B) A Theorem About Perpendicular Lines

We have seen that perpendicularity is one of the fundamental concepts of geometry. Perpendicularity has led to: right angles, complementary angles, right triangles, and altitudes. So far, perpendicular lines have generally been *given* information. We are now ready for methods to *prove* that lines are perpendicular:

Theorem 21

T **If two lines (or rays or segments) intersect to form congruent adjacent angles, then the lines are perpendicular.**

Given: $\angle 1 \cong \angle 2$
Prove: $m \perp p$

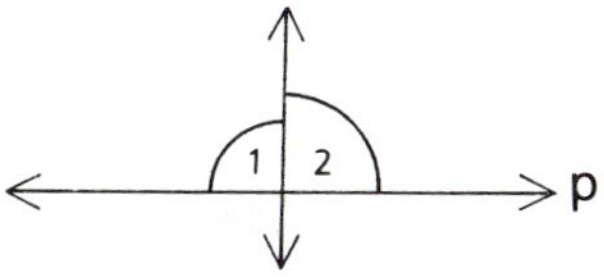

Proof of Theorem 21:

Since ∠1 and ∠2 form a straight angle, (line p), they are supplementary. Thus, $\angle 1 + \angle 2 = 180°$. Since $\angle 1 \cong \angle 2$, we may *substitute* and get $\angle 1 + \angle 1 = 180°$, or $\angle 1 = 90°$. Thus ∠1 is a right angle, and $p \perp m$.

Section 4.3 Sample Problems

1 Given: $\overline{AB} \cong \overline{AC}$
$\overline{BD} \cong \overline{CD}$

Conclusion: $\overline{AD}$ is an altitude

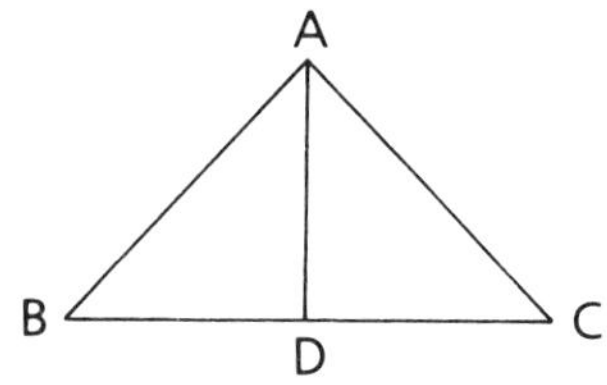

1 $\overline{AB} \cong \overline{AC}$	1 Given
2 $\overline{BD} \cong \overline{CD}$	2 Given
3 $\overline{AD} \cong \overline{AD}$	3 Reflexive Property
4 $\triangle ABD \cong \triangle ACD$	4 SSS (1, 2, 3)
5 $\angle ADB \cong \angle ADC$	5 CPCTC
6 $\overleftrightarrow{AD} \perp \overleftrightarrow{BC}$	6 If two lines intersect to form $\cong$ adjacent $\angle$s, then the lines are $\perp$
7 $\overline{AD}$ is an altitude	7 If a segment is drawn from a vertex of a $\triangle$ and is $\perp$ to the opposite side, it is an altitude of the $\triangle$

2 Given: $\overline{AB} \cong \overline{AD}$, $\overline{BC} \cong \overline{CD}$

Prove: $\overleftrightarrow{AC}$ is the $\perp$ bisector of $\overline{BD}$

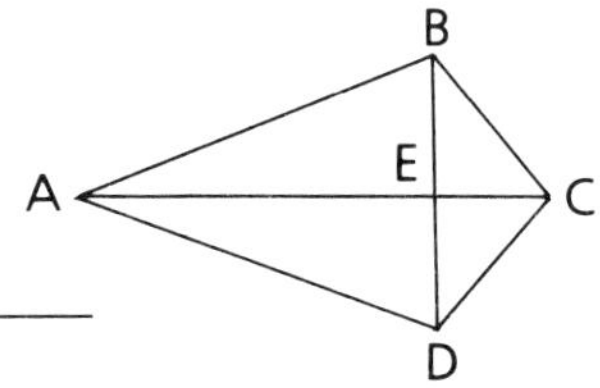

1 $\overline{AB} \cong \overline{AD}$	1 Given
2 $\overline{BC} \cong \overline{CD}$	2 Given
DETOUR [3 $\overline{AC} \cong \overline{AC}$	3 Reflexive Property
4 $\triangle ABC \cong \triangle ADC$]	4 SSS (1, 2, 3)
5 $\angle BAC \cong \angle DAC$	5 CPCTC
6 $\overline{AE} \cong \overline{AE}$	6 Reflexive Property
7 $\triangle ABE \cong \triangle ADE$	7 SAS (1, 5, 6)
8 $\overline{BE} \cong \overline{ED}$	8 CPCTC
9 $\overleftrightarrow{AC}$ bisects $\overline{BD}$	9 If a line divides a segment into two $\cong$ segments, it bisects the segment
10 $\angle AEB \cong \angle AED$	10 CPCTC (step 7)
11 $\overleftrightarrow{AC} \perp \overleftrightarrow{BD}$	11 If two lines intersect to form $\cong$ adjacent $\angle$s, the lines are $\perp$
12 $\overleftrightarrow{AC}$ is the $\perp$ bisector of $\overline{BD}$	12 Combination of steps 9 and 11

Section 4.3 Problem Set A

1 Why does *Theorem 21 not* apply to the figures below?

a

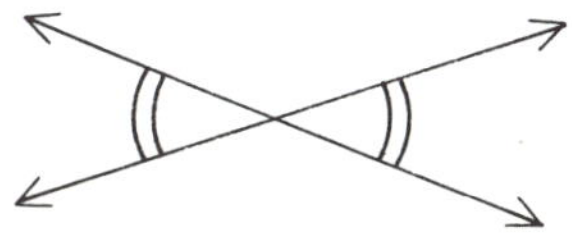

b

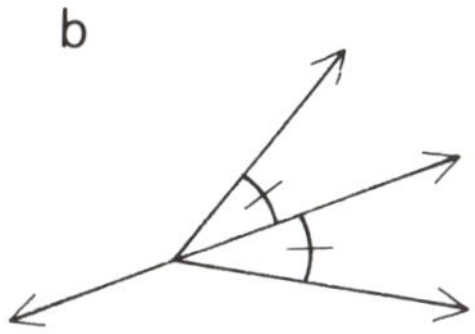

2 Given: $\odot$ P

S is the midpt. of $\overline{QR}$

Prove: $\overleftrightarrow{PS} \perp \overleftrightarrow{QR}$

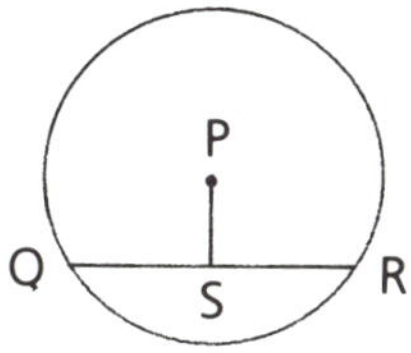

3 Prove: The angle bisector of the vertex angle of an isosceles triangle is perpendicular to the base.

4 Given: $\overline{AB} \cong \overline{BC} \cong \overline{CD} \cong \overline{AD}$

(Such a figure is a rhombus.)

Conclusion: $\overline{AC} \perp \overline{BD}$

(Hint: Detour)

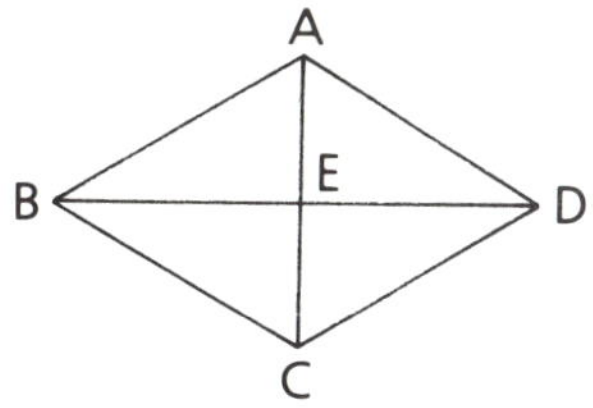

5 Given: $\overrightarrow{XR}$ bisects $\angle YXZ$

$\angle Y \cong \angle Z$

Conclusion: $\overline{XR}$ is an altitude

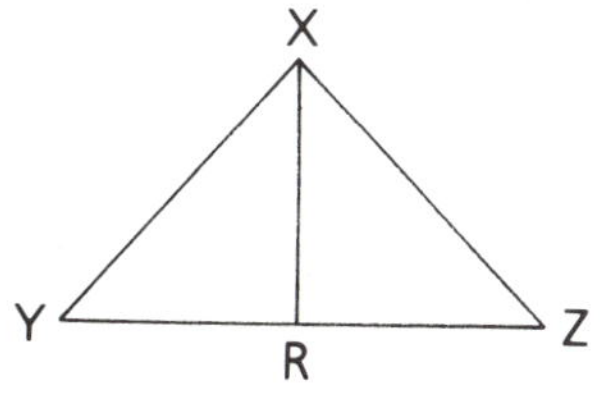

Section 4.3 Problem Set B

6 Given: $\odot$ O

$\angle B \cong \angle C$

Conclusion: $\overline{AO} \perp \overline{BC}$

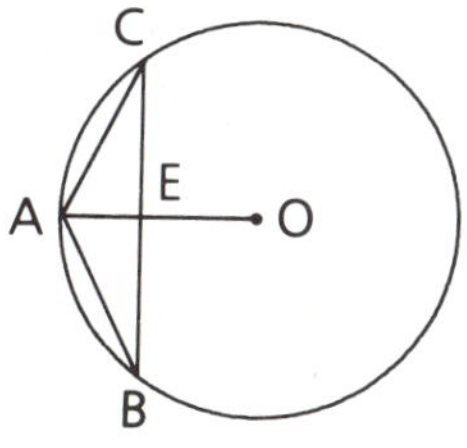

7 Given: $\overline{EX} \cong \overline{EY}$

$\overline{XP} \cong \overline{PY}$

Prove: $\overleftrightarrow{EZ} \perp \overleftrightarrow{XY}$

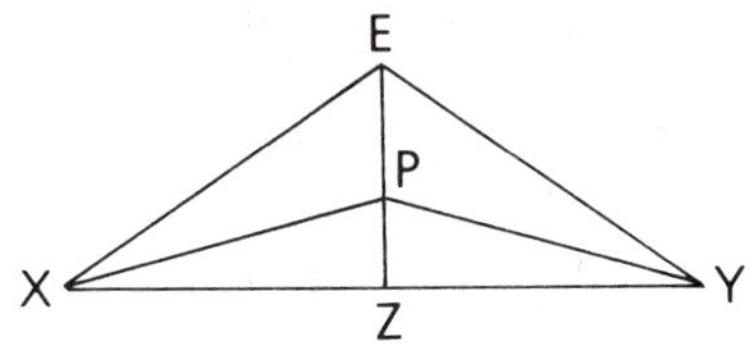

8 Prove that the median to the base of an isosceles triangle is also an altitude to the base.

9 Given: $\overleftrightarrow{PR}$ bisects $\overline{QS}$

$\angle RQT \cong \angle RST$

Prove: $\overleftrightarrow{QS} \perp \overleftrightarrow{PR}$

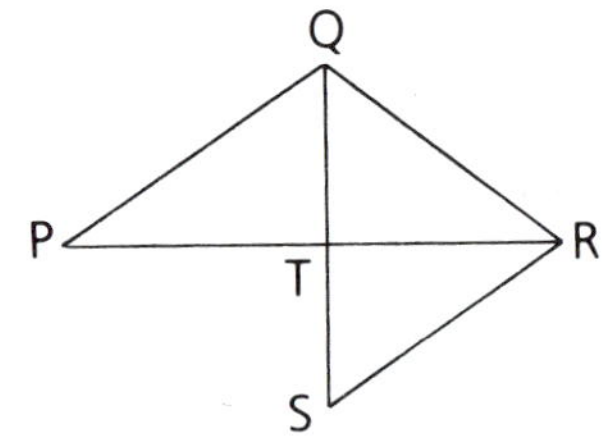

10 Prove that if two circles intersect at two points A and B, then the line joining the centers is perpendicular to $\overline{AB}$.

Section 4.3 Problem Set C

11 Is $b \perp a$?

Justify your answer.

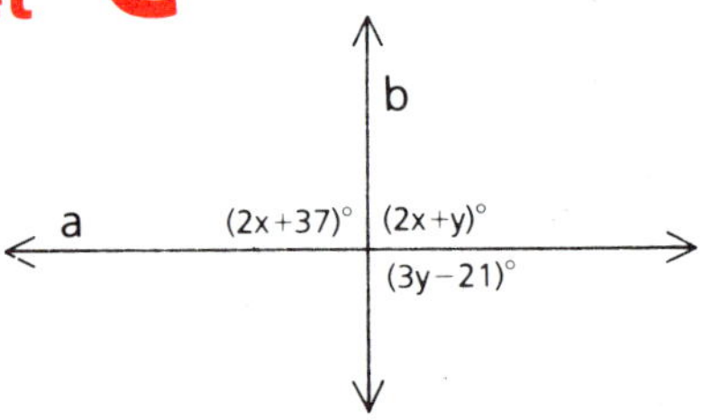

12 Prove that the line drawn from the vertex angle of an isosceles triangle through the intersection of the medians to the legs is perpendicular to the base.

13 The ratio of the complements of two angles is 3:2 and the ratio of their supplements is 9:8. Find the two original angles.

14 To the nearest second, when is the next time after 7:00 that the hands of a clock form a right angle?

Section 4.4 The Equidistance Theorems

D **The DISTANCE between two objects is the length of the shortest path joining them.**

The **DISTANCE** between points R and S is the length of $\overline{RS}$, or RS.

If two points P and Q are both the same distance from a third point, X, then X is said to be **EQUIDISTANT** from P and Q.

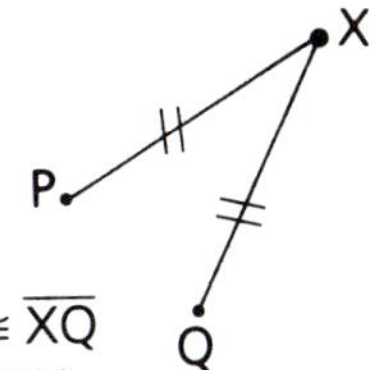

$\overline{PX} \cong \overline{XQ}$
means
X is equidistant from P and Q

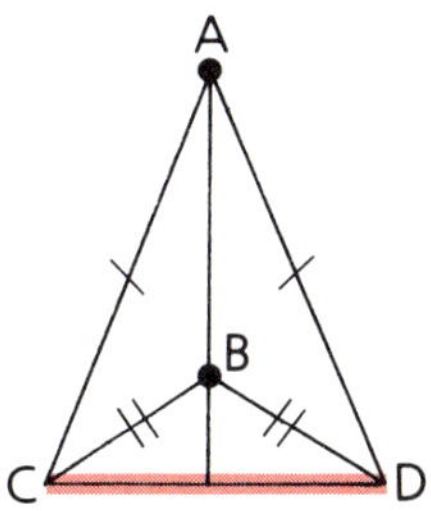

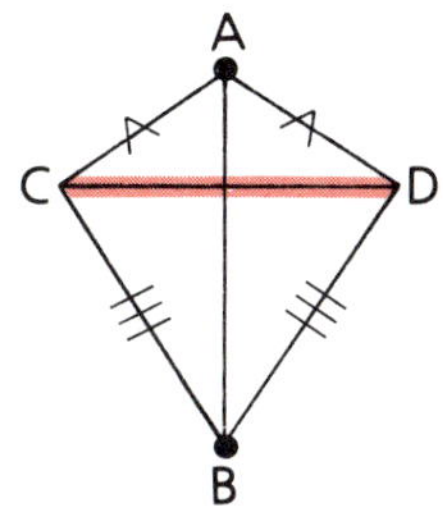

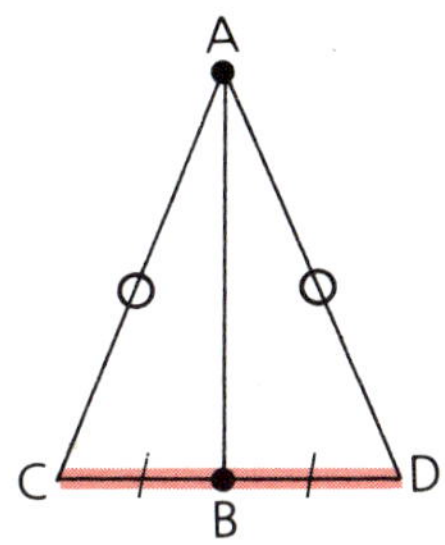

You should recall many problems with diagrams resembling those above. They all have something in common. In each diagram, points A and B are each **EQUIDISTANT** from the endpoints C and D of $\overline{CD}$. In each case, you could prove that $\overleftrightarrow{AB} \perp \overleftrightarrow{CD}$ and $\overleftrightarrow{AB}$ bisects $\overline{DC}$ just by using the theorem that follows this formal definition of a $\perp$ bisector.

D **The PERPENDICULAR BISECTOR of a segment is the *line* that bisects and is perpendicular to the segment.**

Theorem 22

T **If two points are each equidistant from the endpoints of a segment, then the two points determine the perpendicular bisector of that segment.**

Given: $\overline{PA} \cong \overline{PB}$
$\overline{QA} \cong \overline{QB}$
Prove: $\overleftrightarrow{PQ}$ is the $\perp$ bisector of $\overline{AB}$

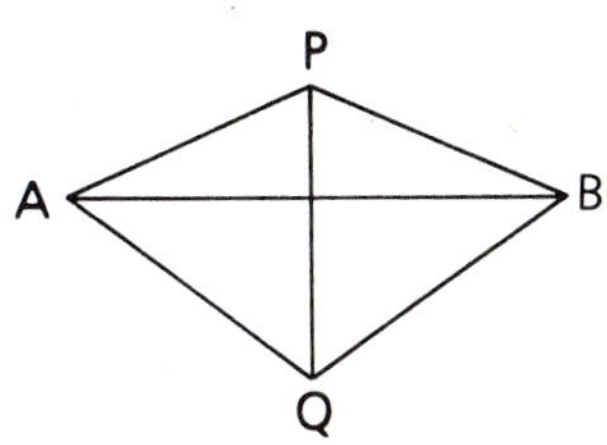

For the proof of Theorem 22, see Sample Problem 2 in section 4.3.

Theorem 23

T **If a point is on the perpendicular bisector of a segment, then it is equidistant from the endpoints of that segment.**

Given: $\overleftrightarrow{PQ}$ is the $\perp$ bisector of $\overline{AB}$

Prove: $\overline{PA} \cong \overline{PB}$

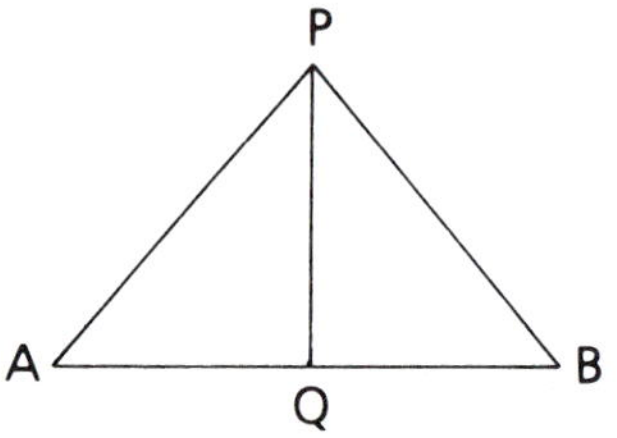

The proof of this theorem is easily done by using the definition of perpendicular bisector and some congruent triangles.

Section 4.4 Sample Problems

1 **Given:** $\angle 1 \cong \angle 2$
$\angle 3 \cong \angle 4$

Prove: $\overleftrightarrow{AE} \perp$ **bis** $\overline{BD}$

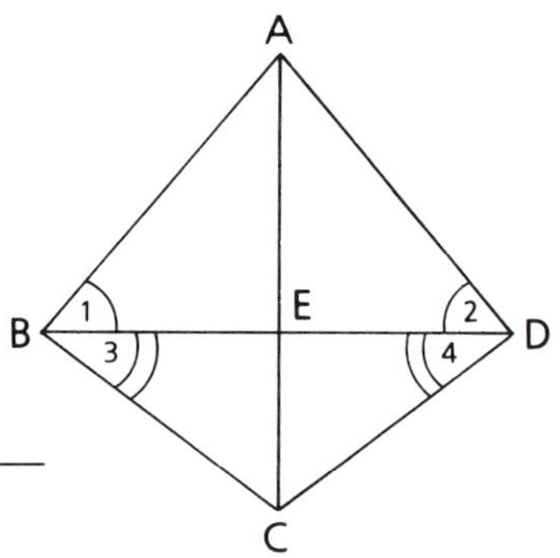

1 $\angle 1 \cong \angle 2$	1 Given
2 $\overline{AB} \cong \overline{AD}$	2 If △ then △
3 $\angle 3 \cong \angle 4$	3 Given
4 $\overline{BC} \cong \overline{CD}$	4 Same as step 2
5 $\overleftrightarrow{AE} \perp$ bis $\overline{BD}$	5 If two points are each equidistant from the endpoints of a segment, they determine the $\perp$ bisector of the segment

Note: We proved that A and C were each equidistant from B and D. Why did we *not* need to use E?

2 Prove: The line joining the vertex of an isosceles triangle to the midpoint of the base is perpendicular to the base.

Given: $\triangle ABC$ is isosceles with
$\overline{AB} \cong \overline{AC}$
E is the midpoint of $\overline{BC}$

Prove: $\overleftrightarrow{AE} \perp \overline{BC}$

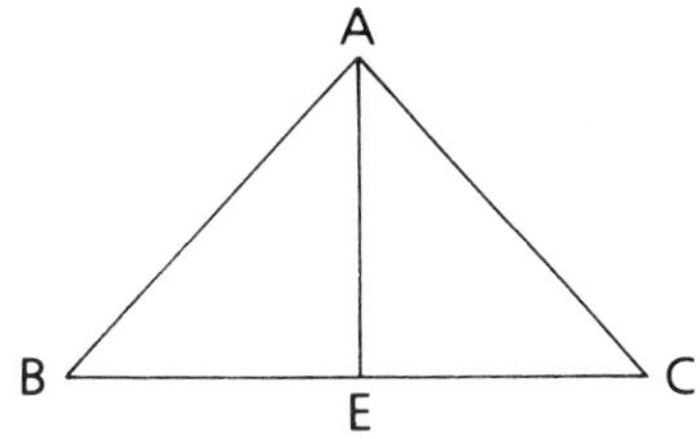

1 $\triangle ABC$ is isosceles with $\overline{AB} \cong \overline{AC}$	1 Given
2 E is midpt. of $\overline{BC}$	2 Given
3 $\overline{BE} \cong \overline{EC}$	3 The midpoint of a segment divides the segment into two $\cong$ segments
4 $\overleftrightarrow{AE} \perp \overline{BC}$	4 Two points each equidistant from the endpoints of a segment determine the $\perp$ bisector of the segment

3 Given: $\overline{AB} \cong \overline{AD}$
$\overline{BC} \cong \overline{CD}$

Conclusion: $\overline{BE} \cong \overline{ED}$

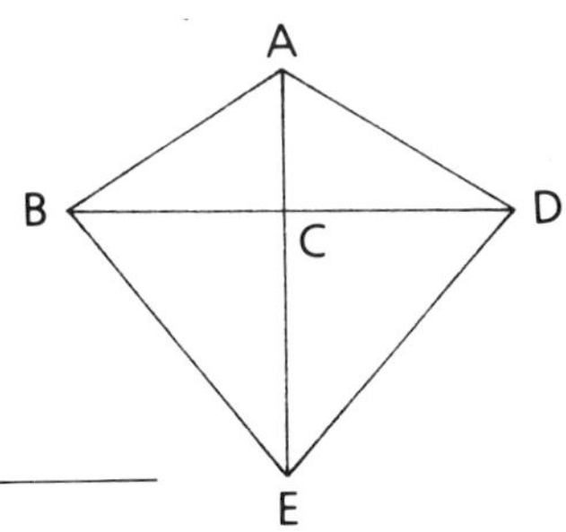

1 $\overline{AB} \cong \overline{AD}$	1 Given
2 $\overline{BC} \cong \overline{CD}$	2 Given
3 $\overleftrightarrow{AC} \perp$ bis $\overline{BD}$	3 Two points each equidistant from the endpoints of a segment determine the $\perp$ bisector of the segment
4 $\overline{BE} \cong \overline{ED}$	4 A point on the $\perp$ bisector of a segment is equidistant from the endpoints of the segment

These sample problems could have been done without Theorems 22 and 23, but would have been harder and longer. How much longer?

Section 4.4 Problem Set **A**

As you do these proofs, see if the equidistance theorems apply; they can save you a lot of work.

1 Given: $\odot$ O, M is the midpt. of $\overline{AB}$

Conclusion: $\overline{OM} \perp \overline{AB}$

(Hint: Introduce two segments.)

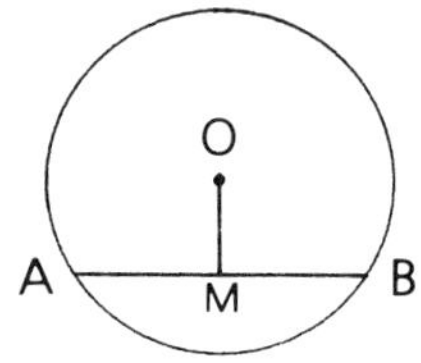

2 Given: $\overleftrightarrow{WZ}$ ⊥ bis $\overline{XY}$

Prove: △WXY is isosceles

(This problem can be done in 3 steps if you use Theorem 23.)

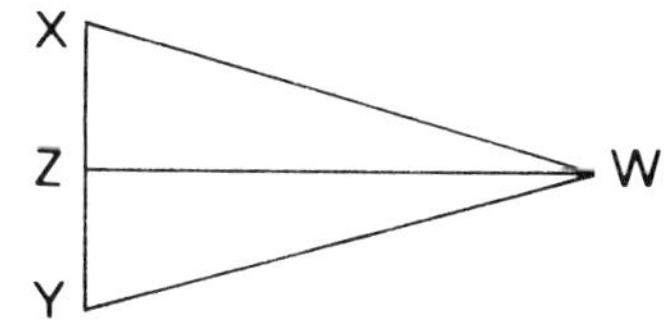

3 Given: $\odot$O, $\overline{AB} \cong \overline{AC}$

Conclusion: $\overleftrightarrow{AD}$ ⊥ bis $\overline{BC}$

(Hint: Show that A and O are each equidistant from B and C.)

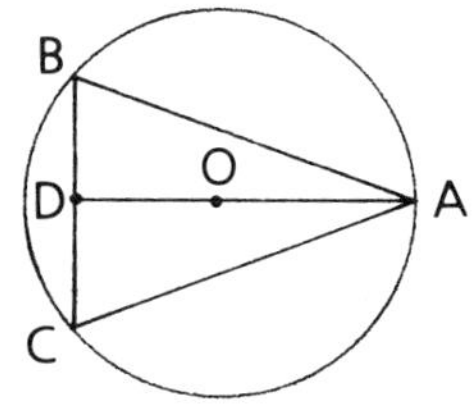

4 Given: ⓢ P and Q

Prove: $\overleftrightarrow{PQ}$ ⊥ bis $\overline{RS}$

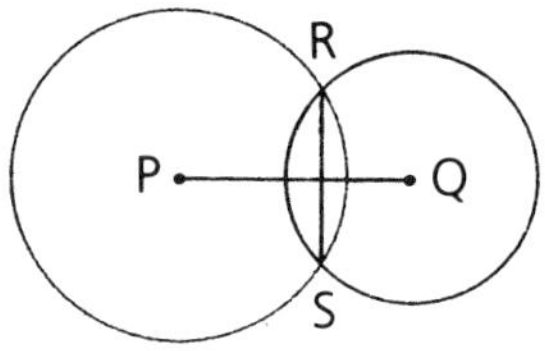

5 Given: $\overleftrightarrow{AD}$ ⊥ bis $\overline{BC}$

Prove: △ABE ≅ △ACE

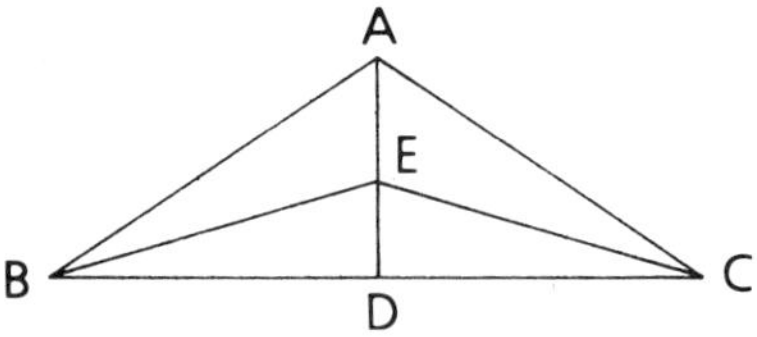

6 Given: $\overleftrightarrow{AG}$ ⊥ bis $\overline{BC}$ and $\overline{DE}$

Conclusion: $\overline{BD} \cong \overline{CE}$

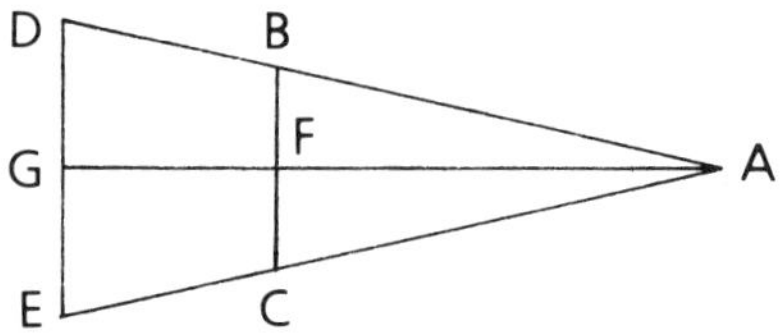

Section 4.4 Problem Set B

Remember: the equidistance theorems will help you write a concise proof.

7 Draw isosceles $\triangle$ PQR with P the vertex. Draw the bisectors of the base angles. They intersect at S. Prove that $\overleftrightarrow{PS} \perp \overline{QR}$.
(Hint: Use Theorem 22.)

8 Given: $\overline{AB} \cong \overline{BC}$
$\overline{AE} \cong \overline{EC}$

Prove: $\overline{AD} \cong \overline{DC}$
(Hint: This can be done in 4 steps.)

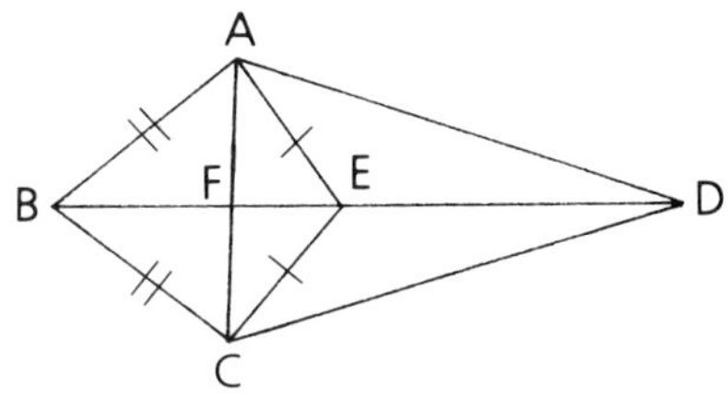

9 Given: $\overline{WY}$ and $\overline{XZ}$ are $\perp$ bis of *each other*

Prove: $\overline{WX} \cong \overline{XY} \cong \overline{YZ} \cong \overline{ZW}$
(Just information: WXYZ is a RHOMBUS.)

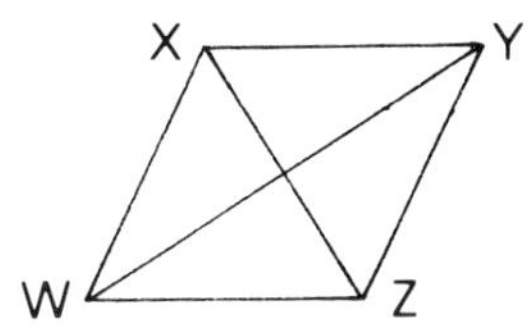

10 Given: WXYZ is a kite, $\overline{WX} \cong \overline{WZ}$, $\overline{XY} \cong \overline{YZ}$

Prove: $\triangle$WPZ is a right triangle

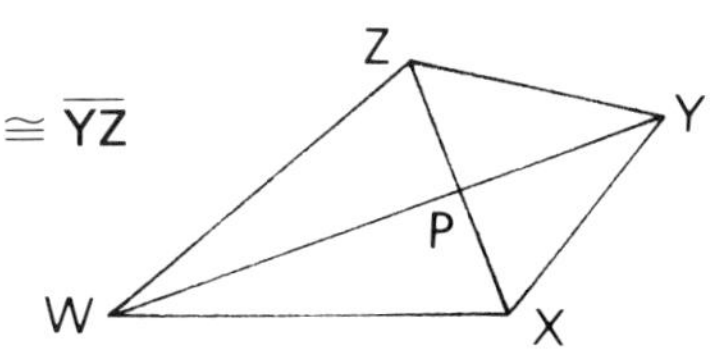

11 Given: $\angle$ADC and $\angle$ABC are right $\angle$s, $\overline{AB} \cong \overline{AD}$

Conclusion: $\overleftrightarrow{AC} \perp$ bis $\overline{BD}$

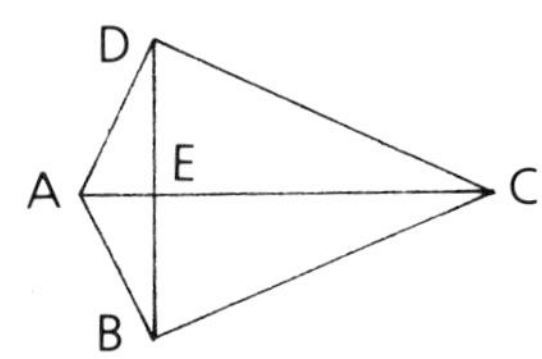

12 The median to the base of an isosceles triangle is also an altitude. (Prove this *without using congruent triangles*.)

13 Given: F is the midpt. of $\overline{BC}$
$\overline{DB} \cong \overline{EC}$
$\overline{DB} \perp \overline{DF}$
$\overline{EC} \perp \overline{EF}$

Conclusion: $\overleftrightarrow{AF} \perp \overleftrightarrow{BC}$

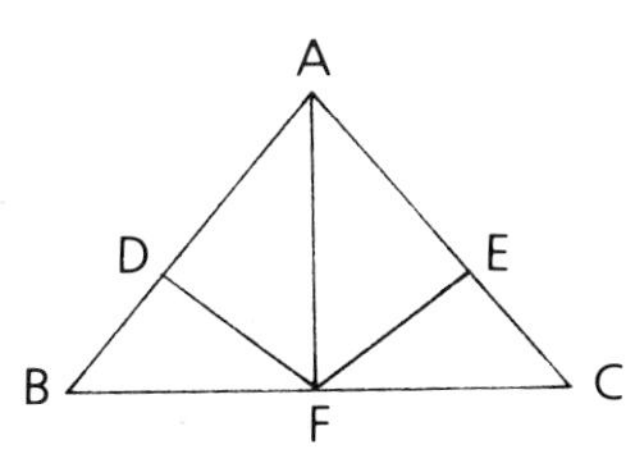

Section 4.4 Problem Set C

14 Given: $\overline{AB} \cong \overline{BC}$
$\overline{AE} \cong \overline{EC}$

Conclusion: $\overline{AD} \cong \overline{DC}$

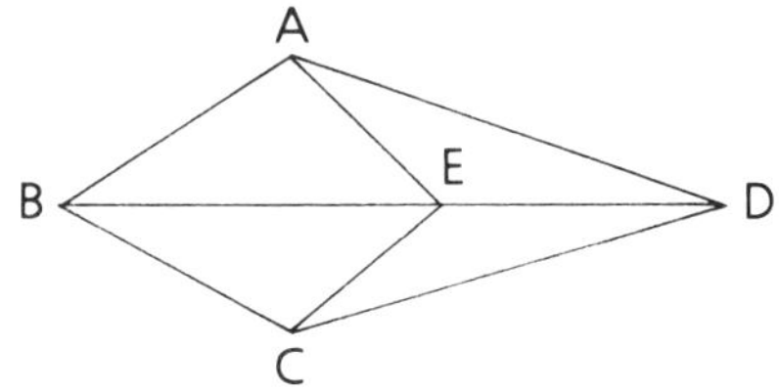

15 Given: ABCDE is equilateral and equiangular
F is the midpt. of $\overline{AE}$

Prove: $\overleftrightarrow{FC} \perp$ bis $\overline{BD}$

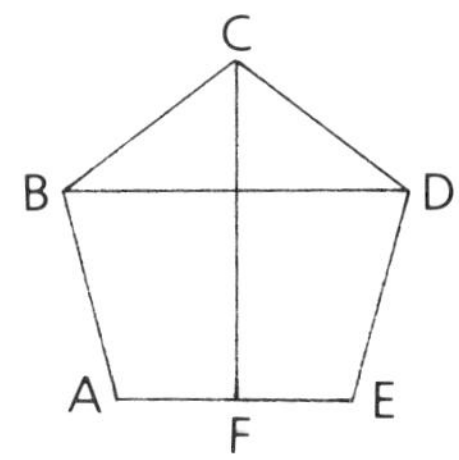

16 A four-sided figure with two distinct pairs of adjacent sides congruent is called a KITE. The two segments joining opposite vertices are its diagonals. Prove that one of these diagonals is the perpendicular bisector of the other diagonal.

17 Prove that if each of the three altitudes of a triangle bisects the side to which it is drawn, then the triangle is equilateral.

18 a If two of the points A, B, C, D, E, and M are chosen at random, what is the probability that the two chosen points determine the perpendicular bisector of $\overline{AB}$?

b If three of the six points are chosen at random, what is the probability that the points are collinear?

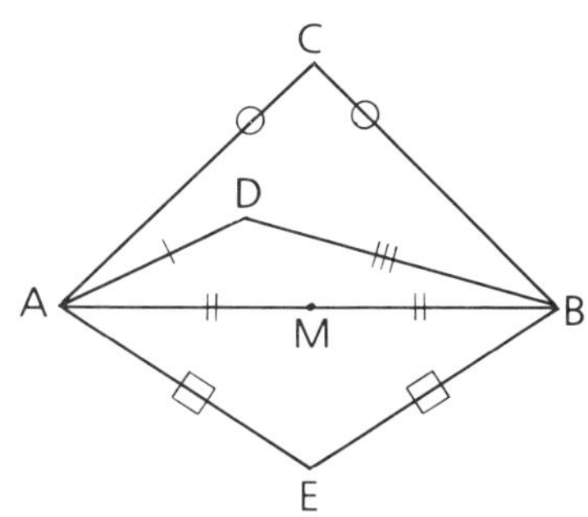

19 Find the angle formed between the minute hand and the second hand of a clock at 20 seconds after 6:01.

Section 4.5 Lies On or Passes Through

Theorem 23 stated that any point that lies on the perpendicular bisector of a segment will be equidistant from the endpoints of that segment. The reverse statement is also true.

Theorem 24

T **If a point is equidistant from the endpoints of a segment, then it *lies on* the perpendicular bisector of that segment. (LIES ON THEOREM)**

Given: $\overleftrightarrow{PM} \perp$ bis $\overline{BC}$

$\overline{AB} \cong \overline{AC}$

Prove: A lies on $\overleftrightarrow{PM}$ (or $\overleftrightarrow{PM}$ passes through A)

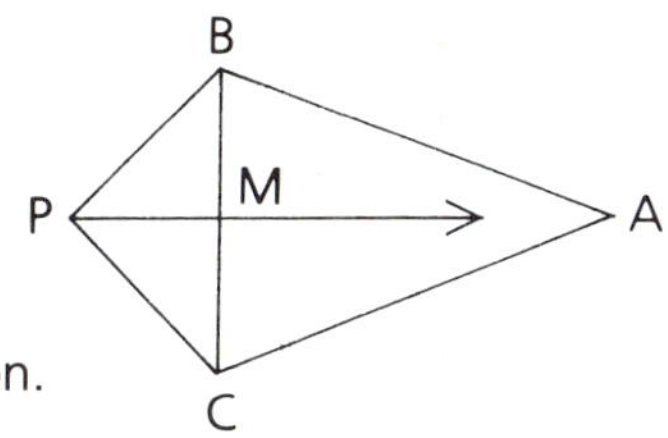

The proof is called for in problem 15 of this section.

The uses for this theorem may not be at all obvious. Read on and you will see some applications.

This is the first theorem we have presented that will enable you to conclude that *a point lies on a line* (or that *a line passes through a given point*). For such proofs, there is a strict procedure to be followed:

STRICT PROCEDURE

SP 1 Prove that the *line in question* is the perpendicular bisector of *some segment* in the figure. (Temporarily *ignore* the point in question.)

SP 2 Prove that the *point in question* is equidistant from the endpoints of the *same segment* used in SP 1.

SP 3 Apply Theorem 24 to conclude that the *point* lies on the *line.*

Section 4.5 Sample Problems

1 Prove that the perpendicular bisector of a chord of a circle must pass through the center of the circle.

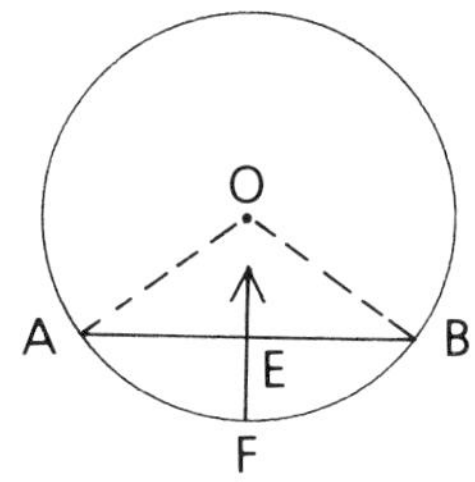

Given: $\odot$ O with chord $\overline{AB}$
$\overleftrightarrow{EF} \perp$ bis $\overline{AB}$

Conclusion: $\overleftrightarrow{FE}$ passes through O

	1 $\odot$ O	1 Given
SP 1	2 $\overleftrightarrow{FE} \perp$ bis $\overline{AB}$	2 Given
	3 Draw $\overline{OA}$ and $\overline{OB}$	3 Two points determine a line
SP 2	4 $\overline{OA} \cong \overline{OB}$	4 All radii of a circle are $\cong$
SP 3	5 $\overleftrightarrow{FE}$ passes through O	5 A point equidistant from the endpoints of a segment lies on the $\perp$ bisector of that segment *(Lies On Theorem)*

Notice that in the above problem, we were trying to prove that $\overleftrightarrow{EF}$ passes through point O and could not assume that this fact was known. Therefore, we could not draw $\overleftrightarrow{EF}$ through O. Observe how that was handled here and in the following diagrams.

2 Given: $\overleftrightarrow{AC} \perp \overleftrightarrow{BD}$
$\overline{AB} \cong \overline{AD}$
$\angle 1 \cong \angle 2$

Prove: P lies on $\overleftrightarrow{AC}$ *

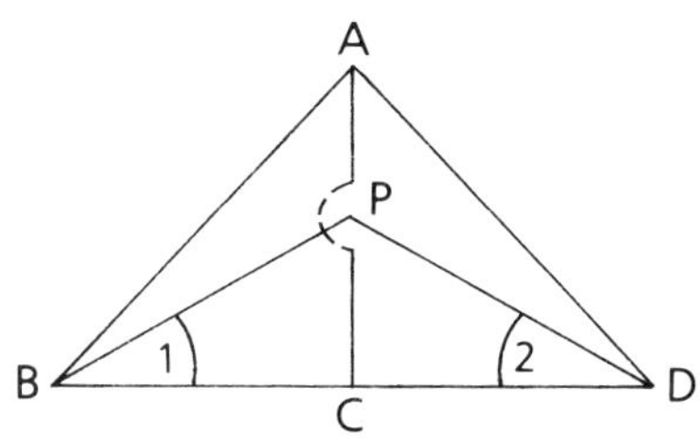

1 $\overleftrightarrow{AC} \perp \overleftrightarrow{BD}$	1 Given
2 $\angle ACB$, $\angle ACD$ are right angles	2 If two lines are perpendicular, then they form right angles
3 $\overline{AB} \cong \overline{AD}$	3 Given
4 $\overline{AC} \cong \overline{AC}$	4 Reflexive Property

* Notice that in the proof, P is not ever mentioned until after $\overleftrightarrow{AC}$ is proved to be the $\perp$ bisector of $\overline{BD}$.

	Statements	Reasons
	5 $\triangle ABC \cong \triangle ADC$	5 HL (2, 3, 4)
	6 $\overline{BC} \cong \overline{CD}$	6 CPCTC
SP 1	7 $\overleftrightarrow{AC} \perp$ bis $\overline{BD}$	7 Two points each equidistant from the endpoints of a segment determine the $\perp$ bisector of the segment
	8 $\angle 1 \cong \angle 2$	8 Given
SP 2	9 $\overline{BP} \cong \overline{PD}$	9 If △ then △
SP 3	10 P lies on $\overleftrightarrow{AC}$	10 *Lies On Theorem*

Section 4.5 Problem Set A

Study the STRICT PROCEDURE before attempting these problems.

1 **Answer the questions, copy the proof, and fill in the missing statements and reasons.**

Given: $\triangle ABC$ **is isosceles with base** $\overline{BC}$
$\triangle BPC$ **is isosceles with base** $\overline{BC}$
M is the midpt. of $\overline{BC}$

Prove: $\overleftrightarrow{PM}$ **passes through A**

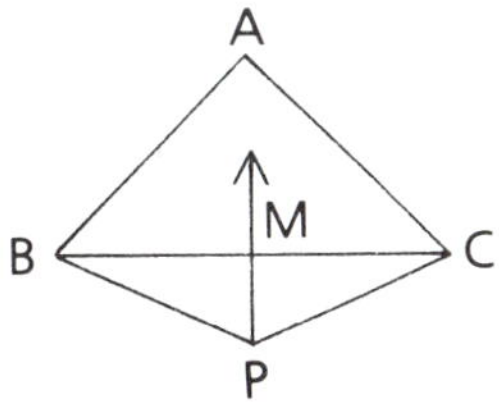

Questions: **a** **What is the** ***line in question?***
b **What must be proved about this line?**
c **What is the** ***point in question?***
d **What must be proved about this point?**

	Statements	Reasons
	1 $\triangle BPC$ is isosceles, base $\overline{BC}$	1 Given
	2 $\overline{BP} \cong \overline{PC}$	2 ____________
	3 M is the midpt. of $\overline{BC}$	3 Given
	4 ____________	4 ____________
SP 1	5 $\overleftrightarrow{PM} \perp$ bis $\overline{BC}$	5 ____________
	6 $\triangle ABC$ is isosceles, base $\overline{BC}$	6 Given
SP 2	7 ____________	7 ____________
SP 3	8 $\overleftrightarrow{PM}$ passes through A	8 ____________

2 Copy the proof, filling in the missing reasons.
Given: $\overline{AB} \cong \overline{AF}$, and $\overline{BC} \cong \overline{FE}$
$\overleftrightarrow{DG} \perp$ bis $\overline{CE}$

Prove: A lies on $\overleftrightarrow{DG}$

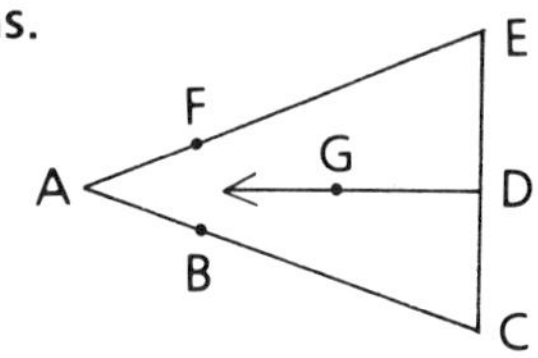

	Statements	Reasons
SP 1	1 $\overleftrightarrow{DG} \perp$ bis $\overline{CE}$	1 __ __ __ __ __ __
	2 $\overline{AB} \cong \overline{AF}$	2 __ __ __ __ __ __
	3 $\overline{BC} \cong \overline{FE}$	3 __ __ __ __ __ __
SP 2	4 $\overline{AC} \cong \overline{AE}$	4 __ __ __ __ __ __
SP 3	5 A lies on $\overleftrightarrow{DG}$	5 __ __ __ __ __ __

3 Copy the proof, filling in the missing reasons.
Given: $\odot$ O, $\angle J \cong \angle M$
K is the midpt. of $\overline{JM}$

Prove: $\overleftrightarrow{HO}$ passes through K

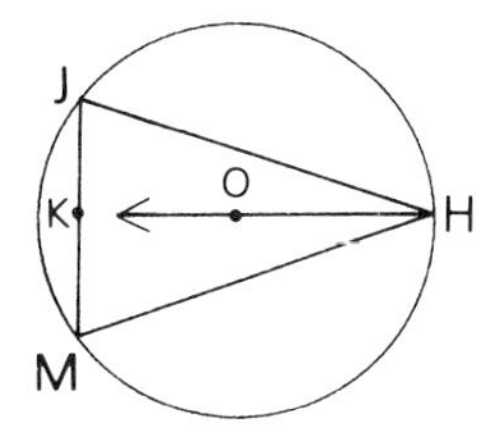

	Statements	Reasons
	1 $\odot$ O	1 __ __ __ __ __ __
	2 Draw $\overline{OJ}$	2 __ __ __ __ __ __
	3 Draw $\overline{OM}$	3 __ __ __ __ __ __
	4 $\overline{OJ} \cong \overline{OM}$	4 __ __ __ __ __ __
	5 $\angle J \cong \angle M$	5 __ __ __ __ __ __
	6 $\overline{HJ} \cong \overline{HM}$	6 __ __ __ __ __ __
SP 1	7 $\overleftrightarrow{HO} \perp$ bis $\overline{JM}$	7 __ __ __ __ __ __
	8 K is the midpt. of $\overline{JM}$	8 __ __ __ __ __ __
SP 2	9 $\overline{JK} \cong \overline{KM}$	9 __ __ __ __ __ __
SP 3	10 $\overleftrightarrow{HO}$ passes through K	10 __ __ __ __ __ __

4 Given: $\overline{AM}$ is a median of $\triangle ABC$
$\overline{AB} \cong \overline{AC}$
$\triangle PBC$ is isosceles, base $\overline{BC}$

Prove: P lies on $\overleftrightarrow{AM}$

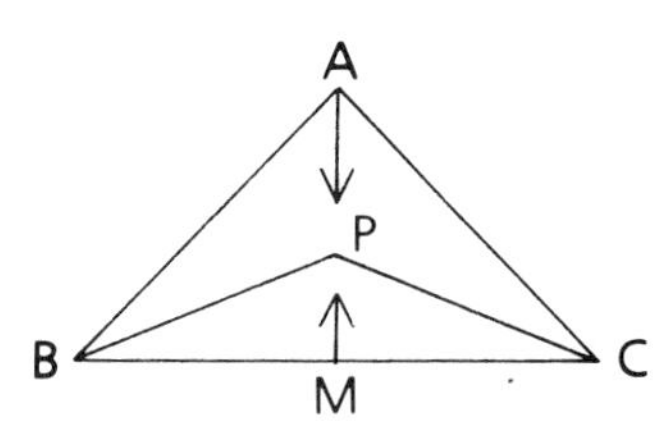

Questions:

a What line must be proved to be the $\perp$ bisector of $\overline{BC}$?

b What must be proved about point P? Remember to ignore point P until you have finished proving part a.

5 Given: ⓢ P and Q
$\overline{XA} \cong \overline{AY}$

Prove: $\overleftrightarrow{PA}$ passes through Q

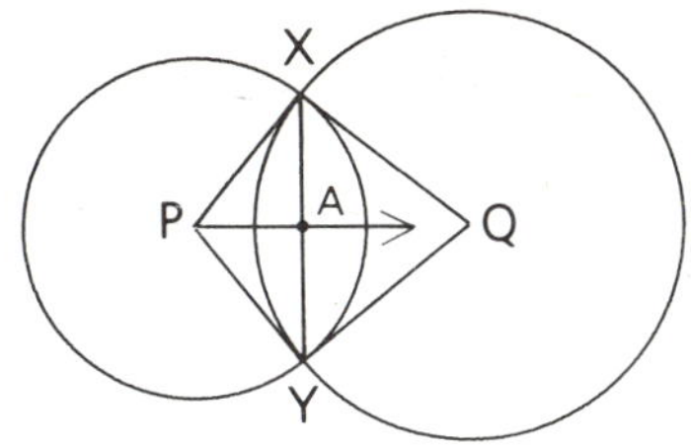

Section 4.5 Problem Set B

6 Given: ⊙ P and ⊙ Q
$\overline{PJ} \perp \overline{XY}$

Prove: $\overleftrightarrow{PJ}$ passes through Q

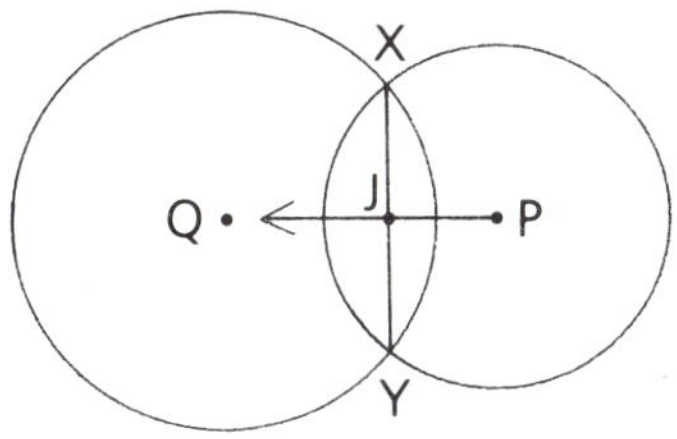

7 Given: $\overline{AB} \cong \overline{AC}$
$\overrightarrow{AD}$ bisects $\angle BAC$
M is the midpt. of $\overline{BC}$

Prove: $\overleftrightarrow{AD}$ passes through M

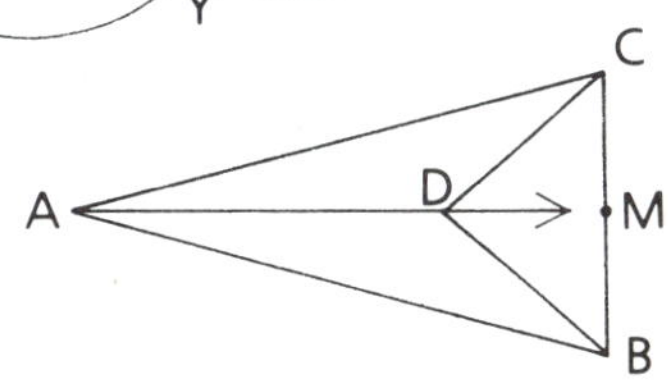

8 Given: $\overrightarrow{AD}$ bisects $\angle BAC$
$\overline{AB} \cong \overline{AC}$
$\angle 1 \cong \angle 2$

Prove: E lies on $\overleftrightarrow{AD}$

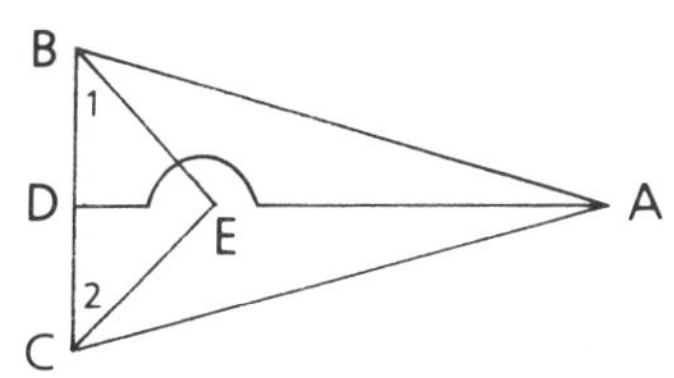

9 Given: $\overleftrightarrow{DB}$ bisects $\angle ABC$ and $\angle ADC$
$\overline{AP} \cong \overline{CP}$

Prove: $\overleftrightarrow{BD}$ passes through P

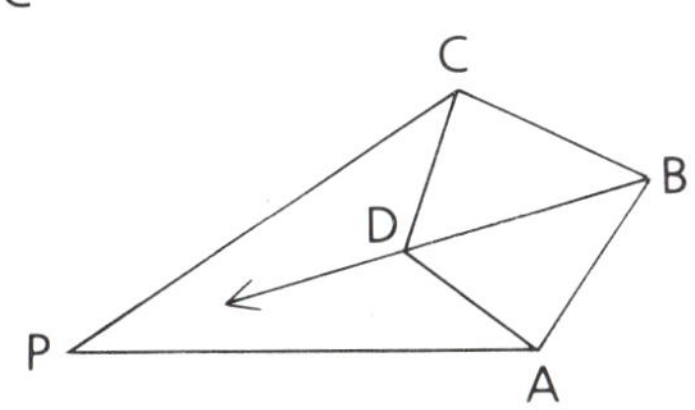

10 Given: ⊙ O, with $\angle B \cong \angle C$
$\overline{AK}$ is the median to $\overline{BC}$

Conclusion: O lies on $\overleftrightarrow{AK}$

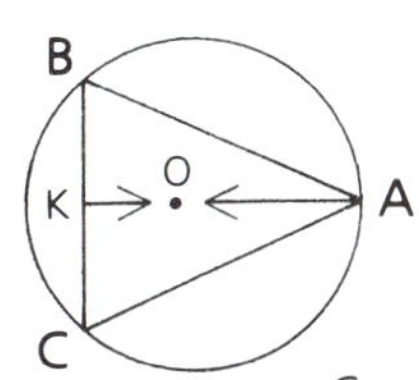

11 Given: $\overline{AB} \cong \overline{BC}$
$\overrightarrow{BD}$ bisects $\angle ABC$

Prove: $\overleftrightarrow{BD}$ passes through P

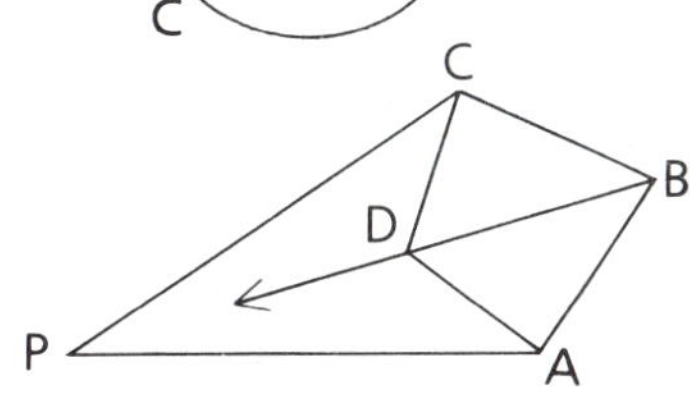

Section 4.5 Problem Set C

12 A polygon is *inscribed* in a circle if all of its vertices lie on the circle. Prove that if an isosceles triangle is inscribed in a circle, then the altitude to the base passes through the center of the circle.

13 Given: $\overleftrightarrow{MN} \perp$ bis $\overline{BC}$

$\overline{AB} \cong \overline{DC}$

$\angle ABC \cong \angle DCB$

Prove: $\overleftrightarrow{MN}$ passes through E

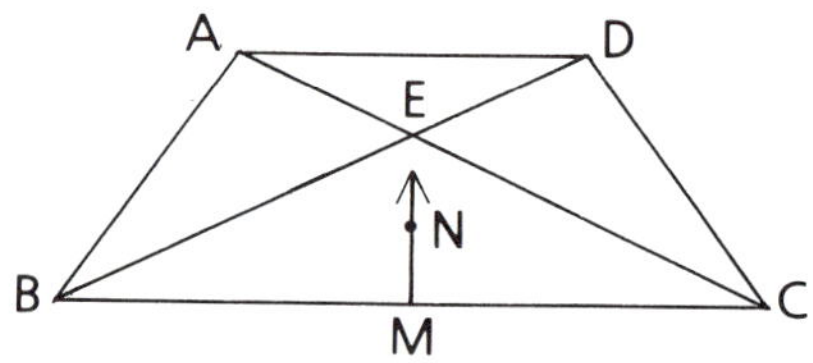

14 Given: $\overleftrightarrow{EG}, \overleftrightarrow{PF}, \overleftrightarrow{PD}$ are the $\perp$ bisectors of the sides of $\triangle ABC$

Prove: $\overrightarrow{EG}$ passes through P (This is essentially the proof that the $\perp$ bisectors of the sides of any $\triangle$ are concurrent.)

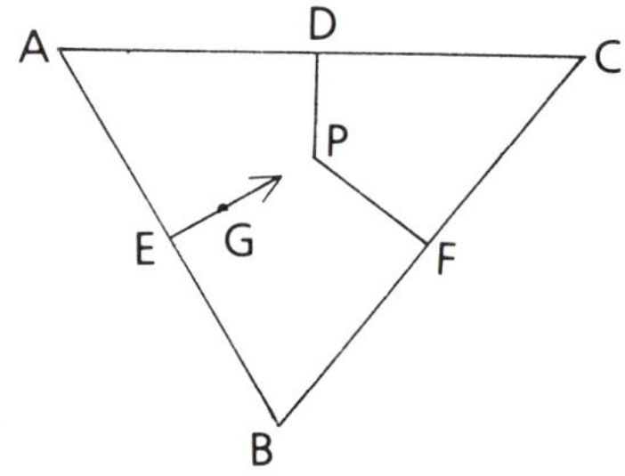

15 Prove Theorem 24. You may use these key steps:

Given: $\overleftrightarrow{PM} \perp$ bis $\overline{BC}$

$\overline{AB} \cong \overline{AC}$

Prove: A lies on $\overleftrightarrow{PM}$

Draw $\overline{AM}$

$\triangle AMB \cong \triangle AMC$ by SSS

$\angle AMB \cong \angle AMC \Rightarrow \angle BMA = 90^\circ$

But $\angle BMP = 90^\circ$ also. $\therefore \angle AMP$ is a straight $\angle$.

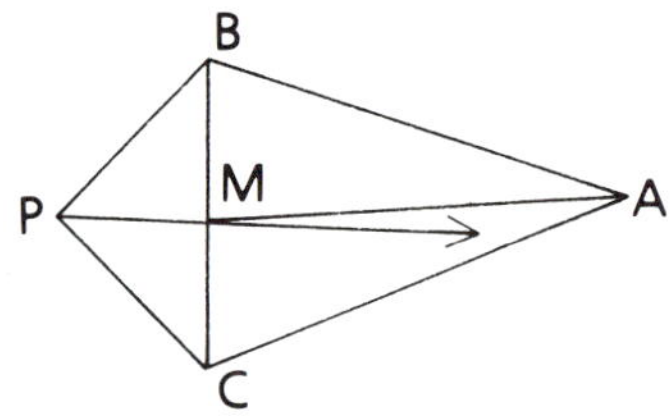

Section 4.5 Problem Set D

16 Prove that the intersection point of the medians to the legs of an isosceles triangle lies on the altitude to the base.

Chapter 4 Study Guide

As you study and review this chapter, be sure that you are able to:

1. Show how to solve DETOUR PROBLEMS such as taking a detour through one pair of triangles in order to prove that another pair is $\cong$. (pp. 167, 168)

2. Solve *The Case of the Missing Diagram* (p. 173)

   a Decide the *given* and *conclusion* from a *Conditional Statement.* (p. 173)
   b Convert a simple sentence to "If . . . , then . . ." form. (pp. 173, 174)

3. Use congruent angles to prove that lines are perpendicular. (p. 177)

4. State the definition of the perpendicular bisector of a segment. (p. 181).

5. Explain what is meant by the distance between two geometric objects. (p. 181)

6. Fill in the left column of a proof to state that a point Q is equidistant from points A and B. (p. 178)

7. Explain when congruent adjacent angles do *not* imply that some lines are perpendicular. (p. 179)

8. State what is true about any point on the perpendicular bisector of a segment. (p. 182)

9. Prove that a line is a perpendicular bisector of a segment without using the definition of a perpendicular bisector or its reverse. (p. 181)

10. Follow the STRICT PROCEDURE in using the LIES-ON Theorem. (p. 187)

Chapter 4 Review Problems A

For **1, 2** copy each problem, filling in the blanks with the correct reasons or statements.

1 **Given:** **P is the midpt. of $\overline{XZ}$**
$\angle 1 \cong \angle 2$

Conclusion: **$\overline{XY} \cong \overline{YZ}$**

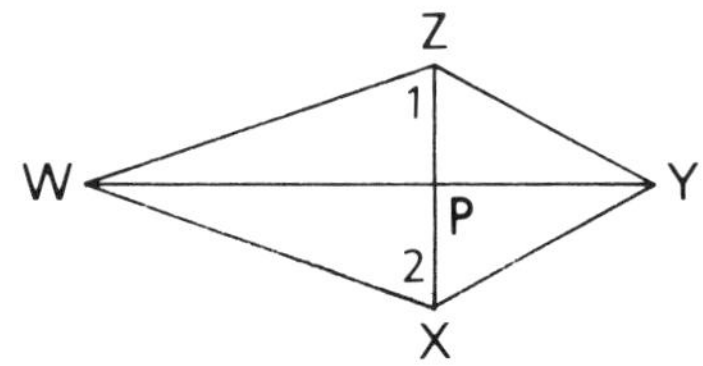

1 P is the midpt. of $\overline{XZ}$ | 1 Given
2 ______________ | 2 ______________
3 $\angle 1 \cong \angle 2$ | 3 Given
4 ______________ | 4 ______________
5 $\overleftrightarrow{WY} \perp$ bis $\overline{XZ}$ | 5 ______________
6 $\overline{XY} \cong \overline{YZ}$ | 6 ______________

2 **Given:** **ⓢ P and Q**
M is the midpt. of $\overline{AB}$

Prove: **$\overleftrightarrow{PQ}$ passes through M**

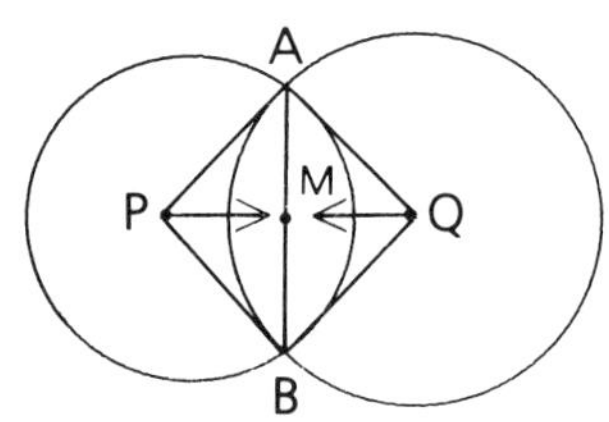

a **What must we prove about line PQ?**

b **What must we prove about point M?**

1 ⓢ P and Q | 1 Given
2 $\overline{PA} \cong \overline{PB}$ | 2 ______________
3 $\overline{AQ} \cong \overline{QB}$ | 3 Reason 2
4 ______________ | 4 ______________
5 M is the midpt. of $\overline{AB}$ | 5 Given
6 ______________ | 6 ______________
7 $\overleftrightarrow{PQ}$ passes through M | 7 ______________

3 **Given:** **$\angle ADB \cong \angle CDB$**
$\overline{AD} \cong \overline{DB}$

Prove: **$\overline{BD}$ is an altitude**

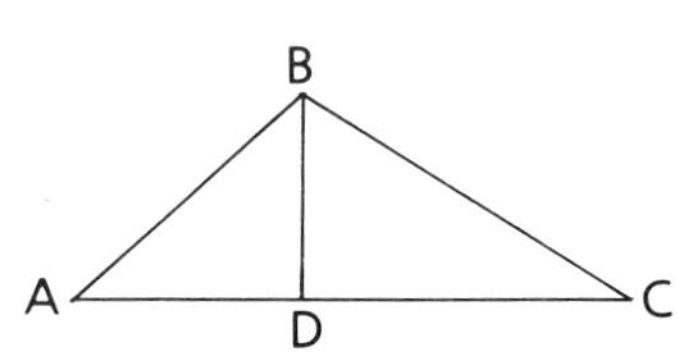

4 Given: $\angle 1 \cong \angle 4$
$\overrightarrow{FC}$ bisects $\angle BFD$

Conclusion: $\overleftrightarrow{CF} \perp \overleftrightarrow{AE}$

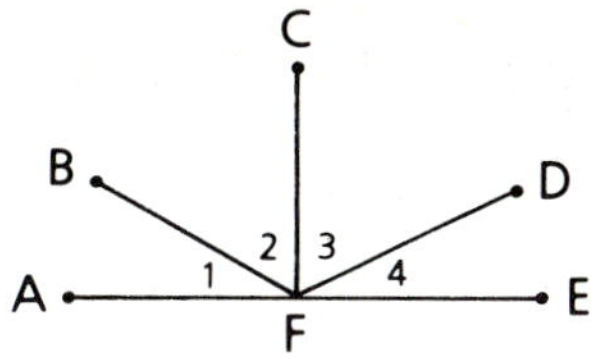

5 Set up the diagram for this problem. Then do the proof.

Given: Two isosceles triangles with the same base

Prove: The line joining the vertices of the vertex $\angle$s of the $\triangle$s is the $\perp$ bisector of the base.

6 Given: $\triangle ABC$ is isosceles, base $\overline{AC}$
$\angle 1 \cong \angle 2$

Conclusion: $\overleftrightarrow{BD} \perp \overleftrightarrow{AC}$

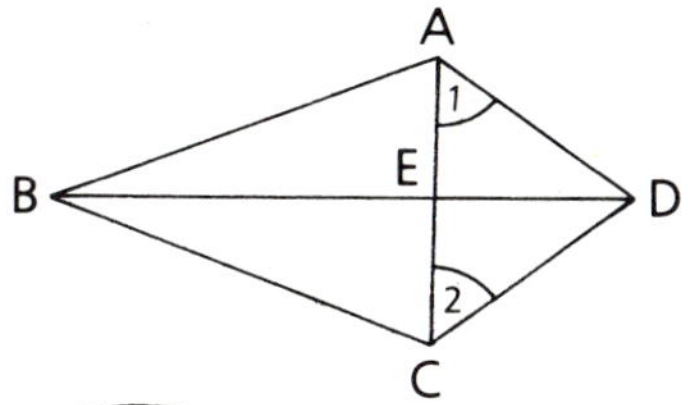

7 Given: $\odot$ O
M is the midpt. of $\overline{AB}$

Conclusion: $\overline{OM} \perp \overline{AB}$

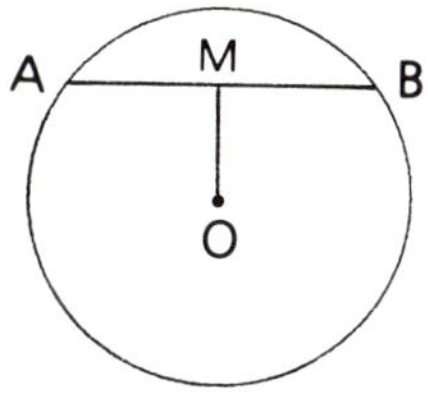

For **8, 9** set up each proof, but *do not prove* it.

8 The bisectors of the base angles of an isosceles $\triangle$ are $\cong$.

9 If two chords of a circle are $\cong$, then the segments joining the midpoints of the chords to the center of the circle are $\cong$. (A *chord* is a segment whose endpoints are on the circle.)

Chapter 4 Review Problems B

10 Given: $\odot$ O
$\angle 1 \cong \angle 2$

Conclusion: $\overline{OY} \perp \overline{WX}$

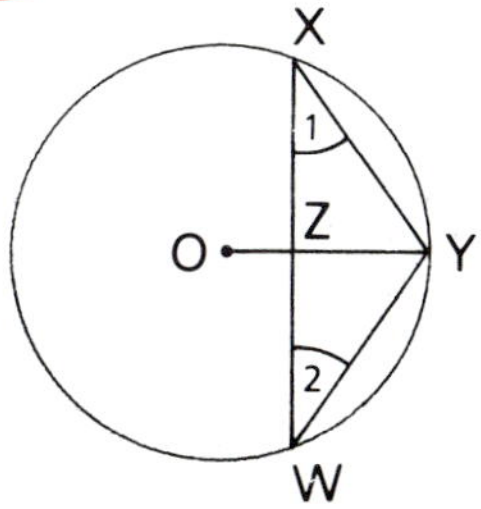

11 Given: $\angle 1 \cong \angle 2 \cong \angle 3 \cong \angle 4$
$\overline{BE} \cong \overline{BF}$
Prove: $\triangle ABE \cong \triangle CBF$

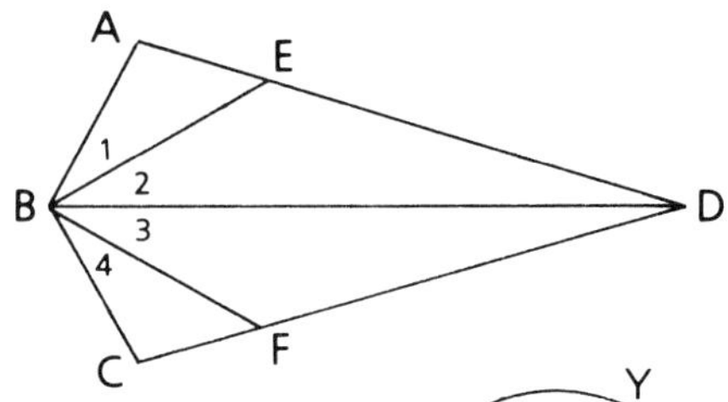

12 Given: $\odot O$
$\overline{DX} \cong \overline{DY}$
Conclusion: $\overleftrightarrow{DZ}$ bisects $\overline{XY}$

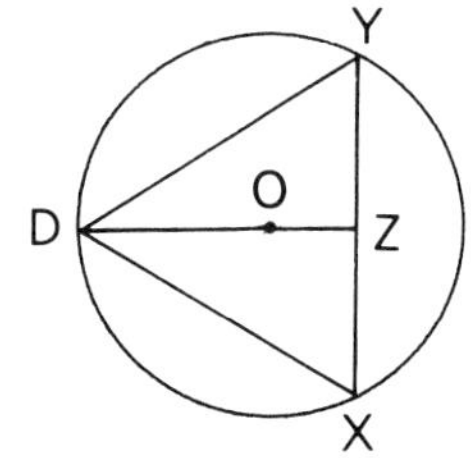

13 Given: $\angle WXY \cong \angle ZYX$
$\overline{WX} \cong \overline{ZY}$
Conclusion: $\overline{WR} \cong \overline{RZ}$

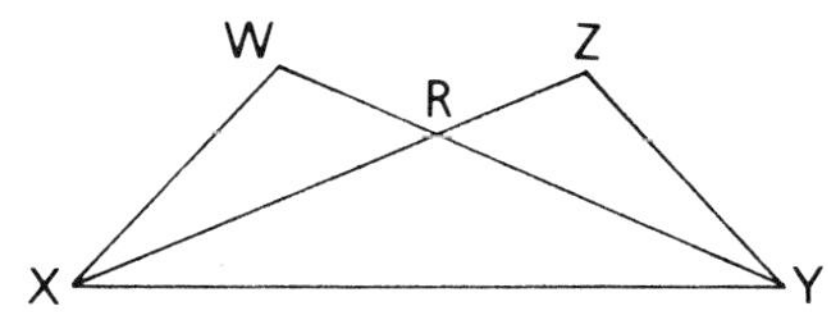

14 Given: $\angle ABC \cong \angle ACB$
$\angle 1 \cong \angle 2$
E is the midpt. of $\overline{BC}$
Prove: $\overleftrightarrow{AD}$ passes through E

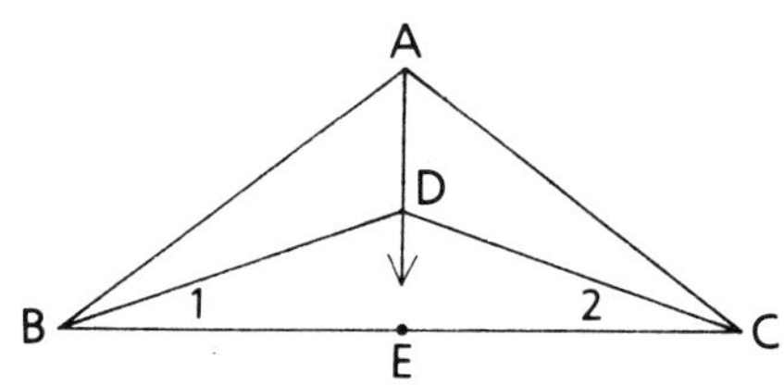

15 Given: $\overline{AB} \cong \overline{AF}$
$\overline{BD} \cong \overline{DF}$
$\angle 1 \cong \angle 2$
Conclusion: $\overline{AD} \perp \overline{CE}$

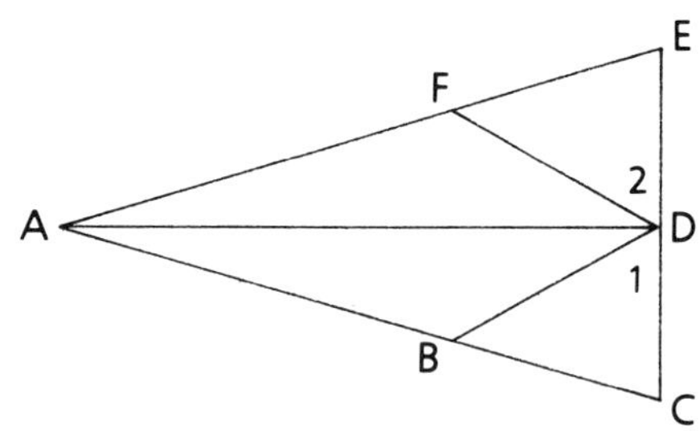

16 Given: $\odot O$
$\overline{PQ} \cong \overline{PR}$
$\overline{PM}$ is the median to $\overline{QR}$
Prove: $\overleftrightarrow{PM}$ passes through O

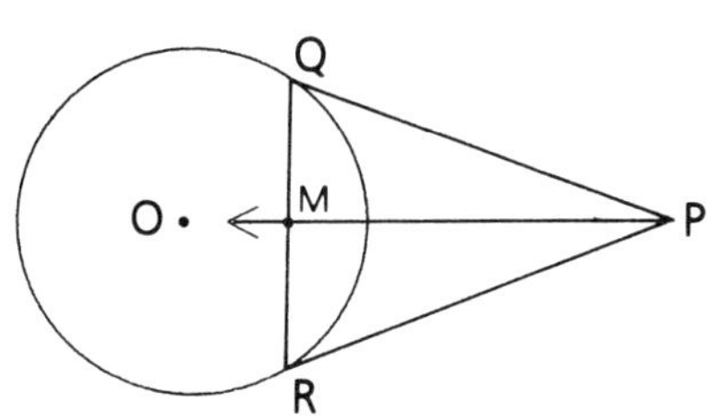

17 Given: $\overleftrightarrow{AD} \perp$ bis $\overline{BC}$

Conclusion: $\angle 1 \cong \angle 2$

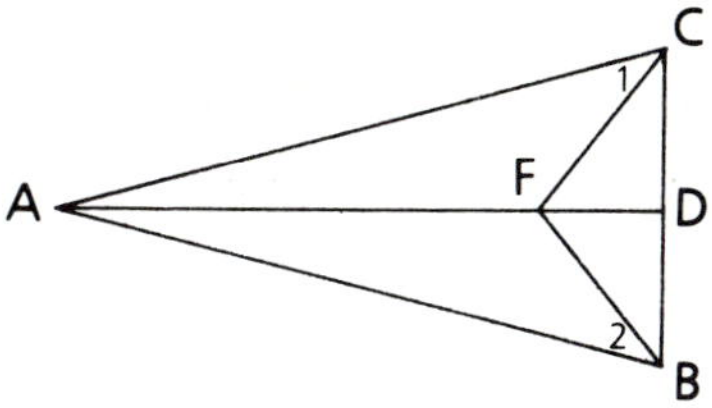

Chapter 4 Review Problems C

18 Given: $\overline{AB} \cong \overline{AF}$

$\overline{BC} \cong \overline{FE}$

Conclusion: $\overline{CD} \cong \overline{DE}$

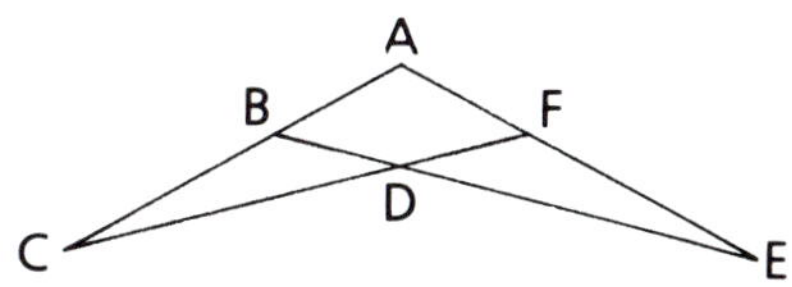

19 Prove: If the bisector of an angle whose vertex lies on a circle passes through the center of the circle, then it will be the perpendicular bisector of the segment joining the points of intersection of the sides of the angle with the circle.

20 Given: $\overline{CD}$ and $\overline{BE}$ are the medians to the legs of isosceles $\triangle ABC$

Prove: $\overleftrightarrow{AG} \perp$ bis $\overline{BC}$

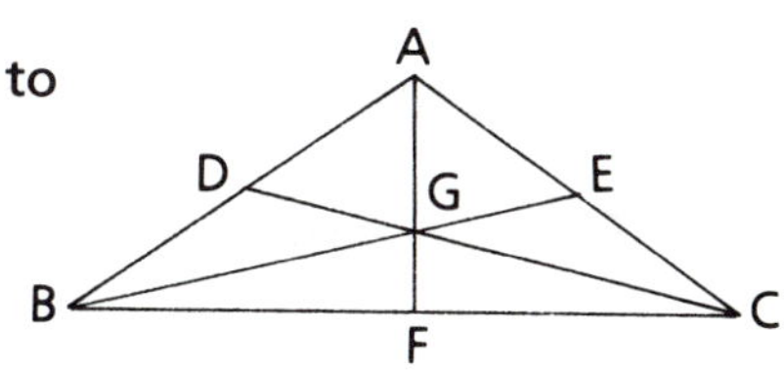

21 Given: $\odot$ O, $\overline{PQ} \cong \overline{PR}$

$\overrightarrow{PS}$ bisects $\angle QPR$

Prove: $\overleftrightarrow{PS}$ passes through the center of the circle

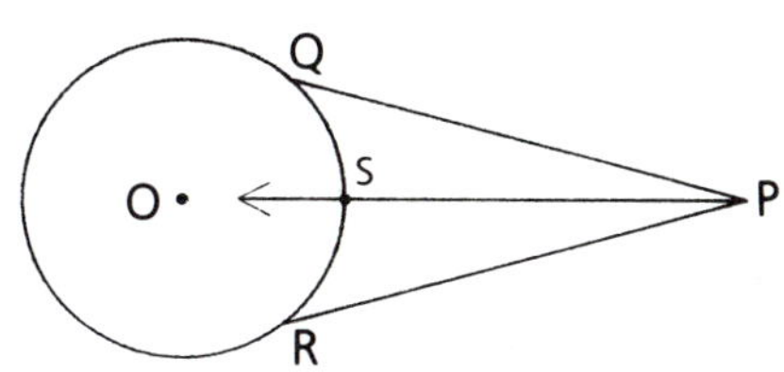

22 Given: $\overline{AB} \cong \overline{AC}$

$\overline{BF} \cong \overline{FC}$

$\angle BAE \cong \angle CAD$

Prove: $\overleftrightarrow{AF} \perp \overleftrightarrow{DE}$

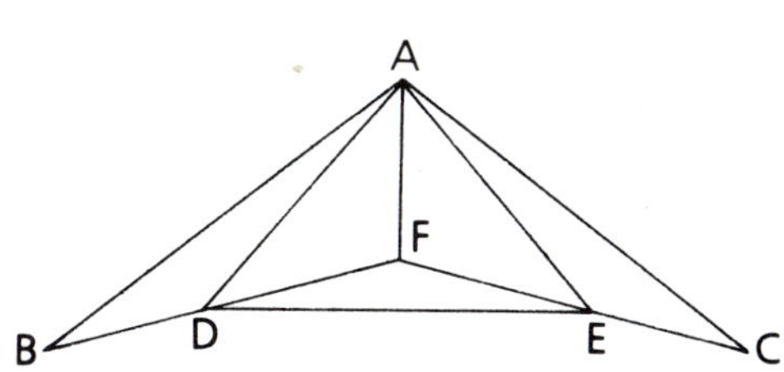

Chapter 5

PARALLEL LINES AND RELATED FIGURES

Section 5.1 Introduction to Parallel Lines

In this section we discuss:

A PLANES, COPLANAR, NON-COPLANAR
B TRANSVERSALS, INTERIOR AND EXTERIOR REGIONS
C ANGLE PAIRS RELATED TO TRANSVERSALS
D PARALLEL LINES

5.1 (A) Planes

In order to explain parallel lines adequately, we must first acquaint you with planes.

D **A PLANE is a surface such that if any two points on the surface are connected by a line, all points of the line will lie on the surface.**

A **PLANE** has only two dimensions—length and width. Both the length and width are infinite. A **PLANE** has no thickness.

D **If points, lines, segments, etc., lie in the same plane, we call them COPLANAR.**
Similarly, points, lines, segments, etc., that do not all lie in the same plane are called NON-COPLANAR.

There is more about **PLANES** in Chapter 6.

5.1 (B) Transversals, Interior and Exterior Regions

In the figure, line t is a **TRANSVERSAL** for lines a and b.

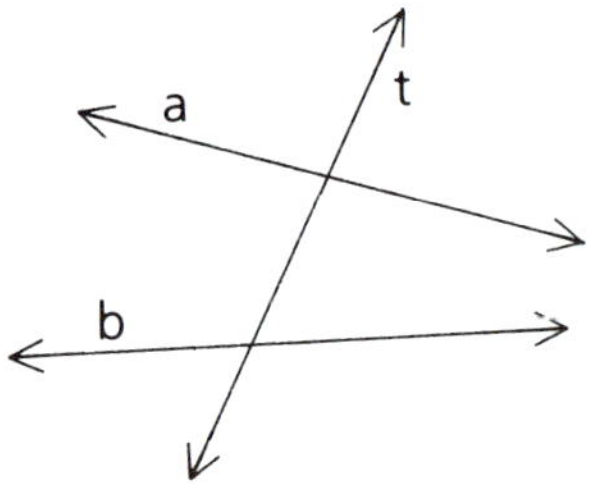

D **In a plane a TRANSVERSAL is a line that intersects two other lines in two distinct points.**

The region between lines d and e is the **INTERIOR REGION.**

The region outside lines d and e is the **EXTERIOR REGION.**

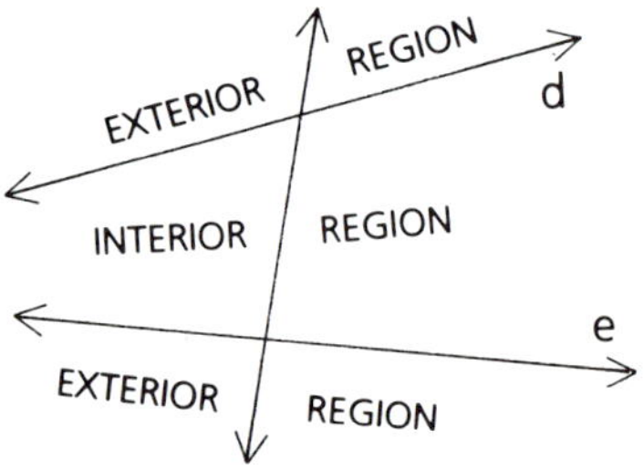

Lines f and g, being cut by **TRANSVERSAL** h, provide another example of regions.

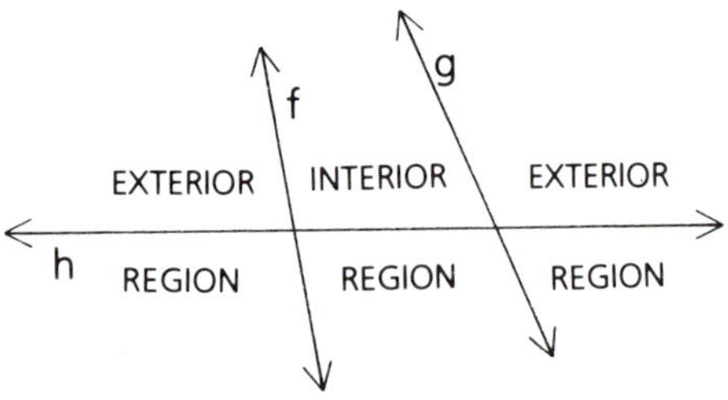

5.1(C) Angle Pairs Related To Transversals

$\overleftrightarrow{AB}$ and $\overleftrightarrow{CD}$ are cut by transversal $\overleftrightarrow{EF}$

The two pairs of ALTERNATE INTERIOR ANGLES are 2 and 7, 6 and 3.

The two pairs of ALTERNATE EXTERIOR ANGLES are 1 and 8, 5 and 4.

The four pairs of CORRESPONDING ANGLES are 1 and 3, 2 and 4, 5 and 7, 6 and 8.

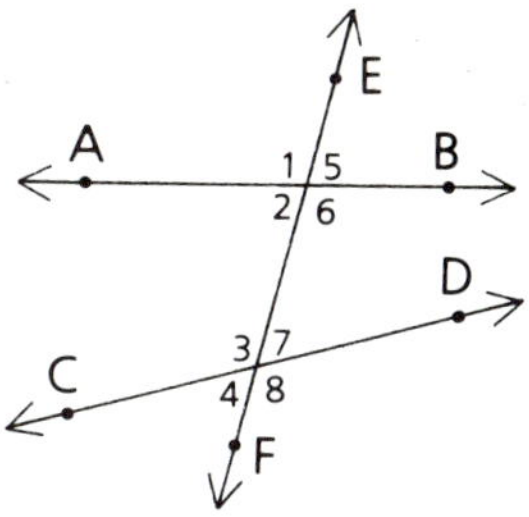

$\overleftrightarrow{GH}$ and $\overleftrightarrow{JK}$ are cut by transversal $\overleftrightarrow{MO}$.

The ALTERNATE INTERIOR ANGLES are b and g, f and c.

The ALTERNATE EXTERIOR ANGLES are a and h, e and d.

The CORRESPONDING ANGLES are a and c, b and d, e and g, f and h.

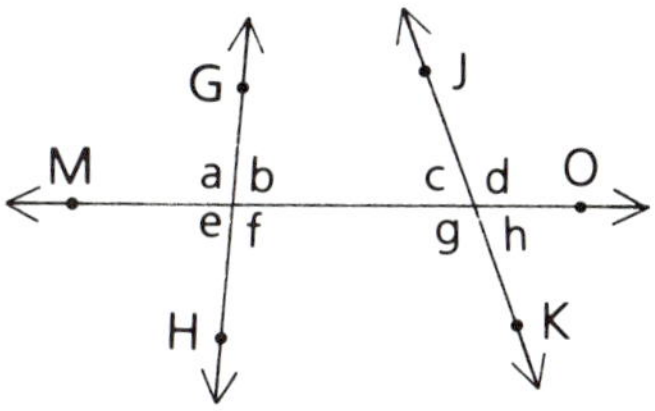

D **ALTERNATE INTERIOR ANGLES are two angles formed by a transversal intersecting two lines. Both angles must lie in the interior region, must lie on alternate sides of the transversal, and must have different vertices.**

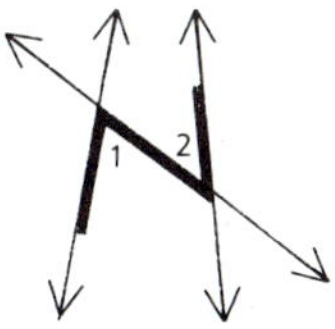

Look for N or Z.

D **ALTERNATE EXTERIOR ANGLES are two angles formed by a transversal intersecting two lines. Both angles must lie in the exterior region, must lie on alternate sides of the transversal, and must have different vertices.**

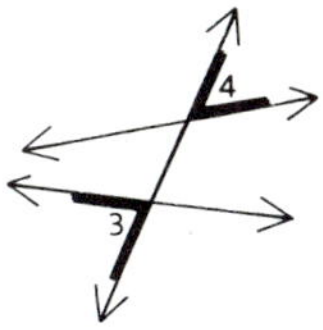

D **CORRESPONDING ANGLES are two angles formed by a transversal intersecting two lines. One angle must lie in the interior region, and the other must lie in the exterior region. Both angles lie on the same side of the transversal but have different vertices.**

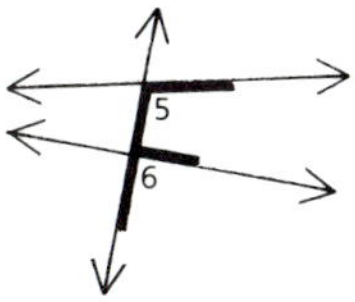

Look for F.

It is important to be able to recognize these angles when figures are more complicated and only segments are showing.

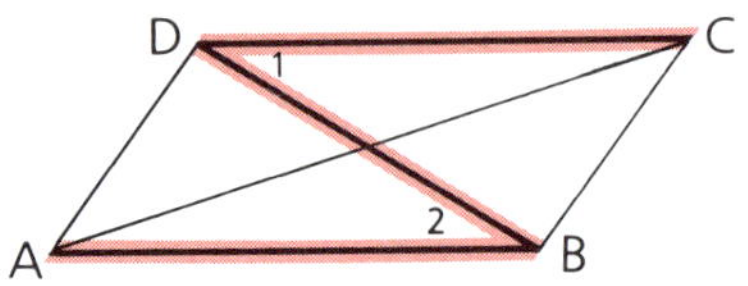

$\angle 1$ and $\angle 2$ are **ALTERNATE INTERIOR ANGLES** for $\overleftrightarrow{BD}$ as the transversal of $\overleftrightarrow{CD}$ and $\overleftrightarrow{AB}$.

Note: $\angle 1$ and $\angle 2$ are *not* **ALTERNATE INTERIOR ANGLES** for $\overleftrightarrow{BD}$ as the transversal of $\overleftrightarrow{AD}$ and $\overleftrightarrow{BC}$. ($\overrightarrow{DA}$ and $\overrightarrow{BC}$ are *not* sides of $\angle 1$ and $\angle 2$.)

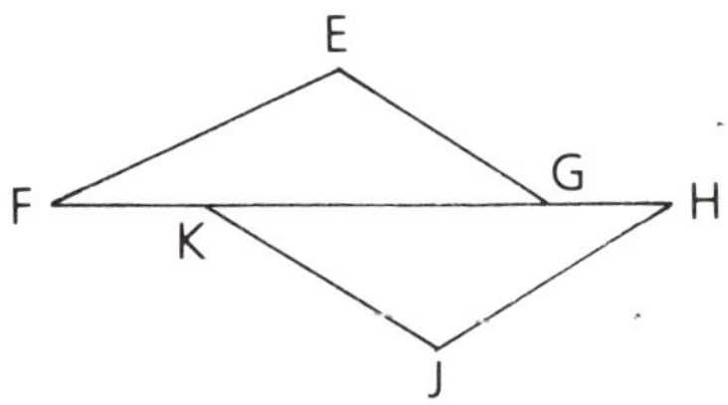

$\angle$F and $\angle$H are **ALTERNATE INTERIOR ANGLES.** Can you name the lines and their transversal?

$\angle$s FKJ and EGH are **ALTERNATE EXTERIOR ANGLES.** Can you name the lines and their transversal?

5.1 (D) Parallel Lines

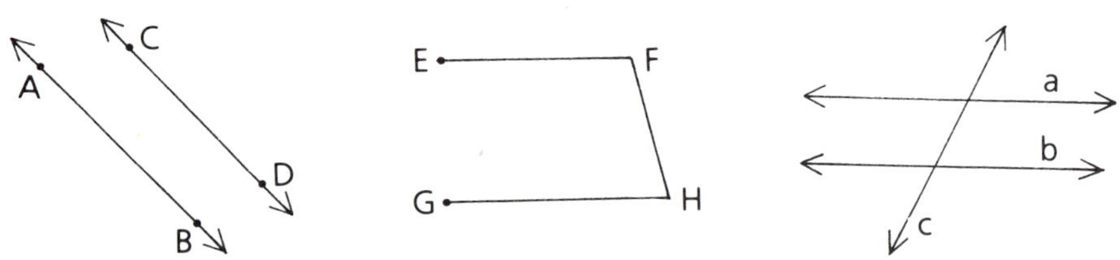

Above are three examples of **PARALLEL (||) LINES.** We write $\overleftrightarrow{AB} \parallel \overleftrightarrow{CD}$, $\overline{EF} \parallel \overline{GH}$, and a || b.

D **PARALLEL LINES are two coplanar lines that do NOT INTERSECT.**

We shall also refer to parallel segments or rays if they are parts of parallel lines. For example, we may say that in the previous diagram $\overline{AB} \parallel \overline{CD}$.

There are many lines that do not intersect and are *not* parallel. To be **PARALLEL,** lines *must* be coplanar. In Chapter 6, lines that are non-coplanar and do not intersect are given a name.

Section 5.1 Problem Set

1 a Name all pairs of alternate interior angles.

b Name all pairs of alternate exterior angles.

c Name all pairs of corresponding angles.

d Name all pairs of interior angles on the same side of the transversal.

e Name all pairs of exterior angles on the same side of the transversal.

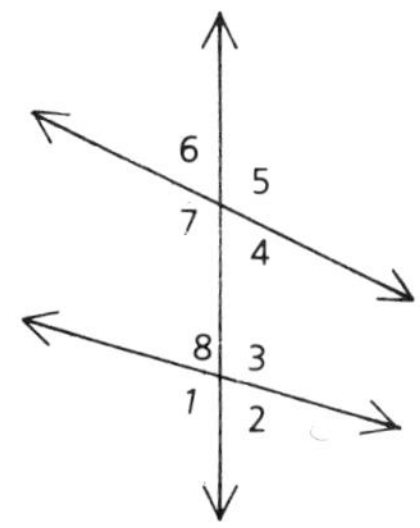

2 a Which line is the transversal?

b Name all pairs of alternate interior angles.

c Name all pairs of alternate exterior angles.

d Name all pairs of corresponding angles.

e Name all pairs of interior angles on the same side of the transversal.

f Name all pairs of exterior angles on the same side of the transversal.

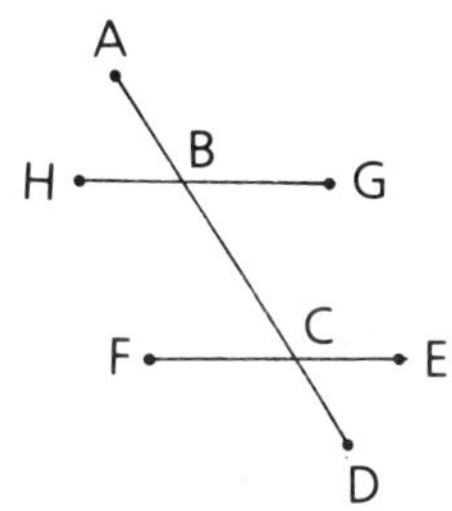

3 a For which pair of lines are ∠s 1 and 4 a pair of alternate interior angles?

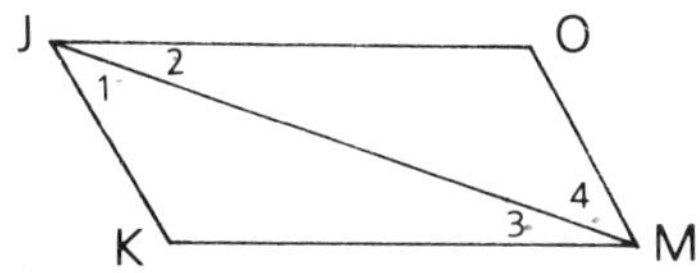

b For which pair of lines are ∠s 2 and 3 a pair of alternate interior angles?

c How many transversals of $\overleftrightarrow{JO}$ and $\overleftrightarrow{KM}$ are shown?

Section 5.2 Theorems Derived from Parallel Lines

In this section we present the following concepts related to parallel lines:

A PAI

B A GENERALIZATION ABOUT ANGLES AND PARALLEL LINES

C FIVE THEOREMS ABOUT PARALLEL LINES

5.2 (A) PAI

We assume the following:

P **If two parallel lines are cut by a transversal, each pair of alternate interior angles are congruent. (PAI)**

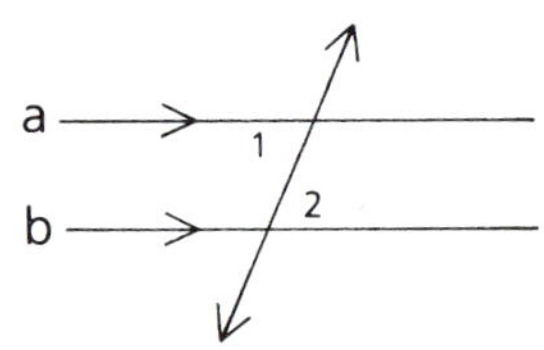

Given: line a || line b

Prove: $\angle 1 \cong \angle 2$

Notice the special tick marks (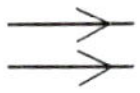) to designate parallel lines.

5.2 (B) A Generalization about Angles and Parallel Lines

Now we can prove the following:

IF TWO PARALLEL LINES ARE CUT BY A TRANSVERSAL, THEN ANY PAIR OF ANGLES FORMED ARE EITHER CONGRUENT OR SUPPLEMENTARY.

The proof is done algebraically by letting x be the measure of any one of the angles. In each diagram, a || b.

Diagram 1

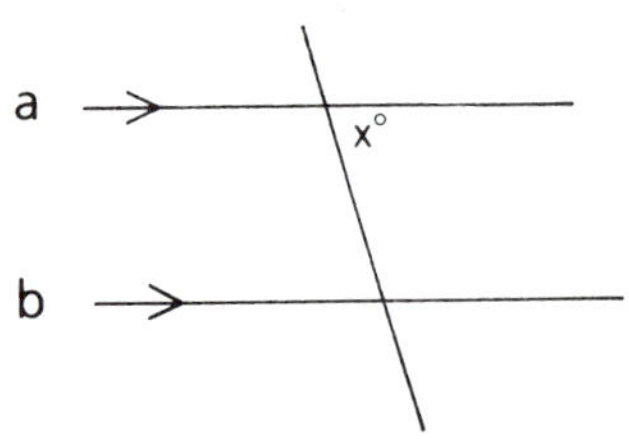

Diagram 2

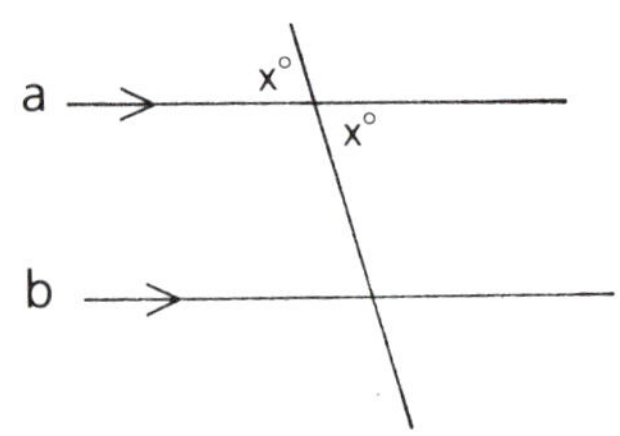

due to vertical $\angle$s

Diagram 3

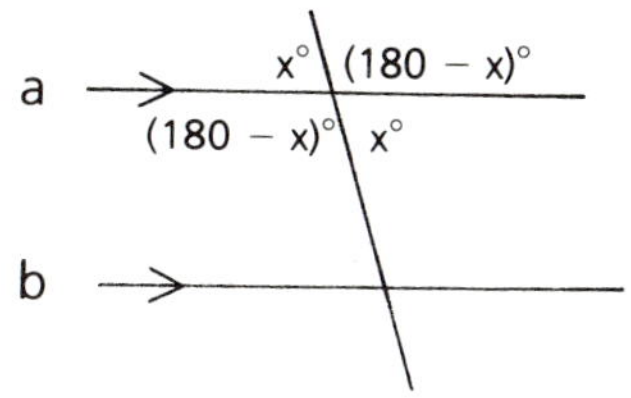

due to straight ∠s

Diagram 4

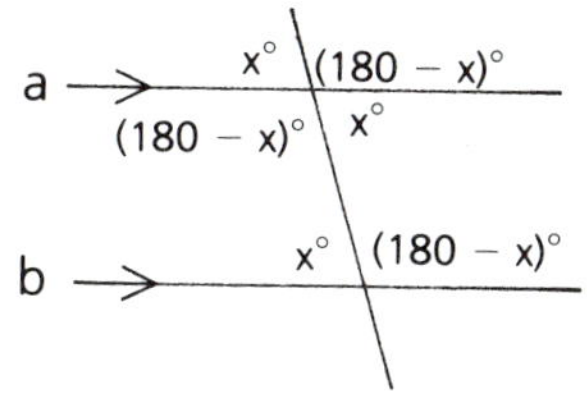

due to PAI

Diagram 5

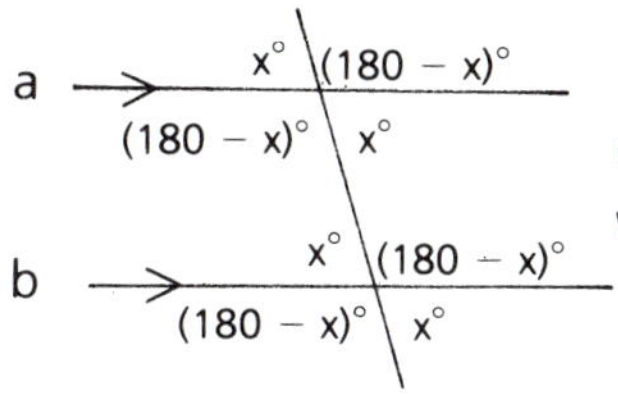

due to vertical ∠s

5.2 (C) Five Theorems about Parallel Lines

Diagram 5 above is the basis for each of the following five theorems.

Theorem 25

T **If two parallel lines are cut by a transversal, each pair of alternate exterior angles are congruent. (PAE)**

Given: a || b

Prove: ∠1 ≅ ∠8

Proof: See Diagram 5

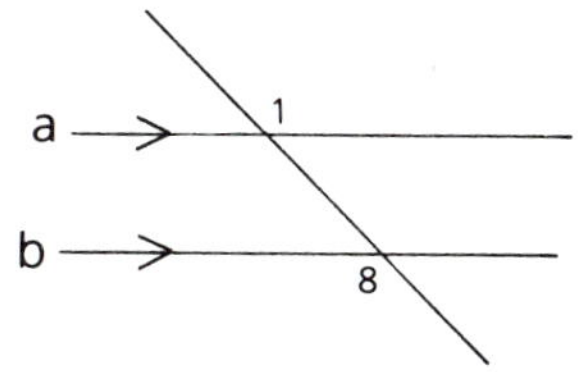

Theorem 26

T **If two parallel lines are cut by a transversal, each pair of corresponding angles are congruent. (PCA)**

Given: a || b

Prove: ∠1 ≅ ∠5

Proof: See Diagram 5

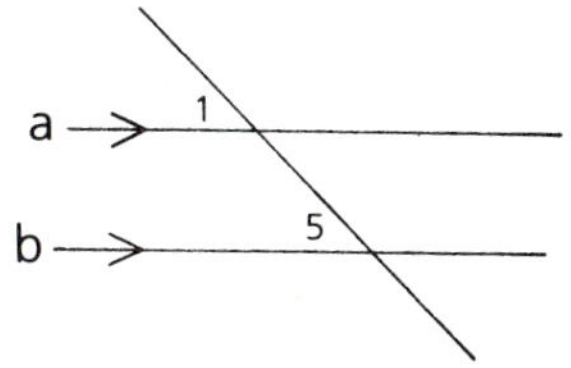

Theorem 27

T **If two parallel lines are cut by a transversal, each pair of interior angles on the same side of the transversal are supplementary.**

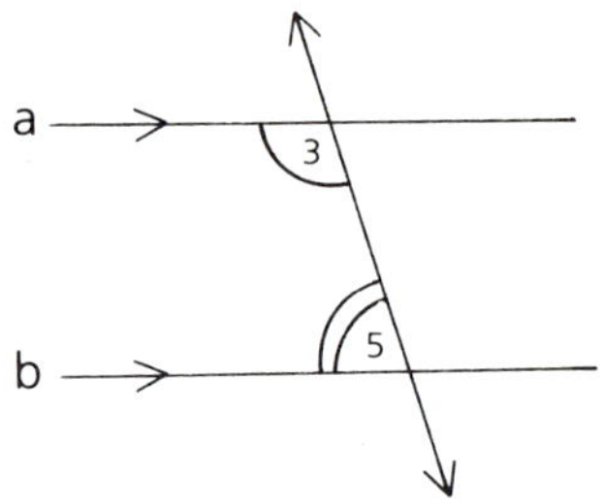

Given: a || b

Prove: $\angle 3$ is supp. to $\angle 5$

Proof: See Diagram 5

Theorem 28

T **If two parallel lines are cut by a transversal, each pair of exterior angles on the same side of the transversal are supplementary.**

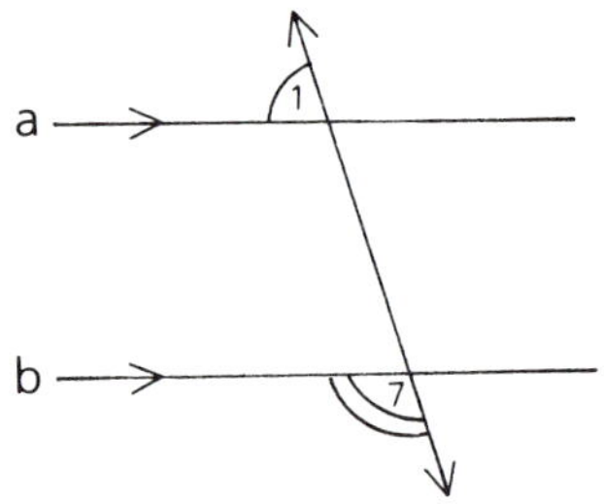

Given: a || b

Prove: $\angle 1$ is supp. to $\angle 7$

Proof: See Diagram 5

Theorem 29

T **In a plane, if a line is perpendicular to one of two parallel lines, it is perpendicular to the other.**

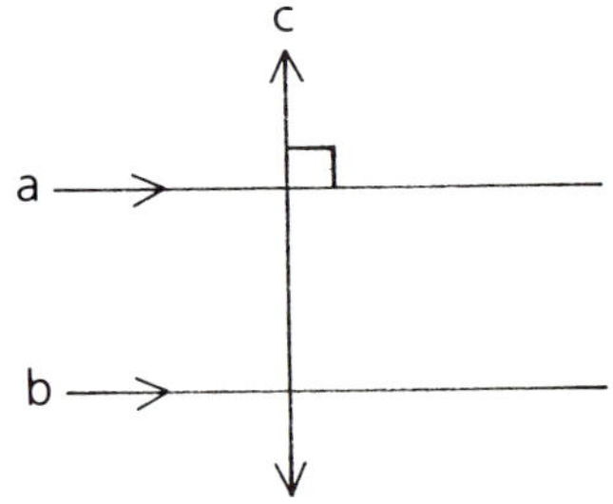

Given: a || b
c $\perp$ a

Prove: c $\perp$ b

Proof: See Diagram 5 and let $x = 90$

In summary, if two **PARALLEL** lines are cut by a transversal, then:

each pair of **ALTERNATE INTERIOR** angles are **CONGRUENT.**

each pair of **ALTERNATE EXTERIOR** angles are **CONGRUENT.**

each pair of **CORRESPONDING** angles are **CONGRUENT.**

each pair of **INTERIOR** angles on the **SAME SIDE** of the transversal are **SUPPLEMENTARY.**

each pair of **EXTERIOR** angles on the **SAME SIDE** of the transversal are **SUPPLEMENTARY.**

Section 5.2 Sample Problems

1 If c ∥ d, find m∠1.

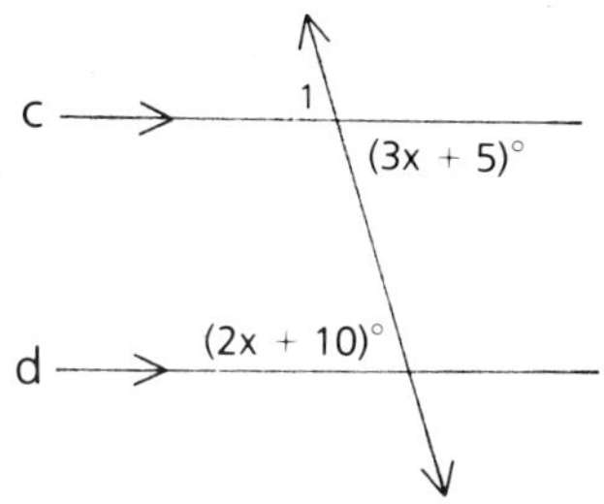

Solution:

From PAI, $3x + 5 = 2x + 10$

$$x + 5 = 10$$
$$x = 5$$
$$(3x + 5)^\circ = 20^\circ$$

Because of vertical angles, $\angle 1 = 20^\circ$

2 Given: $\overline{FA} \parallel \overline{DE}$

$\overline{FA} \cong \overline{DE}$

$\overline{AB} \cong \overline{CD}$

Prove: $\angle F \cong \angle E$

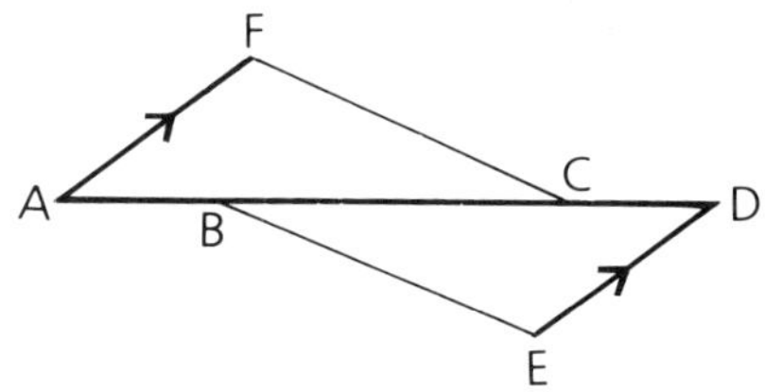

1 $\overline{FA} \parallel \overline{DE}$	1 Given
2 $\angle A \cong \angle D$	2 PAI
3 $\overline{FA} \cong \overline{DE}$	3 Given
4 $\overline{AB} \cong \overline{CD}$	4 Given
5 $\overline{AC} \cong \overline{BD}$	5 Addition Property ($\overline{BC}$ to step 4)
6 $\triangle FAC \cong \triangle EDB$	6 SAS (3, 2, 5)
7 $\angle F \cong \angle E$	7 CPCTC

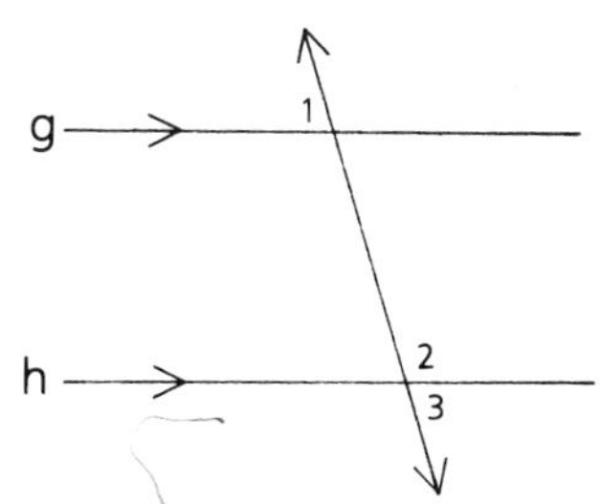

3 Given: g ∥ h

Prove: ∠1 is supp. to ∠2

1 g ∥ h	1 Given
2 ∠2 is supp. to ∠3	2 If two angles form a straight angle, they are supplementary
3 ∠1 ≅ ∠3	3 PAE
4 ∠1 is supp. to ∠2	4 Substitution

Section 5.2 Problem Set A

1 Given: $a \parallel b$
30° angle as shown
Complete the diagram with the measures of the seven remaining angles.

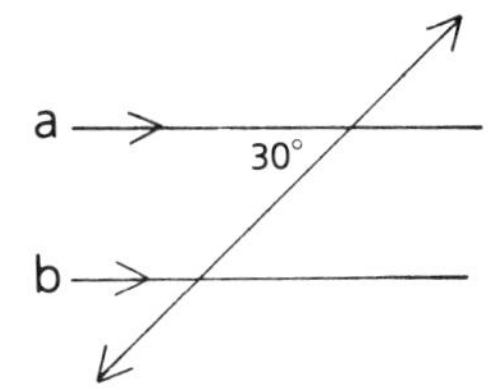

2 Given: $\overline{AB} \cong \overline{DC}$
$\overline{AB} \parallel \overline{DC}$
Conclusion: $\overline{AD} \cong \overline{BC}$

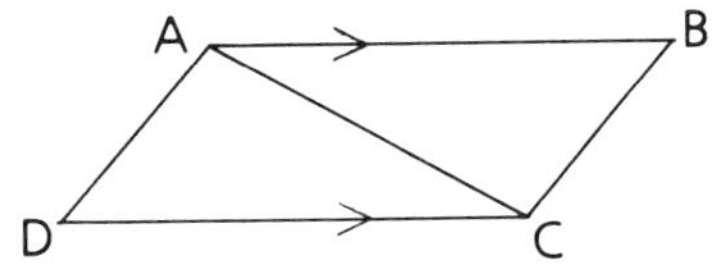

3 Given: $\overline{EF} \parallel \overline{GH}$
$\overline{EF} \cong \overline{GH}$
Conclusion: $\overline{EJ} \cong \overline{JH}$

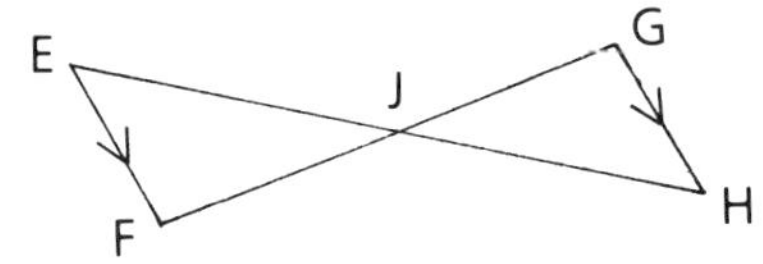

4 Given: $g \parallel h$
$g \parallel j$
Conclusion: $\angle 2 \cong \angle 3$

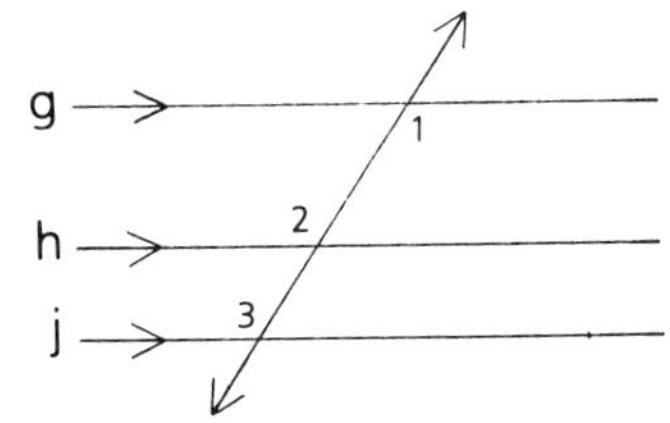

5 Find $m\angle 1$

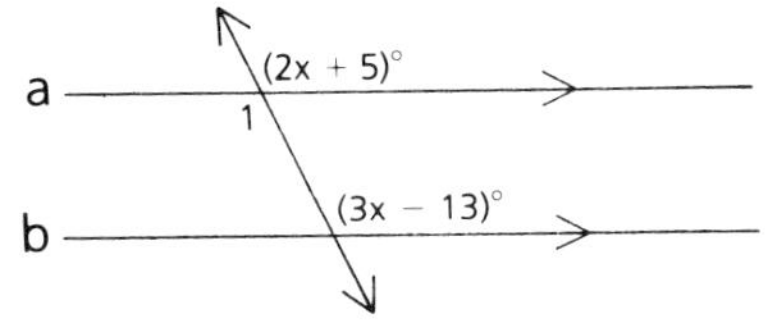

6 $c \parallel d$
Find $m\angle 2$

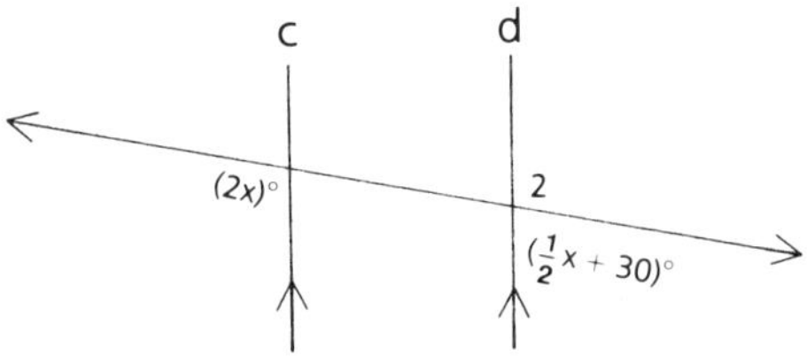

7 Is $e \parallel f$?

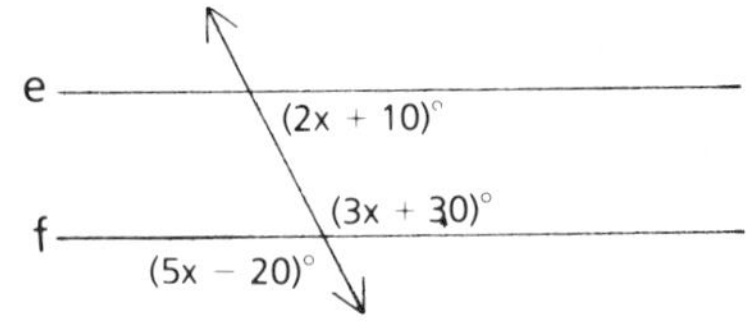

8 $\overline{ST} \parallel \overline{XW}$
$\overrightarrow{ST}$ bisects $\angle VSW$
Find: $m\angle X$
$m\angle W$

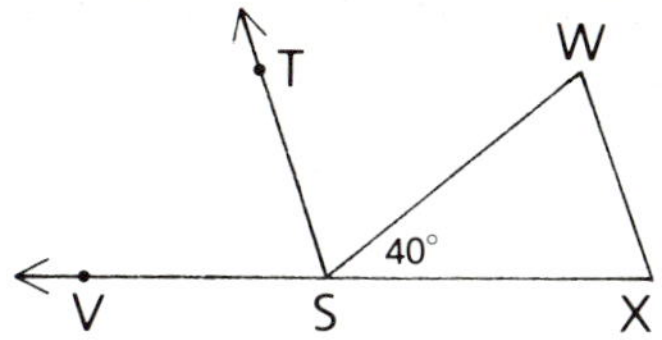

Section 5.2 Problem Set B

9 Given: $\overline{AB} \parallel \overline{CD}$
$\overrightarrow{BC} \parallel \overrightarrow{DE}$
Conclusion: $\angle B \cong \angle D$

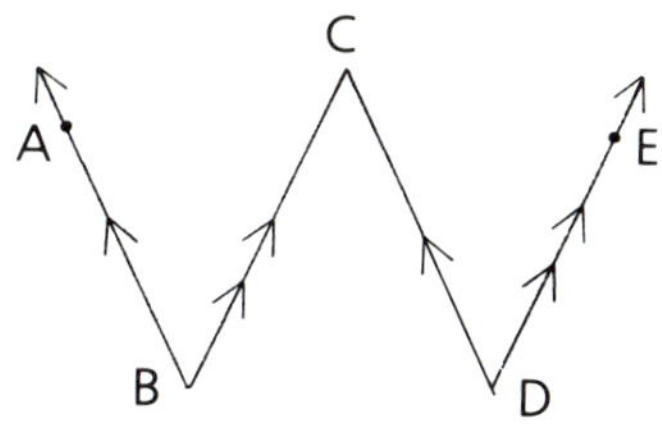

10 Given: $\overleftrightarrow{FH} \parallel \overleftrightarrow{JM}$
$\angle 1 \cong \angle 2$
$\overline{FH} \cong \overline{JM}$
Prove: $\overline{GJ} \cong \overline{HK}$

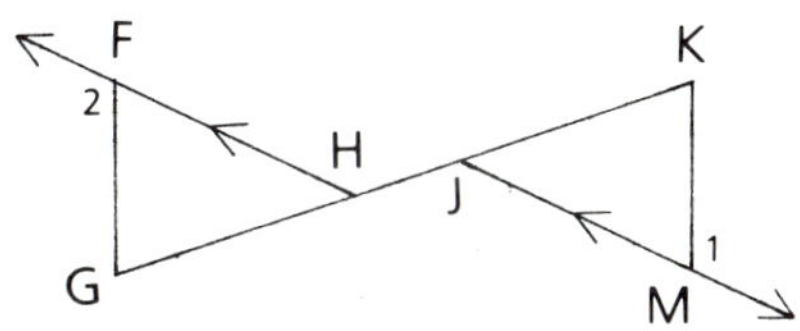

11 Given: $\odot O$
$\overline{NR} \parallel \overline{PS}$
Prove: $\triangle OSP$ is isosceles

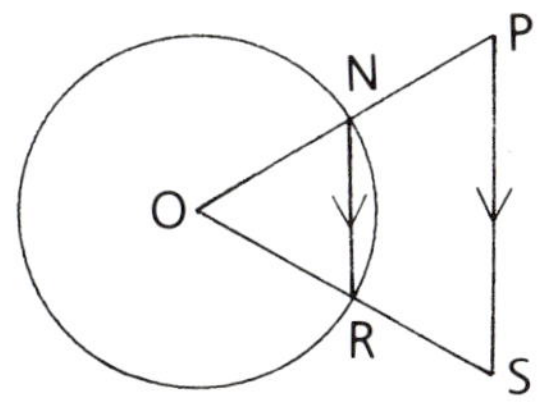

12 Given: $\angle ADC \cong \angle CBA$
$\overleftrightarrow{DC} \parallel \overleftrightarrow{AB}$
Conclusion: $\overline{DC} \cong \overline{AB}$

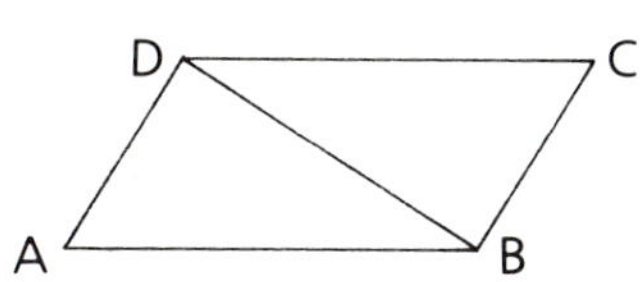

13 Given: $\overline{EH} \parallel \overline{FG}$
$\overline{EH} \cong \overline{FG}$
Prove: $\overline{HF}$ and $\overline{EG}$ bisect each other

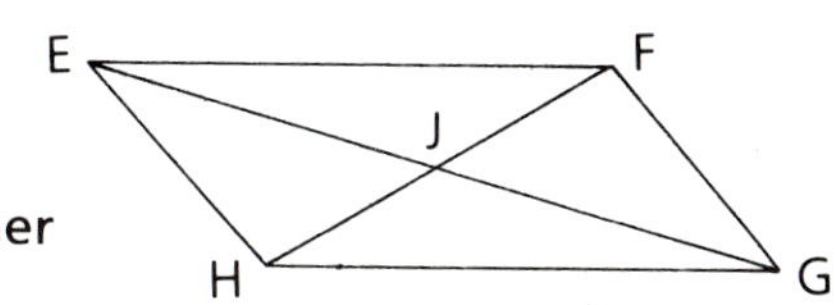

14 Given: $\overline{AD} \parallel \overline{BC}$

Name all pairs of angles that must be congruent.

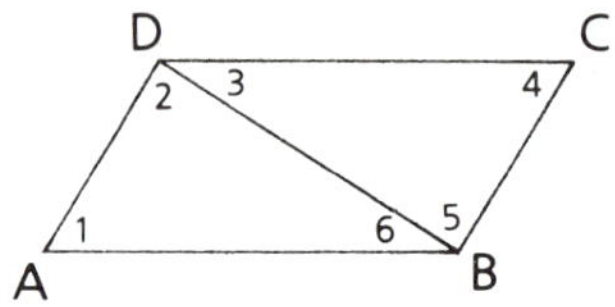

15 If $\overleftrightarrow{DA} \parallel \overleftrightarrow{BC}$, is $\triangle ABC$ equilateral?

Find: $m\angle DAB$

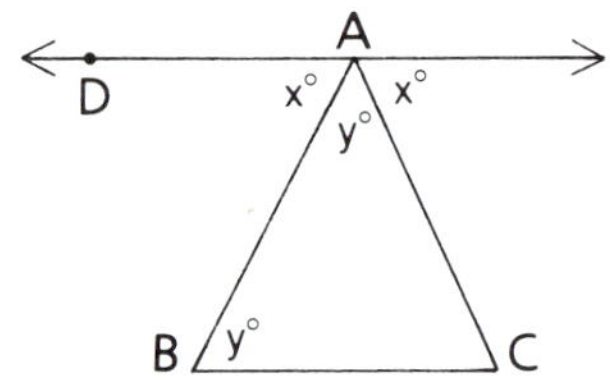

16 Prove that the opposite sides of a parallelogram* are congruent.

17 Prove that the opposite angles of a parallelogram* are congruent.

18 $\overleftrightarrow{JK} \parallel \overleftrightarrow{OM}$, $\overline{JO} \cong \overline{KM}$

$\overleftrightarrow{JO} \parallel \overleftrightarrow{KN}$, $MN = 6$,

$JO = 3x + 4$, $KN = 2x + 7$

Find the perimeter of $\triangle KNM$

(Hint: Use problem 16)

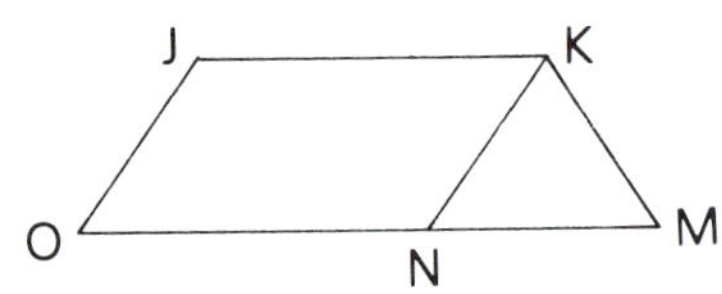

Section 5.2 Problem Set C

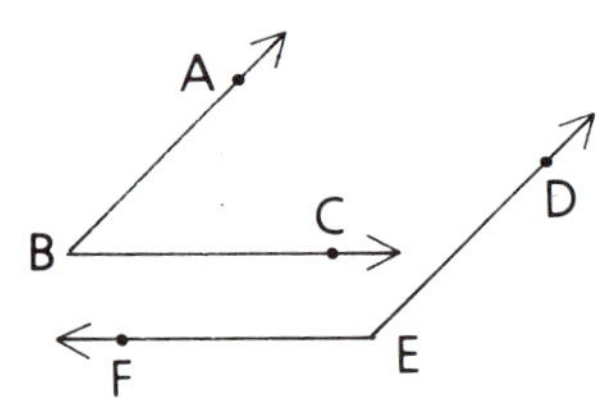

19 Write a paragraph proof that if $\overline{AB} \parallel \overline{DE}$ and $\overline{BC} \parallel \overline{FE}$, then $\angle B$ is supp. to $\angle E$.

20 If two parallel lines are cut by a transversal, eight angles are formed (not counting the straight angles).

a How many *pairs* of angles are formed?

b If a pair of angles is chosen at random, what is the probability that it will either be a pair of alternate interior angles or a pair of alternate exterior angles or a pair of corresponding angles?

c If a pair of angles is chosen at random, what is the probability that the angles are supplementary?

* A parallelogram is a four-sided figure in which *both* pairs of opposite sides are parallel.

Section 5.3 Ways To Prove That Lines Are Parallel

In this section we discuss:

A THE PARALLEL POSTULATE

B WAYS TO PROVE THAT LINES ARE PARALLEL

5.3 (A) The Parallel Postulate

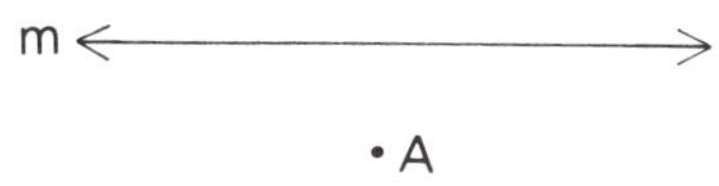

In the figure above, how many lines parallel to line m do you believe can be drawn through point A?

Most people believe that the answer is *exactly one* as shown in the figure below:

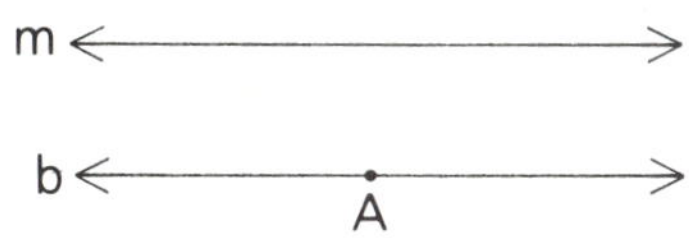

P Through a given point not on a given line, there exists exactly one line parallel to the given line. This assumption is THE PARALLEL POSTULATE.

The geometry of this book is called Euclidean geometry because its structure, including the Parallel Postulate, is essentially that of the Greek mathematician, Euclid (300 B.C.). For centuries, eager mathematicians argued that the Parallel Postulate could be a theorem, provable from the definitions and other postulates. But nobody could prove it. Some even began to think it might be false and replaced the Parallel Postulate with one of their own.

Thus, Janos Bolyai (1802–1860) and Nicolai Ivanovich Lobachevsky (1792–1856) each developed a non-Euclidean geometry using this postulate:

> Through a given point not on a given line, there are *many lines* in the same plane that are parallel to (do not intersect) the given line.

In contrast, Bernard Riemann (1826–1866) developed a non-Euclidean geometry based on this postulate:

> Through a given point not on a given line, there are *no lines* that are parallel to the given line.

In each case, the geometry developed is logically valid, because its conclusions are the same whatever approach is taken within the system. For instance, you have seen classmates make different, but correct, proofs of the same conclusion. What would you think of a structure in which that does not happen?

What is the true geometry of the universe? At present, all we know is that, according to Albert Einstein, objects traveling at the speed of light behave differently. Astronomers and nuclear physicists, who work with such objects, say that the geometry of the universe that would be seen from those objects is probably more non-Euclidean than Euclidean. However, when dealing with automobiles, satellites, and baseballs, astronomers and nuclear physicists use Euclidean geometry.

5.3 (B) Ways To Prove That Lines Are Parallel

Theorem 30

T **If two lines are cut by a transversal such that a pair of alternate interior angles are congruent, the lines are parallel. (AIP)**

Given: $\angle 3 \cong \angle 6$

Prove: $a \parallel b$

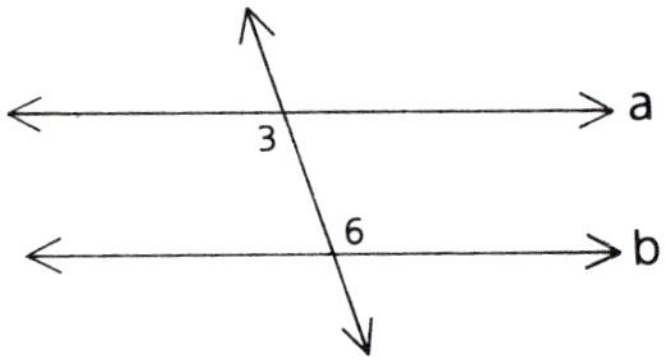

Theorem 31

T **If two lines are cut by a transversal such that a pair of alternate exterior angles are congruent, the lines are parallel. (AEP)**

Given: $\angle 1 \cong \angle 8$

Prove: $a \parallel b$

This can be proved by use of AIP.

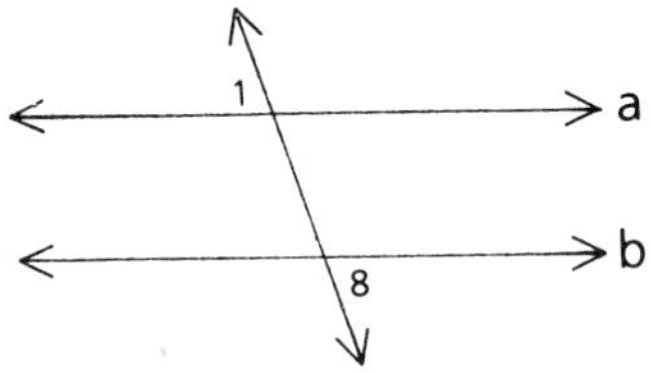

Theorem 32

T **If two lines are cut by a transversal such that a pair of corresponding angles are congruent, the lines are parallel. (CAP)**

Given: $\angle 2 \cong \angle 6$

Prove: $a \parallel b$

This can be proved by use of AIP.

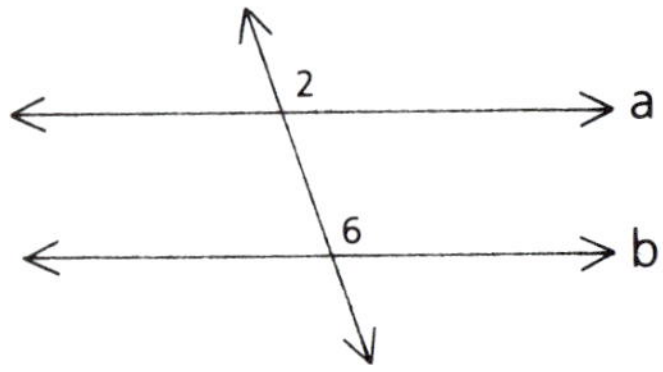

Theorem 33

T **If two lines are cut by a transversal such that a pair of interior angles on the same side of the transversal are supplementary, the lines are parallel.**

Given: ∠3 is supp. ∠5

Prove: a || b

This can be proved by use of AIP.

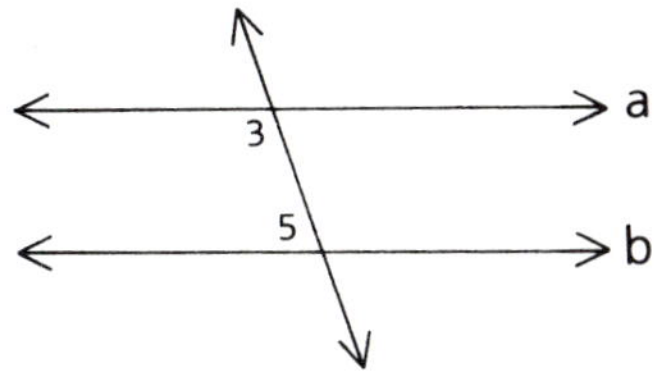

Theorem 34

T **If two lines are cut by a transversal such that a pair of exterior angles on the same side of the transversal are supplementary, the lines are parallel.**

Given: ∠1 is supp. ∠7

Prove: a || b

This can be proved by use of AIP.

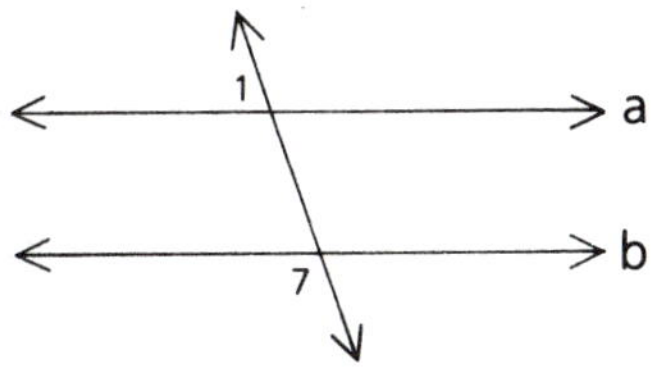

Theorem 35

T **In a plane if two lines are perpendicular to a third line, then they are parallel.**

Given: a ⊥ c, and b ⊥ c

Prove: a || b

This can be proved by use of AIP.

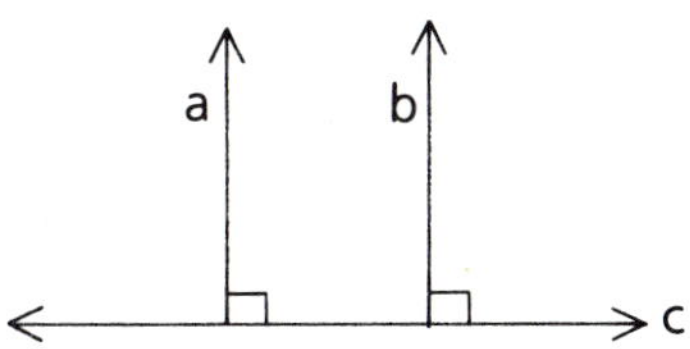

Theorem 36

T **If two lines are each parallel to a third line, they are parallel to each other. (TRANSITIVE PROPERTY OF PARALLEL LINES)**

Given: a || b, and b || c

Prove: a || c

By use of **PAI** and **AIP** you can prove that **THEOREM 36** is true when all three lines lie in the same plane. It can be shown that the theorem holds for lines in space.

Section 5.3 Sample Problems

1 Prove THEOREM 35

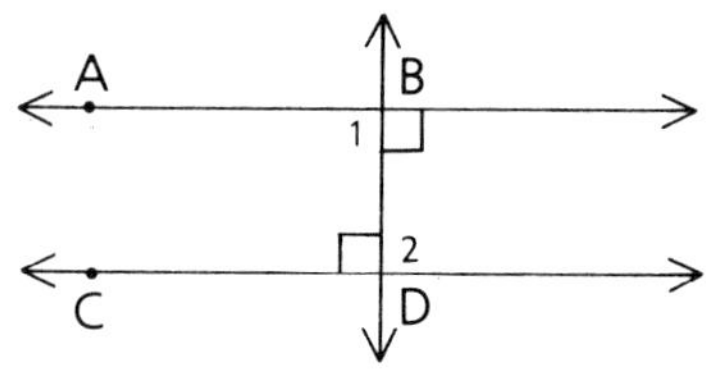

Given: $\overleftrightarrow{AB} \perp \overleftrightarrow{BD}$ and $\overleftrightarrow{CD} \perp \overleftrightarrow{BD}$

Prove: $\overleftrightarrow{AB} \parallel \overleftrightarrow{CD}$

1 $\overleftrightarrow{BD} \perp \overleftrightarrow{AB}$	1 Given
2 $\angle 1$ is a right $\angle$	2 $\perp$ lines form right angles
3 $\overleftrightarrow{BD} \perp \overleftrightarrow{CD}$	3 Given
4 $\angle 2$ is a right $\angle$	4 Same as 2
5 $\angle 1 \cong \angle 2$	5 Right angles are $\cong$
6 $\overleftrightarrow{AB} \parallel \overleftrightarrow{CD}$	6 AIP

2 Given: $\overline{JM}$ and $\overline{PO}$ bisect each other

Conclusion: $\overline{JP} \parallel \overline{OM}$

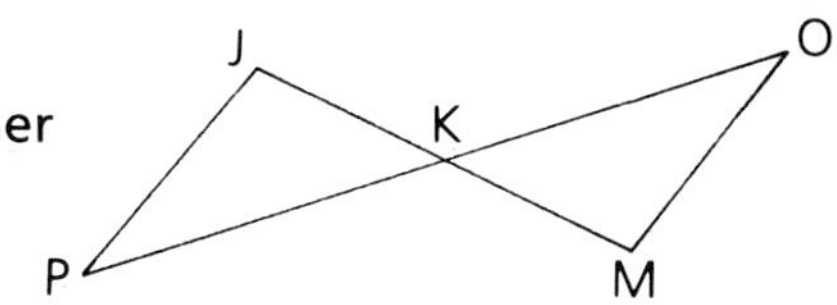

1 $\overline{JM}$ bisects $\overline{PO}$	1 Given
2 $\overline{PK} \cong \overline{OK}$	2 If a segment is bisected, it is divided into two $\cong$ segments
3 $\overline{PO}$ bisects $\overline{JM}$	3 Given
4 $\overline{JK} \cong \overline{MK}$	4 Same as 2
5 $\angle JKP \cong \angle MKO$	5 Vertical angles are congruent
6 $\triangle JKP \cong \triangle MKO$	6 SAS (2, 5, 4)
7 $\angle J \cong \angle M$	7 CPCTC
8 $\overline{JP} \parallel \overline{OM}$	8 AIP

3 The CROOK PROBLEM

If a $\parallel$ b, find m$\angle 1$

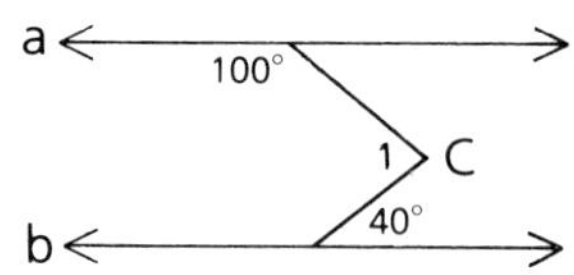

Solution:

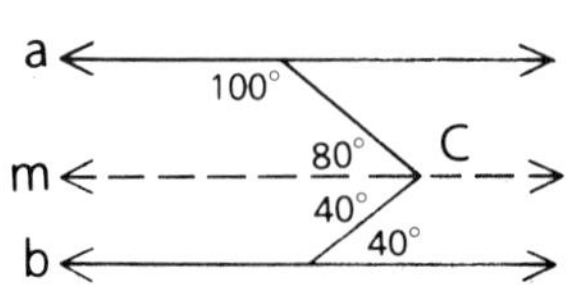

Using the PARALLEL POSTULATE, draw m $\parallel$ a. Use of three theorems about $\parallel$ lines leads to $\angle 1 = 120°$.

4 Given: Figure ABCD* with $\overline{AD} \parallel \overline{BC}$
$\overline{AB} \cong \overline{DC}$, and $\overline{AB} \nparallel \overline{DC}$

Prove: $\angle B \cong \angle C$

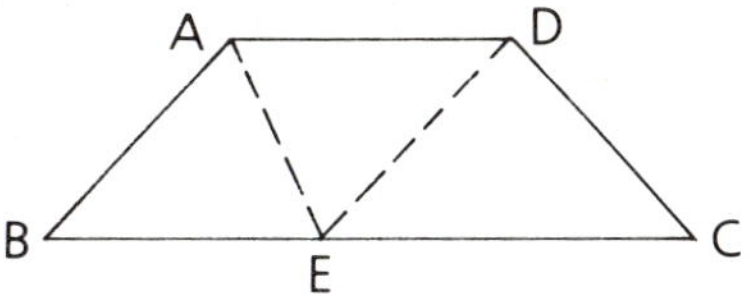

1 Figure ABCD with $\overline{AD} \parallel \overline{BC}$	1 Given
2 $\overline{AB} \nparallel \overline{DC}$	2 Given
3 Draw $\overline{DE} \parallel \overline{AB}$	3 THE PARALLEL POSTULATE
4 Draw $\overline{AE}$	4 Two points determine a line
5 $\angle DAE \cong \angle BEA$	5 PAI
6 $\angle BAE \cong \angle DEA$	6 PAI
7 $\overline{AE} \cong \overline{AE}$	7 Reflexive Property
8 $\triangle AEB \cong \triangle EAD$	8 ASA (5, 7, 6)
9 $\overline{AB} \cong \overline{DE}$	9 CPCTC
10 $\overline{AB} \cong \overline{DC}$	10 Given
11 $\overline{DE} \cong \overline{DC}$	11 Transitive Property
12 $\angle DEC \cong \angle C$	12 If $\triangle$ then $\triangle$
13 $\angle B \cong \angle DEC$	13 PCA
14 $\angle B \cong \angle C$	14 Transitive Property

Section 5.3 Problem Set A

1 Given: $\angle MOP$ is a right angle
$\overline{RP} \perp \overline{OP}$

Conclusion: $\overline{MO} \parallel \overline{RP}$

2 Given: $\overline{EA} \parallel \overline{DB}$, and $\overline{EA} \cong \overline{DB}$
B is the midpt. of $\overline{AC}$

Prove: $\overline{EB} \parallel \overline{DC}$

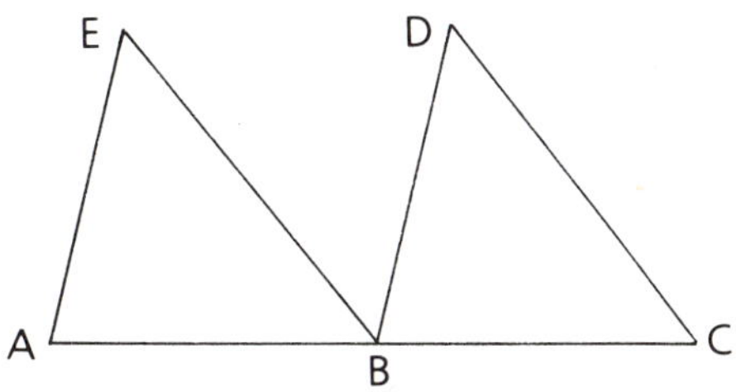

*Note: Figure ABCD is called an isosceles trapezoid.

3 Prove AEP.

4 Given: $\odot$ O

Conclusion: $\overline{ST} \parallel \overline{VX}$

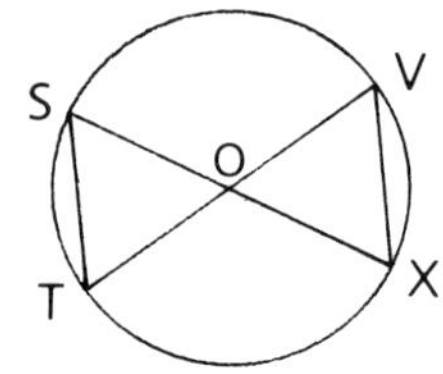

5 Given: a ∥ c

$\angle 1 \cong \angle 2$

Conclusion: b ∥ c

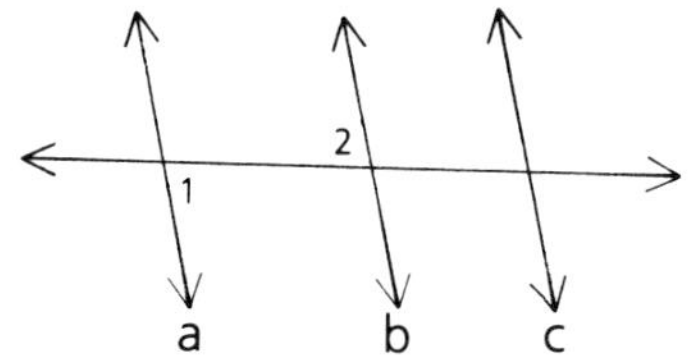

6 Given: $\angle A \cong \angle C$

$\overline{DC} \parallel \overline{AB}$

Prove: $\overline{DA} \parallel \overline{CB}$

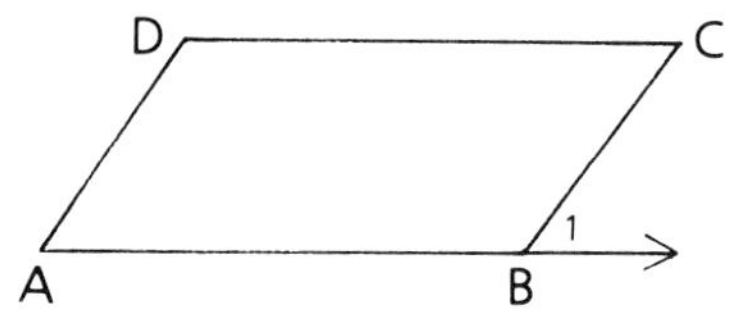

7 Given: $\angle 5 \cong \angle 6$

$\overline{RS} \parallel \overline{NP}$

Prove: $\triangle$NPR is isosceles

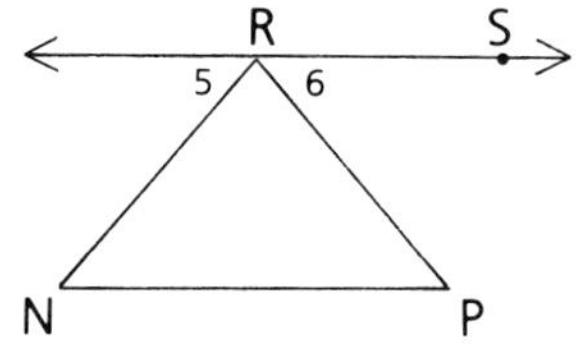

8 Given: $\overline{TE} \parallel \overline{XW}$

$\overline{TE} \cong \overline{XW}$

Conclusion: $\overline{TX} \parallel \overline{EW}$

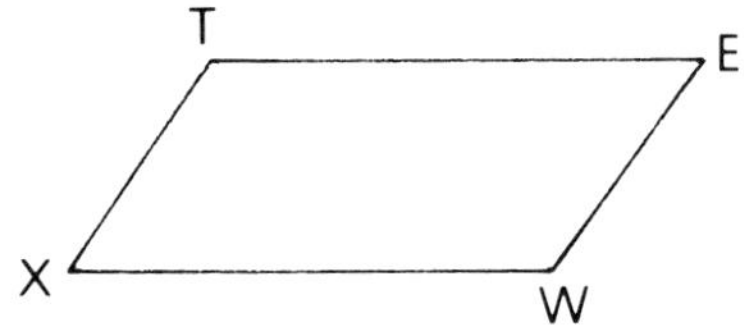

(Hint: Draw an auxiliary segment and prove that some $\triangle$s are $\cong$.)

9 Given: $\overline{AE} \parallel \overline{FC}$, $\odot$ O

Conclusion: $\overline{OD} \cong \overline{OB}$

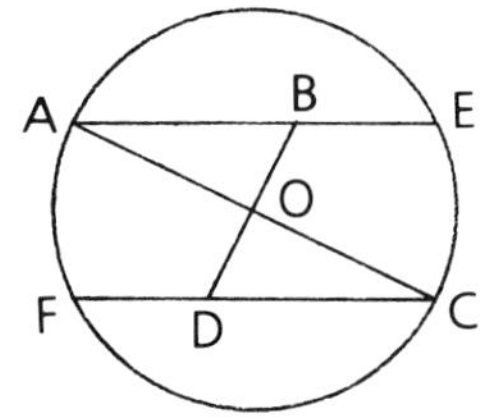

10 Given: $\overline{GO} \cong \overline{JM}$, and $\overline{OH} \cong \overline{MK}$

$\overline{KM} \parallel \overline{OH}$

Prove: $\overline{GH} \parallel \overline{KJ}$

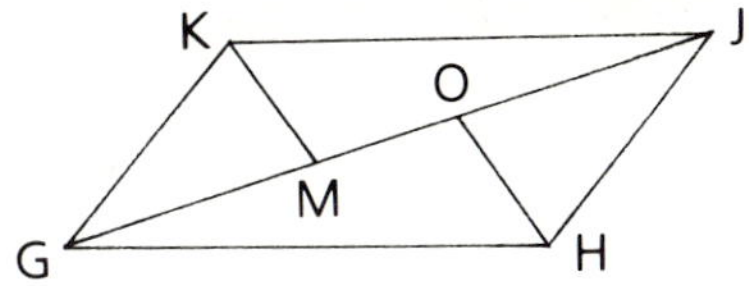

11 $\angle N = 80^\circ$

Find: $m\angle S$, $m\angle R$, $m\angle P$

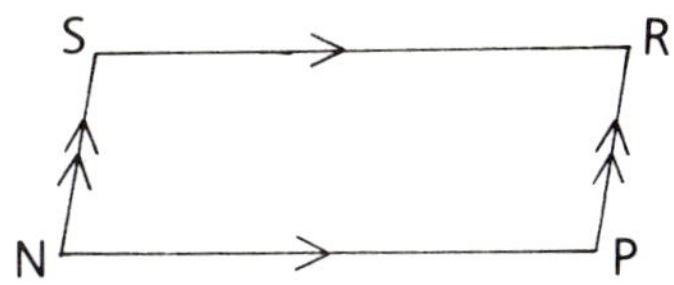

12 $\angle 4 = 64^\circ 26'$

$\angle 1 = 90^\circ$

Find: $m\angle 5$, $m\angle 2$, $m\angle 3$

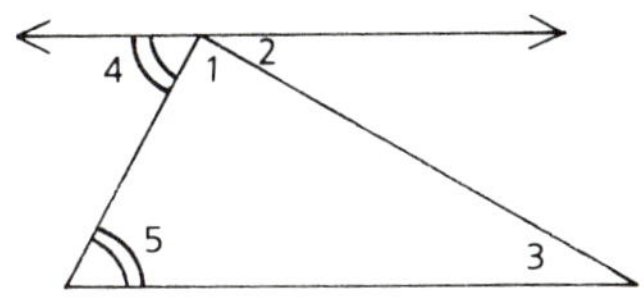

13 A Crook Problem

If $f \parallel g$, find $m\angle 8$

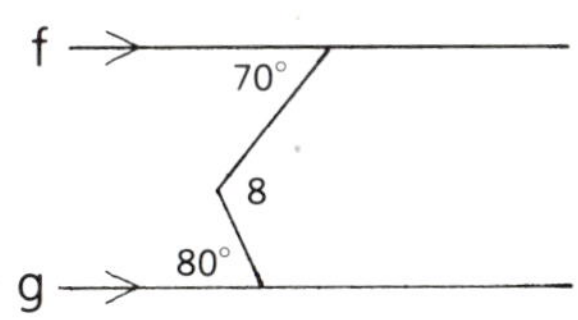

Section 5.3 Problem Set B

14 Given: $\angle C$ is supp.to $\angle D$

Prove: $\angle A$ is supp. to $\angle B$

15 Given: $\overline{RS} \cong \overline{AT}$

$\overline{RS} \perp \overline{ST}$, and $\overline{AT} \perp \overline{ST}$

Conclusion: $\triangle WST$ is isosceles

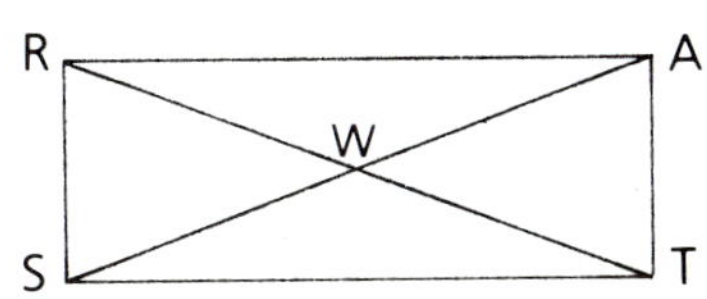

16 Given: $\overline{CY} \cong \overline{AY}$
$\overrightarrow{YZ} \parallel \overline{CA}$

Prove: $\overrightarrow{YZ}$ bisects $\angle AYB$

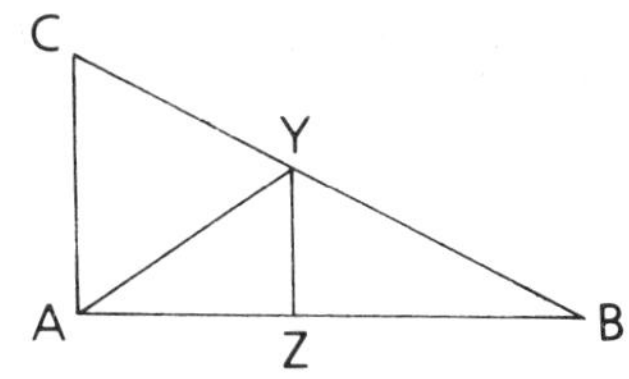

17 Given: $\overline{ED} \cong \overline{HD}$
$\angle DEH \cong \angle DFG$

Conclusion: $\overline{EG} \cong \overline{HF}$

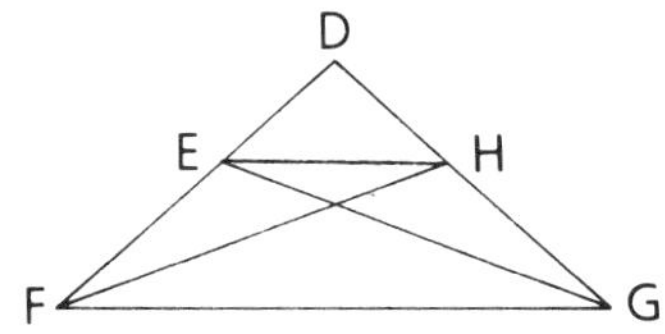

18 Given: $\triangle JRK \cong \triangle OMP$
$\overline{RO} \parallel \overline{JM}$

Prove: $\overleftrightarrow{RK} \parallel \overleftrightarrow{PM}$

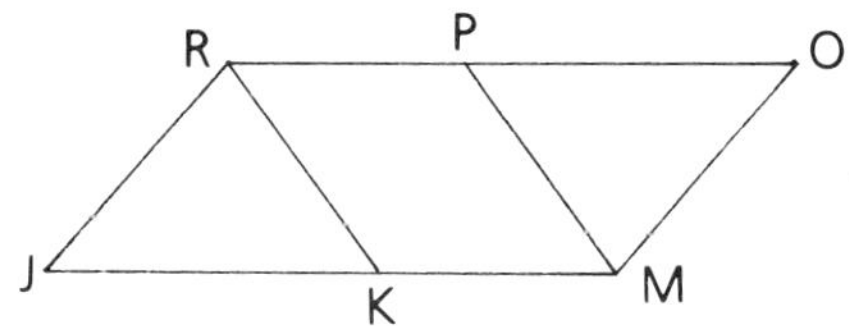

19 Prove that bisectors of a pair of alternate exterior angles formed with a transversal of parallel lines are parallel.

20 $\overline{EJ} \parallel \overline{FH}$, and $\overrightarrow{FH}$ bis $\angle JFG$
$EJ = y + 6$
$EF = 0.8y + 12$
$JF = 2y - 12$
Find the perimeter of $\triangle JEF$

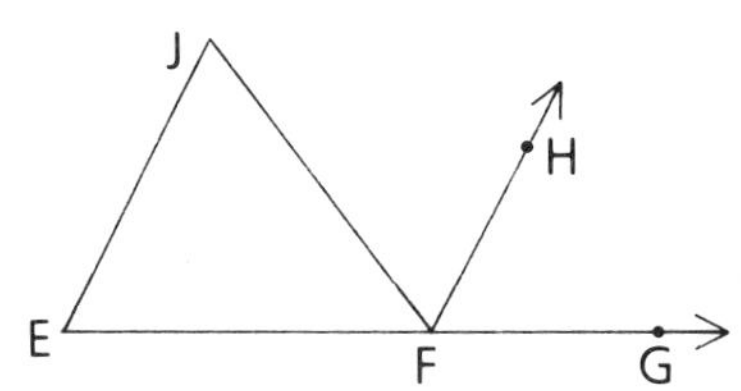

21 $a \parallel b$
$\angle 1 = (x + 3y)^\circ$
$\angle 2 = (2x + 30)^\circ$
$\angle 3 = (5y + 20)^\circ$
Find $m\angle 1$

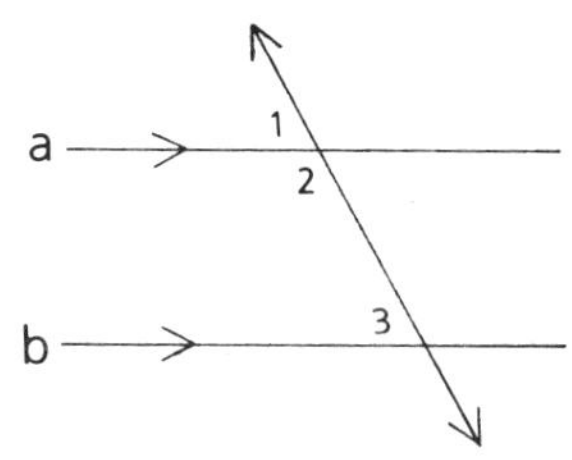

22 $\angle K = 72^\circ$
$\angle KMP \cong \angle OPM$, and $\angle OMP \cong \angle KPM$
$\angle KMP = (x + 4)^\circ$
$\angle OMP = (2x + 32)^\circ$
Find $m\angle OMP$

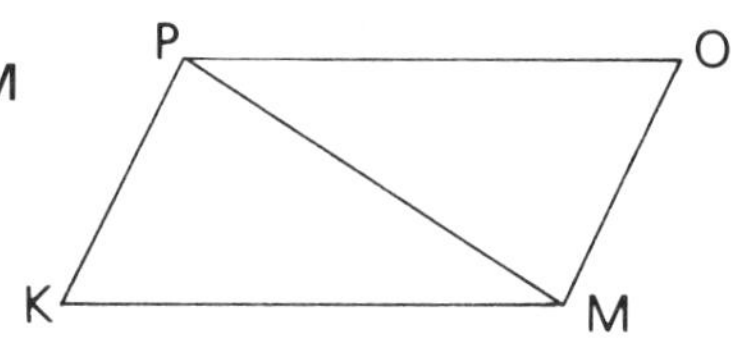

Section 5.3 Problem Set C

23 Given: $\odot$ O
$\overline{DC} \parallel \overline{AB}$

Prove: $\overline{AD} \cong \overline{BC}$

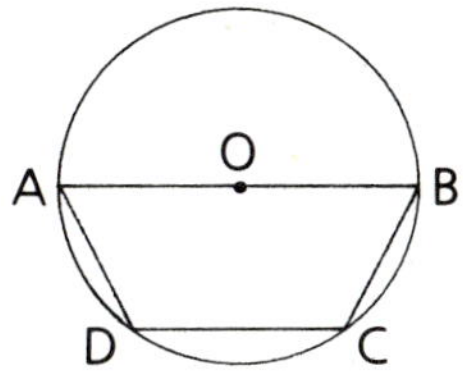

24 Given: $\overline{EF} \perp \overline{FG}$ and $\overline{EF} \perp \overline{EH}$
$\overline{HG} \perp \overline{FG}$ and $\overline{HG} \perp \overline{HE}$
$\overleftrightarrow{KM}$ is $\perp$ bis $\overline{FG}$

Prove: $\overleftrightarrow{KM}$ passes through J

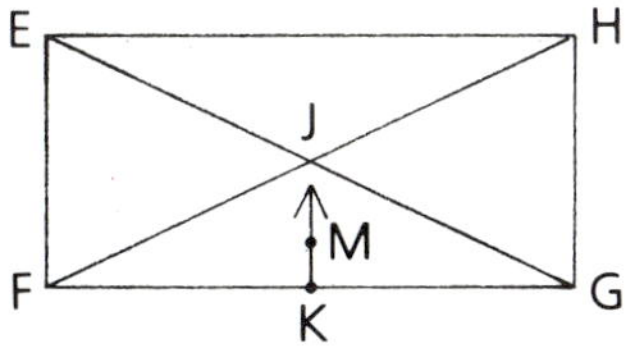

25 Given: $a \parallel b$
$\angle 1 = (x^2 + 4y)^\circ$
$\angle 2 = (x + 4y + 6)^\circ$
$\angle 3 = (5y + 15x)^\circ$

Find: $m\angle 1$

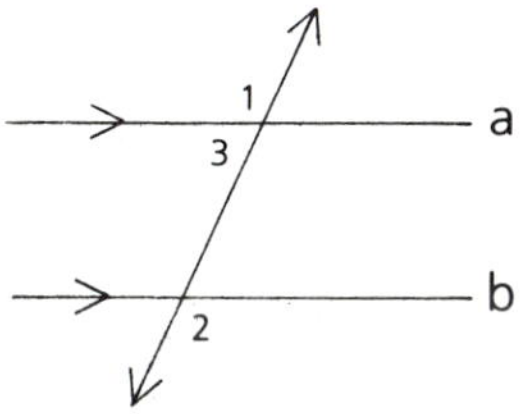

Section 5.3 Problem Set D

26 Given: $\overleftrightarrow{BE} \parallel \overleftrightarrow{CD}$, and $\overline{BE} \cong \overline{EF}$
B is the midpt. of $\overline{AC}$
E is the midpt. of $\overline{AD}$

Prove: $BE = \frac{1}{2}CD$

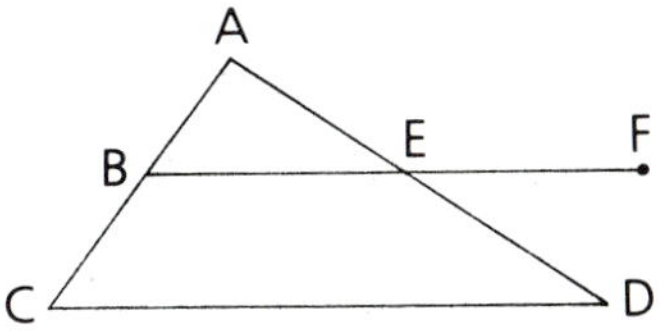

27 Write a paragraph proof that shows that the sum of the three angles of a $\triangle$ is 180°. (Hint: Draw a $\triangle$ and use the $\parallel$ postulate.)

Section 5.4 Four Sided Polygons

This section discusses—

A POLYGONS

B NAMING POLYGONS

C CONVEX POLYGONS

D DIAGONALS OF POLYGONS

E QUADRILATERALS

5.4 (A) Polygons

POLYGONS are coplanar figures. The following are examples of polygons:

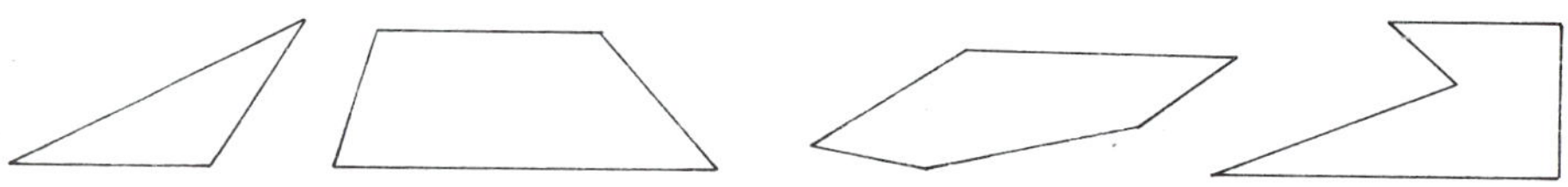

The following are examples of figures that are *not* polygons:

EFGH is *not* a polygon, because a polygon consists entirely of segments.

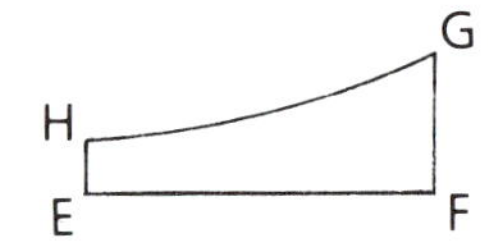

ABCDE is *not* a polygon, because in a polygon consecutive, and only consecutive, sides intersect and only at endpoints.

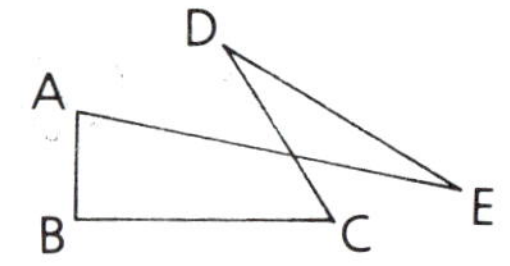

PKMO, PKMOR, and **POR** *are* polygons, *but* **PKMOPRO** is *not,* because each vertex must belong to *exactly* 2 sides. (Vertex P belongs to 3 sides in PKMOPRO.)

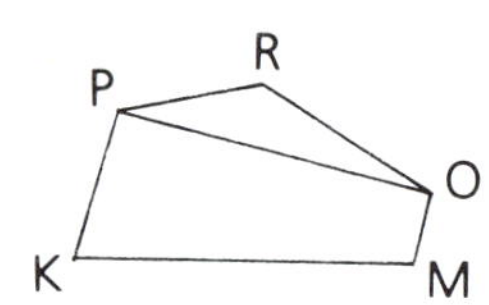

SVTY *is* a polygon, *but* **SVTXY** is *not,* because consecutive sides must be *non*-collinear.

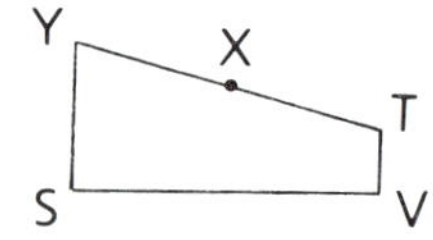

Why is **PLAN** *not* a polygon?

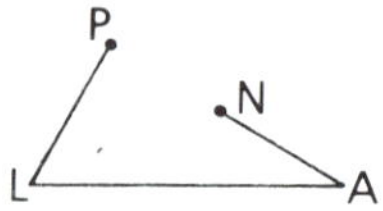

5.4 (B) Naming Polygons

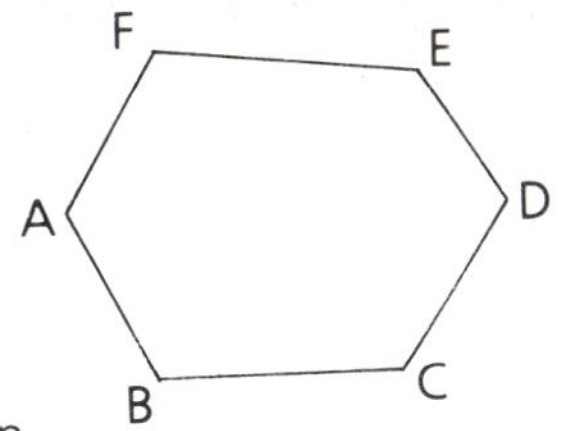

We name a polygon by starting at any vertex and then proceeding either clockwise or counterclockwise. If we start at A, we can call this polygon ABCDEF or AFEDCB. Can you start at B and name the polygon in two different ways?

5.4 (C) Convex Polygons

D **A CONVEX polygon is a polygon such that each interior angle has less than 180°.**

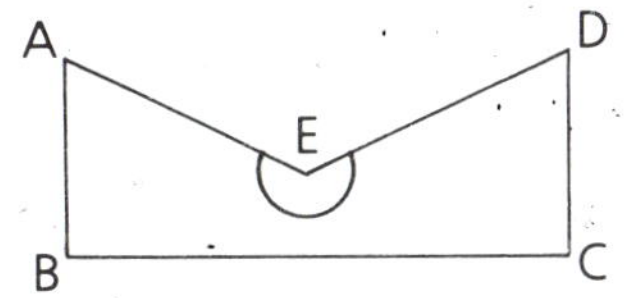

Polygon ABCDE is *not* convex because the angle that lies in the interior of the polygon at E has more than 180°. Hence, the polygon is *not* convex.

For the rest of the text, unless we expressly state otherwise, assume all polygons are convex.

5.4 (D) Diagonals

In each of the following figures, the dotted segments are DIAGONALS of the polygon.

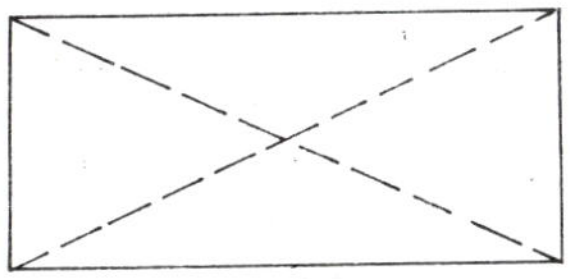

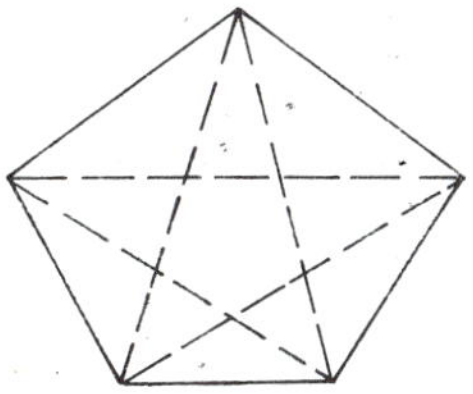

D **A DIAGONAL of a polygon is any segment that connects two non-consecutive (non-adjacent) vertices of the polygon.**

5.4 (E) Quadrilaterals

A QUADRILATERAL is a four-sided polygon.

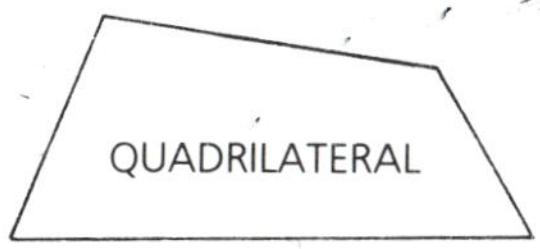

The following are examples of special quadrilaterals.

A **PARALLELOGRAM** is a quadrilateral in which both pairs of opposite sides are parallel.

A **RECTANGLE** is a *parallelogram* in which at least one angle is a right angle.

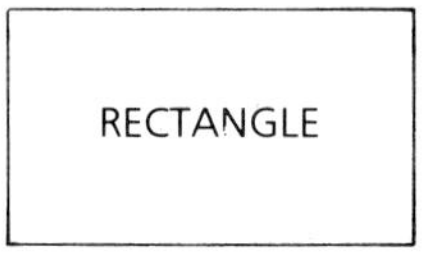

A **RHOMBUS** is a *parallelogram* in which at least two consecutive sides are ≅.

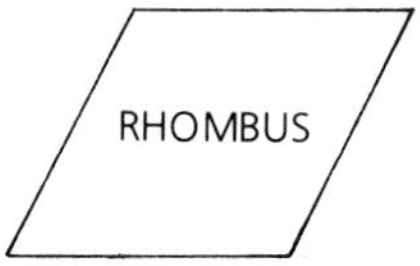

A **KITE** is a quadrilateral with two distinct pairs of congruent consecutive sides.

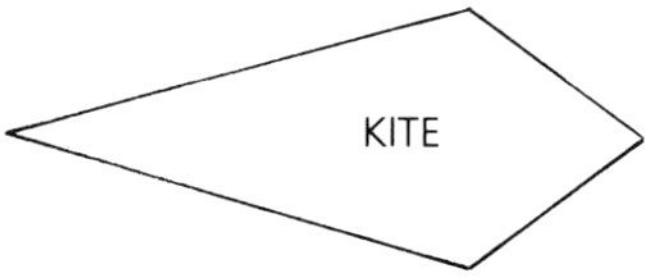

A **SQUARE** is a *parallelogram* that is both a rectangle and a rhombus.

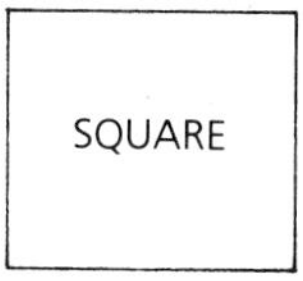

A **TRAPEZOID** is a quadrilateral with exactly one pair of parallel sides. The parallel sides are called BASES of the trapezoid.

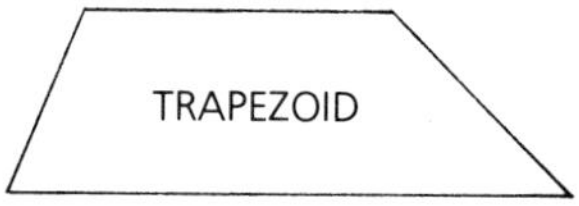

An **ISOSCELES TRAPEZOID** is a trapezoid in which the non-parallel sides (LEGS) are congruent. In the figure, ∠s A and B are called the LOWER BASE ANGLES, and ∠s C and D are called the UPPER BASE ANGLES.

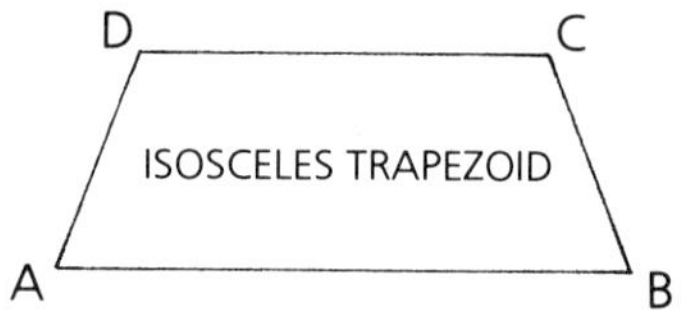

We have given the meaning (definition) of each of the above figures in as simple a manner as possible. Each special quadrilateral will have further properties associated with it. Those properties are discussed in the next section.

Section 5.4 Sample Problems

1 Can an isosceles trapezoid be a parallelogram?

Ans: No, an isosceles trapezoid has only one pair of parallel sides whereas a parallelogram has two pairs of parallel sides.

2 How many rectangles do there appear to be in this figure?

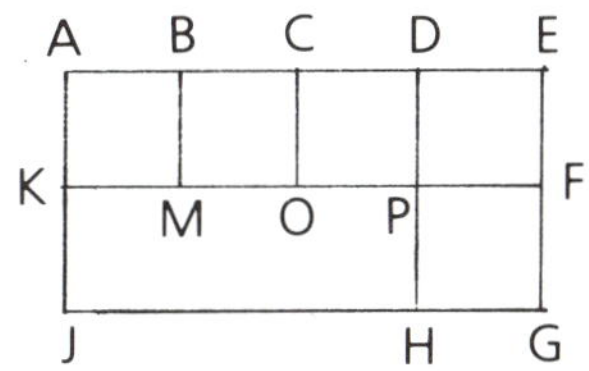

Solution:

ABMK	ACOK	ADPK	AEFK	ADHJ	AEGJ
BCOM	BDPM	BEFM	KFGJ		
CDPO	CEFO	KPHJ			
DEFP	DEGH				
PFGH				Ans:	16

Section 5.4 Problem Set A

1 If the statement is always true, write A; if sometimes true, write S; if never true, write N (A, S, or N).

- **a** A square is a rhombus.
- **b** A rhombus is a square.
- **c** A kite is a parallelogram.
- **d** A rectangle is a polygon.
- **e** A polygon has the same number of vertices as sides.
- **f** A parallelogram has three diagonals.
- **g** A trapezoid has three bases.

2 In the isosceles trapezoid shown, $\overline{ST} \parallel \overline{RV}$.

Name: a the bases

b the diagonals

c the legs

d the lower base angles

e the upper base angles

f all pairs of congruent alternate interior ∠s

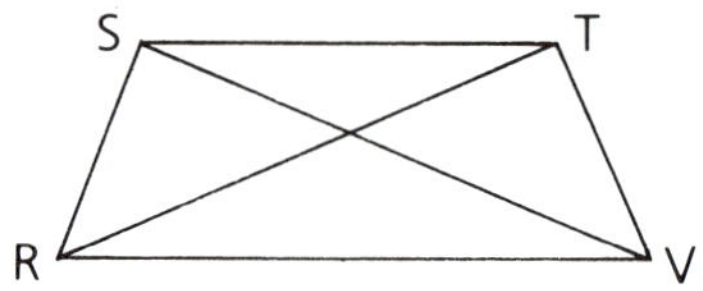

3 a Is GEF a polygon?

b Is ABCDEFGE a polygon?

c Is AGEDCB a polygon?

d Name a polygon that is not convex.

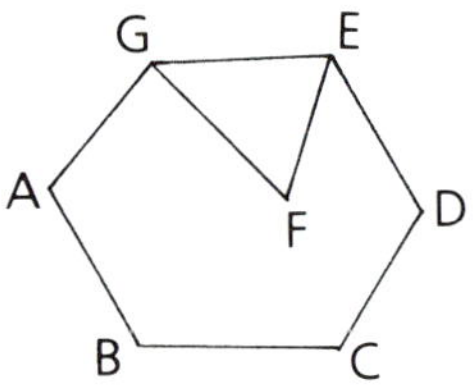

4 Judging by the *appearance* of each figure, give the most descriptive name:

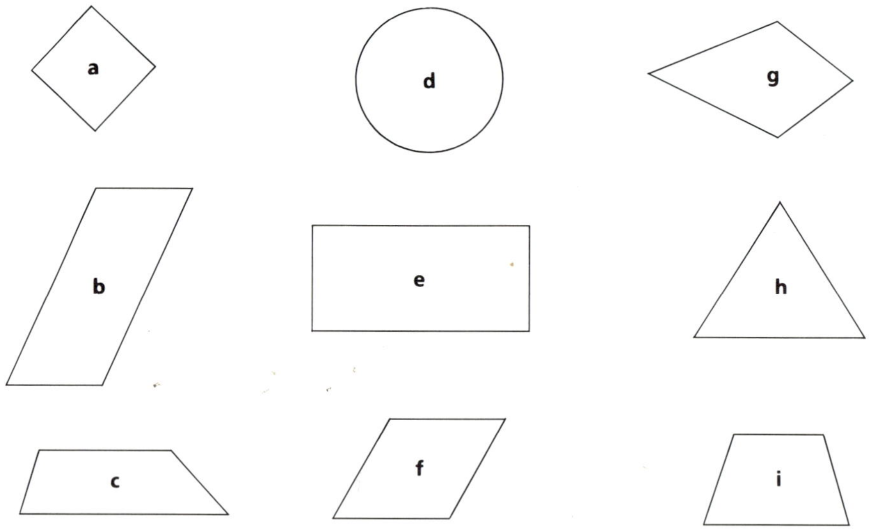

5 Prove that in a parallelogram each pair of consecutive angles are supplementary.

6 Prove that all of the angles of a rectangle are right angles.

Section 5.4 Problem Set B

7 How many rectangles are shown in the figure to the right, where all of the angles are right angles?

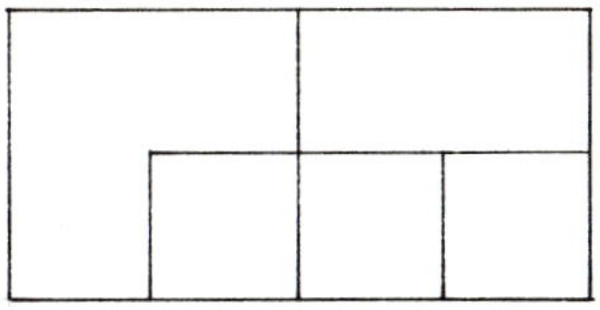

8 Given: ABCD is a kite
AB = x + 3
BC = x + 4
CD = 2x − 1
AD = 3x − y

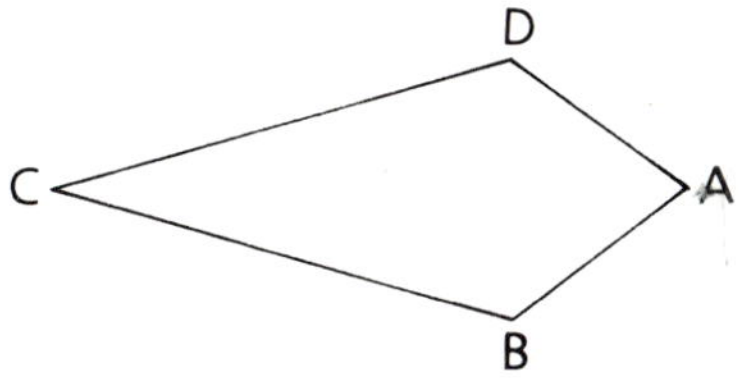

a Solve for x and y.

b What is the perimeter of the kite?

c Is it possible for $\overline{AC}$ to be 19 units long? Why or why not?

9 a PQRS is a kite and also a rectangle. What else do we know about PQRS?

b Draw a quadrilateral that is not convex and still satisfies the definition of a kite.

10 Prove that in a parallelogram each pair of opposite sides is congruent.

11 Prove that the diagonals of a rectangle are congruent.

12 Prove that all of the sides of a rhombus are congruent.

Section 5.4 Problem Set C

13 Given the seven shapes on page 222, what is the probability that if two are picked at random, they will both have a pair of congruent opposite sides?

Section 5.5 Properties of Quadrilaterals

This section presents PROPERTIES of:

A PARALLELOGRAMS
B RECTANGLES
C RHOMBUSES
D SQUARES
E KITES
F ISOSCELES TRAPEZOIDS

We shall simply list some of the properties of the special quadrilaterals. You should be able to prove many of them. Read the section carefully and learn these properties.

5.5 (A) Properties of Parallelograms

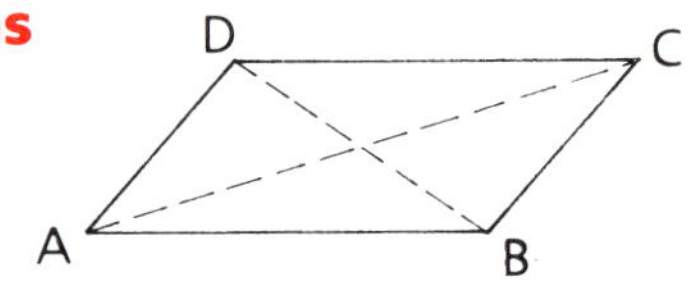

In a parallelogram—

D 1 the opposite sides are parallel (by definition). ($\overline{DC} \parallel \overline{AB}$, $\overline{AD} \parallel \overline{BC}$)

T 2 the opposite sides are congruent. ($\overline{DC} \cong \overline{AB}$, $\overline{AD} \cong \overline{BC}$)

T 3 the opposite angles are congruent. ($\angle DAB \cong \angle DCB$, $\angle ABC \cong \angle ADC$)

T 4 the diagonals bisect each other. ($\overline{AC}$ and $\overline{BD}$ bisect each other)

T 5 any pair of consecutive angles are supplementary. ($\angle DAB$ is supp. to $\angle ABC$, etc.)

5.5 (B) Properties of Rectangles

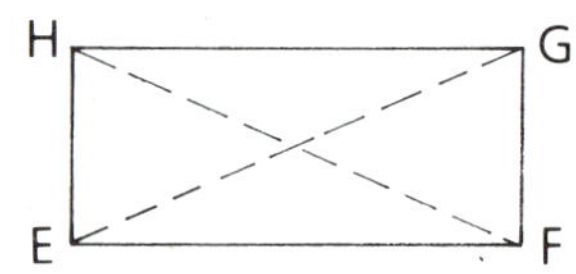

In a rectangle—

1 all the properties of a parallelogram apply (by definition).

T 2 all angles are right angles. ($\angle E$ is a right angle, etc.)

T 3 the diagonals are congruent. ($\overline{EG} \cong \overline{HF}$)

5.5 (C) Properties of Rhombuses

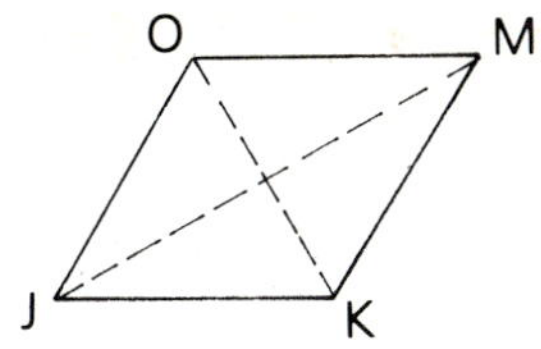

In a rhombus—

1 all the properties of a parallelogram apply (by definition).

T 2 all sides are $\cong$ (a rhombus is equilateral). ($\overline{JK} \cong \overline{KM} \cong \overline{MO} \cong \overline{OJ}$)

T 3 the diagonals bisect the angles of the polygon. ($\overline{JM}$ bisects $\angle$s OJK and OMK, $\overline{OK}$ bisects $\angle$s JOM and JKM)

T 4 the diagonals are perpendicular bisectors of each other. ($\overline{JM} \perp$ bis $\overline{OK}$, $\overline{OK} \perp$ bis $\overline{JM}$)

5.5 (D) Properties of Squares

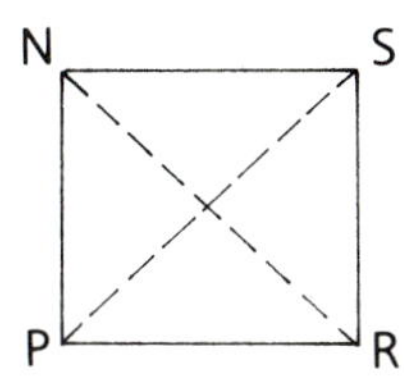

In a square—

1 all the properties of a rectangle apply (by definition).

2 all the properties of a rhombus apply (by definition).

T 3 the diagonals form four isosceles right triangles. ($45°–45°–90°$ ⚠)

5.5 (E) Properties of Kites

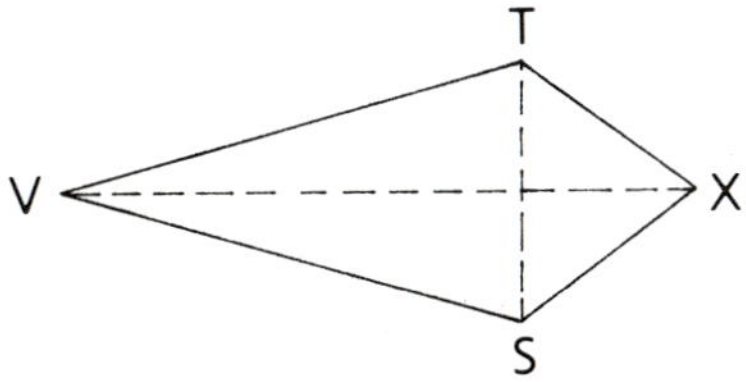

In a kite—

D 1 two distinct pairs of consecutive sides are congruent (by definition). ($\overline{TV} \cong \overline{VS}$, $\overline{TX} \cong \overline{XS}$)

T 2 one of the diagonals is the perpendicular bisector of the other diagonal. ($\overline{VX} \perp$ bis $\overline{TS}$)

3 if the kite is also a rhombus or a square, it inherits those properties as well.

5.5 (F) Properties of Isosceles Trapezoids

In an isosceles trapezoid—

1 **the legs are congruent (by definition). ($\overline{AD} \cong \overline{BC}$)**

2 **the bases are parallel (by definition of a trapezoid). ($\overline{AB} \parallel \overline{DC}$)**

T 3 **the lower base angles are congruent. ($\angle A \cong \angle B$)**

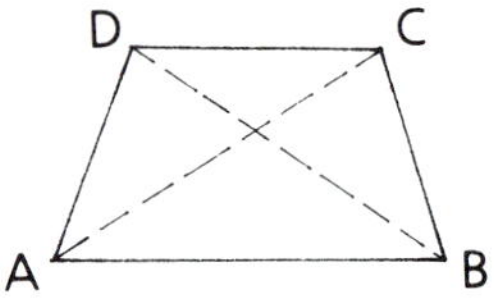

T 4 **the upper base angles are congruent. ($\angle D \cong \angle C$)**

T 5 **the diagonals are congruent. ($\overline{AC} \cong \overline{DB}$)**

T 6 **any lower base angle is supplementary to any upper base angle. ($\angle A$ is supp. to $\angle C$, etc.)**

In the problems that follow, you will be asked to prove some of these properties. You may use any prior property to help in the proof of a later property. For example, if you are asked to prove property 5 of parallelograms, you may use properties 1–4 to help you in the proof.

Section 5.5 Sample Problems

1 **Given:** ABCD is a ▱ (parallelogram)
$\angle GHA \cong \angle FEC$
$\overline{HB} \cong \overline{DE}$

Conclusion: $\overline{GH} \cong \overline{EF}$

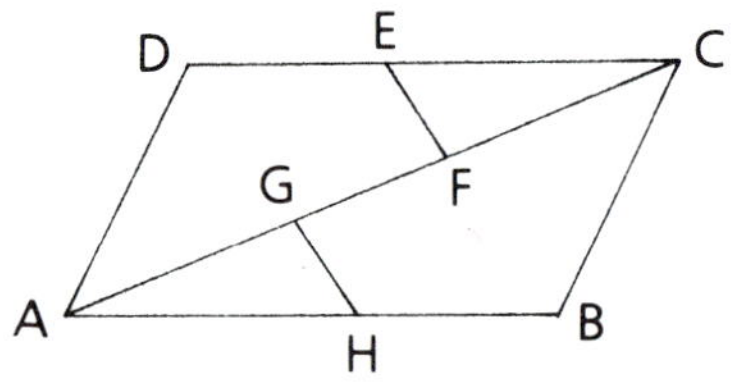

1 ABCD is a ▱	1 Given
2 $\overline{DC} \parallel \overline{AB}$	2 The opposite sides of a ▱ are ∥
3 $\angle ECF \cong \angle HAG$	3 PAI
4 $\overline{AB} \cong \overline{DC}$	4 The opposite sides of a ▱ are ≅
5 $\overline{HB} \cong \overline{DE}$	5 Given
6 $\overline{HA} \cong \overline{EC}$	6 Subtraction Property
7 $\angle GHA \cong \angle FEC$	7 Given
8 $\triangle GAH \cong \triangle FCE$	8 ASA (3, 6, 7)
9 $\overline{GH} \cong \overline{EF}$	9 CPCTC

2 Given: VRZA is a ▱

$AV = 2x - 4$

$VR = 3y + 5$

$RZ = \frac{1}{2}x + 8$

$ZA = y + 12$

Find the perimeter of VRZA

Solution: The opposite sides of a ▱ are congruent, so—

$$2x - 4 = \frac{1}{2}x + 8$$
$$1\tfrac{1}{2}x - 4 = 8$$
$$1\tfrac{1}{2}x = 12$$
$$x = 8$$

AV = 12 and RZ = 12

and

$$3y + 5 = y + 12$$
$$2y + 5 = 12$$
$$2y = 7$$
$$y = 3\tfrac{1}{2}$$

$VR = 15\frac{1}{2}$ and $ZA = 15\frac{1}{2}$

Adding up the four sides, we find that the perimeter is 55.

3 **Prove Property 4 of Parallelograms.**

Given: ▱ ABCD

Prove: $\overline{AC}$ and $\overline{BD}$ bisect each other

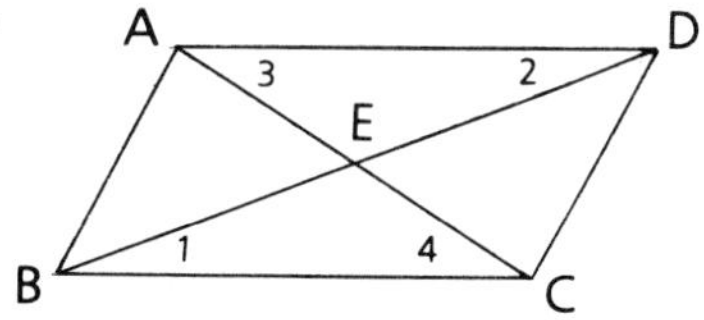

1 ▱ ABCD	1 Given
2 $\overline{AD} \parallel \overline{BC}$	2 The opposite sides of a ▱ are ∥
3 $\angle 1 \cong \angle 2$	3 PAI
4 $\angle 3 \cong \angle 4$	4 PAI
5 $\overline{AD} \cong \overline{BC}$	5 The opposite sides of a ▱ are ≅
6 $\triangle BEC \cong \triangle DEA$	6 ASA (3, 5, 4)
7 $\overline{BE} \cong \overline{DE}$	7 CPCTC
8 $\overline{AE} \cong \overline{EC}$	8 CPCTC
9 $\overline{AC}$ and $\overline{BD}$ bisect each other	9 If two segments divide each other into ≅ segments, they bisect each other

4 **Prove Property 3 of Isosceles Trapezoids.**

(This was proved on page 215, Sample Problem 4.)

Section 5.5 Problem Set A

1 Given: ▱ ABCD (ABCD is a ▱)

Conclusion: $\triangle ABC \cong \triangle CDA$

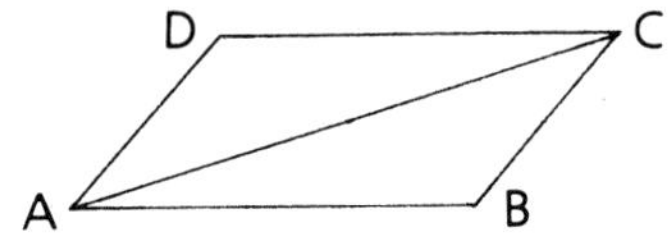

2 Given: ▱ EFHJ

$\angle 1 \cong \angle 2$

Conclusion: $\overline{KH} \cong \overline{EG}$

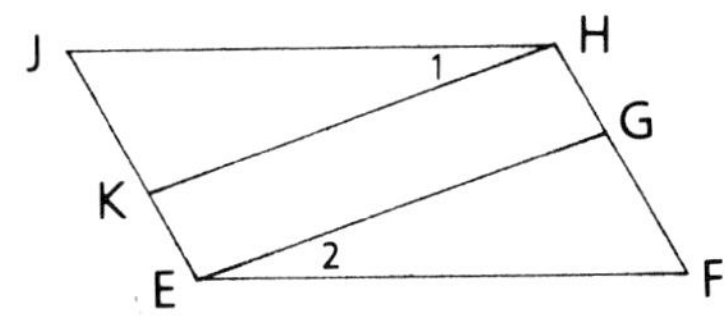

1 ▱ EFHJ	1 ______
2 $\angle J \cong \angle F$	2 ______
3 $\overline{JH} \cong \overline{EF}$	3 ______
4 $\angle 1 \cong \angle 2$	4 ______
5 $\triangle KJH \cong \triangle GFE$	5 ______
6 $\overline{KH} \cong \overline{EG}$	6 ______

3 Given: Rectangle MPRS

$\overline{MO} \cong \overline{PO}$

Prove: $\triangle ROS$ is isosceles

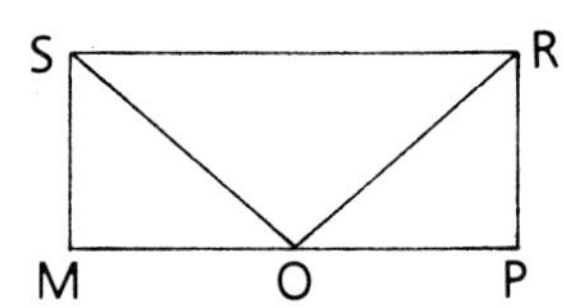

4 Given: ▱ ABCD

$\overline{AE} \cong \overline{CF}$

Conclusion: $\overline{DE} \cong \overline{BF}$

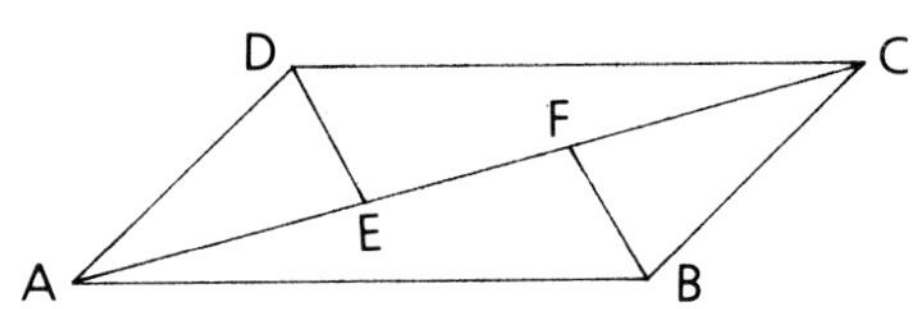

5 Given: ▱ WSTV

$WS = x + 5$

$WV = x + 9$

$VT = 2x + 1$

Find the perimeter of WSTV

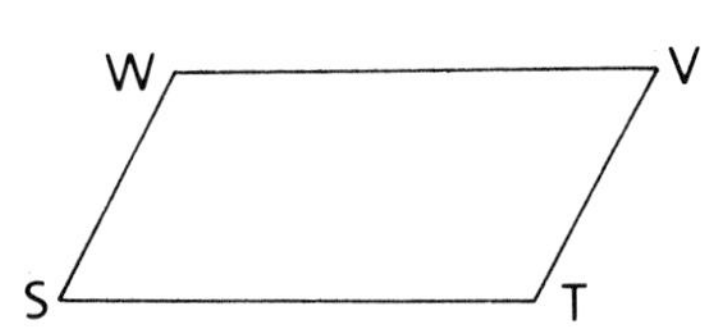

6 Given: ▱ ABCD

$\angle A = (x)^\circ$

$\angle D = (3x - 4)^\circ$

Find: $m\angle D$, and $m\angle C$

7 Given: EFGH is an isosceles trapezoid, legs $\overline{HE}$ and $\overline{GF}$
$EJ = x + 5$
$JG = 2x - 1$
$HF = 13$

Find: EJ, JG, and HJ

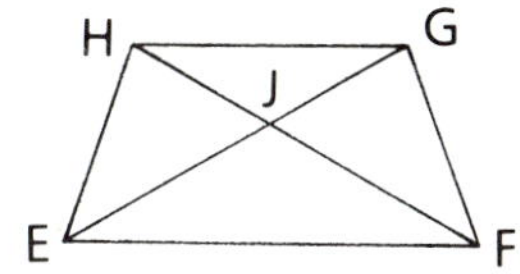

8 Prove Property 3 of Parallelograms.

9 Prove Property 4 of Rhombuses.

10 Prove Property 5 of Isosceles Trapezoids.

Section 5.5 Problem Set B

11 Given: ABCD is a ▱
$\overline{AF} \cong \overline{CE}$

Prove: $\overline{DF} \parallel \overline{EB}$

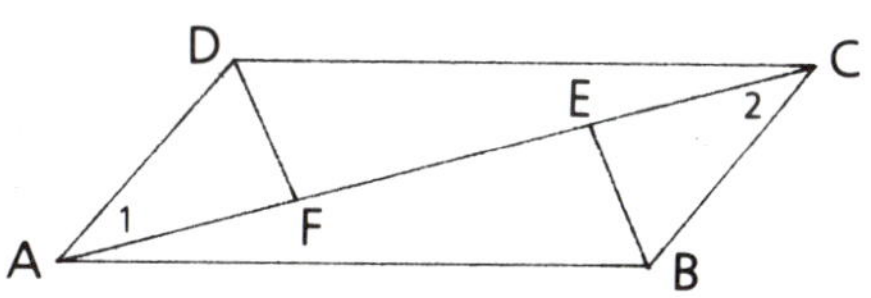

12 Given: PHJM is a rectangle
$\overline{PG} \cong \overline{MK}$

Prove: $\triangle$OGK is isosceles

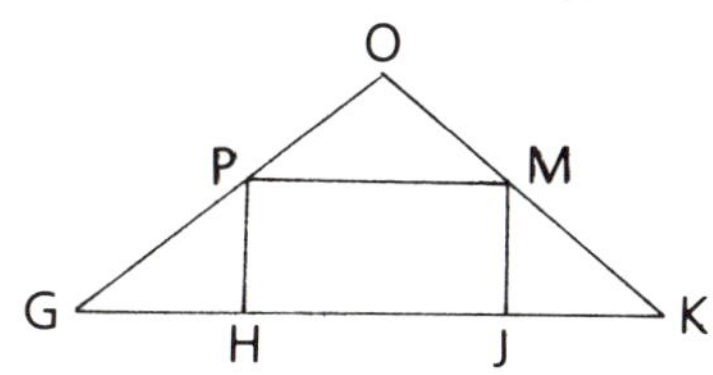

13 Given: VRST is an isosceles trapezoid with legs $\overline{VR}$ and $\overline{TS}$

Prove: $\triangle$ARS is isosceles

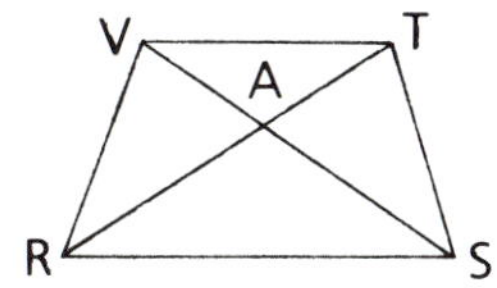

14 The diagonals of a rhombus divide the rhombus into four $\cong$ $\triangle$s.

15 Given: FGKM is a rectangle
GHJK is a rectangle
$\overline{MK} \cong \overline{KJ}$

Prove: a $\overline{FK} \cong \overline{GJ}$
b $\overline{FK} \parallel \overline{GJ}$

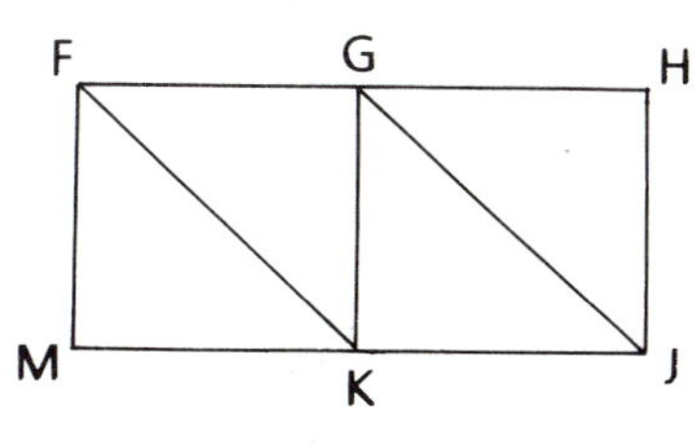

16 ▱ KMOP
$\angle M = (x + 3y)^\circ$
$\angle O = (x - 4)^\circ$
$\angle P = (4y - 8)^\circ$
Find $m\angle K$

17 ABCD is an isosceles trapezoid with upper base $\overline{AD}$
$BE = x + 7,\ CE = y - 3$
$AE = x + 5,\ BD = y + 4$
Find AC

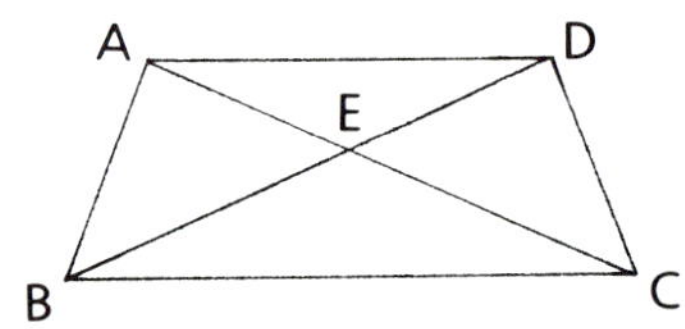

Section 5.5 Problem Set C

18 Given: BDEF is a ▱
$\overline{FE} \perp \overline{CF}$
$\overline{BC} \cong \overline{EH}$
Prove: $\triangle ACH$ is isosceles

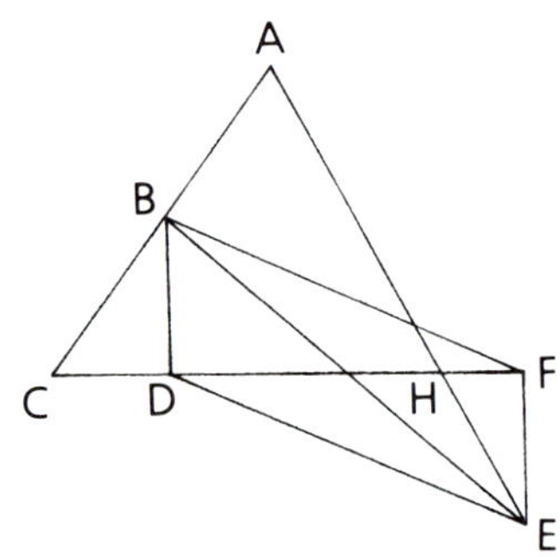

19 Given: $m \parallel n$
a Solve for a in terms of x and y.
b If $a > 90$, what must be true of $y - x$?

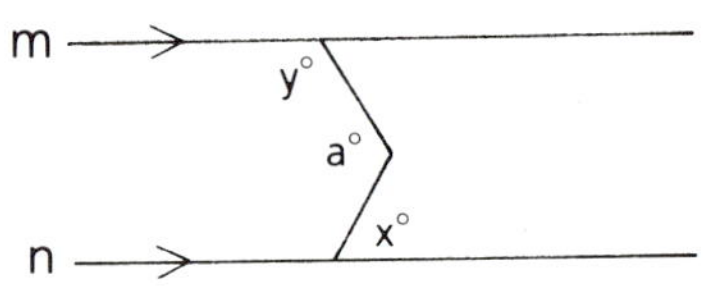

20 In the space figure to the right,
▱ ABCD ≅ ▱ EFGH
Prove: $\overline{HF} \cong \overline{DB}$

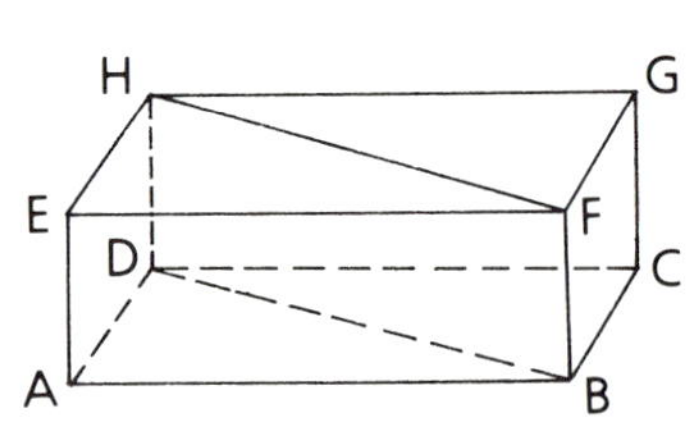

21 Prove: The perpendicular bisector of the *upper* base of an isosceles trapezoid passes through the point of intersection of the bisectors of the *lower* base angles.

Section 5.6 Proving That Figures Are Special Quadrilaterals

In this section, we present ways to prove that a quadrilateral is a—

A PARALLELOGRAM

B RECTANGLE

C RHOMBUS

D SQUARE

E ISOSCELES TRAPEZOID

5.6(A) Proving That a Quadrilateral Is a Parallelogram

Any one of the following methods might be used to prove that quadrilateral ABCD is a parallelogram.

D 1 **If *both* pairs of opposite sides of a quadrilateral are parallel, then the quadrilateral is a parallelogram (reverse of the definition).**

T 2 **If *both* pairs of the opposite sides of a quadrilateral are congruent, then the quadrilateral is a parallelogram.**

T 3 **If two sides of a quadrilateral are both parallel and congruent, then the quadrilateral is a parallelogram.**

T 4 **If the diagonals of a quadrilateral bisect each other, then the quadrilateral is a parallelogram.**

T 5 **If *both* pairs of opposite angles of a quadrilateral are congruent, then the quadrilateral is a parallelogram.***

*Note: The proof of this theorem is difficult at this time and will be called for in Section 7.3, problem 15, page 296.

5.6 (B) Proving That a Quadrilateral Is a Rectangle

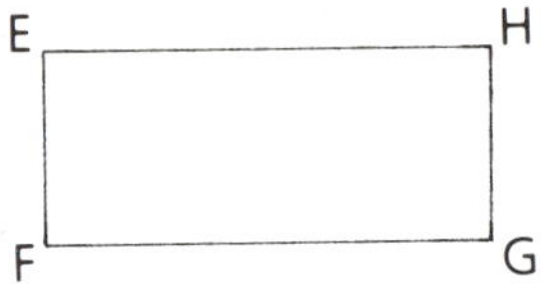

You could prove that quadrilateral EFGH is a rectangle by first showing that the quadrilateral is a *parallelogram.* Then use either of the following methods to complete the proof.

D 1 If a parallelogram contains at least one right angle, then it is a rectangle (reverse of the definition).

T 2 If the diagonals of a parallelogram are congruent, then the parallelogram is a rectangle.

5.6 (C) Proving That a Quadrilateral Is a Rhombus

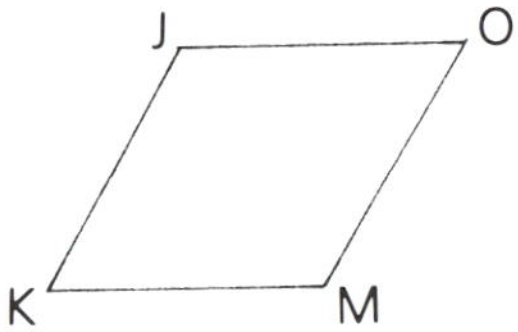

To prove that quadrilateral KMOJ is a rhombus, you may first show that it is a *parallelogram.* Then finish with either of the following methods.

D 1 If a parallelogram contains a consecutive pair of sides that are congruent, then it is a rhombus (reverse of the definition).

T 2 If either diagonal of a parallelogram bisects two angles of the polygon, then the parallelogram is a rhombus.

You could also prove that a quadrilateral is a rhombus without first showing that it is a parallelogram, if you use the next method:

T 3 If the diagonals of a quadrilateral are perpendicular bisectors of each other, then the quadrilateral is a rhombus.

5.6 (D) Proving That a Quadrilateral Is a Square

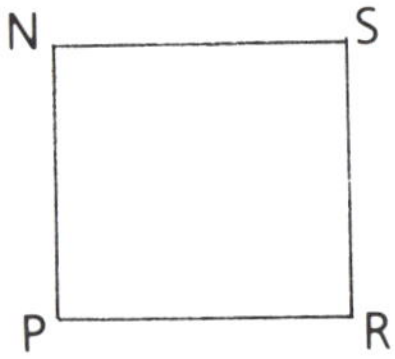

The following method could be used to prove that NPRS is a square.

D 1 If a quadrilateral is both a rectangle and a rhombus, then it is a square (reverse of the definition).

5.6 (E) Proving That a Trapezoid Is Isosceles

Any one of the following methods might be used to prove that a *trapezoid* is isosceles.

D 1 **If the non-parallel sides of a trapezoid are congruent, then it is isosceles (reverse of the definition).**

T 2 **If the lower or the upper base angles of a trapezoid are congruent, then it is isosceles.**

T 3 **If the diagonals of a trapezoid are congruent, then it is isosceles.**

Section 5.6 Sample Problems

1 Given: ACDF is a ▱
$\angle AFB \cong \angle ECD$
Prove: FBCE is a ▱

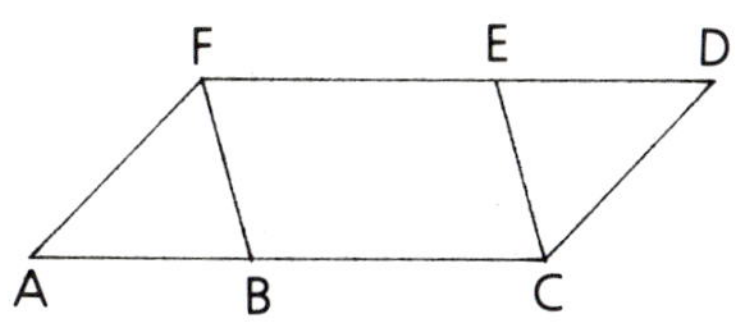

1 ACDF is a ▱	1 Given
2 $\angle A \cong \angle D$	2 The opposite angles of a ▱ are $\cong$
3 $\overline{AF} \cong \overline{DC}$	3 The opposite sides of a ▱ are $\cong$
4 $\angle AFB \cong \angle ECD$	4 Given
5 $\triangle AFB \cong \triangle DCE$	5 ASA (2, 3, 4)
6 $\overline{FB} \cong \overline{EC}$	6 CPCTC
7 $\overline{AB} \cong \overline{ED}$	7 CPCTC
8 $\overline{AC} \cong \overline{FD}$	8 Same as 3
9 $\overline{BC} \cong \overline{FE}$	9 Subraction Property
10 FBCE is a ▱	10 If both pairs of opposite sides of a quadrilateral are $\cong$, it is a ▱

2 Given: GJMO is a ▱

$\overline{OH} \perp \overline{GK}$

$\overline{MK}$ is an altitude of $\triangle$MKJ

Prove: OHKM is a rectangle

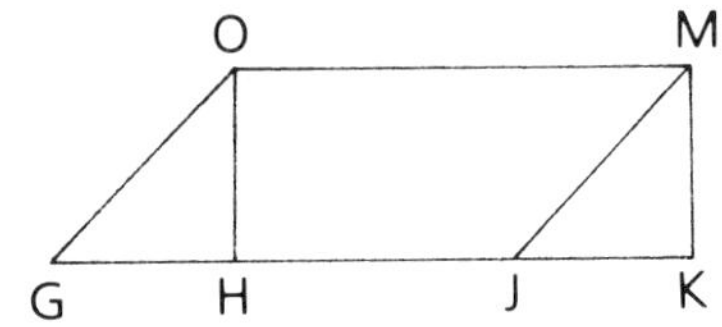

1 GJMO is a ▱	1 Given
2 $\overleftrightarrow{OM} \parallel \overleftrightarrow{GK}$	2 The opposite sides of a ▱ are $\parallel$
3 $\overline{OH} \perp \overline{GK}$	3 Given
4 $\overline{MK}$ is alt. of $\triangle$ MKJ	4 Given
5 $\overline{MK} \perp \overline{GK}$	5 An altitude of a $\triangle$ is $\perp$ to the side to which it is drawn
6 $\overline{OH} \parallel \overline{MK}$	6 In a plane, if two lines are $\perp$ to a third line, they are $\parallel$ to each other
7 OHKM is a ▱	7 If both pairs of opposite sides are $\parallel$, then a quadrilateral is a ▱
8 $\angle$OHK is a right $\angle$	8 $\perp$ segments form a right $\angle$
9 OHKM is a rectangle	9 If a ▱ contains at least one right angle, it is a rectangle

3 Given: NRTW is a ▱

$\overline{NX} \cong \overline{TS}$

$\overline{WV} \cong \overline{PR}$

Prove: XPSV is a ▱

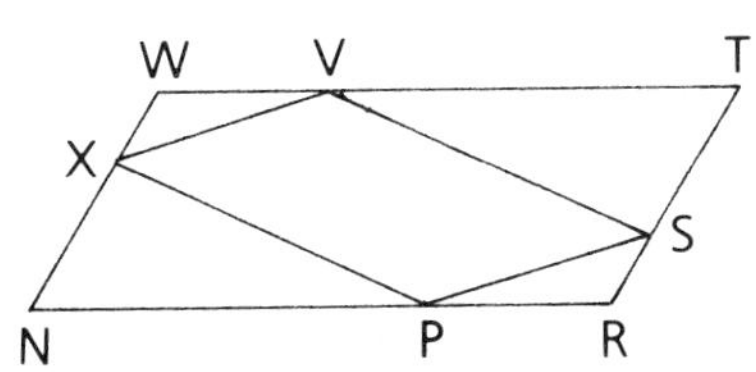

1 NRTW is a ▱	1 Given
2 $\angle N \cong \angle T$	2 The opposite $\angle$s of a ▱ are $\cong$
3 $\overline{NX} \cong \overline{TS}$	3 Given
4 $\overline{NR} \cong \overline{WT}$	4 The opposite sides of a ▱ are $\cong$
5 $\overline{WV} \cong \overline{PR}$	5 Given
6 $\overline{NP} \cong \overline{VT}$	6 Subtraction Property
7 $\triangle NXP \cong \triangle TSV$	7 SAS (3, 2, 6)
8 $\overline{XP} \cong \overline{VS}$	8 CPCTC
9 In a similar manner, $\triangle WXV \cong \triangle RSP$, $\overline{XV} \cong \overline{PS}$	9 Steps 1–8
10 XPSV is a ▱	10 If both pairs of opposite sides are $\cong$, a quadrilateral is a ▱

4 Prove that if either diagonal bisects two angles of a ▱, the ▱ is a rhombus (Method 2 of proving that a quadrilateral is a rhombus).

Given: ABCD is a ▱
$\overleftrightarrow{BD}$ bisects $\angle ADC$ and $\angle ABC$

Prove: ABCD is a rhombus

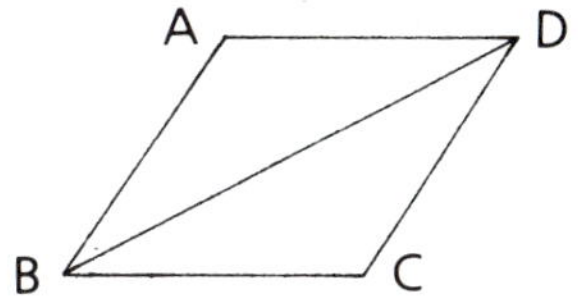

1 ABCD is a ▱	1 Given
2 $\angle ADC \cong \angle ABC$	2 The opposite angles of a ▱ are $\cong$
3 $\overleftrightarrow{BD}$ bis $\angle ADC$ and $\angle ABC$	3 Given
4 $\angle ABD \cong \angle ADB$	4 Division Property
5 $\overline{AB} \cong \overline{AD}$	5 If $\triangle$ then $\triangle$
6 ABCD is a rhombus	6 If a ▱ contains a consecutive pair of sides that are $\cong$, it is a rhombus

Section 5.6 Problem Set A

1 For each of the following sets of GIVEN information, state the method you would use to PROVE that ABCD is a parallelogram.

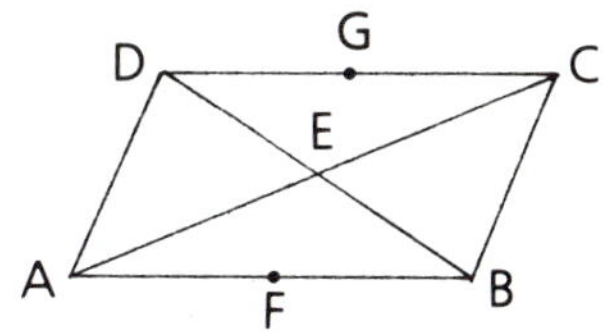

a $\overline{AB} \parallel \overline{DC}$
$\overline{AD} \parallel \overline{BC}$

b $\overline{AF} \cong \overline{DG}$
$\overline{FB} \cong \overline{GC}$
$\overline{DC} \parallel \overline{AB}$

c E is the midpt. of $\overline{AC}$
E is the midpt. of $\overline{BD}$

d $\angle DAC \cong \angle BCE$
$\overline{AD} \cong \overline{BC}$

e $\overline{BC} \cong \overline{GC}$
$\overline{AD} \cong \overline{DG}$
G is the midpt. of $\overline{DC}$
$\overline{DC} \cong \overline{AB}$

f $\overline{AF} \cong \overline{DG}$
G is the midpt. of $\overline{DC}$
F is the midpt. of $\overline{AB}$
$\angle CAB \cong \angle ECD$

2 Given: RKMP is a ▱
$\angle JRK \cong \angle PMO$
Prove: RJMO is a ▱

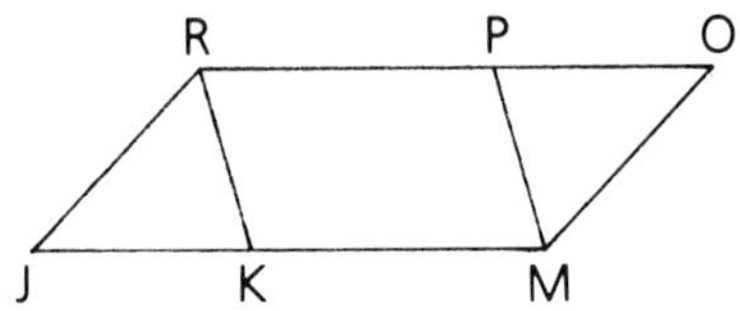

1 RKMP is a ▱	1 ______
2 $\overleftrightarrow{RO} \parallel \overleftrightarrow{JM}$	2 ______
3 $\overline{RK} \cong \overline{PM}$	3 ______
4 $\angle RKM \cong \angle MPR$	4 ______
5 $\angle JKR$ is supp. $\angle RKM$	5 ______
6 $\angle OPM$ is supp. $\angle MPR$	6 ______
7 $\angle JKR \cong \angle OPM$	7 ______
8 $\angle JRK \cong \angle PMO$	8 ______
9 $\triangle JRK \cong \triangle OMP$	9 ______
10 $\overline{JK} \cong \overline{PO}$	10 ______
11 $\overline{RP} \cong \overline{KM}$	11 ______
12 $\overline{RO} \cong \overline{JM}$	12 ______
13 RJMO is a ▱	13 ______

3 Given: $\angle XRV \cong \angle RST$
$\angle RSV \cong \angle TVS$
Conclusion: RSTV is a ▱

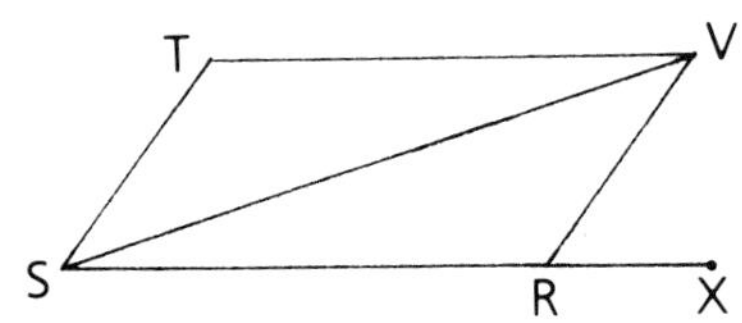

4 Given: ⊙ O
Conclusion: SMPR is a ▱

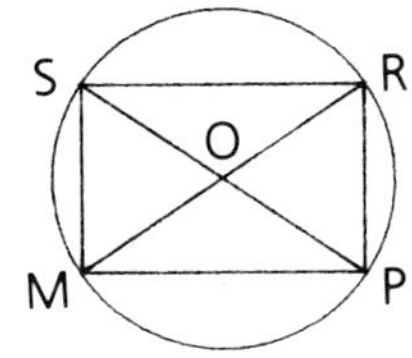

5 Given: YTWX is a ▱
$\overline{YP} \perp \overline{TW}$
$\overline{ZW} \perp \overline{TY}$
$\overline{TP} \cong \overline{TZ}$
Conclusion: TWXY is a rhombus

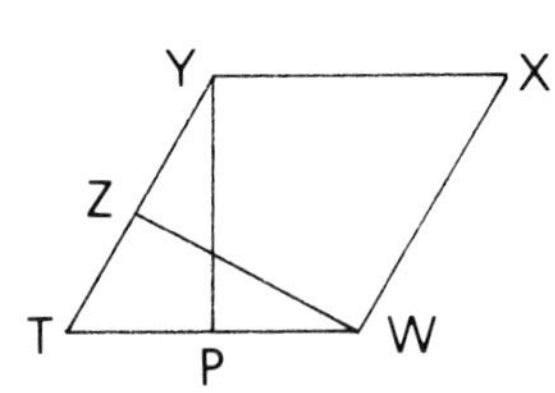

6 Given: $\overline{CD} \parallel \overline{AB}$
$\angle EDA \cong \angle CBF$
Prove: ABCD is a parallelogram

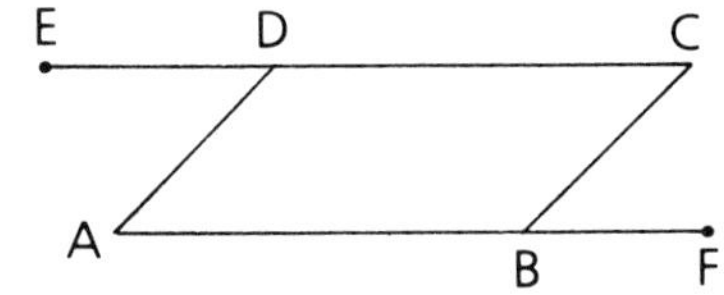

Section 5.6 Problem Set C

20 Given: $\overline{AC} \cong \overline{AD}$, and $\angle 1 \cong \angle 2$
$\overleftrightarrow{BE} \parallel \overleftrightarrow{CD}$, and $\overline{FC} \nparallel \overline{DG}$

Prove by Method 1:
FCDG is an isosceles trapezoid.

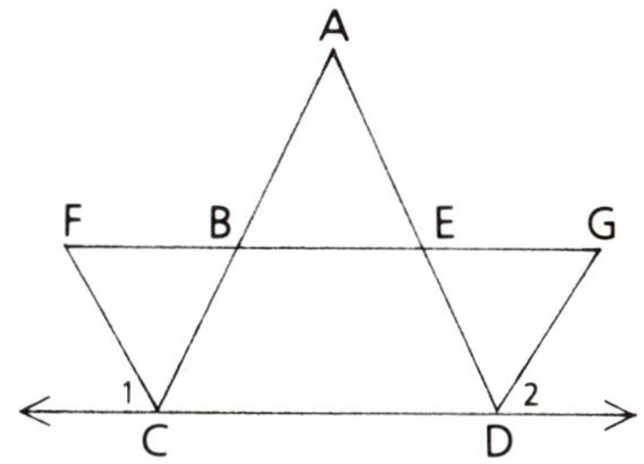

21 Given: △KOR is equilateral
KOPR is a ▱
KMOR is a ▱

Prove: △JMP is equilateral

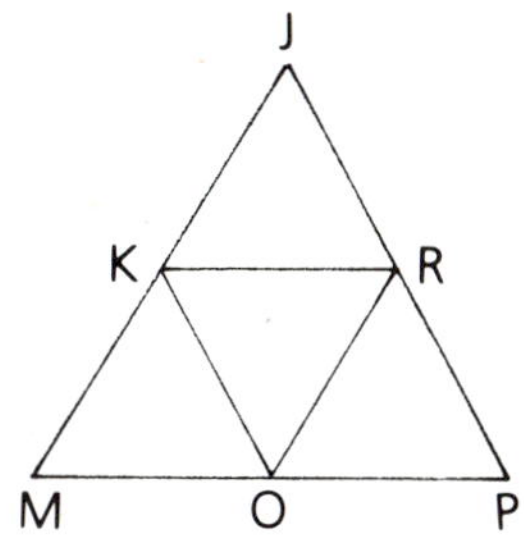

22 Prove that if the diagonals are ≅, a ▱ is a rectangle.
(Method 2 of proving a ▱ is a rectangle)

Section 5.6 Problem Set D

23 A rectangle and a parallelogram that is not a rectangle are in a box.

a If two of the eight angles are selected at random, what is the probability that the angles are congruent?

b A man offers to let you have two tries at getting a pair of ≅ angles. In other words, you would draw a pair of angles at random, then replace the pair, and then draw a pair again. The man is willing to bet you \$20 that you won't have a ≅ pair either time. Should you take the bet?

24 In trapezoid ABCD, $\overline{AC} \cong \overline{BD}$
$AD = x + 5$, $DC = 2x - 1$
$BC = 3x - 3$, $AB = 4x - 4$
State which two legs are parallel and find the perimeter.

Chapter 5 Study Guide

As you study and review this chapter, be sure that you are able to:

1 Explain the meaning of each line, mark, connection, and location of this Quadrilateral Family Tree. (p. 222)

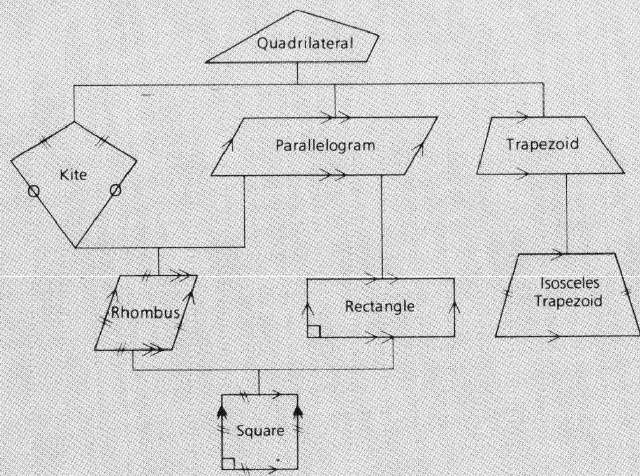

2 Check every property that MUST be true for each polygon (pp. 226–228):

Quadrilateral **Parallelogram** **Rhombus** **Trapezoid**

Kite **Rectangle** **Square** **Isos. Trapezoid**

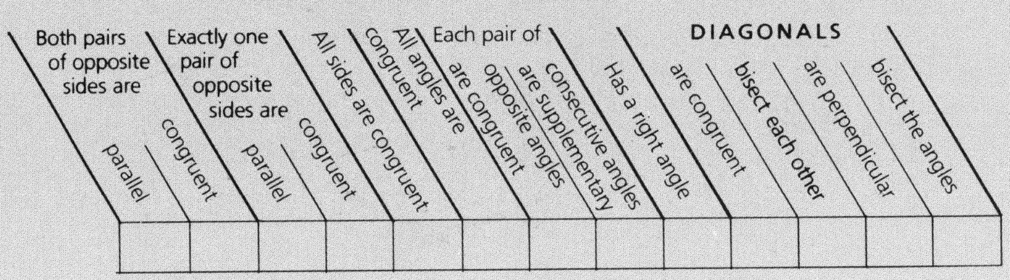

3 Give an example of each pair of angles (pp. 201, 206):

a alternate exterior angles
b alternate interior angles
c corresponding angles
d exterior angles on the same side of the transversal
e interior angles on the same side of the transversal

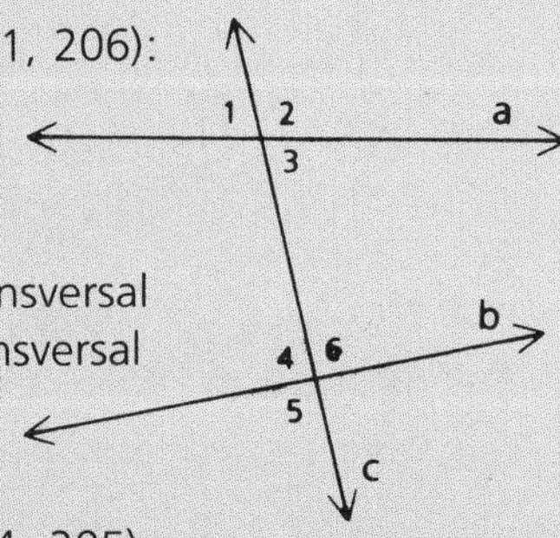

4 State which pairs of angles are congruent and which pairs are supplementary if a ∥ b. (pp. 204–205)

5 Explain the meaning of PAI, PCA, PAE, AIP, CAP, AEP. (pp. 204–205, 212)

6 Describe the PARALLEL POSTULATE. (p. 211)

7 Solve CROOK problems. (p. 214)

8 State the methods for proving that a quadrilateral is each of the following (pp. 233–235):

a a kite
b a parallelogram
c a rectangle
d a rhombus
e a square
f a trapezoid
g an isosceles trapezoid

The Royal Road: Geometry on a Sphere

Until recent times, people treated geometry as a description of the world. The terms *point* and *line* were undefined, but we all could describe **models** for points and for lines.

Recently, some mathematicians have looked at other models for geometry. Consider the surface of a sphere. A *point* is a location on the surface. You will find that the shortest distance between two points, which should determine a *line*, is a portion of a "great circle" of the sphere.

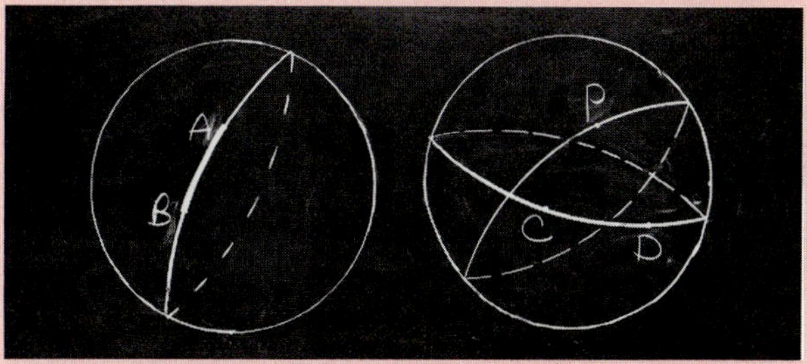

Most of the postulates of geometry are satisfied by this model. However, one postulate is not satisfied. Page 211 of the text presents the postulate that *through a given point not on a given line, there exists exactly one line parallel to the given line*. On a sphere, every "line" must meet every other "line"—every line through point P must meet line $\overleftrightarrow{CD}$.

Geometry on a sphere has many theorems that are different from theorems of Euclidean geometry. Draw three "lines" on a sphere that form a triangle. Can you find the sum of the measures of the three angles? Is that sum the same for triangles that are different sizes? For different shapes?

Chapter 5 Review Problems A

1 a Write out AIP in full b Write out PCA in full

2 ABCD is a ▱
AB = 2x + 6
BC = 8
CD = x + 8
Find the perimeter of ABCD

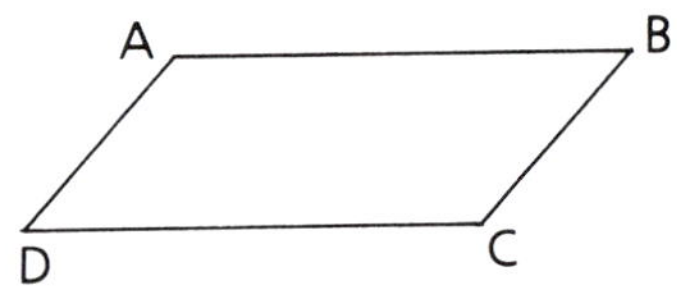

3 EFGH is a square
∠E = (x + 60)°
EF = x + 1
Find the perimeter of EFGH

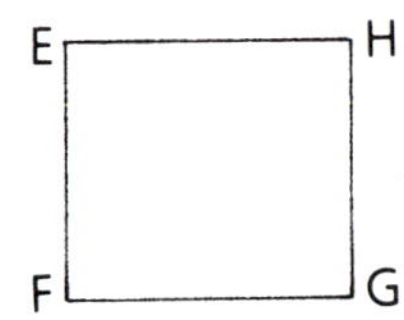

4 JKMP is a rectangle
PK = .2x
JM = x − 12
Find PK

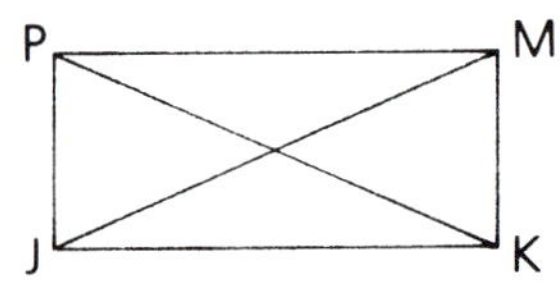

5 RSTV is a ▱
∠R = 70°
Find: m∠S, m∠T, m∠V

6 In a ▱, the measure of one of the angles is twice that of another. Are these opposite angles or consecutive angles? Find the measure of each angle of the ▱.

7 In each of these diagrams, is m ∥ p?

a

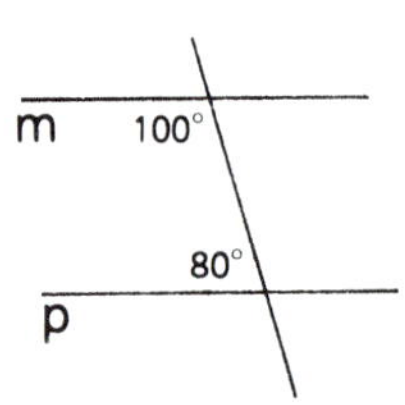

b

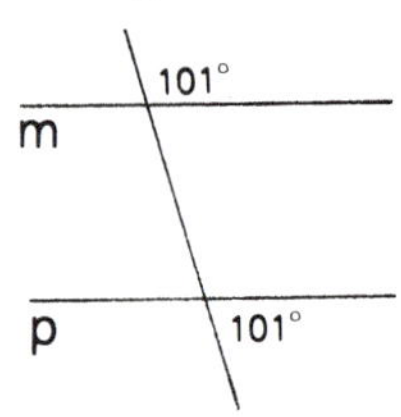

c

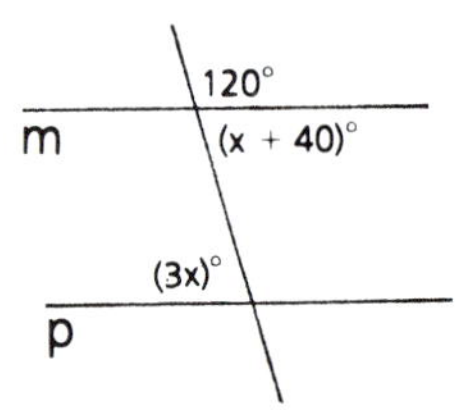

8 Name five properties of a parallelogram.

9 Given: $\overline{AB} \cong \overline{CD}$, and $\overline{AG} \cong \overline{BE}$
$\overline{AG} \parallel \overline{BE}$

Conclusion: $\overline{GC} \parallel \overline{ED}$

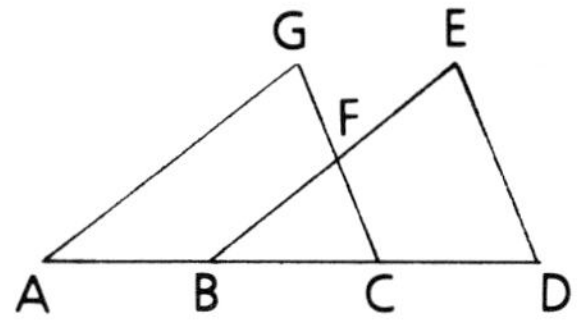

10 Given: HJKM is a ▱
$\angle JHP \cong \angle MKO$

Conclusion: $\overline{MP} \cong \overline{JO}$

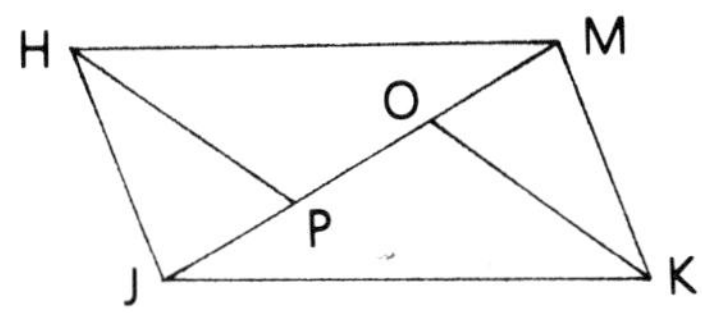

11 Given: RTBX is a ▱
$\overline{RS} \cong \overline{WB}$

Conclusion: $\overline{XT}$ bisects $\overline{SW}$

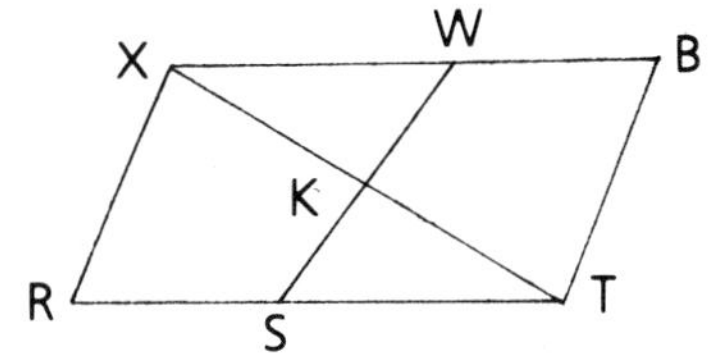

12 Given: $\triangle DEH \cong \triangle BGF$
$\triangle AEF \cong \triangle CGH$

Prove: **a** EFGH is a ▱
b ABCD is a ▱

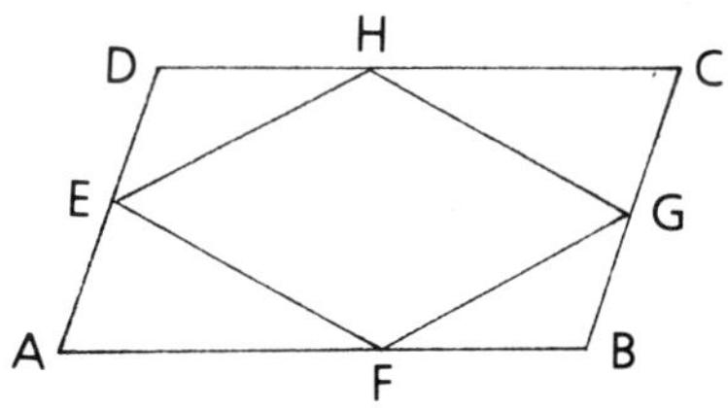

13 Given: $\overline{KR}$ is a median to $\overline{JO}$
$\overline{RP} \cong \overline{KM}$
$\overline{RM} \cong \overline{KM}$

Prove: JMOP is a ▱

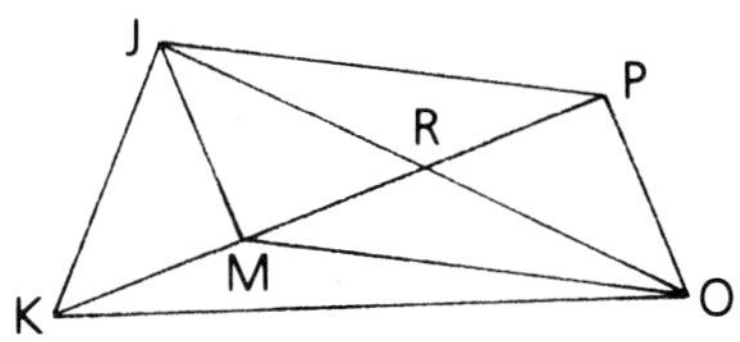

14 **a** How many squares appear to be in the figure to the right?
b How many rectangles?

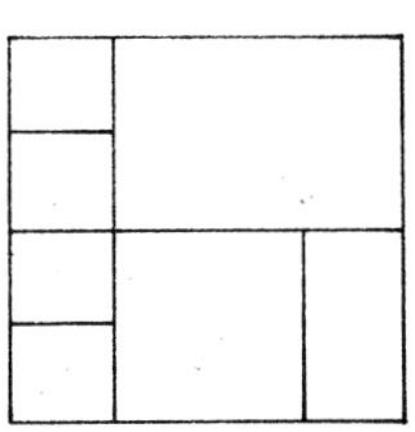

15 In ▱ ABCD, $\angle A = (2x + 6)^\circ$, $\angle B = (x + 24)^\circ$. Find m$\angle$C.

Chapter 5 Review Problems B

16 If the statement is always true, write A; if sometimes true, write S; if never true, write N (A, S, or N).

a If the diagonals of a quadrilateral are congruent, the quadrilateral is an isosceles trapezoid.

b If the diagonals of a quadrilateral divide each angle into two 45 degree angles, the quadrilateral is a square.

c If a parallelogram is equilateral, it is equiangular.

d If two of the angles of a trapezoid are congruent, the trapezoid is isosceles.

e The perpendicular bisector of the lower base of a non-isosceles trapezoid passes through the point of intersection of the diagonals.

17 Prove: If the consecutive midpoints of a parallelogram are joined, the resulting figure is a parallelogram.

18 Prove: If, in a triangle, the bisector of an exterior angle formed by extending one of the sides is parallel to a side of the triangle, the triangle is isosceles.

19 Given: EFGH is a ▱
$\overline{AE} \cong \overline{BF} \cong \overline{CG} \cong \overline{DH}$

Prove: ABCD is a ▱

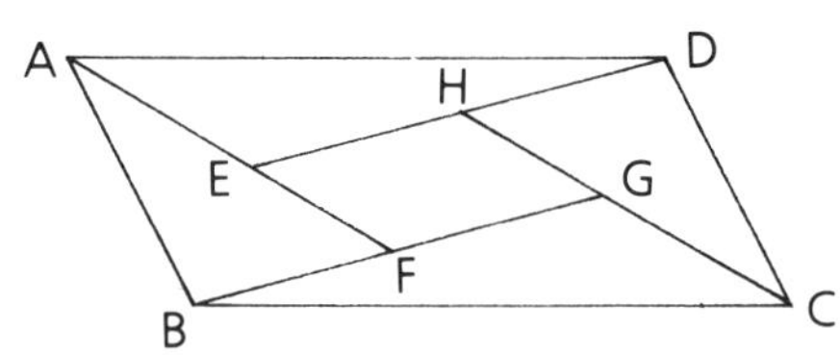

20 Given: $\angle 1 \cong \angle 2$
$\overline{PS} \cong \overline{KS}$
P is the midpt. of $\overline{RO}$
K is the midpt. of $\overline{JM}$

Prove: RJMO is a ▱

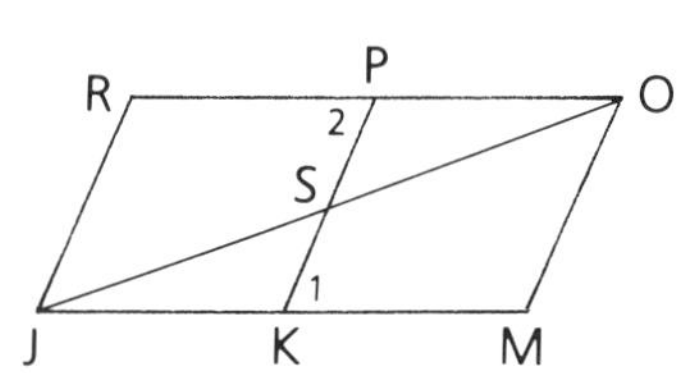

21 Find x.

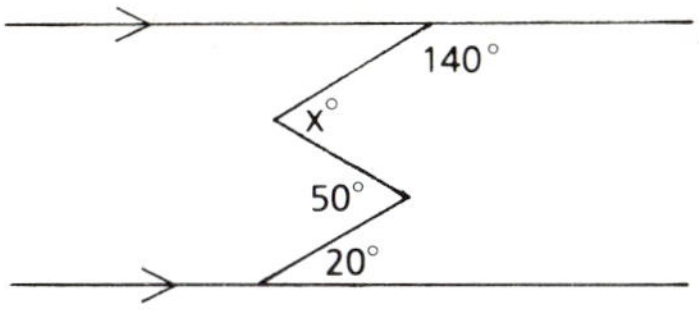

22 Given: $\triangle TWX$ is isosceles with base $\overline{WX}$
$\overline{RY} \parallel \overline{WX}$

Prove: RWXY is an isosceles trapezoid

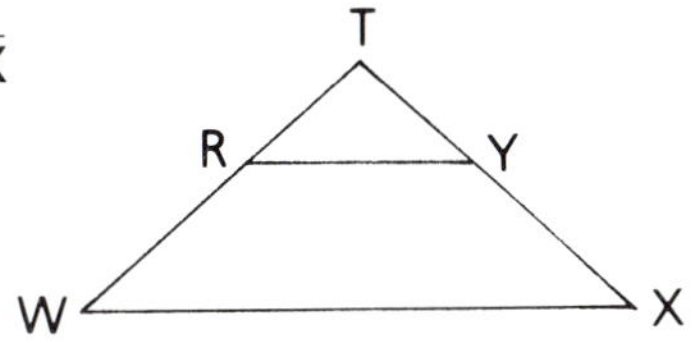

23 If two of the five labeled angles are chosen at random, what is the probability that they are supplementary?

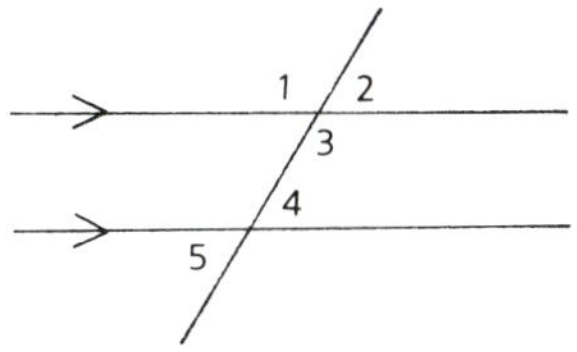

Chapter 5 Review Problems C

24 Given: $\overline{AB} \cong \overline{DC}$
$\overline{AB}$ and $\overline{DC} \perp \overline{BC}$

Prove: $\triangle DEC$ is isosceles

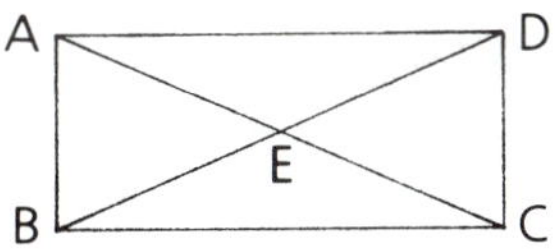

25 Given: $\triangle AED$ and $\triangle BEC$ are isosceles with congruent bases $\overline{AD}$ and $\overline{BC}$

Prove: ABCD is a rectangle

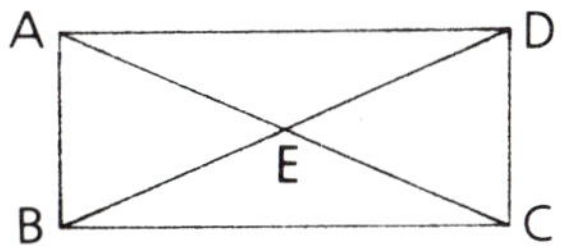

Chapter 5 Review Problems D

26 Given: eight segments with lengths of 6, 6, 6, 9, 9, 5, 5, and 7. If four of the eight segments are selected at random, what is the probability that they could be joined together to form a parallelogram?

Chapter 6

LINES AND PLANES IN SPACE

Section 6.1 Introductory Ideas

In this section the following will be discussed:

A PLANES, NOTATION, COPLANAR, NON-COPLANAR, FOOT

B METHODS OF DETERMINING A PLANE

C FURTHER ASSUMPTIONS CONCERNING LINES AND PLANES

6.1 (A) Planes, Notation, Coplanar, Non-Coplanar, Foot

Recall from Chapter 5 Section 5.1(A) the definition of a plane:

D **A PLANE is a surface such that if any two points on the surface are connected by a line, all points of the line will lie on the surface.**

Because a surface has no thickness, a plane must be "flat" if it is to contain the straight lines through all pairs of points. It must also be infinitely long and wide. Thus, a plane has just two dimensions, length and width.

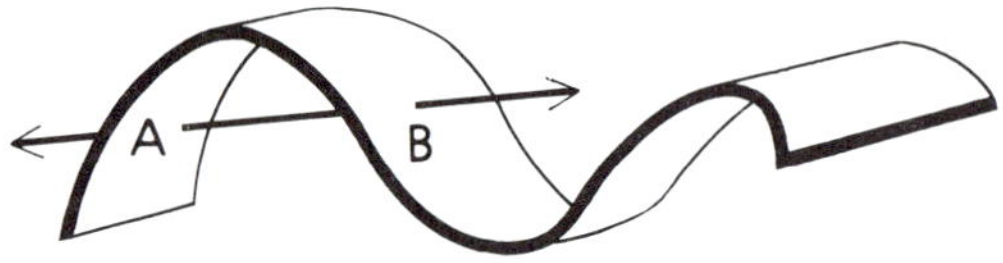

A surface that is not a plane

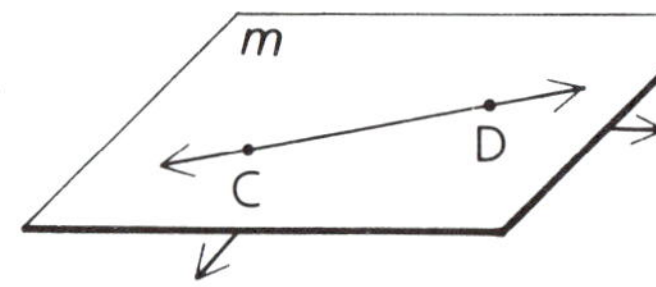

Plane surface

To give the appearance of a plane surface, a plane is frequently drawn as shown in the diagram above, right. In this case, the plane is made to appear to be horizontal, with heavily darkened edges closest to you. A plane can be named by placing a single lower-case letter in one of the corners.

It is most important to understand that while our *picture* of a plane has edges and corners, an actual plane has neither, and should be thought of as infinite in length and width.

You may recall the following definitions from Section 5.1(A):

D **If points, lines, segments, etc., lie in the same plane, we call them COPLANAR. Similarly, points, lines, segments, etc., which do not all lie in the same plane are called NON-COPLANAR.**

In the diagram below,
$\overleftrightarrow{AB}$ and $\overleftrightarrow{ST}$ lie in plane m. $\overleftrightarrow{RP}$ does not lie in the plane,
but intersects m at V.

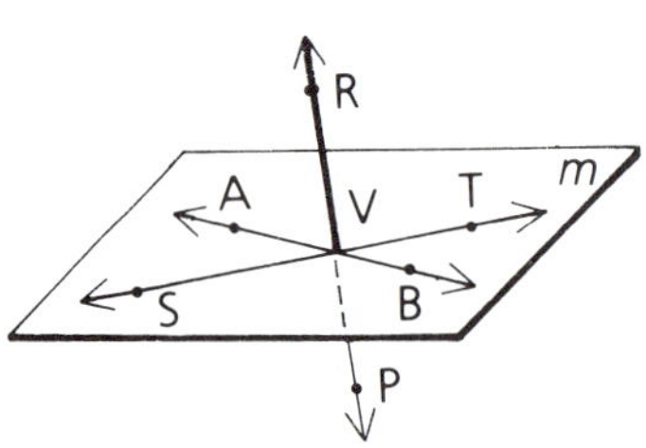

A, B, S, T, and V are COPLANAR points.
$\overleftrightarrow{AB}$ and $\overleftrightarrow{ST}$ are COPLANAR lines.
$\overline{AB}$ and $\overline{ST}$ are COPLANAR segments.

A, B, S, T, and R are NON-COPLANAR points.
$\overleftrightarrow{AB}$, $\overleftrightarrow{ST}$, and $\overleftrightarrow{RP}$ are NON-COPLANAR lines.
$\overline{AB}$, $\overline{ST}$, and $\overline{RP}$ are NON-COPLANAR segments.

D **The point of intersection of a line and a plane is called the FOOT of the line.**

In the diagram above, V is the foot of $\overleftrightarrow{RP}$ in plane m.

6.1 (B) Methods of Determining a Plane

In Chapter 3 we learned the following:

P **Two points determine a line.**

We would now like to find conditions under which a *plane* is determined.

One point will obviously not determine a plane since there are infinitely many planes passing through a single point.

The diagram to the right shows that two points will *not* determine a unique plane. It shows two different planes, *m* and *n*, each of which contains both points A and B. The same diagram shows that three points, A, B, and C, will not determine a plane if the three points are collinear.

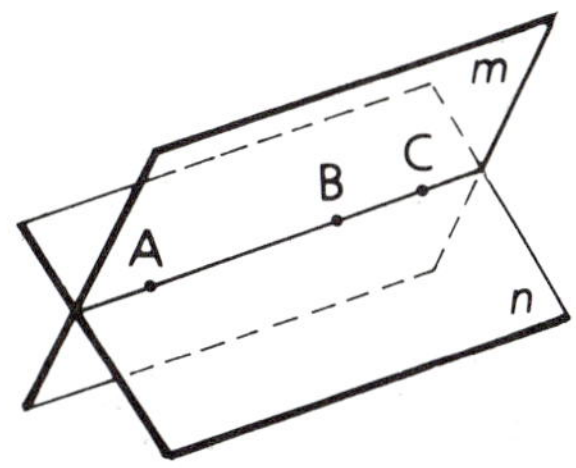

However, if the three points are non-collinear, they will determine a plane:

There is one and only one plane that contains the three non-collinear points A, B, and C. This plane can be named either plane ABC or plane *k*.

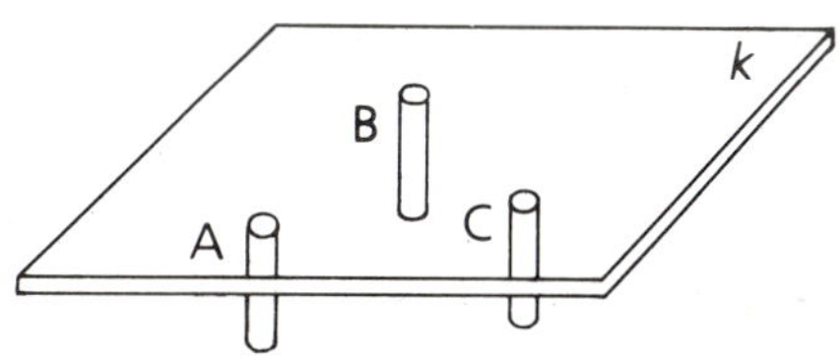

The above discussion is formalized in the following postulate:

P **Three non-collinear points determine a plane.**

There are other methods of determining a plane. Following are three stated as theorems.

Theorem 37

T **A line and a point not on the line determine a plane.**

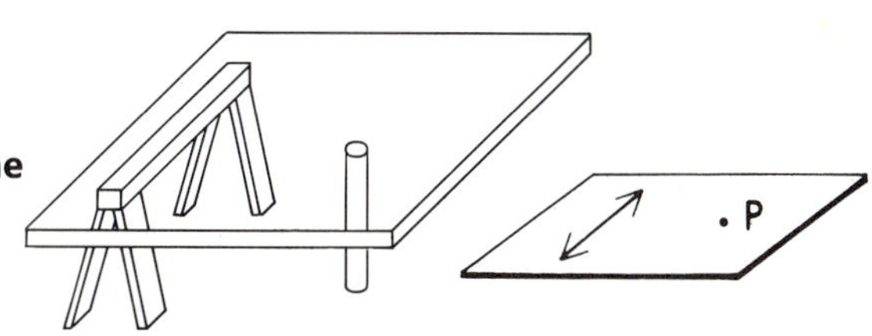

Theorem 38

T **Two intersecting lines determine a plane.**

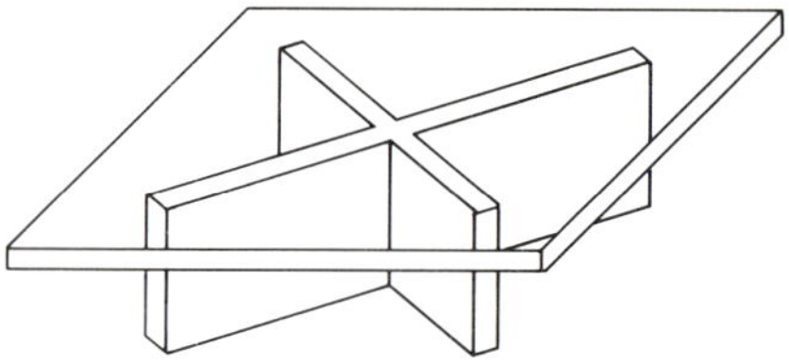

The proofs of Theorems 37 and 38 are left to the student in Problem Set B.

Theorem 39

T **Two parallel lines determine a plane.**

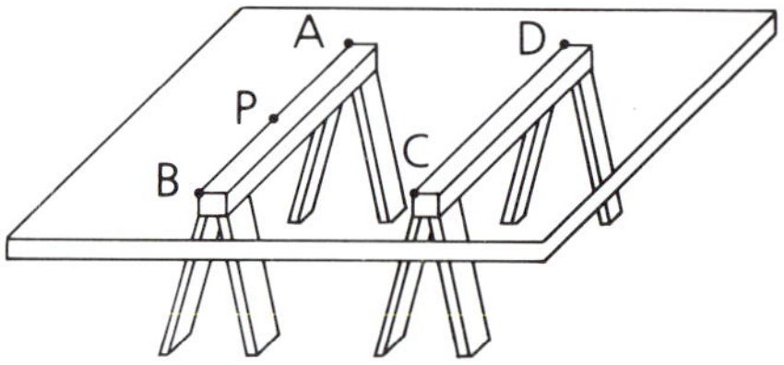

Proof of Theorem 39: If $\overleftrightarrow{AB}$ and $\overleftrightarrow{CD}$ are parallel, then from the definition of parallel lines they lie in one plane. We need to show that they lie in *only one* plane. If P is any point on $\overleftrightarrow{AB}$, then by Theorem 37 there is only one plane containing P and $\overleftrightarrow{CD}$. Thus, there is only one plane that contains $\overleftrightarrow{AB}$ and $\overleftrightarrow{CD}$, because every plane containing $\overleftrightarrow{AB}$ contains P.

Section 6.1 (C) Further Assumptions Concerning Lines and Planes

We shall assume the following two postulates:

P **If a line intersects a plane not containing it, then the intersection is exactly one point.**

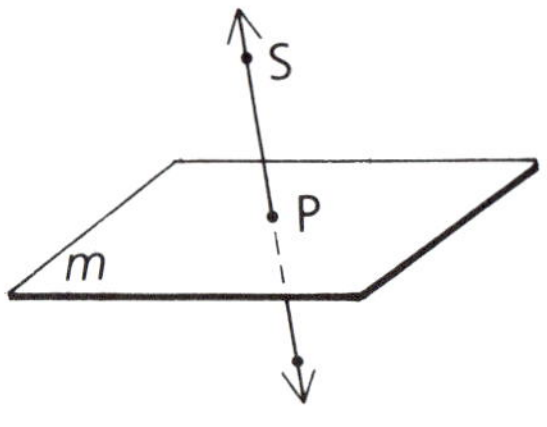

P **If two planes intersect, their intersection is exactly one line.**

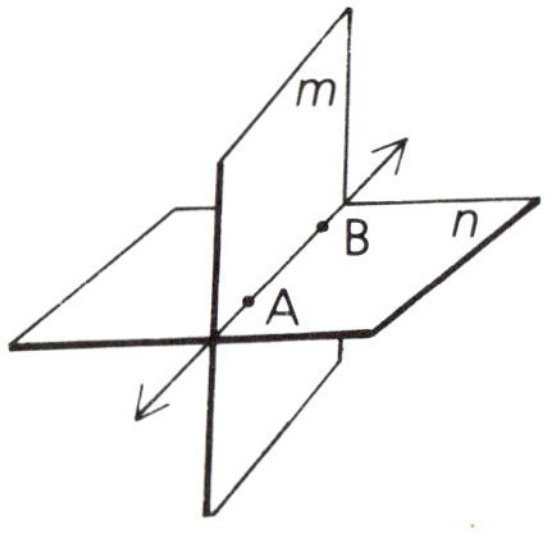

Section 6.1 Sample Problems

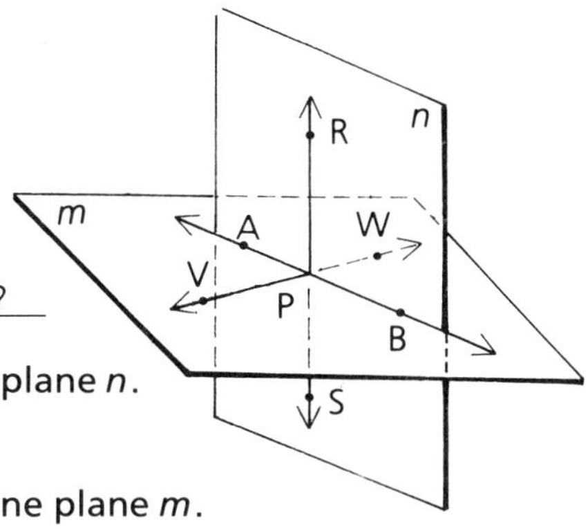

1 **a** $m \cap n =$ __?__

b A, B, V determine plane __?__

c Name the foot of $\overleftrightarrow{RS}$ in m.

d $\overleftrightarrow{AB}$ and $\overleftrightarrow{RS}$ determine plane __?__

e $\overleftrightarrow{AB}$ and point __?__ determine plane n.

f Does W lie in plane n?

g Line AB and line __?__ determine plane m.

h A, B, V and __?__ are coplanar points.

i A, B, V and __?__ are non-coplanar points.

j If R and S lie in plane n, what can be said concerning $\overleftrightarrow{RS}$?

Answers

a $\overleftrightarrow{AB}$
b m
c P
d n
e R or S
f No
g VW
h W or P
i R or S
j $\overleftrightarrow{RS}$ lies in plane n

Note: There are other planes besides the two you see in the diagram. For example, the non-collinear points R, P, and V determine a plane.

2 Given: A, B, C lie in plane m
$\overline{PB} \perp \overline{AB}$
$\overline{PB} \perp \overline{BC}$
$\overline{AB} \cong \overline{BC}$

Prove: $\angle APB \cong \angle CPB$

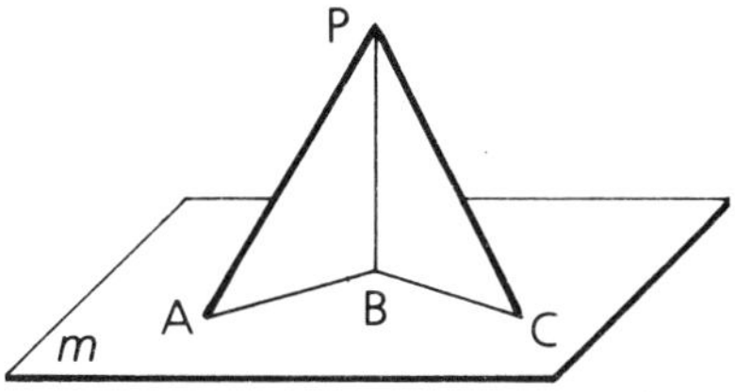

1 $\overline{PB} \perp \overline{AB}$, $\overline{PB} \perp \overline{BC}$	1 Given
2 $\angle PBA$ and $\angle PBC$ are right $\angle$s	2 $\perp$ lines form right $\angle$s
3 $\angle PBA \cong \angle PBC$	3 Right angles are congruent
4 $\overline{AB} \cong \overline{BC}$	4 Given
5 $\overline{PB} \cong \overline{PB}$	5 Reflexive
6 $\triangle PBA \cong \triangle PBC$	6 SAS (4, 3, 5)
7 $\angle APB \cong \angle CPB$	7 CPCTC

Section 6.1 Problem Set A

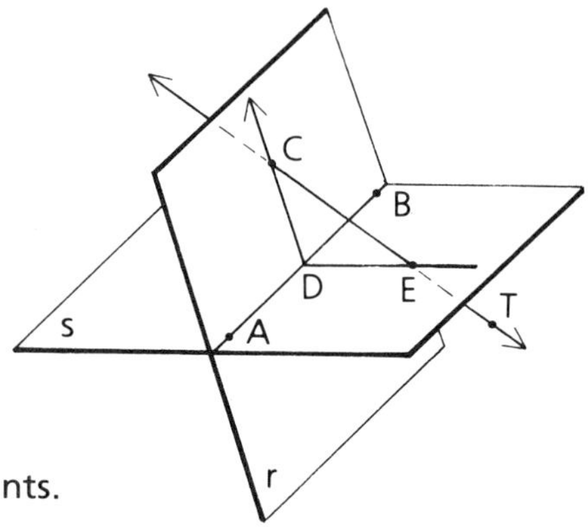

1 a $r \cap s =$ ___?___

b $\overleftrightarrow{AB} \cap s =$ ___?___

c Name three collinear points.

d Name four non-coplanar points.

e Points A, B, E determine what plane?

f $\overleftrightarrow{AB}$ and $\overleftrightarrow{ED}$ determine what plane?

g Name the foot of $\overleftrightarrow{TC}$ with respect to plane s.

h Name the foot of $\overleftrightarrow{TC}$ with respect to plane r.

i Do $\overleftrightarrow{CD}$ and $\overleftrightarrow{ED}$ determine a plane?

j If $\overleftrightarrow{CD} \perp \overleftrightarrow{AB}$, name the right angles formed.

2 Consider some spherical object, such as an orange or a globe. If two points are marked on it and a straight line is drawn through the two points, does the line lie on the surface? Is it possible to draw a straight line that will lie entirely on the surface?

3 Choose any two points on a cylindrical surface, such as the curved surface of a tin can. Does the line connecting the two points *always* lie on the surface? Does it *ever* lie on the surface?

4 Make a free-hand sketch of each of the following: (a) a horizontal plane, (b) a vertical plane, and (c) two intersecting planes.

5 A three-legged milking stool will not rock, even if the legs are of different lengths. Many four-legged stools wobble. Explain.

6 What theorem or assumption of this chapter provides the *best* explanation for the following: When you saw a board, the edge of the cut is a straight line.

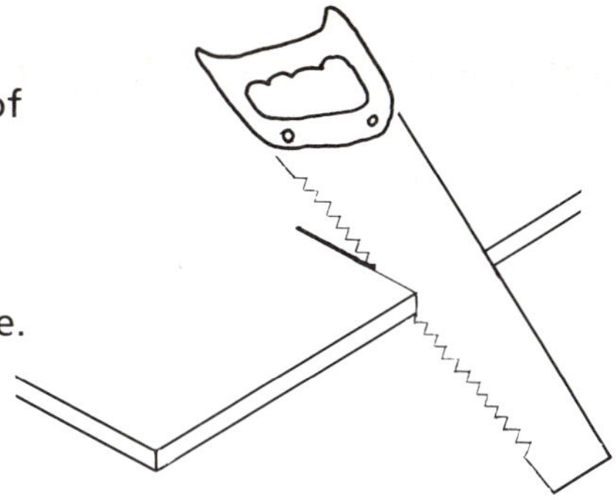

Section 6.1 Problem Set **B**

7 Given: A, P, B lie in plane *m* as shown
$\overleftrightarrow{CP} \perp \overleftrightarrow{AP}$ and $\overleftrightarrow{PB}$
$\overline{PA} \cong \overline{PB}$

Prove: $\overline{CA} \cong \overline{CB}$

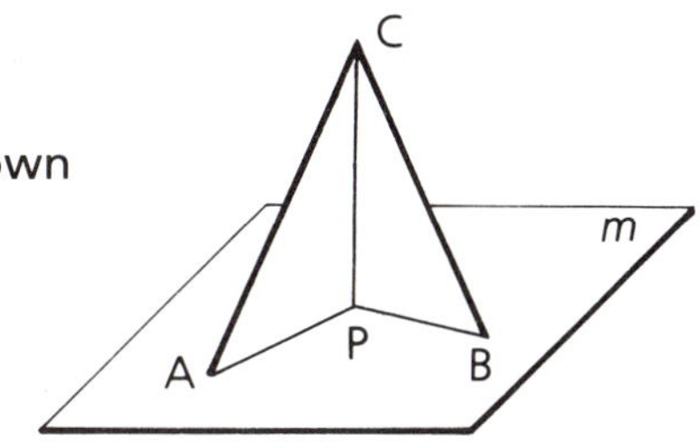

8 Given: ⊙ O lies in plane *p*
$\overleftrightarrow{VO} \perp \overleftrightarrow{OS}$
$\overleftrightarrow{VO} \perp \overleftrightarrow{OT}$

Prove: $\angle VSO \cong \angle VTO$

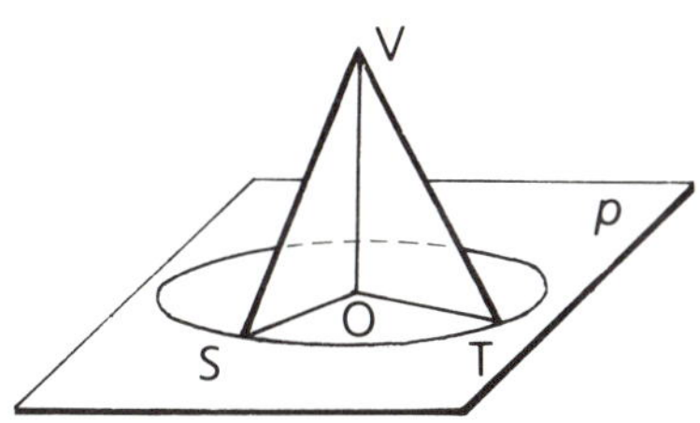

9 Prove Theorem 37: A line and a point not on the line determine a plane. (Use a paragraph form.)

10 Prove Theorem 38: Two intersecting lines determine a plane. (Use a paragraph form.)

11 Can you hold two pencils so that they do not intersect and are not parallel? Are they coplanar? Lines that do not intersect and that are not coplanar are called SKEW lines.

12 Cut a quadrilateral out of paper and fold it along a diagonal as shown in the figure. Is every four-sided figure a plane figure?

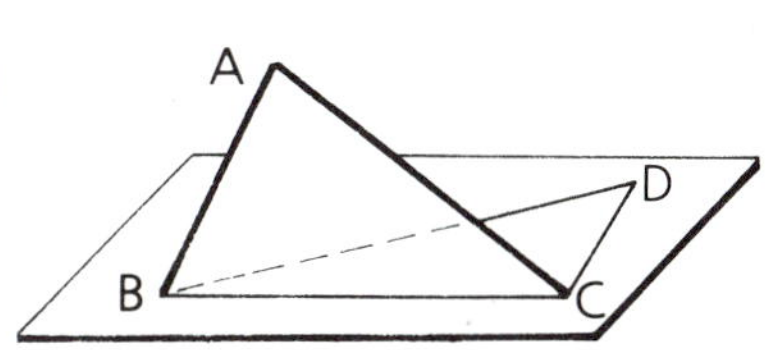

13 If two points in *space* are equidistant from the endpoints of a segment, will the line joining them be the perpendicular bisector of the segment? Explain.

14 Given: Planes *m* and *n* intersect in $\overleftrightarrow{RS}$
m contains R, S, and V
n contains R, S, and T
$\overline{TS} \cong \overline{VR}$
$\overline{TR} \perp \overline{RS}$
$\overline{VS} \perp \overline{RS}$
Prove: $\overline{TR} \cong \overline{VS}$

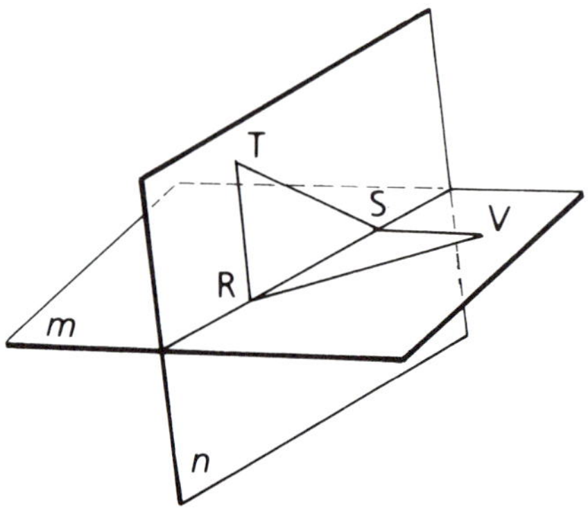

Section 6.1 Problem Set C

15 The figure to the right is a square pyramid. How many planes are determined by its vertices? There are more than 5. Name them.

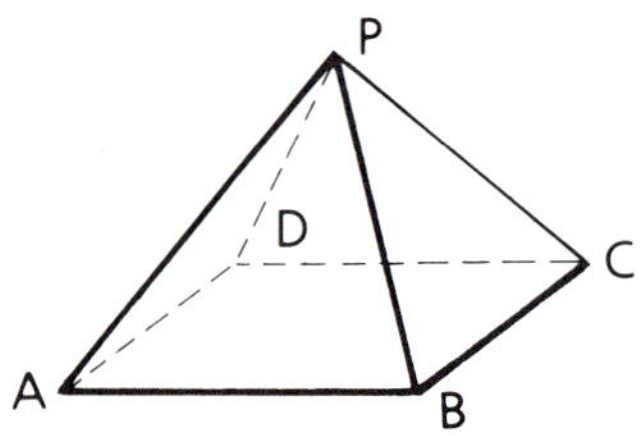

16 Given: A, B, and X lie in plane *m*
X is on $\overline{AB}$; P and Q are above *m*
B is equidistant from P and Q
A is equidistant from P and Q
Prove: X is equidistant from P and Q

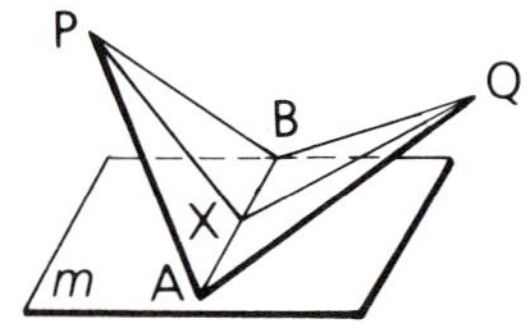

17 Given: $\triangle ABC \cong \triangle DBC$
Prove: $\triangle AXD$ is isosceles

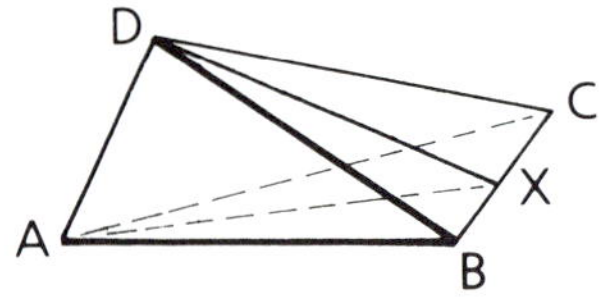

18 Given: A, B, C, D lie in plane *m*
$\overleftrightarrow{ST}$ intersects *m* at B
D is any point on $\overline{AC}$
$\overleftrightarrow{ST} \perp \overleftrightarrow{AB}$ and $\overleftrightarrow{BC}$
$\overline{SB} \cong \overline{TB}$
Prove: $\overline{ST} \perp \overleftrightarrow{BD}$

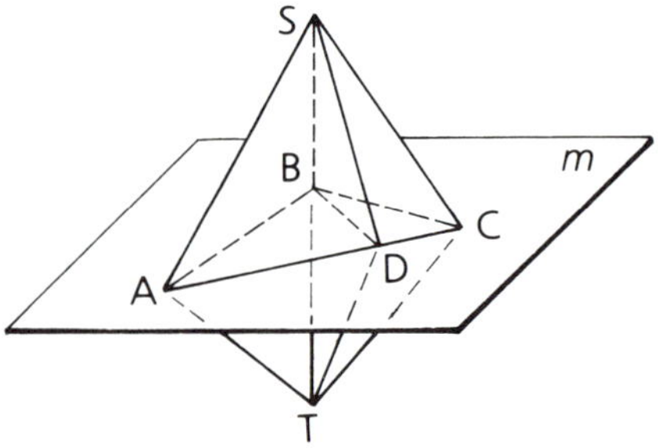

Section 6.2 Perpendicularity Between a Line and Plane

In this section we discuss the following concepts:

A DEFINITION OF A LINE PERPENDICULAR TO A PLANE

B THE BASIC THEOREM CONCERNING PERPENDICULARITY BETWEEN A LINE AND A PLANE

6.2 (A) Definition of a Line Perpendicular to a Plane

D **A LINE PERPENDICULAR to a PLANE is a line perpendicular to *every* line in the plane that passes through its FOOT.**

Observe that we now have two kinds of perpendicularity:

1 between two lines: $\overleftrightarrow{AB} \perp \overleftrightarrow{BD}$

2 between a line and a plane: $\overleftrightarrow{AB} \perp m$

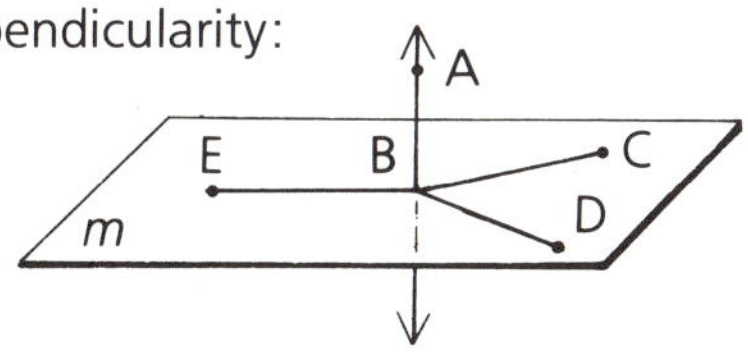

The definition above is a very powerful statement, because of the word *every*. If, in the diagram we are given the fact that $\overleftrightarrow{AB} \perp m$, then we know *three* conclusions:

1 $\overleftrightarrow{AB} \perp \overleftrightarrow{BC}$ 2 $\overleftrightarrow{AB} \perp \overleftrightarrow{BD}$ 3 $\overleftrightarrow{AB} \perp \overleftrightarrow{BE}$

6.2 (B) The Basic Theorem Concerning Perpendicularity Between a Line and a Plane

We have seen in the previous sub-section that the definition of a line perpendicular to a plane will be very useful when it is *given* that a line is perpendicular to a plane. Now, we wish to reverse the situation. How do we *prove* that a line is indeed perpendicular to a plane? To apply the definition in reverse we would have to show that the line is perpendicular to *every* line in the plane that passes through its foot. To consider the infinitely many lines one by one would be an endless process.

If a line is perpendicular to only *one* line that lies in the plane and passes through its foot, will it be perpendicular to the plane? Or, must it be perpendicular to *two, three,* or *four* lines in order to be perpendicular to the plane? The following theorem answers that question.

Theorem 40

T **If a line is perpendicular to TWO distinct lines that lie in a plane and that pass through its foot, then it is perpendicular to the plane.**

Given: $\overleftrightarrow{BF}$ and $\overleftrightarrow{CF}$ lie in plane m
$\overleftrightarrow{AF} \perp \overleftrightarrow{FB}$
$\overleftrightarrow{AF} \perp \overleftrightarrow{FC}$

Prove: $\overleftrightarrow{AF} \perp m$

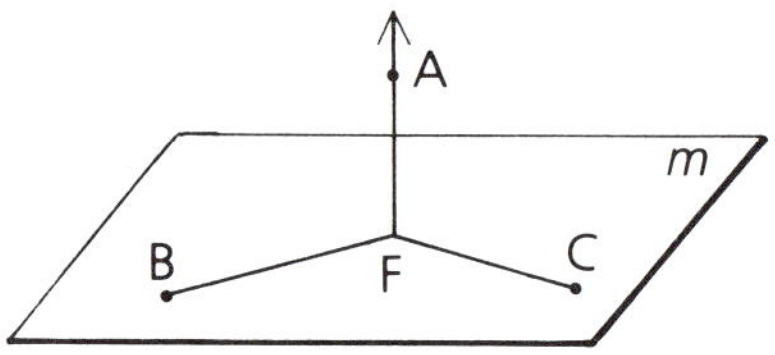

The proof is left as a challenge. You may have already done part of it as problem 18 of Section 6.1.

Section 6.2 Sample Problems

1 **If $\angle$STR is a right $\angle$, can you conclude that $\overleftrightarrow{ST} \perp m$? Why, or why not?**

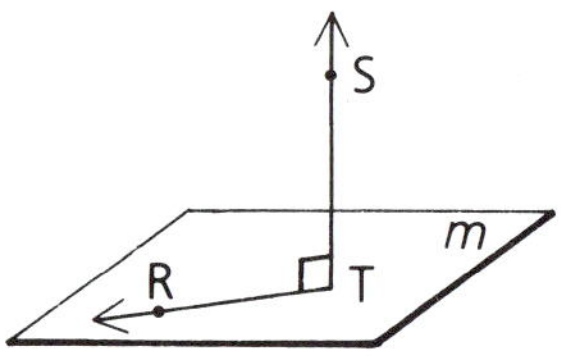

Ans: No. To be $\perp$ to plane m, $\overleftrightarrow{ST}$ must be $\perp$ to at least *two* lines that lie in m and pass through T, the foot.

2 **Given: $\overline{PF} \perp k$**
$\overline{PG} \cong \overline{PH}$
Prove: $\angle G \cong \angle H$

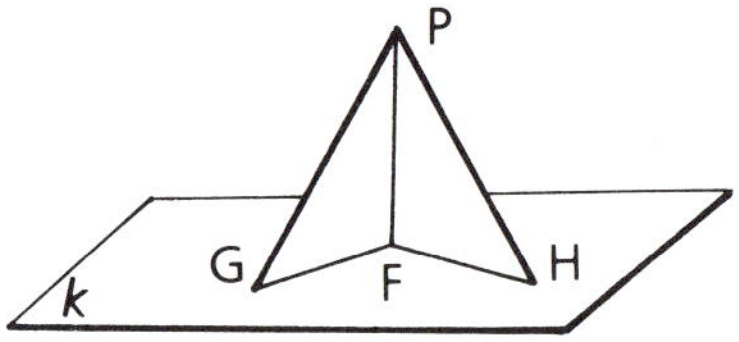

1 $\overline{PF} \perp k$	1 Given
2 $\overline{PF} \perp \overline{FG}$ $\overline{PF} \perp \overline{FH}$	2 If a line is $\perp$ to a plane, it is $\perp$ to every line in the plane that passes through its foot
3 $\angle$PFG is a right $\angle$ $\angle$PFH is a right $\angle$	3 $\perp$ lines form right $\angle$s
4 $\overline{PG} \cong \overline{PH}$	4 Given
5 $\overline{PF} \cong \overline{PF}$	5 Reflexive
6 $\triangle PFG \cong \triangle PFH$	6 HL (3, 4, 5)
7 $\angle G \cong \angle H$	7 CPCTC

3 Given: B, C, D, and E lie in plane n

$\overline{AB} \perp n$

$\overleftrightarrow{BE} \perp$ bis $\overline{CD}$

Prove: $\triangle ADC$ is isosceles

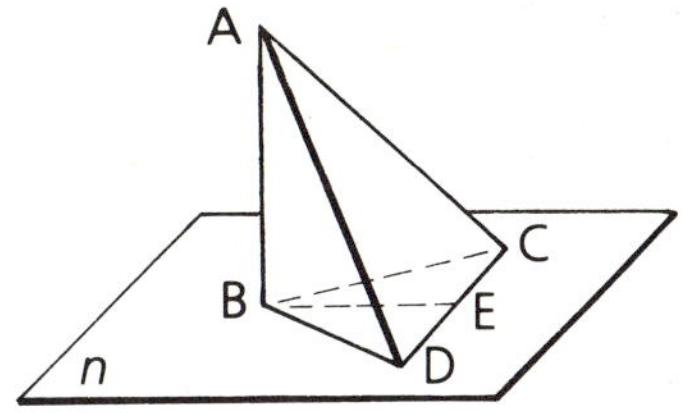

1 $\overline{AB} \perp n$	1 Given
2 $\overline{AB} \perp \overline{BD}$ $\overline{AB} \perp \overline{BC}$	2 If a line is $\perp$ to a plane, it is $\perp$ to every line in the plane that passes through its foot
3 $\angle ABC$ is a right $\angle$ $\angle ABD$ is a right $\angle$	3 Perpendicular lines meet to form right angles
4 $\angle ABC \cong \angle ABD$	4 All right angles are $\cong$
5 $\overleftrightarrow{BE} \perp$ bis $\overline{CD}$	5 Given
6 $\overline{BC} \cong \overline{BD}$	6 If a point is on the $\perp$ bis of a segment, then it is equidistant from the endpoints
7 $\overline{AB} \cong \overline{AB}$	7 Reflexive
8 $\triangle ABC \cong \triangle ABD$	8 SAS (6, 4, 7)
9 $\overline{AD} \cong \overline{AC}$	9 CPCTC
10 $\triangle ADC$ is isosceles	10 A $\triangle$ with two $\cong$ sides is isosceles

Section 6.2 Problem Set A

1 ▱ABCD is a square that lies in plane t.

$\overleftrightarrow{PF} \perp t$

How many right angles are there in the figure?

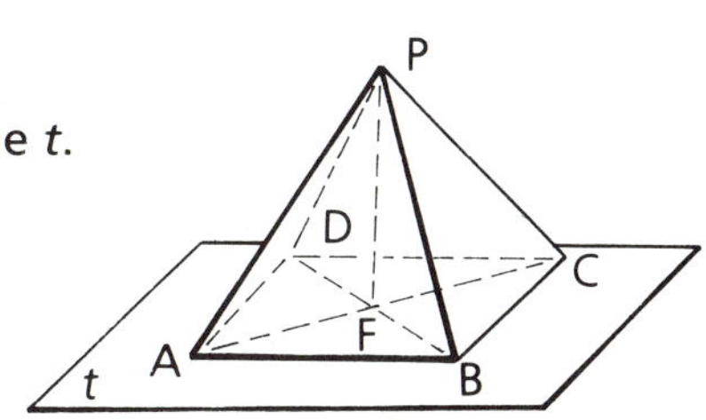

2 Given: $\overleftrightarrow{PB} \perp m$

$\angle APB \cong \angle CPB$

Prove: $\overline{AB} \cong \overline{CB}$

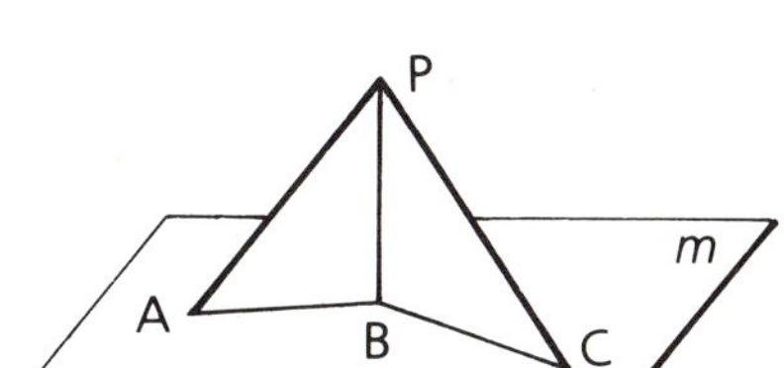

3 Given: ⊙ O lies in plane n

$\overline{RO} \perp n$

Prove: $\overline{RS} \cong \overline{RT}$

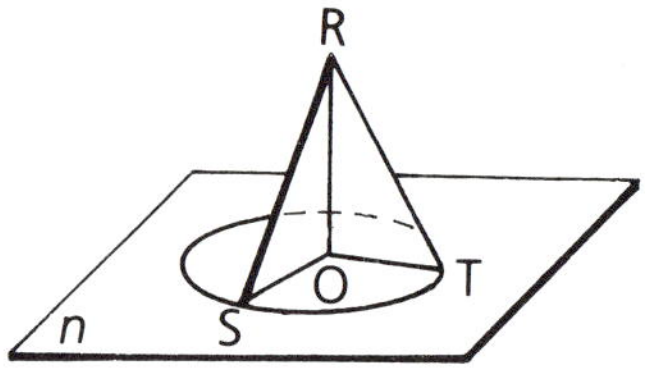

4 Given: $\overleftrightarrow{TS} \perp m$

$\overline{PV}$ bisects $\overline{TS}$

Prove: $\overrightarrow{PV}$ bisects $\angle TPS$

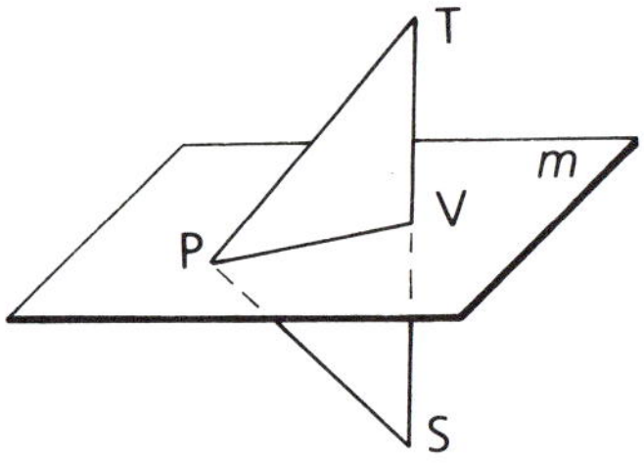

5 Given: $\overleftrightarrow{AB}$ and $\overleftrightarrow{CD}$ lie in plane s

$\overleftrightarrow{PT} \perp s$

$\overline{PC} \cong \overline{PD}$

$\overline{PA} \cong \overline{PB}$

Prove: T is the midpt. of $\overline{AB}$ and $\overline{CD}$

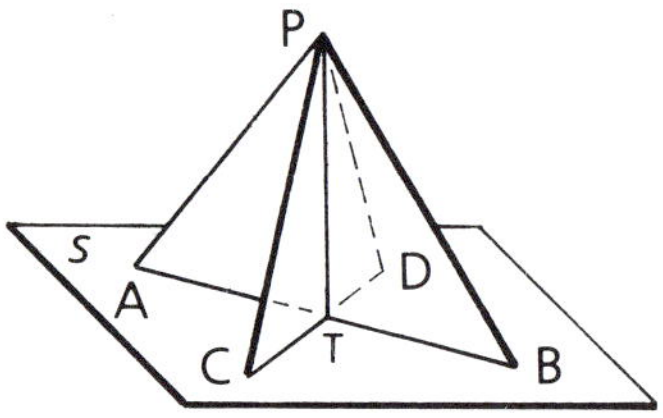

Section 6.2 Problem Set B

6 Given: Q, R, S, T lie in plane m

$\angle PQR$ and $\angle PQT$ are right $\angle$s

Prove: $\angle PQS$ is a right $\angle$

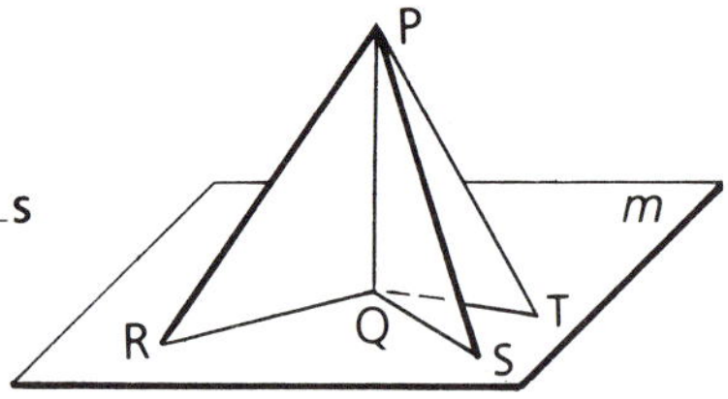

7 If $\overline{AB} \perp \overline{BD}$,

$m\angle ABD = \frac{2}{3}x + 56$ and

$m\angle ABC = 2x - 10$,

is $\overline{AB} \perp m$?

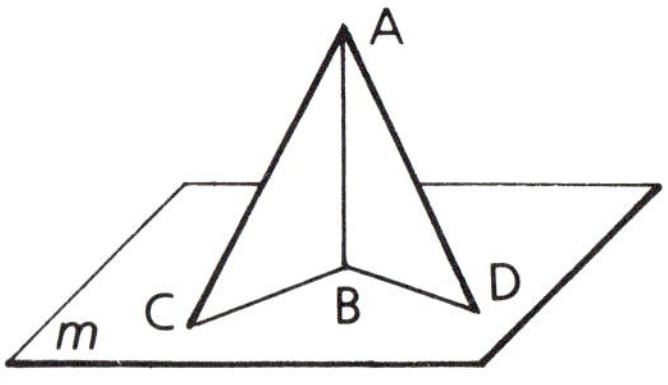

8 Given: $\overline{AB} \perp m$

Equilateral $\triangle DBC$ lies in plane m

Prove: $\triangle ACD$ is isosceles

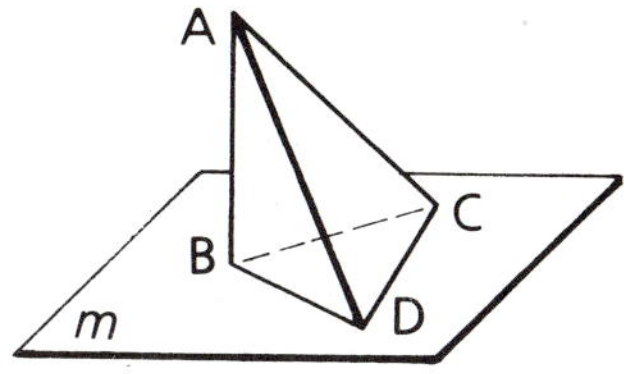

9 Given: $\overline{PA} \perp s$

P is equidistant from B and C

Prove: A is equidistant from B and C

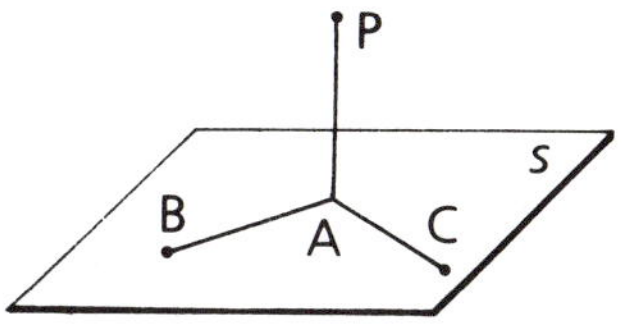

10 Given: $\overline{AB} \perp n$

$\overline{CD} \perp n$

$\overline{AC}$ bisects $\overline{BD}$

Prove: $\overleftrightarrow{BD}$ bisects $\overline{AC}$

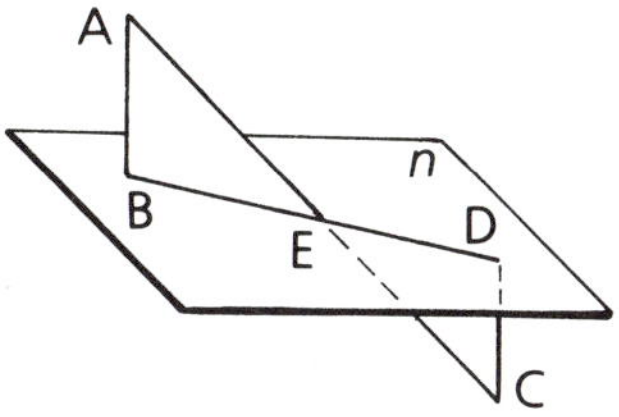

11 Given: $\overleftrightarrow{PB} \perp m$

D is the midpt. of $\overline{AC}$

$\triangle PAC$ is isosceles with base $\overline{AC}$

Prove: $\overleftrightarrow{BD} \perp$ bis $\overline{AC}$

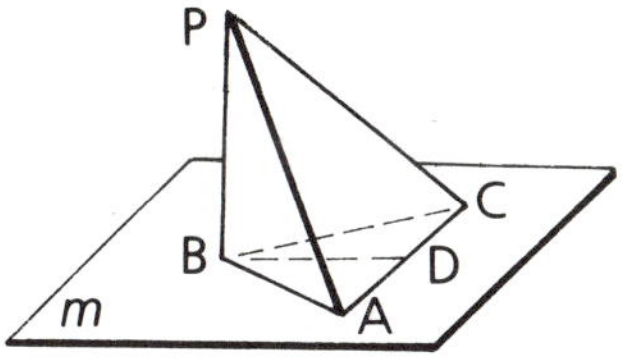

12 From any point on a line perpendicular to a plane, two lines are drawn oblique to the plane. If the foot of the perpendicular is equidistant from the feet of the oblique lines, prove that the oblique segments are congruent.

Section 6.2 Problem Set C

13 Given: $\overline{EF} \perp \overline{CF}$

$\overline{CE} \cong \overline{DE}$

$\angle FCD \cong \angle FDC$

Prove: $\overline{EF} \perp m$

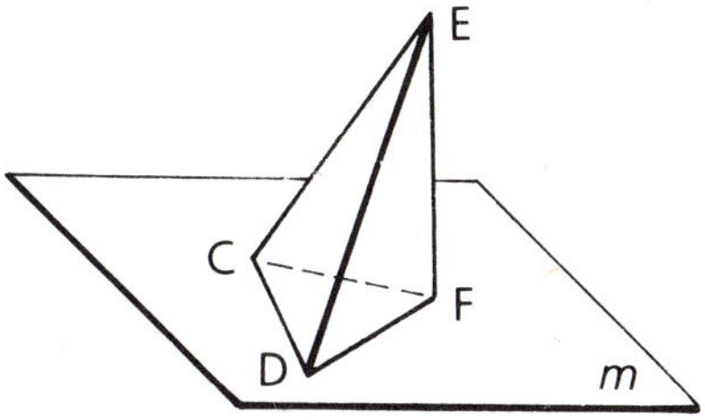

14 Given: $\overleftrightarrow{AD}$ and $\overleftrightarrow{BC}$ intersect at E
$\overleftrightarrow{AC} \perp m$ and n
$\overleftrightarrow{BD} \perp m$ and n

Prove: $\overline{AD} \cong \overline{BC}$

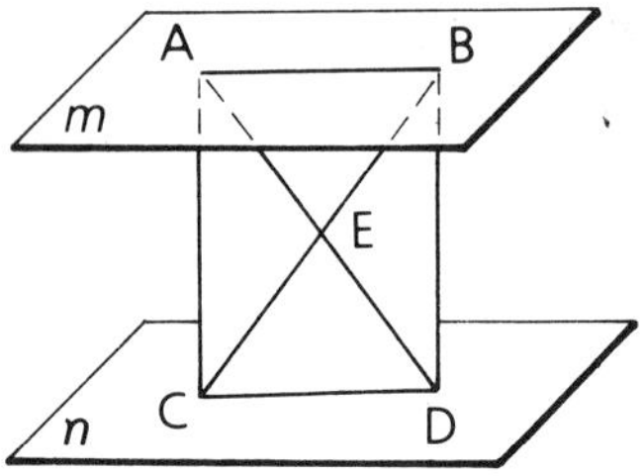

15 Given: A, B, C, D lie in m
$\overleftrightarrow{ED} \perp \overleftrightarrow{BC}$
$\overleftrightarrow{AD} \perp$ bis $\overline{BC}$

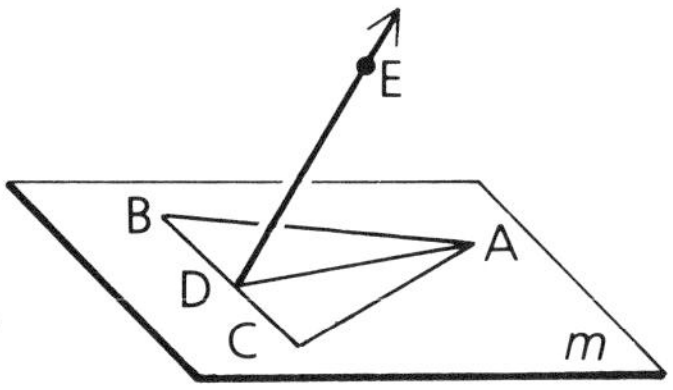

a Which segment is $\perp$ to which plane?

b How many planes are determined in this figure?

16 Given: ABCD is an isosceles trapezoid in plane t
$\overline{BC} \parallel \overline{AD}$
$\overline{PF} \perp t$
$\overline{PF}$ bisects $\overline{AD}$

Prove: $\triangle PAB \cong \triangle PDC$

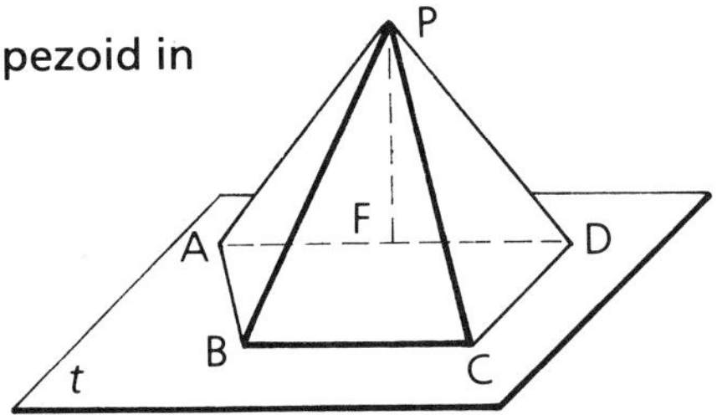

17 Prove that if a line is $\perp$ to the plane of a $\odot$ at its center, any point on that line is equidistant from any two points of the $\odot$.

18 Given: $\overleftrightarrow{SX} \perp m$
$\overleftrightarrow{SX} \perp n$
$\overline{TP} \cong \overline{TR}$

Prove: $\triangle SZW$ is isosceles

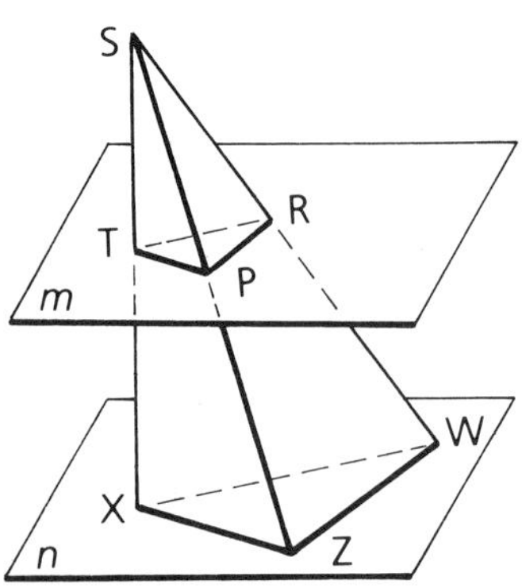

Section 6.3 Basic Facts About Parallel Planes

In this section we discuss:

A LINES PARALLEL TO PLANES, PLANES PARALLEL TO PLANES, SKEW LINES
B PROPERTIES RELATING PARALLEL LINES AND PLANES

6.3 (A) Lines Parallel to Planes, Planes Parallel to Planes

Since we developed the notion of a line being parallel to another line in Chapter 5, it seems logical now to investigate the possibilities of a line being parallel to a plane and of two planes being parallel to each other.

D **A LINE and a PLANE that do *not* intersect are PARALLEL.**

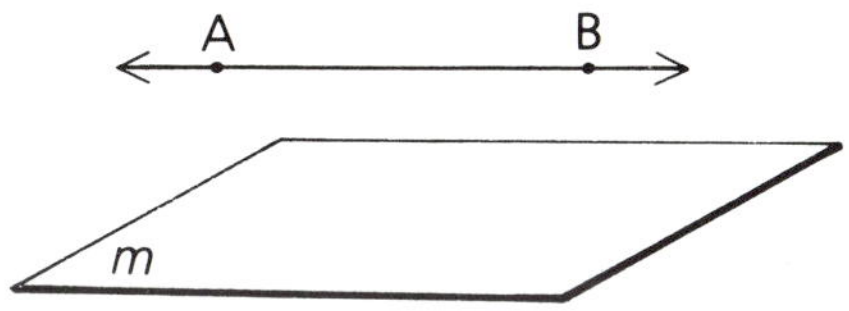

D **PARALLEL PLANES are planes that do *not* intersect.**

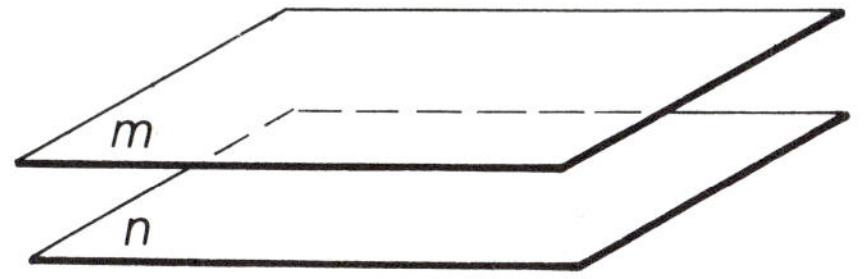

The diagram at the right shows two lines located in two parallel planes. Although the planes are parallel, the lines are not parallel, because A, B, C, and D do not determine a plane.
Such lines are called SKEW.

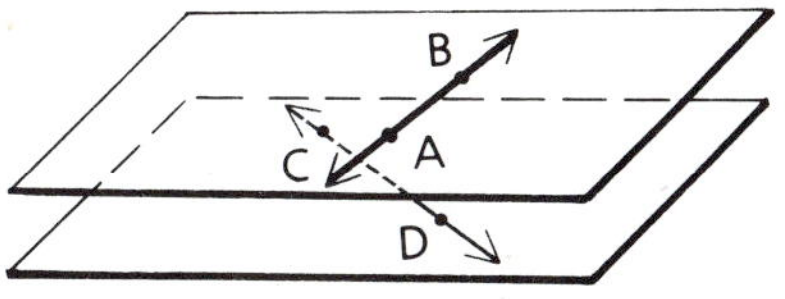

D **SKEW LINES are two lines that are *not* coplanar.**

We shall see that the development of parallels in space is very similar to parallelism in a plane. There are, however, notable differences. For example, there are no skew *planes.* Planes are either intersecting or parallel.

The following theorem is basic to the understanding of parallels in space:

Theorem 41

T **If a plane intersects two parallel planes, the lines of intersection are parallel.**

Given: $m \parallel n$
plane s intersects m and n in lines $\overleftrightarrow{AB}$ and $\overleftrightarrow{CD}$

Prove: $\overleftrightarrow{AB} \parallel \overleftrightarrow{CD}$

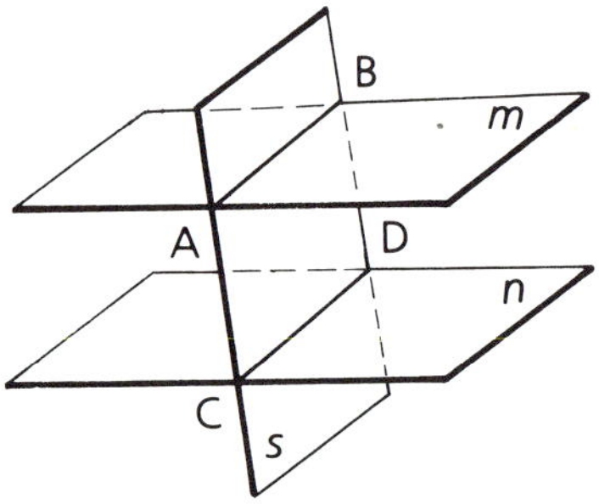

Proof of Theorem 41: We know $\overleftrightarrow{AB}$ and $\overleftrightarrow{CD}$ are coplanar since they both lie in s. Also, they cannot intersect each other, because they lie in planes m and n respectively, which, being parallel, have no intersection. Thus, $\overleftrightarrow{AB} \parallel \overleftrightarrow{CD}$ by the definition of parallel lines.

6.3 (B) Properties Relating Parallel Lines and Planes

There are numerous properties relating lines and planes in space, many of which are similar to theorems about parallel lines in Chapter 5. We present some of these properties without their proofs:

1 If two planes are perpendicular to the same line, then they are parallel to each other.

2 If a line is perpendicular to one of two parallel planes, it is perpendicular to the other plane as well.

3 If two planes are parallel to the same plane, then they are parallel to each other.

4 If two lines are perpendicular to the same plane, then they are parallel to each other.

5 If a plane is perpendicular to one of two parallel lines, it is perpendicular to the other line as well.

6 If two lines are parallel to the same line, then they are parallel to each other.

Section 6.3 Sample Problem

1 Given: $m \parallel n$
$\overleftrightarrow{AB}$ lies in m
$\overleftrightarrow{CD}$ lies in n
$\overleftrightarrow{AC} \parallel \overleftrightarrow{BD}$

Prove: $\overline{AD}$ bisects $\overline{BC}$

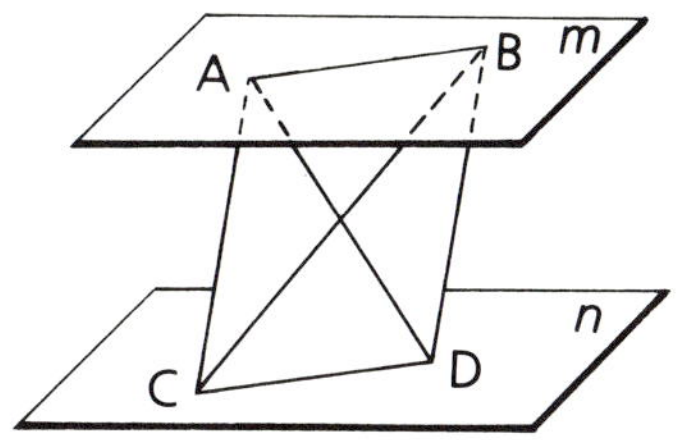

Statements	Reasons
1 $m \parallel n$	1 Given
2 $\overleftrightarrow{AB}$ lies in m $\overleftrightarrow{CD}$ lies in n	2 Given
3 $\overleftrightarrow{AC} \parallel \overleftrightarrow{BD}$	3 Given
4 $\overleftrightarrow{AC}$ and $\overleftrightarrow{BD}$ determine a plane, ACDB	4 Two $\parallel$ lines determine a plane
5 $\overleftrightarrow{AB} \parallel \overleftrightarrow{CD}$	5 If a plane intersects two $\parallel$ planes, the lines of intersection are $\parallel$
*6 ACDB is a ▱	6 If both pairs of opposite sides of a quadrilateral are $\parallel$, it is a ▱
7 $\overline{AD}$ bisects $\overline{BC}$	7 The diagonals of a ▱ bisect each other

* Before making statement 6, it was essential to have shown that ABDC is a **COPLANAR** figure. See problem 12 of Section 6.1 for a four-sided figure that is **NON-COPLANAR**.

Section 6.3 Problem Set A

1 Answer True or False for each of the five statements.

a If a plane contains one of two skew lines, it can contain the other.

b If a line and a plane never meet, they are parallel.

c If two parallel lines lie in different planes, the planes are parallel.

d If a line is perpendicular to two planes, the planes are parallel.

e If a plane and a line not in the plane are each perpendicular to the same line, then they are parallel to each other.

2 Given: $r \parallel s$
$s \parallel t$
$\overleftrightarrow{AE} \parallel \overleftrightarrow{BF}$

Prove: **a** $r \parallel t$
b ABFE is a coplanar figure
c $\overleftrightarrow{AB} \parallel \overleftrightarrow{EF}$
d $\overline{AB} \cong \overline{EF}$

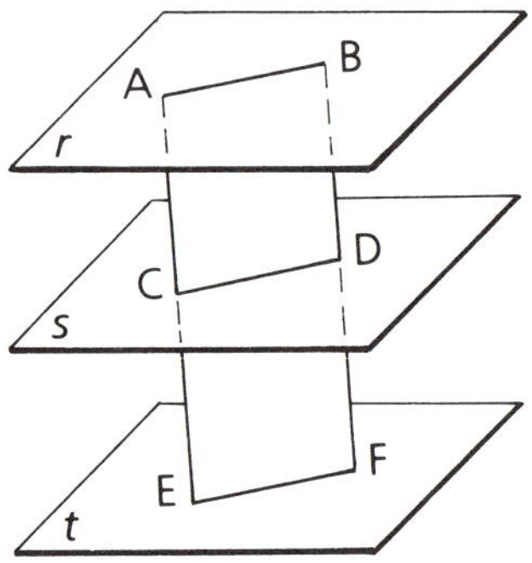

3 Given: $\overline{GJ}$ and $\overline{KH}$ bisect each other at P

a Is GHJK a coplanar figure?
b Is $\overline{GH} \parallel \overline{KJ}$?
c Is $\overline{GH} \cong \overline{KJ}$?
d Is $e \parallel f$?
e What is the most descriptive name for GHJK?

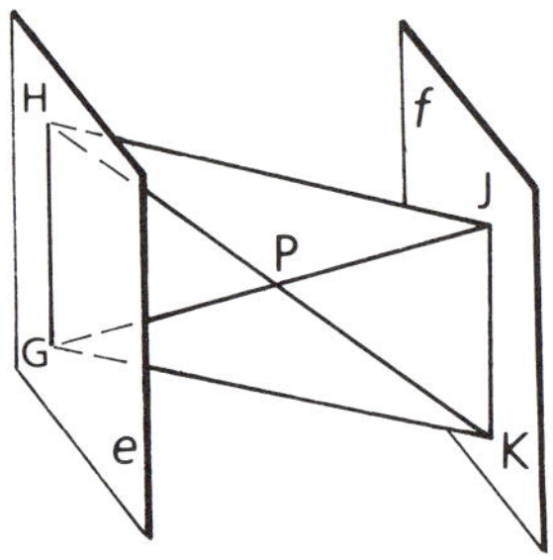

Section 6.3 Problem Set B

4 Given: $m \parallel n$
$\overleftrightarrow{AB} \parallel \overleftrightarrow{CD}$

Prove: $\overline{AB} \cong \overline{CD}$

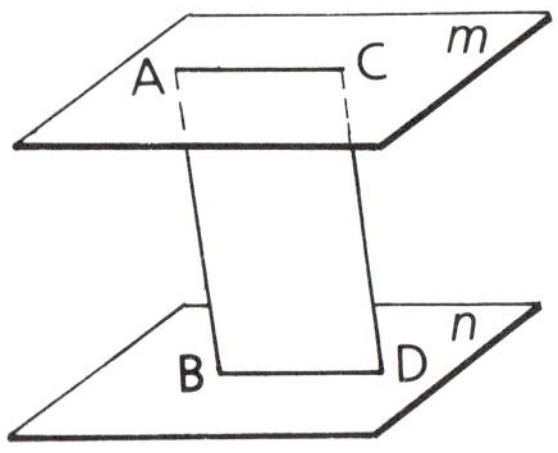

5 Given: $e \parallel f$
$\overline{RT} \cap \overline{VS} = \{P\}$
$\overline{RS} \cong \overline{VT}$

Prove: $\overline{RV} \cong \overline{ST}$

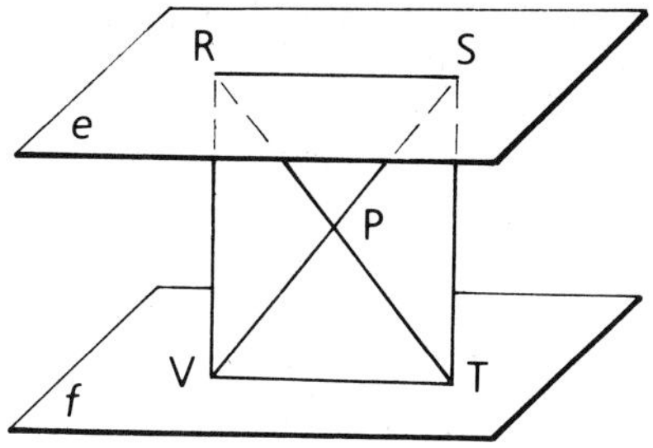

6 When we use a slide projector, the slide is parallel to the screen.

a Prove that a segment on the slide will be parallel to its image on the screen.

b Prove that the angles marked 1 and 2 in the diagram are congruent.

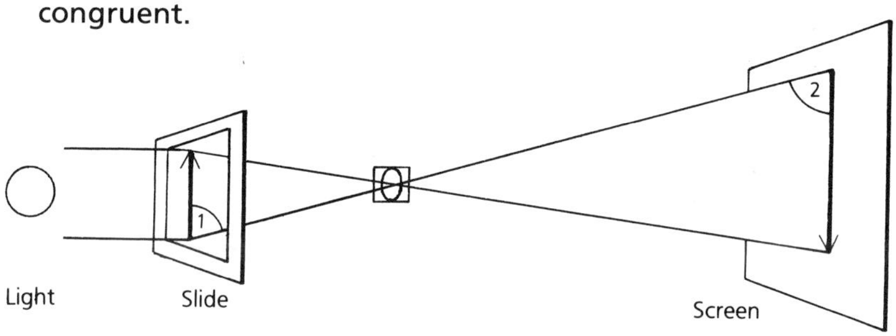

7 Given: $f \parallel g$
RTW is an isosceles $\triangle$ with base $\overline{TW}$

Prove: $\triangle$RSV is isosceles

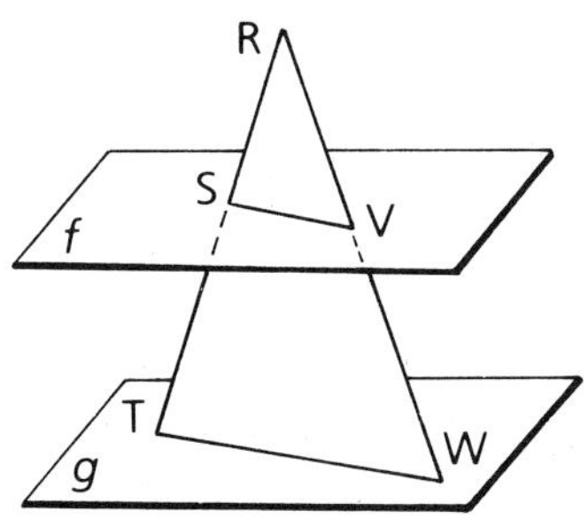

8 Given: $m \parallel n$
$p \parallel n$
$\overline{AD}$ bisects $\overline{BC}$

Prove: $\overline{BC}$ bisects $\overline{AD}$

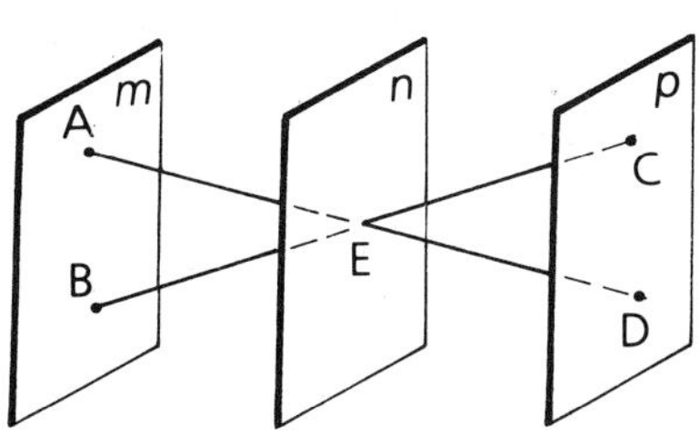

Section 6.3 Problem Set C

9 Given: $m \parallel n$
$\overline{BD} \cong \overline{CE}$
Prove: $\triangle ADE$ is isosceles

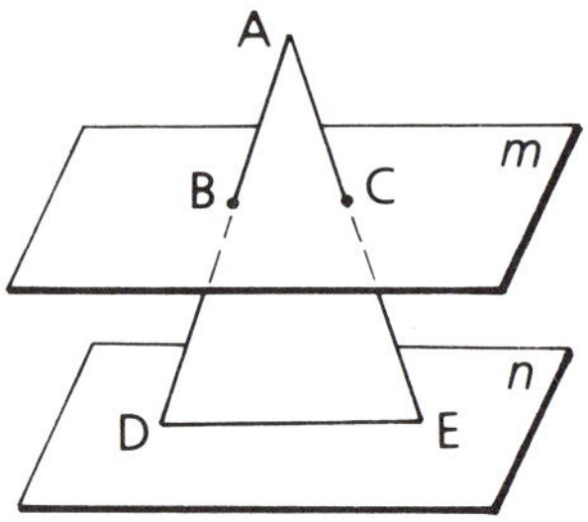

10 Given: $p \parallel q$
$\overleftrightarrow{AD}$ and $\overleftrightarrow{CF} \parallel \overleftrightarrow{BE}$
Prove: $\angle BAC \cong \angle EDF$

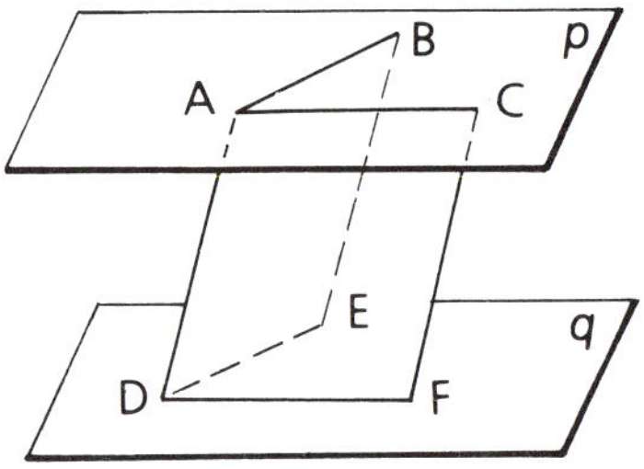

The Royal Road: Geometry in a Circle

One non-Euclidean model of geometry on a sphere was presented on page 243. Another model defines "point" as a location in or on a given circle and "line" as part of another circle that meets the given circle at "right angles."

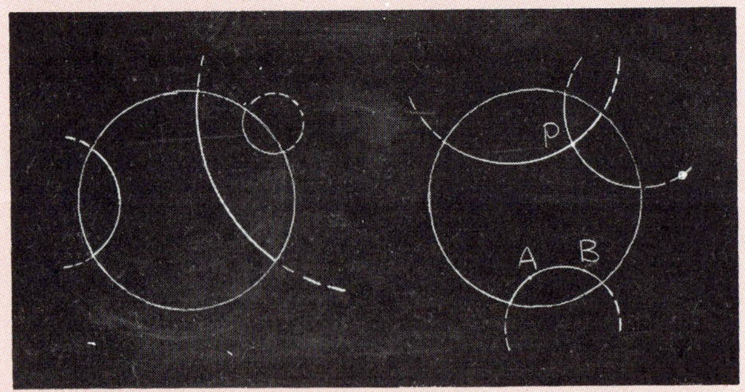

Notice that there are many "lines" through point P that do not intersect "line" $\overleftrightarrow{AB}$. Can you find any rules for the sum of the measures of the angles of a triangle?

Chapter 6 Study Guide

As you study and review this chapter, be sure that you are able to:

1 Define a plane. (p. 249)

2 Draw a representation of a plane and name it in two different ways. (p. 251)

3 Define *foot of a line* and use the phrase correctly. (p. 250)

4 Explain how *skew* lines and *parallel* lines differ. (p. 263)

5 Comment with authority on a claim by Dr. Right that a plane *m* and a plane *n* are skew planes. (p. 264)

6 Draw a figure that shows coplanar points and one that shows non-coplanar points. (p. 250)

7 Name the four ways to determine a plane. (pp. 251–252)

8 Describe the relationship between a line perpendicular TO a plane and certain lines IN the plane. (p. 257)

9 Describe what must be proved in order to claim that a line is perpendicular to a plane. (p. 258)

10 Describe the relationship between a line parallel TO a plane and certain lines IN the plane. (p. 263)

11 List the conditions that will assure that a line in one plane will be parallel to a line in a parallel plane. (p. 263)

12 Describe the kind of problem for which it is essential to prove that a figure is coplanar. (p. 265)

Chapter 6 Review Problems A

1 True-False. Be prepared to defend your answer.

a Two lines must be either intersecting or parallel.

b In a plane, two lines perpendicular to the same line are $\parallel$.

c In space, two lines perpendicular to the same line are $\parallel$.

d If a line is $\perp$ to a plane, then it is $\perp$ to every line in the plane.

e It is possible for two planes to intersect in one point.

f If a line is $\perp$ to a line of a plane, it is $\perp$ to the plane.

g Two lines perpendicular to the same line are $\parallel$.

h A triangle is a plane figure.

i A line that is $\perp$ to a horizontal line is vertical.

j Three parallel lines must be coplanar.

k Every four-sided figure is a plane figure.

2 Given: $\overline{PB} \perp m$
$\overline{PA} \cong \overline{PC}$

Prove: $\triangle ABC$ is isosceles

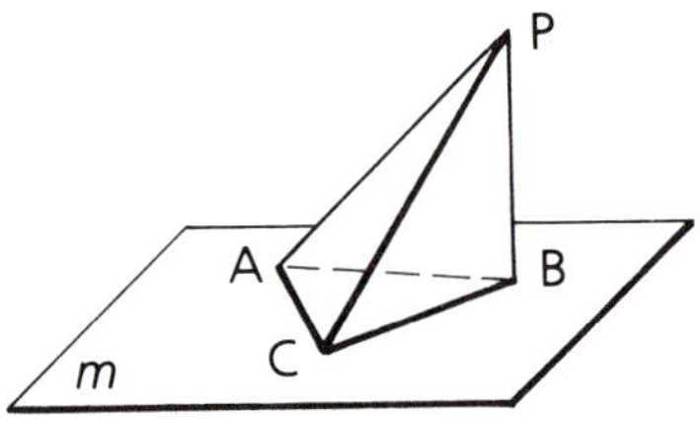

3 Given: $\overline{AB} \cong \overline{AC}$
$\angle DAB \cong \angle DAC$

Prove: $\overline{DB} \cong \overline{DC}$

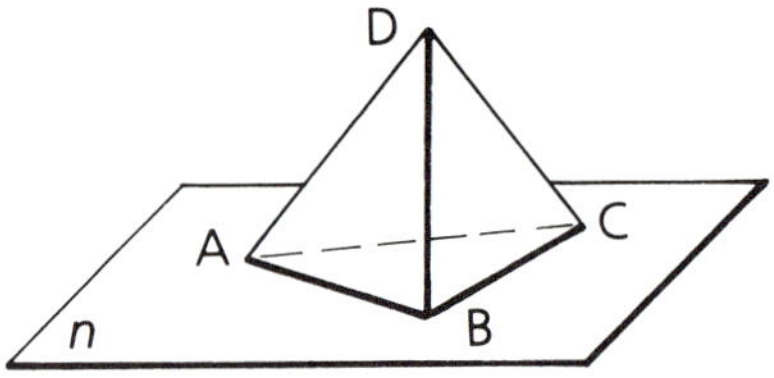

Chapter 6 Review Problems B

4 How many planes are determined by subsets of four non-coplanar points, no three of which are collinear?

5 From the top of a flag pole 48 feet in height, two ropes, each 60 feet long, reach two points on the ground, each of which is 36 feet from the pole. If the ground is assumed to be level, is the pole perpendicular to the ground?

6 a At a given point on a line, how many lines can be drawn $\perp$ to the given line?

b At a given point on a plane, how many lines can be drawn $\perp$ to the plane?

In 7–9 choose the correct answers by letter.

7 $\overline{AB}$ is a segment in plane m, and $\overleftrightarrow{XY}$ is a perpendicular bisector of $\overline{AB}$. Which of the following *must* be true?

a $\overleftrightarrow{XY} \parallel m$
b $\overleftrightarrow{XY} \perp m$
c If P is on $\overleftrightarrow{XY}$, then $\overline{PA} \cong \overline{PB}$
d If $\overline{PA} \cong \overline{PB}$, then P is on $\overleftrightarrow{XY}$
e $\overleftrightarrow{XY}$ and $\overleftrightarrow{AB}$ are skew

8 $\odot$ P lies in plane m. If A and B are points on $\odot$ P, then which of the following is (are) true if $\overleftrightarrow{QP} \perp m$?

I $\angle APQ \cong \angle BPQ$
II $\overline{AP} \cong \overline{PB}$
III $\overline{QP} \perp \overline{AB}$

a I only?
b II only?
c III only?
d I and II only?
e II and III only?

9 $\overleftrightarrow{AB}$ is parallel to plane m and perpendicular to plane r. $\overleftrightarrow{CD}$ lies in r. Which of the following must be true?

a $r \perp m$
b $r \parallel m$
c $\overleftrightarrow{CD} \perp m$
d $\overleftrightarrow{AB} \parallel \overleftrightarrow{CD}$
e $\overleftrightarrow{AB}$ and $\overleftrightarrow{CD}$ are skew

10 Given: $\triangle BDC$ is isosceles with $\overline{BD} \cong \overline{CD}$

$\angle ADB \cong \angle ADC$

Prove: $\triangle BAC$ is isosceles

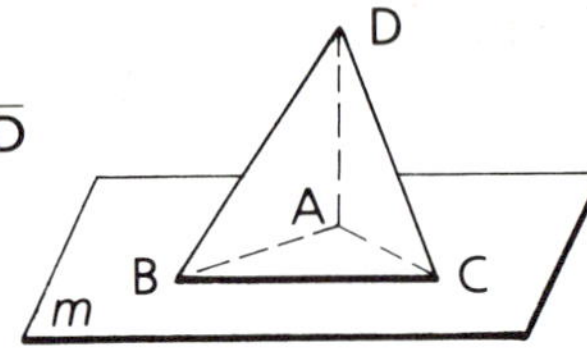

11 Given: $\overline{BP} \perp \overline{PQ}$

$\overline{AP} \perp \overline{PQ}$

A and B are equidistant from P

Prove: $\angle ABQ \cong \angle BAQ$

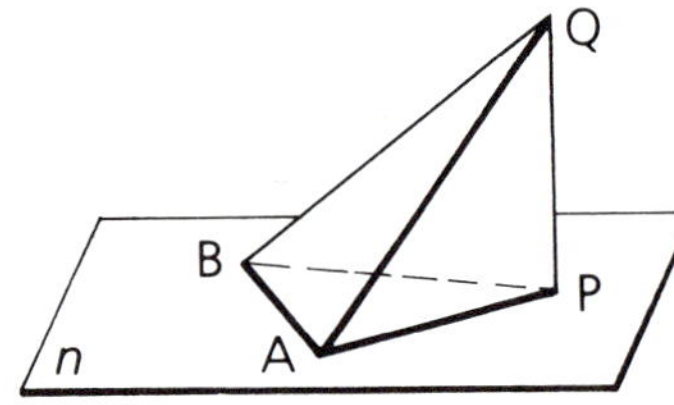

12 A line is drawn perpendicular to the plane of a square at the point of intersection of the diagonals. Prove that any point on the perpendicular is equidistant from the vertices of the square.

Chapter 6 Review Problems C

13 Given: $\overleftrightarrow{PR} \perp m$

$\angle PAB \cong \angle PBA$

Prove: $\angle PAR \cong \angle PBR$

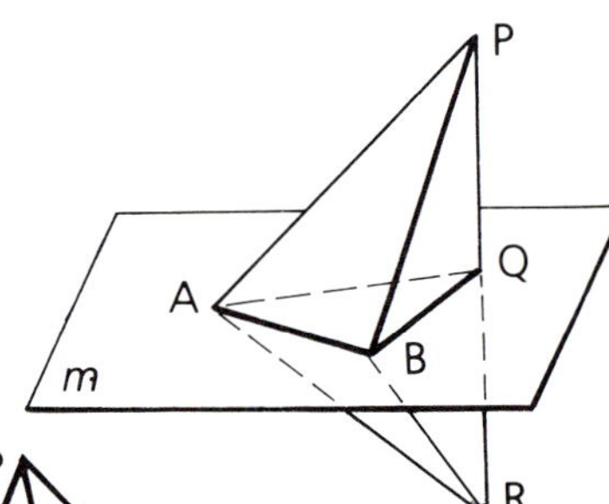

14 Given: $\triangle ABC$ lies in n

$\overline{PA} \cong \overline{PC}$

$\overline{AB} \cong \overline{BC}$

T and S are midpoints

Prove: $\overline{RT} \cong \overline{RS}$

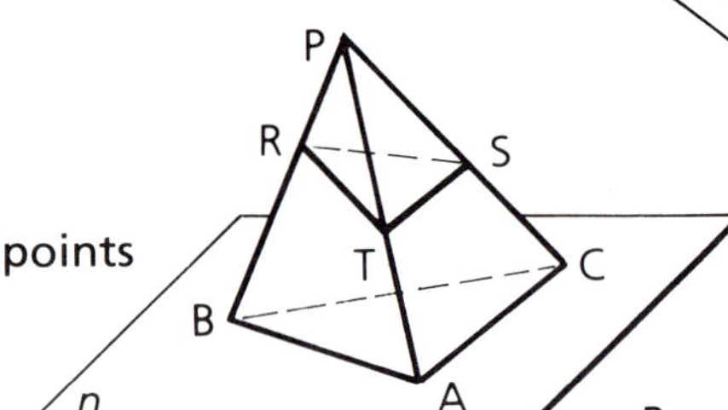

15 Given: $\overline{PC} \cong \overline{QC}$

A is the midpoint of $\overline{PQ}$

$\angle PCB \cong \angle QCB$

Prove: $\overleftrightarrow{BA} \perp \overleftrightarrow{PQ}$

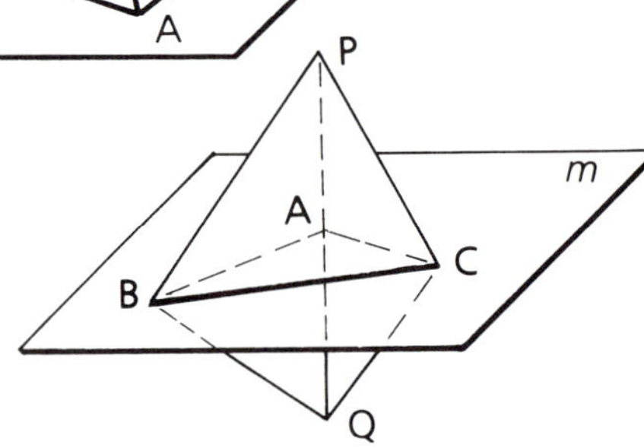

Chapter 1—6 Cumulative Review Problems A

1 Write the most descriptive name for each figure:

- **a** A four-sided figure in which the diagonals are perpendicular bisectors of each other.
- **b** A four-sided figure in which the diagonals bisect each other.
- **c** A triangle in which there is an hypotenuse.
- **d** A four-sided figure in which the diagonals are congruent and all sides are congruent.

2 Find the angle formed by the hands of a clock at 9:30.

3 If one of two supplementary angles is 16° less than three times the other, find the measure of the larger.

4 Given: $\angle OMS \cong \angle OPS$
$\angle SMR \cong \angle SPR$
S is midpt. of $\overline{MP}$
Prove: $\overleftrightarrow{OR}$ passes through S

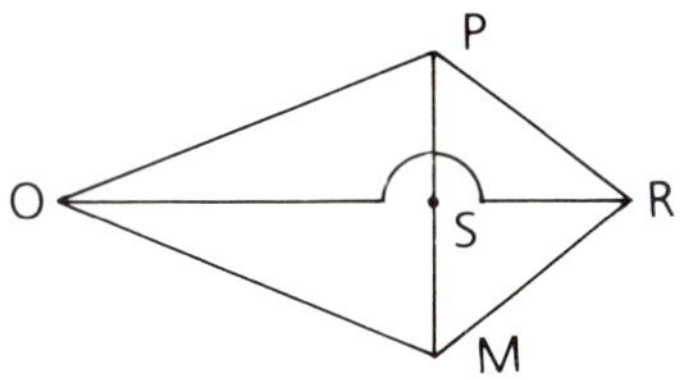

5 Given: TVAX is a rectangle
Conclusion: $\angle TXV \cong \angle VAT$

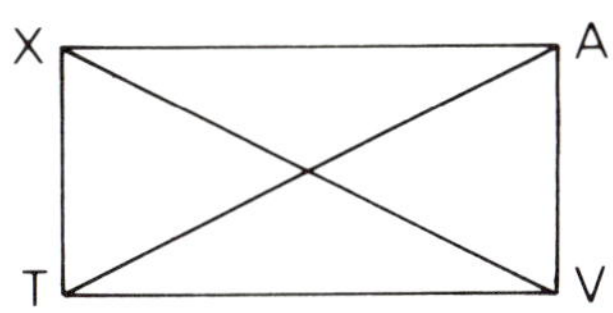

6 Two consecutive angles of a parallelogram are in the ratio of 7 to 5. Find the measure of the larger.

7 Given: NPRS is a ▱ with diagonals $\overline{SP}$ and $\overline{NR}$ intersecting at O
$\overline{TO} \perp$ plane of ▱ NPRS
Prove: $\triangle STP$ is isosceles

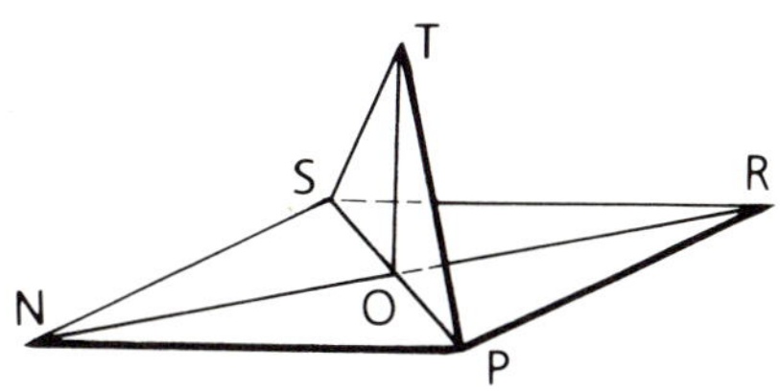

8 Find: $m\angle 1$

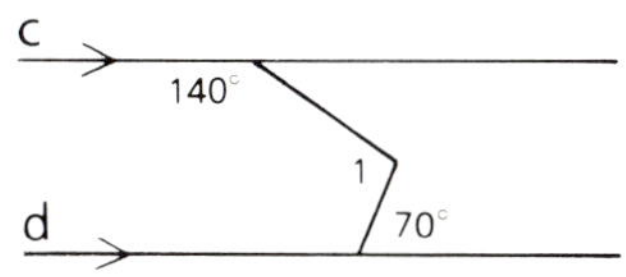

9 Given: $\odot$ P

M is the midpoint of $\overline{AB}$

Prove: $\overleftrightarrow{PQ} \perp \overleftrightarrow{AB}$

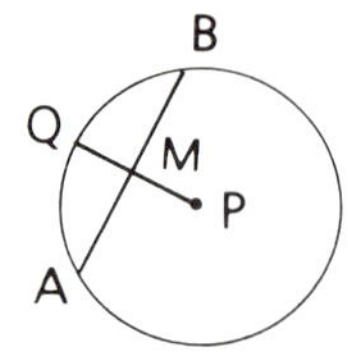

10 Answer a–g Always, Sometimes, or Never (A, S, or N)

a If a triangle is obtuse, it is isosceles.

b The bisector of the vertex angle of a scalene triangle is perpendicular to the base.

c If one of the diagonals of a quadrilateral is the perpendicular bisector of the other, the quadrilateral is a kite.

d If A, B, C, and D are non-coplanar, $\overline{AB} \perp \overline{BC}$ and $\overline{AB} \perp \overline{BD}$, then $\overline{AB}$ is perpendicular to the plane determined by B, C, and D.

e Two parallel lines determine a plane.

f Planes that contain two skew lines are parallel.

g Supplements of complementary angles are congruent.

11 FGHJ is a $\square$

FG = x + 5, GH = 2x + 3

$\angle G = 40°$, $\angle J = (4x + 12)°$

Find: a $m\angle F$

b perimeter of FGHJ

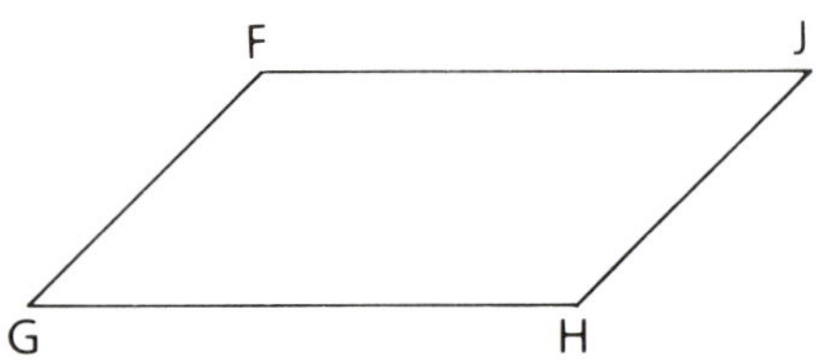

Chapter 1–6 Cumulative Review Problems B

12 Given: $\overline{AB} \cong \overline{AC}$

$\overline{BD} \cong \overline{DC}$

Conclusion: $\angle B \cong \angle C$

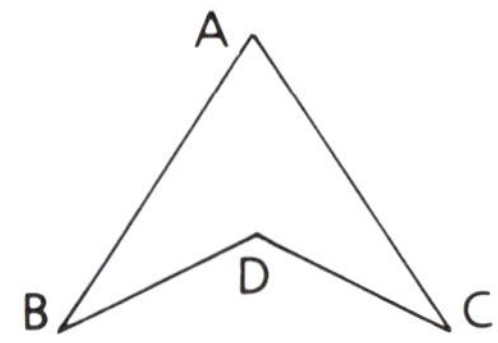

13 Given: $\overline{GH} \cong \overline{GK}$

$\overline{HM} \cong \overline{KM}$

Conclusion: HMKJ is a kite

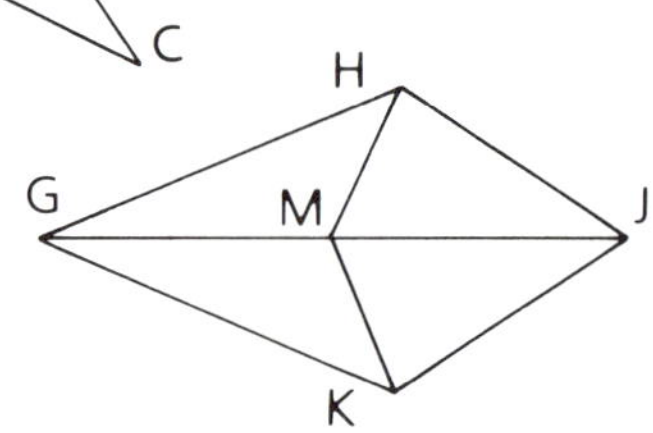

14 ABCD is a ▱
$\angle A = (3x + y)^\circ$
$\angle D = (5x + 10)^\circ$
$\angle C = (5y + 20)^\circ$
Find: $m\angle B$

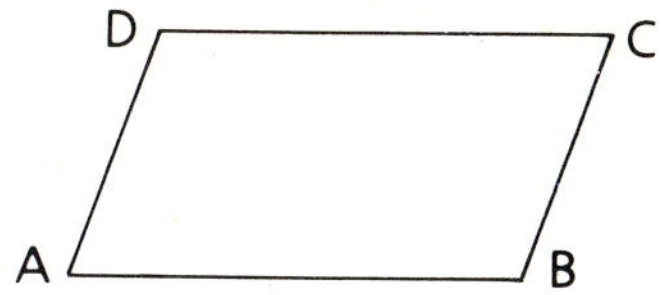

15 Given: A, B, C, and D lie in ***m***
FBCE is a ▱
$\overline{FE} \parallel \overline{AD}$
$\overline{AD} \cong \overline{BC}$
Prove: ABCD is a ▱

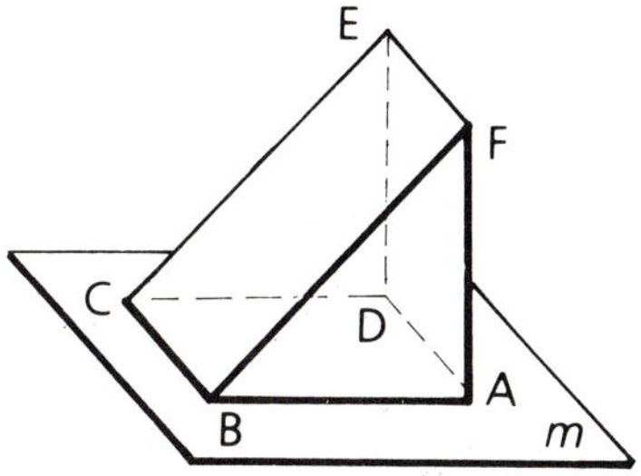

16 If segments drawn from the midpoint of one side of a triangle are perpendicular to the other two sides, and are congruent, then the triangle is isosceles.

17 The measure of the supplement of an angle exceeds three times the measure of the complement of the angle by 12°. Find the measure of half of the supplement.

Chapter 1–6 Cumulative Review Problems C

18 Given: $\overline{AC} \cong \overline{BD}$
$\overline{AB} \cong \overline{CD}$
Prove: $\angle B \cong \angle C$

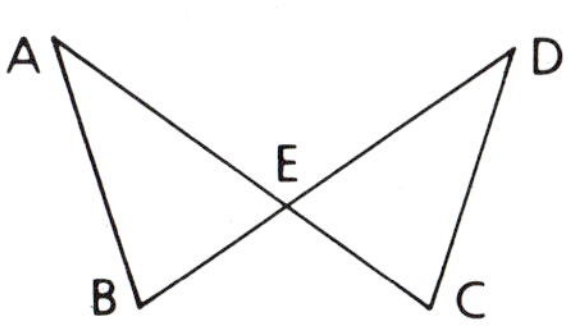

19 Given: ⊙ A lies in ***m***
$\overline{PA} \perp$ ***m***
$\overline{PD} \cong \overline{PE}$
Prove: $\overline{BE} \cong \overline{CD}$

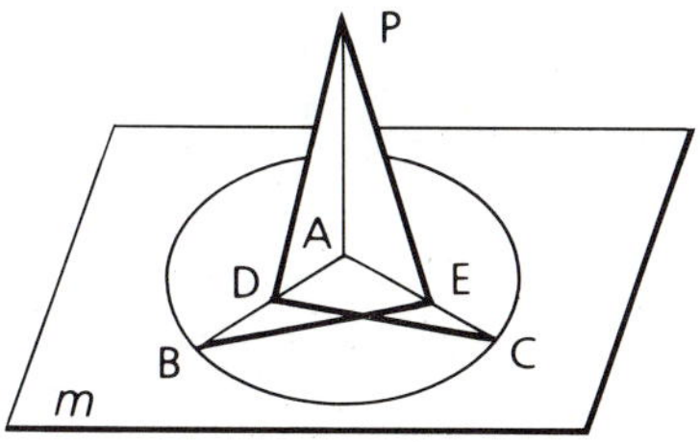

InterSection: Shaping Up

The ideas of geometry can be applied to the human figure. Artists and athletes are two groups of people who study the shapes, sizes and proportions of the body.

Geometry Works: For Artists

Leonardo Da Vinci was an Italian Renaissance artist who lived from 1452 to 1519. Many of his sketches for paintings and sculpture show how he used geometric shapes and proportions to analyze and plan his works.

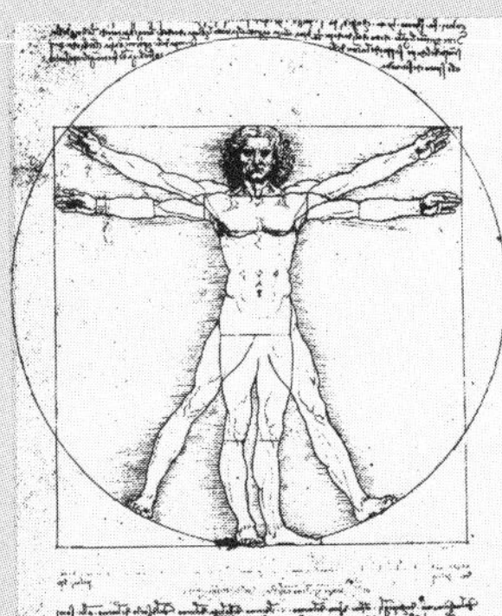

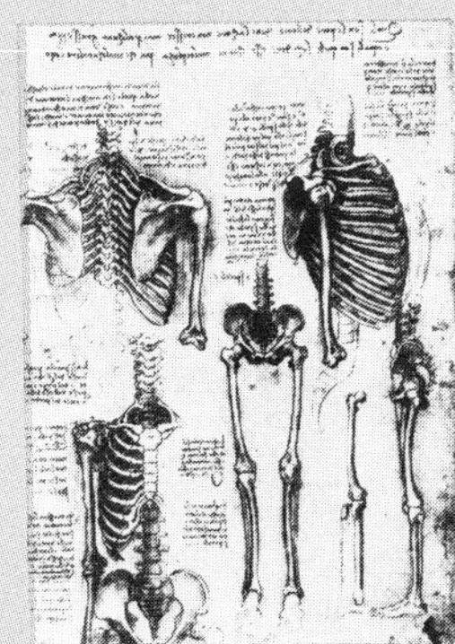

Leonardo also studied anatomy to be aware of every bone, muscle, tendon, ligament, and fold of skin in the body. His figures seem almost alive!

Geometry Works: For Athletes

Elena wants to be a world-class gymnast. Her parents have offered the garage as a practice room for the uneven parallel bars, provided that there is enough room to perform the exercises.

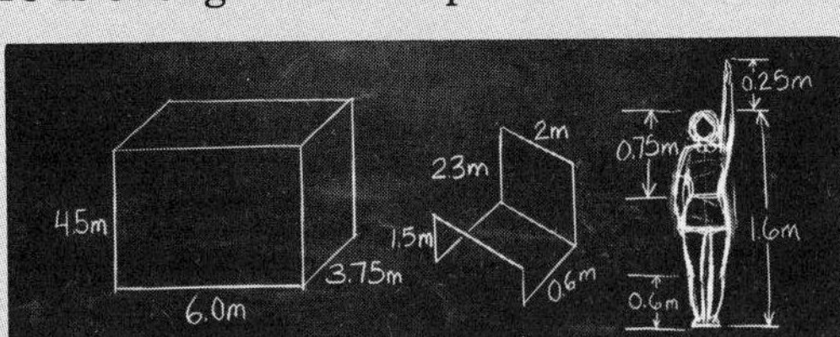

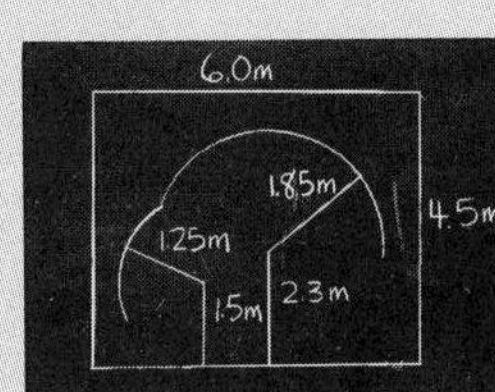

Elena took measurements of the garage, the apparatus, and her own body. When she went to her parents, she told them that there was enough room in the garage, but she would have to practice her dismounts at the gymnasium. Do you agree?

Projects

1 Pole vaulters Roy and Hector disagreed: would their 5.5-meter poles make a right-angle turn in the narrow six-foot wide hallway? Roy drew a diagram and said that the poles would be too long.

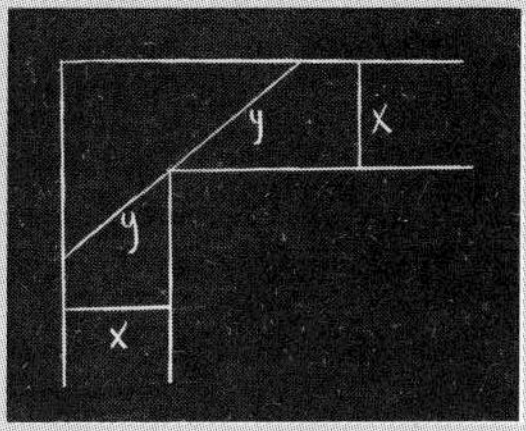

a Find values for x and y in meters.
b The poles were too long; by how much?
c Hector showed Roy that the poles really would make the turn. How did he do it?

2 Leah is a jogger. Her daily routine is to run the half-mile (eight blocks) from her house to a quarter-mile track, run twenty laps, and run home. She averages ten miles per hour on the track and six miles per hour on the streets.

a How far does she usually run?
b How long is she usually out of the house?
c When the track is being used, she runs the same total distance in a square path on the streets. How many blocks does she run in each direction? If she leaves her house at 3:20, what time would she be expected home?

3 Five judges are used in a diving contest. The top mark and the bottom mark are dropped, and the contestant's score is the average of the three remaining scores.

```
 5   INPUT A(1),A(2),A(3),A(4),A(5)
25   FOR N = 1 TO 4
30   FOR K = N + 1 TO 5
35   IF A(N) < = A(K) THEN 51
40   LET W = A(N)
45   LET A(N) = A(K)
50   LET A(K) = W
51   NEXT K
52   NEXT N
56   PRINT A(1);A(2);A(3);A(4);A(5)
60   PRINT "SCORE IS:"; (A(2) + A(3) + A(4)) / 3
70   END
```

a For each dive, the final score is multiplied by the "degree of difficulty" of the dive. Extend the program to accept the degree of difficulty as input and print the contestants' final score.
b Modify the program to calculate the final score based on only the scores of the five judges and the degree of difficulty.

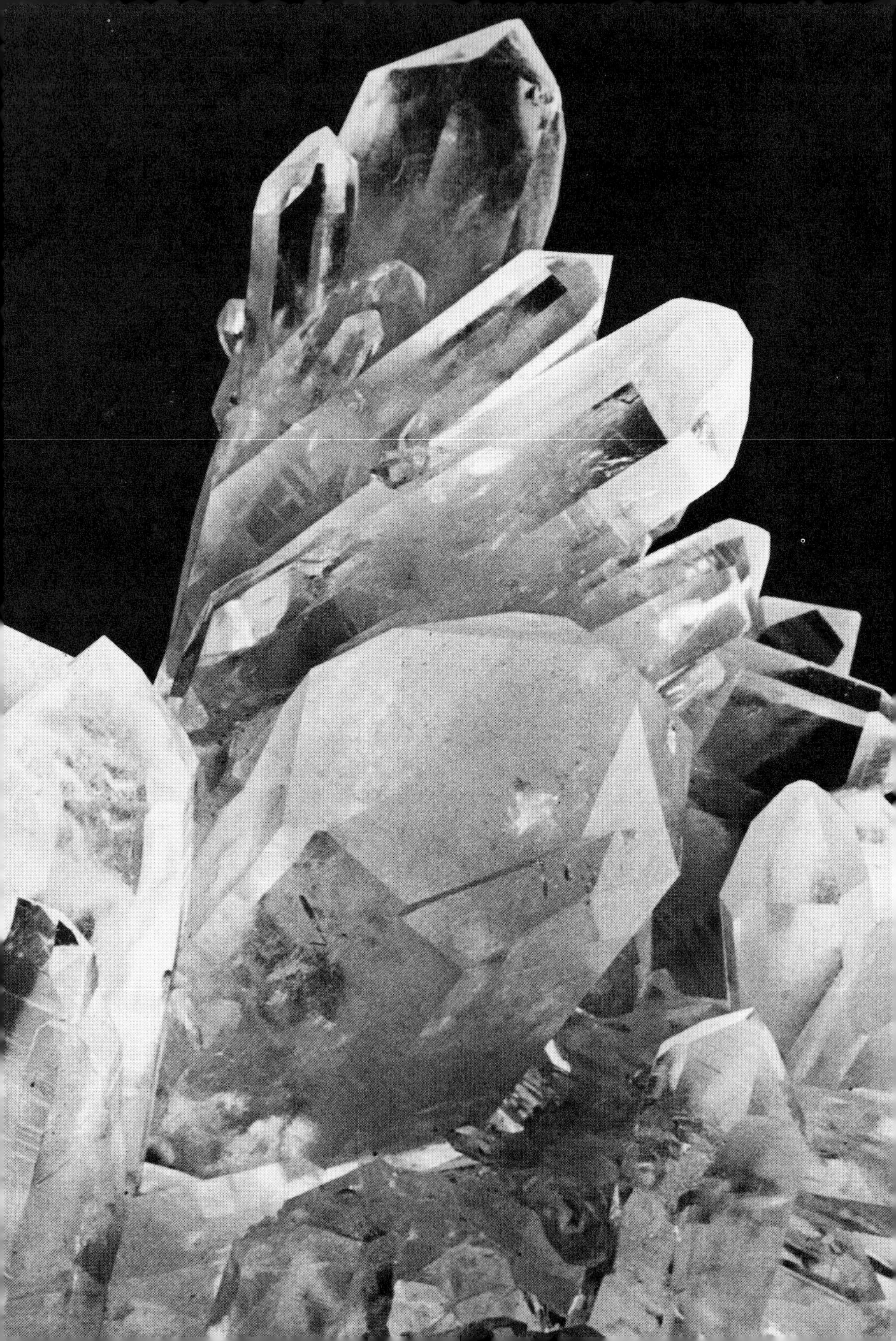

Chapter 7

POLYGONS

Section 7.1 Triangle Application Theorems

The topics of this section are:

A THE SUM OF THE MEASURES OF THE ANGLES OF A TRIANGLE

B AN EXTERIOR ANGLE OF A POLYGON

C A THEOREM ABOUT AN EXTERIOR ANGLE OF A TRIANGLE

D THE MIDLINE THEOREM

Many students have been taught in elementary school that the sum of the measures of the angles of a triangle is 180. Indeed, that theorem is one of the most useful and widely applied.

Theorem 42

T **The sum of the measures of the three angles of a triangle is 180.**

Given: $\triangle ABC$

Prove: $m\angle A + m\angle B + m\angle C = 180$

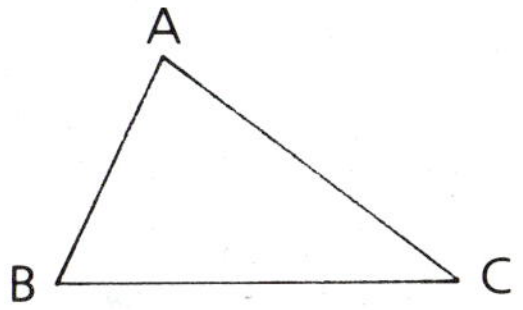

Proof of Theorem 42: By the **PARALLEL POSTULATE** there exists exactly one line through point A parallel to $\overleftrightarrow{BC}$, so the figure can look like this:

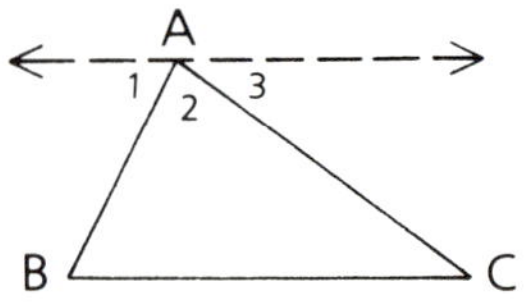

Because of the straight angle, we know that
$\angle 1 + \angle 2 + \angle 3 = 180°$.
Since $\angle 1 \cong \angle B$ and $\angle 3 \cong \angle C$ by PAI,
we may substitute to obtain $\angle B + \angle 2 + \angle C = 180°$.
Hence, $m\angle A + m\angle B + m\angle C = 180$.

Before proving the next theorem, it is necessary to introduce the meaning of an exterior angle of a polygon. In each of the figures below, $\angle 1$ is an exterior angle of a polygon.

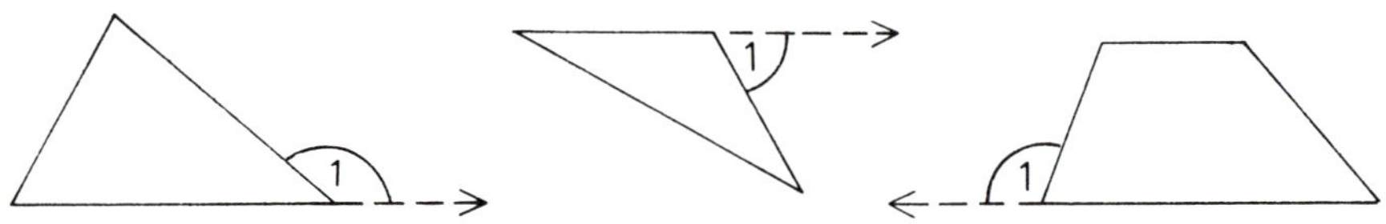

You can see that an exterior angle of a polygon is formed by extending one of the sides of the polygon through a vertex. The definition puts that idea in a form much more useful in proofs and problems.

D **An EXTERIOR ANGLE OF A POLYGON is an angle that is adjacent and supplementary to an interior angle of the polygon.**

The next theorem is a property of triangles only.

Theorem 43

T **The measure of an exterior angle of a *triangle* is equal to the sum of the measures of the remote interior angles.**

Given: $\triangle DEF$ with exterior angle 1 at F

Prove: $m\angle 1 = m\angle D + m\angle E$

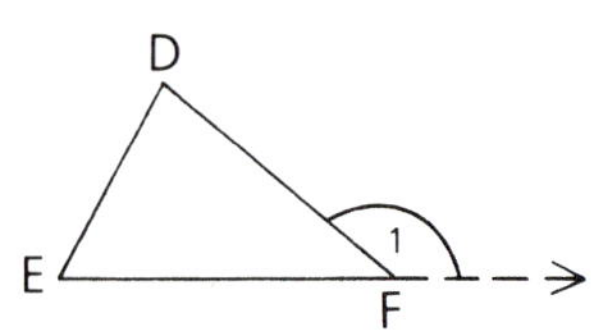

Do you see how the above definition and Theorem 42 are the keys to a proof of Theorem 43?

The following theorem could have been presented in the chapter on parallelograms, but we choose to include it now.

Theorem 44

T **A segment joining the midpoints of two sides of a triangle is parallel to the third side, and its length is one-half the length of the third side. (MIDLINE THEOREM)**

Given: H is a midpoint
M is a midpoint

Prove: **a** $\overline{HM} \parallel \overline{JK}$
b $HM = \frac{1}{2}JK$

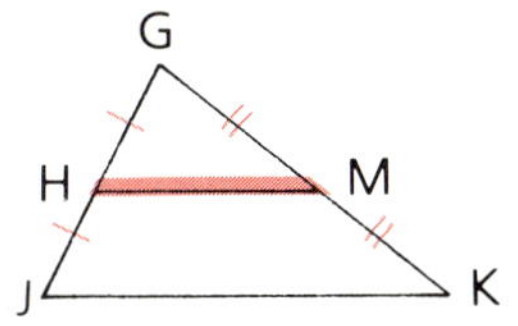

Proof of Theorem 44: Extend $\overleftrightarrow{HM}$ through M to a point P, such that $\overline{MP} \cong \overline{HM}$. P is now established, so P and K determine $\overleftrightarrow{PK}$.

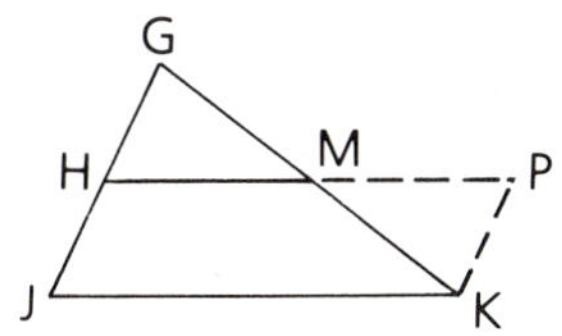

$\overline{GM} \cong \overline{KM}$ (given midpoint).
$\angle GMH \cong \angle KMP$ (vertical angles).
Thus, $\triangle GMH \cong \triangle KMP$ by SAS.
$\angle G = \angle PKM$ by CPCTC, so $\overleftrightarrow{PK} \parallel \overleftrightarrow{HJ}$ by AIP.
Also, $\overline{GH} \cong \overline{PK}$ by CPCTC, and
$\overline{GH} \cong \overline{HJ}$ (given midpoint). By transitivity, then, $\overline{PK} \cong \overline{HJ}$.

Two sides, $\overline{PK}$ and $\overline{HJ}$, are now parallel and congruent, so PKJH is a parallelogram. Therefore, $\overleftrightarrow{HP} \parallel \overleftrightarrow{JK}$.

Opposite sides of a parallelogram are congruent, so $HP = JK$. Also, since we made $MP = HM$, $HM = \frac{1}{2}HP$. Then, substitution yields $HM = \frac{1}{2}JK$.

Section 7.1 Sample Problems

1 **Diagram as marked**
Find: x, y, and z

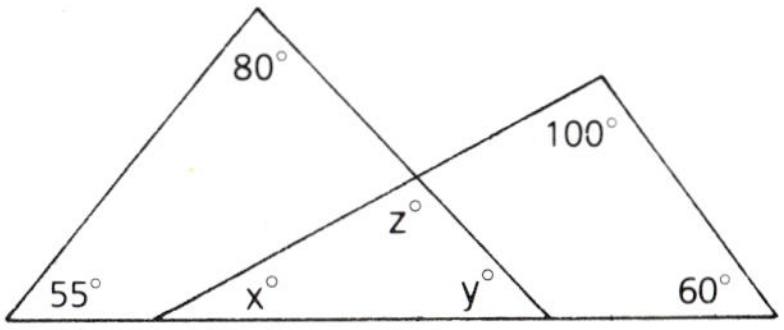

Solution:

Since the sum of the measures of the angles of a triangle is 180,

$x + 100 + 60 = 180$	$55 + 80 + y = 180$	$x + y + z = 180$
$x + 160 = 180$	$135 + y = 180$	$20 + 45 + z = 180$
$x = 20$	$y = 45$	$z = 115$

Substitution Substitution

2 The measures of the three angles of a triangle are in the ratio of 3:4:5. Find the measure of the largest angle.

Solution: Let the measures of the three angles be 3x, 4x, and 5x. Since the sum of the measures of the three angles of a triangle is 180,

$$\begin{aligned} 3x + 4x + 5x &= 180 \\ 12x &= 180 \\ x &= 15 \\ 5x &= 75 \end{aligned}$$

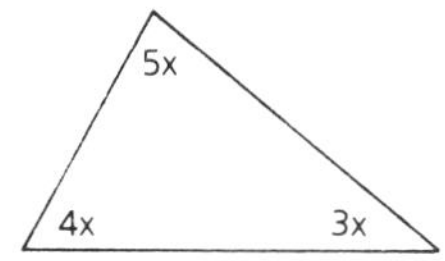

Therefore, the largest angle of the triangle is a 75° angle.

3 If one of the angles of a triangle is 80°, find the measure of the angle formed by the bisectors of the other two angles.

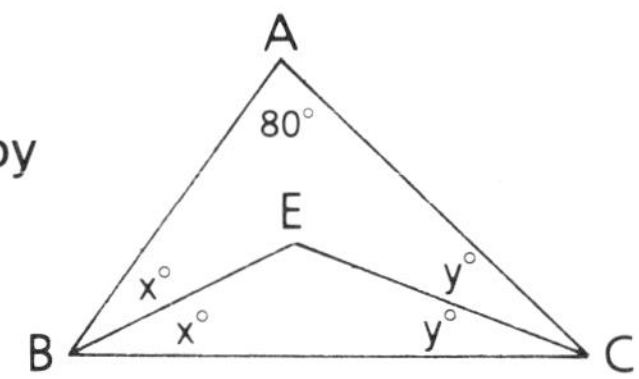

Solution:
The bisectors, $\overrightarrow{BE}$ and $\overrightarrow{CE}$ meet at E so we want to find m∠E.
Let ∠ABC = (2x)°, and ∠ACB = (2y)°

In △ABC

$$\begin{aligned} 2x + 2y + 80 &= 180 \\ 2x + 2y &= 100 \\ x + y &= 50 \end{aligned}$$

In △EBC

$$\begin{aligned} x + y + m\angle E &= 180 \\ \text{Substitution:}\quad 50 + m\angle E &= 180 \\ m\angle E &= 130 \end{aligned}$$

4 ∠1 = 150°, and the measure of ∠D is twice that of ∠E. Find the measure of each angle of the triangle.

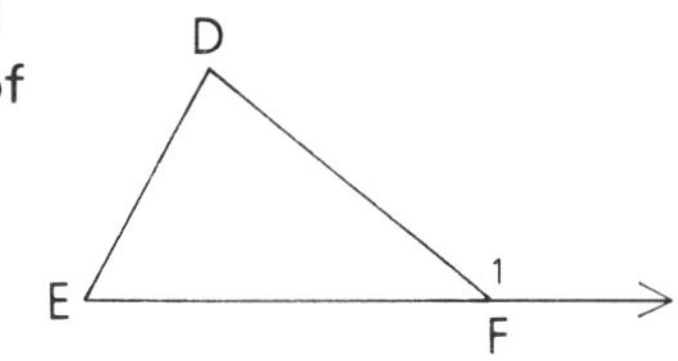

Solution:
Let ∠E = x° and ∠D = (2x)°.

Since the measure of an exterior angle of a triangle is equal to the sum of the measures of the remote interior angles,

$$\begin{aligned} 150 &= x + 2x \\ 150 &= 3x \\ 50 &= x \end{aligned}$$

Hence, ∠E = 50°, ∠D = 100°, and ∠DFE = 30°

Section 7.1 Problem Set **A**

1 Diagram as marked
Find: m∠B

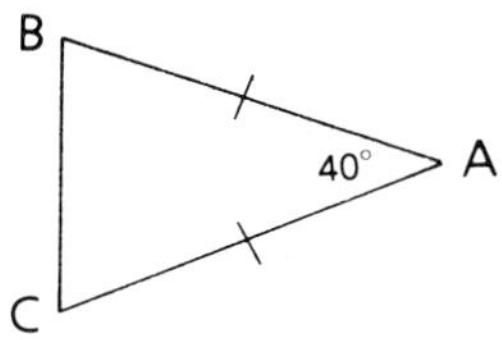

2 ∠1 = 130°
∠7 = 70°
Find the measures of ∠2, ∠3, ∠4, ∠5, ∠6

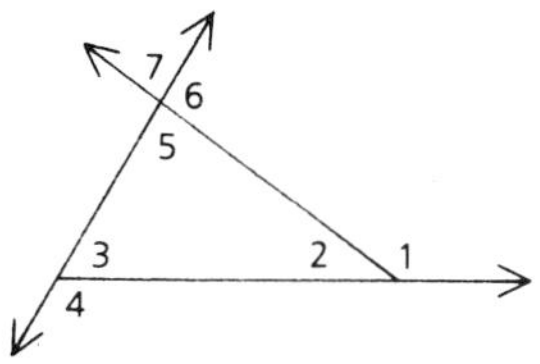

3 ∠CAB = 80°
∠CBA = 60°
$\overline{AE}$ and $\overline{BD}$ are altitudes
Find: m∠C and m∠AFB

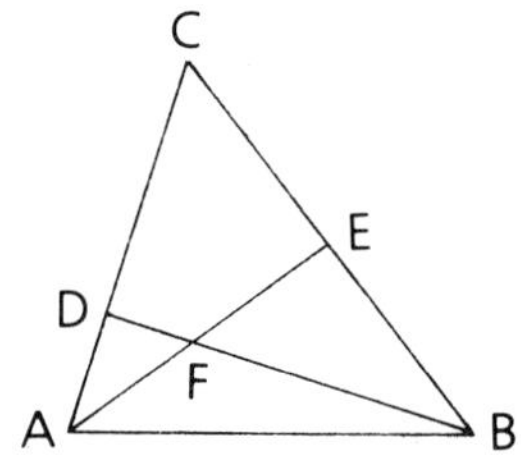

4 In the diagram as marked,
if m∠G = 50, find m∠M.
Hint: See sample problem 3

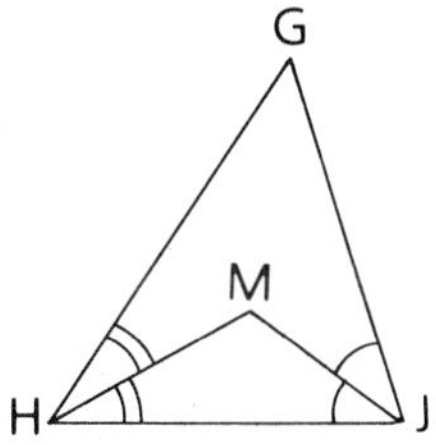

5 The measures of the three angles of a triangle are in the ratio of 4:5:6. Find the measure of each.

6 The vertex angle of an isosceles △ is twice as large as one of the base angles. Find the measure of the vertex angle.

7 ∠ORS = (4x + 6)°
∠P = (x + 24)°
∠O = (2x + 4)°
Find: m∠O

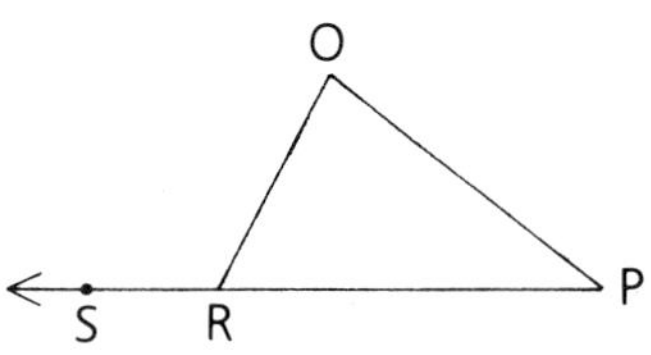

8 In the diagram as marked,
if WX = 18, find AY.

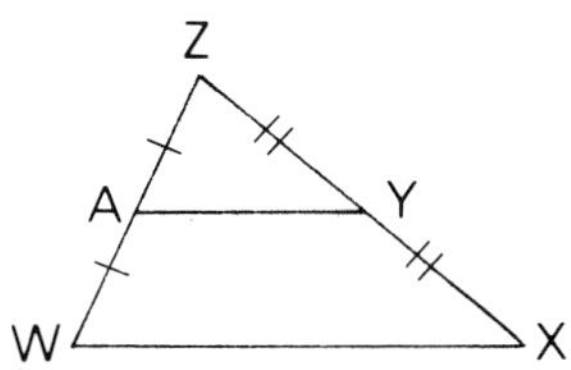

9 Diagram as marked
G and J are midpoints
Find: m∠H, m∠HGJ, and m∠HJG

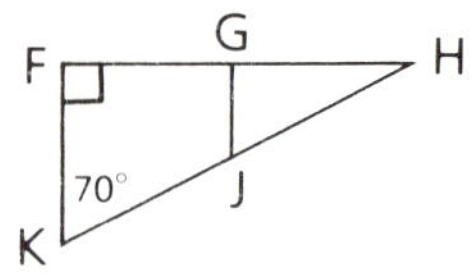

10 ∠P = 10°
$\overrightarrow{RO}$ bisects ∠MRP
Find: m∠ORP, m∠MOR

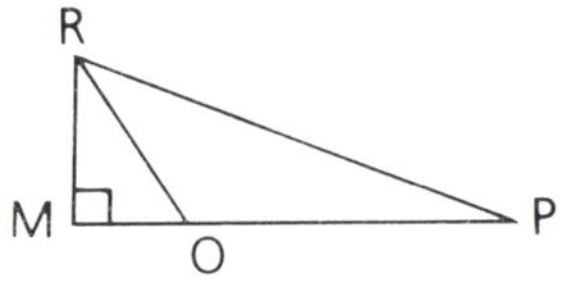

11 Answer a–e Always, Sometimes, or Never (A, S, or N)

a The acute angles of a right triangle are complementary.

b The supplement of one of the angles of a △ is equal in measure to the sum of the other two angles of the △.

c A triangle contains two obtuse angles.

d If one of the angles of an isosceles △ is 60°, the △ is equilateral.

e If the sides of one triangle are doubled to form another triangle, each angle of the second triangle is twice as large as the corresponding angle of the first triangle.

Section 7.1 Problem Set B

12 Prove, in paragraph form, that the acute angles of a right triangle are complementary.

13 Prove, in paragraph form, that if a right triangle is isosceles, it must be specifically a 45°–45°–90° triangle.

14 The first and second angles of a triangle have measures in the ratio of 2:3 respectively. If the third angle is 4 degrees larger than the second, find the measure of an exterior angle at the third vertex.

15 ∠A = 30°, $\overline{AB} \cong \overline{AC}$
$\overrightarrow{CD}$ bisects ∠ACB
$\overrightarrow{BD}$ is *one* of the trisectors of ∠ABC
Find: m∠D

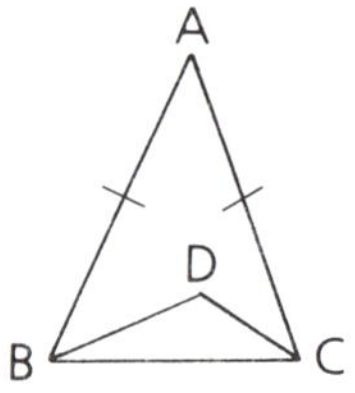

16 EFGH is a rectangle
FH = 20
J, K, M, and O are midpoints

a Find the perimeter of JKMO
b What is the most descriptive name for JKMO?

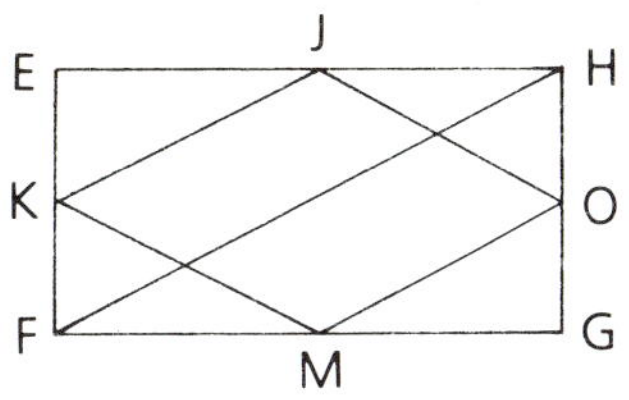

17 $\angle PST = (x + 3y)^\circ$
$\angle P = 45^\circ$, $\angle R = (2y)^\circ$
$\angle PSR = (5x)^\circ$
Find $m\angle PST$

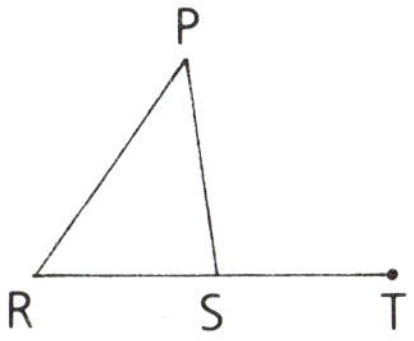

Section 7.1 Problem Set C

18 Prove that the midpoint of the hypotenuse of a right triangle is equidistant from all three vertices. (Hint: see the method used to prove the MIDLINE THEOREM, page 281.)

19 Prove: If the midpoints of a quadrilateral are joined in order, the figure formed is a parallelogram.

20 Given: $\overline{AB} \cong \overline{AC}$
$\overline{AE} \cong \overline{DE} \cong \overline{DB} \cong \overline{BC}$
Find: $m\angle A$

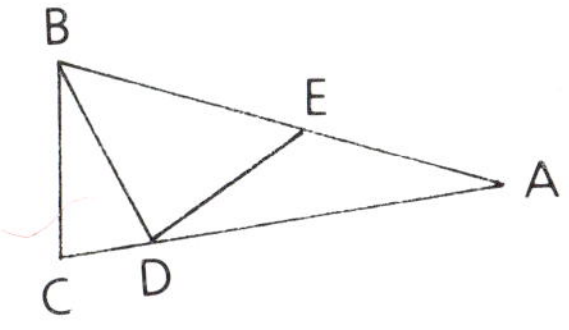

21 Given: $\angle E = 70^\circ$
$\overrightarrow{JH}$ and $\overrightarrow{FH}$ bisect the exterior angles of $\triangle JEF$ at J and F

a Find: $m\angle H$
b Can you find a formula for $m\angle H$ in terms of $m\angle E$?

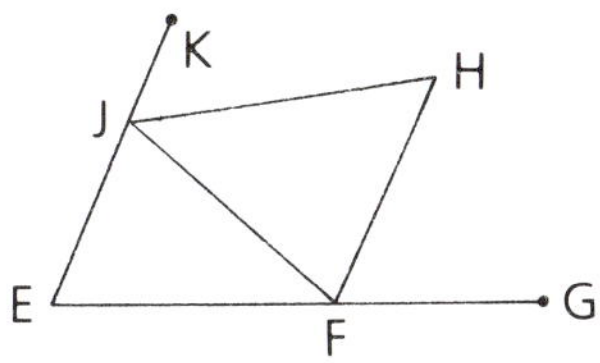

22 Show: $a + e_1 + c_1 = d + e_2 + b_2$

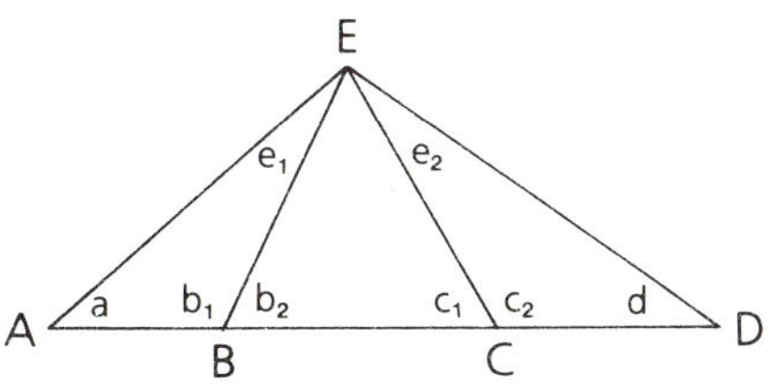

Section 7.2 Two Proof-Oriented Triangle Theorems

In this section we present the following theorems:

A NO CHOICE THEOREM

B AAS

Theorem 45

T **If two angles of one triangle are congruent to two angles of a second triangle, then the third angles are congruent. (We refer to this theorem as the NO CHOICE THEOREM, since the third angles have no choice except to be congruent.)**

Given $\angle A \cong \angle D$

$\angle B \cong \angle E$

Conclusion: $\angle C \cong \angle F$

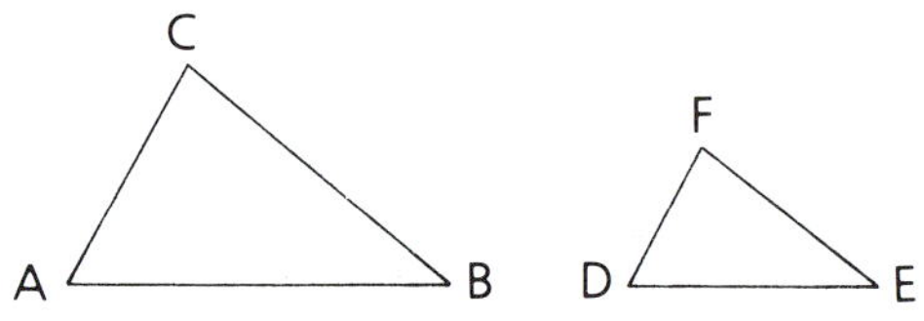

Proof of Theorem 45: Since the sum of the angles in each triangle is 180°, the sums may be set equal. If we then apply the Subtraction Property, we see that $\angle C$ and $\angle F$ have no choice but to be $\cong$.

Note: Having congruent triangles is not a necessary condition for use of the NO CHOICE THEOREM.

Theorem 46

T **If there exists a correspondence between the vertices of two triangles such that two angles and a non-included side are congruent respectively to the corresponding parts of a second triangle, the triangles are congruent. (AAS)**

Given: $\angle G \cong \angle K$

$\angle H \cong \angle M$

$\overline{JH} \cong \overline{OM}$

Prove: $\triangle GHJ \cong \triangle KMO$

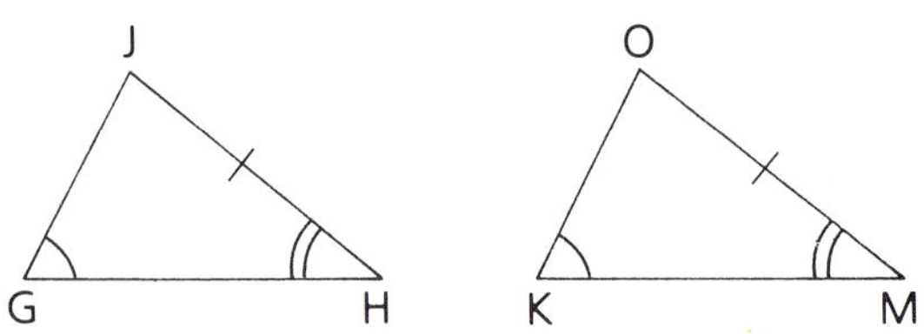

Proof of Theorem 46:

1 $\angle G \cong \angle K$	1 Given
2 $\angle H \cong \angle M$	2 Given
3 $\angle J \cong \angle O$	3 No Choice Theorem
4 $\overline{JH} \cong \overline{OM}$	4 Given
5 $\triangle GHJ \cong \triangle KMO$	5 ASA (2, 4, 3)

Section 7.2 Sample Problems

1 **Given:** $\angle A \cong \angle D$

Prove: $\angle E \cong \angle C$

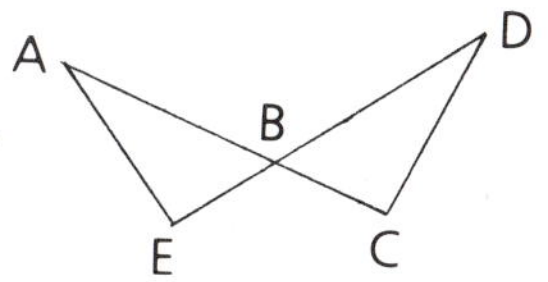

1 $\angle A \cong \angle D$	1 Given
2 $\angle ABE \cong \angle DBC$	2 Vertical angles are congruent
3 $\angle E \cong \angle C$	3 No Choice Theorem

2 **Given:** $\angle N \cong \angle R$, $\angle NTR \cong \angle P$
$\overline{TO} \perp \overline{NP}$, $\overline{TS} \perp \overline{PR}$
$\overline{TO} \cong \overline{TS}$

Prove: NPRT is a rhombus

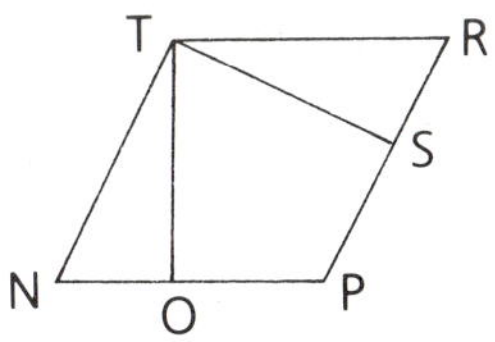

1 $\angle N \cong \angle R$	1 Given
2 $\angle NTR \cong \angle P$	2 Given
3 NPRT is a ▱	3 If both pairs of opposite ∠s of a quadrilateral are ≅, it is a ▱
4 $\overline{TO} \perp \overline{NP}$	4 Given
5 ∠TON is a right ∠	5 ⊥ segments form right ∠s
6 $\overline{TS} \perp \overline{PR}$	6 Given
7 ∠TSR is a right ∠	7 Same as 5
8 $\angle TON \cong \angle TSR$	8 Right angles are congruent
9 $\overline{TO} \cong \overline{TS}$	9 Given
10 $\triangle TON \cong \triangle TSR$	10 AAS (1, 8, 9)
11 $\overline{TN} \cong \overline{TR}$	11 CPCTC
12 NPRT is a rhombus	12 If two consecutive sides of a ▱ are ≅, it is a rhombus

Section 7.2 Problem Set A

1 Given: $\overline{JM} \perp \overline{GM}$
$\overline{GK} \perp \overline{KJ}$
Conclusion: $\angle G \cong \angle J$

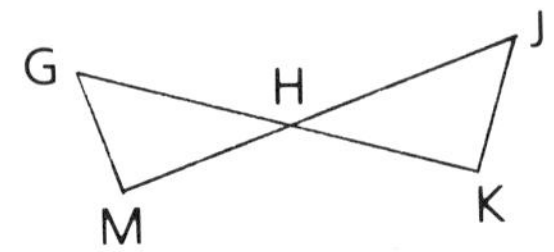

2 Given: $\overline{CB} \perp \overline{AB}$
$\overleftrightarrow{DE} \parallel \overleftrightarrow{AB}$
$\angle CDE = 40^\circ$
Find: $m\angle A$, $m\angle C$, and $m\angle CED$

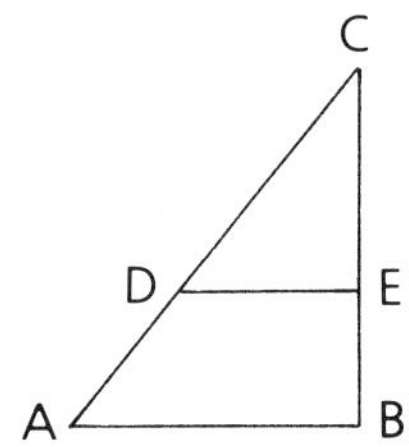

3 Given: $\overline{PD}$ and $\overline{PC}$ lie in plane *m*
$\overline{BP} \perp m$
$\angle C \cong \angle D$
Prove: $\angle PBC \cong \angle PBD$

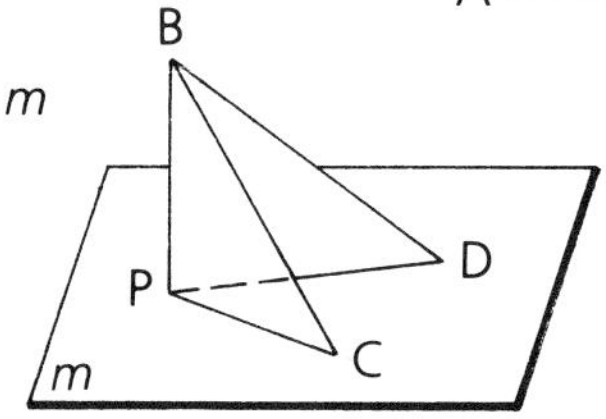

4 Given: $\overline{MR} \perp \overline{KP}$
$\overline{KO} \perp \overline{PM}$
$\angle RKM \cong \angle OMK$
Prove: $\triangle RKM \cong \triangle OMK$

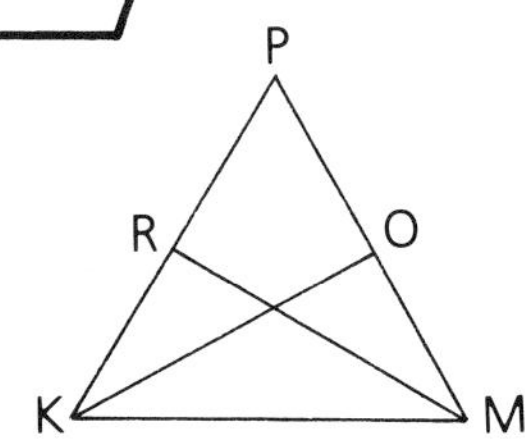

5 Given: $\odot$ O
$\angle SOV \cong \angle TOW$
$\angle WSO \cong \angle VTO$
Prove: $\overline{SO} \cong \overline{TO}$

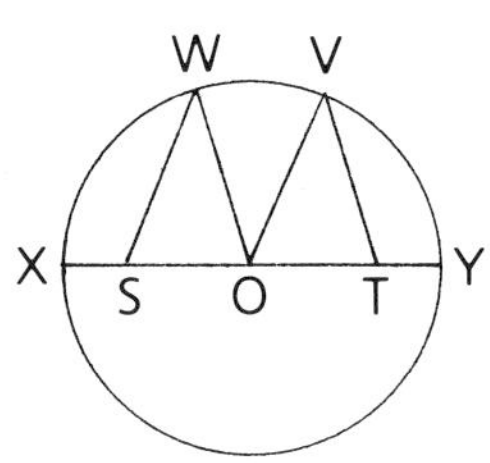

6 Given: GJKM is a rhombus
$\overline{OJ} \perp \overline{GM}$
$\overline{MH} \perp \overline{GJ}$
Conclusion: $\overline{MH} \cong \overline{JO}$

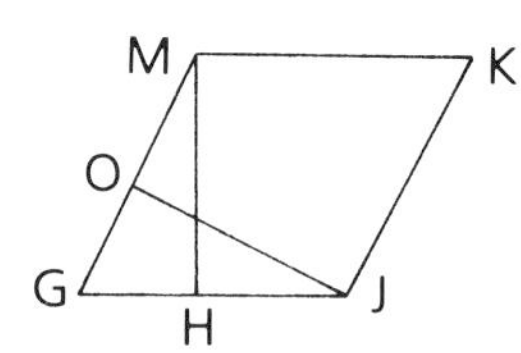

7 Given: $\angle A \cong \angle X$
$\angle AVZ \cong \angle XYB$
$\angle ZVB \cong \angle YBX$
Prove: VBYZ is a ▱

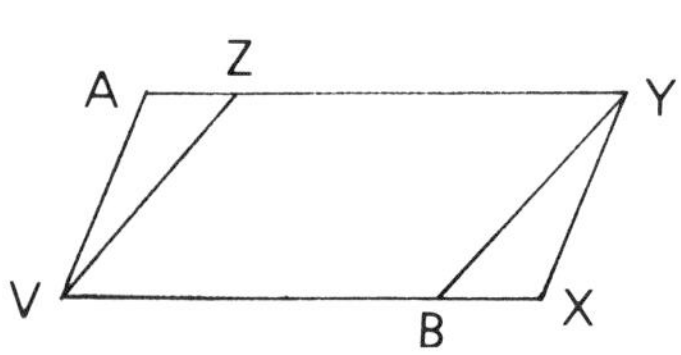

8 The measures of the angles of a △ are in the ratio of 3:4:8. Find the measure of the supplement of the largest angle.

9 Triangle as marked
Find: m∠1

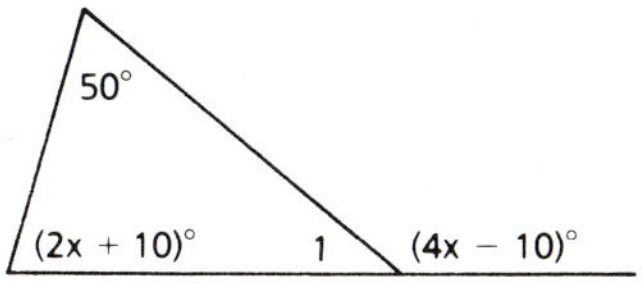

Section 7.2 Problem Set B

10 Prove that the altitude to the base of an isosceles triangle is also a median to the base.

11 Prove that segments drawn from the midpoint of the base of an isosceles triangle and perpendicular to the legs are congruent if they terminate at the legs.

12 Given: OHJM is an isosceles trapezoid with bases $\overline{HJ}$ and $\overline{OM}$
∠HPJ ≅ ∠JKH

Prove: **a** △HRJ is isosceles

b The ⊥ bis of $\overline{HJ}$ passes through R

c $\overline{HP} \cong \overline{JK}$

d R is equidistant from O and M

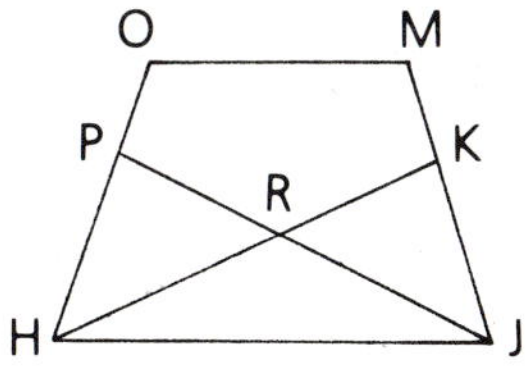

13 Given: $\overleftrightarrow{AC} \parallel \overleftrightarrow{XY}$
$\overleftrightarrow{AB} \parallel \overleftrightarrow{CY}$
∠ZAC ≅ ∠XAB

Prove: ∠X ≅ ∠Z

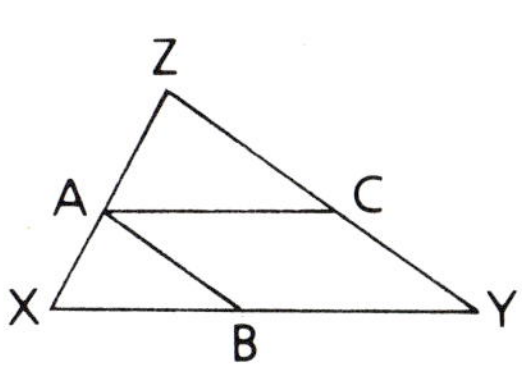

14 If the perimeter of △DEF is 145, find the perimeter of △GHJ.

Can you state a generalization based on this problem?

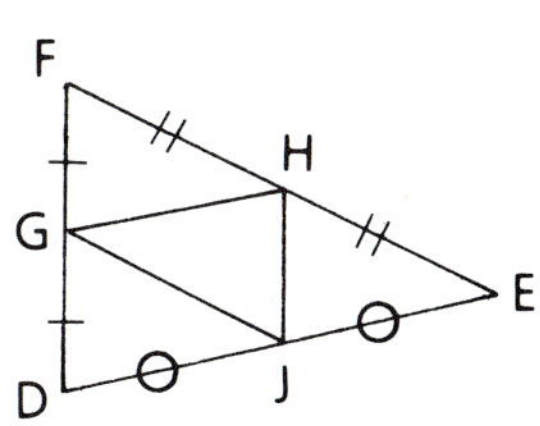

15 Prove that HL is true.

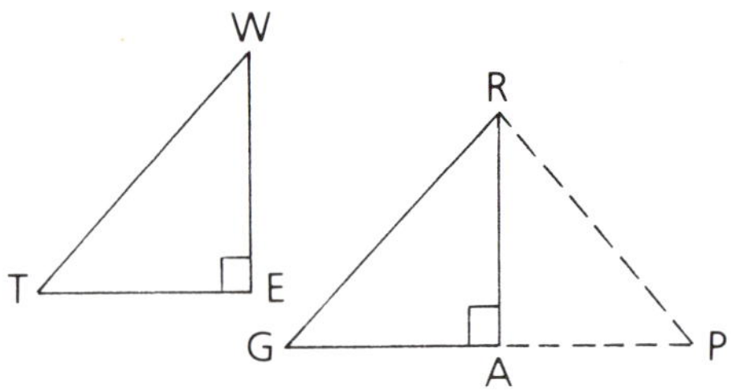

Given: $\overline{TW} \cong \overline{GR}$
$\overline{WE} \cong \overline{AR}$
$\angle E$ and $\angle A$ are rt. $\angle$s

Conclusion: $\triangle WET \cong \triangle RAG$

Hint: Extend $\overrightarrow{GA}$ to P, so that $\overline{AP} \cong \overline{ET}$. Prove that $\triangle WET \cong \triangle RAP$ by SAS. Prove that $\triangle RGP$ is isosceles. Prove that $\triangle RAG \cong \triangle RAP$ by AAS. Then consider the significance of two triangles being congruent to $\triangle RAP$.

Section 7.2 Problem Set C

16 Give the most descriptive name to the figure formed by connecting consecutive midpoints of each of the following figures. Be prepared to defend your answer in each case.

a Rhombus **c** Square **e** Parallelogram **g** Isosceles trapezoid
b Kite **d** Rectangle **f** Quadrilateral

17 Given: $\overline{EF}$ is the median to $\overline{AC}$
$\angle CBD \cong \angle ADB$
$\overline{CD}$ is the base of isosceles $\triangle FDC$

Prove: ABCD is a rectangle

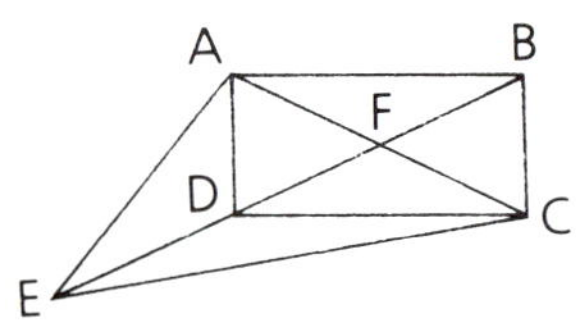

18 Given: P, T, and R lie in plane f
$\angle TNR \cong \angle TSR$, $\overline{NS} \perp f$
$\angle TNP \cong \angle TSP$

Conclusion: $\triangle NPR \cong \triangle SPR$

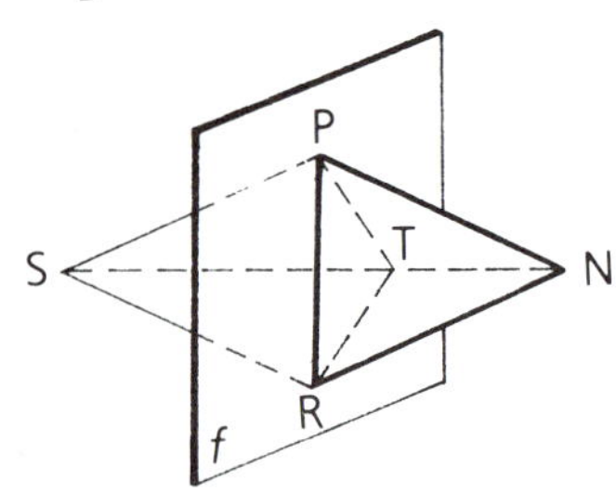

Section 7.2 Problem Set D

19 $\triangle RST$ is equiangular
$\overline{GH} \cong \overline{KH}$

Solve for a in terms of b and c

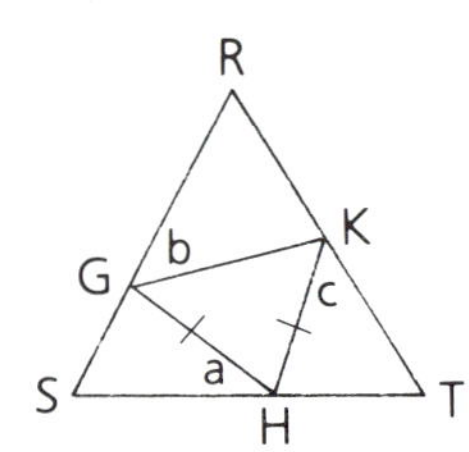

The Royal Road: Computers as Random Number Generators

The random number generator of a computer can be used to **simulate**, or represent, many problems in probability. For example, the probability that each dart will land in a red region of this dartboard is .75. What is the probability that at least two of the next three darts will land in the red? The following computer program simulates one trial of three darts:

```
10   LET S = 0
20   FOR I = 1 TO 3
30   X = INT (RND(1)*100)
40   IF X < 75 THEN S = S + 1
50   NEXT
60   IF S > = 2 THEN 90
70   PRINT "FAILURE"
80   GOTO 100
90   PRINT "SUCCESS"
100  END
```

1 Repeat this trial 10 times. The probability that at least two darts will land in the red is the number of "successes" divided by 10, the number of trials.
2 Modify the program to perform any given number of trials.

Section 7.3 Formulas Involving Polygons

A polygon with three sides can be called a 3-gon; similarly, a polygon with 7 sides can be called a 7-gon.

Most of the polygons we shall encounter have special names as given on the following chart:

Number of Sides (or Vertices)	Name of Polygon	Number of Sides (or Vertices)	Name of Polygon
3	Triangle	8	Octagon
4	Quadrilateral	9	Nonagon
5	Pentagon	10	Decagon
6	Hexagon	12	Duodecagon
7	Heptagon	15	Pentadecagon
		n	n-gon

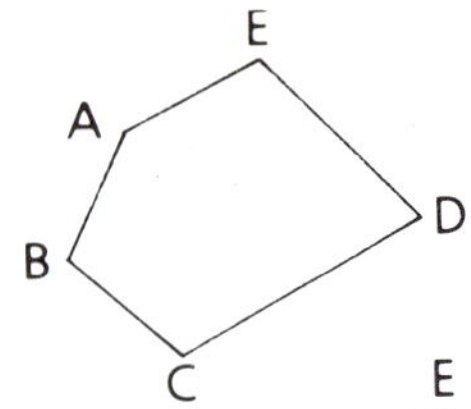

What is the sum of the measures of the five angles in the figure? To answer that question, start at any vertex and draw diagonals. Three triangles have been formed. By adding the measures of the angles of the three triangles, we can obtain the sum of the measures of the five original angles. Hence, figure ABCDE contains 3 (180), so 540 is the sum of the measures of its five angles.

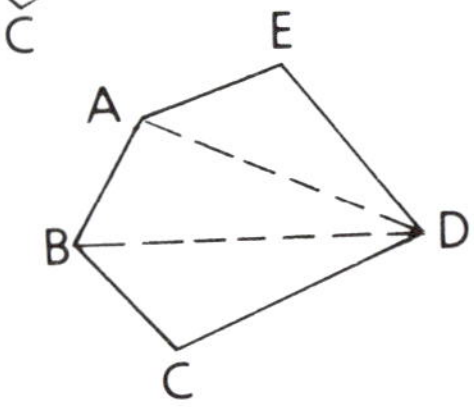

We follow a similar process with the next figure:

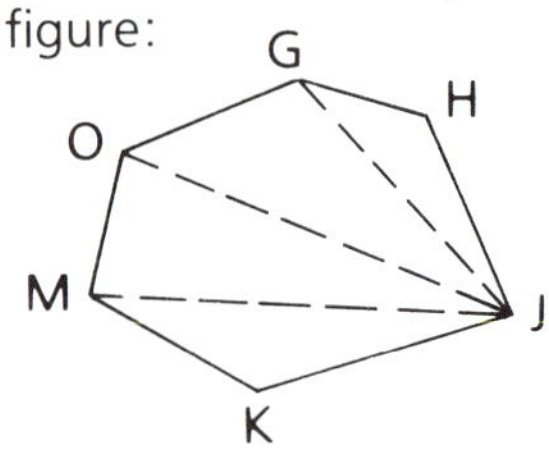

Since there are four triangles, the sum of the measures of the angles of figure GHJKMO is 4 (180) or 720 .

These two examples suggest the following theorem, which we present without formal proof:

Theorem 47

T **The sum of the measures of the angles of a polygon with n sides is given by the formula, $S_i = (n-2)180$.**

On occasion, we *may* refer to the angles of a polygon as the *interior* angles of the polygon.

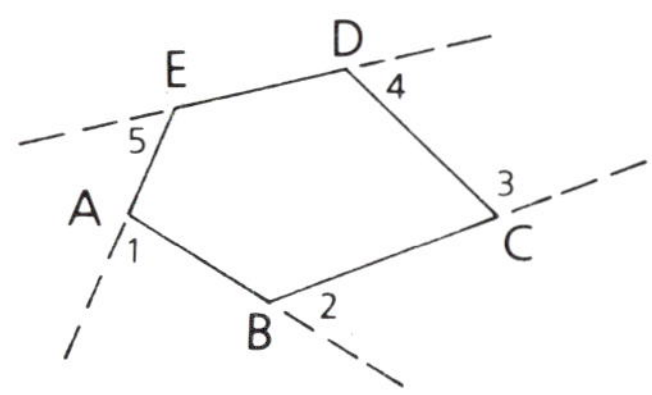

In the figure above, we have formed an exterior angle at each vertex by extending one of the sides of the polygon.

At vertex A, $m\angle 1 + m\angle EAB = 180$. In a similar manner we add each exterior angle to its adjacent interior angle, getting 180 at each vertex. Since there are five vertices, the total is 5 (180) or 900 .

According to Theorem 47, the sum of the measures of the angles of polygon ABCDE is 540 .

Since $900 - 540 = 360$, we may conclude that $m\angle 1 + m\angle 2 + m\angle 3 + m\angle 4 + m\angle 5 = 360$.

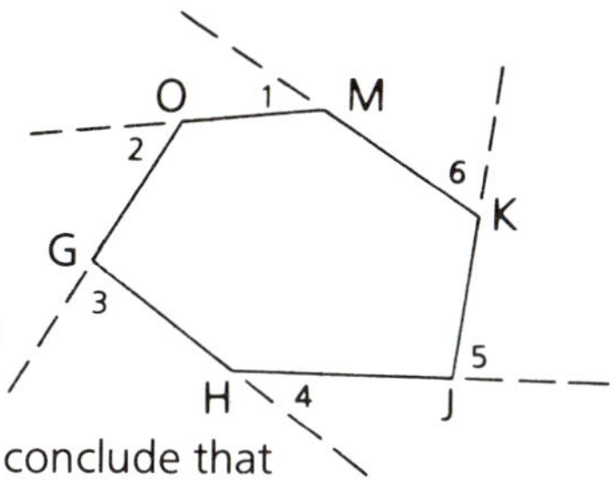

What is the sum of the measures of exterior angles 1, 2, 3, 4, 5, and 6 in this figure?

Again, we can calculate a total of 180 at each of six vertices for a total of 6 (180) or 1080 .

Theorem 47 tells us that the sum of the measures of the angles of polygon GHJKMO is 720 .

Because 1080 − 720 = 360, we may once again conclude that $m\angle 1 + m\angle 2 + m\angle 3 + m\angle 4 + m\angle 5 + m\angle 6 = 360$.

These two examples suggest the next theorem, which we present without formal proof:

Theorem 48

T **Taking one exterior angle at each vertex, the sum of the measures of the exterior angles of a polygon is given by the formula, $S_e = 360$.**

The following theorem is presented without proof. Problem 16 in Problem Set C asks you to explain this formula.

Theorem 49

T **The number d of diagonals that can be drawn in a polygon of *n* sides is given by the formula, $d = \frac{n(n-3)}{2}$.**

Section 7.3 Sample Problems

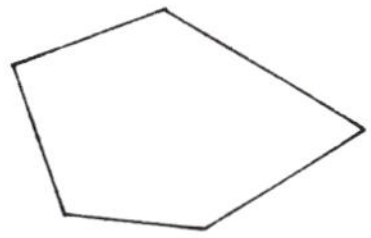

1 Find the sum of the measures of the angles of the figure to the right.

Solution: The figure has 5 sides and 5 vertices.

The formula is $S_i = (n - 2)180$.

Substituting 5 for n,

$S_i = (5 - 2)180$, or 540

2 Find the number of diagonals that can be drawn in a pentadecagon.

Solution: The formula is $d = \frac{n(n-3)}{2}$

Substituting 15 for n, $d = \frac{15(15-3)}{2}$

$d = \frac{15(12)}{2}$, or 90

3 What is the name of a polygon if the sum of the measures of its angles is 1080?

Solution: The formula is $S_i = (n-2)180$

Substituting 1080 for S_i $1080 = (n-2)180$

$1080 = 180n - 360$

$1440 = 180n$

$8 = n$

Since there are 8 sides, the polygon is an octagon.

Section 7.3 Problem Set A

1 Find the sum of the measures of the angles of a:

a quadrilateral
b heptagon
c octagon
d duodecagon
e 93-gon

2 $m\angle A = 160$, $m\angle B = 50$
$m\angle C = 140$, $m\angle D = 150$
Find: $m\angle E$

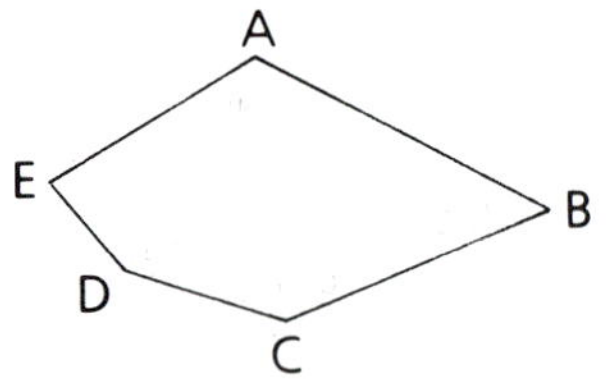

3 $m\angle F = 110$
$m\angle G = 80$
$m\angle H = 74$
Find: $m\angle 1$

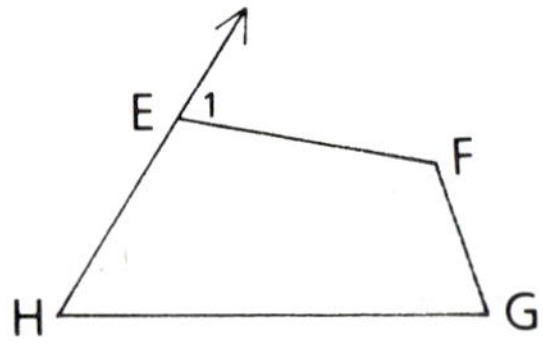

4 Given: K is a midpoint
P is a midpoint
$m\angle M = 70$
$m\angle JKP = y + 15$
$m\angle JPK = y - 10$

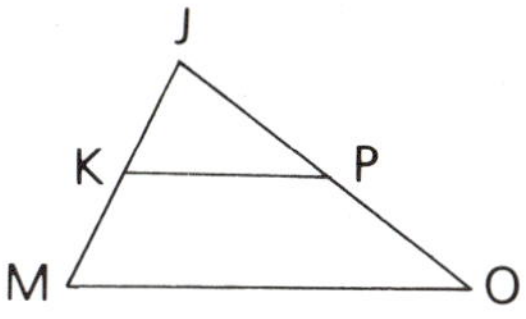

Find: **a** $m\angle JKP$ **b** $m\angle JPK$ **c** $m\angle J$

5 How many diagonals can be drawn in each figure below?

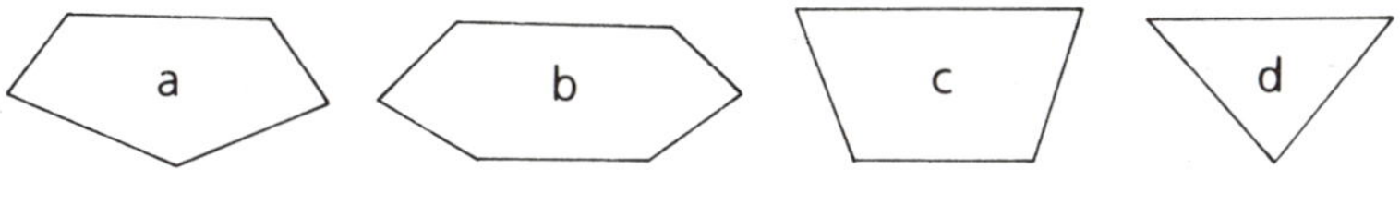

6 Find the sum of the measures of the exterior angles, one per vertex, of each of these polygons:

a triangle **b** heptagon **c** nonagon **d** 1984-gon

7 What is the fewest number of sides a polygon can have?

Section 7.3 Problem Set B

8 On a clock a segment is drawn connecting the mark representing the 12 to the mark representing the 1, another segment from the mark for the 1 to the mark with the 2, etc., all the way around the clock.

a What is the sum of the measures of the angles of the polygon formed?

b What is the sum of the measures of the exterior angles, one per vertex, of the polygon?

9 Prove: Corresponding altitudes of $\cong$ triangles are $\cong$.

10 How many sides does each of the following polygons have if the sum of the measures of its angles is:

a 900 **c** 2880 **e** 436

b 1440 **d** $180x - 720$ **f** 6 right angles

11 a In what polygon is the sum of the measures of the exterior angles, one per vertex, equal to the sum of the measures of the angles?

b In what polygon is the sum of the measures of the angles of the polygon equal to twice the sum of the measures of the exterior angles, one per vertex?

12 If the sum of the measures of the angles of a polygon is increased by 900°, how many sides will have been added to the polygon?

13 What are the names of the polygons that contain the following numbers of diagonals? a 14 b 35 c 209

14 Answer a–d Always, Sometimes, or Never (A, S, or N)

a As the number of sides of a polygon increases, the number of exterior angles increases.

b As the number of sides of a polygon increases, the sum of the measures of the exterior angles increases.

c The sum of the lengths of the diagonals of a polygon is more than the perimeter of the polygon.

d If consecutive midpoints of the sides of a polygon are joined, the sum of the measures of the angles of the original polygon is equal to the sum of the measures of the angles of the midpoint-connecting polygon.

Section 7.3 Problem Set C

15 In Chapter 5, we noted that one of the ways to prove that a quadrilateral is a parallelogram is to prove that both pairs of opposite angles are congruent. Until this chapter, the proof of that method would have been extremely long and involved. Let's see if you can prove it now.

Given: $\angle B \cong \angle D$

$\angle A \cong \angle C$

Prove: ABCD is a ▱

(Hint: Let $m\angle B = x$ and $m\angle C = y$)

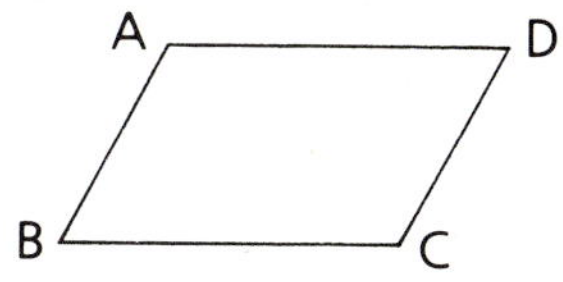

16 Explain why each of the three ingredients in the formula of Theorem 49 (the n, the n − 3, and the 2) is needed.

17 We stated that in this text a polygon will be a CONVEX polygon and that angles greater than 180° will not be considered. Let us ignore that rule for this problem.

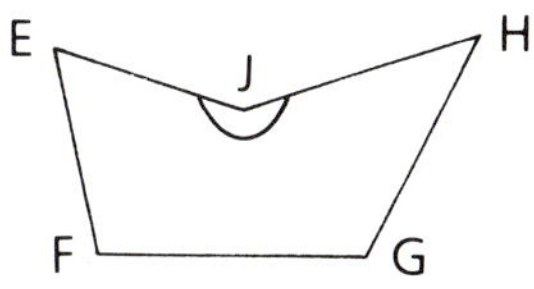

- **a** If we consider the non-convex polygon EFGHJ, the angle *of the polygon* at J is greater than 180°.

  Can you demonstrate that the sum of the measures of the angles of the non-convex polygon EFGHJ is 540 ?

- **b** Can you demonstrate that the sum of the measures of the angles of the non-convex octagon is 1080?

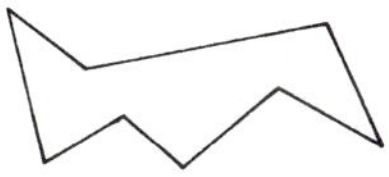

- **c** Do you believe that the sum of the measures of the angles of a non-convex polygon of n sides is (n − 2)180?
- **d** Is the sum of the measures of the exterior angles, one per vertex, of a non-convex polygon equal to 360? Explain.

18 Seven of the angles of a decagon have measures whose sum is 1220. Of the remaining three angles, exactly two are complementary and exactly two are supplementary. Find the measures of the three remaining angles.

Section 7.3 Problem Set D

19 Find the set of polygons for which the number of diagonals is greater than the sum of the measures of the angles.

Section 7.4 Regular Polygons

The topics of this section are:

A REGULAR POLYGONS
B A SPECIAL FORMULA FOR EQUIANGULAR POLYGONS

7.4(A) Regular Polygons

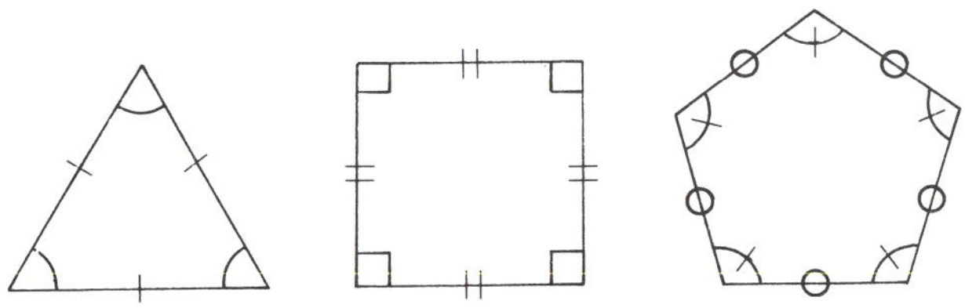

Above are shown three examples of **REGULAR POLYGONS.**

D **A REGULAR POLYGON is a polygon that is both equilateral and equiangular.**

A **REGULAR POLYGON** of three sides is usually called an equilateral (or equiangular) triangle. A **REGULAR POLYGON** of four sides is a square. But a **REGULAR POLYGON** of seven sides is a regular heptagon, and a **REGULAR POLYGON** of 34 sides is a regular 34-gon.

7.4(B) A Special Formula for Equiangular Polygons

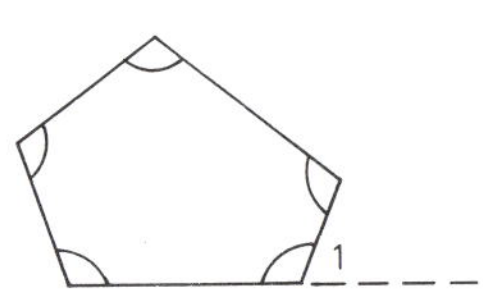

Can you find $m\angle 1$ in the *equiangular* pentagon above? In Section 7.3(B), we learned that the sum of the measures of the exterior angles, one per vertex, of any polygon is 360.

Since each of the five exterior angles has the same measure, we can work the problem above by dividing 360 by 5.

$$m\angle 1 = \frac{360}{5} = 72$$

This problem suggests the next theorem, which we present without formal proof.

Theorem 50

T **The measure of each exterior angle, E, of an equiangular polygon of n sides is given by the formula,** $E = \frac{360}{n}$.

You will see several applications of this theorem in the problems that follow.

Section 7.4 Sample Problems

1 How many degrees are there in each exterior angle of an equiangular heptagon?

Solution: Using $E = \frac{360}{n}$, we have

$$E = \frac{360}{7}, \text{ or } 51\frac{3}{7}$$

2 If each exterior angle of a polygon is 18°, how many sides does the polygon have?

Solution: The formula is $E = \frac{360}{n}$

By substitution, $18 = \frac{360}{n}$

$$18n = 360$$
$$n = 20$$

3 If each angle of a polygon is 108°, how many sides does the polygon have?

Solution: First, find the measure of an exterior ∠ by noting that an ∠ of a polygon and an adjacent exterior ∠ are supp.

Thus, $180 - 108 = 72$

By formula, $E = \frac{360}{n}$

By substitution, $72 = \frac{360}{n}$

$$72n = 360$$
$$n = 5$$

4 Find the measure of each angle of a regular octagon.

Solution: By formula, $E = \frac{360}{n}$

By substitution, $E = \frac{360}{8} = 45$

Thus, the measure of each *interior* angle is $180 - 45 = 135$.

5 Find the measure of each exterior angle of an equilateral quadrilateral.

Solution: An equilateral quadrilateral is not necessarily equiangular, so there is no answer.
The question is undetermined.

Section 7.4 Problem Set A

1 Find the measure of each *exterior* angle of the following equiangular polygons:

a triangle **b** quadrilateral **c** octagon **d** pentadecagon **e** 23-gon

2 Find the measure of each angle of the following equiangular polygons:

a pentagon **b** hexagon **c** nonagon **d** duodecagon **e** 21-gon

3 How many sides does an equiangular polygon have if each *exterior* angle is:

a 60° **b** 40° **c** 36° **d** 2° **e** $7\frac{1}{2}^\circ$

4 How many sides does an equiangular polygon have if each angle is:

a 144° **b** 120° **c** 156° **d** 162° **e** $172\frac{4}{5}^\circ$

5 Given: PENTA is a regular pentagon

Prove: $\triangle PNT$ is isosceles

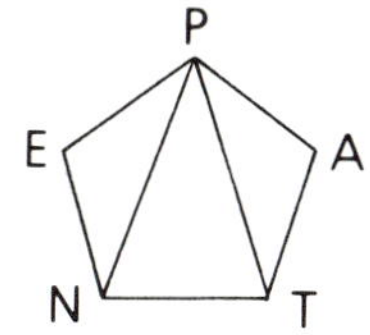

6 Given: A Stop Sign

Is $\triangle NTE$ scalene, isosceles, equilateral, or undetermined?

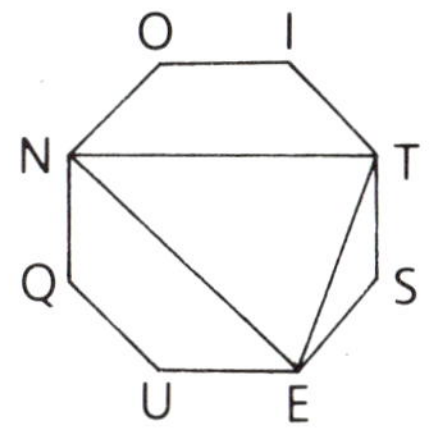

7 In an equiangular polygon the measure of each exterior angle is 25% of the measure of each interior angle. What is the name of the polygon?

Section 7.4 Problem Set B

8 a Prove: The perpendicular bisector of a side of a regular pentagon passes through the opposite vertex.

b Can you generalize about the perpendicular bisectors of the sides of regular polygons?

9 Given: $\overline{AB} \cong \overline{AD}$

$\overline{FC} \perp \overline{BD}$

Conclusion: $\triangle AEF$ is isosceles

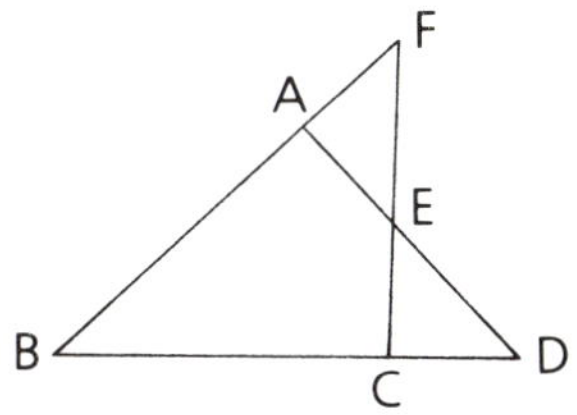

10 The sum of the measures of the angles of a regular polygon is 5040. Find the measure of each angle.

11 What is the name of the polygon whose sum of the angle measures is nine times that of the measure of each exterior angle of a regular hexagon?

12 What is the name of an equiangular polygon if the ratio of the measure of an interior angle to the measure of an exterior angle is 7 to 2?

13 Answers a–f Always, Sometimes, or Never (A, S, or N)

a As the number of sides of an equiangular polygon doubles, the measure of each exterior angle is halved.

b The measure of an exterior angle of a decagon is larger than the measure of an exterior angle of a quadrilateral.

c A regular polygon is equilateral.

d An equilateral polygon is regular.

e If the midpoints of the sides of a scalene quadrilateral are joined in order, the figure formed is equilateral.

f If the midpoints of the sides of a rhombus are joined in order, the figure formed is equilateral, but not equiangular.

Section 7.4 Problem Set C

14 Given: ABCDEF is a regular hexagon

Prove: ACDF is a rectangle

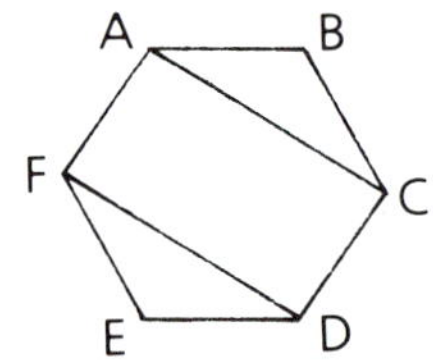

15 Given: $\overline{RO} \perp$ plane GHJ

O, M, K are coplanar

GHJKMP is a regular hexagon

$\overrightarrow{HO}$ bisects $\angle GHJ$

$\overline{RH} \cong \overline{RJ}$

Prove: $\triangle HOJ$ is regular

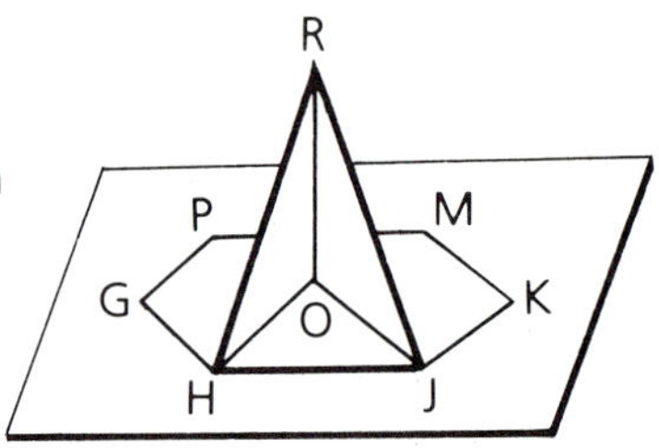

16 Given: $105 < m\angle T < 145$

An equiangular polygon can be drawn with $\angle T$ as one of the angles

Find: The set of possibilities for $m\angle V$

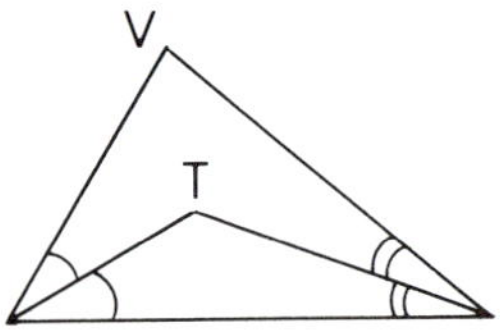

17 We shall call the figure to the right a regular semioctagon. (What do you think that means?)

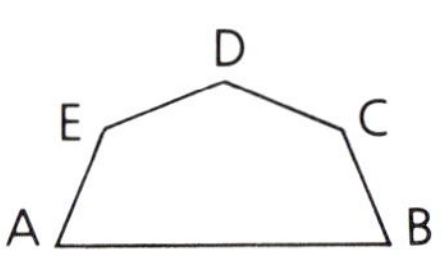

If $m\angle E = 3x + 3y + 9$ and $m\angle A = 2x + y - 4\frac{1}{2}$, solve for x and y.

Chapter 7 Study Guide

As you study and review this chapter, be sure that you are able to:

1. Prove the "second most important" theorem of Geometry, *The sum of the measures of the angles of a triangle is 180.* (p. 279)

2. Describe two ways that an *exterior* angle of a triangle is related to the *interior* angles of that triangle. (p. 280)

3. Name all five ways to prove that two triangles are $\cong$. (pp. 110, 111, 150, 286)

4. State an important fact about the acute angles of a right triangle. (p. 284)

5. Explain why the NO CHOICE theorem has such a name. (p. 286)

6. State both conclusions of the MIDLINE theorem. (p. 281)

7. Describe the difference between a hexagon and a pentagon. (p. 291) How many other polygons can you name?

8. Find the sum of the measures of the angles of an octagon. (pp. 292, 299, 300)

   a If the octagon is REGULAR, find the measure of each exterior angle and then each interior angle.
   b Explain how you could answer those questions for a polygon of *any* number of sides.

9. Use the formulas of this chapter to decide whether there is a polygon whose angles add up to 2000°. (p. 292)

10. State whether a decagon has twice as many diagonals as a pentagon. (p. 293)

11. Describe each of the following symbols and the formulas in which they appear: S_i, S_e, E, n, d (pp. 292, 293, 299)

Chapter 7 Review Problems A

1 Given: $\angle DBC \cong \angle E$

Conclusion: $\angle A \cong \angle BDC$

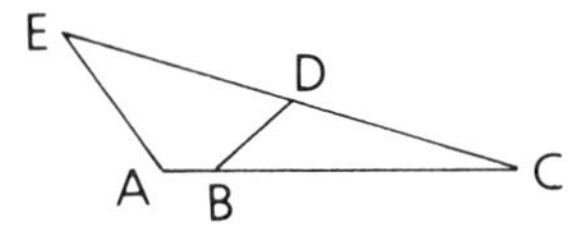

2 Given: $\odot O$

$\angle W \cong \angle V$

$\angle ORW \cong \angle OSV$

Prove: $\overline{PR} \cong \overline{ST}$

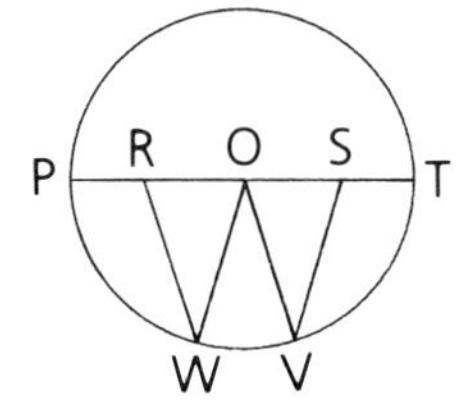

3 Given: $\overline{AC} \cong \overline{AE}$

$\angle CBD \cong \angle EFD$

Prove: $\angle BDC \cong \angle FDE$

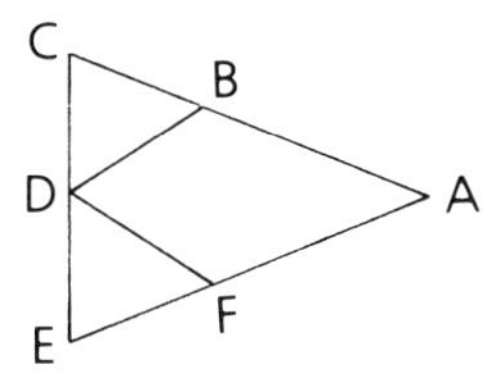

4 Given: $\angle S \cong \angle ROP$

$\angle ROS \cong \angle P$

Prove: $\angle SRO \cong \angle PRO$

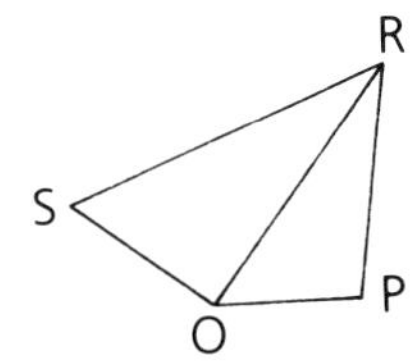

Why can you not work by AAS to prove that the $\triangle$s are $\cong$?

5 Given: $\overline{SV}$ lies in plane *m*

$\overline{VX}$ lies in plane *m*

$\angle S \cong \angle X$

$\overleftrightarrow{TV} \perp$ plane *m*

Prove: $\overline{TS} \cong \overline{TX}$

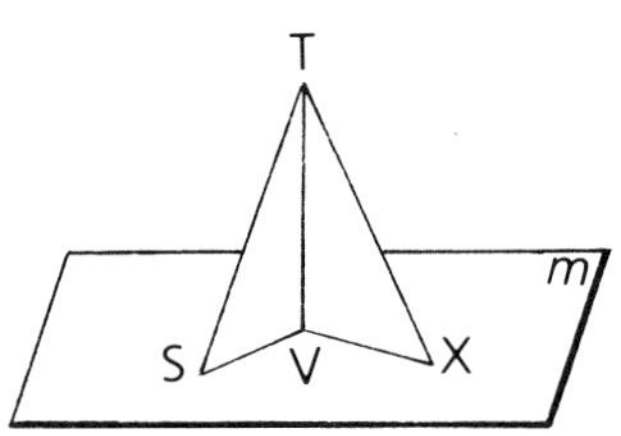

6 The measures of three of the angles of a quadrilateral are respectively 40, 70, and 130. What is the measure of the fourth angle?

7 The measures of the angles of a triangle have the ratio of 1:2:3. Find the measure of half of the largest angle.

8 Diagram as marked

Find: $m\angle 1$

$m\angle 2$

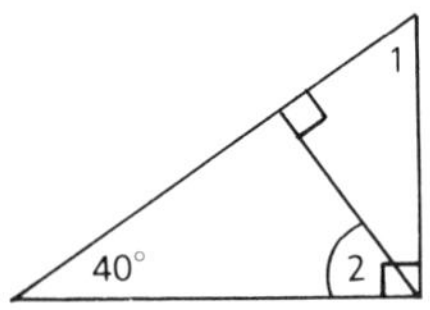

9 Diagram as marked
Find: m∠YZA

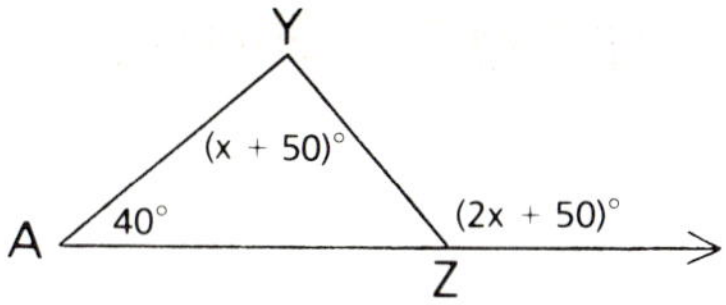

10 Given: C is the midpt. of $\overline{BD}$
E is the midpt. of $\overline{BF}$
DF = 12
m∠D = 80, m∠B = 60
Find: CE, m∠BCE, m∠BEC

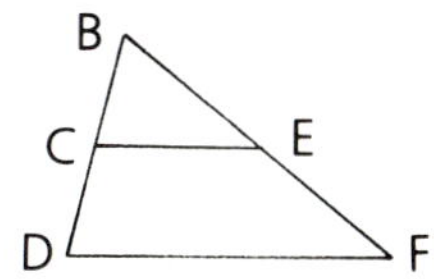

11 Find m∠3 of the diagram as marked

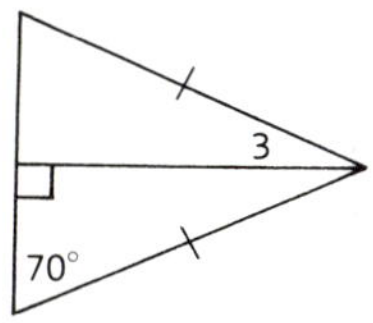

12 If the measure of an exterior angle of a regular polygon is 15, how many sides does the polygon have?

13 If a polygon contains 33 sides, find:

a the sum of the measures of the angles of the polygon.

b the sum of the measures of the exterior angles, one per vertex, of the polygon.

14 The sum of the measures of the angles of a polygon is 1620. Find the number of sides of the polygon.

15 Find the number of diagonals in a pentadecagon.

16 The measure of each exterior angle of an equiangular polygon is twice that of each interior angle. What is the name of the polygon?

Chapter 7 Review Problems B

17 Prove that any two diagonals of a regular pentagon are congruent. Is this statement true for regular polygons in general?

18 Given: $\triangle ABC$ is isosceles with base $\overline{BC}$
$\overrightarrow{BE}$ bisects $\angle ABC$
$\overrightarrow{CE}$ bisects $\angle FCD$
$\angle A = 50^\circ$

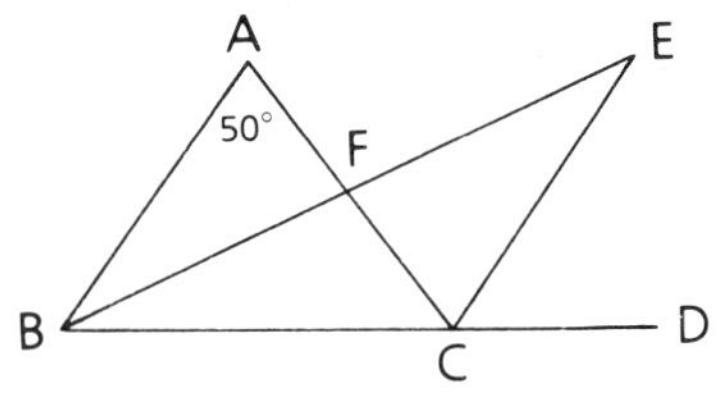

Find: **a** $m\angle ABF$ **b** $m\angle BCE$ **c** $m\angle E$

19 The measure of one of the angles of a right triangle is five times that of a second angle of the triangle. What are the possibilities for the measure of the second largest angle?

20 Given: $\overline{AB} \cong \overline{AC}$
$\angle DBC \cong \angle DCA$
$m\angle A = 50$

Find: $m\angle BDC$

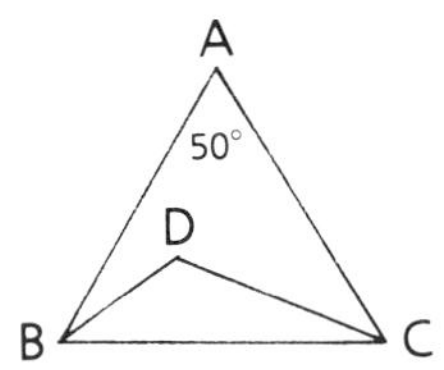

21 Answer a–d Always, Sometimes, or Never (A,S,or N)

a An equiangular triangle is isosceles.

b The number of diagonals in a polygon is the same as the number of sides.

c An exterior angle of a triangle is larger in measure than any angle of a triangle.

d One of the base angles of an isosceles triangle has a measure larger than one of the exterior angles of the triangle.

22 Five of the angles of an "octagon" have a total measure of 540. What conclusion can you draw about the "octagon"?

23 An ARITHMETIC PROGRESSION is a sequence of terms in which the difference between any two consecutive terms is always the same. For example: 1, 5, 9, 13 is an ARITHMETIC PROGRESSION because the difference between any two consecutive terms is 4.

Do the numbers of diagonals in a triangle, quadrilateral, pentagon, and hexagon form an ARITHMETIC PROGRESSION?

24 The measure of an angle of an equiangular polygon exceeds four times the measure of an exterior angle by 30. What is the name of the polygon?

Chapter 7 Review Problems C

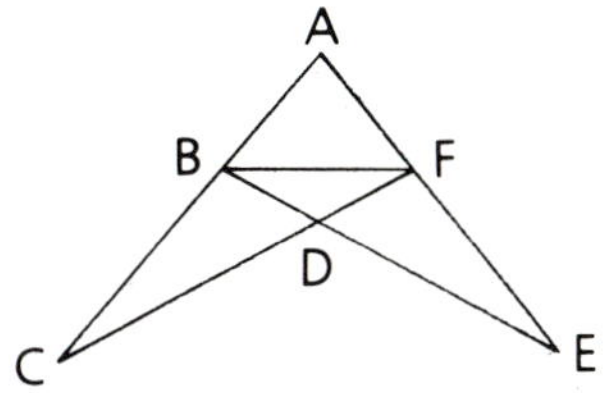

25 Given: $\overline{BC} \cong \overline{FE}$

$\angle C \cong \angle E$

Prove: $\triangle ABF$ is isosceles

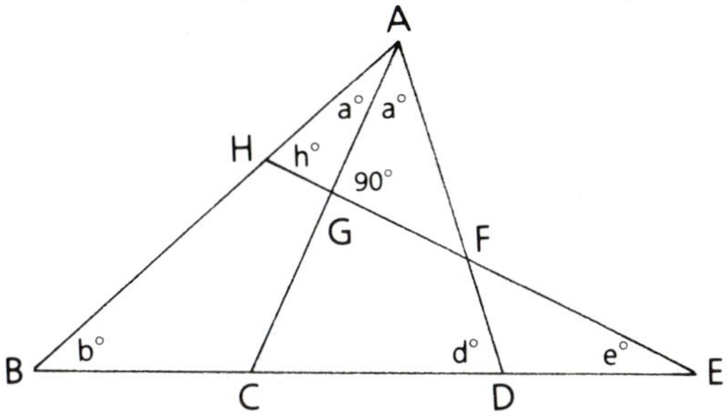

26 Show that $h = \frac{1}{2}(b + d)$

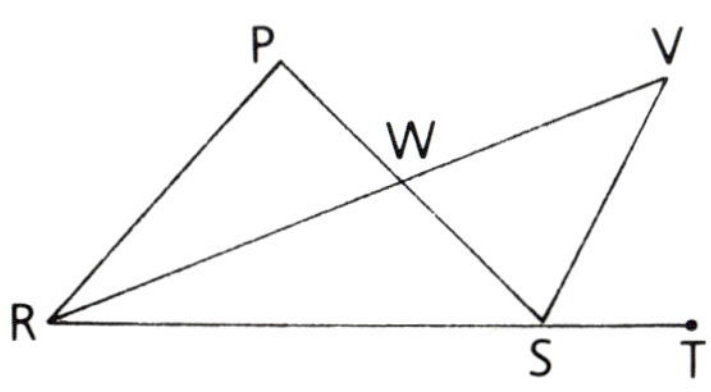

27 Given: $PR = PS$

$\overrightarrow{RV}$ bisects $\angle PRS$

$\overrightarrow{SV}$ bisects $\angle PST$

Prove: $m\angle V = \frac{1}{2}m\angle P$

(Hint: Let $m\angle P = 4x$)

28 In a drawer there is a regular triangle, a regular quadrilateral, a regular pentagon, and a regular hexagon. The drawer is opened, and an angle is selected at random. What is the probability that the measure of the angle is an integral multiple of 30?

Chapter 7 Review Problem D

29 Show that the number of diagonals in a polygon is never the same as the sum of the measures of the exterior angles, one per vertex, of a polygon.

Chapter 8

SIMILAR POLYGONS

Section 8.1 Ratio and Proportion

This section is about the following algebraic concepts:

A RATIO, PROPORTION, TERMS, MEANS, EXTREMES, MEAN PROPORTIONAL

B MEANS-EXTREMES PRODUCTS THEOREM

C MEANS-EXTREMES RATIO THEOREM

D EQUIVALENT FORMS

8.1 (A) Ratio, Proportion, Terms, Means, Extremes, Mean Proportional

D **A RATIO is the quotient of two numbers.**

The ratio of 5 meters to 3 meters is written in any of the following ways:

$$\frac{5}{3} \qquad 5:3 \qquad 5 \text{ to } 3 \qquad 5 \div 3$$

Notice that the 5 (or first number) is the numerator and the 3 (or second number) is the denominator.

Unless otherwise specified, a ratio is given in lowest terms. For example, the ratio of 15 to 6 is

$\frac{15}{6}$ before reduction to lowest terms, or $\frac{5}{2}$ when reduced.

D **A PROPORTION is an equation in which two or more ratios are set equal.**

Here are three examples:

$$\frac{1}{2} = \frac{5}{10} \qquad \frac{3}{5} = \frac{9}{15} = \frac{x}{y} = \frac{6}{10} \qquad \frac{a}{b} = \frac{c}{d}$$

Most PROPORTIONS consist only of two ratios such as:

$$\frac{a}{b} = \frac{c}{d} \quad \text{or} \quad a:b = c:d$$

In the above expressions of the same proportion:

a is called the FIRST TERM — c is called the THIRD TERM

b is called the SECOND TERM — d is called the FOURTH TERM

In a proportion consisting of four terms:

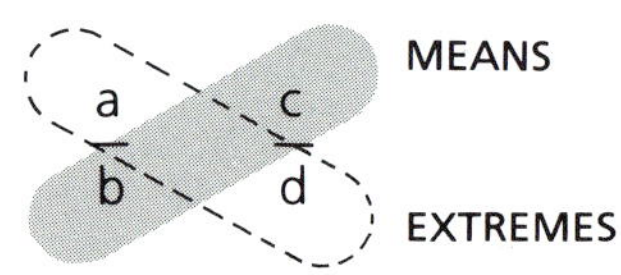

the first and fourth terms are called the EXTREMES

the second and third terms are called the MEANS

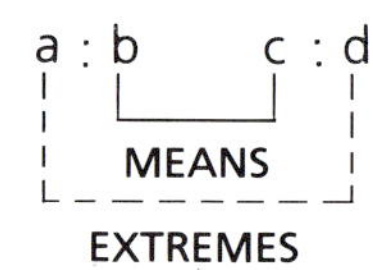

In a MEAN PROPORTION the MEANS are the same. $\frac{1}{4} = \frac{4}{16} \qquad \frac{a}{x} = \frac{x}{r}$

D **If the means are equal, either mean is called a MEAN PROPORTIONAL between the extremes.**

In the first example above, 4 is a MEAN PROPORTIONAL between 1 and 16.

Can you name a MEAN PROPORTIONAL in the second example?

In some textbooks the MEAN PROPORTIONAL is called the GEOMETRIC MEAN. We shall use either name.

8.1 (B) Means-Extremes Products Theorem

Theorem 51

T **In a proportion, the product of the means is equal to the product of the extremes. (MEANS-EXTREMES PRODUCTS THEOREM)**

Given: $\frac{a}{b} = \frac{c}{d}$ Prove: $ad = bc$

Proof of Theorem 51:

1 $\frac{a}{b} = \frac{c}{d}$	1 Given
2 $bd\left(\frac{a}{b}\right) = bd\left(\frac{c}{d}\right)$	2 Multiplication Property (We multiplied each side by bd.)
3 $ad = bc$	3 Algebra

8.1 (C) Means-Extremes Ratio Theorem

The reverse of Theorem 51 is also true and quite useful:

Theorem 52

T **If the product of a pair of non-zero numbers is equal to the product of a pair of non-zero numbers, then either pair of numbers may be made the extremes, and the other pair, the means. (MEANS-EXTREMES RATIO THEOREM)**

Given: $ad = bc$, where a, b, c, d $\neq 0$ Prove: $\frac{a}{b} = \frac{c}{d}$

1 $ad = bc$	1 Given
2 $\frac{ad}{bd} = \frac{bc}{bd}$	2 Division Property (We divided each side by bd.)
3 $\frac{a}{b} = \frac{c}{d}$	3 Algebra

8.1 (D) Equivalent Forms

The MEANS-EXTREMES PRODUCTS THEOREM is useful in determining if two or more proportions are EQUIVALENT FORMS.

■ Example 1: If a, b, c, d $\neq 0$ and $\frac{a}{b} = \frac{c}{d}$, does $\frac{a}{c} = \frac{b}{d}$?

Solution: Use the MEANS-EXTREMES PRODUCTS THEOREM.

$$\frac{a}{b} = \frac{c}{d} \qquad\qquad \frac{a}{c} = \frac{b}{d}$$

$$ad = bc \qquad\qquad ad = cb$$

$$ad = bc$$

Notice that we obtained $ad = bc$ in each case. Therefore, the two proportions are EQUIVALENT FORMS.

Example 2: If $\frac{x}{y} = \frac{a}{b}$, does $\frac{x - 2y}{y} = \frac{a - 2b}{b}$?

Solution:

$$\frac{x}{y} = \frac{a}{b} \qquad \frac{x - 2y}{y} = \frac{a - 2b}{b}$$
$$xb = ya \qquad (x - 2y)b = (a - 2b)y$$
$$xb = ay \qquad xb - 2by = ay - 2by$$
$$xb = ay$$

Again, the MEANS-EXTREMES PRODUCTS THEOREM reveals that the two proportions are EQUIVALENT FORMS.

Section 8.1 Sample Problems

1 If $\frac{3}{x} = \frac{7}{14}$, solve for x.

Solution:

$\frac{3}{x} = \frac{7}{14}$

$\frac{3}{x} = \frac{1}{2}$ Reduced $\frac{7}{14}$ to $\frac{1}{2}$

$1 \cdot x = 3 \cdot 2$ Means-Extremes Products Theorem

$x = 6$

2 Find the fourth term (sometimes called the fourth proportional) if the first three terms of a proportion are 2, 3, and 4.

Solution:

$$\frac{2}{3} = \frac{4}{x}$$
$$2x = 3 \cdot 4$$
$$x = 6$$

3 If 3x = 4y, find the ratio of x to y.

Solution:

$3x = 4y$

$\frac{x}{y} = \frac{4}{3}$ Means-Extremes Ratio Theorem (x and 3 are the extremes, y and 4 are the means.)

4 Find the mean proportional(s) between 4 and 16.

Solution: $\frac{4}{x} = \frac{x}{16}$

$$x \cdot x = 4 \cdot 16$$
$$x^2 = 64$$
$$x = \pm 8$$

Note: There are two mean proportionals (or geometric means) in this problem. In certain geometry problems, we reject one of such algebraic answers. For example, a length cannot be -8.

5 Show that $\frac{a}{b} = \frac{c}{d}$ and $\frac{a+b}{b} = \frac{c+d}{d}$ are equivalent forms.

Solution: Start with the first proportion and add 1 to each side.

$$\frac{a}{b} + 1 = \frac{c}{d} + 1$$
$$\frac{a}{b} + \frac{b}{b} = \frac{c}{d} + \frac{d}{d}$$
$$\frac{a+b}{b} = \frac{c+d}{d}$$

These steps could also be reversed. Thus the proportions are equivalent.

Obtaining the same products equations $\Rightarrow$ equivalent forms.

Section 8.1 Problem Set A

1 If $\frac{3}{4} = \frac{9}{12}$, what is the third term?

2 If $\frac{7}{10} = \frac{14}{20}$, name the means, and then name the extremes.

3 If $\frac{p}{q} = \frac{r}{s}$ does $\frac{r}{p} = \frac{s}{q}$?

4 Solve for x: **a** $\frac{3}{x} = \frac{12}{16}$ **b** $\frac{x}{18} = \frac{3}{7}$ **c** $\frac{7}{x-4} = \frac{3}{5}$

5 Find the fourth proportional to these three terms:

a 1, 2, 3 **b** $\frac{1}{2}$, 3, 4, **c** a, b, 5

6 Find the ratio of x to y if:

a $2x = 3y$ b $6(y + 3) = 2(x + 9)$ c $\frac{3}{x+5} = \frac{9}{y+15}$

7 What is the ratio of the number of diagonals in a pentagon to the measure of each exterior angle of a regular decagon?

8 What is the ratio of the measure of a side of a square to its perimeter?

9 If the ratio of the measures of a pair of sides of a ▱ is 2:3 while the ratio of the measures of the diagonals is 1:1, what is the most descriptive name of the ▱?

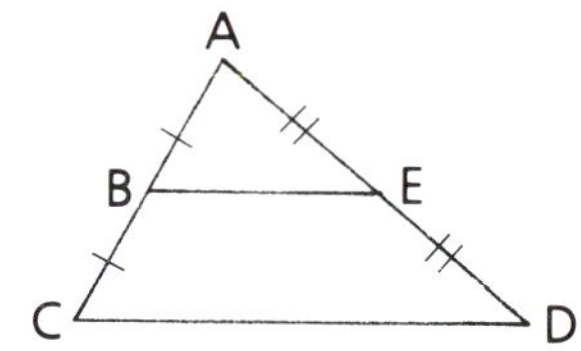

10 a What is the ratio of AB to BC?

b What is AB:AC?

11 Find the mean proportional(s) between each pair of extremes:

a 4 and 25 b 3 and 5 c a and b

12 A 60-meter steel pole is cut into two parts that have the ratio of 11 to 4. How much longer is the larger part than the smaller?

13 The ratio of the measures of the sides of a quadrilateral is 2:3:5:7. If the perimeter is 68, find the length of each side.

Section 8.1 Problem Set B

14 Find the positive geometric mean between 6 and 12.

15 If 4 is a mean proportional between 6 and a number, what is the number?

16 The ratio of the measure of the supplement of an angle to that of the complement of the angle is 5:2. Find the measure of the supplement.

17 If $\frac{x-5}{4} = \frac{c}{3}$, does $\frac{x-1}{4} = \frac{c+3}{3}$? (Use Sample Problem 5 as a theorem.)

18 If $x(a+b) = y(c+d)$, find the ratio of x to y.

19 If $ex - fy = gx + hy$, find the ratio of x to y.

20 Reduce the ratio $\frac{x^2 - 7x + 12}{x^2 - 16}$ to lowest terms.

21 In the figure to the right, F, H, K, and O are midpoints, $\frac{OF}{OK} = \frac{3}{4}$, and $MG + EJ = 42$. Find: OK.

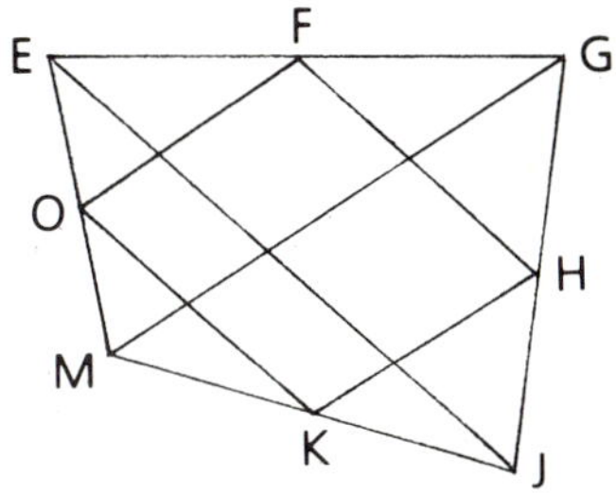

Section 8.1 Problem Set C

22 Show that no polygon exists such that the ratio of the number of diagonals to the sum of the measures of the angles of the polygon is 1 to 18.

23 If $\frac{a}{b} = \frac{c}{d}$, show that $\frac{a-b}{b} = \frac{c-d}{d}$

24 If $\frac{a}{b} = \frac{c}{d}$, show that $\frac{a+b}{a-b} = \frac{c+d}{c-d}$

25 In the figure below, P is said to "divide" $\overline{AB}$ *externally* into two segments $\overline{AP}$ and $\overline{PB}$. If $AB = 30$ and $\frac{AP}{AB} = \frac{5}{2}$, find AP.

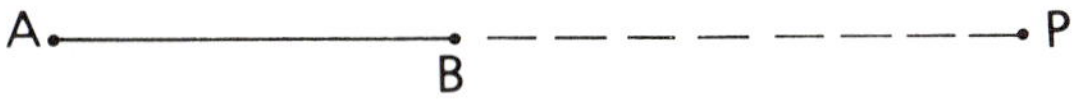

Section 8.1 Problem Set D

26 Given the four numbers: 1, 2, 4, and 8. Two ratios are formed at random using the four numbers. What is the probability that the ratios are equal?

Section 8.2 Similar Figures

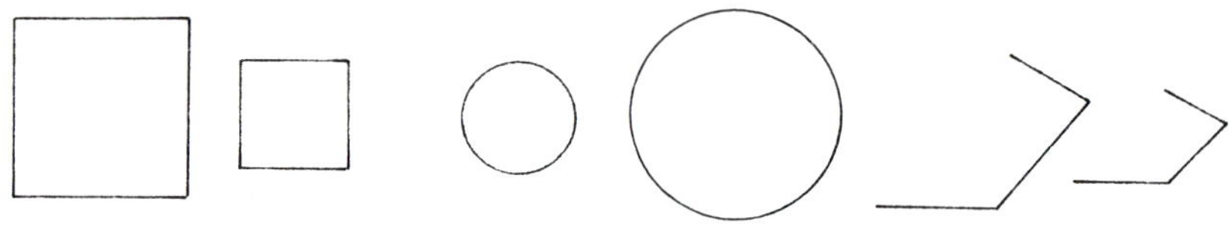

1 2 3

Above are shown three examples of **SIMILAR FIGURES. SIMILAR FIGURES** are figures that have the same *shape,* but *not necessarily* the same *size.*

Real life is full of examples of **SIMILAR FIGURES.** If you look through a pair of binoculars, order an enlargement of a photograph, make a scale drawing, or study a blueprint, you are dealing with similar figures.

Except for a few isolated problems, we shall limit our study of similar figures this year to **SIMILAR POLYGONS.**

D **SIMILAR POLYGONS are polygons in which:**

1 the ratios of the measures of corresponding sides are equal, and

2 corresponding angles are congruent.

The following is an example of two similar (~) triangles:

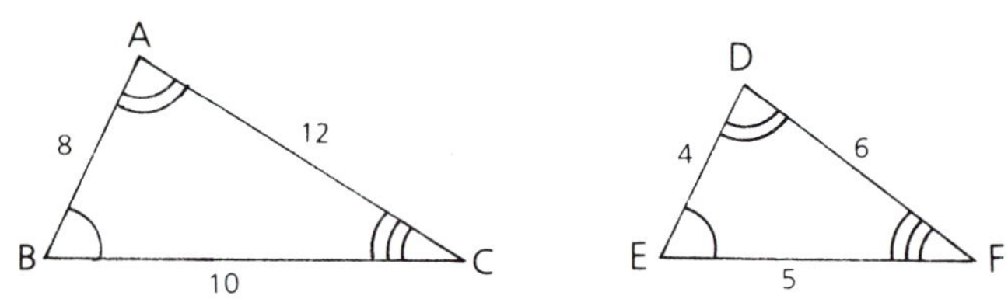

We say that $\triangle ABC \sim \triangle DEF$ ($\triangle ABC$ is similar to $\triangle DEF$), which means that $A \longleftrightarrow D$, $B \longleftrightarrow E$, and $C \longleftrightarrow F$.

As you can see—

1 the ratios of the measures of all pairs of corresponding sides are equal.

$$\left(\text{Notice: } \frac{AB}{DE} = \frac{2}{1} \quad \frac{AC}{DF} = \frac{2}{1} \quad \frac{BC}{EF} = \frac{2}{1}\right)$$

2 each pair of corresponding angles is congruent.

(Notice: $\angle B \cong \angle E$, $\angle A \cong \angle D$, $\angle C \cong \angle F$)

Consider this example:

ABCD $\sim$ EFGH with measures as shown.

a Find FG, GH, EH.

b Find the ratio of perimeters.

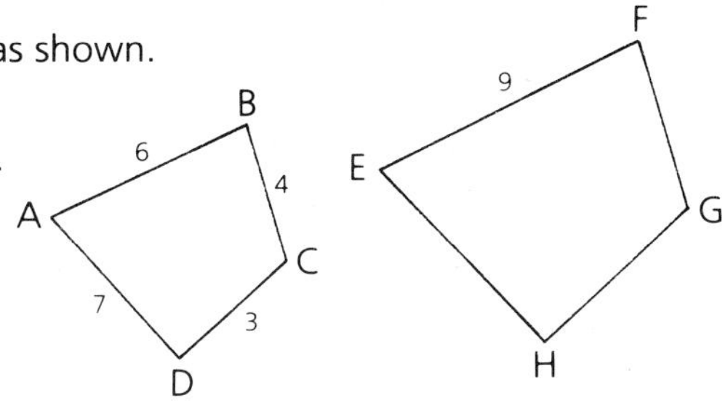

Solutions:

a Since the quadrilaterals are similar, the ratios of the measures of corresponding sides are equal. We begin with one ratio of measures of corresponding sides, preferably one we can simplify.

Thus, we choose: $\frac{AB}{EF} = \frac{6}{9} = \frac{2}{3}$

$$\frac{AB}{EF} = \frac{BC}{FG} \qquad \frac{2}{3} = \frac{4}{FG} \qquad 2(FG) = 12 \qquad FG = 6$$

$$\frac{AB}{EF} = \frac{CD}{GH} \qquad \frac{2}{3} = \frac{3}{GH} \qquad 2(GH) = 9 \qquad GH = 4\tfrac{1}{2}$$

$$AB:EF = AD:EH \qquad 2:3 = 7:EH \qquad 2(EH) = 21 \qquad EH = 10\tfrac{1}{2}$$

b Perimeter of ABCD $= 6 + 4 + 3 + 7 = 20$

Perimeter of EFGH $= 9 + 6 + 4\tfrac{1}{2} + 10\tfrac{1}{2} = 30$

$$\frac{P_{ABCD}}{P_{EFGH}} = \frac{20}{30} = \frac{2}{3}$$

Notice that the ratio of perimeters came out equal to the ratio of sides. This suggests the following theorem.

Theorem 53

T **The ratio of the perimeters of two similar figures equals the ratio of any pair of corresponding sides.**

A proof of Theorem 53 for similar triangles is called for in problem 15, Problem Set C.

Section 8.2 Sample Problems

1 $\triangle JHK \sim \triangle POM$, $\angle H = 90^\circ$, $\angle J = 40^\circ$, $m\angle M = x + 5$, and $m\angle O = \frac{1}{2}y$. Solve for x and y.

Solution: First, draw two triangles, JHK and POM, so that $\angle H = 90^\circ$, $\angle J = 40^\circ$, and the corresponding angles are $\cong$.

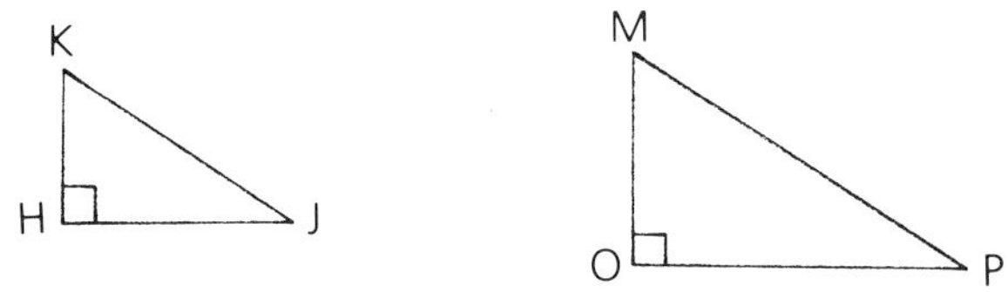

$\angle J$ comp $\angle K$
$\angle J = 40^\circ$
$\Rightarrow \angle K = 50^\circ$

$\angle K \cong \angle M$

$50 = x + 5$

$45 = x$

$\angle H \cong \angle O$

$90 = \frac{1}{2}y$

$180 = y$

2 If $\triangle ABC \sim \triangle XYZ$, show that $\frac{AB + BC}{AC} = \frac{XY + YZ}{XZ}$.

1 $\triangle ABC \sim \triangle XYZ$	1 Given
2 $\frac{AB}{XY} = \frac{AC}{XZ}$	2 If two $\triangle$s are similar, the ratios of the measures of corresponding sides are equal
3 $\frac{BC}{YZ} = \frac{AC}{XZ}$	3 Same as 2
4 $(AB)(XZ) = (AC)(XY)$	4 Means-extremes products theorem (from step 2)
5 $(BC)(XZ) = (AC)(YZ)$	5 Same as 4 (from step 3)
6 $(AB)(XZ) + (BC)(XZ) = (AC)(XY) + (AC)(YZ)$	6 Addition Property of Equality
7 $(XZ)(AB + BC) = (AC)(XY + YZ)$	7 Factoring
8 $\frac{AB + BC}{AC} = \frac{XY + YZ}{XZ}$	8 Means-Extremes Ratio Theorem

You may wonder how we decided what to do in Sample Problem 2.

The truth is we worked BACKWARDS. We started with step 8, did step 7, then step 6, etc. Then we rewrote the proof in proper order. Of course, some problems require more ingenuity than that, but remember the hint of working BACKWARDS. It made our thinking much easier here.

Section 8.2 Problem Set A

1 Which pairs of figures *appear* to be similar?

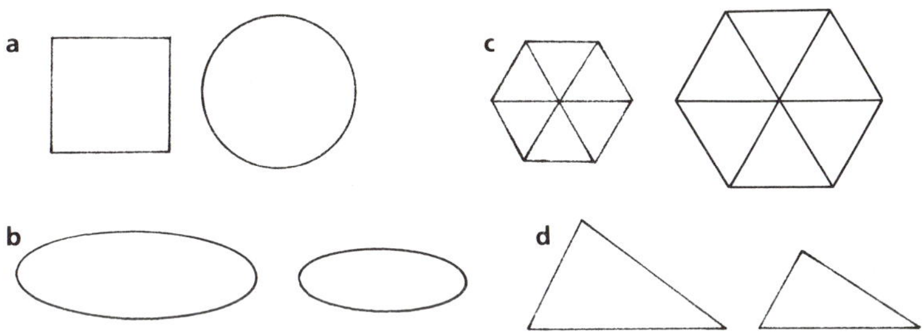

2 Which pairs of triangles can be *proved* to be similar?

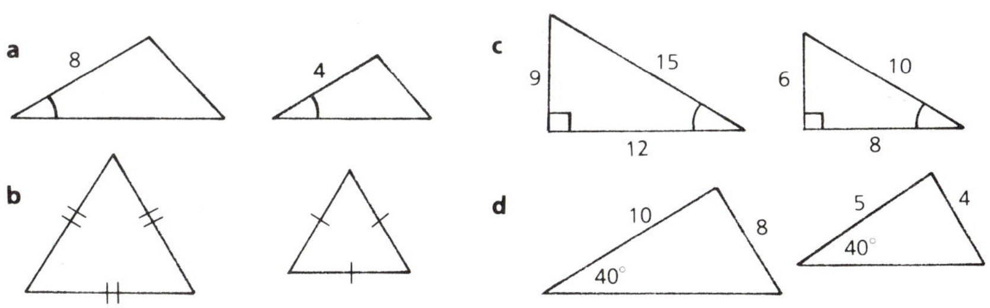

3 Given: $\triangle NPR \sim \triangle STV$

$\angle P = 90^\circ$

$\angle R = 60^\circ$

Find: $m\angle T$, $m\angle S$

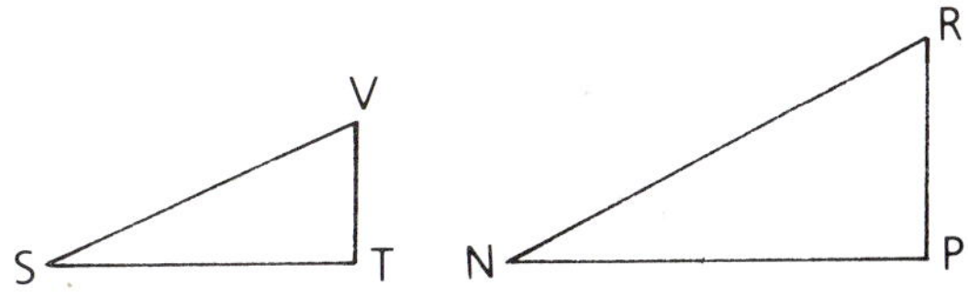

4 $\triangle ABC \sim \triangle DEF$

Lengths as shown

Find: EF

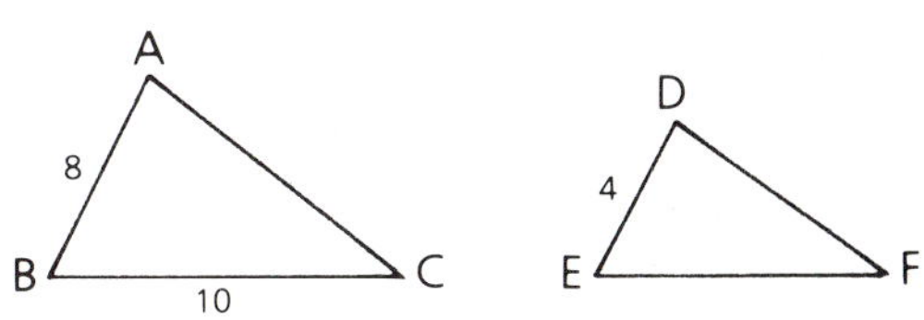

5 $\odot$ O, $\odot$ P, $\triangle AOB \sim \triangle RPS$

$OA = 2,\ AB = 3,\ PR = 6$

Find: PS, RS

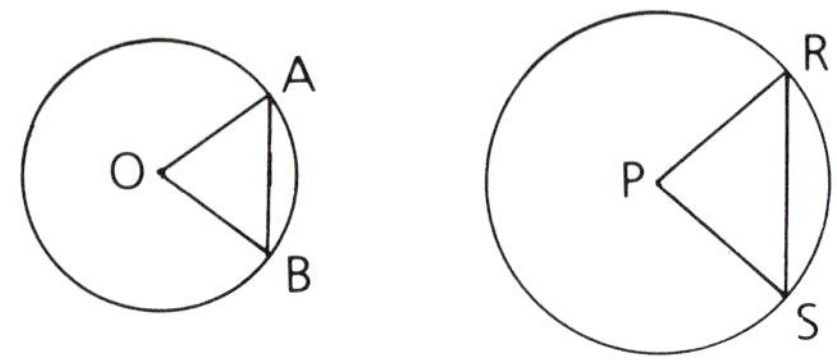

6 Find the mean proportionals between each pair of extremes:

a 4 and 25 **b** 2 and 5

7 If $3x = 5y$, find the ratio of x to y.

Section 8.2 Problem Set B

8 $\triangle ABC \sim \triangle TRS$, $\angle A = 70°$, and $\angle C = 53°$

Find: $m\angle T$, $m\angle R$, $m\angle S$

9 $\triangle SVT \sim \triangle WYX$

Measures of segments as shown

Find: WY

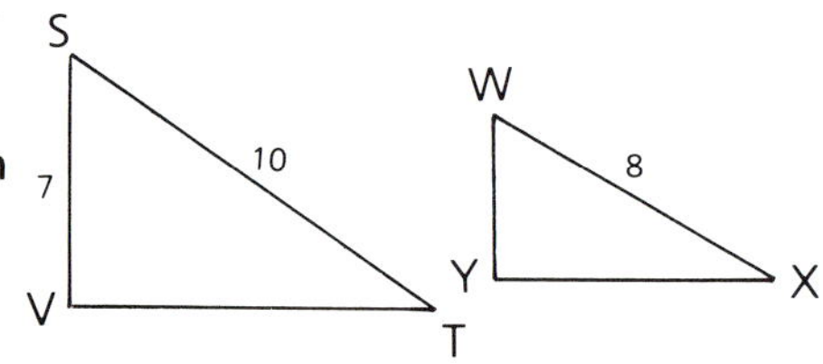

10 Quad ABCD ~ Quad HGFE

Lengths as shown

Find: **a** Ratio of lengths of corresponding sides

b EF

c Perimeter of EFGH

d Ratio of perimeters

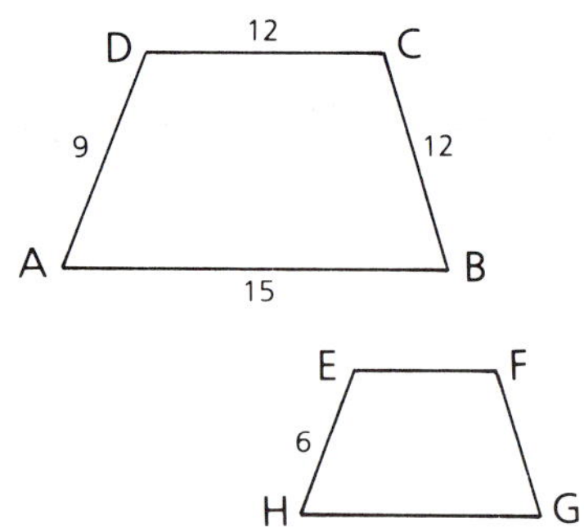

11 $\triangle KJM \sim \triangle OPR$

Angle measures as shown

Find: $\dfrac{x + y + z}{2}$

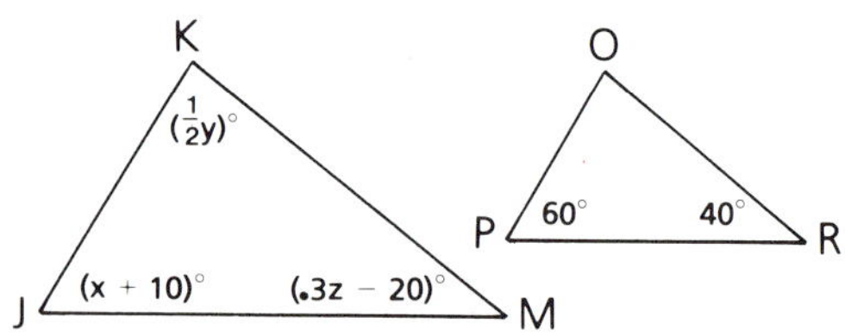

12 Find the ratio of the fourth proportional of 1, 2, and 3 to the fourth proportional of 4, 5, and 6.

13 If $\frac{8}{2x - 3y} = \frac{7}{6x - 4y}$, find the ratio of x to y.

Section 8.2 Problem Set C

14 $\triangle ABC \sim \triangle DEF$
$m\angle A = 50$
$m\angle D = 2x + 5y$
$m\angle F = 5x + y$
$m\angle B = 102 - x$
Find: $m\angle F$

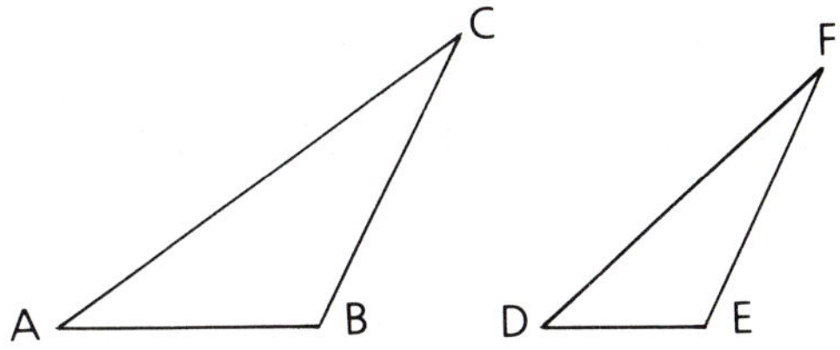

15 If $\triangle GHJ \sim \triangle RST$, show that $\frac{HJ}{ST}$ is equal to the ratio of the perimeters. That is, prove that Theorem 53 is true for triangles.

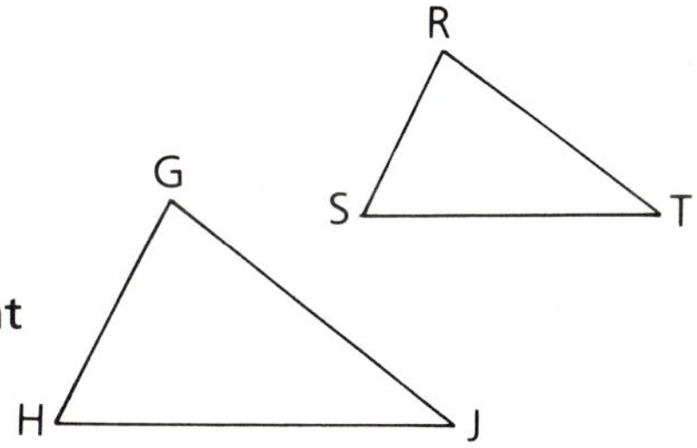

Section 8.2 Problem Set D

16 Show that if the sides of a triangle form an ARITHMETIC PROGRESSION, the sides of any similar triangle form an ARITHMETIC PROGRESSION.

Reminder: An ARITHMETIC PROGRESSION is a sequence of terms in which the difference between any two consecutive terms is constant.

Hint: Let the sides of the given triangle be: x, $x + a$, $x + 2a$
Let the sides of any other similar triangle be: y, $y + b$, $y + c$
Solve for b and c.

Section 8.3 Methods for Proving That Triangles Are Similar

In this section we present ways to prove that triangles are similar. We start by accepting one method as a postulate.

P **If there exists a correspondence between the vertices of two triangles such that three angles of one triangle are congruent respectively to the corresponding angles of a second triangle, then the two triangles are similar. (AAA)**

The following three theorems will be used in proofs much as SSS, SAS, ASA, HL, and AAS were used in proofs for congruency. The first method, AA, is the most frequently used of the three methods.

Theorem 54

T **If there exists a correspondence between the vertices of two triangles such that two angles of one triangle are congruent to the corresponding angles of the other, then the triangles are similar. (AA)**

Given: $\angle A \cong \angle D$

$\angle B \cong \angle E$

Conclusion: $\triangle ABC \sim \triangle DEF$

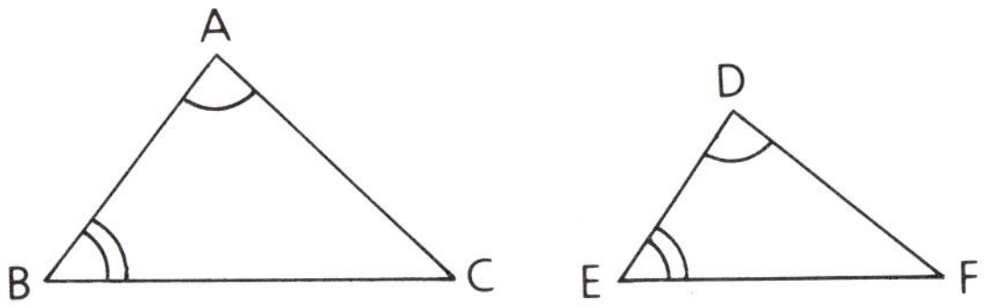

The proof of Theorem 54 follows from the NO CHOICE theorem (p. 286).

We also present, without proof, two additional methods to prove that two triangles are similar. You will discover, however, that AA is most frequently used of the three methods.

Theorem 55

T **If there exists a correspondence between the vertices of two triangles such that the ratios of the measures of corresponding sides are equal, then the triangles are similar. (SSS~)**

Given: $\dfrac{AB}{DE} = \dfrac{BC}{EF} = \dfrac{AC}{DF}$

Prove: $\triangle ABC \sim \triangle DEF$

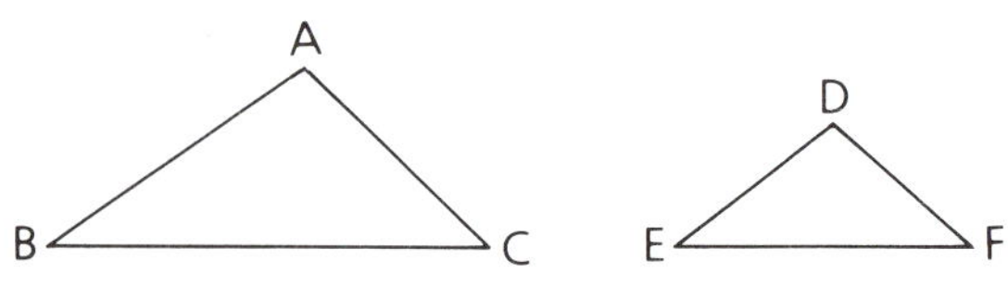

Theorem 56

T **If there exists a correspondence between the vertices of two triangles such that the ratios of the measures of two pairs of corresponding sides are equal and the included angles are congruent, then the triangles are similar. (SAS$\sim$)**

Given: $\frac{AB}{DE} = \frac{BC}{EF}$
$\angle B \cong \angle E$

Prove: $\triangle ABC \sim \triangle DEF$

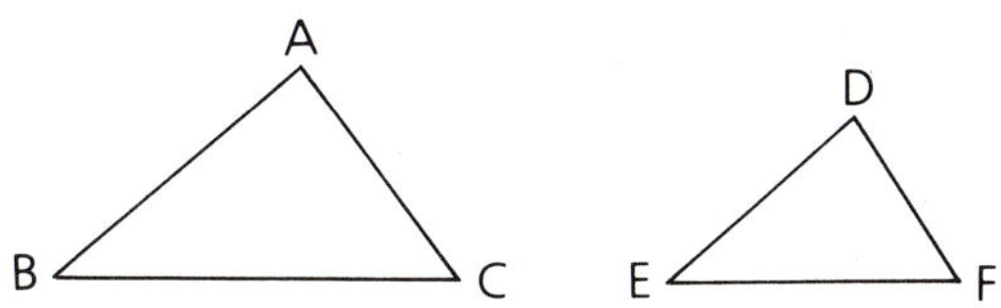

Section 8.3 Sample Problems

1 Given: ABCD is a ▱

Prove: $\triangle BFE \sim \triangle CFD$

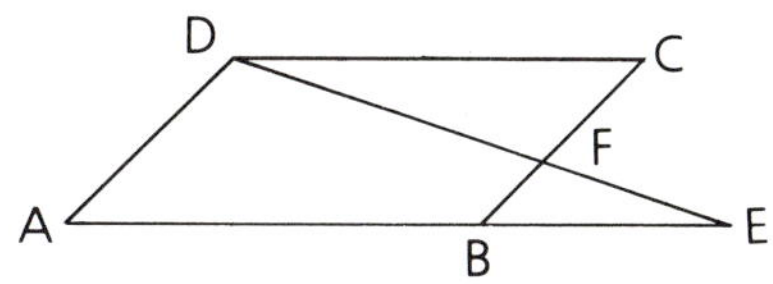

1 ABCD is a ▱	1 Given
2 $\overleftrightarrow{AB} \parallel \overleftrightarrow{DC}$	2 Opposite sides of a ▱ are $\parallel$
3 $\angle CDF \cong \angle E$	3 PAI
4 $\angle DFC \cong \angle EFB$	4 Vertical angles are congruent
5 $\triangle BFE \sim \triangle CFD$	5 AA (3, 4)

2 Given: $\angle MET \cong \angle RST$
$\overline{ST} \perp \overline{MR}$

Conclusion: $\triangle MET \sim \triangle RST$

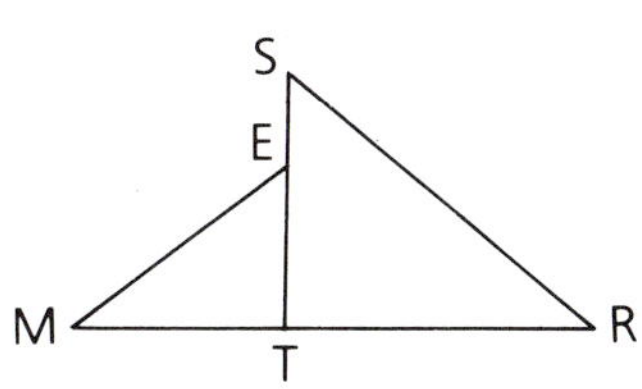

1 $\angle MET \cong \angle RST$	1 Given
2 $\overline{ST} \perp \overline{MR}$	2 Given
3 $\angle STM$ is a right $\angle$ $\angle STR$ is a right $\angle$	3 $\perp$ lines form right $\angle$s
4 $\angle STM \cong \angle STR$	4 Right $\angle$s are $\cong$
5 $\triangle MET \sim \triangle RST$	5 AA (1, 4)

3 Given: $\overline{KH}$ is the altitude to hypotenuse $\overline{GJ}$ of $\triangle GHJ$

Prove: $\triangle KHJ \sim \triangle HGJ$

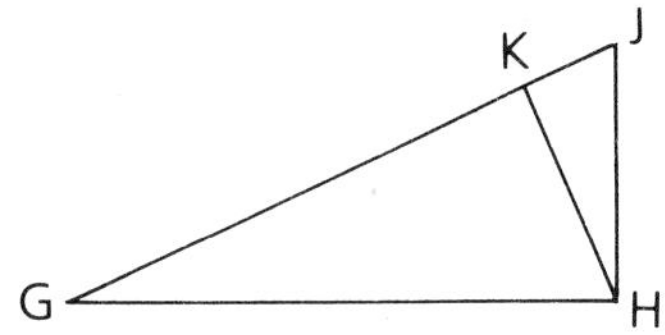

1 $\overline{KH}$ is the altitude to hypotenuse $\overline{GJ}$ of $\triangle GHJ$	1 Given
2 $\angle HKJ$ is a right angle	2 An altitude of a $\triangle$ is drawn from a vertex and forms right $\angle$s with the opposite side
3 $\angle JHG$ is a right angle	3 The hypotenuse is opposite the right $\angle$
4 $\angle HKJ \cong \angle JHG$	4 Right $\angle$s are $\cong$
5 $\angle J \cong \angle J$	5 Reflexive Property
6 $\triangle KHJ \sim \triangle HGJ$	6 AA (4, 5)

4 Prove: If from two points, one on each leg of an isosceles triangle, perpendiculars are dropped to the base, the triangles formed are similar.

Given: $\triangle MTR$ is isosceles
$\overline{MT}$ and $\overline{RT}$ are legs
$\overline{VO} \perp \overline{MR}$, $\overline{SP} \perp \overline{MR}$

Prove: $\triangle MOV \sim \triangle RPS$

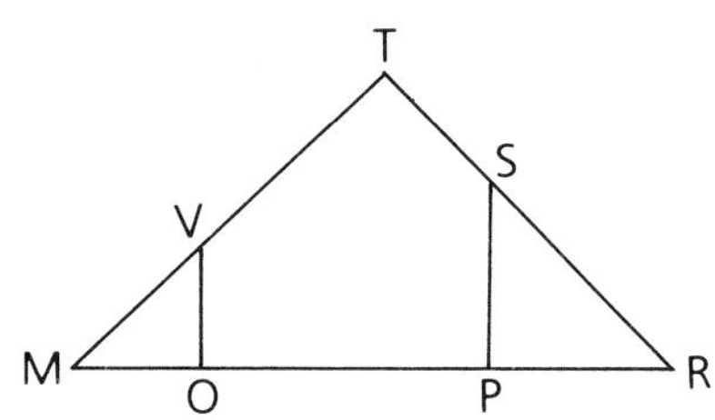

1 $\triangle MTR$ is isosceles with legs $\overline{MT}$ and $\overline{RT}$	1 Given
2 $\angle M \cong \angle R$	2 An isosceles $\triangle$ has $\cong$ legs, and $\triangle \Rightarrow \triangle$
3 $\overline{VO} \perp \overline{MR}$	3 Given
4 $\angle VOM$ is a right angle	4 $\perp$ lines form right $\angle$s
5 $\overline{SP} \perp \overline{MR}$	5 Given
6 $\angle SPR$ is a right angle	6 Reason 4
7 $\angle VOM \cong \angle SPR$	7 Right angles are $\cong$
8 $\triangle MOV \sim \triangle RPS$	8 AA (2, 7)

Section 8.3 Problem Set A

1 Given: $\angle A \cong \angle D$

$\angle 2 \cong \angle 4$

Prove: $\triangle ABC \sim \triangle DEF$

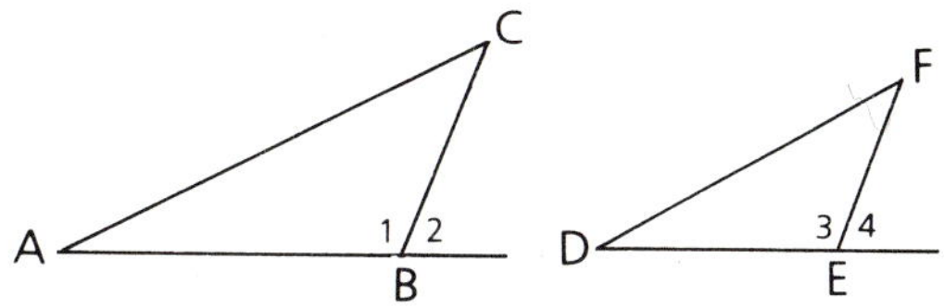

2 Draw a triangle GJK with $\overleftrightarrow{HM} \parallel \overline{JK}$. Make H a point on $\overline{GJ}$ and M a point on $\overline{GK}$. Prove that $\triangle GHM \sim \triangle GJK$.

3 Given: NPRV is a ▱

Conclusion: $\triangle NWO \sim \triangle SWT$

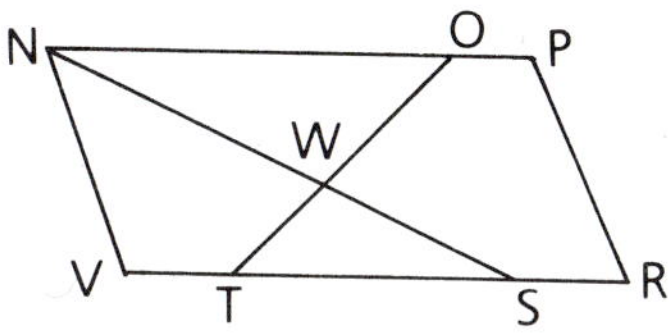

4 Given: $\overline{AC} \cong \overline{AE}$

$\angle CBD \cong \angle EFD$

Prove: $\triangle BCD \sim \triangle FED$

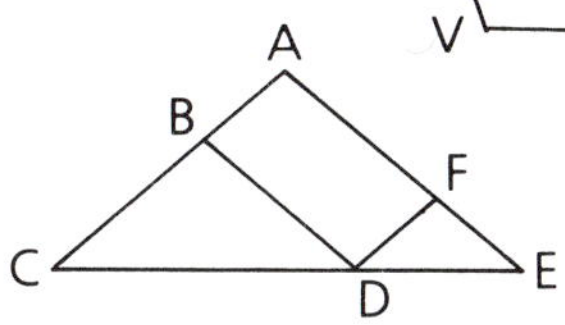

5 Given: TVPX is a trapezoid, bases $\overline{TV}$ and $\overline{XP}$

Conclusion: $\triangle TVY \sim \triangle PXY$

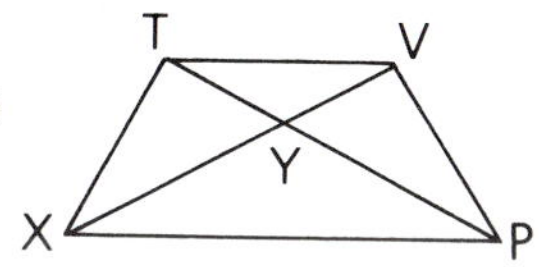

6 Prove that if $\triangle ABC \cong \triangle EDF$, then $\triangle ABC \sim \triangle EDF$.

7 Given: $\angle G$ is a right $\angle$

$\angle K$ is a right $\angle$

$HJ = \frac{1}{2}MO$

Prove: $\triangle GHJ \sim \triangle KMO$

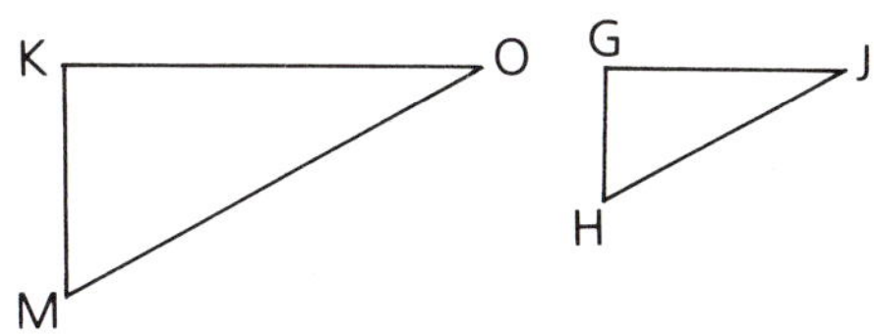

Section 8.3 Problem Set B

8 Given: $\overline{SP}$ is the altitude from S to $\overline{NR}$

$\overline{RT}$ is the altitude from R to $\overline{NS}$

Conclusion: $\triangle NRT \sim \triangle NSP$

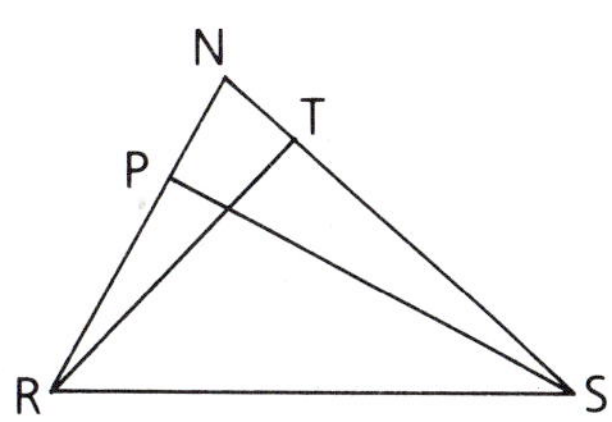

9 Prove: If an acute angle of one right $\triangle$ is congruent to an acute angle of another right $\triangle$, the $\triangle$s are similar.

10 Prove: If the vertex ∠ of one isosceles △ is congruent to the vertex ∠ of a second isosceles △, the Ⓢ are similar.

11 Given: $\overline{AE} \perp \overline{FD}$
$\overline{FC}$ is the altitude to $\overline{BD}$
$\overline{BD}$ is the base of isosceles △ FBD

Prove: △FBC ~ △ADE

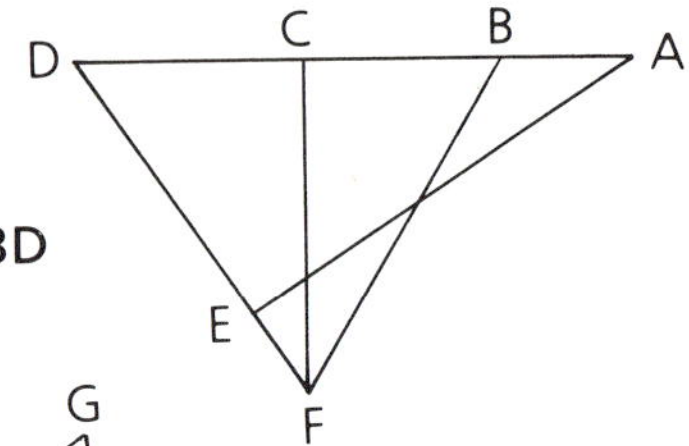

12 Given: $\frac{GJ}{HK} = \frac{GK}{GM}$
∠1 ≅ ∠G

Conclusion: $\overleftrightarrow{HM} \parallel \overleftrightarrow{JK}$

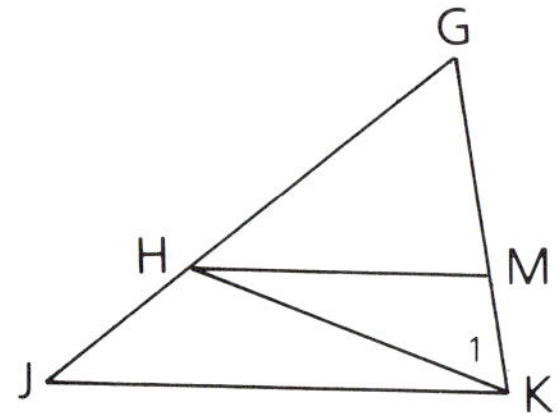

13 Answer a–d Always, Sometimes, or Never (A, S, or N)

a Two right triangles are similar.

b An obtuse triangle is similar to an acute triangle.

c If two triangles are similar, the perimeters have the same ratio as the measures of any pair of corresponding sides.

d Two equilateral polygons are similar.

14 Prove that if you double each side of a triangle, the resulting triangle is similar to the original.

Section 8.3 Problem Set C

15 Prove the transitive property of similar triangles. That is, if two triangles are similar to a third triangle, they are similar to each other. Do you think the transitive property would hold for similar polygons?

16 If two of the six triangles below are selected at random, what is the probability that the two triangles selected are similar?

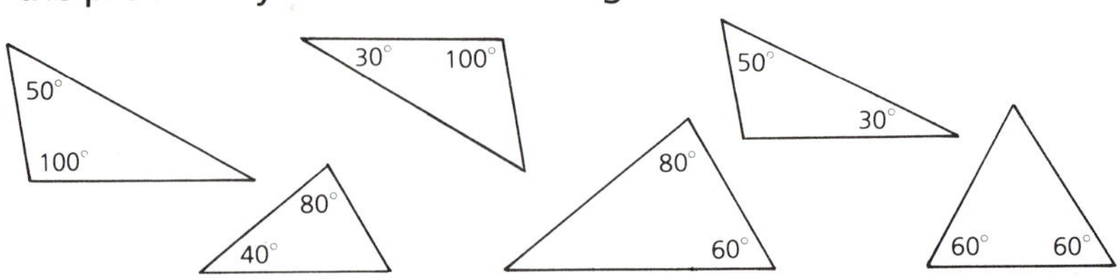

Section 8.4 Ratios and Products from Similar Triangles

After we prove that two triangles are congruent, we use CPCTC. After we prove that two triangles are similar, we use the definition of similar triangles:

1 Corresponding angles of similar triangles are congruent.

2 Corresponding sides of similar triangles are proportional. (The ratios of the measures of corresponding sides are equal.)

If need be, after using the second part of the definition, we use the **MEANS-EXTREMES PRODUCTS THEOREM** to get equal products.

■ Example 1 Given: $\triangle ABC \sim \triangle DEF$

Prove: $\angle A \cong \angle D$

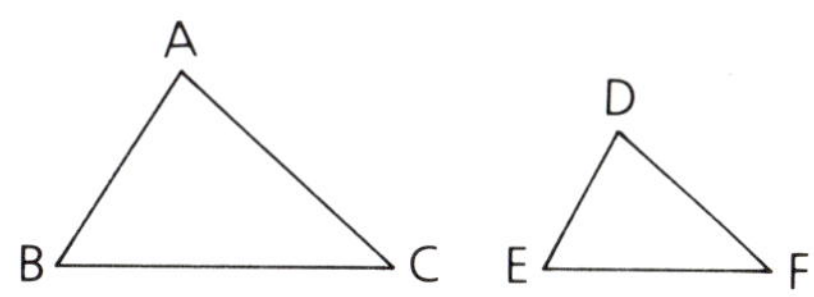

1 $\triangle ABC \sim \triangle DEF$	1 Given
2 $\angle A \cong \angle D$	2 Corresponding $\angle$s of $\sim$ $\triangle$s are $\cong$

■ Example 2 Given: $\triangle ABC \sim \triangle DEF$

Prove: $\frac{AB}{DE} = \frac{AC}{DF}$

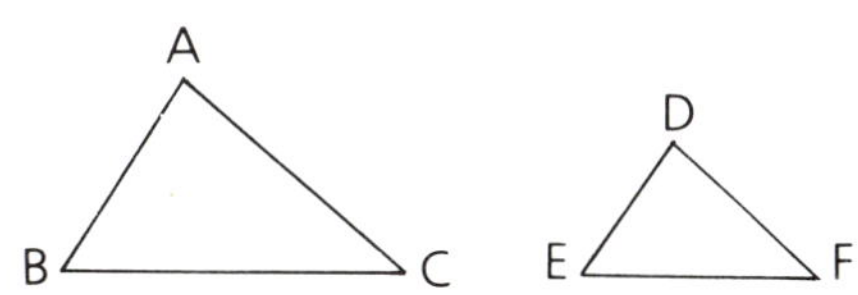

1 $\triangle ABC \sim \triangle DEF$	1 Given
2 $\frac{AB}{DE} = \frac{AC}{DF}$	2 Corresponding sides of $\sim$ $\triangle$s are proportional

Note: We may also claim $\frac{AB}{AC} = \frac{DE}{DF}$ as it is **EQUIVALENT** to $\frac{AB}{DE} = \frac{AC}{DF}$.

■ Example 3 Given: $\triangle ABC \sim \triangle DEF$

Prove: $AB \cdot DF = AC \cdot DE$

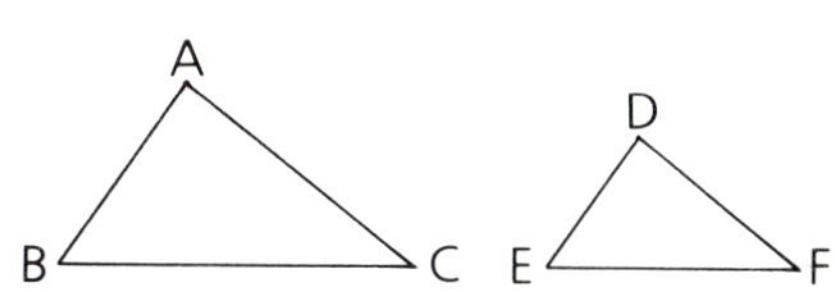

1 $\triangle ABC \sim \triangle DEF$	1 Given
2 $\frac{AB}{DE} = \frac{AC}{DF}$ or $\frac{AB}{AC} = \frac{DE}{DF}$	2 Corresponding sides of $\sim$ $\triangle$s are proportional
3 $AB \cdot DF = AC \cdot DE$	3 Means-Extremes Products Theorem

The sample problems that follow illustrate more difficult problems and the SHADOW PROBLEM.

Section 8.4 Sample Problems

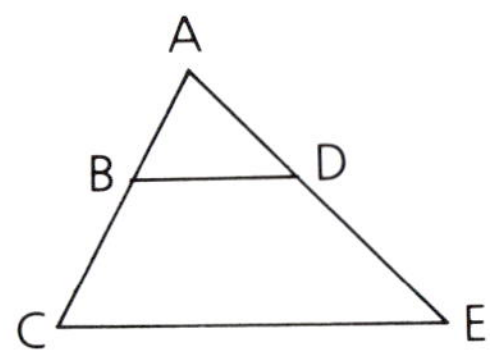

1 Given: $\overleftrightarrow{BD} \parallel \overleftrightarrow{CE}$

Prove: $\dfrac{AB}{AC} = \dfrac{BD}{CE}$

1 $\overleftrightarrow{BD} \parallel \overleftrightarrow{CE}$	1 Given
2 $\angle ABD \cong \angle C$	2 PCA
3 $\angle ADB \cong \angle E$	3 PCA
4 $\triangle ABD \sim \triangle ACE$	4 AA (2, 3)
5 $\dfrac{AB}{AC} = \dfrac{BD}{CE}$	5 Corresponding sides of ~ $\triangle$s are proportional

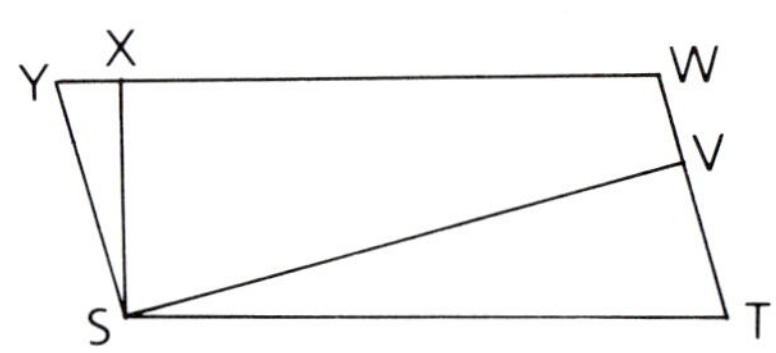

2 Given: $\square$ YSTW

$\overline{SX} \perp \overline{YW}$

$\overline{SV} \perp \overline{WT}$

Prove: $SX \cdot YW = SV \cdot WT$

1 $\square$ YSTW	1 Given
2 $\angle Y \cong \angle T$	2 Opposite $\angle$s of a $\square$ are $\cong$
3 $\overline{SX} \perp \overline{YW}$	3 Given
4 $\angle SXY$ is a right $\angle$	4 $\perp$ segments form right $\angle$s
5 $\overline{SV} \perp \overline{WT}$	5 Given
6 $\angle SVT$ is a right $\angle$	6 Same as 4
7 $\angle SXY \cong \angle SVT$	7 Right $\angle$s are $\cong$
8 $\triangle SXY \sim \triangle SVT$	8 AA (2, 7)
9 $\dfrac{SX}{SV} = \dfrac{SY}{ST}$	9 Corresponding sides of ~ $\triangle$s are proportional
10 $SX \cdot ST = SV \cdot SY$	10 Means-Extremes Products Theorem
11 $\overline{ST} \cong \overline{YW}$	11 The opposite sides of a $\square$ are $\cong$
12 $\overline{SY} \cong \overline{WT}$	12 Same as 11
13 $SX \cdot YW = SV \cdot WT$	13 Substitution (11 and 12 in 10)

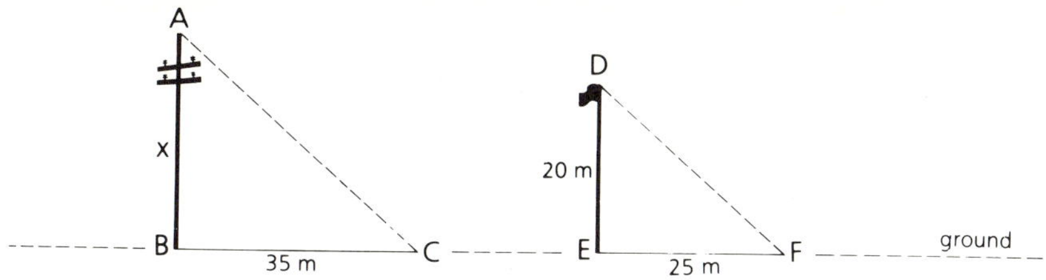

3 (THE SHADOW PROBLEM.) While strolling one morning to get a little sun, Judy noticed that a 20-meter flagpole cast a 25-meter shadow. Nearby was a telephone pole that cast a 35-meter shadow. How tall was the telephone pole?

Because the sun is very far from us, its rays are nearly parallel.

Solution: $\triangle ABC \sim \triangle DEF$ by, AA, so

$$\frac{x}{35} = \frac{20}{25}$$

$$\frac{x}{35} = \frac{4}{5}$$

$$5x = 140$$

$$x = 28\text{m}$$

The Royal Road: Euclid's Proposition 2, Book VI

If a line is parallel to one side of a triangle, it divides the other two sides proportionally.

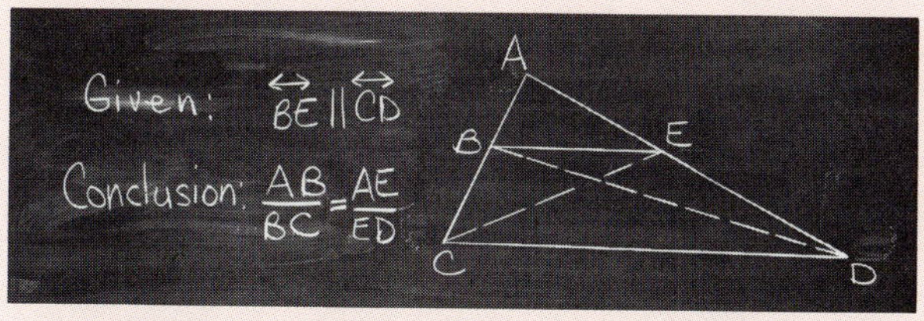

One way to prove this property is to use the AA Theorem for similar triangles (see page 333). In Euclid's organization of the *Elements,* he did not yet have the AA Theorem. He did have the following result: if two triangles have congruent altitudes, their areas are proportional to their bases.

Notice that $\triangle BEA$ and $\triangle BEC$ have the same altitude (from E). Similarly, $\triangle EBA$ and $\triangle EBD$ have the same altitude. Finally, $\triangle BEC$ and $\triangle EBD$ have the same altitude and the same base. Can you show the following:

$$\frac{AB}{BC} = \frac{\text{area of } \triangle BEA}{\text{area of } \triangle BEC} = \frac{\text{area of } \triangle EBA}{\text{area of } \triangle EBD} = \frac{AE}{ED}$$

Section 8.4 Problem Set A

1 Given: $\angle C \cong \angle F$
$\overline{AB} \perp \overline{BC}$
$\overline{DE} \perp \overline{EF}$

Prove: $\frac{AB}{BC} = \frac{DE}{EF}$

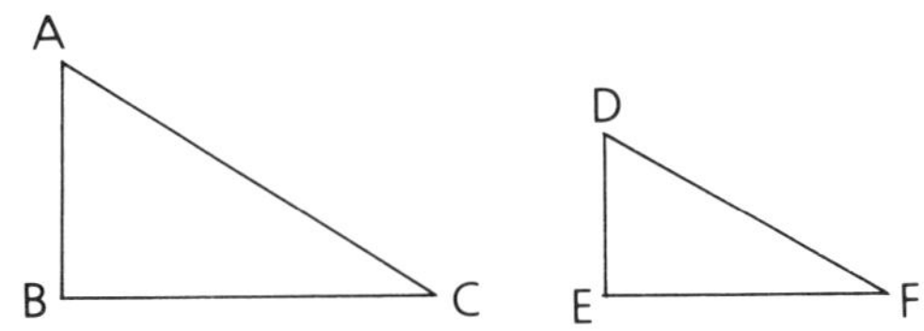

2 Draw a triangle GJK with $\overleftrightarrow{HM} \parallel \overline{JK}$. Make H a point on $\overline{GJ}$ and M a point on $\overline{GK}$. Prove that $\frac{GH}{GJ} = \frac{GM}{GK}$.

3 Given: $\angle X \cong \angle ZBA$

Conclusion: $\frac{AZ}{AB} = \frac{ZY}{XY}$

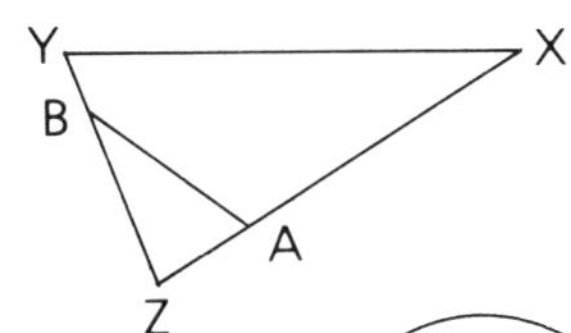

4 Given: $\angle D \cong \angle G$

Conclusion: $\frac{CD}{FG} = \frac{DE}{EG}$

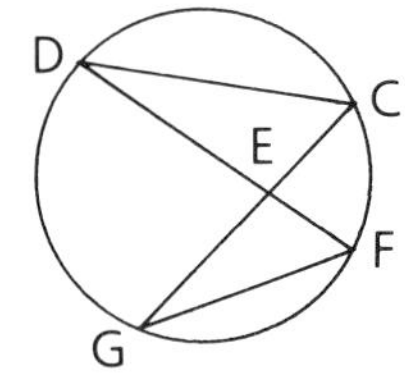

5 Given: $\angle HJK$ is a right $\angle$
$\overline{JM}$ is an altitude

Prove: $\frac{JM}{MK} = \frac{HJ}{JK}$

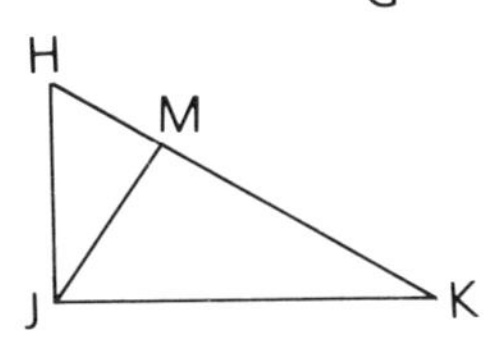

6 Given: $\triangle NOP \sim \triangle RST$

Prove: $NO \cdot RT = RS \cdot NP$

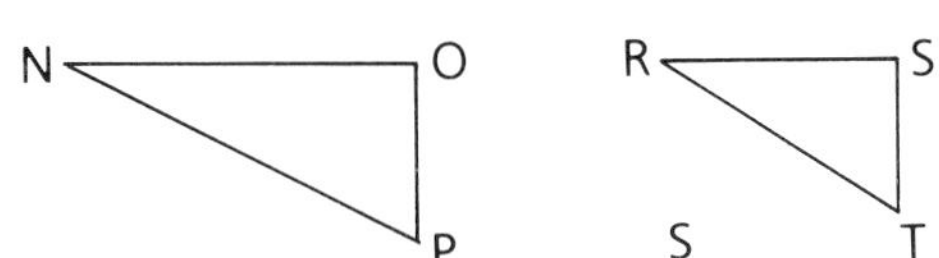

7 Given: $\overleftrightarrow{WZ} \parallel \overleftrightarrow{XY}$

Conclusion: $WS \cdot XY = XS \cdot WZ$

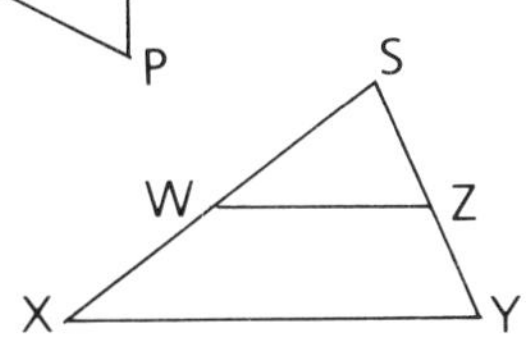

8 $\triangle ABC \sim \triangle DEF$

Find: AC, EF

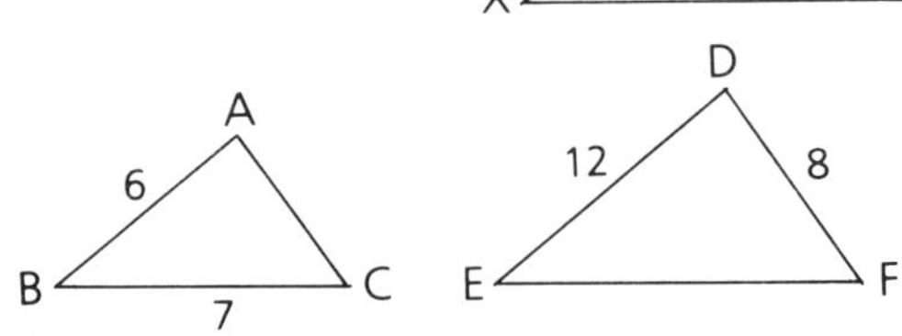

9 Given: GJKL ~ MOPR

Find: OP, PR, MR

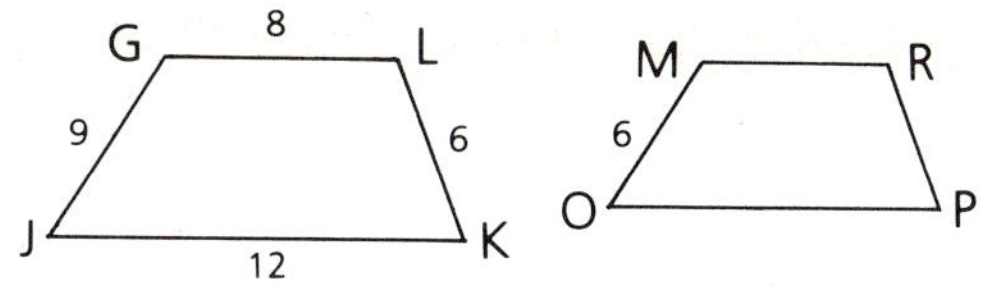

10 (A SHADOW PROBLEM.) Mannertink observed that a tree was casting a 30-meter shadow. A nearby flagpole was casting a 24-meter shadow. If the flagpole was 20 meters high, how tall was the tree?

11 If two similar kites have perimeters of 21 and 28, what is the ratio of the measures of two corresponding sides?

Section 8.4 Problem Set B

12 Given: ▱ ACEG

$\angle ABH \cong \angle EFD$

Prove: $AH \cdot FD = HB \cdot ED$

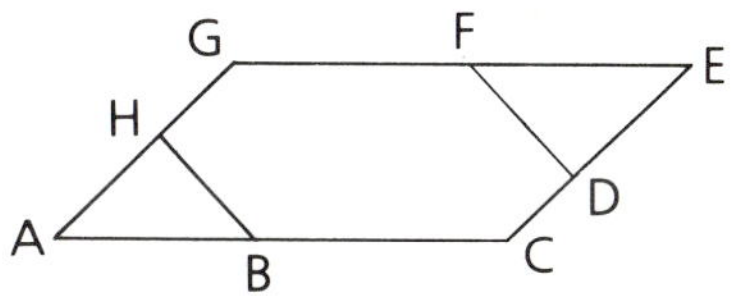

13 Prove: Corresponding altitudes of similar triangles are proportional to any pair of corresponding sides of the similar triangles.

14 Prove: Corresponding angle bisectors of similar triangles are proportional to any pair of corresponding sides of the similar triangles.

15 Given: $m \parallel p \parallel r$

J lies in m

$\overline{KT}$ lies in p

$\overline{OS}$ lies in r

Prove: $\frac{JK}{JO} = \frac{JT}{JS}$

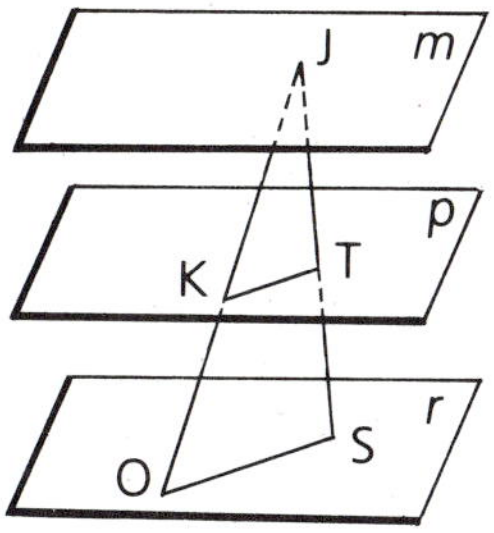

16 Given: Trapezoid ABCD, bases $\overline{AB}$ and $\overline{CD}$

Prove: AE • CD = EC • AB

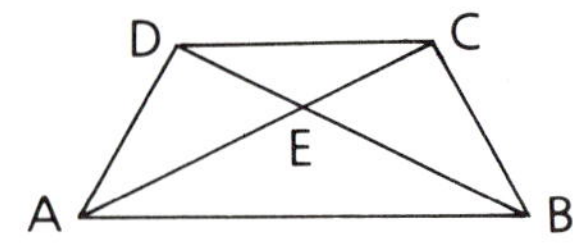

17 Given: $\angle M \cong \angle S$

MP = 8

PR = 6

SP = 7

Find: PO

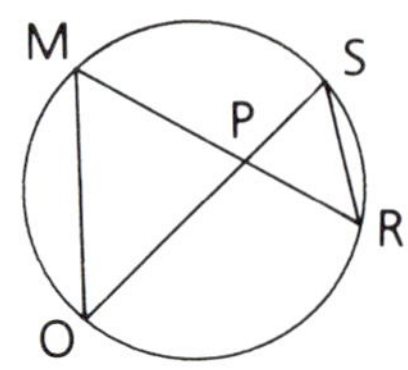

18 If $\triangle TVK \sim \triangle XZY$, TV = 8, VK = 9, TK = 10, ZY = 4, find XY.

19 Given: $\overleftrightarrow{BE} \parallel \overleftrightarrow{CD}$

AB = 6, BC = 2, and BE = 9

Find: CD

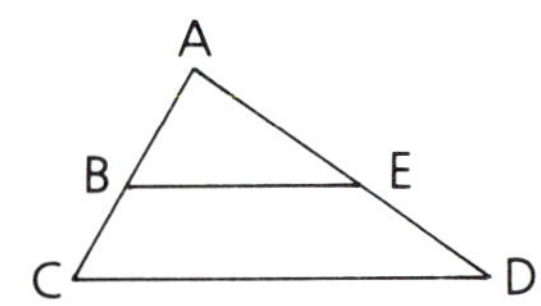

20 Quadrilateral ABCD is similar to quadrilateral RSTV. If the respective perimeters are 10 and 30, and AB = 4, find RS.

Section 8.4 Problem Set C

21 Given: $\angle ACB$ is a right $\angle$

$\overline{CD}$ is an altitude

Prove: $(CD)^2 = (AD)(DB)$

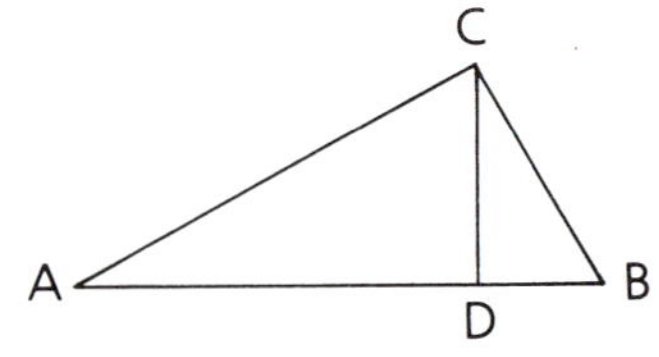

22 Given: EFGK is a ▱

MJ = 4

JH = 5

Find: EM

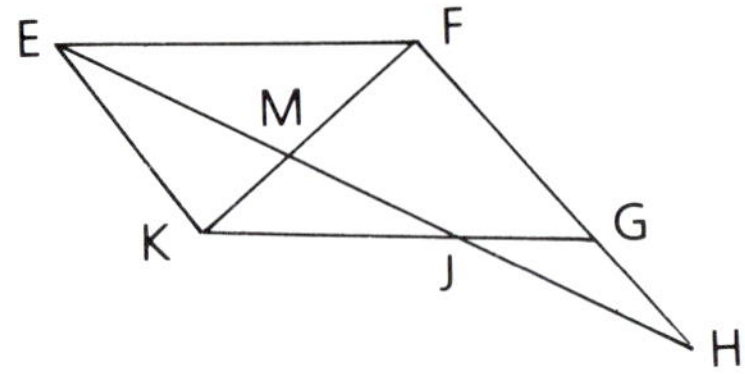

23 Given: $\angle S \cong \angle R$

OP = 6

PR = 8

OT = 4

Find: ST

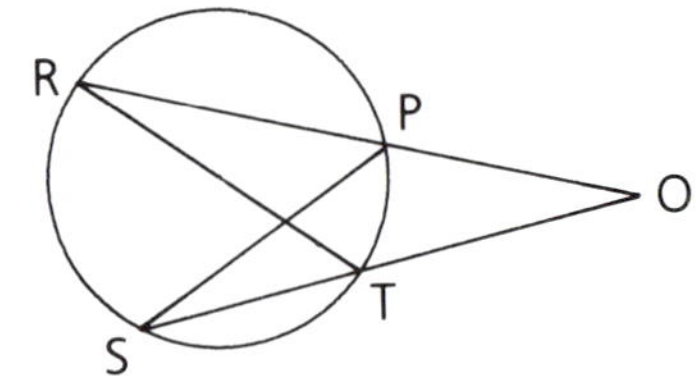

Section 8.5 Three Application Theorems Involving Proportions

In this section we present three application theorems involving proportions.

Theorem 57

T **If a line is parallel to one side of a triangle and intersects the other two sides, it divides them proportionally.**

Given: $\overleftrightarrow{BE} \parallel \overleftrightarrow{CD}$

Prove: $\frac{AB}{BC} = \frac{AE}{ED}$

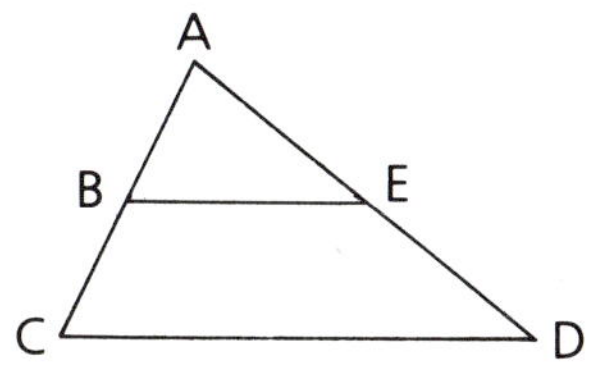

Proof of Theorem 57: Here is a sequence of hints. In a proportion obtained from similar triangles, substitute a sum for AC and a sum for AD. Then use means-extremes products, algebraic simplification, and the means-extremes ratio theorem.

Theorem 58

T **If three or more parallel lines intersect two transversals, they divide the transversals proportionally.**

Given: $\overleftrightarrow{AB} \parallel \overleftrightarrow{CD} \parallel \overleftrightarrow{EF}$

Conclusion: $\frac{AC}{CE} = \frac{BD}{DF}$

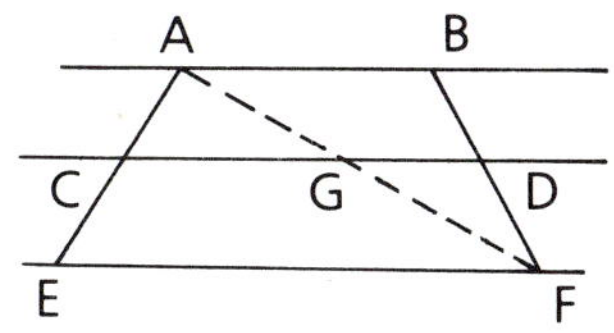

If you wish to prove this theorem, use auxiliary segment AF and think about two opportunities to use Theorem 57. You may also find it a challenge to prove Theorem 58 for transversals of four parallel lines.

It is sometimes handy to be able to cite the *special case* of THEOREM 58 stated here:

If parallel lines cut off (intercept) congruent segments on one transversal, they intercept congruent segments on any transversal.

Do you know the ratio of lengths in such special cases?

Theorem 59

T **If a ray bisects an angle of a triangle, it divides the opposite side into segments that are proportional to the other two sides. (ANGLE BISECTOR THEOREM)**

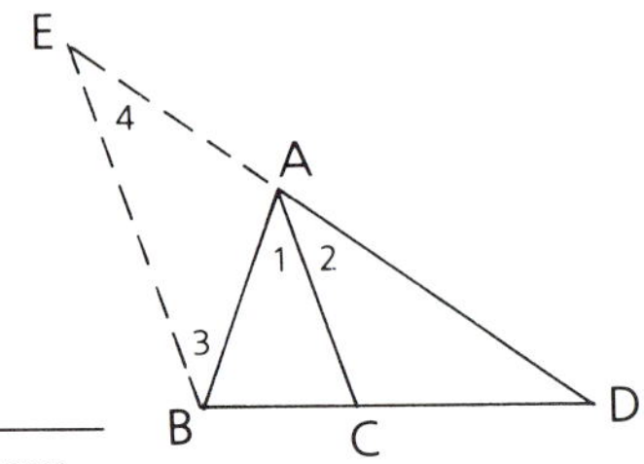

Given: $\triangle ABD$
$\overrightarrow{AC}$ bisects $\angle BAD$

Prove: $\frac{BC}{CD} = \frac{AB}{AD}$

1 $\triangle ABD$	1 Given
2 $\overrightarrow{AC}$ bisects $\angle BAD$	2 Given
3 $\angle 1 \cong \angle 2$	3 If a ray bisects an $\angle$, it divides the $\angle$ into two $\cong$ $\angle$s
4 Draw the line through B that is $\parallel$ to $\overline{AC}$	4 The PARALLEL POSTULATE
5 Extend $\overleftrightarrow{DA}$ to intersect the $\parallel$ line, at some point E	5 A line can be extended as far as desired
6 $\frac{BC}{CD} = \frac{EA}{AD}$	6 If a line is $\parallel$ to one side of a $\triangle$ and intersects the other two sides, it divides them proportionally
7 $\angle 1 \cong \angle 3$	7 PAI
8 $\angle 2 \cong \angle 4$	8 PCA
9 $\angle 3 \cong \angle 4$	9 Transitive Property (3, 7, 8)
10 $\overline{EA} \cong \overline{AB}$	10 If $\triangle$ then $\triangle$
11 $\frac{BC}{CD} = \frac{AB}{AD}$	11 Substitution (10 in 6)

Section 8.5 Sample Problems

1 Given: $\overleftrightarrow{BE} \parallel \overleftrightarrow{CD}$
Lengths as shown

Find: **a** ED
b CD

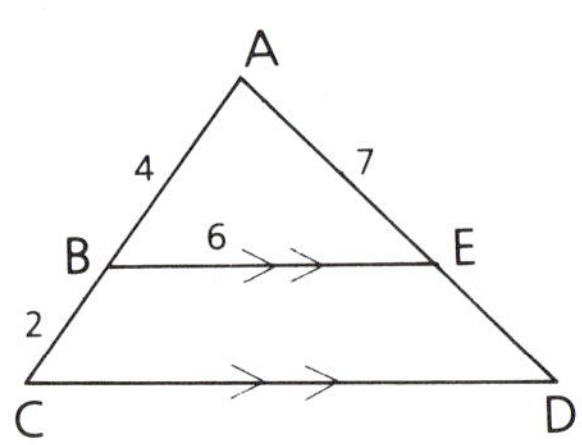

Be alert. When you see this type of figure, two concepts may be involved, Theorem 57 or Similar Triangles.

Solutions:

a Using Theorem 57

$$\frac{AB}{BC} = \frac{AE}{ED}$$
$$\frac{4}{2} = \frac{7}{x}$$
$$\frac{2}{1} = \frac{7}{x}$$
$$2x = 7$$
$$x = 3\tfrac{1}{2}$$

b Using Similar Triangles (necessary when the parallel segments are involved)

$$\frac{AB}{AC} = \frac{BE}{CD}$$
$$\frac{4}{4+2} = \frac{6}{y}$$
$$\frac{2}{3} = \frac{6}{y}$$
$$2y = 18$$
$$y = 9$$

2 Given: $a \parallel b \parallel c \parallel d$
Lengths as shown
$KP = 24$

Find: KM

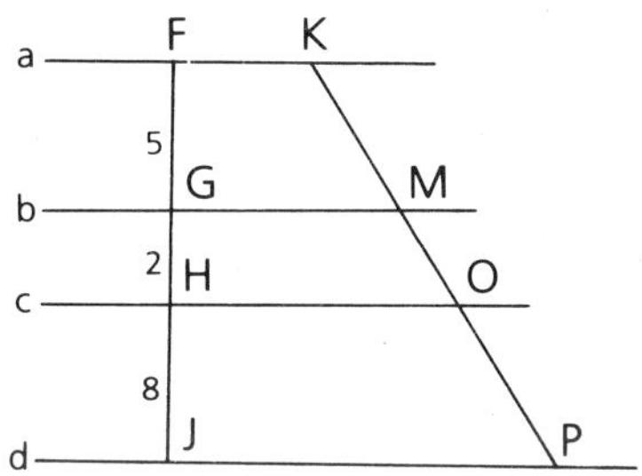

Solution: According to Theorem 58, the ratio KM:MO:OP is 5:2:8. Therefore, we let $KM = 5x$, $MO = 2x$, and $OP = 8x$

$$\text{Then, } KP = 5x + 2x + 8x = 24$$
$$15x = 24$$
$$x = \frac{24}{15} = \frac{8}{5}, \text{ so } KM = 5\left(\frac{8}{5}\right) = 8$$

3 Given: $\angle RVS \cong \angle SVT$
Lengths as shown

Find: ST

Solution:

Using Theorem 59

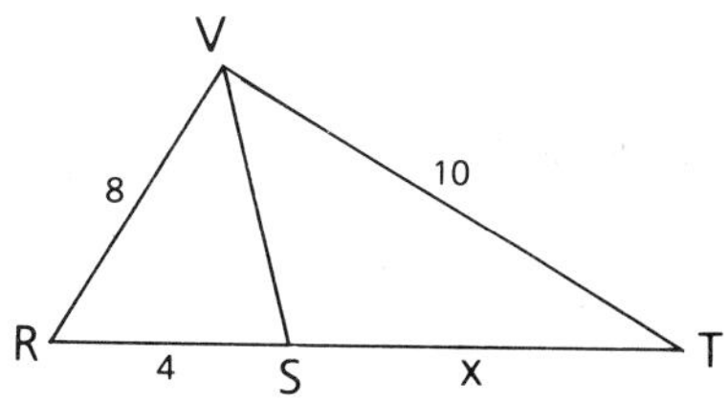

$$\frac{VR}{VT} = \frac{RS}{ST}$$
$$\frac{8}{10} = \frac{4}{x}$$
$$\frac{4}{5} = \frac{4}{x}$$
$$4x = 20$$
$$x = 5$$

4 Given: $\overleftrightarrow{XA} \parallel \overleftrightarrow{YZ}$

$\angle XAY \cong \angle XYA$

Conclusion: $\dfrac{WX}{XA} = \dfrac{WA}{AZ}$

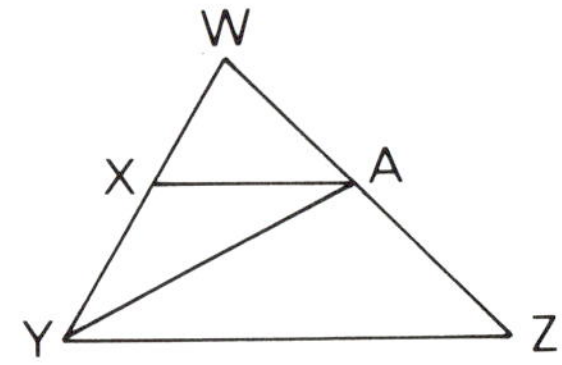

1 $\overleftrightarrow{XA} \parallel \overleftrightarrow{YZ}$	1 Given
2 $\dfrac{WX}{XY} = \dfrac{WA}{AZ}$	2 If a line is $\parallel$ to a side of a △ and intersects the other two sides, it divides them proportionally
3 $\angle XAY \cong \angle XYA$	3 Given
4 $\overline{XA} \cong \overline{XY}$	4 If △ then △
5 $\dfrac{WX}{XA} = \dfrac{WA}{AZ}$	5 Substitution (4 in 2)

5 Given: $\overleftrightarrow{DH} \parallel \overleftrightarrow{BC}$

$\overleftrightarrow{HF} \parallel \overleftrightarrow{BG}$

Prove: $\dfrac{CD}{DE} = \dfrac{GF}{FE}$

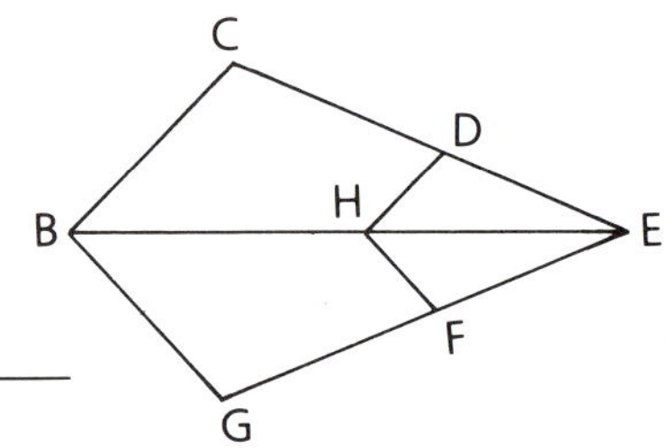

1 $\overleftrightarrow{DH} \parallel \overleftrightarrow{BC}$	1 Given
2 $\dfrac{CD}{DE} = \dfrac{BH}{HE}$	2 If a line is $\parallel$ to a side of a △ and intersects the other two sides, it divides them proportionally
3 $\overleftrightarrow{HF} \parallel \overleftrightarrow{BG}$	3 Given
4 $\dfrac{BH}{HE} = \dfrac{GF}{FE}$	4 Same as 2
5 $\dfrac{CD}{DE} = \dfrac{GF}{FE}$	5 Transitive Property of Equality (If two ratios are equal to the same ratio, they are equal to each other.)

Section 8.5 Problem Set A

For **1–3** see Sample Problem 1

1 Given: $\overline{BE} \parallel \overline{CD}$

Lengths as shown

Find: **a** ED

b CD

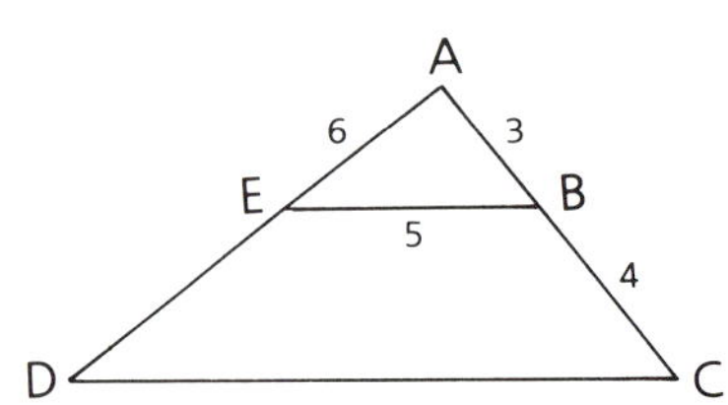

2 Solve for x and y as shown

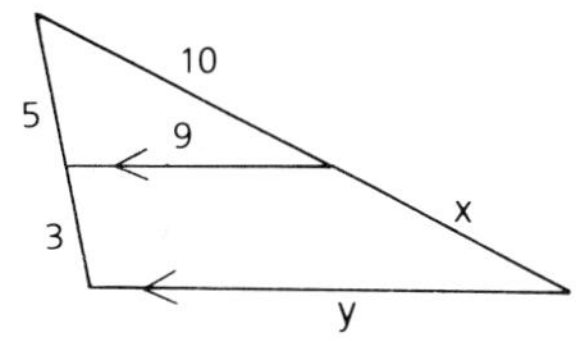

3 Solve for p and q as shown

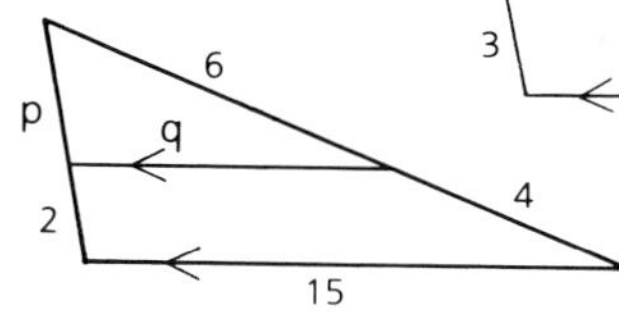

For **4, 5** see Sample Problem 2

4 Given: a ∥ b ∥ c ∥ d
Lengths as shown
KP = 15

Find: KM, MO, OP

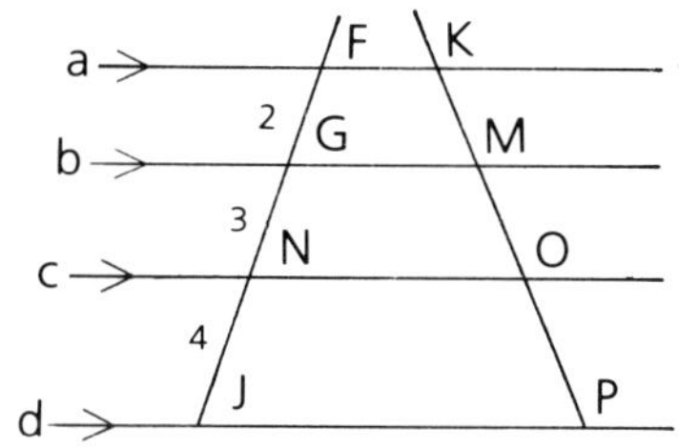

5 Solve for x of the diagram as marked

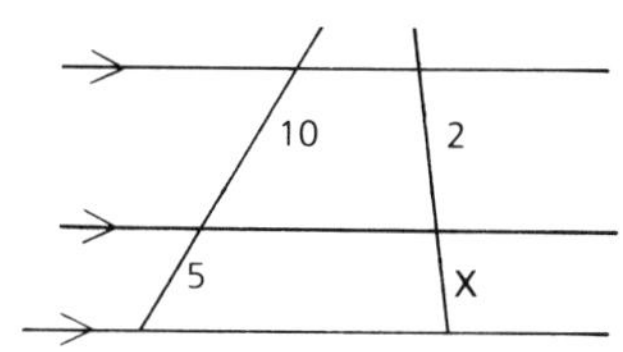

For **6, 7** see Sample Problem 3

6 Given: ∠RVS ≅ ∠SVT
Lengths as shown

Find: ST

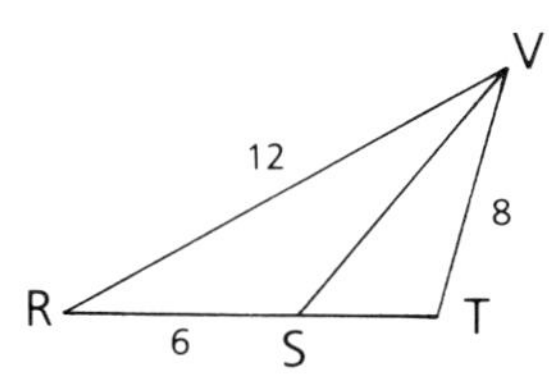

7 Given: Diagram as marked
Solve for x

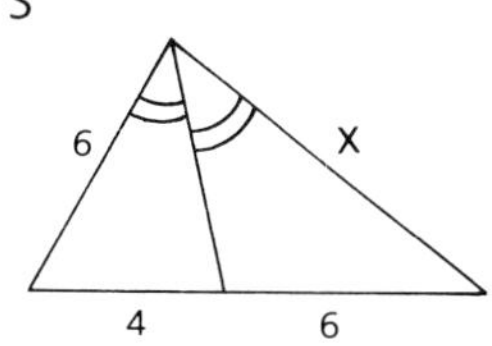

8 A 60-meter tower casts a 50-meter shadow while one-half block away a telephone pole casts a 20-meter shadow. How tall is the telephone pole?

9 Given: ∠J ≅ ∠MKO
MK = 12, KO = 8
MO = 10, JK = 3

Find: PO, JP

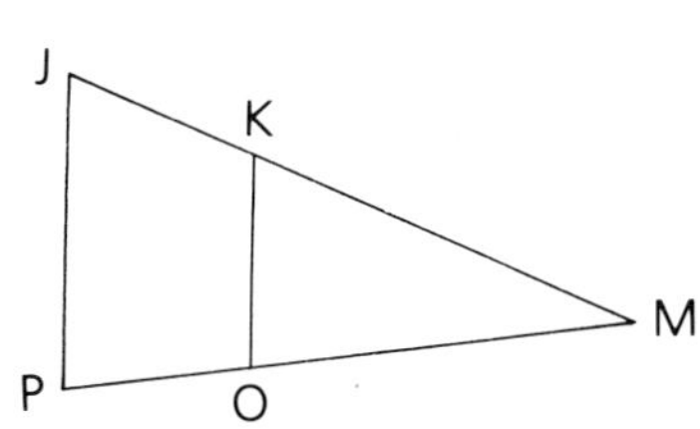

10 Given: $\overleftrightarrow{SV} \parallel \overleftrightarrow{RW}$

$RW = 15,\ RS = 10$

$ST = 3,\ WV = 8$

Find: $SV,\ VT$

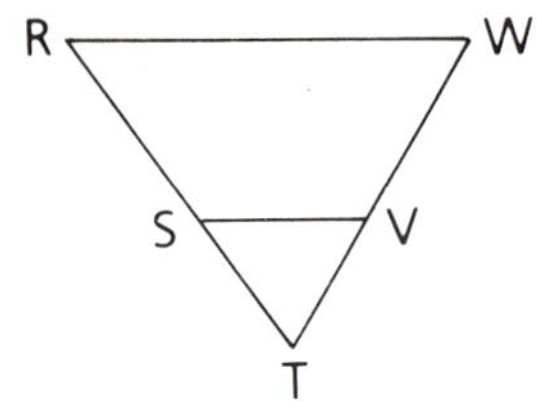

11 Given: $\overleftrightarrow{CD} \parallel \overleftrightarrow{BE}$

$AC = 18,\ AB = 12$

$AE = 10,\ CD = 24$

Find: $BC,\ ED,\ BE$

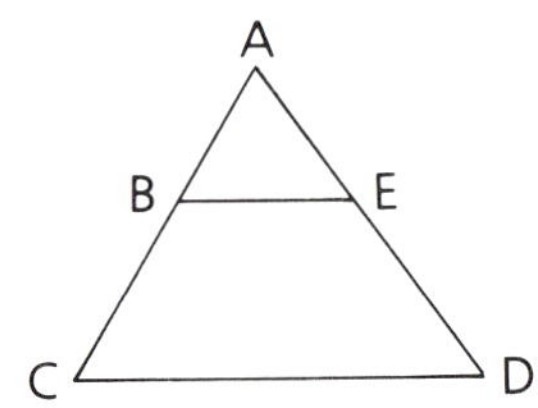

12 Given: $\overrightarrow{GJ}$ bisects $\angle FGH$

$FG = 10,\ GH = 8$

$FJ = 7$

Find: JH

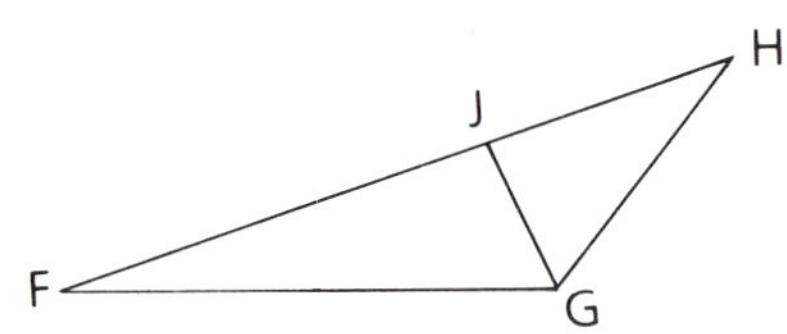

13 Given: $r \parallel s \parallel t$

$WV = 3$

$WX = 8$

$QY = 9$

Find: $QZ,\ ZY$

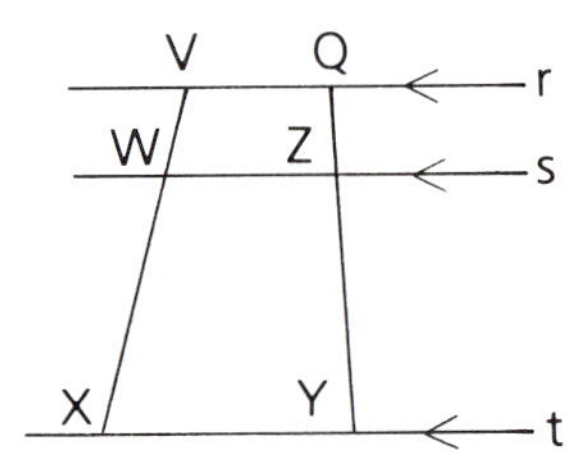

Section 8.5 Problem Set B

14 Given: $\angle 1 \cong \angle 2$

Conclusion: $\dfrac{KM}{JK} = \dfrac{MO}{OP}$

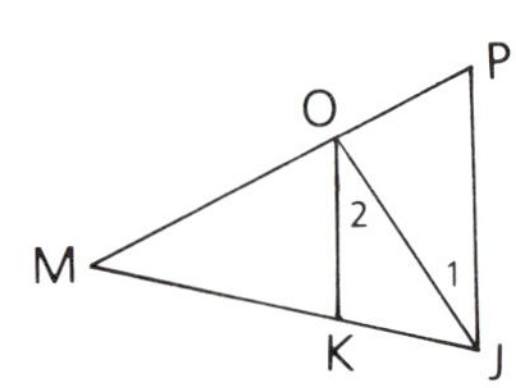

15 Given: $\angle 3 \cong \angle 5$

Prove: $\dfrac{RV}{VT} = \dfrac{RS}{ST}$

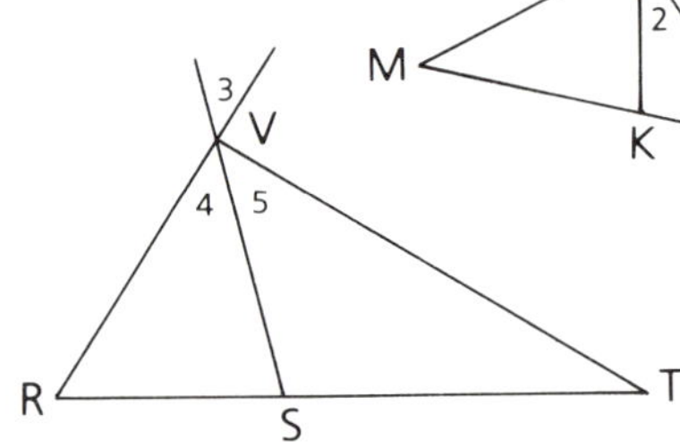

16 Given: WYZB is a trapezoid with bases $\overline{WB}$ and $\overline{YZ}$

$\overleftrightarrow{XA} \parallel \overleftrightarrow{YZ}$

Prove: $\frac{WX}{XY} = \frac{WC}{CZ} = \frac{BA}{AZ}$

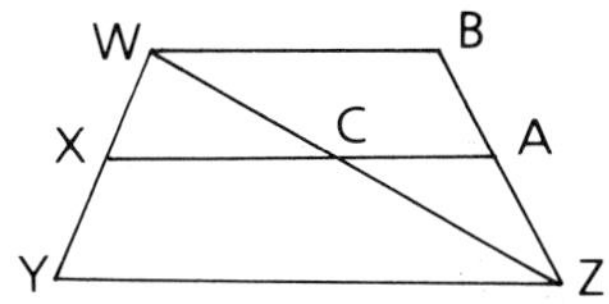

17 Given: $\overleftrightarrow{BE} \parallel \overleftrightarrow{CD}$

$AB = 4x,\ BC = x$

$AD = 8x,\ BE = 5x$

Find: AE, CD (in terms of x)

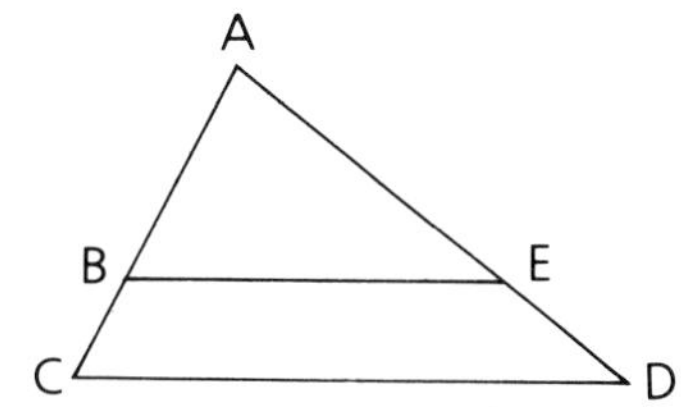

18 a One side of a △ is 4 cm longer than another side. The ray bisecting the ∠ between these sides divides the opposite side into 5 cm and 3 cm segments. Find the perimeter of the △.

b If, in part *a*, one side of the triangle were x cm longer than the second side and the other information were unchanged, find the perimeter in terms of x.

19 Given: $\overleftrightarrow{GK} \parallel \overleftrightarrow{HJ}$

Lengths as shown

Find the perimeter of △HJF

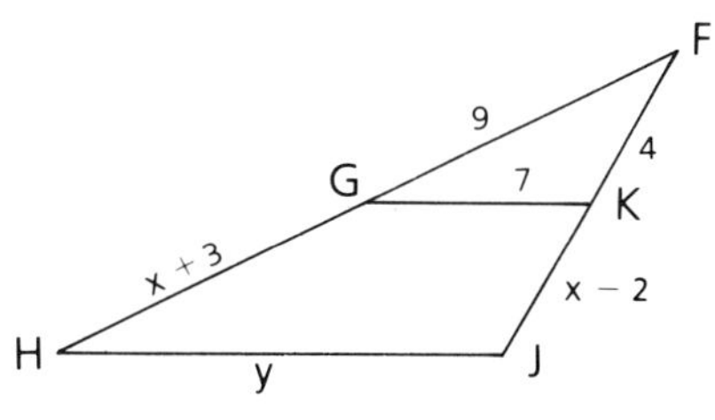

20 Given: $TV = 12,\ \overleftrightarrow{VS} \parallel \overleftrightarrow{MR}$

$VM = 8,\ TS = 15$

$SR = TW = TX$

Find: XP

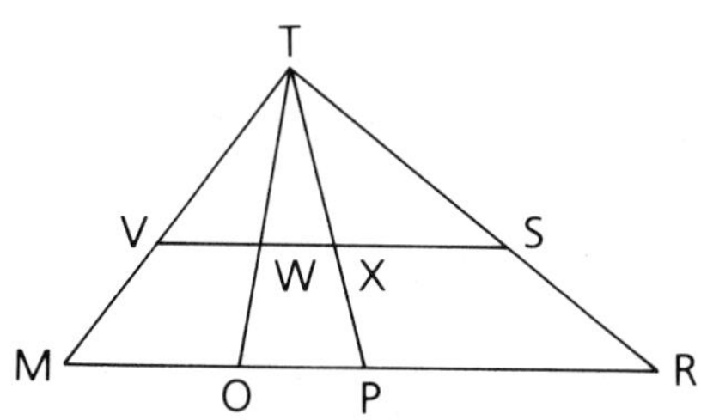

21 Given: $\overleftrightarrow{AC} \perp \overleftrightarrow{BE}$

$\angle 1 \cong \angle 4$

Conclusion: $\frac{BD}{BF} = \frac{DE}{EF}$

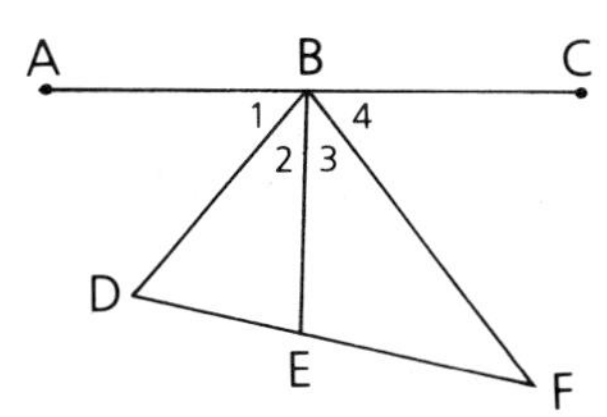

22 Given: $\overleftrightarrow{GD} \parallel \overleftrightarrow{FE}$

$\overleftrightarrow{BD} \parallel \overleftrightarrow{CE}$

Prove: $\frac{AB}{AC} = \frac{AG}{AF}$

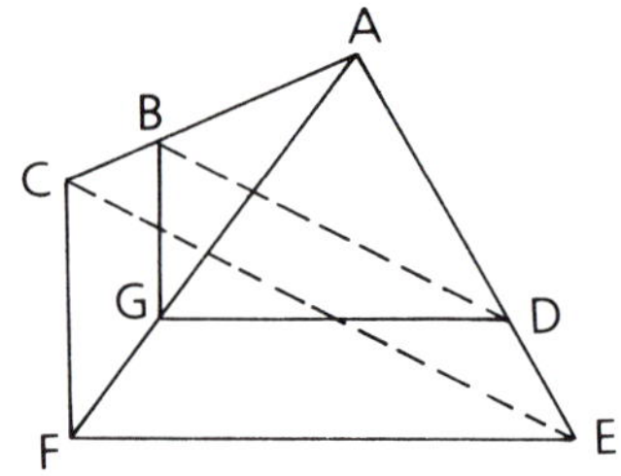

23 Prove: If a line bisects one side of a triangle and is parallel to a second side, it bisects the third side.

24 In order to measure the width ($\overline{GH}$) of the pond on his Texas ranch, Gonzaldiatrez used his transit to sight a 50° angle at H to a point M. Then using his transit at M, he noted a point K, which gave him a 50° angle at M. After completing $\overline{GK}$, he then measured $\overline{HJ}$ as 150 m, $\overline{JM}$ as 50 m, and $\overline{MK}$ as 40 m. How long was $\overline{GH}$, and why did Gonzaldiatrez's method work?

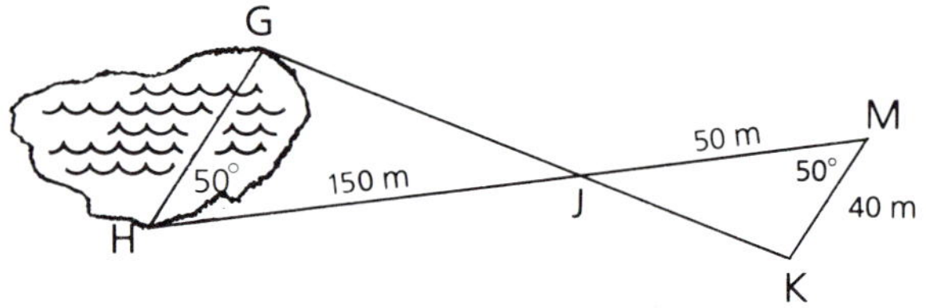

Section 8.5 Problem Set C

25 Given: $\overleftrightarrow{GK} \parallel \overleftrightarrow{HJ}$

Lengths as shown

Find the perimeter of △HJF

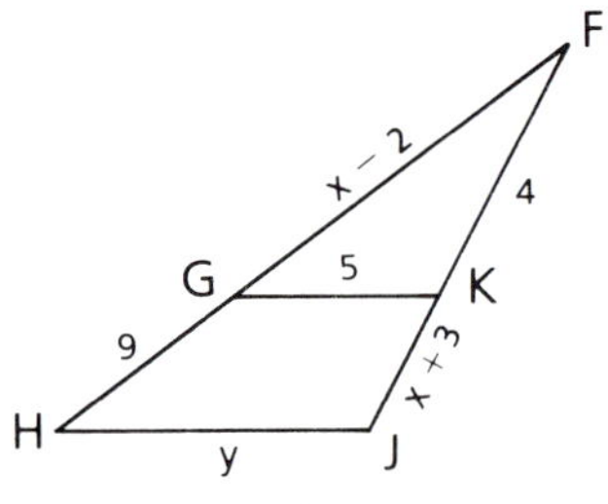

26 Prove: If a line divides two sides of a triangle proportionally, it is parallel to the third side.

27 If two poles are respectively 10 m and 70 m tall and are 100 m apart, find the height of the intersection of the lines joining the top of each pole to the foot of the opposite pole.

28 Given: $\overrightarrow{RW}$ bisects $\angle SRT$
$\overrightarrow{TV}$ bisects $\angle RTS$
$RV = 4,\ SV = 5$
$SW = 6,\ WT = 7$

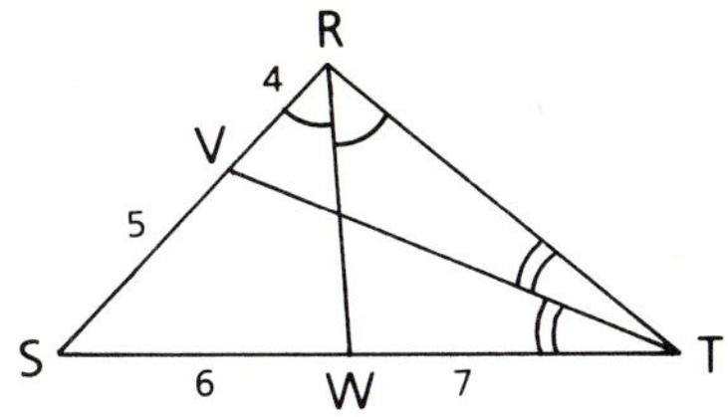

Show that the given information is impossible.

29 *Suppose that Theorem 56* (SAS ~) and *Theorem 57* (If a line is ∥ to one side of a △ and intersects the other two sides, it divides them proportionally) *were postulates.* Prove that AA is true.

Given: $\angle A \cong \angle D$
$\angle B \cong \angle E$

Prove: $\triangle ABC \sim \triangle DEF$

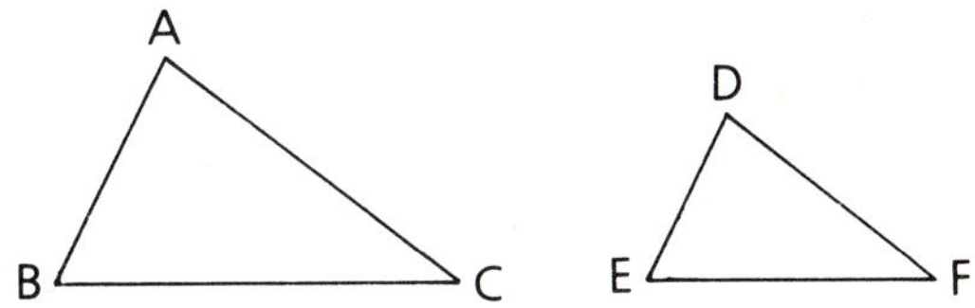

Key steps of a proof:

Introduce R and S such that $\overline{AR} \cong \overline{DE}$ and $\overline{AS} \cong \overline{DF}$.

$\triangle ARS \cong \triangle DEF$

$\overline{RS} \parallel \overline{BC}$

$\frac{AR}{AB} = \frac{AS}{AC}$

$\triangle ARS \sim \triangle ABC$

But $\triangle ARS \cong \triangle DEF$

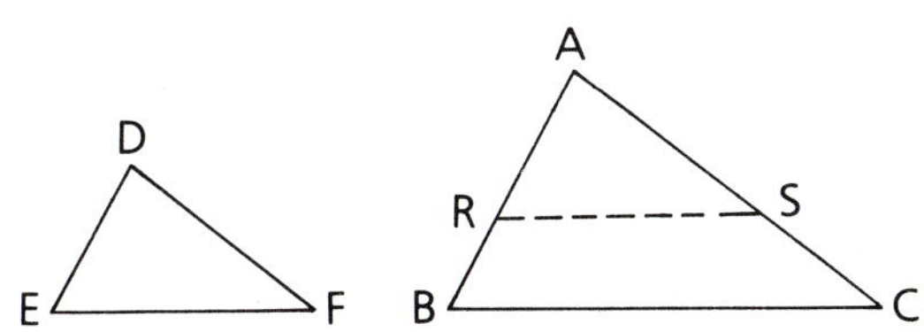

Section 8.5 Problem Set D

30 Prove: If two transversals intersect three parallel planes, the planes divide the transversals proportionally.

31 In problem 30, if more than three parallel planes were involved, would the transversals still be divided proportionally?

Chapter 8 Study Guide

As you study and review this chapter, be sure you are able to:

1 Describe the meaning *and* write a sample algebra problem for each of the following terms (pp. 309–316):

a Ratio
b Proportion
c Terms of a proportion
d Mean proportional (geometric mean)
e Fourth proportional
f Equivalent forms
g Similar polygons

2 Describe when and how to use the Means-Extremes Product Theorem and the Means-Extremes Ratios Theorem. (p. 310, 311)

3 Describe the three ways to prove that two triangles are similar. (pp. 322, 323) Which is used most often?

4 State whether corresponding sides of similar triangles are proportional. (p. 316) Do the corresponding sides have the same ratio as corresponding altitudes? (p. 311) corresponding medians? (p. 331) corresponding angle bisectors? perimeters? (p. 317)

5 Set up and solve a Shadow Problem. (p. 329)

6 **a** State the transitive property of similar polygons. (p. 326)
b State what is true about corresponding angles of similar polygons. (p. 316)

7 **a** Set up a proportion involving a, b, c, d, and tell why the proportion is true. (p. 334)

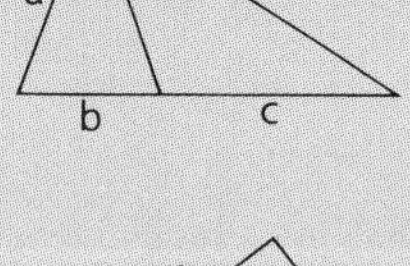

b Write a proportion for f, g, h, j. For f, g, k, m. (Careful!) Why are you correct? (pp. 327, 333)

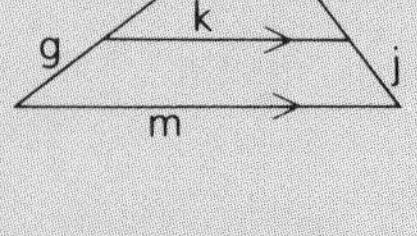

c Describe why $a:b:c = d:e:f$ and whether $\frac{a}{d} = \frac{b}{e} = \frac{c}{f}$ (p. 333)

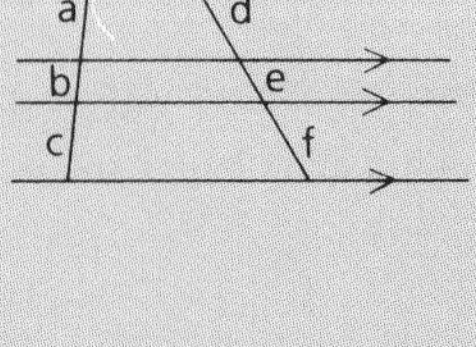

Chapter 8 Review Problems A

1 In the proportion $\frac{a}{b} = \frac{c}{d}$ __?__ and __?__ are the means, and __?__ and __?__ are the extremes.

2 Find the fourth proportional to 4, 6, and 8.

3 Find the mean proportionals between 5 and 20.

4 Find the geometric means between 3 and 6.

5 If $9x = 4y$, find the ratio of x to y.

6 Given: $\triangle ABC \sim \triangle DEF$
Lengths as shown
Find: DF, EF

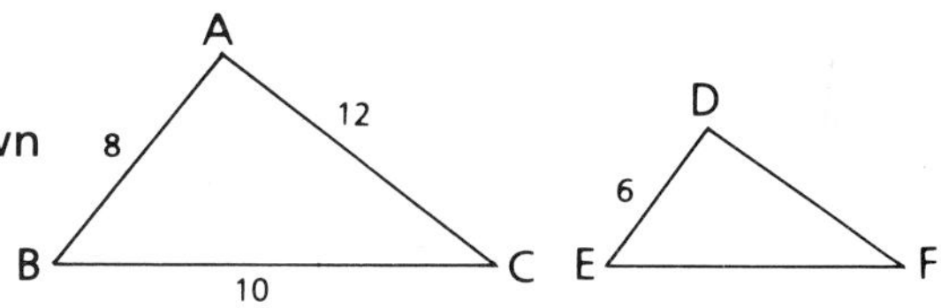

7 Pentagon ABCDE is similar to pentagon $A'B'C'D'E'$. The perimeters are respectively 24 and 30. If $AB = 6$, find $A'B'$.

8 If $\frac{GH}{HJ} = \frac{3}{4}$ and $GJ = 56$, find HJ.

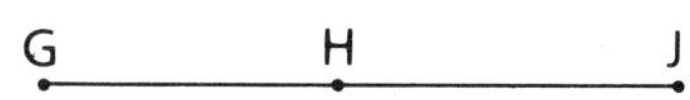

9 If $\frac{r}{3x} = \frac{a}{2b}$, solve for x in terms of a, b, and r.

10 A 100 m tall cable television antenna casts an 80 m shadow. A nearby telephone pole casts a 16 m shadow. Find the height of the telephone pole.

11 $\overleftrightarrow{MR} \parallel \overleftrightarrow{OP}$
Lengths as shown
Find: RP, OP

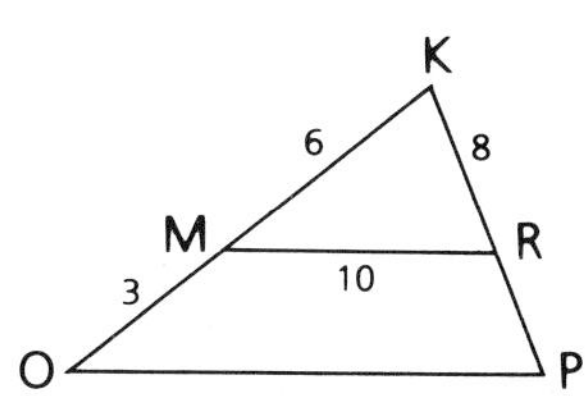

12 $\overleftrightarrow{TP} \parallel \overleftrightarrow{VW}$
Lengths as shown
Find: ST, TV, PT

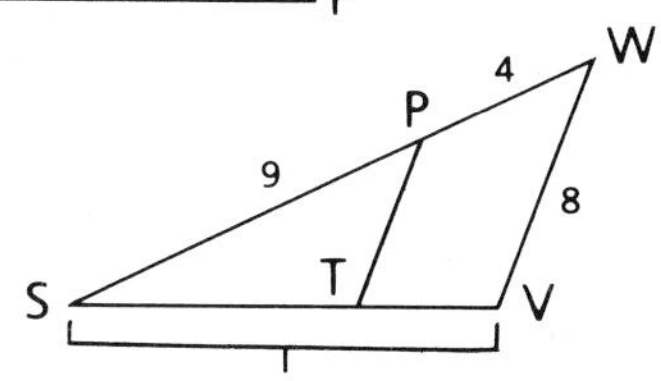

13 Given: $\overrightarrow{CD}$ bisects $\angle ACB$
Lengths as shown
Find: AD

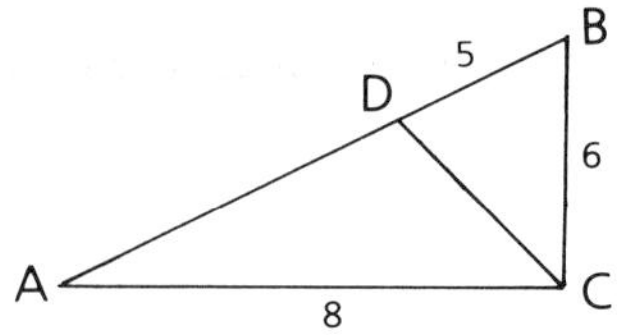

14 Given: Lengths as shown
Solve for x

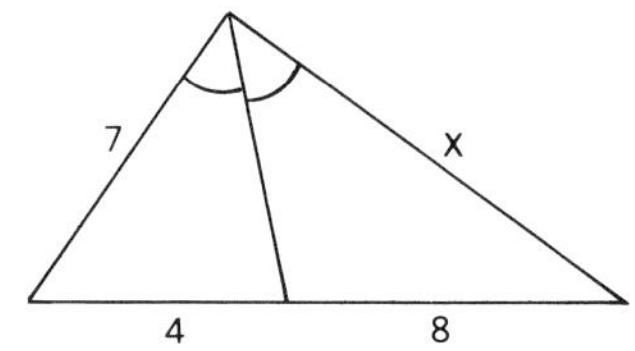

15 $a \parallel b \parallel c$
Lengths as shown
Solve for y

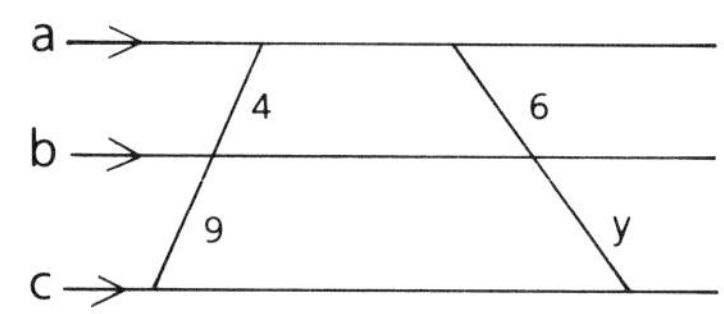

16 $\overleftrightarrow{EF} \parallel \overleftrightarrow{GO} \parallel \overleftrightarrow{HM} \parallel \overleftrightarrow{JK}$
$FG = 2,\ GH = 8$
$HJ = 5,\ EM = 6$
Find: EO, EK

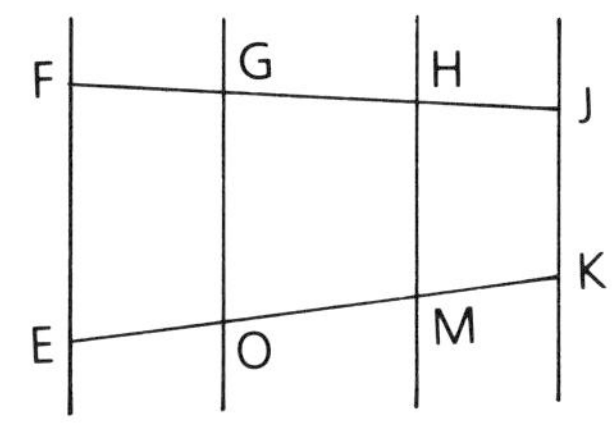

17 Given: $\overline{OS} \parallel \overline{PR}$
$\angle 1 \cong \angle 2$
Prove: $\dfrac{MO}{OP} = \dfrac{MT}{SR}$

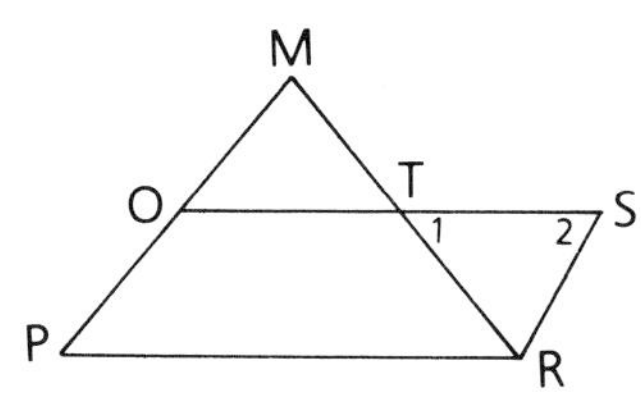

18 Given: BRXZ is a ▱
$\angle 3 \cong \angle 4$
Prove: (RC) (ZA) = (ZY) (RW)

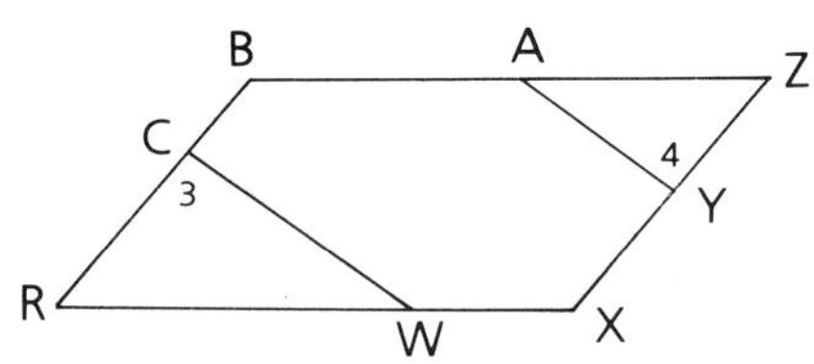

19 Given: $\angle JEF \cong \angle 6$
Conclusion: $\dfrac{EF}{EH} = \dfrac{FG}{GH}$

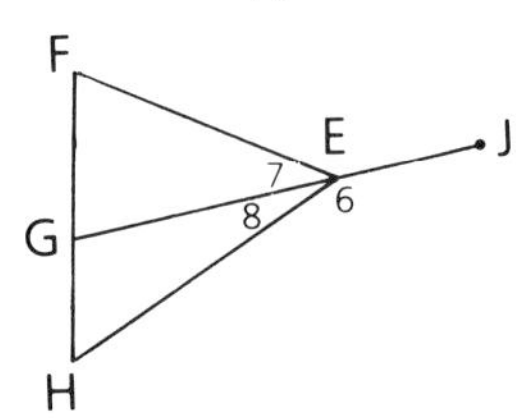

Chapter 8 Review Problems B

20 Answer a–h Always, Sometimes, or Never (A, S, or N)

a Two isosceles triangles are similar if a base angle of one is congruent to a base angle of another.

b Two isosceles triangles are similar if the vertex angle of one is congruent to the vertex angle of another.

c An equilateral triangle is similar to a scalene triangle.

d If two sides of one triangle are proportional to two sides of a second triangle, the triangles are similar.

e In $\triangle ABC$, $\angle A = 40°$, $AB = 6$, $BC = 8$.
In $\triangle RST$, $RS = 12$, $ST = 16$, $\angle R = 80°$.
Therefore, $\triangle ABC \sim \triangle RST$.

f If a line is one of the trisectors of one side of a triangle and is parallel to a second side, then it is one of the trisectors of the third side.

g Two right triangles are similar if the legs of one are proportional to the legs of the other.

h If the ratio of the measures of a pair of corresponding sides of two polygons is 3:4, then the ratio of the perimeters is 5:6.

21 Given: ABDF is a ▱
Conclusion: $\triangle CBD \sim \triangle DFE$

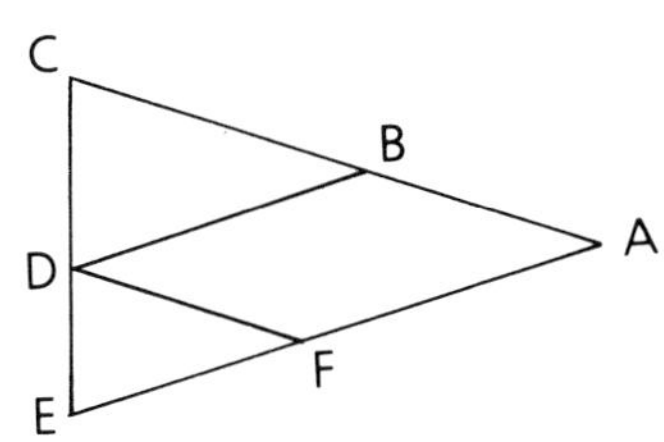

22 Given: $\overleftrightarrow{HR} \parallel \overleftrightarrow{GM}$
Prove: $\frac{PR}{OM} = \frac{PH}{OG}$

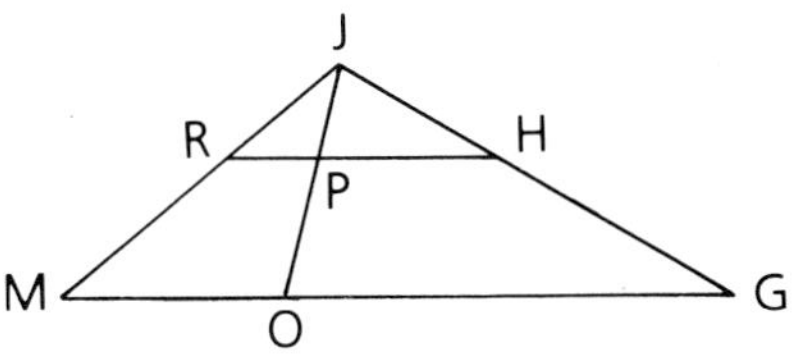

23 Prove: Diagonals of a trapezoid divide themselves proportionally.

24 If 78 is divided into three parts proportional to 3, 5, and 7, the sum of the smallest and largest parts is ___?___.

25 If $\frac{7}{x + 4y} = \frac{9}{2x - y}$, find the ratio of x to y.

26 One side of a △ is 4 cm shorter than a second side. The ray bisecting the ∠ between these sides divides the opposite side into 4 cm and 6 cm segments. Find the perimeter of the △.

27 $NP = 5x - 21$, $PR = 5$
$NT = x$, $TS = 8$, $\overleftrightarrow{PT} \parallel \overleftrightarrow{RS}$
Find: $NR + NS$

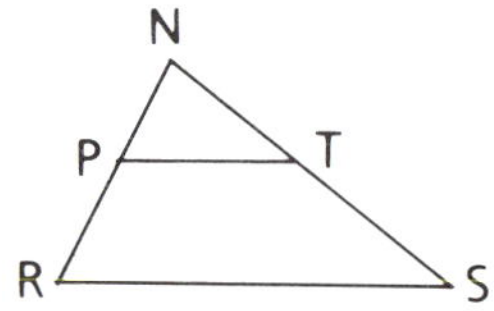

28 The diagram to the right shows a part of the town of Oola, La. First, Second, Third, and Fourth Streets are each perpendicular to Elmwood Avenue. If the total frontage on Sandwick Court is 400 meters, find the length of each block of Sandwick Court.

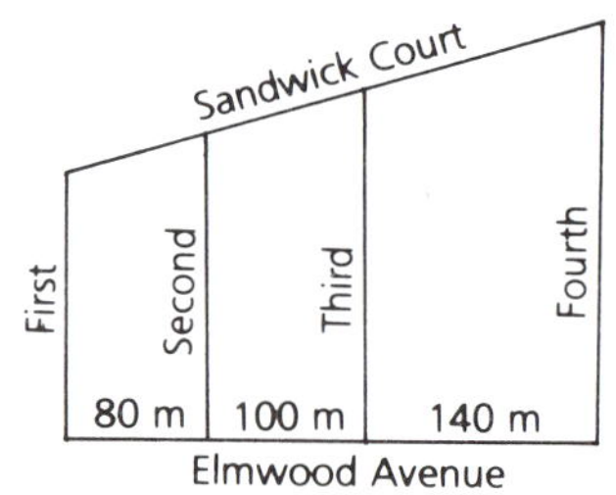

Chapter 8 Review Problems C

29 The sum of four numbers is 771. The ratio of the first to the second is 2:3. The ratio of the second to the third is 5:4. The ratio of the third to the fourth is 5:6. Find the second.

30 Given: $\overline{GB} \perp \overline{AC}$, $\overline{HD} \perp \overline{EC}$
$\overline{JF} \perp \overline{AE}$
$AB = 8$
$BC = 12$
$EC = 15$
$AE = 10$
$GH = 5$
Find: GJ, HJ

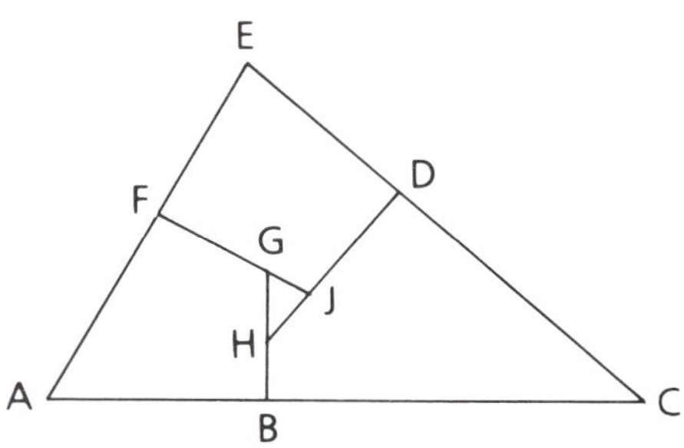

31 Prove: If an altitude is drawn to the hypotenuse of a right triangle, then the product of the measures of the altitude and the hypotenuse is equal to the product of the measures of the legs of the right triangle.

32 Filbert knows that the two triangles ABC and XYZ are similar but he cannot remember what the correct correspondence of letters should be. He guesses $\triangle ABC \sim \triangle XYZ$.

a What is the probability that his guess is correct?

b If Filbert finds out that the triangles are isosceles, then what will the probability be?

c If the triangles are equilateral, what are his chances of guessing a correct correspondence?

Chapter 8 Review Problems D

33 Given: $a + b + c + d = 70$

$b = ga + 7$

$c = 2a$

Find: g

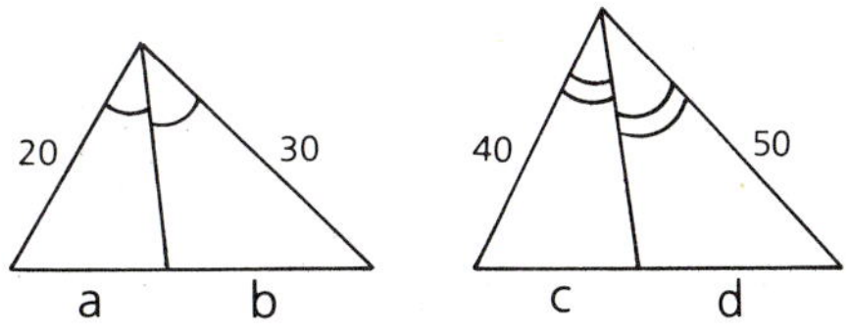

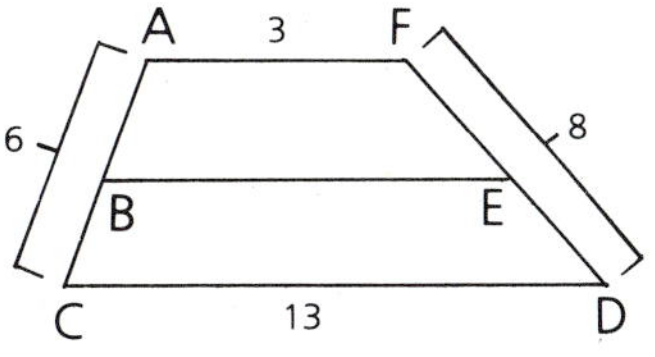

34 Given: ACDF is a trapezoid

$\overline{BE} \parallel \overline{CD} \parallel \overline{AF}$, $AF = 3$

$CD = 13$, $AC = 6$, $FD = 8$

Perimeter ABEF = Perimeter BCDE

Find the ratio of FE to ED

Chapter 9

THE PYTHAGOREAN THEOREM

Section 9.1 Review of Radicals and Quadratic Equations

Some of the problems in the next three chapters will involve radicals and quadratic equations. Although you have already completed a course in algebra, you may have forgotten some of the techniques. Read carefully the following sample problems, which review those two concepts.

Section 9.1 Sample Problems

1 Simplify: $\sqrt{48}$

Solution: $\sqrt{48} = \sqrt{16 \cdot 3}$ (Note: 16 is a perfect square.)

$$= \sqrt{16} \cdot \sqrt{3}$$
$$= 4\sqrt{3}$$

2 Simplify: $\sqrt{18} + \sqrt{32} + \sqrt{75}$

Solution:
$$\begin{aligned}\sqrt{18} + \sqrt{32} + \sqrt{75} &= \sqrt{9 \cdot 2} + \sqrt{16 \cdot 2} + \sqrt{25 \cdot 3}\\ &= 3\sqrt{2} + 4\sqrt{2} + 5\sqrt{3}\\ &= 5\sqrt{3} + 7\sqrt{2}\end{aligned}$$

3 Simplify: $\sqrt{\frac{5}{3}}$

Solution:
$$\begin{aligned}\sqrt{\frac{5}{3}} &= \frac{\sqrt{5}}{\sqrt{3}}\\ &= \frac{\sqrt{5}}{\sqrt{3}} \cdot \frac{\sqrt{3}}{\sqrt{3}}\\ &= \frac{\sqrt{15}}{3}, \text{ or } \frac{1}{3}\sqrt{15}\end{aligned}$$

(The two answers are equivalent simplifications.)

4 Solve for x: $x^2 + 9 = 25$

Method 1:
$$\begin{aligned}x^2 + 9 &= 25\\ x^2 &= 16\\ x &= \pm 4\end{aligned}$$

Method 2: (factoring)
$$\begin{aligned}x^2 + 9 &= 25\\ x^2 - 16 &= 0\\ (x - 4)(x + 4) &= 0\\ x - 4 &= 0, \text{ or } x + 4 = 0\\ x &= 4, \text{ or } x = -4\end{aligned}$$

5 Solve for x: $(3\sqrt{5})^2 + (3\sqrt{2})^2 = x^2$

Solution:
$$\begin{aligned}(3\sqrt{5})^2 + (3\sqrt{2})^2 &= x^2\\ 9 \cdot 5 + 9 \cdot 2 &= x^2\\ 45 + 18 &= x^2\\ 63 &= x^2\\ \pm\sqrt{63} &= x\\ \pm\sqrt{9 \cdot 7} &= x\\ \pm 3\sqrt{7} &= x\end{aligned}$$

6 Solve for x: **a** $x^2 - 10x = -16$ **b** $x^2 + 5x = 0$

Solutions:

a
$$\begin{aligned}x^2 - 10x &= -16\\ x^2 - 10x + 16 &= 0\\ (x - 8)(x - 2) &= 0\\ x - 8 = 0 &\text{ or } x - 2 = 0\\ x = 8 &\text{ or } x = 2\end{aligned}$$

b
$$\begin{aligned}x^2 + 5x &= 0\\ x(x + 5) &= 0\\ x = 0 &\text{ or } x + 5 = 0\\ x = 0 &\text{ or } x = -5\end{aligned}$$

The Royal Road: Euclid's Proof of the Pythagorean Theorem

This text uses properties of similar triangles to show that the square of the measure of the hypotenuse of a right triangle is equal to the sum of the squares of the measures of the legs (see page 359). Euclid's proof dealt directly with squares and their areas.

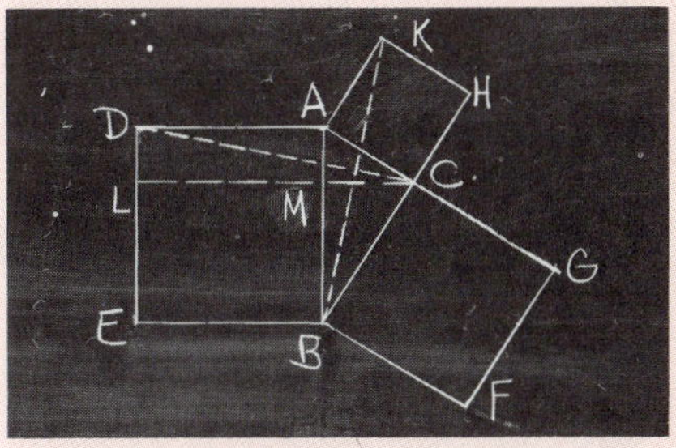

He used the following steps to show that the area of rectangle AMLD equals the area of square ACHK. First, he showed $\triangle KAB \cong \triangle CAD$. Then, he showed that the area of square ACHK is twice that of $\triangle KAB$ and the area of rectangle AMLD is twice that of $\triangle CAD$. Thus square ACHK and rectangle AMLD have the same area. Similarly, square BCGF has the same area as rectangle BELM.

Can you write a complete version of Euclid's proof?

Section 9.1 Problem Set A

1 Simplify:
- a $\sqrt{4}$
- b $\sqrt{27}$
- c $\sqrt{72}$
- d $\sqrt{32}$
- e $\sqrt{98}$
- f $\sqrt{200}$
- g $\sqrt{20}$
- h $\sqrt{24}$

2 Simplify:
- a $5\sqrt{18}$
- b $\sqrt{4 + 9}$
- c $\sqrt{3^2 + 4^2}$
- d $\sqrt{5^2 + 12^2}$
- e $\frac{1}{6}\sqrt{48}$
- f $\sqrt{49 \cdot 3}$

3 Simplify:
- a $\frac{1}{\sqrt{2}}$
- b $\frac{1}{\sqrt{5}}$
- c $\frac{4}{\sqrt{2}}$
- d $\frac{6}{\sqrt{3}}$

4 Simplify:
- a $4\sqrt{3} + 7\sqrt{3}$
- b $7\sqrt{2} + \sqrt{3} + 6\sqrt{3} + \sqrt{2}$
- c $\sqrt{12} + \sqrt{27}$
- d $\sqrt{72} + \sqrt{75} - \sqrt{48}$

5 Solve for x:

a $x^2 = 25$
b $x^2 = 144$
c $x^2 = 169$
d $x^2 = \frac{1}{4}$
e $x^2 = 12$
f $x^2 = 18$

6 Solve for x:

a $x^2 + 16 = 25$
b $x^2 + 6^2 = 100$
c $12^2 + x^2 = 13^2$
d $x^2 + (3\sqrt{3})^2 = 36$
e $(\sqrt{5})^2 + (\sqrt{11})^2 = x^2$
f $x^2 = (5\sqrt{3})^2 + (\sqrt{5})^2$

7 Solve for x:

a $x^2 - 5x - 6 = 0$
b $x^2 + 4x - 12 = 0$
c $x^2 - 8x + 15 = 0$
d $x^2 - 18 - 3x = 0$
e $x^2 - 36 = 9x$
f $-x^2 + 5x + 36 = 0$

8 Solve for x:

a $x^2 - 4x = 0$
b $x^2 = 10x$
c $x^2 - 2x = 11x$
d $5x = x^2 - 3x$

Section 9.1 Problem Set B

9 Solve for x:

a $3x^2 + 5x - 7 = x^2 + 8x + 28$
b $12x^2 - 15 = -11x$
c $8x^2 - 7x + 9 = 2x^2 + 6x + 7$

10 Solve for x: $\frac{7}{x + 1} = \frac{2x + 4}{3x - 3}$

Section 9.1 Problem Set C

11 Simplify:

a $\sqrt{h^2}$ if h represents a negative number
b $\sqrt{(x - 3)^2}$ if $x < 3$
c $\sqrt{p^2q^2}$ if p and q both represent negative numbers
d $\sqrt{x^3y^2}$ if $x > 0$ and $y < 0$

Section 9.2 Altitude on Hypotenuse Theorems

Given: $\triangle ABC$ with altitude $\overline{CD}$ drawn to the hypotenuse.

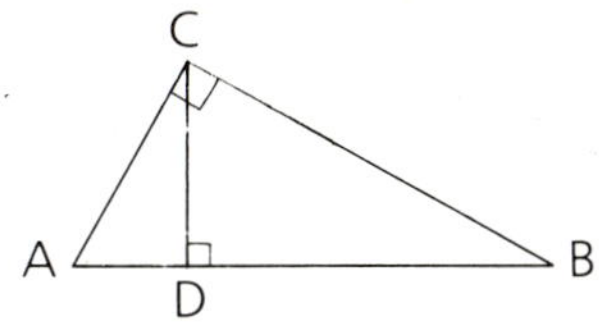

Three similar triangles are formed:

$$\triangle ABC \sim \triangle ACD \sim \triangle CBD$$

- $\triangle ABC \sim \triangle ACD$ by AA, and we notice that $\frac{AB}{AC} = \frac{AC}{AD}$ or $(AC)^2 = (AB) \cdot (AD)$, so AC is the mean proportional between AB and AD.

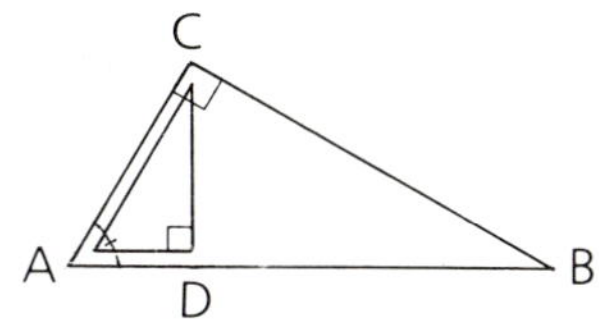

- $\triangle ABC \sim \triangle CBD$ by AA, and we notice that $\frac{AB}{CB} = \frac{CB}{DB}$ or $(CB)^2 = (AB) \cdot (DB)$, so CB is the mean proportional between AB and DB.

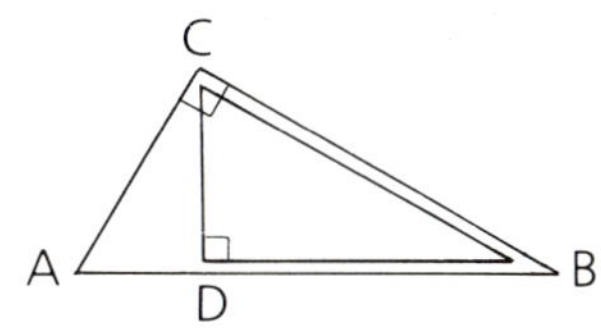

- $\triangle ACD \sim \triangle CBD$ by transitivity of similar triangles, and we notice that $\frac{AD}{CD} = \frac{CD}{DB}$ or $(CD)^2 = (AD) \cdot (DB)$, so CD is the mean proportional between AD and DB.

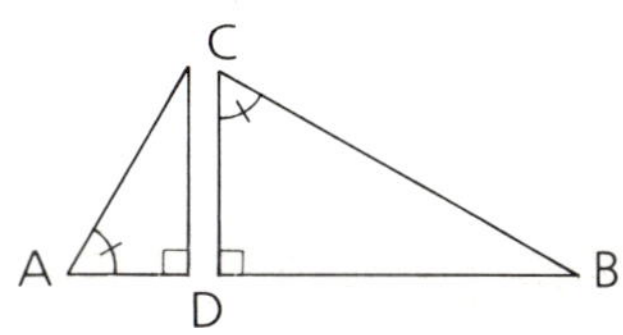

These examples prove three closely related theorems, presented next as one theorem.

Theorem 60 a, b, c

T **If the altitude is drawn to the hypotenuse of a right triangle, then:**

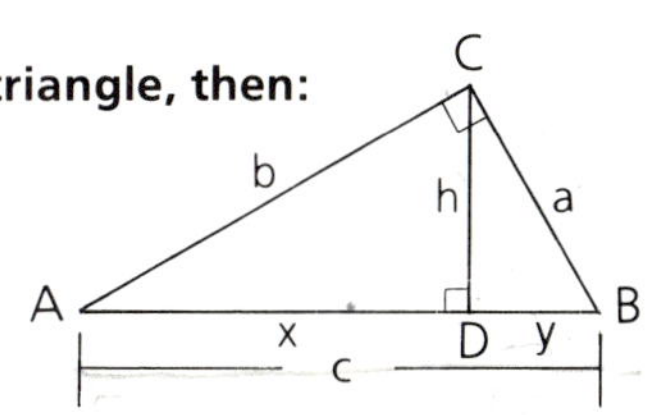

a the two triangles formed are similar to the given right triangle and to each other. $\triangle ADC \sim \triangle ACB \sim \triangle CDB$

b the altitude to the hypotenuse is the mean proportional between the segments of the hypotenuse. $\frac{x}{h} = \frac{h}{y}$ **or** $h^2 = xy$

c either leg of the given right triangle is the mean proportional between the hypotenuse of the given right triangle and the segment of the hypotenuse adjacent to that leg (i.e., the projection of that leg on the hypotenuse).

$\frac{y}{a} = \frac{a}{c}$ **or** $a^2 = yc$ **and** $\frac{x}{b} = \frac{b}{c}$ **or** $b^2 = xc$

To summarize Theorems 60b and 60c,

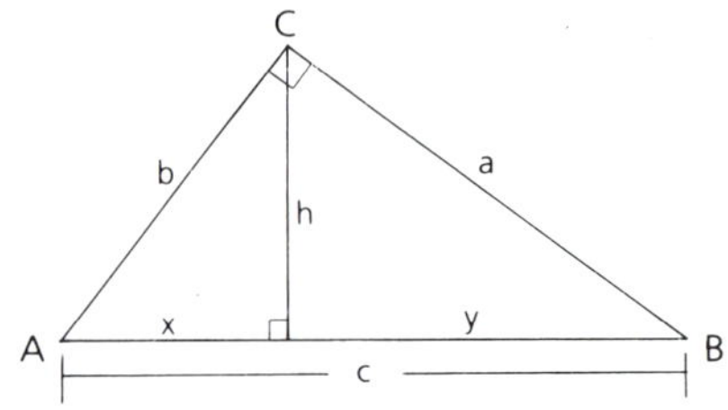

$h^2 = x \cdot y$
$b^2 = x \cdot c$
$a^2 = y \cdot c$

Section 9.2 Sample Problems

1 **a** If AD = 3 and DB = 9, find CD.

Solution:

$$\begin{aligned} CD^2 &= AD \cdot DB \\ x^2 &= 3 \cdot 9 \\ x &= \pm\sqrt{3}\sqrt{9} \\ x &= \pm 3\sqrt{3} \\ CD &= 3\sqrt{3} \end{aligned}$$

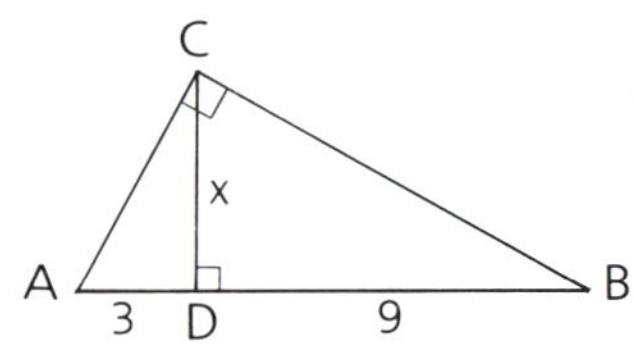

(CD cannot be negative, so reject $-3\sqrt{3}$.)

b If AD = 3 and DB = 9, find AC.

Solution:

$$\begin{aligned} AC^2 &= AD \cdot AB \\ x^2 &= 3 \cdot 12 \\ x^2 &= 36 \\ x &= \pm 6 \\ AC &= 6 \end{aligned}$$

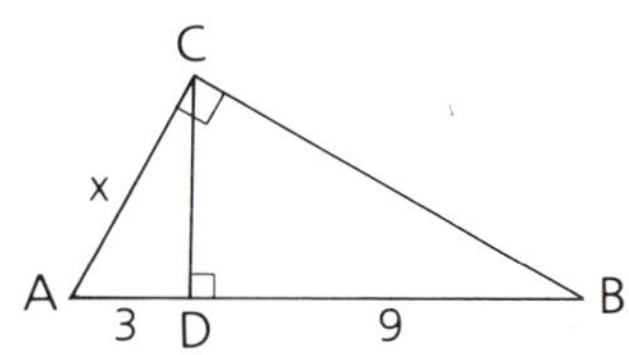

(Reject -6, since AC cannot be negative.)

c If DB = 21 and AC = 10, find AD.

Solution:

$$\begin{aligned} AC^2 &= AD \cdot AB \\ 10^2 &= x(x + 21) \\ x(x + 21) &= 10 \cdot 10 \\ x^2 + 21x &= 100 \\ x^2 + 21x - 100 &= 0 \\ (x + 25)(x - 4) &= 0 \end{aligned}$$

$x + 25 = 0$ or $x - 4 = 0$

$x = -25$ or $x = 4$

AD = 4, reject -25

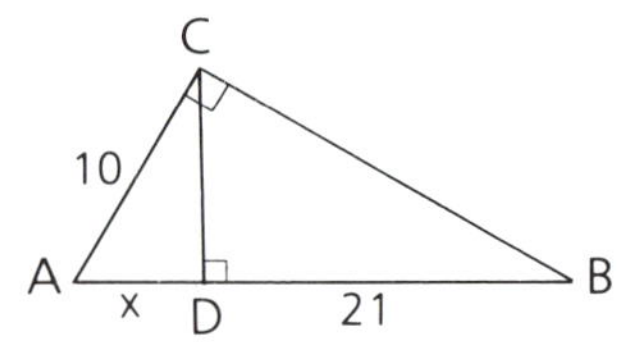

2 Given: $\overline{PK} \perp \overline{JM}$, $\overline{RK} \perp \overline{JP}$, $\overline{KO} \perp \overline{PM}$

Prove: (PO) (PM) = (PR) (PJ)

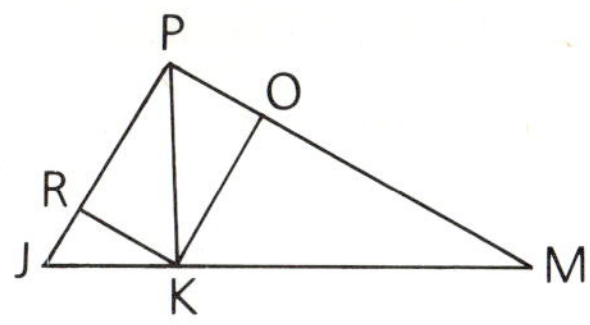

1 $\overline{PK} \perp \overline{JM}$	1 Given
2 $\angle$PKJ is a right $\angle$	2 $\perp$ segments form right $\angle$s
3 $\angle$PKM is a right $\angle$	3 Same as 2
4 $\overline{RK} \perp \overline{JP}$	4 Given
5 $\overline{RK}$ is an altitude	5 A segment drawn from a vertex of a $\triangle$ $\perp$ to the opposite side is an altitude
6 $(PK)^2 = (PR)(PJ)$	6 If the altitude is drawn to the hypotenuse of a right $\triangle$, then either leg of the given right $\triangle$ is the mean proportional between the hypotenuse of the given right $\triangle$ and the segment adjacent to that leg on the hypotenuse
7 Similarly, $(PK)^2 = (PO)(PM)$	7 Reasons 1–6
8 (PO) (PM) = (PR) (PJ)	8 Transitive Property

Section 9.2 Problem Set A

1 **a** If EH = 7 and HG = 3, find HF.

b If EH = 7 and HG = 4, find EF.

c If GF = 6 and EG = 9, find HG.

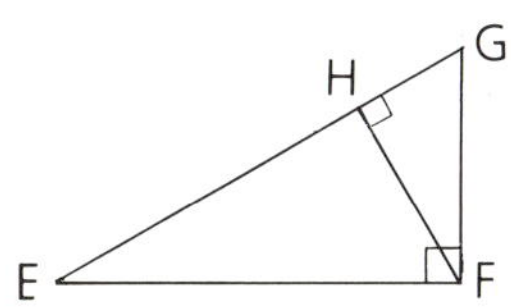

2 Given: $\overline{AC} \perp \overline{CB}$ and $\overline{CD} \perp \overline{AB}$

a If AD = 4 and BD = 9, find CD.

b If AD = 4 and AB = 16, find AC.

c If BD = 6 and AB = 8, find BC.

d If CD = 8 and BD = 16, find AD.

e If AD = 3 and BD = 24, find AC.

f If BC = 8 and BD = 20, find AB.

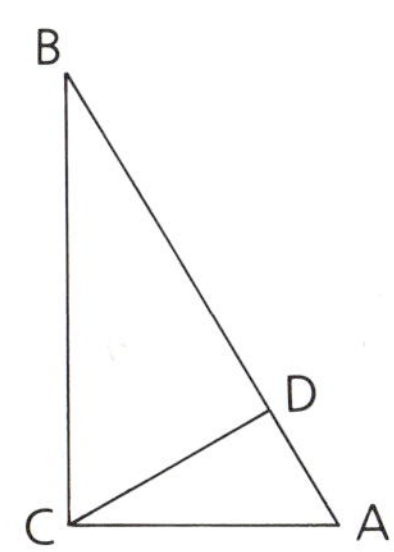

3 Given: $\angle JOM = 90°$, $\overline{OK}$ is an altitude

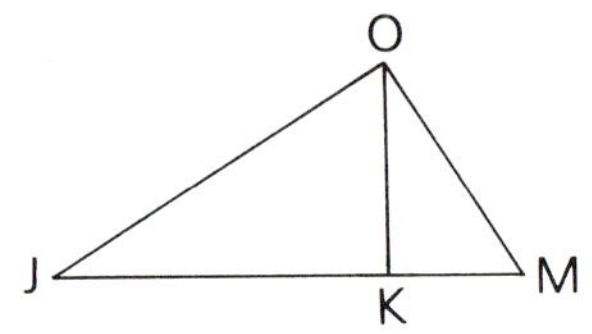

a If $JK = 12$ and $KM = 5$, find OK.

b If $OK = 3\sqrt{5}$ and $JK = 9$, find KM.

c If $JO = 3\sqrt{2}$ and $JK = 3$, find JM.

d If $KM = 5$ and $JK = 6$, find OM.

4 a Find 2x b Find $\frac{1}{2}y$ c Find $z + 8$

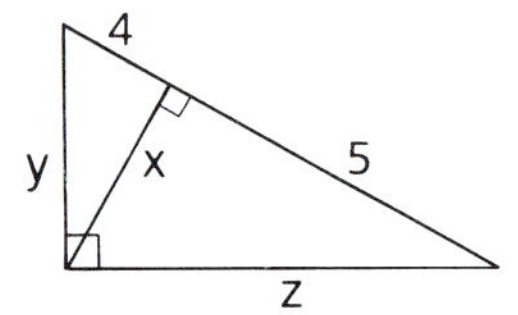

5 a Find a.

b Find ab.

c Find $a + b + c$.

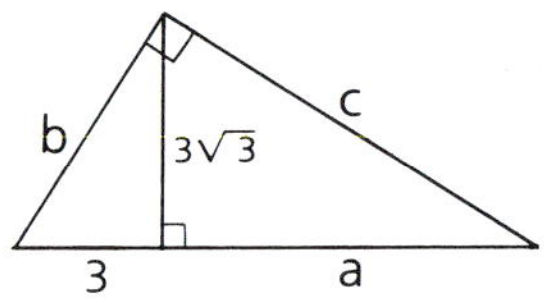

6 Given: $\overline{RT}$ is an altitude, $\angle PRS$ is a right $\angle$

Conclusion: $\frac{PR}{RS} = \frac{RT}{ST}$

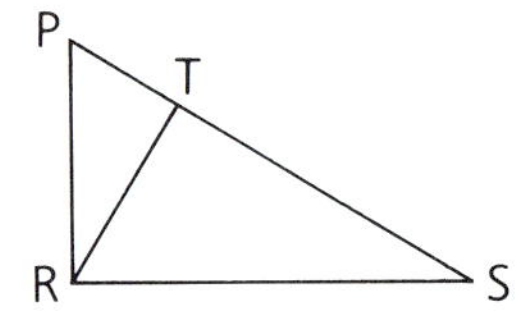

7 Given: $\overline{SY}$ is an altitude, $\angle VSX$ is a right $\angle$

Prove: $XY \cdot SV = XS \cdot YS$

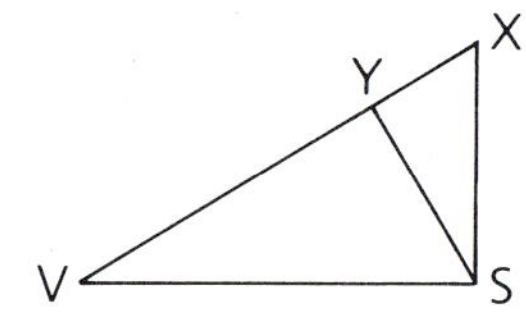

Section 9.2 Problem Set B

8 As Slarpy stood at B, the foot of a six meter pole, he asked Carpy how far it was across the pond from B to C. Carpy got his carpenter's square and climbed the pole. Using his line of sight, he set up the figure shown. When Slarpy found AB = 3 m, Carpy knew the answer. What was it?

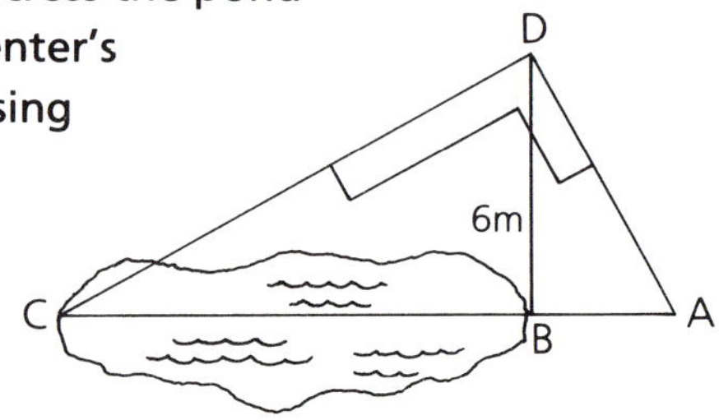

9 a If $HG = 4$ and $EF = 3\sqrt{5}$, find EH.

b If $GF = 6$ and $EH = 9$, find EG.

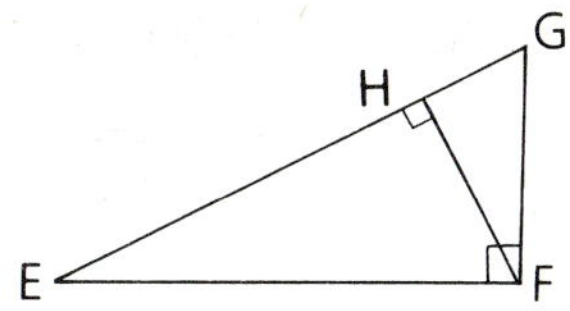

10 Given: ⊙ O, $\overline{CD} \perp \overline{AB}$

$\angle ACB$ is a right $\angle$

Conclusions: a $\frac{AD}{CD} = \frac{CD}{BD}$

b $\frac{AD}{ED} = \frac{ED}{BD}$

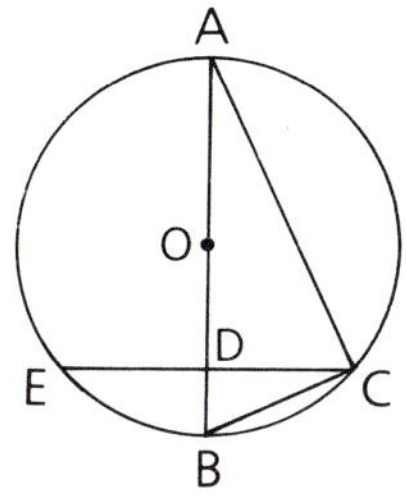

11 a If $AD = 7$ and $AB = 11$, find CD.

b If $CD = 8$ and $AD = 6$, find AB.

c If $AB = 12$ and $AD = 4$, find BC.

d If $AC = 7$ and $AB = 12$, find BD.

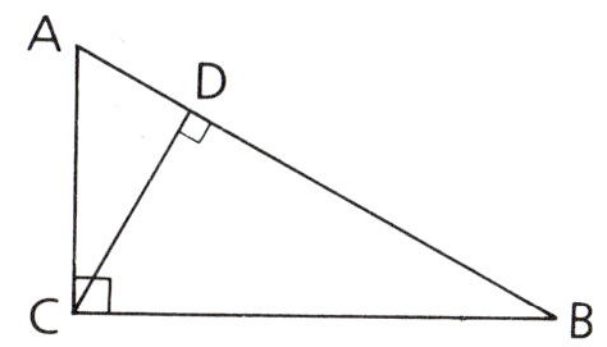

12 $\overline{CD}$ is the altitude to hypotenuse $\overline{AB}$. If the four lengths AD, CD, CD, and BD are written down at random to form two ratios, what is the probability that the ratios are equal?

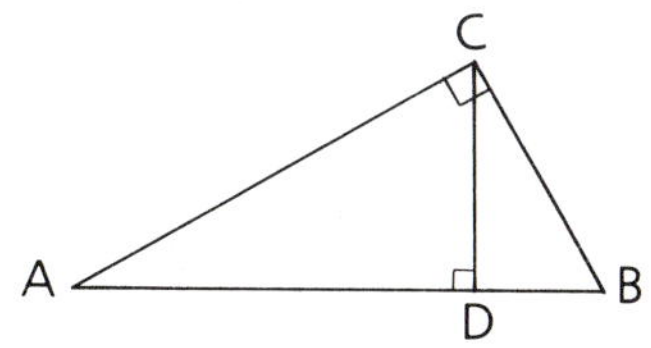

Section 9.2 Problem Set C

13 If $\sqrt{5} \approx 2.236$, find DE to the nearest tenth.

(The symbol $\approx$ means *approximately equals*.)

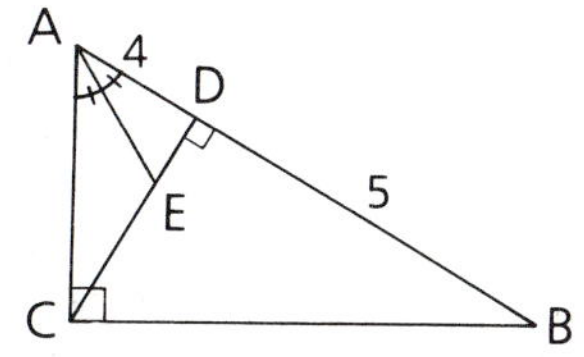

14 Prove: The product of the measures of the legs of a right triangle is equal to the product of the measures of the hypotenuse and the altitude to the hypotenuse.

15 Given: $\overline{AD} \perp \overline{CD}$
$\overline{BD} \perp \overline{AC}$
$BC = 5, AD = 6$

Find: BD

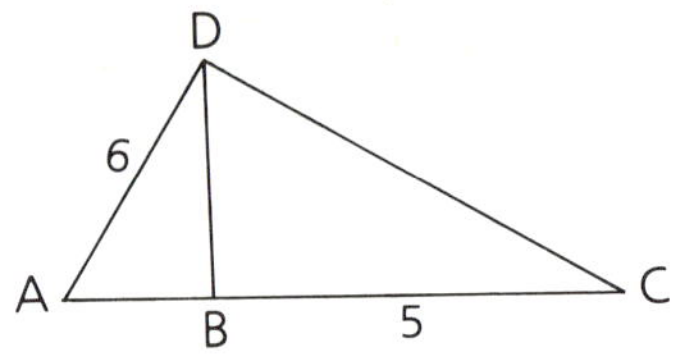

16 Given: $\overline{FG} \perp \overline{GH}$
$\angle 1$ is comp. to $\angle 3$

Prove: $\frac{JH}{GH} = \frac{GH}{HF}$

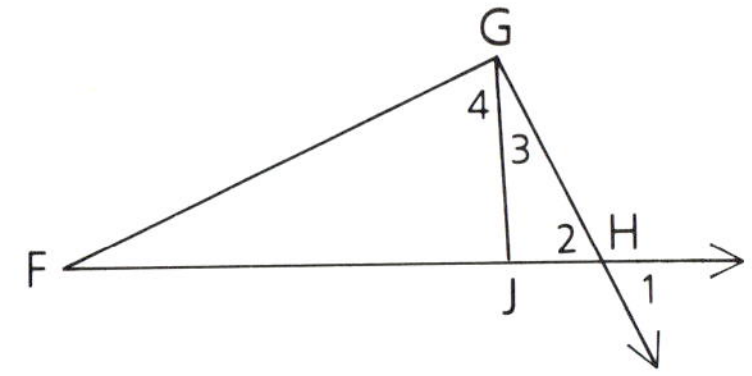

Section 9.2 Problem Set D

17 Given: HKMO is a rectangle
$\overline{PK} \perp \overline{HM}$
$\overline{PJ} \perp \overline{HK}$

Prove: $ab = fx$

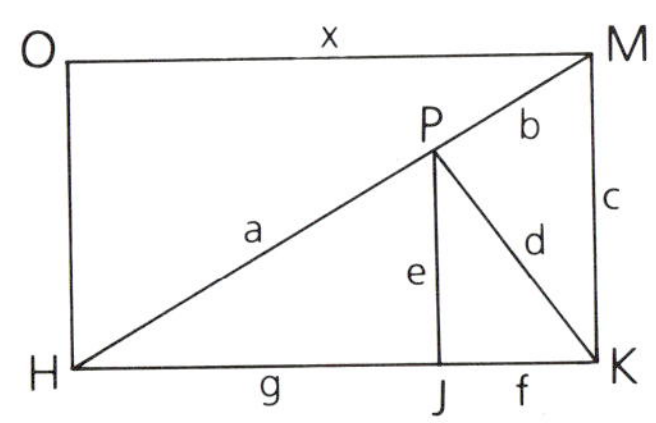

18 In the figure, CD is the mean proportional (or GEOMETRIC MEAN) between AD and BD.

For any two numbers a and b, the ARITHMETIC MEAN is $\frac{1}{2}(a + b)$.

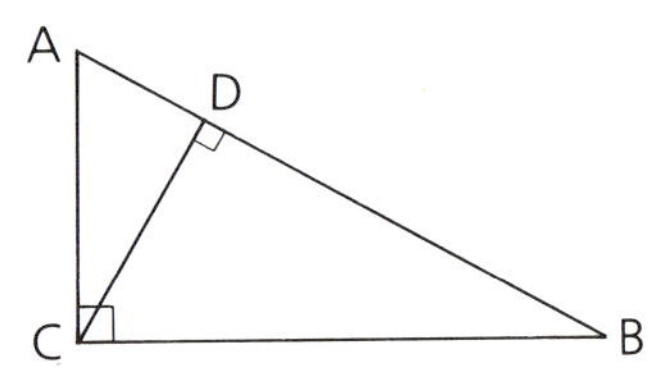

a Find the ARITHMETIC MEAN and the GEOMETRIC MEAN for AD and BD, if:

(1) $AD = 2, BD = 8$
(2) $AD = 3, BD = 12$
(3) $AD = 4, BD = 25$

b Given two positive numbers a and b, prove that the arithmetic mean, $\frac{1}{2}(a + b)$, is always greater than or equal to the positive geometric mean, $\sqrt{ab}$.

Section 9.3 Geometry's Most Elegant Theorem

As the plays of Shakespeare are to literature, as the Constitution is to America, so is the **PYTHAGOREAN THEOREM** to geometry. First, it is basic, for it is the rule for solving right triangles. Second, it is widely applied, because every polygon is made up of right triangles formed by sides of the polygon, diagonals, and altitudes. Third, it enables many ideas (and objects) to fit together very simply. Indeed, it is *elegant* in concept and extremely *powerful.*

The **PYTHAGOREAN THEOREM** was known to the ancient Egyptians and Greeks. The first proof is attributed to Pythagoras, a Greek mathematician who lived about 500 B.C. There are now over 300 proofs* of the theorem, and a book has been published consisting solely of such proofs.

Theorem 61

T **The square of the measure of the hypotenuse of a right triangle is equal to the sum of the squares of the measures of the legs. (The PYTHAGOREAN THEOREM)**

Proof of the Pythagorean Theorem:

Given: $\triangle ACB$ is a right $\triangle$ with right $\angle ACB$

Prove: $a^2 + b^2 = c^2$

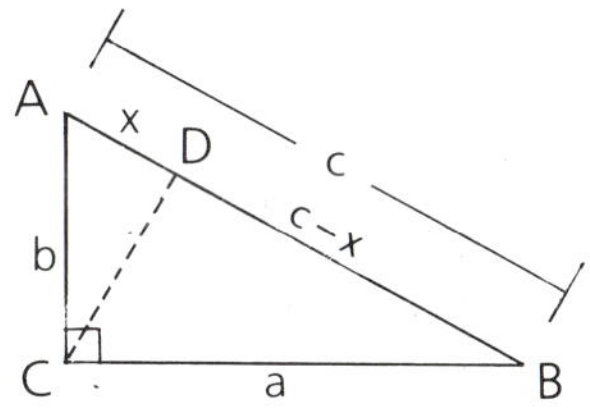

1 $\angle ACB$ is a right $\angle$	1 Given
2 Draw $\overline{CD}$, making it $\perp$ to $\overline{AB}$	2 From a point outside a line, only one $\perp$ can be drawn to the line
3 $\overline{CD}$ is an altitude	3 A segment drawn from a vertex of a $\triangle$ and $\perp$ to the opposite side is an altitude
4 $a^2 = (c - x)c$	4 Altitude on hypotenuse $\Longrightarrow$ $\text{leg}^2 = (\text{adjacent seg.}) \cdot (\text{hypotenuse})$
5 $a^2 = c^2 - cx$	5 Distributive property
6 $b^2 = xc$	6 Same as 4
7 $a^2 + b^2 = c^2 - cx + cx$	7 Addition Property
8 $a^2 + b^2 = c^2$	8 Algebraic Properties

To this day, one of the simplest ways to know that two lines are perpendicular is to find out if they form a right angle in a triangle. For that, we need the **CONVERSE** of the **PYTHAGOREAN THEOREM,** given next.

* Different sets of postulates and theorems lead to different proofs.

Theorem 62

T **If the square of the measure of one side of a triangle equals the sum of the squares of the measures of the other two sides, then the angle opposite the longest side is a right angle.**

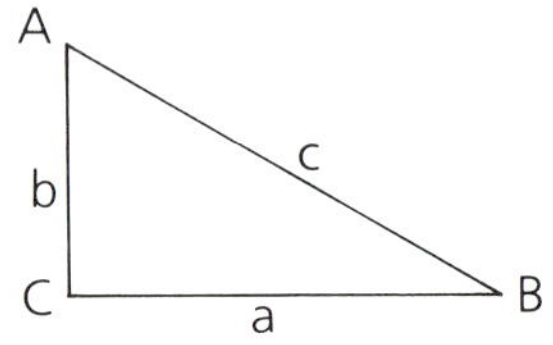

Given: $a^2 + b^2 = c^2$

Prove: $\triangle ACB$ is a right $\triangle$
and $\angle C$ is the right $\angle$

The proof of Theorem 62 is called for in problem 21 of this section.

If in the above diagram, we had increased c while keeping a and b the same, $\angle C$ would have been larger. Try it. Thus, a valuable extension of Theorem 62 can be found:

If c is the length of the longest side of a triangle, and:

1 if $a^2 + b^2 > c^2$, then it is an *acute* triangle.

2 if $a^2 + b^2 = c^2$, then it is a *right* triangle.

3 if $a^2 + b^2 < c^2$, then it is an *obtuse* triangle.

Section 9.3 Sample Problems

1 Solve for x.

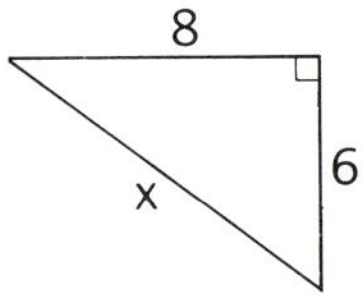

Solution: From the Pythagorean Theorem

$$6^2 + 8^2 = x^2$$
$$36 + 64 = x^2$$
$$100 = x^2$$
$$\pm 10 = x, \text{ reject } -10$$
$$x = 10$$

2 Find the perimeter of the rectangle shown.

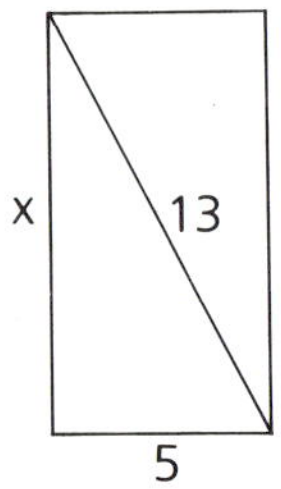

Solution: From the Pythagorean Theorem

$$x^2 + 5^2 = 13^2$$
$$x^2 + 25 = 169$$
$$x^2 = 144$$
$$x = \pm 12, \text{ reject } -12$$
$$\text{Perimeter} = 5 + 12 + 5 + 12 = 34$$

3 Find the perimeter of a rhombus with diagonals of 6 and 10.

Solution: Remember that the diagonals of a rhombus are perpendicular bisectors of each other.

$$3^2 + 5^2 = x^2$$
$$9 + 25 = x^2$$
$$34 = x^2$$
$$\pm\sqrt{34} = x, \quad \text{reject } -\sqrt{34}$$

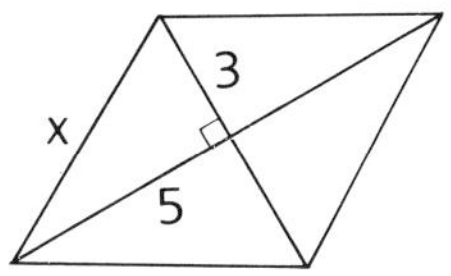

Since all sides of a rhombus are congruent, the perimeter is $4\sqrt{34}$.

4 Nadia skips 3 m north, 2 m east, 4 m north, 13 m east, and one m north. How far is Nadia from where she started?

Solution: Since Nadia started at S and ended at E, we are looking for the hypotenuse of $\triangle$SAE. She has gone a total of 8 m north and 15 m east, so:

$$8^2 + 15^2 = x^2$$
$$64 + 225 = x^2$$
$$289 = x^2$$
$$\pm 17 = x, \text{ reject } -17$$
$$SE = 17 \text{ meters}$$

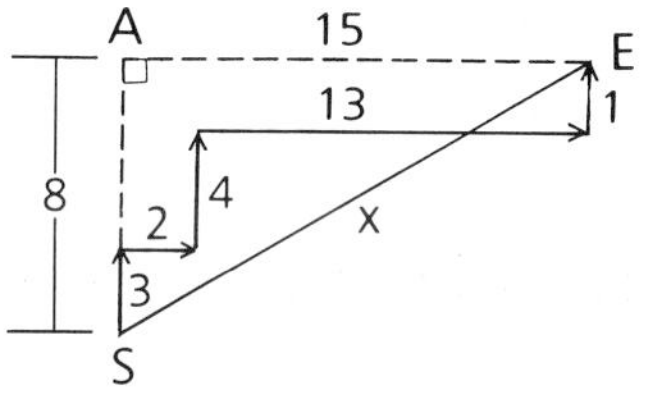

5 Find the altitude of an isosceles trapezoid whose sides have lengths of 10, 30, 10, and 20.

Solution: An altitude of a trapezoid is a segment, such as $\overline{AE}$, perpendicular to both bases. We often draw two altitudes, such as $\overline{AE}$ and $\overline{BD}$ to obtain a rectangle, AEDB. Thus, ED = 20. Right $\triangle$AEF is congruent to right $\triangle$BDC, and FE = DC = $\frac{1}{2}(30 - 20) = 5$.

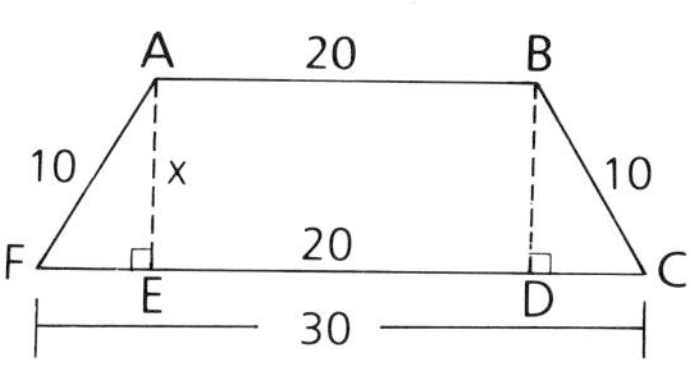

In $\triangle$AEF, $x^2 + 5^2 = 10^2$
$$x^2 + 25 = 100$$
$$x^2 = 75$$
$$x = \pm\sqrt{75} = \pm\sqrt{25 \cdot 3} = \pm 5\sqrt{3}, \text{ reject } -5\sqrt{3}$$
$$\text{The altitude} = 5\sqrt{3}$$

6 Classify the triangle shown.

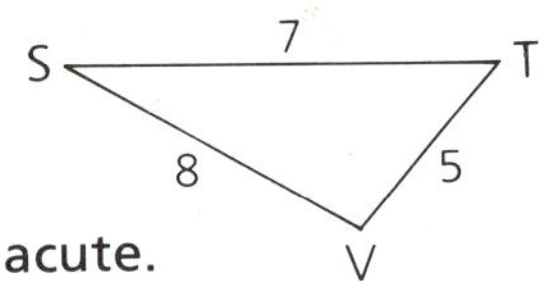

Solution: If $5^2 + 7^2 > 8^2$, the triangle is acute.
If $5^2 + 7^2 = 8^2$, the triangle is right.
If $5^2 + 7^2 < 8^2$, the triangle is obtuse.

$$5^2 + 7^2 \ ? \ 8^2$$
$$25 + 49 \ ? \ 64$$
$$74 > 64$$ Therefore, the triangle is acute.

Section 9.3 Problem Set A

1 Solve for the missing letter.

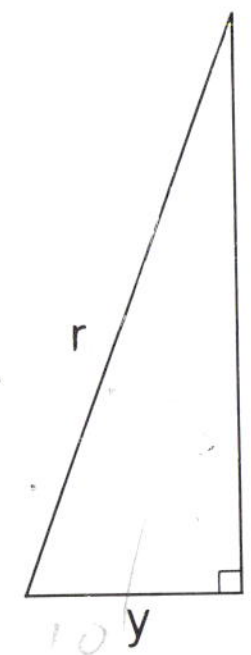

a $x = 4, \quad y = 5$
b $x = 15, \quad r = 17$
c $y = 9, \quad r = 15$
d $x = 12, \quad r = 13$
e $x = 5, \quad y = 5\sqrt{3}$
f $x = 5, \quad r = \sqrt{29}$
g $x = 2\sqrt{5}, \quad r = \sqrt{38}$

2 Find the length of the diagonal of a square with perimeter 12 cm.

3 Find the perimeter of a rhombus with diagonals 12 and 16 km.

4 Find the perimeter of a rectangle whose diagonal is 17 mm long and whose base is 15 mm long.

5 Given: $\overline{JG}$ is the altitude to base $\overline{FH}$ of isosceles triangle JFH. FJ = 15, FH = 24

Find: JG

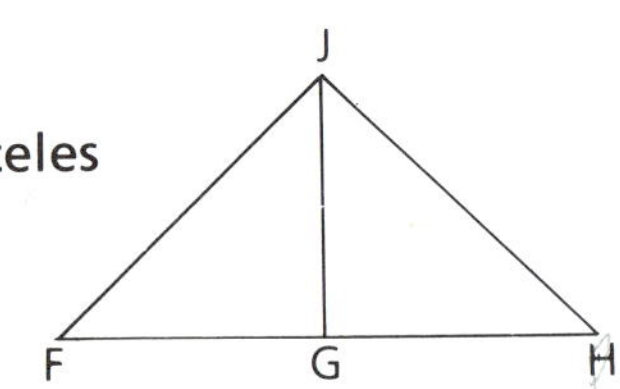

6 $\overline{PM}$ is an altitude of equilateral triangle PKO. If PK = 4, find PM.

7 Find the missing length in the trapezoid.

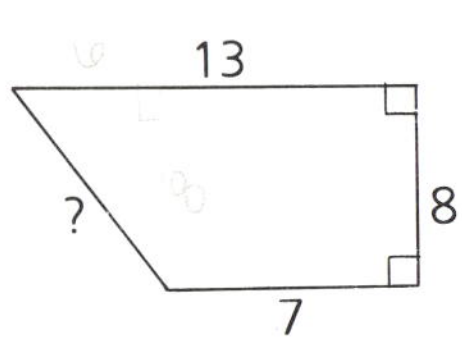

8 How far is the foot of the ladder from the wall?

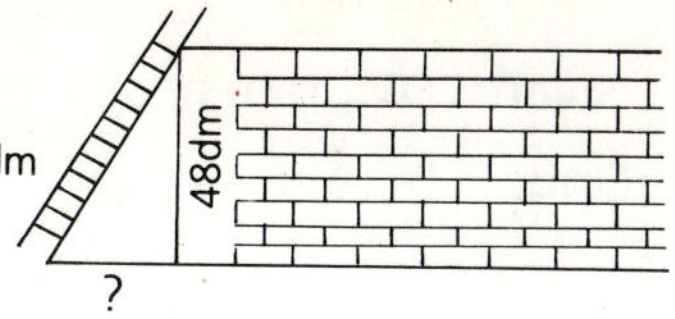

Section 9.3 Problem Set B

9 Find the missing length in terms of the variable provided:

a $AC = x$, $BC = y$, $AB = ?$

b $AC = 2$, $BC = x$, $AB = ?$

c $AC = 3a$, $BC = 4a$, $AB = ?$

d $AB = 13c$, $AC = 5c$, $BC = ?$

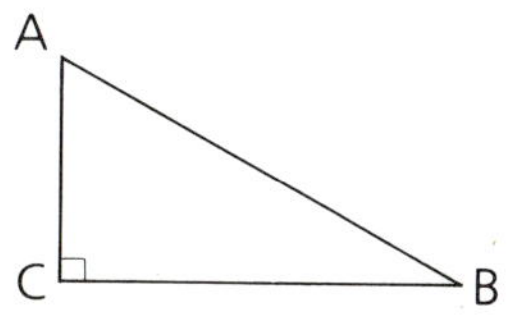

10 $\angle ACB$ is a right angle and $\overline{CD} \perp \overline{AB}$.

a $AD = 7$, $BD = 4$, find CD.

b $CD = 8$, $DB = 6$, find CB.

c $BC = 8$, $BD = 2$, find AB.

d $AC = 21$, $AB = 29$, find CB.

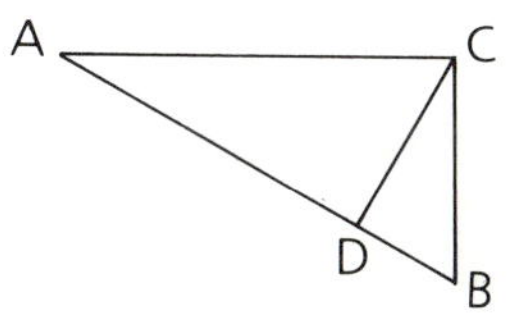

11 Al Capone walked 2 km north, 6 km west, then 4 km north, and 2 km west. If Big Al decides to go "straight," how far must he walk across the fields to his starting point?

12 A piece broke off rectangle ABDF, leaving trapezoid ACDF. If $BD = 16$, $BC = 7$, $FD = 24$, and E is the midpoint of $\overline{FD}$, what is the perimeter of $\triangle ACE$?

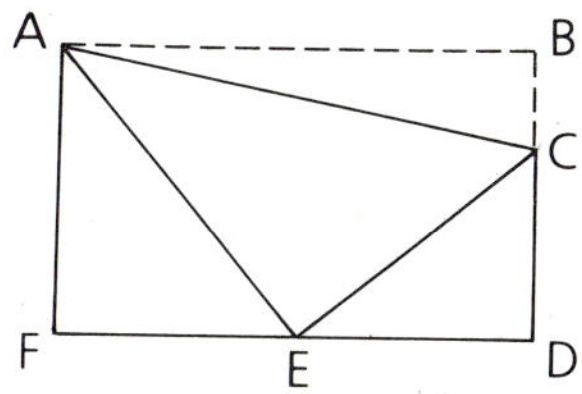

13 Find the altitude (length of segments perpendicular to both bases) of the isosceles trapezoid shown.

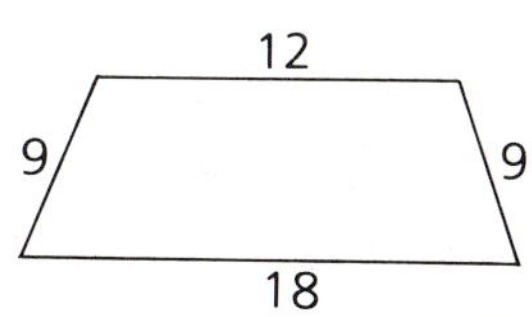

14 Given: Diagram as shown

Find: CD

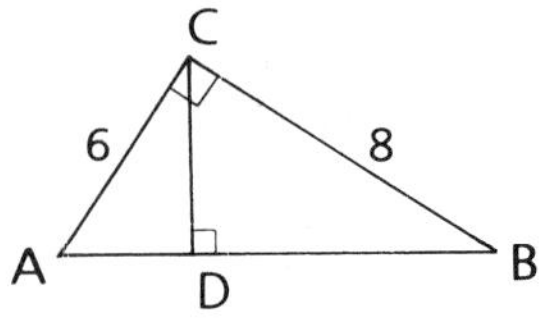

15 Solve for x in the partial spiral to the right.

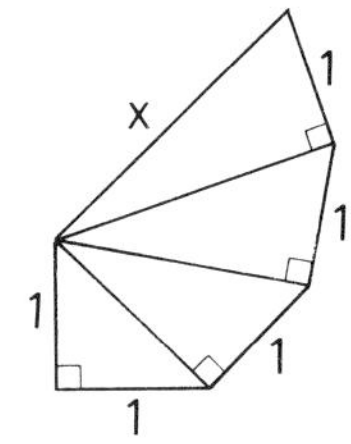

16 If the perimeter of a rhombus is $8\sqrt{5}$, and one diagonal has a length of $4\sqrt{2}$, find the length of the other diagonal.

17 Woody Woodpecker pecked at a 17 m wooden pole until it cracked and the upper part fell, reaching a point on the ground 10 m from the foot of the pole. Since the pole had not completely broken off, Woody pecked away where it had cracked. How far was Woody above the ground?

18 Find the perimeter of an isosceles right △ with a 6 cm hypotenuse.

19 The diagonals of a rhombus have a length ratio of 2:1. If the perimeter is 20, find the sum of the lengths of the diagonals.

20 Classify the triangles.

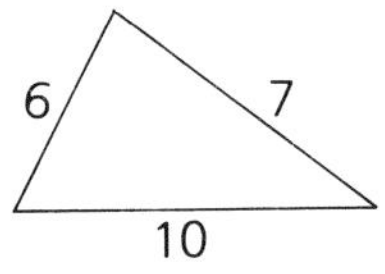

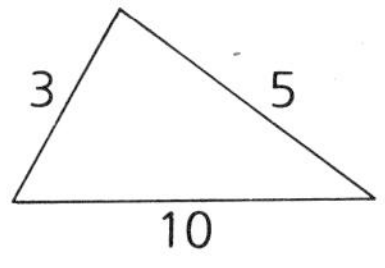

21 Prove Theorem 62

Given: $a^2 + b^2 = c^2$

Prove: ∠C is a right ∠

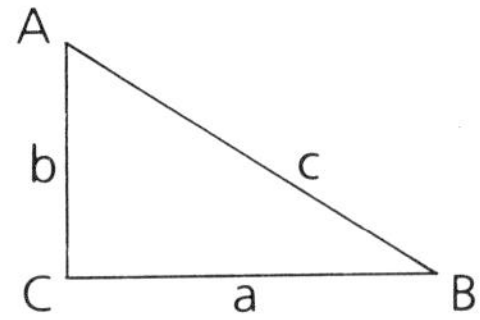

Hint: At some point D, draw a right ∠. Make one side *a* units long, the other *b* units long. Prove that △ACB ≅ △EDF.

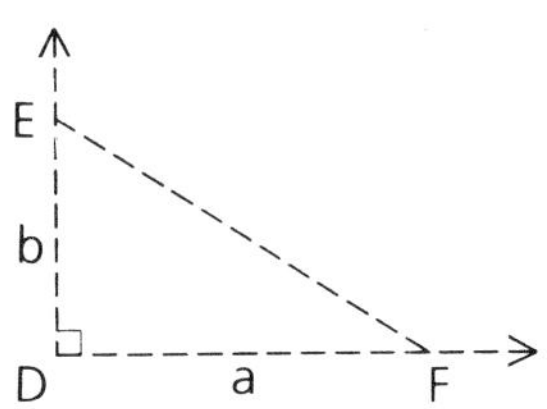

Section 9.3 Problem Set C

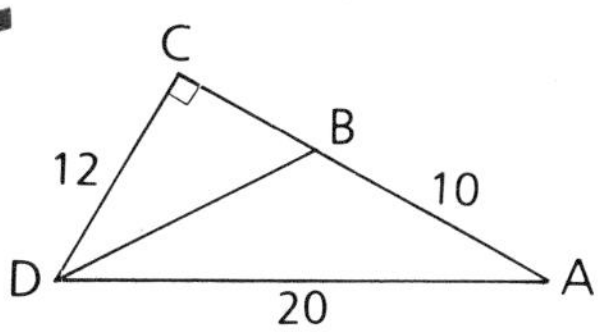

22 Find the perimeter of $\triangle DBC$.

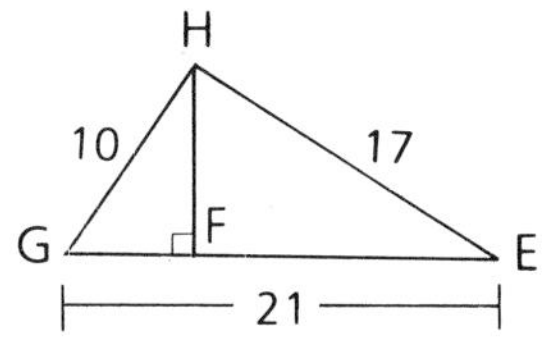

23 a Find HF.

b Is $\triangle EHF \sim \triangle HGF$?

24 The perimeter of an isosceles triangle is 32 and the length of the altitude to the base is 8. Find the length of a leg.

25 A ladder 25 units long (JO) is leaning against the wall, reaching a point 20 units above the ground (MO). The ladder is then moved so that $JK = 2(PO)$. Find KM.

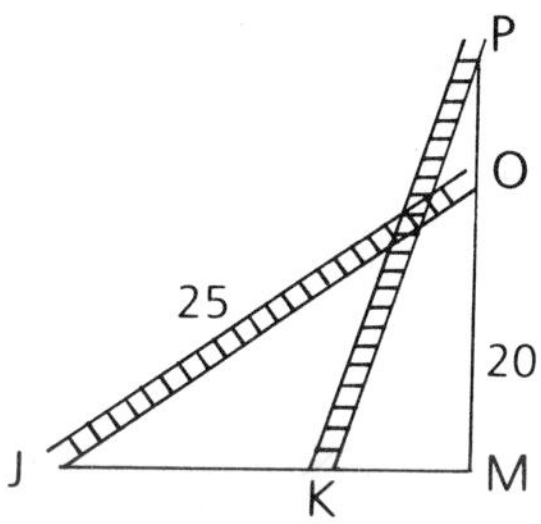

26 The medians of a right triangle that are drawn from the vertices of the acute angles have lengths of $2\sqrt{13}$ and $\sqrt{73}$. Find the length of the hypotenuse.

27 The diagonals of an isosceles trapezoid are each 17, the altitude is 8, and the upper base is 9. Find the perimeter of the trapezoid.

28 a Show that if the lengths of one leg of a right triangle and the hypotenuse are consecutive integers, then the square of the length of the second leg is equal to the sum of the lengths of the first leg and the hypotenuse.

b Show by counter-example that the converse of part *a* is not necessarily true. Note: The converse is, "If the square of the length of one of the legs of a right triangle is equal to the sum of the lengths of the other leg and the hypotenuse, then the lengths of the second leg and the hypotenuse are consecutive integers."

Section 9.3 Problem Set D

29 The legs of a right triangle have lengths of 3 m and 4 m. A point is taken on the hypotenuse 2 m from the intersection of the hypotenuse with the longer leg. How far is the point from the vertex of the right angle?

30 RSTV is an isosceles trapezoid with RS = 9, RV = 12, ST = 18.

Find the length of the perpendicular segment from T to $\overline{SW}$.

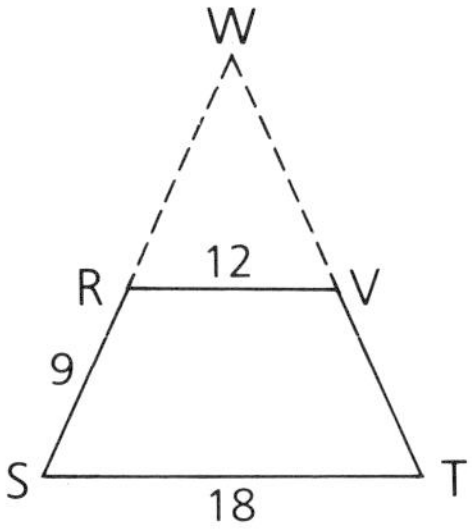

31 Given: $\overline{PR} \perp \overline{RT}$, PT = 25, PR = 15
PS = ST + 12

Find: SR

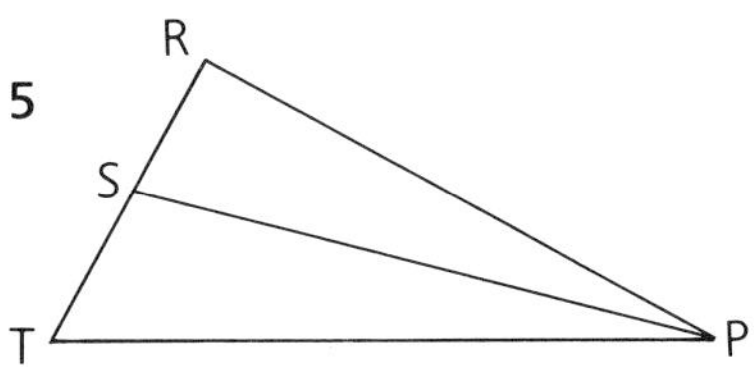

32 Abigail Adventuresome took a shortcut along the diagonal of a rectangular field and saved a distance equal to $\frac{1}{3}$ of the length of the longer side. Find the ratio of the length of the shorter side of the rectangle to that of the longer side.

33 a Given: P is any point in the interior of the rectangle ABCD.

Show: $(BP)^2 + (PD)^2 = (AP)^2 + (CP)^2$

b Is the result the same when P is in the exterior of the rectangle?

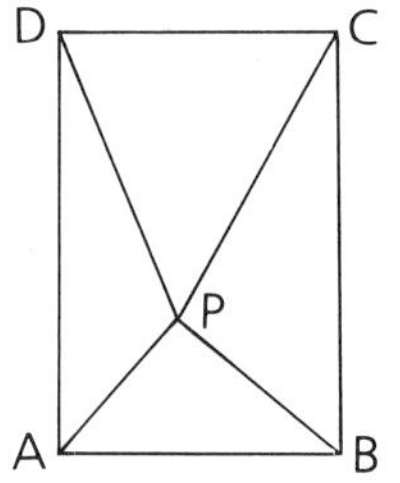

Section 9.4 Families of Right Triangles

In this section we cover the following concepts:

A PYTHAGOREAN TRIPLES

B THE PRINCIPLE OF THE REDUCED TRIANGLE

9.4 (A) Pythagorean Triples

In this section we consider some combinations of whole numbers that satisfy the Pythagorean Theorem. Knowing combinations is not essential, but knowing some of them can save you appreciable time and effort.

D **Any three whole numbers that satisfy the equation $a^2 + b^2 = c^2$ form a PYTHAGOREAN TRIPLE.**

Below is a set of right triangles you have encountered many times in this chapter. Do you see how the triangles are related?

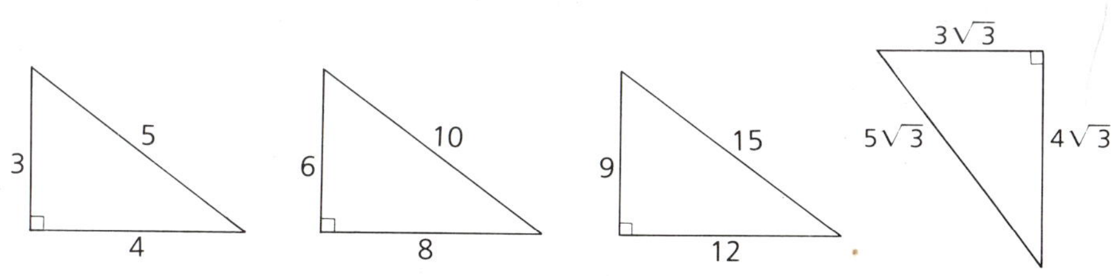

These four triangles are all members of the (3, 4, 5) family. For example, the triple (6, 8, 10) is (3•2, 4•2, 5•2).

The last triangle $(3\sqrt{3}, 4\sqrt{3}, 5\sqrt{3})$ is also a member of the (3, 4, 5) family but the numbers are *not* a **PYTHAGOREAN TRIPLE** because they are *not* whole numbers.

Other common families are:

(5, 12, 13) of which (15, 36, 39) is another member

(7, 24, 25) of which (14, 48, 50) is another member

(8, 15, 17) of which $(4, 7\frac{1}{2}, 8\frac{1}{2})$ is another member

There are infinitely many families, including the (9, 40, 41), the (11, 60, 61), the (20, 21, 29), and the (12, 35, 37), but they are not used very often.

9.4 (B) The Principle of the Reduced Triangle

■ Given: the right triangle shown

Find: x

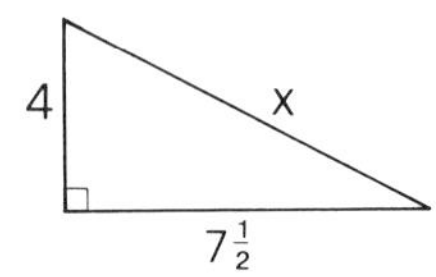

The fraction may complicate our work and we may not wish to complete the long calculation, $(4)^2 + (7\frac{1}{2})^2 = x^2$.

Our alternative is to find a more easily recognized member of the same family. Multiply each side by the common denominator, 2. Clearly, the family is (8, 15, 17).

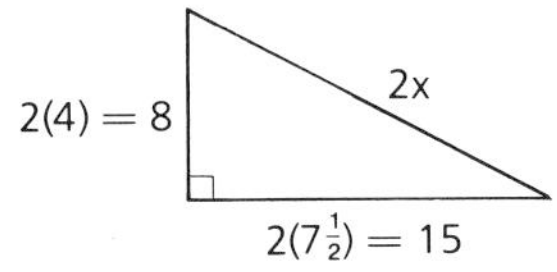

Thus, $2x = 17$ and $x = 8\frac{1}{2}$ (in the original triangle).

> **PRINCIPLE OF THE REDUCED TRIANGLE**
>
> **1 Reduce the difficulty of the problem by multiplying or dividing the three lengths by the same number to obtain a similar, but simpler triangle in the same family.**
>
> **2 Solve for the missing side of this easier triangle.**
>
> **3 Convert back to the original problem.**

The next example shows that the method may save time even if the "reduced" triangle is not a proper Pythagorean Triple.

■ Solve for x.

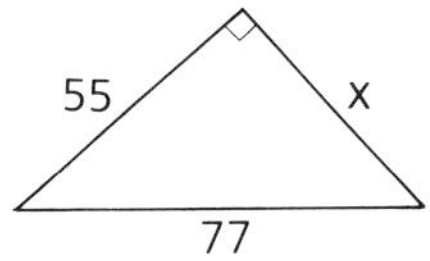

First, notice that both 55 and 77 are multiples of 11. Then *reduce the problem* to the easier problem shown.

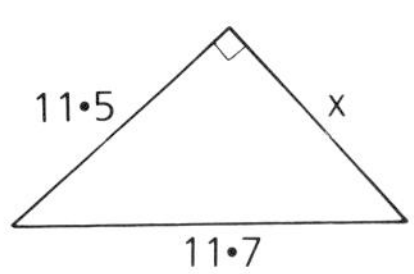

is in the family,

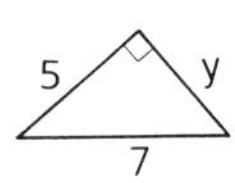

where

$$5^2 + y^2 = 7^2$$
$$25 + y^2 = 49$$
$$y^2 = 24$$
$$y = \pm 2\sqrt{6}$$

Thus, $x = 11 \cdot 2\sqrt{6} = 22\sqrt{6}$

Section 9.4 Sample Problems

1 Solve for AB.

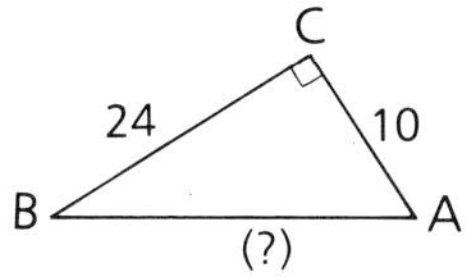

Method One:

(10, 24, ?) belongs to the (5, 12, 13) family

Notice:
$$10 = 5 \cdot 2$$
$$24 = 12 \cdot 2$$
$$\text{so } AB = 13 \cdot 2 = 26$$

Method Two:

$$\begin{aligned} 10^2 + 24^2 &= (AB)^2 \\ 100 + 576 &= (AB)^2 \\ 676 &= (AB)^2 \\ \pm\sqrt{676} &= AB, \text{ reject the } -\sqrt{676} \\ 26 &= AB \end{aligned}$$

2 Find x.

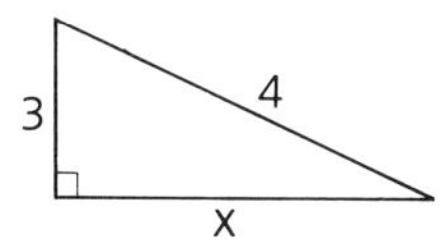

Solution: You may think 5 is the answer, but in (3, 4, 5) the 5 *must* be the hypotenuse. So we are stuck with the long way.

$$\begin{aligned} 3^2 + x^2 &= 4^2 \\ x^2 &= 7 \\ x &= \pm\sqrt{7}, \qquad \{\sqrt{7}\} \end{aligned}$$

3 Find the hypotenuse of the right triangle.

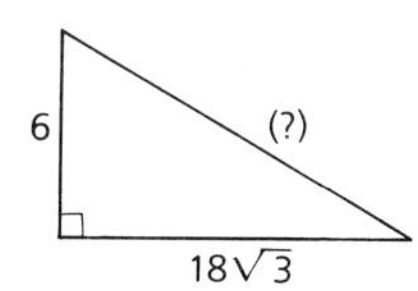

Method One:

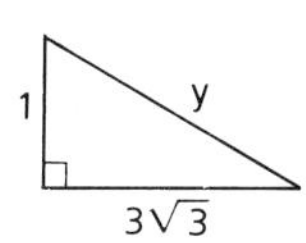

REDUCED TRIANGLE PRINCIPLE

Divide each given length by 6 to obtain the reduced similar triangle.

$$\begin{aligned} 1^2 + (3\sqrt{3})^2 &= y^2 \\ 1 + 27 &= y^2 \\ \pm\sqrt{28} &= y \\ \pm 2\sqrt{7} &= y, \text{ reject } -2\sqrt{7} \end{aligned}$$

Now multiply by 6 to convert back to the original triangle:

$$x = 6(2\sqrt{7}) = 12\sqrt{7} \quad \text{Ans.}$$

Method Two:

PYTHAGOREAN THEOREM

$$\begin{aligned} 6^2 + (18\sqrt{3})^2 &= c^2 \\ 36 + 972 &= c^2 \\ 1008 &= c^2 \\ \sqrt{1008} &= c \\ \pm\sqrt{144 \cdot 7} &= c \end{aligned}$$

(Would *you* have discovered those factors?)

Reject negative root

$$12\sqrt{7} = c$$

Section 9.4 Problem Set A

1–5 Find the missing side in each triangle:

1 (3, 4, 5)

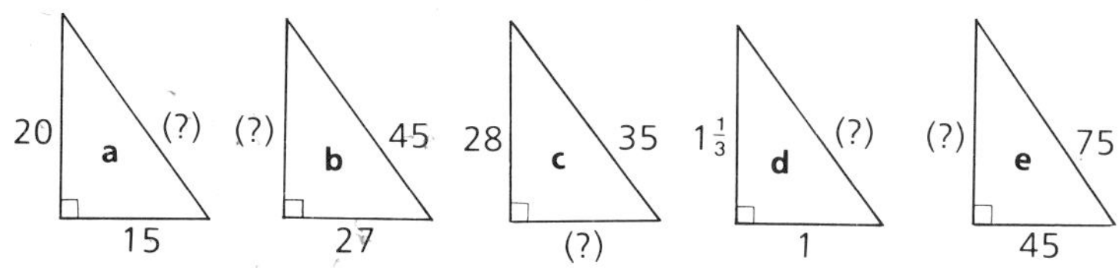

2 (5, 12, 13)

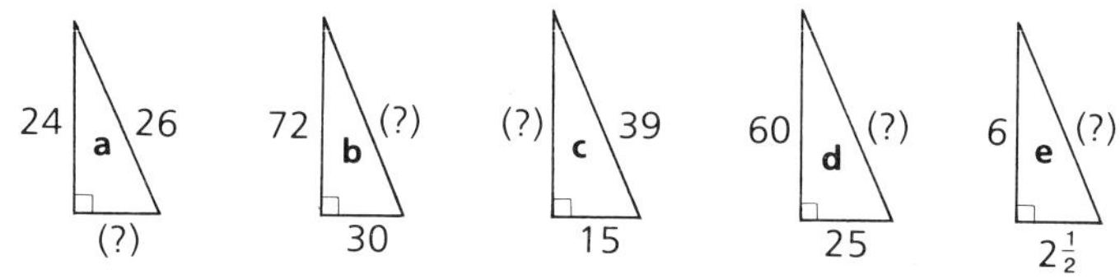

3 (7, 24, 25)

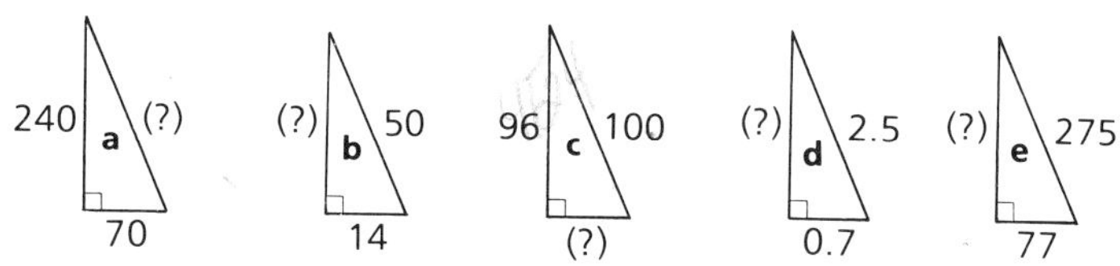

4 (8, 15, 17)

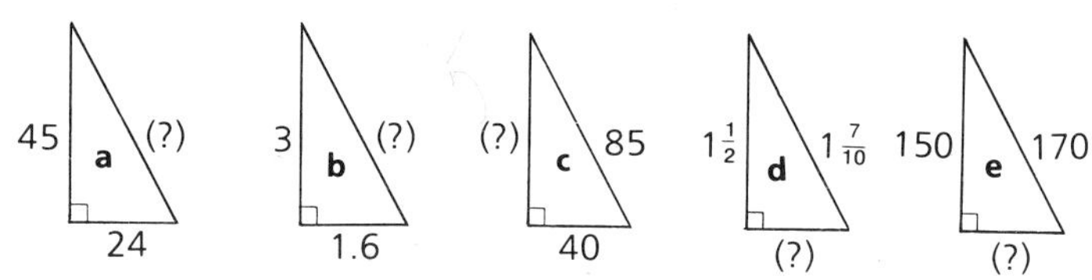

5 (MIXED)

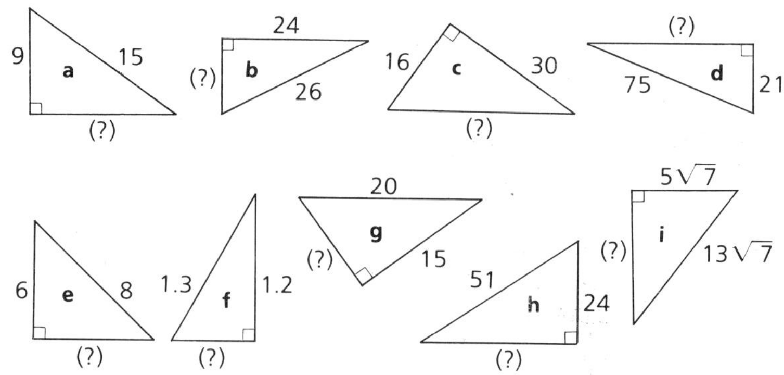

6 Find the diagonal of a rectangle whose sides are 20 and 48.

7 Find the perimeter of an isosceles triangle whose base is 16 dm and whose height is 15 dm.

8 Find the length of the upper base of the isosceles trapezoid.

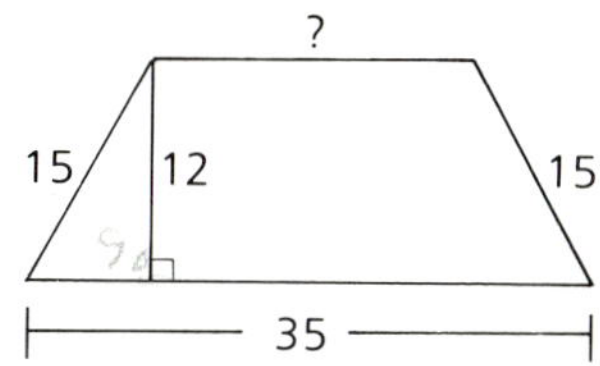

9 Use the reduced triangle principle to find the missing side.

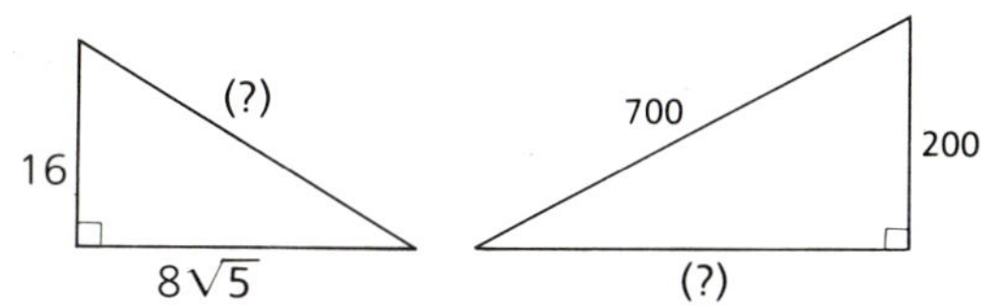

10 Find QD.

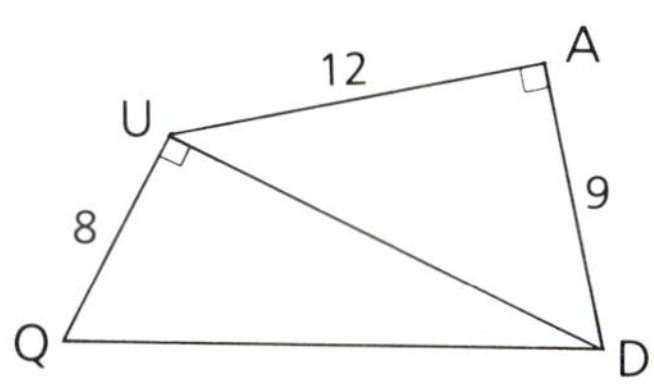

11 Find AD and BC.

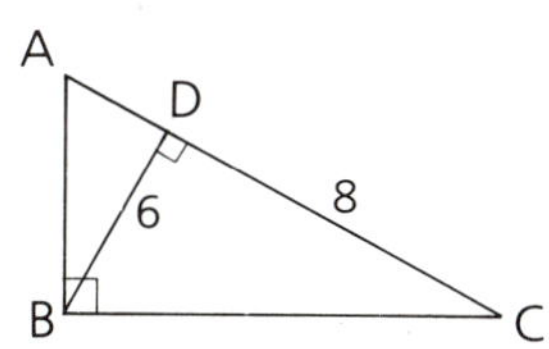

12 RHOM is a rhombus with diagonals RO = 48, and HM = 14. Find the perimeter of the rhombus.

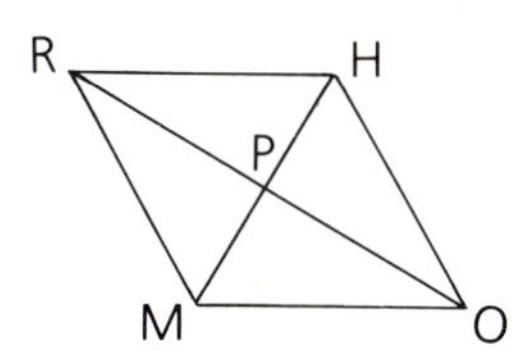

Section 9.4 Problem Set B

13 Mary and Larry left the riding stable at 10 A.M. Mary trotted south at 10 kph while Larry galloped east at 16 kph. How far apart were they at 11:30?

14 Find the missing side of each triangle:

a

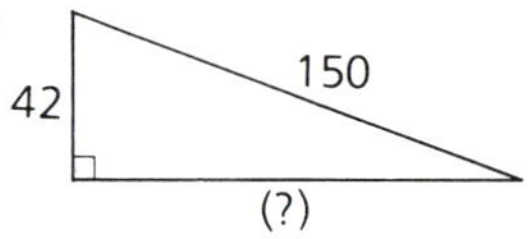

b

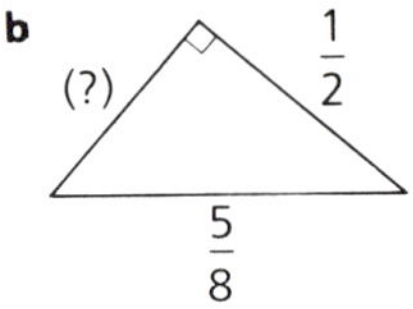

c

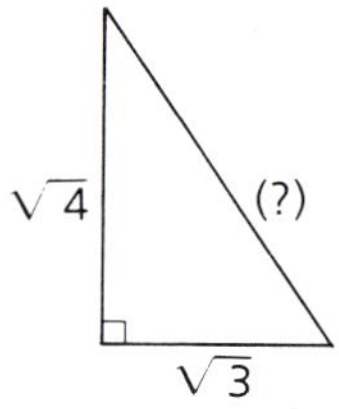

15 **a** Find x.

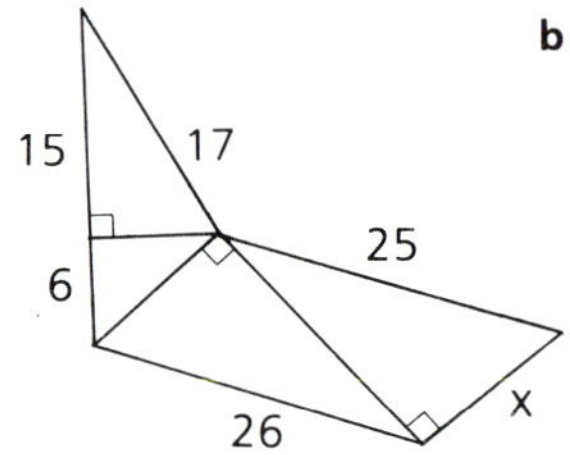

b Find x and y.

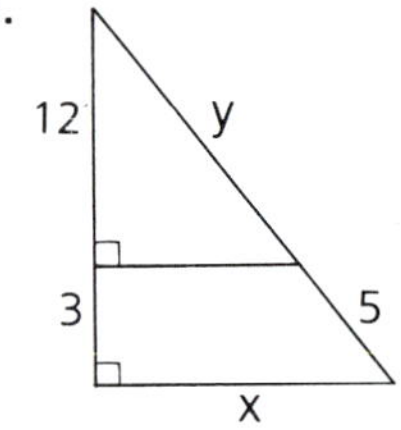

16 Find the altitude of a trapezoid with sides 2, 41, 20, and 41.

17 A model rocket shot up to a point 20 m above ground, hitting a smokestack, and then dropped straight down to a point 11 meters from its launch site. Find the total distance traveled from launch to touchdown.

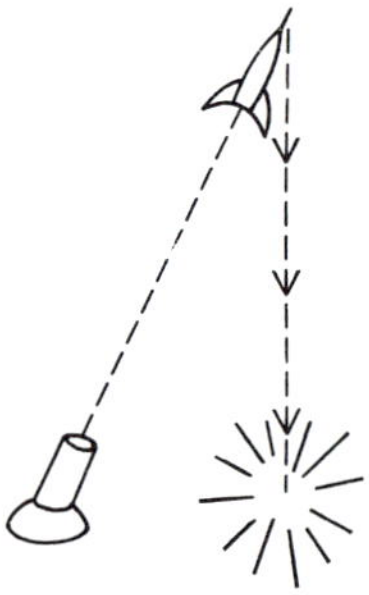

18 A submarine travels an evasive course trying to outrun a destroyer. It travels 1 km north, then 1 km west, then 1 km north, then 1 km west, etc., until it travels a total of 41 km. How many kilometers is the sub from the point at which it started?

Section 9.4 Problem Set C

19 Each of the following triples is a method for generating numbers that will represent the sides of a right triangle. Prove that each of these does indeed generate triples.

a Rule of Pythagoras:
(n is any odd number)

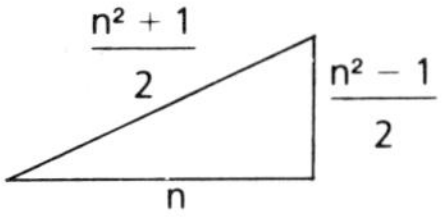

b Rule of Plato:
(m is any even number)

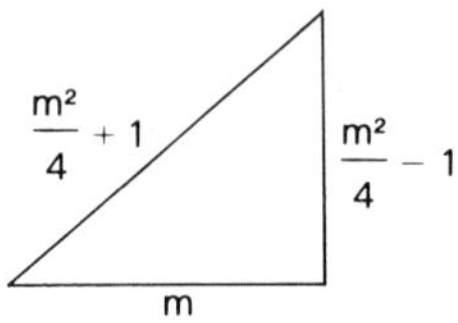

c Rule of Euclid:
(m, n are both odd or both even)

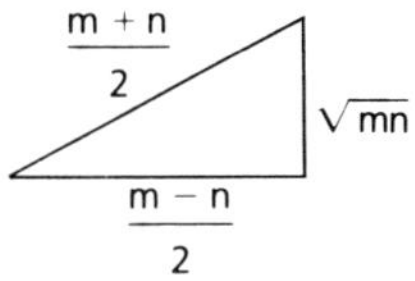

d Rule of Masères:
(m, n are any two integers)

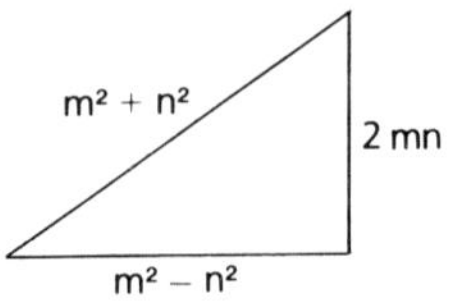

20 Show that the only right triangle in which the lengths of the sides are consecutive integers is the 3, 4, 5 triangle.

21 If a 650 cm ladder is placed against a building at a certain angle, it just reaches a point on the building that is 520 cm above the ground. If the ladder is moved to reach a point 80 cm higher up, how much closer will the foot of the ladder be to the building?

22 The lengths of the legs of a right triangle are x and $3x + y$. The length of the hypotenuse is $4x - y$. Find the ratio of x to y.

23 Six slips of paper, each containing one of the numbers 3, 4, 5, 6, 8, or 10 in one-to-one correspondence, are placed in a hat, and then two of the slips are drawn at random.

a What is the probability that the slips drawn will be the lengths of two of the sides of a triangle of the (3, 4, 5) family?

b What is the probability that the slips drawn will be the lengths of a leg and hypotenuse of a triangle of the (3, 4, 5) family?

Section 9.5 Special Right Triangles

A 30°–60°–90° TRIANGLES

B 45°–45°–90° TRIANGLES

9.5 (A) 30°–60°–90° Triangles

Theorem 63

T **In a triangle whose angles have the measures 30, 60, and 90, the lengths of the sides opposite these angles are respectively proportional to x, $x\sqrt{3}$, and 2x. (30°–60°–90° TRIANGLE THEOREM)**

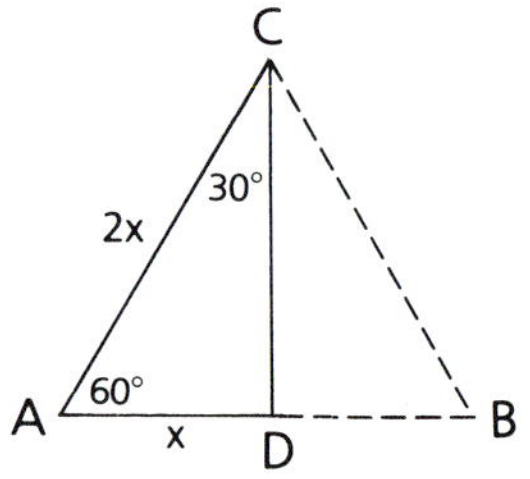

Given: $\triangle ABC$ is equilateral

$\overrightarrow{CD}$ bisects $\angle ACB$

Prove: $AD:DC:AC = x:x\sqrt{3}:2x$

Proof of Theorem 63: Since $\triangle ABC$ is equilateral, $\angle ACD = 30°$, $\angle A = 60°$, $\angle ADC = 90°$, and $AD = \frac{1}{2}AC$.

By the Pythagorean Theorem in $\triangle ADC$,

$$\begin{aligned} x^2 + (DC)^2 &= (2x)^2 \\ x^2 + (DC)^2 &= 4x^2 \\ (DC)^2 &= 3x^2 \\ DC &= x\sqrt{3} \\ \text{Thus, } AD:DC:AC &= x:x\sqrt{3}:2x \end{aligned}$$

9.5 (B) 45°–45°–90° Triangles

Theorem 64

T **In a triangle whose angles have the respective measures of 45, 45, and 90, the lengths of the sides opposite these angles are respectively proportional to x, x, and $x\sqrt{2}$. (45°–45°–90° TRIANGLE THEOREM)**

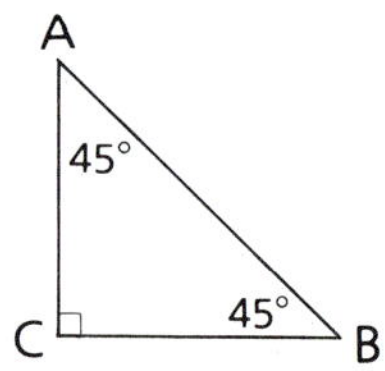

Given: $\triangle ACB$ with $\angle A = 45°$, $\angle B = 45°$

Prove: $AC:CB:AB = x:x:x\sqrt{2}$

The proof of this theorem is left to you.

You will see 30°–60°–90° and 45°–45°–90° triangles frequently here and in higher mathematics courses. Their ratios are worth memorizing now.

Six Common Families of Right Triangles

- $30^\circ-60^\circ-90^\circ \Longleftrightarrow (x, x\sqrt{3}, 2x)$
- $(3, 4, 5)$
- $(7, 24, 25)$
- $45^\circ-45^\circ-90^\circ \Longleftrightarrow (x, x, x\sqrt{2})$
- $(5, 12, 13)$
- $(8, 15, 17)$

Section 9.5 Sample Problems

Problems 1 and 2 are examples of $30^\circ-60^\circ-90^\circ$ triangles. In each, start by placing x on the side opposite (across from) the 30° angle, $x\sqrt{3}$ on the side opposite the 60° angle, and 2x on the hypotenuse.

1 *Type: Hypotenuse Known (2x)*
Find BC and AC.

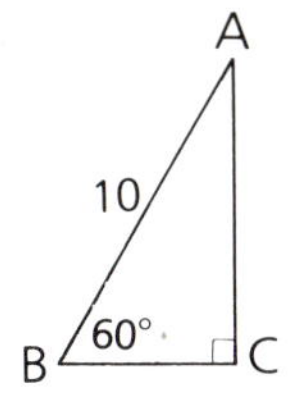

Solution:
Place x, $x\sqrt{3}$, and 2x on a copy of the diagram.

$$\text{For AB, } 2x = 10$$
$$x = 5$$
$$\text{Hence, BC} = 5$$
$$\text{AC} = 5\sqrt{3}$$

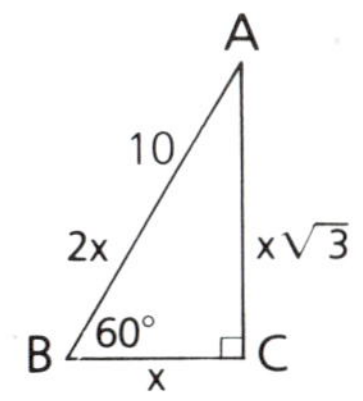

2 *Type: Longer Leg Known ($x\sqrt{3}$)*
Find JK and HK.

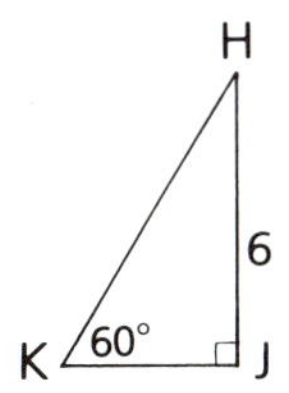

Solution:
Place x, $x\sqrt{3}$, and 2x on the figure as shown.

$$\text{For HJ, } x\sqrt{3} = 6$$
$$x = \frac{6}{\sqrt{3}} = \frac{6}{\sqrt{3}} \cdot \frac{\sqrt{3}}{\sqrt{3}}$$
$$x = \frac{6\sqrt{3}}{3} = 2\sqrt{3}$$
$$\text{Hence, JK} = 2\sqrt{3}$$
$$\text{HK} = 2(2\sqrt{3}) = 4\sqrt{3}$$

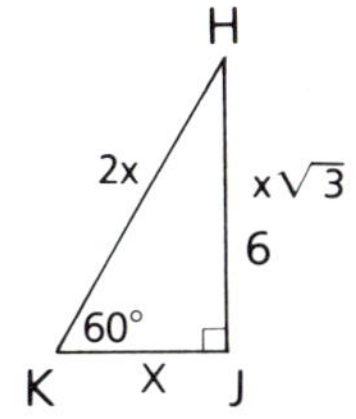

Problems 3 and 4 are examples of 45°–45°–90° triangles. In each, start by placing x on each leg and $x\sqrt{2}$ on the hypotenuse.

3 *Type: Leg Known (x)*

MOPR is a square.

Find MP.

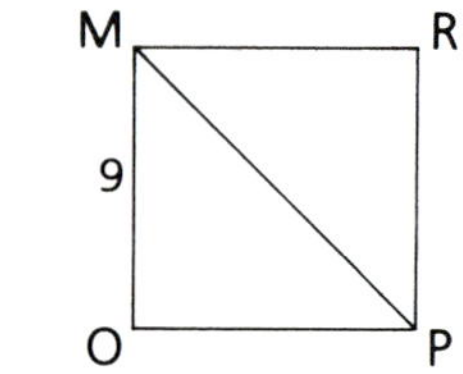

Solution: A diagonal divides a square into two 45°–45°–90° triangles. Place x, x, and $x\sqrt{2}$ as shown.

For MO, $x = 9$

Hence, $MP = 9\sqrt{2}$

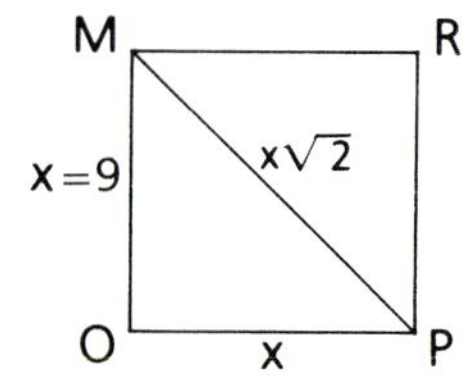

4 *Type: Hypotenuse Known ($x\sqrt{2}$)*

Find ST and TV.

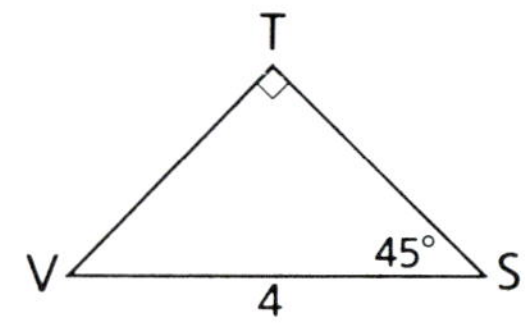

Solution: Place x, x, and $x\sqrt{2}$ as shown below.

For SV, $x\sqrt{2} = 4$

$$x = \frac{4}{\sqrt{2}} \cdot \frac{\sqrt{2}}{\sqrt{2}}$$

$$x = \frac{4\sqrt{2}}{2} = 2\sqrt{2}$$

Hence, $ST = TV = 2\sqrt{2}$

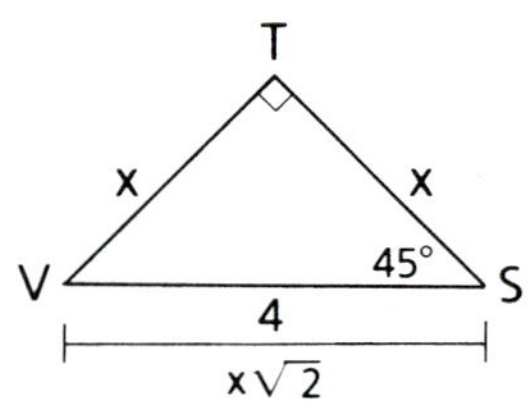

Section 9.5 Problem Set A

1 Find the two missing sides in each 30°–60°–90° triangle. Try to do them in your head.

a

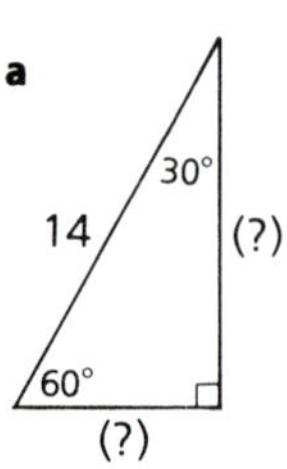

b

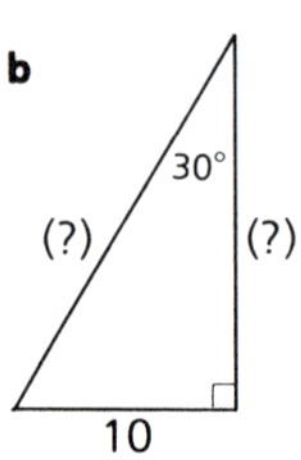

c

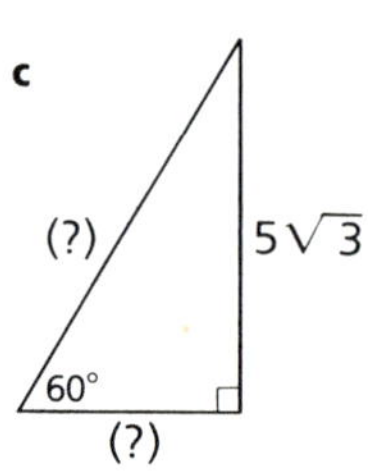

d

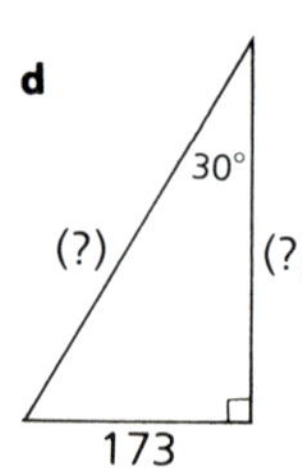

e

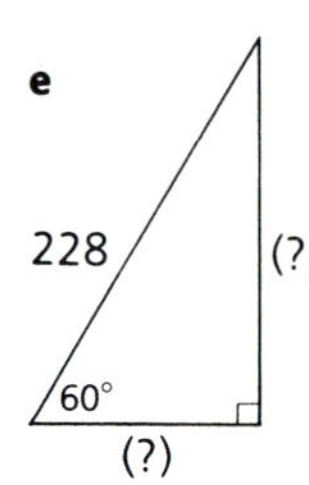

2 The following are a bit harder and you may want to put x, $x\sqrt{3}$, and 2x on the proper sides as shown in the Sample Problems. Find the two missing sides.

a

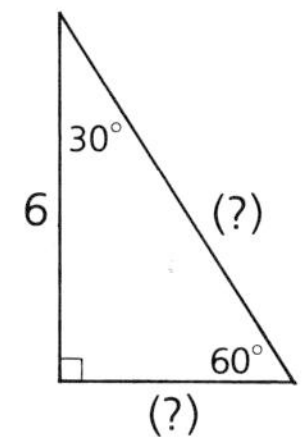

b

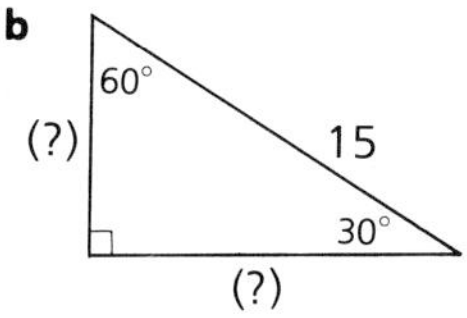

c

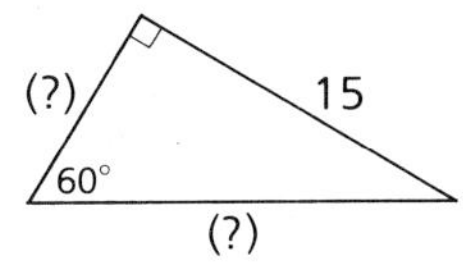

d

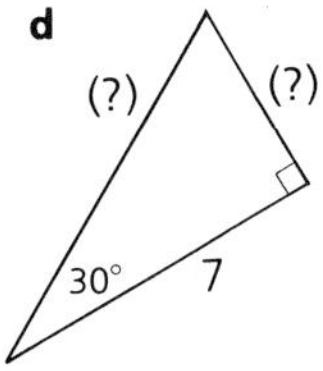

3 Solve for the variable in each of these equilateral triangles:

a

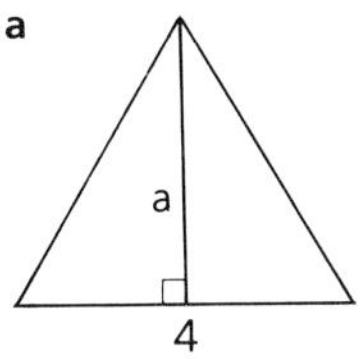

b

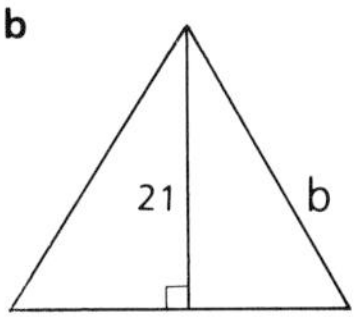

c

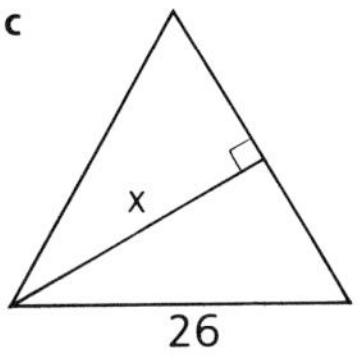

4 Solve for the variable in each of these 45°–45°–90° triangles:

a

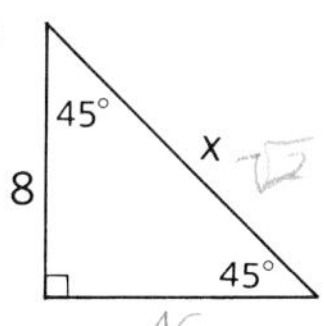

b

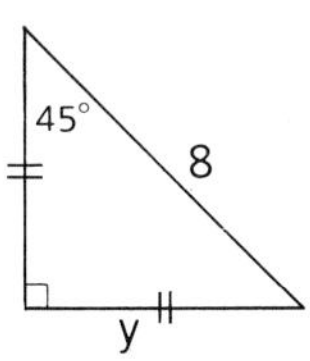

c

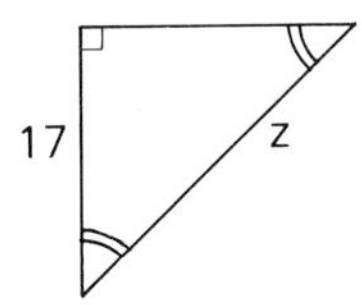

d

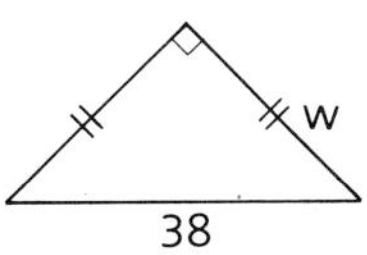

5 The perimeter of a square is 44. Find the length of a diagonal.

6 Find the altitude of an equilateral triangle if a side is 6 mm long.

7 Find the length of the diagonal of the rectangle.

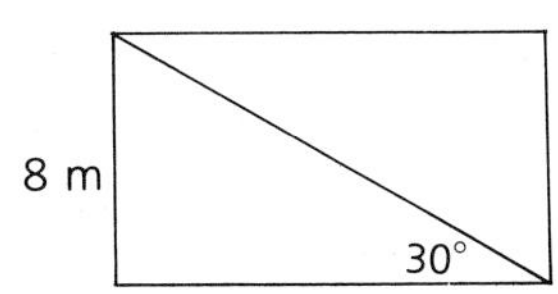

8 Given: $\overline{AC} \perp \overline{BC}$, $\overline{CD} \perp \overline{AB}$
$\angle B = 30°$, $BC = 8\sqrt{3}$

Find: CD

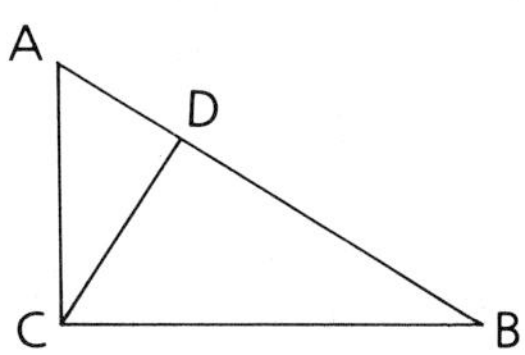

9 TRWX is a kite, $\overline{TR} \cong \overline{WR}$ and $\overline{TX} \cong \overline{XW}$
RY = 5, TW = 10, YX = 12.
Find: a TR
b WX

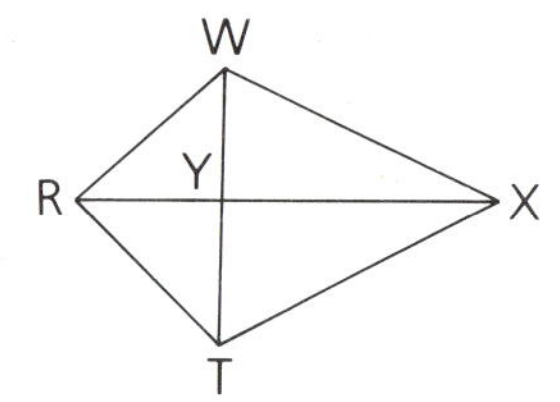

10 a In a 30°–60°–90° triangle, find the ratio of the longer leg to the hypotenuse.

b In a 45°–45°–90° triangle, find the ratio of one of the legs to the hypotenuse.

11 Plato is alleged to have said the 30°–60°–90° triangle was the most beautiful right triangle in the world. Grunts Giraffe, sophomore student at Animal High, is alleged to have said the 30°–60°–90° triangle didn't look very pretty to him. Who was Plato, and what do you think he meant by *beautiful?*

Section 9.5 Problem Set B

12 Show that in a 30°–60°–90° triangle the altitude on the hypotenuse divides the hypotenuse in the ratio of 1:3.
Hint: Let DB = x.
Then CD = $x\sqrt{3}$.
Now solve for AD.

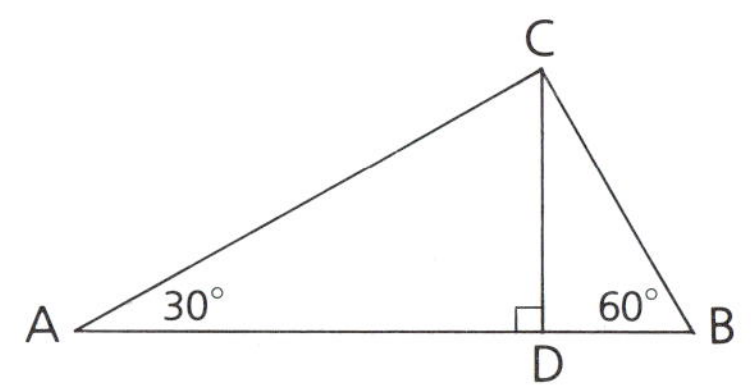

13 Find the perimeter of the isosceles trapezoid EFGH.
Hint: Drop altitudes of the trapezoid from E and H.

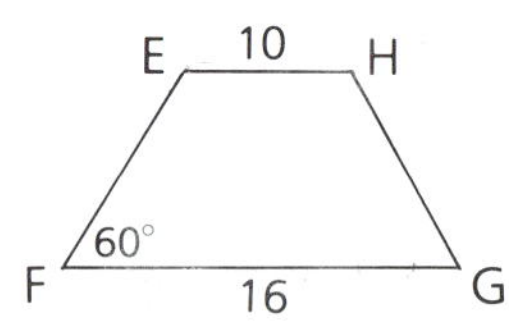

14 $\overline{PK}$ is an altitude of isosceles trapezoid JMOP.
PK = 6, PO = 8, ∠J = 45°.
Find the perimeter of JMOP.

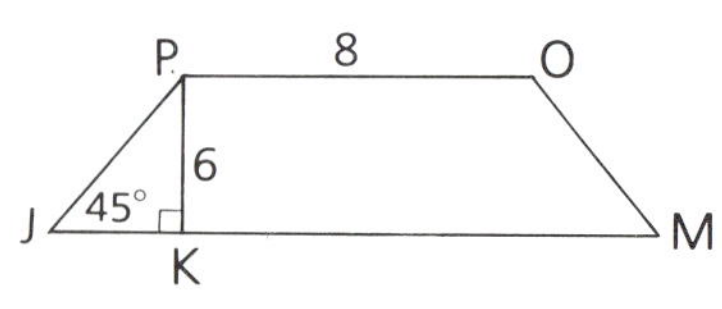

15 Using the figure, find:

a VS b ST c VT

d the ratio of the perimeter of $\triangle VSR$ to the perimeter of $\triangle VRT$.

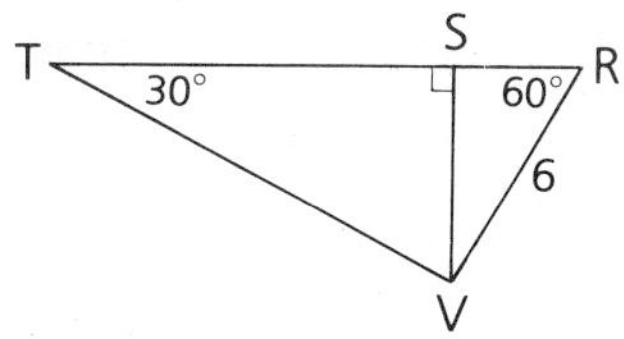

16 One of the angles of a rhombus has a measure of 120. If the perimeter of the rhombus is 24, find the length of each diagonal.

17 The perimeters of two 30°–60°–90° triangles are in the ratio of 1:2. If the hypotenuse of the larger triangle is 20 centimeters, find the length of the longer leg of the smaller triangle.

18 Find the perimeter of the trapezoid.

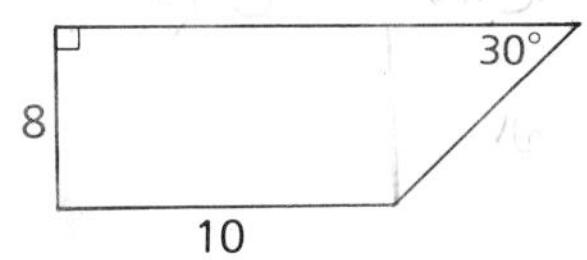

19 Any regular hexagon can be divided into 6 equilateral triangles by drawing in the three diagonals shown. Find the span of a regular hexagon with sides 12 dm long.

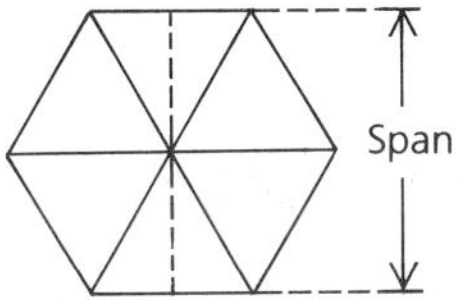

20 Any regular octagon can be divided into rectangles and right triangles. Here, a side of the central square is 6 units long.

a Find the perimeter of the octagon.

b Find the span of the octagon.

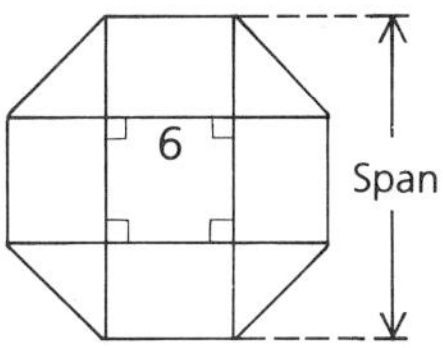

21 Find the altitude to the base of the isosceles triangle shown.

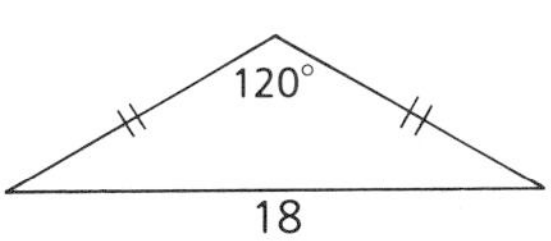

22 Find x, y, and z.

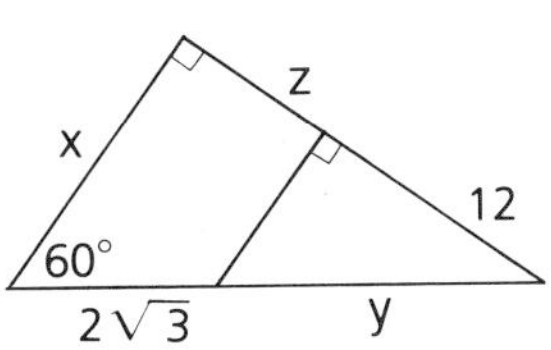

Section 9.5 Problem Set C

23 Find x and y.

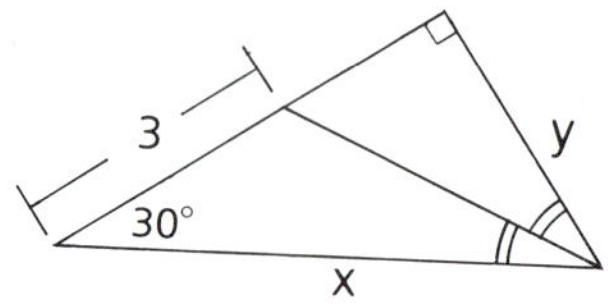

24 ABCD is a trapezoid, $\overline{DC} \parallel \overline{AB}$
$AB = AD = 4$
$\angle A = 60°$, $\angle C = 45°$
Find: **a** DC
b BC

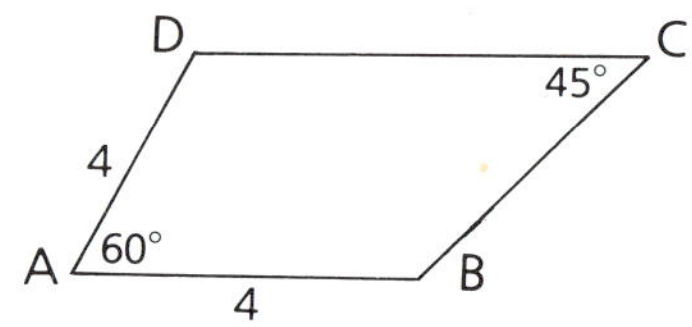

Section 9.5 Problem Set D

25 Given: $\angle ACB$ is a right angle
$\overrightarrow{CD}$ and $\overrightarrow{CE}$ trisect $\angle ACB$
$AC = 5$, $BC = 12$
Find: CE
Hint: Draw a perpendicular from E to $\overline{CB}$.

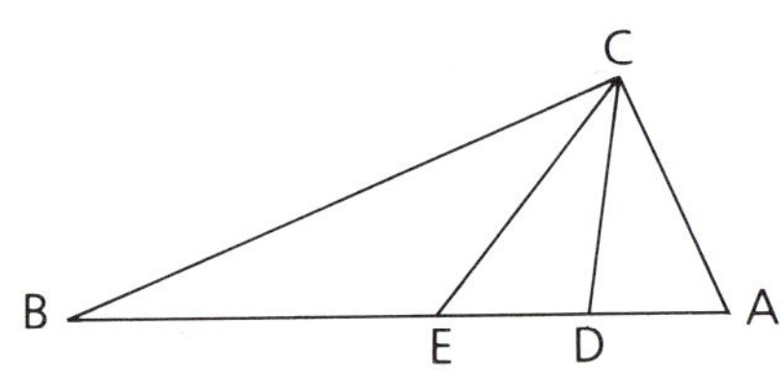

26 A Sample Problem: In solving probability problems, a "tree diagram" is sometimes helpful. Consider the following problem.

A bag contains 7 red marbles, 2 blue marbles, and a white marble. A lady reaches into the bag and draws two marbles.

a What is the probability that she has drawn two red marbles?
b What is the probability that she has drawn one or more red marbles?

Solutions:

a The tree diagram on the next page shows that the probability of drawing Red and then another Red is
$\frac{7}{10} \cdot \frac{2}{3} = \frac{7}{15}$.

$$\text{So RR} = \frac{7}{15}.$$

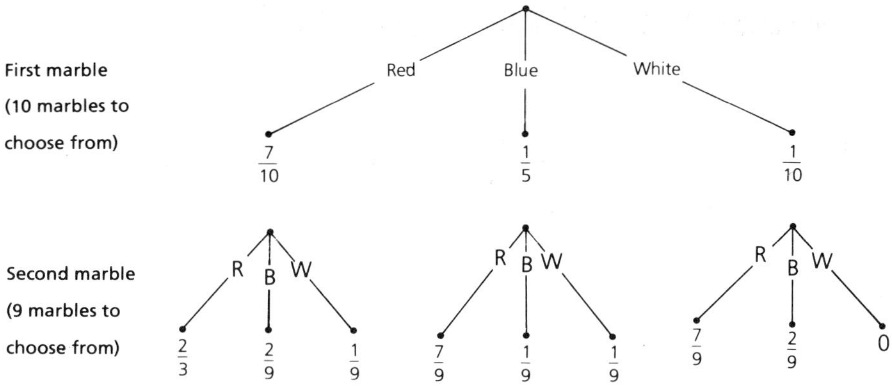

Similarly,

$$WB = \frac{1}{10} \cdot \frac{2}{9} = \frac{1}{45} \qquad RB = \frac{7}{10} \cdot \frac{2}{9} = \frac{7}{45} \qquad BW = \frac{1}{5} \cdot \frac{1}{9} = \frac{1}{45}$$

$$WW = \frac{1}{10} \cdot 0 = 0 \qquad RW = \frac{7}{10} \cdot \frac{1}{9} = \frac{7}{90} \qquad WR = \frac{1}{10} \cdot \frac{7}{9} = \frac{7}{90}$$

b Then by adding the probabilities of RR, RB, RW, BR, and WR, the probability of drawing one or more Red marbles is $\frac{14}{15}$.

27 As in problem 26, use a tree diagram to solve this one.

A bag contains 8 right triangles. Five are members of the (3, 4, 5) family, and two are 30°–60°–90° triangles. A puppy falls over the bag, and two triangles fall out on the floor.

a What is the probability that both are members of the (3, 4, 5) family?

b What is the probability that at least one of the triangles is a member of the (3, 4, 5) family?

c What is the probability that one is a member of the (3, 4, 5) family and the other is a 30°–60°–90° triangle?

28 The numbers 4, 5, 6, 8, 10, 12, 13, $4\sqrt{3}$ are placed in a bag, and two are drawn at random. What is the probability that the two numbers drawn are the lengths of two of the sides of:

a a triangle of the (3, 4, 5) family?

b a triangle of the (5, 12, 13) family?

c a 30°–60°–90° triangle?

d one of the families in a, b, or c?

Section 9.6 The Pythagorean Theorem in Space Figures

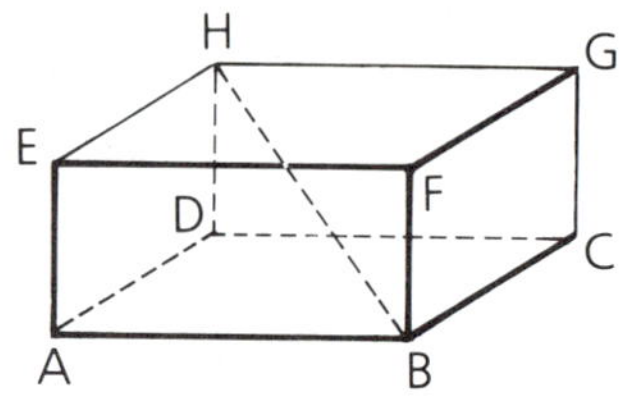

RECTANGULAR SOLID

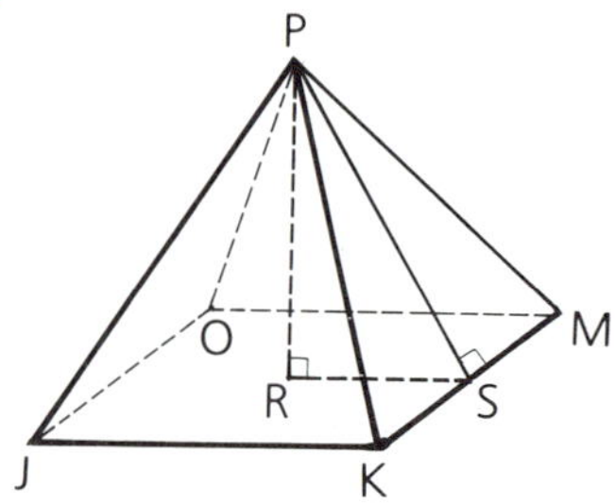

REGULAR SQUARE PYRAMID

Many of the problems in Section 9.6 will involve one of the two figures above.

In the **RECTANGULAR SOLID:**

ABFE is one of its 6 rectangular *faces.*

$\overline{AB}$ is one of its 12 *edges.*

$\overline{HB}$ is one of the four *diagonals of the solid.* The others are $\overline{AG}$, $\overline{CE}$, and $\overline{DF}$.

In the **REGULAR SQUARE PYRAMID:**

JKMO is a *square,* and it is called the *base.* P is the *vertex.*

$\overline{PR}$ is the *altitude* of the pyramid and is perpendicular to the base at its center.

$\overline{PS}$ is called a *slant height,* and is perpendicular to a side of the base.

(A **CUBE** is a rectangular solid with all edges congruent.)

Section 9.6 Sample Problems

1 The dimensions of a rectangular solid are 3, 5, and 7. Find the diagonal.

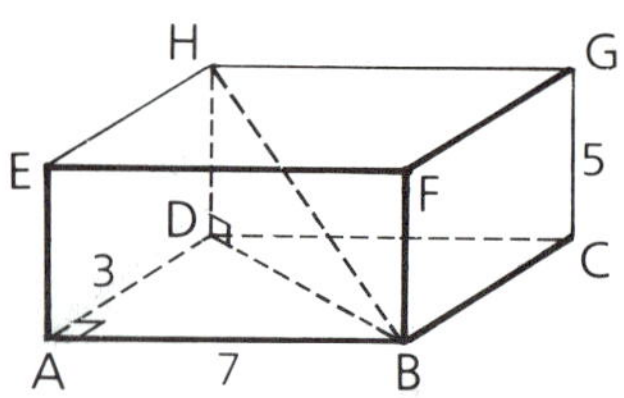

Solution: It does not matter which edges are given the lengths 3, 5, or 7.
Let AD = 3, AB = 7, and HD = 5.

Use the Pythagorean Theorem twice:

In $\triangle$ABD,

$$3^2 + 7^2 = (DB)^2$$
$$9 + 49 = (DB)^2$$
$$\sqrt{58} = DB$$

In $\triangle$HDB,

$$5^2 + (\sqrt{58})^2 = (HB)^2$$
$$25 + 58 = (HB)^2$$
$$\sqrt{83} = HB \quad \text{Ans.}$$

2 Given: Regular square pyramid with altitude $\overline{PR}$, slant height $\overline{PS}$, Perimeter of JKMO = 40, PK = 13

Find: **a** JK **b** PS **c** PR

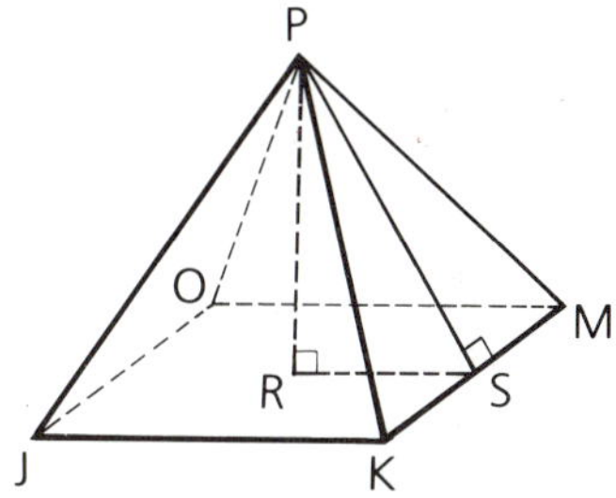

Solutions:

a $JK = \frac{1}{4}(40) = 10$

b The slant height of a regular pyramid is the $\perp$ bis of $\overline{MK}$.

In right $\triangle PSK$, $(SK)^2 + (PS)^2 = (PK)^2$

$$5^2 + (PS)^2 = 13^2$$

$$PS = 12$$

c The altitude of a regular pyramid is perpendicular to the base at its center. Thus, $RS = \frac{1}{2}JK = 5$.

In right $\triangle PRS$, $(RS)^2 + (PR)^2 = (PS)^2$

$$5^2 + (PR)^2 = (12)^2$$

$$25 + (PR)^2 = 144$$

$$PR = \sqrt{119}$$

Section 9.6 Problem Set A

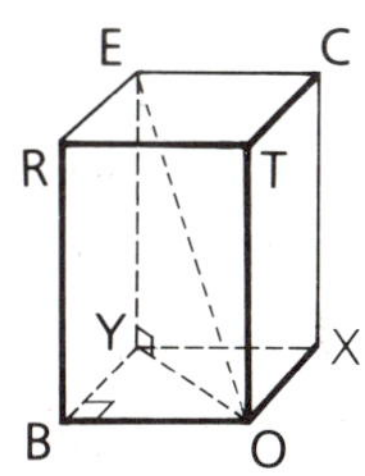

1 Given: A rectangular box with edges BY = 3, OB = 4, and EY = 12

Find: **a** YO, a diagonal of face BOXY
b EO, a diagonal of the box

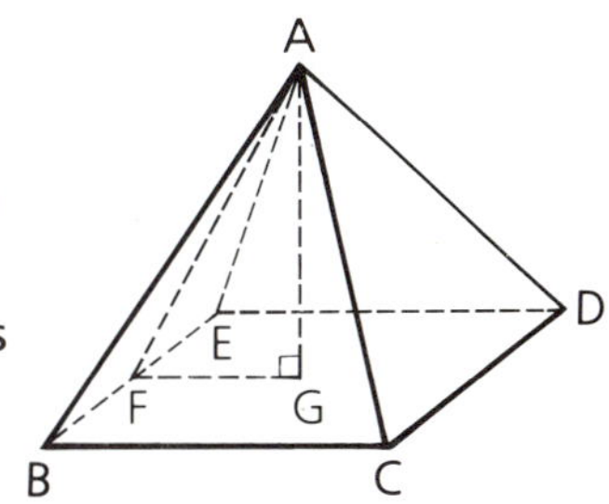

2 Given: Regular square pyramid ABCDE with slant height $\overline{AF}$, altitude $\overline{AG}$, and base BCDE. The perimeter of BCDE is 40 and $\angle AFG = 60°$.

Find: The altitude and the slant height

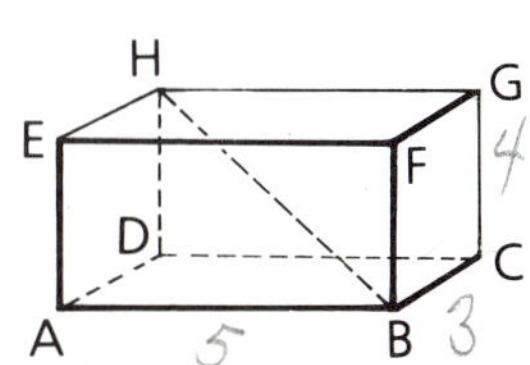

3 Given: A rectangular solid with edges GC = 8, HG = 12, and BC = 9

Find: **a** HB, a diagonal of the solid

b AG, another diagonal of the solid

4 Find the diagonal of a rectangular solid whose dimensions are 3, 4, and 5.

5 Given: A regular square pyramid with altitude $\overline{PY}$ and slant height $\overline{PR}$

ID = 14, PY = 24

Find:

a AD

b YR

c PR

d the perimeter of base AMID

e a *diagonal of the base* (not shown in the diagram)

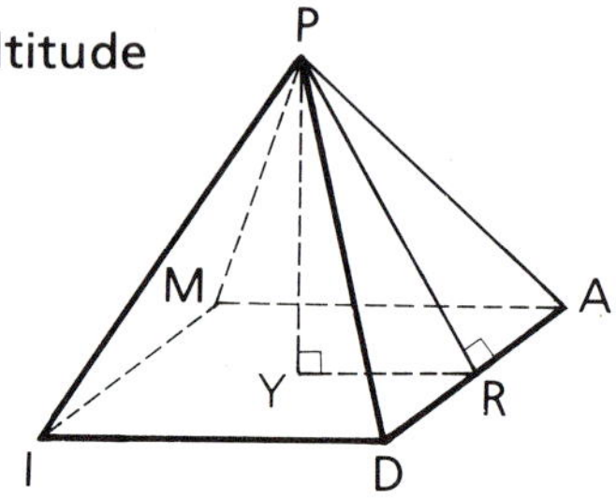

6 Find the slant height of a regular square pyramid if the altitude is 12, and one of the sides of the square base is 10.

Section 9.6 Problem Set B

7 ABCDEFGH is a rectangular solid.

a If the face diagonal CH = 17, edge GH = 8, and edge FG = 6, how long is solid diagonal $\overline{AG}$?

b If solid diagonal AG = 50, edge AE = 40, and edge EF = 3, how long is edge $\overline{FG}$?

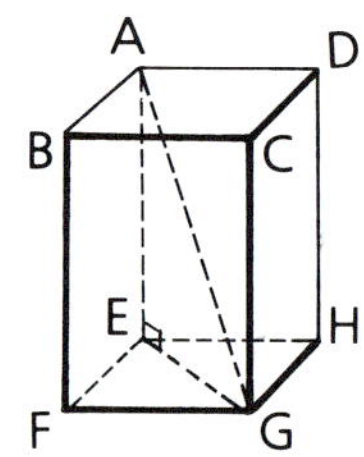

8 PADIM is a regular square pyramid with slant height PR = 10, and base diagonals $12\sqrt{2}$ long.

a Find ID.

b Find the altitude of the pyramid.

c Find RD.

d Find PD (length of a lateral edge).

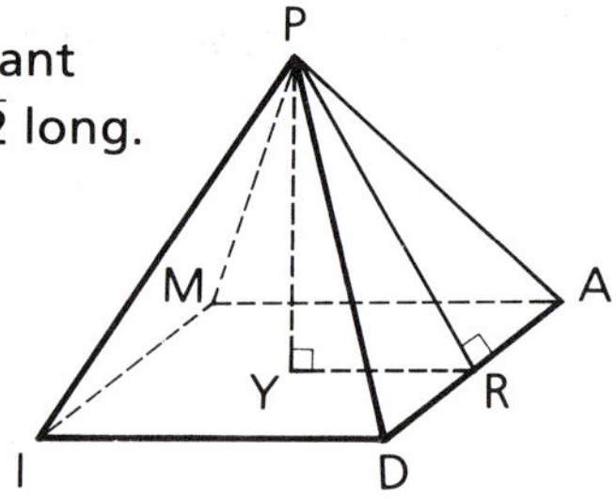

9 Find the diagonal of a cube if each edge is 2.

10 Find the diagonal of a cube if the perimeter of a face is 20.

11 Find the edge of a cube whose diagonal is $7\sqrt{3}$.

12 The perimeter of the base of a regular square pyramid is 24. If the slant height is 5, find the altitude.

Section 9.6 Problem Set C

13 In the cube, find the measure of the diagonal in terms of x if:

a $AB = x$ b $AC = x$

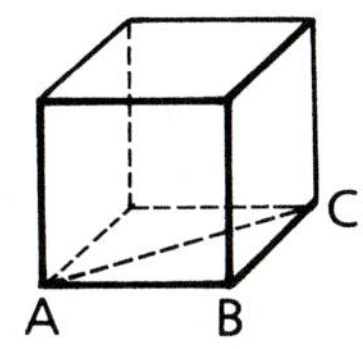

14 The three dimensions of a rectangular solid are a, b, and c. Find a formula for the length of a diagonal of a rectangular solid.

15 The dimensions of a rectangular solid are in the ratio 3:4:5. If the diagonal is $200\sqrt{2}$, find the three dimensions.

16 The face diagonals of a rectangular box are 2, 3, and 6. Find the diagonal of the box.

17 A pyramid is formed by assembling a square 6 cm to a side and four equilateral triangles. Find the altitude and the slant height.

Section 9.6 Problem Set D

18 The strongest rectangular beam to be cut from a circular log is one in which the cross-section is such that perpendicular segments from opposite vertices to the diagonal joining the other two vertices trisect the diagonal. If $AB = a$, $BC = b$, $CE = x$, $DE = y$, show that $\frac{b}{a} = \frac{\sqrt{2}}{1}$

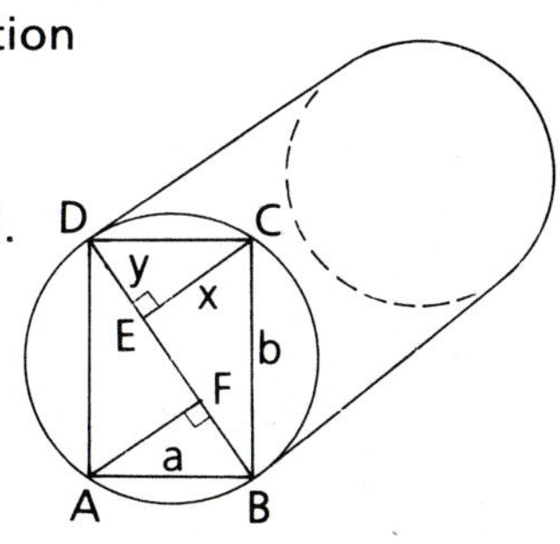

Section 9.7 Introduction to Trigonometry

This section presents the three basic trigonometric ratios: SINE, COSINE, and TANGENT. The ideas of Similar Triangles and the Pythagorean Theorem can be used to develop the **TRIGONOMETRY OF THE RIGHT TRIANGLES.**

Consider the following 30°-60°-90° triangles:

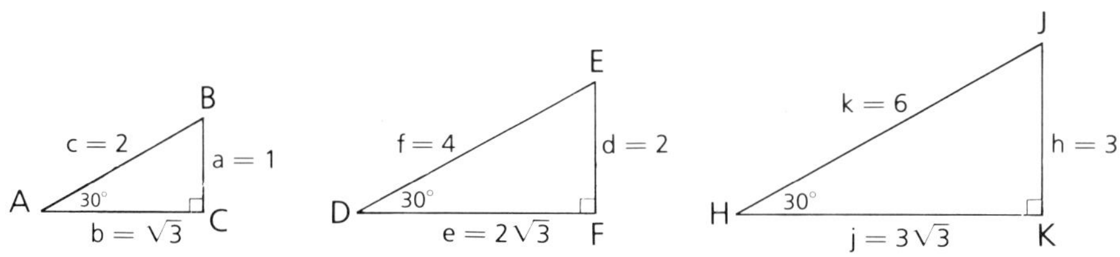

Compare the length of the leg opposite the 30° angle to the length of the hypotenuse in each triangle.

In △ABC, $\frac{a}{c} = \frac{1}{2} = .5$ In △DEF, $\frac{d}{f} = \frac{2}{4} = .5$ In △HJK, $\frac{h}{k} = \frac{3}{6} = .5$

If you think about similar triangles, you will see that in every 30°-60°-90° triangle $\frac{\text{leg opposite } 30^\circ \angle}{\text{hypotenuse}} = \frac{1}{2}$

For each triangle above, verify that $\frac{\text{leg adjacent to } 30^\circ \angle}{\text{hypotenuse}} = \frac{\sqrt{3}}{2}$

For each triangle above, find the ratio $\frac{\text{leg opposite } 30^\circ \angle}{\text{leg adjacent to } 30^\circ \angle}$

In △ABC and △DEF,

$\frac{a}{c} = \frac{d}{f} = \frac{6}{10} = \frac{3}{5}$

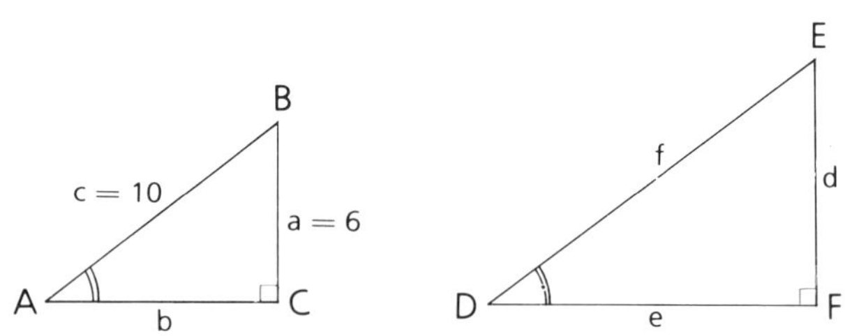

Engineers and scientists have found it convenient to formalize these relationships by naming the ratios of sides. Memorize the three basic ratios presented on the next page.

D THREE TRIGONOMETRIC RATIOS

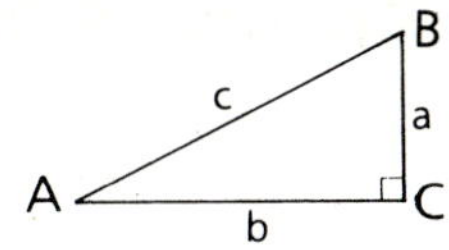

$$\text{sine } \angle A = \sin \angle A = \frac{\text{opposite leg}}{\text{hypotenuse}}$$

$$\text{cosine } \angle A = \cos \angle A = \frac{\text{adjacent leg}}{\text{hypotenuse}}$$

$$\text{tangent } \angle A = \tan \angle A = \frac{\text{opposite leg}}{\text{adjacent leg}}$$

$\sin \angle A = \frac{a}{c}$ $\quad \sin \angle B = \frac{b}{c}$

$\cos \angle A = \frac{b}{c}$ $\quad \cos \angle B = \frac{a}{c}$

$\tan \angle A = \frac{a}{b}$ $\quad \tan \angle B = \frac{b}{a}$

Section 9.7 Sample Problems

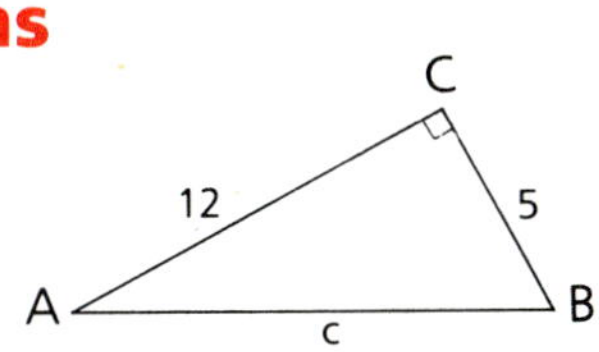

1 Find: **a** $\cos \angle A$
b $\tan \angle B$

Solution: By the Pythagorean Theorem, $c = 13$.

Thus, **a** $\cos \angle A = \dfrac{\text{adjacent leg}}{\text{hypotenuse}} = \dfrac{12}{13}$

b $\tan \angle B = \dfrac{\text{leg opposite } \angle B}{\text{leg adjacent to } \angle B} = \dfrac{12}{5}$

2 Find the three trigonometric ratios for $\angle A$ and $\angle B$

Solution: $\sin \angle A = \frac{3}{5}$ $\quad \sin \angle B = \frac{4}{5}$

$\cos \angle A = \frac{4}{5}$ $\quad \cos \angle B = \frac{3}{5}$

$\tan \angle A = \frac{3}{4}$ $\quad \tan \angle B = \frac{4}{3}$

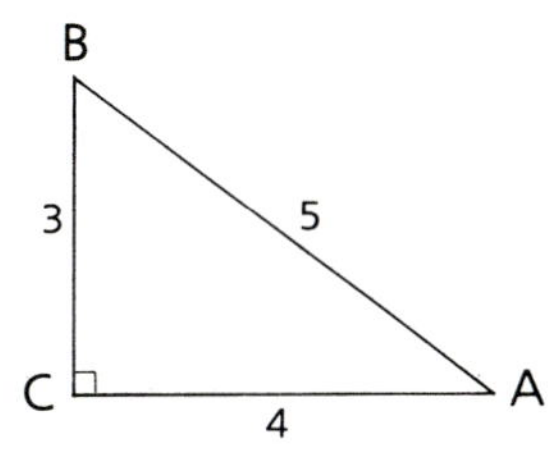

3 $\triangle ABC$ is an isosceles triangle as marked.
Find $\sin \angle C$.

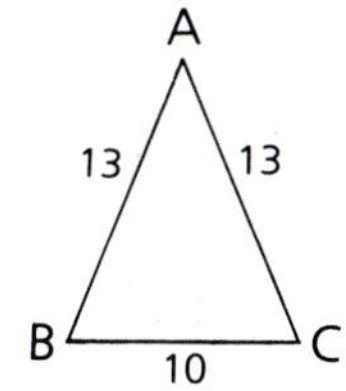

Solution: We must have a right triangle, so draw the altitude to the base.
Thus, in $\triangle ADC$ $\sin \angle C = \frac{12}{13}$.

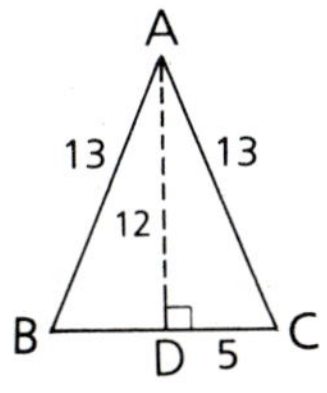

4 Use the fact that $\tan 40^\circ \approx .8391$ to find the height of the tree to the nearest foot.

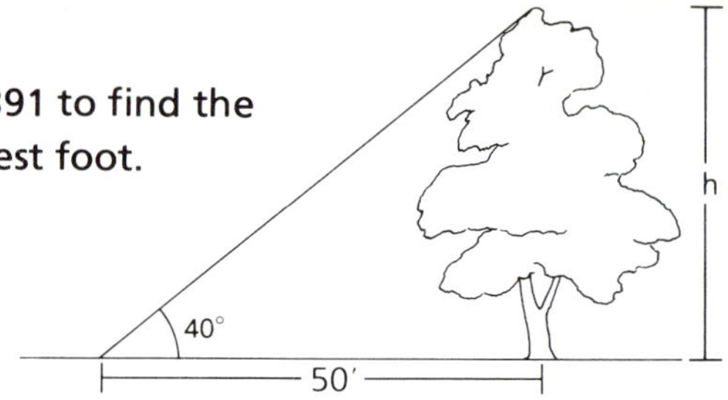

Solution:

$$\tan 40^\circ = \frac{h}{50}$$
$$.8391 \approx \frac{h}{50}$$
$$h \approx 41.955$$
$$\approx 42 \text{ ft}$$

Section 9.7 Problem Set A

1 Find: **a** $\sin \angle A$ **b** $\cos \angle A$ **c** $\tan \angle A$ **d** $\sin \angle B$ **e** $\cos \angle B$ **f** $\tan \angle B$

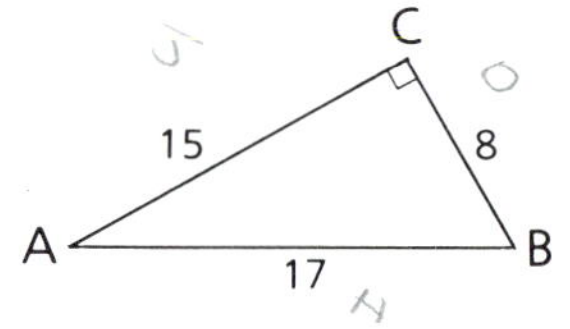

2 Find: **a** $\sin 30^\circ$ **b** $\cos 30^\circ$ **c** $\tan 30^\circ$ **d** $\sin 60^\circ$ **e** $\cos 60^\circ$ **f** $\tan 60^\circ$

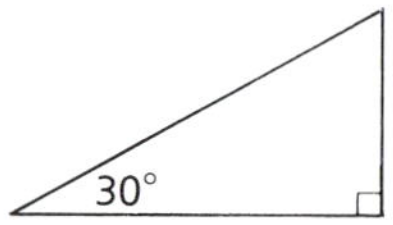

3 Find: **a** $\sin 45^\circ$ **b** $\cos 45^\circ$ **c** $\tan 45^\circ$

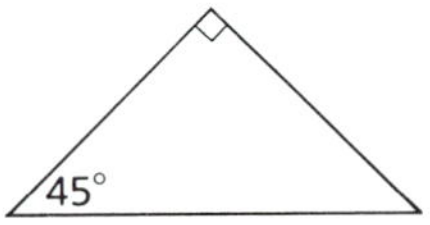

4 Find: **a** $\cos \angle H$ **b** $\tan \angle K$

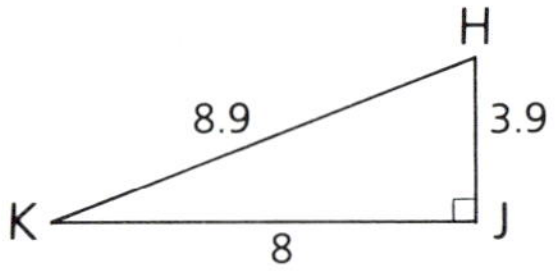

5 If $\tan \angle M = \frac{3}{4}$, find $\cos \angle M$. (Hint: Start by drawing the triangle.)

6 Using the figure as marked, find:

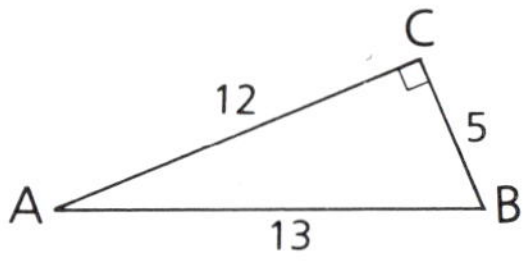

a $\frac{5}{12} = \tan \angle$ __?__ **b** $\frac{12}{13} = \cos \angle$ __?__ **c** $\frac{5}{13} = \sin \angle$ __?__

7 Find: **a** BC **b** $\sin \angle A$ **c** $\tan \angle B$

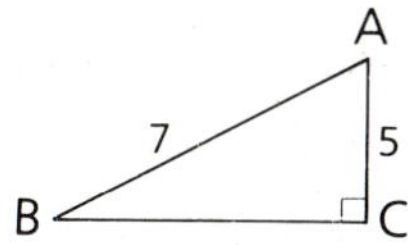

8 Given: RECT is a rectangle
ET=26, RT=24
Find: **a** $\sin \angle RET$ **b** $\cos \angle RET$

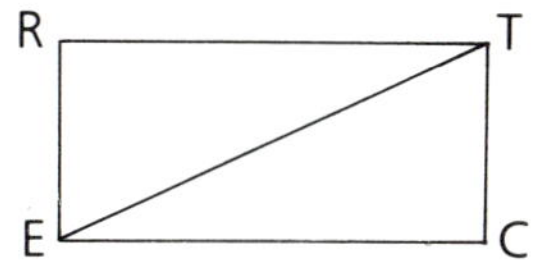

Section 9.7 Problem Set B

9 Using the given figures,
find: **a** $\cos \angle A$
b $\sin \angle E$
c $\sin \angle DFG$

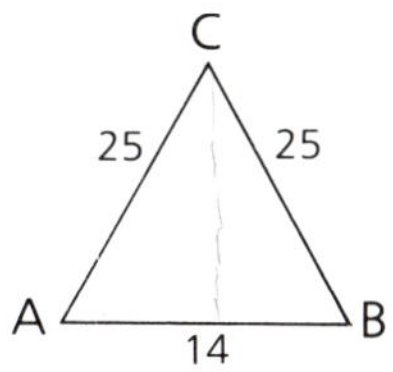

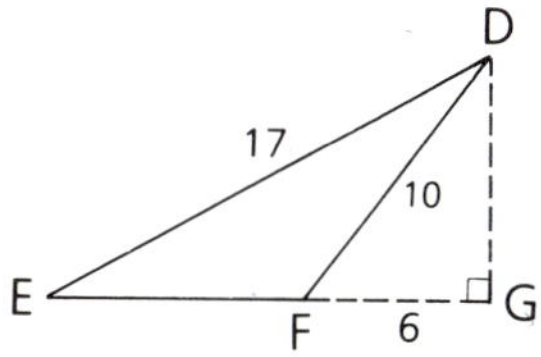

10 Use the fact that $\sin 40° \approx .6428$ to find the height of the kite to the nearest meter.

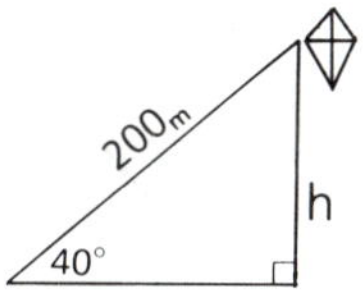

11 **a** If $\tan \angle A = 1$, find $m\angle A$.
b If $\sin \angle P = .5$, find $m\angle P$.

12 Given: $\sin \angle P = \frac{3}{5}$ and PQ = 10
Find: $\cos \angle P$

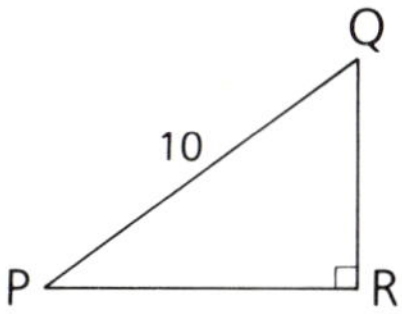

13 Using the figure,
find: **a** $\tan \angle ACD$
b $\sin \angle A$

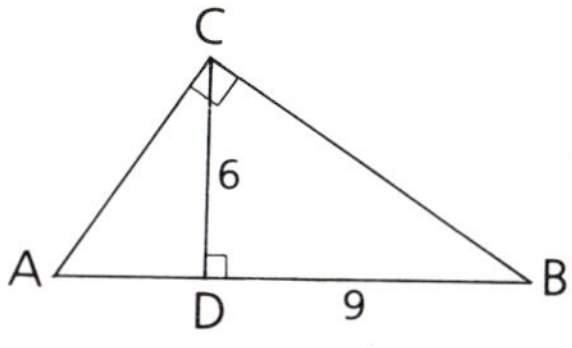

14 Given: RHOM is a rhombus
RO = 18, HM = 24
Find: **a** $\cos \angle BRM$ **b** $\tan \angle BHO$

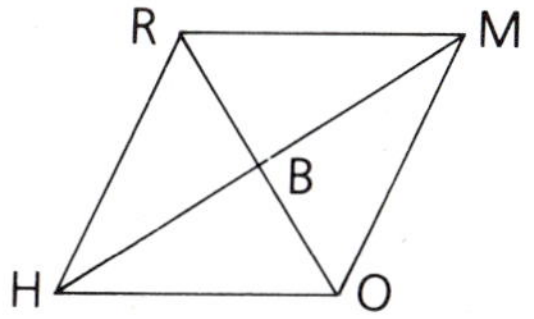

15 Given a trapezoid with sides 5, 10, 17 and 10, find the sine of one of the acute angles.

16 Answer a–c Always, Sometimes, or Never (A, S, or N)
Given $\triangle ABC$ with $\angle C = 90°$:

a $\sin \angle A = \cos \angle B$ b $\sin \angle A = \tan \angle A$ c $\sin \angle A = \cos \angle A$

17 If $\triangle EQU$ is equilateral and $\triangle RAT$ is a right triangle with $RA = 2$, $RT = 1$, and $\angle T = 90°$, show $\sin \angle E = \cos \angle A$.

Section 9.7 Problem Set C

18 Use the definitions of the trigonometric ratios to verify the following relationships, given $\triangle ABC$ where $\angle C = 90°$:

a $(\sin \angle A)^2 + (\cos \angle A)^2 = 1$

b $\dfrac{a}{\sin \angle A} = \dfrac{b}{\sin \angle B}$

c $\dfrac{\sin \angle A}{\cos \angle A} = \tan \angle A$

d $\sin \angle A = \cos (90° - \angle A)$

19 Rhombus PQRS has a perimeter of 60 and one diagonal of 15. Find the *two* answers for $\sin \angle PQS$.

20 In $\triangle ABC$ as marked, two sides are picked at random to form a ratio. What is the probability the ratio is the tangent of $\angle A$?

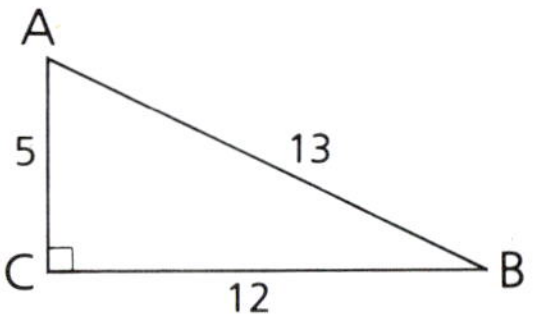

21 If $\triangle PQR$ is a right triangle, what is the probability $\tan \angle R$ is *not* a trigonometric ratio?

22 Given: KITE is a kite with segments as marked.
Find: $\tan \angle KEI$

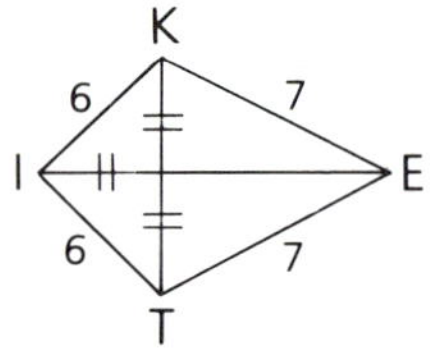

Section 9.8 Trigonometric Ratios

Trigonometry is used to solve triangles other than the $30° - 60° - 90°$ and the $45° - 45° - 90°$. The TABLE OF TRIGONOMETRIC RATIOS shows four place decimal approximations of the ratios for other angles: for instance, $\sin 23° \approx .3907$, and the angle whose tangent is 1.5399 is approximately 57°.

m ∠A	sin ∠A	cos ∠A	tan ∠A
1°	.0175	.9998	.0175
2°	.0349	.9994	.0349
3°	.0523	.9986	.0524
4°	.0698	.9976	.0699
5°	.0872	.9962	.0875
6°	.1045	.9945	.1051
7°	.1219	.9925	.1228
8°	.1392	.9903	.1405
9°	.1564	.9877	.1584
10°	.1736	.9848	.1763
11°	.1908	.9816	.1944
12°	.2079	.9781	.2126
13°	.2250	.9744	.2309
14°	.2419	.9703	.2493
15°	.2588	.9659	.2679
16°	.2756	.9613	.2867
17°	.2924	.9563	.3057
18°	.3090	.9511	.3249
19°	.3256	.9455	.3443
20°	.3420	.9397	.3640
21°	.3584	.9336	.3839
22°	.3746	.9272	.4040
23°	.3907	.9205	.4245
24°	.4067	.9135	.4452
25°	.4226	.9063	.4663
26°	.4384	.8988	.4877
27°	.4540	.8910	.5095
28°	.4695	.8829	.5317
29°	.4848	.8746	.5543
30°	.5000	.8660	.5774
31°	.5150	.8572	.6009
32°	.5299	.8480	.6249
33°	.5446	.8387	.6494
34°	.5592	.8290	.6745
35°	.5736	.8192	.7002
36°	.5878	.8090	.7265
37°	.6018	.7986	.7536
38°	.6157	.7880	.7813
39°	.6293	.7771	.8098
40°	.6428	.7660	.8391
41°	.6561	.7547	.8693
42°	.6691	.7431	.9004
43°	.6820	.7314	.9325
44°	.6947	.7193	.9657
45°	.7071	.7071	1.0000
46°	.7193	.6947	1.0355
47°	.7314	.6820	1.0724
48°	.7431	.6691	1.1106
49°	.7547	.6561	1.1504
50°	.7660	.6428	1.1918
51°	.7771	.6293	1.2349
52°	.7880	.6157	1.2799
53°	.7986	.6018	1.3270
54°	.8090	.5878	1.3764
55°	.8192	.5736	1.4281
56°	.8290	.5592	1.4826
57°	.8387	.5446	1.5399
58°	.8480	.5299	1.6003
59°	.8572	.5150	1.6643
60°	.8660	.5000	1.7321
61°	.8746	.4848	1.8040
62°	.8829	.4695	1.8807
63°	.8910	.4540	1.9626
64°	.8988	.4384	2.0503
65°	.9063	.4226	2.1445
66°	.9135	.4067	2.2460
67°	.9205	.3907	2.3559
68°	.9272	.3746	2.4751
69°	.9336	.3584	2.6051
70°	.9397	.3420	2.7475
71°	.9455	.3256	2.9042
72°	.9511	.3090	3.0777
73°	.9563	.2924	3.2709
74°	.9613	.2756	3.4874
75°	.9659	.2588	3.7321
76°	.9703	.2419	4.0108
77°	.9744	.2250	4.3315
78°	.9781	.2079	4.7046
79°	.9816	.1908	5.1446
80°	.9848	.1736	5.6713
81°	.9877	.1564	6.3138
82°	.9903	.1392	7.1154
83°	.9925	.1219	8.1443
84°	.9945	.1045	9.5144
85°	.9962	.0872	11.4301
86°	.9976	.0698	14.3007
87°	.9986	.0523	19.0811
88°	.9994	.0349	28.6363
89°	.9998	.0175	57.2900

For some applications of trigonometry, you need to know the meanings of ANGLE OF ELEVATION and ANGLE OF DEPRESSION.

If an observer at a point P sights upward at an object at A, the angle the line of sight $\overleftrightarrow{PA}$ makes with the horizontal $\overleftrightarrow{PH}$ is called the ANGLE OF ELEVATION.

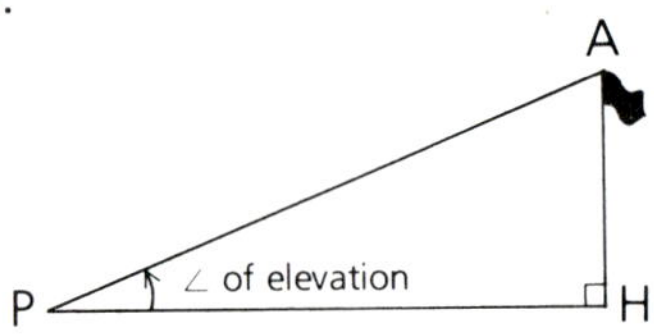

If an observer at a point P sights downward at an object at B, the angle the line of sight $\overleftrightarrow{PB}$ makes with the horizontal $\overleftrightarrow{PH}$ is called the ANGLE OF DEPRESSION.

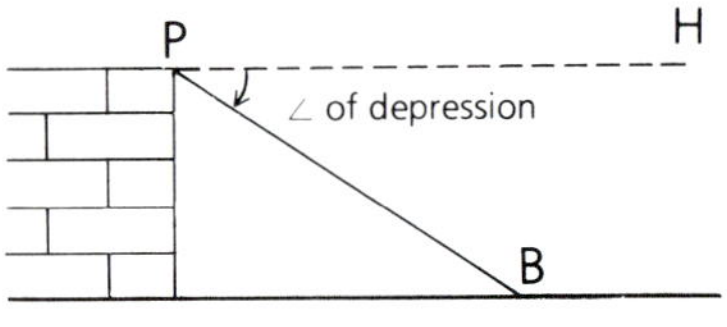

(Notice: An angle of elevation or depression is the angle between the LINE OF SIGHT and the HORIZONTAL. *Do not use the vertical.*)

Section 9.8 Sample Problems

1 Given: Right $\triangle DEF$ as shown

Find: **a** $m\angle D$ to the nearest degree

b e to the nearest tenth

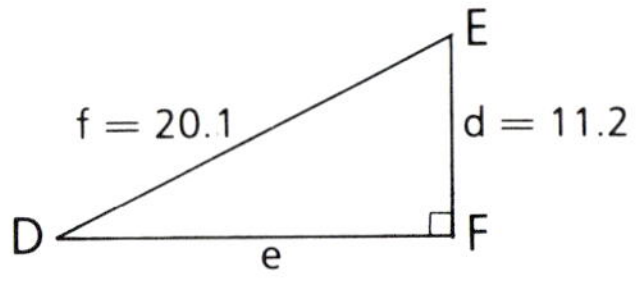

Solutions: **a** $\sin \angle D = \dfrac{11.2}{20.1}$

$\sin \angle D \approx .5572$

The number nearest to .5572 in the sin column of the Table is $\angle D \approx 34°$.

b Using the result from part *a*

$\cos 34° \approx \dfrac{e}{20.1}$

$.8290 \approx \dfrac{e}{20.1}$

$16.7 \approx e$

2 As viewed from a cliff 360 m above the level of water, the angle of depression of a ship is 28°. How far is the ship from shore?

Solution: Start by drawing a diagram.

By PAI, $\angle CSH = 28°$

Thus, $\tan 28° = \dfrac{360}{x}$

$.5317 \approx \dfrac{360}{x}$

$x \approx 677$ m

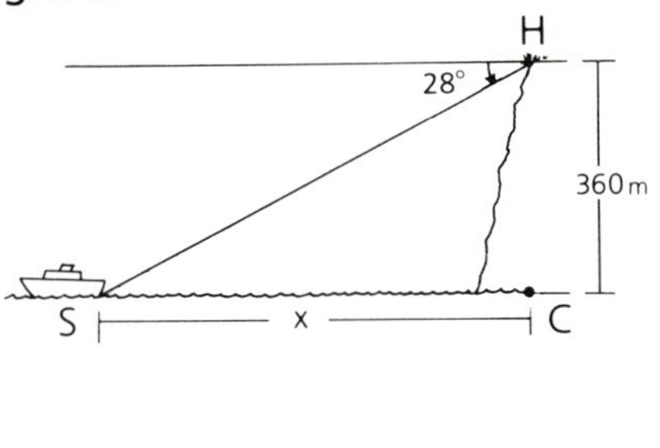

Section 9.8 Problem Set A

1 Find each of the following from the Table of Trigonometric ratios.

a $\sin 21^\circ$ **b** $\tan 52^\circ$ **c** $\cos 5^\circ$ **d** $\tan 45^\circ$ **e** $\sin 60^\circ$

2 Using the Trig Table, find $m\angle A$ in each case:

a $\sin \angle A = .4067$ **b** $\tan \angle A = 3.4874$ **c** $\cos \angle A = .7071$

3 Without using the Trig Table, find $m\angle A$ in each case:

a $\tan \angle A = 1$ **b** $\sin \angle A = \frac{1}{2}$ **c** $\sin \angle A = \frac{\sqrt{3}}{2}$

4 In each case, find x to the nearest integer:

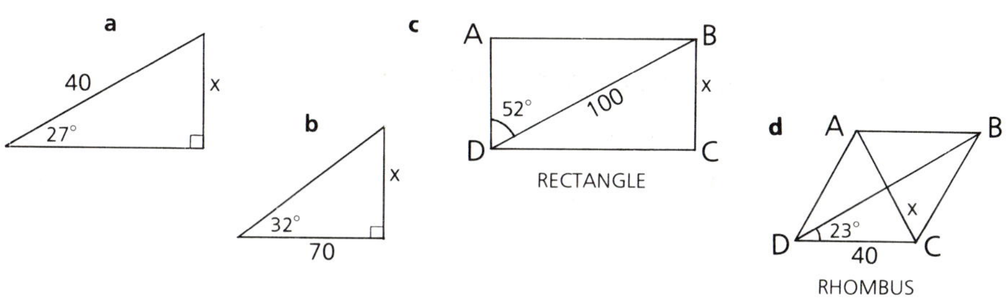

5 Find the height of isosceles trapezoid ABCD as marked.

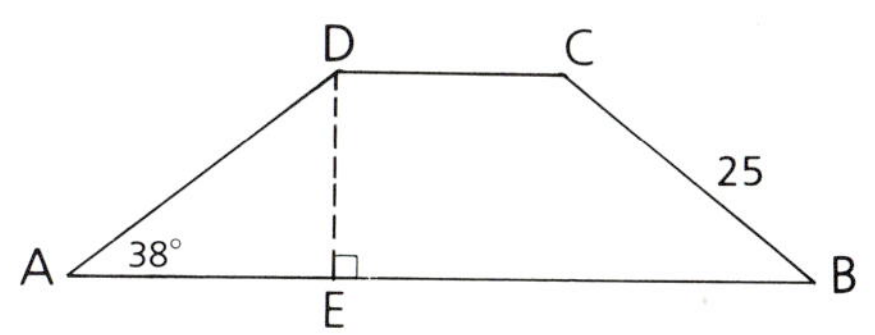

Section 9.8 Problem Set B

6 Solve each equation for x to the nearest integer.

a $\sin 25^\circ = \frac{x}{40}$ **b** $\cos 73^\circ = \frac{35}{x}$ **c** $\sin x = \frac{29}{30}$

7 A department store escalator is 80 feet long. If it rises 32 feet vertically, find the angle it makes with the floor.

8 Find the measure of the sides and angles of each right triangle.

a

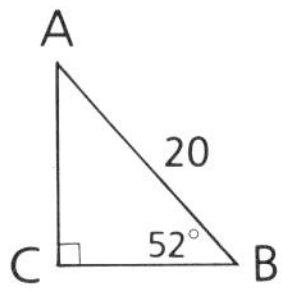

b

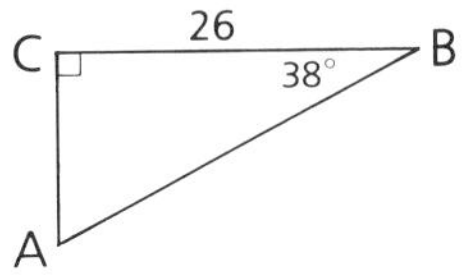

c

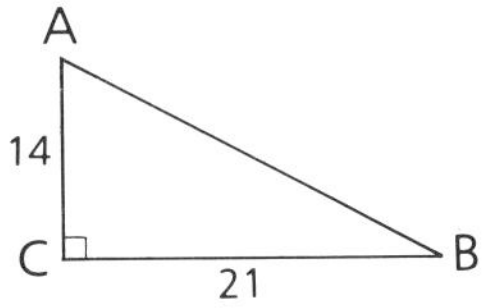

9 Find, to the nearest degree, the angles of a 3-4-5 triangle.

10 A sonar operator on a cruiser detects a submarine at a distance of 500 m and an angle of depression of 37°. How deep is the sub?

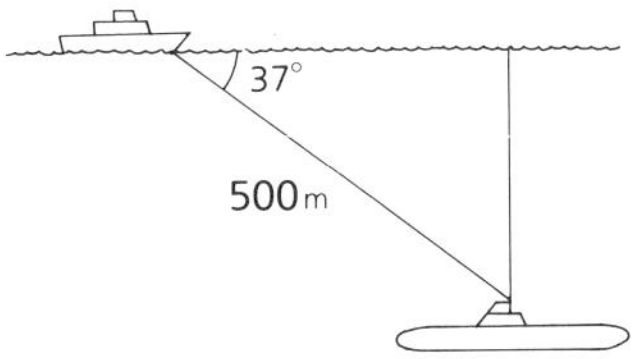

11 The legs of an isosceles triangle are each 18. The base is 14.

Find: **a** the base angles to the nearest degree
b the exact length of the altitude to the base.

12 One diagonal of a rhombus makes an angle of 27° with a side of the rhombus. If each side of the rhombus has a length of 6.2″, find the length of each diagonal to the nearest tenth of an inch.

13 Find the perimeter of trapezoid ABCD in which $\overline{CD} \parallel \overline{AB}$, $\cos \angle A = \frac{1}{2}$, and AD = DC = CB = 2.

14 Find the length of the apothem of a regular pentagon that has a perimeter of 50 cm.

15 Two buildings are 100 dm apart across a street. A sun bather at point P finds the angle of elevation of the roof of the taller building to be 25°, and the angle of depression of its base to be 30°. Find the height of the taller building, to the nearest dm.

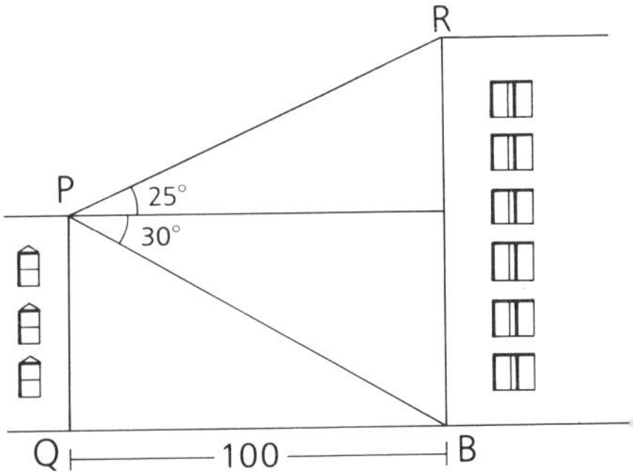

Section 9.8 Problem Set C

16 An observer on a cliff 1000 dm above sea level sights two ships due east. The angles of depression of the ships are 47° and 32°. Find, to the nearest dm, the distance between the ships.

17 Each side of the base of a regular square pyramid is 20 and the altitude is 35.

Find: a PT b BP c ∠PTF d ∠PBF

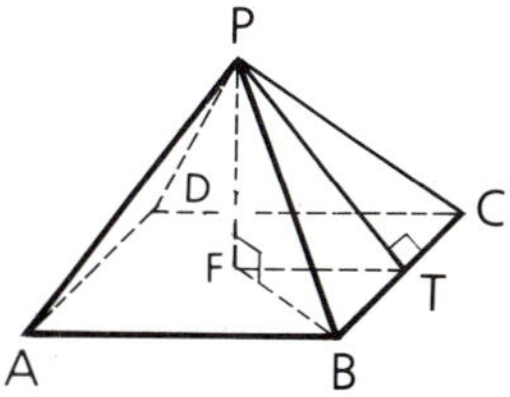

18 Find the height, PB, of a mountain, whose base and peak are inaccessible. At point A the angle of elevation of the peak is 30°; one km closer to the mountain, at point C, the angle of elevation is 35°.

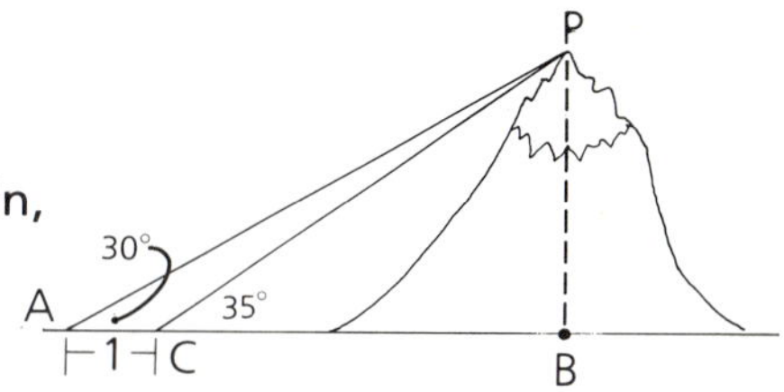

19 Find the angle formed by:

a a diagonal of a cube and a diagonal of a face of the cube.
b two face diagonals that intersect at a vertex of a cube.

Section 9.8 Problem Set D

20 Prove that $c^2 = a^2 + b^2 - 2ab(\text{Cos}\angle C)$ is true for any acute $\triangle ABC$. (The Law of Cosines)

21 Given: Diagram as shown

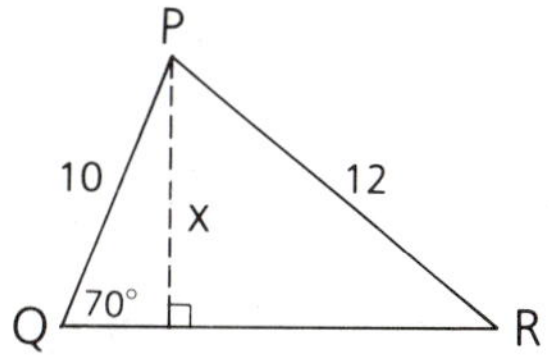

Find: a m∠R to the nearest degree
b QR to the nearest integer
c Show that $\frac{PR}{\sin\angle Q} = \frac{PQ}{\sin\angle R}$
d Generalize the result of part c for the sides and angles of any acute triangle. (The Law of Sines)

Chapter 9 Study Guide

As you study and review this chapter, be sure that you are able to:

1 Complete these expressions (p. 354) and state what theorem applies. (p. 353)

$a^2 = (\ \ )(\ \ )$
$b^2 = (\ \ )(\ \ )$
$h^2 = (\ \ )(\ \ )$

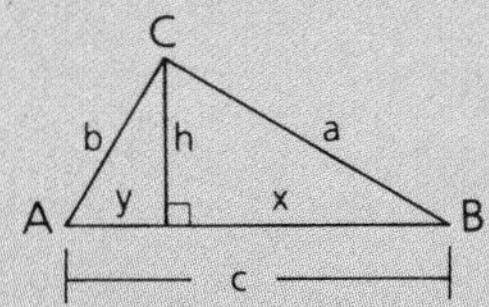

2 Apply the Pythagorean Theorem to right triangles and test a triangle to tell whether it is acute, right, or obtuse. (pp. 359–360)

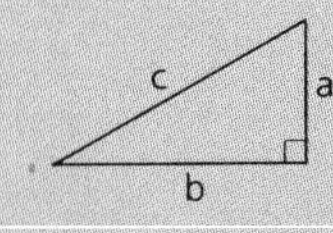

3 **a** List some families of common Pythagorean Triples (p. 367):

3,	4,	5	5,	12,	13	7,	24,	25	8,	15,	17
6,	8,	10	10,	24,	26	14,	48,	50	16,	30,	34
⋮	⋮	⋮	⋮	⋮	⋮	⋮	⋮	⋮	⋮	⋮	⋮

b Be aware of other families such as (9, 40, 41), (11, 60, 61), (12, 35, 37), (13, 84, 85), (16, 63, 65), and (20, 21, 29). (p. 367)

4 Use the Principle of the Reduced Triangle to find the missing side of each right triangle. (p. 368)

a

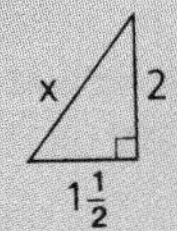

b

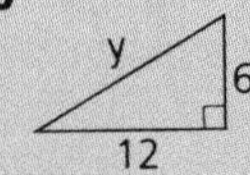

c

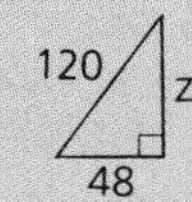

5 **a** Use the pattern of a 30°–60°–90° triangle. (pp. 374, 375)

Is the leg opposite the 30° angle equal to half the hypotenuse?

6 Describe the pattern for a 45°–45°–90° triangle. (pp. 374, 376)

7 Use right triangle techniques in applied problems:

a altitudes (pp. 354, 361)
b zig-zag problems (p. 361)
c diagonals of rectangles (pp. 371, 377)
d ladder problems (pp. 363, 365)

8 Apply the Pythagorean Theorem to rectangular solids and to pyramids. (p. 382)

9 Apply the three trigonometric ratios. (p. 387)

Chapter 9 Review Problems A

1 a If HG = 4 and EG = 6, find GF.

b Find EH, if GH = 4 and GF = 12.

c Find HF, if EF = $2\sqrt{5}$ and GF = 4.

d If EH = 2 and EF = 3, find HF.

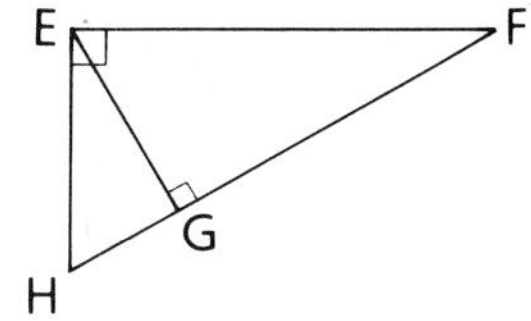

2 Identify the family of each of these special right triangles:

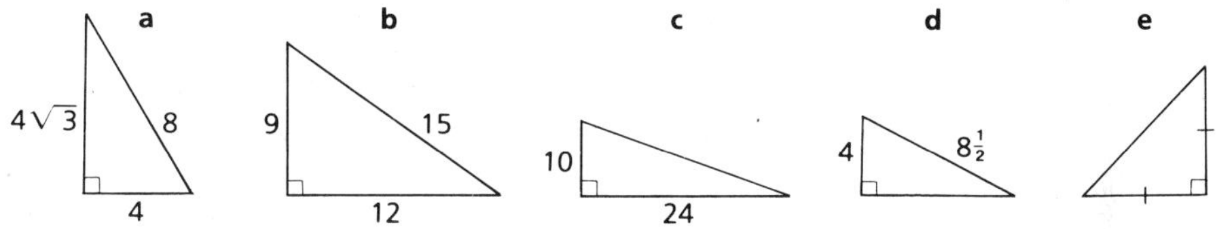

3 Find the missing lengths:

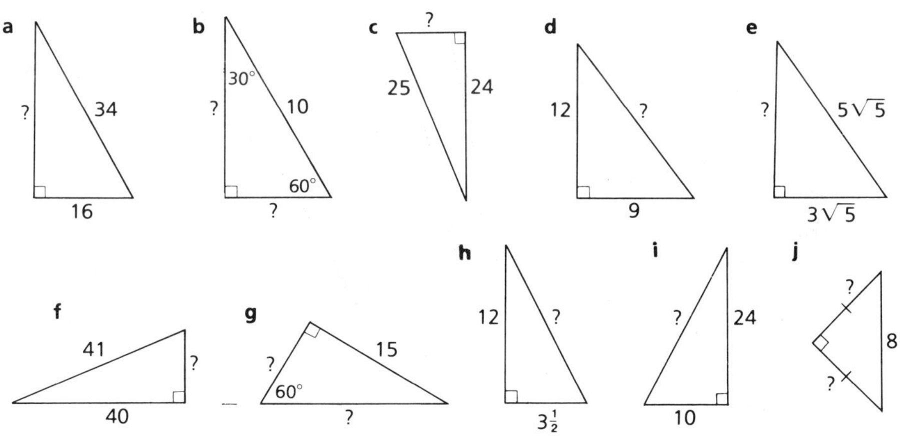

4 AE = 6, BE = 8

Find the perimeter of the rhombus shown.

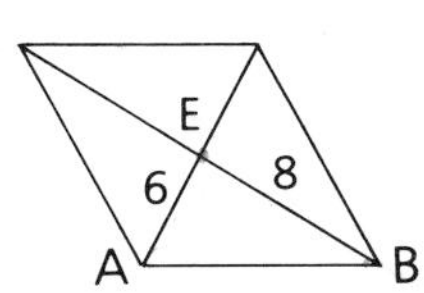

5 Find the altitude of the triangle shown.

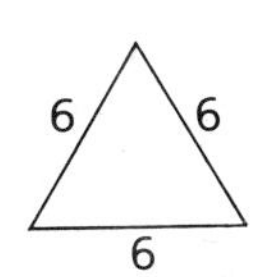

6 Vail skied 2 km north, 2 km west, 1 km north, and 2 km west. How far was she from her starting point?

7 A twenty-five foot ladder just reaches a point on a wall twenty-four feet above the ground. How far is the foot of the ladder from the wall?

8 Find the altitude to the base of an isosceles triangle whose sides have lengths of 8, 6, and 8.

9 If the altitude of an equilateral triangle is $8\sqrt{3}$, find the perimeter of the triangle.

10 What is the length of a diagonal of a 2 by 5 rectangle?

11 In the trapezoid shown, find RS.

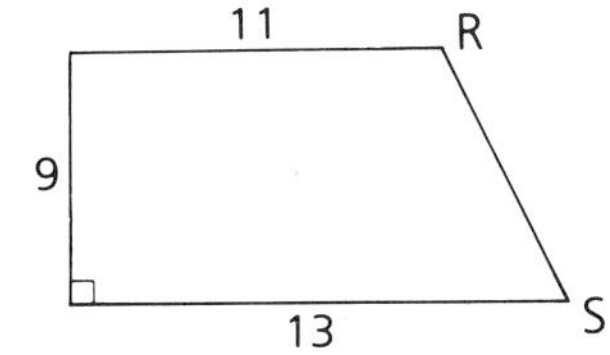

12 Given: TVWX is an isosceles trapezoid
$TX = 8$, $VW = 12$, $\angle V = 30^\circ$

Find: TV and TZ

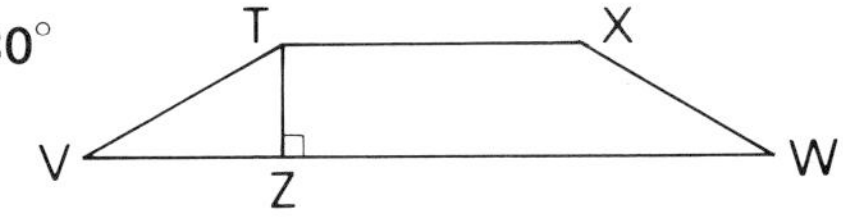

13 Find the diagonal of a rectangular solid whose dimensions are 4, 3, and 12.

14 Given: The regular square pyramid shown
$PR = 20$, $PS = 25$

Find: the perimeter of the base JKMO

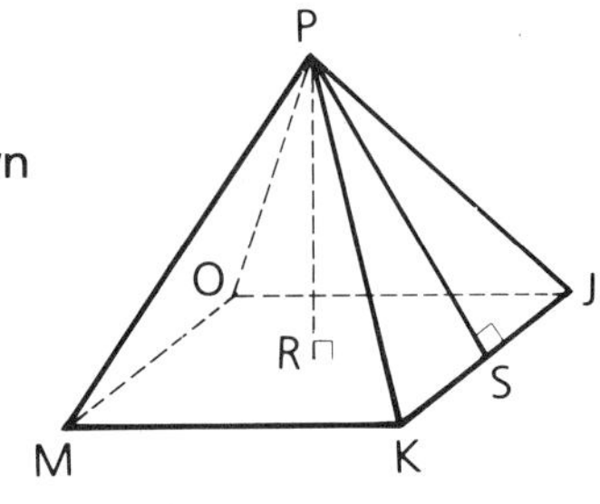

15 In the rectangular solid shown, find AG if:

DC = 12, CG = 7, and AD = 4

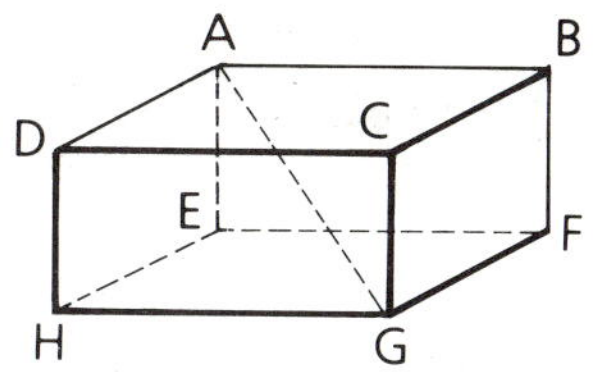

16 Given: $\overline{AC} \perp \overline{CB}$, $\overline{DE} \parallel \overline{CB}$

AC = 15, AB = 17, DE = 4

Find: **a** CB **c** AE **e** DC

b AD **d** EB

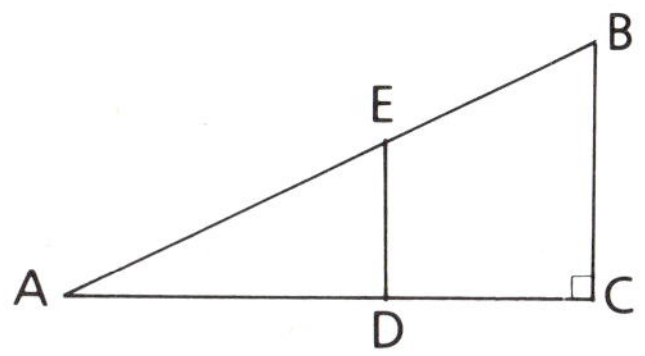

Chapter 9 Review Problems B

17 Two boats leave the harbor at 9:00 A.M. Boat A sails north at 20 kilometers per hour. Boat B sails west at 15 kph. How far apart are the two boats at noon?

18 **a** Find x.

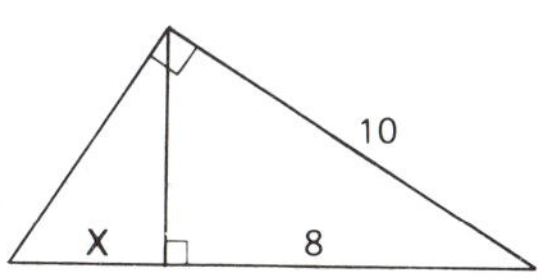

b Find y.

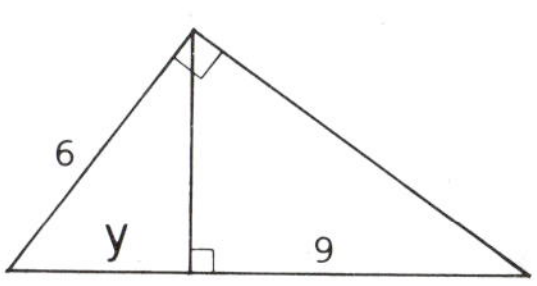

19 A boy standing on the shore of a lake 1 mile wide wants to reach the "Golden Arches" 3 miles down the shore on the opposite side of the lake. If he swims at 2 mph and walks at 4 mph, is it quicker for him to swim directly across the lake and then walk to the "Golden Arches," or to swim directly to the "Golden Arches"?

20 A boat is tied to a pier by a 25′ rope. The pier is 15′ above the boat. If 8 feet of rope is pulled in, how many feet does the boat move forward?

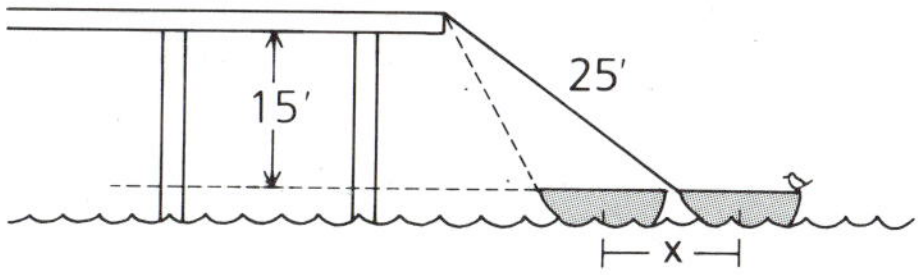

21 Find x.

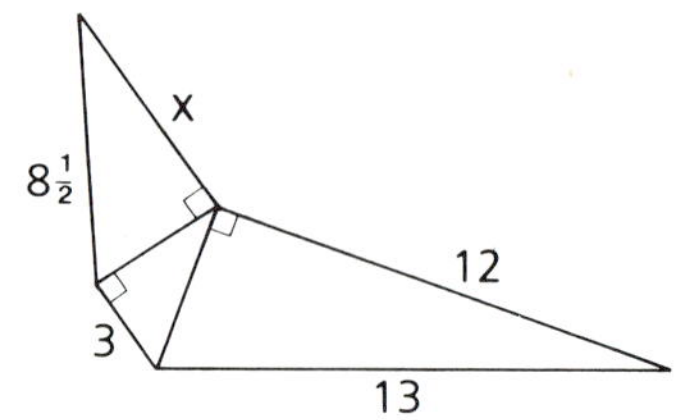

22 Follow the treasure map of Captain Zig Zag to see how far the treasure is from the old stump.

23 Given: Kite KITE with right ∠s KIT and KET
KP = 9, TP = 4

Find: **a** IE

b the perimeter of KITE

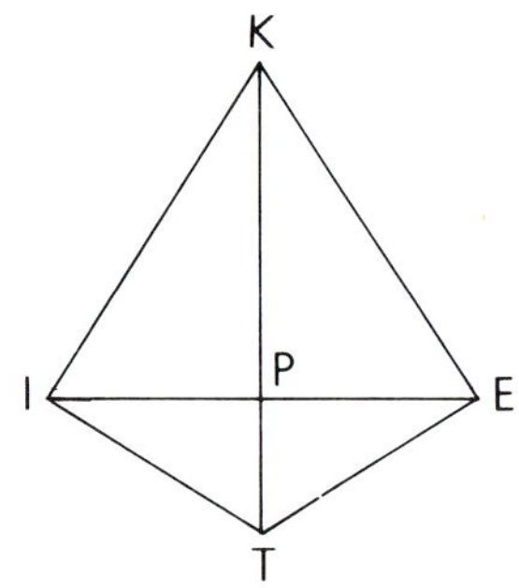

Chapter 9 Review Problems C

24 Given: ∠C is a right angle
E is the midpoint of $\overline{AC}$
F is the midpoint of $\overline{BC}$
$AF = \sqrt{41}$, $BE = 2\sqrt{26}$

Find: AB

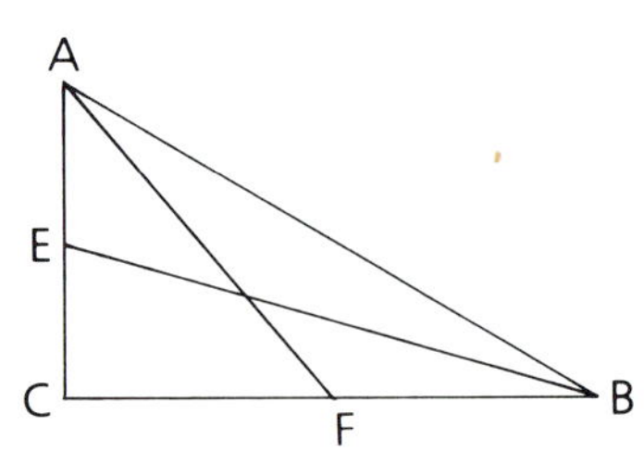

25 The altitude to the hypotenuse divides the hypotenuse of a right △ in the ratio 4:1. What is the ratio of the legs of the right △?

26 A 12 m rope is used to form a triangle the lengths of whose sides are integers. If one of the possible triangles is selected at random, what is the probability that the triangle is a right triangle?

Chapters 1–9 Cumulative Review Problems A

1 A pair of consecutive angles of a ▱ are in a ratio of 5:3. Find the measure of the smaller angle.

2 Find x.

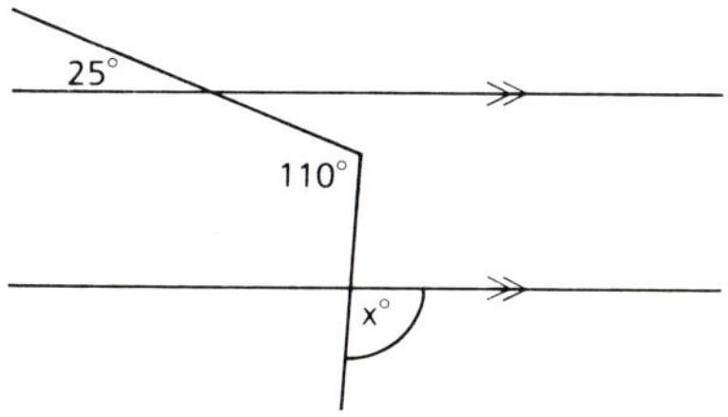

3 a Find the sum of the measures of the angles of a nonagon.

b If each angle of a regular polygon is a 168° angle, how many sides does the polygon have?

c How many diagonals does a heptagon have?

4 a Find x.

b Is △ABC isosceles?

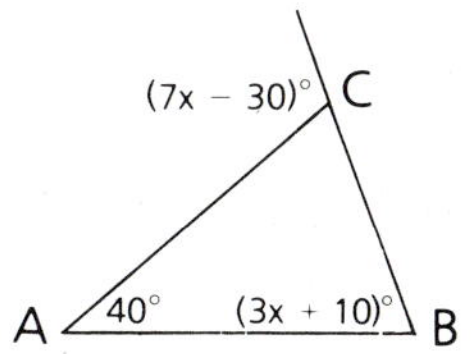

5 A boy 180 cm tall casts a 150 cm shadow. A nearby flagpole casts a shadow 12 m long. What is the length of the flagpole?

6 Given: $\angle ABC = 60°$, $\angle ACB = 70°$

Find: $\angle BFC$

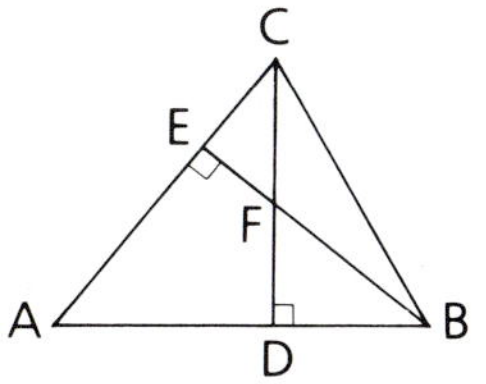

7 Find the perimeter of a rhombus whose diagonals are 10 and 24.

8 a Find BC.

b Find AB.

c Is △DEC acute, right, or obtuse?

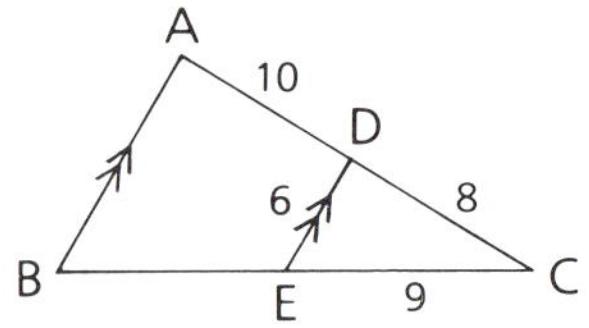

9 a Find the mean proportionals between $\frac{1}{4}$ and 49.

b Solve for y: $\frac{5}{5-y} = \frac{10}{y-10}$

10 Is a ∥ b?

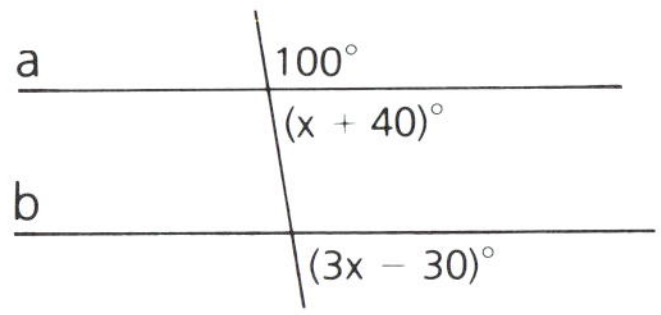

11 Given: $\overline{EB} \cong \overline{DF}$, $\overline{AG} \cong \overline{GC}$

$\angle EAG \cong \angle FCG$

Prove: ABCD is a parallelogram

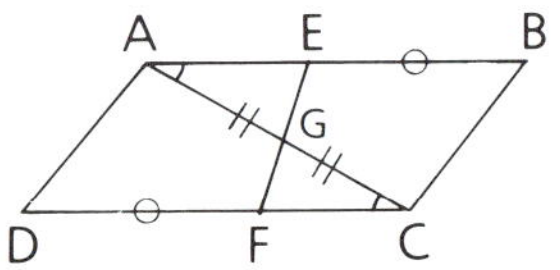

12 Given: ⊙ Q lies in plane *m*

$\overline{PQ} \perp m$

Prove: $\angle R \cong \angle S$

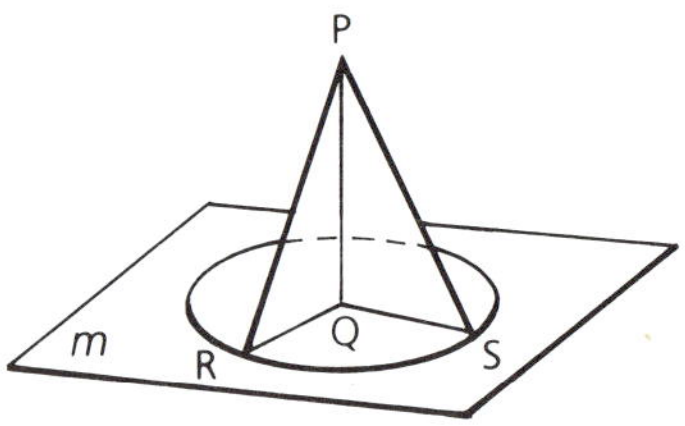

Chapters 1–9 Cumulative Review Problems B

13 Find the angle formed by the hands of a clock at:

a 11:50 b 12:01

14 The sum of an angle and four times its complement is 20° more than the supplement of the angle. Find the angle's complement.

15 The sum of the angles of an equiangular polygon is 3960°. Find the measure of each exterior angle.

16 Find AB.

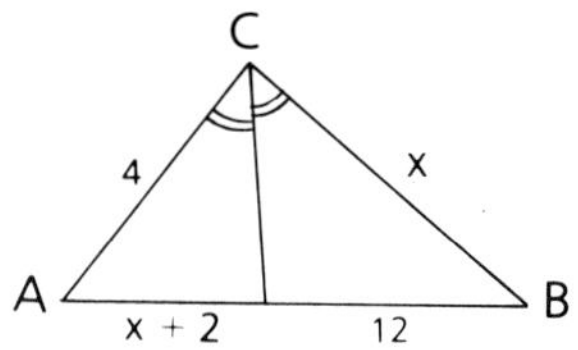

17 Given: Ⓢ P and Q intersect at A and B
M is the midpoint of $\overline{AB}$
Prove: $\overleftrightarrow{PM}$ passes through Q

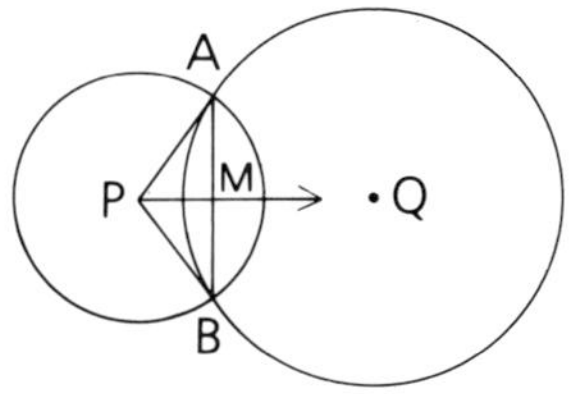

18 Given: $\angle A = 120°$
$\overrightarrow{BD}$ and $\overrightarrow{BE}$ trisect $\angle ABC$
$\overrightarrow{CD}$ and $\overrightarrow{CE}$ trisect $\angle ACB$

a Find $m\angle D$ and $m\angle E$

b Do $m\angle A$, $m\angle D$, and $m\angle E$ form an arithmetic progression?

(See Chapter 7, Review Problem B23)

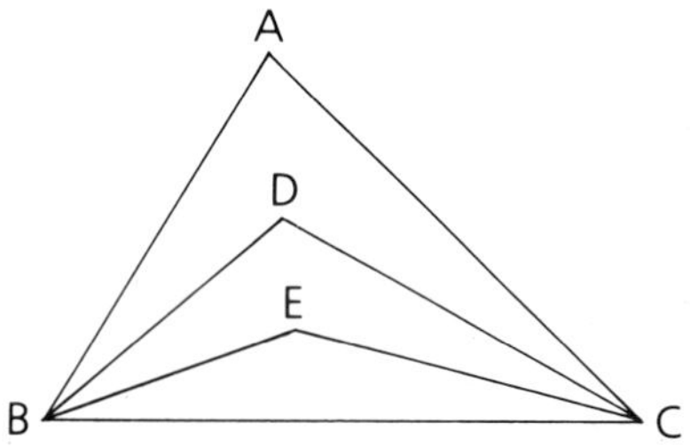

19 Given: $\overline{AB}$ lies in m, $\overline{CD}$ lies in n
$m \parallel n$
$\overline{AC}$ intersects $\overline{BD}$ at P
$\overline{AD} \perp n$, and $\overline{BC} \perp n$
Prove: $\overline{AC} \cong \overline{DB}$

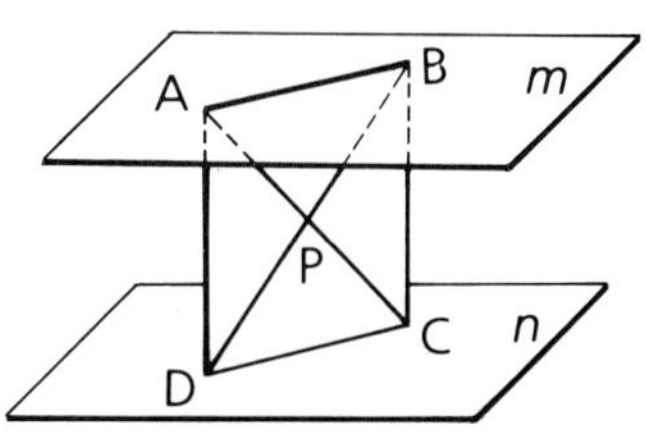

20 The shorter diagonal of a regular hexagon is 6 units. Find the length of the longer diagonal.

Chapters 1–9 Cumulative Review Problems C

21 Given: $\overrightarrow{BD}$ and $\overrightarrow{BE}$ trisect $\angle ABC$
$DE = 4$, $EC = 5$, $BD = 12$
Find: **a** BC **b** BE **c** AD

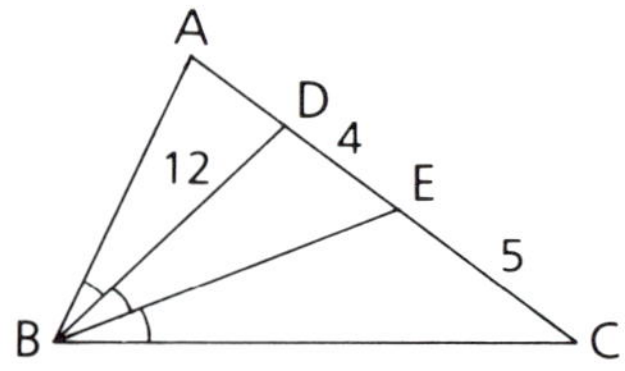

22 Given: $\angle A \cong \angle E$
$\overline{FA} \cong \overline{FE}$
Prove: $\overrightarrow{CF}$ bisects $\angle BCD$

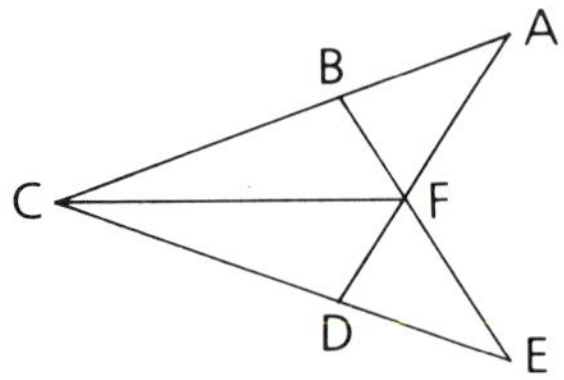

23 Given: The cube as shown with J the midpoint of $\overline{HF}$
$AB = 6$
Find: AJ

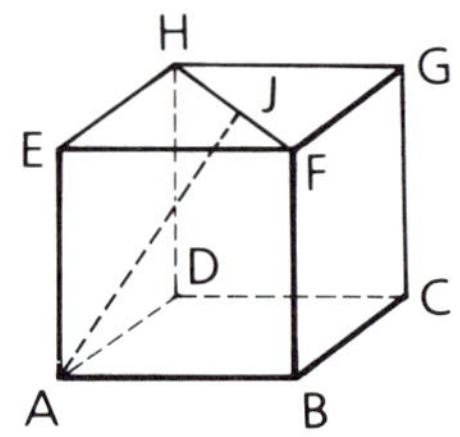

24 Given: $\overline{AD} \cong \overline{DC}$
$\angle B + 50° = \angle DAB$
Find: $m\angle CAB$

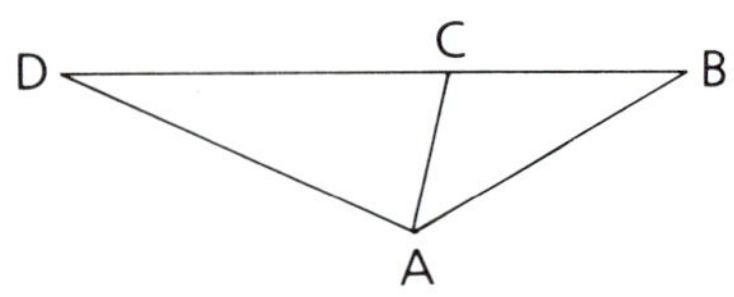

25 Find the sum of the measures of angles A, B, C, D, and E.
Does your answer depend on knowing whether any polygons are equilateral or equiangular?

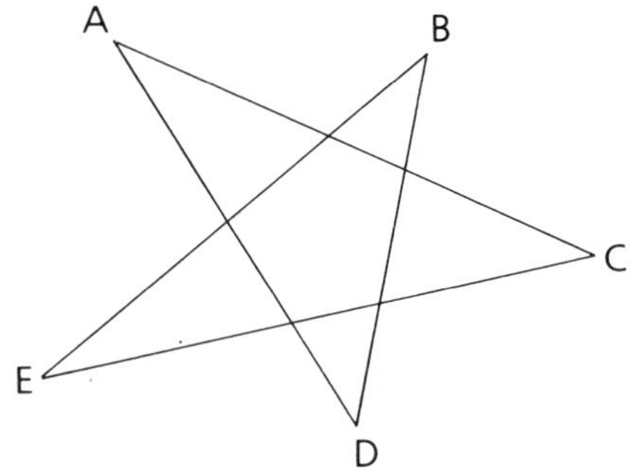

26 What is the probability that a diagonal chosen at random in a regular decagon will be one of the shortest diagonals?

InterSection: Shape and Perception

When the eye sees a picture in two dimensions, the mind often interprets that picture as objects in three dimensions. Figures that aid three-dimensional perception are called **perspective drawings**. Figures (or objects) that fool three-dimensional perception are called **illusions**.

Geometry Works: For Perspective Drawing

In perspective drawing, shapes viewed from the front are not changed, but shapes viewed at an angle are distorted. Also, the apparent size of an object is reduced as the object is further from the viewer.

These diagrams illustrate two examples of perspective drawing. In the figure at the left, the shapes that are perpendicular to the line of sight (from the eye to the object) are not changed, but all the lines that are parallel to the line of sight appear to meet at a single **vanishing point**. In the figure at the right, no surfaces are perpendicular to the line of sight. The vertical lines remain vertical, but other sets of parallel lines meet at one of two vanishing points.

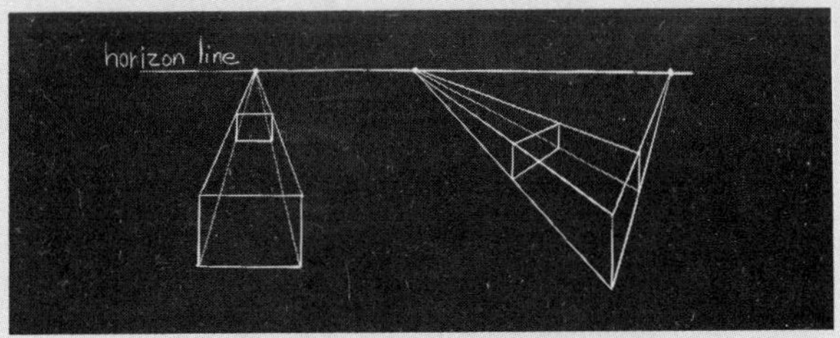

In both cases, surfaces that are further from the viewer seem smaller.

The **horizon line** in perspective drawing represents the level of the viewer's eyes. In a drawing that has many horizontal lines, each set of parallel lines will have a vanishing point, and all the vanishing points will lie on the horizon line.

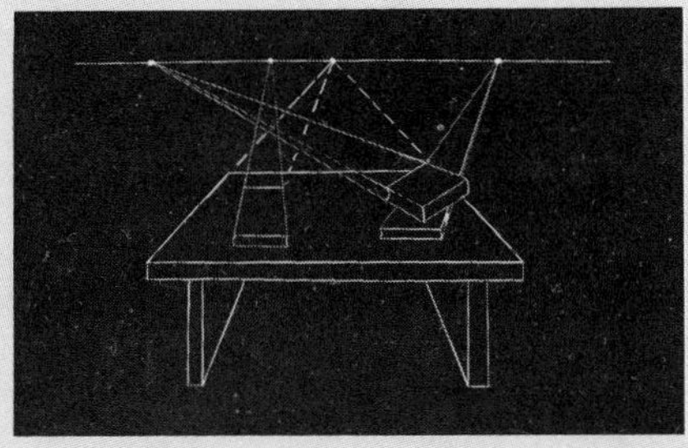

Geometry Works: For Illusionists

With a puff of smoke, an illusionist can present the appearance of a disappearance! For example, a very real elephant is standing in front of a black curtain. With a puff of smoke, the elephant is not there! One method used is to fool the depth perception of the audience. Suppose you viewed the illusion from above:

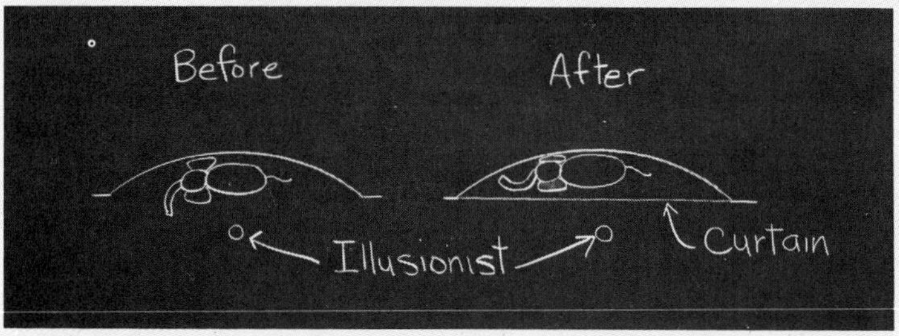

The puff of smoke hides the act of dropping a curtain in front of the elephant. The audience's perception was a straight black curtain with an elephant in front. As the smoke clears, the new curtain is perceived as the original, straight curtain. The elephant vanished!

Another type of illusion can be performed with a piece of wood cut in the shape of an isosceles trapezoid. Attach the trapezoid to a spindle on a slowly rotating turntable:

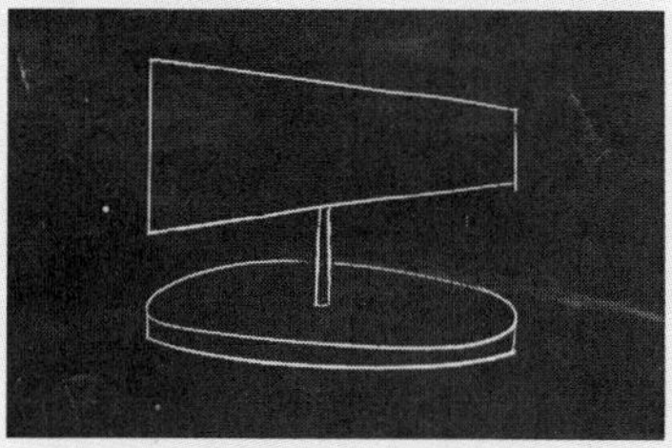

The eye is likely to interpret the trapezoid as a rectangle viewed from an angle or "in perspective." The eye will interpret the change of shape as a figure moving back and forth rather than as a rotating trapezoid. Compare the top view with the view of the perceiver.

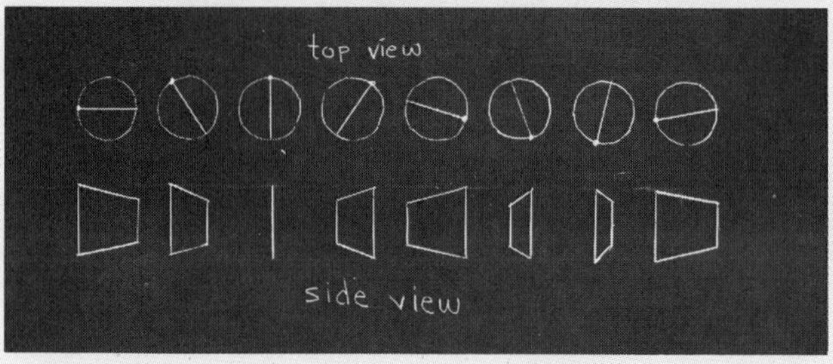

Projects

1 Discuss the following illusions. What does the mind "see"? Why was it fooled?

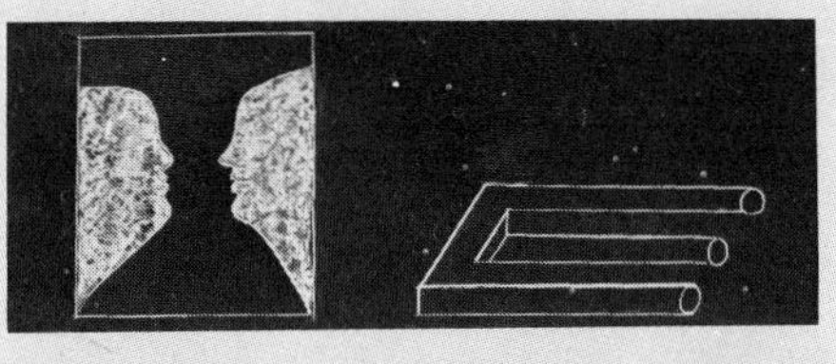

2 Draw a desk using one point perspective and two point perspective. Describe the location of the viewer for each drawing.

3 Draw another desk using two point perspective so the horizon line is lower than the top of the desk. Describe the location of the viewer for your drawing.

4 How does each drawing try to mislead the viewer?

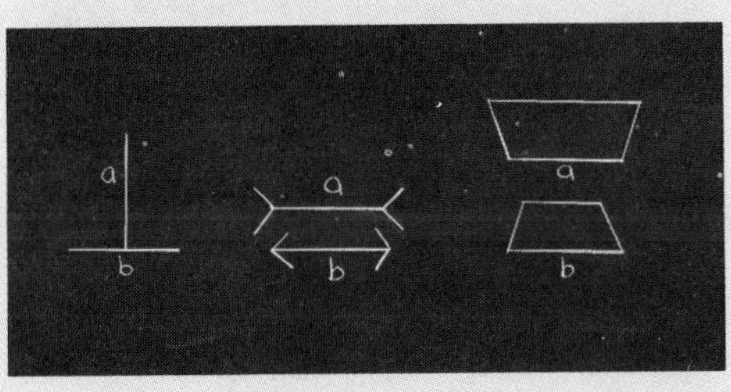

5 The following program calculates the apparent area of the trapezoid on page 406.

```
5 INPUT H,A,B
10 PRINT "AREA IS:"; ___?___
```

a Complete the program to calculate the apparent area of the trapezoid.

b The apparent "height" of the isosceles trapezoid can be calculated using trigonometry.

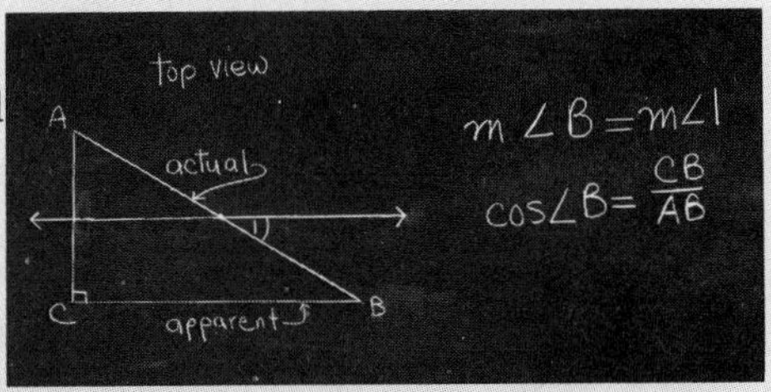

Extend the program to calculate the apparent area of the trapezoid based on the lengths of the parallel sides and the measure of ∠1.

Chapter 10

CIRCLES

Section 10.1 The Circle

This section examines:

A BASIC PROPERTIES AND DEFINITIONS

B CHORDS AND DIAMETERS

C RELATIONSHIPS BETWEEN A RADIUS AND A CHORD

10.1 (A) Basic Properties and Definitions

D **A CIRCLE is the set of all points in a plane that are a given distance from a given point in the plane. That point is called the CENTER, and a segment joining the center to a point on the circle is a RADIUS (plural: RADII).**

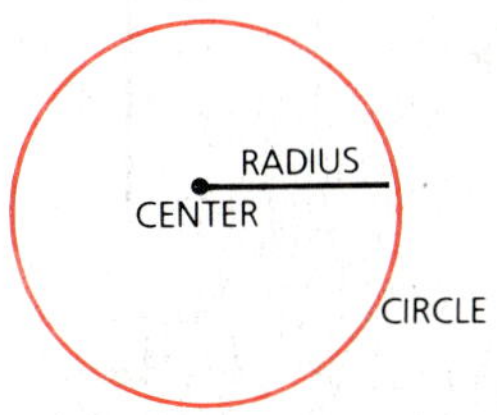

D **Two coplanar circles with the same center are called CONCENTRIC circles.**

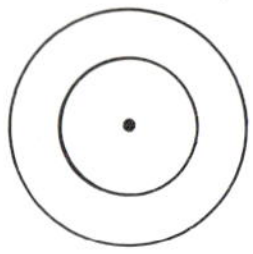

From the definition of a CIRCLE we have the following property:

All radii of a circle are congruent.

All circles do have the same *shape* but their *sizes* are determined by the measures of their radii.

D **Two circles are *congruent* whenever a radius of one of them equals a radius of the other.**

D **A point is inside (in the INTERIOR of) a circle, if its distance from the center is less than the radius.**

Points O and A are in the interior of ⊙ O.

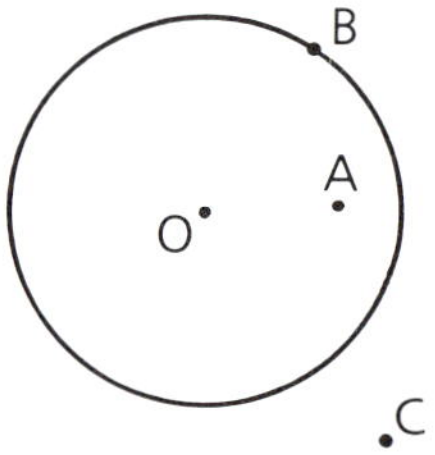

D **A point is outside (in the EXTERIOR of) a circle, if its distance from the center is greater than the radius.**

Point C is in the exterior of ⊙ O.

D **A point is ON a circle if its distance from the center is equal to the length of a radius.**

Point B is on ⊙ O.

10.1 (B) Chords and Diameters

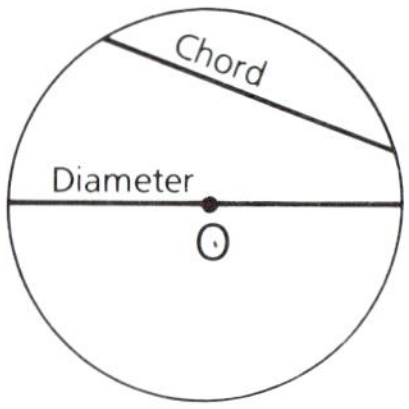

D **A CHORD of a circle is a segment joining any two points of the circle.**

D **A DIAMETER of a circle is a *chord* that passes through the center of the circle.**

10.1 (C) Relationships Between a Radius and a Chord

D **The DISTANCE FROM THE CENTER OF A CIRCLE TO A CHORD is the measure of the *perpendicular* segment from the center to the chord.**

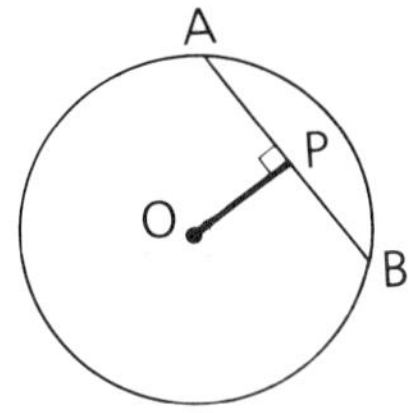

OP is the distance from O to chord $\overline{AB}$.

A radius and a chord can be related in three ways:

Theorem 65

T **If a radius is perpendicular to a chord, then it bisects the chord.**

Given: $\odot$ O
$\overline{OD} \perp \overline{AB}$
Prove: $\overline{OD}$ bisects $\overline{AB}$

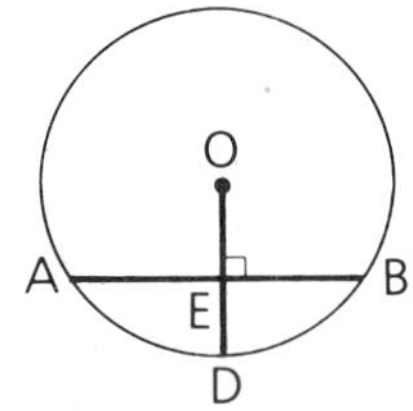

Theorem 66

T **If a radius of a circle bisects a chord *that is not a diameter,* then it is perpendicular to that chord.**

Given: $\odot$ O
$\overline{OH}$ bisects $\overline{EF}$
Prove: $\overline{OH} \perp \overline{EF}$

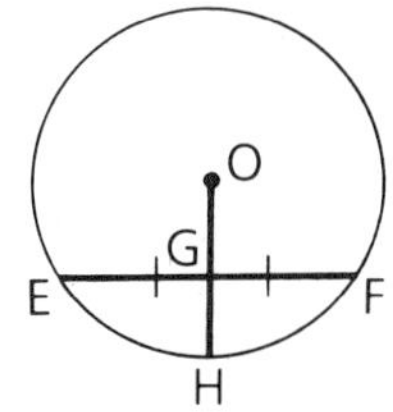

Theorem 67

T **The perpendicular bisector of a chord passes through the center of the circle.**

Given: $\overleftrightarrow{PQ}$ is the $\perp$ bisector of $\overline{CD}$
Prove: $\overleftrightarrow{PQ}$ passes through O.

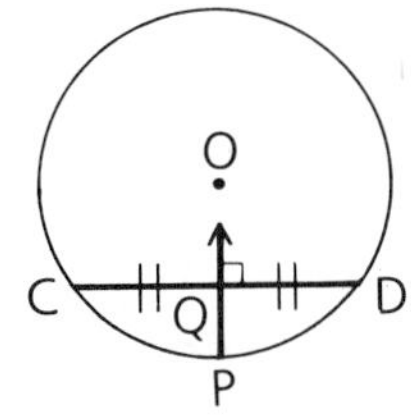

The proofs of Theorems 65–67 are called for in the problem sets.

Section 10.1 Sample Problems

1 Given: $\odot$ Q, $\overline{PR} \perp \overline{ST}$

Prove: $\overline{PS} \cong \overline{PT}$

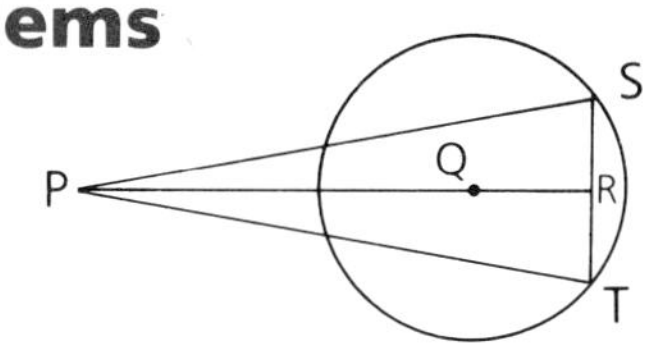

Statements	Reasons
1 $\odot$ Q, $\overline{PR} \perp \overline{ST}$	1 Given
2 $\overline{PR}$ bisects $\overline{ST}$	2 If a radius is $\perp$ to a chord, it bisects the chord ($\overline{QR}$ is part of a radius)
3 $\overline{PR}$ $\perp$ bis of $\overline{ST}$	3 Combine steps 1 and 2.
4 $\overline{PS} \cong \overline{PT}$	4 If a point is on the $\perp$ bis of a segment, then it is equidistant from the endpoints.

2 The radius of circle O is 13 mm.
The length of chord $\overline{PQ}$ is 10 mm.

Find the distance from chord $\overline{PQ}$ to the center, O.

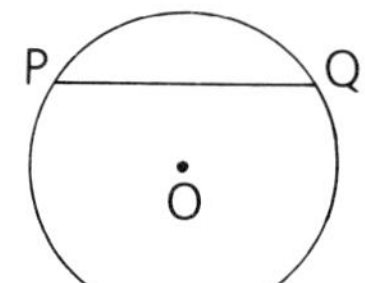

Solution: Draw the distance segment $\overline{OR} \perp \overline{PQ}$.
Draw radius $\overline{OP}$ to complete a right △.

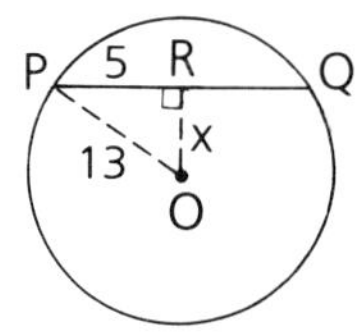

Since a radius $\perp$ to a chord bisects the chord, $PR = \frac{1}{2}PQ = \frac{1}{2}(10) = 5$.

Using the Pythagorean Theorem, $x^2 + 5^2 = 13^2$

$OR = 12$

3 Given: △ABC is isosceles, $\overline{AB} \cong \overline{AC}$
$\overline{BC} \parallel \overline{PQ}$, Ⓢ P and Q

Prove: $\odot$ P $\cong$ $\odot$ Q

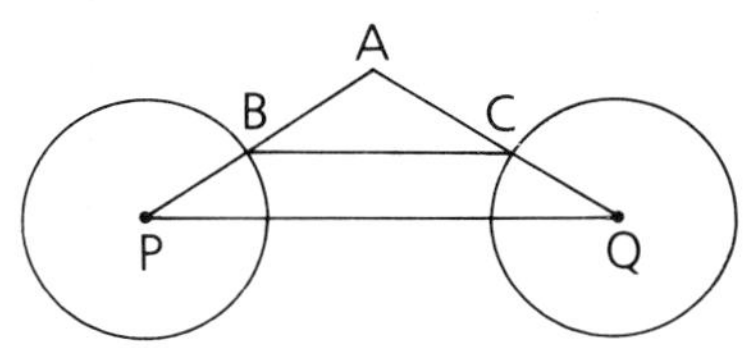

Statements	Reasons
1 △ABC is isosceles, $\overline{AB} \cong \overline{AC}$	1 Given
2 $\overline{BC} \parallel \overline{PQ}$, $\odot$ P and $\odot$ Q	2 Given
3 $\angle ABC \cong \angle P$, $\angle ACB \cong \angle Q$	3 PCA
4 $\angle ABC \cong \angle ACB$	4 If △ then △
5 $\angle P \cong \angle Q$	5 Transitive property
6 $\overline{AP} \cong \overline{AQ}$	6 If △ then △
7 $\overline{PB} \cong \overline{CQ}$	7 Subtraction (1 from 6)
8 $\odot$ P $\cong$ $\odot$Q	8 Ⓢ with $\cong$ radii are $\cong$

Section 10.1 Problem Set A

1 Given: $\odot$ O, chord $\overline{AB}$

Prove: **a** $\triangle AOB$ is isosceles

b $\angle A \cong \angle B$

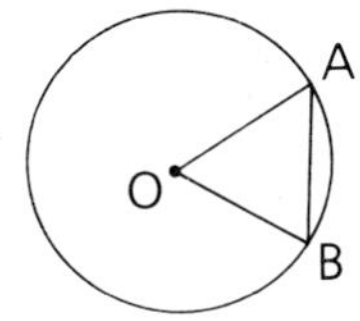

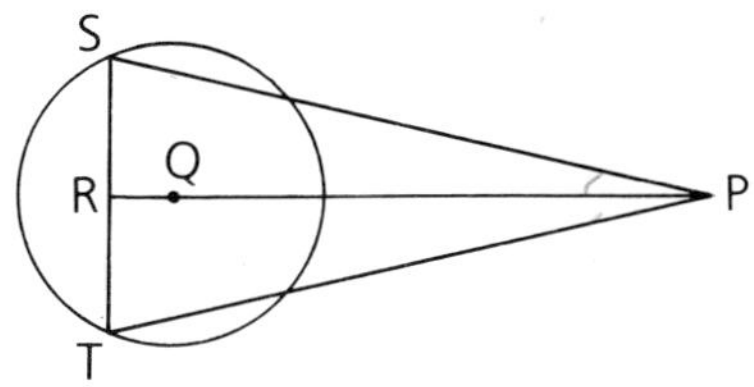

2 Given: $\overline{PR} \perp \overline{ST}$ in $\odot$ Q

Prove: $\angle S \cong \angle T$

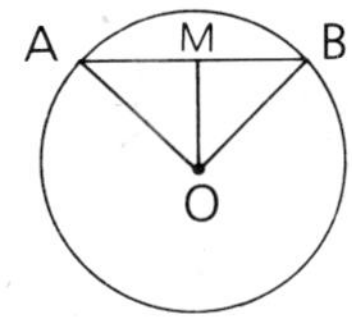

3 Given: $\odot$ O, $\overline{OM}$ is a median

Conclusion: $\overline{OM}$ is an altitude of $\triangle BOA$

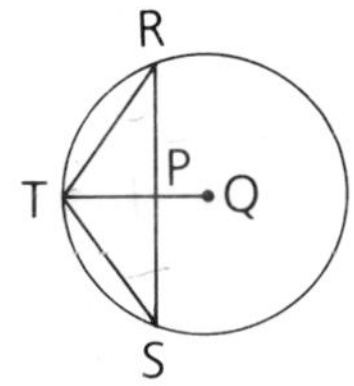

4 Given: $\odot$ Q, $\overline{QT} \perp \overline{RS}$

Prove: $\overrightarrow{TQ}$ bisects $\angle RTS$

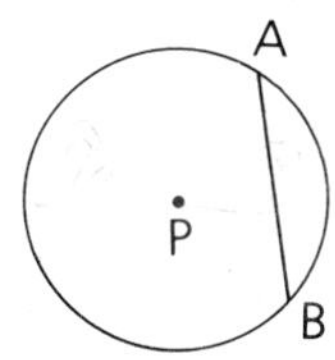

5 Chord $\overline{AB}$ is 12 mm long.
The radius of $\odot$ P is 10 mm.
Find the distance from $\overline{AB}$ to P.

6 Find the length of a chord that is 15 cm from the center of a circle with a radius of 17 cm.

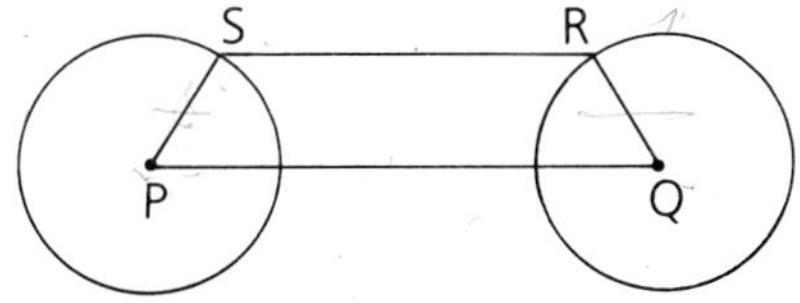

7 Given: PQRS is an isosceles trapezoid with $\overleftrightarrow{SR} \parallel \overleftrightarrow{PQ}$

Conclusion: $\odot P \cong \odot Q$

8 Prove Theorem 65.

9 Given: $\odot A \cong \odot B$

$\overleftrightarrow{AD} \parallel \overleftrightarrow{BC}$

Prove: ABCD is a $\square$

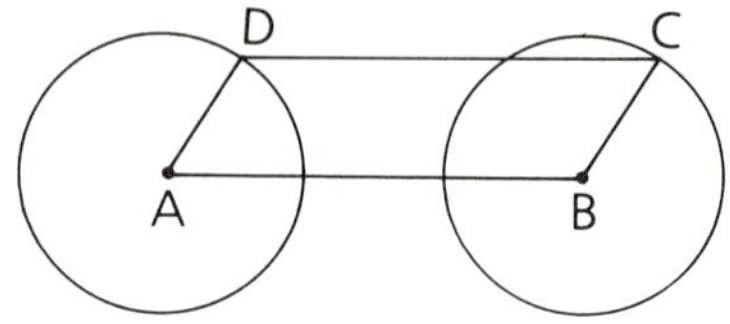

Section 10.1 Problem Set B

10 Given: $\odot O$

$\overleftrightarrow{OR}$ bisects $\overline{PQ}$

Prove: $\overrightarrow{RO}$ bisects $\angle PRQ$

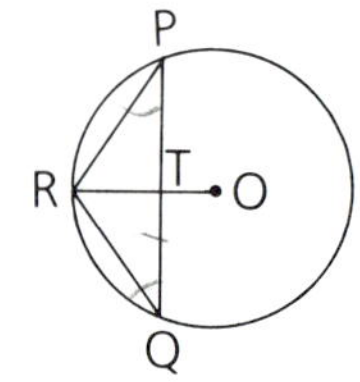

11 Find the distance from the center of a circle to a chord 30 m long, if the *diameter* of the circle is 34 m.

12 Find the radius of a circle if a 24 cm chord is 9 cm from the center.

13 Given: Ⓢ A and B intersect as shown

$\overline{DE} \parallel \overline{FC}$, $\angle ADE \cong \angle FCB$

$\overline{DE} \cong \overline{FC}$

Prove: $\odot A \cong \odot B$

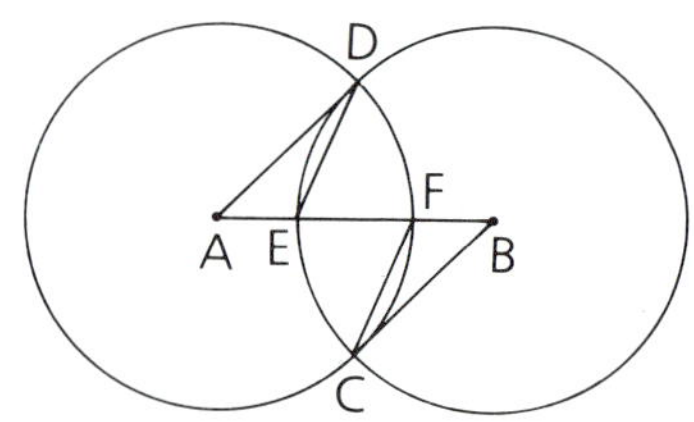

14 Two circles intersect and have a common chord 24 cm long. The centers of the circles are 21 cm apart. The radius of one circle is 13 cm. Find the radius of the other circle.

15 Given: $\odot P$

$\overleftrightarrow{QT} \parallel \overleftrightarrow{RS}$

Conclusion: $\overline{QR} \cong \overline{TS}$

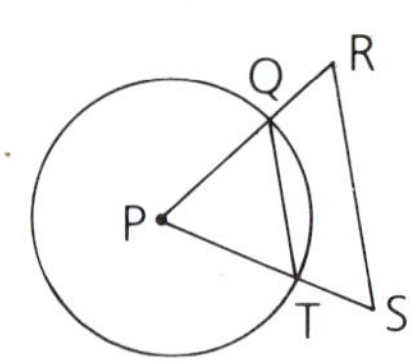

16 Set up and prove: **a** Theorem 66 **b** Theorem 67

17 Given: Chords $\overline{AB} \cong \overline{CD}$ in $\odot P$

Distance segments $\overline{PQ}$ and $\overline{PR}$ are $\perp$ to the chords

Prove: $\overline{PQ} \cong \overline{PR}$

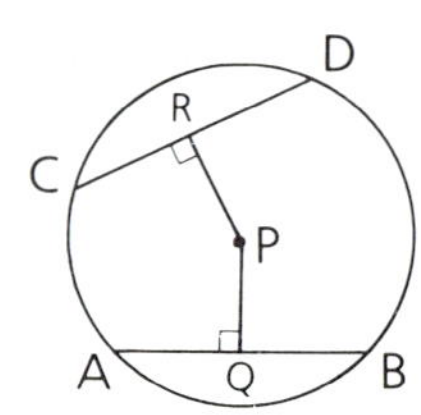

18 Given: $\odot$ P
Z is the midpt. of $\overline{WX}$
$\triangle WAX$ is isosceles, base $\overline{WX}$
Prove: $\overrightarrow{AZ}$ passes through P

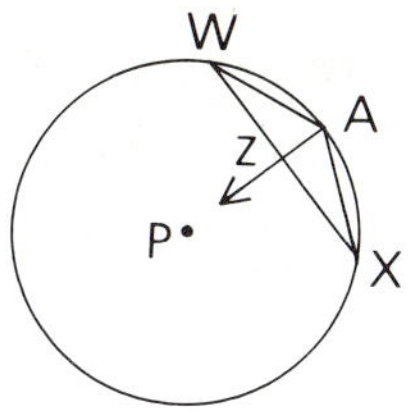

Section 10.1 Problem Set C

19 Given: Two concentric circles with center P
Line m intersects the circles at A, B, C, D
Conclusion: $\overline{AB} \cong \overline{CD}$

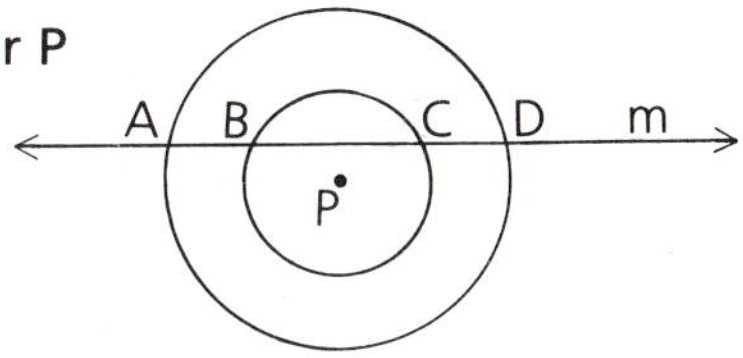

20 Given: $\odot$ P, $\overline{WX} \cong \overline{YZ}$
Prove: $\overline{WQ} \cong \overline{ZR}$

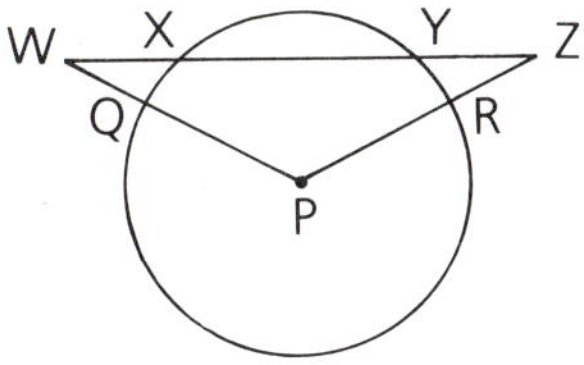

21 Given: $\overline{AB}$ is a diameter of $\odot$ O
$\overleftrightarrow{AC} \parallel \overleftrightarrow{BD}$
Conclusion: $\overline{AC} \cong \overline{BD}$

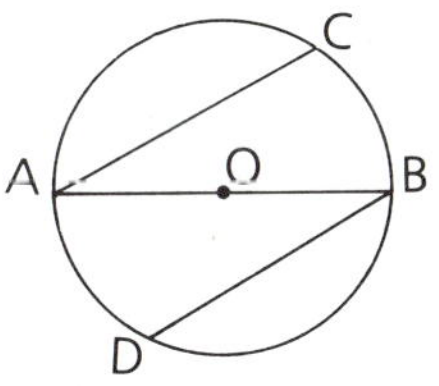

22 Find the radius of a circle in which a 48 cm chord is 8 cm closer to the center than a 40 cm chord.

23 In circle O: PQ = 4, RQ = 10, PO = 15.
Find PS (the distance from P to the $\odot$O).

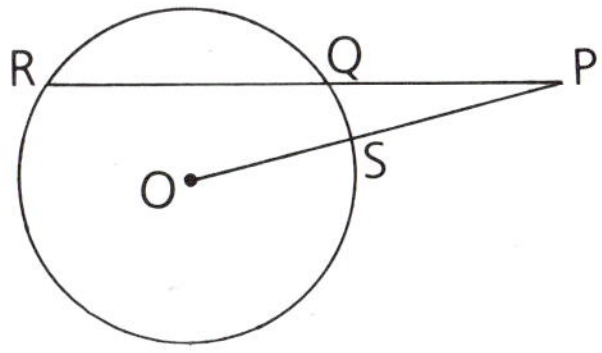

24 An isosceles triangle, whose legs are each 13, is inscribed in a circle. If the altitude to the base of the triangle is 5, find the radius of the circle.

25 Two circles intersect and have a common chord. The radii of the circles are 13 and 15. The distance between their centers is 14. Find the length of their common chord.

Section 10.2 Congruent Chords

What is the longest chord in a circle?
Is there a shortest chord?

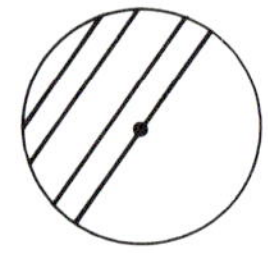

If two chords are the same distance from the center of a circle, what can we conclude?

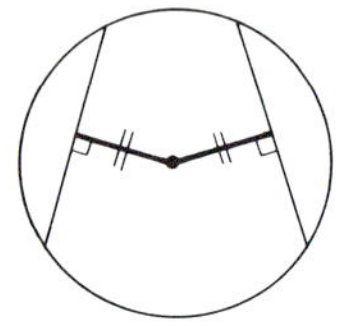

Theorem 68

T **If two chords of a circle are equidistant from the center, then they are congruent.**

Given: $\odot$ P with $\overline{PX} \perp \overline{AB}$,
$\overline{PY} \perp \overline{CD}$, $\overline{PX} \cong \overline{PY}$

Prove: $\overline{AB} \cong \overline{CD}$

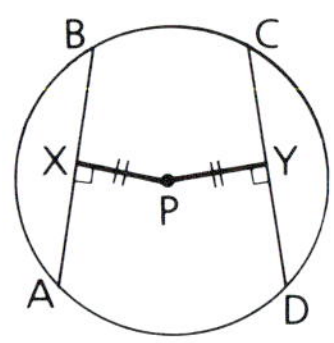

The proof of Theorem 68 is left for you to do.
Use four congruent triangles.
The converse of Theorem 68 can also be proved:

Theorem 69

T **If two chords of a circle are congruent, then they are equidistant from the center of the circle.**

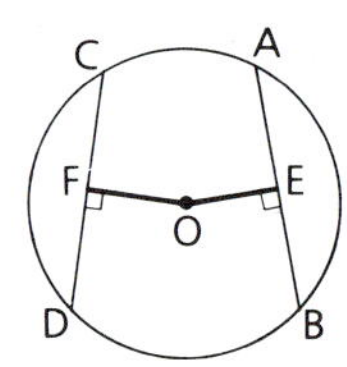

Given: $\odot$ O with $\overline{AB} \cong \overline{CD}$
$\overline{OE} \perp \overline{AB}$, $\overline{OF} \perp \overline{CD}$

Prove: $\overline{OE} \cong \overline{OF}$

You proved this theorem if you proved problem 10.1, 17.

Section 10.2 Sample Problems

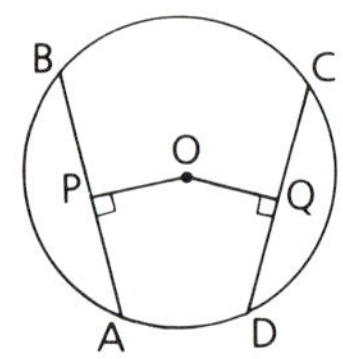

1 Given: $\overline{AB} \cong \overline{CD}$ in circle O
$OP = 12x - 5$, $OQ = 4x + 19$

Find: OP

Solution: Since $\overline{AB} \cong \overline{CD}$, then $OP = OQ$

$$12x - 5 = 4x + 19$$
$$x = 3$$

Thus $OP = 12(3) - 5 = 31$

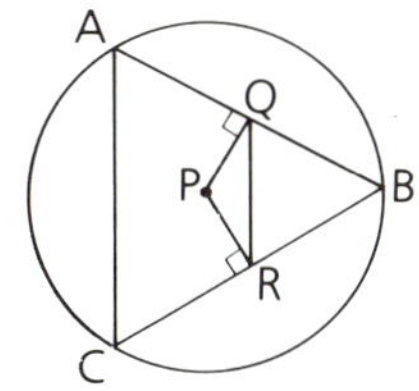

2 Given: $\triangle ABC$ is isosceles, base $\overline{AC}$
$\odot P$, $\overline{PQ} \perp \overline{AB}$, $\overline{PR} \perp \overline{CB}$

Prove: $\triangle PQR$ is isosceles

1 $\odot P$, $\overline{PQ} \perp \overline{AB}$, $\overline{PR} \perp \overline{CB}$	1 Given
2 $\triangle ABC$ is isosceles, base $\overline{AC}$	2 Given
3 $\overline{AB} \cong \overline{BC}$	3 An isosceles $\triangle$ has two $\cong$ sides
4 $\overline{PQ} \cong \overline{PR}$	4 If two chords of a circle are $\cong$, then they are equidistant from the center
5 $\triangle PQR$ is isosceles	5 A $\triangle$ with two $\cong$ sides is isosceles

Why do you think it was necessary to be given $\overline{PQ} \perp \overline{AB}$ and $\overline{PR} \perp \overline{CB}$, even though they did not *seem* to play an active role in the proof?

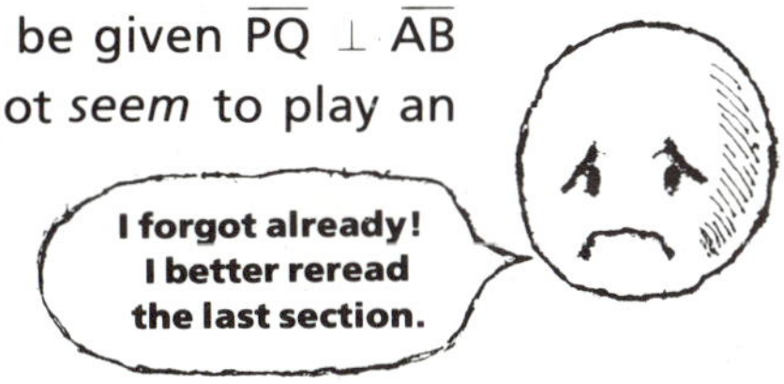

Section 10.2 Problem Set A

1 In a circle, chord $\overline{AB}$ is 325 cm long and chord $\overline{CD}$ is $3\frac{1}{4}$ meters long. Which is closer to the center?

2 Given: $\odot P$ with $\overline{PQ} \cong \overline{PR}$
$AB = 6x + 14$
$CD = 4 - 4x$

Find: AB

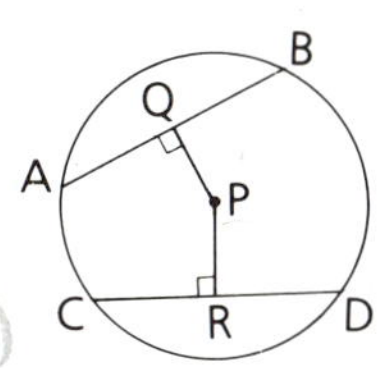

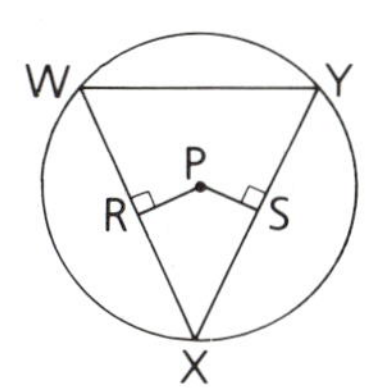

3 Given: $\odot P$, $\overline{PR} \perp \overline{WX}$
$\overline{PS} \perp \overline{XY}$, $\overline{PR} \cong \overline{PS}$

Conclusion: $\angle W \cong \angle Y$

4 Given: Equilateral $\triangle$ ABC is inscribed in $\odot$ Q

Conclusion: $\overline{AB}$, $\overline{BC}$, and $\overline{CA}$ are equidistant from the center

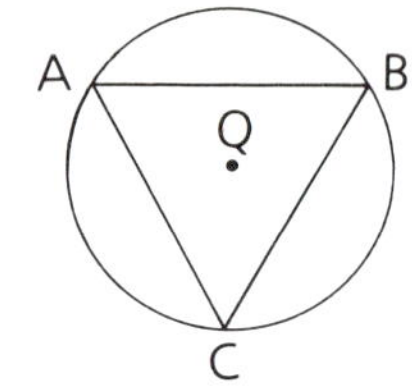

5 Given: $\odot$ P

P is the midpoint of $\overline{MN}$

$\overline{MN} \perp \overline{AD}$ and $\overline{BC}$

Conclusion: ABCD is a $\square$

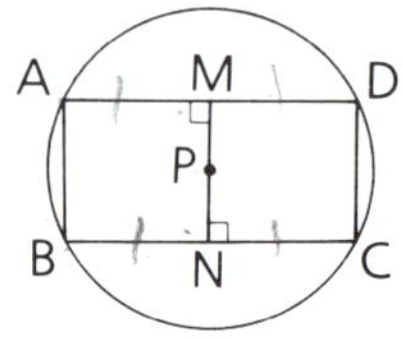

6 A fly is sitting at the midpoint of a wooden chord of a circular wheel. The wheel has a radius of 10 cm and the chord has a length of 12 cm.

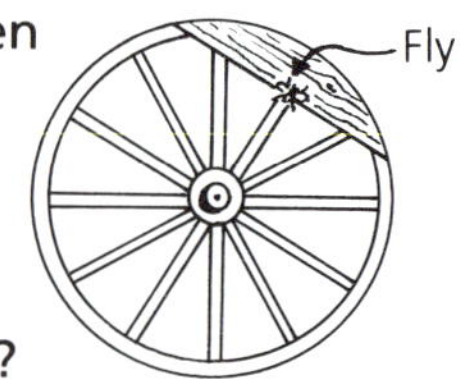

a How far from the hub (center) is the fly?

b The wheel is spun. What is the path of the fly?

Section 10.2 Problem Set B

7 Given: Circle Q, $\overline{PS} \perp \overline{RT}$

$\overline{MQ} \perp \overline{RP}$, $\overline{NQ} \perp \overline{PT}$

Conclusion: $\overline{MQ} \cong \overline{QN}$

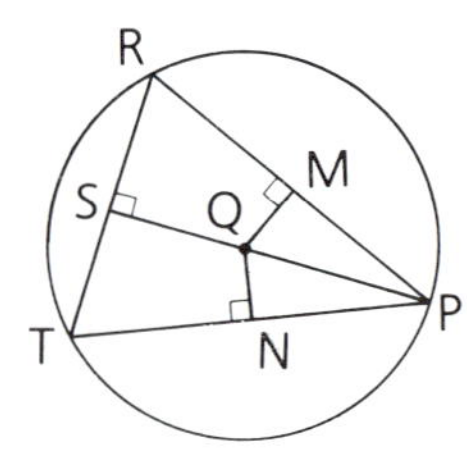

8 Given: Circle F, $\overline{FE} \perp \overline{BC}$

$\overline{FD} \perp \overline{AB}$

$\overrightarrow{BF}$ bisects $\angle$ABC

Prove: $\overline{BC} \cong \overline{BA}$

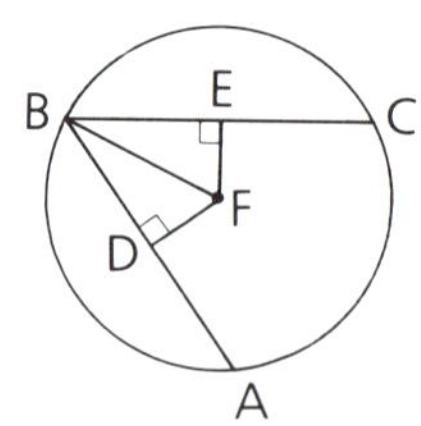

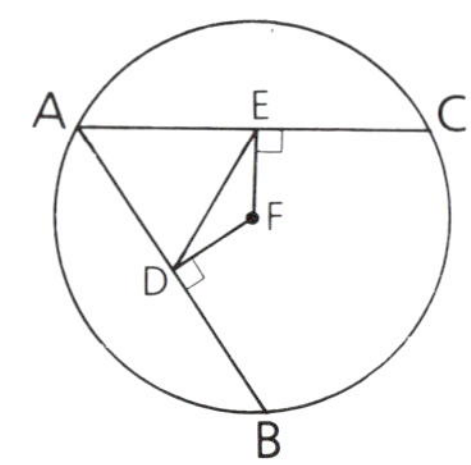

9 Given: Circle F, $\overline{AB} \cong \overline{AC}$

$\overline{DF} \perp \overline{AB}$, $\overline{EF} \perp \overline{AC}$

Prove: $\triangle$ADE is isosceles

10 In circle O, $PB = 3x - 17$,
$CD = 15 - x$,
$OQ = OP = 3$

Find: **a** AB **b** the radius of $\odot$ O

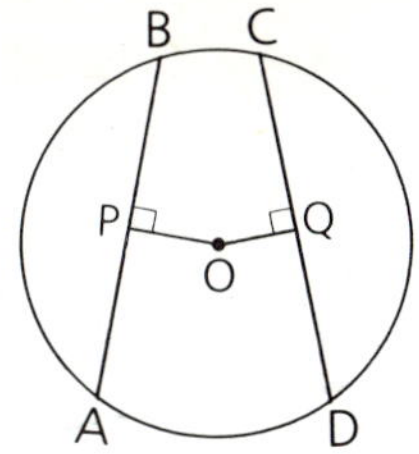

11 Given: Circle Q with $\overline{QM} \cong \overline{QN}$
$\overline{QM} \perp \overline{PT}$, $\overline{QN} \perp \overline{UP}$

Conclusion: $\overline{PR} \perp \overline{TU}$

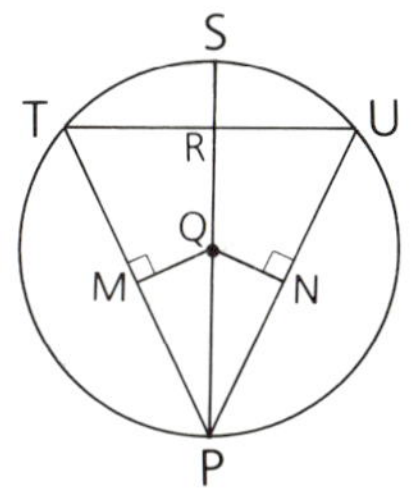

Section 10.2 Problem Set C

12 Given: Circle F with $\overline{DB} \cong \overline{EC}$
$\angle FDB$ and $\angle FEC$ are right $\angle$s

Prove: $\overleftrightarrow{AF} \perp$ bis $\overline{DE}$

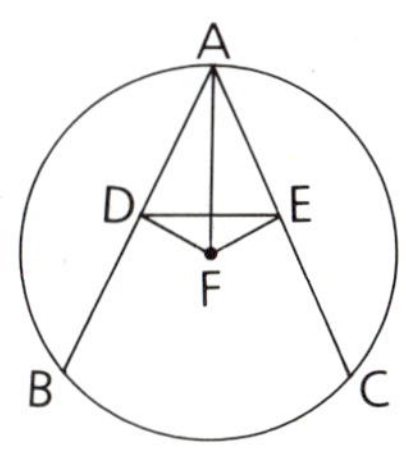

13 Given: $\odot$ O with $\overline{PQ} \cong \overline{TS}$

Prove: $\overline{RQ} \cong \overline{RS}$

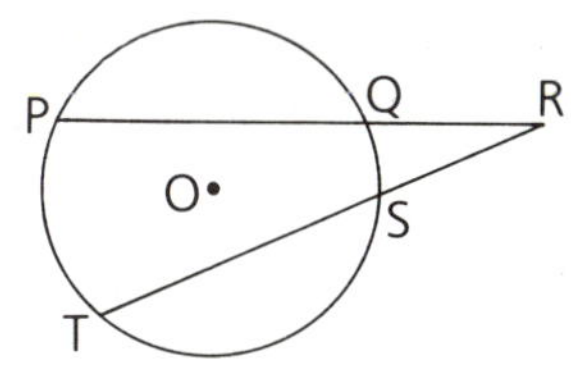

14 $\triangle ABC$ is isosceles with $\overline{AB} \cong \overline{AC}$
$\odot E$, $\overline{AD} \perp \overline{BC}$, $\overline{EF} \perp \overline{AC}$,
$AF = 6$, $ED = 1$.

Find the radius of the circle and the perimeter of $\triangle ABC$.

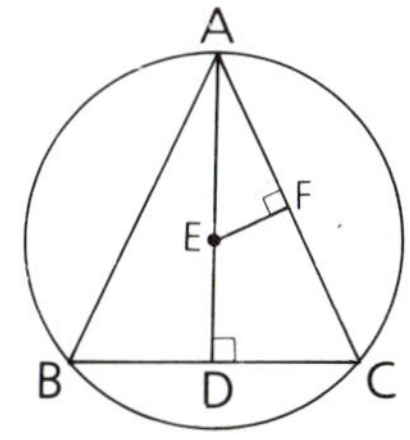

15 Two chords intersect inside a circle. If a diameter drawn to the intersection point bisects the angle formed by the chords, then the chords are congruent. (Hint: Prove that the chords are equidistant from the center of the circle.)

16 In paragraph form, prove that the midpoints of all congruent chords of a circle form another circle concentric with the given circle.

Section 10.3 Arcs of a Circle

A TYPES OF ARCS
B THE MEASURE OF AN ARC
C CONGRUENT ARCS
D RELATING CONGRUENT ARCS, CHORDS, AND CENTRAL ANGLES

10.3 (A) Types of Arcs

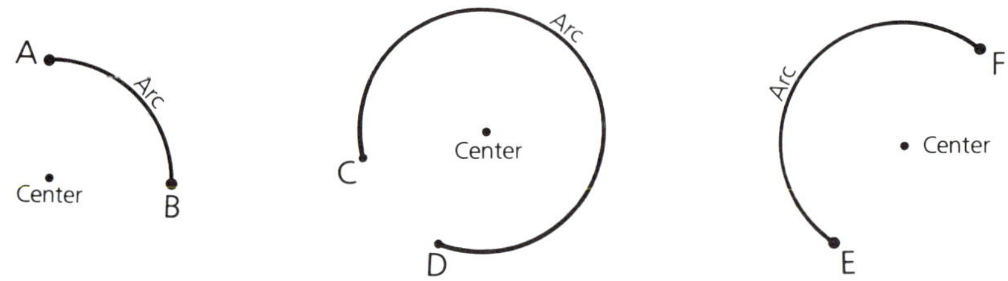

D **An ARC consists of two points on a circle and all points on the circle needed to connect the points by a single path.**

D **The CENTER of an arc is the center of the *circle* of which the arc is a part.**

D **A CENTRAL ANGLE is an angle whose vertex is at the center of a circle.**

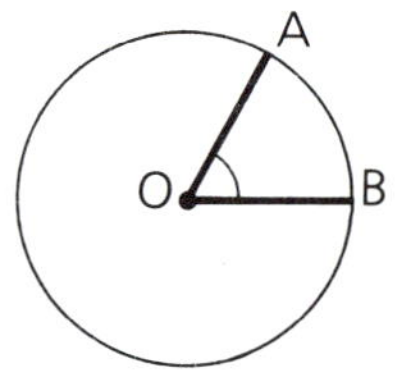

Radii $\overline{OA}$ and $\overline{OB}$ determine central $\angle AOB$.

D **A MINOR ARC is an arc whose points are on or between the sides of a central angle.**

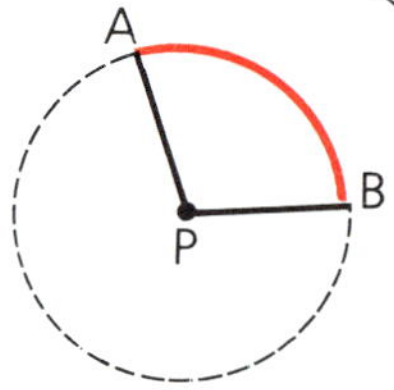

Central $\angle APB$ determines minor arc AB.

D **A MAJOR ARC is an arc whose points are on or outside of a central angle.**

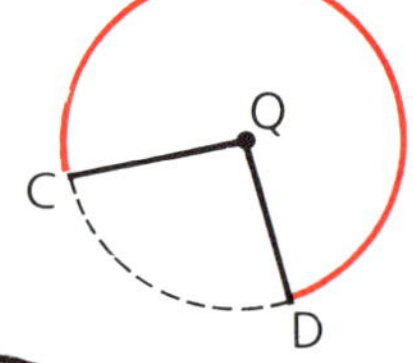

Central $\angle CQD$ determines major arc CD.

D **A SEMICIRCLE is an arc of a circle, whose endpoints are the endpoints of a diameter.**

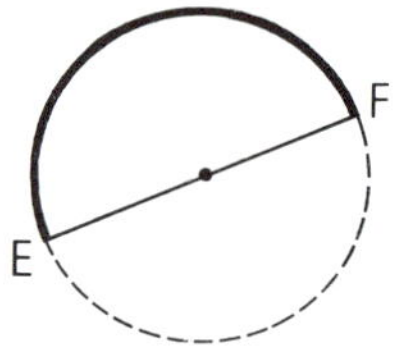

Arc EF is a semicircle.

To label arcs, the symbol "$\frown$" is used in the same way that the symbol "—" is used to denote *segment*. There is a chance for confusion due to the fact that every two points on a circle will be the endpoints of *two* arcs, one major and one minor, or the arcs may be semicircles. In either case, writing $\overset{\frown}{AB}$ would not tell which arc was being discussed. Therefore, when an arc AB is known to be a *minor* arc it will be denoted $\overset{\frown}{AB}$, and in all other cases a third point, X, will be placed between the endpoints, A and B. Thus, $\overset{\frown}{AXB}$ means the arc from A to B by way of X.

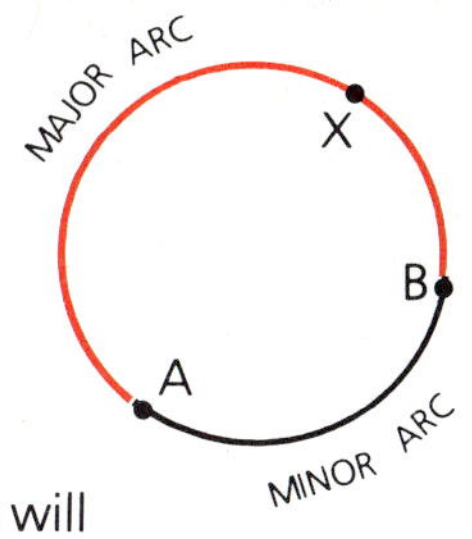

10.3 (B) The Measure of an Arc

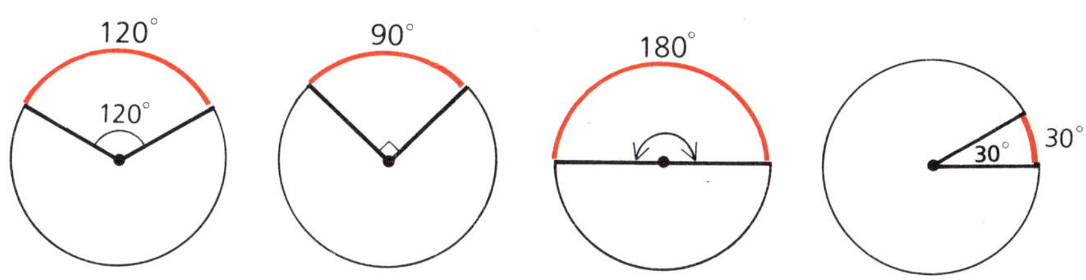

D **The MEASURE OF A MINOR ARC OR SEMICIRCLE is the same as the measure of the central angle that intercepts the arc.**

D **The MEASURE OF A MAJOR ARC in degrees is 360 minus the measure of the minor arc with the same endpoints.**

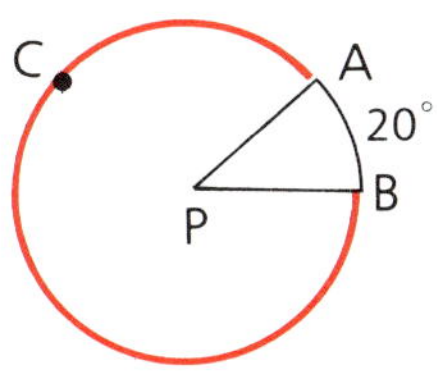

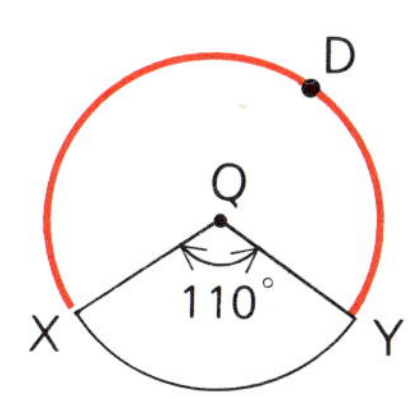

Given: $m\overset{\frown}{AB} = 20$

Find: $m\overset{\frown}{ACB}$

Solution:

$m\overset{\frown}{ACB} = 360 - 20$

$= 340$

Given: $m\angle XQY = 110$

Find: m(major arc XDY)

Solution:

The minor arc XY is 110°

so major arc XDY $= 360° - 110°$

$m(\overset{\frown}{XDY}) = 250$

10.3 (C) Congruent Arcs

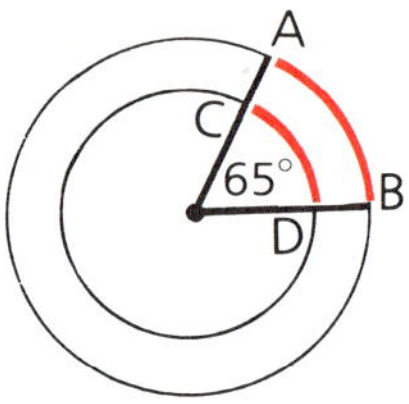

Two arcs that have the same degree measure are not necessarily congruent arcs. In the concentric circles shown, $\overset{\frown}{AB}$ and $\overset{\frown}{CD}$ are each 65°, but $\overset{\frown}{AB}$ and $\overset{\frown}{CD}$ are *not* congruent.

Under what conditions will two arcs be congruent?

D **Two ARCS are CONGRUENT whenever they have the same measure *and* are in the *same* or *congruent* circles.**

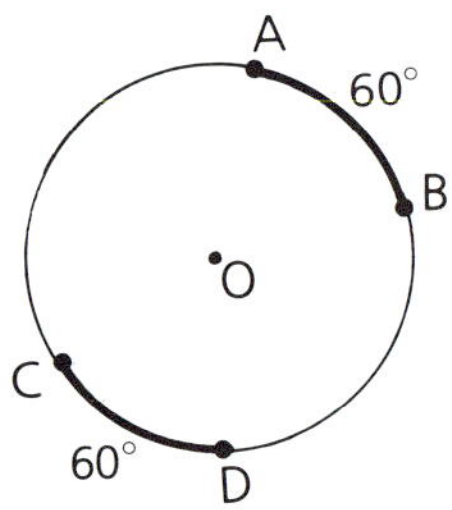

We may conclude $\overset{\frown}{AB} \cong \overset{\frown}{CD}$

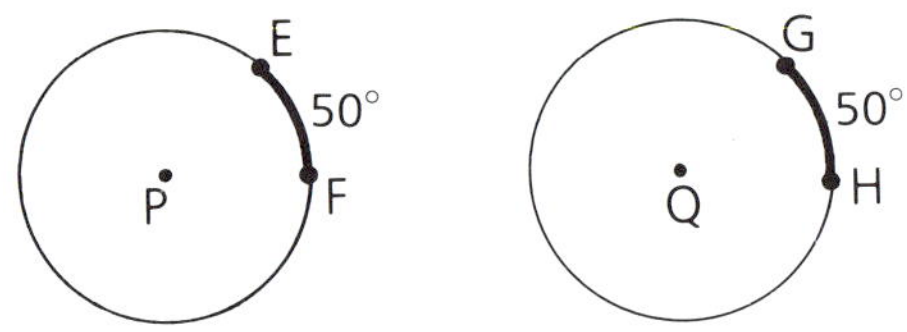

Given: $\odot P \cong \odot Q$

We may conclude $\overset{\frown}{EF} \cong \overset{\frown}{GH}$

10.3 (D) Relating Congruent Arcs, Chords, and Central Angles

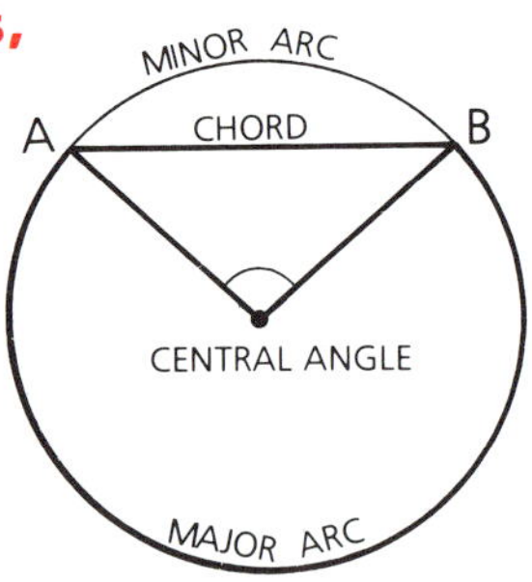

In the diagram, points A and B determine one central angle, one chord, and two arcs (one major and one minor).

You can readily prove the following theorems:

Theorem 70

T **If two central angles of a circle (or of congruent circles) are congruent, then their intercepted arcs are congruent.**

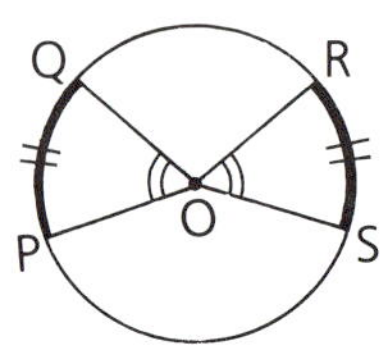

Theorem 71

T **If two arcs of a circle (or of congruent circles) are congruent, then their central angles are congruent.**

Theorem 72

T **If two central angles of a circle (or of congruent circles) are congruent, then their chords are congruent.**

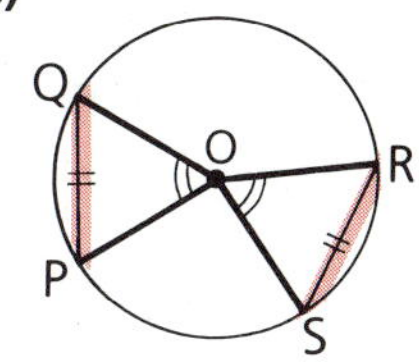

Theorem 73

T **If two chords of a circle (or of congruent circles) are congruent, then their central angles are congruent.**

Theorem 74

T **If two arcs of a circle (or of congruent circles) are congruent, then their corresponding chords are congruent.**

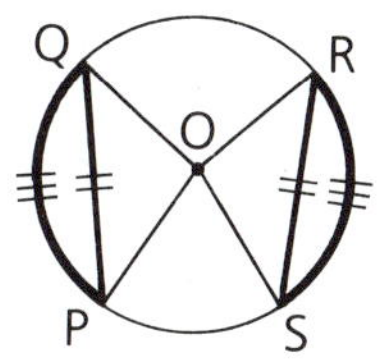

Theorem 75

T **If two chords of a circle (or of congruent circles) are congruent, then their corresponding arcs are congruent.**

To summarize, in the same circle or in congruent circles,

CONGRUENT CHORDS ⟺ CONGRUENT ARCS ⟺ CONGRUENT CENTRAL ANGLES

Section 10.3 Sample Problems

1 Given: ⊙ B

D is the midpt. of $\overset{\frown}{AC}$

Conclusion: $\overrightarrow{BD}$ bisects ∠ABC

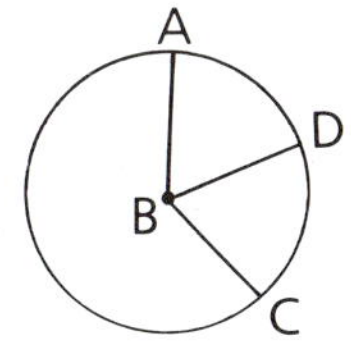

1 ⊙ B, D is the midpt. of $\overset{\frown}{AC}$	1 Given
2 $\overset{\frown}{AD} \cong \overset{\frown}{DC}$	2 The midpoint of an arc divides the arc into two congruent arcs
3 ∠ABD ≅ ∠DBC	3 If two arcs of a circle are ≅, then their central ∠s are ≅
4 $\overrightarrow{BD}$ bisects ∠ABC	4 If a ray divides an ∠ into two ≅ ∠s, then the ray bisects the ∠

2 Circle O, $m\widehat{AB} = 102$

Find $m\angle A$ and $m\angle B$ in $\triangle AOB$

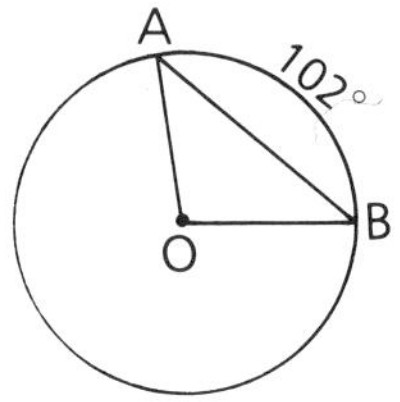

Solution: $\widehat{AB} = 102°$, therefore, $\angle AOB = 102°$.

The sum of the angles of a triangle is 180°, so

$$\begin{aligned} \angle AOB + \angle A + \angle B &= 180° \\ 102 + \angle A + \angle B &= 180° \\ \angle A + \angle B &= 78°. \end{aligned}$$

But, $\overline{OA} \cong \overline{OB}$ so that $\angle A \cong \angle B$.

Hence, $m\angle A = 39$ and $m\angle B = 39$.

3 a What fractional part of a circle is an arc of 36°? of 200°?

b Find the measure of an arc that is $\frac{7}{12}$ of its circle.

Solutions:

a $36°$ is $\frac{36}{360}$ or $\frac{1}{10}$ of a ⊙

$200°$ is $\frac{200}{360}$ or $\frac{5}{9}$ of a ⊙

b There are 360° in a whole ⊙.

$\frac{7}{12}$ of $360 = \frac{7}{12} \cdot \frac{360}{1} = 210$

4 Given: ⓢ P and Q, $\angle P \cong \angle Q$,

$\overline{AR} \cong \overline{RD}$

Prove: $\widehat{AB} \cong \widehat{CD}$

(First prove that ⊙ P ≅ ⊙ Q.)

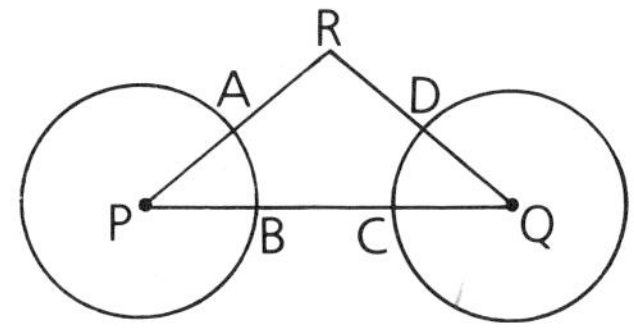

1 ⓢ P and Q	1 Given
2 $\angle P \cong \angle Q$	2 Given
3 $\overline{RP} \cong \overline{RQ}$	3 If $\triangle$ then $\triangle$
4 $\overline{AR} \cong \overline{RD}$	4 Given
5 $\overline{AP} \cong \overline{DQ}$	5 Subtraction property
6 ⊙ P ≅ ⊙ Q	6 ⓢ with ≅ radii are ≅
7 $\widehat{AB} \cong \widehat{CD}$	7 If two central ∠s of *congruent* ⓢ are ≅, then their intercepted arcs are ≅

Section 10.3 Problem Set A

1 Match each item in the left column with its correct name from the right column.

1	$\overset{\frown}{QRS}$	a	Radius
2	$\overline{QS}$	b	Diameter
3	$\overset{\frown}{RQS}$	c	Chord
4	$\overset{\frown}{RS}$	d	Minor Arc
5	$\overline{RS}$	e	Major Arc
6	$\angle RPQ$	f	Semicircle
7	$\overline{PS}$	g	Central Angle

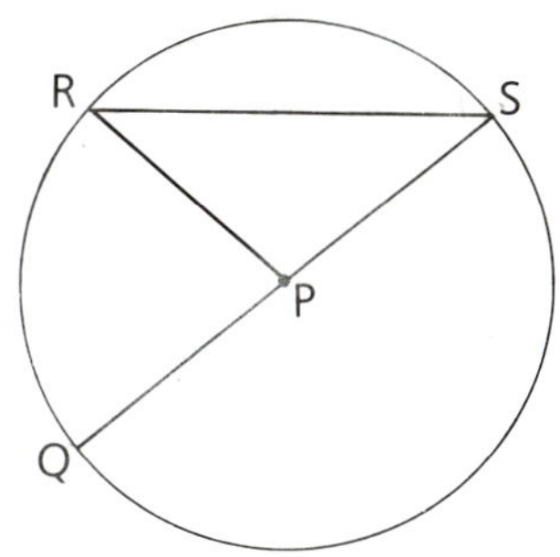

2 Given: Two concentric circles with center O, $\angle BOC$ is acute.

a Name a major arc of the smaller circle.

b Name a minor arc of the larger circle.

c What is $m\overset{\frown}{BC} + m\overset{\frown}{PQ}$?

d Which is greater $m\overset{\frown}{BC}$ or $m\overset{\frown}{PQ}$?

e Is $\overset{\frown}{BC} \cong \overset{\frown}{QR}$?

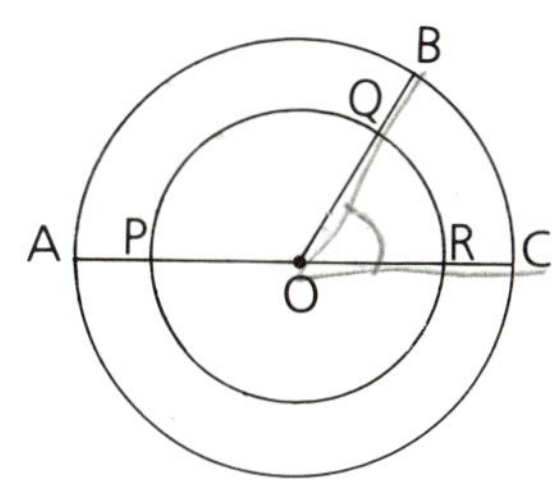

3 In circle E, find each of the following:

a $m\overset{\frown}{BC}$ c $m\overset{\frown}{ACD}$ e $m\overset{\frown}{ADC}$

b $m\overset{\frown}{AD}$ d $m\overset{\frown}{BAD}$

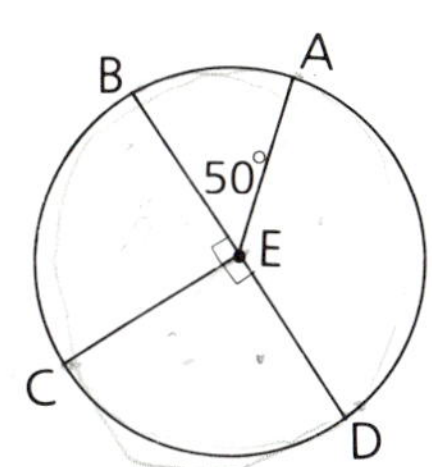

4 In circle Q, $\angle A = 25°$.
Find $m\overset{\frown}{AB}$.

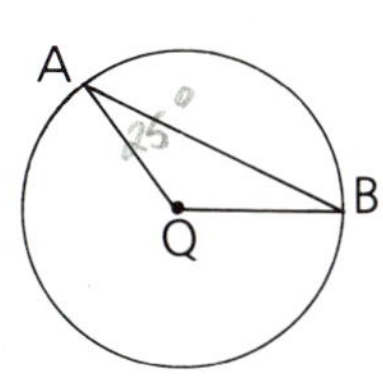

5 Given: $\odot$ P
$\overset{\frown}{WY} \cong \overset{\frown}{XZ}$

Conclusion: $\overline{WX} \cong \overline{YZ}$

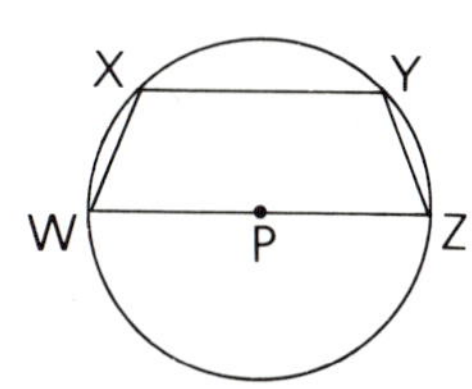

6 Given: $\odot$ D, $\angle B \cong \angle C$

Conclusion: $\overset{\frown}{AB} \cong \overset{\frown}{AC}$

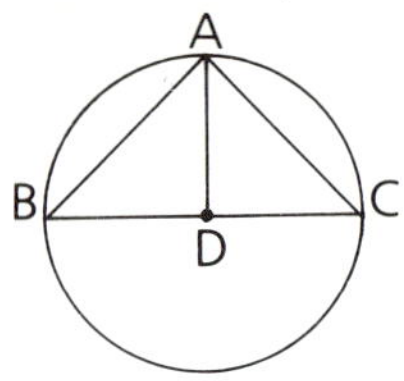

7 Given: $\overline{AB} \cong \overline{CD}$

Conclusion: $\overset{\frown}{AC} \cong \overset{\frown}{BD}$

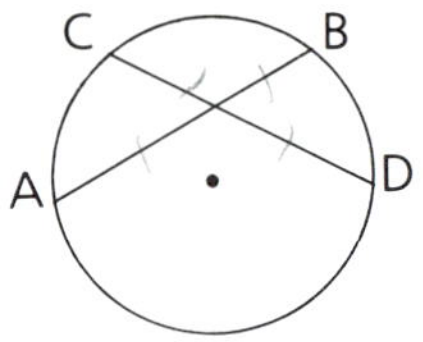

8 Given: $\odot$ E

$\overline{AB} \cong \overline{CD}$

Prove: $\overline{BD} \cong \overline{AC}$

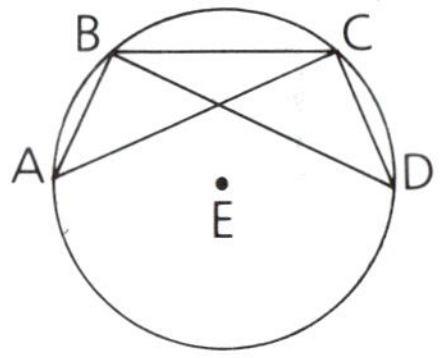

9 What fractional part of a circle is an arc that measures:

a 8° c 120° e 240° g 300°

b 72° d 144° f 270° h 1°

10 Find the measure of an arc that is:

a $\frac{3}{5}$ of its circle c 70% of its circle

b $\frac{5}{9}$ of its circle d $\frac{1}{40}$ of its circle

11 Prove this extension of Theorem 65: If a radius is $\perp$ to a chord, it bisects the arc of the chord. Use Theorem 70, p. 422.

Section 10.3 Problem Set B

12 $\overline{AD}$ is a diameter of $\odot$ E

$m\overset{\frown}{AB} = 9x + 30$

$m\overset{\frown}{CD} = 54 - x$

C is the midpoint of $\overset{\frown}{BD}$

Find $m\angle AEC$

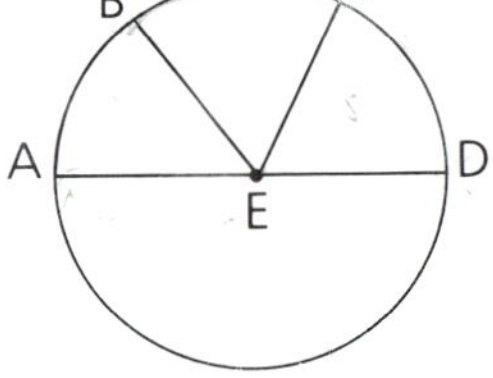

13 Find the length of a chord that cuts off an arc of measure 60 in a circle of radius 12.

14 $\overline{AB}$ is a chord of circle E, and C is the midpoint of $\overset{\frown}{AB}$.
Prove $\overleftrightarrow{EC}$ is the perpendicular bisector of chord $\overline{AB}$.

15 Given: $\odot$ Q, B is the midpt. of $\overset{\frown}{AC}$
Conclusion: $\angle A \cong \angle C$

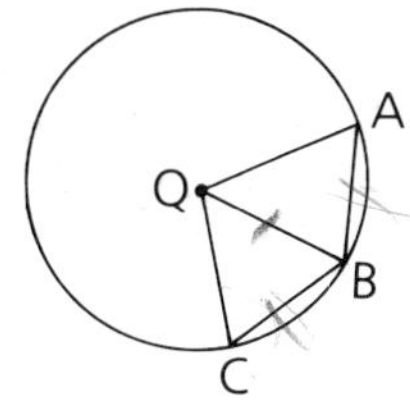

16 Given: $\odot$ O
$\overline{OP} \perp \overline{AC}$, $\overline{OQ} \perp \overline{BD}$
$\overline{OP} \cong \overline{OQ}$
Conclusion: $\overset{\frown}{AB} \cong \overset{\frown}{CD}$

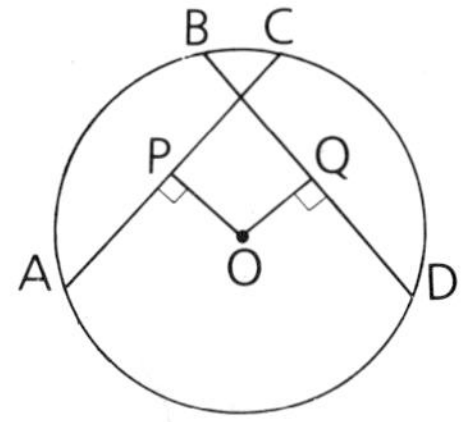

17 Given: $\odot$ B $\cong$ $\odot$ D
$\overset{\frown}{AE} \cong \overset{\frown}{CE}$
Prove: ABCD is a ▱

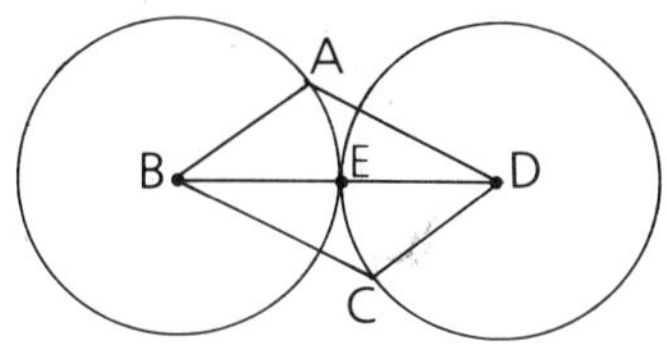

18 A polygon is inscribed in a $\odot$ if all its vertices lie on the $\odot$.
Find the measure of the arc cut off by a side of each of these regular inscribed polygons:

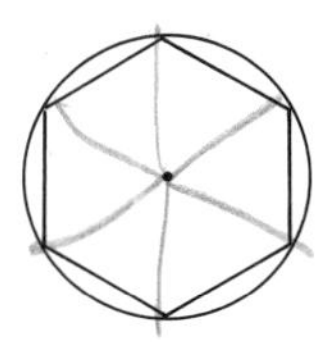

a Hexagon **b** Pentagon **c** Octagon

19 Given: $\odot$ P
C is the midpt. of $\overset{\frown}{AB}$
D is the midpt. of $\overline{AB}$
Prove: $\overleftrightarrow{CD}$ passes through P

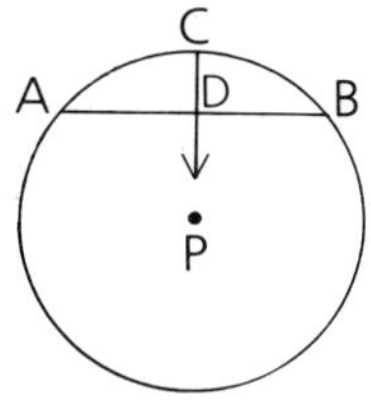

20 Given: $\odot$ P $\cong$ $\odot$ Q
$\overline{BC} \cong \overline{CD}$
Conclusion: $\angle A \cong \angle E$

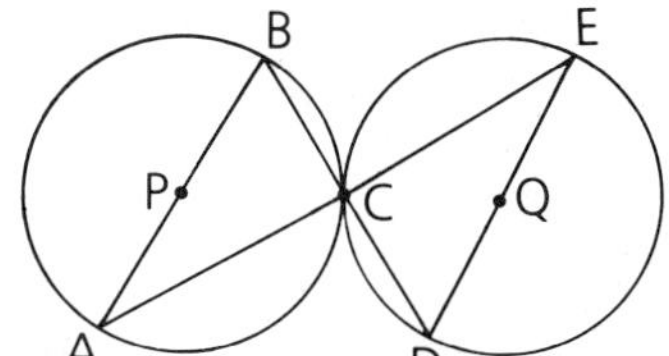

Section 10.3 Problem Set C

21 Given: $\odot E$

$\overline{AB} \cong \overline{CD}$

Conclusion: $\overset{\frown}{FB} \cong \overset{\frown}{CG}$

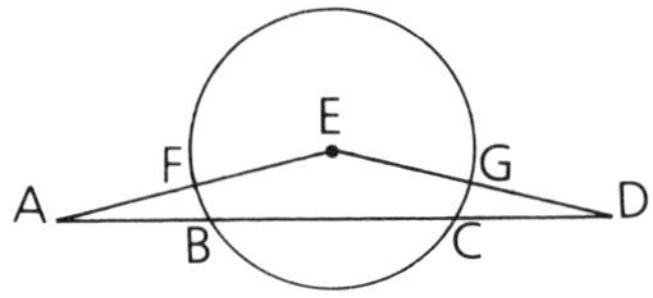

22 From point Q on circle P, another arc is drawn that contains point P. Find the measure of the arc AQB that is cut off.

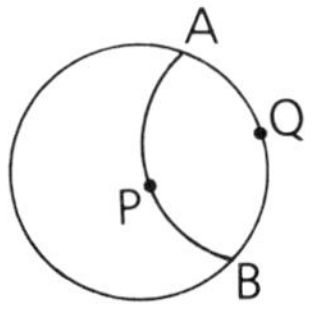

23 If there are "n" points on a given circle, find a formula:

a for the number of chords that can be drawn between pairs of these points.

b to tell how many arcs are formed (including major, minor and semi-circle). Hint: draw circles and count arcs for n = 1, 2, 3 . . . until you see a number pattern.

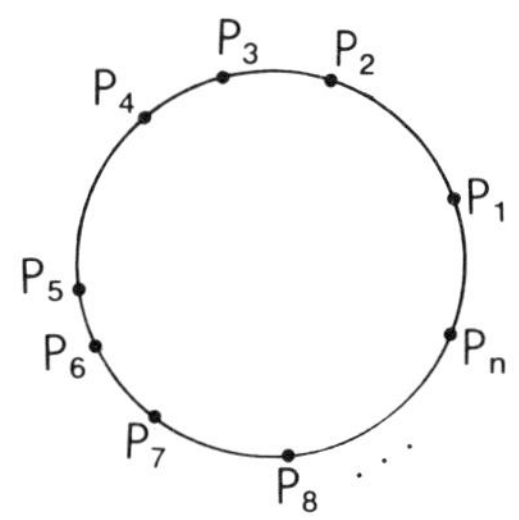

c for the measure of an arc formed by a side of a *regular* n-gon inscribed in the circle.

24 $\odot P \cong \odot Q$

$XY = 8$

$RP = QS = 1$

Find PQ

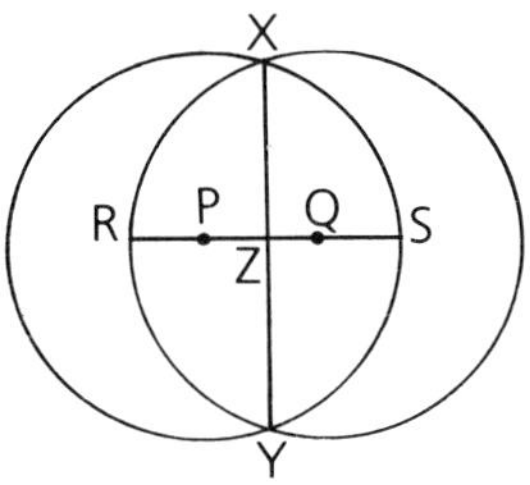

25 Prove that if an equilateral polygon is inscribed in a circle, then it is equiangular.

26 If a diameter bisects two chords of a circle, then the chords are parallel.

a Under what conditions is the above statement true?

b Prove that the statement is true under those conditions.

Section 10.4 Secants and Tangents

10.4 (A) Secant and Tangent Lines

D **A SECANT is a line that intersects a circle at exactly two points. (Every secant contains a chord of the circle.)**

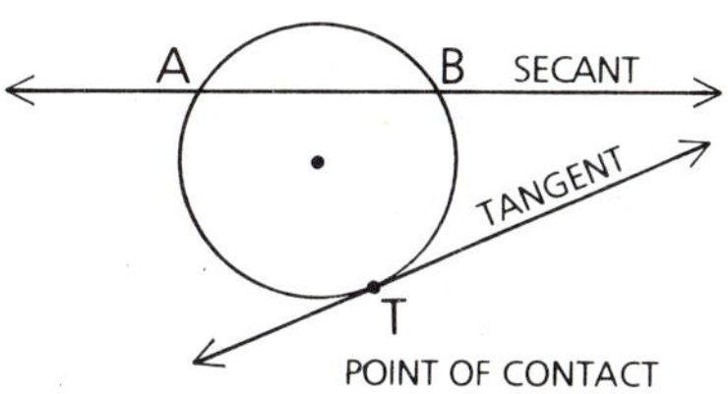

D **A TANGENT is a line that intersects a circle at exactly one point. This point is called the POINT OF TANGENCY or POINT OF CONTACT.**

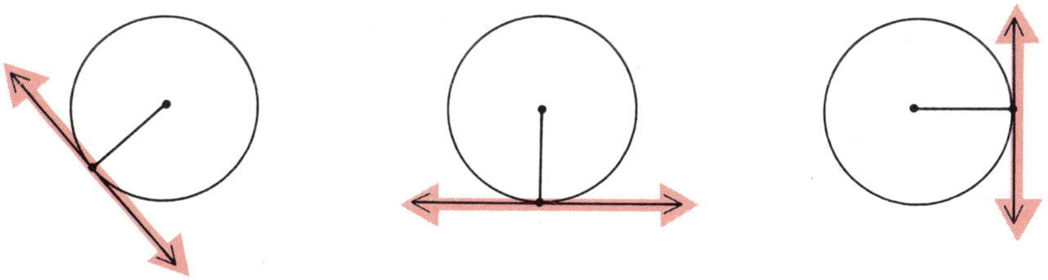

The diagrams above suggest the following postulates about **TANGENTS**:

P **A tangent line is perpendicular to the radius drawn to the point of contact.**

P **If a line is perpendicular to a radius at its outer endpoint, then it is tangent to the circle.**

10.4 (B) Secant Segments and Tangent Segments

D **A TANGENT SEGMENT is that part of a tangent line between the point of contact and a point outside the circle.**

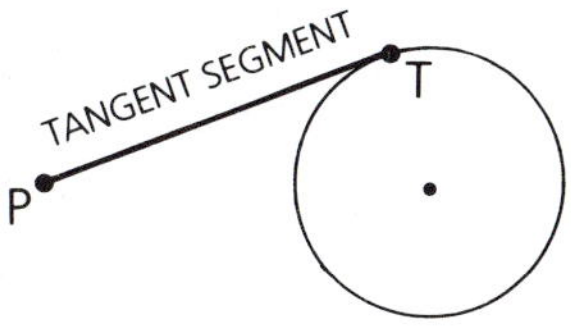

D **A SECANT SEGMENT is the part of a *secant line* that joins a point outside the circle to the *further* intersection point of the secant and the circle.**

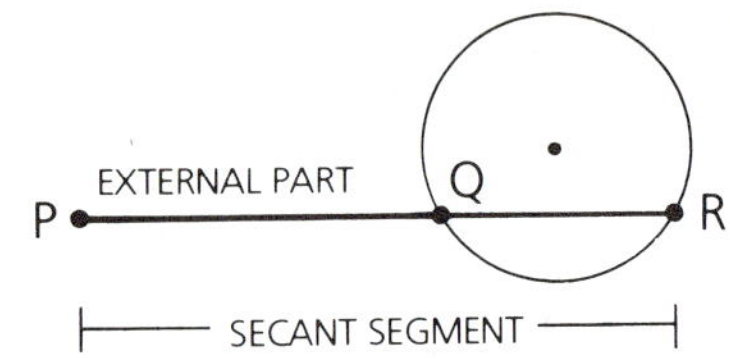

D **The EXTERNAL PART of a secant segment is the part of a secant line that joins the outside point to the closer intersection point.**

Theorem 76

T **If two tangent segments are drawn to a circle from an external point, then those segments are congruent. (TWO TANGENT THEOREM)**

Given: $\odot$ O
$\overline{PX}$ and $\overline{PY}$ are tangent segments

Prove: $\overline{PX} \cong \overline{PY}$

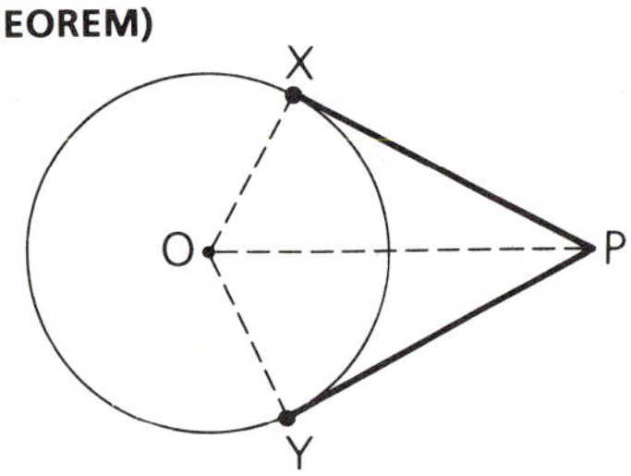

The Two Tangent Theorem is easily proved with congruent triangles.

More theorems of secant segments and tangent segments are in Section 10.8.

10.4 (C) Tangent Circles

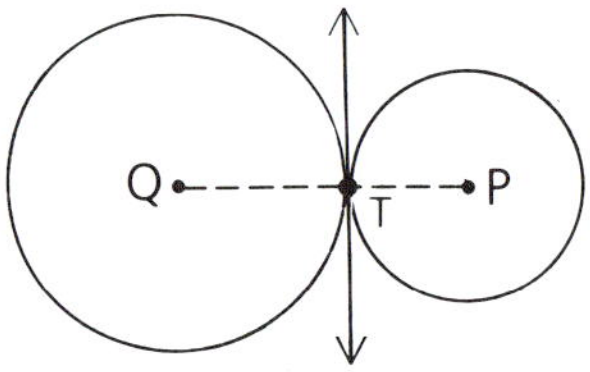

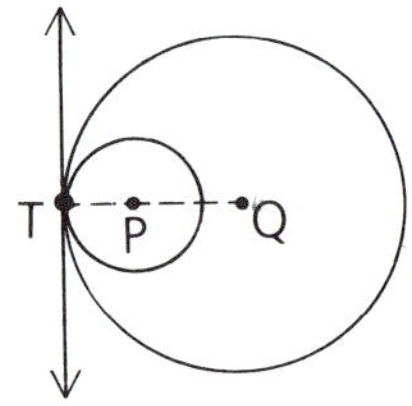

D **TANGENT CIRCLES are two circles that intersect each other at exactly one point.**

D **EXTERNALLY TANGENT CIRCLES are two tangent circles such that one circle lies *outside* the other. (See the left figure, above.)**

D **INTERNALLY TANGENT CIRCLES are two tangent circles such that one circle lies *inside* the other. (See the right figure, above.)**

Notice that in each case the tangent circles have *one common tangent* at their point of contact. Also the *point of contact lies on* the LINE OF CENTERS, $\overleftrightarrow{PQ}$.

10.4 (D) Common Tangents

$\overleftrightarrow{PQ}$ is the line of centers.

$\overleftrightarrow{XY}$ is a common internal tangent.

$\overleftrightarrow{AB}$ is a common external tangent.

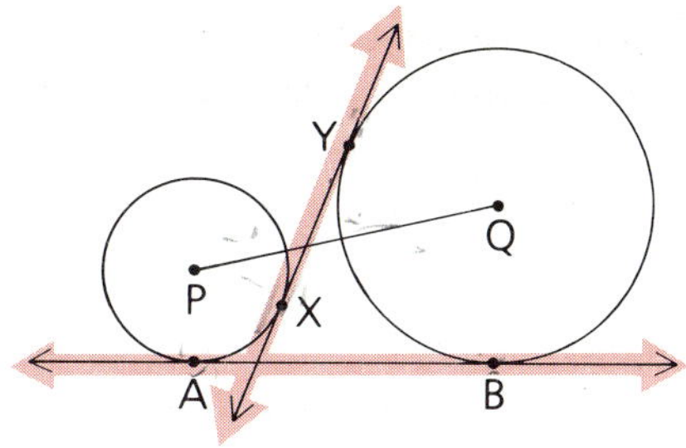

D **A COMMON TANGENT is one line tangent to two circles (not necessarily at the same point). Moreover, a tangent is a COMMON INTERNAL TANGENT if the tangent lies between the circles (crosses the *segment* joining the centers), and is a COMMON EXTERNAL TANGENT if the tangent is not between the circles (does not intersect the *segment* joining the centers).**

Section 10.4 Sample Problems

1 Given: $\overline{XY}$ is a common internal tangent to circles P and Q at X and Y.
$\overline{XS}$ is tangent to ⊙ P at S
$\overline{YT}$ is tangent to ⊙ Q at T

Conclusion: $\overline{XS} \cong \overline{YT}$

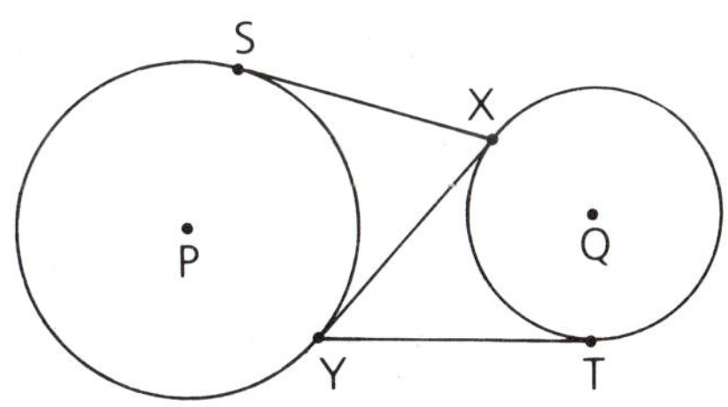

1 $\overline{XS}$ tangent to ⊙ P $\overline{YT}$ tangent to ⊙ Q	1 Given
2 $\overline{XY}$ tangent to ⊙ P and ⊙ Q	2 Given
3 $\overline{XS} \cong \overline{XY}$	3 Two Tangent Theorem
4 $\overline{XY} \cong \overline{YT}$	4 Same as 3
5 $\overline{XS} \cong \overline{YT}$	5 Transitive property

2 $\overleftrightarrow{TP}$ is tangent to circle O at T.
The radius of circle O is 8 mm.
The tangent segment, $\overline{TP}$, is 6 mm long.

Find the length of $\overline{OP}$.

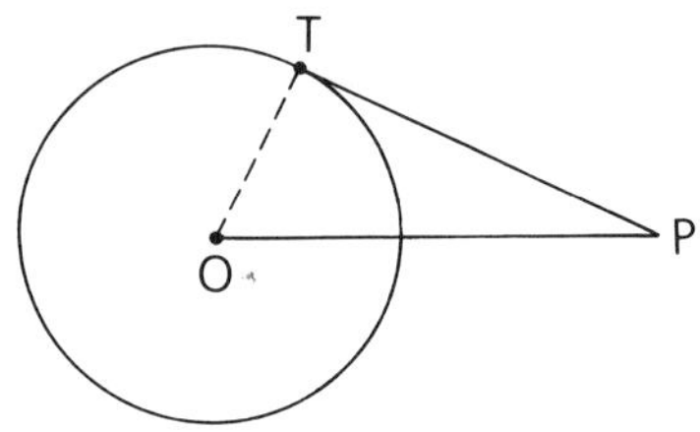

Solution: Draw radius $\overline{OT}$, which will form a right triangle, OTP.

$$\begin{aligned} TP^2 + TO^2 &= OP^2 \\ 6^2 + 8^2 &= OP^2 \\ \pm 10 &= OP, \text{ reject } -10 \\ \text{Thus, } OP &= 10 \text{ mm} \end{aligned}$$

3 A circle of radius 8 cm is externally tangent to a circle of radius 18 cm. Find the length of a common external tangent.

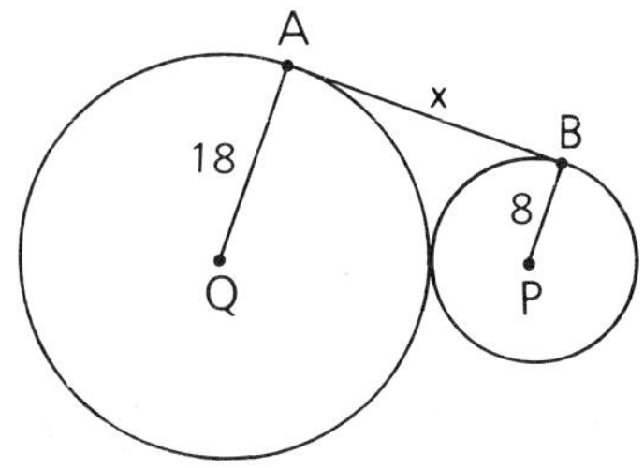

There is a standard procedure for solving problems involving common tangents (both internal and external).

COMMON TANGENT PROCEDURE

1 **Draw the segment of centers.**

2 **Draw the radii to the points of contact.**

3 **Through the center of the smaller circle draw a line parallel to the common tangent.**

4 **Observe that this line will intersect the radius of the larger circle (extended if necessary) to form a rectangle and a right triangle.**

5 **Use the Pythagorean Theorem and properties of a rectangle.**

Solution: For $\triangle RPQ$

$$\begin{aligned} QR^2 + RP^2 &= PQ^2 \\ 10^2 + RP^2 &= 26^2 \\ RP &= \pm 24 \\ \text{thus, } AB &= 24 \text{ cm} \end{aligned}$$

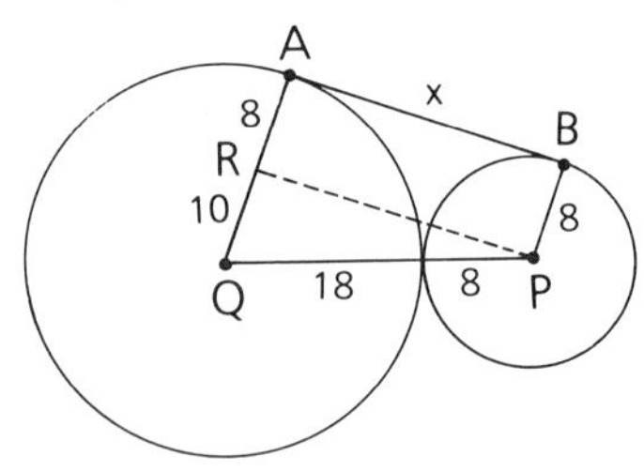

4 A WALK-AROUND PROBLEM

Given: Each side of quadrilateral ABCD is tangent to the circle. AB = 10, BC = 15, AD = 18

Find: CD

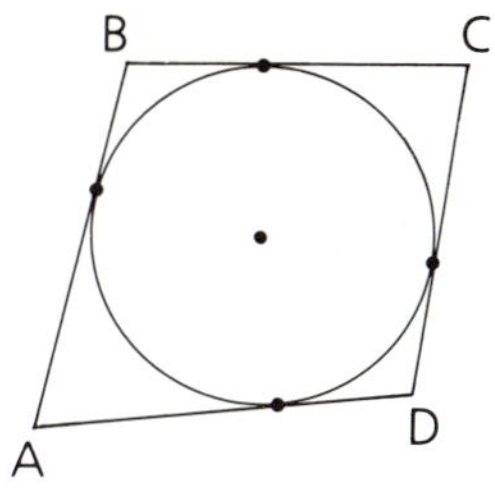

Solution:

Let BE = x and walk around the figure using the given information and the Two Tangent Theorem.

$$
\begin{aligned}
CD &= 15 - x + 18 - (10 - x) \\
&= 15 - x + 18 - 10 + x \\
&= 23
\end{aligned}
$$

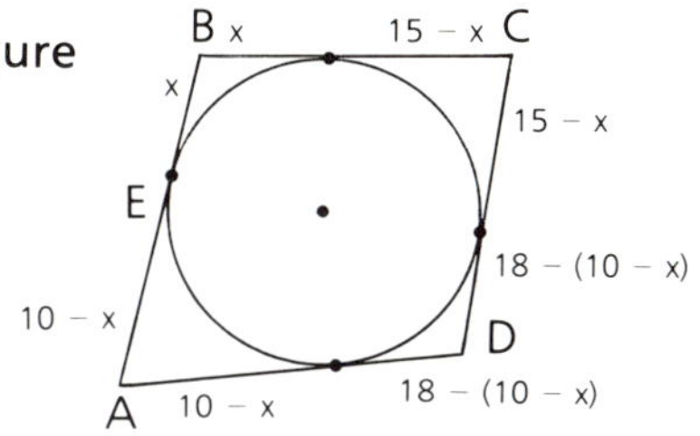

See 13, 17, 18, 24 for other types of Walk-around Problems.

Section 10.4 Problem Set A

1 The radius of ⊙A is 8 cm. Tangent segment $\overline{BC}$ is 15 cm long.

Find the length of $\overline{AC}$.

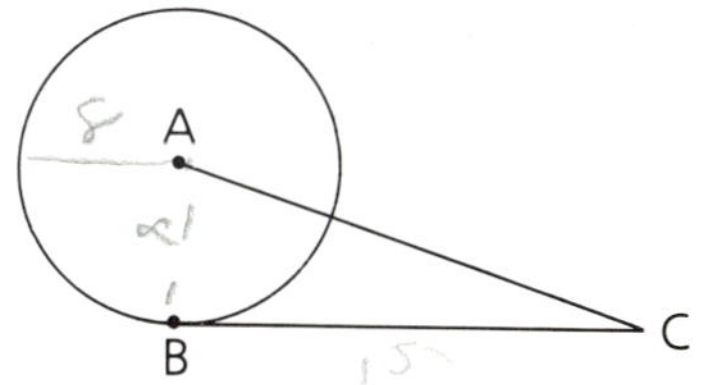

2 Concentric circles with radii 8 and 10 have center P. $\overline{XY}$ is a tangent to the inner ⊙ and is a chord of the outer ⊙.

Find $\overline{XY}$. (Hint: Draw $\overline{PX}$ and $\overline{PY}$.)

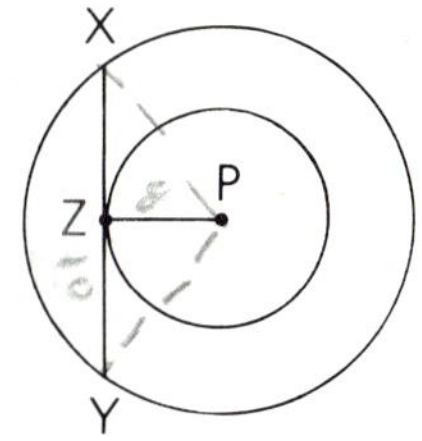

3 Given: $\overline{PR}$ and $\overline{PQ}$ are tangents to ⊙ O at R and Q

Prove: $\overrightarrow{PO}$ bisects ∠RPQ (Hint: Draw $\overline{RO}$ and $\overline{OQ}$.)

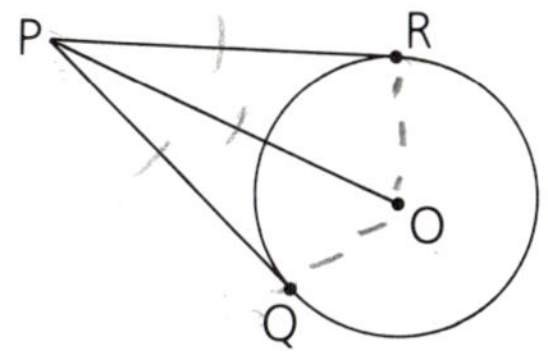

4 Given: $\overline{AC}$ is a diameter of ⊙ B s, m are tangents to the ⊙ at A and C

Conclusion: s ∥ m

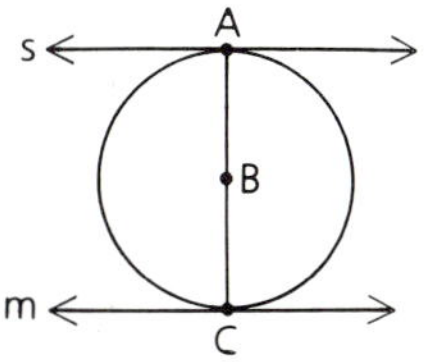

5 For each pair of circles illustrated below, tell how many common internal tangents and how many common external tangents can be drawn.

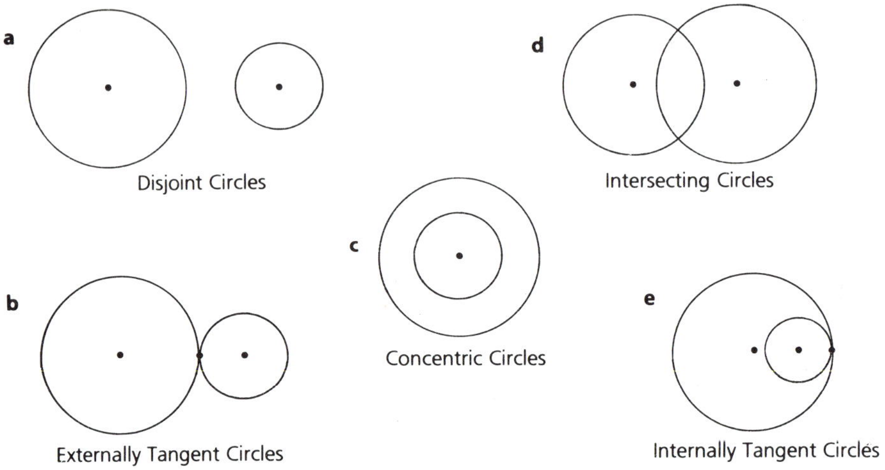

6 Given: $\overline{CE}$ is a common internal tangent to circles A and B at C and E

Prove: **a** $\angle A \cong \angle B$

b $\dfrac{AD}{BD} = \dfrac{CD}{DE}$

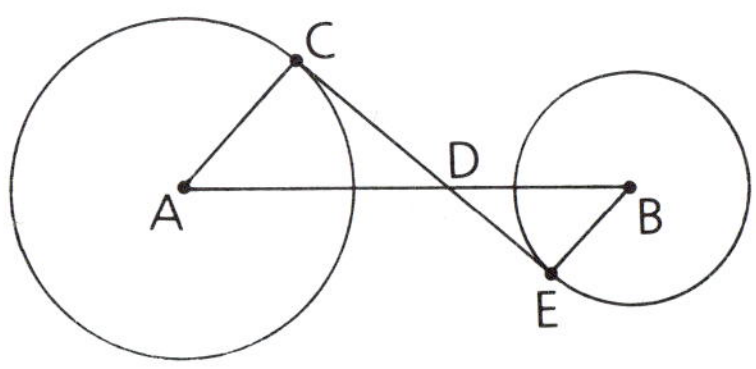

7 Given: $\overline{QR}$, $\overline{QS}$ are tangent to ⊙ P at points R and S

Prove: $\overline{PQ} \perp \overline{RS}$

(This can be done in just a few steps.)

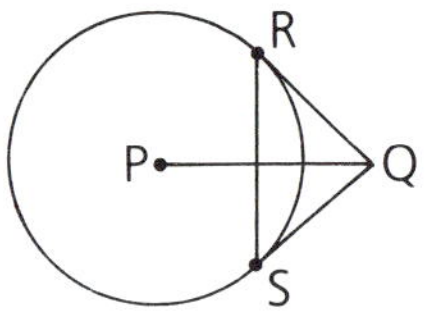

8 Given: $\overline{PW}$, $\overline{PZ}$ are common tangents to Ⓢ A and B at W, X, Y, and Z

Prove: $\overline{WX} \cong \overline{YZ}$

(Note: No auxiliary lines are needed.)

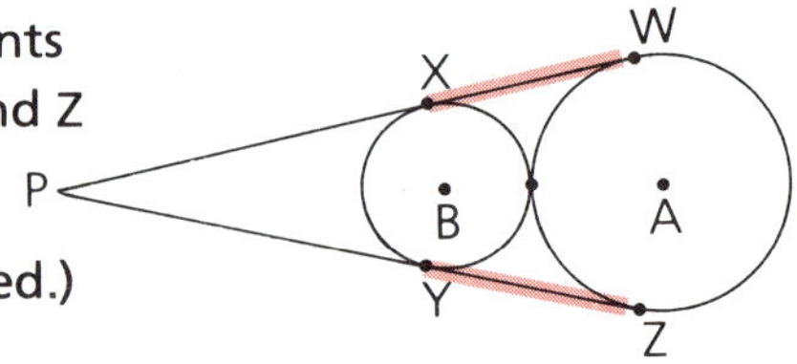

This is part of the proof of the property:

The common external tangent segments of two circles are congruent.

Section 10.4 Problem Set B

9 Two concentric circles have radii 3 and 7. Find the length of a chord of the larger circle that is tangent to the smaller circle. (Hint: See problem 2 for a diagram.)

10 The centers of two circles of radii 10 and 5 cm are 13 cm apart.

a Find the length of a common external tangent. (See the COMMON TANGENT PROCEDURE.)

b Are the circles intersecting or disjoint?

11 The centers of two circles with radii 3 and 5 are 10 units apart. Find the length of a common internal tangent. (See the COMMON TANGENT PROCEDURE.)

12 Given: $\overline{PT}$ is tangent to Ⓢ Q and R at points S and T

Conclusion: $\frac{PQ}{PR} = \frac{SQ}{TR}$

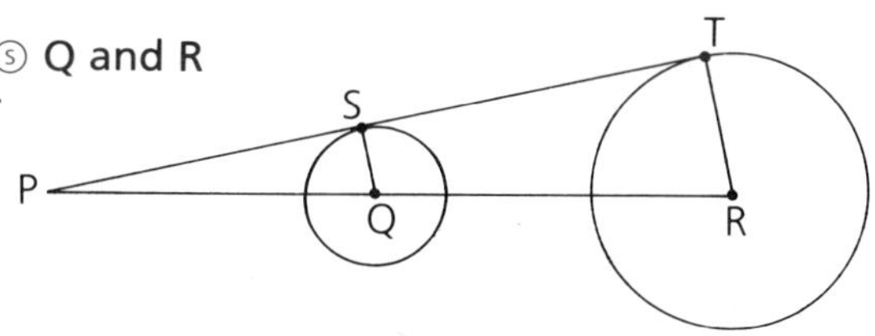

13 Given: Three tangent Ⓢ, A, B, and C
AB = 8, BC = 13, AC = 11

Find: the radii of the three Ⓢ
(A Walk-around problem)

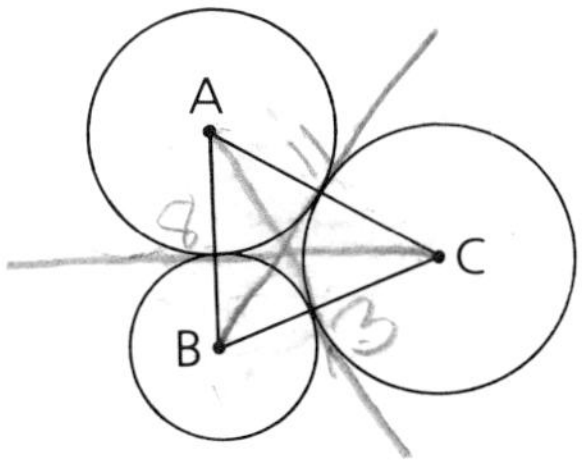

14 The radius of ⊙ O is 10. The secant segment $\overline{PX}$ measures 21 and is 8 units from the center of the ⊙.

a Find the external part, PY, of the secant.

b Find the distance, OP, from point P to the center of the circle.

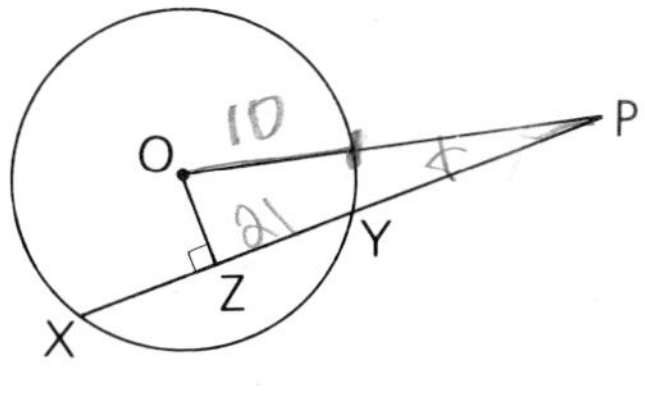

15 Given: $\triangle ABC$ is isosceles with base $\overline{BC}$

Conclusion: $\overline{BR} \cong \overline{RC}$

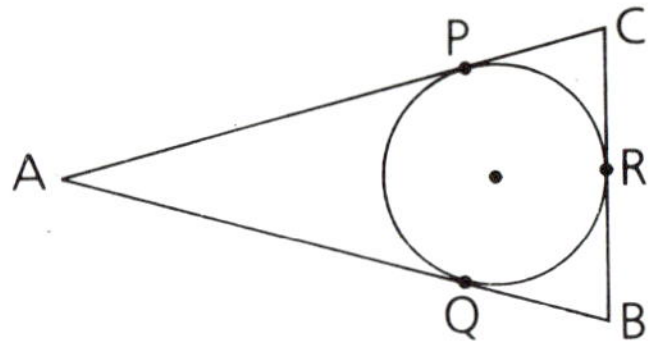

16 Given a ⊙ with center O and points A, B, and C on the ⊙. Segments are drawn connecting each two of the four points. If one of these segments is drawn at random, what is the probability that:

a it is a radius?

b its length equals that of a radius?

Section 10.4 Problem Set C

17 Given: Quadrilateral WXYZ is circumscribed about ⊙ O (sides tangent to the ⊙)

Prove: $m\overline{XY} + m\overline{WZ} = m\overline{WX} + m\overline{YZ}$

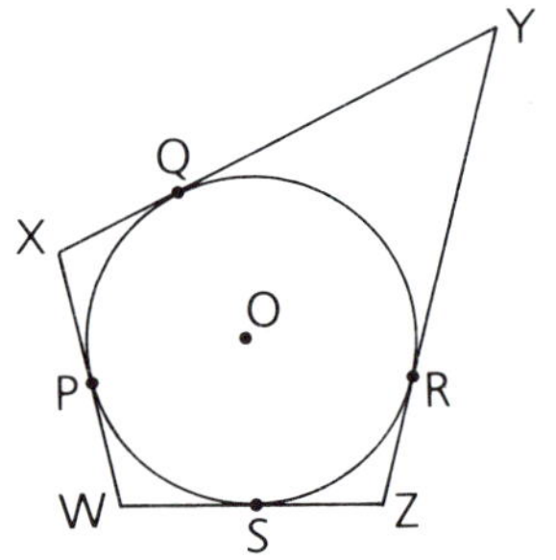

18 Find the perimeter of right triangle WXY, if the radius of the circle is 4 and WY = 20.

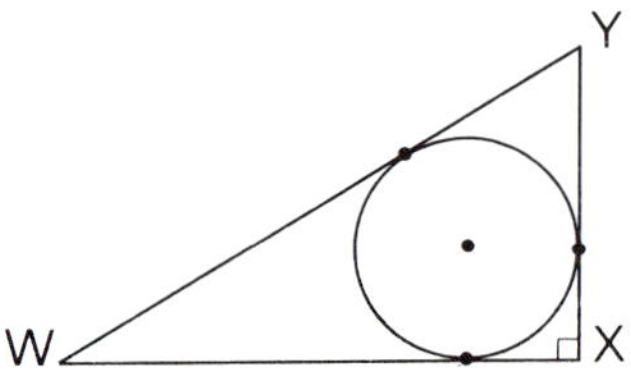

19 B is 34 mm from the center of circle O, which has radius 16 mm.
$\overline{BP}$, $\overline{BR}$, and $\overline{AC}$ are tangent segments.
Find the perimeter of $\triangle ABC$.

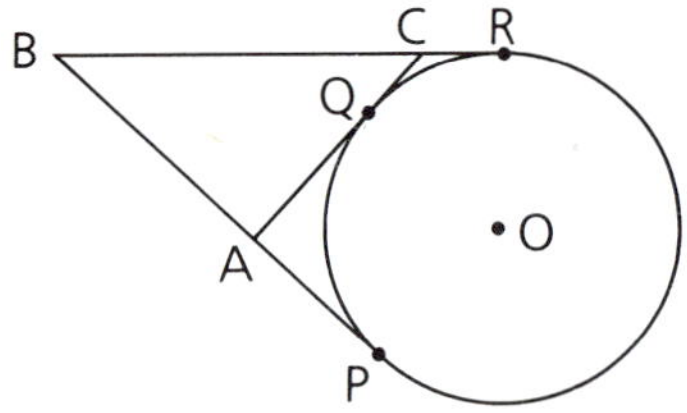

20 Given two concentric circles with center E.
AB = 40, CD = 24, $\overline{CD} \perp \overline{AE}$.
$\overline{AB}$ is tangent at C.
Find AF.

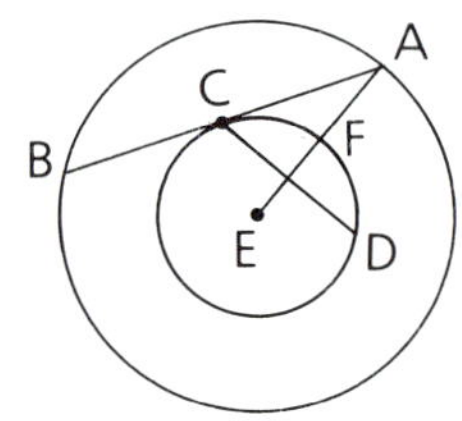

21 $\overline{BC}$ is tangent to ⊙ A at B, and $\overline{BD} \cong \overline{BA}$.

Explain why $\overline{BD}$ bisects $\overline{AC}$.

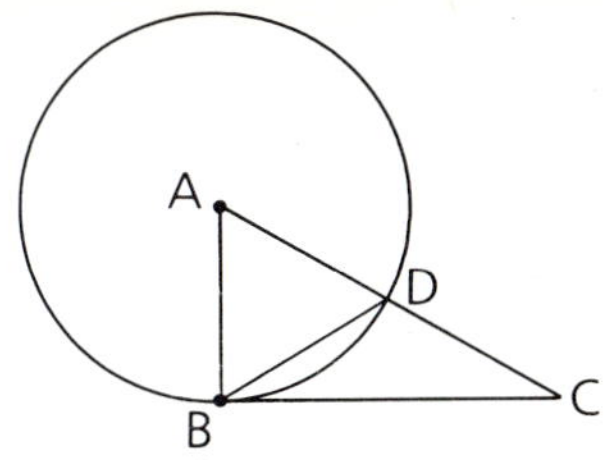

22 Given: Ⓢ E and F with $\overline{AC}$ tangent at B and C
DE = 10, FB = 4

Find: AB

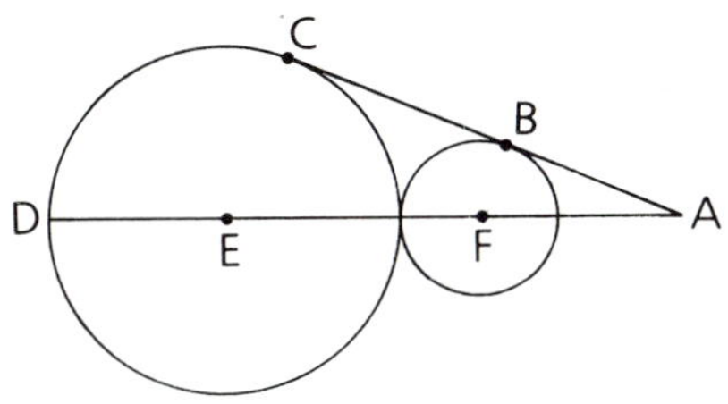

Section 10.4 Problem Set D

23 Two circles are externally tangent at P. A secant line is drawn through P and intersects the circles at A and B, respectively. Prove: The tangents to the circles at A and B are parallel.

24 Given: Three tangent Ⓢ A, B, and C
BC = a, AC = b, and AB = c

a Find the radius of ⊙ A in terms of a, b, and c.

b Can three such circles be found for *any* triangle? Explain why or why not. (Assume that the sum of the lengths of two sides of a triangle is greater than the third side.)

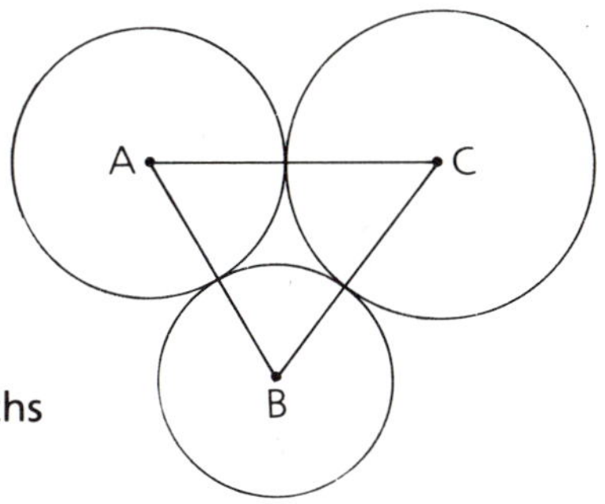

25 Given: ⊙ P ≅ ⊙ Q
$\overset{\frown}{TS} \cong \overset{\frown}{RU}$

Prove: PRQS is a ▱

Hint: 1 Draw a perpendicular from S to $\overline{PR}$ and from R to $\overline{SQ}$.

2 Draw $\overline{SR}$.

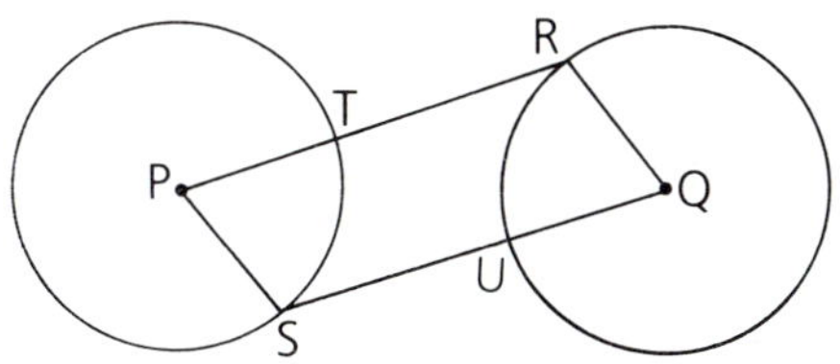

Section 10.5 Angles Related to a Circle

The measure of an angle whose sides intersect a circle is determined by the measure of its intercepted arcs.

The location of the VERTEX of each angle is the key to remembering how to compute the measure of the angle.

An angle may be placed in one of these categories:

A VERTEX AT THE CENTER OF A CIRCLE

B VERTEX ON A CIRCLE

C VERTEX INSIDE BUT NOT AT THE CENTER OF A CIRCLE

D VERTEX OUTSIDE OF A CIRCLE

10.5(A) Vertex at the Center of a Circle

An angle with its vertex at the center of a circle is a *central angle,* already defined to be *equal* in measure to its *intercepted arc* (Section 10.3A and 10.3B).

In $\odot$ O, $\widehat{AB} = 50^\circ$

Find m$\angle$AOB

m$\angle$AOB = 50

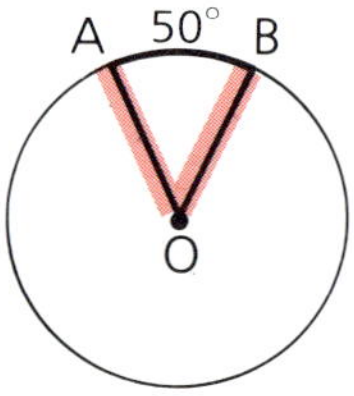

10.5(B) Vertex On the Circle

The two types of angles whose vertices are on a circle are shown below:

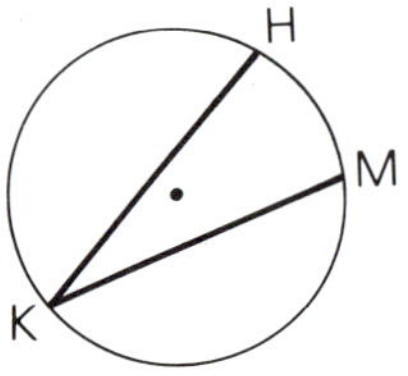

$\angle$HKM is an INSCRIBED ANGLE

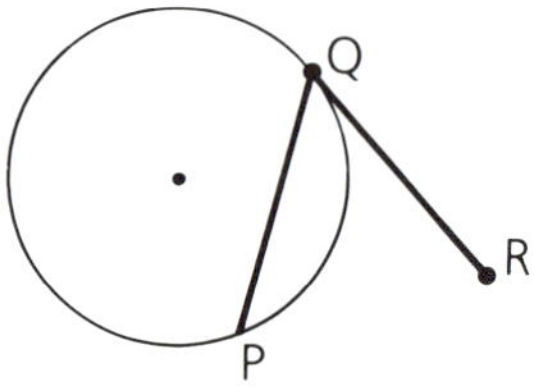

$\angle$PQR is a TANGENT-CHORD ANGLE

D **An INSCRIBED ANGLE is an angle whose vertex is *on* a circle and whose sides are determined by two chords.**

D **A TANGENT-CHORD ANGLE is an angle whose vertex is *on* a circle and whose sides are determined by a tangent and a chord that intersect at the point of contact.**

Theorem 77

T **The measure of an angle whose VERTEX is ON a circle (an INSCRIBED ANGLE or a TANGENT-CHORD ANGLE) is *half* of the measure of its intercepted arc.**

The proof of Theorem 77 for inscribed angles is unusual because three cases must be proved. In each case some key information is shown for proof that

$m\angle BAC = \frac{1}{2}\, m\widehat{BC}$.

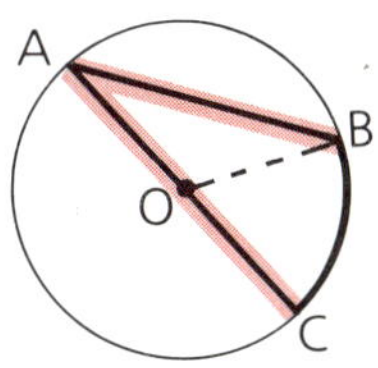

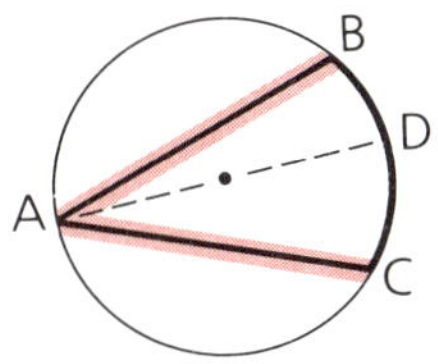

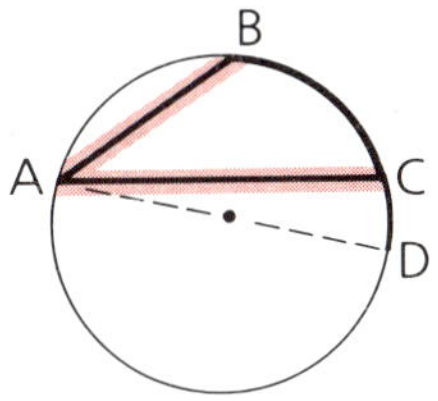

CASE 1

The center lies on a side of the angle.

1 $m\angle BOC = m\widehat{BC}$

2 $\angle BOC = \angle BAC + \angle ABO$
or $m\angle BOC = 2(m\angle BAC)$

CASE 2

The center lies inside the angle

1 Use Case 1 twice

2 Add $\angle$s and arcs

CASE 3

The center lies outside the angle

1 Use Case 1 twice

2 Subtract $\angle$s and arcs

Proof of Theorem 77 for a Tangent-Chord Angle is called for in Problem 37.

■ Examples

$m\widehat{AC} = 112$

Find $m\angle B$

$m\angle B = \frac{1}{2}\, m\widehat{AC}$

$= \frac{1}{2} \cdot 112$

$= 56$

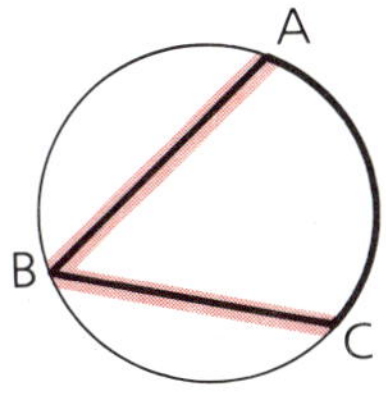

$\overline{FE}$ is tangent at E

$m\widehat{DE} = 80$

Find $m\angle DEF$

$m\angle DEF = \frac{1}{2}\, m\widehat{DE}$

$= \frac{1}{2} \cdot 80$

$= 40$

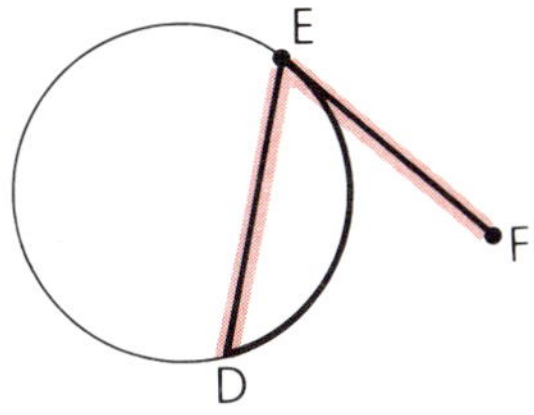

10.5(C) Vertex Inside But Not at the Center of a Circle

D **A CHORD-CHORD angle is an angle formed by two chords that intersect INSIDE a circle, but not at the center.**

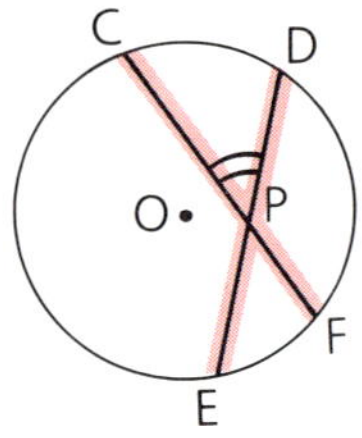

$\angle CPD$ is one of four CHORD-CHORD angles formed by chords $\overline{CF}$ and $\overline{DE}$ in circle O.

Theorem 78

T **The measure of a CHORD-CHORD ANGLE equals *half* of the *sum* of the measures of the arcs cut off by the chord-chord angle and its vertical angle.**

Notice that *half of the sum* is the same as *the average.*

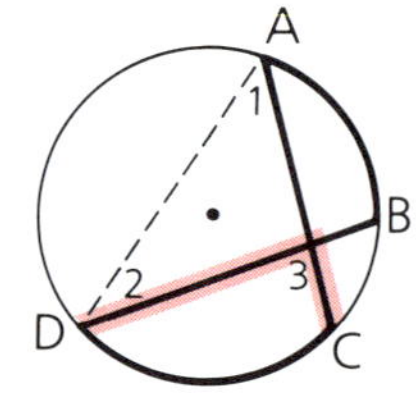

Given: $\angle 3$ is a chord-chord angle

Prove: $m\angle 3 = \frac{1}{2}(m\widehat{AB} + m\widehat{CD})$

Here are some key steps in a proof of Theorem 78.

1 $m\angle 3 = m\angle 1 + m\angle 2$

2 $m\angle 3 = \frac{1}{2}m\widehat{CD} + \frac{1}{2}m\widehat{AB}$

10.5(D) Vertex Outside of a Circle

There are three types of angles with their vertices outside of a circle.

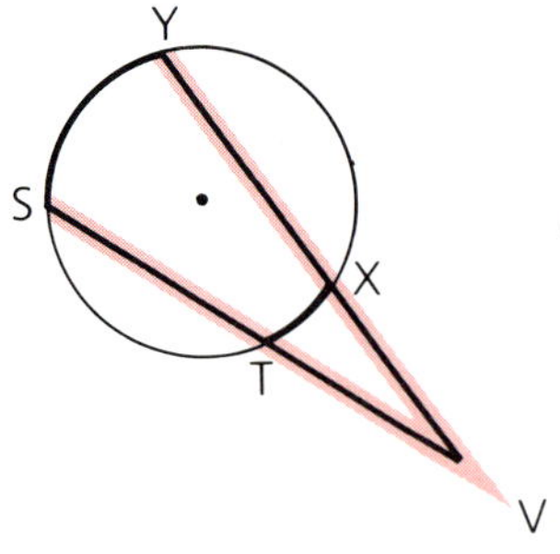

$\angle V$ is a SECANT-SECANT ANGLE

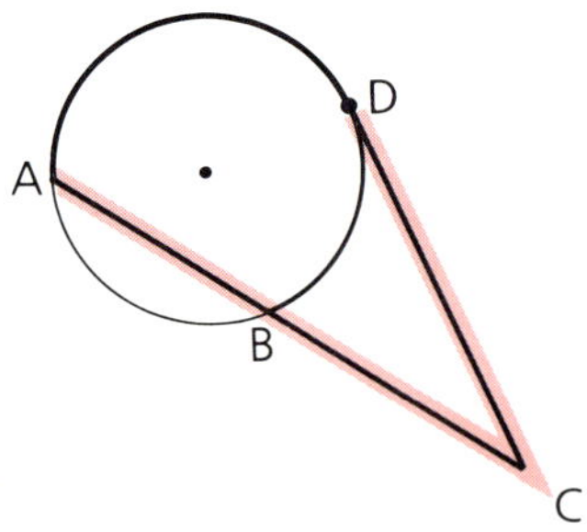

$\angle C$ is a SECANT-TANGENT ANGLE

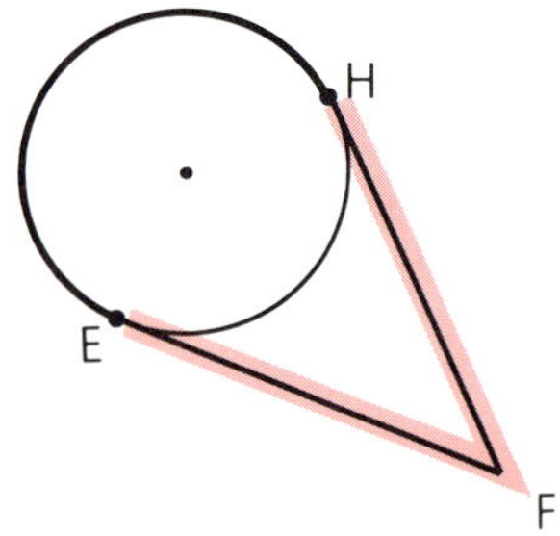

$\angle F$ is a TANGENT-TANGENT ANGLE

D **A SECANT-SECANT ANGLE is an angle whose vertex is outside a circle and whose sides are determined by two secants.**

D **A SECANT-TANGENT ANGLE is an angle whose vertex is outside a circle and whose sides are determined by a secant and a tangent.**

D **A TANGENT-TANGENT ANGLE is an angle whose vertex is outside a circle and whose sides are determined by two tangents.**

Theorem 79

T **The measure of an angle whose *vertex* is OUTSIDE a circle (SECANT-SECANT ANGLE, SECANT-TANGENT ANGLE, or TANGENT-TANGENT ANGLE) is *half* of the DIFFERENCE of the two intercepted arcs.**

Key steps for a proof of Theorem 79 for Secant-Secant angles follow:

Prove: $m\angle E = \frac{1}{2}(m\widehat{SC} - m\widehat{AB})$

1 $m\angle 3 = m\angle E + m\angle 2$; solve for $m\angle E$

2 $m\angle 2 = \frac{1}{2}m\widehat{AB}$, $m\angle 3 = \frac{1}{2}m\widehat{SC}$

3 Substitute and simplify.

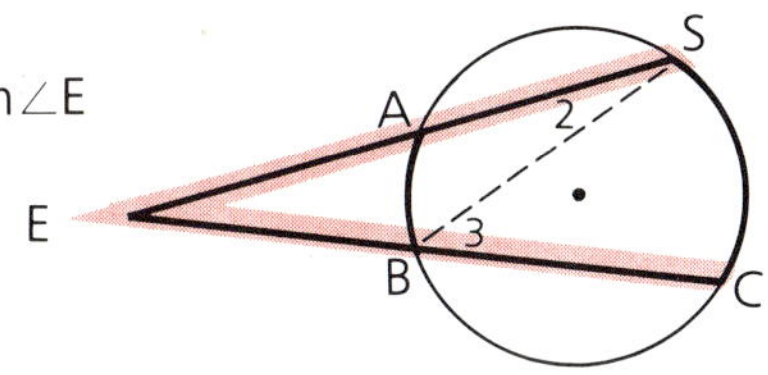

Examples:

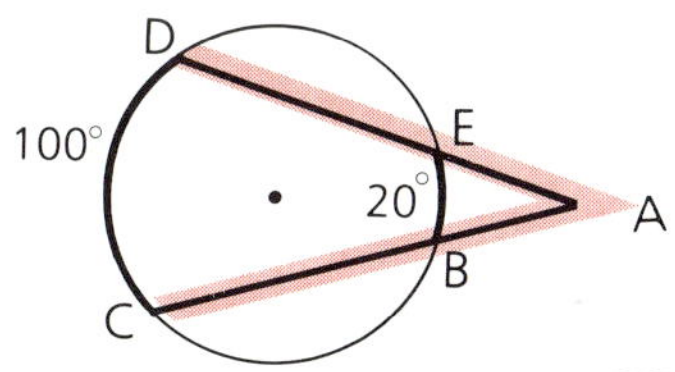

Find $m\angle A$

$m\angle A = \frac{1}{2}(m\widehat{CD} - m\widehat{BE})$

$= \frac{1}{2}(100 - 20)$

$= 40$

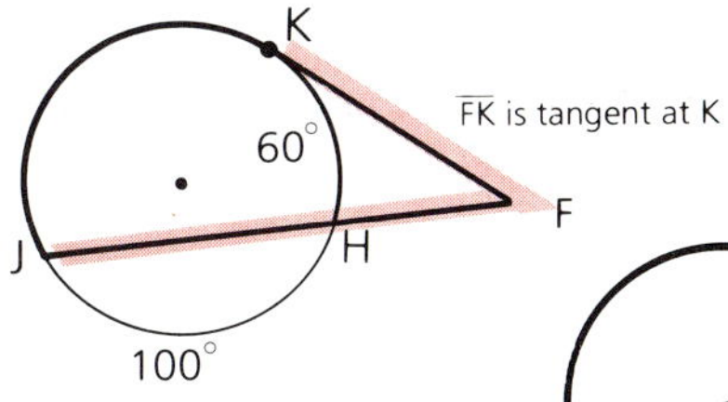

Find $m\angle F$

$m\widehat{JK} = 360 - 100 - 60$

$= 200$

$m\angle F = \frac{1}{2}(m\widehat{JK} - m\widehat{HK})$

$= \frac{1}{2}(200 - 60)$

$= 70$

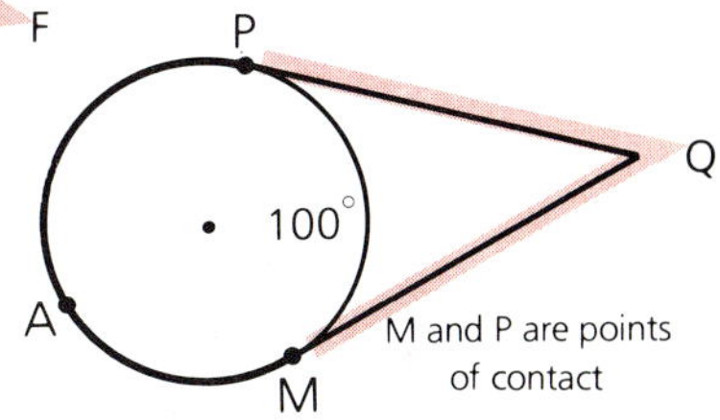

Find $m\angle Q$

$m\widehat{MAP} = 360 - 100 = 260$

$m\angle Q = \frac{1}{2}(m\widehat{MAP} - m\widehat{MP})$

$= \frac{1}{2}(260 - 100)$

$= 80$

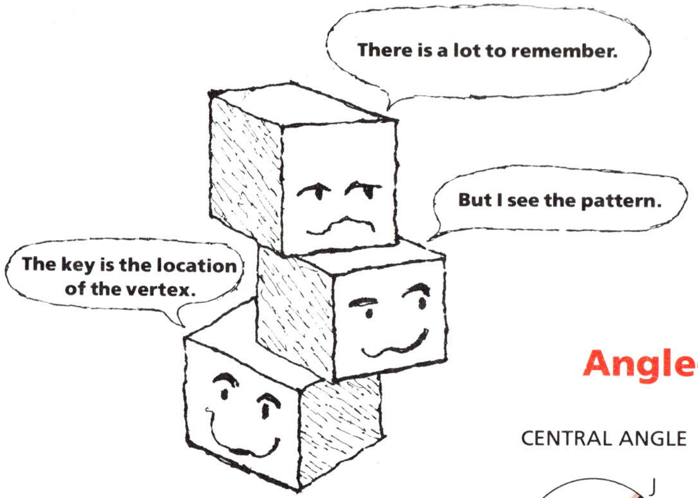

Angle-Arc Summary

CENTRAL ANGLE

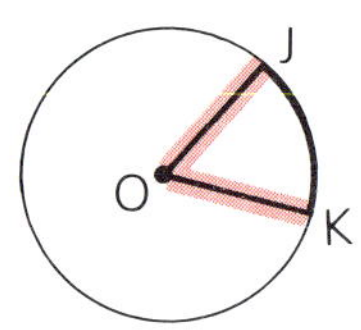

$m\angle KOJ = m\widehat{JK}$

VERTEX AT CENTER ⟹ EQUAL

CHORD-CHORD ANGLE

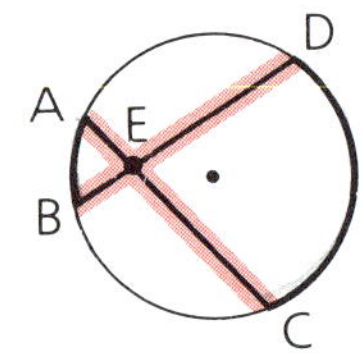

$m\angle DEC = \frac{1}{2}(m\widehat{AB} + m\widehat{CD})$

VERTEX INSIDE ⟹ HALF THE SUM

INSCRIBED ANGLE

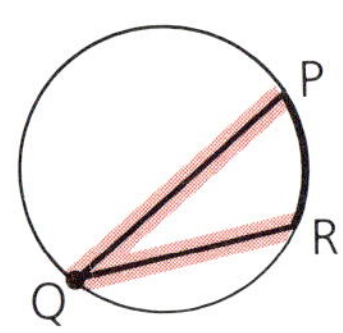

$m\angle Q = \frac{1}{2}(m\widehat{PR})$

TANGENT-CHORD ANGLE

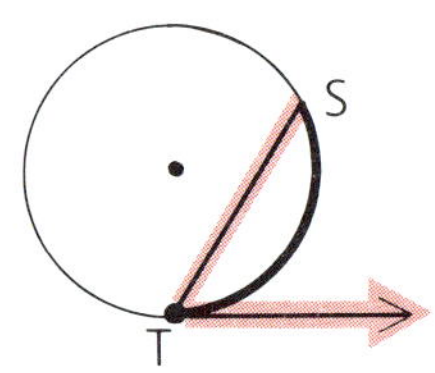

$m\angle T = \frac{1}{2}(m\widehat{ST})$

VERTEX ON CIRCLE ⟹ HALF THE ARC

SECANT-SECANT ANGLE

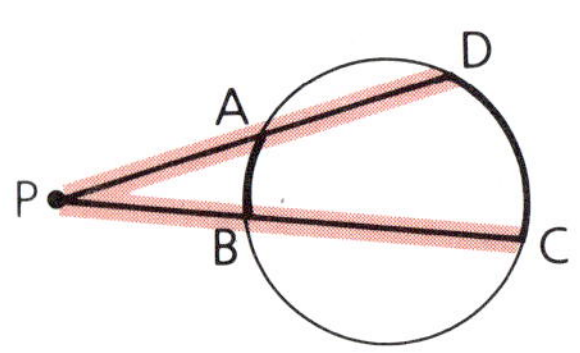

$m\angle P = \frac{1}{2}(m\widehat{CD} - m\widehat{AB})$

TANGENT-TANGENT ANGLE

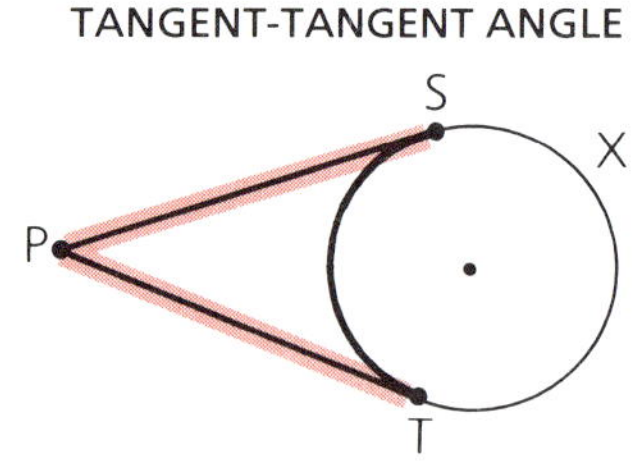

$m\angle P = \frac{1}{2}(m\widehat{SXT} - m\widehat{ST})$

SECANT-TANGENT ANGLE

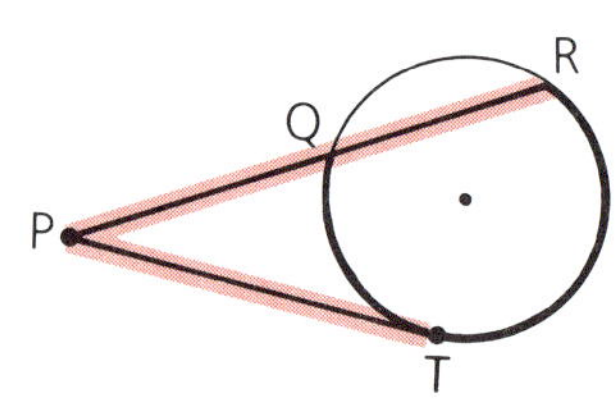

$m\angle P = \frac{1}{2}(m\widehat{RT} - m\widehat{QT})$

VERTEX OUTSIDE CIRCLE ⟹ HALF THE DIFFERENCE

Section 10.5 Sample Problems

1 Find x, y, and z.

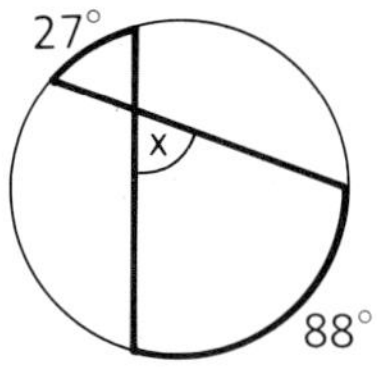

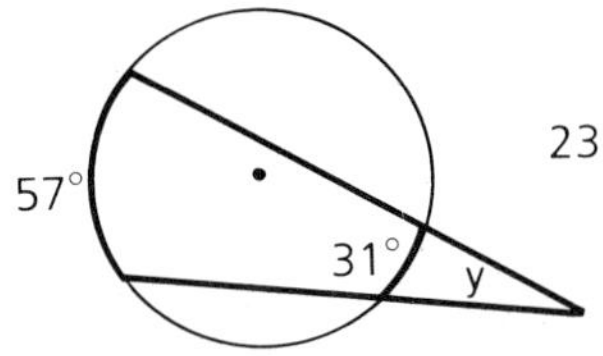

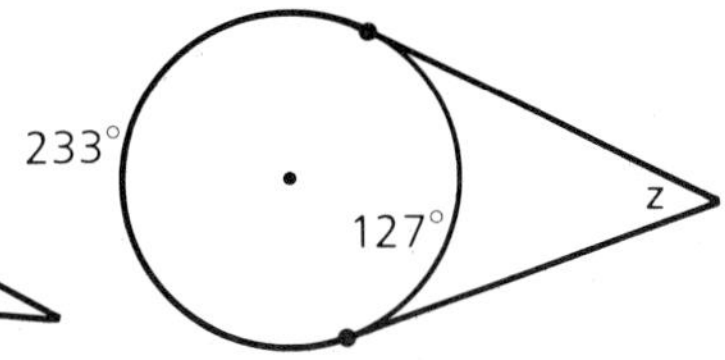

Solutions:

$x = \frac{1}{2}(88 + 27)$

$x = 57\frac{1}{2}°$

$y = \frac{1}{2}(57 - 31)$

$y = 13°$

$z = \frac{1}{2}(233 - 127)$

$z = 53°$

2 **a** Find y. **b** Find z. **c** Find a.

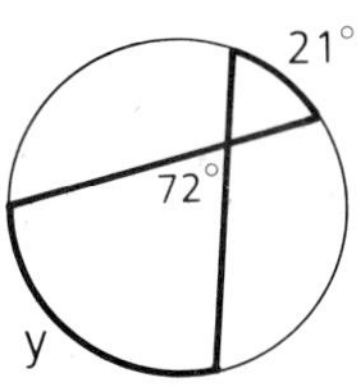

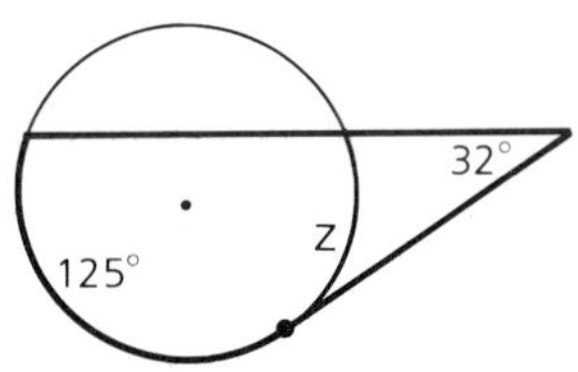

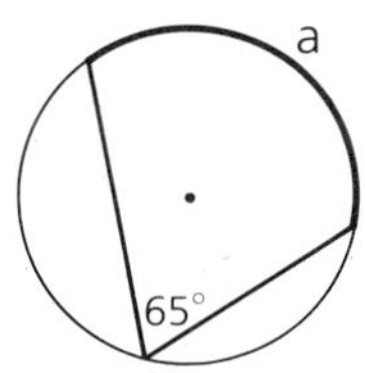

Solutions:

$$\begin{aligned} \frac{1}{2}(21 + y) &= 72 \\ 21 + y &= 144 \\ y &= 123° \end{aligned}$$

$$\begin{aligned} \frac{1}{2}(125 - z) &= 32 \\ 125 - z &= 64 \\ z &= 61° \end{aligned}$$

$$\begin{aligned} \frac{1}{2}(a) &= 65 \\ a &= 130° \end{aligned}$$

3 $\overline{AB}$ is a diameter of ⊙ P.
$\widehat{BD} = 20°$, $\widehat{DE} = 104°$.
Find m∠C.

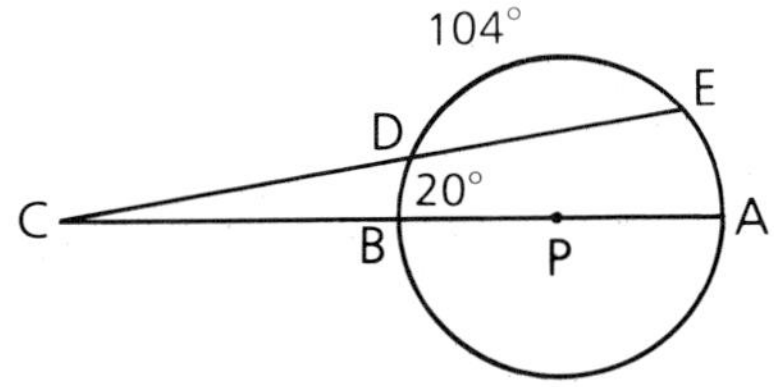

Solution: First find m$\widehat{EA}$.
m$\widehat{AEB}$ = 180, so m$\widehat{EA}$ = 180 − (104 + 20) = 56

$m\angle C = \frac{1}{2}(m\widehat{EA} - m\widehat{DB}) = \frac{1}{2}(56 - 20) = 18$

4 Find y.

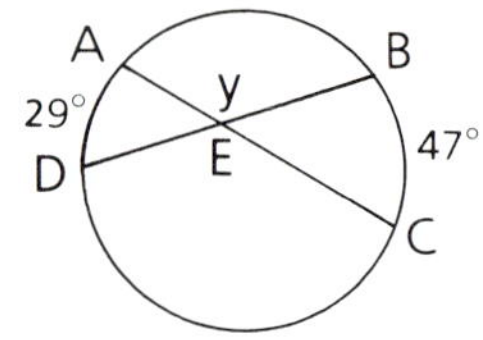

Solution: Solve for $\angle BEC$ first.

$\angle BEC = \frac{1}{2}(29 + 47) = 38°$

Then, $y = 180 - \angle BEC = 142°$

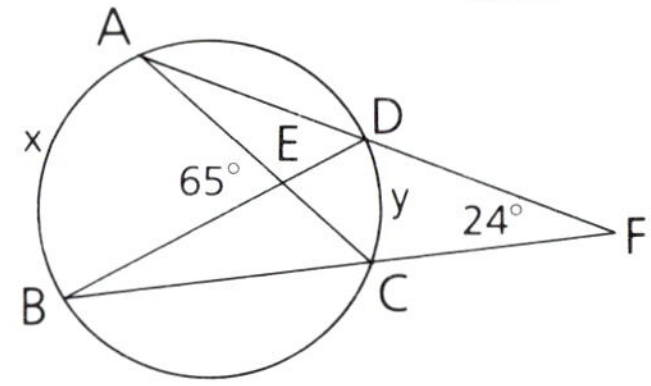

5 Find $m\widehat{AB}$ and $m\widehat{CD}$.

Let $m\widehat{AB} = x$, and $m\widehat{CD} = y$. Then $\frac{1}{2}(x + y) = 65$ and $\frac{1}{2}(x - y) = 24$

$x + y = 130 \qquad x - y = 48$

By the addition method.

$$\begin{bmatrix} x + y = 130 \\ x - y = 48 \end{bmatrix}$$

$$2x = 178$$
$$x = 89$$
$$89 + y = 130$$
$$y = 41$$

Thus, $m\widehat{AB} = 89$ and $m\widehat{CD} = 41$

Section 10.5 Problem Set A

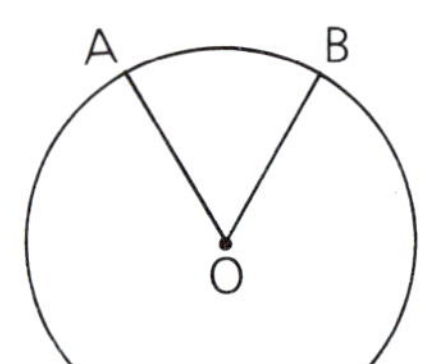

1 VERTEX at CENTER

Given: Circle O, $\widehat{AB} = 62°$

Find: $m\angle O$

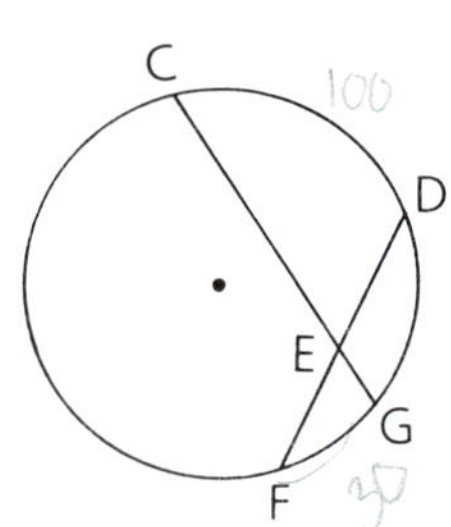

2 VERTEX INSIDE

Given: $\widehat{CD} = 100°$, $\widehat{FG} = 30°$

Find: $m\angle CED$

3 VERTEX ON

a Given: $\widehat{AC} = 70°$

Find: $m\angle B$

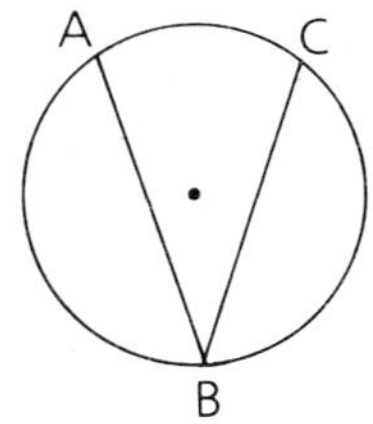

b $\overline{DE}$ is tangent at E

$\widehat{EF} = 150°$

Find: $m\angle DEF$

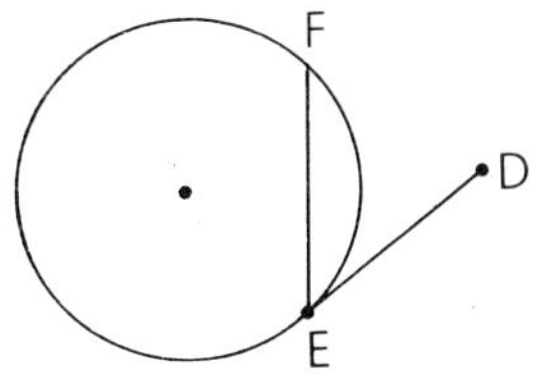

4 VERTEX OUTSIDE

a

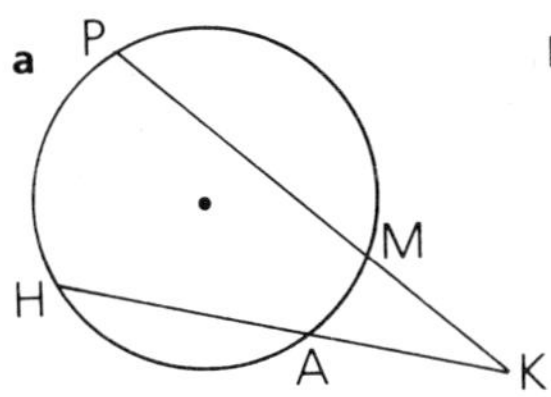

b

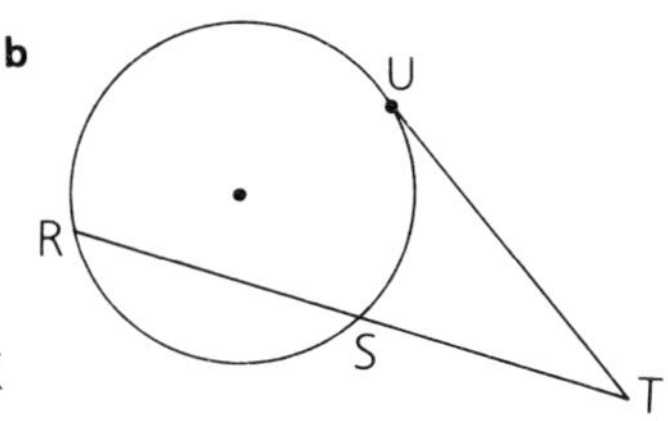

c

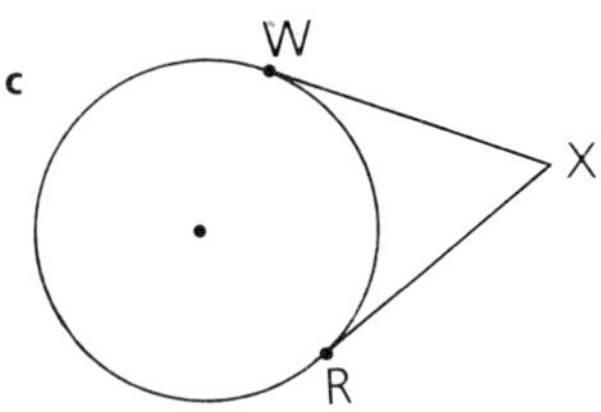

a Given: $\widehat{HP} = 120°$

$\widehat{AM} = 36°$

Find: $m\angle K$

b Given: $\overline{TU}$ is tangent at U

$\widehat{RU} = 160°$

$\widehat{SU} = 60°$

Find: $m\angle T$

c Given: W and R are points of contact

$\widehat{WR} = 140°$

Find: $m\angle X$

5 Find the measure of each angle or arc that is labeled with a letter:

a

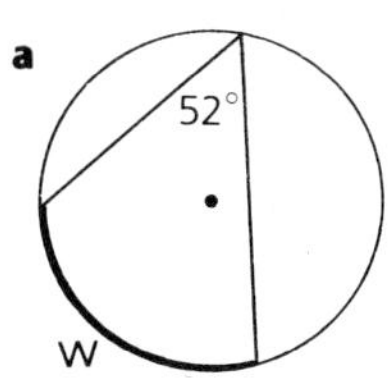

b

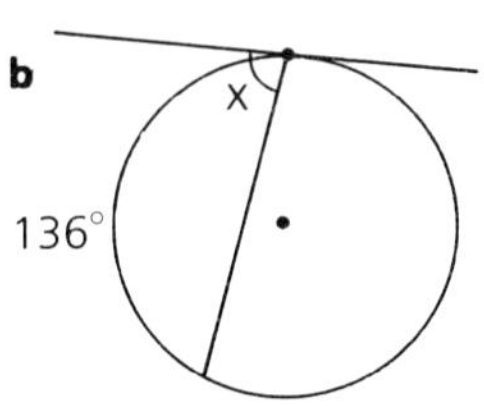

c

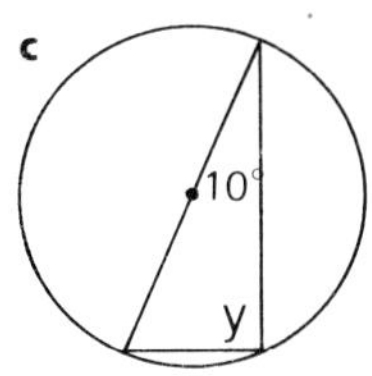

d

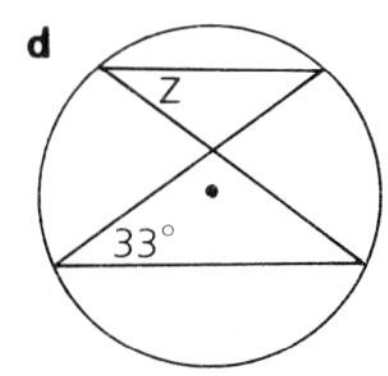

6 Find the measure of each angle or arc that is labeled with a letter:

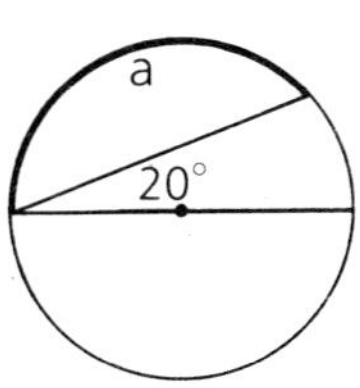

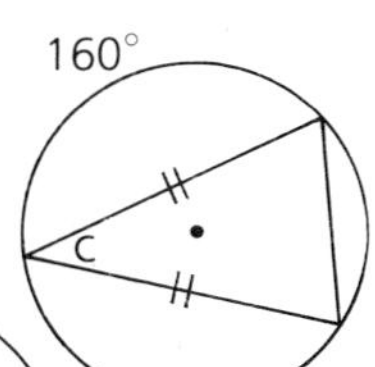

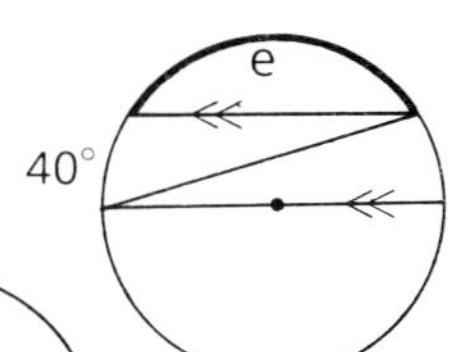

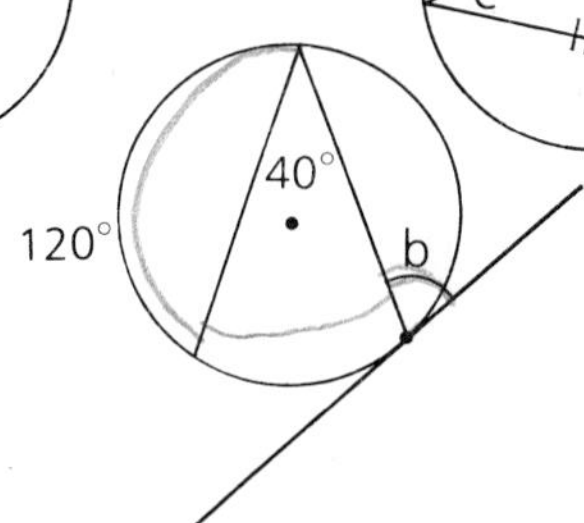

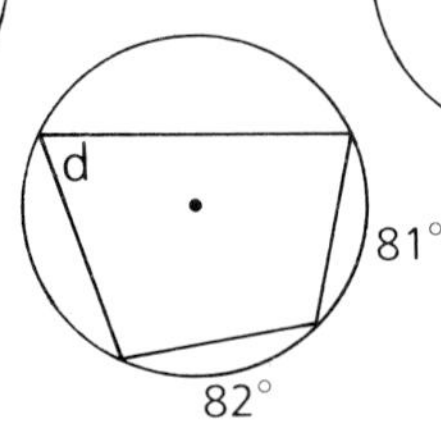

7 $\overset{\frown}{AB} = 108^\circ$, $\overset{\frown}{CD} = 62^\circ$

Find $\angle AXB$ and $\angle Y$

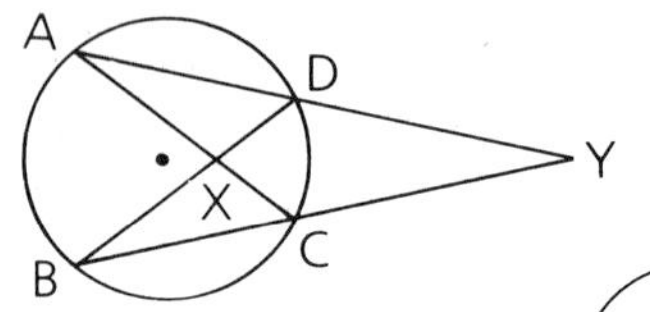

8 $\overset{\frown}{TP} = 170^\circ$, $\overset{\frown}{PQ} = 135^\circ$

Find $\angle R$

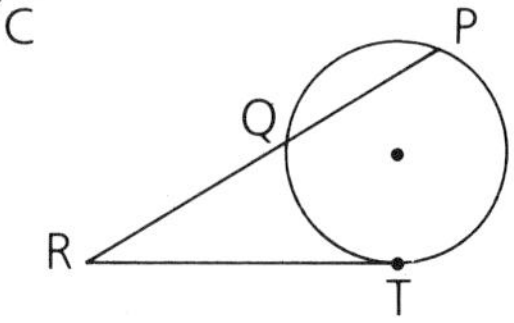

9 $\angle AEB = 30^\circ$

$\overset{\frown}{AB} = 50^\circ$

Find $\overset{\frown}{CD}$

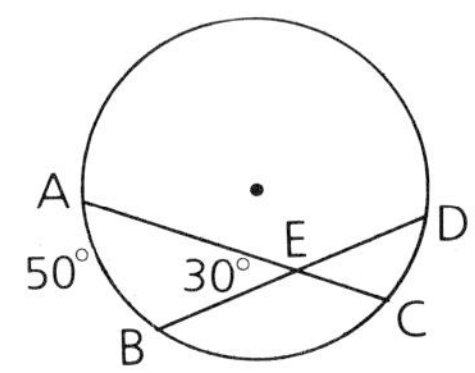

10 $\angle P = 17^\circ$

$\overset{\frown}{TQ} = 42^\circ$

Find $\overset{\frown}{SR}$

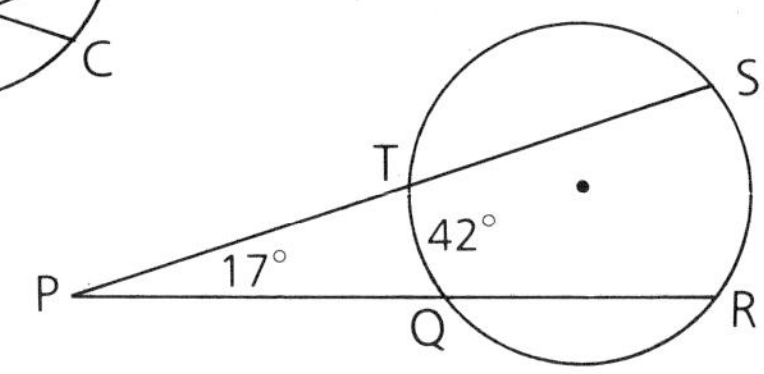

11 $\angle Y = 22^\circ$, $\overset{\frown}{WZ} = 125^\circ$

$\overleftrightarrow{YZ}$ is tangent at Z

Find $\overset{\frown}{XZ}$

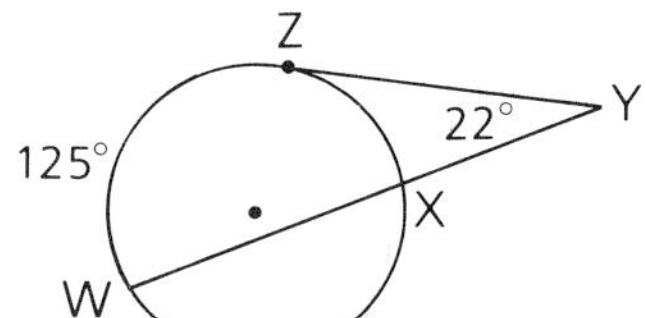

12 $\overset{\frown}{AB} = 85^\circ$

$\overset{\frown}{CD} = 25^\circ$

Find $\angle AED$

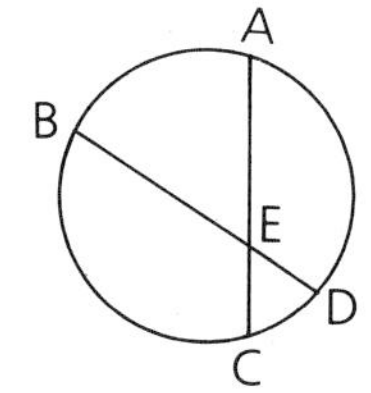

13 $\overset{\frown}{ST} = 85^\circ$, $\overset{\frown}{SQ} = 95^\circ$

$\overset{\frown}{TR} = 175^\circ$

Find $\angle P$

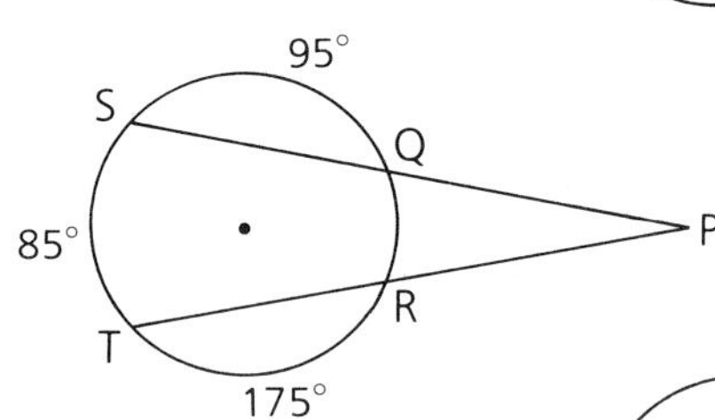

14 $\overline{WY}$ is a diameter of circle E

$\overset{\frown}{WX} = 50^\circ$, $\angle XPY = 120^\circ$

Find $\overset{\frown}{WZ}$

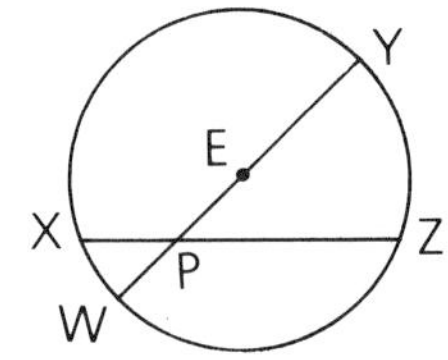

15 A circle is divided into three arcs in the ratio of 3:4:5. A tangent-chord angle intercepts the largest of the three arcs. Find the measure of the tangent-chord angle.

16 An inscribed angle intercepts an arc that is $\frac{1}{9}$ of the circle. Find the measure of the inscribed angle.

17 A secant-secant angle intercepts arcs that are respectively $\frac{2}{3}$ and $\frac{1}{4}$ of the circle. Find the measure of the secant-secant angle.

Section 10.5 Problem Set B

18 $\overline{QS}$ is a diameter of $\odot$ O
$\overset{\frown}{PQ} = 115°$, $\angle RPS = 36°$

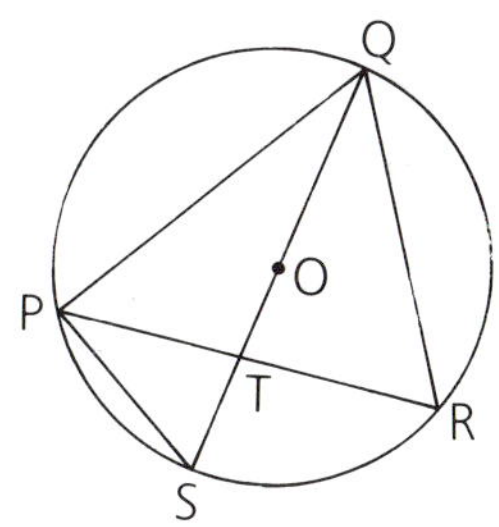

Find:

a $\angle R$	**e** $\angle QPR$	**i** $\overset{\frown}{PRQ}$
b $\angle S$	**f** $\angle QPS$	**j** $\overset{\frown}{RSP}$
c $\overset{\frown}{SR}$	**g** $\angle QTP$	
d $\overset{\frown}{QR}$	**h** $\angle QTR$	

19 Find the measure of each arc or angle labeled with a letter:

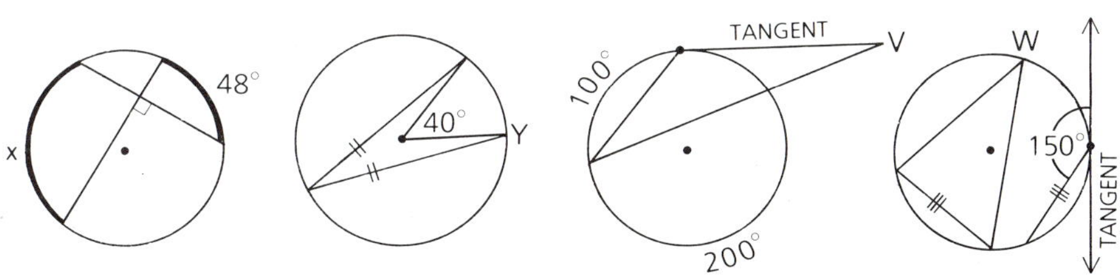

20 Given concentric circles at O,
$\overline{AB}$ tangent to the inner circle.
$\overset{\frown}{BC} = 84°$.

Find the measures of
$\angle A$, $\overset{\frown}{DE}$, and $\overset{\frown}{DF}$.

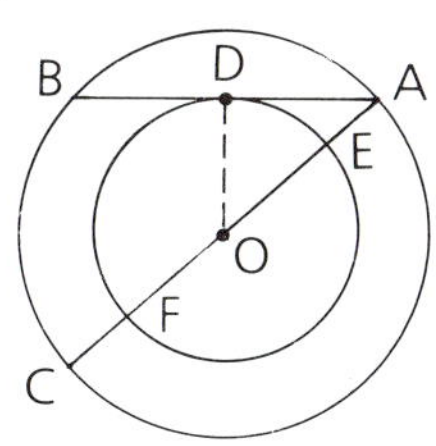

21 $m\angle P = 60$, $m\overset{\frown}{PSR} = 128$
Find $m\angle Q$, $m\angle R$, $m\angle S$

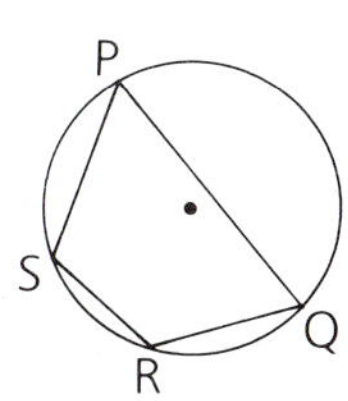

22 $\widehat{AB} = 92°$

$\angle AEB = 82°$

Find $\widehat{AD}$

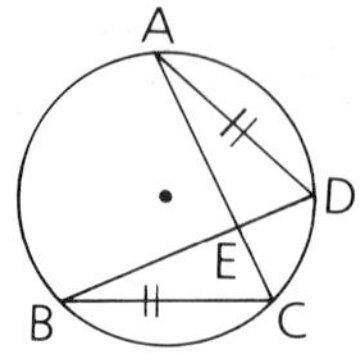

23 $\angle AFE = 89°$

$\angle C = 15°$

Find $\widehat{AE}$ and $\widehat{BD}$

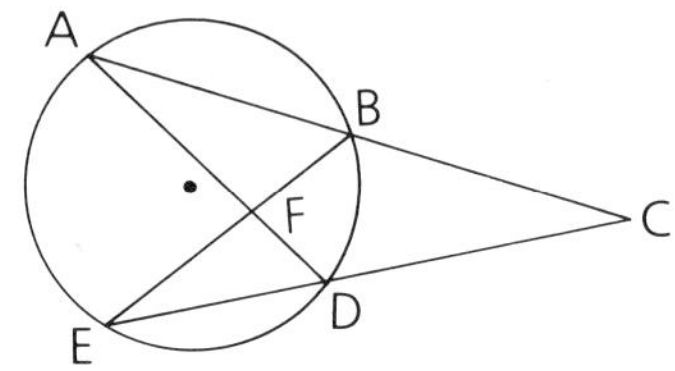

24 The major arc cut off by two tangents to a circle from an outside point is five thirds of the minor arc. Find the angle formed by the tangents.

25 $\widehat{SY} = 112°$

$\widehat{DC} = 87°$

Find $\widehat{AB}$

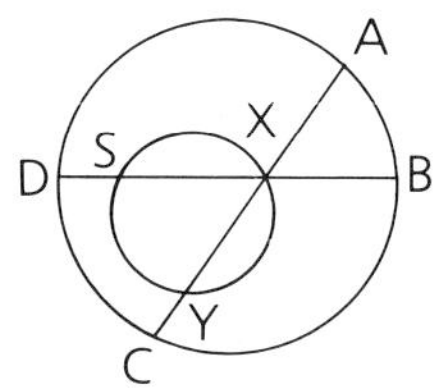

26 $\widehat{DC} = (5x + 6)°$, $\widehat{AB} = (2x)°$

$\angle AEB = 94°$

Find $\widehat{AB}$

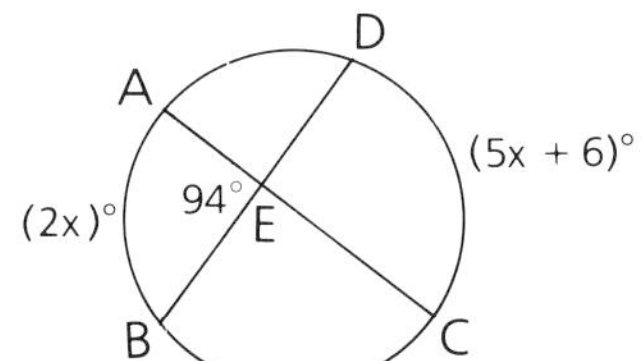

27 A secant-secant angle intercepts arcs that are respectively $\frac{3}{5}$ and $\frac{3}{8}$ of the circle. Find the measure of a chord-chord angle that intercepts those arcs.

28 $\triangle ABC$ is inscribed in a circle (all sides are chords), $AB = 12$, $AC = 6$, and $BC = 6\sqrt{3}$. Find m $\widehat{BC}$.

29 **a** An angle is inscribed in a circle and intercepts an arc of 140°. Find the measure of the inscribed angle.

b An angle is inscribed in a 140° arc (the vertex is on the arc and the sides contain the end points of the arc). Find the measure of the inscribed angle.

30 Prove Theorem 79 for Tangent-Tangent angles. Hint: Draw a secant through the vertex, use the theorem for tangent-secant angles, and do some adding.

Section 10.5 Problem Set C

31 Given: $\overleftrightarrow{AB} \parallel \overleftrightarrow{CD}$

$\overleftrightarrow{DC}$ is tangent to $\odot$ O at T

Conclusion: $\overset{\frown}{AT} \cong \overset{\frown}{BT}$

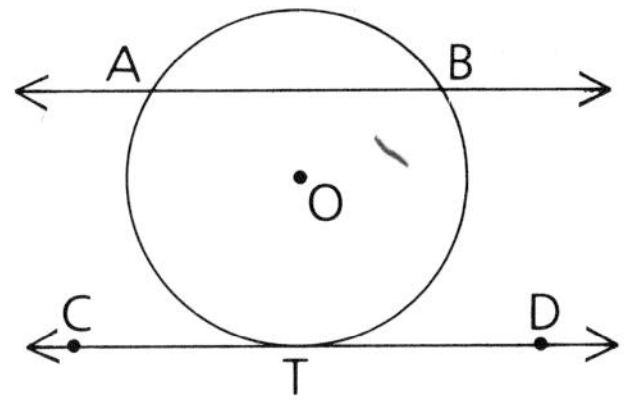

32 A quadrilateral ABCD is inscribed in a circle. Its diagonals intersect at X. If $\overset{\frown}{AB} = 100^\circ$, $\overset{\frown}{BC} = 50^\circ$, and $\overline{AD} \cong \overline{BD}$, find m$\angle$DXC.

33 $\overline{WZ} \cong \overline{XY}$

$\overset{\frown}{WXY} = 200^\circ$

Find $\angle$P

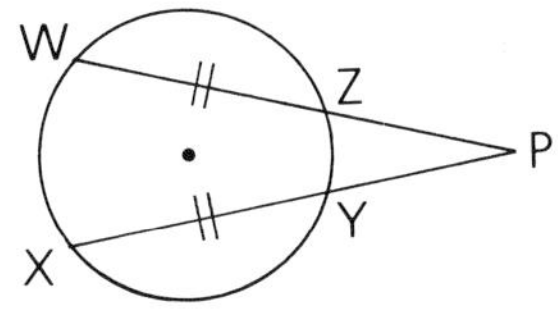

34 A secant and a tangent to a circle intersect to form an angle of 38°. If the measures of the arcs intercepted by the secant and the tangent are in a ratio of 2:1, find the measure of the third arc.

35 Given: Ⓢ P and Q are tangent internally at T. Diameter $\overline{NS}$ of $\odot$ Q is tangent to $\odot$ P at A. m$\overset{\frown}{MR}$ = 42, $\overline{TM}$ passes through A

Find: m$\overset{\frown}{NM}$

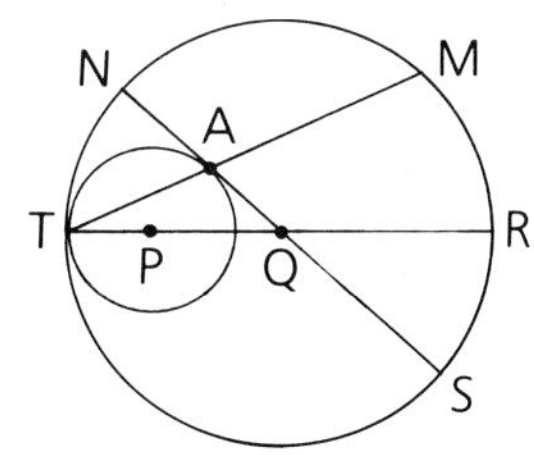

36 The two circles intersect at A and B. $\angle AXB = 70^\circ$, $\overset{\frown}{CD} = 20^\circ$, $\overset{\frown}{EF} = 160^\circ$

Find the *difference* between the measures of $\overset{\frown}{AB}$ of the small circle and $\overset{\frown}{AB}$ of the larger circle.

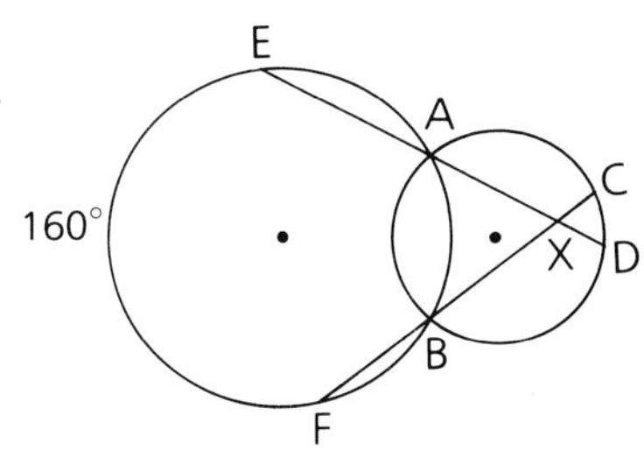

37 Prove Theorem 77 for tangent-chord angles. Some key steps to prove that m$\angle 2 = \frac{1}{2}\overset{\frown}{BC}$ are:

Draw $\overline{CD} \parallel \overleftrightarrow{BA}$

$\angle 1 \cong \angle 2$, m$\angle 1 = \frac{1}{2}\overset{\frown}{DB}$

Use Problem 31 and transitivity.

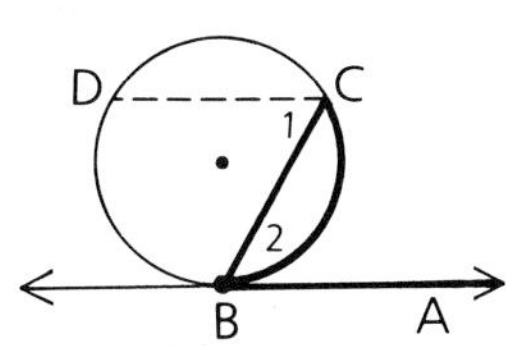

Section 10.6 More Angle-Arc Theorems

In this section we shall discuss:

A ANGLES THAT INTERCEPT THE SAME ARC OR CONGRUENT ARCS
B ANGLE INSCRIBED IN A SEMICIRCLE
C A SPECIAL THEOREM ABOUT TANGENT-TANGENT ANGLES

10.6 (A) Angles That Intercept the Same Arc or Congruent Arcs

Our knowledge of the relationship between angles and their intercepted arcs leads easily to the next three theorems.

Theorem 80

T **If two inscribed or tangent-chord angles intercept the *same arc,* then they are congruent.**

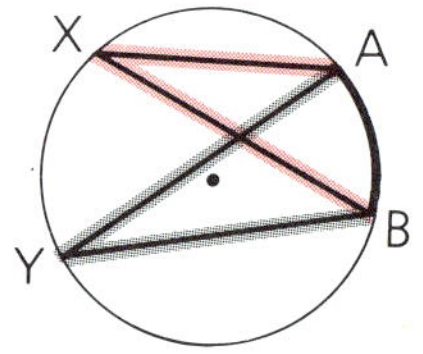

Given: X and Y are inscribed angles intercepting arc AB

Conclusion: $\angle X \cong \angle Y$

Proof: $m\angle X = \frac{1}{2}m\widehat{AB}$ and $m\angle Y = \frac{1}{2}m\widehat{AB}$ since the measure of an inscribed angle is one half of the measure of its arc.
Thus, $m\angle X = m\angle Y$, by the transitive property. Therefore, $\angle X \cong \angle Y$.

In the same way, we could prove Theorem 80 for two tangent-chord angles and for the combination of a tangent-chord angle and an inscribed angle.

Theorem 81

T **If two inscribed or tangent-chord angles intercept CONGRUENT ARCS, then they are congruent.**

■ Example: If we are given $\overleftrightarrow{ED}$ as the tangent at D and $\widehat{AB} \cong \widehat{CD}$, we may conclude that $\angle P \cong \angle CDE$.

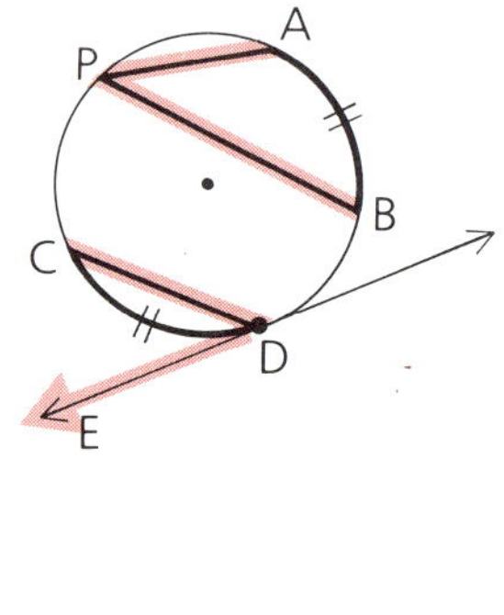

10.6 (B) Angle Inscribed in a Semicircle

Theorem 82

T **An angle inscribed in a semicircle is a right angle.**

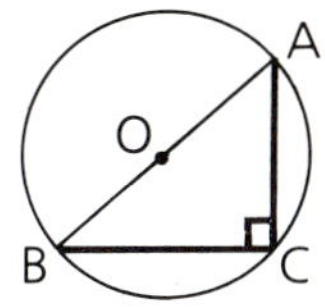

Given: $\overline{AB}$ is a diameter of $\odot$ O

Prove: $\angle C$ is a right angle

Proof: Since the measure of an inscribed angle is half of the measure of its intercepted arc, and a semicircle is 180°, $\angle C = 90°$.

10.6 (C) A Special Theorem About Tangent-Tangent Angles

Theorem 83

T **The sum of the measures of a TANGENT-TANGENT angle and its minor arc is 180.**

Given: $\overline{PT}$, $\overline{PS}$ are tangent to circle O

Prove: $m\angle P + m\widehat{TS} = 180$

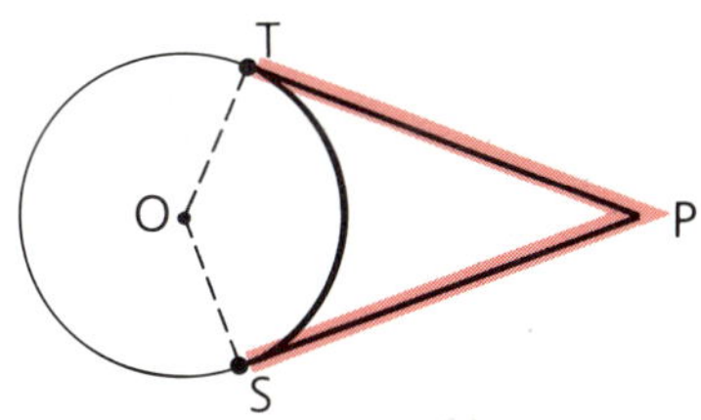

Proof: Since tangents are perpendicular to the radii drawn to the point of contact, $\angle T$ and $\angle S$ are right angles (90° each).

The sum of the measures of the angles in any quadrilateral is 360°.
Thus $\angle O + \angle T + \angle P + \angle S = 360°$.

Since $\angle T + \angle S = 180°$, then $\angle P + \angle O = 180°$
But, central $\angle O$ measures $\widehat{TS}$. Thus, $m\angle P + m\widehat{TS} = 180$.

■ Example: $\overleftrightarrow{PT}$ and $\overleftrightarrow{PS}$ are tangents at T and S
Find $m\angle P$

Solution:
$$m\angle P + m\widehat{TS} = 180$$
$$m\angle P + 130 = 180$$
$$m\angle P = 50$$

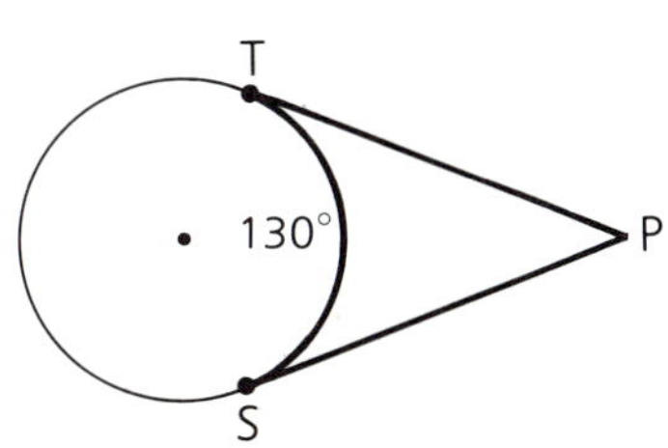

Section 10.6 Sample Problems

1 Given: $\odot O$

Conclusions: $\triangle LVE \sim \triangle NSE$

$EV \cdot EN = EL \cdot SE$

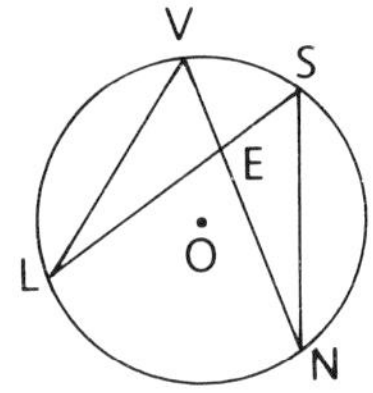

1 $\odot O$	1 Given
2 $\angle V \cong \angle S$	2 If two inscribed angles intercept the same arc, they are congruent
3 $\angle L \cong \angle N$	3 Same as 2
4 $\triangle LVE \sim \triangle NSE$	4 AA (2, 3)
5 $\frac{EV}{SE} = \frac{EL}{EN}$	5 Ratios of corresponding sides of $\sim$ $\triangle$s are $=$
6 $EV \cdot EN = EL \cdot SE$	6 Means-extremes products theorem

2 In circle O, $\overline{BC}$ is a diameter and the radius of the circle is 20.5 mm. Chord $\overline{AC}$ has a length of 40 mm. Find AB.

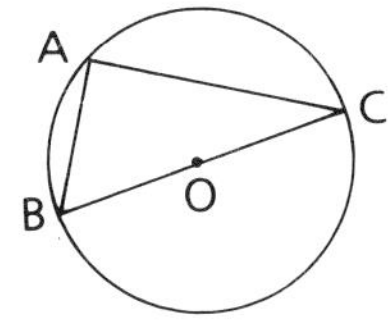

Solution: Since $\angle A$ is inscribed in a semicircle, it is a right angle.

Applying the Pythagorean Theorem, $AB^2 + AC^2 = BC^2$

$$AB^2 + 40^2 = (41)^2$$

$$AB = 9 \text{ mm}$$

3 Given: $\overleftrightarrow{AB} \parallel \overleftrightarrow{CD}$ in $\odot O$

$\overleftrightarrow{AB}$ is tangent at B

Prove: $\angle C \cong \angle BDC$

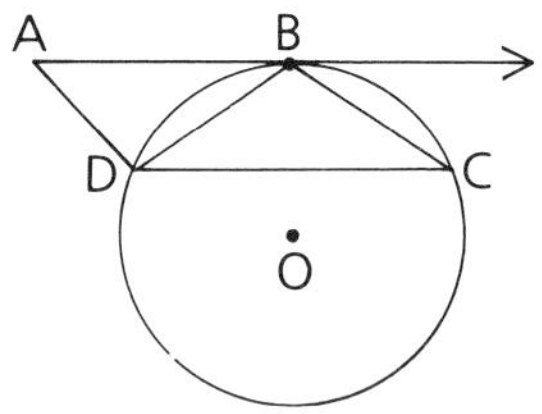

1 $\overleftrightarrow{AB}$ tangent to $\odot O$	1 Given
2 $\overleftrightarrow{AB} \parallel \overleftrightarrow{CD}$	2 Given
3 $\angle ABD \cong \angle BDC$	3 PAI
4 $\angle C \cong \angle ABD$	4 If an inscribed $\angle$ and a tangent-chord $\angle$ cut off the same arc, they are $\cong$
5 $\angle C \quad \angle BDC$	5 Transitive property

Section 10.6 Problem Set A

1 Given: X is the midpt. of $\overset{\frown}{WY}$

Prove: $\overrightarrow{ZX}$ bisects $\angle WZY$

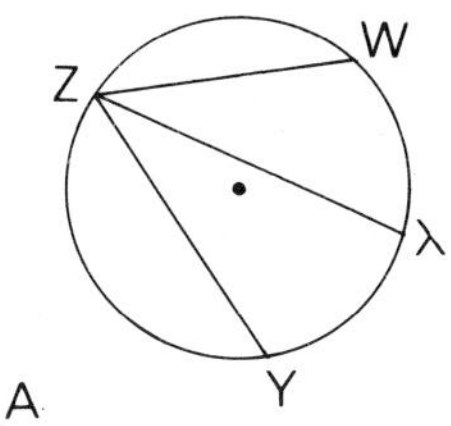

2 Given: $\odot$ E with diameter $\overline{AC}$
$\overline{BC} \cong \overline{CD}$

Conclusion: $\triangle ABC \cong \triangle ADC$

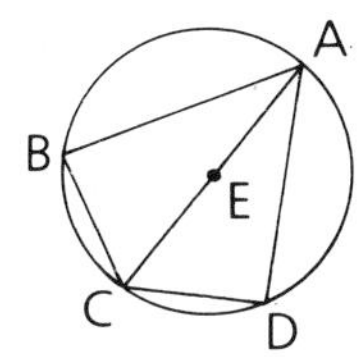

3 In $\odot$ P, $\overline{BC}$ is a diameter.
AC = 12 mm, BA = 16 mm.

Find the length of the radius.

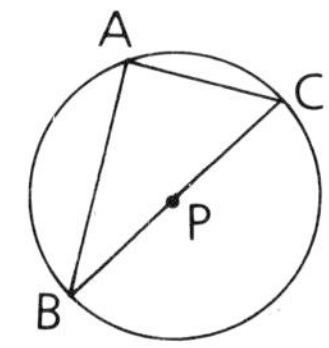

4 $\overline{PQ}$, $\overline{PR}$ are tangent segments, $\overset{\frown}{QR} = 163°$.

Find: **a** $\angle P$

b $\angle PQR$

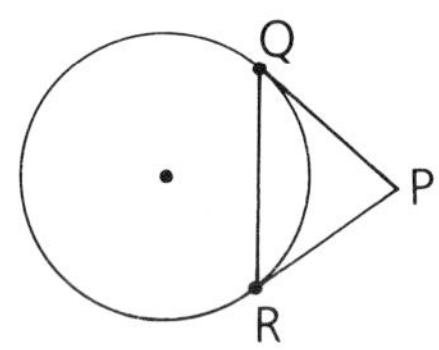

5 A, B, C are points of contact
$\overset{\frown}{AB} = 145°$, $\angle Y = 48°$

Find $\angle Z$

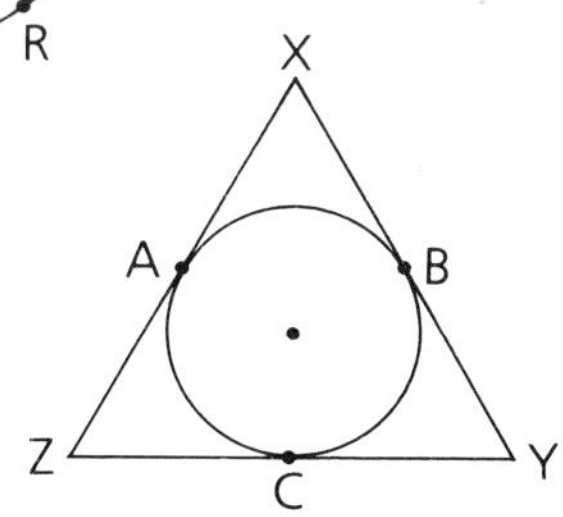

6 Given: $\overset{\frown}{BC} \cong \overset{\frown}{ED}$, AB = 8
BC = 4, CD = 9

a Is $\overline{BE} \parallel \overline{CD}$?

b Find BE.

c Is $\triangle ACD$ scalene?

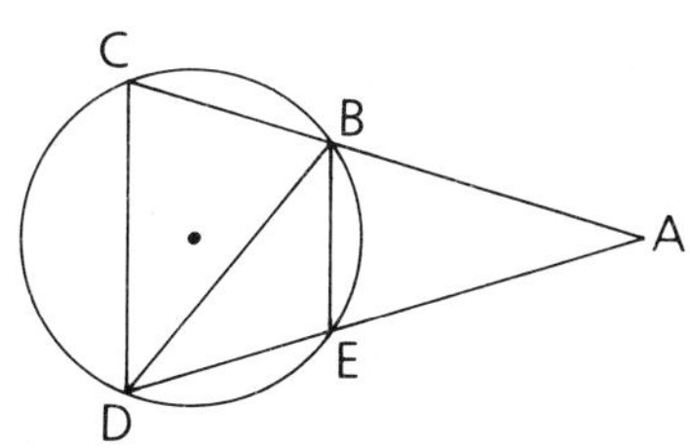

7 $\overleftrightarrow{PY}$ and $\overleftrightarrow{QW}$ are tangents.
$\overset{\frown}{WZ} = 126°$, $\overset{\frown}{XY} = 40°$.

Find $\overset{\frown}{PQ}$.

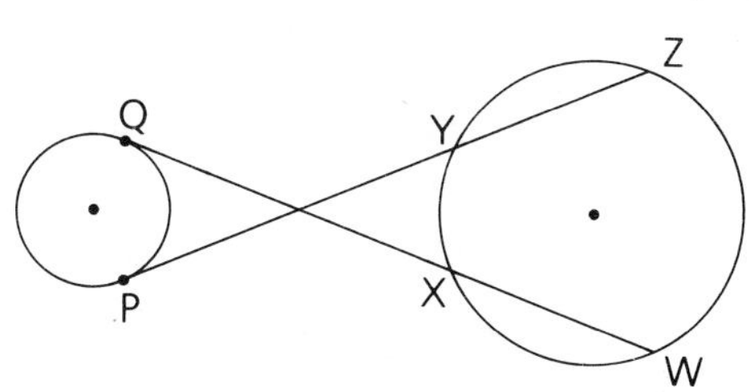

8 Given: $\triangle ABC$ inscribed in a circle, $\overset{\frown}{AC} \cong \overset{\frown}{AB}$

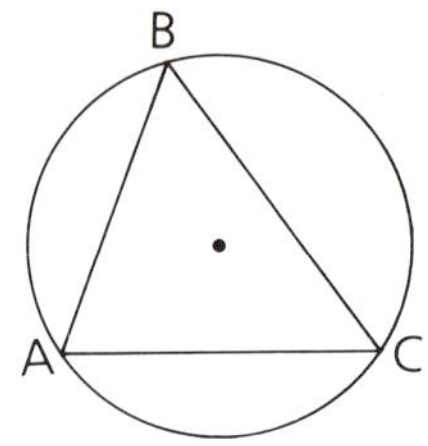

Tell whether each of the following:
must be true, could be true,
or cannot be true.

a $\overline{AB} \cong \overline{AC}$

b $\overline{AC} \cong \overline{BC}$

c $\overline{AB}$ and $\overline{AC}$ are equidistant from the center of the circle

d $\angle B \cong \angle C$

e $\angle BAC$ is a right angle

f $\angle ABC$ is a right angle

9 **a** Given: Circle O, $AP = 4$
$\angle A = 45^\circ$

Find: AB

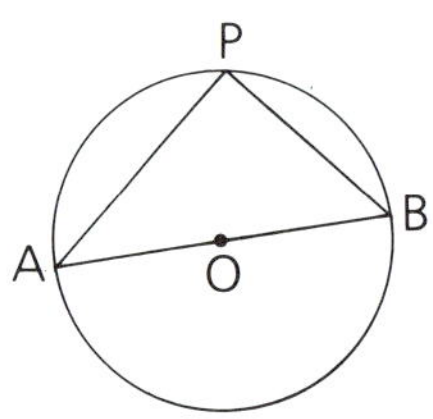

b Given: Circle Q, $ED = \frac{1}{2}CD$
$CE = 8\sqrt{3}$

Find: CD and $m\angle D$

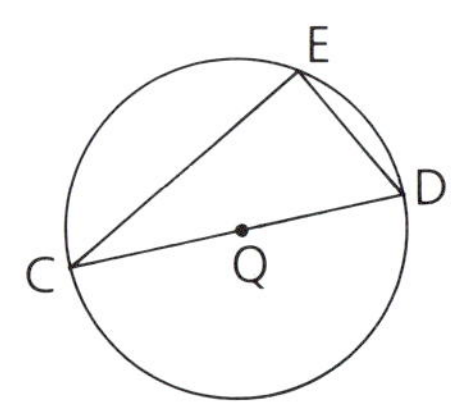

10 $\overline{AB}$ is a diameter of $\odot$ P.
$CB = 1.5$ m, $CA = 2$ m.
Find the radius of $\odot$ P.

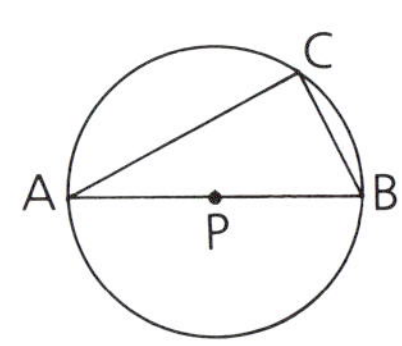

11 In $\odot$ Z, $\overset{\frown}{WX} = 120^\circ$ and the radius $= 6$ cm

Find: **a** AX

b the perimeter of $\triangle WAX$

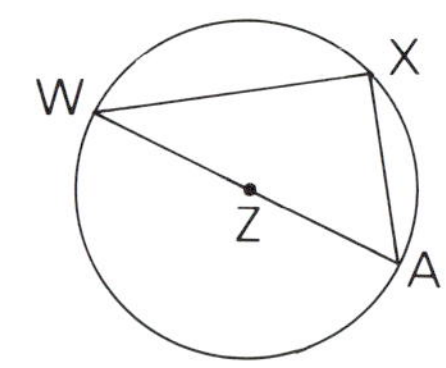

Section 10.6 Problem Set B

12 A rectangle with dimensions 18 by 24 is inscribed in a circle. Find the radius of that circle.

13 A square is inscribed in a circle of radius 10. Find the length of a side of the square.

14 Quadrilateral ABCD is inscribed in circle O. AB = 12, BC = 16, CD = 10, and $\angle ABC$ is a right angle.

Find the measure of $\overline{AD}$ in simplified radical form.

15 Circles O and P are tangent at F.
$\overline{AC}$ and $\overline{CE}$ are tangent to $\odot$ P at B and D.
$\widehat{DFB} = 223°$.
Find $\widehat{AE}$.

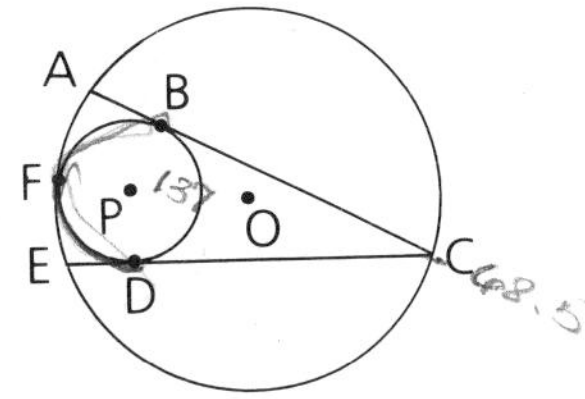

16 $\angle S = 88°$, $\widehat{QT} = 104°$
$\widehat{ST} = 94°$, tangent $\overline{PQ}$
Find: **a** $\angle P$
b $\angle STQ$

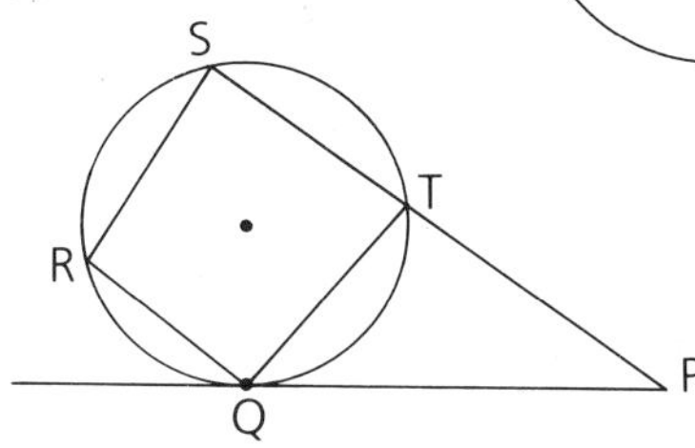

17 Given: $\widehat{BC} \cong \widehat{CD}$
Conclusion: $\triangle ABC \sim \triangle AED$

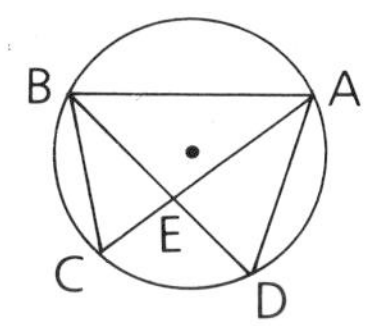

18 Given: $\overleftrightarrow{AC}$ is tangent at A
$\angle APR$ and $\angle AQR$ are right $\angle$s
R is the midpoint of $\widehat{AB}$
Conclusion: $\overline{PR} \cong \overline{RQ}$
(Hint: Draw $\overline{AR}$)

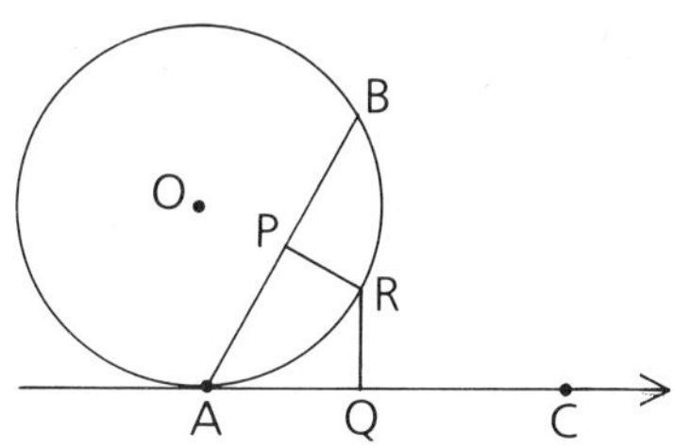

19 Given: $\triangle WXZ$ is isosceles
with $\overline{WX} \cong \overline{WZ}$
$\overline{WZ}$ is a diameter of $\odot$ O
Prove: Y is the midpoint of $\overline{XZ}$
(Hint: Draw $\overline{WY}$)

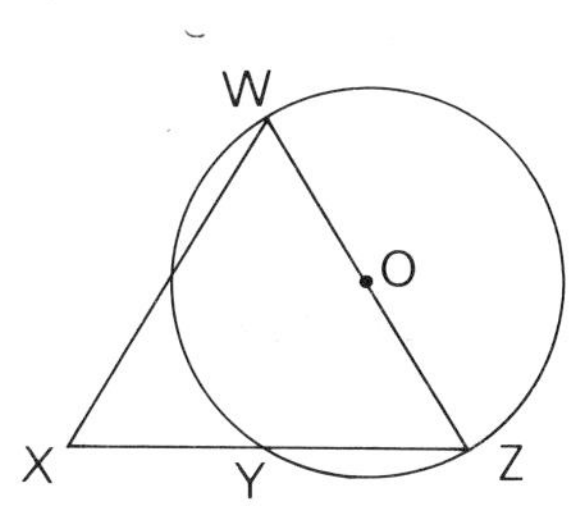

20 Given: $\overline{AC}$ is tangent to $\odot$ O at A
Conclusion: $\triangle ADC \sim \triangle BDA$

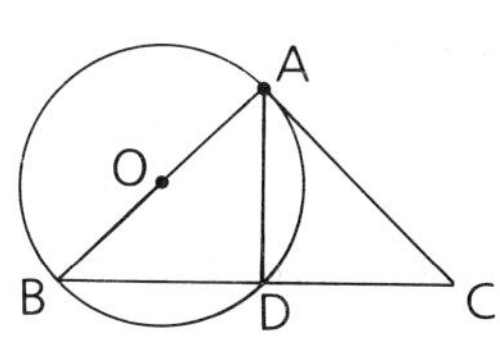

Section 10.6 Problem Set C

21 Given: Circle P
WXYZ is a ▱
Conclusion: △VXY is isosceles

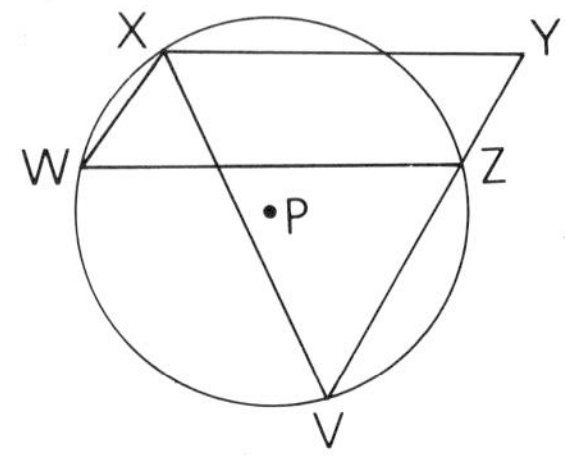

22 Given: ⊙ O with chords $\overline{AC}$ and $\overline{BD}$ intersecting at E
Prove: **a** $m\widehat{AB} + m\widehat{CD} = 2m\angle CED$
b $AE \cdot EC = BE \cdot ED$

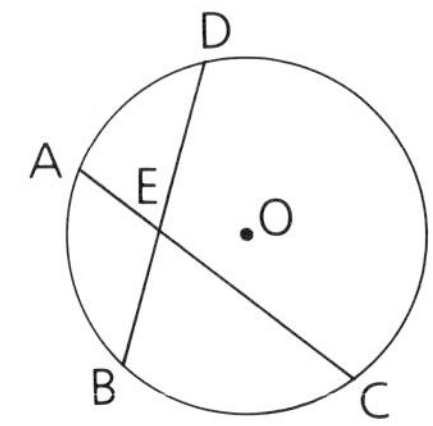

23 $\overline{AB}$ is a diameter of ⊙ P.
$QR = 6$, $AB = 13$, $\overline{QR} \perp \overline{AB}$.
Find RB.

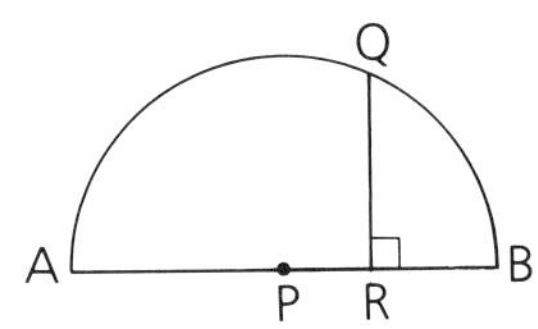

24 Given: Semicircle with center Q
$\overline{QB}$ bisects $\widehat{AC}$
$\overline{QD}$ bisects $\widehat{CE}$
Prove: ∠BQD is a right ∠

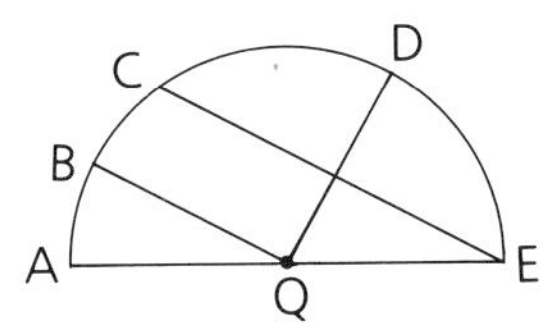

25 Given: △ABC is inscribed in ⊙ P
$\overline{AE}$ and $\overline{CD}$ are chords such that $\overline{AE} \perp \overline{BC}$ and $\overline{CD} \perp \overline{AB}$
Prove: $\widehat{BD} \cong \widehat{BE}$

26 Two circles are tangent internally so that the center of the larger circle lies on the smaller circle. Prove that any chord that has one endpoint at the point of tangency is bisected by the smaller circle.

27 Given: ⊙ A is tangent to ⊙ B at R
$\overline{PT}$ is a common external tangent at P and T, $\angle Q = 43^\circ$
Find: ∠S

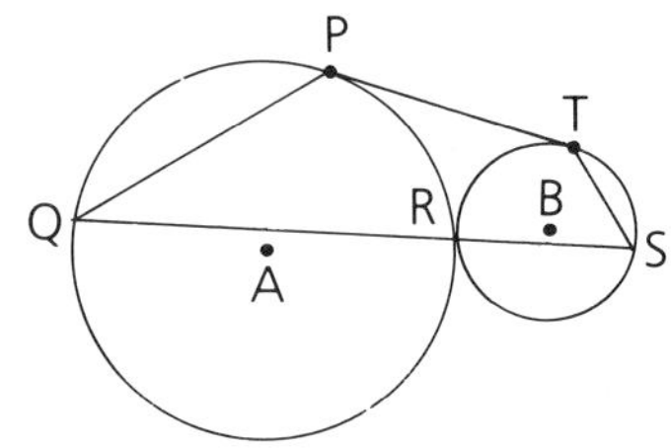

Section 10.7 Inscribed and Circumscribed Polygons

A INSCRIBED AND CIRCUMSCRIBED POLYGONS
B A THEOREM ABOUT INSCRIBED QUADRILATERALS
C THE STORY OF THE PLAIN OLD PARALLELOGRAM

10.7 (A) Inscribed and Circumscribed Polygons

Triangle ABC is **INSCRIBED IN** circle O.

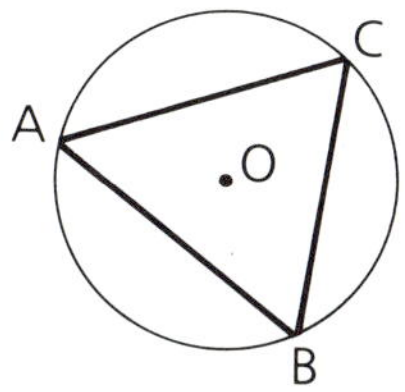

D **A polygon is INSCRIBED IN a circle if all of its vertices lie on the circle.**

Polygon **PQRST** is **CIRCUMSCRIBED ABOUT** circle F.

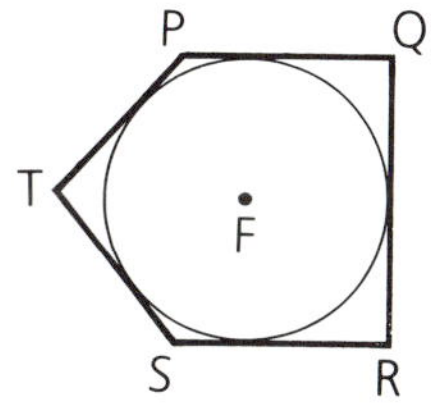

D **A polygon is CIRCUMSCRIBED ABOUT a circle if each of its sides is tangent to the circle.**

We can also speak of circles being circumscribed about a polygon or inscribed in a polygon.

The diagram illustrates that
Quadrilateral ABCD is **INSCRIBED IN** ⊙ O, and
Circle O is **CIRCUMSCRIBED ABOUT** ABCD
have the same meaning.

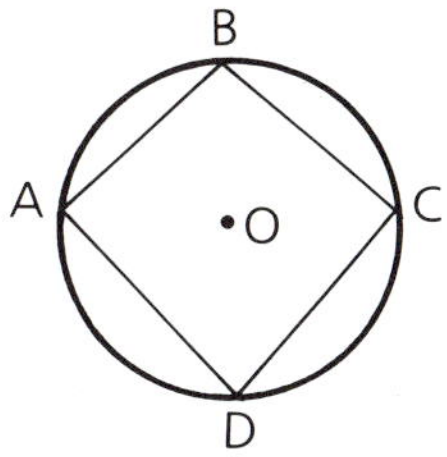

D **The center of a CIRCUMSCRIBED circle is the CIRCUMCENTER of the polygon.**
O is a circumcenter.

Hexagon PQRSTU is **CIRCUMSCRIBED ABOUT** circle F.
Circle F is **INSCRIBED IN** Hexagon PQRSTU.

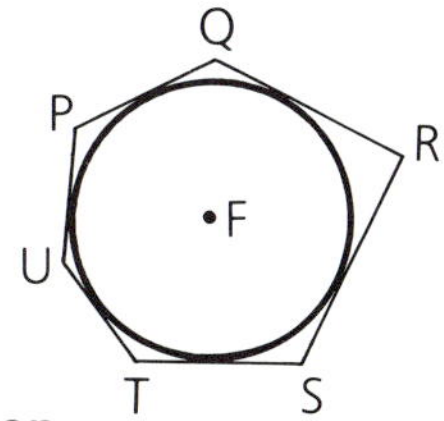

D **The center of an INSCRIBED circle is the INCENTER of the polygon.**
F is an incenter.

10.7 (B) A Theorem About Inscribed Quadrilaterals

Theorem 84

T **If a quadrilateral is inscribed in a circle, the opposite angles are supplementary.**

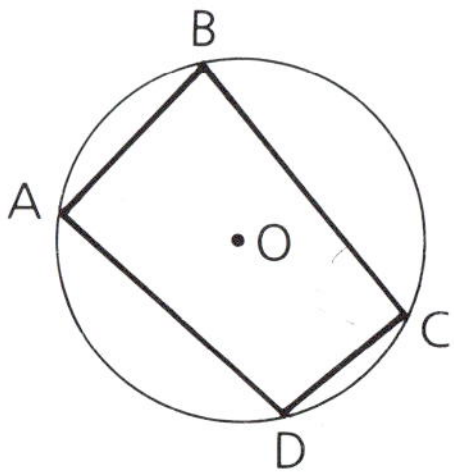

Given: Quadrilateral ABCD inscribed in circle O

Prove: $\angle A$ is supp. to $\angle C$
$\angle B$ is supp. to $\angle D$

Proof: $\angle A$, $\angle B$, $\angle C$, and $\angle D$ are inscribed angles, so

$$m\angle A = \frac{1}{2}m\overset{\frown}{BCD}, \text{ and } m\angle C = \frac{1}{2}m\overset{\frown}{BAD}$$

$$\begin{aligned} \text{Thus, } m\angle A + m\angle C &= \frac{1}{2}m\overset{\frown}{BCD} + \frac{1}{2}m\overset{\frown}{BAD} \\ &= \frac{1}{2}(m\overset{\frown}{BCD} + m\overset{\frown}{BAD}) \\ &= \frac{1}{2}(360), \text{ because } \overset{\frown}{BCD} \cup \overset{\frown}{BAD} = \text{whole } \odot \\ &= 180 \end{aligned}$$

Thus, $\angle A$ is supplementary to $\angle C$, and similarly, $\angle B$ is supplementary to $\angle D$.

10.7 (C) The Story of the Plain Old Parallelogram

Once upon a time there was a plain old parallelogram who looked like this:

One day a new family moved into town. The new family was called the circle family.

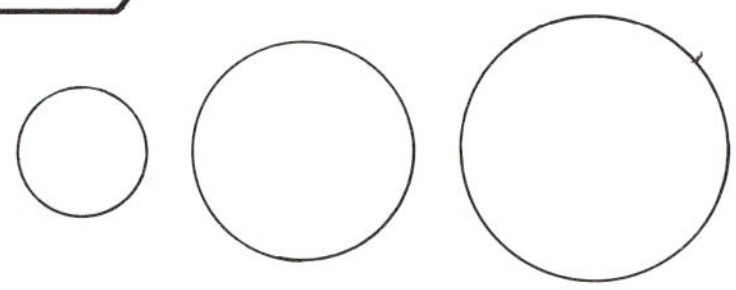

Well, like many polygons, the plain old parallelogram, called P.O.P. for short, felt rather bored and decided to make friends with the circle family. So P.O.P. called on the circles, and they invited him in. P.O.P. went in, or at least part of him did. Here's what happened.

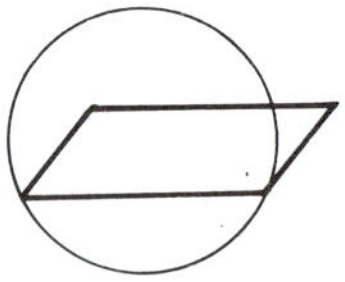

The circles were most displeased. They told P.O.P. he just didn't fit in. P.O.P. saw that he didn't, but he tried again.

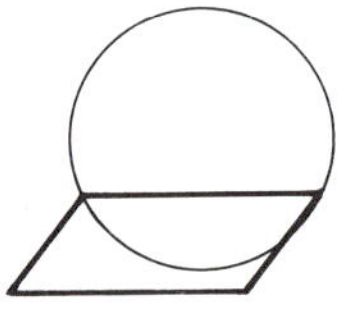

No matter how hard P.O.P. tried, he just couldn't fit in. So he went to the hospital for an operation. Some called it plastic surgery. When he left the hospital, he looked like this:

So the new P.O.P., called Recty for short, visited the circle family again. Here's what happened.

You see, Recty fit in just perfectly. And so Recty lived happily, if not forever after, at least until the next story.

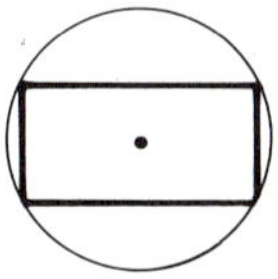

Now what is the point of this story about P.O.P. and Recty? The principle is this:

Theorem 85

T **If a parallelogram is inscribed in a circle, it must be a rectangle.**

Many conclusions follow from Theorem 85.

If ABCD is an inscribed parallelogram, then:

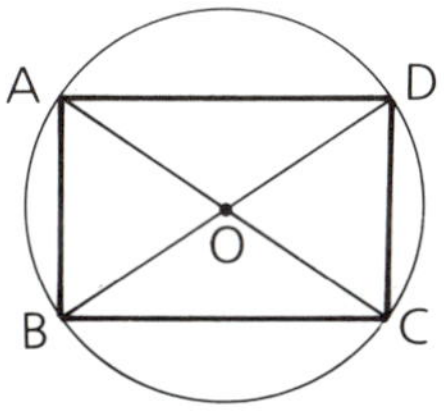

1 $\overline{BD}$ and $\overline{AC}$ are diameters.
2 O is the center of the circle.
3 $\overline{OA}$, $\overline{OB}$, $\overline{OC}$, $\overline{OD}$ are radii.
4 $(AB)^2 + (BC)^2 = (AC)^2$, etc.

Section 10.7 Sample Problems

1 **Given: Quadrilateral ABCD inscribed in ⊙ O**

Prove: ∠B ≅ ∠ADE

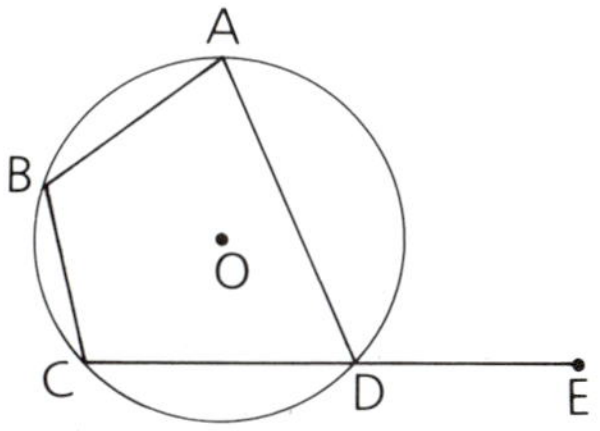

1 ABCD inscribed in ⊙ O	1 Given
2 ∠B is supp. to ∠ADC	2 If a quadrilateral is inscribed in a ⊙, the opposite ∠s are supp.
3 ∠ADC supp. to ∠ADE	3 Two ∠s forming a straight ∠ are supp.
4 ∠B ≅ ∠ADE	4 Two ∠s supp. to the same ∠ are ≅

2 Parallelogram ABCD is inscribed in a circle and the diagonals intersect at E.

a Draw the figure.

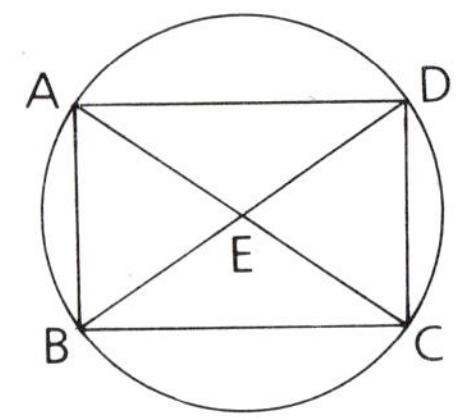

b What is true about ▱ ABCD?

Solution: A ▱ inscribed in a ⊙ must be a rectangle. So ABCD is a rectangle.

c What is $\overline{BD}$?

Solution: ∠BCD is an inscribed right ∠, so $\frac{1}{2}m\widehat{BAD} = 90$, making $\widehat{BAD} = 180^\circ$, a semicircle. Thus chord $\overline{BD}$ is a diameter.

d If AB = 5 and BC = 6, find AC.

Solution: Since △ABC is a right △,
$$(AB)^2 + (BC)^2 = (AC)^2$$
$$5^2 + 6^2 = (AC)^2$$
$$\sqrt{61} = AC$$

Section 10.7 Problem Set A

1 ∠A = 104°, ∠B = 67°

Find ∠D and ∠C

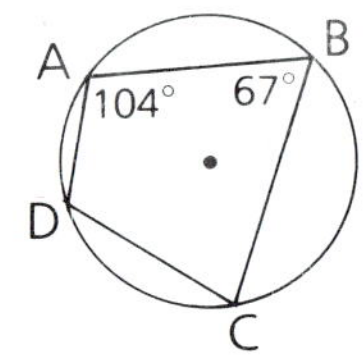

2 $\widehat{PS} = 110^\circ$, $\widehat{PQ} = 100^\circ$

Find m∠R and m∠P

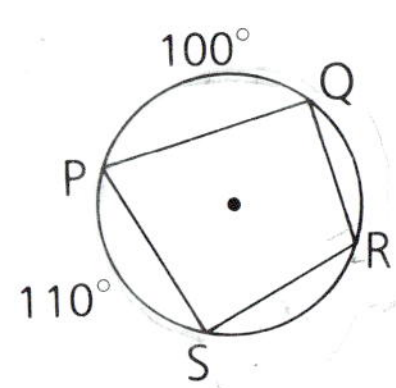

3 ∠A = 110°, $\overline{BC} \cong \overline{CD}$, ∠D = 95°

Find: a ∠C c ∠B

b $\widehat{BC}$ d $\widehat{AB}$

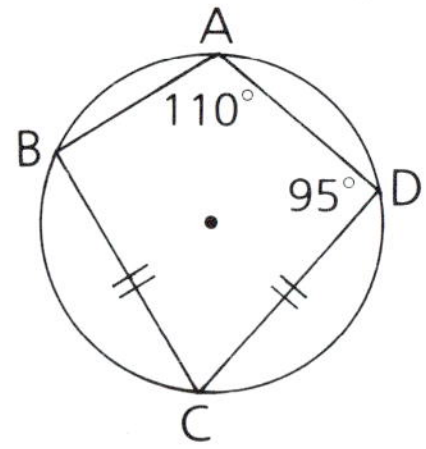

4 Given: ⊙ O

Prove: ∠Q ≅ ∠PST

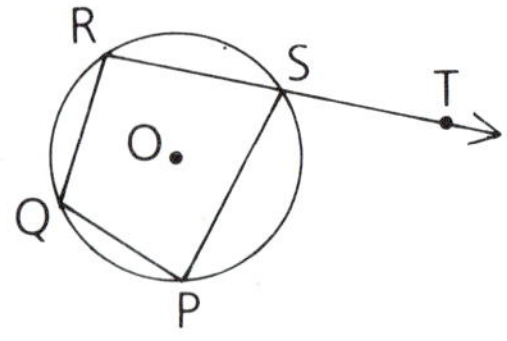

5 Can a parallelogram with a 100° angle be inscribed in a circle?

6 Given: PQRST is a regular pentagon
ABCDEF is a regular hexagon

Find: **a** $m\widehat{PQ}$ **d** $m\widehat{BD}$

b $m\widehat{RT}$ **e** $m\widehat{DEA}$

c $m\widehat{AB}$

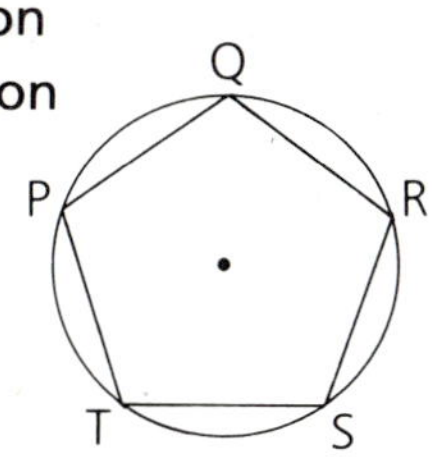

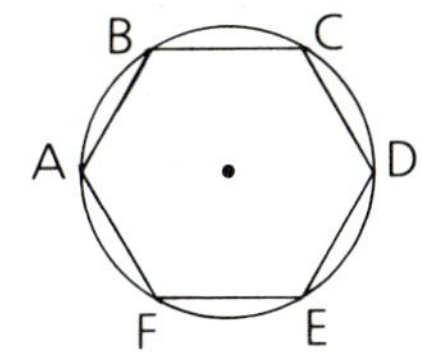

7 If a rhombus is inscribed in a circle, what must be true about the rhombus?

8 If a trapezoid is inscribed in a circle, what must be true about the trapezoid?

Section 10.7 Problem Set B

9 Given: $\angle B = 115°$, $\widehat{AD} = 60°$
$\overline{BC} \parallel \overline{EF}$

Find: **a** $\angle ADC$ **c** $\angle C$

b $\angle CDF$ **d** $\angle A$

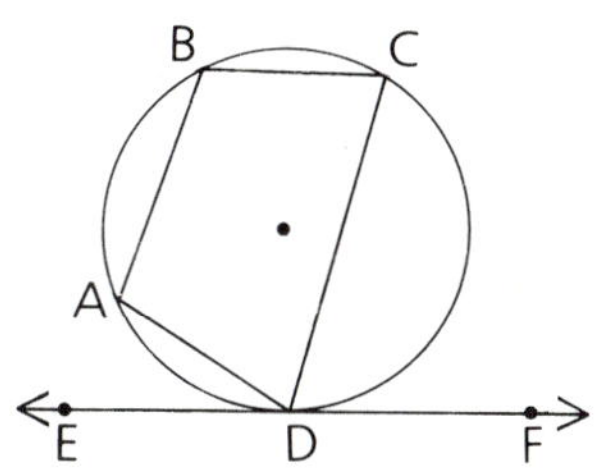

10 PQ = 15, QR = 20, RS = 7
$\angle Q$ is a right angle
Find PS

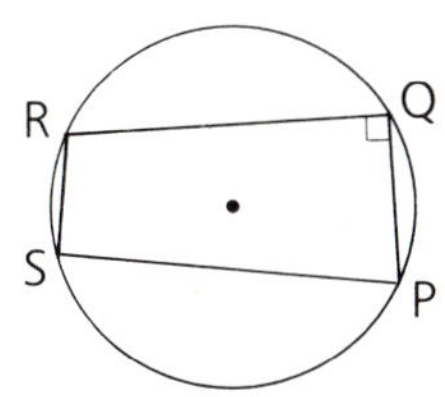

11 Trapezoid WXYZ is circumscribed about circle O. $\angle X$ and $\angle Y$ are right $\angle$s, XW = 16, YZ = 7.
Find the perimeter of WXYZ.

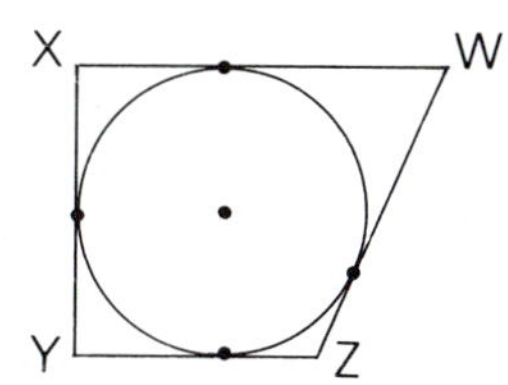

12 Prove that if one pair of opposite angles in a quadrilateral is supplementary, then the other pair is also supplementary. (Use a paragraph proof.)

13 Prove: A trapezoid inscribed in a circle is isosceles.

14 Parallelogram RECT is inscribed in circle O. If RE = 6 and EC = 8, find the perimeter of $\triangle$ECO.

15 Parallelogram ABCD is inscribed in circle P. If AB = 12, CB = 16, how far is chord $\overline{AB}$ from P?

16 Given: $\odot$ O, EFGH is a $\square$
$\widehat{HG} = 120^\circ$, OJ = 6

Find: the perimeter of EFGH

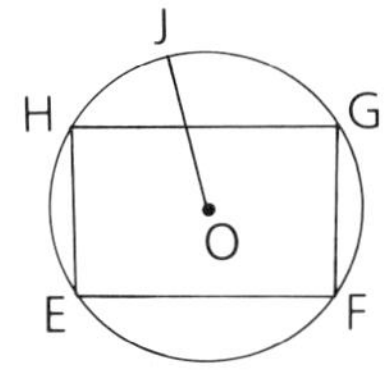

17 A quadrilateral can be inscribed in a circle if, and only if, a pair of opposite angles is supplementary.

Which of the following quadrilaterals can be inscribed in a circle?

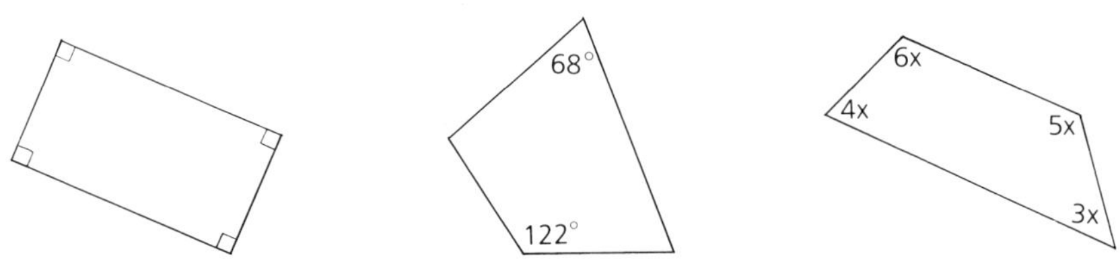

18 Prove: Any isosceles trapezoid can be inscribed in a circle. (Hint: See problem 17.)

19 Equilateral triangle PQR is inscribed in one circle and circumscribed about another circle. The circles are concentric.

a If the radius of the smaller circle is 10, find the radius of the larger circle.

b In general, for an equilateral triangle, what is the ratio of the radius of the inscribed circle to the radius of the circumscribed circle?

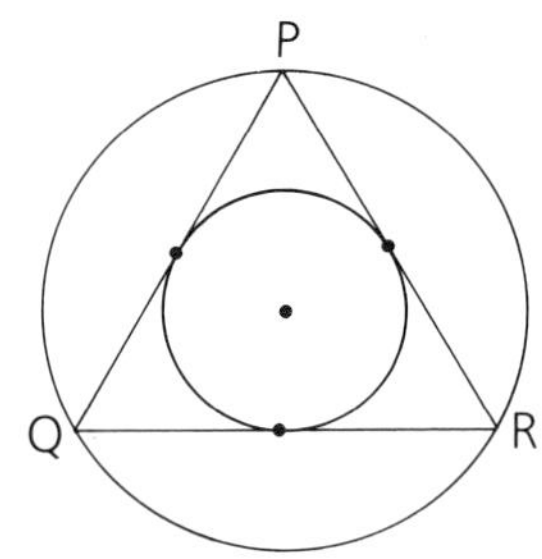

20 ABCD is a kite with $\overline{AB} \cong \overline{BC}$, $\overline{AD} \cong \overline{CD}$, and m$\angle$B = 120. The radius of the circle is 3.

Find: the perimeter of ABCD.

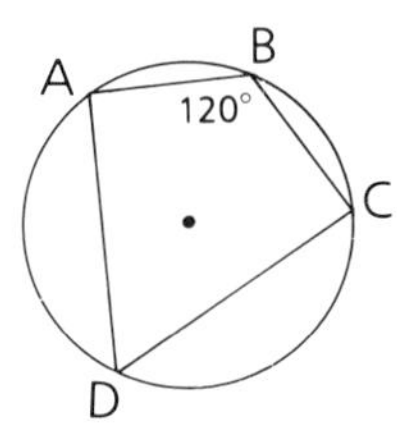

Section 10.7 Problem Set C

21 Given: $\angle A$ and $\angle C$ are right $\angle$s
$\overline{AB} \cong \overline{AD}$, $\overline{BC} \cong \overline{DC}$
Prove: ABCD is a square

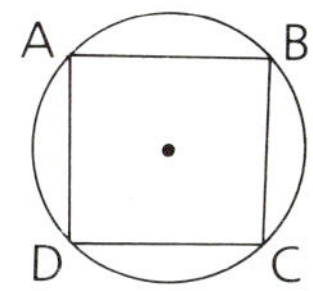

22 A set of points are CONCYCLIC if they all lie on the same circle. Prove that the vertices of any triangle are concyclic.

23 Are the vertices of each of the following figures concyclic? Answer Always, Sometimes, or Never.

a a rectangle
b a parallelogram
c a rhombus
d a non-isosceles trapezoid
e an equilateral polygon
f an equiangular polygon

24 ⊙ P is inscribed in trapezoid WXYZ. $\angle W$ and $\angle X$ are right $\angle$s. The radius of ⊙ P is 5, YZ = 14. Find the perimeter of WXYZ.

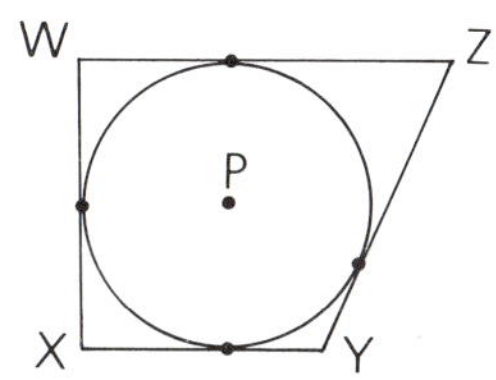

25 $\triangle XYZ$ is a right triangle with M the midpoint of hypotenuse $\overline{XZ}$. $\overline{XY} \cong \overline{MY}$. Find $m\angle ZXY$.

26 A circle is inscribed in a triangle with sides 8, 10, and 12. The point of tangency of the circle on the side 8 units long divides that side in a ratio x:y, $x < y$. Find that ratio.

Section 10.7 Problem Set D

27 Determine the conditions under which an equiangular polygon inscribed in a circle will be equilateral. Prove your conjecture.

28 Discuss the location of the center of the circumscribed circle of each of the following types of triangles:
a right b acute c obtuse

Section 10.8 The Power Theorems

Theorem 86

T **If two chords of a circle intersect inside the circle, then the product of the measures of the segments of one chord is equal to the product of the measures of the segments of the other chord. (CHORD-CHORD POWER THEOREM)**

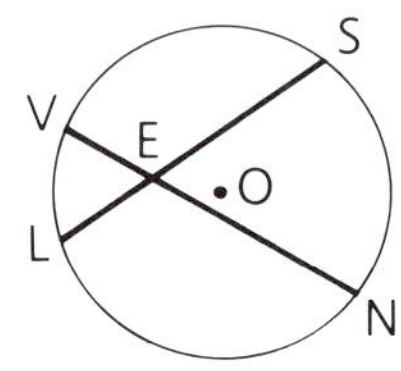

Given: Chords $\overline{VN}$ and $\overline{LS}$ intersect at E inside circle O

Prove: EV • EN = EL • SE

Theorem 86 was proved in Section 10.6, Sample Problem 1.

Theorem 87

T **If a tangent and secant segment are drawn from an external point to a circle, then the square of the measure of the tangent segment is equal to the product of the measures of the entire secant segment and its external part. (TANGENT-SECANT POWER THEOREM)**

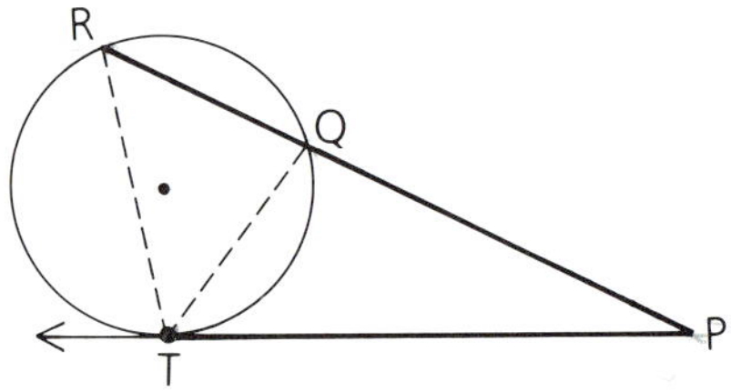

Given: $\overline{PR}$ is a secant segment, and $\overline{PT}$ is a tangent segment to the ⊙

Prove: $(TP)^2 = (PR)(PQ)$

Proof: Similar triangles are formed by drawing $\overline{TQ}$ and $\overline{TR}$. $\angle PTQ \cong \angle R$ (Why?), and $\angle P \cong \angle P$, so $\triangle PTR \sim \triangle PQT$.

Thus, $\frac{TP}{PR} = \frac{PQ}{TP}$, and $(TP)^2 = (PQ)(PR)$

Theorem 88

T **If two secant segments are drawn from a point outside a circle, then the product of the measures of one secant segment and its external part is equal to the product of the measures of the other secant segment and its external part. (SECANT-SECANT POWER THEOREM.)**

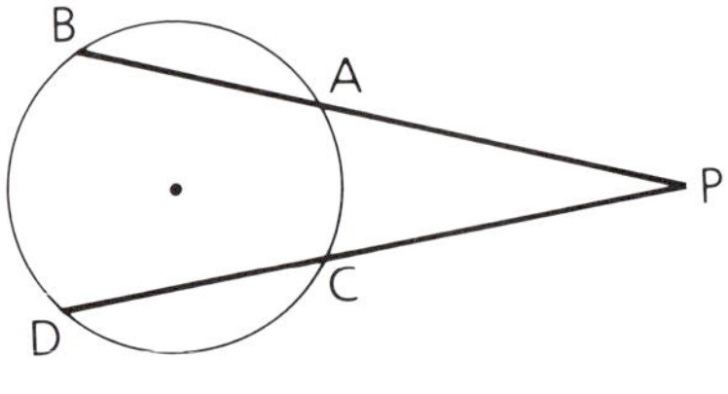

Given: Secant segments $\overline{PB}$ and $\overline{PD}$

Prove: PB • PA = PD • PC

The Royal Road: The Pentagram

If you draw all the diagonals of a regular pentagon, you will find another regular pentagon, and inside that, still another pentagon, and so on. The figure formed by the diagonals of each of the pentagons is called a **pentagram**.

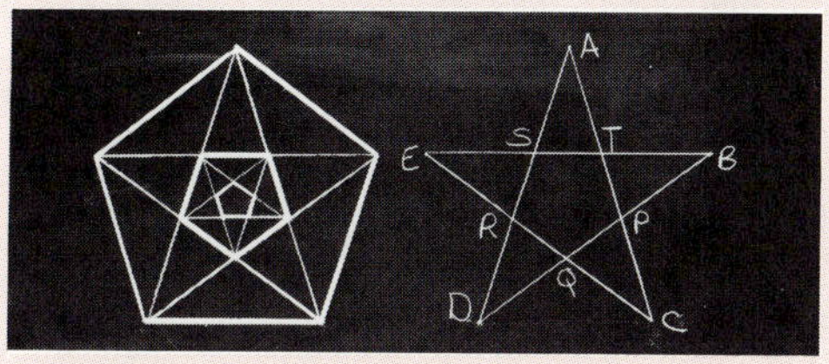

Every angle in a pentagram is 36°, 72°, or 108°, and there are many similar triangles that can be formed.

One of the properties of a pentagram is that a point of intersection, say point P, divides side $\overline{AC}$ such that the ratio of the whole length of $\overline{AC}$ to a portion of $\overline{AC}$ is the same as the ratio between the portions of $\overline{AC}$:

$$\frac{AC}{AP} = \frac{AP}{PC}$$

To prove this, first show that $\overline{DP}$ is in isosceles $\triangle DPA$ and in isosceles $\triangle DPC$, so that $AP = DP = DC$. Next, show $\triangle ACD \sim \triangle DPC$, so that:

$$\frac{AC}{DP} = \frac{CD}{PC}$$

Finally, substitute values for DP and CD in the proportion.

Section 10.8 Sample Problems

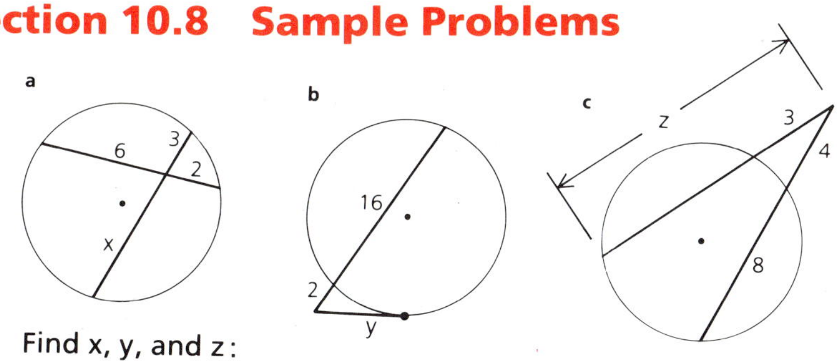

1 Find x, y, and z:

Solutions:

By the CHORD-CHORD POWER THEOREM

$6 \cdot 2 = 3 \cdot x$

$4 = x$

By the TANGENT-SECANT POWER THEOREM

$y^2 = 2 \cdot 18$

$y = \pm 6$, reject -6

$y = 6$

By the SECANT-SECANT POWER THEOREM

$4 \cdot (8 + 4) = 3 \cdot z$

$4 \cdot 12 = 3z$

$16 = z$

2 Tangent segment PT = 8 cm. The radius of the circle is 6 cm. Find the distance from P to the circle.

Solution: Draw secant from P through center R. PT = 8 and QR = RS = 6. Let PQ = x, the distance from P to the ⊙.

By the TANGENT-SECANT POWER THEOREM:

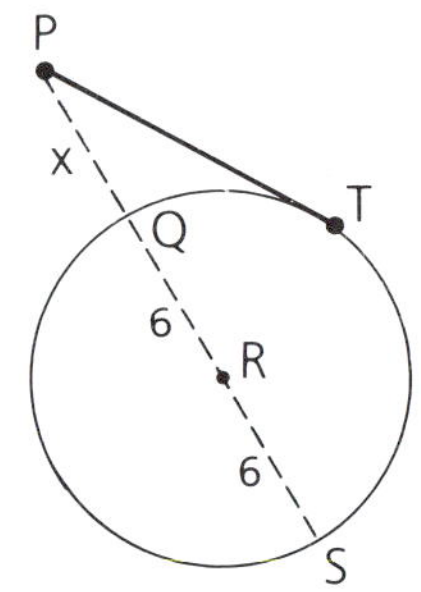

$$\begin{aligned} PQ \cdot PS &= PT^2 \\ x \cdot (x + 12) &= 8^2 \\ x^2 + 12x &= 64 \\ x^2 + 12x - 64 &= 0 \\ (x - 4)(x + 16) &= 0 \end{aligned}$$

$$x - 4 = 0 \quad \text{or} \quad x + 16 = 0$$

$$x = 4 \quad \text{or} \quad x = -16$$

Thus, PQ is 4 cm

Section 10.8 Problem Set A

1 Solve for x, y, and z.

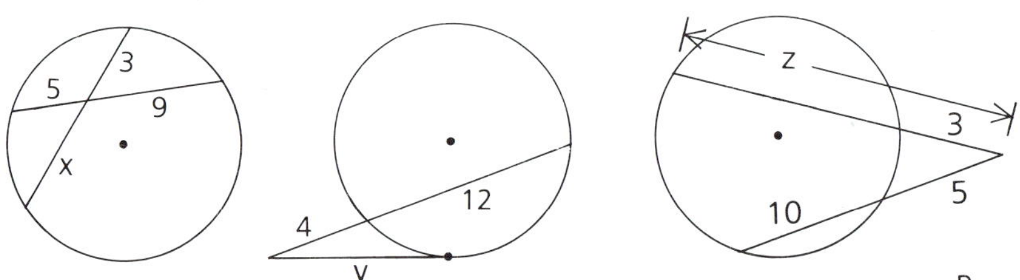

2 T is the midpt. of $\overline{QS}$ in ⊙ O, PT = 8, QS = 40.

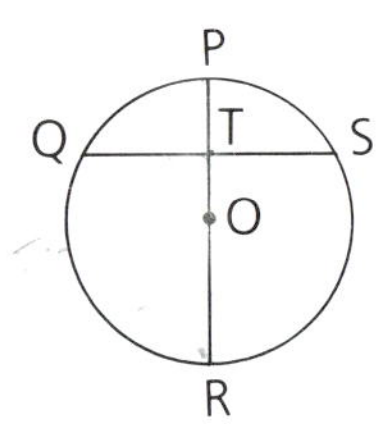

a Find TR.

b Find the diameter of ⊙ O.

3 a If TR = 10 and QR = 5, find PR.

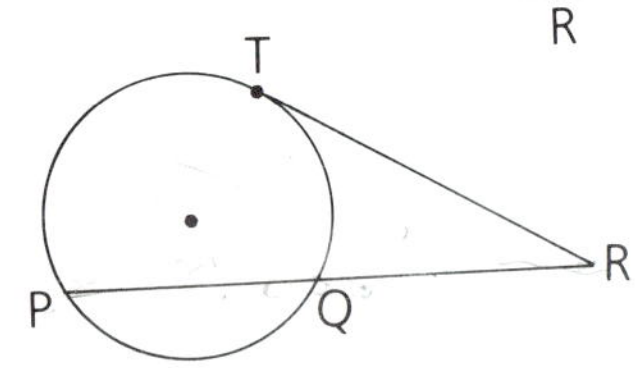

b If TR = 10 and QR = 4, find PQ.

c If TR = 10 and PR = 50, find PQ.

4 a If AE = 6, AB = 10, and CE = 3, find ED.

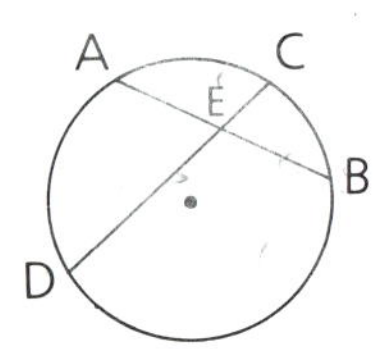

b If AE = 8, AB = 14, and ED = 16, find DC.

c If CE = 2, ED = 18, and $\overline{AE} \cong \overline{EB}$, find AB.

5 Find the radius of ⊙ P.

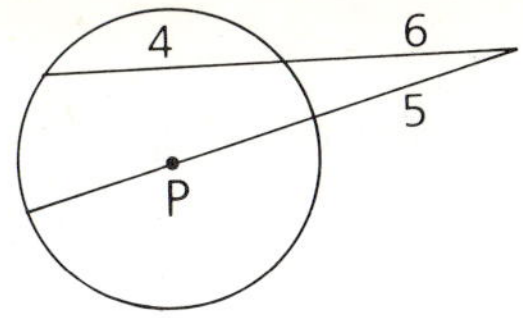

6 AP = 3, PQ = 5, QB = 7,
CP = 2, QD = 14.

Find PD and EQ.

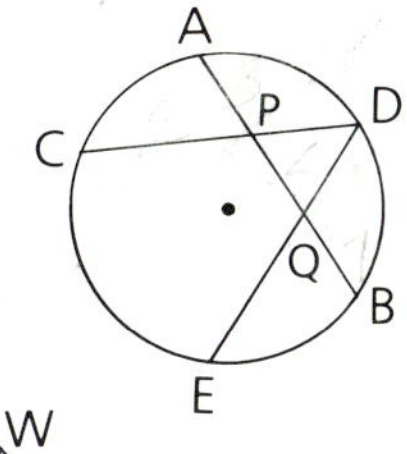

7 TZ = 6, YZ = 4, SX = 3, WX = 1

Find XT (Hint: Find SZ)

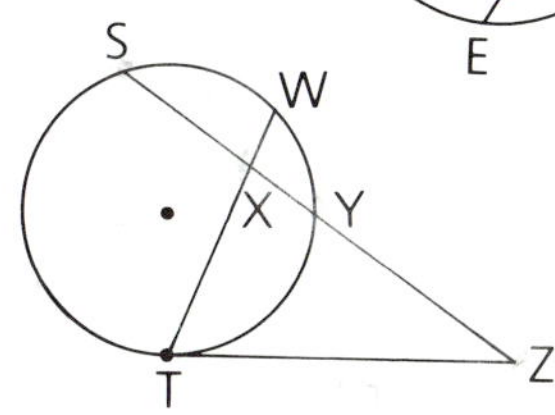

Section 10.8 Problem Set B

8 a Find y.

b Is the triangle acute, right, or obtuse?

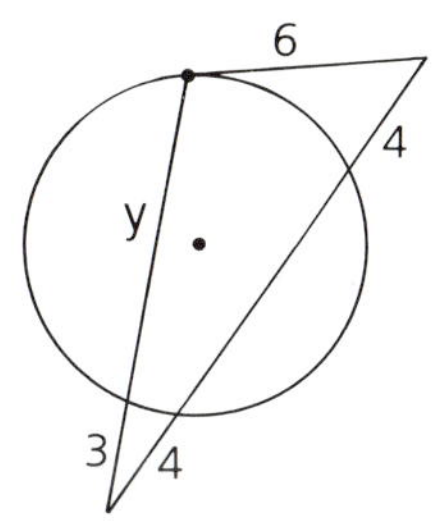

9 AB = 7, CD = 5, ED = 2

Find AE

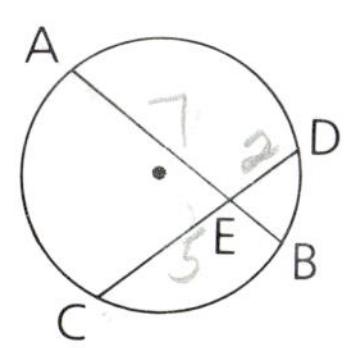

10 PT = 2, QR = 3

Find PQ

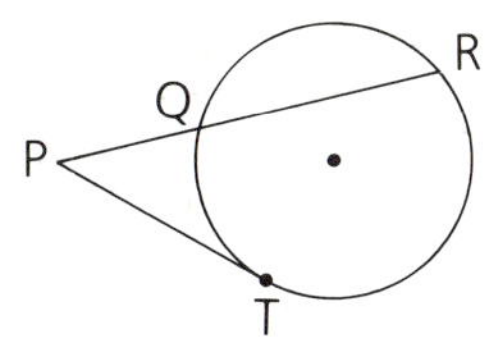

11 Solve for x.

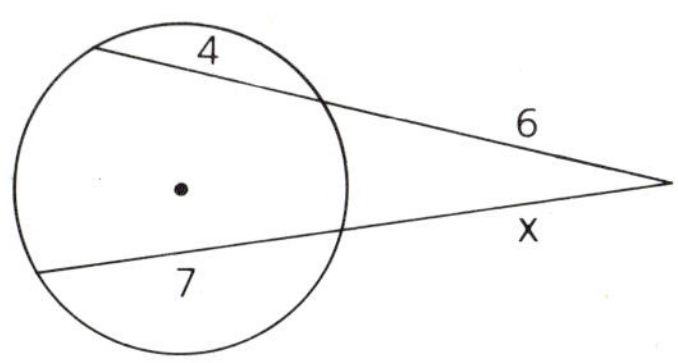

12 $\overline{AB}$ is diameter of ⊙ O.
$\overline{CD}$ tangent at D, CD = 6, BC = 4.

Find the radius of the ⊙.

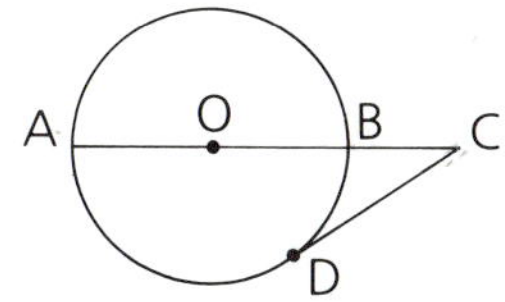

13 An arch supports a pipeline across a river 20 m wide. Midway, the suspending cable is 5 m long. Find the radius of the arch.

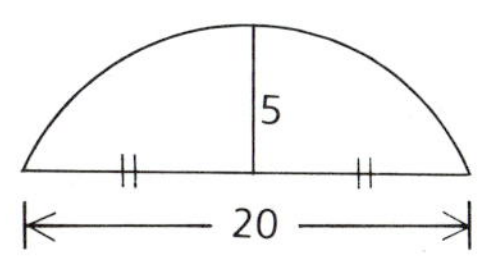

14 The diameter of the earth is approximately 8,000 miles. Heavenly Helen, in a space ship 100 miles above earth, sights Earthy Ernest coming over the horizon. Approximate how far apart Helen and Ernest are.

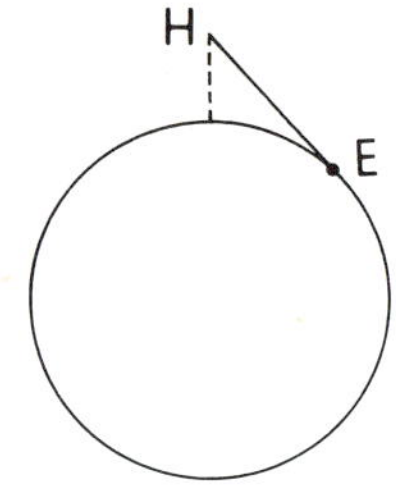

Section 10.8 Problem Set C

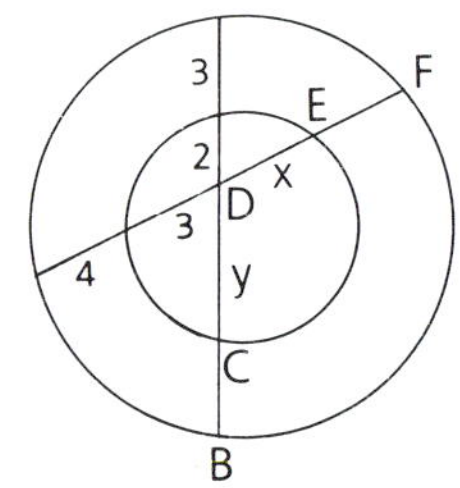

15 Given concentric circles as shown.

Find DE and DC.

16 The radius of each circle is 3. Triangle WXY is equilateral.

a Find WY.

b Find the ratio of the perimeters of △ABC to △PQR to △WXY.

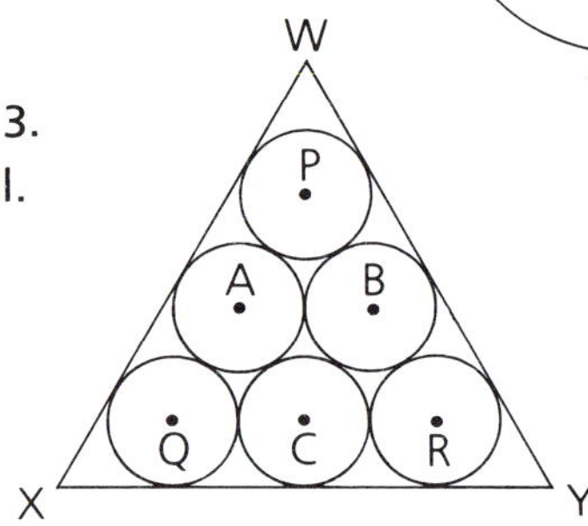

17 a Find x.

b What restrictions must be placed on y in this problem?

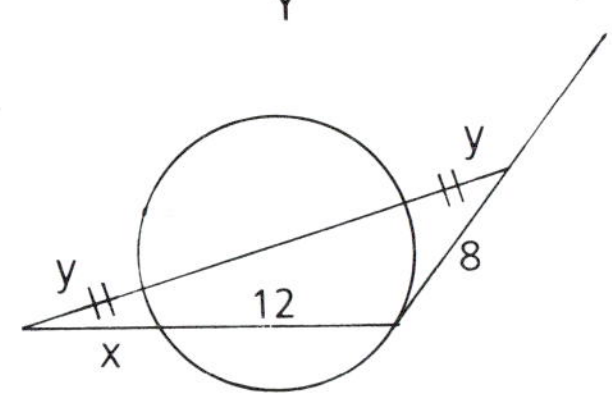

18 Tangent $\overline{AT} = 12$, $AB = 8$, $\overline{AT} \perp \overline{AB}$

a Find the diameter of the circle.

b How far is the circle from point A?

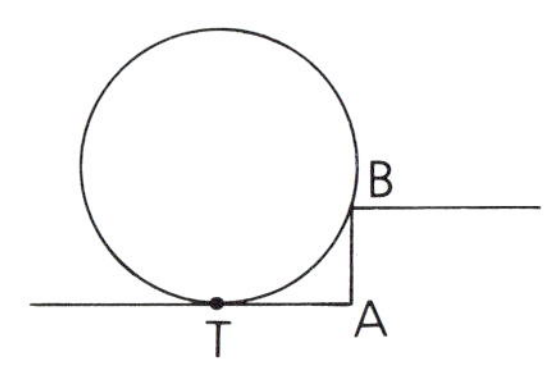

Section 10.8 Problem Set D

19 a Assume $AB = 5$. Then solving for y will yield $5\sqrt{5}$. Why is this answer impossible?

b What set of values for AB will allow a valid solution for y?

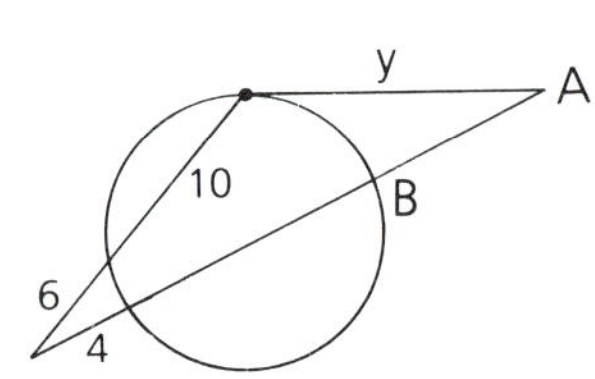

Section 10.9 Circumference and Arc Length

In this section we discuss:

A CIRCUMFERENCE OF A CIRCLE

B LENGTH OF AN ARC

10.9(A) Circumference of a Circle

D **The CIRCUMFERENCE of a circle is its perimeter.**

The formula for the circumference C of a circle of diameter d is a result of the fact that, regardless of the circle's size, the ratio of the circumference to the diameter is always the same number. This number is given the special letter π (Greek letter "Pi"). Its *approximate* value is 3.14 or $\frac{22}{7}$.

P $\mathbf{C = \pi d} \qquad \pi \approx \mathbf{3.14} \approx \mathbf{\frac{22}{7}}$

■ Example: Find the circumference of a circle whose radius is 5.

$d = 2 \cdot$ radius, so $d = 2 \cdot 5 = 10$
$C = \pi d$
$C = \pi(10) = 10\pi$

Leave the answer in terms of π, unless you are asked to approximate the answer. Then use 3.14 or $\frac{22}{7}$ in place of π. In the example above, you would continue with
$C \approx 10\,(3.14) \approx 31.4$

10.9(B) Length of an Arc

The length of an arc is a linear measurement similar to the length of a line segment. Arc length depends upon the circumference of the circle and the *degree measure* of the arc. The units of arc length are feet, meters, centimeters, etc.

■ Example: Find the length of a 40° arc of a circle of 18 cm radius.

Solution: $C = \pi \cdot 36$ or 36π

The 40° arc is $\frac{40}{360}$ or $\frac{1}{9}$ of the circle.

$$\text{Arc Length AB} = \frac{1}{9}(\text{Circumference})$$
$$= \frac{1}{9}(36\pi) = 4\pi$$

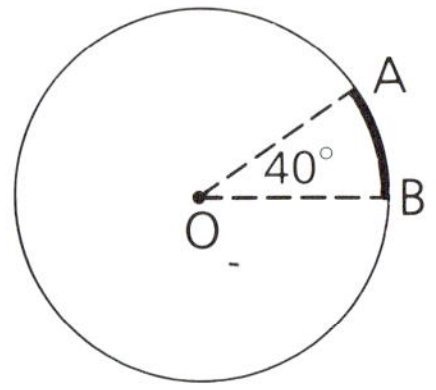

Theorem 89

T **The LENGTH OF AN ARC is equal to the *circumference* times the *fractional part* of the circle determined by the arc.**

> Length of $\widehat{PQ} = \left(\frac{m\widehat{PQ}}{360}\right)\pi d$
> d = diameter
> $\widehat{PQ}$ is measured in degrees

Section 10.9 Sample Problems

1 Find the radius of a circle whose circumference is 50π.

Solution:
$$C = \pi d$$
$$50\pi = \pi d$$
$$50 = d\text{ , so } r = 25$$

2 For a circle of 12 cm radius, find the length of:

a 30° arc **b** 105° arc

Solutions:

a length of arc $= \frac{30°}{360°}(24\pi)$
$= 2\pi$ cm

b length of arc $= \frac{105°}{360°}(24\pi)$
$= 7\pi$ cm

3 The outside diameter of a bicycle tire is 70 cm.

a How far will the bicycle travel if the wheel rotates 1000 times? (Approximate the answer in meters.)

b How many revolutions will the tire make if the bicycle travels 15 meters? (Approximate to the nearest tenth of a revolution.)

Solutions: The distance covered in each revolution is the circumference of the tire: $C = \pi d = 70\pi \approx 220$ cm per revolution.

a In 1000 revolutions, distance $\approx$ (1,000 rev.) (220 cm per rev.)

$\approx$ 220,000 cm, or 2200 m

b The bicycle moves approximately 220 cm, or 2.2 m per rev. Let x be the number of revolutions.

Total distance = (number of rev.) (distance per rev.)

15 m $\approx$ (x) (2.2)

6.8 $\approx$ x, so the wheel made approx. 6.8 rev.

Section 10.9 Problem Set A

1 Find the circumference (in both π and approximate forms)

a of a circle whose diameter is 21 mm (use $\pi \approx \frac{22}{7}$).

b of a circle whose radius is 6 mm (use $\pi \approx 3.14$).

2 Find the radius of a circle whose circumference is:

a 56π b 314.0 (use $\pi \approx 3.14$) c 17π d 88 (use $\pi \approx \frac{22}{7}$)

3 Find the length of each arc in a circle of radius 10.

a 72° arc b 90° arc c 60° arc d semicircle

4 A bicycle has wheels of 30 cm in diameter. Find the approximate distance that the bicycle moves forward during:

a 1 revolution b 10 revolutions c 1000 revolutions

5 Find the complete perimeter of each shape. Leave answers in terms of π and whole numbers.

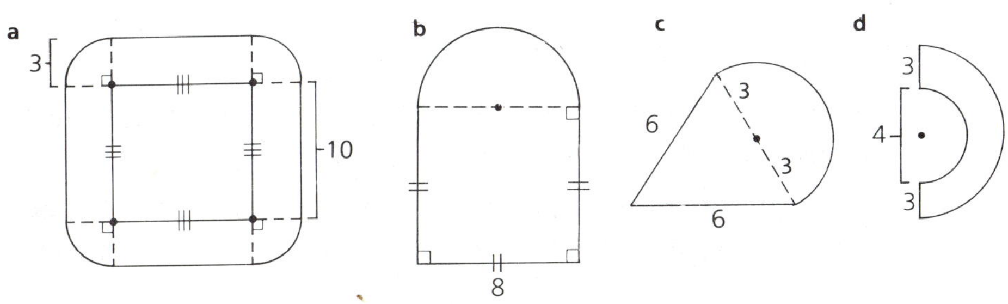

6 a Find the length of $\widehat{AB}$.

b Find the perimeter of sector AOB. (The shaded region is a sector.)

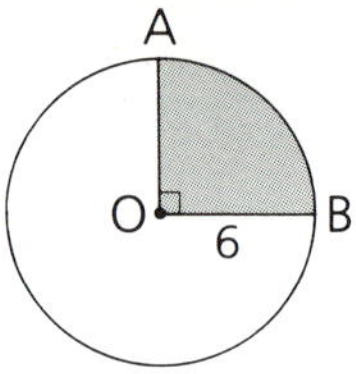

7 Find the length of fencing needed to surround the race track.

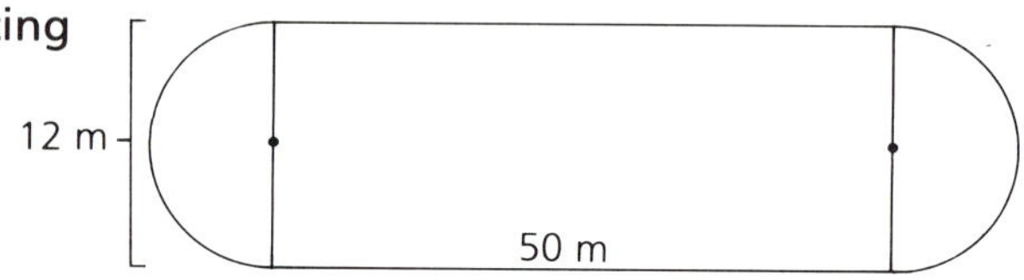

8 Radius of ⊙ O = 10 mm, length of $\widehat{AB} = 4\pi$ mm.

a Find the circumference of ⊙ O.

b Find the measure of $\widehat{AB}$.

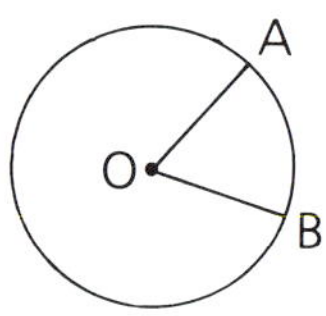

Section 10.9 Problem Set B

9 Given arcs $\widehat{AB}$ mounted on equilateral triangles as shown. Find the length of each arc $\widehat{AB}$. In each case $\overline{OA}$ is a radius of $\widehat{AB}$.

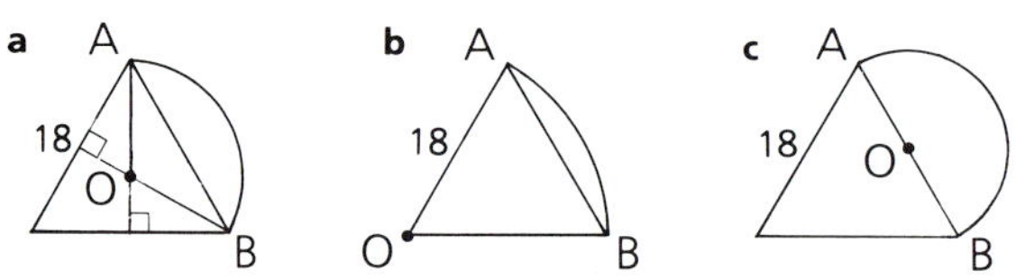

10 100 turns of thread are on this circular spool. The diameter is 4 cm. Find the approximate length of thread.

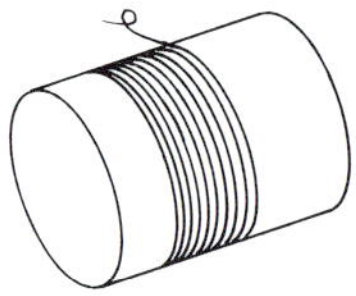

11 Awful Kanaufil plans to ride his cycle on a single loop track. There is 100 m of straight track before the loop, 20 m after. The loop has a radius of 15 m. What is the total length of the track he must ride? (Use $\pi \approx 3.14$.)

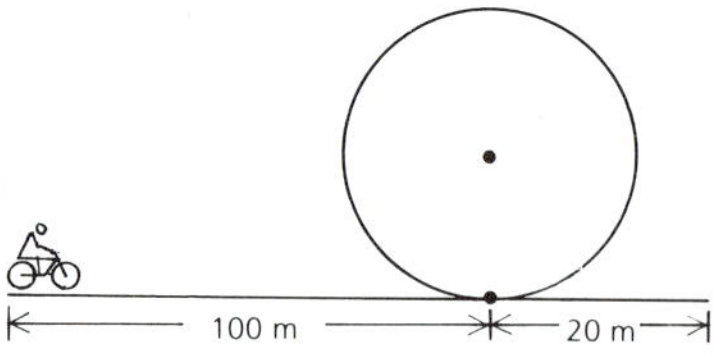

12 Find the outer perimeter of the figure. It is composed of semicircles mounted on the sides of a rectangle.

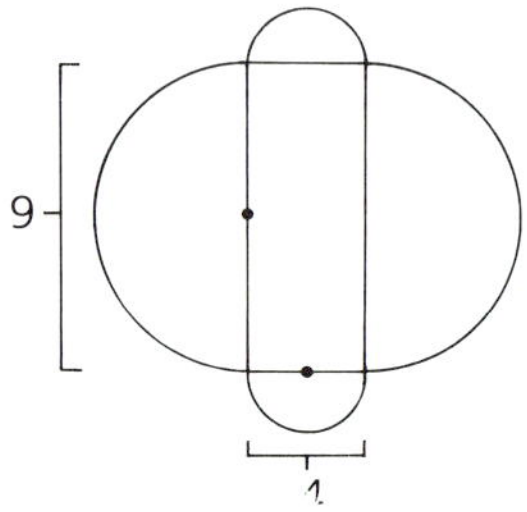

13 Sandy skated on a rink as shown. How far did she travel going once around in the outside lane? In the inside lane? (Use $\pi \approx 3.14$.)

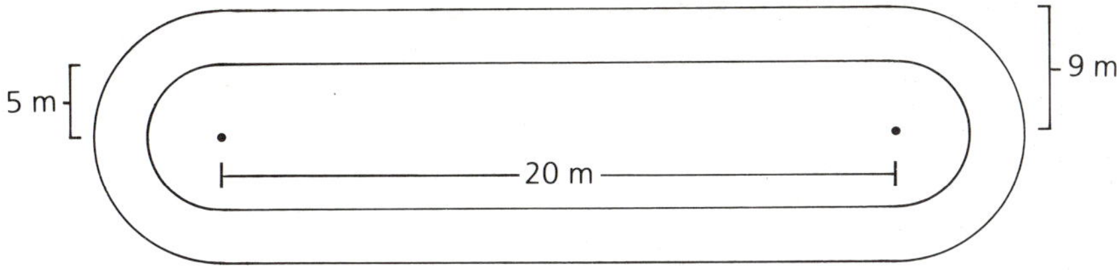

14 Given circle O of radius 6 and point P outside it. A belt wrapped tightly around the circle forms a right angle at P. Find the length of the belt.

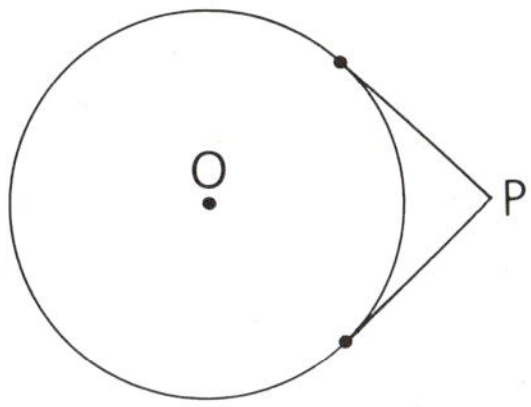

Section 10.9 Problem Set C

15 A circular garbage can is wedged into a rectangular corner. The can has a diameter of 48 cm.

a Find the distance from the corner point to the can, PA.

b Find the distance from the corner point to the point of contact of the can with the wall, PB.

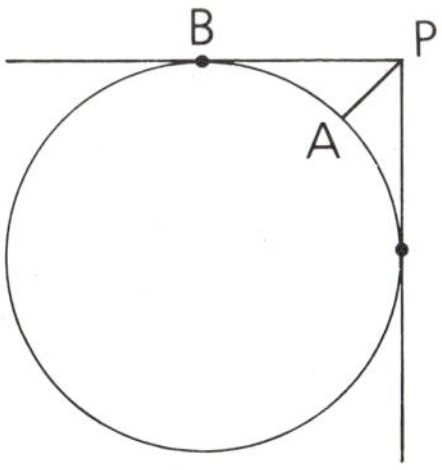

Section 10.9 Problem Set D

16 Two pulleys are connected by a belt. The radii of the pulleys are 3 cm and 15 cm and the distance between centers is 24 cm.

Find the total length of belt needed to connect the pulleys.

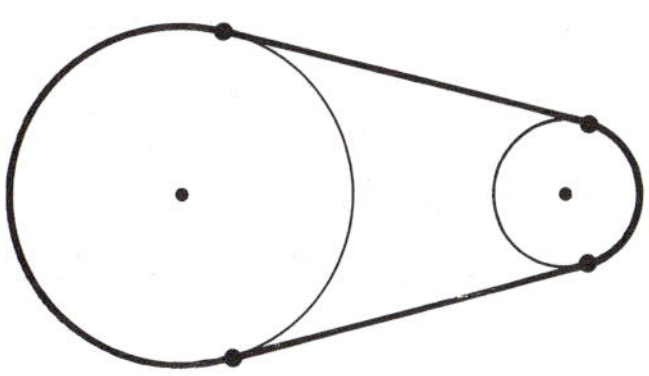

Chapter 10 Study Guide

As you study and review this chapter, be sure that you are able to:

1 Use the following vocabulary:

circle: center, exterior of, interior of, radius, diameter, chord (pp. 409–410), circumference (p. 469), pi (p. 469)

circles: congruent (p. 410), tangent—2 types, (p. 430), disjoint (p. 434), intersecting (p. 434), line of centers (p. 430), concentric (pp. 410, 434)

arcs: minor, major, semicircle (p. 420), measure of (p. 421), length of (p. 470), congruent (p. 422), midpoint of (p. 423), intercepted (pp. 421, 422, 438, 442, 449)

distances: chord to center (p. 411), point to circle (p. 466)

polygons and circles: either one inscribed or circumscribed about the other, incenter, circumcenter (p. 457)

lines and circles: secant, tangent (pp. 429, 430)

2 Apply the following theorems or postulates:

a Radius $\perp$ chord (p. 411) or tangent (p. 429)
b $\cong$ chords (pp. 416, 423)
c Two-tangent (p. 430)
d Power theorems, 3 kinds (pp. 464–465)
e $\cong$ chords $\Longleftrightarrow$ $\cong$ arcs $\Longleftrightarrow$ $\cong$ central angles (p. 423)

3 Describe how to measure these circle-related angles (p. 442):

a vertex inside (2 kinds)
b vertex on (2 kinds)
c vertex outside (3 kinds)

4 Apply special angle formulas in these situations:

a angles intercepting the same or congruent arcs (p. 450)
b angles inscribed in a semicircle (p. 451)
c tangent-tangent angle and its minor arc (p. 451)
d opposite angles of an inscribed quadrilateral (p. 458)

5 Solve the following types of problems:

a distance from chord to center (p. 412)
b radius perpendicular to tangent (p. 432)
c common tangent problems (p. 432)
d Walk-around (p. 433)
e POP (pp. 458–459)
f circumference (p. 470)
g arc lengths (p. 470)

Chapter 10 Review Problems A

1 Find x in each circle.

a

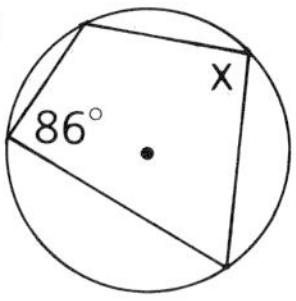

b

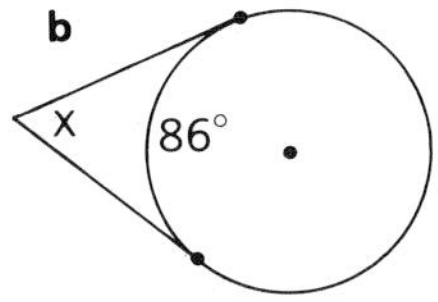

c

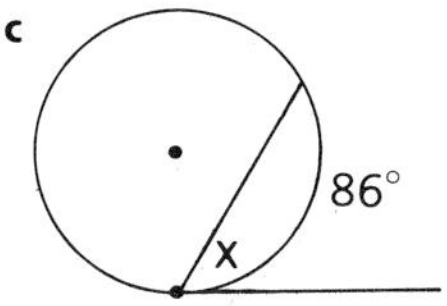

2 $\widehat{AB} = 98°$, $\widehat{CD} = 34°$.
Find x and y.

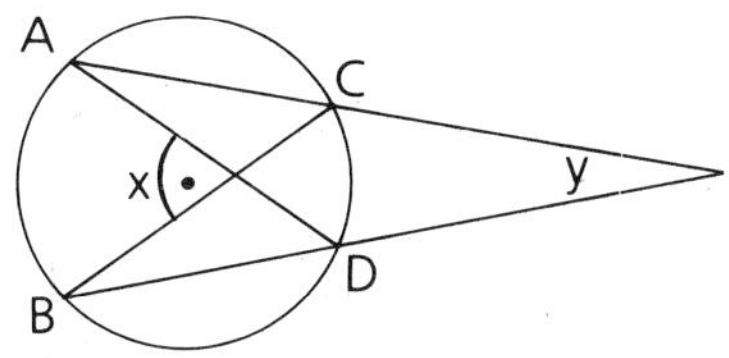

3 **a** Find BD.

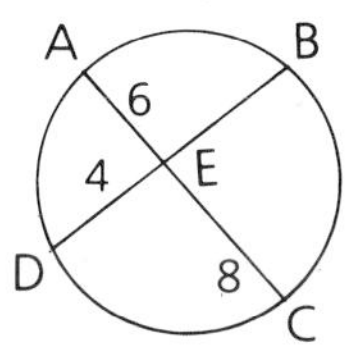

b Find PT.

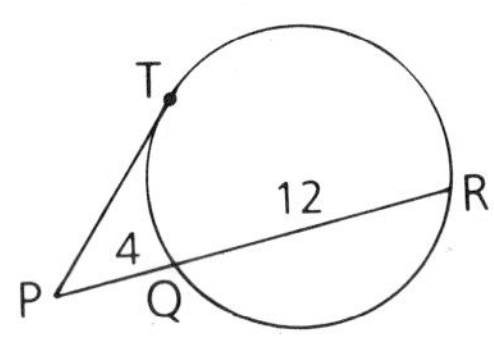

c Find WX.

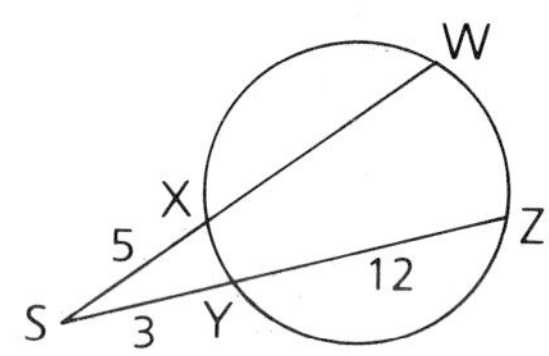

4 Find the radius of each circle.

a

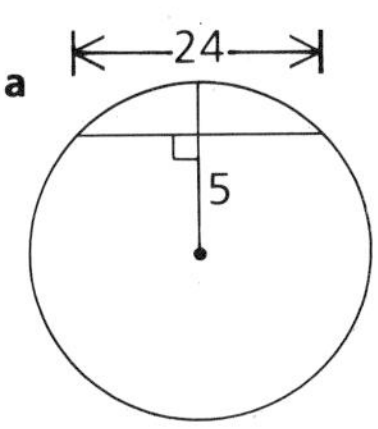

b

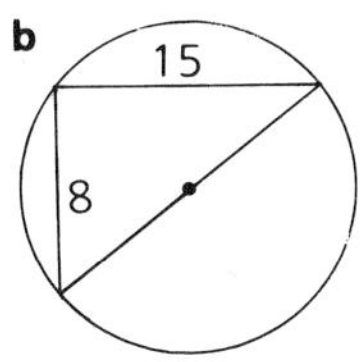

c

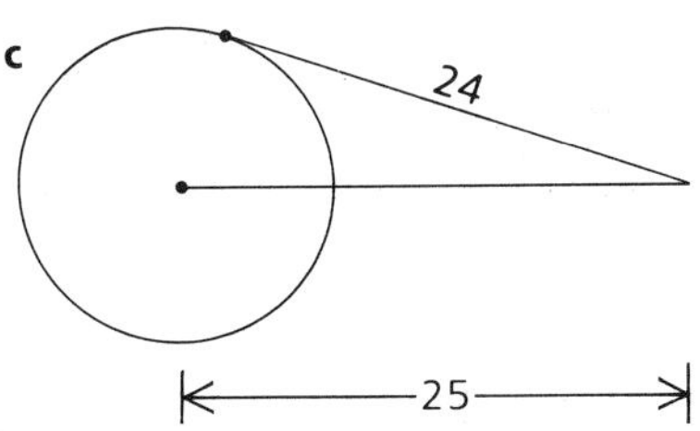

5 Given concentric circles at O with $\overline{PZ}$ and $\overline{PY}$ tangent to the inner circle at W and X. $\widehat{YZ} = 110°$.
Find the measure of $\widehat{WX}$.

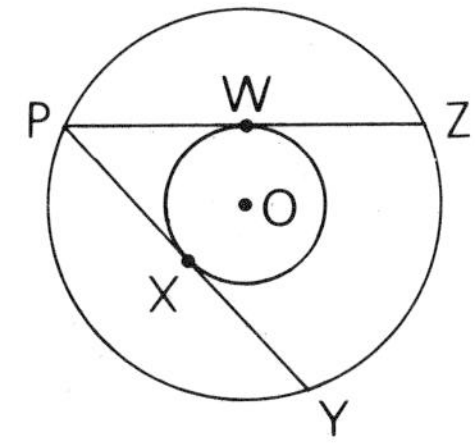

6 $\triangle ABC$ is isosceles with base $\overline{AB}$
$\angle DAC = 70°$, $\widehat{BC} = 160°$.
Find $\widehat{AB}$ and $\widehat{AD}$.

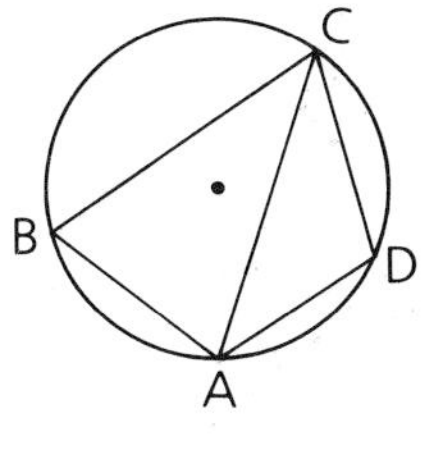

7 XOY is a sector of $\odot$ O with radius OY = 6 cm and central $\angle XOY = 45°$.

Find: **a** the length of $\overset{\frown}{XY}$

b the perimeter of sector XOY.

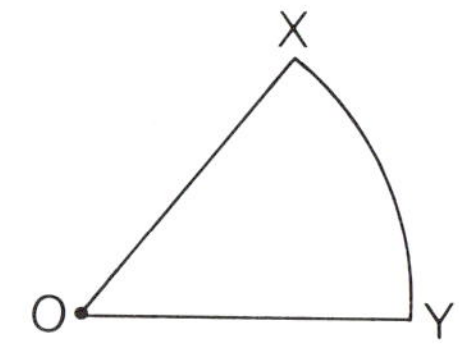

8 Circles A, B, C are tangent. AB = 7, BC = 10, CA = 11.

a Find the radius of $\odot$ A.

b Which circle is the largest?

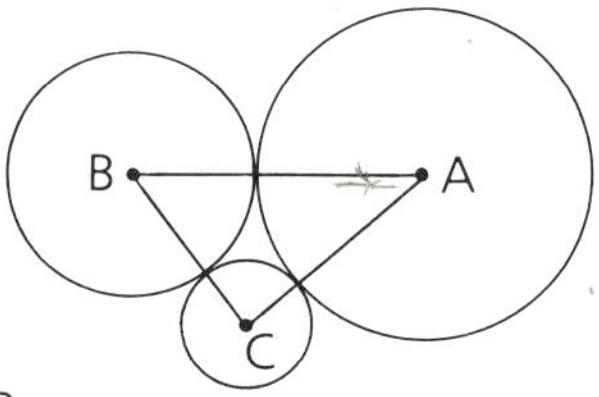

9 Given: $\odot$ O, $\overline{OM} \perp \overline{AB}$

Prove: $\overrightarrow{OM}$ bisects $\angle AOB$

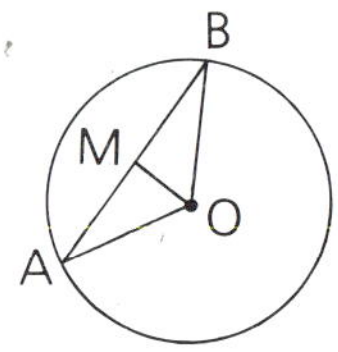

10 Given: $\overline{OP} \perp \overline{WX}$, $\overline{OQ} \perp \overline{YZ}$

$\triangle OPQ$ is isosceles with base $\overline{PQ}$

Conclusion: $\overset{\frown}{WX} \cong \overset{\frown}{YZ}$

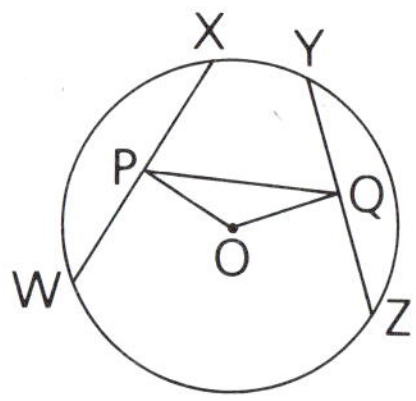

11 Given: $\overline{ZX}$ and $\overline{ZY}$ are tangent at X and Y

Prove: $\overline{WZ}$ bisects $\overset{\frown}{XY}$

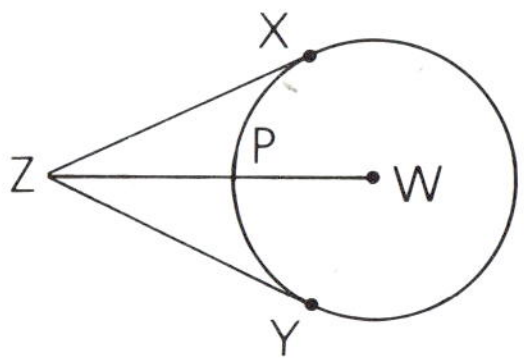

Chapter 10 Review Problems B

12 How far from the center of a circle of radius 15 is a chord of 18?

13 TP = 8, PQ = 6

Find RQ

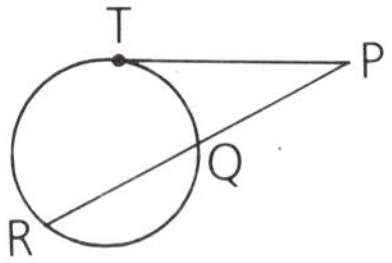

14 Given: Ⓢ O and P externally tangent

OA = 8, PB = 2

Find: common external tangent, AB

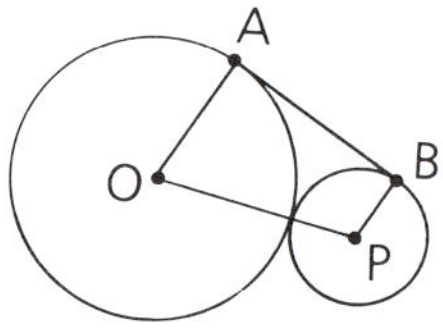

15 Jim knows that ⊙ O is inscribed in isosceles △ ABC. He forgets the congruent sides, but remembers that AB = 14 and the perimeter is 38.

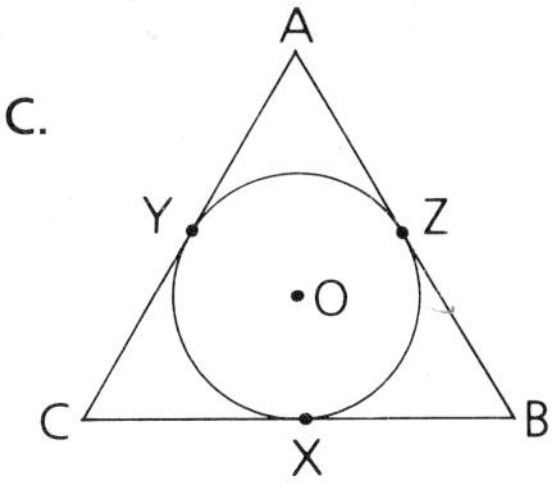

a Find XC.

b What are the three possible lengths of $\overline{BX}$?

16 A quadrilateral is inscribed in a circle. The vertices divide the circle into 4 arcs in a ratio 1:2:5:4. Find the angles of the quadrilateral.

17 $\widehat{AB} = 30°$, $\widehat{BC} = 40°$, $\widehat{CD} = 50°$

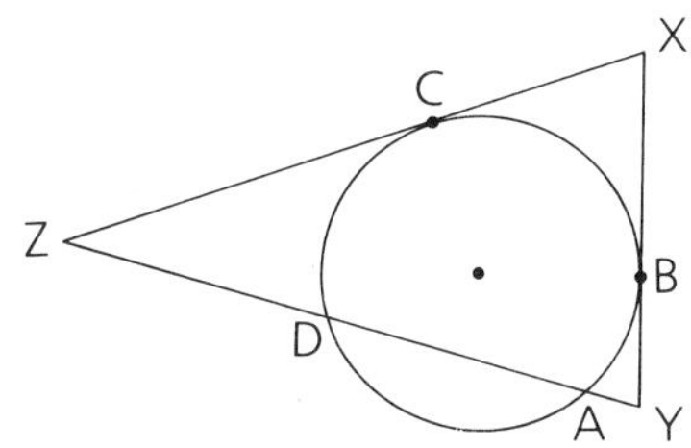

Find: a ∠X

b ∠Y

c ∠Z

18 Tangent segment $\overline{TP}$ is 15. PQ = 5.

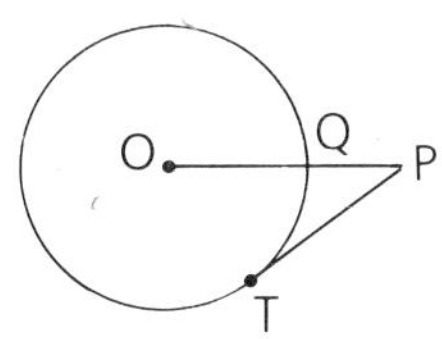

Find the radius of ⊙ O.

19 $m\widehat{AD} + m\widehat{BC} = 200$

$m\angle P = 30$

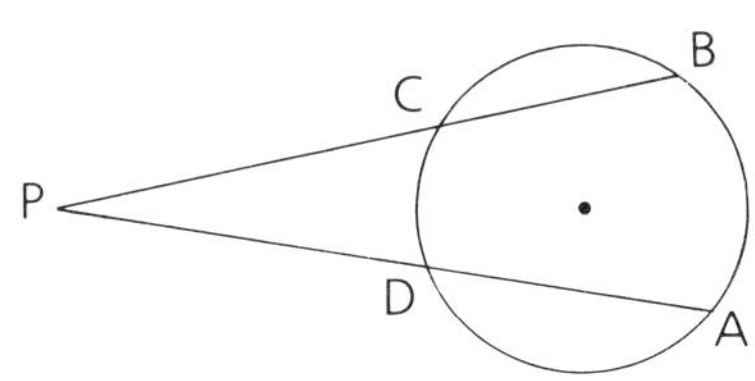

Find $m\widehat{AB}$ and $m\widehat{CD}$.

20 Given: ⊙ F with $\overline{EG} \perp \overline{AB}$

$\overline{EC} \cong \overline{ED}$

Prove: $\overline{AD}$ and $\overline{BC}$ are equidistant from F

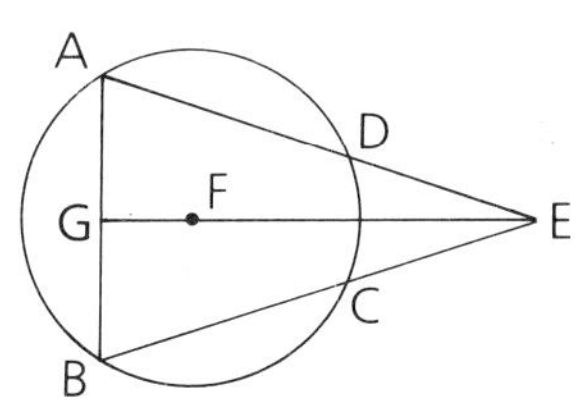

21 WXYZ is a parallelogram. $\overline{WZ}$ and $\overline{YZ}$ are tangent segments.

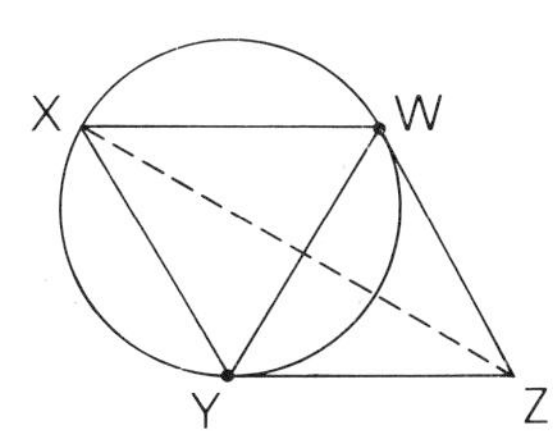

a Show that WXYZ is a rhombus.

b Find m∠Z.

c If WY = 15, find the perimeter of WXYZ.

d If WY = 15, find XZ.

22 Find x and y.

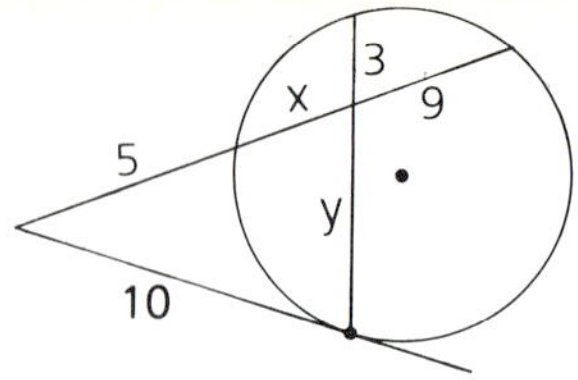

23 Given: WXYZ is a square
P is on $\widehat{WZ}$

Prove: $\overrightarrow{PX}$ and $\overrightarrow{PY}$ trisect $\angle WPZ$

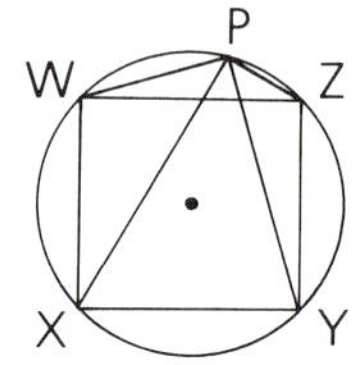

Chapter 10 Review Problems C

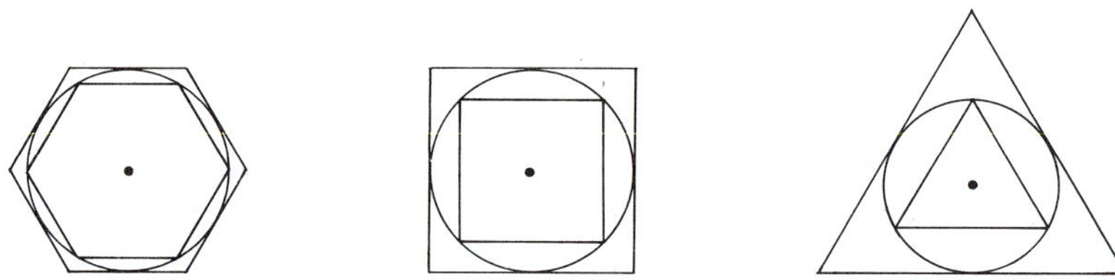

24 Each circle above is inscribed in a regular polygon and is circumscribed about another regular polygon.

a If the length of a side of each outer polygon is 12, find the length of a side of the inner polygon.

b Find the ratio of the sides of the small polygon to the sides of the larger polygon.

25 a An inscribed triangle divides a circle into three arcs in a ratio a:b:c. Find the ratio of the angles of the triangle.

b An inscribed quadrilateral divides a circle into four arcs in a ratio a:b:c:d. Find the ratio of the angles of the quadrilateral.

26 Given: $\overline{WZ}$ is a diameter of the ⊙

Show: $m\angle P = \dfrac{m\widehat{WX} + m\widehat{YZ}}{2}$

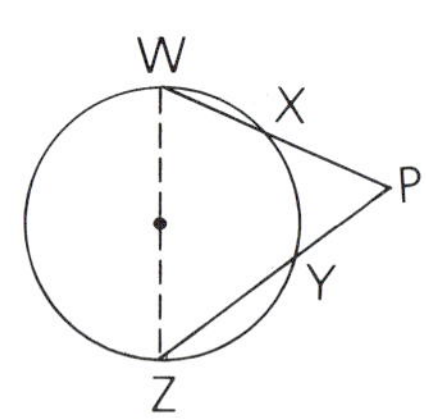

27 Given: Ⓢ P and Q are tangent internally at T

Prove: AC:CT = BD:DT

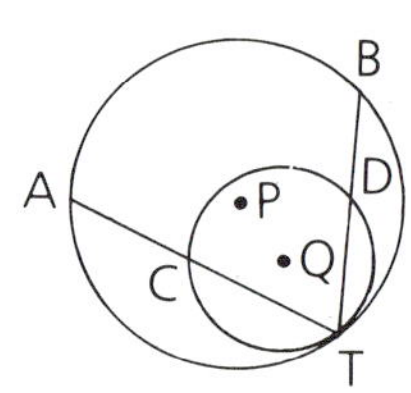

28 $\widehat{AQ} \cong \widehat{RB}$.
$\overline{PR}$ divides major and minor arcs AB in ratio of $\widehat{AQ}:\widehat{QB} = 4:3$ and $\widehat{AR}:\widehat{RB} = 7:5$.
Find the ratio of $\angle APQ : \angle BPQ$.

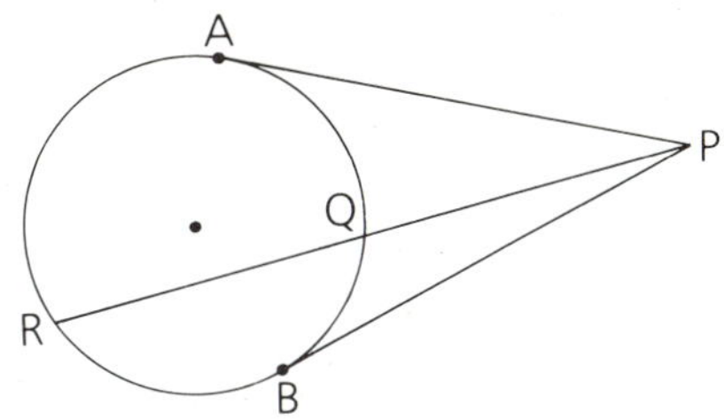

29 Given: $\overline{PA}$, $\overline{PB}$, $\overline{PC}$, $\overline{PD}$, and $\overline{PE}$ are such that three of the segments are secant segments to circle O, and the remaining two are tangent segments to circle O.

If two of the segments are selected at random, what is the probability that a secant-tangent angle is formed?

Chapter 10 Review Problems D

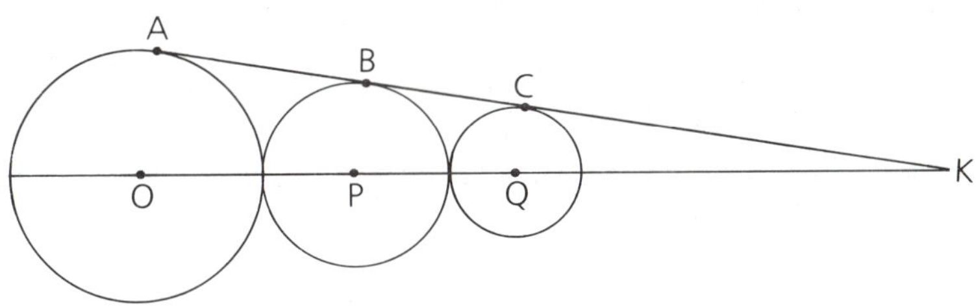

30 Given: ⊙ O is tangent to ⊙ P, and ⊙ P is tangent to ⊙ Q.
P is between O and Q.
$\overline{AC}$ is a common tangent to ⓢ O, P, and Q at A, B and C.

a If AB:BC = 2:1, find the ratio of radii of ⊙ O and ⊙ Q.

b If AB:BC = x:y, find the ratio of radii of ⊙ O and ⊙ Q.

c If the centers are not collinear, will the same result occur?

31 Given: $\overline{AB}$ is a diameter of ⊙ O, and $\overleftrightarrow{PQ}$ is tangent at B
$\overline{AP}$ and $\overline{AQ}$ intersect the circle at X and Y respectively

Show: $\angle AXY \cong \angle AQP$

32 If a triangle is inscribed in a circle, what is the probability that the triangle is obtuse?

Chapter 11

AREA

Section 11.1 The Concept of Area

This section discusses:

A THE MEANING OF AREA

B THE AREAS OF RECTANGLES AND SQUARES

C BASIC PROPERTIES OF AREA

11.1 (A) The Meaning of Area

When we measure the length of a line segment, such standard units as meter, yard, mile, centimeter, kilometer, and so on are used. These are often called **LINEAR UNITS** because they measure length.

The standard units of area are **SQUARE UNITS,** such as square meter, square yard, and square mile. A **SQUARE METER** is the area of a square whose sides are each one meter in length.

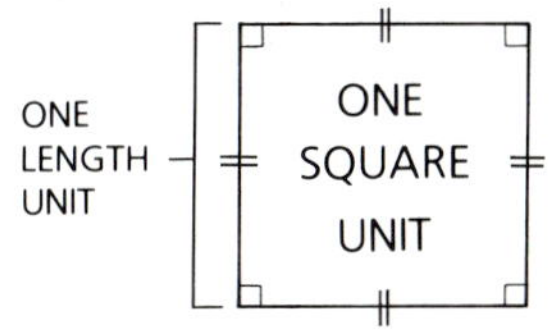

D **The AREA of a closed region is a number. The AREA is the number of "square units" contained within the boundary of the region.**

We can estimate the area of a region by observing the number of square units it contains.

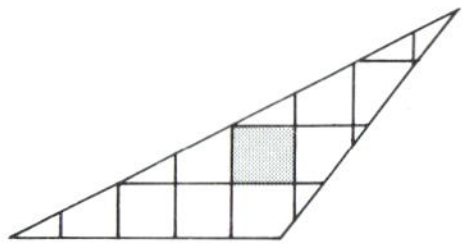

Estimated Area = 10 sq units

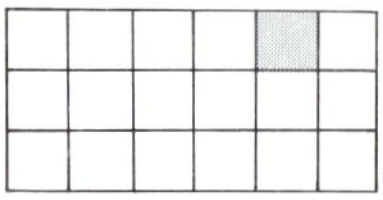

Estimated Area = 18 sq units

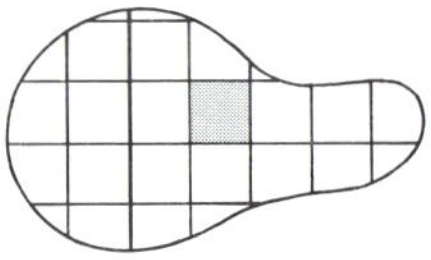

Estimated Area = 19 sq units

Counting is not the easiest nor the best way to find the number of square units in a region. We will develop formulas for computing the areas of regions bounded by the common geometrical figures. Such regions are usually named by their boundaries, as in "area of a rectangle."

11.1 (B) The Areas of Rectangles and Squares

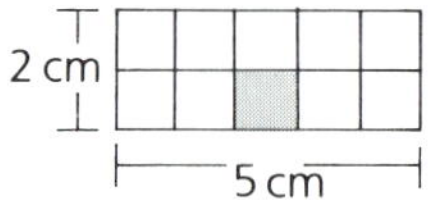

Area = 10 sq cm

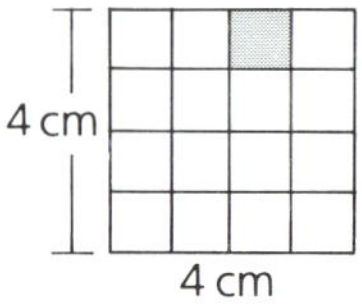

Area = 16 sq cm

In the figures above, there are two ways to find the areas:

1 The number of square units can be counted individually, or

2 The number of square units can be computed by multiplying the number of columns by the number of rows (*height*, or altitude).

The second method suggests the following formula, which may be used to compute areas even when the lengths are fractions or irrational numbers.

P **The AREA OF A RECTANGLE equals the product of the *base* times the *height* for that base.**

$$A_{rect} = b \cdot h$$

where b = base
h = height

In a square the base and height are equal, so:

Theorem 90

T **The AREA OF A SQUARE is equal to the square of a side.**

$$A_{sq} = s^2$$

where s = length of a side

11.1 (C) Basic Properties of Area

We make three basic assumptions about area:

P **1 Every closed region has an area.**

P **2 If two closed figures are congruent, then their areas are equal.**

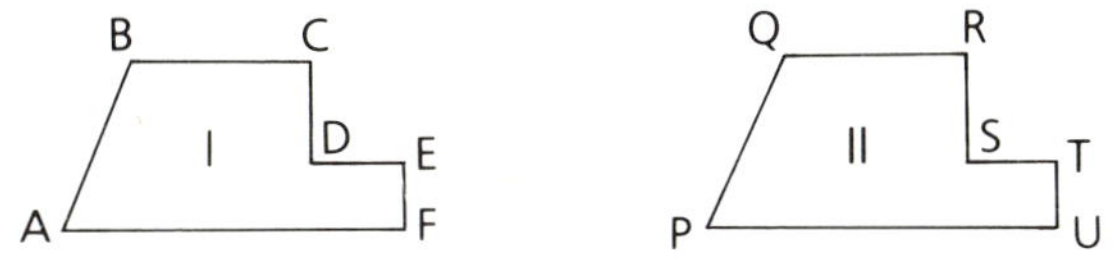

If ABCDEF ≅ PQRSTU, then Area of Region I = Area of Region II

P **3 If two closed regions intersect *only along a common boundary,* then the area of their union is equal to the sum of their individual areas.**

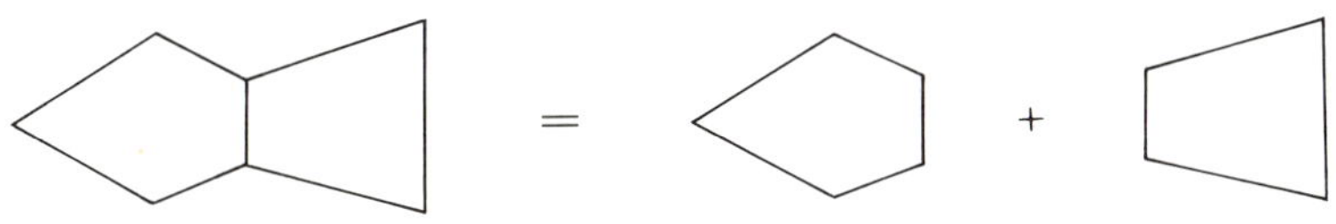

Section 11.1 Sample Problems

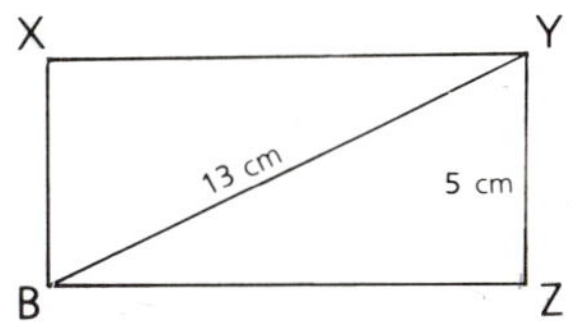

1 Find the area of the rectangle.

Solution: $A_{rect} = bh$

We need to find base, BZ.

△BZY is a right △ of the (5, 12, 13) family,

so BZ = 12.

$A_{rect} = (12)(5) = 60\text{cm}^2$

2 Given the area of a rectangle is 20 sq dm and the altitude is 5 dm, find the base.

Solution: Let x be the number of dm in the base.

$$\begin{aligned} A_{rect} &= bh \\ 20 &= (x)(5) \\ 4 &= x \\ \text{Base} &= 4 \text{ dm} \end{aligned}$$

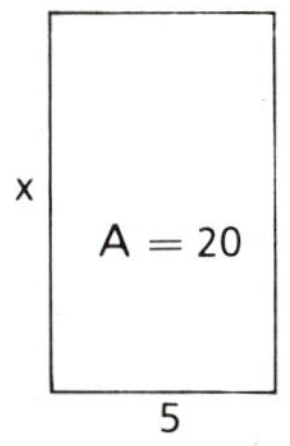

3 Find the area of the shaded region.

Solution: We show two methods. One uses subtraction and the other uses addition.

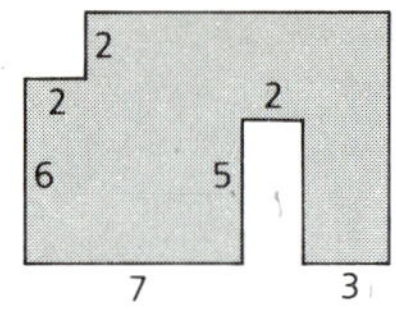

Method One

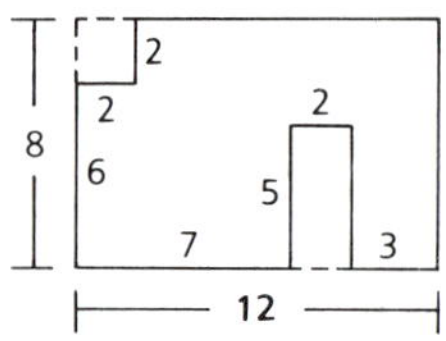

Area of large rectangle $= 12 \cdot 8 = 96$

Area of square $= 2^2 = 4$

Area of small rectangle $= 2 \cdot 5 = 10$

Shaded area $= 96 - 4 - 10 = 82$

Method Two

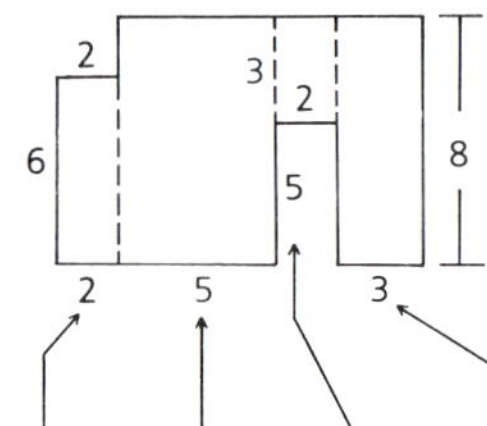

$A = 2 \cdot 6 + 5 \cdot 8 + 2 \cdot 3 + 3 \cdot 8$

Shaded area $= 82$

Section 11.1 Problem Set A

1 Find the area of a rectangle whose length and width are 12.5 cm and 6 cm respectively.

2 Find the area of each figure below. (Assume right angles.)

a

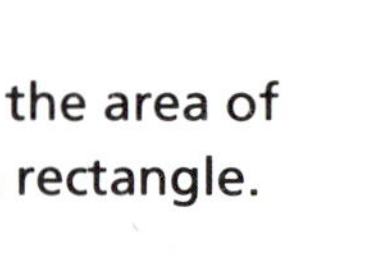

b

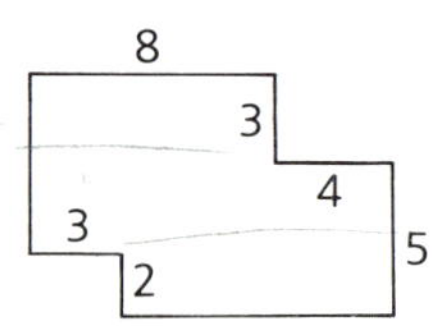

3 Find the area of each rectangle.

a

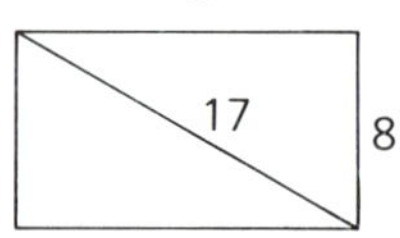

b

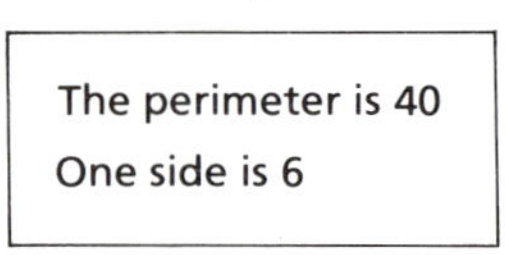

4 The area of a rectangle is 48 sq mm and the altitude is 6 mm.

Find: **a** the base **b** the length of the diagonal

5 **a** Find the area of a square whose side is 12.

b Find the area of a square whose diagonal is 10.

c Find the side of a square whose area is 49.

d Find the perimeter of a square whose area is 81.

e Find the area of a square whose perimeter is 36.

6 Find the area of each shaded region. (Assume right angles.)

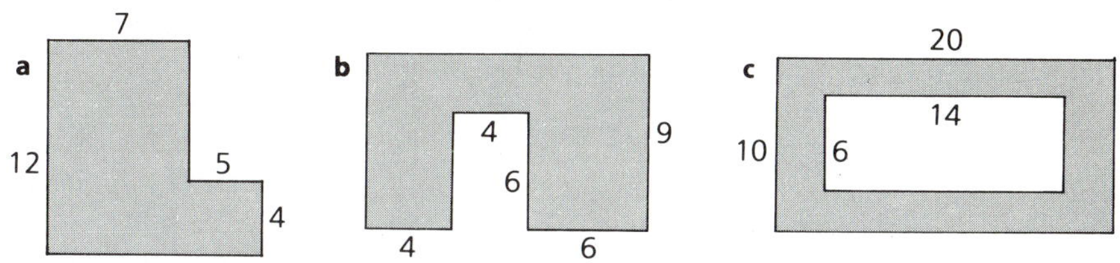

7 The diagonal of a rectangle is $\sqrt{29}$ and the base is 2.

Find: **a** the area **b** the semiperimeter

Section 11.1 Problem Set B

8 The cross section of a steel "I" beam is shown. Assume right angles and symmetry from appearances. Find the area of the cross section.

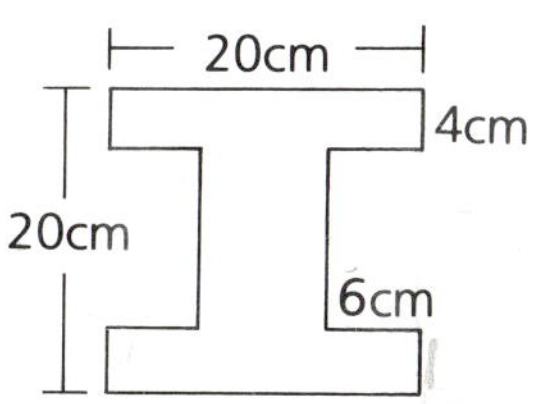

9 Each rectangular garden below has an area of 100.

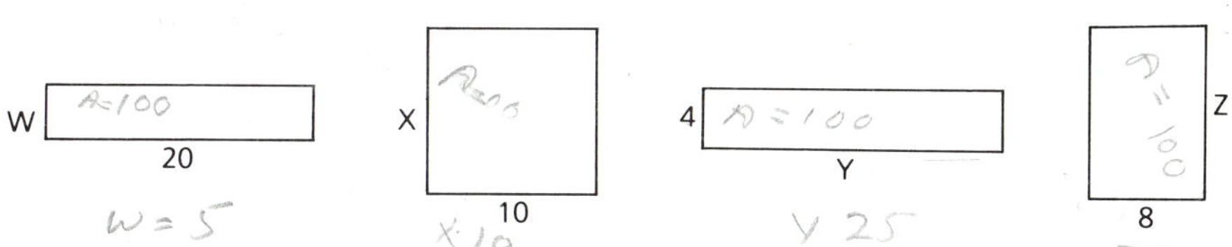

a Find the missing dimension in each.

b What length of fencing is needed to surround each?

c Which figure has the *least* perimeter?

d What do you think must be true about a rectangle if you wish to obtain the maximum area with the smallest perimeter?

10 A rectangular picture measures 12 cm by 30 cm. It is mounted in a frame 2 cm wide. Find the area of the frame.

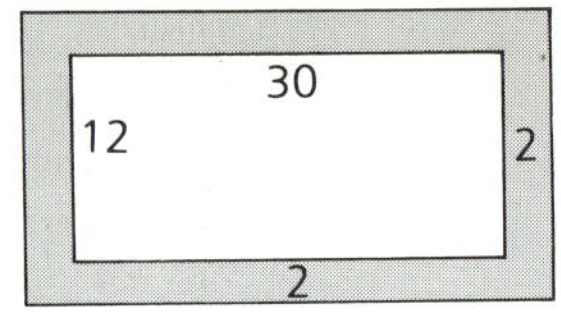

11 The sides of a rectangle are in a ratio 3:5 and the area is 135 sq. meters. Find the dimensions of the rectangle.

12 The area of square ABCD is 64 sq units. MNOP is formed by joining the midpoints of the sides of ABCD. Find the area and perimeter of MNOP.

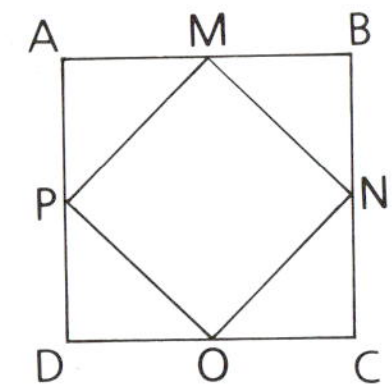

13 The area of an 8 by 10 rectangular picture and its white border of uniform width is 120. How wide is the border?

14 The dimensions of a rectangle of area 72 are whole numbers. List all the rectangles that could be drawn. Choose any two. What is the probability that each has a perimeter greater than 40?

Section 11.1 Problem Set C

15 A rectangle is formed inside a regular hexagon as shown. Each side of the hexagon is 12. Find the area of the rectangle.

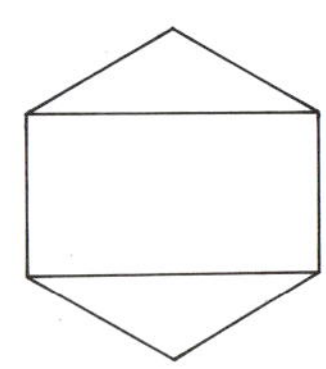

16 An American flag has dimensions 65 by 39. Each short stripe has a length 39. What fractional part of the flag is red?

17 In a series of squares, the first square has sides 1 unit long. The diagonal of the second square equals a side of the first. The diagonal of the third equals each side of the second square, etc. What is the sum of the areas of the infinite sequence of squares?

Section 11.2 Areas of Parallelograms and Triangles

A THE AREA OF A PARALLELOGRAM

B THE AREA OF A TRIANGLE

11.2 (A) The Area of a Parallelogram

Many areas may be found by a "cut and paste" method. For example, to find the area of a parallelogram of base b and altitude h, we may do this:

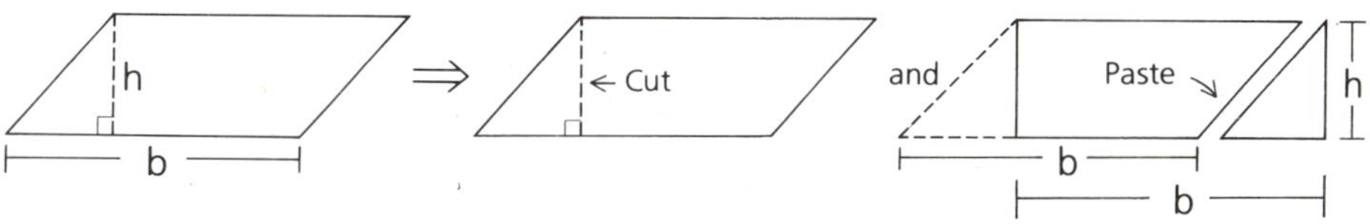

Thus,
the AREA of 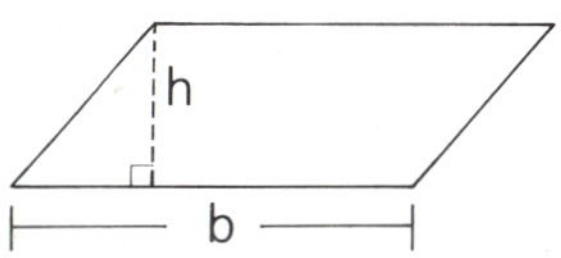

= the AREA of 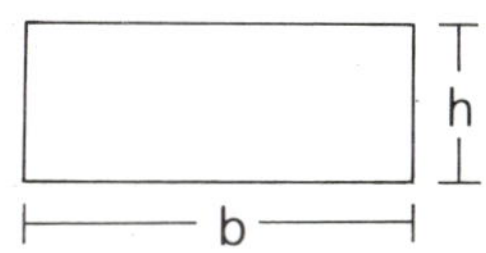

Theorem 91

$A_{\square} = \mathbf{b \cdot h}$
where b = base
h = height

T **The AREA OF A PARALLELOGRAM is equal to the product of the *base* times the *height*.**

Formal area proofs are often based on the "cut and paste" method. For instance, the key steps in a proof of Theorem 91 could be those below.

Given: ▱ PACT
$\overline{RT}$ is an altitude to $\overline{PA}$

Prove: $A_{PACT} = (PA)(RT)$

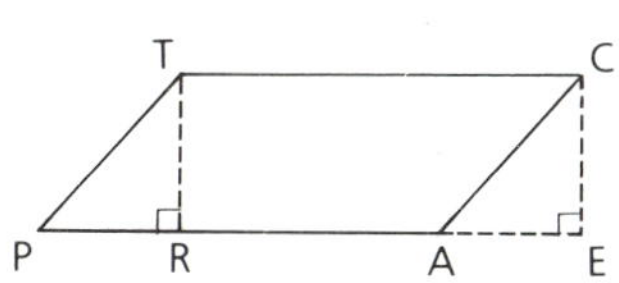

Key steps:
1 Extend $\overleftrightarrow{PA}$ and draw altitude $\overline{CE}$ to $\overleftrightarrow{PA}$; RECT is a rectangle
2 $A_{PRT} = A_{AEC}$ because $\triangle PRT \cong \triangle AEC$ by HL
3 $A_{PACT} = A_{RECT}$ since $A_{CART} + A_{PRT} = A_{CART} + A_{ACE}$
4 $A_{RECT} = (TC)(RT)$ (Why?)
5 $A_{PACT} = (PA)(RT)$ because $PA = TC$

11.2 (B) The Area of a Triangle

The area of any triangle can be shown to be equal to one-half of the area of a parallelogram with the same base and height. For example,

the area of 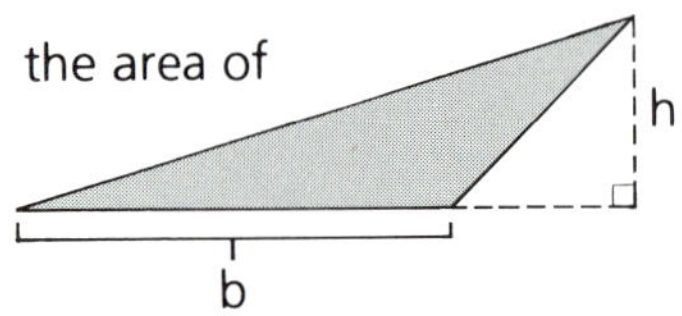

$= \left(\frac{1}{2}\right)$ area of 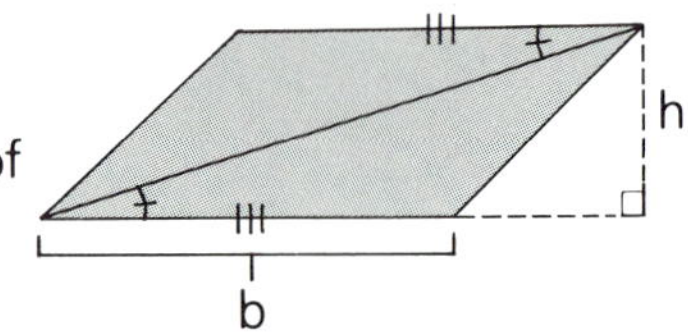

Theorem 92

T **The AREA OF A TRIANGLE equals one-half the product of *base times* the *height* (or altitude) for that base.**

$$\mathbf{A_{\triangle} = \tfrac{1}{2}\, b \cdot h}$$

where $b =$ base
$h =$ altitude

Section 11.2 Sample Problems

1 Find the area of each triangle below.

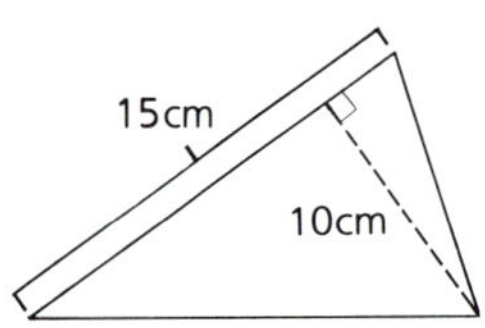

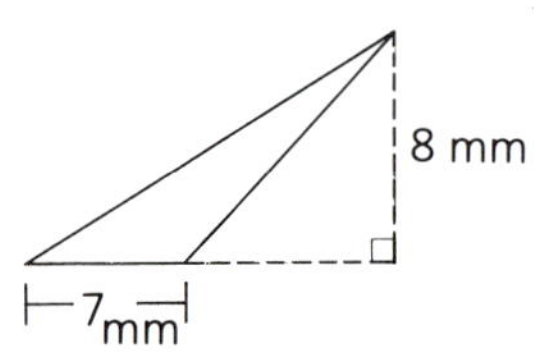

Solutions:

a $A_{\triangle} = \frac{1}{2} bh$

$= \frac{1}{2}(15)(10)$

$= 75$ sq cm

Note: The base of a triangle is not always on the bottom. The 10 cm altitude is the altitude associated with the 15 cm long base.

b $A_{\triangle} = \frac{1}{2} bh$

$= \frac{1}{2}(7)(8)$

$= 28$ sq mm

Note: The altitude of a triangle is not always inside the triangle.

2 Find the base of a triangle with altitude 15 and area 60.

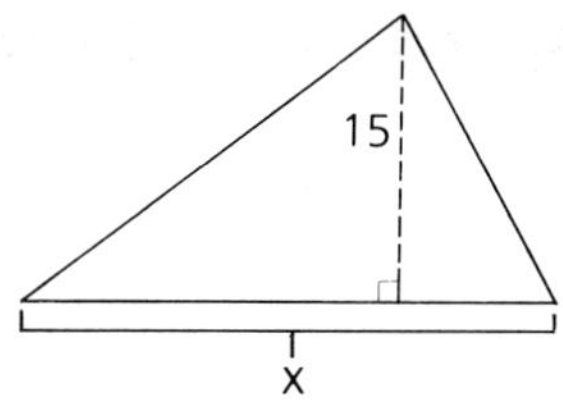

Solution: Let x be the base

Then, $A_{\triangle} = \frac{1}{2}bh$

$$60 = \tfrac{1}{2}(x)(15)$$
$$8 = x$$

3 Find the area of a parallelogram whose sides are 14 and 6, and whose acute angle is 60°.

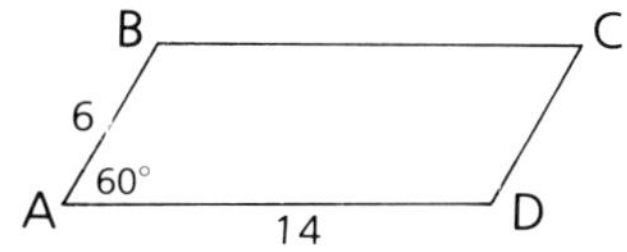

Solution: You may use 14 as the base, but first find the height for that base. When altitude $\overline{BE}$ is drawn, a 30°-60°-90° triangle is formed,

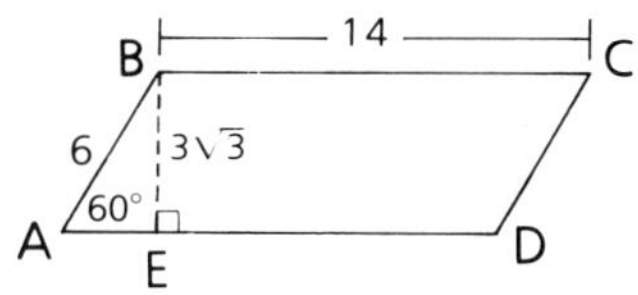

$$\text{so } h = 3\sqrt{3}$$
$$\text{Thus } A_{\square} = bh$$
$$= (14)(3\sqrt{3}) = 42\sqrt{3}$$

4 Find the area of trapezoid WXYZ.

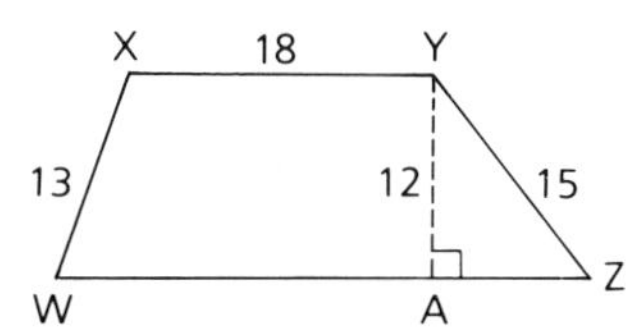

Solution: Copy the diagram. Use the "divide and conquer" method. By drawing another altitude, $\overline{XB}$, two right triangles and a rectangle are formed.

Find the areas of these figures and add them together.

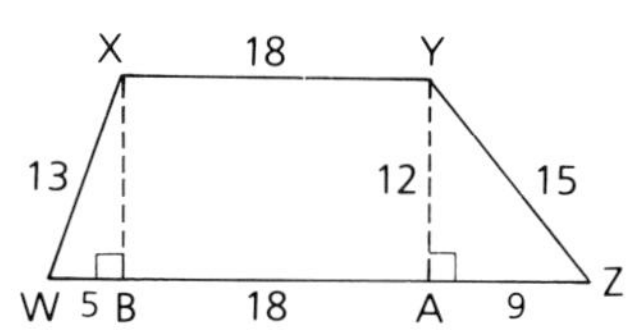

In $\triangle$WBX, WB = 5 by a Pythagorean triple.
Similarly, in $\triangle$YAZ, AZ = 9.

$A_{\triangle WBX} = \frac{1}{2}bh$	$A_{rect} = bh$	$A_{\triangle YAZ} = \frac{1}{2}bh$
$= \frac{1}{2}(5)(12)$	$= (18)(12)$	$= \frac{1}{2}(9)(12)$
$= 30$	$= 216$	$= 54$

The sum of the three areas, 300, is the area of the trapezoid.

Section 11.2 Problem Set A

1 Find the area of each triangle.

a

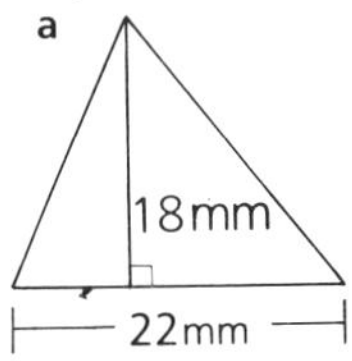

b

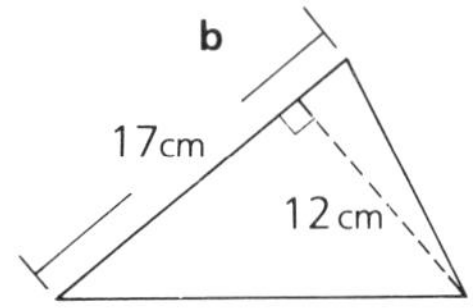

c

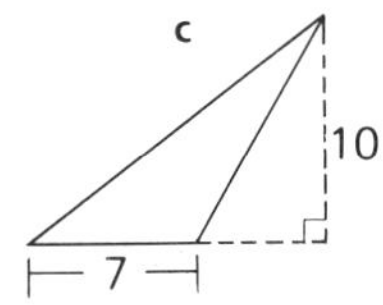

2 Find the area of the triangle.

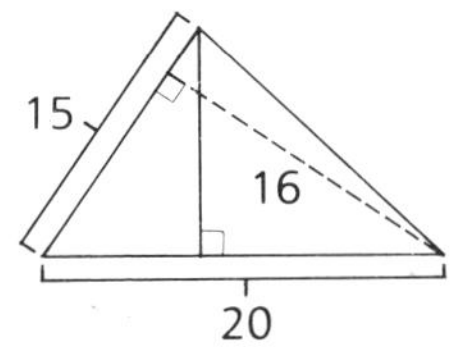

3 Find the total area of each. In each figure the triangle is mounted on a rectangle.

a

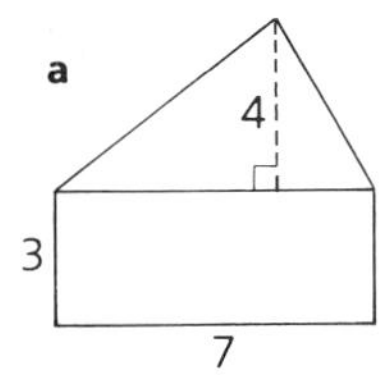

b

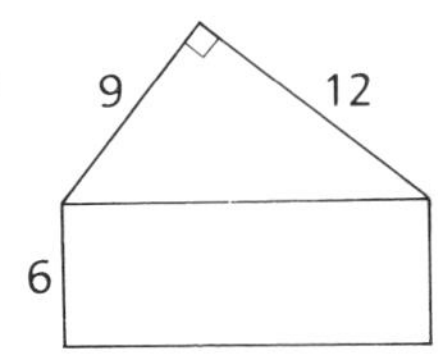

4 Find the altitude of a triangle if the base is 7 and the area is 21.

5 Find the area of an isosceles triangle with sides 10, 10, and 16.

6 Find the area of a parallelogram of base 17 and height 11.

7 Find the base of a parallelogram of height 3 and area 42.

8 Find the area of each obtuse triangle.

a

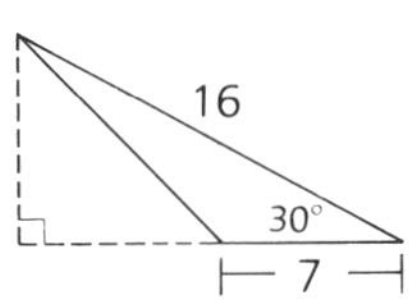

b

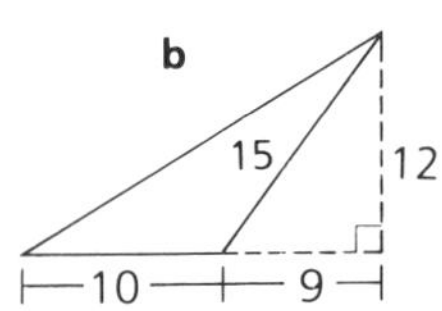

9 Find the area of each triangle.

a

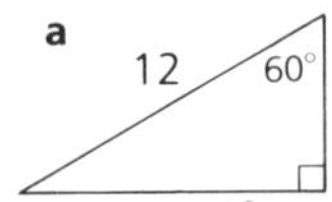

b

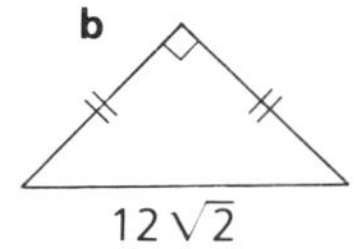

c

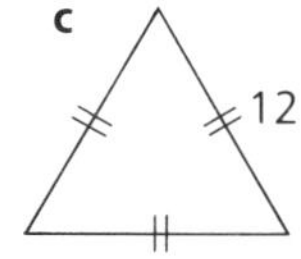

10 Find the area of each parallelogram.

a

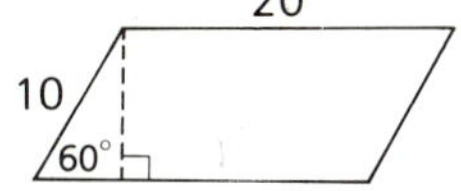

b

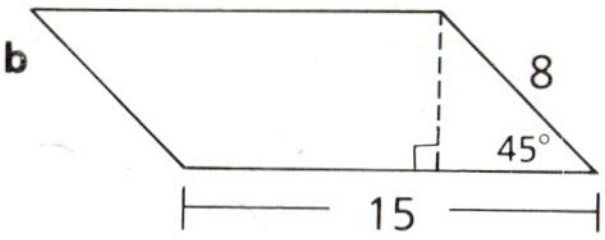

11 Find the area of each trapezoid by dividing it into a rectangle and triangle(s).

a

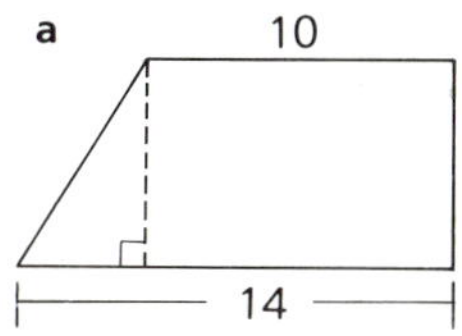

b

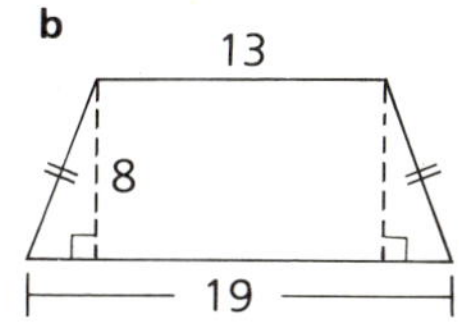

Section 11.2 Problem Set B

12 A triangle has the same area as a 6 by 8 rectangle. The base of the triangle is 8. Find the altitude of the triangle.

13 Lines $\overleftrightarrow{CF}$ and $\overleftrightarrow{AB}$ are parallel and 10 mm apart. Several triangles have been drawn below with base $\overline{AB}$ and a vertex on $\overleftrightarrow{CF}$. Which triangle has the largest area? Explain.

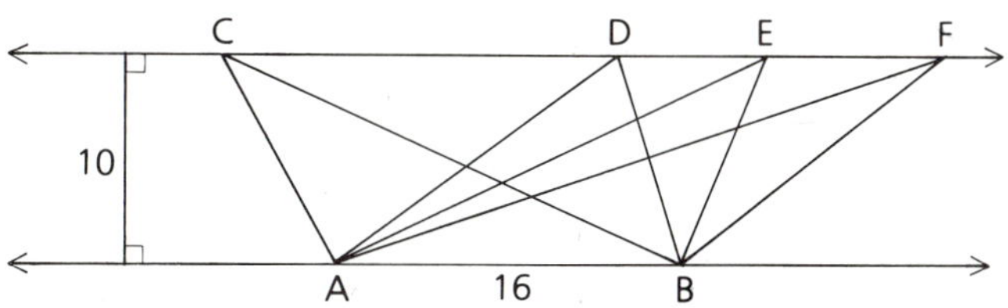

14 Find the area of the shaded region.

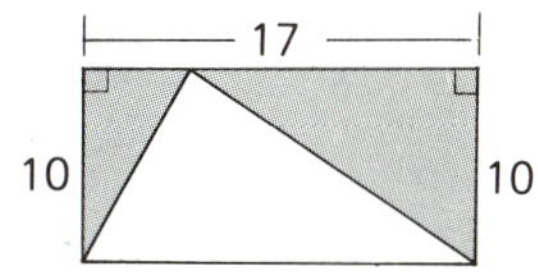

15 Find the area of the shaded triangular region.

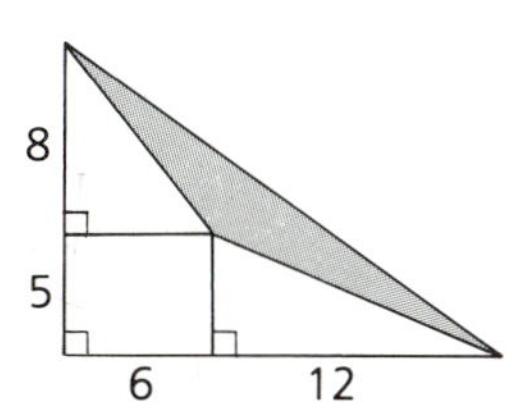

16 QT = 12, PR = 15
PS = 10

Find: **a** the area of $\triangle PQR$

b RQ

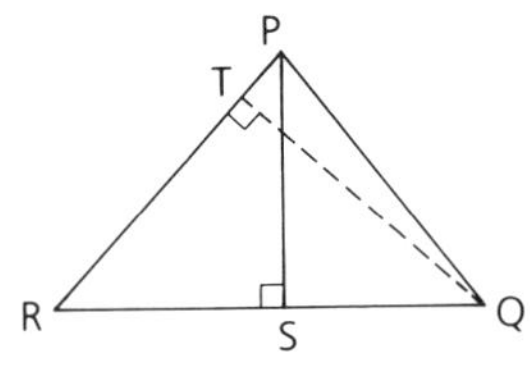

17 In a triangle a base and its altitude are in a ratio 3:2. The area is 48. Find the base and altitude.

18 **a** Find the area of a triangle whose sides are 25, 25, and 14.

b Find the area of a right triangle whose legs are 9 and 40.

c Find the area of an isosceles right triangle with hypotenuse 18.

19 Find the area of an equilateral triangle of 45 meters perimeter.

20 Find the area of each parallelogram as marked.

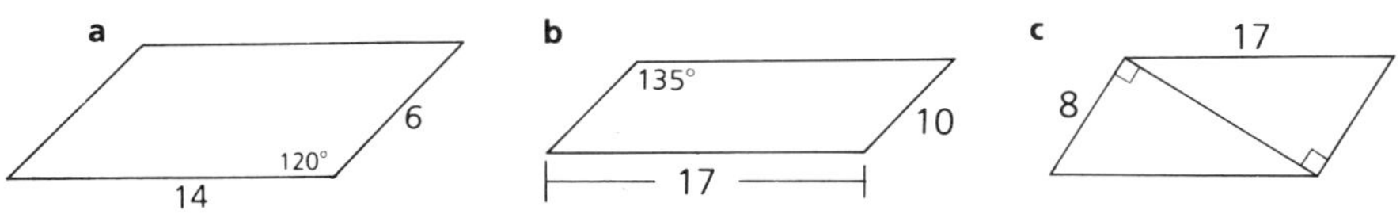

21 Find the area of each trapezoid below by dividing it into other figures (rectangles and triangles or parallelograms and triangles).

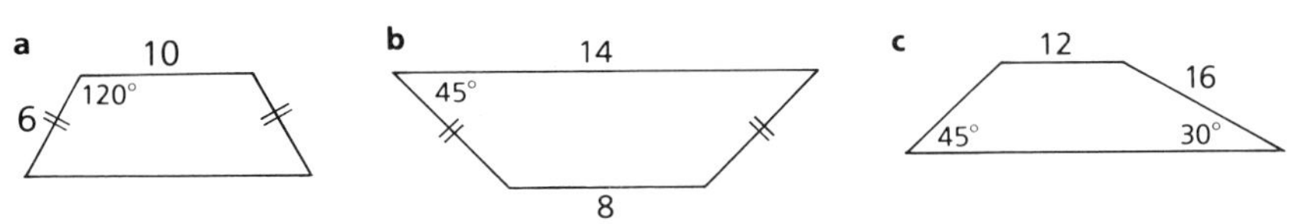

22 Find the area of a trapezoid whose sides are 10, 12, 10, and 28.

23 The hypotenuse of a right triangle is 50 and one leg is 14.
Find: **a** the area of the triangle

b the altitude to the hypotenuse

24 **a** Find m∠A in ▱ AXYZ.

b Find AX.

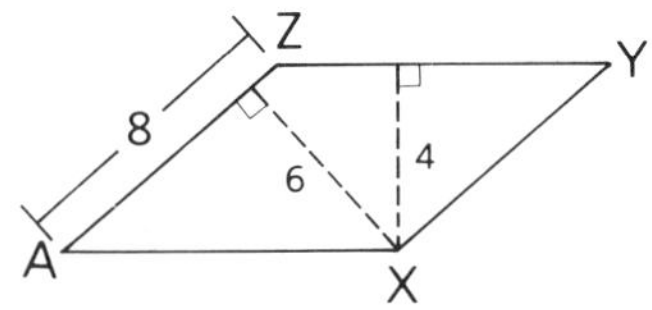

Section 11.2 Problem Set C

25 If the diagonals of a rhombus are 10 and 24, find the area and the perimeter.

26 **a** The area of an equilateral triangle is $9\sqrt{3}$. Find the length of one side.

b Find a formula for the area of an equilateral triangle with sides s units long.

27 Find the area of the triangle.

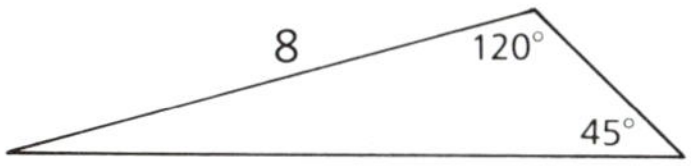

28 Find the area of the parallelogram.

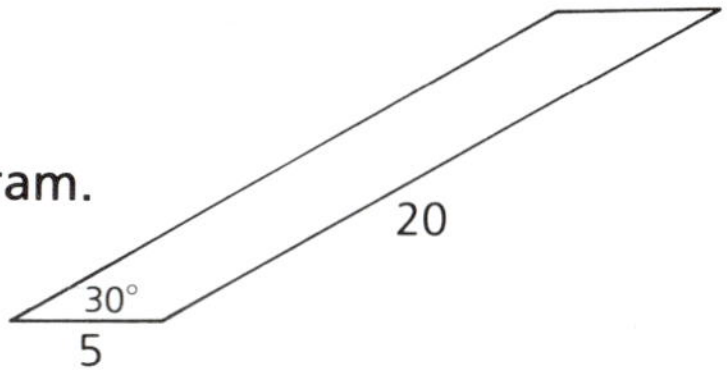

29 What is the name of the parallelogram that yields the largest area for a given fixed perimeter?

30 The diagonals of a kite are 10 and 24. Find the area.

31 The perimeter of the parallelogram is 154.

Find the area.

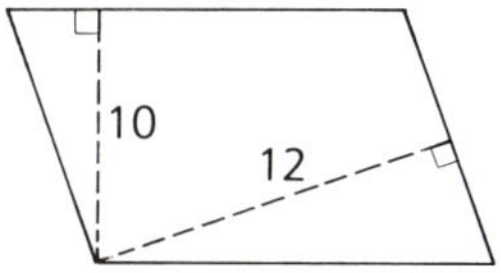

32 Let P be any point in the interior of rectangle ABCD. Four triangles are formed by joining P to each vertex.

a Demonstrate that $A_{\triangle APD} + A_{\triangle BPC} = A_{\triangle APB} + A_{\triangle PCD}$

b Does this work if ABCD is a parallelogram?

c Does this work if ABCD is a trapezoid?

Section 11.3 The Area of a Trapezoid

In this section we discover some special formulas for finding:

A THE AREA OF A TRAPEZOID

B THE MEDIAN OF A TRAPEZOID

11.3 (A) The Area of a Trapezoid

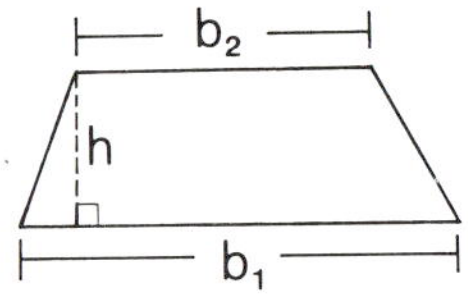

You learned that the area of a trapezoid may be found by dividing the trapezoid into simpler shapes, such as triangles, rectangles, and parallelograms (divide and *conquer*).

A trapezoid has its own formula for computing the area:

Theorem 93

T **The AREA OF A TRAPEZOID equals one half of the product of the *height* times the *sum* of *the bases.***

$$A_{trap} = \frac{h}{2}(b_1 + b_2)$$

where b_1 = one base
b_2 = the other base
h = height

11.3 (B) The Median of a Trapezoid

D **The line segment joining the midpoints of the non-parallel sides of a trapezoid is called the MEDIAN of the trapezoid.**

We can use the Midline Theorem to find out what happens when the midpoints of the nonparallel sides of a trapezoid are joined.

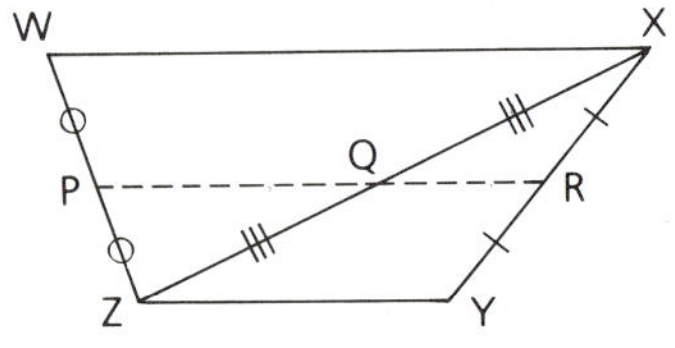

In trapezoid WXYZ, the midpoints of the sides of $\triangle WXZ$ and $\triangle XYZ$ are P, Q, and R. P, Q, and R are collinear, because $\overline{PQ}$ and $\overline{QR}$ share Q and are both parallel to $\overline{WX}$ and $\overline{ZY}$.

$\overline{PR}$ is called the MEDIAN of Trapezoid WXYZ.

By the MIDLINE THEOREM, $PQ = \frac{1}{2}WX$ and $QR = \frac{1}{2}YZ$

Thus, $PR = PQ + QR = \frac{1}{2}WX + \frac{1}{2}YZ = \frac{1}{2}(WX + YZ)$

Theorem 94

T **The measure of the MEDIAN OF A TRAPEZOID equals the *average* of the *bases.***

$$\mathbf{M = \tfrac{1}{2}(b_1 + b_2)}$$

b_1 = one base
b_2 = other base

You can now easily prove a shorter form of Theorem 93.

The area of a trapezoid is the product of the median by the height.

$$\mathbf{A_{trap} = M \cdot h}$$

M = median
h = height

Section 11.3 Sample Problems

1 Find the area of trapezoid WXYZ,
given: height = 7
lower base = 18
upper base = 12

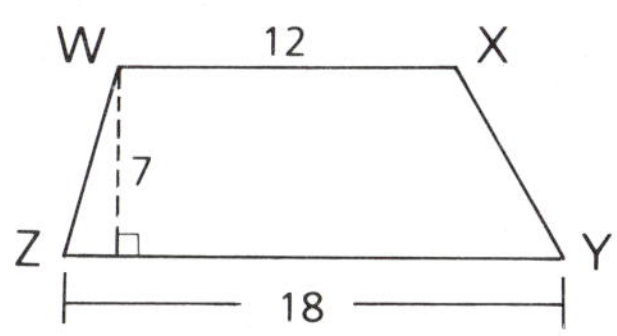

Solution: $A_{trap} = \frac{1}{2}h(b_1 + b_2)$

$= \frac{1}{2}(7)(18 + 12) = 105$

2 Find the smaller base of a trapezoid, if the area is 52, the altitude is 8, and the larger base is 10.

Solution: Let x be the length of the smaller base.

$A_{trap} = \frac{1}{2}h(b_1 + b_2)$

$52 = \frac{1}{2}(8)(10 + x)$

$52 = 4(10 + x)$

$3 = x$ Ans.

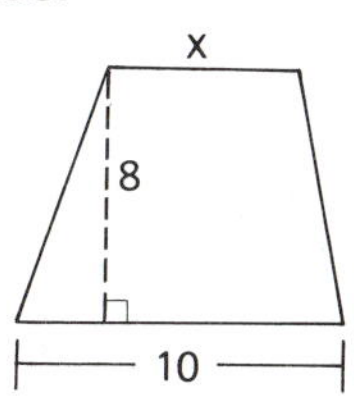

3 The height of a trapezoid is 12. The bases are 6 and 14.

a Find the median. **b** Find the area.

Solutions:

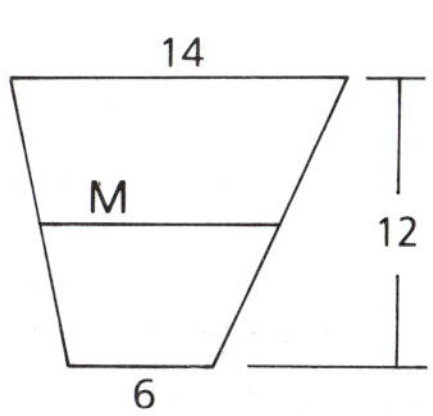

a $M = \frac{1}{2}(b_1 + b_2)$

$= \frac{1}{2}(14 + 6)$

$= 10$

b $A_{trap} = Mh$

$= (10)(12)$

$= 120$

Section 11.3 Problem Set A

1 A trapezoid has bases 15 and 11 and height 8.

Find: **a** the area **b** the median

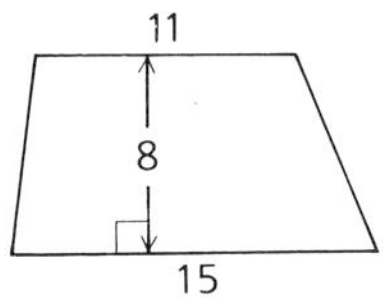

2 Find the area of each trapezoid.

a

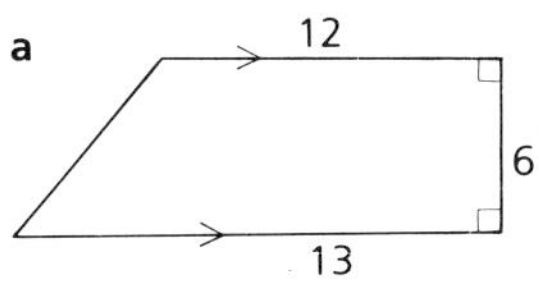

c

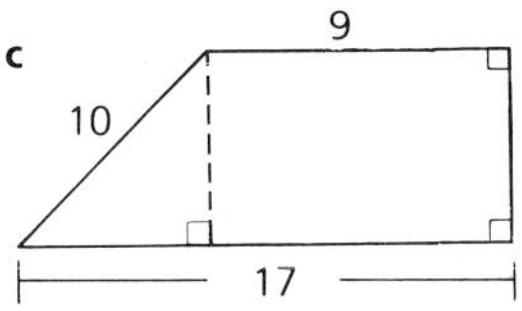

b

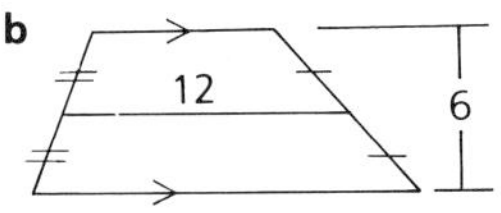

d

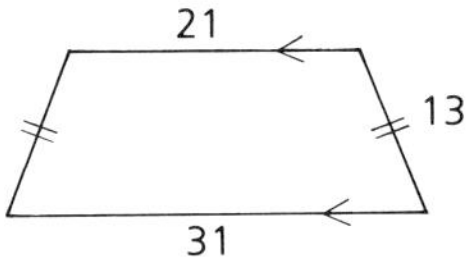

3 Given a trapezoid with bases 6 and 15 and height 7.

Find: **a** the median **b** the area

4 The bases of a trapezoid are 8 and 22 and the area is 135.

Find the height.

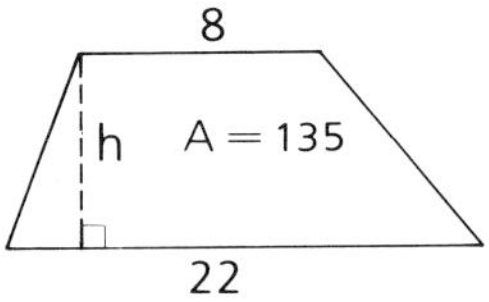

5 The height of a trapezoid is 10 and the area is 130. If one base is 15, find the other base.

6 A straight wire stretches between the tops of two poles whose heights are 30 feet and 14 feet. Find the height of a third pole, which is to be placed halfway between the original poles to support the wire.

Assume the poles are ⊥ to the ground. (Do you see a trapezoid and its median?)

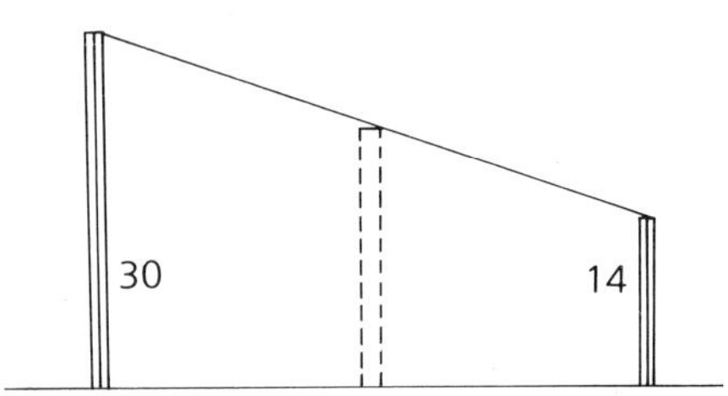

Section 11.3 Problem Set B

7 Find the total area of each region.

a

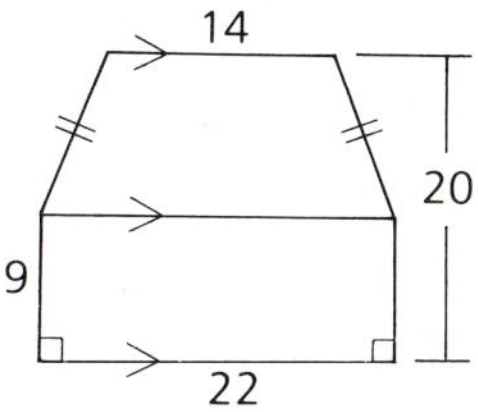

b

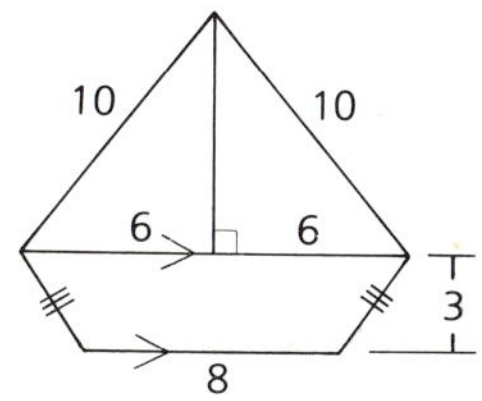

8 Find the total area of each figure.

a

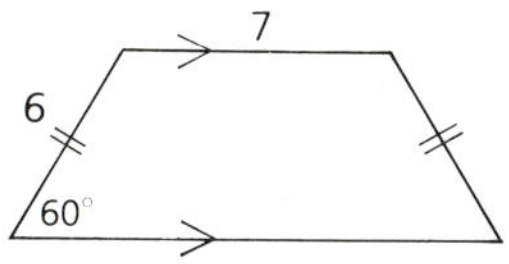

b

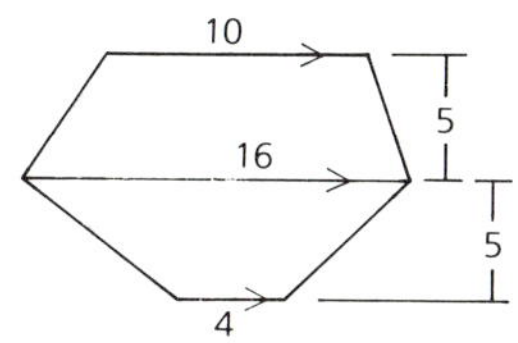

9 Find the lower base of a trapezoid whose upper base is 10 and whose median is 17.

10 The area of triangle PQS is 25.
The median of trapezoid PQRS is 14.
Base RS = 18.

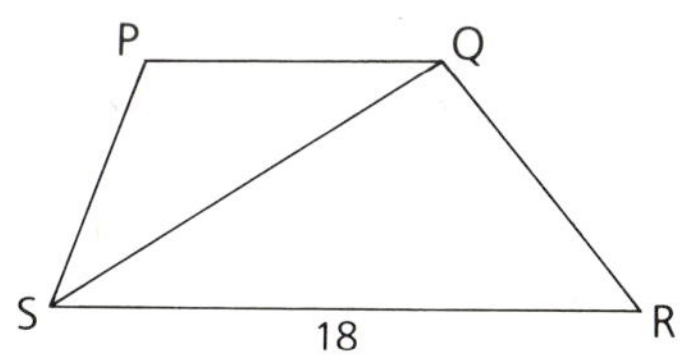

Find: **a** the length of base $\overline{PQ}$

b the height to base $\overline{PQ}$ of $\triangle PQS$

c the height of trapezoid PQRS

d the area of trapezoid PQRS

11 Find the area of the figure shown, which was formed by cutting two identical isosceles trapezoids out of a square.

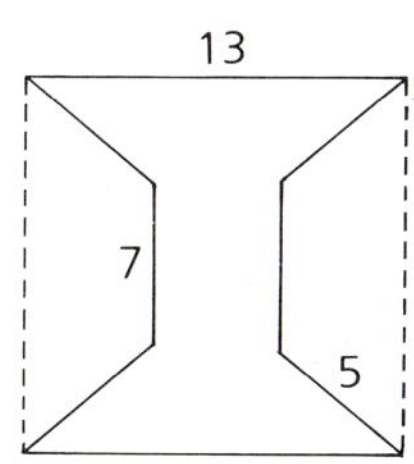

12 The perimeter of a trapezoid is 35. The non-parallel sides are 7 and 8. Find the area if the height is 5.

13 The consecutive sides of an isosceles trapezoid are in a ratio 2:5:10:5 and the perimeter is 44. Find the area of the trapezoid.

Section 11.3 Problem Set C

14 The region shown is composed of four regions of equal height. The triangle and the trapezoid are isosceles, and each side of the trapezoid is parallel to a side of the triangle.

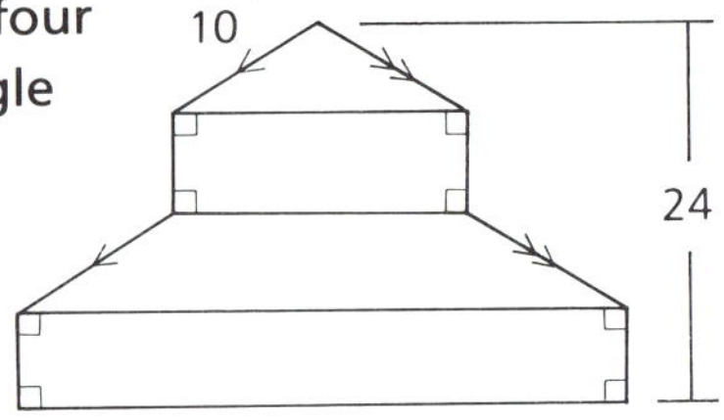

Find the total area.

15 An isosceles triangle is folded to put its vertex on the midpoint of the base, thus forming a trapezoid of 12 square units of area. Find the area of the original triangle.

16 The sides of a trapezoid are in the ratio of 2:5:8:5. The area is 245. Find the altitude and the perimeter of the trapezoid.

17 Find the area of each trapezoid:

a

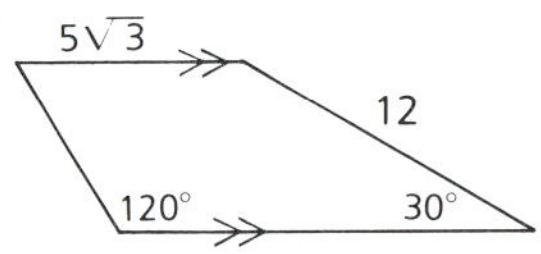

b

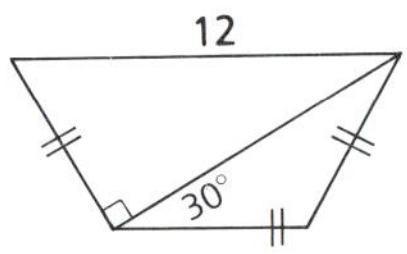

18 In trapezoid ABCD, X and Y are midpoints of sides, and P and Q are midpoints of diagonals. Derive and verify a formula for PQ.

(Hint: See the proof of Theorem 94.)

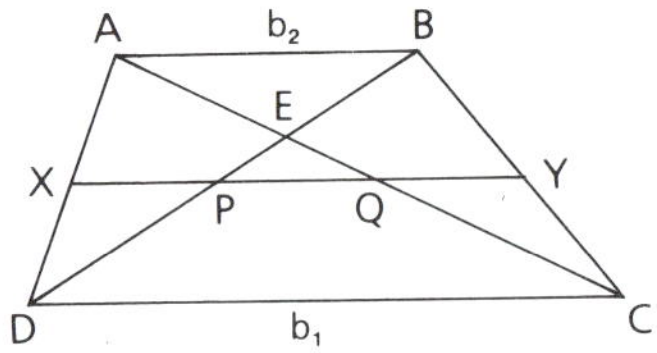

19 Prove that the area of a trapezoid is $\frac{1}{2}h(b_1 + b_2)$ by these methods:

a Draw a diagonal and use the two triangles formed.

b Draw altitudes and use the rectangle and triangles formed.

Section 11.4 The Area of a Kite and Related Figures

Remember that in a kite the diagonals are perpendicular.

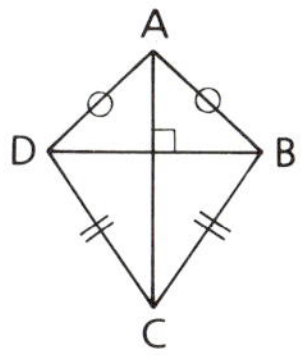

Also, a kite can be divided into two isosceles triangles with a common base.

Then,

$A_{kite} = A_{\triangle ABD} + A_{\triangle DBC}$

$= \frac{1}{2}(BD)(AE) + \frac{1}{2}(BD)(EC)$

$= \frac{1}{2}(BD)(AE + EC)$

$A_{kite} = \frac{1}{2}(BD)(AC)$

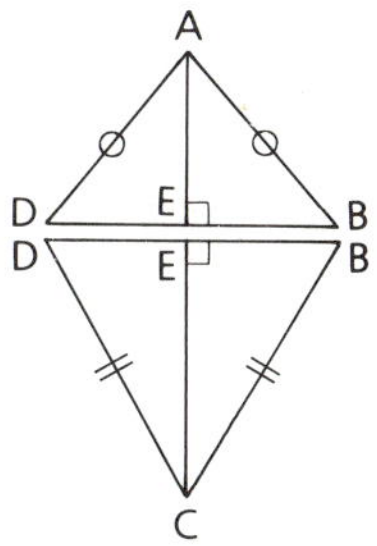

We notice that $\overline{BD}$ and $\overline{AC}$ are the diagonals and that we have just proved the following formula:

Theorem 95

T **The AREA OF A KITE equals half the product of its diagonals.**

$A_{kite} = \frac{1}{2} d_1 \cdot d_2$
d_1 = one diagonal d_2 = other diagonal

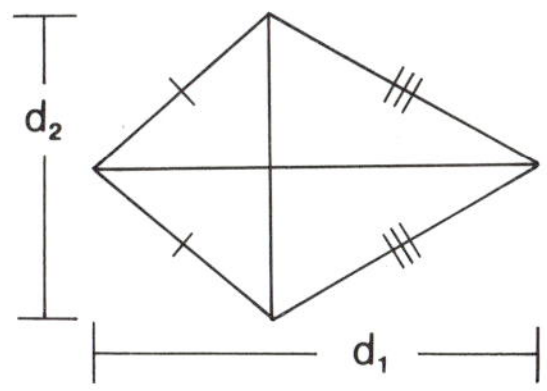

This formula can be applied to any kite, including the special cases of a rhombus and a square.

Section 11.4 Sample Problems

1 Find the area of a kite with diagonals 9 and 14.

Solution:

$A_{kite} = \frac{1}{2} d_1 \cdot d_2$

$= \frac{1}{2}(14)(9) = 63$

AC = 9
BD = 14

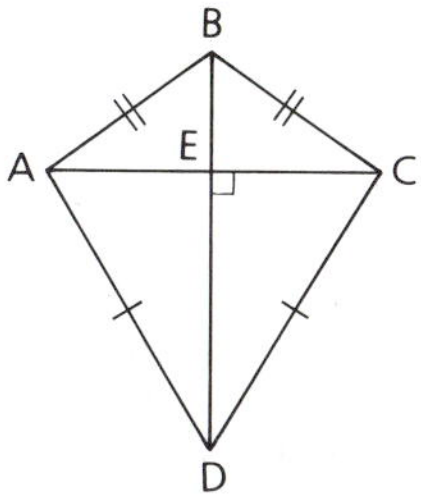

2 Find the area of a rhombus whose perimeter is 20 and whose longer diagonal is 8.

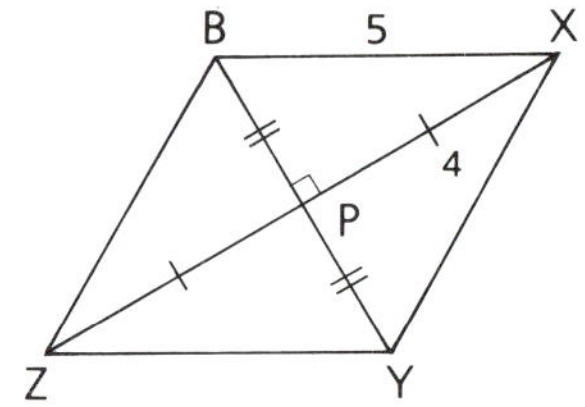

Solution:

A rhombus is a ▱ so the diagonals bisect each other. It is also a kite, so the diagonals are $\perp$ to each other.

Thus, XZ = 8, and XP = 4.

The perimeter = 20, so XB = 5. (Why?)

$\triangle$BPX is a right triangle.

Thus, BP = 3, and BY = 6.

$$A_{kite} = \tfrac{1}{2} d_1 \cdot d_2$$
$$= \tfrac{1}{2}(6)(8) = 24$$

Section 11.4 Problem Set A

1 Find the area of a kite with diagonals 6 and 20.

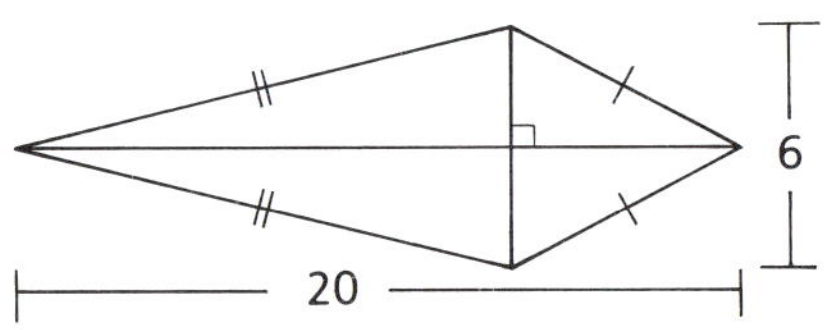

2 Find the area of each kite.

a

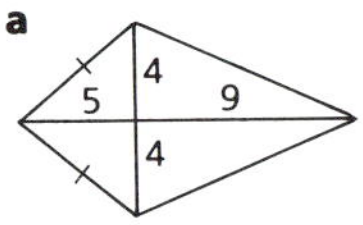

b

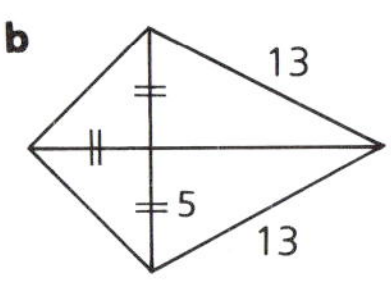

c

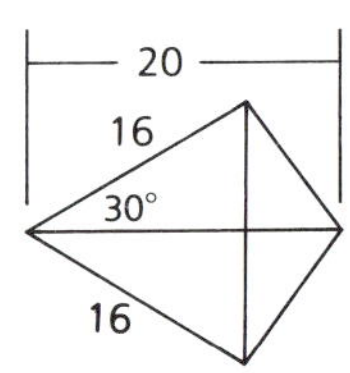

3 The area of a kite is 20. The longer diagonal is 8. Find the shorter diagonal.

Section 11.4 Problem Set B

4 Find the area of the kite shown.

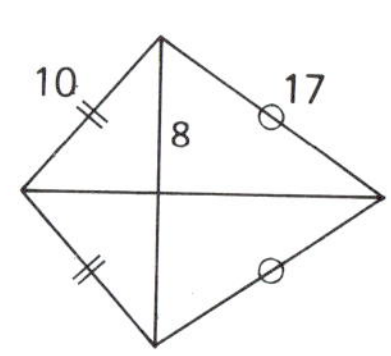

5 Find the area of each rhombus.

a

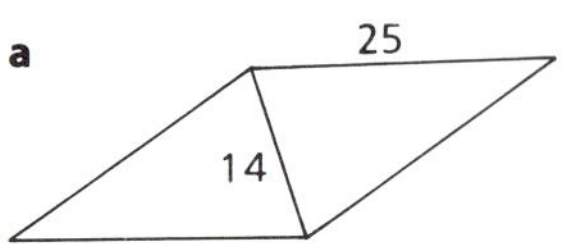

b

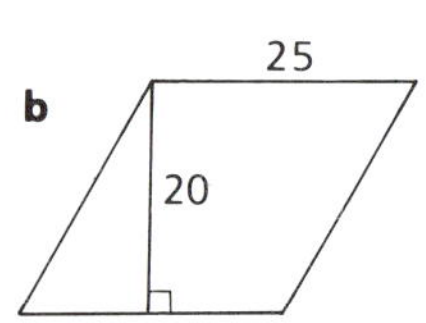

6 ABCD is a kite.
∠BAD is a right ∠.
BD = 10, BC = 13
Find the area of ABCD.

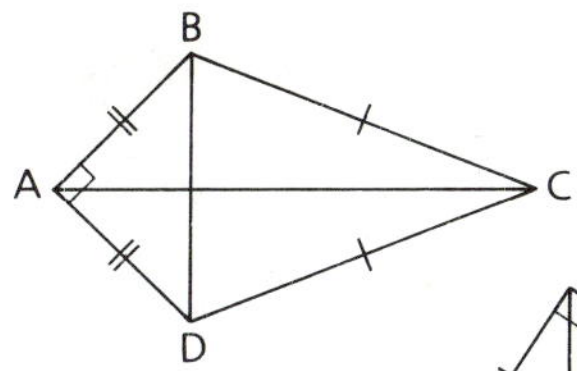

7 Find the area of the kite as marked.

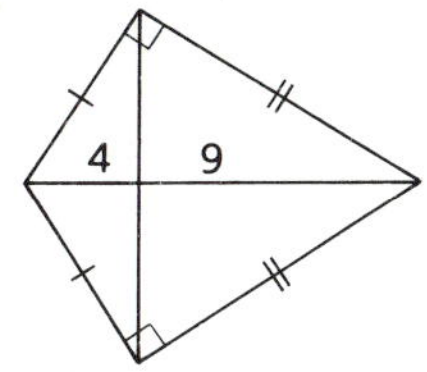

8 Find the area of a rhombus with a perimeter of 40 and an angle of 60°.

Section 11.4 Problem Set C

9 Given a rhombus with diagonals 18 and 24. Find the height.

10 The kite formula applies to *any* quadrilateral whose diagonals are perpendicular.

Prove that the area of any quadrilateral with perpendicular diagonals equals half of the product of the diagonals.

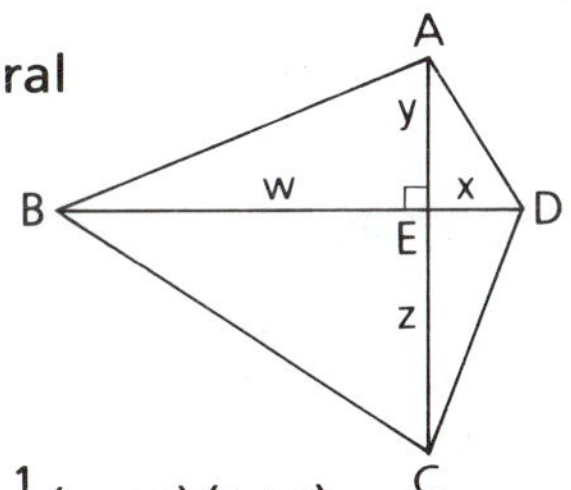

Hint: Use w, x, y, and z as marked to show that $A = \frac{1}{2}(w + x)(y + z)$.

11 Observe the figure at the right.
It resembles a kite and yet it is not convex (it is "dented in").
Does the kite formula still hold?
(That is, is $A = \frac{1}{2}xy$?)

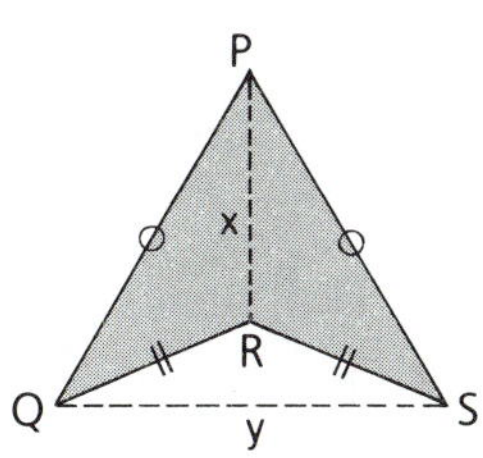

12 In rectangle ABCD, X and Y are midpoints of $\overline{AB}$ and $\overline{CD}$, and $\overline{PD} \cong \overline{QC}$.

a Compare the area of quadrilateral XQYP to the area of ABCD.

b Prove your conjecture.

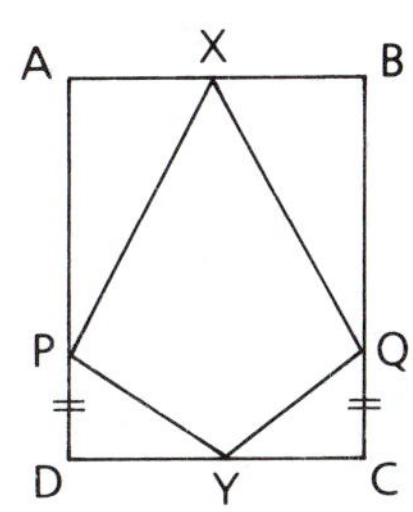

Section 11.5 Areas of Regular Polygons

In this section we discuss:

A THE AREA OF AN EQUILATERAL TRIANGLE

B THE AREA OF A REGULAR POLYGON

11.5(A) The Area of an Equilateral Triangle

Equilateral triangles are used so frequently that a special formula for their areas will be useful.

Remember that the altitude of an equilateral triangle divides it into two 30°-60°-90° right triangles.

Thus, if $WY = s$, then $ZY = \frac{s}{2}$, and $WZ = \frac{s}{2}\sqrt{3}$

Therefore, $A_{WXY} = \frac{1}{2}bh$

$$A_{WXY} = \frac{1}{2}(s)\left(\frac{s}{2}\sqrt{3}\right) = \frac{s^2}{4}\sqrt{3}$$

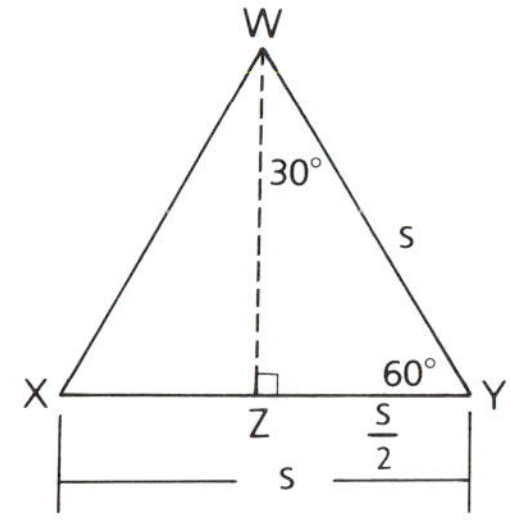

Theorem 96

T **The AREA OF AN EQUILATERAL TRIANGLE is the product of one-fourth of the *square of a side* times the square root of three.**

$$A_{\triangle} = \frac{s^2}{4}\sqrt{3}$$

s = common length of the sides.

11.5(B) The Area of a Regular Polygon

A regular polygon has all interior angles congruent and all sides congruent.

In regular polygon PENTA

O is the CENTER

$\overline{OA}$ is the RADIUS

$\overline{OM}$ is called the APOTHEM

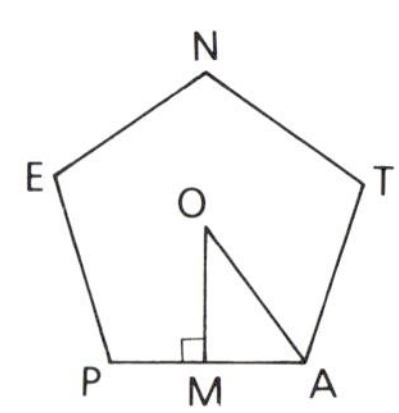

D **A RADIUS OF A REGULAR POLYGON is a segment joining the center to any vertex.**

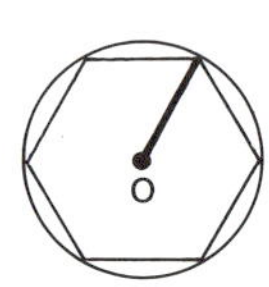

D **An APOTHEM OF A REGULAR POLYGON is a segment joining the center to the midpoint of any side.**

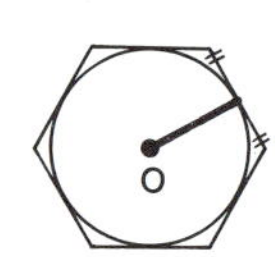

Here are some important observations:

All apothems of a regular polygon are congruent.

Only regular polygons have apothems.

An apothem is a radius of the inscribed circle.

An apothem is the perpendicular bisector of a side.

A radius of a regular polygon is a radius of the circumscribed circle.

A radius of a regular polygon bisects an angle of the polygon.

If all of the radii are drawn in a regular polygon, the polygon will be divided into congruent isosceles triangles. If you use some algebra on the sum of the areas of those isosceles triangles, you can derive the next formula. What is your choice for an altitude of each triangle?

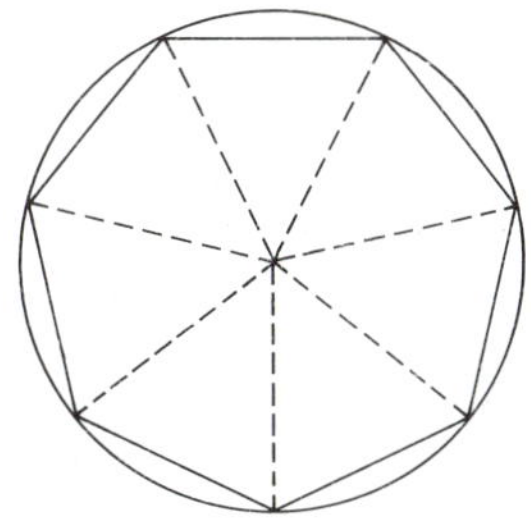

Theorem 97

T **The AREA OF A REGULAR POLYGON equals one-half the product of the *apothem* times the *perimeter*.**

$$A_{\text{reg. poly}} = \tfrac{1}{2}\mathbf{a} \cdot \mathbf{p}$$

a = length of an apothem
p = perimeter

Section 11.5 Sample Problems

1 A regular polygon has a perimeter of 40 and an apothem of 5. Find the area.

Solution: $A_{\text{reg. poly.}} = \tfrac{1}{2}a \cdot p$

$$= \frac{1}{2}(5)(40) = 100$$

2 An equilateral triangle has a side 10 cm long. Find the area.

Solution: $A_{\triangle} = \dfrac{s^2}{4}\sqrt{3}$

$$= \frac{(10)^2}{4}\sqrt{3}$$

$$= 25\sqrt{3}\ \text{cm}^2$$

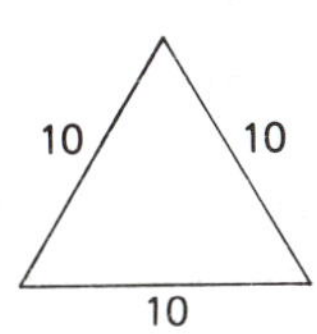

3 A circle of radius 6 is inscribed in an equilateral triangle. Find the area of the triangle.

Solution: Notice that $\overline{OP}$ is an apothem 6 units long.

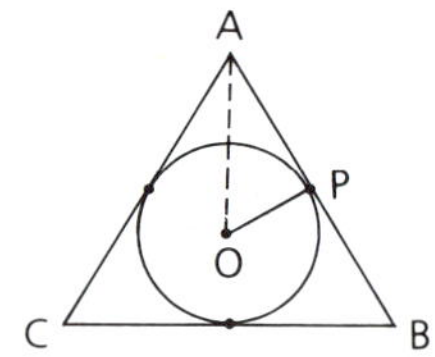

Observe that AOP is a 30°–60°–90° triangle.

Thus, OA = 12, AP = $6\sqrt{3}$, and the perimeter of $\triangle$ABC is $36\sqrt{3}$.

An equilateral triangle is a regular polygon, so

$A_{\triangle} = \frac{1}{2}a \cdot p$

$= \frac{1}{2}(6)(36\sqrt{3}) = 108\sqrt{3}$

4 Find the area of a regular hexagon with sides 18 units long.

Solution: AF = 18, so AP = 9.

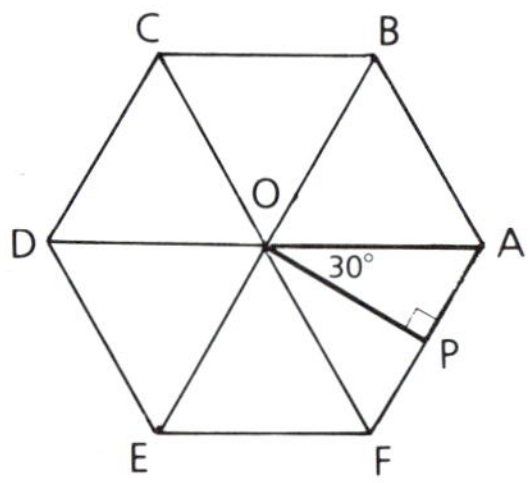

Observe that OPA is a 30°–60°–90° $\triangle$ so that OP = $9\sqrt{3}$ = apothem.

Perimeter = 6 (18) = 108

$A = \frac{1}{2}a \cdot p$

$= \frac{1}{2}(9\sqrt{3})(108) = 486\sqrt{3}$

Section 11.5 Problem Set A

1 The perimeter of a regular polygon is 24 and the apothem is 3. Find the area.

2 Find the area of an equilateral triangle with a *side* of:

a 6 **b** 7 **c** 8 **d** $2\sqrt{3}$

3 Find the area of an equilateral triangle with an *apothem* of:

a 6 **b** 4 **c** 3 **d** $2\sqrt{3}$

4 Find the area of a regular hexagon whose:

a side is 6 **c** apothem is 6

b side is 8 **d** apothem is 8

5 The radius of a regular hexagon is 12.

Find: a the length of one side

b the apothem

c the area

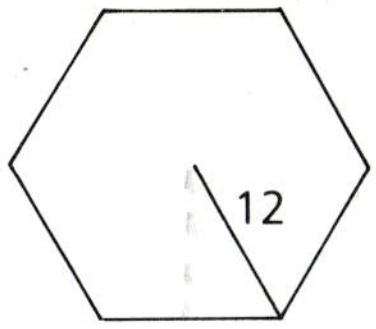

6 Find the area of a square whose:

a apothem is 5
b apothem is 12
c side is 7
d diagonal is 10
e radius is 6
f perimeter is 12

7 Find the apothem of a square whose area is 36 mm^2.

8 Find the side of an equilateral triangle whose area is $9\sqrt{3}$ km^2.

9 Find the area of a square if the radius of its inscribed circle is 9.

10 Find the area of an equilateral triangle if the radius of its inscribed circle is 3.

11 Find the area of a regular hexagon if the radius of its inscribed circle is 12.

Section 11.5 Problem Set B

12 Find the area of:

a an equilateral triangle whose side is 9

b a square whose apothem is $7\frac{1}{2}$

c a regular hexagon whose side is 7.

13 Find the length of one side and the apothem of:

a a square whose area is 121

b an equilateral triangle whose area is $36\sqrt{3}$ sq meters

c a regular hexagon whose perimeter is 24 cm.

14 Find the perimeter of a regular polygon whose area is 64 and whose apothem is 4.

15 A circle of radius 12 is circumscribed about each regular polygon below. Find the area of each polygon.

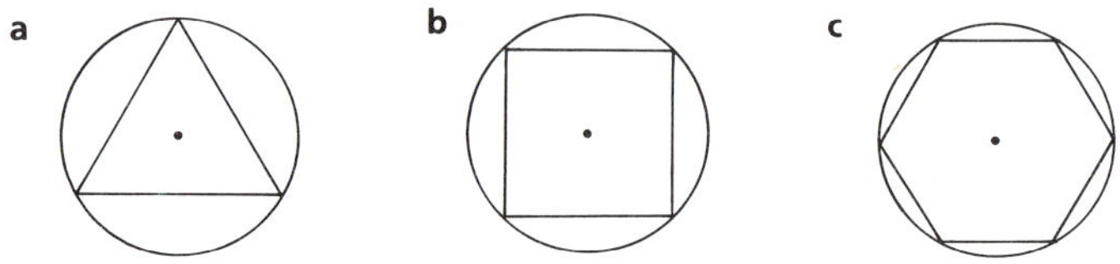

16 A circle is inscribed in one regular hexagon and circumscribed about another. If the circle has a radius of 6, find the ratio of the areas of the two hexagons.

17 Find the area of the shaded region. (Assume regular polygons.)

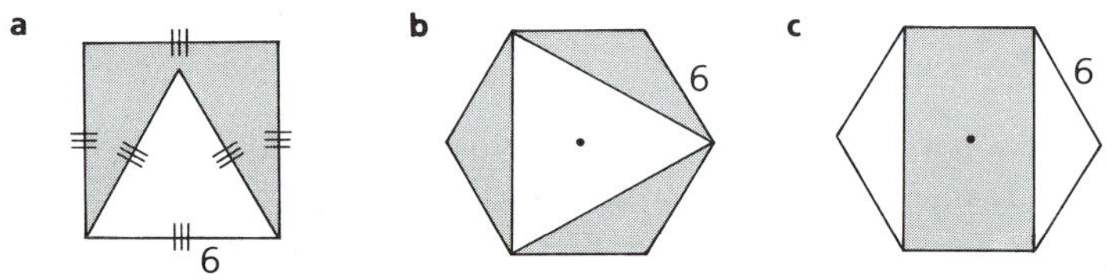

18 Given a scalene triangle, an equilateral triangle, a kite, a square, a regular polygon, and a regular hexagon. If two of the six figures are selected at random, what is the probability that both have apothems?

Section 11.5 Problem Set C

19 **a** The span s of a regular hexagon is 30. Find the area.

b Find the span of a regular hexagon with an area of $32\sqrt{3}$.

c Find a formula for the area of a regular hexagon with a given span s.

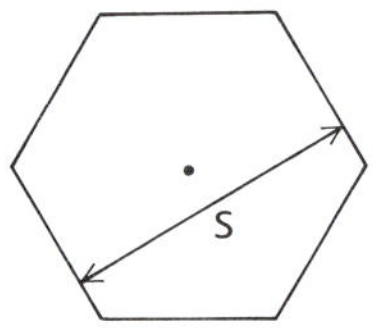

20 a Find the apothem of the octagon.

b Find the area of the regular octagon.

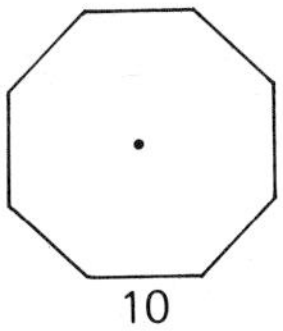

21 A square is formed by joining the mid-points of alternate sides of a regular octagon. A side of the octagon is 10.

Find: a the area of the square

b the area of the shaded region.

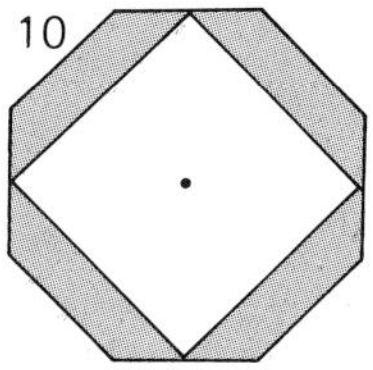

22 A square and a regular hexagon are inscribed in the same circle. Find:

a the ratio of a side of the square to a side of the hexagon

b the ratio of the area of the square to the area of the hexagon.

23 Given a set of four nested regular hexagons, each one unit larger in radius than the next smaller. Find the total area of the shaded regions.

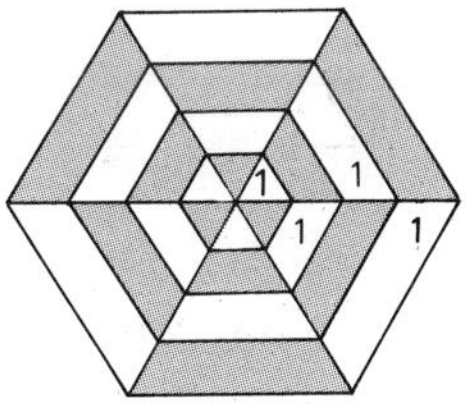

24 A square is inscribed in an equilateral triangle as shown. Find the area of the shaded region.

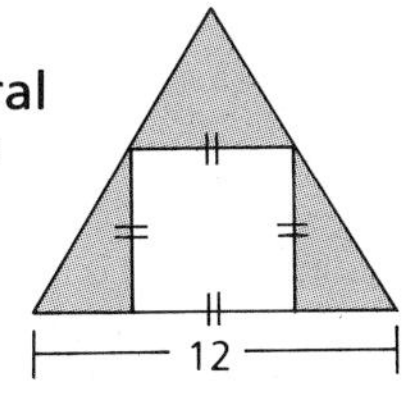

Section 11.6 The Areas of Circles, Sectors, and Segments

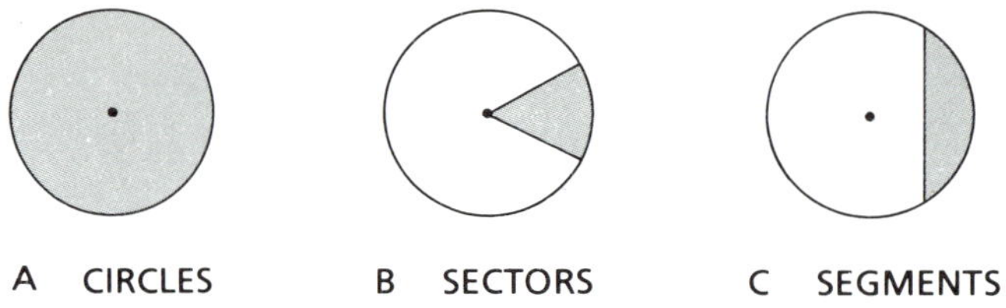

11.6 (A) The Area of a Circle

You may already know the area formula for a circle:

$$A_{\odot} = \pi r^2$$

r = radius

P **The area of a circle is the product of π times the square of the radius.**

11.6 (B) The Area of a Sector

D **A SECTOR OF A CIRCLE is a region bounded by two radii and an arc of the circle.**

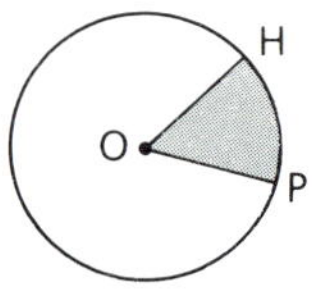

sector HOP

Just as the arc length of $\widehat{AB}$ was a fractional part of the circumference, the area of a sector is a fractional part of the area of the circle.

Theorem 98

T **The AREA OF A SECTOR of a circle is the area of the circle times the fractional part of the circle determined by the arc.**

$$A_{\text{sector HOP}} = \left(\frac{m\widehat{HP}}{360}\right)\pi r^2$$

r = radius

$\widehat{HP}$ is measured in degrees

11.6 (C) The Area of a Segment

D **A SEGMENT OF A CIRCLE is a region bounded by a chord of the circle and its arc.**

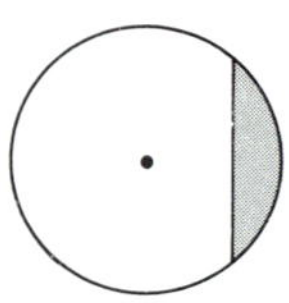

You may see from the figure above what to do to find the area of a segment (shaded). Sample problem 4 illustrates the procedure in detail.

Section 11.6 Sample Problems

1 Find the area of a circle whose diameter is 10.

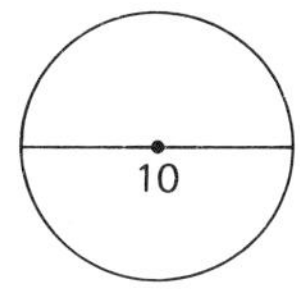

Solution: $A_{\odot} = \pi\ r^2$

but $r = \frac{1}{2}$ (diameter) $= 5$

$A_{\odot} = \pi\ (5^2) = 25\ \pi$ sq units.

2 Find the circumference of a circle whose area is $49\ \pi$.

Solution: Find the radius from the given area.

$A_{\odot} = \pi r^2$	$C = 2\pi r$
$49\pi = \pi r^2$	$C = 2\pi(7) = 14\ \pi$ Ans.
$7 = r$	

3 Find the area of a sector with a radius of 12 and a 45° arc.

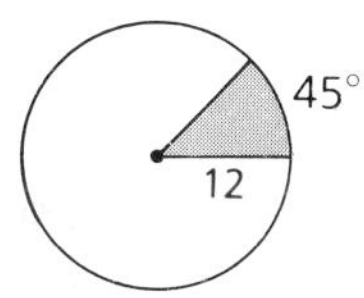

Solution: $A_{sector} = \dfrac{arc^\circ}{circle^\circ} \cdot A.$

$= \dfrac{45}{360} \cdot \pi(12)^2 = 18\pi$ sq units

4 The measure of the arc of the segment is 90.
The radius is 10. Find the area of the segment.

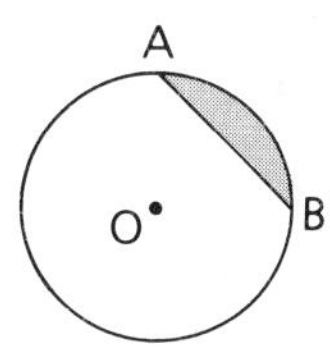

Solution: Draw radii to the endpoints of AB, forming sector AOB

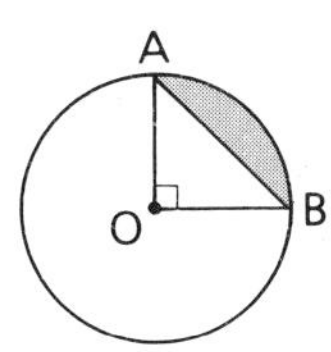

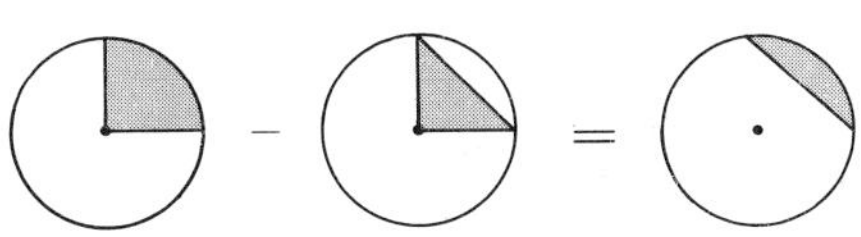

Area of Sector − Area of $\triangle$ = Area of Segment

$\left(\dfrac{m\widehat{AB}}{360}\right)\pi r^2 \quad - \quad \frac{1}{2}bh \quad =$

$\dfrac{90}{360}\ \pi(10)^2 \quad - \frac{1}{2}\ (10)\ (10) = 25\pi - 50 =$ Area of Segment

Section 11.6 Problem Set A

1 Find the area and circumference of a circle whose radius is:

a 1 b 8 c 15

2 Find the radius of a circle whose area is:

a 16π b 169π

3 Find the circumference of a circle whose area is $100\,\pi$ sq cm.

4 Find the area of a circle whose circumference is $18\,\pi$ dm.

5 Find the area of each shaded sector.

a
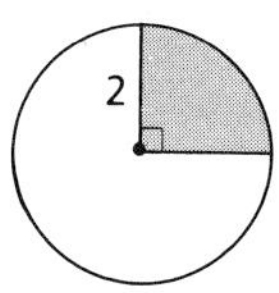

b
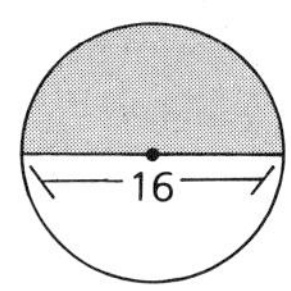

c
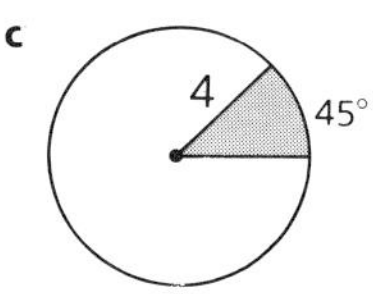

d
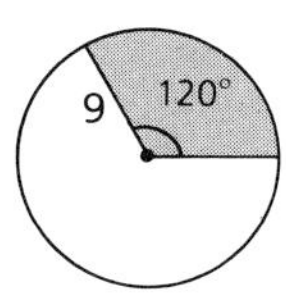

e
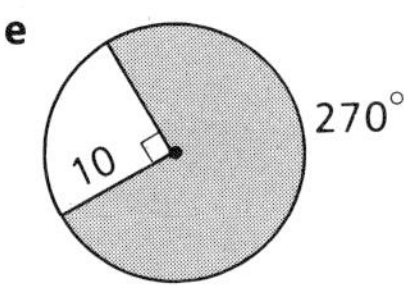

6 Two sprinklers are used to water the rectangular lawn as shown. Each sprinkler is capable of spraying a circular region of 3m radius. Find:

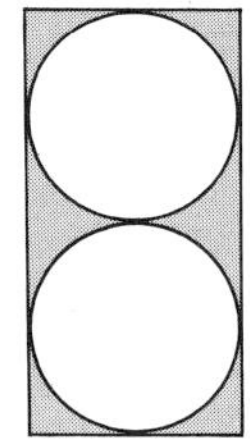

a the total area of lawn that is watered
b the area of the whole lawn
c the area of lawn *not* watered (shaded).

7 Find the total area of the region shown.

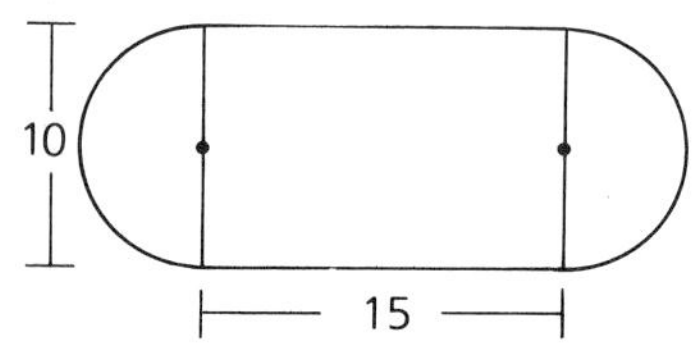

Section 11.6 Problem Set B

8 Find the radius of a circle whose area is: **a** 24π **b** 36

9 Find the area of each sector.

a

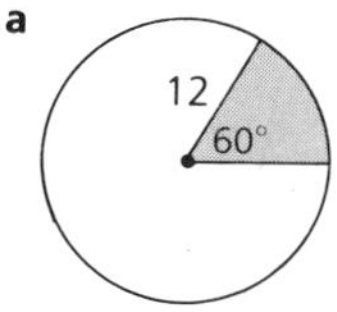

b

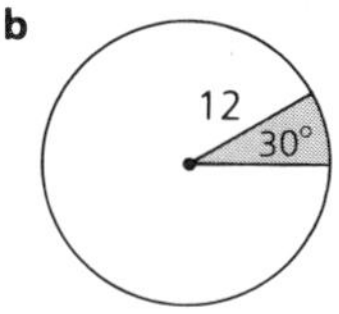

c

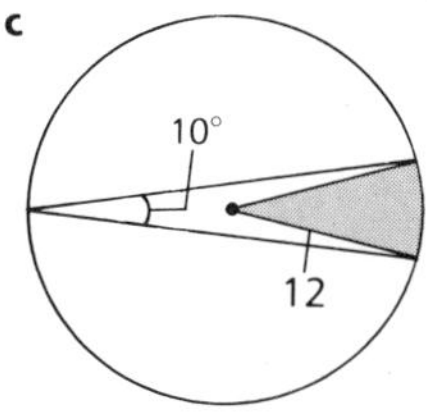

10 Find the degree measure of the arc of a sector whose area is 24π, given that the area of the circle is 60π.

11 Find the area of each segment.

a

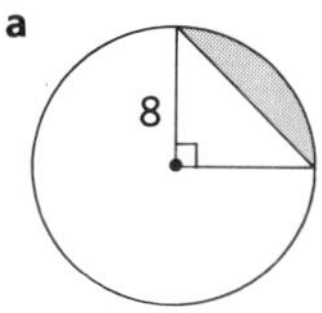

b

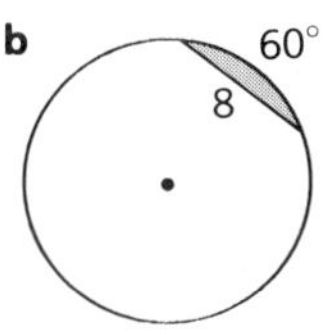

c

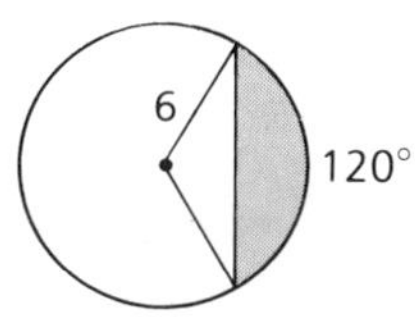

12 a Find the area of a "washer" if the inner radius is 3 and the outer radius is 5. (Such a figure is called an ANNULUS.)

b If the inner circle has a radius r and the outer circle has a radius R, derive the formula for the area of any annulus.

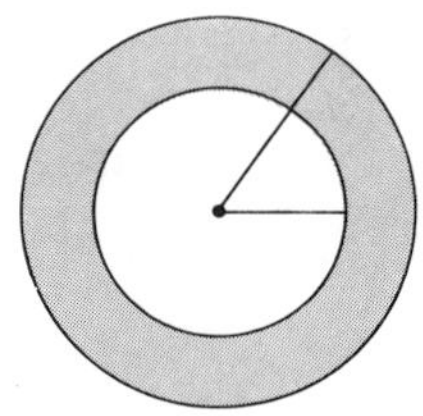

13 Find the area of the shaded region if:

a $x = 6$ **b** $x = 10$ c $x = 7$

What observation is possible?

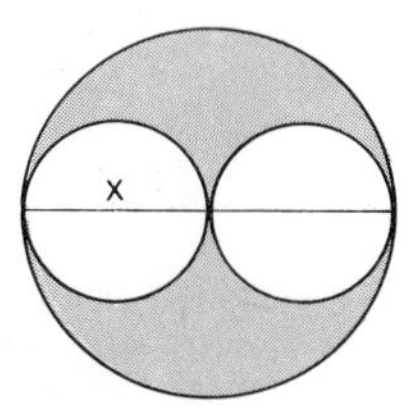

14 Find the area of the shaded region.

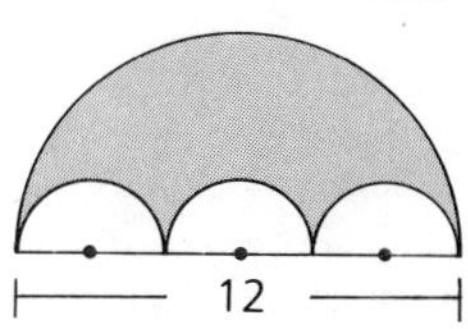

15 Find the shaded area. (Assume regular polygons.)

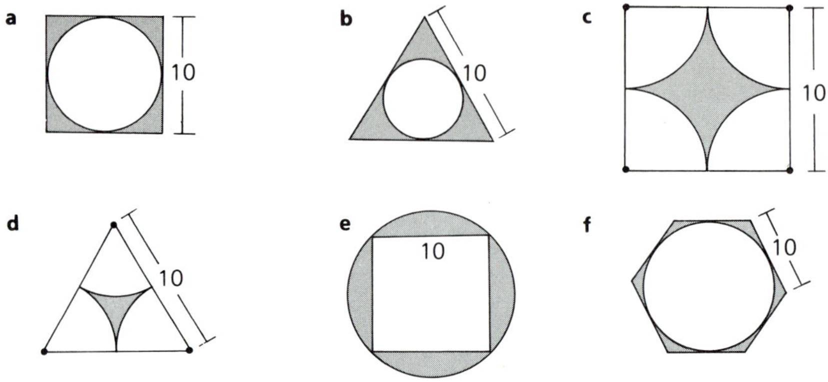

16 A target has a bulls-eye of 5 cm diameter and each band has a $2\frac{1}{2}$ cm width. Find:

a the total area covered by dark bands

b the total area covered by light bands.

What is the probability that if you hit the target, you will get a bulls-eye? (Assume that no skill is involved.)

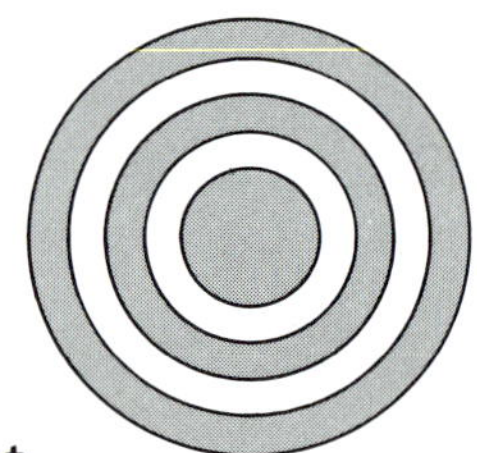

17 In the square grid each square is 2 cm wide.

Find the area of the region bounded by the circular arcs.

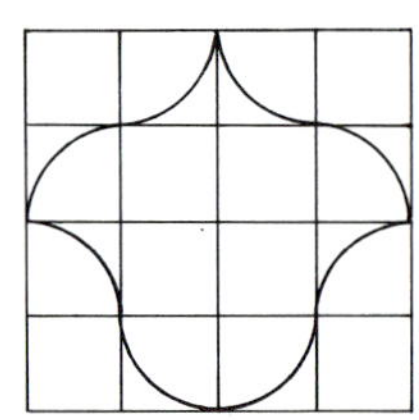

Section 11.6 Problem Set C

18 Find the area of each shaded region.

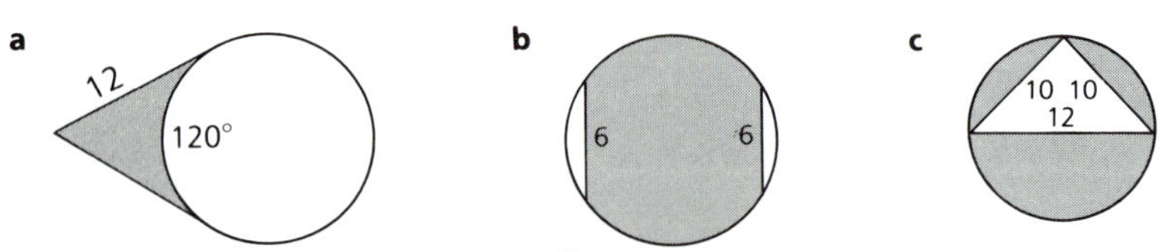

diameter = 12

19 The Wankel Rotary automotive engine is based on the geometric shape shown. Each arc has the opposite vertex of the equilateral triangle as its center.

Find: **a** the area

b the perimeter

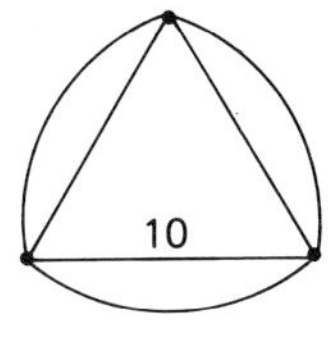

20 Find the areas of the shaded regions.

a

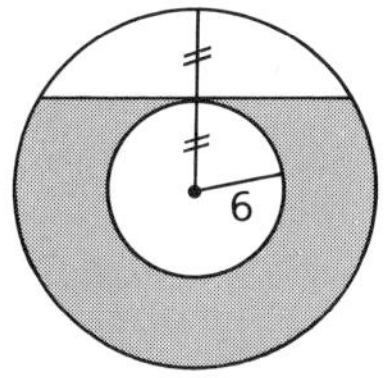

b

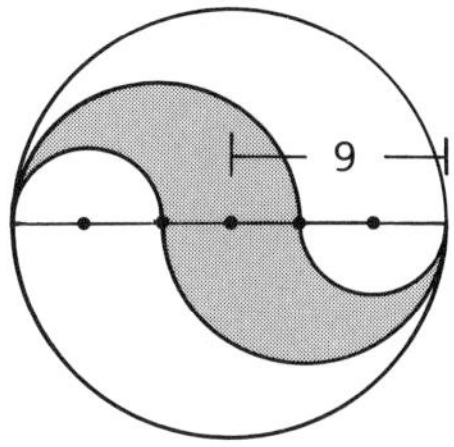

c

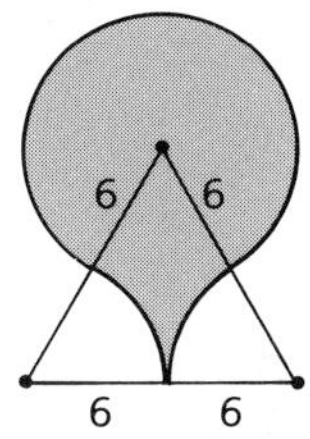

21 Three arcs are drawn from the midpoints of the sides of a triangle and meet at the vertices as shown. Find:

a the area of the shaded regions, which are called the LUNES OF HIPPOCRATES.

b the area of the triangle.

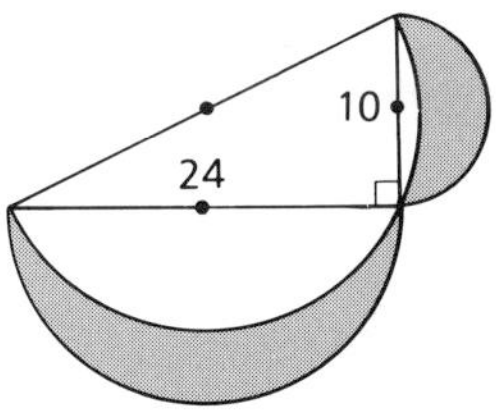

22 Find the area of the shaded region if the circle has a radius of 2.

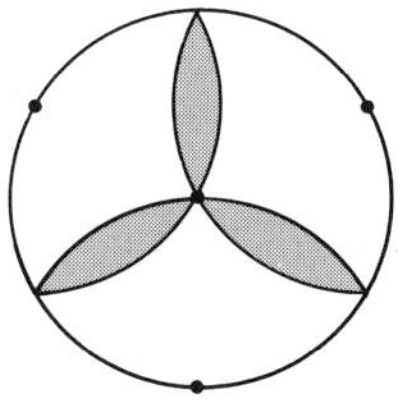

23 A circle is inscribed in a rhombus.

Find the area of the shaded region if the diagonals of the rhombus are 30 and 40.

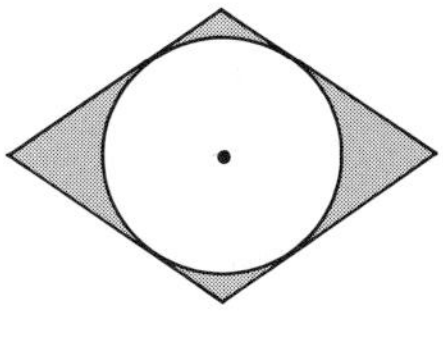

Section 11.7 Ratios of Areas

The topics of this section are two methods of finding the ratio of the areas of figures:

A COMPUTING THE AREAS

B SIMILAR FIGURES

11.7(A) Computing the Areas

■ Example 1

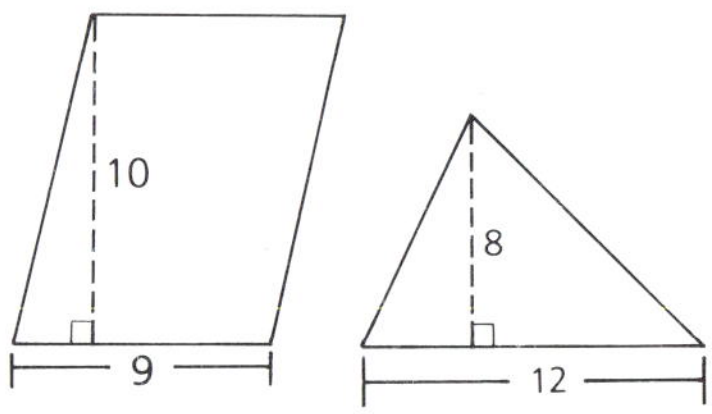

Find the ratio of the area of the parallelogram to the area of the triangle.

Solution by COMPUTING THE AREAS:

$$\frac{A_{\square}}{A_{\triangle}} = \frac{b_1h_1}{\frac{1}{2}b_2h_2} = \frac{9 \cdot 10}{\frac{1}{2} \cdot 12 \cdot 8} = \frac{90}{48} = \frac{15}{8} \text{ or } 15{:}8.$$

■ Example 2

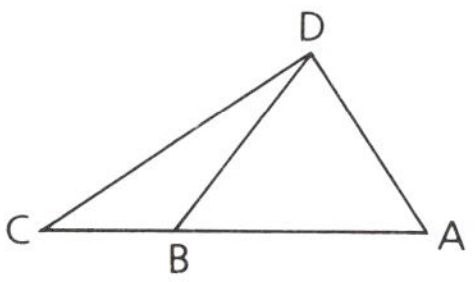

In the diagram AB = 5 and BC = 2.
Find the ratio of the areas of △ABD and △CBD.

Solution by COMPUTING THE AREAS:

Notice that the height of △ABD is the same as the height of △CBD and is labeled by the letter h.

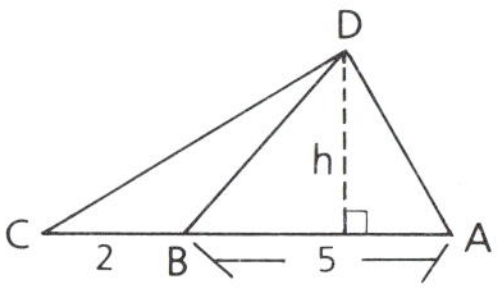

$$\frac{A_{\triangle ABD}}{A_{\triangle CBD}} = \frac{\frac{1}{2}b_1h_1}{\frac{1}{2}b_2h_2} = \frac{(5)\,(h)}{(2)\,(h)} = \frac{5}{2}$$

11.7(B) Similar Figures

As you know, if two triangles are similar, the ratio of any pair of their corresponding altitudes, medians, or angle bisectors equals the ratio of their corresponding sides. Application of that fact in an example leads to an interesting formula.

■ Example 3 Given that △PQR ~ △WXY, find the ratio of their areas.

Notice that the ratio of the corresponding sides is $\frac{3}{2}$.

The ratio of the areas is:

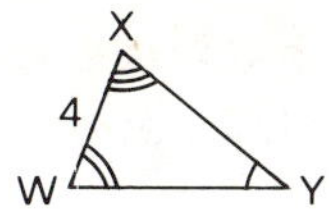

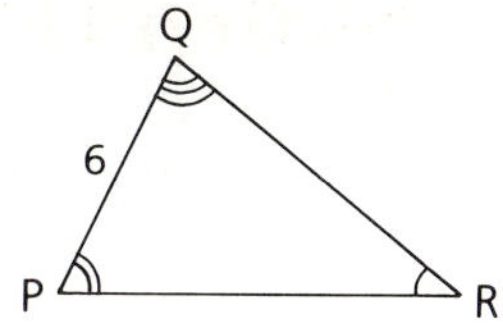

$$\frac{A_{\triangle PQR}}{A_{\triangle WXY}} = \frac{\frac{1}{2}b_1h_1}{\frac{1}{2}b_2h_2} = \frac{b_1}{b_2} \cdot \frac{h_1}{h_2}$$

but $\frac{b_1}{b_2} = \frac{3}{2}$ and $\frac{h_1}{h_2} = \frac{3}{2}$, so $\frac{A_{\triangle PQR}}{A_{\triangle WXY}} = \frac{3}{2} \cdot \frac{3}{2} = \left(\frac{3}{2}\right)^2 = \frac{9}{4}$.

Notice that $\frac{9}{4}$ is the SQUARE of $\frac{3}{2}$.

The example shows the key steps of a proof about the areas of similar triangles. Because convex polygons can be divided into triangles, you may suspect that similar polygons have the same characteristic. They do.

Theorem 99

T **If two figures are similar, then the *ratio of their areas* equals the *square of the ratio* of corresponding segments.**
(SIMILAR FIGURES THEOREM)

$\frac{A_1}{A_2} = \left(\frac{s_1}{s_2}\right)^2$
A_1 and A_2 are areas
s_1 and s_2 are corresponding segments.

Corresponding segments can be sides, altitudes, medians, diagonals, radii, etc.

■ Example 4 Given the similar pentagons shown, find the ratio of their areas.

Using the Similar Figures Theorem

$\frac{A_I}{A_{II}} = \left(\frac{s_1}{s_2}\right)^2$, and $\frac{s_1}{s_2} = \frac{12}{9} = \frac{4}{3}$

so $\frac{A_I}{A_{II}} = \left(\frac{4}{3}\right)^2 = \frac{16}{9}$, or 16:9

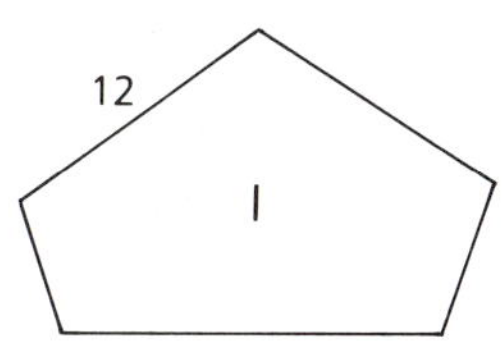

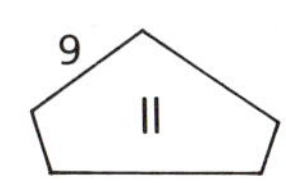

Section 11.7 Sample Problems

1 $\triangle ABC \sim \triangle DEF$ **(Note the correspondence.)**
Find the ratio of the areas of the two triangles.

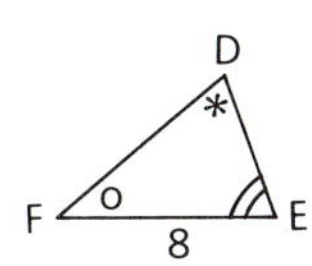

Solution: Use the Similar Figures Theorem

$$\frac{A_1}{A_2} = \left(\frac{s_1}{s_2}\right)^2$$

$$\frac{A_1}{A_2} = \left(\frac{12}{8}\right)^2 = \left(\frac{3}{2}\right)^2 = \frac{9}{4}$$

Note: $\frac{9}{4}$ is the SQUARE of $\frac{3}{2}$.

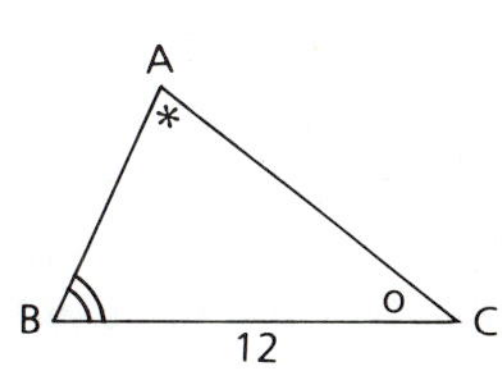

2 If the ratio of the areas of two similar parallelograms is 49:121, find the ratio of their bases.

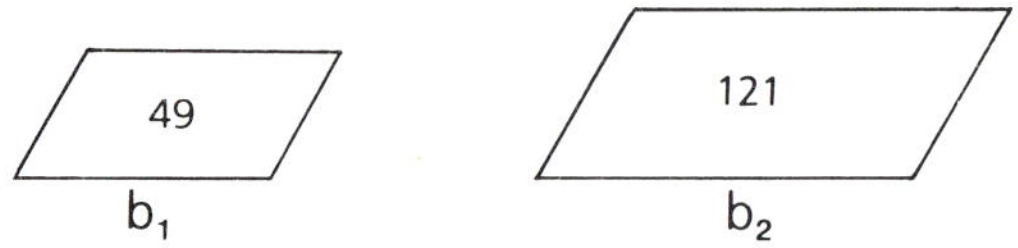

Solution: The Similar Figures Theorem is indicated.

$$\frac{A_1}{A_2} = \left(\frac{b_1}{b_2}\right)^2$$

$$\frac{49}{121} = \left(\frac{b_1}{b_2}\right)^2$$

Thus, $\frac{7}{11} = \frac{b_1}{b_2}$

Note: $\frac{7}{11}$ is the SQUARE ROOT of $\frac{49}{121}$

3 $\overline{AM}$ is the median of $\triangle ABC$.

Compare the areas by their ratio,

$A_{\triangle ABM} : A_{\triangle ACM}$

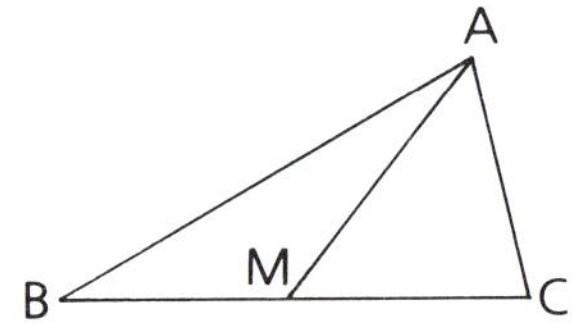

Solution: The Similar Figures Theorem does not apply.

a Compare the altitudes from A. The same!

Call this common altitude x.

b Compare the bases. BM = MC, because $\overline{AM}$ is a MEDIAN.

Let y represent BM and MC.

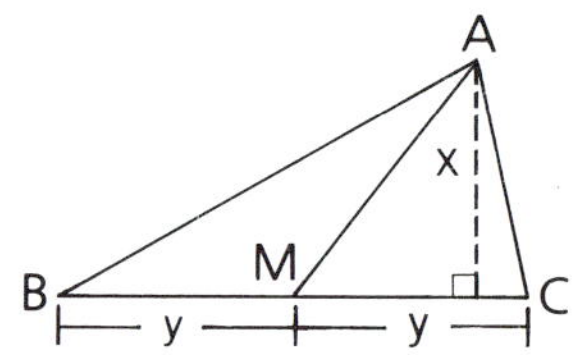

c $\frac{A_{\triangle ABM}}{A_{\triangle ACM}} = \frac{\frac{1}{2}b_1h_1}{\frac{1}{2}b_2h_2} = \frac{xy}{xy} = 1$ or 1:1

Since the triangles have equal bases and equal heights, their areas are equal.

This problem may be stated as a theorem:

Theorem 100

T **A median of a triangle divides the triangle into two triangles with equal areas.**

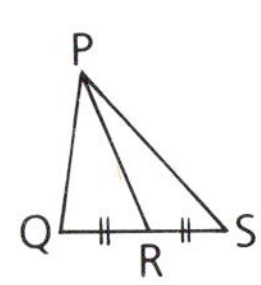

$\Longrightarrow$ $\mathbf{A}_{\triangle PQR} = \mathbf{A}_{\triangle PRS}$

Section 11.7 Problem Set A

1 By computing the areas, find the ratio of the areas for each pair of figures shown.

a

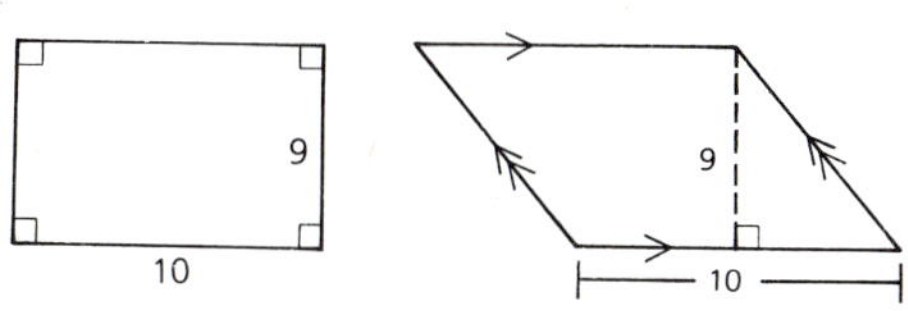

b

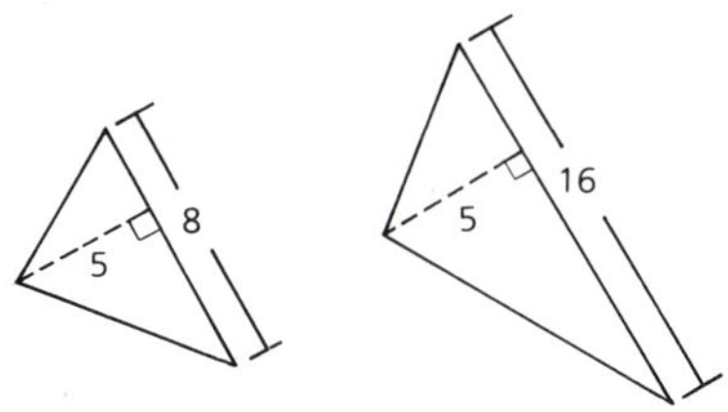

c

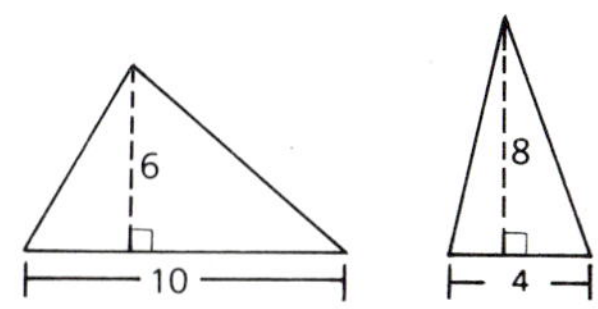

d

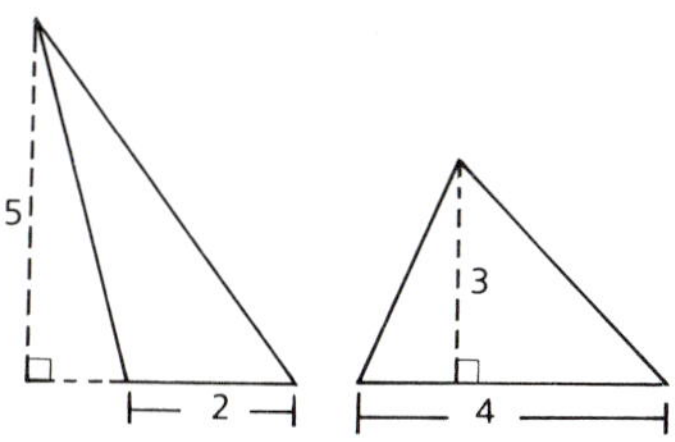

2 By using the Similar Figures Theorem, find the ratio of the areas of each pair of similar figures.

a

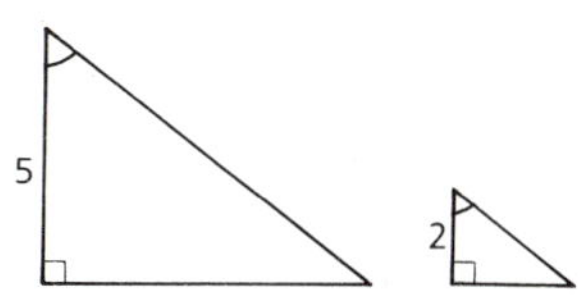

b

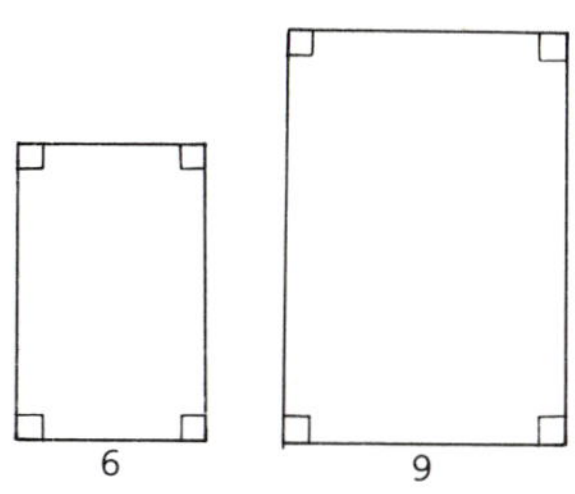

c

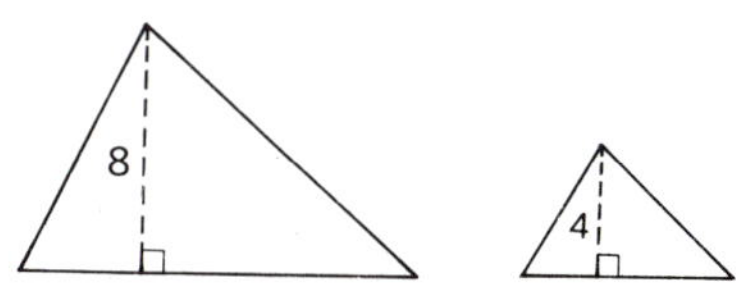

d

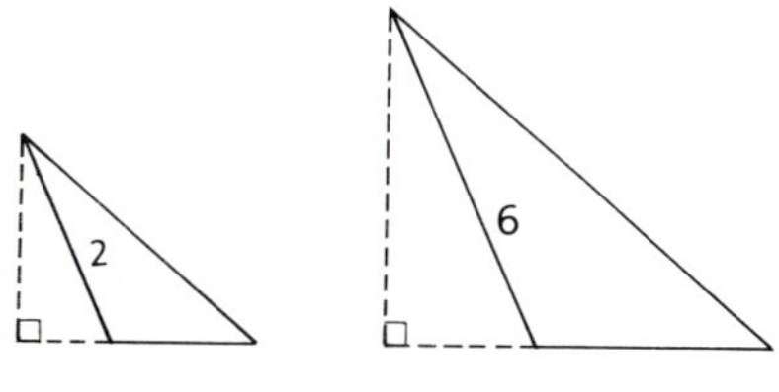

3 Given: $\overline{PM}$ is a median

Find: **a** $A_{\triangle PQM} : A_{\triangle PRM}$

b $A_{\triangle PQM} : A_{\triangle PQR}$

c QR:MR

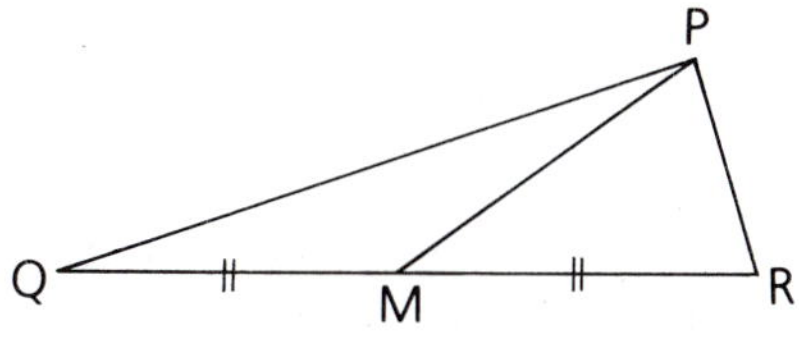

4 A pair of corresponding sides of two similar triangles are 4 and 9 respectively. Find the ratio of the areas.

5 If the ratio of the areas of two similar polygons is 9:16, find the ratio of a pair of corresponding altitudes.

6 Gladys Gardenia has a square garden, 3 m on a side. She wishes to make it exactly twice as large. Gladys decides to double the length and double the width. Does she succeed?

7 Find the ratio of the areas of the regular hexagons.

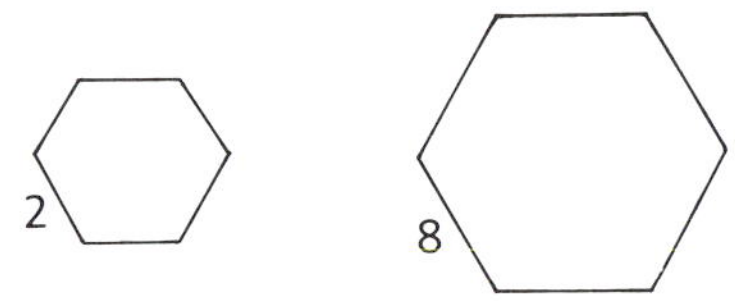

8 Find the ratio of the areas of the triangles.

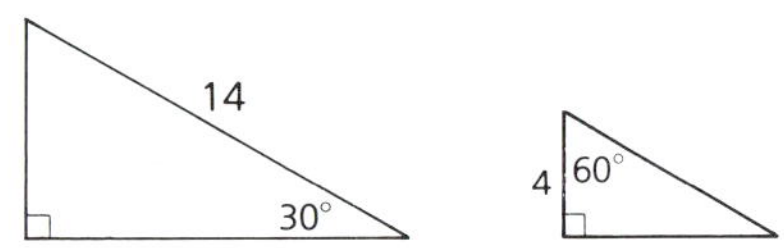

Section 11.7 Problem Set B

9 Find the ratio of area I to area II in each problem. Each figure is a parallelogram, rectangle, or triangle as shown.

a

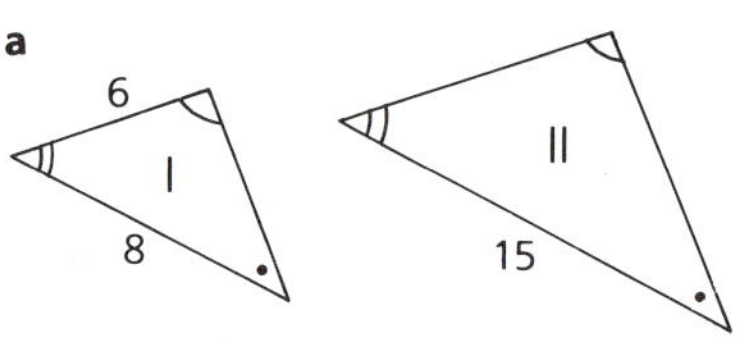

b

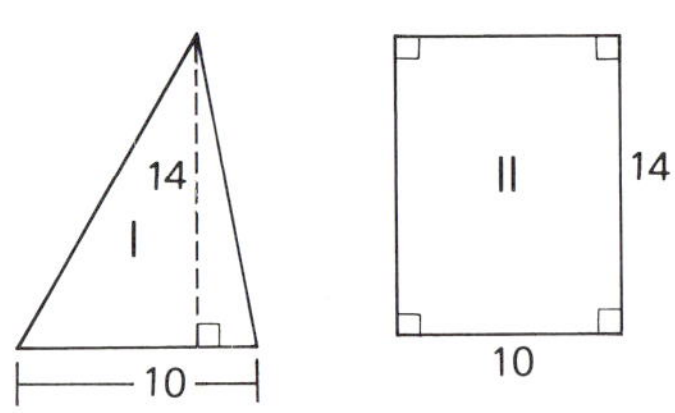

c

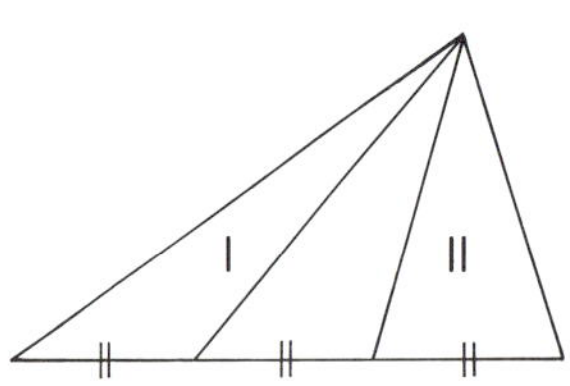

d

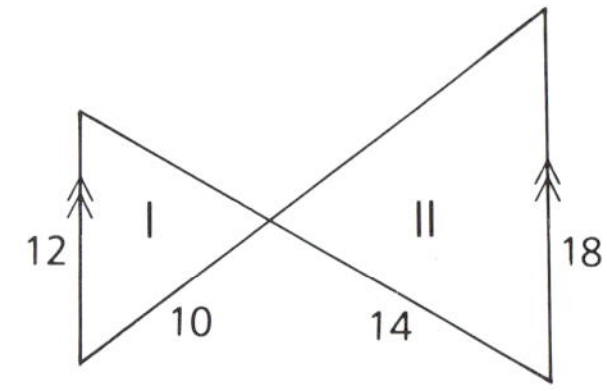

10 Find the ratio of the areas of the shaded triangle to the whole triangle in each case.

a

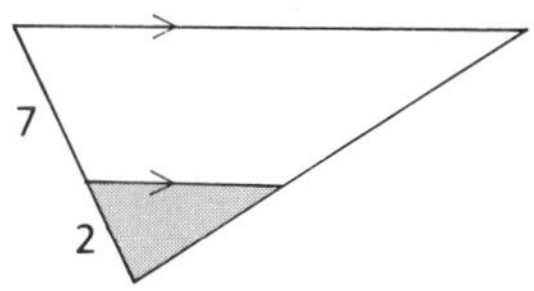

b

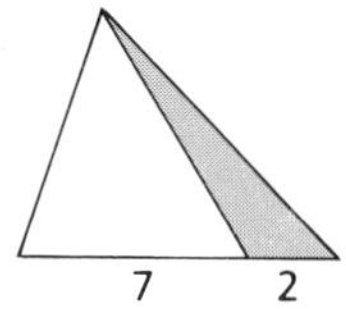

c

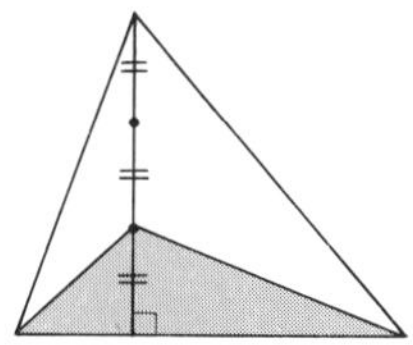

d

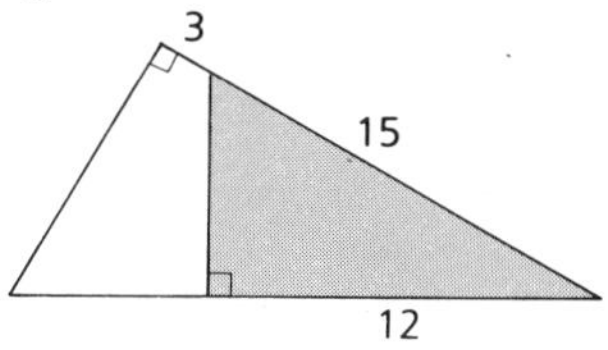

11 Find the ratio of the areas of the two triangles.

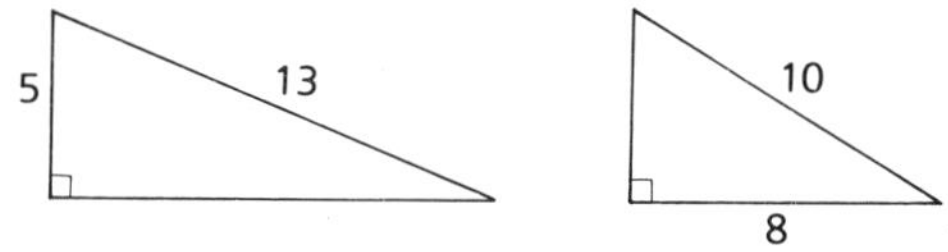

12 The ratio of the areas of two similar pentagons is 8:18.

Find: **a** the ratio of their corresponding sides

b the ratio of their perimeters.

13 The ratio of the corresponding medians of two similar triangles is 5:2. Find the area of the larger triangle if the smaller triangle has an area of 40.

14 One triangle has sides 13, 13, and 10. A second triangle has sides 12, 20, and 16. Find the ratio of their areas.

15 Find the ratio of the areas of two circles if their radii are 4 and 9.

16 Find the ratio of the areas of two equilateral triangles with sides 6 and 8.

17 Find $\frac{A\triangle ACD}{A\triangle BCD}$

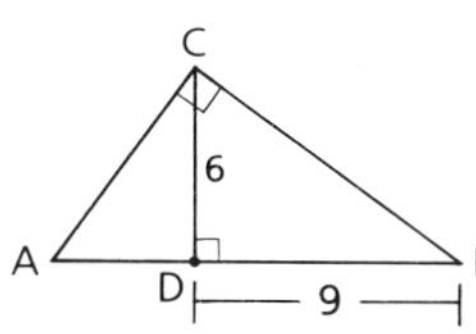

Section 11.7 Problem Set C

18 In trapezoid WXYZ, find the ratio of the areas of:

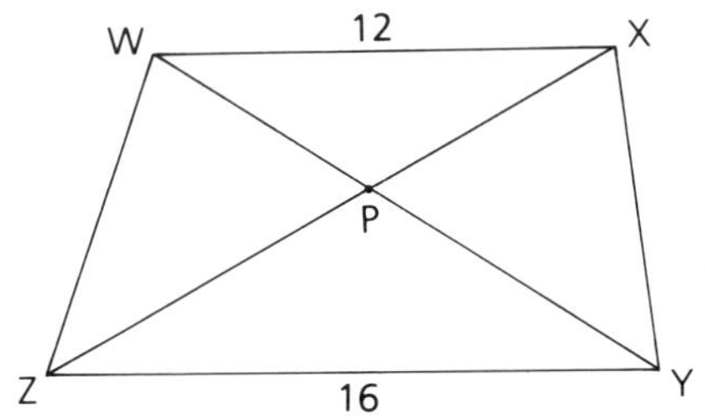

a $\triangle$WYZ to $\triangle$XYZ

b $\triangle$WXZ to $\triangle$WXY

c $\triangle$WPZ to $\triangle$XPY

d $\triangle$WPX to $\triangle$ZPY

e $\triangle$WPX to $\triangle$XPY

19 Given: $\triangle$ABC is a right $\triangle$
E and F are midpoints
D, H, and G divide $\overline{AC}$ into 4 $\cong$ segments

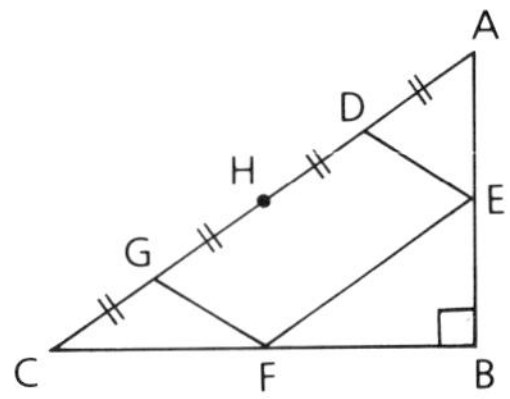

Find: a the ratio of the areas of $\triangle$ABC to $\triangle$EBF

b the ratio of the areas of $\triangle$ABC to $\triangle$GFC

c the ratio of the areas of $\triangle$ADE to $\triangle$GFC

d the ratio of the areas of ▱DEFG to $\triangle$ABC

e the perimeter of ▱DEFG if AC = 20.

20 If the midpoints of the sides of a quadrilateral are joined in order, another quadrilateral is formed. Find the ratio of the area of the larger quadrilateral to that of the smaller quadrilateral.

21 Given: Trapezoid ABCD

Find: The ratio of the areas of $\triangle$I to $\triangle$II.

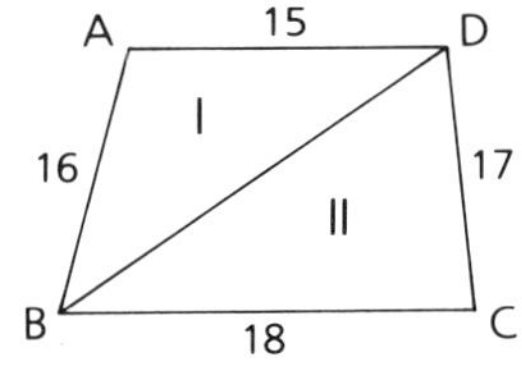

22 In ▱PQRS, T is a midpoint.

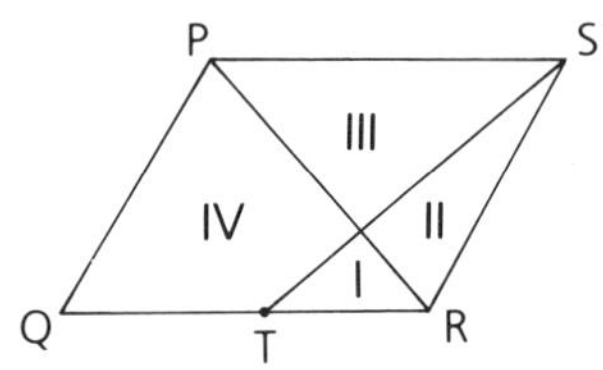

a If the area of PQRS is 60, find the areas of regions I, II, III, and IV.

b If T divides $\overline{QR}$ such that $\frac{QT}{TR} = \frac{x}{y}$, find the ratio of the areas of region I to ▱ PQRS.

Section 11.8 Surface Area

A SURFACE AREAS OF PRISMS
B SURFACE AREAS OF PYRAMIDS
C SURFACE AREAS OF CIRCULAR SOLIDS

11.8 (A) Surface Areas of Prisms

Solids with flat faces are called **POLYHEDRA** (meaning "many faces"). The faces are polygons, and the lines where they intersect are called edges.

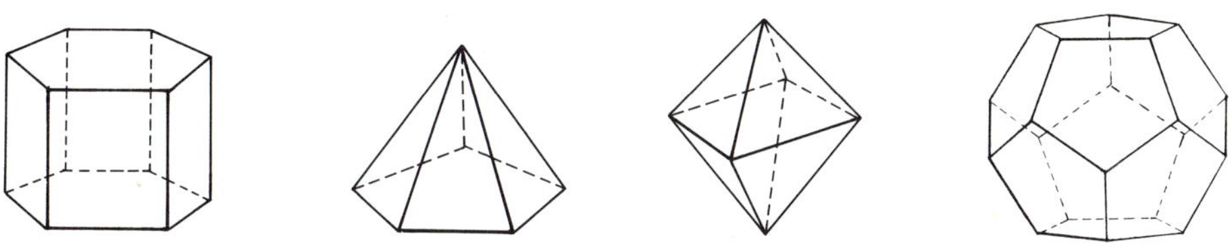

One familiar type of polyhedron is the prism. Here are three examples:

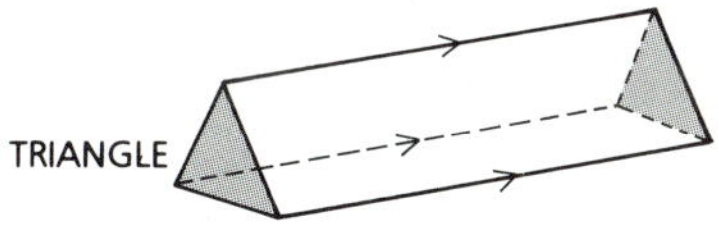

Triangular Prism

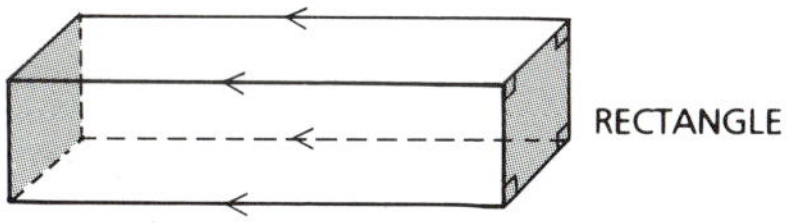

Rectangular Prism

Every **PRISM** has two *congruent, parallel faces* (shaded in the examples) and a set of *parallel edges* that connect corresponding vertices of the two parallel faces.

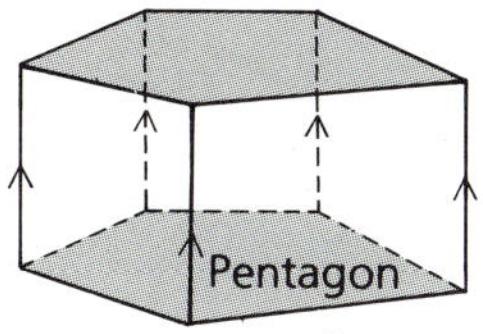

Pentagonal Prism

The two parallel and congruent faces are **BASES**. The parallel edges of a prism joining the vertices of the bases are called **LATERAL EDGES**. The faces of the prism that are not bases are called **LATERAL FACES**. The lateral faces of all prisms are parallelograms! Therefore, we name prisms by their bases, as in hexagonal prism.

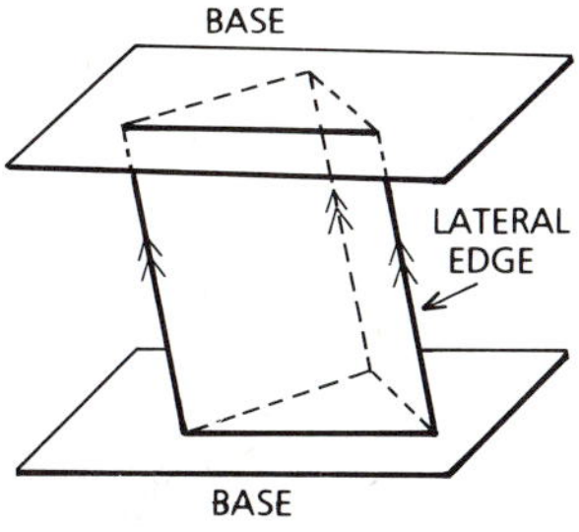

D **The LATERAL SURFACE AREA OF A PRISM is the sum of the areas of the lateral faces.**

D The TOTAL SURFACE AREA OF A PRISM is the sum of the lateral area and the areas of the two bases.

If the lateral edges are *perpendicular* to the bases, then the lateral faces will be rectangles. Why? When this happens, we put the word *right* in front of the name of the prism. In this book, the word "box" will refer to a right prism.

■ Examples:

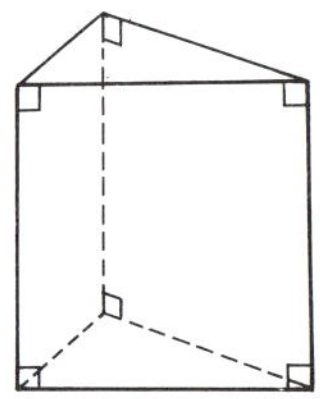

Right Triangular Prism
(The base is not necessarily a right triangle.)

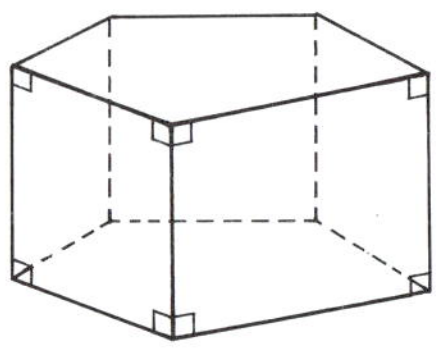

Right Pentagonal Prism

Section 11.8(A) Sample Problem

1 Given the right triangular prism as shown, find:

a Lateral Area (L.A.)

b Total Area (T.A.)

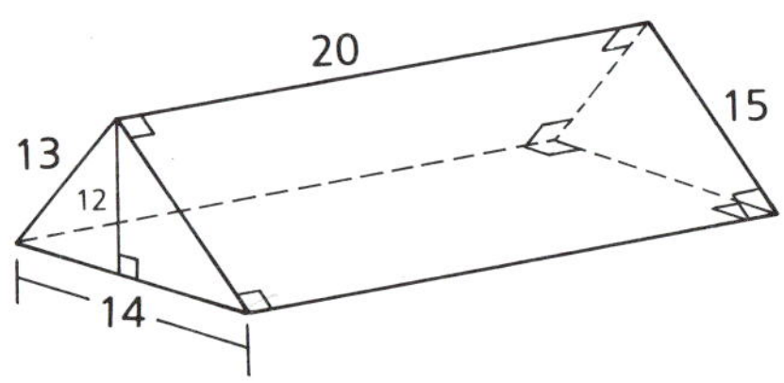

Solution: The right triangular prism can be divided into two triangles (the parallel bases) and three rectangles (the lateral faces).

a Lateral Area:

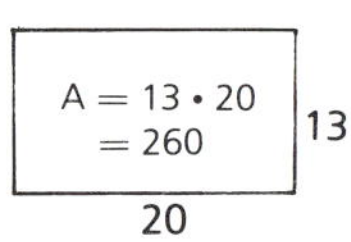

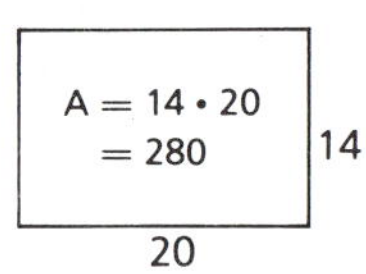

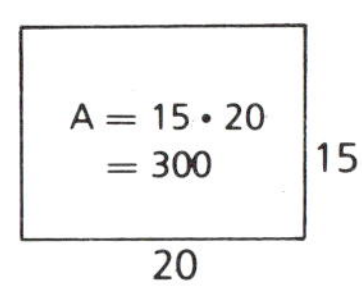

L.A. = 260 + 280 + 300 = 840

b Total Area:

T.A. = L.A. +

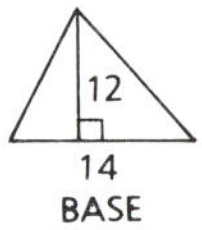

\+

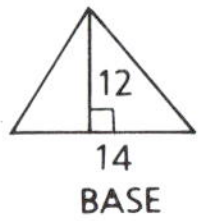

but the area of each base $= \frac{1}{2} b \cdot h$

$= \frac{1}{2}(14)(12) = 84$

Thus, T.A. = 840 + 84 + 84 = 1008

Section 11.8(A) Problem Set A

1 Find the total surface area of a right rectangular prism with the following dimensions:

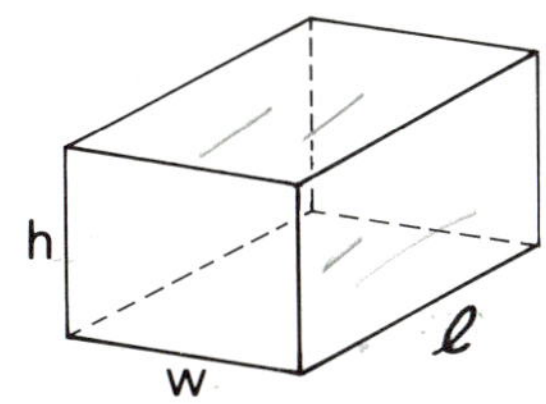

a $\ell = 15\text{cm}$, $w = 5\text{cm}$, $h = 10\text{cm}$

b $\ell = 12\text{mm}$, $w = 7\text{mm}$, $h = 3\text{mm}$

c $\ell = 18$, $w = 9$, $h = 9$

2 Find the *lateral area* of the right triangular prism with these edges:

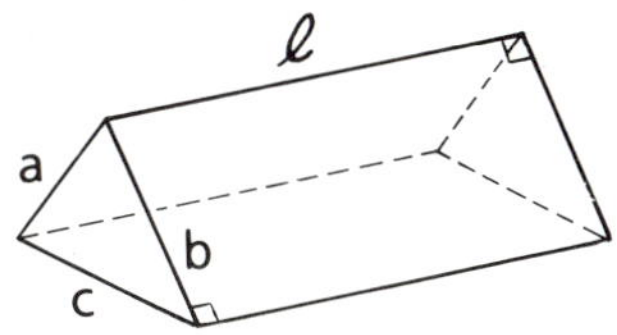

a $\ell = 10$, $a = 3$, $b = 5$, $c = 7$

b $\ell = 14$, $a = 2$, $b = 3$, $c = 4$

3 A right triangular prism has bases that are isosceles △s.

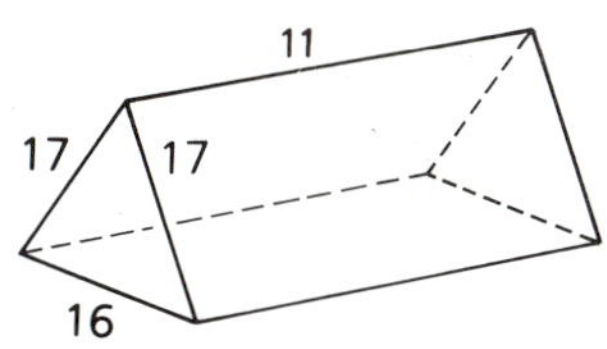

Find: a the lateral area

b the area of one base

c the total area.

4 Find the total surface area of the right equilateral triangular prism with these edges:

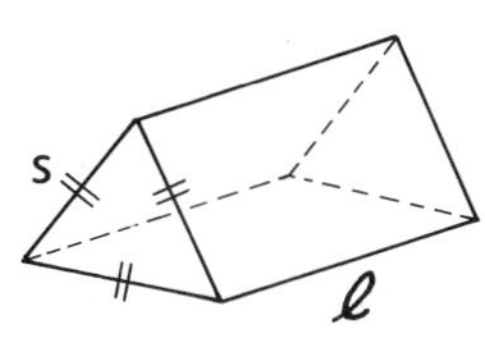

a $s = 6$, $\ell = 5$

b $s = 12$, $\ell = 10$

5 A cube is a rectangular prism with each face a square. Find the total surface area of a cube whose edge is:

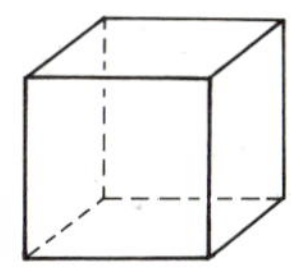

a 5 b 7

Section 11.8(A) Problem Set B

6 Find the lateral area and the total area for each prism.

a Right Square Prism

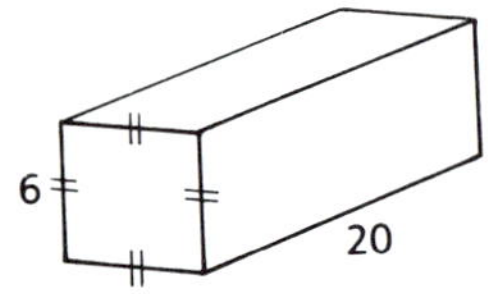

b Right Triangular Prism

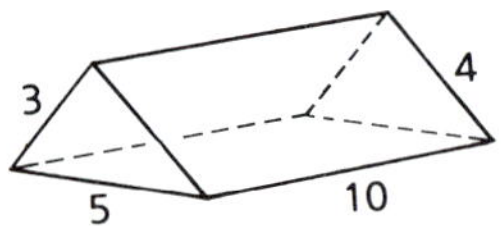

c Right Isosceles Triangular Prism

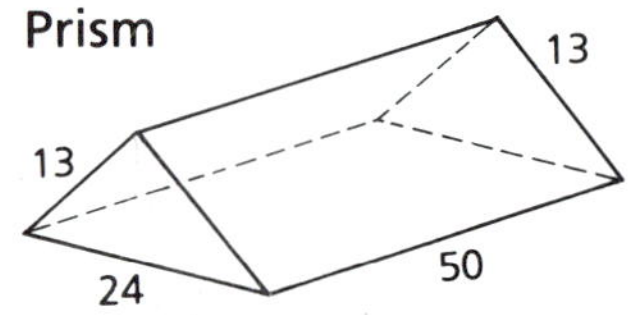

d Regular Hexagonal Prism

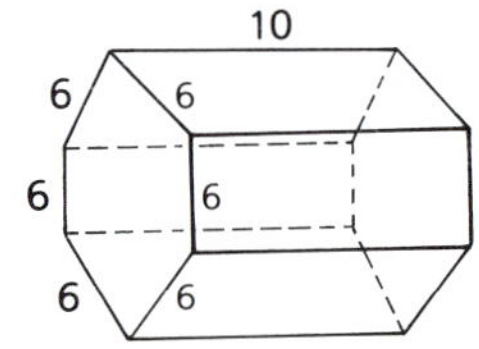

7 Find the total area of cardboard needed to construct an *open* box as shown. (Remember, a box is a right prism.)

a

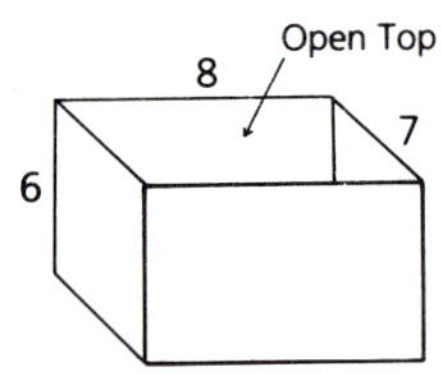

b

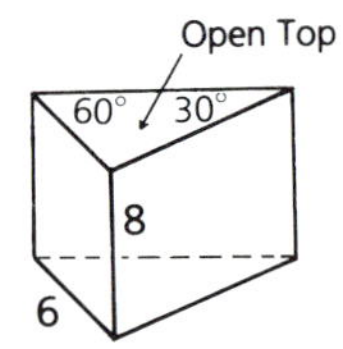

8 Find the total area of the right prism.

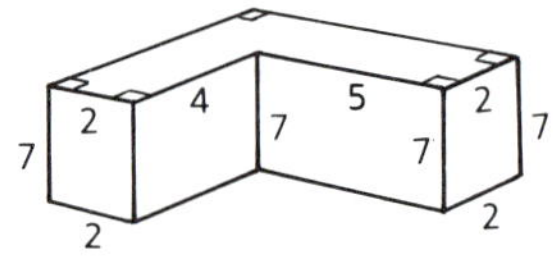

Section 11.8(A) Problem Set C

9 Find the lateral area and total area of the right prism shown.

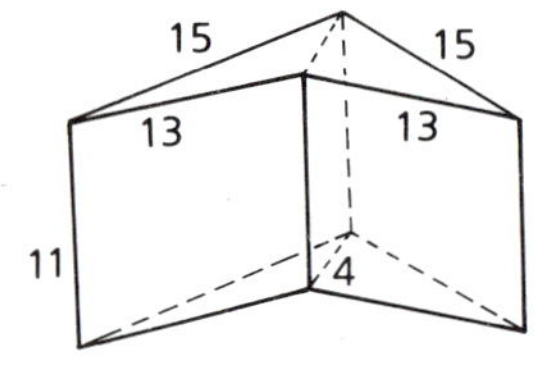

10 The perimeter of the scalene base of a pentagonal right prism is 17, and the lateral edge is 10. Find the lateral area.

11 A 6-inch cube painted on the outside is cut into 27 smaller cubes.

a How many of the cubes have 6 faces painted? 5? 4? 3? 2? 1? None?

b What is the probability that if one cube is picked at random, it will have at least 2 painted faces?

c What is the total area of the unpainted surfaces?

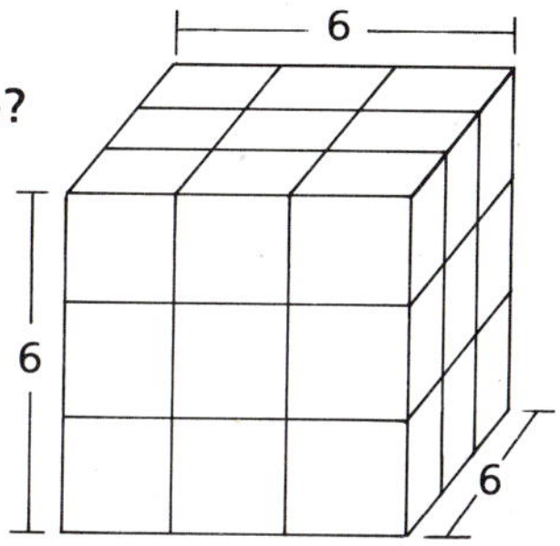

11.8 (B) Surface Areas of Pyramids

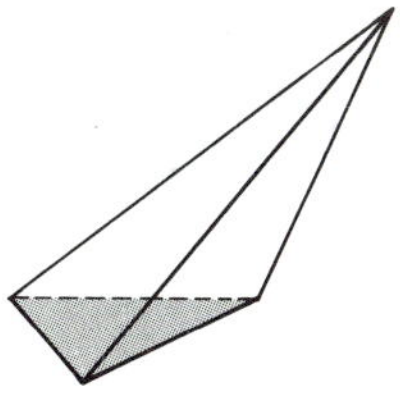

Triangular Pyramid

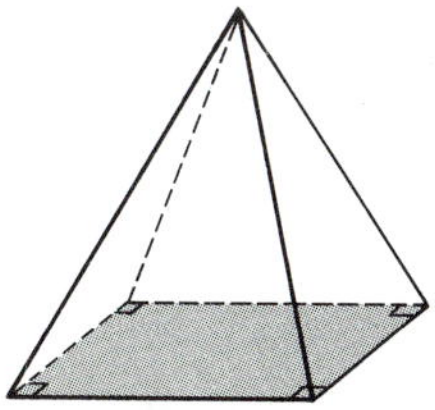

Rectangular Pyramid

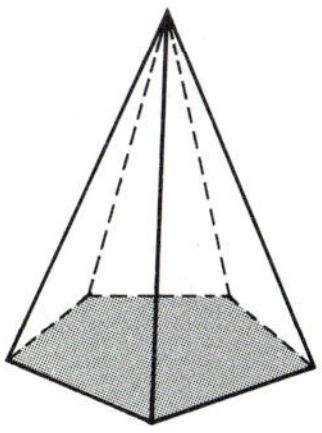

Pentagonal Pyramid

A **PYRAMID** has only one **BASE**. Its **LATERAL EDGES** are not parallel but meet at a single point called the **VERTEX**. The base may be any type of polygon, but the lateral faces will always be triangles. The diagram above illustrates three types of pyramids. Notice that each pyramid is named by its base.

A **REGULAR PYRAMID** has a regular polygon for a base and also has congruent lateral edges. Thus, the lateral faces of a regular pyramid are congruent isosceles triangles.

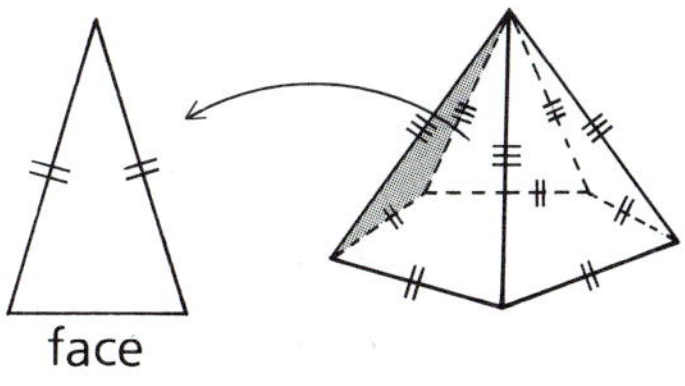

The **ALTITUDE** in a regular pyramid is perpendicular to the base. The foot of the altitude of a regular pyramid is the center of the base.
The **SLANT HEIGHT** is the height of a lateral face.

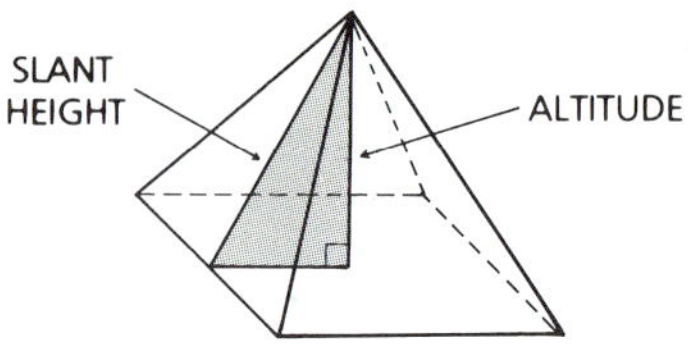

The altitude and a slant height determine a right triangle.

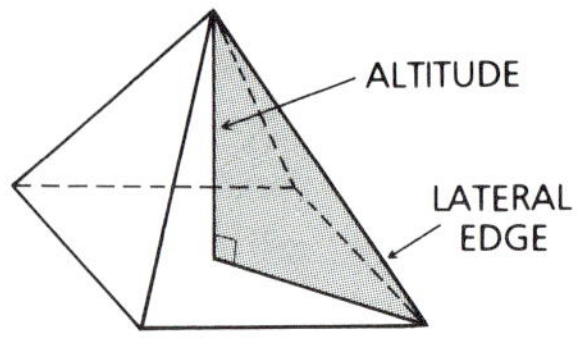

The altitude and a lateral edge determine a right triangle.

Section 11.8(B) Sample Problems

1 Given: The regular pyramid shown at the right

Find: **a** Lateral Area (L.A.)

b Total Area (T.A.)

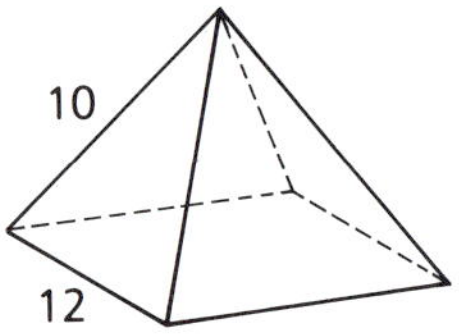

Solutions:

a The lateral area is the sum of the areas of four congruent isosceles triangles.

A Pythagorean triple shows the slant height to be 8.

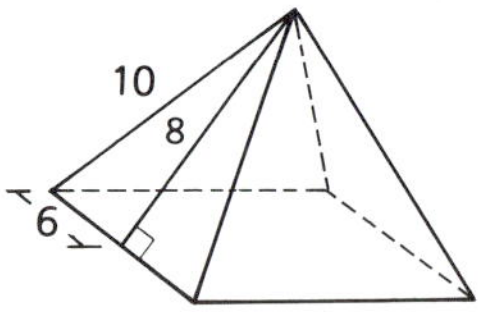

For each lateral face,

$A_{\triangle} = \frac{1}{2}bh$

$= \frac{1}{2}(12)(8) = 48$

L. A. $= 4(48) = 192$

b The total area is equal to the lateral area plus the area of the base. $A_{sq} = s^2$

$= (12)^2 = 144$

So, T.A. $= 192 + 144 = 336$

2 The base of rectangular pyramid ABCDE is 10 by 18. The altitude is 12. The lateral edges are ≅.

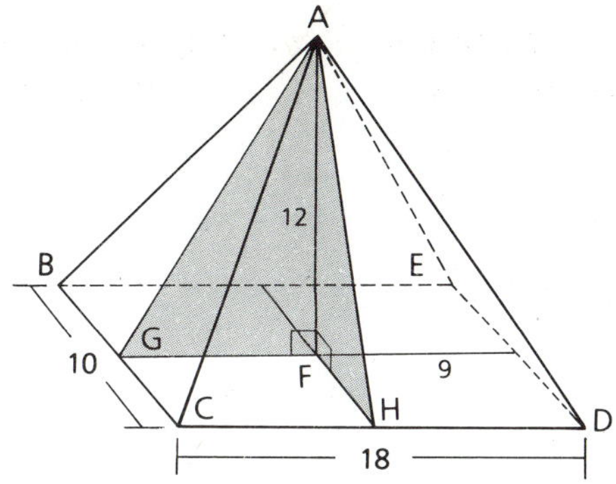

a Why is ABCDE *not* a regular pyramid?

b Find the total surface area.

Solutions:

a The base is not regular, so ABCDE is not regular.

b AH and AG are the heights of the lateral faces.
AH = 13, from △AFH and the Pythagorean Theorem
AG=15, from △AFG and the Pythogorean Theorem
There are five faces:

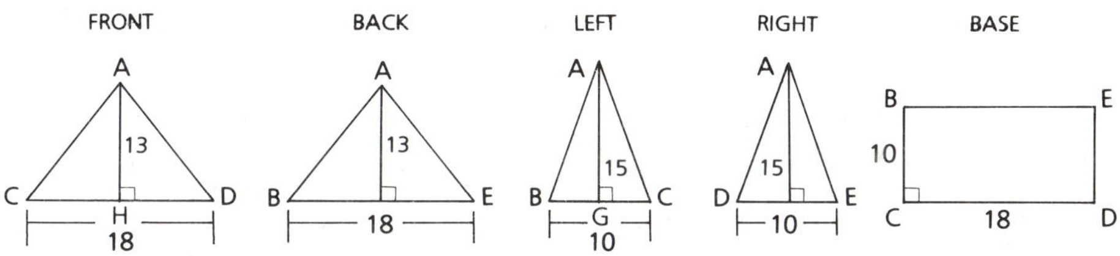

$$\tfrac{1}{2}(18)(13) + \tfrac{1}{2}(18)(13) + \tfrac{1}{2}(10)(15) + \tfrac{1}{2}(10)(15) + (18)(10)$$

$$117 + 117 + 75 + 75 + 180$$

Total Surface Area = 564

Section 11.8(B) Problem Set A

1 Given a regular pyramid with a square base as shown, find:

a the area of each lateral face

b the lateral area

c the total area.

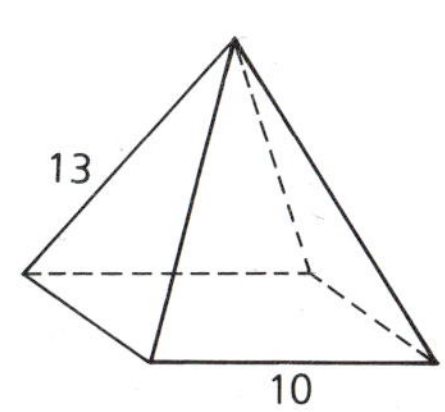

2 Given a regular pyramid with a triangular base, what is:

a the area of each lateral face?

b the area of the base?

c the total area?

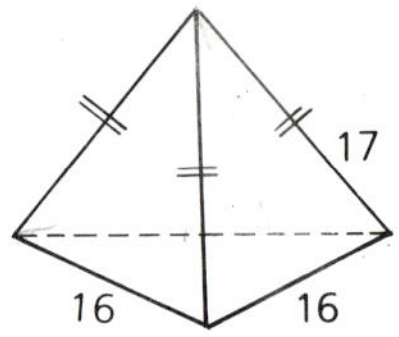

3 The pyramid shown has a rectangular base, and its lateral edges are congruent.

a Why is this pyramid *not* regular?

b What is the lateral area?

c What is the total area?

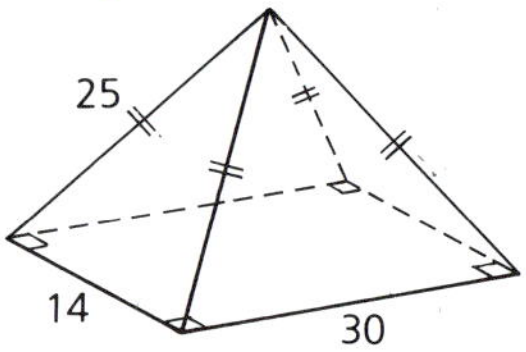

4 The solid at the right is a combination of a prism and a regular pyramid.

a Is ABCD a face of the solid?

b How many faces does this solid have?

c Find the total area.

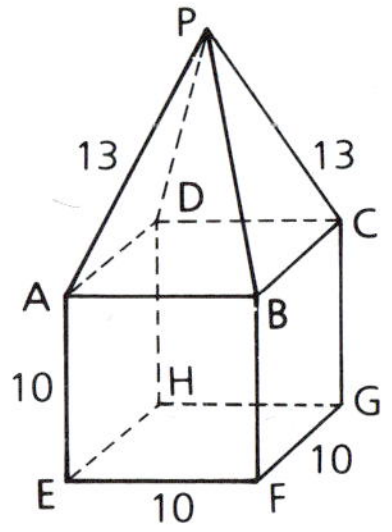

5 PRXYZ is a regular pyramid. The midpoints of the lateral edges are joined to form a square, ABCD. PR = 10, and RX = 12.

Find: a the lateral area of PRXYZ

b the lateral area of pyramid PABCD

c the area of square ABCD

d the area of square RXYZ

e the ratio of the area, ABCD to RXYZ

f the area of trapezoid ABXR.

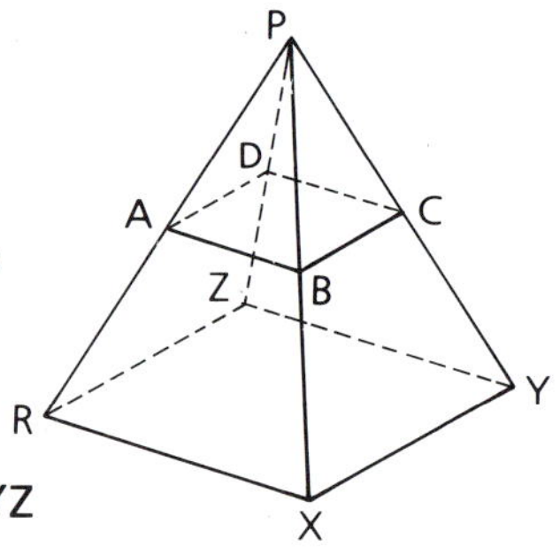

Section 11.8 (B) Problem Set B

6 A regular pyramid has a slant height of 8. The area of its square base is 25. Find the total area.

7 A regular pyramid has a slant height of 12 and a lateral edge of 15.

Find: a the perimeter of the base

b the lateral area

c the area of the base

d the total area.

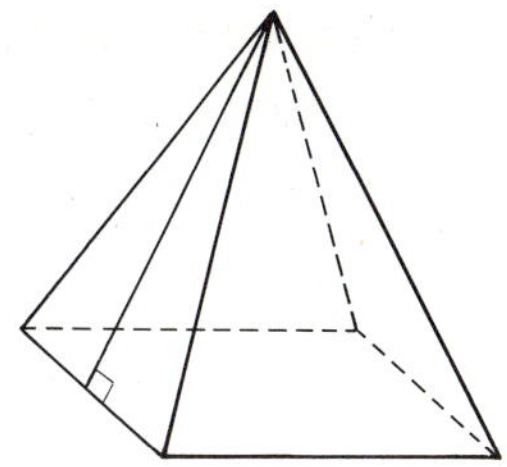

8 PABCD is a regular square pyramid.

a If the sides of the base are 14 each, and the altitude PQ is 24, find the lateral and total areas.

b If the slant heights are each 17 and the altitude PQ is 15, find the lateral and total areas.

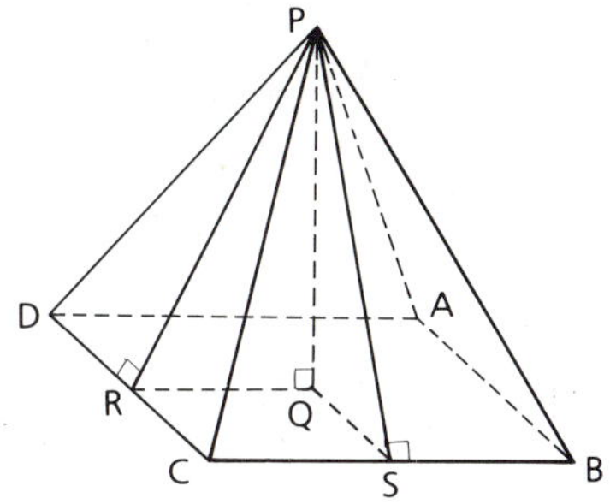

9 Suppose that the pyramid of problem 8 is not regular, but that the base is rectangular and the lateral edges are congruent.

a If altitude PQ = 8, CD = 12, and BC = 30,

find: 1 PR, the slant height of face PCD

2 PS, the slant height of face PBC

3 the lateral area and the total area.

b If the lateral edges are each 25, and the base is 24 by 30, find the height, PQ, of the pyramid.

Section 11.8 (B) Problem Set C

10 Each lateral edge of a regular square pyramid is 3 and the height of the pyramid is 1.

Find: a a diagonal of the base

b the slant height

c the area of the base

d the lateral area.

11 A *regular tetrahedron* ("four faces") is a pyramid, all four of whose faces are equilateral triangles.

Find: **a** the total surface area if an edge is 6

b the height of the pyramid.

12 A *regular octahedron* is a solid with 8 faces, and each face is an equilateral triangle. If each edge is 6 mm, find:

a the total surface area

b the distance from C to E

c the distance from A to B

d the shape of quadrilateral ACBE.

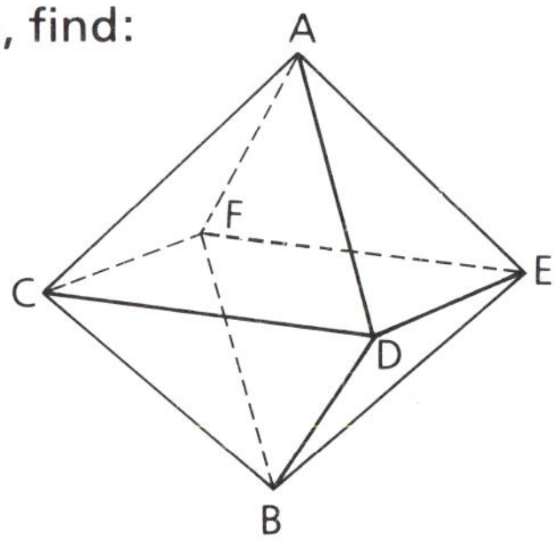

13 A *regular hexahedron* is a solid that does not have triangular faces. What is the common name for a regular hexahedron?

11.8 (C) Surface Area of Circular Solids

Consider the following three solids that are based on the circle:

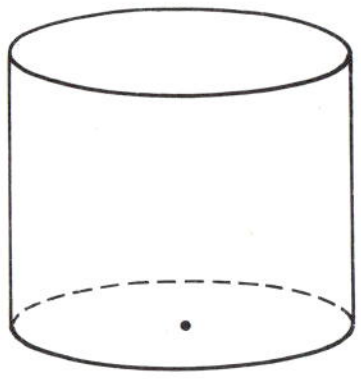
Cylinder

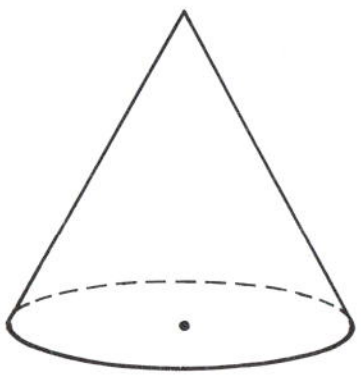
Cone

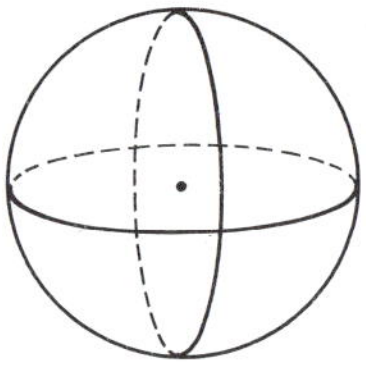
Sphere

A **CYLINDER** resembles a prism, because it has two congruent parallel bases. The bases, however, are circles. In this text, "cylinder" will mean only a right circular cylinder; that is, the line containing the centers of the bases is perpendicular to each base.

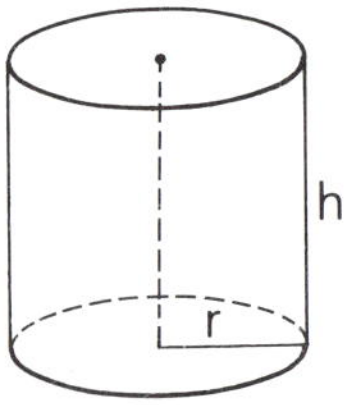

Cylinder

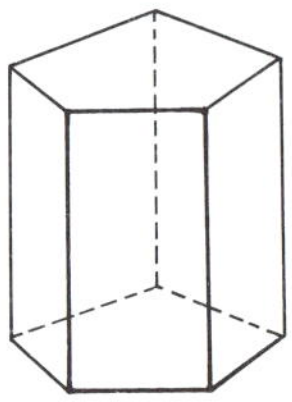
Prism

The lateral area can be determined by thinking of a cylinder as a "can" and the lateral area as the "label" on the can. If we cut the label and spread it out, we see that it is a rectangle. The height of the rectangle is the same as the height of the can. The base of the rectangle is the circumference of the can.

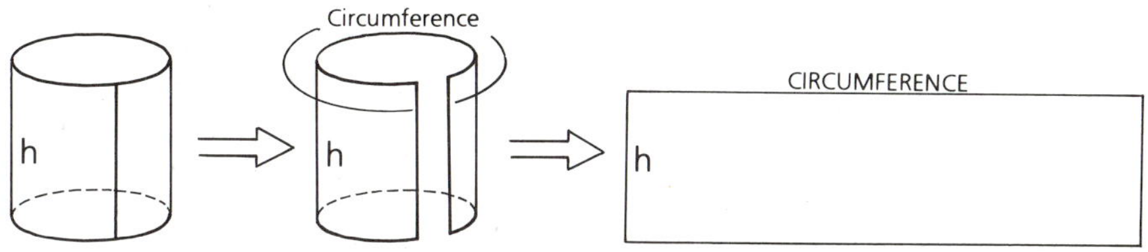

Theorem 101

T **The LATERAL AREA OF A CYLINDER is the product of the height by the circumference of the base.**

$$\text{L.A.} = C \cdot h$$
$$\text{L.A.} = 2\pi r \cdot h$$

C = circumference
h = height
r = radius

D **The TOTAL AREA OF A CYLINDER is the lateral area and the areas of the two bases.**

$$\text{T.A.} = \text{L.A.} + A_{Base1} + A_{Base2}$$

A CONE resembles a pyramid except that its base is a circle. In a pyramid the slant height and the lateral edge are different, while in a cone they are the same.

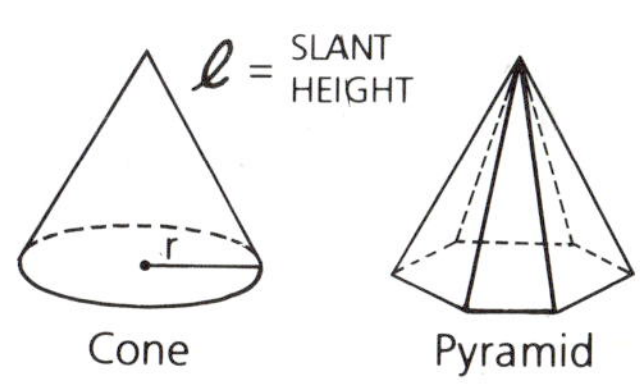

In this text the word "CONE" will mean right circular cone only. That is, the altitude passes through the center of the circular base.

Theorem 102

T **The LATERAL AREA OF A RIGHT CONE is one-half the product of the slant height by the circumference of the base.**

$$\text{L.A.} = \tfrac{1}{2} C \cdot \ell = \pi r \ell$$

C = circumference of base
r = radius of base
ℓ = slant height

D **The TOTAL AREA OF A CONE is the sum of the lateral area and the area of the base.**

$$\text{T.A.}_{\text{cone}} = \text{L.A.} + A_{Base}$$

A **SPHERE** is a special figure with a special surface area formula. (There are no lateral edges and there is no lateral area.) The proof of the formula requires the concept of limits and will not be given here.

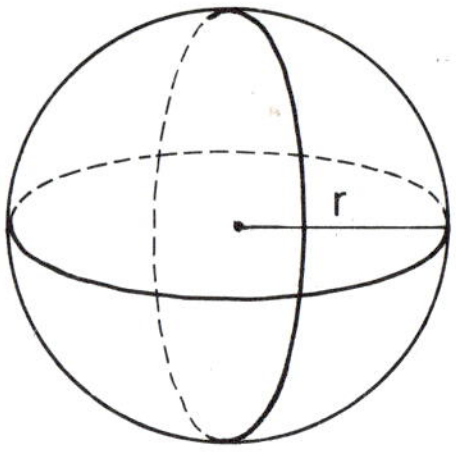

P **Total Area of a Sphere** $= 4\pi r^2$

Section 11.8 (C) Sample Problem

Find the total area of each figure below:

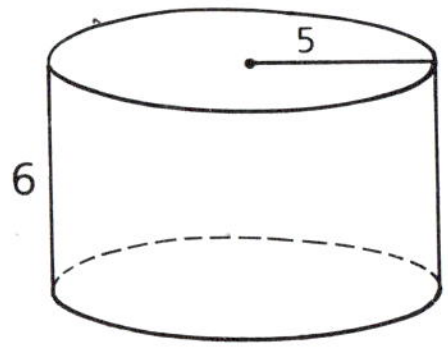

a CYLINDER

T.A. = L.A. + $\text{Base}_1 + \text{Base}_2$

$= 2\pi rh + \pi r^2 + \pi r^2$

$= 2\pi\,(5)\,(6) + \pi(5)^2 + \pi(5)^2 = 110\pi$

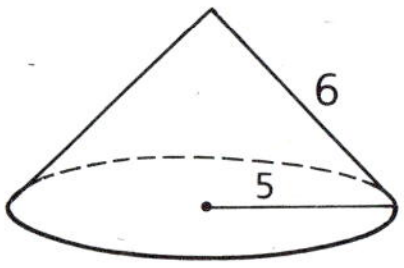

b CONE

T.A. = L.A. + Base

$= \pi r \ell + \pi r^2$

T.A. $= \pi(5)\,(6) + \pi(5)^2 = 55\pi$

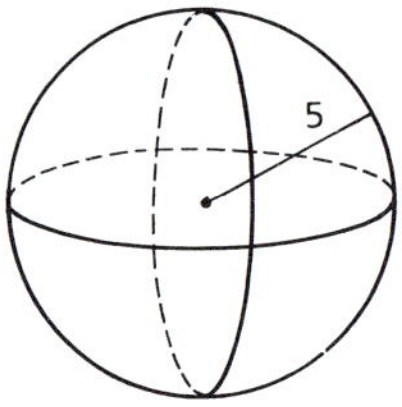

c SPHERE

T.A. $= 4\pi r^2 = 4\pi(5)^2 = 4\pi(25) = 100\pi$

Section 11.8 (C) Problem Set A

1 Find the total area of a sphere with a;

a radius of 7 **c** diameter of 6

b radius of 3 **d** diameter of 5

2 For each figure, find the lateral area and the total area.

a

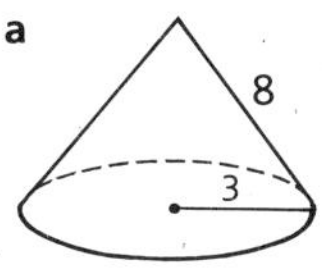

b

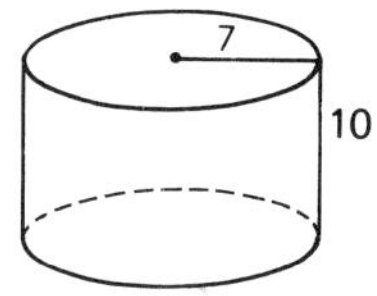

c

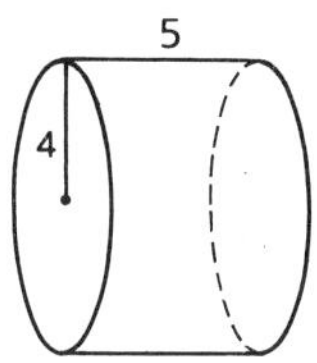

d

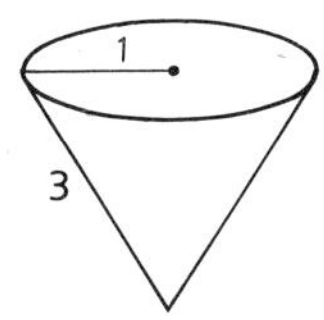

3 Find the radius of a sphere whose surface area is 144π.

4 Find the total area of each solid.

(Be sure you include only outside surfaces and be certain you do not miss any!)

a

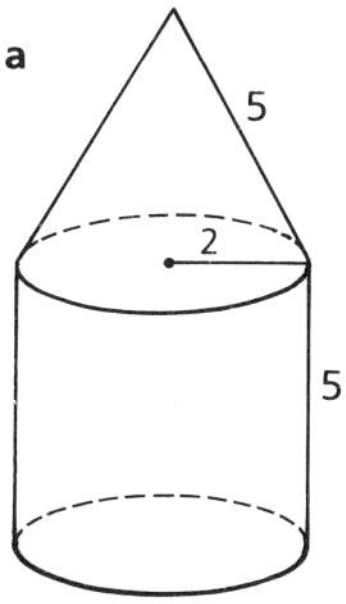

b

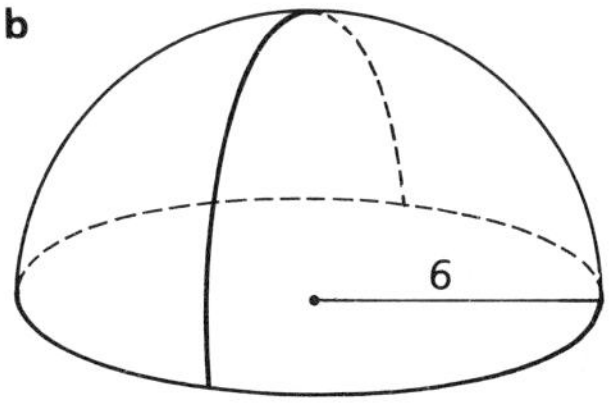

This is a hemisphere ("half sphere").
The T.A. includes the area of the circular base

Section 11.8 (C) Problem Set B

5 Find the total surface area, including the flat sides, of a half cylinder of radius 5 and lateral edge 2.

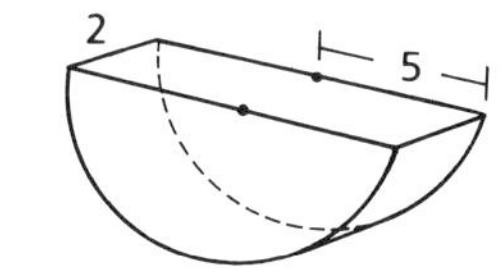

6 Find the total area of each solid.

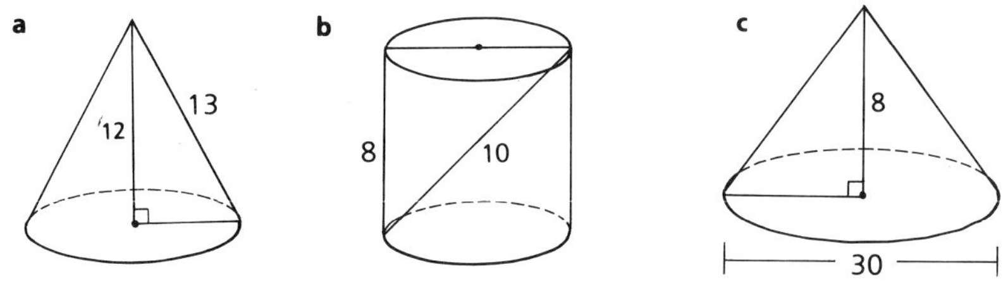

7 Find the number of liter cans of paint needed to paint a tower, as shown, that is *10 meters in total height.* The paint in a one-liter can will cover 10 square meters of area. (Use $\pi \approx 3.14$.) (First find the total painted area.)

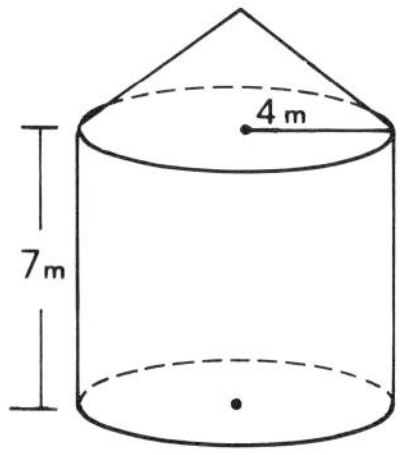

8 What size label (length and width) will just fit on a can 8 cm in diameter and 14 cm high?

9 Find the total area of the solid.

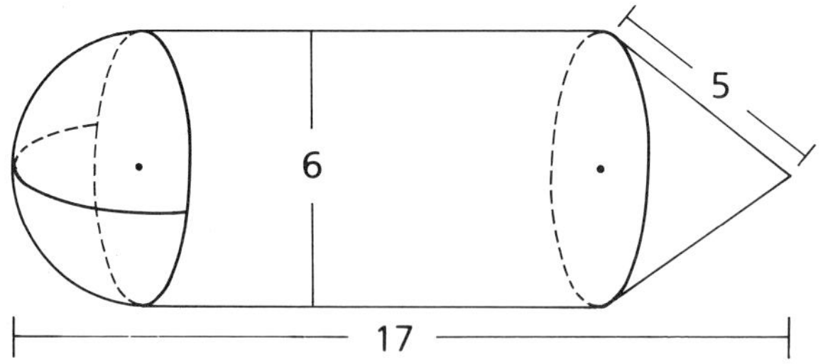

Section 11.8 (C) Problem Set C

10 Find the total surface area of the solid shown, including the surface inside the hole.

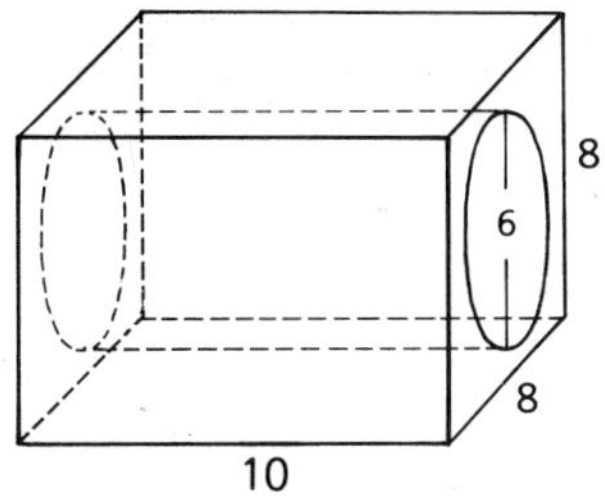

11 The solid at the right is a portion of a cone called the *frustum.* Find its total area if the radii of the top and bottom circles are 4 and 8 respectively, and the slant height is 5.

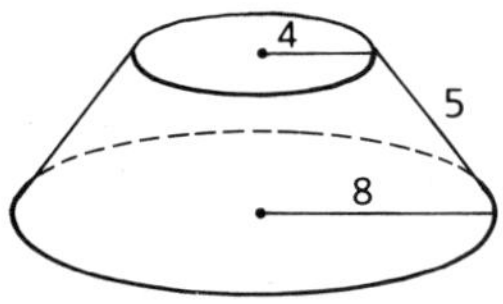

12 A *surface of rotation* is generated by revolving a shape about a fixed line called the *axis of rotation.*

For example:

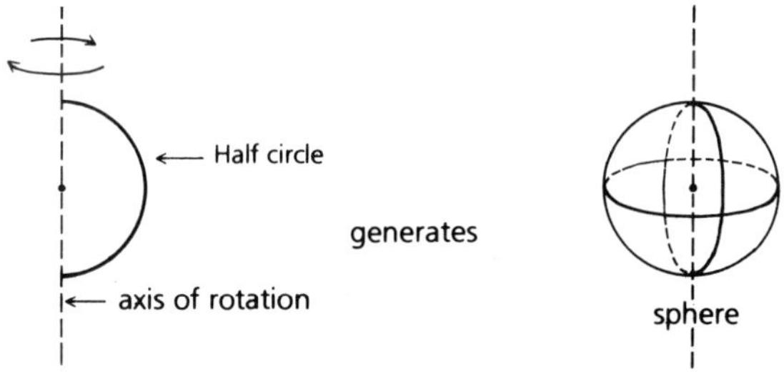

Find each surface of rotation below and compute the total surface area of the solid generated.

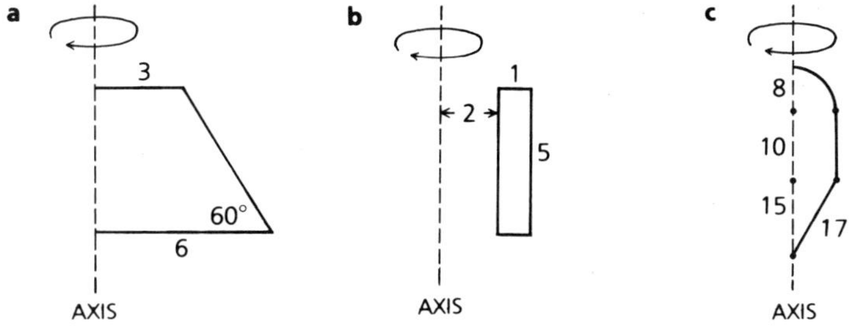

Section 11.9 Hero and Brahmagupta

A useful formula for finding the area of a triangle was derived about 2000 years ago by the mathematician, Hero of Alexandria.

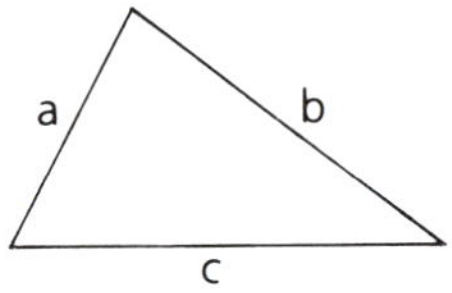

HERO'S FORMULA: $A_{\triangle} = \sqrt{s(s-a)(s-b)(s-c)}$,
where a, b, c are the lengths of the sides of the triangle and s = semiperimeter = $\frac{a+b+c}{2}$.

A Hindu mathematician, Brahmagupta, in about A.D. 628 recorded a formula for the area of an inscribed quadrilateral. This formula holds only for quadrilaterals that can be inscribed in a circle (a quadrilateral with *concyclic* vertices).

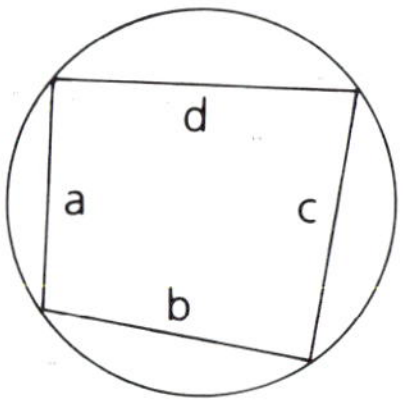

BRAHMAGUPTA'S FORMULA: $A_{\text{inscribed quad.}} = \sqrt{(s-a)(s-b)(s-c)(s-d)}$,
where a, b, c, d are the sides of an inscribed quadrilateral and s = semiperimeter = $\frac{a+b+c+d}{2}$.

Section 11.9 Sample Problems

1 Find the area of a triangle with sides 3, 6, and 7.

Solution: Using Hero's Formula

$A_{\triangle} = \sqrt{s(s-a)(s-b)(s-c)}$, $\quad s = \frac{a+b+c}{2} = \frac{3+6+7}{2} = 8$

Thus, $A = \sqrt{8(8-3)(8-6)(8-7)}$
$= \sqrt{8(5)(2)(1)}$
$= \sqrt{16 \cdot 5} = 4\sqrt{5}$

2 Find the area of the inscribed quadrilateral with sides 2, 7, 6, and 9.

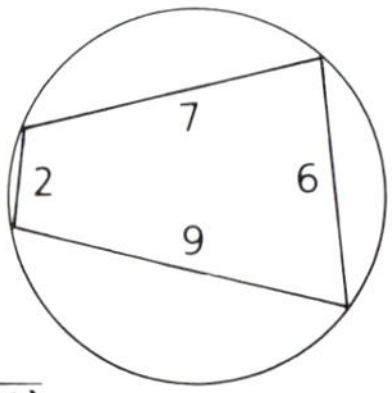

Solution: Using Brahmagupta's Formula

$A_{\text{inscribed quad.}} = \sqrt{(s-a)(s-b)(s-c)(s-d)}$,

$s = \frac{2+7+6+9}{2} = 12$

Thus, $A = \sqrt{(12-2)(12-7)(12-6)(12-9)}$
$\sqrt{10 \cdot 5 \cdot 6 \cdot 3} = \sqrt{900} = 30$

The Royal Road: Hero's Formula

Hero's Formula (page 536) expresses the area of a triangle in terms of the lengths of its sides:

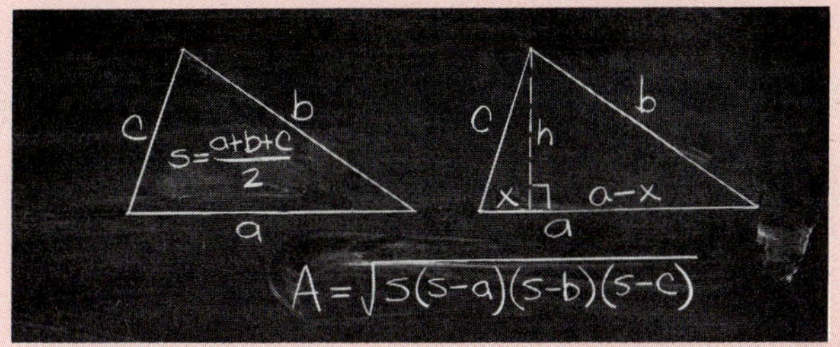

Can you derive the formula? First, use the Pythagorean Theorem to establish that $h^2 = c^2 - x^2 = b^2 - (a - x)^2$ and thus:

$$x = \frac{a^2 + c^2 - b^2}{2a}$$

Substitute that value for x into $h^2 = c^2 - x^2 = (c - x)(c + x)$ and show:

$$h^2 = \frac{(b - a + c)(b + a - c)(a + c - b)(a + c + b)}{4a^2}$$

Use $2s = a + b + c$ and show that $4a^2h^2 = 16s(s - a)(s - b)(s - c)$. Finally, find an expression for $\frac{1}{2}ah$, the area of the triangle.

Section 11.9 Problem Set A

1 Use Hero's Formula to find the area of each triangle whose sides are given below.

a 3, 4, 5 **c** 5, 6, 9 **e** 8, 15, 17

b 3, 3, 4 **d** 3, 7, 8 **f** 13, 14, 15

2 Use Hero's Formula to find the area of an equilateral triangle with 8 units to a side.

3 Use Brahmagupta's Formula to find the area of an inscribed quadrilateral whose sides are:

a 5, 7, 4, 10 **b** 2, 4, 5, 9 **c** 3, 5, 9, 5 **d** 1, 5, 9, 11

Section 11.9 Problem Set B

4 Use Hero's Formula to find the area of a triangle with these sides:

a 2, 5, and 7 b 4, 6, and 12

What explanation can you give for the results in parts a and b?

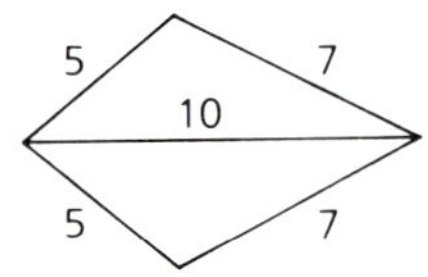

5 Find the area of the figure.

6 Find the area of each.

a

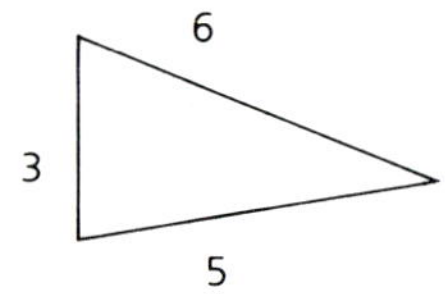

b

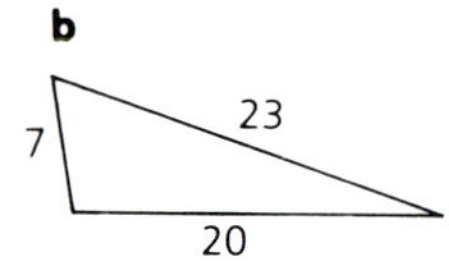

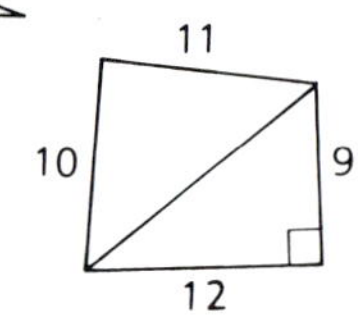

7 Verify that the area of this quadrilateral is $12\sqrt{21} + 54$.

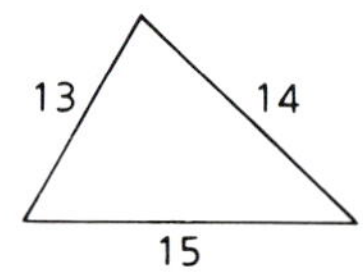

8 Find the three altitudes of the triangle at the right. (Hint: Use Hero's Formula to find the area, and then use $A = \frac{1}{2}bh$ to find each altitude.)

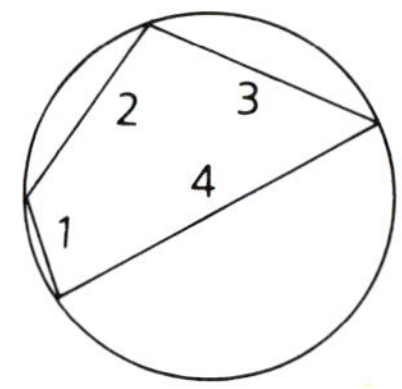

9 Find the area of the quadrilateral.

Section 11.9 Problem Set C

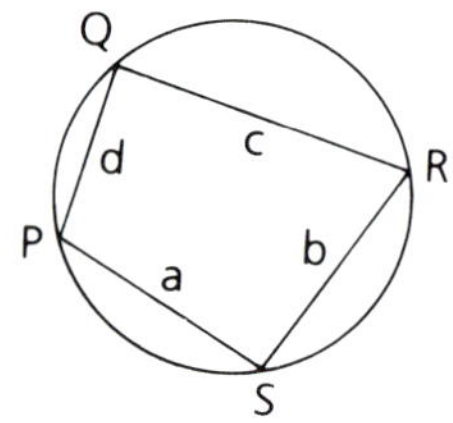

10 a As $\overline{PQ}$ gets smaller and smaller, what happens to the figure PQRS?

b What happens to Brahmagupta's Formula if P and Q become the same point?

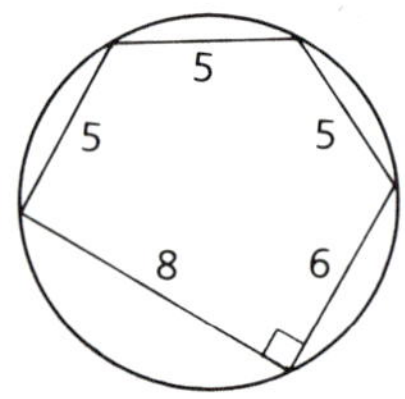

11 Find the area of the pentagon.

Chapter 11 Study Guide

As you study and review this chapter, be sure that you are able to:

1 Use the following vocabulary:

Polygon: median of a trapezoid (p. 494), regular polygon—apothem, radius (p. 502)

Circle: sector, segment of (p. 508)

Surfaces of a solid: lateral area (p. 521), total area (p. 522), sphere (p. 532). cylinder (p. 531), cone (p. 531)

Polyhedron: faces, edges (p. 521)

Prism: bases (how related?), lateral edges (how related?), lateral faces (what figure?), names of prisms, altitude (pp. 521–522)

Right Prism: lateral faces (what figure?), lateral edges (relation to bases?) (p. 522)

Pyramid: vertex, altitude, lateral faces (what figure?), slant height, names of pyramids (pp. 525–526)

Regular Pyramid: slant height, base, altitude (where is its foot?), lateral faces (what figure?) (pp. 525–526)

2 Use similarity in problems that involve ratios of areas. (p. 515)

a Describe what to do when figures are not similar. (p. 514)
b State the ratio into which a median divides the area of a triangle. (p. 516)

3 Use the "Divide and Conquer" method. (p. 489)

4 Substitute correctly in these area formulas:

Basic Area Formulas
rectangle (p. 482)
parallelogram (p. 487)
triangle (p. 488)
circle (p. 508)
surface of cylinder (p. 531)
surface of cone (p. 531)
surface of sphere (p. 532)
Hero's formula (p. 536)
Brahmagupta's formula (p. 536)

Formulas that Save Time
kite, rhombus, or square (p. 499)
equilateral triangle (p. 502)
regular polygon (p. 503)
trapezoid (pp. 494 and 495)

Combinations of Other Areas
surface of a prism (p. 522)
surface of a pyramid (p. 526)
segment of a circle (p. 509)

Chapter 11 Review Problems A

1 Find the areas of these quadrilaterals:

a a rectangle with base 12 and height 7

b a triangle with base 12 and height 7

c a parallelogram with base 15 and height 5

d a trapezoid with bases 3 and 10, and height 8

e a kite with diagonals 5 and 8

f a trapezoid with median 4 and height 2.

2 Find the areas of these rhombuses.

a The base is 9 and height is 7.

b The diagonals are 6 and 11.

3 Find the area of each shaded region.

a
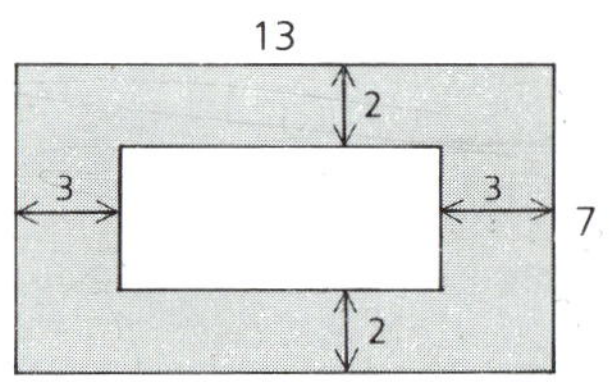

b
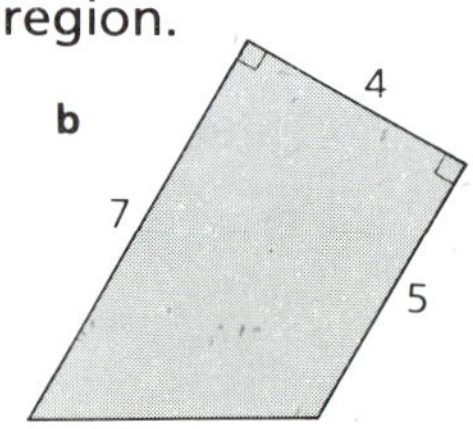

c
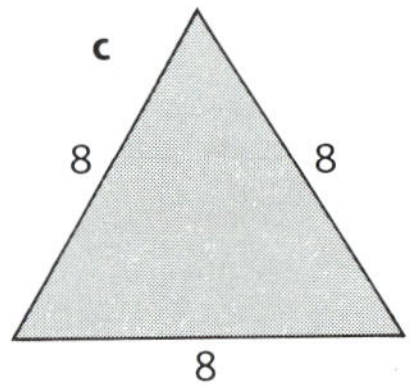

4 A rectangular driveway is to be paved. The driveway is 20 meters long and 4 meters wide. The cost will be $15 per square meter. What is the total cost to pave the driveway?

5 Find the area of a parallelogram with sides 12 and 8 and included angle 60°.

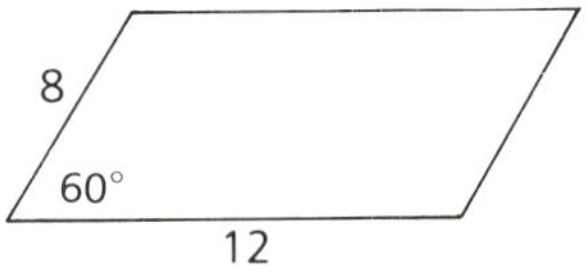

6 Find the area of an isosceles trapezoid with sides 8, 20, 40, and 20.

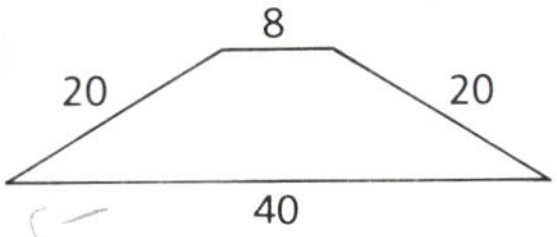

7 Find the area of the triangle shown at the right.

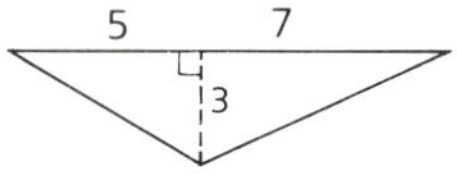

8 John has two sticks 90 cm and 50 cm long to use in making a kite. What will be the cost of the kite, if the sticks and glue are gifts, and the material costs 3c per sq decimeter?

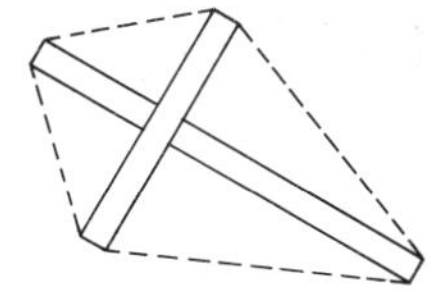

9 The apothem of a regular polygon is 7 and the perimeter is 56. Find the area.

10 Find the area of a circle if the circumference is 16π.

11 Find the area of a square whose semiperimeter is 18 meters.

12 Find the area of a semicircle whose diameter is 14 millimeters.

13 Find the area of each shaded region.

a

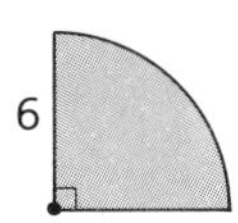

b

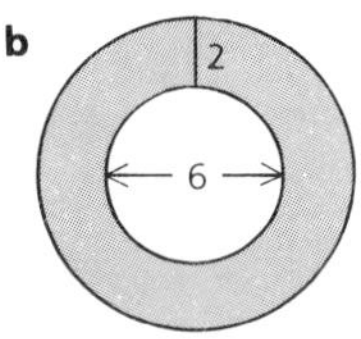

c

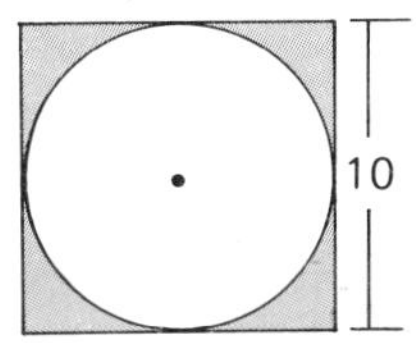

14 Find the area of each sector.

a

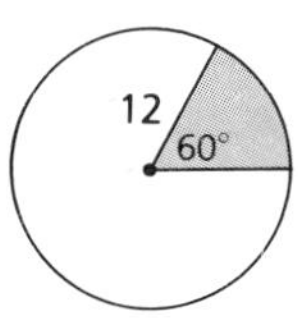

b

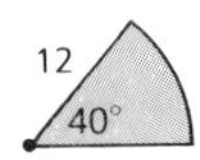

c

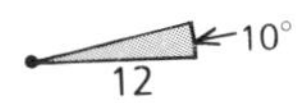

15 Find the ratio of the areas of each pair of figures.

a

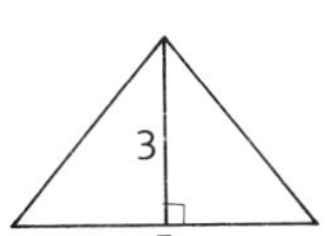

:

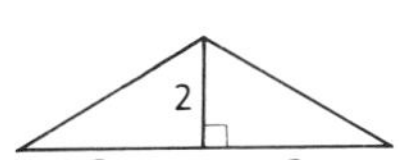

b

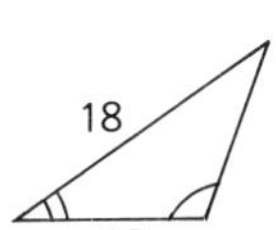

:

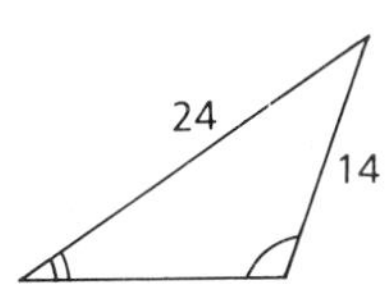

16 Find the lateral area and total area.

a

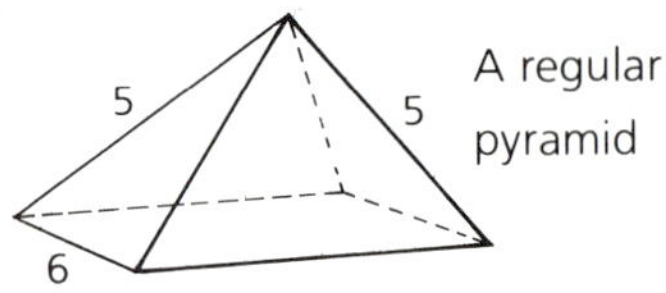

b

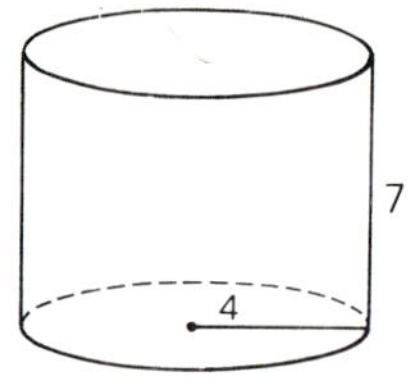

Chapter 11 Review Problems B

17 Find the area of a triangle with sides 41, 41, and 18.

18 Find the area of a parallelogram with sides 6 and 7 and included angle 45°.

19 Find the area of a rhombus whose perimeter is 52 and longer diagonal is 24.

20 Find the area of an equilateral triangle with perimeter 21.

21 Find the area and perimeter of an isosceles trapezoid with lower base 18, upper base 4, and upper base angle 120°.

22 Find the circumference of a circle whose area is: **a** 24π **b** 49

23 **a** The diagonal of a square is 26. Find the area.

b Find the diagonal of a square whose area is 18.

24 Find the area of a regular hexagon whose span is 36.

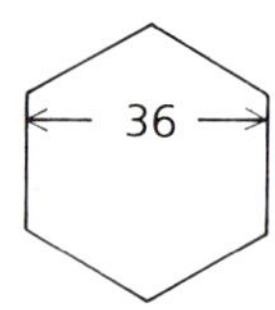

25 Find the area of each shaded region.

a

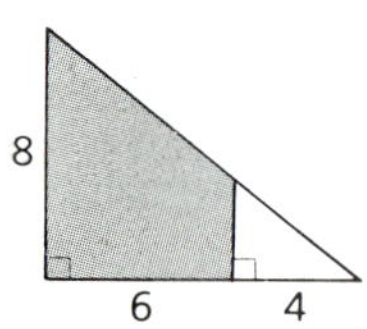

b

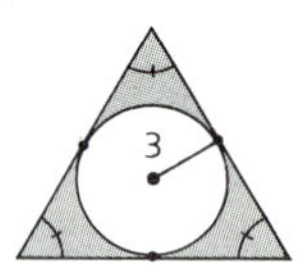

c

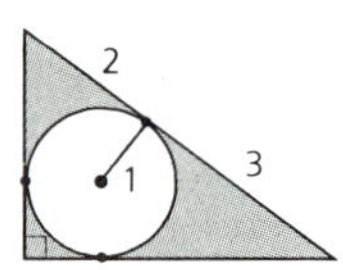

26 Find the ratio of the areas of the whole figure to the shaded region:

a

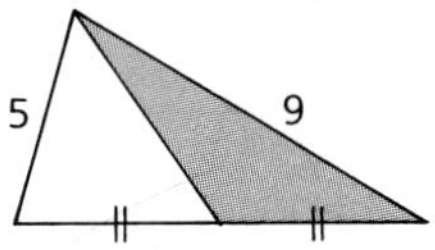

b

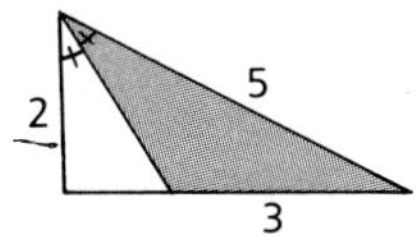

c

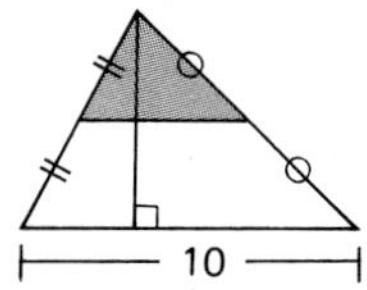

27 Find the area of each segment of a circle shown below:

a

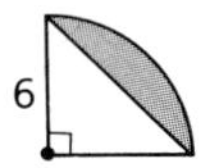

b

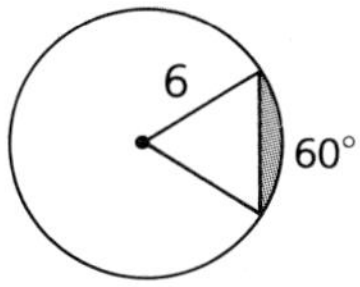

28 Find the ratio of the areas of the two regions I and II:

a

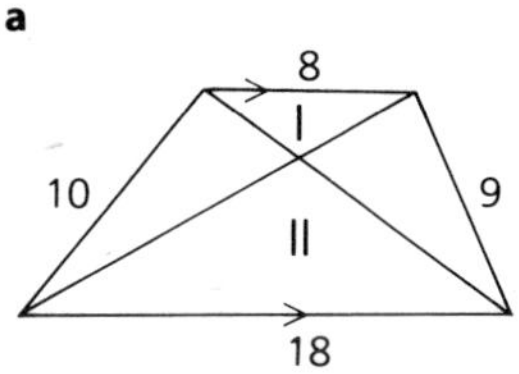

b

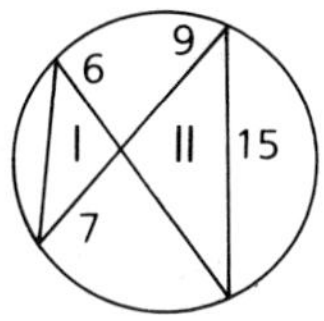

c

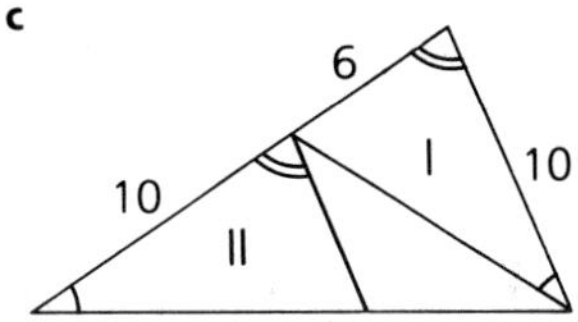

29 Find the lateral area and the total area of these solids.

a

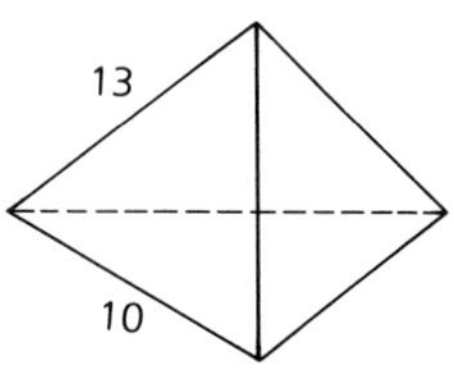

The base is equilateral.

b

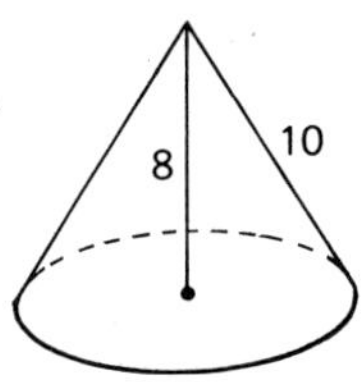

c

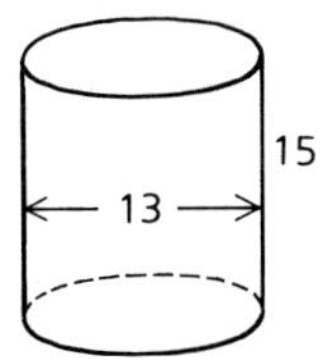

30 Find the total surface area of each solid. (Don't forget the flat faces.)

a

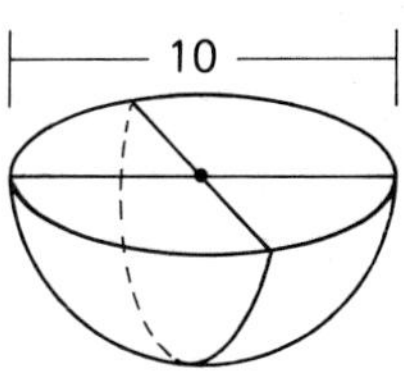

b

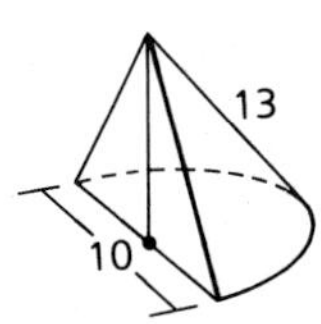

Chapter 11 Review Problems C

31 Find the ratio of the areas of region I to region II.

a

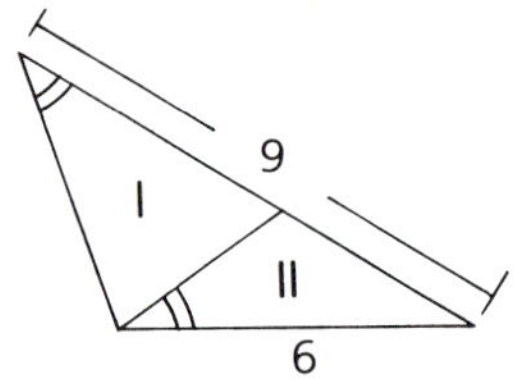

b

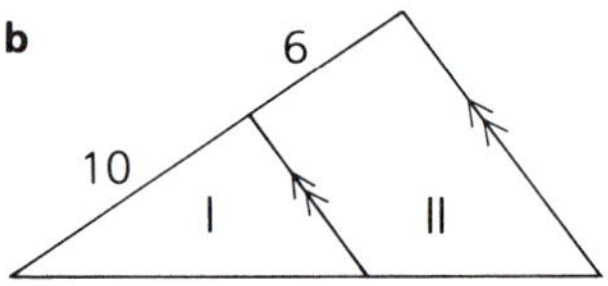

32 Given an isosceles trapezoid with a smaller base of 2 and a perimeter of 70. If the acute base angles are 60°, find the area.

33 A cylinder is cut into 4 equal parts. Find the total area of the part shown.

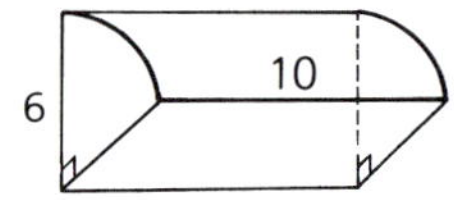

34 Given five circles tangent as shown. If each small circle has radius 3, find the shaded area.

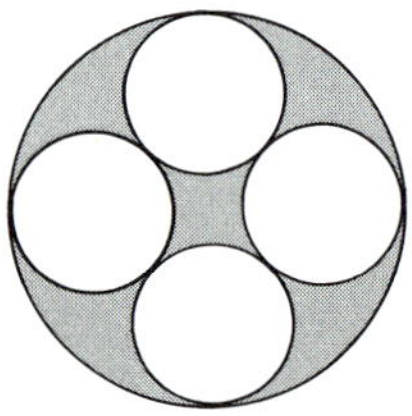

35 Find the shaded areas.

a

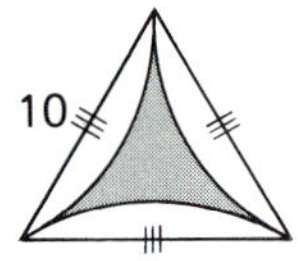

arc radius = 10

b

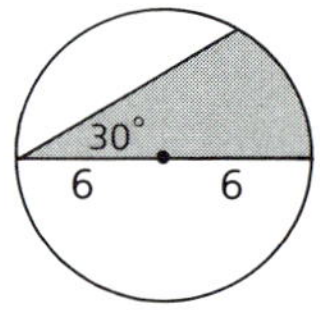

36 Find the area of a trapezoid whose diagonals are each 30 and whose height is 18.

37 AB:BC = 1:1. FE:ED = 1:5.

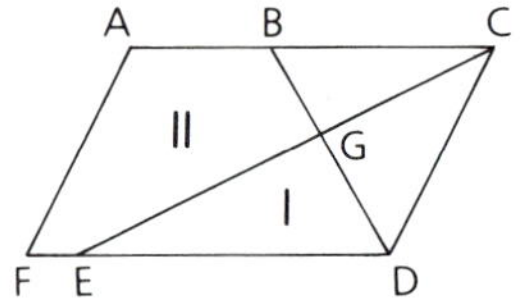

a Find the ratio of the area of Region I to the area of ▱ACDF.

b What is the probability that if a gnat lands in ▱ACDF, it lands in Region II?

38 $\overline{AT}$ is tangent to $\odot$ P at T. AB = 12.

Find the shaded area.

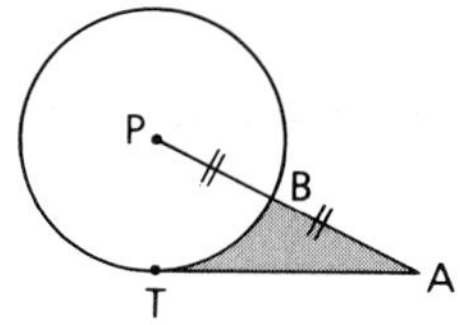

39 Archibold left his horse, Gremilda, tied to the corner of the barn by a 12 meter line. The barn is 8 meters by 10 meters. Find the total grazing area left for Gremilda.

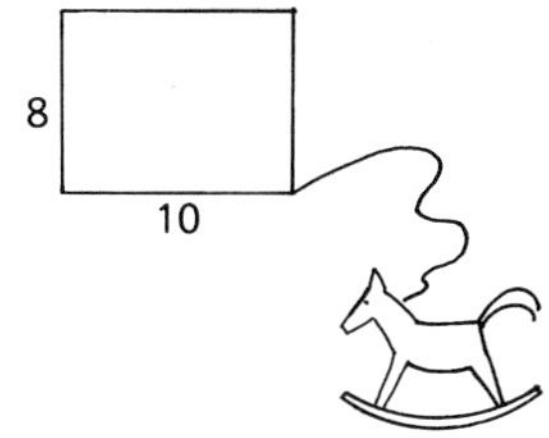

40 The cross has all 12 sides congruent and equal to 4. All angles are right angles.

Find the area of the region within the circle and outside the cross.

41 Find the area of the shaded region.

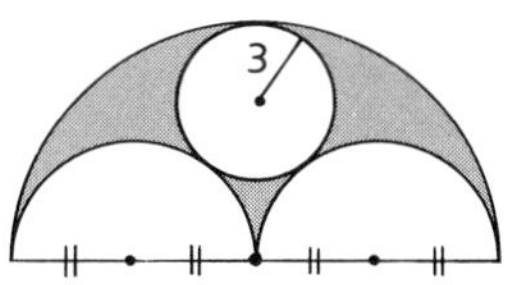

42 Find the area of a trapezoid with sides 12, 17, 40, and 25. The bases are 12 and 40.

43 Buster Bee lives on a honeycomb.

What percentage of the honeycomb is made of wax?

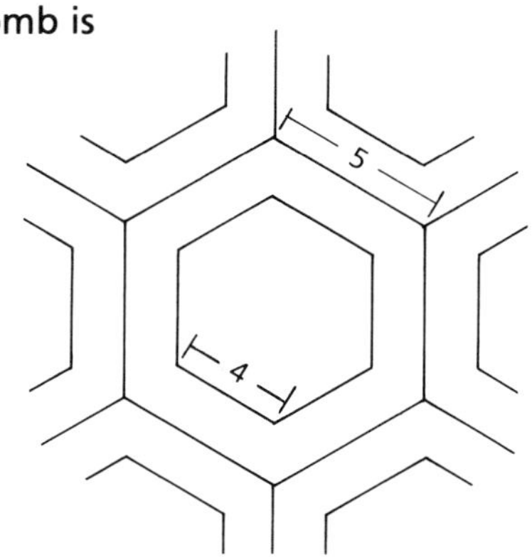

Chapter 12

VOLUMES

Section 12.1 Prisms and Cylinders

In this section the following topics are discussed:

A VOLUME OF A RIGHT RECTANGULAR PRISM
B VOLUME OF A PRISM
C VOLUME OF A CIRCULAR CYLINDER
D CROSS SECTION OF A PRISM OR A CYLINDER

12.1 (A) Volume of a Right Rectangular Prism

The measure of the interior of a solid is called **VOLUME**. Its definition is much like that of area, the measure of the interior of a closed coplanar figure.

D **The VOLUME OF A SOLID is the number of cubic units contained by the solid.**

A cubic unit is the volume of a cube with edges one unit long. A cube is a right rectangular prism with congruent edges, so all faces are squares. In Chapter 11 we used the word "box" for a right prism. Thus, a right rectangular prism is also a rectangular box.

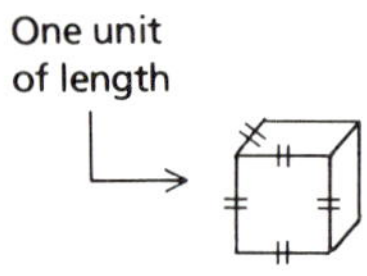

One cubic unit

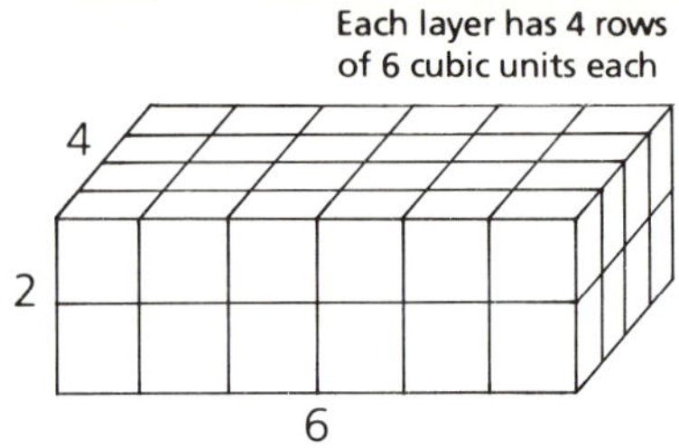

Rectangular box

The rectangular box above contains 48 cubic units. The formula given below not only counts cubic units rapidly, but works with fractional dimensions.

P **The VOLUME OF A RIGHT RECTANGULAR PRISM equals the product of the *length*, *width*, and *height*.**

$$V_{\text{rect. box}} = \ell \cdot w \cdot h$$

ℓ = length
w = width
h = height

Another way to think of the volume of a rectangular prism is to imagine it is a stack of congruent rectangular sheets of paper. The area of each sheet is $\ell \cdot w$ and the height of the stack is h. Since the base of the prism matches each sheet, there is a second formula for the volume of a rectangular box.

$$\begin{aligned} V &= \ell \cdot w \cdot h \\ &= (\ell \cdot w) \cdot h \\ &= (\text{area of sheet}) \cdot h \end{aligned}$$

Theorem 103

T **The VOLUME OF A RIGHT RECTANGULAR PRISM equals the *height* times the *area* of the base.**

$$V_{\text{rect. box}} = \mathcal{B} \cdot h$$

$\mathcal{B}$ = area of base
h = height

12.1 (B) Volume of a Prism

In the same way, you may compute the volume of any PRISM. Assume that each sheet in the stack has the same shape and size as the base.

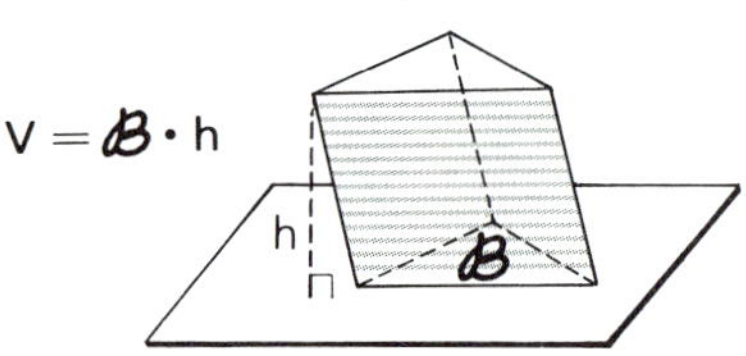

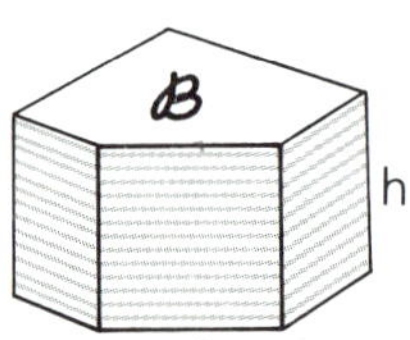

$V = \mathcal{B} \cdot h$

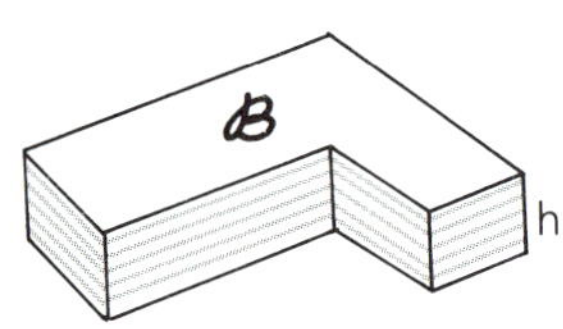

$V = \mathcal{B} \cdot h$

Theorem 104

T **The VOLUME OF ANY PRISM equals the *height* times the *area* of the base.** (Note: the height of a *right prism,* or box, is also its lateral edge.)

$$V_{prism} = \mathcal{B} \cdot h$$

$\mathcal{B}$ = area of base
h = height

12.1 (C) Volume of a Circular Cylinder

A cylinder has the same stacking property as a prism, so the formula for a prism applies to a cylinder. Furthermore, since the base of a circular cylinder is a circle, there is a second, more popular formula:

Theorem 105

T **The VOLUME OF ANY CIRCULAR CYLINDER equals the product of its *base area* by its *height*.**

$$V_{cyl.} = \mathcal{B} \cdot h$$
$$= (\pi r^2)h$$

r = radius of base
h = height of cylinder
$\mathcal{B}$ = area of base

12.1 (D) Cross Section of a Prism or a Cylinder

In dealing with prisms and cylinders, we have stacked congruent sheets, so the area of any of the sheets can be substituted for $\mathcal{B}$. Such a sheet between the bases is called a CROSS SECTION. In general,

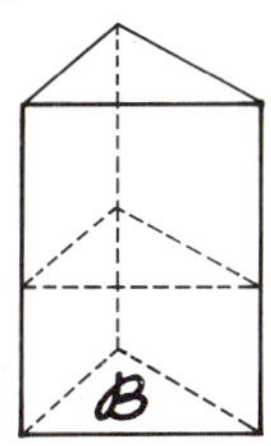

D **A CROSS SECTION is the intersection of a solid with a plane.**

We can now combine Theorems 104 and 105 with the help of the symbol ¢, which represents the area of a cross section parallel to the base.

Theorem 106

T **The VOLUME OF A PRISM OR CYLINDER equals the product of the *cross-section area* by the *height* of the solid.**

$$V_{prism\ or\ cyl.} = ¢ \cdot h$$

¢ = area of cross section
h = height of solid

Section 12.1 Sample Problems

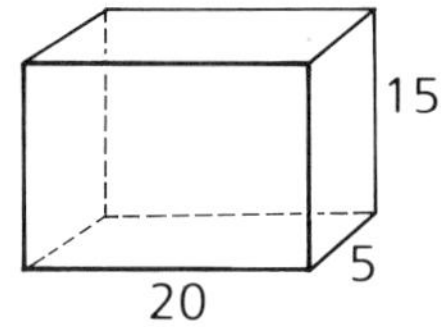

1 Find the volume of the rectangular prism.

Solution:

$V = \ell \cdot w \cdot h$ or $V = \mathcal{B} \cdot h$

$= 20 \cdot 5 \cdot 15$ Using a 5 by 20 face as a base,

$= 1500$ $V = (5 \cdot 20) \cdot 15 = 1500$

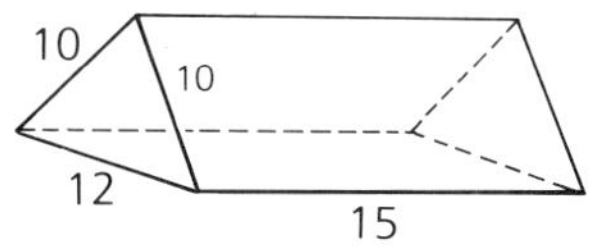

2 Find the volume of the triangular prism.

Solution: Notice that the base of this prism is this triangle.

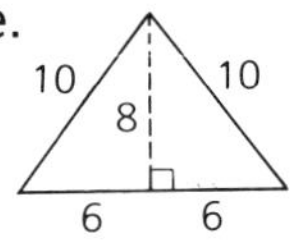

$A_{\triangle} = \frac{1}{2}(12)(8)$
$= 48$

Then, $V = \mathcal{B} \cdot h$
$= 48 \cdot 15$
$= 720$

3 Find the volume of a cylinder with a radius 3 and height 12.

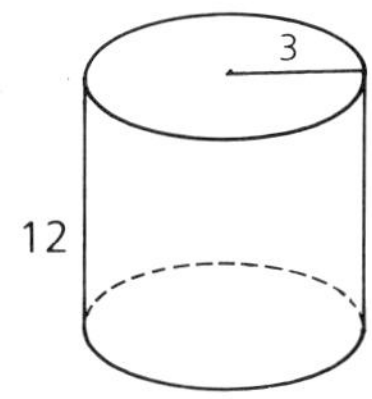

Solution: $V = \mathcal{B} \cdot h$

$V = \pi r^2 h$

$= \pi(3^2)(12)$

$= \pi(9)(12) = 108\pi$

4 Find the volume of the right prism shown. Take the left face as a representative cross section.

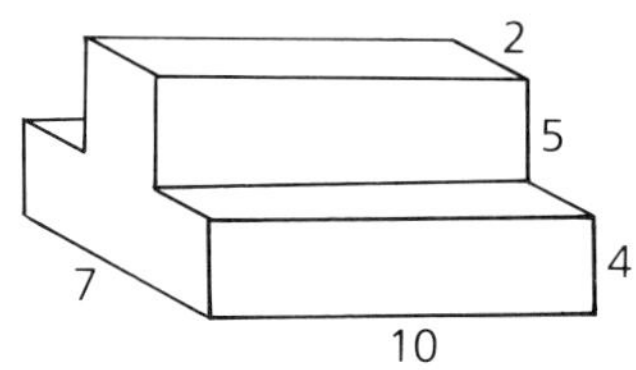

Solutions: Use either of two methods.

a Divide and Conquer

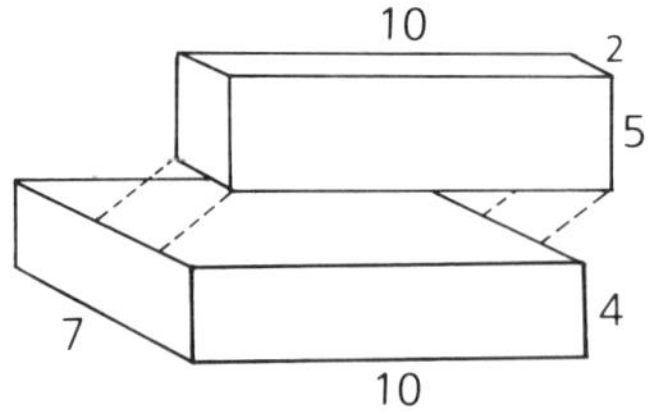

$V_{\text{top box}} = 2 \cdot 5 \cdot 10 = 100$

$V_{\text{bottom box}} = 7 \cdot 10 \cdot 4 = 280$

$V_{\text{solid}} = 280 + 100 = 380$

b Cross Section Times Height

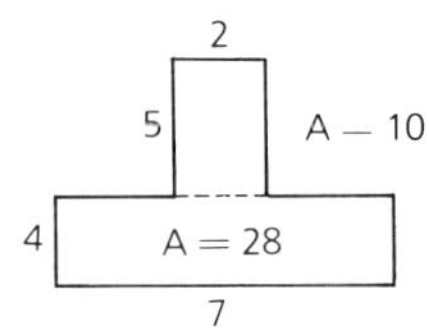

$¢ = 10 + 28 = 38$

$V = ¢ \cdot h$

$= (38)(10) = 380$

The Royal Road: 1:2:3

Archimedes used the following figure to summarize several of his mathematical discoveries:

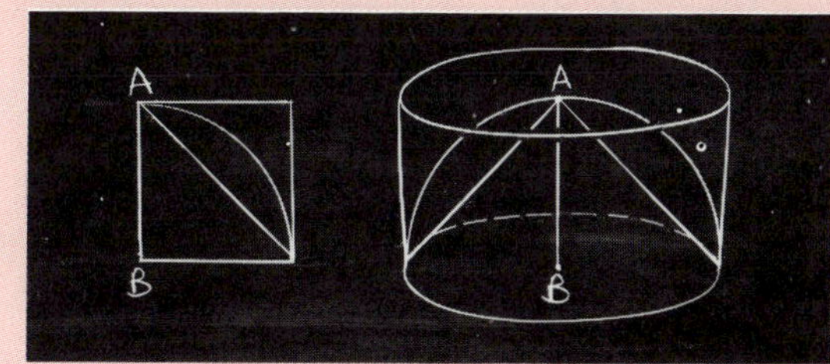

If you "spin" the figure around fixed $\overline{AB}$, you will have a cone inside a hemisphere inside a cylinder. What do you think is the ratio of the three volumes?

Section 12.1 Problem Set A

1 Find the volume of each solid below:

a

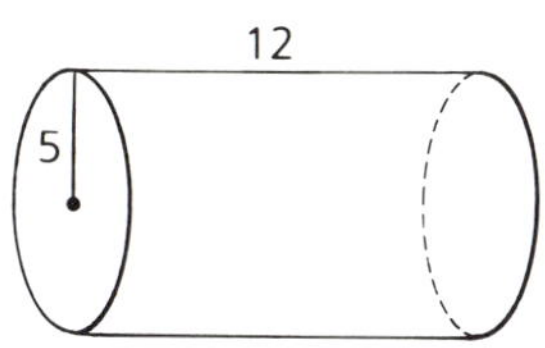

Right circular cylinder

b

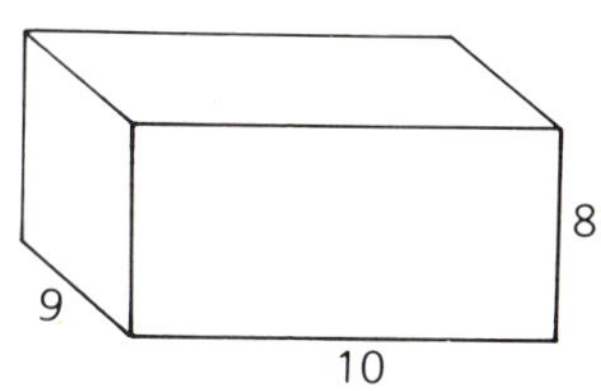

Rectangular solid

2 Find the volume of cement needed to form the concrete pedestal shown: (Leave in π form)

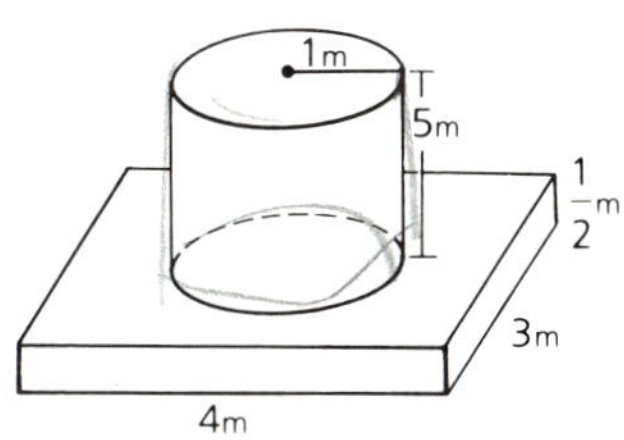

3 The area of the shaded face of the right pentagonal prism is 51. Find the volume.

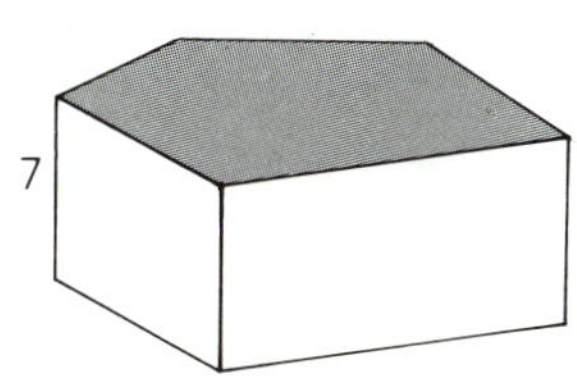

4 Find the volume and the total surface area of the rectangular box as marked.
(Note: Review Chapter 11 for Surface Area)

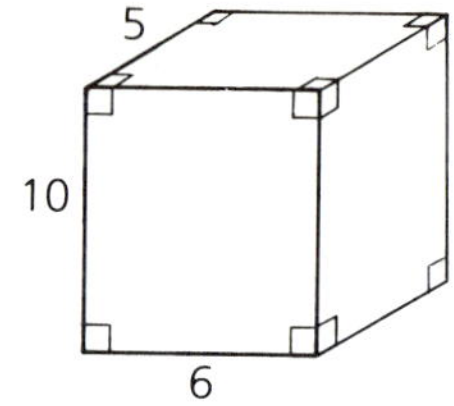

5 Find: a the volume of a cube with each edge 7.

b the volume of a cube with each edge e.

c the edge of a cube with volume 125.

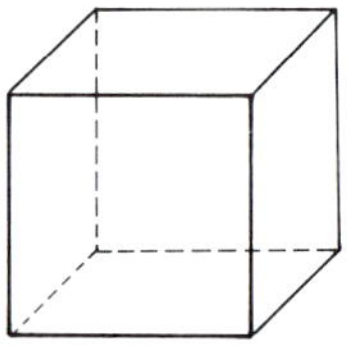

6 Find the lateral edge of a right prism with volume 286 and base area 13.

Section 12.1 Problem Set B

7 Find the volume and the total area.

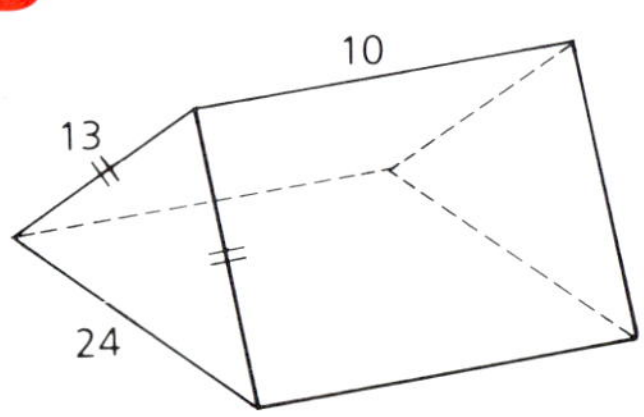

8 For each right cylindrical solid as marked, find the volume and the total area (including flat faces).

a

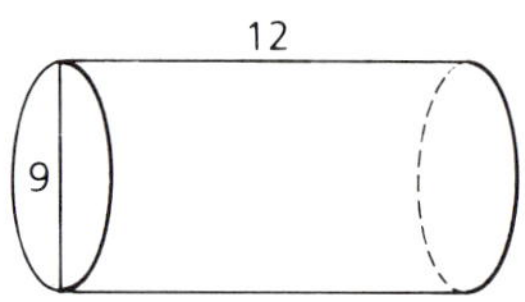

b

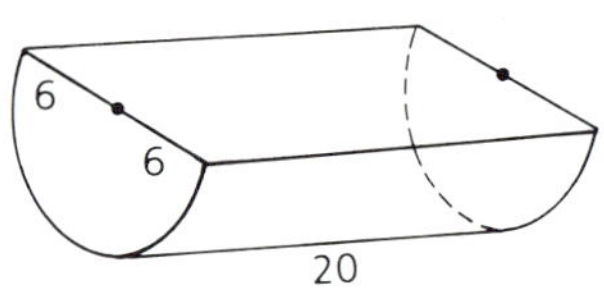

9 Find the volume and surface area of the regular hexagonal right prism.

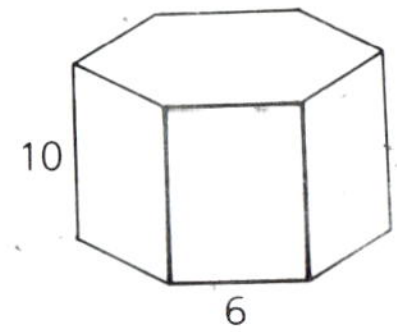

10 When Hilda computed the volume and surface area of a cube, each answer was the same number. Find the length of one side of the cube.

11 Find the volume of a cube in which a face diagonal is 10.

12 A rectangular cake pan has a base 10 cm by 12 cm and a height of 8 cm. If 810 cc of batter is poured into the pan, how far up the side will the batter come? (cc is the abbreviation for cubic centimeters.)

13 A rectangular container is to be formed by folding the cardboard along the dotted lines. Find the volume of this container.

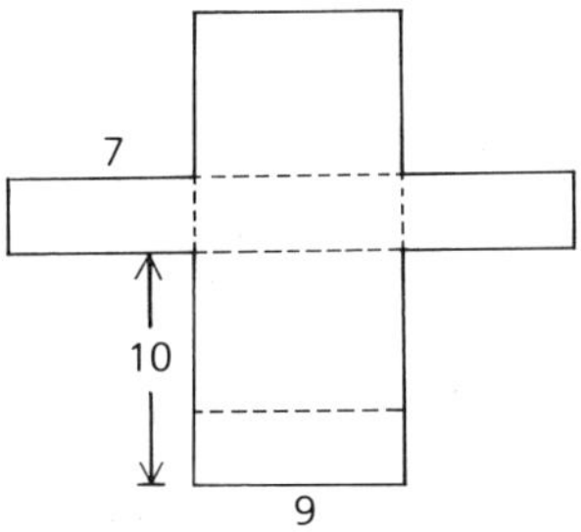

14 The cylindrical glass is full of water, which is poured into the rectangular pan. Will the pan overflow? (Use $\pi \approx 3.14$)

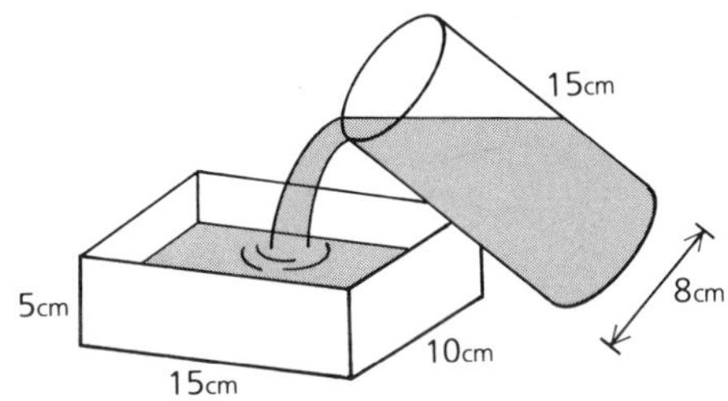

15 Jim's lunch box is in the shape of a half cylinder on a rectangular box.

Find: **a** the total volume it contains

b the total area of metal needed to manufacture it.

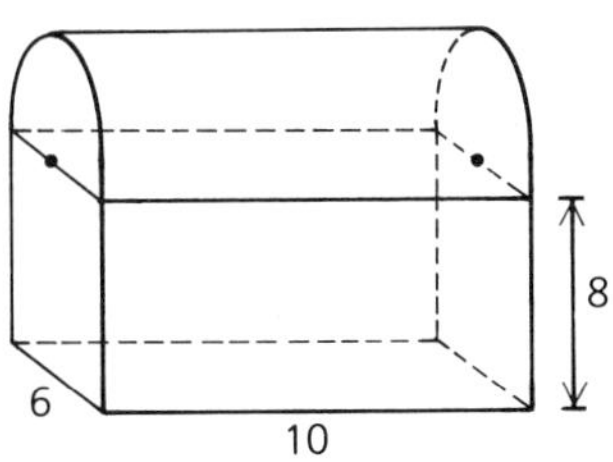

16 A cistern is to be built of cement. The walls and bottom will be 1 ft. thick. The outer height will be 20 ft. The inner diameter is 10 ft.

How much cement is needed for the job?

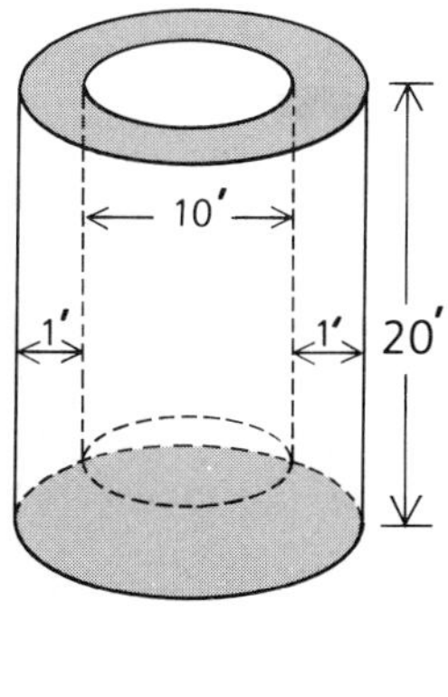

Section 12.1 Problem Set C

17 A wedge of cheese is cut from a cylindrical block. Find the volume and total surface area of this wedge.

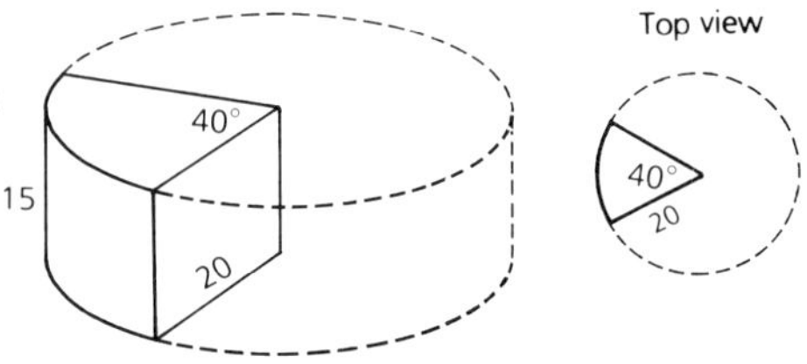

18 An ice-cube manufacturer makes ice cubes with holes in them. Each cube is 4 cm on a side and the hole is 2 cm in diameter. (Use $\pi \approx 3.14$)

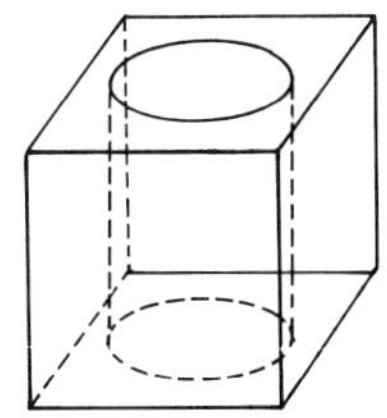

a What is the volume of ice in the cube?

b Ice decreases its volume by 11% when it melts into water. What volume of water will be left when 10 cubes melt?

c What is the total surface area (including inside the hole) of a single cube?

d The manufacturer claims that these cubes cool a drink twice as fast as a regular cube the same size. Verify whether this claim is true by comparison of surface areas. (The ratio of areas = the ratio of cooling speeds.)

19 Find the volume of the solid at the right. (A general cross section is shown.)

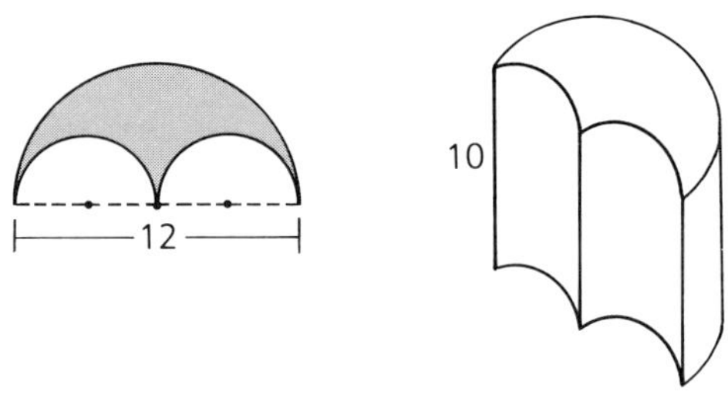

20 A cylinder is cut on a slant as shown.

Find the volume.

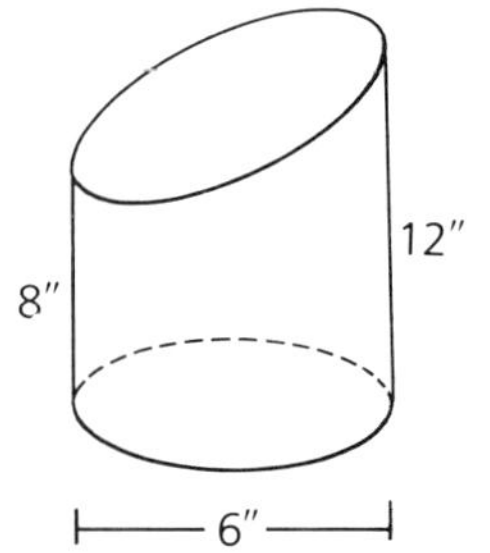

Section 12.2 Pyramids and Cones

This section deals with the volumes of solids that have a flat base and rise to a peak or vertex.

A VOLUME OF A PYRAMID
B VOLUME OF A CIRCULAR CONE
C CROSS SECTION OF A PYRAMID OR CIRCULAR CONE

12.2 (A) Volume of a Pyramid

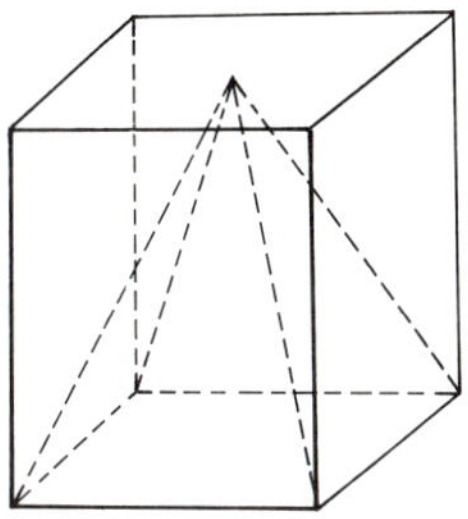

The volume of a pyramid is related to the volume of a prism having the same base and height. At first glance, many people would guess that the volume of the pyramid is half that of the prism.

Wrong! To be correct you would have to guess that the volume of a pyramid is ONE-THIRD of the volume of a prism with the same base and height.

Theorem 107

T **The VOLUME OF A PYRAMID is one-third of the product of the *height* times the *area* of the base.**

$$V_{pyramid} = \frac{1}{3}\mathcal{B} \cdot h$$

$\mathcal{B}$ = area of base
h = height of pyramid

The proof of this formula is complex and will not be shown here.

12.2 (B) Volume of a Circular Cone

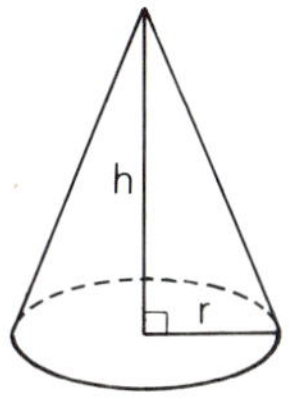

A cone is a close relative of the pyramid.
Instead of a polygonal base,
a cone has a circular base.

Theorem 108

T **The VOLUME OF A CIRCULAR CONE is one-third of the product of the *height* times the *area* of its base.**

$$V_{cone} = \frac{1}{3}\mathcal{B} \cdot h$$
$$= \frac{1}{3}(\pi r^2)h$$

r = radius of base
h = height of cone
$\mathcal{B}$ = area of base

12.2 (C) Cross Section of a Pyramid or Circular Cone

The cross sections of cones and pyramids are not congruent to the base as they are in prisms and cylinders. Observe that the cross section parallel to the base is a figure *similar* to the base in each case.

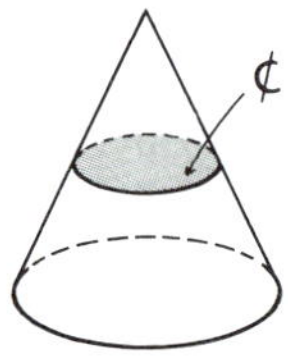

Cross section of a cone

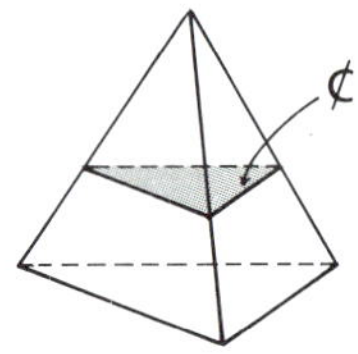

Cross section of a pyramid

The Similar Figures Theorem (p. 515) suggests that the *area* of a cross section is related to the *square* of the distance from the vertex to the cross section.

Theorem 109

T **In a PYRAMID OR CONE, the *ratio* of the area of a cross section to the area of the base equals the *square of the ratio of the distances* from the vertex, respectively.**

$$\frac{¢}{\mathcal{B}} = \left(\frac{\mathbf{k}}{\mathbf{h}}\right)^2$$

¢ = area of cross section
$\mathcal{B}$ = area of base
k = distance, vertex to cross section
h = height of pyramid or cone

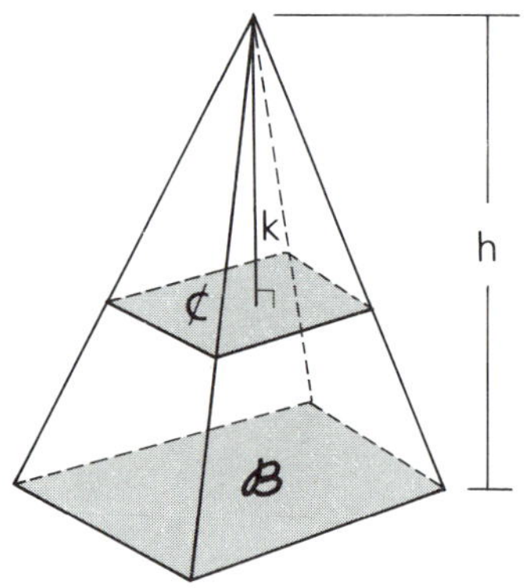

The proof of Theorem 109 is called for in problem 15.

Section 12.2 Sample Problems

1 Find the volume of the pyramid.
The height is 21 cm and the base is an equilateral triangle with 8 cm sides.

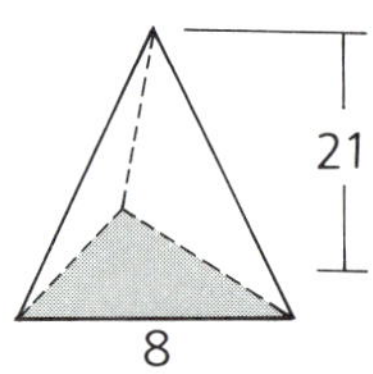

Solution: $V_{pyramid} = \frac{1}{3} \mathcal{B} \cdot h$

Since the base is an equilateral triangle, $= \frac{s^2}{4}\sqrt{3} = 16\sqrt{3}$

Then $V = \frac{1}{3}(16\sqrt{3})(21) = 112\sqrt{3}$

2 Find the volume of a cone with a base radius of 6 and slant height of 10.

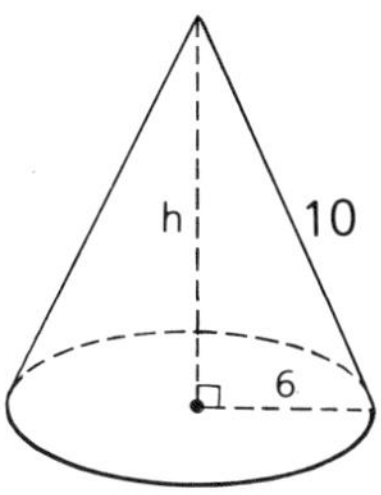

Solution: From the right triangle shown, the height of the cone is 8.

$$V_{cone} = \frac{1}{3} \mathcal{B} \cdot h$$

$$= \frac{1}{3} \pi r^2 h$$

$$= \frac{1}{3}\pi(6^2)(8) = 96\pi$$

3 A pyramid has a base area of 24 sq cm and a height of 12 cm. A cross section is cut 3 cm from the base.

Find: **a** the volume of the upper pyramid (above the cross section).

b the volume of the frustum (below the cross section).

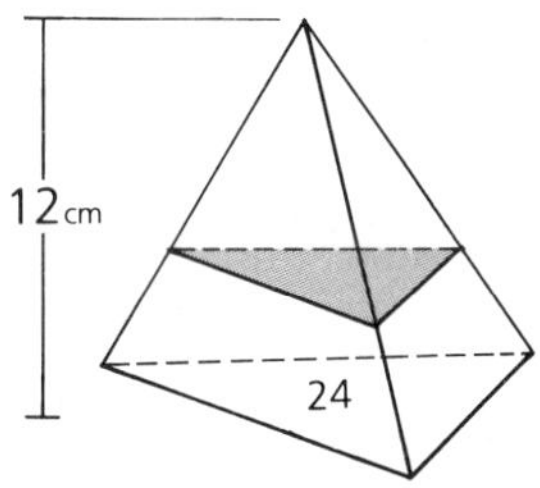

Solutions:

a Since the cross section is 3 cm from the base, the distance k from the peak is 9 cm.

$$\frac{¢}{\mathcal{B}} = \left(\frac{k}{h}\right)^2 \qquad V_{upper\ pyramid} = \frac{1}{3} ¢ \cdot k$$

$$\frac{¢}{24} = \left(\frac{9}{12}\right)^2 \qquad = \frac{1}{3} \cdot \frac{27}{2} \cdot (9)$$

$$¢ = \frac{27}{2} \qquad = 40.5cc$$

b To find the volume of the frustum, subtract the volume of the upper pyramid from the volume of the whole pyramid.

$$V_{frustum} = V_{whole\ pyramid} - V_{upper\ pyramid}$$

$$V = \frac{1}{3}(24)(12) - 40.5$$

$$= 96 - 40.5$$

$$= 55.5 \text{ cc}$$

Section 12.2 Problem Set **A**

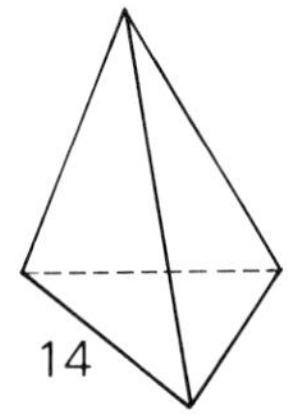

1 Find the volume of a pyramid whose base is an equilateral triangle, if the height of the pyramid is 30, and a side of the base is 14.

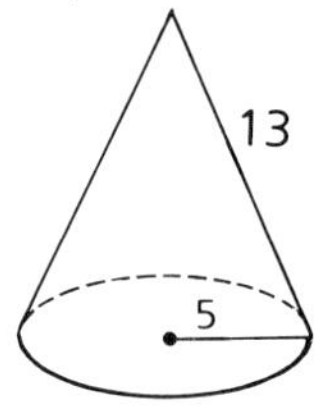

2 Find the volume of a cone with slant height 13 if the radius of the base is 5.

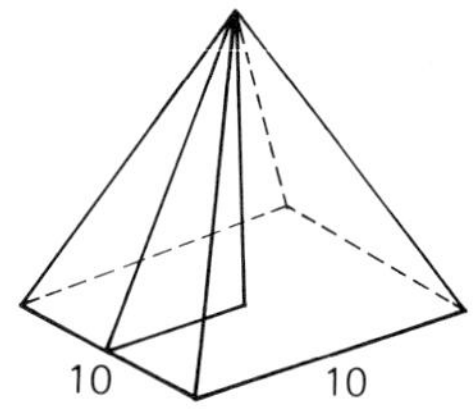

3 Given the pyramid shown with a square base and height 12.

a Find the volume.

b Find the total area.

4 The volume of a pyramid is 42. If the base has an area of 14, find the height.

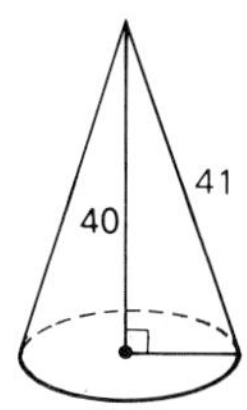

5 Given: The right circular cone as shown

Find: a volume b lateral area c total area

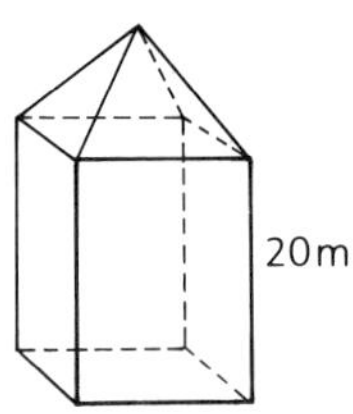

6 A tower has a total height of 24 m. The height of the wall is 20 m. The base is a rectangle with an area of 25 sq m.

Find the total volume of the tower.

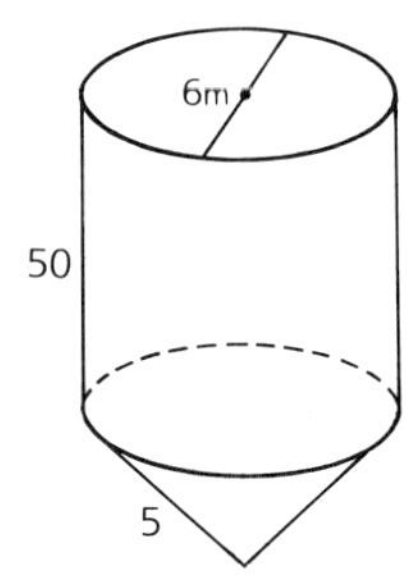

7 A well is dug with a cylindrical wall 50 m deep and with a diameter of 6 m. The bottom is tapered to a cone with slant height 5 m.

Find the volume of water the well could hold.

Section 12.2 Problem Set B

8 Find the volume of a cone with a 60° vertex angle and a slant height of 12.

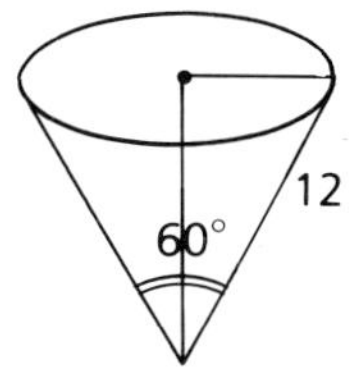

9 A pyramid has a square base with a diagonal of 10. The lateral edges are each 13.

Find the volume of the pyramid.

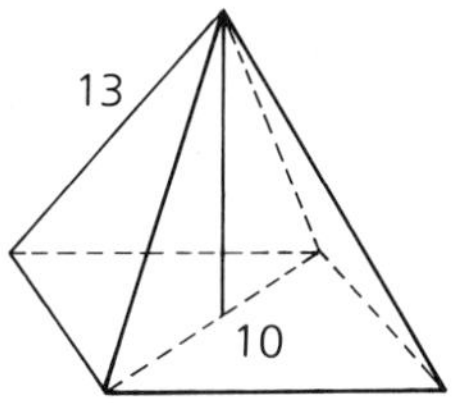

10 Two identical cylinders are shown. One cone is drawn in the left cylinder. Two cones, whose heights add up to the total height, are drawn in the right cylinder. Compare the volume of the cone in the left figure to the sum of the volumes of the cones at the right.

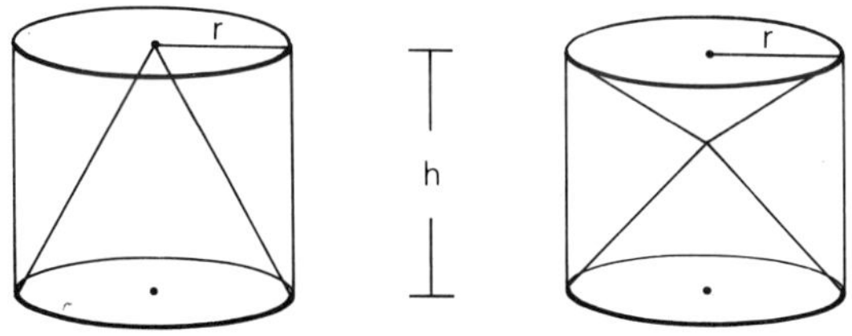

11 Find the volume remaining if the smaller cone is removed from the larger.

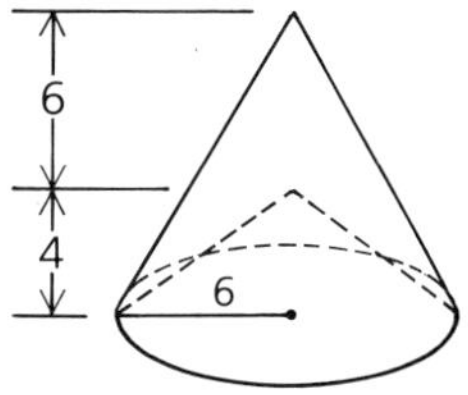

12 A rocket has dimensions shown. If 60% of the volume of the rocket is to contain fuel, what is the volume of that portion of the rocket that can contain other equipment?

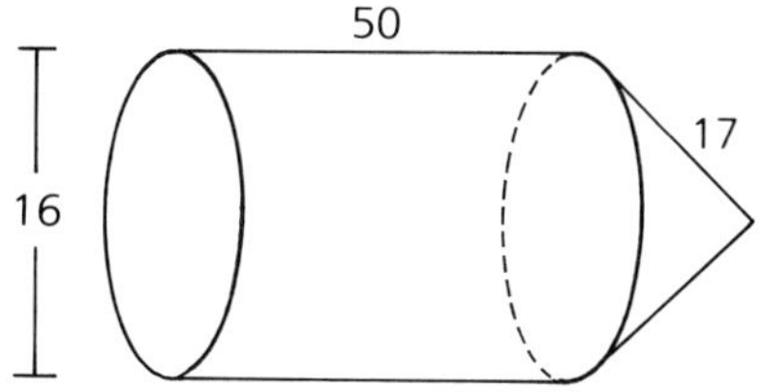

13 A gazebo (garden house) has a pentagonal base with an area of 60 sq m. The total height to the peak is 16 m. The height of the pyramidal roof is 6 m.

Find the total volume.

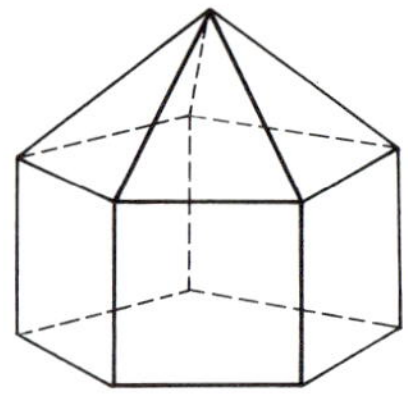

14 Given: The diagram as marked

Find: **a** x

b the radii of the circles

c the volume of the smaller cone

d the volume of the larger cone

e the volume of the frustum

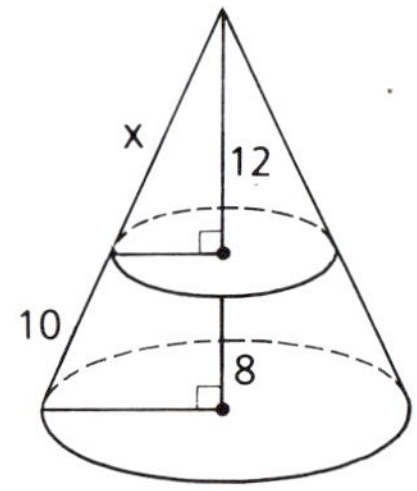

15 Set up and prove Theorem 109. (Hint: First prove that the ratio of corresponding segments of a cross section and a base equals the ratio of h to k.)

Section 12.2 Problem Set C

16 Given: A regular tetrahedron (Each of the 4 faces is an equilateral triangle.)

Find: the total volume if the edge is

a 6 **b** s

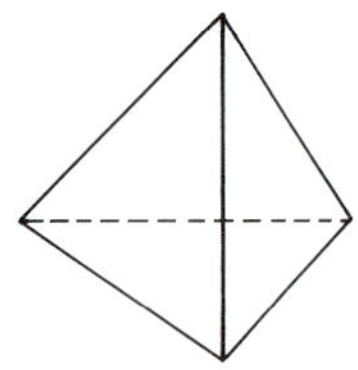

17 A regular octahedron (8 equilateral faces) has an edge of 6.

Find the volume.

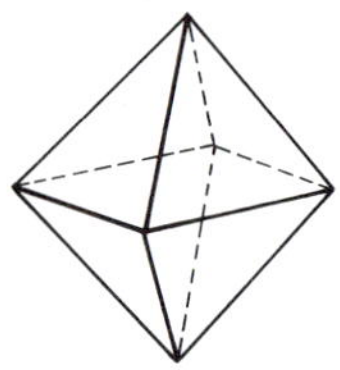

18 Find the volume of the pyramid as marked.

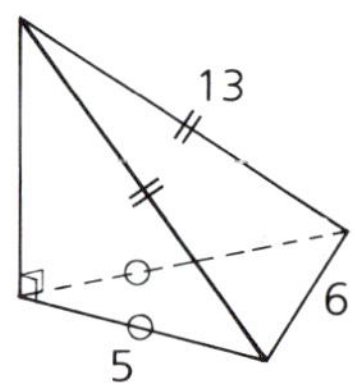

19 Find the volume of the frustum of the cone shown.

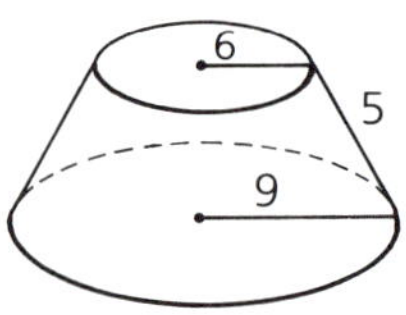

Section 12.3 Spheres

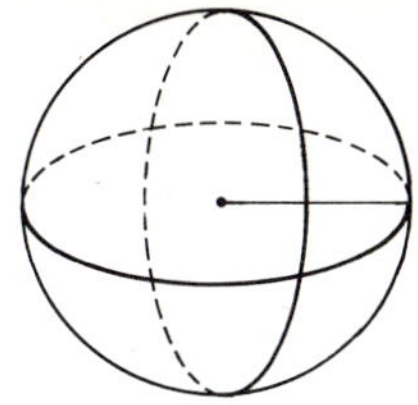

Theorem 110

T **The VOLUME OF A SPHERE is four-thirds of π times the *cube* of the radius.**

$$V_{sphere} = \frac{4}{3}\pi r^3$$

r = radius

Theorem 110 can be derived as an application of *Cavalieri's Principle,* Section 12.3 Problem Set D.

Some of the problems that follow ask for both the volume and the surface area of spheres. As an aid, recall this formula from Chapter 11: Total Surface Area of a Sphere = $4\pi r^2$.

Section 12.3 Sample Problem

1 Find the volume of a hemisphere of radius 6.

Solution: First find the volume of the sphere of radius 6.

$$V_{sphere} = \frac{4}{3}\pi r^3$$
$$= \frac{4}{3}\pi(6)^3 = 288\pi$$

A hemisphere has half the volume of a sphere.

Thus, $V_{hemisphere} = 144\pi$.

Section 12.3 Problem Set A

1 Find the volume of a sphere with:

a radius 3 **b** diameter 18 **c** radius 5

2 Find the volume and surface area of a sphere of radius 6.

3 Find the volume of the grain silo.

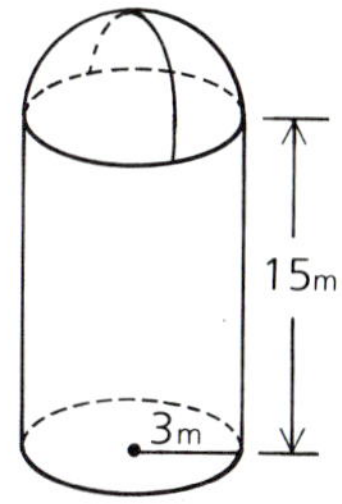

4 A plastic bowl is in the shape of a cylinder with a hemisphere cut out. The dimensions are shown.

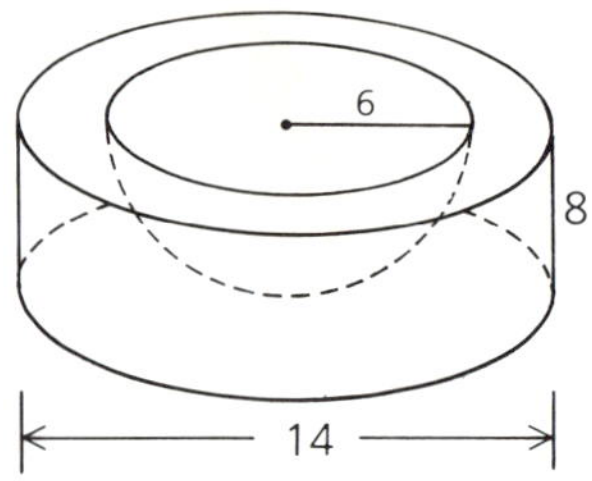

a What is the volume of the cylinder?

b What is the volume of the hemisphere?

c What is the volume of plastic used to make the bowl?

5 What volume of gas (to the nearest cubic foot) is needed to inflate a spherical balloon to a diameter of 10 feet?

Section 12.3 Problem Set B

6 A rubber ball is formed by a rubber shell filled with air. The outer diameter is 48 mm. The inner diameter is 42 mm. Find the volume of rubber used to make the ball.

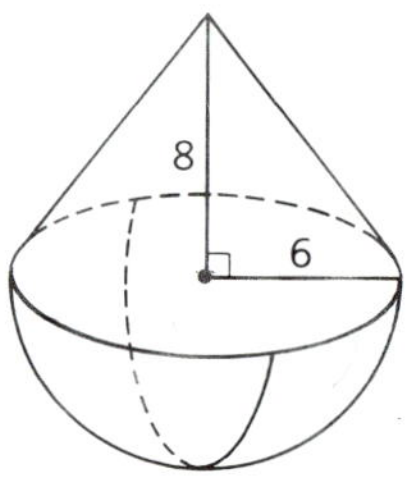

7 Given: a cone and hemisphere as marked

Find: a the total volume of the solid

b the total surface area of the solid

8 A hemispherical dome has a height of 30 meters.

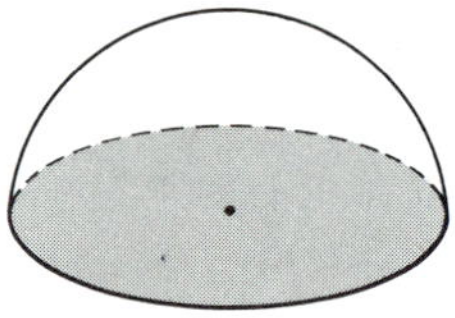

a Find the total volume enclosed.

b Find the area of ground covered by the dome (shaded area).

c How much more paint is needed to paint the dome than to paint the floor?

d To double the area of ground covered by the dome, what must the radius of the dome become?

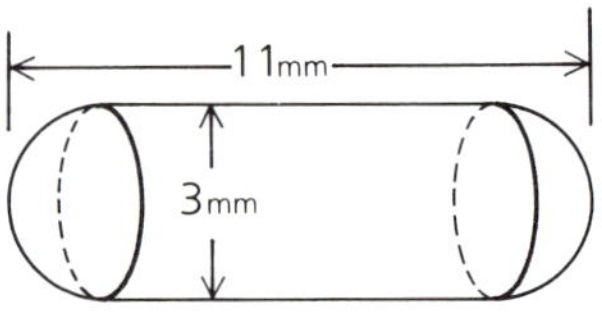

9 A cold capsule is 11 mm long and 3 mm in diameter. Find the volume of medicine it contains.

10 A mini-submarine has the dimensions shown.

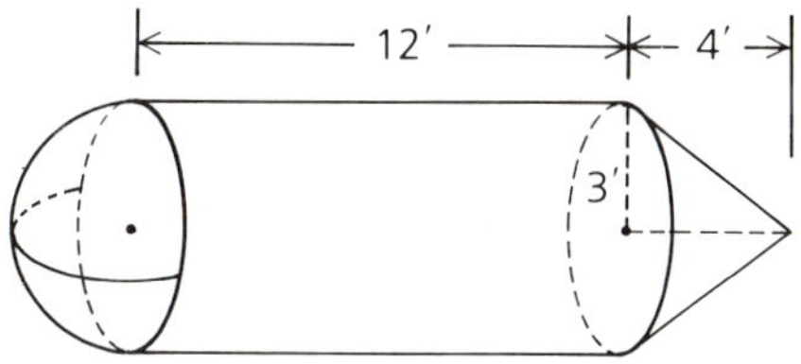

a What is the total volume?

b The surface area is important in computing the pressure that the ship will withstand. What is the total surface area?

11 The radii of two spheres are in a ratio of 2:5.

Find: a the ratio of their volumes

b the ratio of their surface areas.

Section 12.3 Problem Set C

12 An ice-cream cone is 9 cm deep and 4 cm across the top. A single scoop of ice cream 4 cm in diameter is placed on top. If the ice cream melts into the cone, will it overflow? (Justify your answer.)

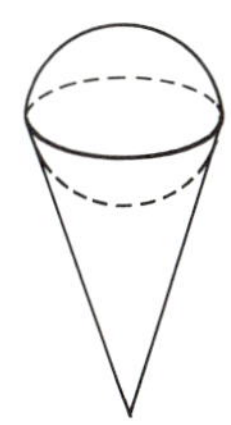

13 The volume of a cube is 1000 cubic meters.

a What is the volume of the largest sphere that can be inscribed inside the cube?

b What is the volume of the smallest sphere that can be circumscribed about the cube?

14 Find the ratio of the volume of a sphere to the volume of the smallest right circular cylinder that can contain it.

15 Compare the volumes of a hemisphere and a cone with congruent bases and equal heights.

Section 12.3 Problem Set D

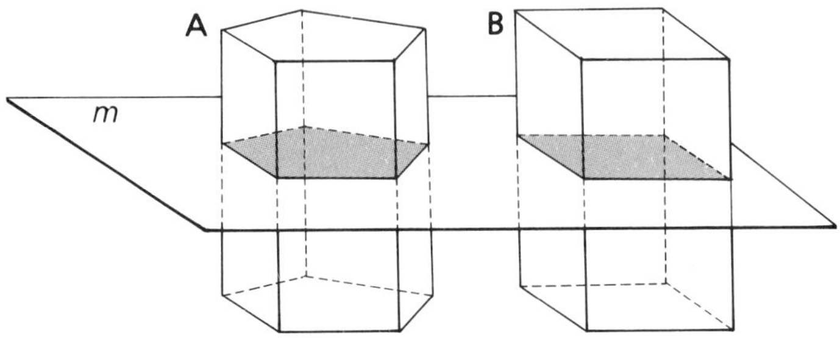

Plane m is a cross-sectional plane that cuts through solids A and B.

Cavalieri's Principle

If two solids A and B can be placed with their bases coplanar and if all planes parallel to the plane of their bases determine cross sections in A of area equal to those in B, then A and B have equal volumes.

16 Show that the volume of cylindrical shell a is equal to the volume of cylinder b by using Cavalieri's Principle.

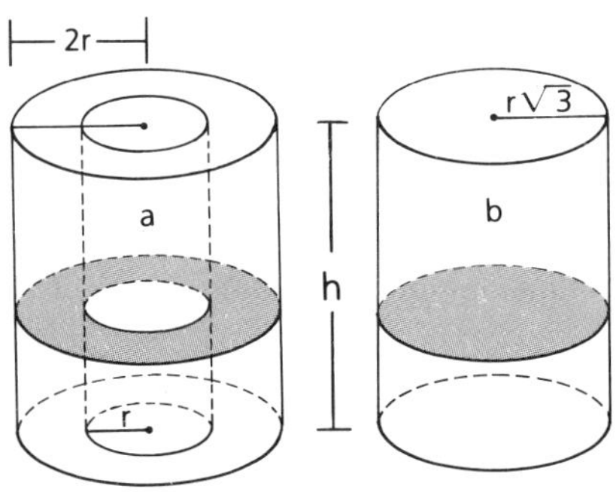

17 Given the figures below:

a compare cross-sectional areas

b use Cavalieri's Principle to derive the formula for the volume of a hemisphere and hence a sphere.

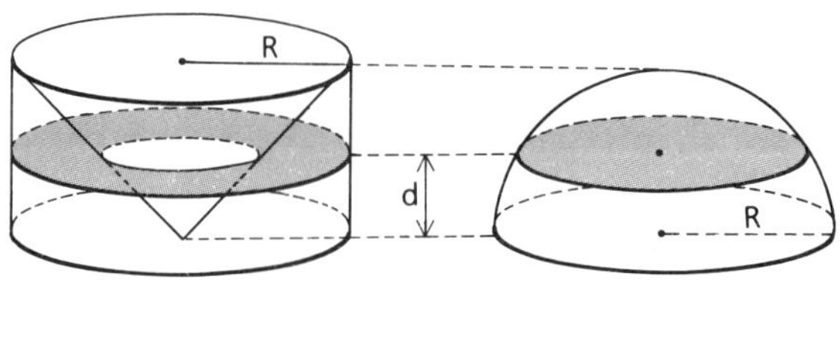

Chapter 12 Study Guide

As you study and review this chapter, be sure that you are able to:

1 Apply the volume formula for each of these solids:

Solids with Congruent Bases	**Pointed Solids**	**Spheres**
rectangular box (p. 548)	pyramid (p. 555)	sphere (p. 561)
prism (p. 549)	cone (p. 555)	
cylinder (p. 549)		

2 Find the total surface area of each solid in part 1.

3 Explain how to use a cross section to find the volume of a solid. (p. 549)

4 Apply the "Divide and Conquer" method for volumes. (p. 550)

5 Identify these parts of a cone: vertex, slant height, altitude and radius. (p. 530)

6 Describe the following ratios involving similar figures: Let s_1 and s_2 be any corresponding linear measures (such as sides, radii, altitudes, etc.)

The ratio of any other corresponding *linear* measures is $\frac{s_1}{s_2}$.

The ratio of any corresponding *area* measures is $\left(\frac{s_1}{s_2}\right)^2$.

The ratio of any corresponding *volume* measures is $\left(\frac{s_1}{s_2}\right)^3$.

7 Use the ratios for similar figures to solve problems such as the following:

a If two spheres have radii of 7 and 4, what is the ratio of their areas? (p. 515) What is the ratio of their volumes? (p. 563)

b If two cubes have total surface areas of 64 and 36, what is the ratio of their edges? (p. 516) What is the ratio of their volumes? (p. 563)

Chapter 12 Review Problems A

1 Find the volume of:

a a cube with side 8

b a box that is 3 by $4\frac{1}{2}$ by 8

c a cylinder with radius 7 and height 2

d a pyramid with height 5 and base area 12

e a prism with height 5 and base area 12

f a sphere with radius 2

2 Find the volume and total surface area of each solid below:

a
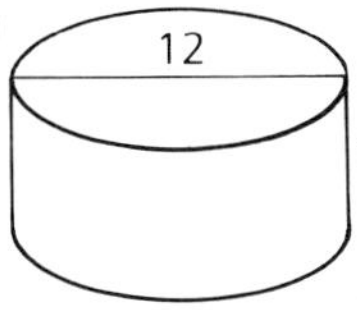

b
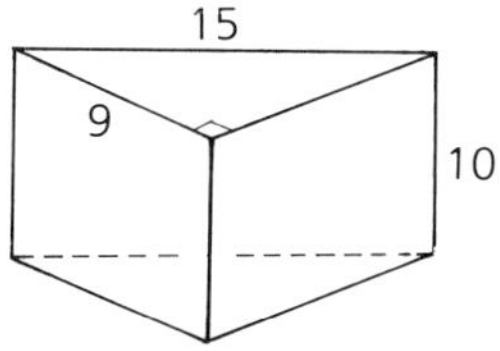

c
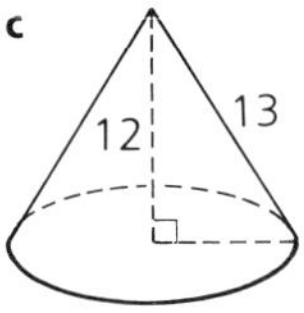

3 Find the volume of the solid that is formed when folds are made along the dotted lines.

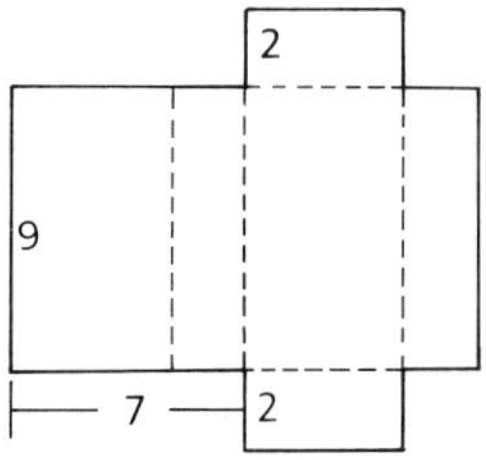

4 Find the height of:

a a box with volume 100, length 15, and width $1\frac{1}{3}$

b a cube with volume 216.

5 Find the volume of a cylindrical glass if its height is 15 cm and a 17 cm straw just fits inside the glass as shown.

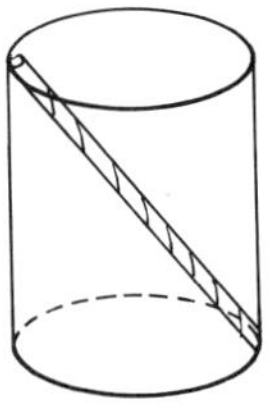

6 A concrete staircase is to be built. Each step is 15 cm high, 25 cm deep, and 1 m wide. The top platform is square. What volume of concrete is needed?

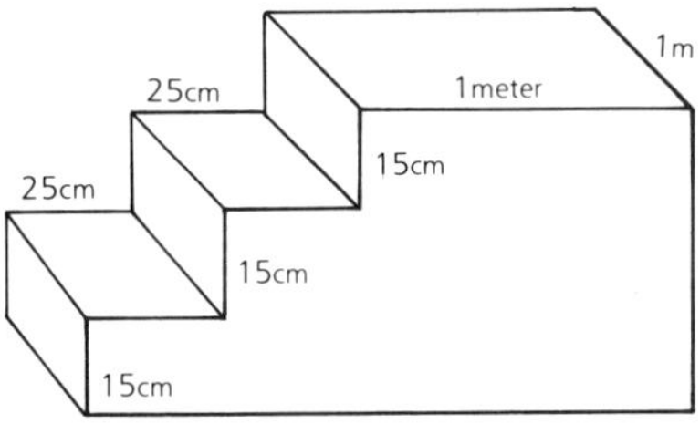

7 Find the volume of a square pyramid with slant height 10 and base 16 by 16.

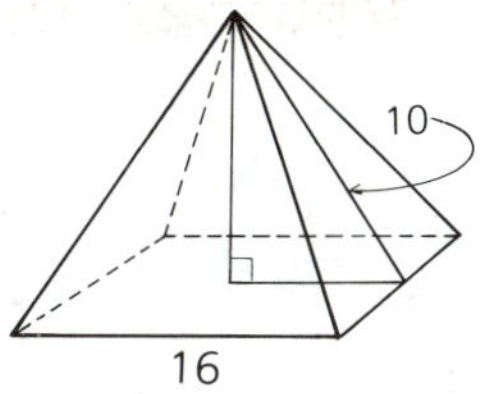

8 Find the volume of a sphere whose surface area is 36π.

Chapter 12 Review Problems B

9 Find the volume of the cylinder formed from the pattern at the right. The area of each circle is 16π. The rectangle has an area of 24π.

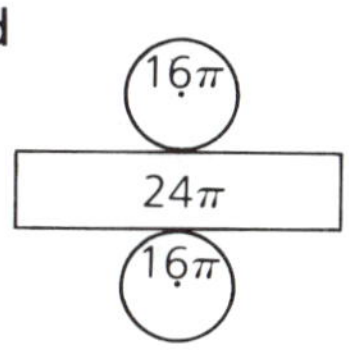

10 A pyramid has a height of 5. The base is a rhombus with diagonals 7 and 6. Find the volume of the pyramid.

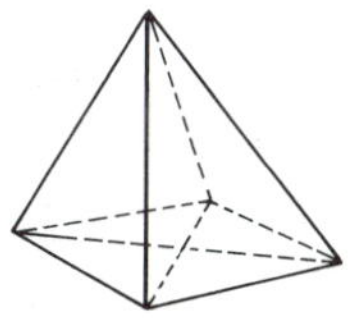

11 A cross section of a hatbox is a regular hexagon with a side 12 cm long. The height of the box is 20 cm. Find the volume.

12 Find the volume of the wedge.

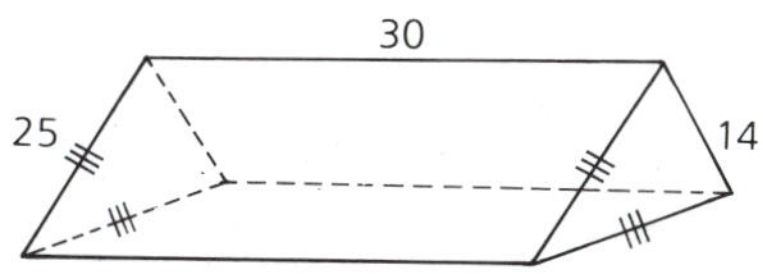

13 A hole of diameter 2 in. is drilled through a block as shown. Find the remaining volume approximated to the nearest whole number, using $\pi \approx 3.14$.

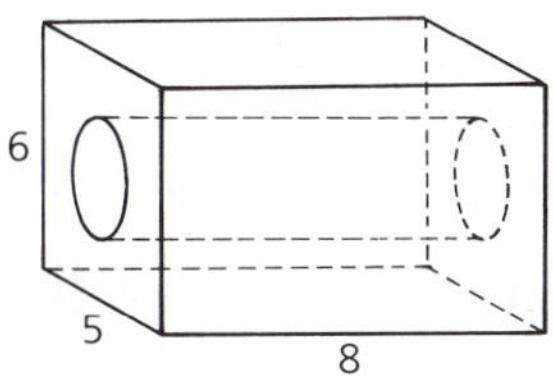

14 Find the volume of Kerry's prism shown at the right.
Hint: If you do it, you will be a HERO.

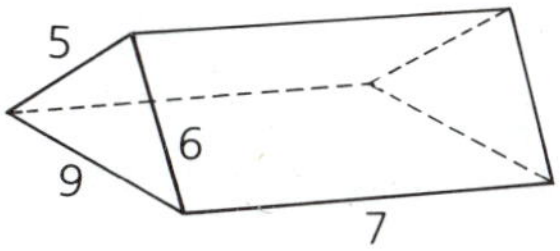

15 Find the total volume of the castle and towers.

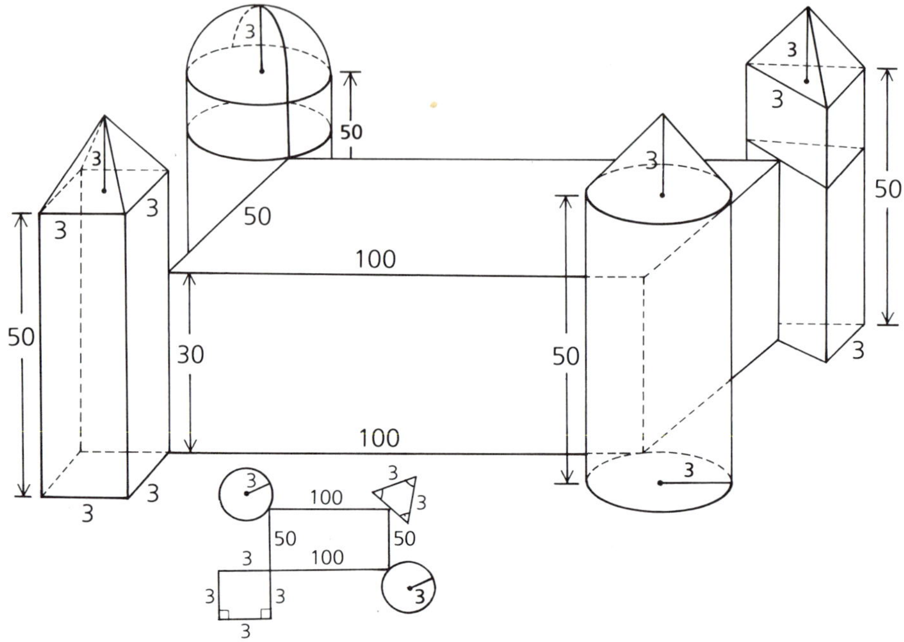

Chapter 12 Review Problems **C**

16 A right cylindrical log was cut parallel to the axis. Find the volume and total surface area.

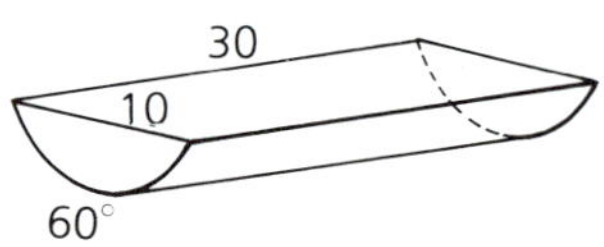

17 A frustum of a cone is shown. Find the volume of this solid.

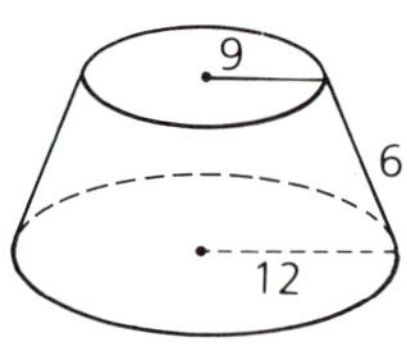

18 Find the volume of each solid generated by rotating the solid line or figure about the dashed line.

a

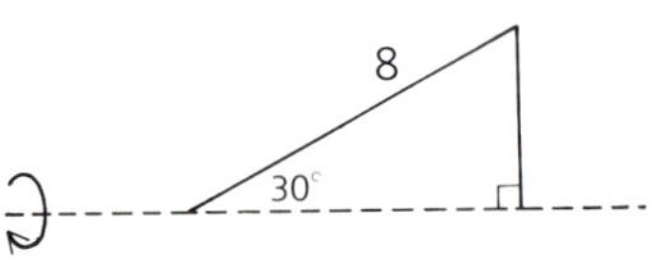

b

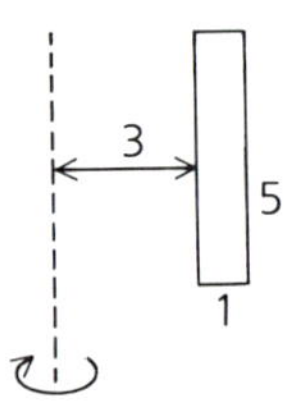

Chapters 1–12 Cumulative Review Problems A

1 The measure of one of the acute angles of a right triangle is 9 times that of the other acute angle. Find the measure of the larger acute angle.

2 The perimeter of $\triangle ABC$ is 28, with AB $= 2x + 3$, BC $= 4x - 5$, and CA $= 8x - 19$. Is $\triangle ABC$ scalene, isosceles, or equilateral?

3 Given: $\overline{BD} \perp \overline{AD}$, $\overline{BD} \perp \overline{BC}$, $\overline{AB} \cong \overline{CD}$

Prove: ABCD is a ▱

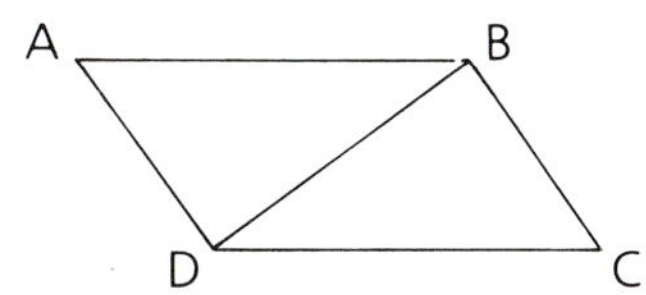

4 Given: $\overline{PQ} \parallel \overline{TR}$

Find: **a** PT

b TR

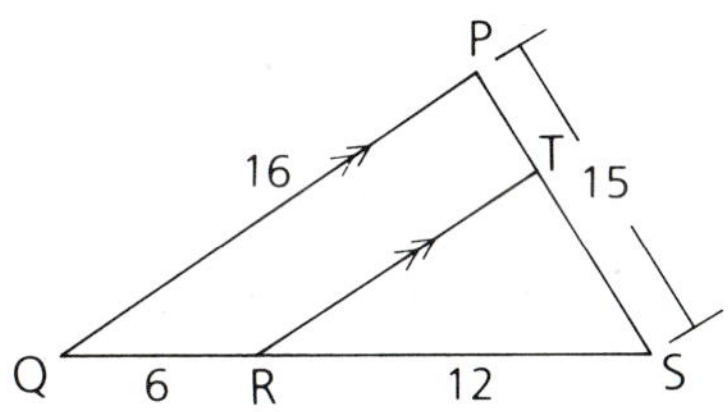

5 Find the value of x:

a

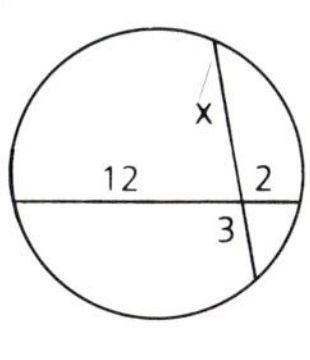

c

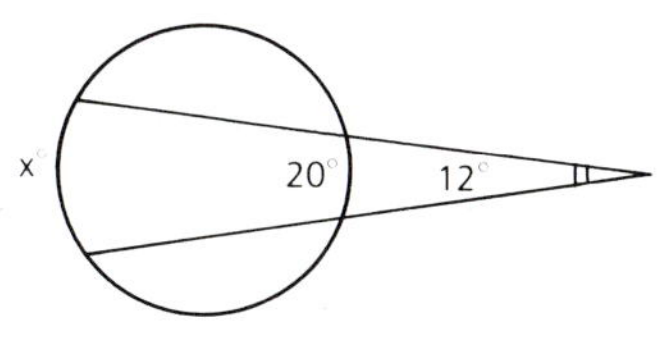

b

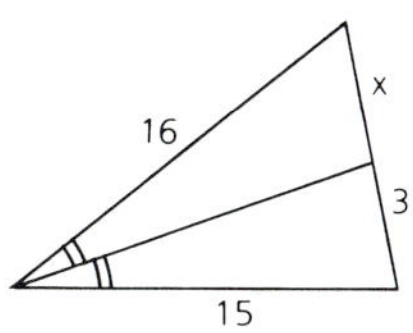

d

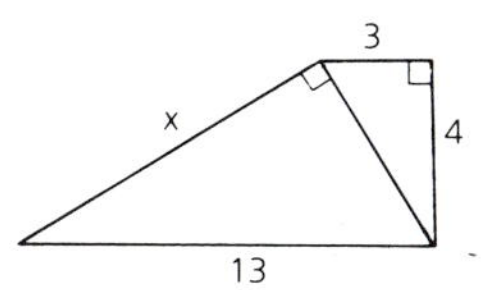

6 A pair of similar triangles has areas 9 and 25 respectively.

a What is the ratio of a pair of corresponding sides?

b What is the ratio of their perimeters?

7 Diagram as marked

Find: **a** WY

b YZ

c XZ

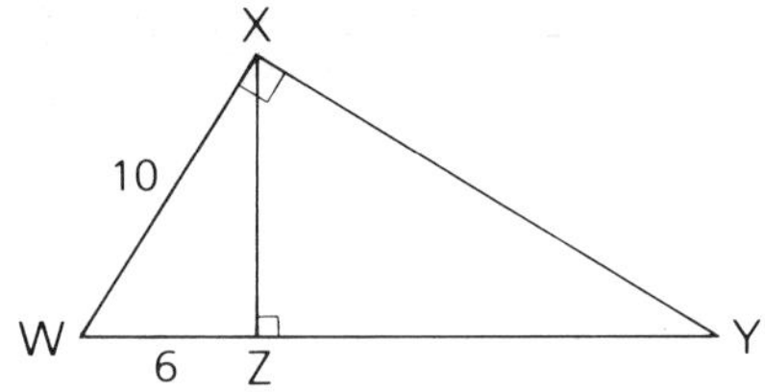

8 Find the area of the trapezoid, triangle, and circle:

a

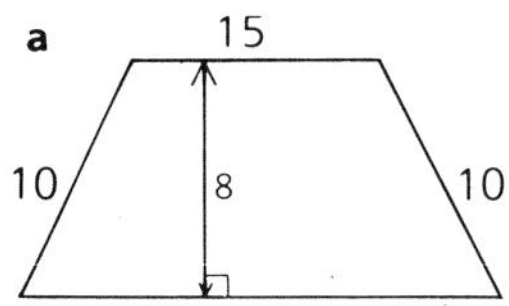

b

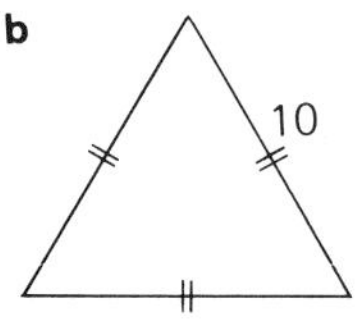

c

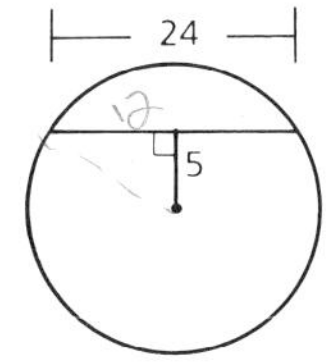

9 SPQR is an isosceles trapezoid.

$\angle S = (x + 40)^\circ$

$\angle Q = (2x - 7)^\circ$

Find $\angle R$.

10 What is the most descriptive name for a quadrilateral where

a the diagonals bisect each other?

b one diagonal is the perpendicular bisector of the other?

c the figure is equiangular?

d all sides are congruent (equilateral)?

11 Find p, q, r, and s.

a

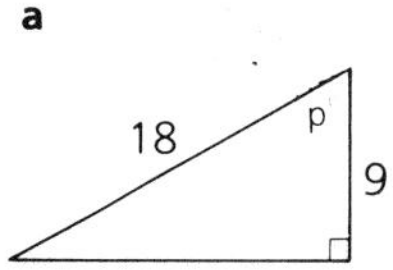

b

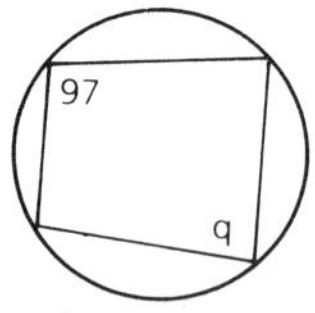

c

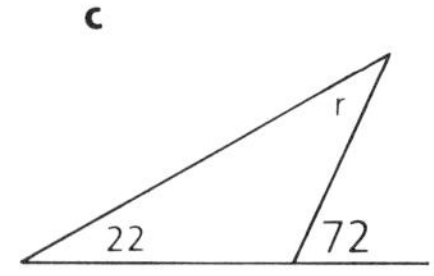

d

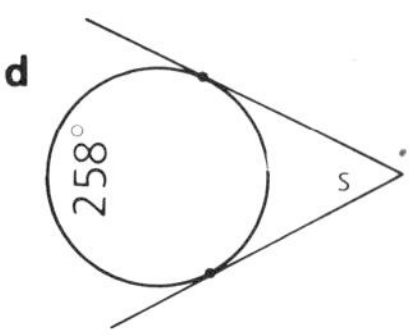

12 **a** Find the 4th proportional in a proportion if the first three terms are 5, 3, and 30.

b Find the mean proportionals between 8 and 18.

13 Given: $\odot$ O with tangent $\overline{CD}$

$CD = 15$

$BC = 9$

Find: **a** AC

b the diameter of the circle

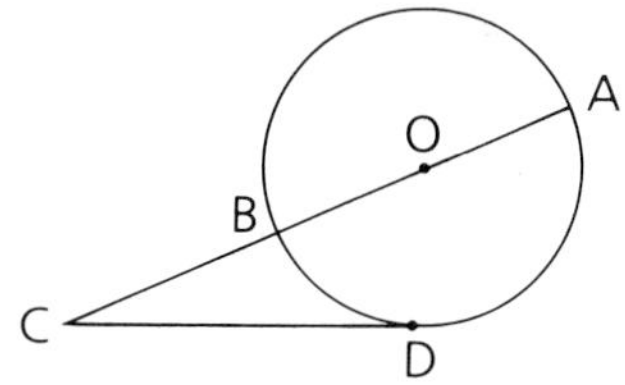

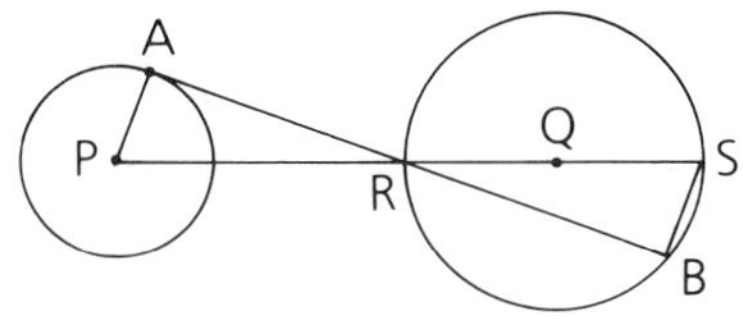

14 Given: $\overline{AR}$ tangent to $\odot$ P

$\overline{RS}$ diameter of $\odot$ Q

Prove: $\triangle PAR \sim \triangle SBR$

15 In $\triangle ABC$, D and E are the midpoints of $\overline{AB}$ and $\overline{AC}$, $DE = 4x$, and $BC = 2x + 48$. Find BC.

16 Find the area of each figure.

a

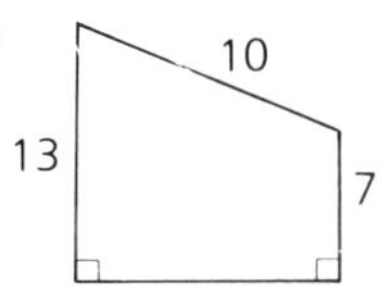

b

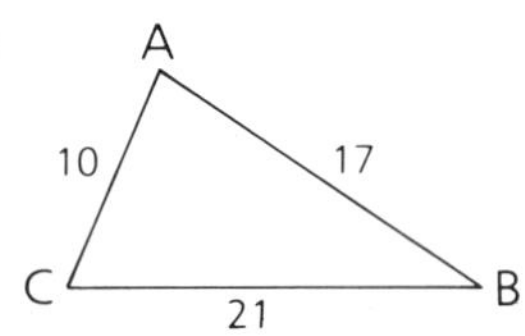

c

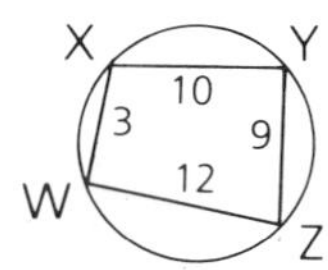

17 Find the number of sides of an equiangular polygon if each interior angle is 170°.

18 The perimeter of an isosceles triangle is 36. One side is 10. What are the possible lengths of the base?

19 Each polygon shown is regular.

Find: **a** the measure of $\angle 1$

b the measure of $\angle 2$

c the measure of $\angle 3$

d the measure of $\angle 4$.

Will a regular pentagon fit at $\angle 5$?

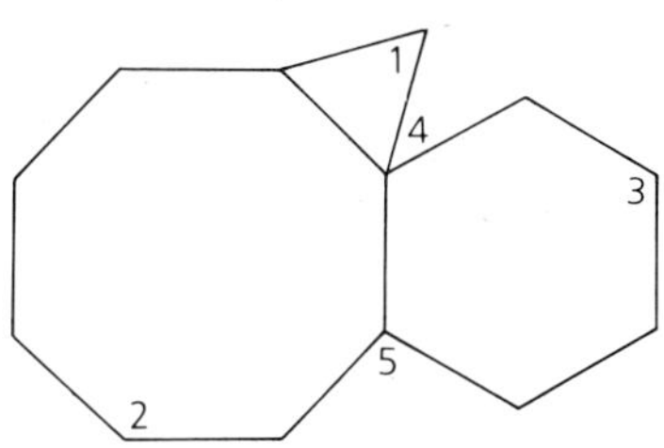

20 Given: Parallelogram as marked

Find: x

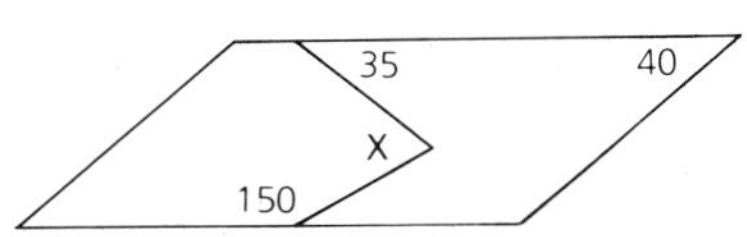

21 Given: WXYZ is an isosceles trapezoid with $\overline{WZ} \cong \overline{XY}$
$\triangle PZY$ is isosceles

Prove: P is midpoint of $\overline{WX}$

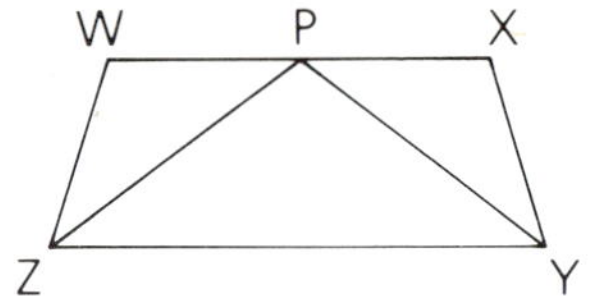

22 Given: $\overleftrightarrow{AC} \perp m$, $\overline{BC} \cong \overline{BA}$

Prove: D is the midpoint of $\overline{AC}$

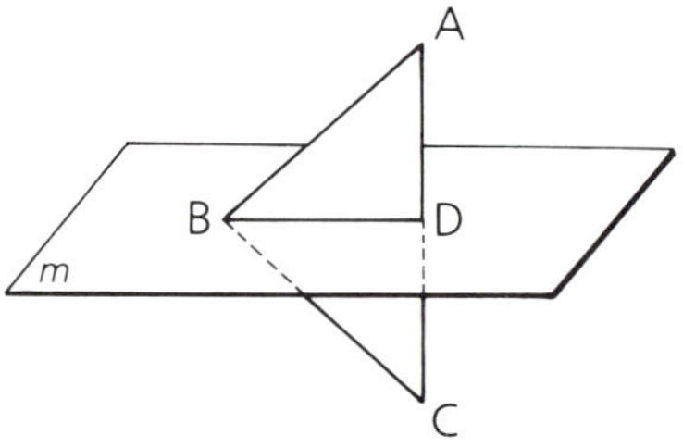

23 Find the length of a 45° arc in a circle whose radius is 8.

Chapters 1–12 Cumulative Review Problems B

24 Find the angle formed by the hands of a clock at:

a 11:30 **b** 2:05 **c** 3:24

25 Given: Rectangle RECT in ⊙ R
RT = 5, TQ = 2

Find: ET

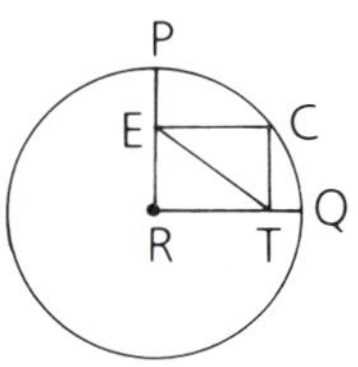

26 Prove that in a quadrilateral if one pair of opposite sides are parallel and one pair of opposite angles are congruent, then the quadrilateral is a parallelogram.

27 A woman walks 20 m west, 100 m south, then another 8 m west, and then 4 m north. How far is she from her starting point?

28 Given: ⊙ O, CB = 9,
$\angle C = 30°$, $\overline{BC} \cong \overline{BD}$,
$\overline{CD}$ tangent

Find: **a** $m\widehat{AD}$ **b** CD **c** radius of ⊙ O

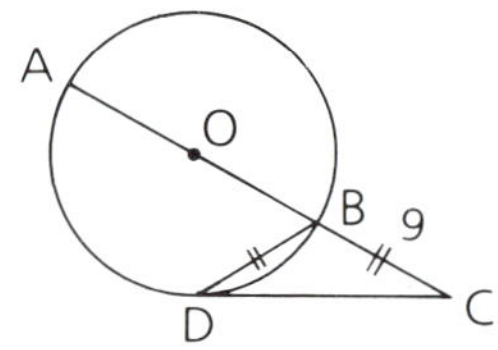

29 Given: $\overline{AB}$ is tangent to circles O and P, AB = 2.4
Radius of $\odot$ O = .7
Radius of $\odot$ P = 1.1

Find: **a** OP
b the distance between the circles

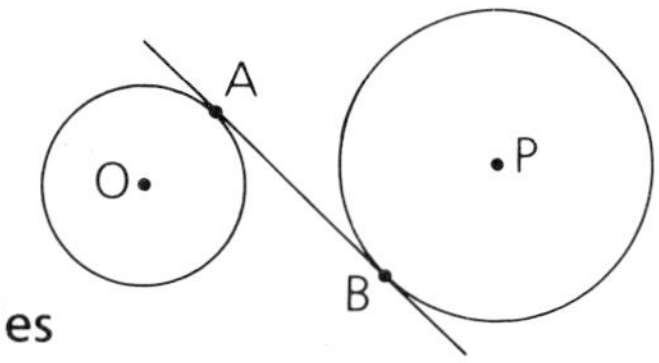

30 Given: Figure as marked
$\overline{PA}$ and $\overline{PD}$ tangent to $\odot$ O

Find: **a** $\widehat{AD}$
b m∠P

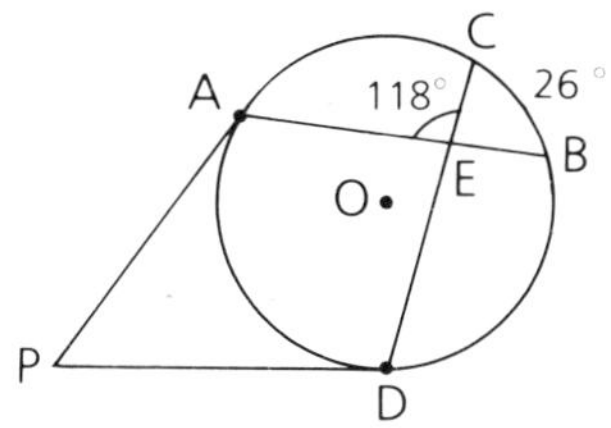

31 Given: Triangle as marked
Find: m∠WYA

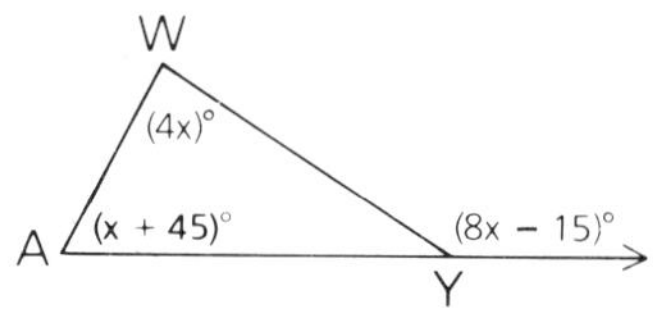

32 A drain pipe has water 18 cm deep. The width of the water surface is 48 cm.

Find the radius of the pipe.

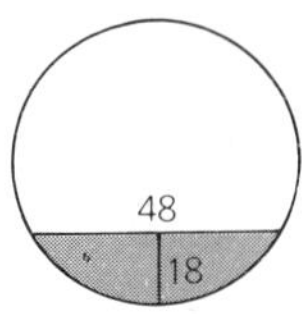

33 Given: Figure as shown
$\overline{RQ}$ is tangent

Find: x

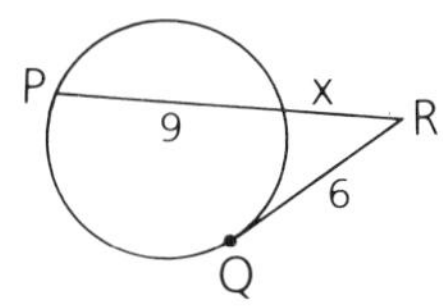

34 Given: ABCD is a ▱
Find: **a** AE
b x:y

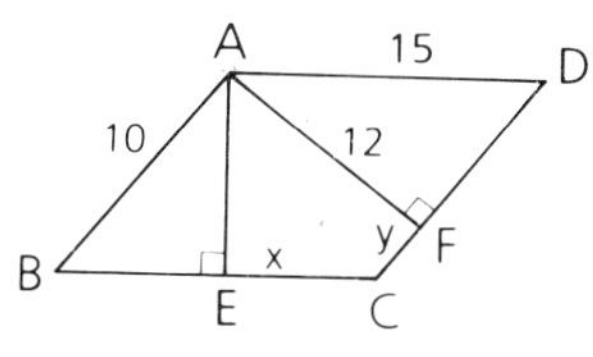

35 Given: $m\widehat{AB}:m\widehat{CD} = 5:2$
$\angle P = 24°$
Find: $\widehat{CD}$

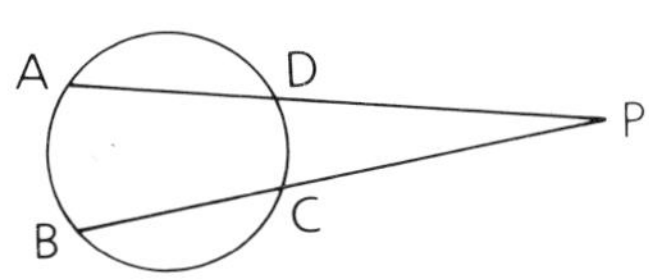

Chapters 1–12 Cumulative Review Problems C

36 A sled dog traveled 6 miles east, 6 miles northeast, then another 6 miles east. How far is he from his starting point? (Leave in simplified radical form)

37 Given: $\overline{BC}$ is the base of isosceles $\triangle ABC$
$\overline{DE}$ is the base of isosceles $\triangle AED$
$\angle BAE = 40^\circ$

Find: $m\angle DEC$

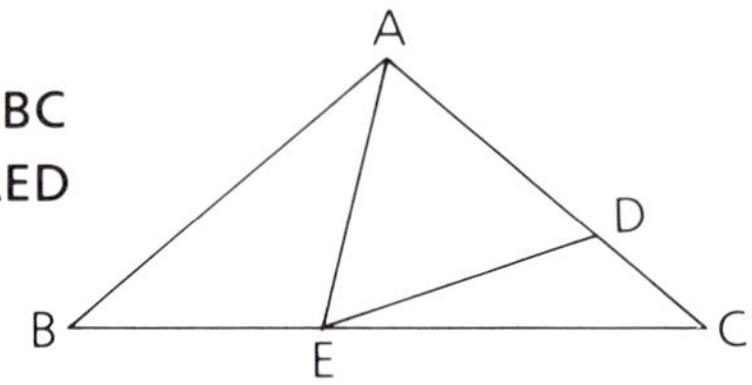

38 Two sides of one triangle are congruent to two sides of a second triangle and the included angles are supplementary. The area of one triangle is 41. Can the area of the second triangle be found?

39 In this set of three semicircles, B is any point on $\overline{AC}$. Prove that the shaded area is equal to π times the product of the radii of the unshaded semicircles.

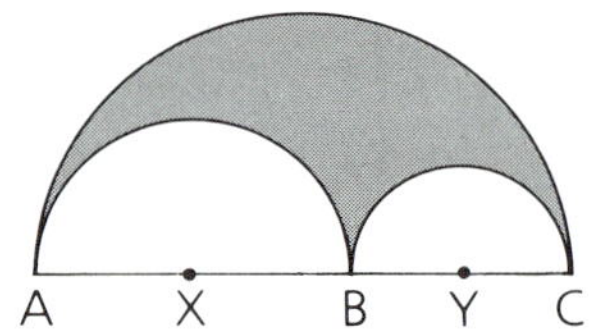

40 Given: QUAD has $\overline{QU} \cong \overline{AD}$, $\angle A$ is supp. to $\angle Q$, $\overline{QD} \not\cong \overline{AU}$

Prove: QUAD is an isosceles trapezoid.

41 The sides of a hexagon are in an arithmetic progression. The perimeter is 30, and the longest side is 7. Find the next-to-longest side.

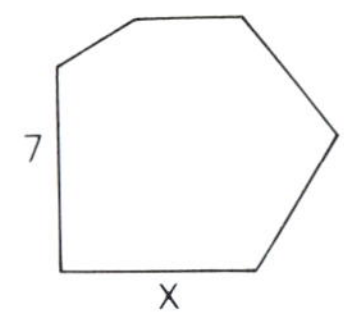

42 Clarence bragged that he ate most of a pizza, but could not remember the diameter. So, on the remaining piece, he made the measurements shown. From the midpoint of the arc to the midpoint of the corresponding chord is 5 cm. The chord measures 30 cm. Find the diameter of the pizza Clarence ate.

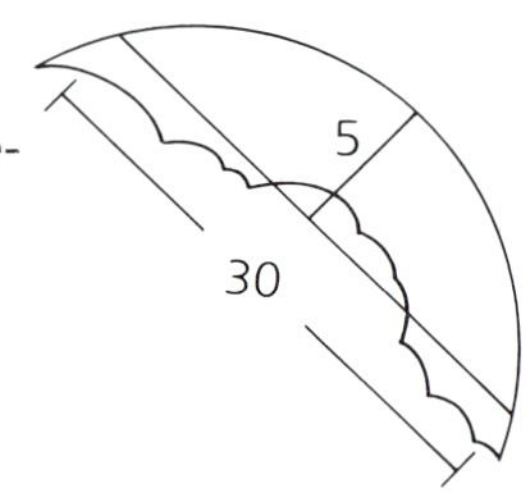

InterSection: Shape-Preserving Changes

Part of geometry is the study of shapes. Geometry can also study movement. For example, think of each point of △ABC being moved to the right. The result of that movement is △A′B′C′ (triangle A-prime, B-prime, C-prime).

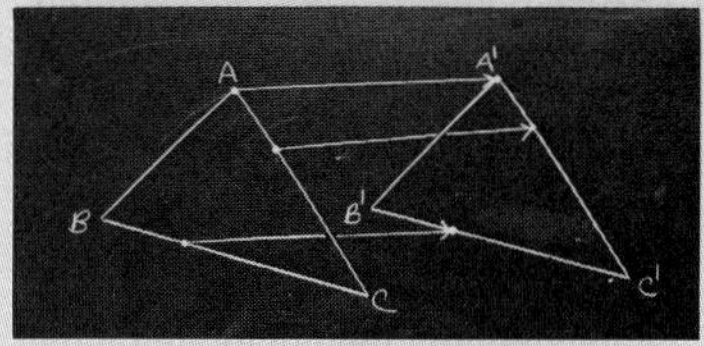

This kind of movement can be called a **translation.**

Another movement is to rotate (turn) △ABC around some fixed point. Rotating the triangle around point C, the result could be △A′B′C.

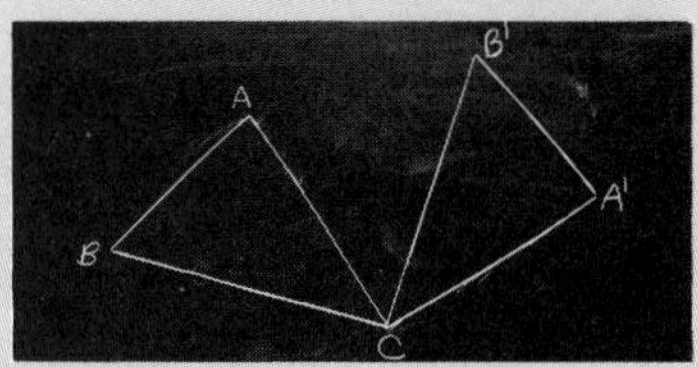

This kind of movement is called a **rotation**.

Another kind of movement is called a **reflection**. For example, the reflection of △ABC about the line l would be △A′B′C.

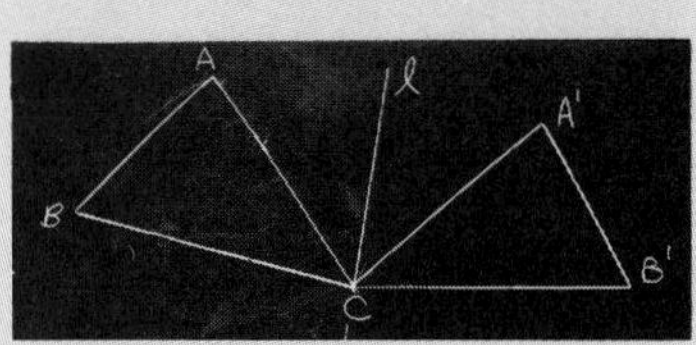

The three movements, translations, rotations, and reflections, are called **geometric transformations**.

The important properties of geometry can be described in terms of transformations. For example, if two triangles are congruent then there is a transformation that will make one triangle "exactly fit" over the other triangle. Another example involves all the points equidistant from two given points. Notice that every point on line l is equidistant from point A and its reflection A′.

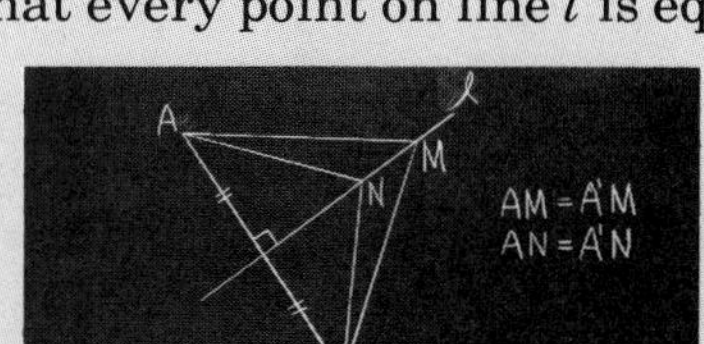

Thus the perpendicular bisector of a segment joining a point and its reflection about a line is that line of reflection.

Geometry Works: For Fencing a Field

The shape of a field is a large acute triangle. How can you separate the field into four triangles using the least amount of fencing?

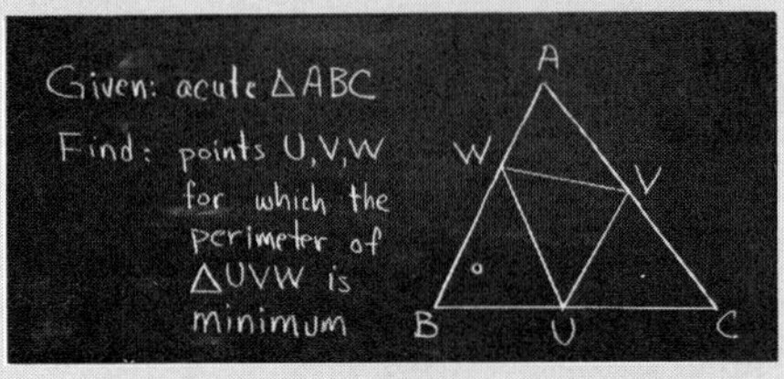

This problem can be solved using some general properties of reflections. The following results hold for any location of points U, V, and W:

Find reflections U′ and U″ (U-prime and U-double prime). Thus U′W = UW and U″V = UV.

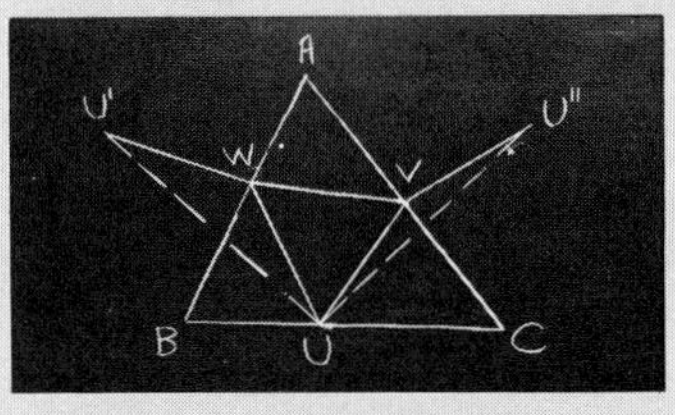

The length of U′WVU″ is equal to the perimeter of △UVW. That length is a minimum when U′WVU″ is straight.

$\overleftrightarrow{AB}$ is the ⊥ bis. of UU′, so AU′ = AU. Similarly, AU″ = AU, so △AU′U″ is isosceles. Another way to make U′U″ the smallest is to make ∠U′AU″ smaller. As ∠A′AU″ gets smaller, AU′ and AU″ must get shorter. AU′ = AU, and AU is shortest when $\overline{AU}$ ⊥ $\overline{BC}$. Thus to find point U, drop a ⊥ from A to $\overline{BC}$.

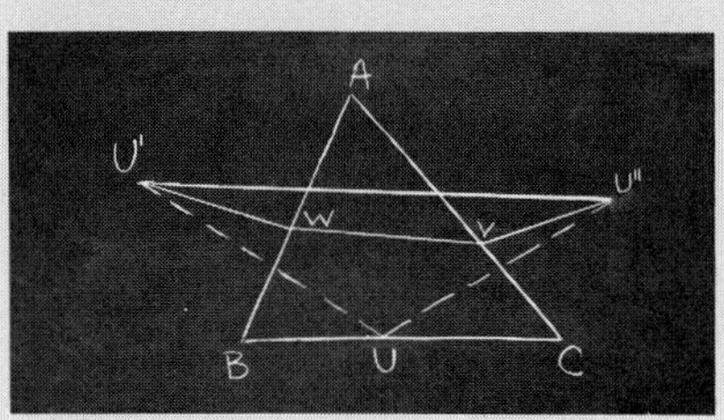

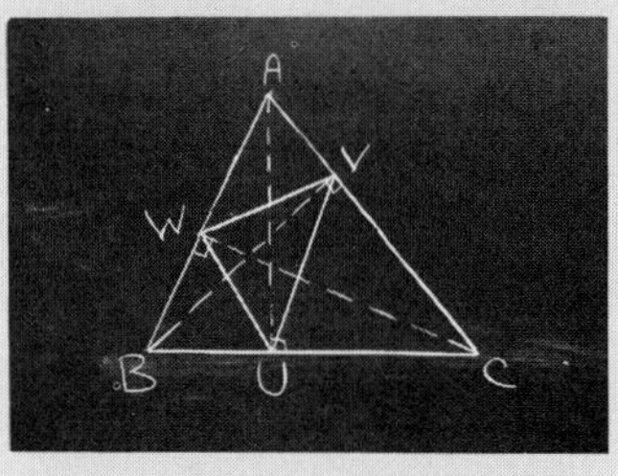

Similarly, find V and W by dropping ⊥s from B and C. △UVW will have the least perimeter of all inscribed triangles.

Projects

1 Folding a piece of paper can represent a reflection about a line. Cut a triangle out of paper. First, find the midpoints of two sides of the triangle. Then fold the triangle on the midline determined by the midpoints.

Fold points A and B to C′. What theorem does this suggest?

2 If a square is folded on a diagonal, it reflects onto itself. Thus the diagonal is a **line of symmetry**. How many lines of symmetry does a square have? a rectangle? a kite? a rhombus? an isosceles trapezoid? a circle?

3 Starting with a rectangular piece of paper ABCD, fold BA onto BA′. What shape is ABA′T? Prove it.

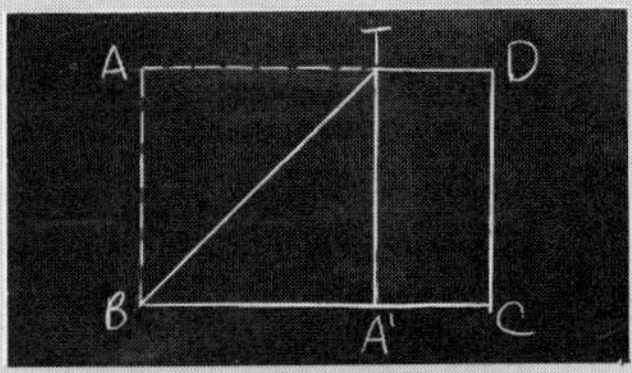

4 The Royal Road, page 715, asks for a figure in which the bisector of ∠BAC meets the ⊥ bisector of $\overline{BC}$ at point P. By folding, tell whether P will be inside or outside the triangle. Is P ever **on** the triangle?

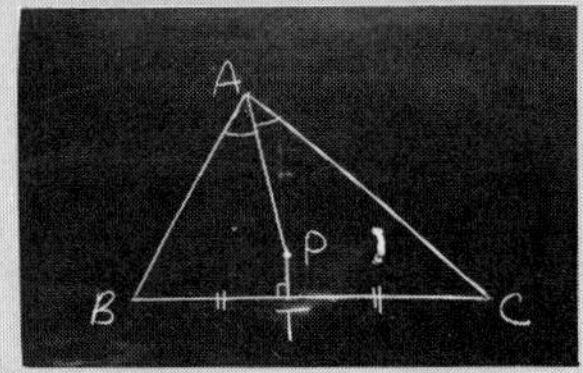

5 This two-step program takes a point (x, y) on a graph and translates it to a point 4 units to the right.

```
10 INPUT X,Y
20 PRINT "THEN NEW POINT IS:("; X+4;",";Y;")"
```

a Extend the program to translate the point N units horizontally, where N is an input value.

b Modify the program to translate the point N units horizontally and M units vertically.

c Add to the program in part **b** to print the distance between the original point and its translation.

d Write a program that takes a point (x,y) and finds the reflection over the line y = x.

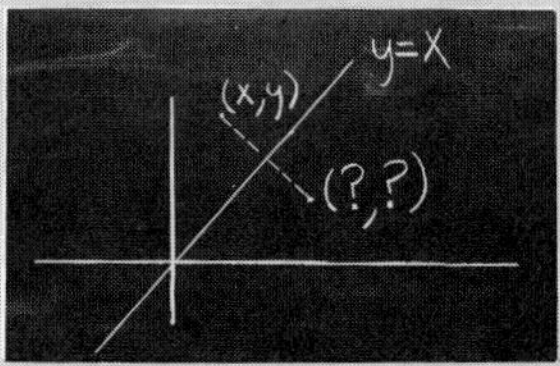

Chapter 13

COORDINATE GEOMETRY

Section 13.1 Introductory Terms

This section reviews the concepts used to graph points on a RECTANGULAR COORDINATE PLANE.

To construct a RECTANGULAR COORDINATE SYSTEM, or CARTESIAN COORDINATE SYSTEM, (named after René Descartes, 1596–1650), draw two perpendicular lines intersecting at point 0 called the ORIGIN.

The horizontal line through the origin is the X-AXIS and the vertical line is the Y-AXIS. The two AXES divide the plane into four QUADRANTS. The plane of the axes is called the real number plane and a unique ordered pair of real numbers, called COORDINATES, locates each point in the plane.

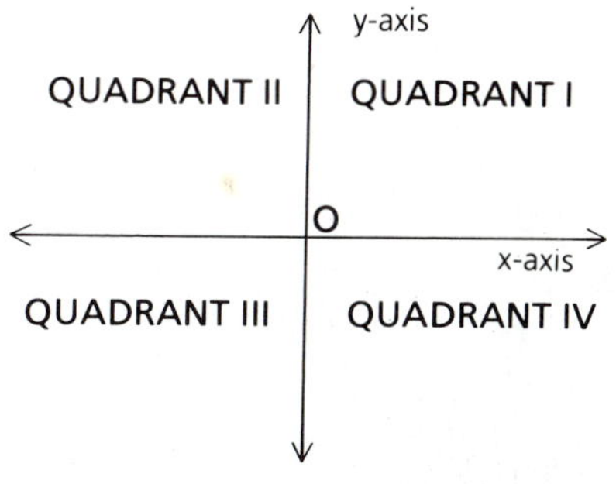

Five points are graphed and labeled with their coordinates. Point A is represented by the ordered pair (3, 2). The **X-COORDINATE,** or **ABSCISSA,** of point A is 3, and the **Y-COORDINATE,** or **ORDINATE,** of A is 2.

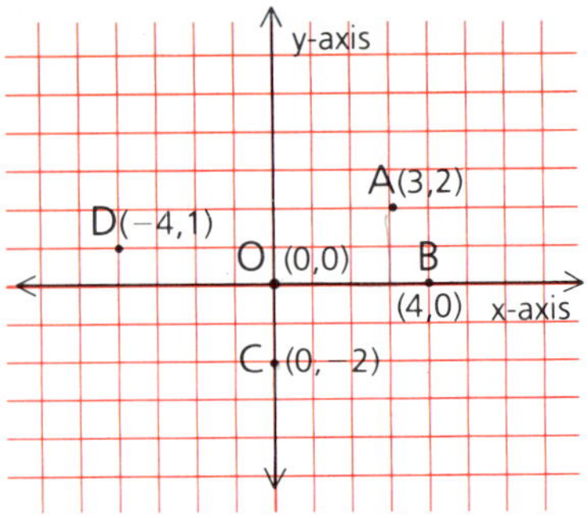

If a point lies *on* the x-axis, it has an **X-INTERCEPT.** The x-intercept of point B(4, 0) is 4. In general, the x-intercept of (x, 0) is x. The **Y-INTERCEPT** of point C(0, −2) is −2. In general, the y-intercept of (0, y) is y. Can you explain why the point A(3, 2) has neither an x-intercept nor a y-intercept?

A Cartesian Coordinate System has many uses because it combines algebra with geometry.

■ Example: To make a graph of the algebraic equation $y = 2x + 1$, first construct a **TABLE OF VALUES.** Do this by choosing values for either x or y and then substitute each value in the equation to find the other member of each ordered pair.

If $x = 1$, then $y = 2(1) + 1 = 3$.

Therefore, (1, 3) lies on the graph of $y = 2x + 1$.

Plot this point and the other points determined from the table. Connect the points with a smooth line. What is the least number of points needed to determine the line for this linear equation?

TABLE OF VALUES

for
$y = 2x + 1$

x	y
1	3
2	5
0	1
−1	−1
−2	−3

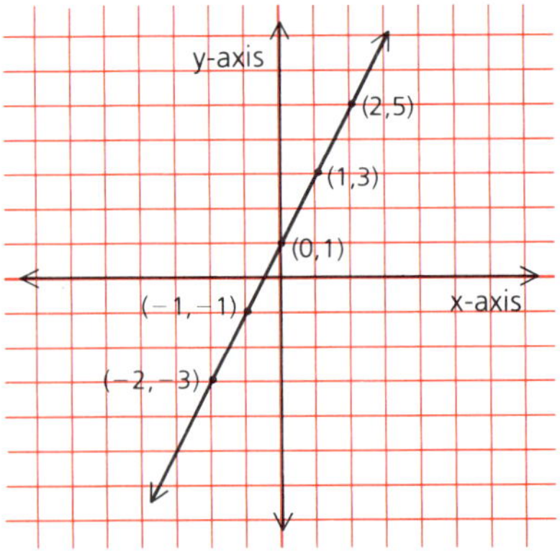

In the graph, the line $y = 2x + 1$ *appears* to cross the x-axis at $(-.5, 0)$. Can we be sure the x-intercept is $-.5$? To determine the exact value of the x-intercept, substitute 0 for y in the equation. Thus,

$$\begin{aligned} y &= 2x + 1 \\ 0 &= 2x + 1 \\ -1 &= 2x \\ -.5 &= x, \text{ the x-intercept} \end{aligned}$$

To determine the y-intercept, substitute 0 for x in the equation $y = 2x + 1$. Therefore, the y-intercept is 1, which agrees with the graph.

In summary:
To find the x-intercept(s), substitute 0 for y.
To find the y-intercept(s), substitute 0 for x.

Section 13.1 Sample Problems

1 Does the point $(-3,7)$ lie on the graph of the line $y = 2x + 10$?

Solution: Test $(-3,7)$ to see if it satisfies the equation.

$$\begin{aligned} y &\stackrel{?}{=} 2x + 10 \\ 7 &\stackrel{?}{=} 2(-3) + 10 \\ 7 &\neq 4 \end{aligned}$$

Thus, the point $(-3,7)$ is not on the line.

2 Find the intercepts of the graph of the line $y = 4x - 2$.

x-intercept (Substitute 0 for y)	*y-intercept* (Substitute 0 for x)
$y = 4x - 2$	$y = 4x - 2$
$0 = 4x - 2$	$y = 4(0) - 2$
$2 = 4x$	$y = 0 - 2$
$.5 = x$	$y = -2$

Thus, the x-intercept is .5, and the y-intercept is -2.
Note: These intercepts mean that the line will pass through the points $(.5,0)$ and $(0,-2)$.

Section 13.1 Problem Set A

1 On a coordinate system, graph the following points and connect them in the order listed: (0,5), (−6, 0), (3, 0), (0, 5), (0, 0), (0, −1), (−7, −1), (−5, −3), (1, −3), (4, −1), and (0, −1).

2 a Name the point whose abscissa is eleven more than its ordinate.

b In which quadrant, if any, is B?

c Which two points lie on a line parallel to the y-axis?

d Does D lie on $y = x + 13$?

e Find AB (the distance between A and B).

f Find BC.

g Find AC by the Pythagorean Theorem.

h Find the area of △ABC.

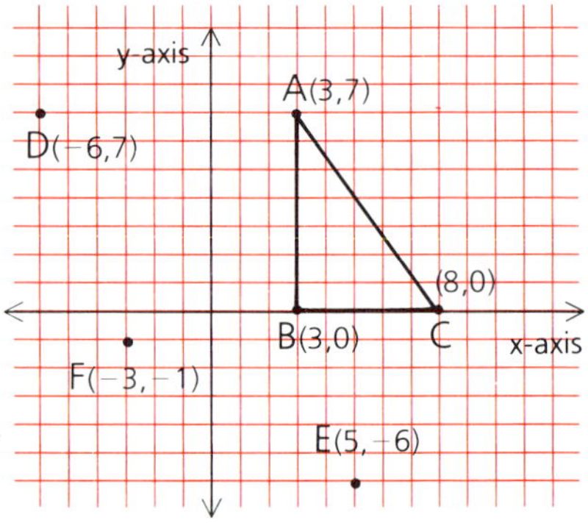

3 In which quadrant are abscissas positive and ordinates negative?

4 Make a table of values and graph the line $y = 2x - 5$.

5 If H is the point (7, 2), find the coordinates of these points:

a 4 units left of H b 8 units above H c 5 units below H

6 A square has vertices at (2, 5), (2, −1), (8, −1) and A. Find the coordinates of A.

7 OABC is a rectangle. OC = 5, OA = 8. Find the coordinates of points O, A, B, and C.

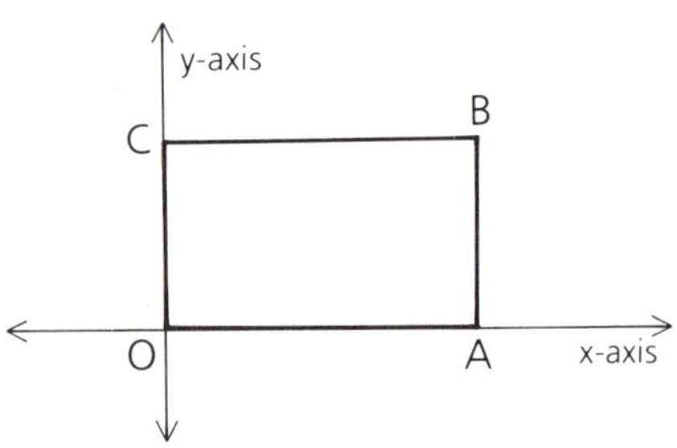

8 Use the diagram in problem 7. If OC = b and OA = a, what are the coordinates of points O, A, B, and C?

Section 13.1 Problem Set B

9 Find the area of the triangle with vertices $(-2, 0)$, $(4, 0)$, and $(2, 3)$.

10 The vertices of a right triangle are $(0, 0)$, $(3, 0)$, and $(3, 4)$.

Find: **a** the lengths of the three sides
b the length of the altitude to the hypotenuse.

11 Given: The isosceles trapezoid as shown

Find: **a** the coordinates of vertex B
b the lengths of the bases
c the length of the median
d the area

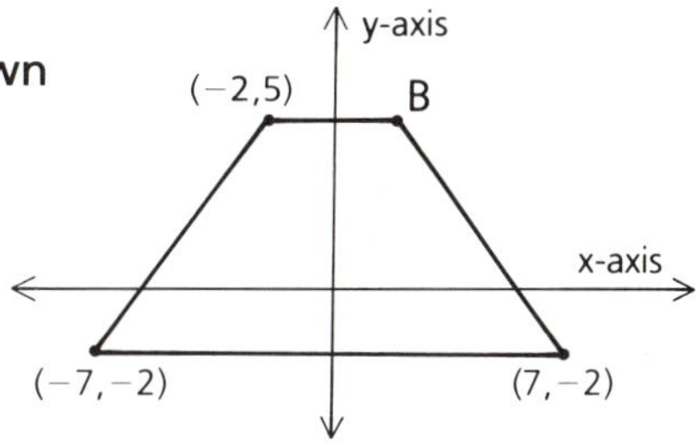

12 Use a table of values and graph the circle $x^2 + y^2 = 25$.

13 Use a table of values and graph the V-shaped figure $y = |x + 2|$.

14 Given 6 points: $(1, 4)$, $(2, 11)$, $(7, 3)$, $(4, \sqrt{3})$, $(-2, 6)$, and $(4, -8)$.
If two of these points are selected at random, what is the probability that both points lie in Quadrant I?

15 A parallelogram has vertices $(-5, -1)$, $(4, -1)$, and $(7, 6)$. Find the fourth vertex if two sides are parallel to the x-axis.

Section 13.1 Problem Set C

16 A point and its REFLECTION over a reflecting line (or point) are endpoints of a segment whose perpendicular bisector (midpoint) is the reflecting line (point). For example, $(-4, 3)$ is the reflection of $(4, 3)$ over the y-axis.
Name the reflection of the point $(-5, 6)$:

a over the x-axis **b** over the y-axis **c** over the origin

17 Find the intercepts and draw the graph of:
a $y = x\sqrt{3} + 6$ **b** $y = x^2 - 8x - 20$

Section 13.2 The Midpoint Formula

■ Example 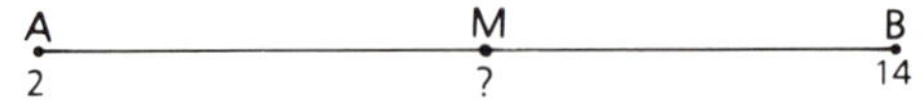

On the number line shown, the coordinate of A is 2 and the coordinate of B is 14. Find the coordinate of M, the midpoint of $\overline{AB}$.

Solution: There are several ways to do this problem, but we choose the AVERAGING PROCESS. To average two numbers, divide their sum by 2. Let the coordinate of M be x_m (read x sub m).

$$x_m = \frac{2 + 14}{2}$$
$$x_m = \frac{16}{2} = 8$$

Check: $AM = 8 - 2 = 6$
$MB = 14 - 8 = 6$
So 8 is the coordinate of midpoint M.

The averaging process may be used to find the coordinates of the midpoint of any segment in the coordinate plane. The proof of the formula is left to you.

Theorem 111 The Midpoint Formula

T **If $A = (x_1,y_1)$ and $B = (x_2,y_2)$, then the midpoint $M(x_m,y_m)$ of $\overline{AB}$ can be found by using the averaging process:**

$$\mathbf{M = (x_m,y_m) = \left(\frac{x_1 + x_2}{2}, \frac{y_1 + y_2}{2}\right)}$$

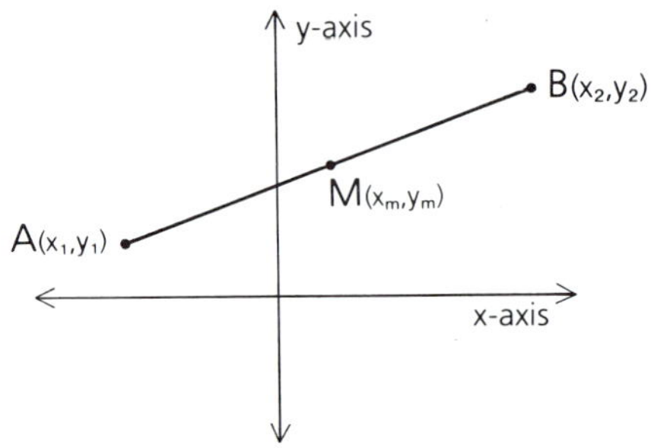

Section 13.2 Sample Problems

1 **Find the coordinates of M, the midpoint of $\overline{AB}$.**

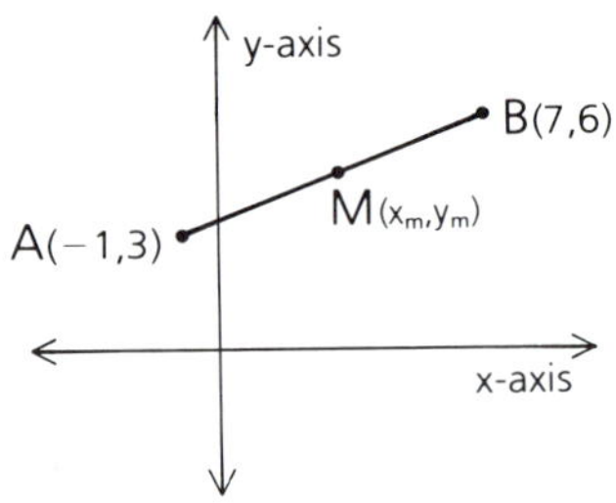

Solution:

Use the Midpoint Formula

$$x_m = \frac{x_1 + x_2}{2} = \frac{-1 + 7}{2}, \quad x_m = 3$$

$$y_m = \frac{y_1 + y_2}{2} = \frac{3 + 6}{2}, \quad y_m = 4\tfrac{1}{2}$$

Thus, $M = (x_m,y_m) = (3, 4\tfrac{1}{2})$

2 THE CASE OF THE MISSING ENDPOINT

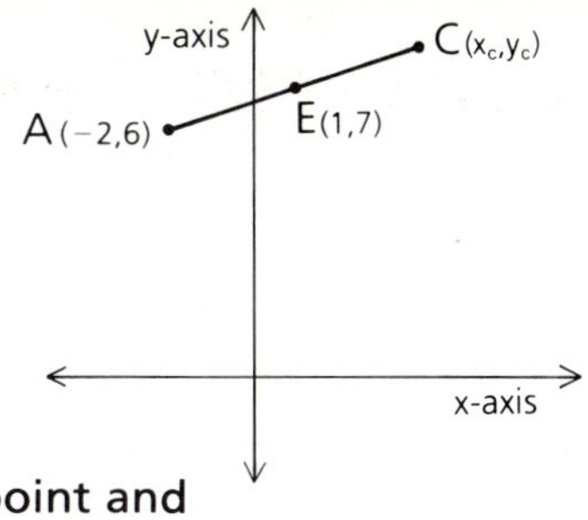

E is the midpoint of $\overline{AC}$.
$A = (-2, 6)$, $E = (1, 7)$
Find the coordinates of C.

Solution: Use the Midpoint Formula

This time, however, we know one endpoint and the midpoint, and are trying to find the other endpoint.

$$x_m = \frac{x_1 + x_2}{2} \qquad y_m = \frac{y_1 + y_2}{2}$$

$$1 = \frac{-2 + x_c}{2} \qquad 7 = \frac{6 + y_c}{2}$$

$$2 = -2 + x_c \qquad 14 = 6 + y_c$$

$$4 = x_c \qquad 8 = y_c$$

Thus, $C = (x_c, y_c) = (4,8)$

Section 13.2 Problem Set A

1 Determine the coordinates of the midpoint of each segment, given the endpoints:

- **a** (4, 0) and (6, 0)
- **b** (−5, 1) and (7, −3)
- **c** (−8, −4) and (−5, 10)
- **d** $(\sqrt{50}, 8)$ and $(\sqrt{18}, -4)$
- **e** $(1\frac{1}{2}, 6)$ and $(0, -4)$
- **f** (−2.7, −4.8) and (1.8, −3.1)

2 Find the midpoint of each side of $\triangle ABC$.

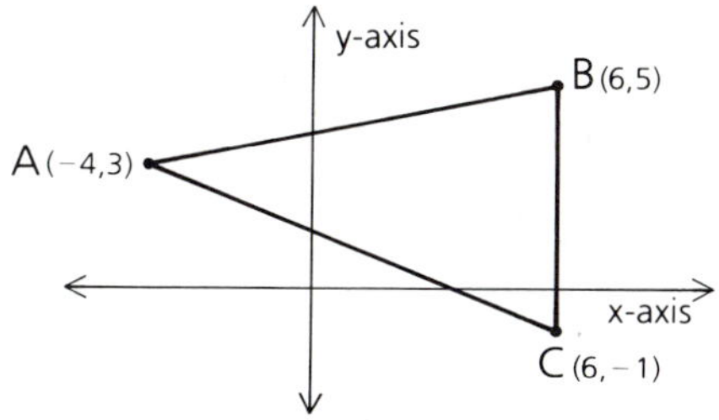

3 Find the midpoint of each side of parallelogram OADC.

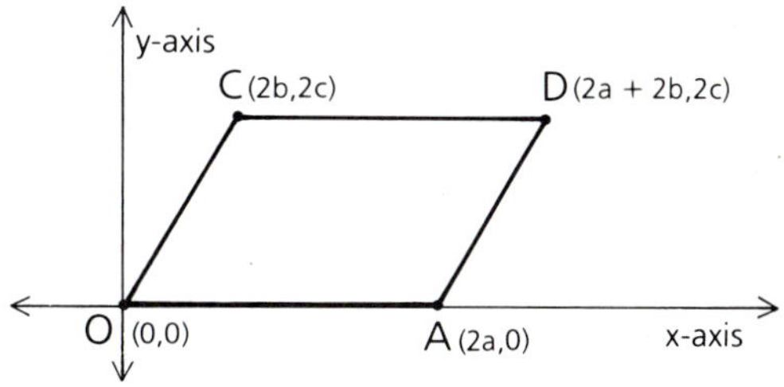

4 A quadrilateral ABCD has vertices A(−1, 5), B(4, 6), C(3, −2), D(−2, 0). Do the diagonals bisect one another?

5 The endpoints of a diameter of a circle are located at (−2, 7) and (6, 21). Find the coordinates of the center of the circle.

Section 13.2 Problem Set B

6 M(2, 5) is the midpoint of $\overline{AB}$. If A = (0, 9), find B.

7 Find the coordinates of the point that is three-fourths of the way from (2, −3) to (6, −11).

8 In △DEF, D = (−2, 6), E = (4, 10), F = (11, 5). Find the coordinates of the point where the median from F intersects $\overline{DE}$.

9 In parallelogram HJKM, H = (−5, 2) and M = (3, 8). The diagonals intersect at P (6, 3). Find the coordinates of J and K.

10 Two vertices of an equilateral triangle are (0, 0) and (4, 0). Find the two possibilities for the coordinates of the remaining vertex.

11 Given △ABC with A = (1, 2), B = (9, 2), and C = (5, 6). Find the length of the median from C to $\overline{AB}$.

12 (7, 6x + 1), (3x + 1, 15), and (9x + 7, 2x − 8) represent the endpoints of $\overline{PR}$ and its midpoint Q, in some order. What are the coordinates of point Q?

13 Find the point that divides the segment from (−11, 3) to (9, 3) in the ratio 2:3.

14 Find the trisection points of the segment joining (−5, 4) and (13, −9).

Section 13.3 Slope of a Line

A DEFINITION OF SLOPE OF A LINE
B PHYSICAL INTERPRETATION OF SLOPE
C SLOPES OF PARALLEL LINES
D SLOPES OF PERPENDICULAR LINES

Section 13.3 (A) Definition of Slope of a Line

D **The SLOPE m of a non-vertical line, segment, or ray containing (x_1,y_1) and (x_2,y_2) is defined by the formula**

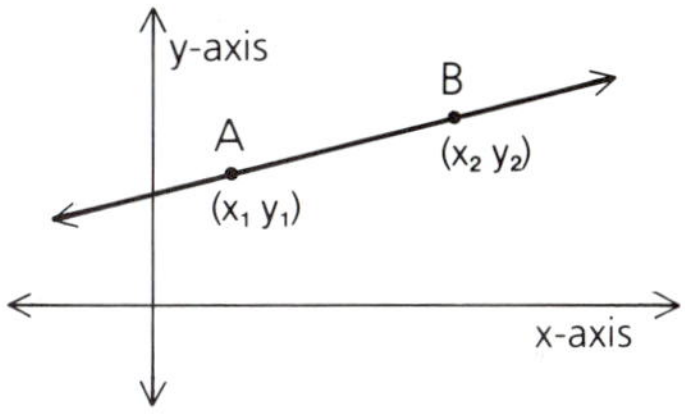

$$m = \frac{y_2 - y_1}{x_2 - x_1} \text{ or } \frac{y_1 - y_2}{x_1 - x_2} \text{ or } \frac{\triangle y}{\triangle x} \text{ or } \frac{\text{rise}}{\text{run}}$$

(Note: In more advanced mathematics courses it is common to replace $y_2 - y_1$ by $\triangle y$ [read delta y] and $x_2 - x_1$ by $\triangle x$ [delta x]. $\triangle y$ may also be regarded as "change in y," meaning *the change of y-coordinates between A and B.* Likewise $\triangle x$ may be read "change in x.")

■ Example: Find the slope of the segment joining (−2, 3) and (6, 5).

$$m = \frac{y_2 - y_1}{x_2 - x_1} \qquad \text{or} \qquad m = \frac{y_1 - y_2}{x_1 - x_2}$$

$$m = \frac{5 - 3}{6 - (-2)} \qquad\qquad = \frac{3 - 5}{-2 - 6}$$

$$m = \frac{2}{8} = \frac{1}{4} \qquad\qquad m = \frac{-2}{-8} = \frac{1}{4}$$

Notice that it does not matter which point is chosen as (x_1, y_1).

When the formula is applied to a vertical line, such as $\overleftrightarrow{CD}$, the denominator is zero. Division by zero is **UNDEFINED,** so a **VERTICAL** line has **NO SLOPE**.

$$m \stackrel{?}{=} \frac{y_2 - y_1}{x_2 - x_1}$$

$$\stackrel{?}{=} \frac{12 - 2}{6 - 6}$$

$$\stackrel{?}{=} \frac{10}{0}, \text{ which is undefined}$$

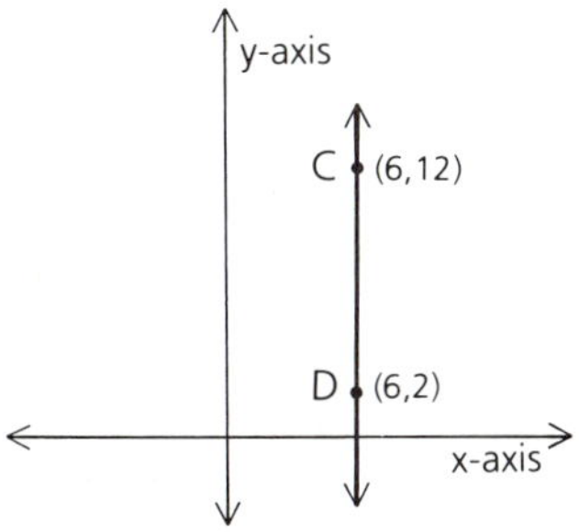

Do not confuse no slope with a slope that is zero. For horizontal lines, $y_2 = y_1$, but $x_2 \neq x_1$. Therefore, the numerator is zero, while the denominator is not. Hence, a **HORIZONTAL** line has **ZERO SLOPE.**

13.3 (B) Physical Interpretation of Slope

A slope is a number that gives a clue to the direction a line is taking. The following diagrams illustrate this notion.

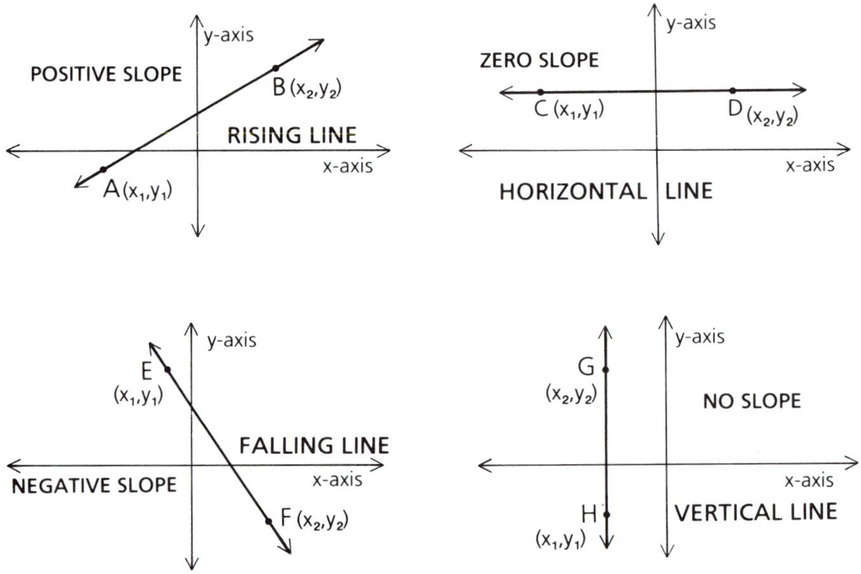

In summary,

RISING LINE	$\Longleftrightarrow$	POSITIVE SLOPE
HORIZONTAL LINE	$\Longleftrightarrow$	ZERO SLOPE
FALLING LINE	$\Longleftrightarrow$	NEGATIVE SLOPE
VERTICAL LINE	$\Longleftrightarrow$	NO SLOPE

Section 13.3 (C) Slopes of Parallel Lines

Since slope gives a clue to direction, parallel lines ought to have equal slopes.
In the coordinate plane, all vertical lines are parallel to the y-axis, and all vertical lines have no slope.
The proof of the following theorem is left to the student.

Theorem 112

T **If two non-vertical lines are parallel, then their slopes are equal.**

Given: $\overleftrightarrow{AB} \parallel \overleftrightarrow{CD}$
Prove: slope $\overleftrightarrow{AB}$ = slope $\overleftrightarrow{CD}$

Hints for Proof:
Draw $\overleftrightarrow{AE}$ and $\overleftrightarrow{CF} \parallel$ x-axis.
Draw $\overleftrightarrow{BE}$ and $\overleftrightarrow{DF} \parallel$ y-axis.
$\triangle ABE \sim \triangle CDF$. (Why?)
$\frac{BE}{AE} = \frac{DF}{CF}$. (Why?)

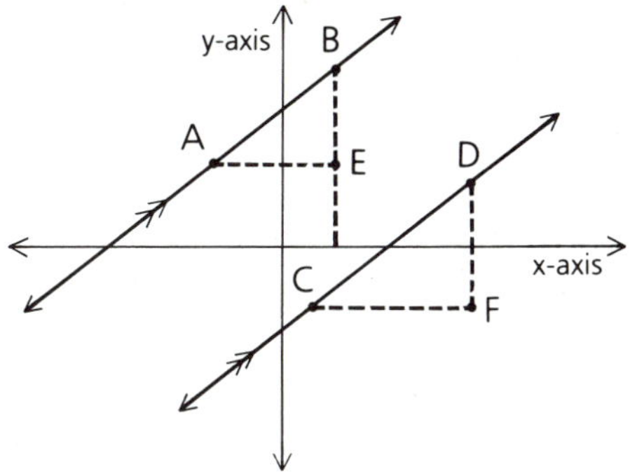

The converse theorem, which reverses the "if" and "then" parts of Theorem 112, is also true.

Theorem 113

T **If the slopes of two non-vertical lines are equal, then the lines are parallel.**

(Outline of a proof: Use the diagram for Theorem 112. If slope $\overrightarrow{AB}$ = slope $\overleftrightarrow{CD}$, the triangles are similar by SAS ~, and ≅ corresponding ∠s make $\overleftrightarrow{AB} \parallel \overleftrightarrow{CD}$.)

Section 13.3 (D) Slopes of Perpendicular Lines

Since the axes are perpendicular, any horizontal line is perpendicular to any vertical line. Examine the diagram and follow the informal discussion to discover what is true about the slopes of other lines that are perpendicular.

Given: $\ell_1 \perp \ell_2$
Draw $\overline{PS} \perp$ x-axis
$\triangle APS \sim \triangle PBS$ (Why?)
So $\frac{y}{c} = \frac{d}{y}$
Since ℓ_2 is rising, slope $\ell_2 = \frac{y}{c}$
Since ℓ_1 is falling, slope $\ell_1 = -\frac{y}{d}$
$-$slope $\ell_1 = \frac{y}{d}$
By substitution
slope $\ell_2 = \frac{1}{-\text{slope } \ell_1}$

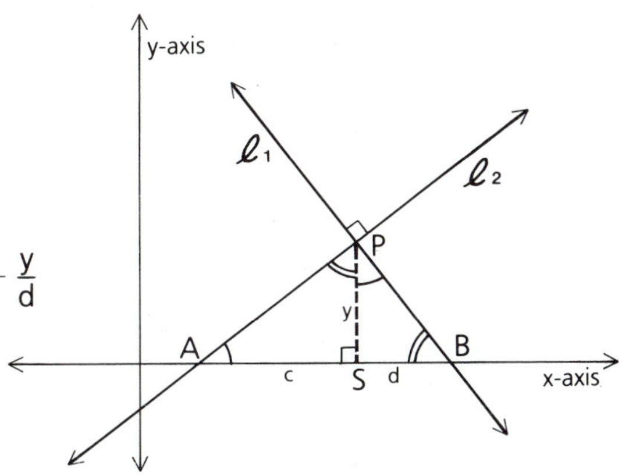

Therefore, the slopes of perpendicular lines are NEGATIVE RECIPROCALS of each other. Similarly, the converse can be proved. In summary:

Theorem 114

T **If two lines are perpendicular, their slopes are NEGATIVE RECIPROCALS (with the exception of a horizontal and a vertical line).**

Theorem 115

T **If the slopes of two lines are NEGATIVE RECIPROCALS, the lines are perpendicular.**

Section 13.3 Sample Problems

1 If A = (4, −6) and B = (−2, −8), find the slope of $\overleftrightarrow{AB}$.

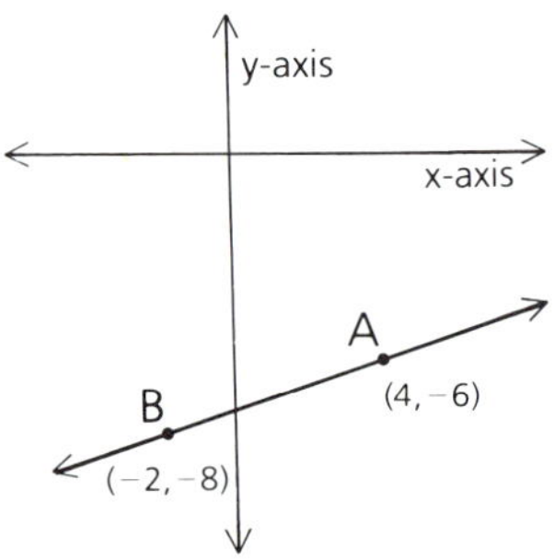

Solution: By the slope formula:

$$m = \frac{y_2 - y_1}{x_2 - x_1} = \frac{-8 - (-6)}{-2 - 4} = \frac{-8 + 6}{-6} = \frac{1}{3}$$

(Note: The line is rising, so the slope is positive. A diagram helps prevent careless errors.)

2 Show that CEF is a right triangle.

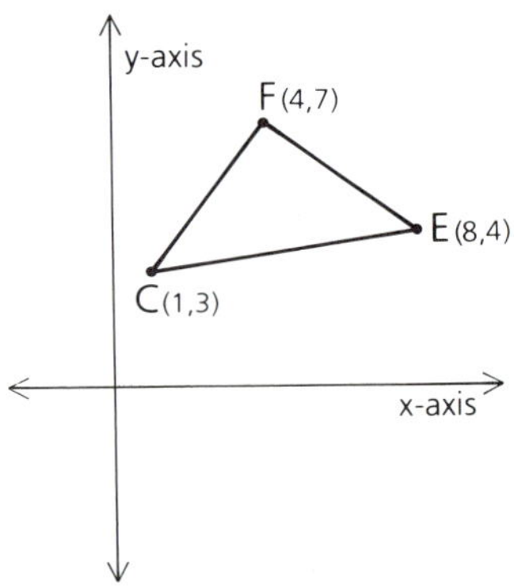

Solution: Find the slopes of the sides.

$$\text{Slope of } \overleftrightarrow{CE} = \frac{\triangle y}{\triangle x} = \frac{4-3}{8-1} = \frac{1}{7}$$

$$\text{Slope of } \overleftrightarrow{FE} = \frac{\triangle y}{\triangle x} = \frac{7-4}{4-8} = \frac{3}{-4} = -\frac{3}{4}$$

$$\text{Slope of } \overleftrightarrow{FC} = \frac{\triangle y}{\triangle x} = \frac{3-7}{1-4} = \frac{-4}{-3} = \frac{4}{3}$$

Since the slopes of $\overleftrightarrow{FE}$ and $\overleftrightarrow{FC}$ are negative recipiocals, $\angle$F is a right angle and $\triangle$CEF is a right triangle.

3 Given: $\triangle ABE$ as shown

Find: **a** the slope of altitude $\overline{AC}$

b Find the slope of median $\overline{AD}$

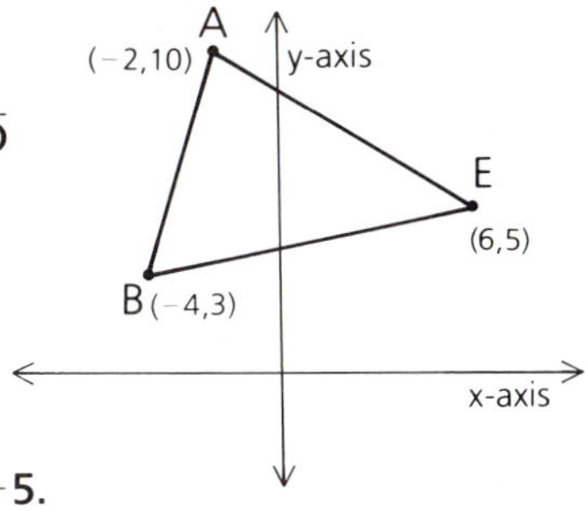

Solutions:

a Slope $\overleftrightarrow{BE} = \frac{\triangle y}{\triangle x} = \frac{5-3}{6-(-4)} = \frac{1}{5}$

Then by the Negative Reciprocal property of $\perp$ lines, slope of $\overleftrightarrow{AC} = -5$.

b By the Midpoint Formula, D = (1, 4). Then by using A(−2, 10) and D(1, 4), slope of $\overleftrightarrow{AD} = \frac{\triangle y}{\triangle x} = \frac{4-10}{1-(-2)} = -2$.

Section 13.3 Problem Set A

1 Find the slope determined by each pair of points.

a (1, 7) and (10,15)

b (−2, 6) and (5, 7)

c (−8, −7) and (−2, 4)

d (5, 4) and (−2, 4)

e $(\sqrt{3}, 7)$ and $(\sqrt{3}, -9)$

f (5a, 6c) and (2a, −9c)

2 $\overleftrightarrow{AB}$ has a slope of $1\frac{2}{3}$. $\overleftrightarrow{CD} \perp \overleftrightarrow{AB}$. What is the slope of $\overleftrightarrow{CD}$?

3 $\overleftrightarrow{EF} \parallel \overleftrightarrow{GH}$. If $\overleftrightarrow{EF}$ has a slope of −4, what is the slope of $\overleftrightarrow{GH}$?

4 Given the diagram as marked.

$\overline{AC}$ is an altitude. $\overline{AD}$ is a median.

Find the slope of:

a $\overleftrightarrow{BE}$ **b** $\overleftrightarrow{AC}$ **c** $\overleftrightarrow{AD}$

d a line through A and parallel to $\overline{BE}$

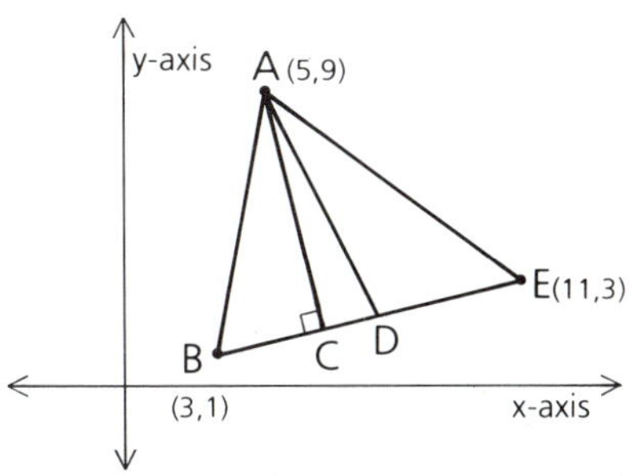

5 F = (2, 8), H = (9, 10), J = (13, 5), and K = (6, 3).

Show that FHJK is a parallelogram.

(Hint: Show that *both* pairs of opposite sides are parallel.)

6 In the diagram as marked,

a find the slope of $\overleftrightarrow{PT}$.

b find the slope of $\overleftrightarrow{TV}$.

c are P, T, and V collinear or non-collinear?

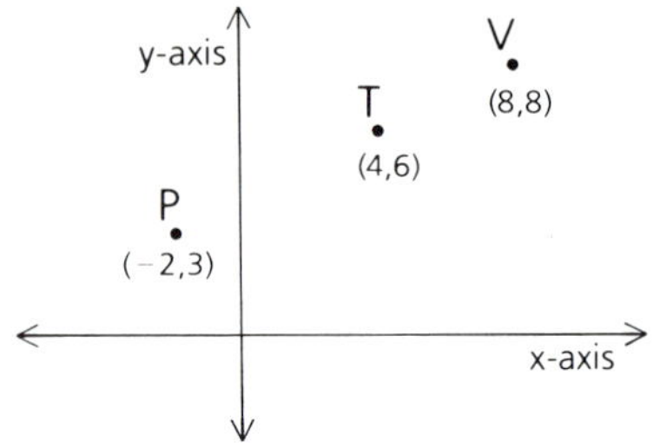

7 $\overleftrightarrow{AB}$ has a slope of $2\frac{1}{2}$. If A = (2, 7) and B = (12, k), solve for k.

8 Given A = (6,11), B = (1, 5), C = (7, 0). Show without graphing that △ABC is a right triangle. Name the hypotenuse.

Section 13.3 Problem Set B

9 Express the slope of the line joining (x_1, y_1) to (x_3, y_3).

10 In △ABC, A = (−6, −2), B = (−4, 5), and C = (8, 1). Find the slope of the perpendicular bisector of $\overline{AB}$.

11 Find the slope of the line represented by: $2x + 3y = 6$.
(Hint: Use a table of values to find two points on the line.)

12 Is (7, 8) on a line passing through (1, 5) if the line has a slope .5?

13 In the diagram, D is one-half the distance from A to B, and E is one-fourth the distance from A to C. Find the slope of $\overleftrightarrow{DE}$.

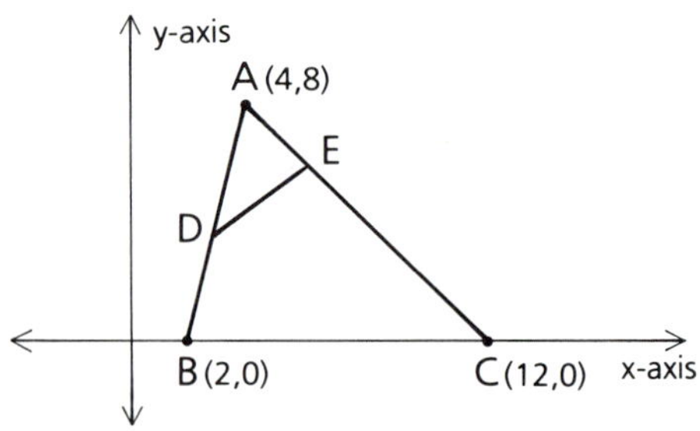

14 a A ray is drawn from the origin so that it bisects Quadrant I. What is the slope of the ray?

b A ray is drawn from the origin so that it bisects Quadrant IV. What is the slope of the ray?

15 a Are $(-6, 5)$, $(1, 7)$, and $(15, 10)$ collinear?

b Are $(74, 20)$, $(50, 16)$, and $(2, 8)$ collinear?

16 In what type of situation are two lines perpendicular if the slopes of the lines are *not* negative reciprocals?

17 In isosceles trapezoid ABCD, $A = (0, 0)$, $B = (14, 0)$, $D = (2, 6)$. If $\overline{AB}$ is the lower base, find

a the coordinates of C

b the area of the isosceles trapezoid.

18 See Section 13.2, problem 3. Show that a figure with such vertices really is a parallelogram.

Section 13.3 Problem Set C

19 A 30°–60°–90° triangle is placed with its 60° vertex at the origin, its shortest side along the x-axis, and the hypotenuse in Quadrant I. Find the slope of the hypotenuse.

20 Show that the quadrilateral formed by joining $(0, 0)$, $(6, 8)$, $(16, 8)$, and $(10, 0)$ is a rhombus.

21 Show that if the midpoints of the sides of any quadrilateral are joined in order, the figure formed is a parallelogram.

Let the quadrilateral have vertices of: $(2a, 2b)$, $(2c, 2d)$, $(2e, 2f)$, $(2g, 2h)$.

22 Three of the four vertices of a parallelogram are $(-4, 5)$, $(0, -2)$, and $(6, 8)$. Find the three possibilities for the coordinates of the remaining vertex.

23 Given five lines with the following slopes:

$\frac{2}{3} \quad -\frac{3}{2} \quad \frac{4}{5} \quad -\frac{5}{4} \quad \frac{2}{7}$

If three of the lines are chosen at random, what is the probability that a right triangle will be formed?

Section 13.4 Equations of Lines

A EQUATIONS FOR NON-VERTICAL LINES
B EQUATIONS FOR HORIZONTAL LINES
C EQUATIONS FOR VERTICAL LINES
D FORMS FOR THE EQUATION OF A LINE

13.4(A) Equations of Non-Vertical Lines

Consider the line with a y-intercept of b and a slope of m. Then (0, b) is one point on the line.
Let (x, y) represent any other point on the line.

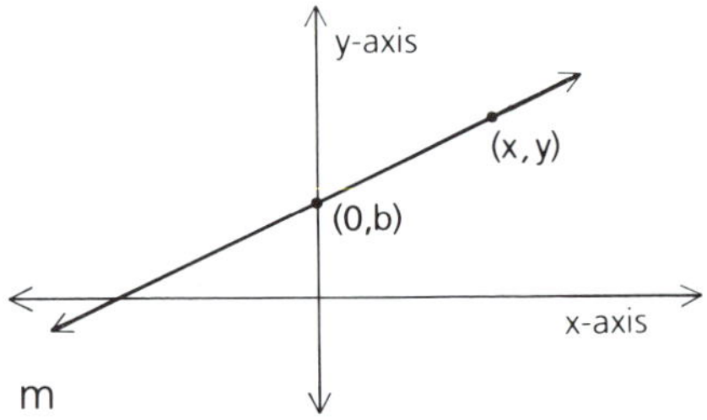

Substituting in the Slope Formula: $\frac{y - b}{x - 0} = m$

$$y - b = mx + o$$
$$y = mx + b$$

Theorem 116

T **Y-FORM, or SLOPE-INTERCEPT FORM, of the equation of a non-vertical line is**

y = mx + b	**where b is the y-intercept of the line** **m is the slope of the line**

■ *Example:*

Use the Y-form to write an equation for the line containing (−1, 4) and (1, 8).

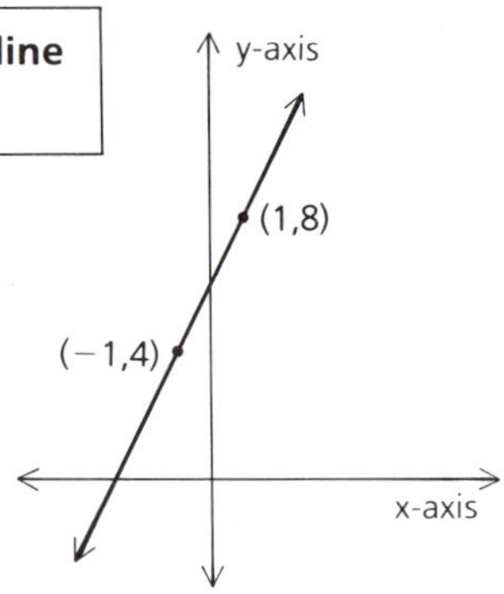

Solution:
First find the slope: $m = \frac{8 - 4}{1 - (-1)} = \frac{4}{2} = 2$

Since the line has a slope, apply the Y-form: $y = mx + b$
Substituting 2 for m and (1, 8) for (x, y): $8 = 2(1) + b$
$6 = b$

Therefore, the equation is $y = 2x + 6$.
You will get the same equation if you use (−1, 4) instead of (1, 8) for (x, y). Try it and see!

13.4(B) Equations of Horizontal Lines

Since a horizontal line is non-vertical, the Y-FORM is used to develop the formula for the equation of any horizontal line.

$\overleftrightarrow{AB}$ is a horizontal line. Every point has the same ordinate. Thus, the y-intercept is 4, that is, $b = 4$. The slope of a horizontal line is zero, so $m = 0$.

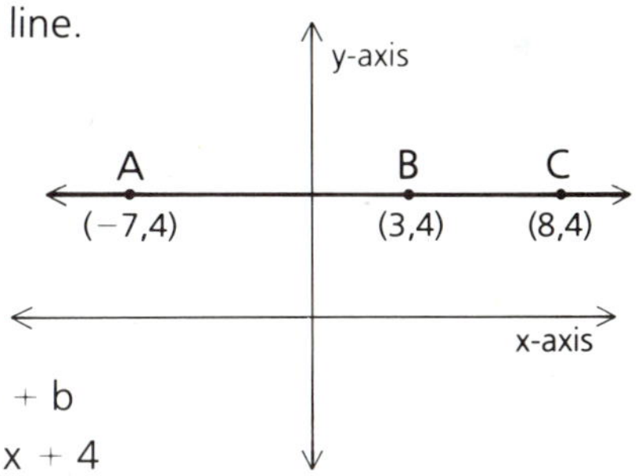

$$\begin{aligned} \text{Y-form} \quad y &= mx + b \\ y &= 0 \cdot x + 4 \\ \text{Equation of } \overleftrightarrow{AB} \text{ is} \quad y &= 4 \end{aligned}$$

In general, every point on a horizontal line has the same y-coordinate b, while the x-coordinate can be any real number. However, since the slope m is zero, x does not appear in the final equation for the line.

Theorem 117

T **The formula for the equation of a HORIZONTAL LINE is**

$y = b$	**where b is the y-coordinate of every point on the line.**

The trick is to *recognize* a horizontal line, which may be disguised in a problem.

■ *Examples* of horizontal lines:

Find the equation of:

	Answers
the line containing (2, 5) and (24, 5)	$y = 5$
the x-axis	$y = 0$
the line $7\frac{1}{2}$ units below the x-axis	$y = -7\frac{1}{2}$
the line perpendicular to the y-axis and passing through $(11, \sqrt{3})$	$y = \sqrt{3}$

13.4 (C) Equations of Vertical Lines

A vertical line has **NO** slope. Therefore the previous formulas cannot apply. However, every point on a vertical line has the same x-coordinate (abscissa), while the y-coordinate may be any real number.

Theorem 118

The formula for the equation of a VERTICAL LINE is

$x = a$	**where a is the x-coordinate of every point on the line.**

The trick is to *recognize* when a line is vertical.

■ *Examples* of vertical lines:

Find the equation of each of these lines:

the line containing $(-1, 6)$ and $(-1, 7)$	the line with an x-intercept of 8 and passing through $(8, 5\sqrt{2})$	the line containing $(-10, 4)$ and perpendicular to line $y = 7$
Ans. $x = -1$	Ans. $x = 8$	Ans. $x = -10$

13.4 (D) Forms for the Equation of a Line

We emphasize the Y-form, but the other forms in the following table may be helpful, especially if you have used them before.

Equations of a Line

FORM	FORMULA	PERMITTED USE
Slope-Intercept (Y-form)	$y = mx + b$ m = slope b = y-intercept	Non-vertical lines only
Point-Slope	$y - y_1 = m(x - x_1)$ m = slope (x_1, y_1) = known point	Non-vertical lines only
Two-Point	$\frac{y - y_1}{x - x_1} = \frac{y_2 - y_1}{x_2 - x_1}$ (x_1, y_1) and (x_2, y_2) are known points	Non-vertical lines only
General Linear	$ax + by + c = 0$ a, b, c are real numbers	Any line
Intercept	$\frac{x}{a} + \frac{y}{b} = 1$ a = x-intercept b = y-intercept	Lines *not passing* through the origin (non-zero intercepts)

Section 13.4 Sample Problems

1 Find the equation of the line containing $(7, -3)$ and $(4, 1)$.

Solution: First find the slope. $m = \frac{1 - (-3)}{4 - 7} = \frac{4}{-3} = -\frac{4}{3}$

Then substitute in the Y-form formula, using either $(7, -3)$ or $(4, 1)$ for (x, y).

$$y = mx + b$$
$$1 = -\frac{4}{3}(4) + b$$
$$\frac{3}{3} = -\frac{16}{3} + b$$
$$\frac{19}{3} = b$$
$$\text{Thus, } y = -\frac{4}{3}x + \frac{19}{3}$$

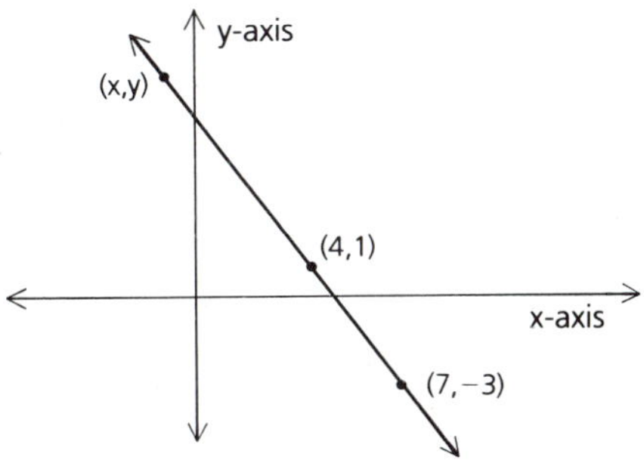

2 Find the equation of the line of slope 3 and x-intercept 5.

Solution: If the line has an x-intercept of 5, it must contain the point (5, 0). Therefore, (5, 0) is substituted for (x, y) and the given slope for m in the Y-form formula.

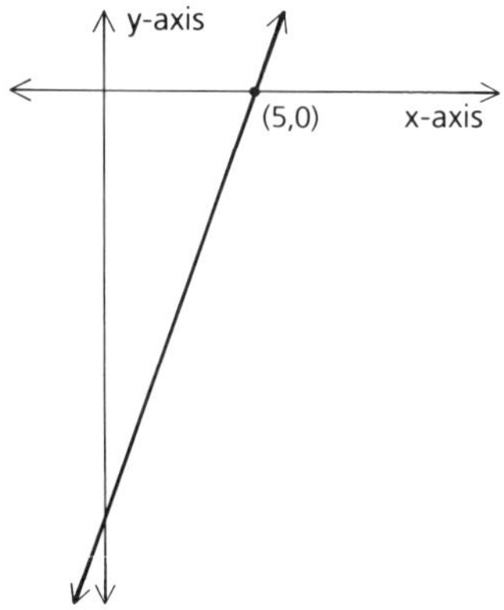

$$\begin{aligned} y &= mx + b \\ 0 &= 3(5) + b \\ 0 &= 15 + b \\ -15 &= b \end{aligned}$$

Thus, $y = 3x - 15$ is the equation of the line.

3 a Find the equation of the line passing through (2, 5) and (17, 5).

Solution: The line is horizontal, so the equation is $y = 5$.

b Find the equation of the line parallel to the y-axis and containing $(-\sqrt{6}, 1)$.

Solution: The line is vertical, so the equation is $x = -\sqrt{6}$.

4 Given $\triangle ABC$ with $A = (-3, 2)$, $B = (9, 4)$, and $C = (5, 12)$.

a Find the equation of the median to $\overline{AB}$.

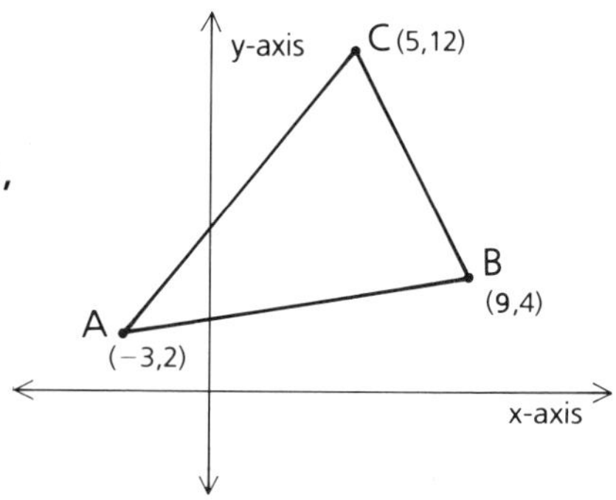

Solution: By calculation, the midpoint of $\overline{AB} = (3, 3)$.
Let $(3, 3) = (x_1, y_1)$ and $(5, 12) = (x_2, y_2)$ in the Two Point Formula.

$$\begin{aligned} \frac{y - y_1}{x - x_1} &= \frac{y_2 - y_1}{x_2 - x_1} \\ \frac{y - 3}{x - 3} &= \frac{12 - 3}{5 - 3} = \frac{9}{2} \\ 2y - 6 &= 9x - 27 \\ 2y &= 9x - 21 \\ y &= \frac{9}{2}x - \frac{21}{2} \end{aligned}$$

(Note: Actually, $y = \frac{9}{2}x - \frac{21}{2}$ is the equation of the *line* containing the median, which is a segment. Unless otherwise stated, *equation of a segment or ray* will mean *equation of the containing line.*)

b Find the equation of the perpendicular bisector of $\overline{AB}$.

Solution: First, the slope of $\overleftrightarrow{AB} = \frac{4 - 2}{9 - (-3)} = \frac{1}{6}$

Since the slopes of two perpendicular lines are negative reciprocals (except for horizontal and vertical lines), the slope of the perpendicular bisector is $m = -6$. Let midpoint $(3, 3) = (x_1, y_1)$ in the Slope Definition or the Point-Slope Formula (below).

$$\begin{aligned} y - y_1 &= m(x - x_1) \\ y - 3 &= -6(x - 3) \\ y &= -6x + 21 \end{aligned}$$

c Find the equation of the altitude to $\overline{AB}$.
Solution: The altitude contains C (5, 12) and has a slope of -6 (it is perpendicular to $\overline{AB}$). Use the Y-form or the Point-Slope Formula or the Slope Definition (below).

By Y-form

$$\begin{aligned} y &= mx + b \\ 12 &= -6(5) + b \\ 42 &= b \\ y &= -6x + 42 \end{aligned}$$

By Slope Definition:

Using (5, 12) and (x, y)

$$\begin{aligned} \frac{y - 12}{x - 5} &= -6 \\ y - 12 &= -6x + 30 \\ y &= -6x + 42 \end{aligned}$$

Section 13.4 Problem Set A

1 Name the slope and y-intercept for each line.

a $y = 3x + 7$

b $y = 4x$

c $y = \frac{1}{2}x - \sqrt{3}$

d $y = 13 - 6x$

e $y = -5x - 6$

f $y = 7$

2 Place each equation in Y-form and give the slope and y-intercept for each line.

a $y - 3x = 1$

b $y + 5x = 2$

c $2x + 3y = 6$

d $7 - (6 - 2x) = 4y$

3 Find the equation of a line 6 units below and parallel to the x-axis.

4 Find the equation of a line perpendicular to the x-axis and passing through (8, 1).

5 Which two of the following three lines are parallel?

a $y = 5x - 1$ b $y = 7x + 2$ c $y = 2 + 5x$

6 Find the Y-form equation of each line.

a y-intercept = 2, slope = 4
b $m = 5$, passes through $(0, -2)$
c parallel to $y = 10x - 6$, and y-intercept = 1
d perpendicular to $2y = x + 16$, and passes through $(0, -5)$
e y-intercept of 2, and perpendicular to the line containing $(-4, 6)$ and $(1, 11)$.

7 In the graph, find

a the slope of $\overleftrightarrow{AB}$
b the equation of $\overleftrightarrow{AB}$
c the slope of $\overleftrightarrow{CD}$
d the equation of $\overleftrightarrow{CD}$

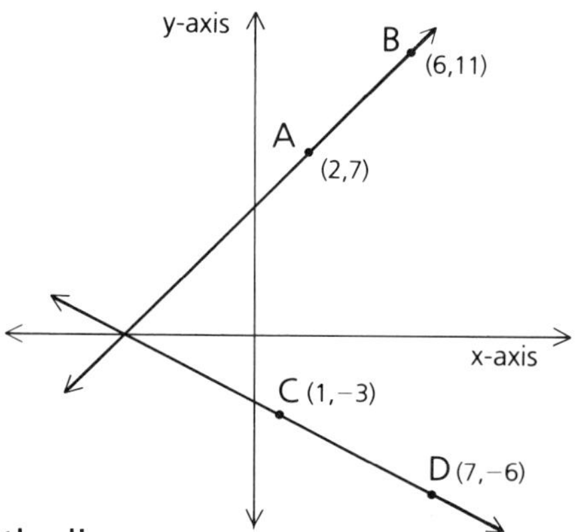

8 Find the Y-form equation of the line.

a containing (2, 1) and (3, 4)
b containing $(-6, 3)$ and $(2, -1)$
c containing (1, 5) and $(-3, 5)$
d having x-intercept of 2 and slope of 7
e having x-intercept of 3 and passing through (1, 8)
f passing through $(-3, 6)$ and $(-3, 10)$
g passing through (8, 7) and perpendicular to $3y = -2x + 24$

Section 13.4 Problem Set B

9 The line $y = 8x - 1$ contains the point $(k, 5)$. Find k.

10 Line $\overleftrightarrow{CD}$ is perpendicular to $2x + 3y = 8$.
If $C = (1, 4)$, find the equation of $\overleftrightarrow{CD}$.

11 Show that $-\frac{a}{b}$ is the slope of the line $ax + by + c = 0$.

12 Show that $-\frac{c}{b}$ is the y-intercept of the line $ax + by + c = 0$.

For **13–17** use $\triangle ABC$ in the diagram.

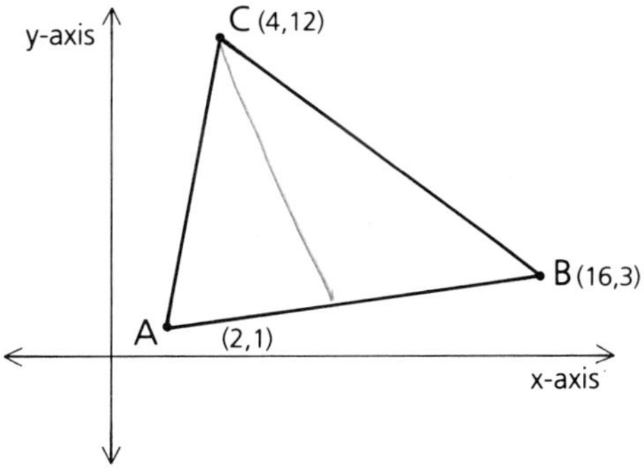

13 Find the equation of a line through C parallel to $\overleftrightarrow{AB}$.

14 Find the equation of the perpendicular bisector of $\overline{AB}$.

15 Find the equation of the altitude from C to $\overline{AB}$.

16 Find the equation of the median from C to $\overline{AB}$.

17 Find the slope of the line joining the midpoints of $\overline{AC}$ and $\overline{BC}$.

18 A line passes through a point 3 units to the left and 2 units above the origin. Find the equation of the line if it is parallel to the: **a** x-axis **b** y-axis

19 If $P = (-2, 5)$ and $R = (0, 9)$, find the equation of the perpendicular bisector of $\overline{PR}$.

Section 13.4 Problem Set C

20 In $\triangle ABC$, $A = (12, -3)$, $B = (r, 9)$, $C = (r, 19)$, and $r \neq 12$. Find the equation of the altitude from A to $\overline{BC}$.

21 Does the point $(12, -3)$ lie on the line whose slope is $-\frac{3}{4}$ and y-intercept is 5? Support your answer.

22 A line has a y-intercept of 2 and makes a 60° angle with the x-axis. Find the two possibilities for the equation of the line.

23 Find the equation of the line whose intercepts are twice those of the line $2x + 5y = 10$.

24 Find the equation of a line through point (p, q) and parallel to the line containing (a, b) and (c, d). Be sure to discuss all special cases.

25 Given $\triangle ABC$ with $A = (0, 0)$, $B = (4, 0)$, $C = (2, 6)$. Show that the medians of $\triangle ABC$ all intersect at $(2, 2)$.

Note: It can be shown that the medians of any triangle are concurrent at a point called the CENTROID of the triangle.

26 Find the center of the circle containing $D = (-3, 5)$, $E = (3, 3)$, and $F = (11, 19)$. Note: The center of this circle is called the CIRCUMCENTER of $\triangle DEF$.

27 Find the reflection of the point $(-9, 7)$ over the reference line $y = x$. (See Section 13.1, problem 16.)

Note: A point and its reflection (image) determine a segment whose perpendicular bisector is the reference line.

28 Find the equation of the reflection of the line $y = \frac{3}{4}x - 1$ over:

a the x-axis b the y-axis c the line $y = x$

Section 13.5 The Distance Formula

In $\triangle AOB$, $AO = 4$ since we can count the 4 spaces from O to A. $OB = 3$ since we can count the 3 spaces from O to B.

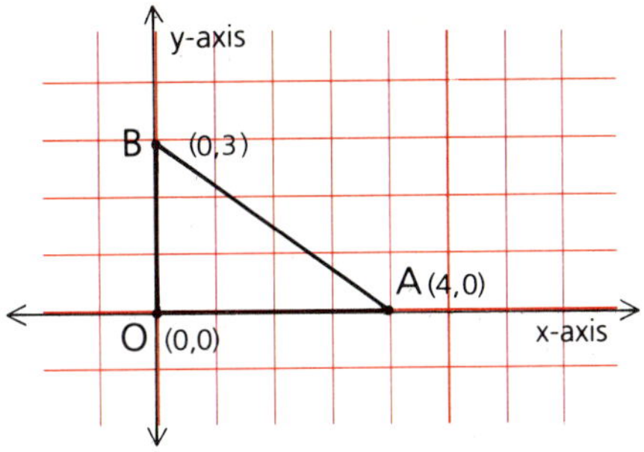

When a segment is either horizontal or vertical, its length is easily computed. To compute the length of $\overline{AB}$, we must find a new method. Since $\triangle AOB$ is a right triangle, we apply the Pythagorean Theorem:

$$\begin{aligned}(OA)^2 + (OB)^2 &= (BA)^2\\ 3^2 + 4^2 &= (BA)^2\\ 25 &= (BA)^2\\ 5 &= BA\end{aligned}$$

To compute any non-vertical, non-horizontal length, we could draw a right triangle and use the Pythagorean Theorem.

$$\begin{aligned}(AB)^2 &= (CA)^2 + (BC)^2\\ (AB)^2 &= (x_2 - x_1)^2 + (y_2 - y_1)^2\\ AB &= \sqrt{(x_2 - x_1)^2 + (y_2 - y_1)^2}\\ \text{or } AB &= \sqrt{(\triangle x)^2 + (\triangle y)^2}\end{aligned}$$

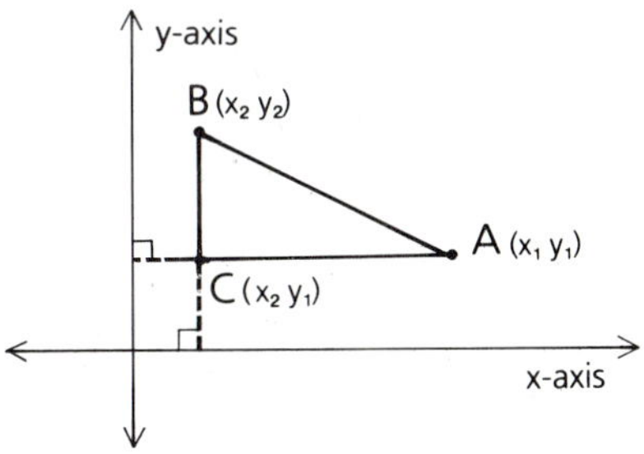

However, it is easier to use **THE DISTANCE FORMULA**, which is derived from the Pythagorean Theorem.

Theorem 119 The Distance Formula

T **If $P = (x_1, y_1)$ and $Q = (x_2, y_2)$ are any two points, then the distance between them is**

$$\begin{aligned}\mathbf{PQ} &= \sqrt{(x_2 - x_1)^2 + (y_2 - y_1)^2}\\ \textbf{or } \mathbf{PQ} &= \sqrt{(\triangle x)^2 + (\triangle y)^2}\end{aligned}$$

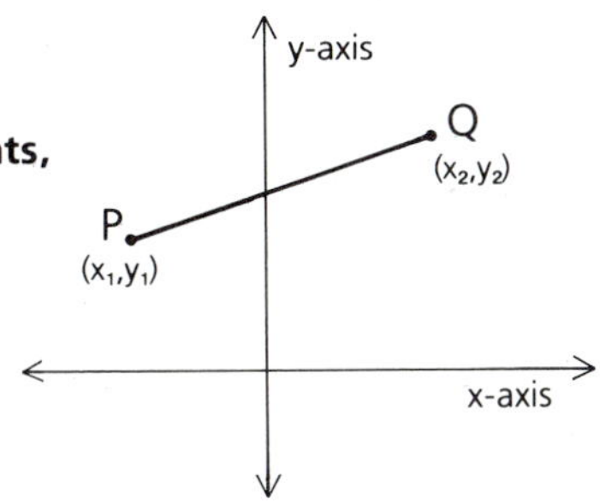

Section 13.5 Sample Problems

1 If A = (2, 3) and B = (7, 15), find AB.

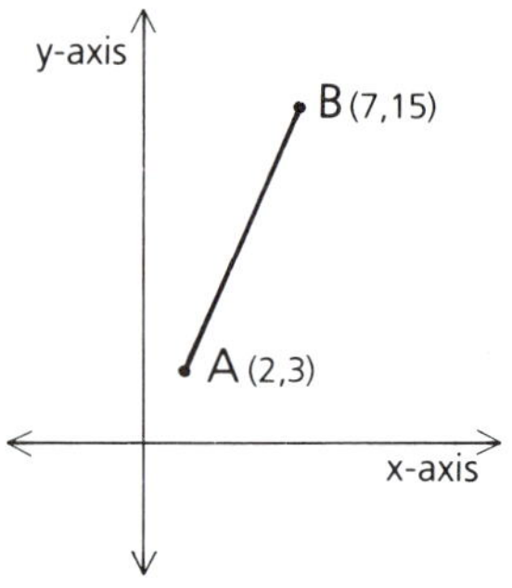

Solution: By the Distance Formula

$$AB = \sqrt{(\triangle x)^2 + (\triangle y)^2}$$
$$= \sqrt{(7 - 2)^2 + (15 - 3)^2}$$
$$= \sqrt{5^2 + 12^2}$$
$$= \sqrt{169}$$
$$AB = 13$$

2 If D = (7, 1), E = (9, −5) and F = (6, −4), find the length of the median from F to $\overline{DE}$.

Solution: By the Midpoint Formula, the midpoint M of $\overline{DE}$ is (8, −2).

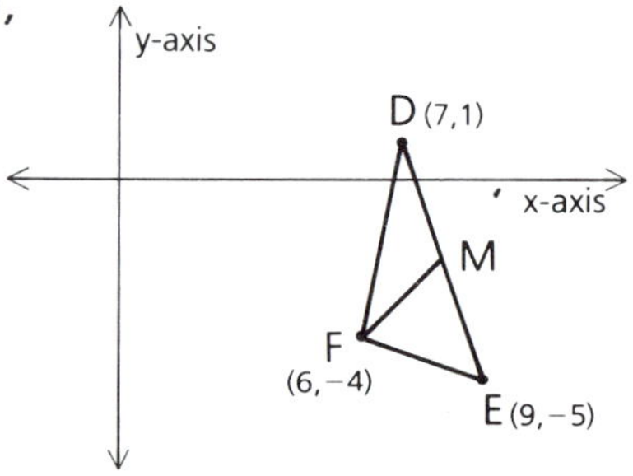

By the Distance Formula

$$FM = \sqrt{(\triangle x)^2 + (\triangle y)^2}$$
$$= \sqrt{(6 - 8)^2 + (-4 - (-2))^2}$$
$$= \sqrt{(-2)^2 + (-2)^2}$$
$$FM = \sqrt{8} = 2\sqrt{2}$$

Section 13.5 Problem Set A

1 Find the distance between each pair of points:

a (4, 0) and (6, 0)

b (2, 3) and (2, −1)

c (4, 1) and (7, 5)

d (−2, −4) and (−8, 4)

e The origin and (2, 5)

f (2, 1) and (6, 3)

2 Find the perimeter of △ABC if A = (2, 6), B = (5, 10), and C = (0, 13).

3 Show that the triangle with vertices at (8, 4), (3, 5), and (4, 10) is a right triangle by using: a the Distance Formula b slopes

4 Using the distance formula show that $\triangle DOG$ is equilateral if $D = (6, 0)$, $O = (0, 0)$, and $G = (3, 3\sqrt{3})$.

5 Find the area of the circle that passes through $(9, -4)$ and whose center is $(-3, 5)$.

6 Given: $\triangle RTV$ as shown

Find: a the length of the median from T

b the length of the segment joining the midpoints of $\overline{RT}$ and $\overline{TV}$

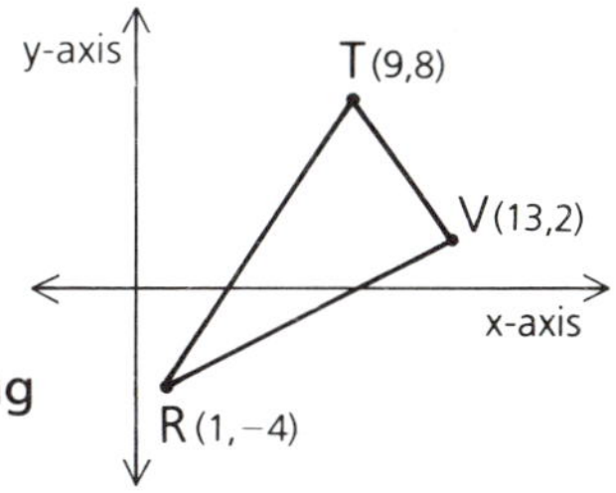

Section 13.5 Problem Set B

7 Show that $(4, -3)$ lies on the perpendicular bisector of the segment joining $(1, 8)$ and $(11, 6)$.

8 Show that $(7, 11)$, $(7, -13)$, and $(14, 4)$ lie on a circle with center at $(2, -1)$.

9 Find the perimeter of a quadrilateral with $A = (2, 1)$, $B = (7, 3)$, $C = (12, 1)$, and $D = (7, -4)$ and give its most descriptive name.

10 Show that a parallelogram whose vertices are $(-1, -3)$, $(2, 1)$, $(3\ -2)$, and $(-2, 0)$ is *not* a rhombus.

11 Show that a triangle with vertices at $(-2, 1)$, $(5, 5)$, and $(-1, -7)$ is isosceles.

12 The vertices of a rectangle are $(0, 0)$, $(8, 0)$, $(0, 6)$, and $(8, 6)$. Find the sum of the lengths of the two diagonals.

13 Show that $(1, 2)$, $(4, 6)$, and $(10, 14)$ are collinear by:

a the Distance Formula (Hint: What is true about the lengths of the three segments formed by three collinear points?)

b using slopes.

14 The point (5, y) is equidistant from (1, 4) and (10, −3). Find y.

Section 13.5 Problem Set C

15 Given: O = (0, 0), A = (2, 3), and B = (3, −2)
Find: the equation of the line that bisects ∠AOB

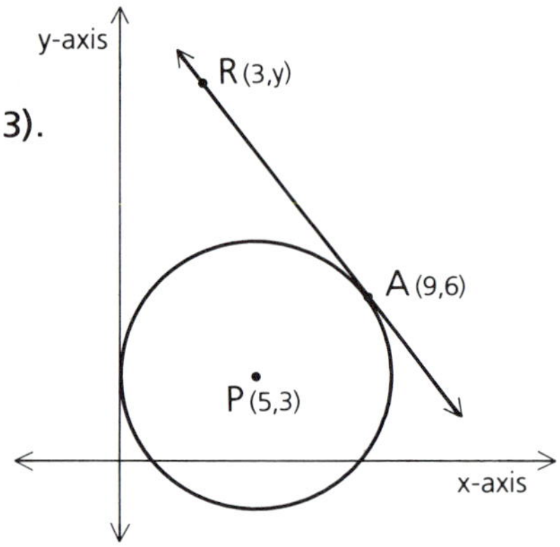

16 Given circle P, whose center is (5, 3). $\overline{RA}$ is tangent to circle P at (9, 6).

a If R = (3, y), solve for y.

b Find the equation of $\overleftrightarrow{RA}$.

c Find PR.

d How far is R from circle P?

17 In isosceles trapezoid ABCD, A = (−2a, 0) and B = (2a, 0), where a > 0. The height of the trapezoid is 2h, and the upper base, $\overline{CD}$, has a length of 4p.

Find: a the coordinates of C and D

b the length of the lower base

c the length of the segment joining the midpoints of $\overline{AD}$ and $\overline{BC}$

d the length of the segment joining the midpoints of the diagonals of the trapezoid.

Section 13.5 Problem Set D

18 Two of the vertices of an equilateral triangle are (2, 1) and (6, 5). Find the possible coordinates for the remaining vertex.

Section 13.6 Systems of Equations

If two linear equations are graphed on the same real number plane, the resulting lines could be:

parallel (a ∥ b)
intersecting (c intersects d)
identical (e, f coincide)

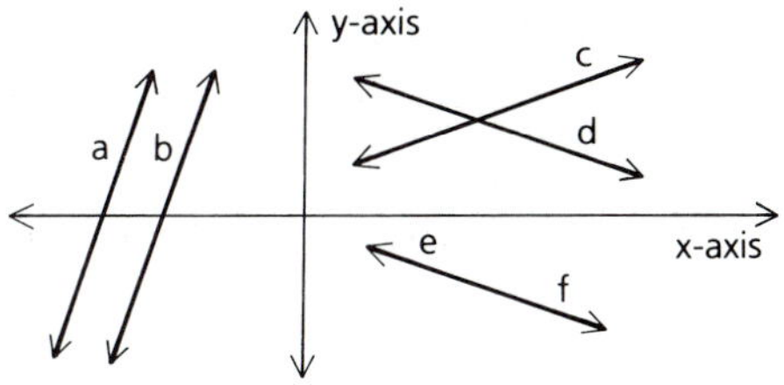

Each pair of lines may be represented by a **SYSTEM OF EQUATIONS**, such as the following:

System	Graph	Intersection
$\begin{cases} y = x + 8 \\ y = x + 13 \end{cases}$	Parallel lines	Empty
$\begin{cases} y = x + 3 \\ y = -2x + 21 \end{cases}$	Intersecting lines	One point
$\begin{cases} y = \frac{1}{2}x - 10 \\ 2x - 4y = 40 \end{cases}$	Identical lines	All points on the line

Most of the problems in this section require solving a system of two linear equations.
The Sample Problems illustrate two methods of solving systems:

ADDITION or SUBTRACTION SUBSTITUTION

Section 13.6 Sample Problems

1 Find the intersection of the two lines: $x = 4$ and $y = 2x + 8$.

Solution by Substitution: Since $x = 4$ from the first equation, substitute 4 for x in the second equation.

$$\begin{aligned} y &= 2x + 8 \\ y &= 2(4) + 8 \\ y &= 8 + 8 \\ y &= 16 \end{aligned}$$

Thus, the intersection is (4, 16)

2 Find the intersection of the following pair of lines:

$$\begin{cases} 8x - 3y = 7 \\ 10x + 4y = 1 \end{cases}$$

Solution by Addition or Subtraction:

first line: $8x - 3y = 7$
second line: $10x + 4y = 1$

$4 \cdot$ first line: $32x - 12y = 28$
$3 \cdot$ second line: $30x + 12y = 3$

sum: $62x + 0 = 31$

$$x = \frac{1}{2}$$

Now substitute $\frac{1}{2}$ for x in the first or second line.

first line:

$$\begin{aligned} 8x - 3y &= 7 \\ 8\left(\frac{1}{2}\right) - 3y &= 7 \\ y &= -1 \end{aligned}$$

Answer: $(\frac{1}{2}, -1)$

3 Find the intersection of the following pair:

$$\begin{cases} y = 3x + 1 \\ 6x - 2y = -2 \end{cases}$$

Solution by Substitution: Substitute $(3x + 1)$ for y in the second equation.

$$\begin{aligned} 6x - 2y &= -2 \\ 6x - 2(3x + 1) &= -2 \\ 6x - 6x - 2 &= -2 \\ -2 &= -2 \end{aligned}$$

Since the above statement is ALWAYS true, the intersection is the entire set of points of the first line, which is therefore identical to the second line. Answer: $\{(x, y) \mid y = 3x + 1\}$

Section 13.6 Problem Set A

1 Determine the point of intersection of each pair of lines:

a $\begin{cases} x + y = 10 \\ x - y = 2 \end{cases}$ **b** $\begin{cases} y = 5 \\ x + y = 7 \end{cases}$ **c** $\begin{cases} y = 2x - 1 \\ y = 4x + 5 \end{cases}$ **d** $\begin{cases} x + 2y = 7 \\ 4x - y = 10 \end{cases}$

2 Determine the intersection of each pair.

a $\begin{cases} x = 4 \\ x^2 + y^2 = 25 \end{cases}$ **b** $\begin{cases} y = 3 \\ |y - 2| = x \end{cases}$

3 Where do the pairs intersect?

a $\begin{cases} x + 2y = 12 \\ \text{x-axis} \end{cases}$ **b** $\begin{cases} y = 3x - 7 \\ 9x - 3y = 21 \end{cases}$

4 Find the points each pair has in common.

a $\begin{cases} 2x + y = 10 \\ 8x + 4y = 17 \end{cases}$ **b** $\begin{cases} y = 4x + 1 \\ \text{The line to the right of the y-axis, parallel to it, and 4 units from it.} \end{cases}$

Section 13.6 Problem Set B

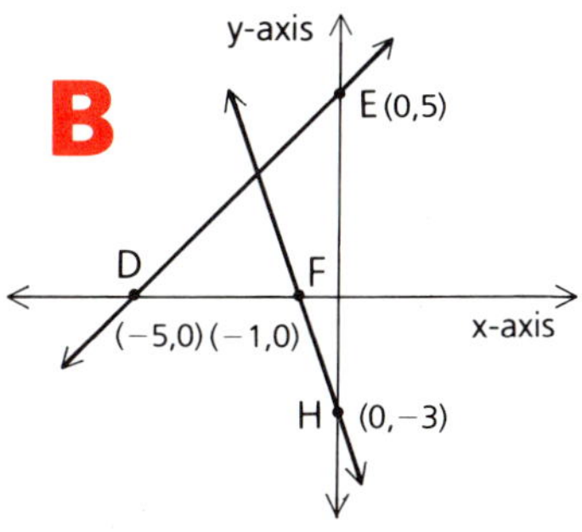

5 Where does $\overleftrightarrow{DE}$ intersect $\overleftrightarrow{FH}$?

6 Find the intersection of the lines $x = a$ and $3x + 2y = 12$.

7 Show that the 3 lines are concurrent (each intersecting the others at the same point).

$$\begin{cases} 2x + 3y = 2 \\ y = 2x - 10 \\ 3x - y = 14 \end{cases}$$

8 The graph of $x^2 + y^2 = 25$ is a circle and will be studied later in this chapter. The graph of $x^2 - y^2 = 7$ is a hyperbola and is normally studied in a later math course. Use one of the methods of solving a system to find the intersection of the circle and the hyperbola.

9 Find the equation of the line containing (2,1) and the intersection of $3x - y = 3$ and $x + 2y = 15$.

10 Find the equation of the line that is parallel to $2x + 3y = 5$ and contains the intersection of $y = 4x + 8$ and $y = x + 5$.

11 Find the point of intersection of the following lines:
$y - 3 = \frac{1}{2}(x - 1)$ and $y + 1 = -\frac{3}{2}(x - 1)$

12 Given line $y = 2x + 1$. Line two contains (5,3) and is parallel to the given line. Line three contains (5, 16) and has the same y-intercept as the given line. Find the intersection of lines two and three.

Section 13.6 Problem Set C

13 Given two intersecting lines $ax + by = c$ and $dx + ey = f$. Find the point of intersection.

14 In $\triangle ABC$, $A = (5, -1)$, $B = (1,1)$, and $C = (5, -11)$. Find the length of the altitude from A to $\overline{BC}$.

15 Find the distance between the parallel lines $y = 2x + 3$ and $y = 2x + 7$. (Hint: Start by choosing a convenient point on one of the lines.)

16 Find the intersection of the V-shaped figure $y = |x - 3|$ with the line $y = 2x + 1$.

17 Find the area of the triangle whose sides lie on the lines $3x + y + 1 = 0$, $x + 4y - 7 = 0$, and $-5x + 2y + 13 = 0$.

18 Find the reflection of the point $(-6, 5)$ over the line $2y - x = 6$. (See Section 13.1, Problem 16.)

Section 13.7 Graphing Inequalities

In this section we discuss the graphs of selected inequalities.

TWO PART PROCEDURE FOR GRAPHING INEQUALITIES

1 Pretend the inequality is an equation. Graph this equation as a *boundary*.

2 In the inequality, *test* points of the various *regions* separated by the boundary. Shade the proper region(s).

In the final graph, the boundary is *dotted* if it is *not* included in a shaded region.

Study the following SAMPLE PROBLEMS closely.

Section 13.7 Sample Problems

1 Graph $y > 2x + 8$

Solution by the Two Part Procedure:

Boundary: Pretend that $y = 2x + 8$

x	y
0	8
1	10
−1	6
−2	4

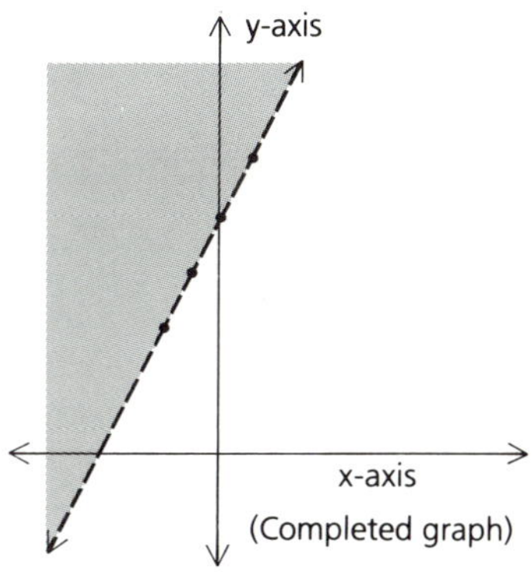

(Completed graph)

The boundary line is *dotted* since there is no equal sign in the original problem.

Test of Regions: In the inequality, test a convenient point not on the boundary line. Use (0, 0) in this case.

$y > 2x + 8$
Is $0 > 2(0) + 8$?
Is $0 > 8$?

Since $0 > 8$ is *false*, *do not shade* the region.

Now test a point such as (−10, 10) in the other region.

Is $10 > 2(-10) + 8$?
$10 > -12$

Since $10 > -12$ is *true, do shade* the region containing (−10, 10).

2 Graph $y \geq |x - 2|$

Solution by the Two Part Procedure:

Boundary: $y = |x - 2|$

x	y
0	2
1	1
2	0
3	1
4	2

A KEY value (for x = 2, y = 0)

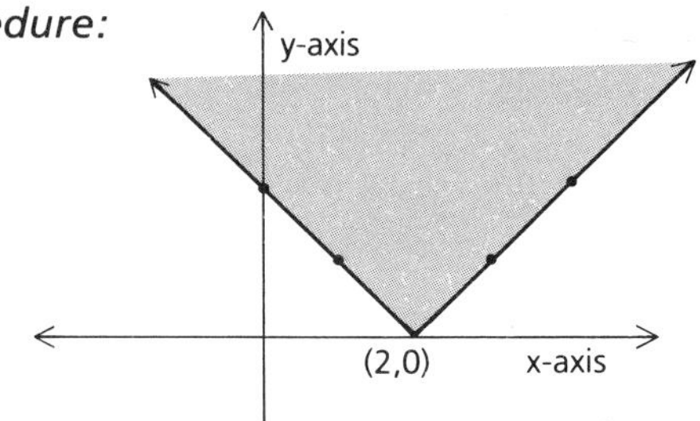

(Completed graph)

The boundary line (a V) is *solid* because there *is* an equal sign in the original problem. Two regions are formed.

Test of Regions: Test (0,0) in the inequality.

$y > |x - 2|$

Is $0 > |0 - 2|$? — $0 > 2$ is *false.*

Is $0 > |-2|$? — Do not shade its region.

Is $0 > 2$? — Further tests confirm shading the other.

3 Show, by graphing, the intersection of this system:

$$\begin{cases} y \leq \frac{2}{5}x + 4 \\ y \geq -\frac{1}{2}x + 4 \\ 2x + y \leq 16 \end{cases}$$

Solution:

Follow the Two Part Procedure three times.

Boundary

$y = \frac{2}{5}x + 4$

x	y
0	4
5	6
10	8
15	10

After testing regions, shade *below* the boundary line.

Boundary

$y = -\frac{1}{2}x + 4$

x	y
0	4
2	3
4	2
6	1

After testing regions, shade *above* the boundary line.

Boundary

$2x + y = 16$

x	y
0	16
1	14
2	12
3	10

After testing regions, shade *below* the boundary line.

The solution consists of the union of the triangle and its interior as shown in the final graph.

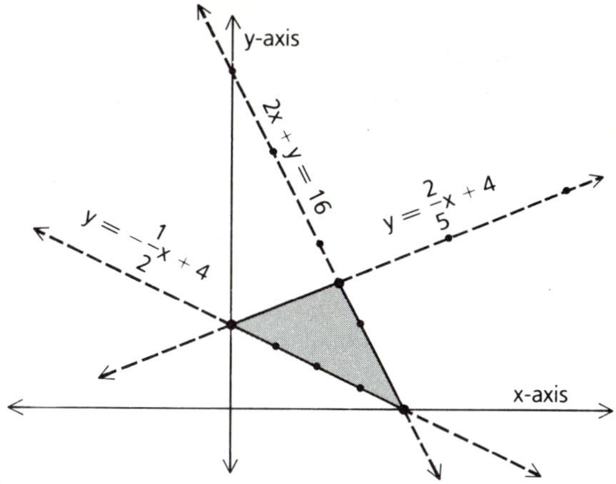

Section 13.7 Problem Set A

1 Graph the inequality for each.

a $2x - 3y < 6$

b $y < \frac{1}{2}x - 1$

c $5x + 2y \geq 10$

d $x < -2$

e $y \geq 2x + 3$

2 What is the inequality represented by each graph below?

a

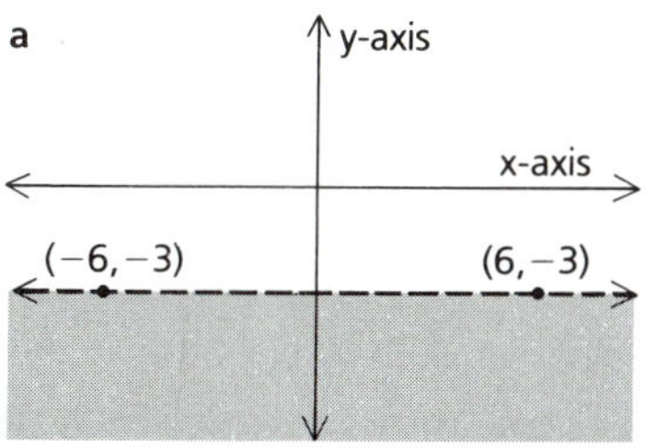

b

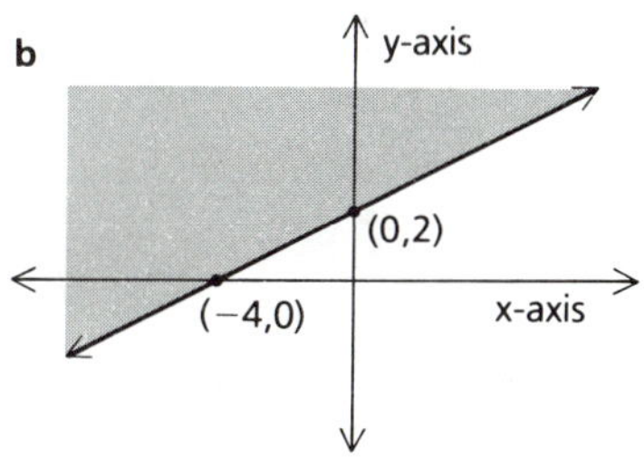

3 Graph each of the following.

a $y \geq |x + 1|$

b $\{(x, y) : x > 2 \text{ or } x < -1\}$

c $\{(x, y) : 5 < y < 7\}$

d $\{(x, y) : |x| < 3\}$

Section 13.7 Problem Set B

4 Determine, by graphing, the intersection of the solution sets of the two inequalities, $y > 2$ and $x + 2y < 6$.

5 Graph the system of inequalities for each of the following.

a $\left[\begin{array}{l} y \geq x + 4 \\ y \leq -2x + 6 \end{array}\right.$

b $\left[\begin{array}{l} x + y \leq 4 \\ 2x - y \leq 6 \\ x \geq 0 \end{array}\right.$

c $\left[\begin{array}{l} x + y > 12 \\ x - y \leq 4 \end{array}\right.$

d $\left[\begin{array}{l} 4y - 3x < 6 \\ y < 3x \\ 2x < 6 - 3y \end{array}\right.$

e $\left[\begin{array}{l} y > |x - 1| \\ x + 3y < 12 \end{array}\right.$

f $\left[\begin{array}{l} y < 2x + 5 \\ 2x - y < 3 \end{array}\right.$

6 Determine the *union* of the solution sets of the inequalities. $x + y > 4$ and $y < 2x - 6$.

Section 13.7 Problem Set C

7 Graph the system of inequalities.

a $\left[\begin{array}{l} y < x^2 + 8 \\ y > -x + 12 \end{array}\right.$

b $\left[\begin{array}{l} x^2 + y^2 \leq 25 \\ y \geq |x| \end{array}\right.$

c $\left[\begin{array}{l} xy < 12 \\ x^2 + y^2 < 16 \end{array}\right.$

8 Graph each of the inequalities.

a $|x + y| \leq 4$

b $|x| + |y| \leq 4$

The Royal Road: Computers for Computation

Many of the techniques of coordinate geometry involve computations. The following computer program computes the midpoint and the length of a segment $\overline{AB}$, given points A and B.

```
120  PRINT "TYPE IN THE COORDINATES OF TWO"
125  PRINT "POINTS (X1,Y1) AND (X2,Y2):"
130  INPUT X1,Y1,X2,Y2
170  PRINT "MIDPOINT: ("(X1 + X2) / 2", "(Y1 + Y2) / 2")"
180  PRINT "LENGTH OF SEGMENT:";SQR ((X2-X1)^2+(Y2-Y1)^2)
```

1 Add to the program so that it provides the slope of $\overleftrightarrow{AB}$ and the equation of $\overleftrightarrow{AB}$. (Will your program deal with vertical lines?)

2 Write a program that provides the equation of the perpendicular bisector of $\overline{AB}$, given points A and B.

Section 13.8 Circles

The equation of a circle is a form of the Distance Formula (Section 13.5) when applied to the length of a radius of the circle.

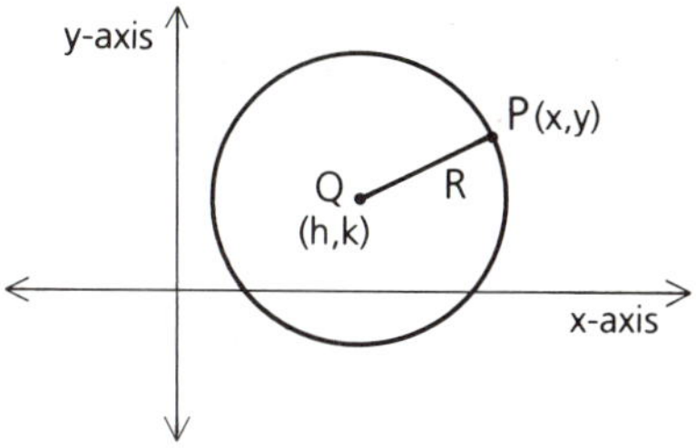

Theorem 120

T **The EQUATION OF A CIRCLE whose center is (h, k) and whose radius is R is**

$$(x - h)^2 + (y - k)^2 = R^2$$

This CIRCLE FORMULA may be used in either of two ways:

■ 1 Find the equation of a circle whose center is (1, 5) and whose radius is 4.

Ans: $(x - 1)^2 + (y - 5)^2 = 16$

■ 2 Find the center and radius of the circle $(x - 2)^2 + (y + 7)^2 = 64$.

Solution: Rewrite the given equation in the same form as the Circle Equation:

$$(x - h)^2 + (y - k)^2 = R^2$$
$$(x - +2)^2 + (y - -7)^2 = 8^2$$

Hence: $h = 2$, $k = -7$, $R = 8$. Answer: center $= (2, -7)$, radius $= 8$.

The next example uses the Circle Formula in the same way as example 2, but the preparation is more complicated.

■ 3 Is $x^2 - 8x + y^2 - 10y = 8$ a circle?

Solution: Use the process of Completing the Square to derive an equivalent equation in the form of the Circle Equation.

$$x^2 - 8x + y^2 - 10y = 8$$
$$x^2 - 8x + \underset{\text{Key Number}}{\mathbf{16}} + y^2 - 10y + \underset{\text{Key Number}}{\mathbf{25}} = 8 + \mathbf{16} + \mathbf{25}$$
$$(x^2 - 8x + 16) + (y^2 - 10y + 25) = 49$$
$$(x - 4)^2 + (y - 5)^2 = 49$$

Yes, the solution set is a circle. The center is (4, 5) and the radius is 7.
The two key numbers, 16 and 25, are found by COMPLETING THE SQUARE.
Notice that 16 is the square of half of (-8),
and 25 is the square of half of (-10).

Section 13.8 Sample Problems

1 Find the equation of the circle with center (0, −2) and radius 3.

Solution: Use the Circle Formula.

$$\begin{aligned}(x-h)^2+(y-k)^2 &= R^2\\ (x-0)^2+(y-(-2))^2 &= 3^2\\ x^2+(y+2)^2 &= 9\end{aligned}$$

2 Find the circumference of the circle: $3x^2+3y^2+6x-18y=15$

Solution:

Divide each side by 3: $x^2+y^2+2x-6y=5$

Rearrange the terms: $x^2+2x+y^2-6y=5$

Complete the Square: $x^2+2x+1+y^2-6y+9 = 5+1+9$

$$(x+1)^2+(y-3)^2 = 15$$

Hence, the radius is $\sqrt{15}$ and the circumference $= 2\pi R = 2\pi\sqrt{15}$.

3 a Discuss: $(x-2)^2+(y+5)^2=0$

Solution: This equation is in the form for a circle with its center at (2, −5) and a *radius of 0.* This is sometimes called a "point circle," a "circle" that has shrunk to the single point (2, −5).

b Discuss: $x^2+(y-4)^2=-25$

Solution: This equation is in the form for a circle with its center at (0,4) and a *radius* of $\sqrt{-25}$. However, $\sqrt{-25}$ is *not* a real number. Hence, it is called an "imaginary circle."

Section 13.8 Problem Set A

1 Determine the equation of each circle.

a Center (0, 0), radius 4

b Center (−2, 1), radius 5

c Center (0, −2), radius $2\sqrt{3}$

d Center (−6, 0), radius $\frac{1}{2}$

2 Graph each circle.

a $x^2 + y^2 = 9$ b $(x - 1)^2 + (y + 2)^2 = 16$

3 For each circle find the center, radius, diameter, circumference, and area.

a $x^2 + y^2 = 36$ c $(x - 3)^2 + (y + 6)^2 = 100$

b $(x + 5)^2 + y^2 = \frac{9}{4}$ d $\frac{(x + 5)^2}{3} + \frac{(y - 2)^2}{3} = 27$

4 Determine the equation of each circle.

(Hint: Find R and use the Circle Equation.)

a

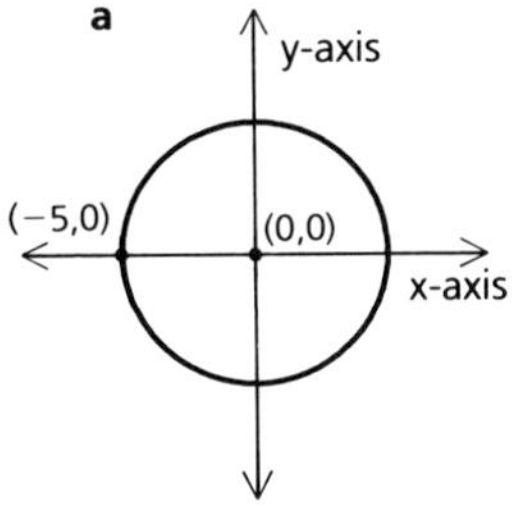

b

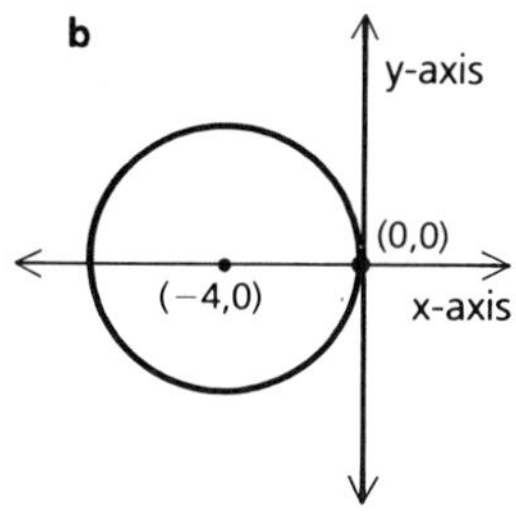

c

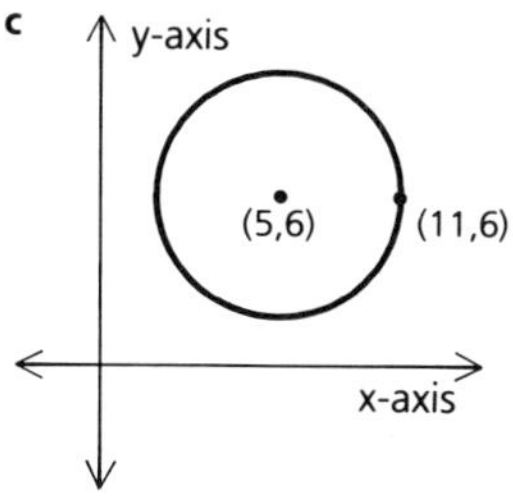

5 Given the circle $(x - 3)^2 + (y + 2)^2 = 17$. Is

a (4, 2) on the circle? b (3, −2) on the circle?

6 a What type of "circle" is $(x - 3)^2 + (y + 1)^2 = 0$?

b What type of "circle" is $(x + 5)^2 + y^2 = -100$?

7 In circle P, the radius is 7.
$\overline{AB}$ is the horizontal diameter,
and $\overline{CD}$ is the vertical diameter.
Find the coordinates of A, B, C, and D.

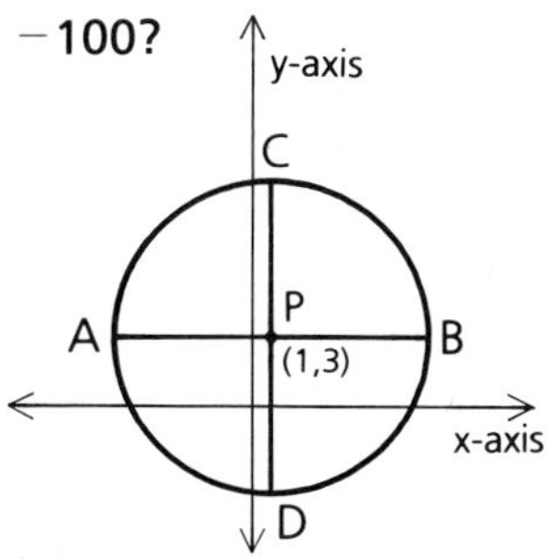

8 Determine the equation of each circle:

a The center is the origin and it passes through (0, −5).

b The endpoints of a diameter are (−2, 1) and (8, 25).

c The center is (−1, 7) and it passes through the origin.

d The center is (2, −3) and it passes through (3, 0).

9 For each circle, tell whether the given point is:

In the interior of the given circle
On the given circle
In the exterior of the given circle

a $(2, 5); x^2 + y^2 = 29$ **c** origin; $(x - 2)^2 + (y + 5)^2 = 16$
b $(3, 0); x^2 + y^2 = 100$ **d** $(-2, 1); x^2 + (y + 6)^2 = 23$

10 Graph the solution of the system.

$$\begin{bmatrix} x^2 + y^2 \geq 9 \\ x^2 + y^2 \leq 25 \end{bmatrix}$$

Section 13.8 Problem Set B

11 Find the center and radius of each circle.

a $x^2 + y^2 - 8y = 9$ **c** $x^2 + 10x + y^2 - 12y = -10$
b $(x + 7)^2 + y^2 + 6y = 27$ **d** $x^2 + y^2 = 8x - 14y + 35$

12 Find the solution to each system.

a $\begin{bmatrix} x^2 + y^2 = 25 \\ x = 3 \end{bmatrix}$ **c** $\begin{bmatrix} x^2 + y^2 = 34 \\ x + y = 8 \end{bmatrix}$

b $\begin{bmatrix} x^2 + y^2 = 25 \\ x^2 - y^2 = 7 \end{bmatrix}$ **d** $\begin{bmatrix} |y| = 6 \\ x^2 + y^2 = 100 \end{bmatrix}$

13 Find the distance between the points of intersection of the circle $x^2 + y^2 = 17$ and the line $x + y = 3$.

14 In circle Q as marked, find

a the equation of the tangent to the circle at (6, 8).

b the circumference of the circle.

c the distance from A to Q.

d the distance from A to the circle.

e the area of the sector shown.

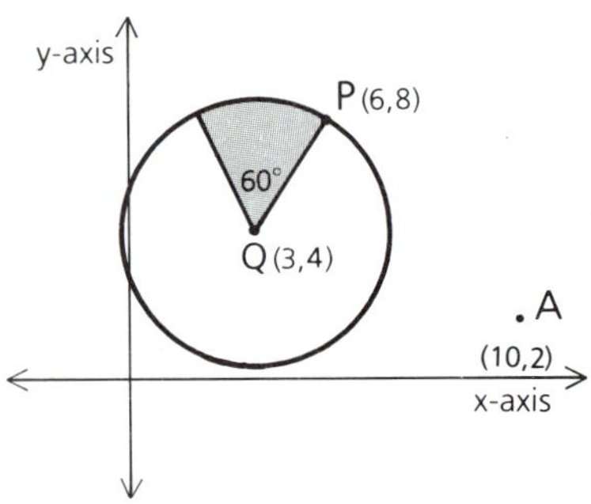

15 Given the circle $(x - 2)^2 + (y + 3)^2 = 61$, find the equation of the tangent to the circle at point $(8, -8)$.

Section 13.8 Problem Set C

16 Find the center and radius of the circle $3x^2 + 12x + 3y^2 - 5y = 2$.

17 Find the area of the circle shown.

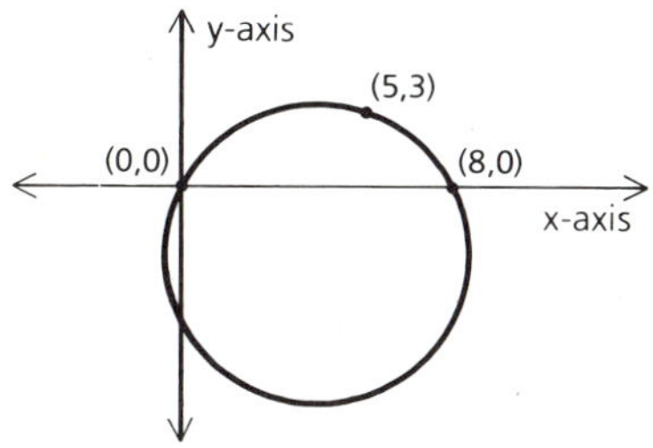

18 Find the equation of the path of a point that moves so that its distance from the point $(3, 0)$ is always twice its distance to the point $(-3, 0)$.

19 Given the circle $x^2 - 12x + y^2 = 28$. A marble at point $(2, 4\sqrt{3})$ was flicked clockwise around the circle until it stopped at the intersection of the circle with the positive x-axis. Find:

a the distance the marble traveled.

b the distance that would have been saved if the marble rolled in a straight line?

Section 13.8 Problem Set D

20 Given: ⊙ P with center at the origin and a radius of 15
⊙ Q with center at $(14, 0)$ and a radius of 13

Find:
a the point(s) of intersection of the circles
b the length of a common external tangent
c the coordinates of the points of tangency of the common external tangent segment that lies in Quadrant I

Section 13.9 Application Review

This section contains a number of applications of coordinate geometry to concepts of previous chapters.

Section 13.9 Problem Set A

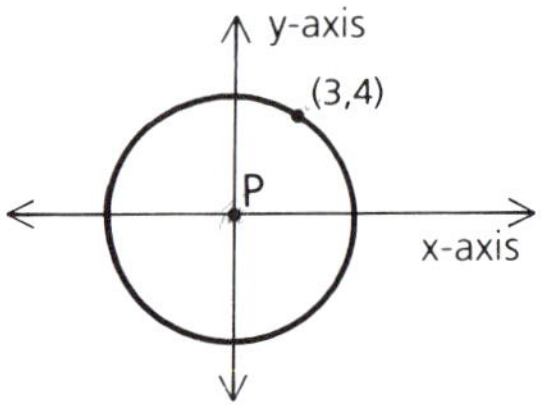

1 a Write the equation of circle P.
b Find the area of the circle.
c Find the circumference.

2 Find the area of the shaded sector.

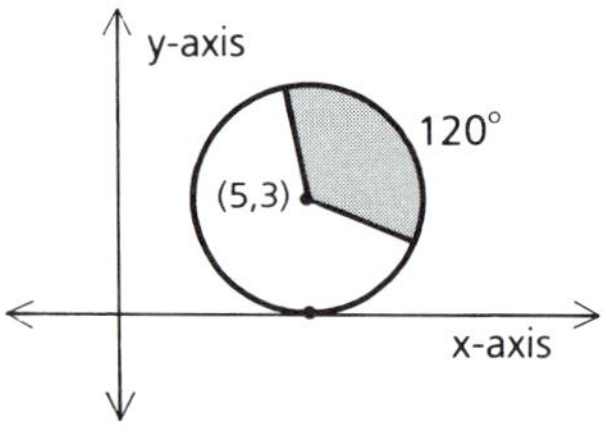

3 Find the area of the shaded region in each problem:

a **ABCD is a square**

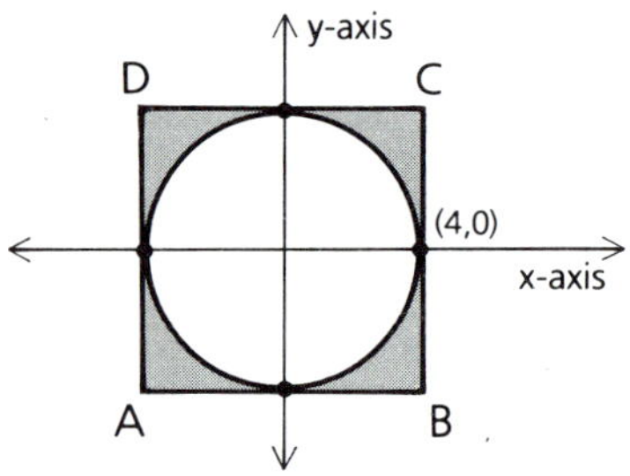

b

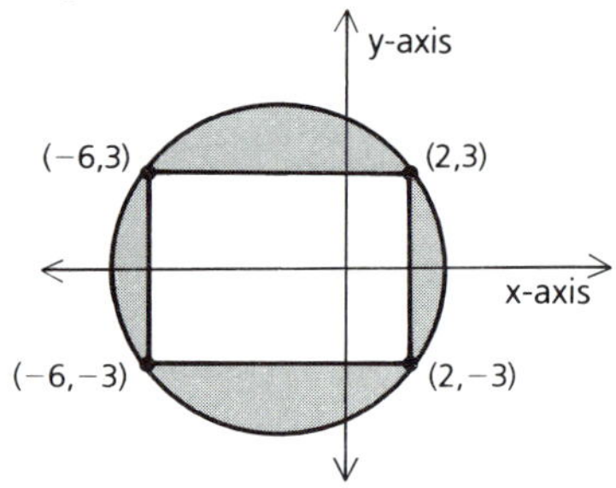

4 Find the area of the square with vertices at (1, 2), (6, 2), (6, 7), and (1, 7).

5 In the figure as marked, M is the midpoint of $\overline{EF}$.

Find: a OM
b EM
c FM

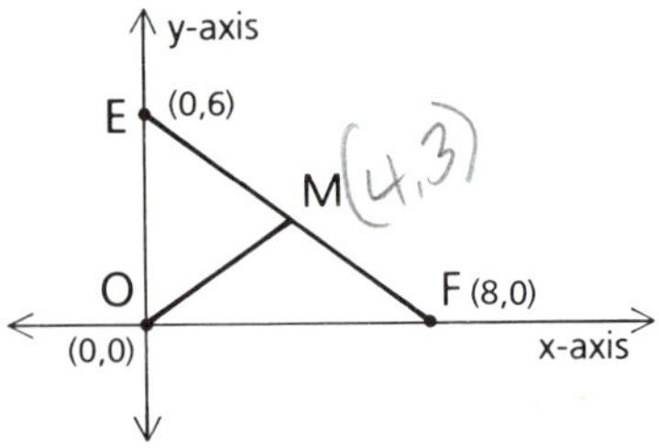

6 In the diagram as marked, $\triangle$ABC is equilateral. Find the coordinates of C.

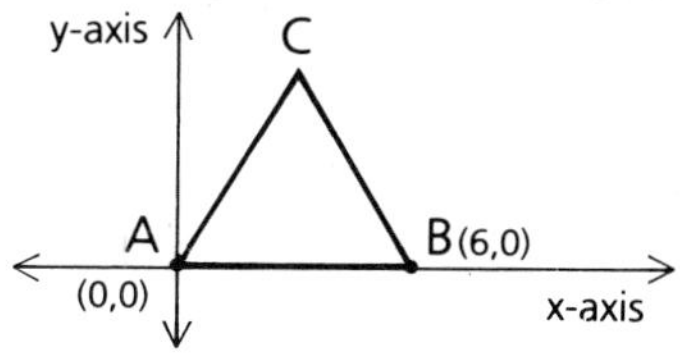

7 In rectangle ABCD, A = (2, 7) and C = (8, 15). Find BD.

8 Find the area of the triangle with vertices at (0, 8), (0, 0), and (3, 0).

9 In the figure as marked, find the slope of the tangent to circle P at point H.

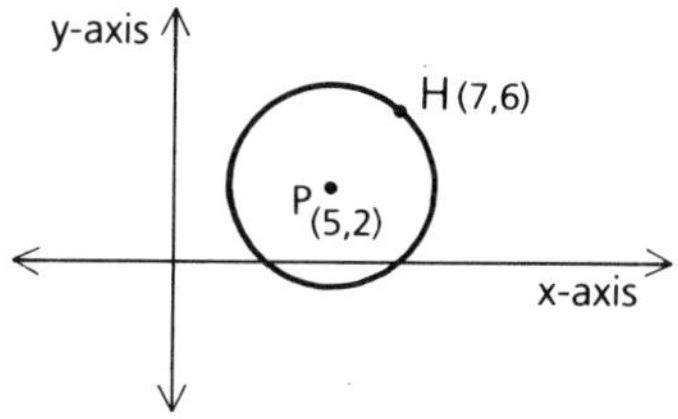

10 In the figure as marked, $\overline{JK}$ is a chord of circle Q, and $\overline{QM} \perp \overline{JK}$. Find QM.

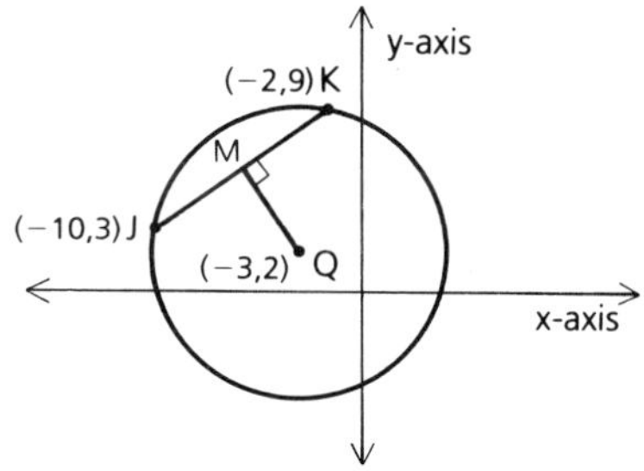

11 In the figure as marked, $\overline{AC}$ and $\overline{DE}$ are chords, A = (8, 0), B = (18, 0), C = (24, 0), and D = (22, −3). Find BE.

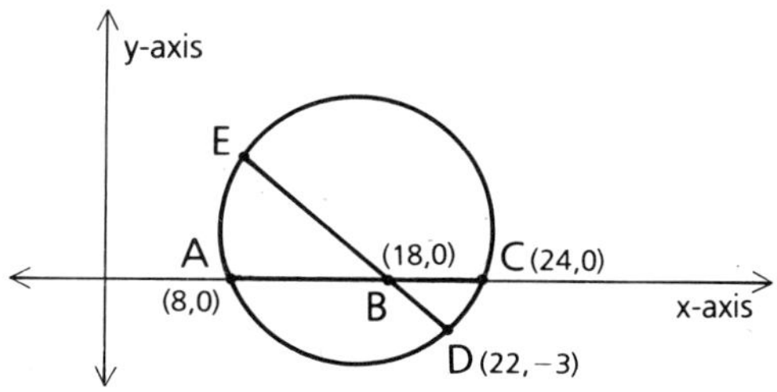

12 Given rectangle HJKM with H = (4, 2), J = (14, 4), and M = (3, 7).
Find: **a** the coordinates of K
b the area of the rectangle

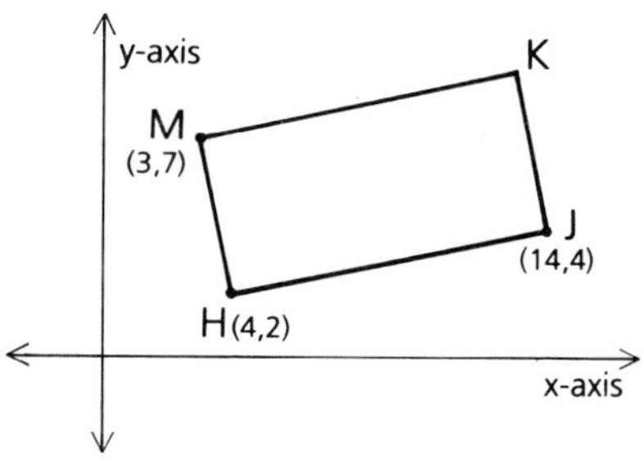

Section 13.9 Problem Set B

13 $\odot$ P is tangent to the x-axis and the y-axis at the points shown.
Find: **a** the equation of the circle
b the area of the shaded region bounded by the circle and the axes

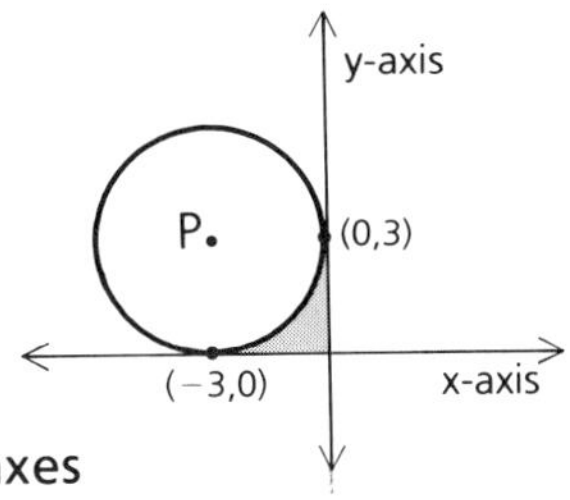

14 In the figure as marked, find the area of $\triangle$ABC.
(THE ENCASEMENT PRINCIPLE)

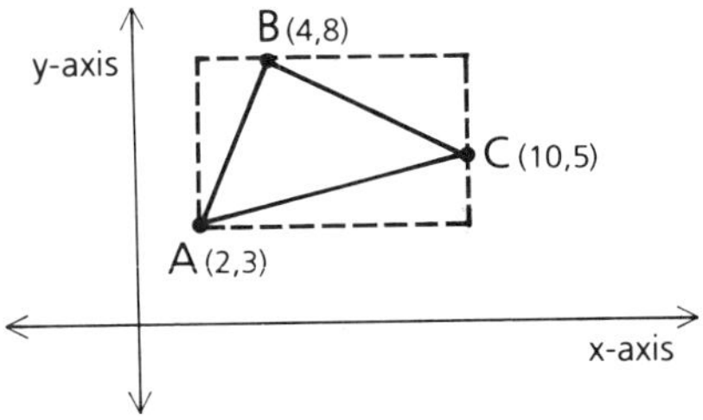

15 In the figure as marked, find the area of $\triangle$DEF.

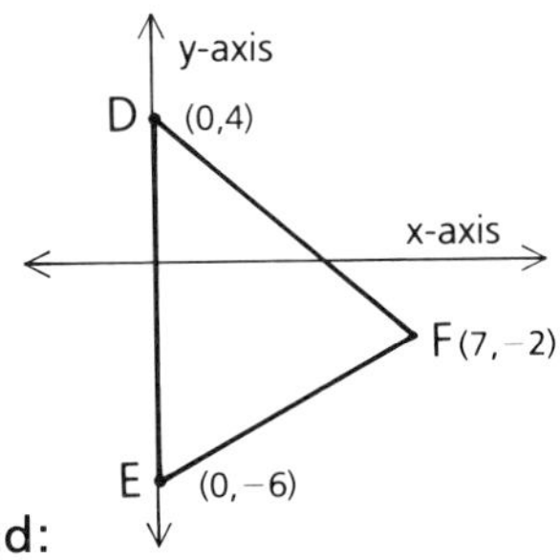

16 (13, 9) is on a circle centered at (7, 1). Find:

- **a** the equation
- **b** the area
- **c** the circumference
- **d** the point on the circle directly opposite (13, 9)
- **e** the equation of the line tangent at (13, 9)
- **f** the distance of (19, 6) from the center
- **g** the distance of (19, 6) from the circle

17 Find the area of the isosceles trapezoid with vertices at (4, 8), (2, 3), (14, 3), (12, 8).

18 $\triangle$ABC is an isosceles right triangle with base $\overline{AB}$. If A = (−3, −2) and B = (−3, 4), find the two possibilities for the coordinates of C.

19 In $\triangle$DEF, D = (1, 2), E = (7, 2), and F = (1, 10). Find the length of the altitude from D to $\overline{EF}$. (Hint: First, find the area of $\triangle$DEF.)

20 Given the circle $(x - 4)^2 + (y + 2)^2 = 50$ with its center at P. Point T lies on chord $\overline{AB}$ such that $\overline{PT}$ is perpendicular to $\overline{AB}$. If A = (11, −1) and B = (5, −9), find: **a** PT **b** m∠TPA

Section 13.9 Problem Set C

21 Given a fixed line segment $\overline{OA}$, as marked. M can lie anywhere on a circle of radius 3 and center O. B moves so that M is always the midpoint of $\overline{AB}$. Find the equation of the circle on which B lies.

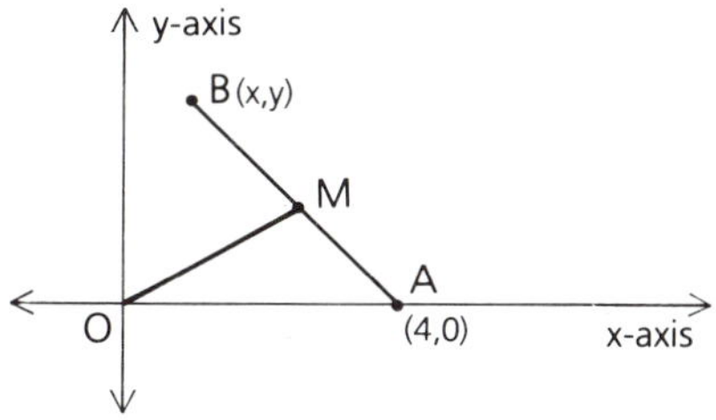

22 △AOB is placed on a graph such that A = (6, 12), B = (21, 3), O = (0, 0). A segment $\overline{CD}$ is drawn parallel to $\overline{OB}$ so that C lies on $\overline{AO}$ and D lies on $\overline{AB}$. If C = (4, 8), find

a the coordinates of D

b the ratio of the area of △ACD to the area of △AOB.

23 Find the distance between the lines $y = 2x - 1$ and $y = 2x + 7$.

24 Given △AOB with A = (6, 0), B = (0, 8), and O = (0, 0). What is the volume of the solid formed:

a if the triangle is rotated around $\overline{OA}$?

b if the triangle is rotated around $\overline{OB}$?

c *and the total surface area* if the triangle is rotated around $\overline{AB}$?

25 Given the circles $(x + 9)^2 + (y - 4)^2 = 52$ and $(x - 12)^2 + (y - 3)^2 = 13$. Find: the length of a

a common internal tangent **b** common external tangent

26 Given a triangle whose vertices have coordinates that are all rational numbers. Prove that the area of the triangle is rational. (Hint: Use the Encasement Principle of Problem 14 in a paragraph proof.)

Section 13.9 Problem Set D

27 A LATTICE POINT is a point whose coordinates are *integers,* including zero. How many lattice points are on the boundary and inside the region bounded by the positive x-axis, the positive y-axis, the circle $x^2 + y^2 = 25$, and the line passing through $(-3, 0)$ and $(0, 2)$?

28 A green billiard ball is located at (3, 1), and a gray ball at (8, 9). Fats Tablechalk wants to strike the green ball so that it bounces off the y-axis and hits the gray ball. At what point on the y-axis should he aim? (The Reflection Principle)

29 The points of $\overline{A_1B_1}$ are "mapped" onto a new coordinate system with shorter units such that A_1B_1 is turned around, and A_1 becomes A_2 and B_1 becomes B_2.
On the new coordinate system, find:

a C_2 if $C_1 = (3\frac{1}{2}, 0)$.

b D_2 if $D_1 = (4, 0)$.

c E_2 (in terms of x_1) if $E_1 = (x_1, 0)$.

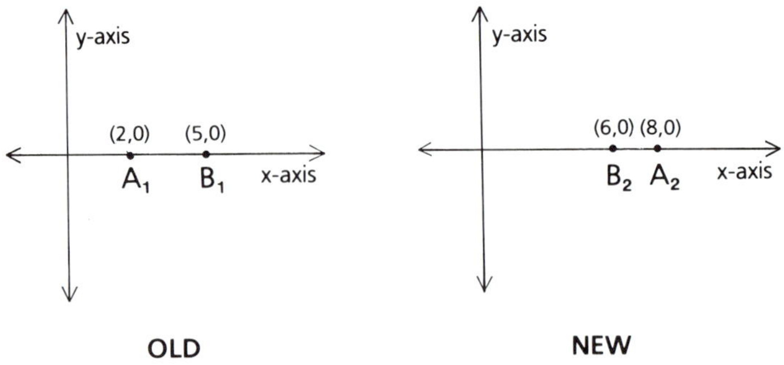

Chapter 13 Study Guide

As you study and review this chapter, be sure that you are able to:

1 Use the following vocabulary (pp. 579–580):

- **a** origin
- **b** x-axis (horizontal)
- **c** y-axis (vertical)
- **d** quadrants
- **e** ordered pair
- **f** abscissa, ordinate
- **g** x-intercept
- **h** y-intercept

2 Apply the following formulas:

- **a** midpoint (p. 584)
- **b** slope (p. 587)
- **c** distance (p. 603)

3 Use the midpoint formula to find an endpoint. (p. 585)

4 Describe the graph of a line with any of these slopes (p. 588):

- **a** zero slope
- **b** positive slope
- **c** negative slope
- **d** no slope

5 Explain how slope is related to parallelism and to perpendicularity. (pp. 589–590)

6 Graph points, given an equation in any one of the following forms:

- **a** horizontal line formula (p. 595)
- **b** vertical line formula (p. 596)
- **c** y-form or slope-intercept form (p. 594)
- **d** two-point formula (p. 598)
- **e** circle formula (p. 615)

7 Find the equation of a graphed line or circle. (pp. 597, 615)

8 Use equations to tell whether two coplanar lines are parallel, intersecting, or identical. (p. 607)

9 Use two methods to solve a system of equations. (p. 608)

10 Apply the Two Part Procedure for inequalities. (p. 611)

11 Write an equation for a circle in standard form by completing the square. (p. 615)

Chapter 13 Review Problems **A**

1 Is the point (7, 5) on the line $2x + 3y = 62$?

2 In which quadrant are both coordinates negative?

3 Is the graph of $y > 2x + 1$ above or below the line $y = 2x + 1$?

4 In each case find the coordinates of point D.

a
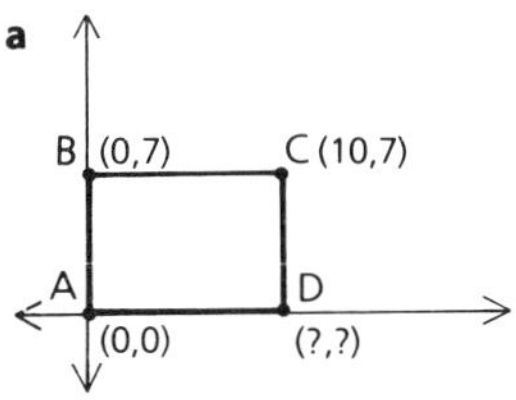

ABCD is a rectangle

b
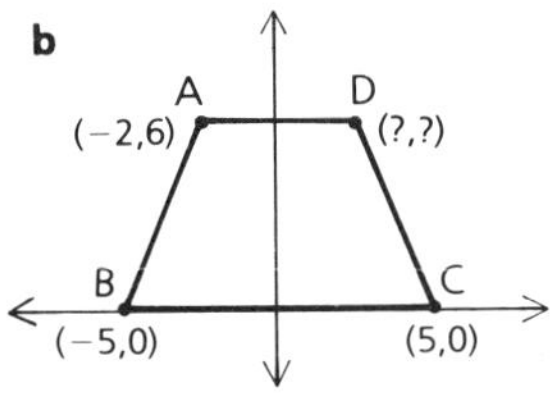

ABCD is an isosceles trapezoid

c
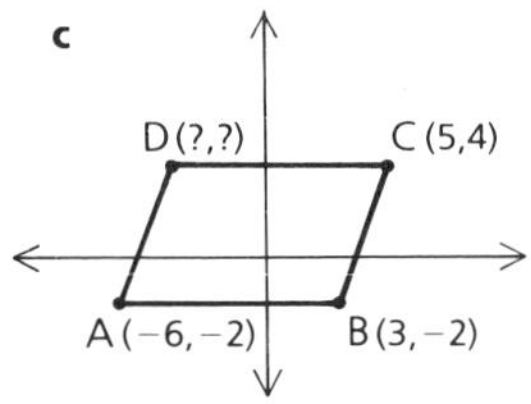

ABCD is a parallelogram

5 Given P (4, −2) and Q(10, 6). Find:
a PQ b the midpoint of $\overline{PQ}$ c the slope of $\overleftrightarrow{PQ}$

6 Using the figure as marked, find:

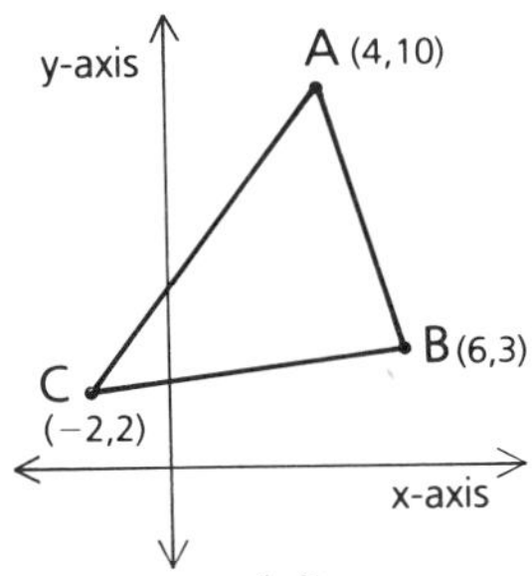

a the slope of $\overline{AC}$
b the midpoint of $\overline{AC}$
c the slope of the median from B
d the length of the median from B
e the slope of the altitude from B
f the slope of a line through A and parallel to $\overleftrightarrow{BC}$
g the slope of the perpendicular bisector of $\overline{AC}$.

7 In each case find the area of the shaded region.

a
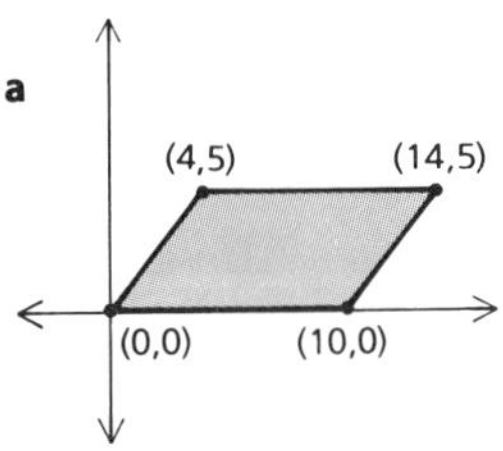

b
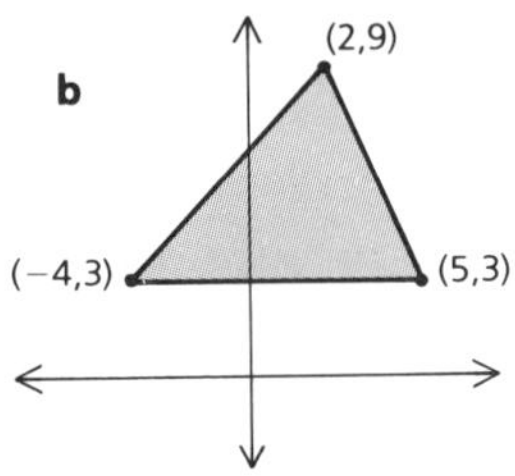

c
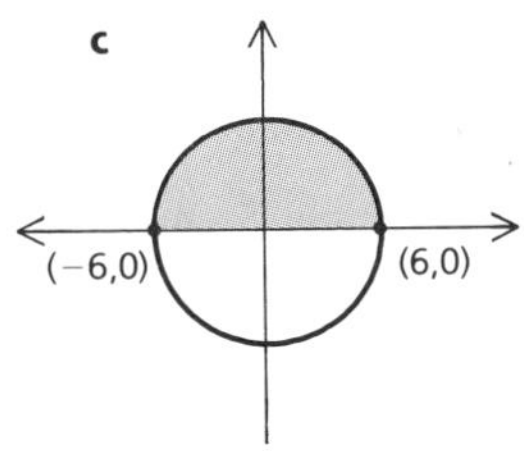

8 Give the equation of each circle:

a center at $(2, -3)$, radius of 4
b center at origin, passing through $(6, 8)$
c endpoints of a diameter are $(0, 0)$ and $(10, 0)$.

9 Find the equation of each line:

a slope of 2, y-intercept of 1
b containing the points $(2, 3)$ and $(2, 7)$
c parallel to and 5 units to the left of the y-axis
d containing the points $(2, 4)$ and $(6, 16)$
e slope of $\frac{1}{2}$, x-intercept of 4
f parallel to $y = 3x + 1$, with the same y-intercept as $y = 2x - 7$
g x-intercept of 6, y-intercept of -3.

10 Find the slope of each line. Are the lines perpendicular, parallel, or neither? a $x + 2y = 10$ b $y = 2x + 3$

11 Are points $(2, 4)$, $(5, 13)$, $(26, 76)$ collinear?

12 Find the slope of $\overleftrightarrow{AB}$.

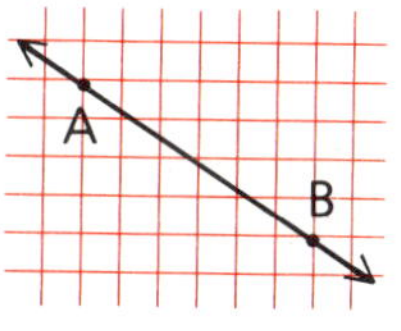

Chapter 13 Review Problems B

13 Find the x-intercept of the line joining $(-2, 3)$ and $(5, 7)$.

14 The points $(2, 1)$, $(4, 0)$, and $(-4, k^2)$ are collinear. Solve for k.

15 $A = (-6, 1)$ and $B = (2, 3)$. If B is the midpoint of $\overline{AC}$, find C.

16 Find the coordinates of the point $\frac{1}{4}$ the way from $(-5, 0)$ to $(7, 8)$.

17 The vertices of $\triangle ABC$ are A $(2, 3)$, B $(12, 5)$, and C $(9, 8)$. Find:

a the length of the median from C to $\overline{AB}$.
b the equation of the median to $\overline{AB}$.
c the equation of the $\perp$ bisector of $\overline{AB}$.
d the equation of the altitude from C to $\overline{AB}$.
e the equation of a line containing C and parallel to $\overleftrightarrow{AB}$.

18 Using the figure as marked, find:

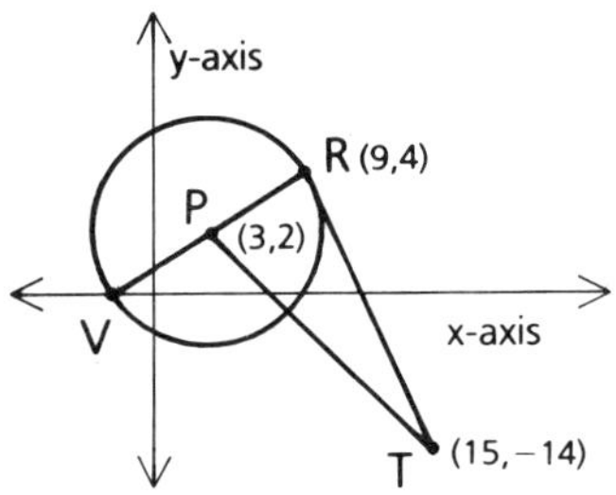

a the equation of circle P
b the area of circle P
c the coordinates of V
d the equation of the tangent $\overleftrightarrow{RT}$
e PT
f how far T is from the circle
g the area of △PRT.

19 Graph each of the following systems of inequalities:

a $\begin{cases} y \geq 2x + 1 \\ x^2 + y^2 \leq 25 \end{cases}$ b $\begin{cases} y \geq 0 \\ x \leq 0 \\ x - 3y \leq 6 \end{cases}$

20 Given five points (2, 1), (6, 4), (5, 17), (−2, −2), (2, 10). If two points are selected at random, what is the probability that:

a both points will lie in Quadrant I?
b both points and one other point will be collinear?

21 Find the intersection of each pair.

a $\begin{cases} y = 4x - 1 \\ y = 2x + 3 \end{cases}$ b $\begin{cases} x - 3y = 10 \\ 2x + y = 13 \end{cases}$ c $\begin{cases} y = 4 \\ (x - 1)^2 + (y - 5)^2 = 17 \end{cases}$

22 Find the center and radius of the circle, $x^2 + 6x + y^2 - 4y = 12$

23 Find the distance between the centers of these circles:

a $(x - 5)^2 + (y - 2)^2 = 29$ b $x^2 + 8x + y^2 = 31$

24 Describe the circle: $x^2 + 2x + y^2 - 6y = -10$

25 The bases of an isosceles trapezoid are parallel to the y-axis. If three vertices are (5, 2), (5, 12), and (−1, 10), find the area.

26 Given quadrilateral PQRS with vertices (3, 1), (15, −3), (9, 7), (5, 7).

a Find the area by using the Encasement Principle.
b What is true about the diagonals?

Chapter 13 Review Problems C

27 Given the line $2x - 5y = 10$. In which quadrant(s) is there a point on the line equidistant from the x and y axes? Find the point(s).

28 Show that the midpoint of the hypotenuse of any right triangle is equidistant from the vertices of the triangle.
(Hint: Use a triangle with vertices at $(0, 0)$, $(2a, 0)$, and $(0, 2b)$.

29 If $A = (0, -17)$, $B = (4, -5)$, and $C = (12, -1)$, find the length of the altitude from C to $\overline{AB}$.

30 Find the distance between the two lines $y = 3x - 8$ and $y = 3x + 2$.

31 A triangle with vertices at $(0, 0)$, $(6, 0)$, and $(0, 6\sqrt{3})$ is rotated around its longest side. Find the volume of the solid formed.

32 If $A = (-8, 5)$ and $B = (7, -3)$, where is the point R that divides $\overline{AB}$ so that $AR:RB = 3:2$?

33 In ▱ PQRS, M, N, and X are midpoints of $\overline{PQ}$, $\overline{PS}$, and $\overline{QR}$ respectively. Find the intersection of $\overline{MN}$ and $\overline{PX}$, if $P = (-8, 1)$, $Q = (0, 5)$, $S = (4, 1)$.

34 How many lattice points are in the intersection of this system:

$$\begin{cases} x > 0 \\ y > 0 \\ y < -|x - 4| + 10 \end{cases}$$

35 Given $A = (2, 10)$ and $C = (8, 4)$. Find point B if it lies on the x-axis and $AB + BC$ is a minimum.

36 Find the "mirror image" of point $(-5, 10)$ when reflected over the:
a x-axis **b** point $(-3, 1)$ **c** line $y = 2x$

37 Find the intersection.

$$\begin{cases} x + y = 16 \\ y = |2x + 10| \end{cases}$$

Chapter 14

LOCUS AND CONSTRUCTIONS

Section 14.1 Locus

D **A LOCUS is a set of points that contains all the points, and only those points, that satisfy specific conditions.**

The word *locus* is derived from the Latin word meaning "place" or "location." The plural form is LOCI.

■ Example 1: Find the locus of points that are one inch from a given point O.

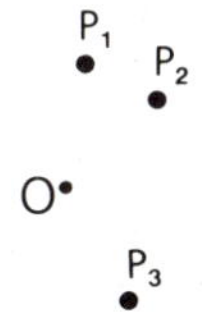

Solution:

Find one point, P_1, that is 1 inch from O.
Find a second such point, P_2.
Continue finding such points until a pattern is formed—in this case, a circle.

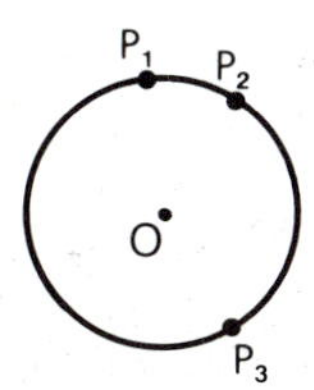

Draw the circle and finish with a verbal description:
The locus of points one inch from a given point O is a circle with O as the center and a radius of one inch.

FOUR STEP PROCEDURE FOR LOCUS PROBLEMS

1. **Find a single point that satisfies the given condition(s).**
2. **Find a second such point, and a third, and another, until a *recognizable pattern* is formed.**
3. **Look outside the pattern for points you may have overlooked. Look within the pattern to exclude points that do not meet the conditions.**
4. **Present the answer by using a diagram and a verbal description of the locus.**

■ Example 2: What is the locus of all points equidistant from the sides of an angle?

Solution:

Step 1 Locate the first point, P_1, so that it is equidistant from the sides of the angle.

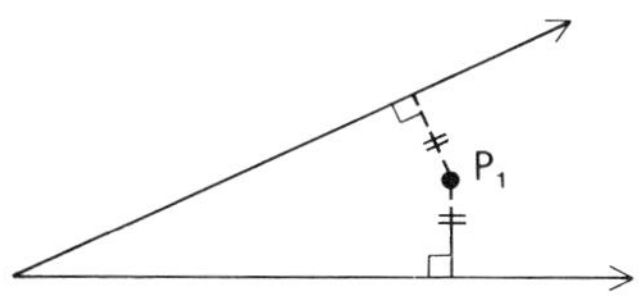

Step 2 Similarly, locate points P_2, P_3, P_4, . . . The pattern appears to be the ray that bisects the angle.

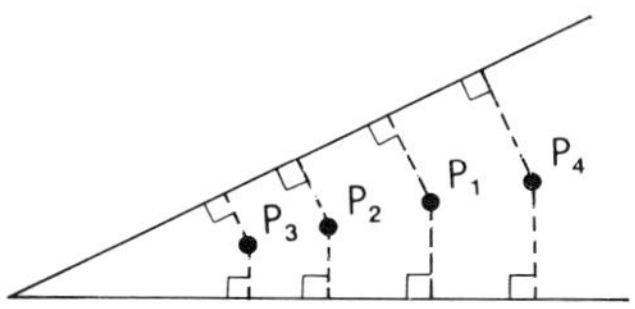

Step 3 By sketching other points outside the pattern, we determine that the *only* points in the locus are on the angle bisector.

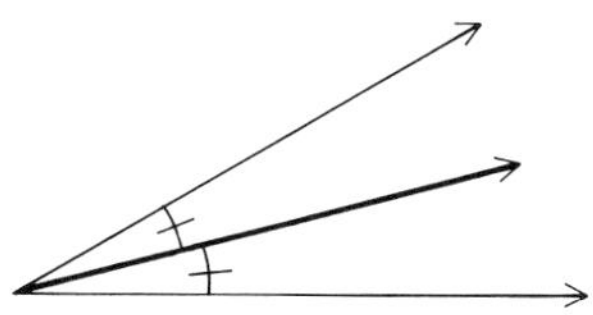

Step 4 Description:
The locus of all points equidistant from the sides of an angle is the bisector of the angle.

The following example uses the FOUR STEP PROCEDURE and draws special attention to the importance of Step 3.

■ Example 3: What is the locus of points 3 cm from a given line ?

Solution:

Step 1 Find a single point P_1 that is 3 cm from ℓ.

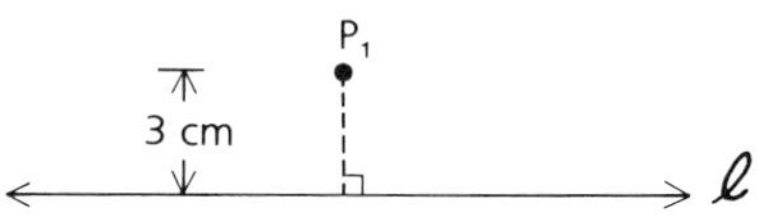

Step 2 Find a second point, a third, and so on until a recognizable pattern is formed.

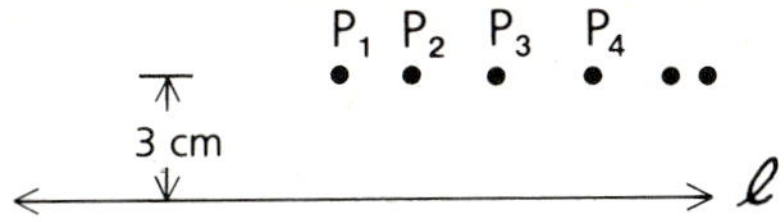

Step 3

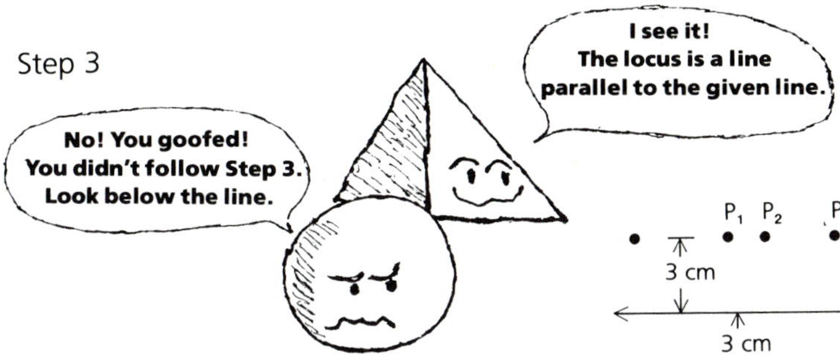

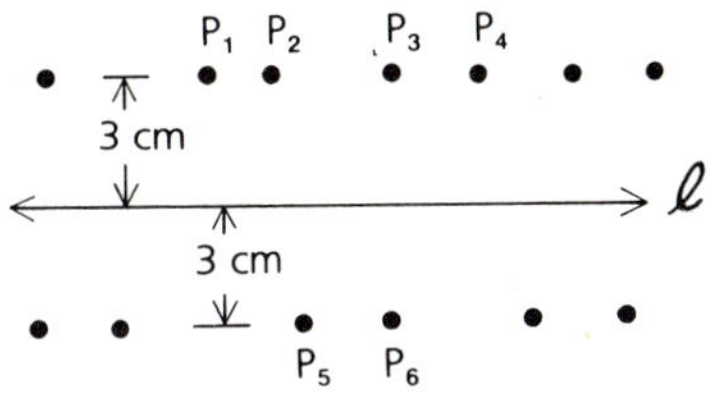

Step 4 Present a diagram and verbal description of the locus. The locus is *two* lines parallel to the given line ℓ, one 3 cm above ℓ and one 3 cm below.

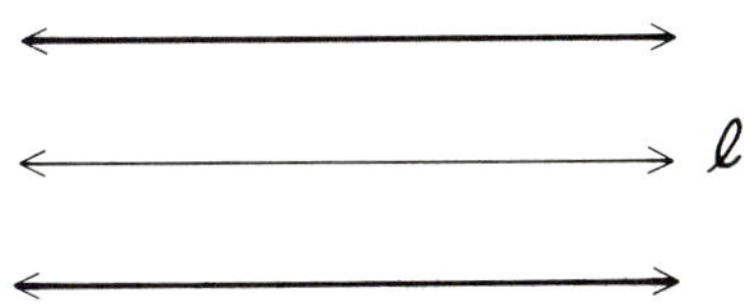

Section 14.1 Sample Problems

Note: Loci in this book are coplanar unless specified otherwise.

1 What is the locus of points 2 inches from a given circle K, whose radius is 5 inches?

Description: The locus of points 2 inches from a given circle K is two circles that are concentric with circle K and have radii of 3 inches and 7 inches.

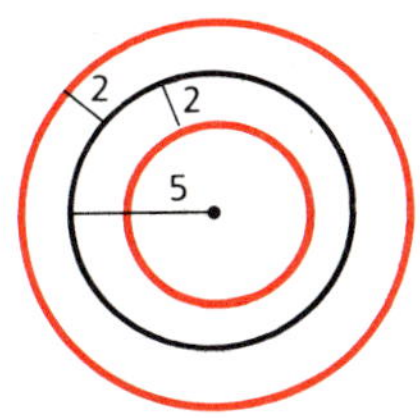

2 Find the locus of points less than 3 cm from a given point A.

Description: The locus of points less than 3 cm from a given point A is the *interior* of a circle with its center at A and a radius of 3 cm.

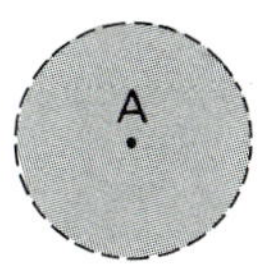

(Note: The circle itself is *not* part of the locus. Do you see why?)

3 What is the locus of the *centers* of all circles having a fixed radius r that are tangent to a given line ℓ ?

Solution: Sketch a few circles tangent to the given line ℓ.

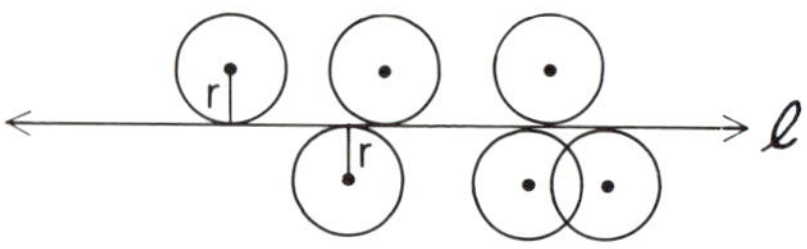

Then consider the pattern of their *centers* only:

Diagram as shown.

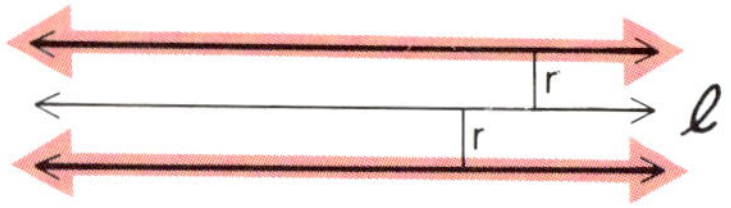

Description: The locus of the *centers* of all circles having a fixed radius r that are tangent to a given line ℓ is two parallel lines on opposite sides of ℓ, each at the distance r from ℓ.

4 Write the equation of the locus of points equidistant from points A and B, where A = (1, 4) and B = (7, 8).

Method One: If (x, y) is any point P on the locus, then PA = PB

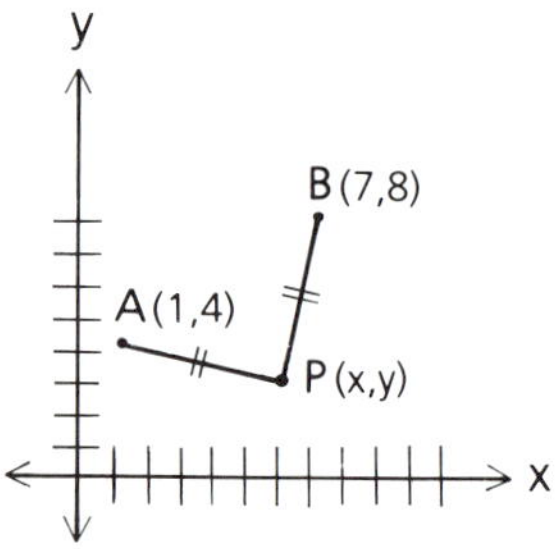

$$\sqrt{(x-1)^2+(y-4)^2} = \sqrt{(x-7)^2+(y-8)^2}$$

$$\text{Squaring, } (x-1)^2+(y-4)^2 = (x-7)^2+(y-8)^2$$

$$x^2-2x+1+y^2-8y+16 = x^2-14x+49+y^2-16y+64$$

$$-2x-8y+17 = -14x-16y+113$$

$$y = -\tfrac{3}{2}x+12$$

Method Two: We know that the locus of all points equidistant from A and B is the perpendicular bisector of $\overline{AB}$.

Midpoint of $\overline{AB}$ = (4, 6) and Slope of $\overline{AB} = \dfrac{8-4}{7-1} = \dfrac{2}{3}$

$\therefore$ slope of perpendicular bisector $= -\frac{3}{2}$

$$\text{Then } y = mx + k$$

$$6 = -\tfrac{3}{2}(4) + k$$

$$12 = k$$

$$\therefore y = -\tfrac{3}{2}x + 12$$

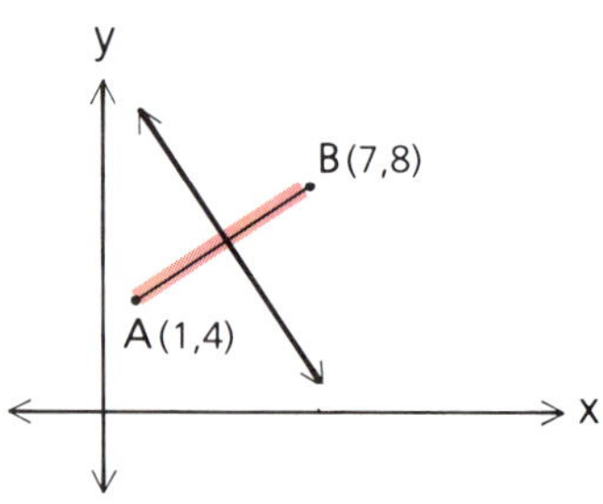

Section 14.1 Problem Set A

Note: Loci in this book are coplanar unless specified otherwise.

For **1–9** draw a sketch and write a verbal description of each locus.

1 What is the locus of points that are 3 cm from a given line, $\overleftrightarrow{AB}$?

2 Find the locus of the midpoints of the radii of a given circle.

3 What is the locus of points equidistant from 2 given points?

4 What is the locus of the center of a dime that rolls around the edge of a quarter?

5 A circle has a radius of 1 ft. Find the locus of points that are 10 in. from the circle.

6 Determine the locus of the centers of all circles tangent to both of two given parallel lines.

7 Find the locus of points equidistant from two given concentric circles. If the radii of the circles are 3 and 8, what is the size of the locus?

8 Write an equation for the locus of points that are 4 units from the origin.

9 Find the locus of points less than or equal to 14 units from a fixed point P.

10 **a** Find the locus of points that are 5 units away from both a and b.

b Find the locus of points that are 4 units away from both a and b.

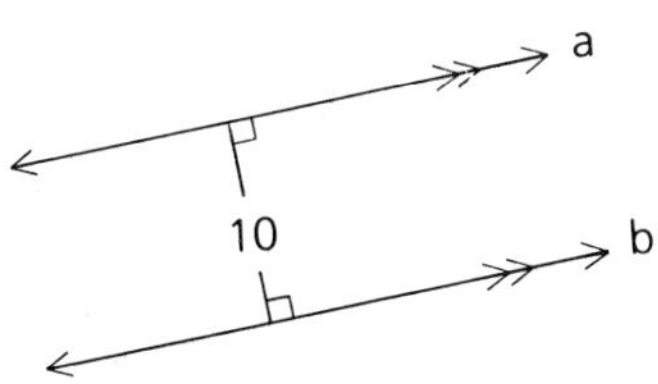

Section 14.1 Problem Set B

11 What is the locus of the midpoints of all chords that can be drawn from a given point of a given circle?

12 What is the locus of the midpoints of all chords congruent to a given chord of a given circle?

13 Determine the locus of the centers of all circles passing through two given points. Find an accurate, simple description of the locus.

14 What is the locus of the midpoints of all segments drawn from one vertex of a triangle to the opposite side of the triangle?

15 What is the locus of points in space that are:
a 5 units from a given point? b 5 units from a given line?

16 Write an equation for the locus of points equidistant from the lines whose equations are $x = -2$ and $x = 7$.

17 Write an equation for the locus of points 6 units from $(-1, 3)$.

18 Find the locus of points that are located 5 units from both the x-axis and the y-axis.

19 Given a circle Q with a radius of 9. Find the locus of points 9 units from the given circle.

20 a Sketch the locus of points 5 units from a segment $\overline{PQ}$.

b Find the area of the locus sketched in part *a*, if $PQ = 6$.

c Sketch the locus of points in space 5 units from segment $\overline{PQ}$.

d Find the volume of the locus sketched in part *c*, if $PQ = 6$.

21 Point P is four units above plane *m*. Find the locus of points that lie in plane *m* and are five units from P.

22 Write an equation for the locus of points equidistant from (3, 5) and (1, −9).

23 Points T and V are fixed. Find the locus of P such that $\overline{PT} \perp \overline{PV}$.

Section 14.1 Problem Set C

24 Write an equation for the locus of points each of which is the vertex of a right angle of a triangle whose hypotenuse is the segment joining (−1, 0) and (1, 0). Describe the set geometrically.

25 a The locus of points equidistant from the vertices of a triangle is the point of intersection of the __?__ of the triangle.

b The locus of points equidistant from the sides of a triangle is the point of intersection of the __?__ of the triangle.

26 Write an equation for the locus of points (x, y) such that the area of the triangle with vertices (x, y), (0, 0), and (3, 0) is 2.

27 Given: P = (−3, 4)

a Sketch the locus of points that are 2 or more units from P and at the same time are 5 or less units from P.

b Describe the locus algebraically.

c Find the area of the locus.

28 A ladder 6 m long leans against a wall. Describe the locus of the midpoint of the ladder in all possible positions. Prove that your answer is correct.

29 Write an equation for the locus of points each of which is twice as far from (−2, 0) as it is from (1, 0).

30 PQRS is a rectangle with $\overline{PQ}$ twice as long as $\overline{QR}$. T is the midpoint of $\overline{RS}$. $\overline{TQ}$ is drawn. Sketch the locus of the midpoints of segments drawn parallel to $\overline{TQ}$ that end in the sides of the rectangle.

Section 14.2 Compound Locus

Many locus problems combine two or more loci in one compound locus.

■ Example: Given points A and B 5 units apart.

The locus of points 3 units from A is a circle shown in Figure 1.

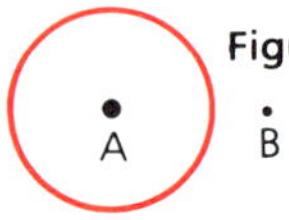

Figure 1

The locus of points 4 units from B is a circle shown in Figure 2.

Figure 2

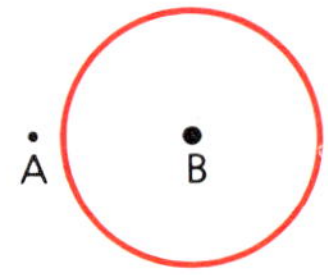

The locus of points that are both 3 units from A and 4 units from B is the two points darkened in Figure 3.

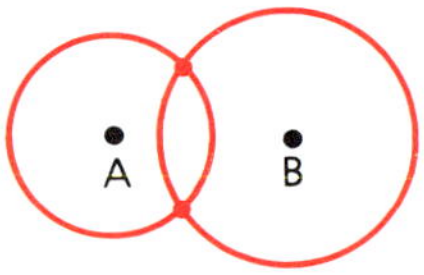

Figure 3

Notice that the **COMPOUND LOCUS** illustrated in Figure 3 is the intersection of the loci of Figure 1 and Figure 2.

	COMPOUND LOCUS PROCEDURE
1	**Solve each part of the compound locus problem separately.**
2	**Intersect the loci in all possible ways.**

Section 14.2 Sample Problems

1 Find the locus of points that are a fixed distance from a given line and lie on a given circle.

Solution: Follow the Compound Locus Procedure.

Step 1 Find each locus individually.

The locus of points that are a fixed distance from a given line is two lines each parallel to the given line.

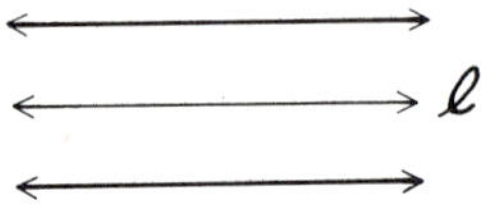

The locus of points that lie on a given circle is simply the circle itself.

Step 2 Intersect the two loci in all possible ways.

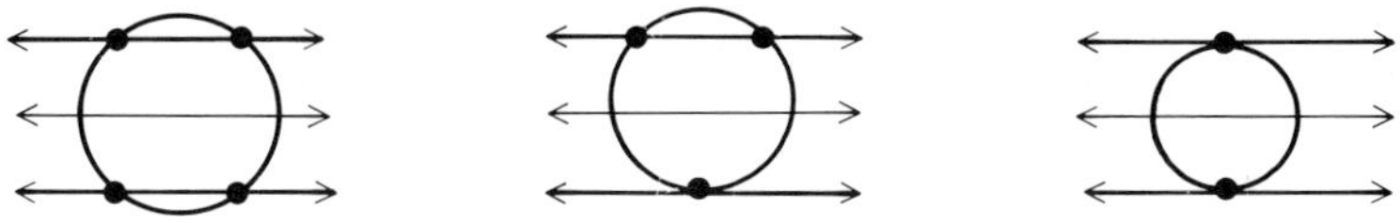

4 points 3 points 2 points

1 point ∅

Thus: The locus of points that are a fixed distance from a given line and lie on a given circle is 4 points, 3 points, 2 points, 1 point, or the empty set.

Important: To solve a compound locus problem, keep any *fixed* distance or *fixed* figure the same in all drawings. Change the size or position of *given* data to show all possible situations.

2 Find the locus of points in space that are a fixed distance from a given plane and a given distance from a fixed point on the plane.
Solution: Follow the Compound Locus Procedure.

Step 1 Find each locus individually.

The locus of points in space that are a fixed distance from a plane is two planes, each parallel to the given plane.

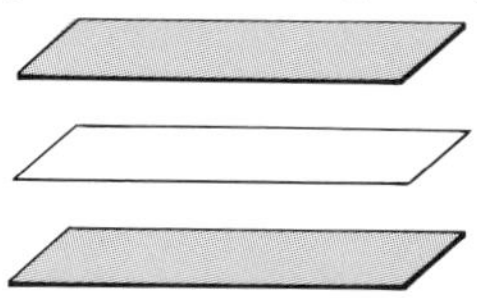

The locus of points in space that are a given distance r from a fixed point P is a sphere with center P and radius r.

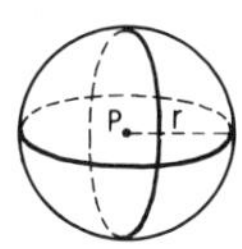

Step 2 Intersect the two loci in all possible ways:

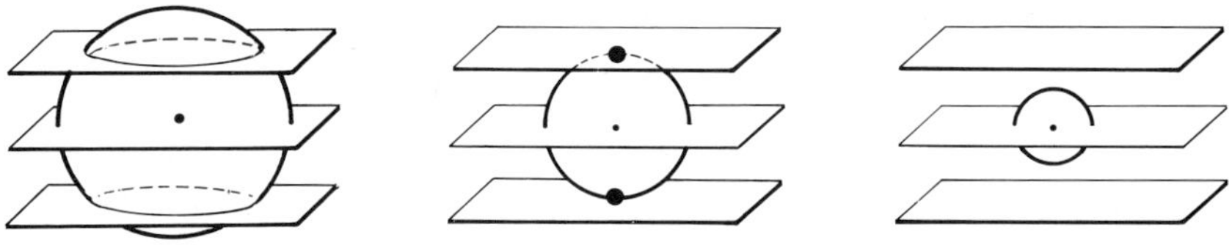

two circles two points ∅

Thus: The locus of points in space that are a fixed distance from a given plane and a given distance from a fixed point on the plane is either two circles, or two points, or the empty set.

Section 14.2 Problem Set A

1 Sketch *all* possible intersections for each.

a Find the locus of points equidistant from two given points and lying on a given circle.

b Given two points, A and B. Find the locus of points that are a given distance from A and another given distance from B.

c Find the locus of points on a graph lying on the line $y = 5$ and on the circle $x^2 + y^2 = r^2$, where $r > 0$.

d Find the locus of points equidistant from two parallel lines and lying on a third line.

e Find the locus of points equidistant from two intersecting lines and a fixed distance from their point of intersection.

f Find the locus of points equidistant from the sides of an angle and equidistant from two parallel lines.

2 Find the locus of points that are 1 cm from a given segment 4 cm long, and 2 cm from the midpoint of the segment.

3 How many points are equidistant from two given parallel lines and equidistant from two fixed points on one of those lines?

4 Given a regular hexagon, find the locus of points that are a given distance from its center and lie on the vertices of the hexagon.

Section 14.2 Problem Set B

5 a What is the locus of points that are less than, or equal to, a fixed distance from a given point and lie on a given line?

b What is the locus of points that are less than a fixed distance from a given point and lie on a given line?

6 Find the locus of points equidistant from two concentric circles and on a diameter of the larger circle.

7 Find all the points on a given line that are a fixed distance from a given circle, and where the distance is less than the radius.

8 Find the locus of points 10 units from the origin of a coordinate system and 6 units from the y axis.

9 Transversal t intersects parallel lines m and n. Find the locus of points equidistant from m and n and 1 unit from t.

10 Given a regular pentagon, find the locus of points that are a given distance from its center and lie on the pentagon.

Section 14.2 Problem Set C

11 Given three points A, B, and C, find the locus of points equidistant from all three points.

12 a In space, find the locus of points that are 3 inches from a given plane and 5 inches from a fixed point on the plane.

b Find the area of the locus.

13 Find all the points equidistant from two given points and at a given distance from a given circle.

14 Find the locus of points in space that are equidistant from two given points and at a given distance from a given line.

15 Find the locus of points that lie on a given square and also lie on a given circle with its center in the interior of the square.

16 Given $\angle A$ and $\angle B$, find the locus of points that are equidistant from the sides of $\angle A$ and the sides of $\angle B$.

17 Find the locus in space of a line segment revolving about its midpoint.

The Royal Road: Constructing a Regular Pentagon

Start with ⊙O and ⊥ diameters $\overline{RS}$ and $\overline{WT}$. Find M, the midpoint of $\overline{OS}$, and find N so that $\overline{MN}$ has the same length as $\overline{WM}$. The length of $\overline{WN}$ is the length of each side of the pentagon.

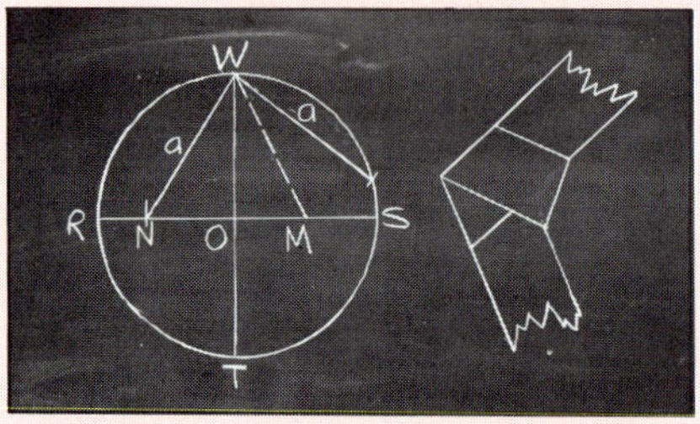

For another "construction," tie a knot in a thin strip of paper!

Section 14.3 The Concurrence Theorems

D **CONCURRENT LINES are lines that intersect in a *single point.***

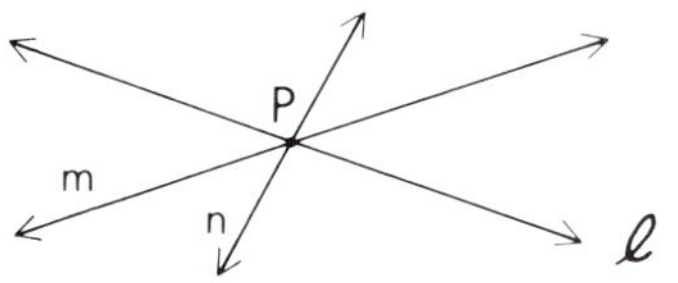

ℓ, m, and n are CONCURRENT at P

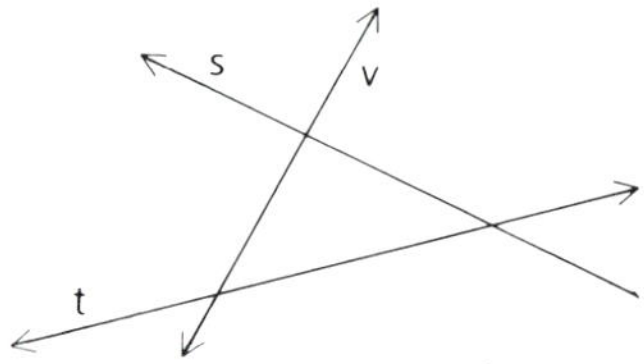

s, t, and v are not CONCURRENT

With this definition and our knowledge of compound locus we can now investigate some theorems of advanced geometry.

Theorem 121

T **The PERPENDICULAR BISECTORS OF THE SIDES OF A TRIANGLE ARE CONCURRENT at a point that is equidistant from the vertices of the triangle. (The point of concurrency of the perpendicular bisectors is the CIRCUMCENTER of the triangle.)**

Given: ℓ is the ⊥ bisector of $\overline{BC}$
m is the ⊥ bisector of $\overline{AC}$
n is the ⊥ bisector of $\overline{AB}$

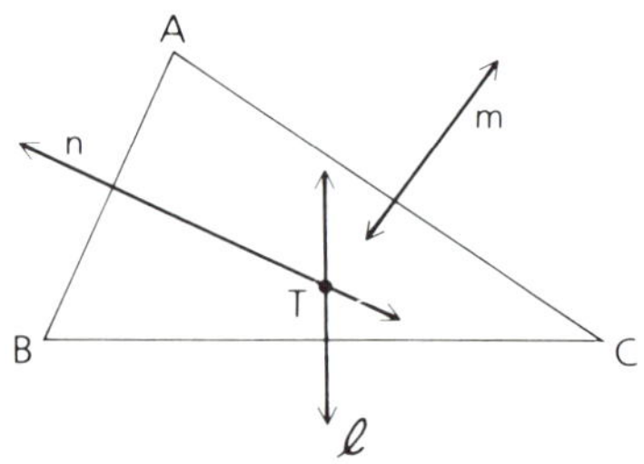

Prove: **a** ℓ, m, and n are concurrent at point T.
b T is equidistant from A, B, and C.

Proof of Theorem 121: Let T be the point of intersection of ℓ and n. (How do we know that ℓ and n intersect?)

We must show that m passes through T.

Because T is on line ℓ, the perpendicular bisector of $\overline{BC}$, T is equidistant from points B and C. Any point on the perpendicular bisector of a line segment is equidistant from the endpoints of that segment.

Similarly, T is equidistant from points A and B because it lies on n, the perpendicular bisector of $\overline{AB}$.

By transitivity, T is equidistant from A and C. Thus, T must lie on m by the Lies-On Theorem.

The bisectors of the angles of the triangle are also concurrent. This statement is formalized in the following theorem, which is presented without proof.

Theorem 122

T **THE BISECTORS OF THE ANGLES OF A TRIANGLE ARE CONCURRENT at a point that is equidistant from the sides of the triangle. (The point of concurrency of the angle bisectors is the INCENTER of the triangle.)**

Given: $\overrightarrow{AP}$ bisects $\angle BAC$
$\overrightarrow{BQ}$ bisects $\angle ABC$
$\overrightarrow{CR}$ bisects $\angle ACB$

Prove: **a** $\overrightarrow{AP}$, $\overrightarrow{BQ}$, and $\overrightarrow{CR}$ are concurrent at point N.
b N is equidistant from $\overline{AB}$, $\overline{BC}$, and $\overline{AC}$.

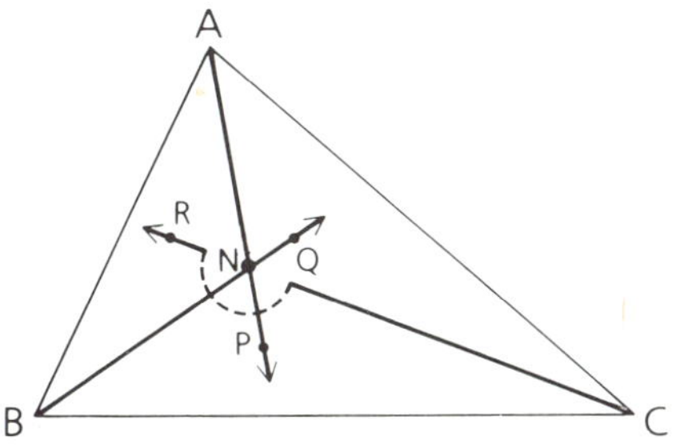

The proof of Theorem 122 uses the fact that the locus of all points equidistant from the sides of an angle is the bisector of the angle (Example 2 of 14.1). The organization of the proof is much like that of Theorem 121.

Theorem 123

T **The lines containing the ALTITUDES OF ANY TRIANGLE ARE CONCURRENT. (The point of concurrency of the altitudes is called the ORTHOCENTER of the triangle.)**

Given: $\overline{AD}$ is altitude to $\overline{BC}$
$\overline{BE}$ is altitude to $\overline{AC}$
$\overline{CF}$ is altitude to $\overline{AB}$

Prove: $\overleftrightarrow{AD}$, $\overleftrightarrow{BE}$, and $\overleftrightarrow{CF}$ are concurrent at O.

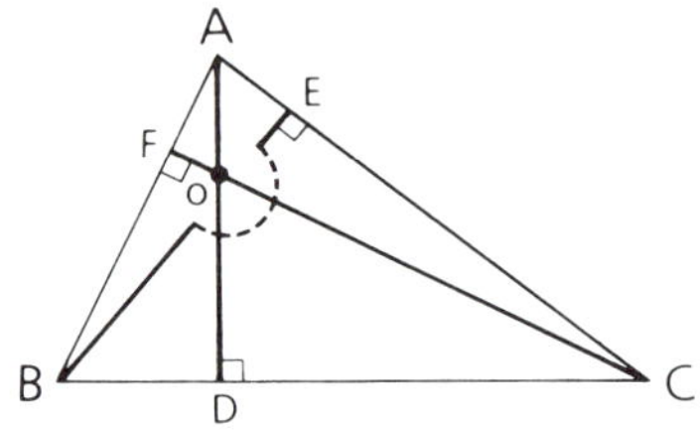

The proof of this theorem is called for in problem 19, Problem Set C.

Note: The ORTHOCENTER of a triangle is *not* always inside the triangle, as you can see in these figures.

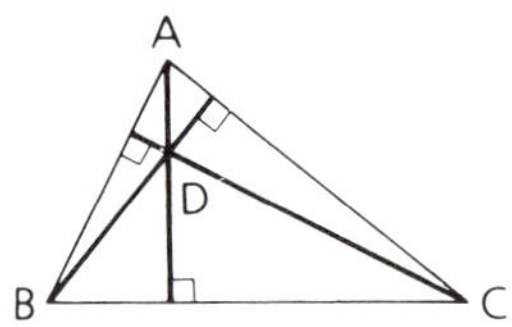

△ABC is an acute △.
D is the orthocenter.

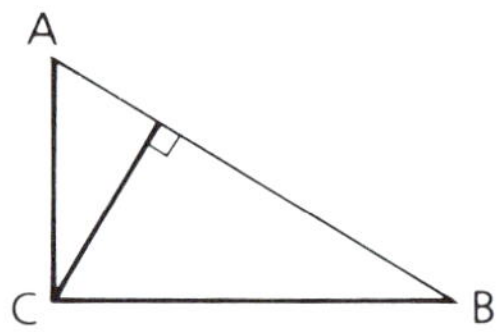

△ABC is a right △.
C is the orthocenter.

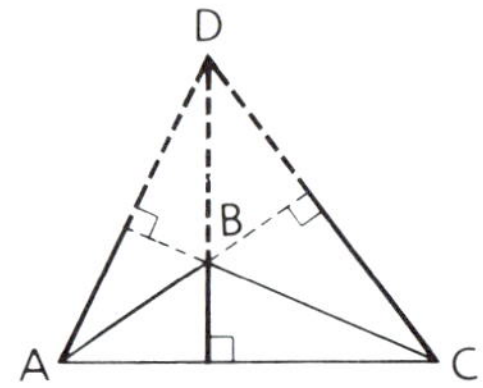

△ABC is an obtuse △.
D is the orthocenter.

Theorem 124

T **The MEDIANS OF ANY TRIANGLE ARE CONCURRENT at a point that is two-thirds of the way from any vertex of the triangle to the midpoint of the opposite side. (The point of concurrency of the medians of a triangle is called the CENTROID of the triangle.)**

Given: Medians $\overline{AM}$, $\overline{BN}$, and $\overline{CP}$

Prove: **a** $\overline{AM}$, $\overline{BN}$, and $\overline{CP}$ are concurrent at T

b $\frac{AT}{AM} = \frac{CT}{CP} = \frac{BT}{BN} = \frac{2}{3}$

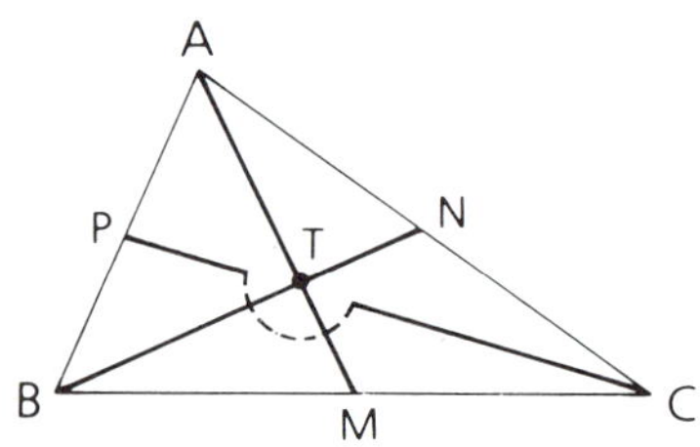

The proof of Theorem 124 is problem 21, Problem Set D.

The centroid of a triangle is important in physics, because it is the CENTER of GRAVITY of the triangle.

Section 14.3 Sample Problems

1 In $\triangle PQR$, medians $\overline{QT}$ and $\overline{PS}$ are concurrent at C.
PC = 4x − 6
CS = x
Find: **a** x **b** PS

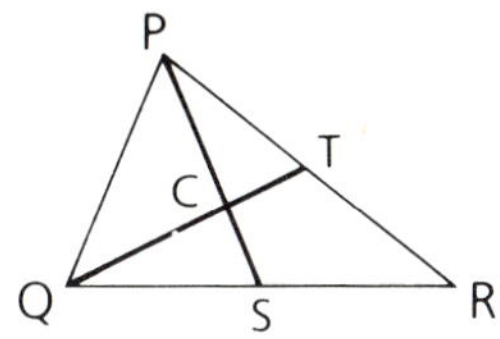

Solutions:

a The medians of a triangle are concurrent at a point that is two-thirds of the way from any vertex of the triangle to the midpoint of the opposite side.

$$\begin{aligned}\text{Thus, } PC &= \tfrac{2}{3}(PS)\\ \text{or } PC &= 2(CS)\\ 4x - 6 &= 2x\\ x &= 3\end{aligned}$$

b
$$\begin{aligned}\therefore PC &= 4x - 6\\ &= 4(3) - 6\\ &= 12 - 6\\ PC &= 6\\ \text{Thus, } PS &= PC + CS\\ &= 6 + 3\\ PS &= 9\end{aligned}$$

Section 14.3 Problem Set A

1 Trace $\triangle ABC$ on a piece of paper. Use a ruler to locate the centroid of $\triangle ABC$.

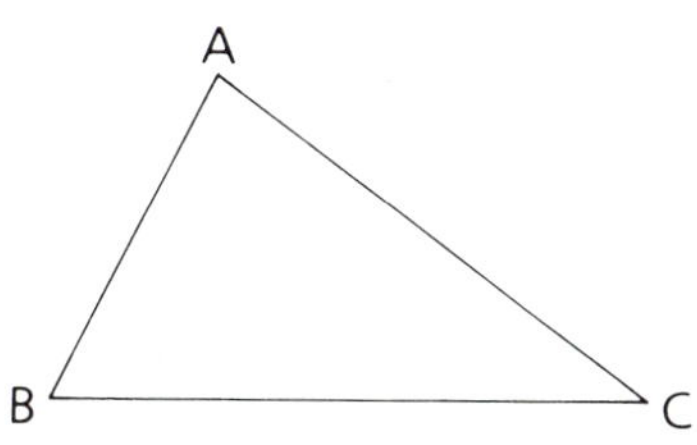

2 Find the orthocenter of right triangle PQR.

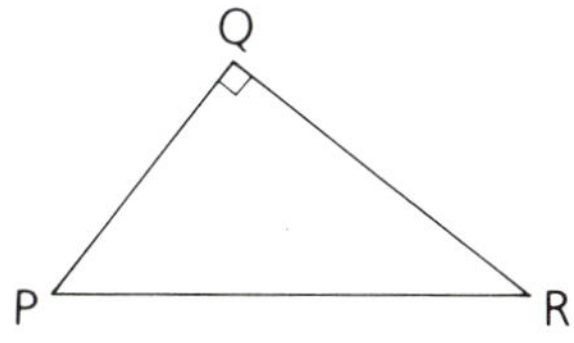

3 Given scalene $\triangle DEF$, explain how to find the locus of points equidistant from $\overline{DE}$, $\overline{EF}$, and $\overline{DF}$.

4 Trace right $\triangle RST$ on a piece of paper.

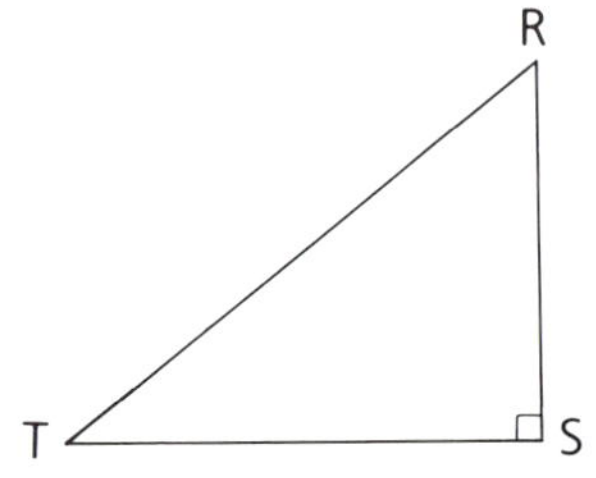

a Use a ruler to estimate where the circumcenter is located.

b Now that you have estimated the location of the circumcenter, can you guess the exact location of the circumcenter of a right triangle?

5 Every triangle has a circumcenter, orthocenter, centroid, and an incenter. Which of the four points will *always* lie in the interior of the triangle?

6 If a triangle is cut from cardboard and the circumcenter, orthocenter, centroid, and incenter are located, upon which point could we balance the triangle?

7 Given $\triangle ABC$ with medians $\overline{AM}$, $\overline{BN}$, and $\overline{CP}$.

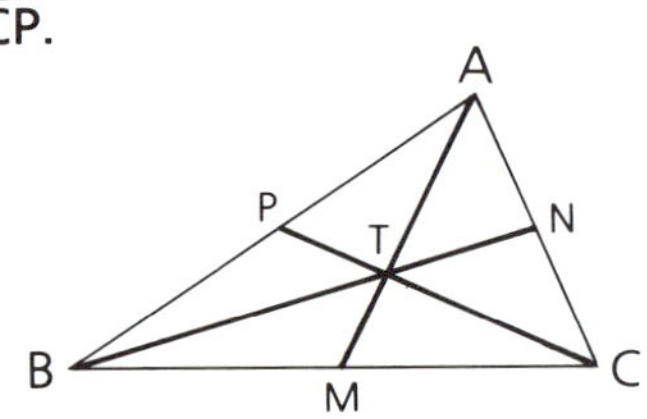

a If $AM = 9$, find AT.
b If $TN = 5$, find BN.
c If $TC = 8$, find PT.
d If $BN = \sqrt{18}$, find TN.

Section 14.3 Problem Set B

8 In what kind of triangle is the orthocenter a vertex of the triangle?

9 In what kind of triangle is the orthocenter the same point as the circumcenter?

10 In what kind of triangle does the centroid lie outside the triangle?

11 Given three non-collinear points. Sketch and describe the locus of points equidistant from all three points.

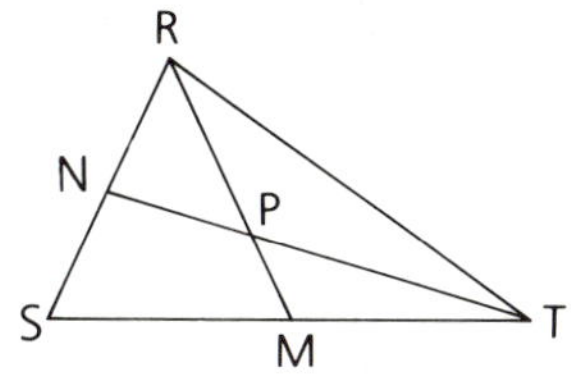

12 Given: $\triangle RST$ with medians $\overline{RM}$ and $\overline{TN}$ intersecting at P
$RP = 2y - x$, $TP = 2y$
$PM = y - 2$, $PN = x + 2$
Find: The longer of the two medians

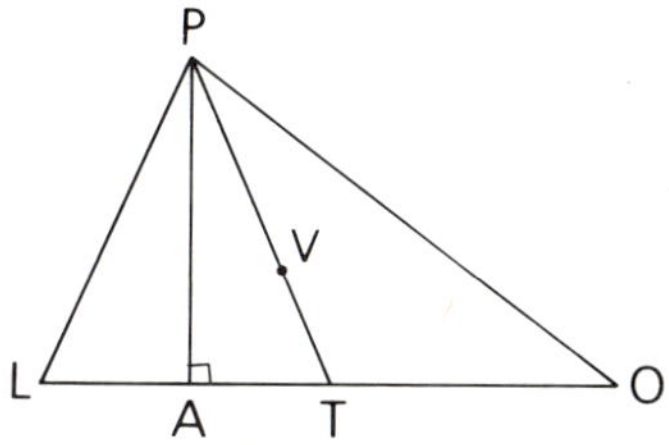

13 Given: $\triangle PLO$ with centroid V
$VT = 6$, $AT = 9$, $OT = 18$
Find: **a** PA
b the area of $\triangle PLO$
c the area of $\triangle POT$
d $m\angle APT$

14 Given $\triangle PQR$, with $P = (0, 0)$, $Q = (5, 12)$, and $R = (10, 0)$. Find the coordinates of its centroid.

Section 14.3 Problem Set C

15 Given $\triangle ABC$ with $A = (1, 3)$, $B = (7, -3)$, and $C = (9, 5)$. Find the circumcenter of the triangle.

16 Given $\triangle RST$ with $R = (-3, 2)$, $S = (4, 5)$, and $T = (7, -2)$. Find the coordinates of the orthocenter of the triangle.

17 Recall that the coordinates of the midpoint of a side of a triangle are found by taking the *average* of the coordinates of the end-points. As an extension of this idea, it can be shown that the coordinates of the centroid of a triangle are found by taking the *average* of the coordinates of the three vertices of the triangle.
Given: $\triangle ABC$ with $A = (-2, 8)$, $B = (-6, -2)$, and $C = (12, 6)$
Find: **a** the coordinates of the centroid of $\triangle ABC$
b the coordinates of the centroid of the triangle formed by joining the midpoints of the sides of $\triangle ABC$

18 Given $\triangle ABC$ with median $\overline{AM}$ and centroid P.

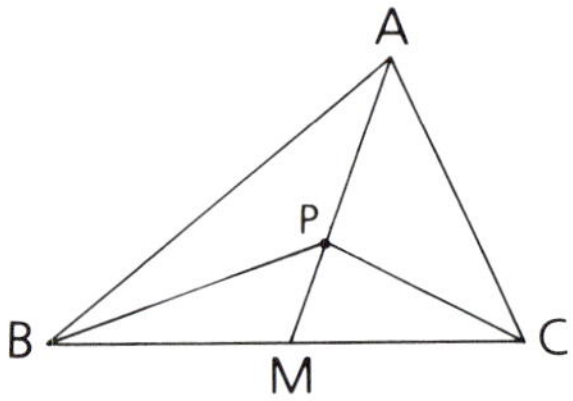

a Using $\overline{BC}$ as the base in each case, prove that the altitude of $\triangle PBC$ is one-third of the altitude of $\triangle ABC$.

b Find the ratio of the area of $\triangle PBC$ to the area of $\triangle ABC$.

19 Given: $\triangle ABC$
Prove Theorem 123: The altitudes of $\triangle ABC$ are concurrent

(Hint: Through each vertex of the triangle, draw a line parallel to the opposite side, obtaining the diagram to the right. Then apply Theorem 121.)

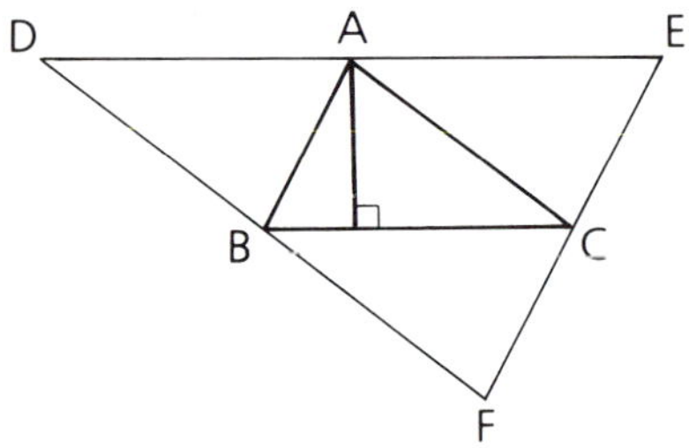

20 The centroid of any triangle is one of the trisection points of each median. Form a triangle by joining the *other* trisection point of *each* median to that of the other two medians.

a Find the ratio of the area of this triangle to the area of the original triangle.

b What is the relationship of this triangle to the triangle formed by joining the midpoints of the sides of the original triangle.

Section 14.3 Problem Set D

21 Given: $\triangle ABC$

Prove Theorem 124: The medians of $\triangle ABC$ are concurrent at a point that is two-thirds the way from any vertex of $\triangle ABC$ to the midpoint of the opposite side.

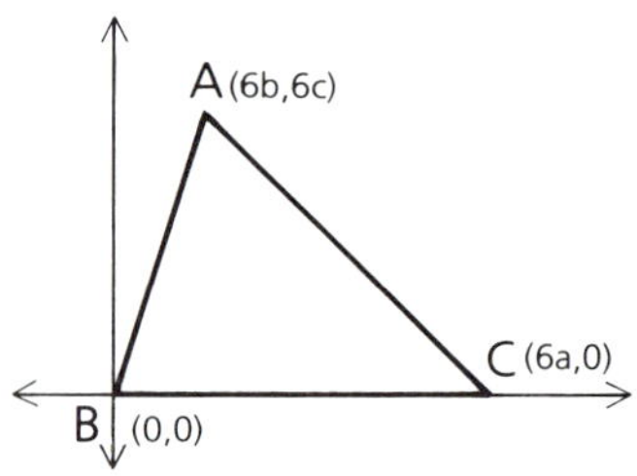

(Hint: Use the coordinates shown at the right and prove the theorem.)

Section 14.4 Basic Constructions

Section 14.4 (A) Introduction to Constructions

A construction is a drawing done with simple tools, and is based on a classical Greek method. Two tools are used:

1 A **COMPASS**, to construct circles or arcs of circles.

2 A **STRAIGHTEDGE**, to draw lines or rays. (A straightedge differs from a ruler only by the absence of marks for measuring distances.)

The method can produce accurate drawings when correctly used. (A sharp pencil and good paper on flat, firm cardboard are necessities.) Admittedly, modern drafting machines can produce more accurate drawings in less time, so our reasons for studying constructions are more like these:

- The tools are simple and portable.
- There is an orderly progression of steps. Nothing is accepted just because the result looks correct.
- Analyzing constructions strengthens knowledge of theorems.
- By limiting equipment and having rules, constructions constitute a game, still enjoyed by most people who learn it. The game has a practical bonus for some, because users of drafting machines also must analyze problems, and the analyses are often the same as those in constructions.

14.4 (B) Shorthand Notation for Constructions

To make step-by-step instructions clear and precise, we use the following notation for constructions:

1 ⊙ (P,PB) represents a CIRCLE drawn with center P, and radius of length PB.

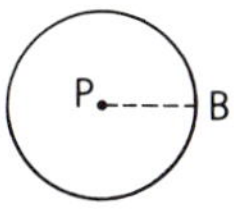

2 arc (P,PB) represents an ARC drawn with center P, and radius of length PB.

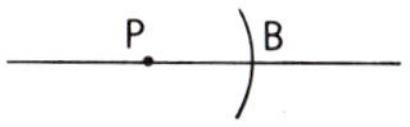

Section 14.4 (C) Six Basic Constructions

Six basic constructions are developed for frequent reference. Because a construction drawing has meaning only in terms of how it grew, we urge you to redraw these constructions, following the instructions step by step.

Construction 1 SEGMENT COPY

Construction of a line segment congruent to a given segment.

Given: $\overline{AB}$ 

Note: In the set-up on your paper, draw a segment $\overline{AB}$ of any length you want.

Construct: A segment $\overline{PQ}$ that is congruent to $\overline{AB}$.

Procedure:

1 Draw a WORKING LINE, w.
2 Let P be any point on w.
3 On the given figure, construct arc (A,AB).
4 Construct arc (P,AB), intersecting w at some point. Call it Q.
5 $\therefore \overline{PQ} \cong \overline{AB}$

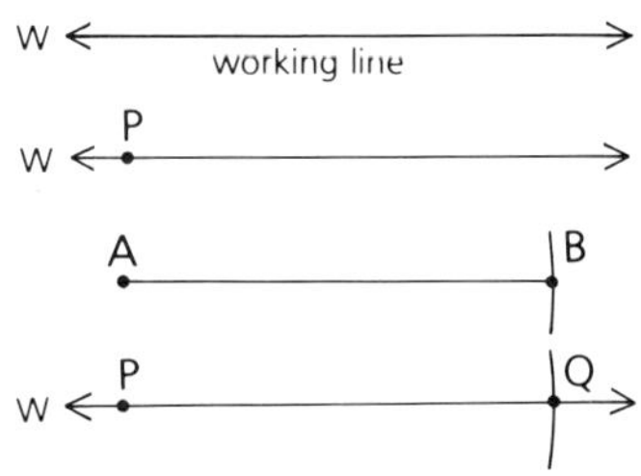

Comment: Lengths are not measured by rulers; they are matched by compass settings.

Your finished paper should look like this:

Given: $\overline{AB}$ 

Construct: $\overline{PQ}$, $\cong$ to $\overline{AB}$ 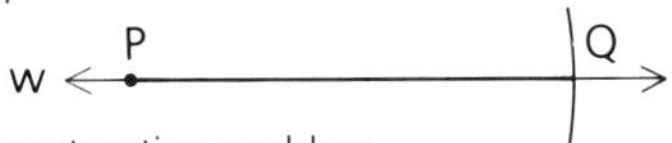 $\therefore \overline{PQ} = \overline{AB}$

Note: Do not erase any arc marks in *any* construction problem.

Construction 2 ANGLE COPY

Construction of an angle congruent to a given angle.

Given: $\angle ABC$

Note: Make your set-up big enough for easy use of the compass, yet not so big that everything won't fit on your paper.

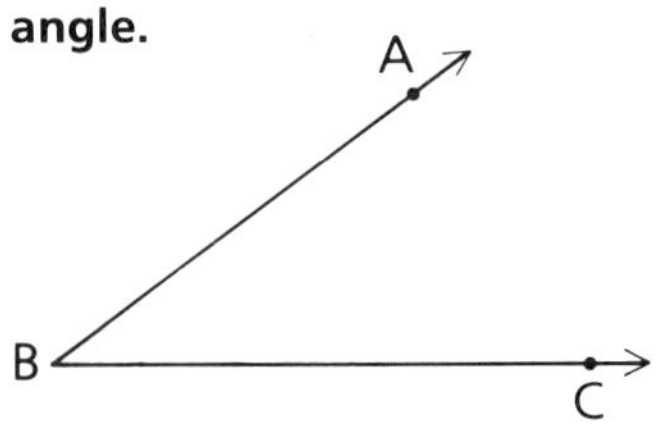

Construct: An angle, ∠PQR, that is congruent to ∠ABC.

Procedure:

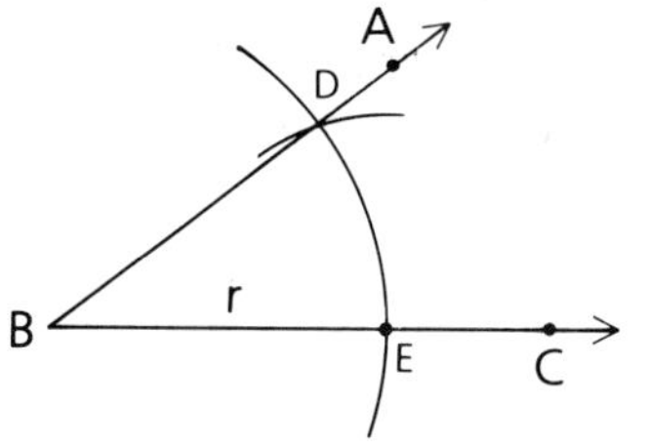

1. On the set-up, use any radius r to construct arc (B,r), intersecting ∠ABC at two points. Call them D and E.
2. Let Q be any point on a working line, w.
3. Construct arc (Q,r) to intersect w at some point, R.
4. Construct arc (E,ED).
5. Construct arc (R,ED) intersecting arc (Q,r) at some point. Call it P.
6. Draw $\overrightarrow{QP}$.
7. ∴ ∠PQR ≅ ∠ABC

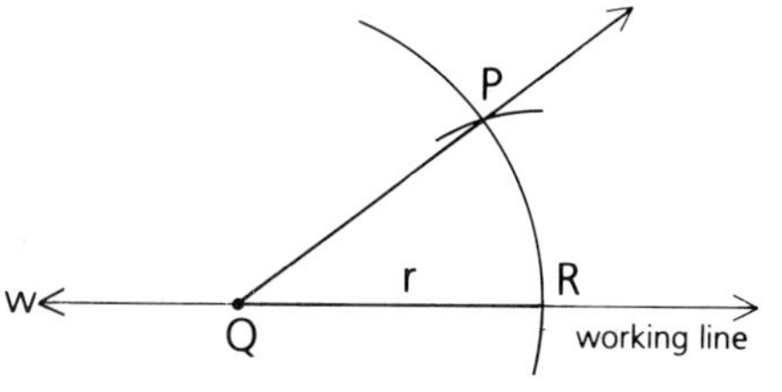

Comment: If we drew $\overline{DE}$ and $\overline{PR}$, we would form △BDE and △QPR. Do you see how SSS is the basis of this construction?

Construction 3 ANGLE BISECTION

Construction of the bisector of a given angle.

Given: ∠ABC
Construct: $\overrightarrow{BP}$, the bisector of ∠ABC

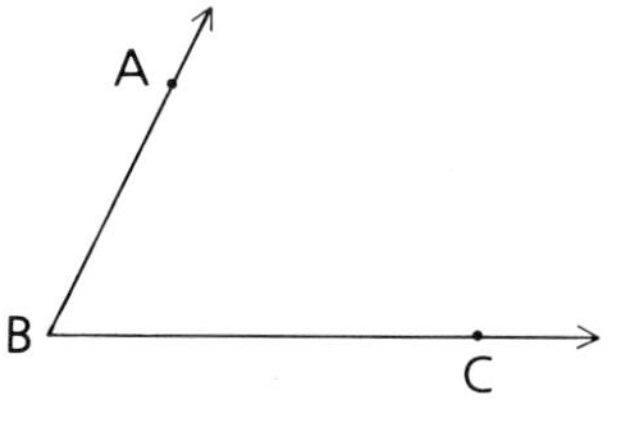

Procedure:

1. Use any radius r to construct arc (B,r), intersecting the sides of ∠ABC at two points, Q and T.
2. Use any radius s (which may or may not be equal to r) to construct arc (Q,s) and arc (T,s), intersecting each other at a point P.
3. Draw $\overrightarrow{BP}$.
4. ∴ $\overrightarrow{BP}$ bisects ∠ABC.

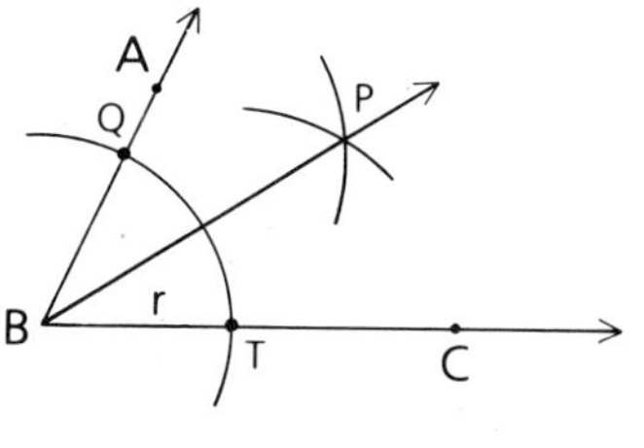

Comment: If we drew $\overline{QP}$ and $\overline{PT}$, each would be s units long. Can you see how SSS is the basis of this construction?

Construction 4 PERPENDICULAR BISECTOR

Construction of the perpendicular bisector of a given line segment.

Given: $\overline{AB}$

Construct: $\overleftrightarrow{PQ}$, the perpendicular bisector of $\overline{AB}$.

Procedure:

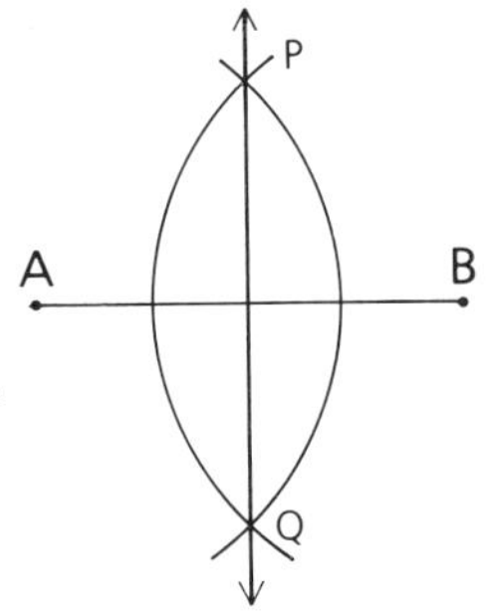

1 Use any radius r that is more than half the length of $\overline{AB}$ to construct arc (A,r).
2 Construct arc (B,r), intersecting arc (A,r) at P and Q.
3 Draw $\overleftrightarrow{PQ}$.
4 $\therefore \overleftrightarrow{PQ}$ is the perpendicular bisector of $\overline{AB}$.

Construction 5 ERECTING A PERPENDICULAR

Construction of a line perpendicular to a given line at a given point on the line.

Given: Line $\overleftrightarrow{AB}$ with point P on the line.

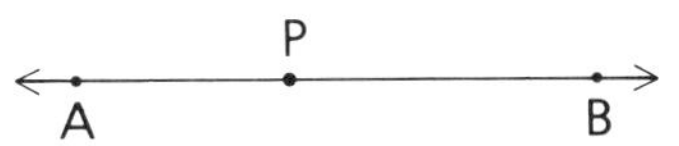

Construct: A line, $\overleftrightarrow{PQ}$, that is perpendicular to $\overleftrightarrow{AB}$ at P.

Procedure:

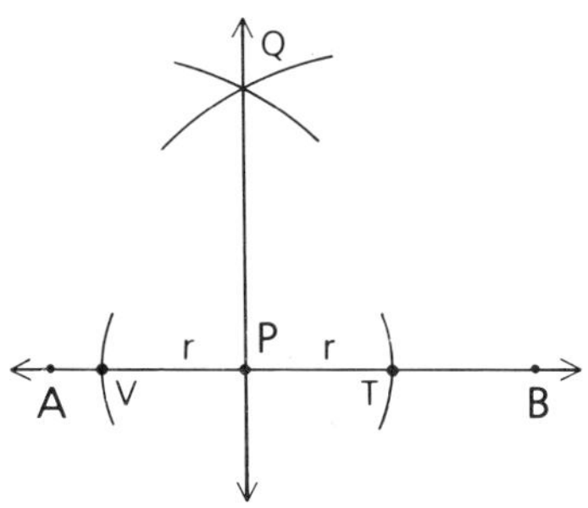

1 Use any radius r to construct arc (P,r), intersecting $\overleftrightarrow{AB}$ at V and T.
2 Use any radius s that is larger than r to construct arc (V,s) and arc (T,s), intersecting each other at a point Q.
3 Draw $\overleftrightarrow{PQ}$.
4 $\therefore \overleftrightarrow{PQ} \perp \overleftrightarrow{AB}$.

Construction 6 DROPPING A PERPENDICULAR

Construction of a line perpendicular to a given line from a given point not on the line.

Given: $\overleftrightarrow{AB}$ and point P not on $\overleftrightarrow{AB}$.

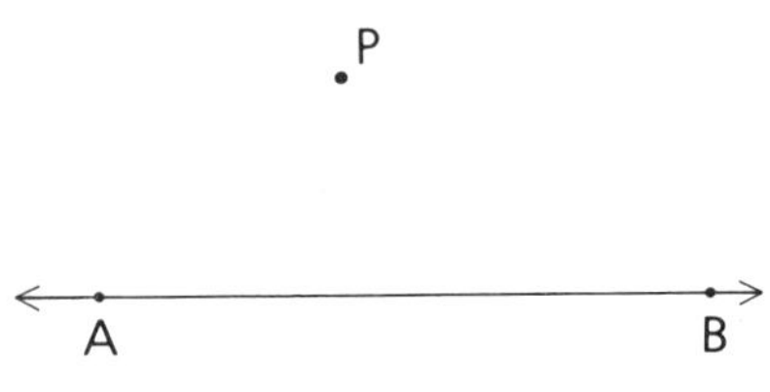

Construct: A line $\overleftrightarrow{PQ}$ through P and perpendicular to $\overleftrightarrow{AB}$.

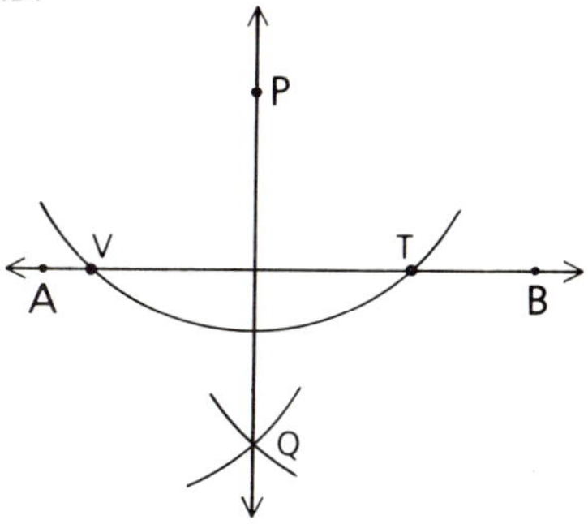

Procedure:

1 Use any radius r to construct arc (P,r) intersecting $\overleftrightarrow{AB}$ at V and T.

2 Use any radius s (which may or may not be equal to r) to construct arc (V,s) and arc (T,s), intersecting each other at a point Q.

3 Draw $\overleftrightarrow{PQ}$.

4 $\therefore \overleftrightarrow{PQ} \perp \overleftrightarrow{AB}$.

Section 14.4 Sample Problems

1 Given: ∠s A and B as marked.

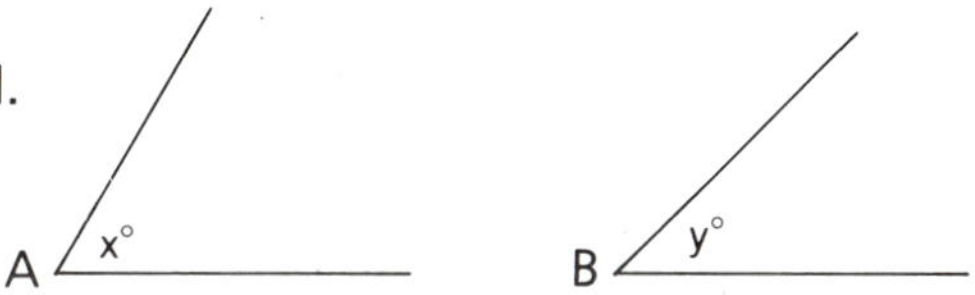

Construct: An angle whose measure is equal to (x + y).

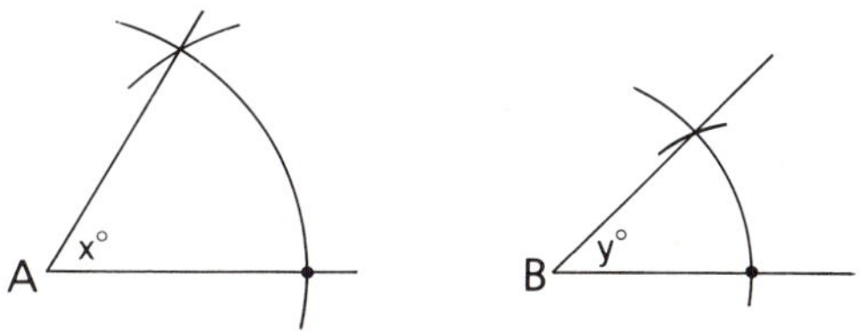

Procedure:

1 On a working line w, use ANGLE COPY to construct ∠VTS ≅ ∠A.

2 With $\overleftrightarrow{TV}$ as a new working line, use ANGLE COPY to construct ∠QTV ≅ ∠B.

3 ∴ ∠QTS is the required angle.

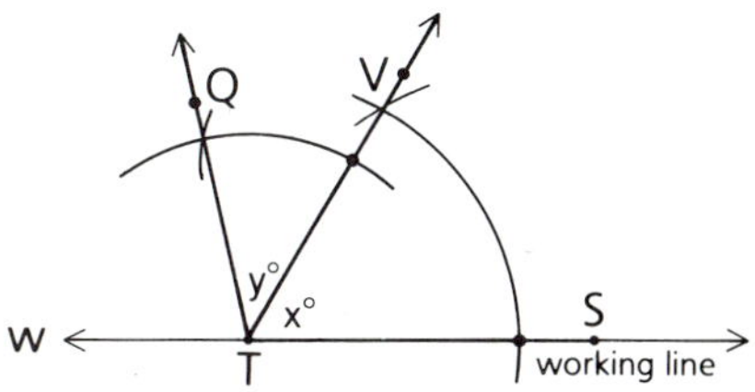

2 Given: ⊙ P and point A on the circle.
Construct: The tangent to ⊙ P at point A.

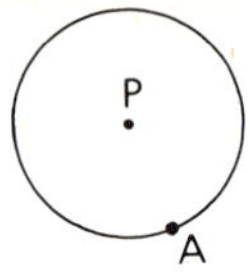

Analysis: Make a free-hand sketch of the required construction. Analyze the geometric relationships between the given figure and the construction to determine the required procedure.

In this problem the sketch will look like this:
Do you see what needs to be done?

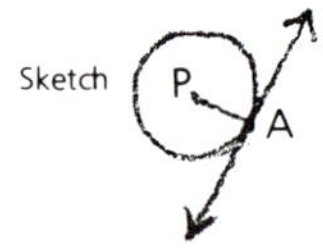

Procedure:

1 Draw $\overrightarrow{PA}$.

2 Construct the $\perp$ to $\overleftrightarrow{PA}$ at A. (see Construction 5)

3 $\therefore$ $\overleftrightarrow{TA}$ is the required tangent.

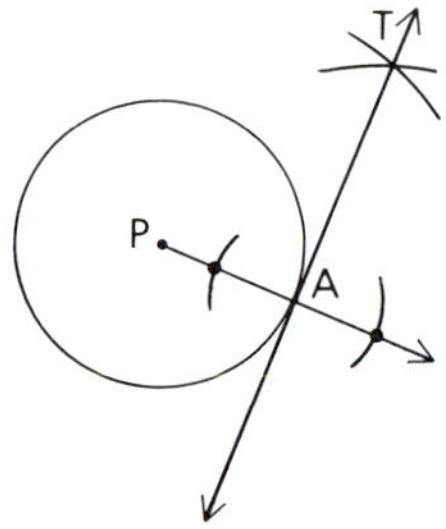

3 Given: ⊙ O and point P outside the circle.

Construct: The tangent to ⊙ O from point P.

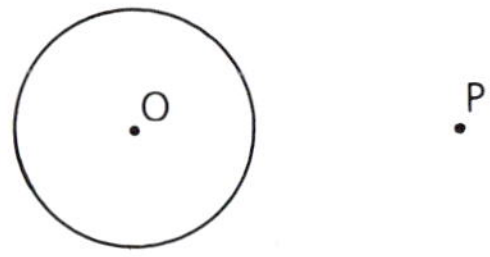

Analysis: At the point of tangency, the radius and the required tangent must form a right angle. See problem 19, Set D.

Procedure:

1 Draw $\overline{OP}$.

2 Find the midpoint M of $\overline{OP}$ by Construction 4.

3 Construct ⊙ (M, MP).

4 Label A and B the intersections of ⊙ O and ⊙ M.

5 Draw $\overline{PA}$.

6 $\therefore$ $\overline{PA}$ is tangent to ⊙ O.

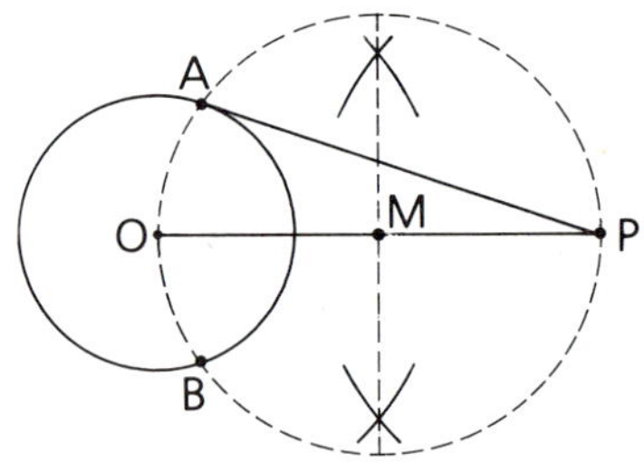

Section 14.4 Problem Set A

1 Construct the locus of points equidistant from two fixed points, A and B.

2 For **a–g** draw two given segments, $\overline{AB}$ and $\overline{CD}$, with $AB > CD$.

- a Construct a segment whose length is the sum of AB and CD.
- b Construct a segment whose length is the difference of AB and CD.
- c Locate the midpoint of $\overline{AB}$ by construction.
- d Construct an equilateral triangle whose sides are congruent to $\overline{CD}$.
- e Construct an isosceles triangle, making its base congruent to $\overline{CD}$ and each leg congruent to $\overline{AB}$.
- f Construct a square whose sides are congruent to $\overline{AB}$.
- g Construct a circle whose diameter is congruent to $\overline{CD}$.

3 For **a–e** draw an acute angle, $\angle ABC$, and an obtuse angle, $\angle WXY$.

- a Construct an angle $\angle FGH$ congruent to $\angle WXY$.
- b Construct the complement of $\angle ABC$.
- c Construct the supplement of $\angle WXY$.
- d Construct an angle whose measure is the difference of $\angle WXY$ and $\angle ABC$.
- e Construct an angle whose measure is double that of $\angle ABC$.

4 Construct the following angles: **a** $90°$ **b** $45°$ **c** $60°$ **d** $75°$

5 Draw an obtuse triangle. Construct the bisector of each angle.

Section 14.4 Problem Set B

6 If a and b are the lengths of two segments and $a < b$, construct a segment whose length is equal to $\frac{1}{2}(b - a)$.

7 Given $\angle A$ and $\angle B$. Construct an angle equal to $\frac{1}{2}(m\angle A + m\angle B)$.

8 Construct an angle whose measure is: **a** 135 **b** $112\frac{1}{2}$ **c** 165

9 Inscribe a square in a given circle. (Hint: Use the diagonals.)

10 Construct the three medians of a given $\triangle PQR$.

11 Construct the three altitudes of an acute $\triangle ABC$.

12 Given circle P with point Q in the interior of the circle, construct a chord of the circle having Q as its midpoint.

13 $\angle A$ is the vertex angle of an isosceles triangle. Find, by construction, one of the base angles of the triangle. (Can you do this without drawing a triangle?)

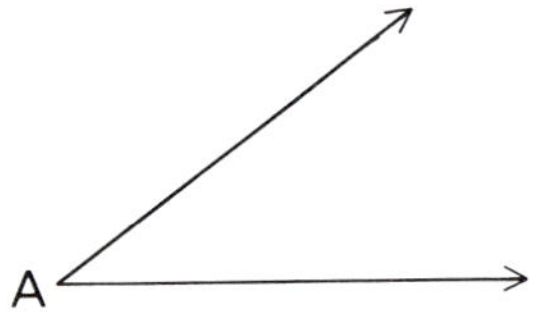

Section 14.4 Problem Set C

14 Draw any triangle. Construct a second triangle similar to the first such that the ratio of the perimeters is 1:2.

15 Construct a square whose diagonal is equal to AB.

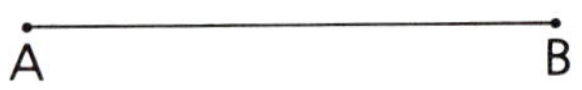

16 Explain how you would construct an angle of:
a $32\frac{1}{2}^\circ$ (if given an angle of 80°) **b** $41\frac{1}{4}^\circ$

17 Construct two parallel lines.

18 Construct a line, $\overleftrightarrow{CD}$, that is parallel to $\overleftrightarrow{AB}$ and tangent to $\odot$ O.

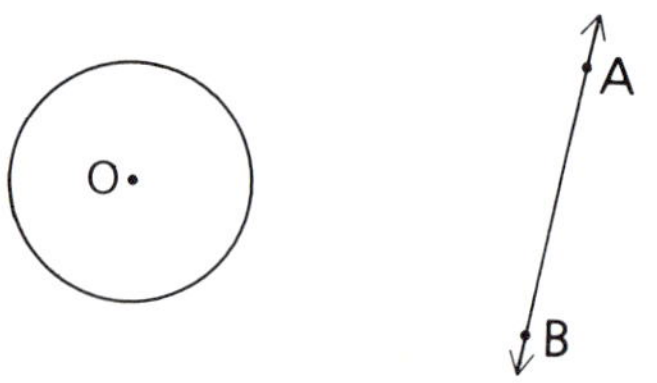

Section 14.4 Problem Set D

19 Write a paragraph proof to show that the construction of a tangent to a circle from an external point, as given in Sample Problem 3, is valid.

Section 14.5 Applications of the Basic Constructions

The six basic constructions may be used to develop more complicated constructions. Four of these are presented in this section. Once you have mastered all ten constructions you will enjoy the challenge of future problem sets.

Construction 7 PARALLELS

Construction of a line parallel to a given line through a point not on the given line.

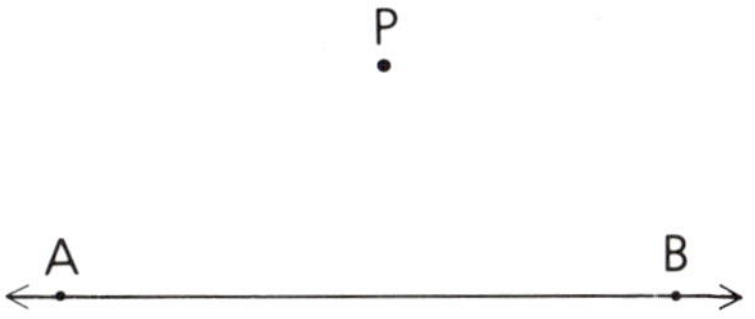

Given: $\overleftrightarrow{AB}$ with point P not on $\overleftrightarrow{AB}$.
Construct: A line, $\overleftrightarrow{PQ}$, through P and parallel to $\overleftrightarrow{AB}$.

Procedure:

1. Draw any line, t, through P, intersecting $\overleftrightarrow{AB}$ at some point. Call it C.
2. Use **ANGLE COPY** to construct $\angle QPR \cong \angle PCB$.
3. $\therefore \overleftrightarrow{PQ} \parallel \overleftrightarrow{AB}$ by CAP.

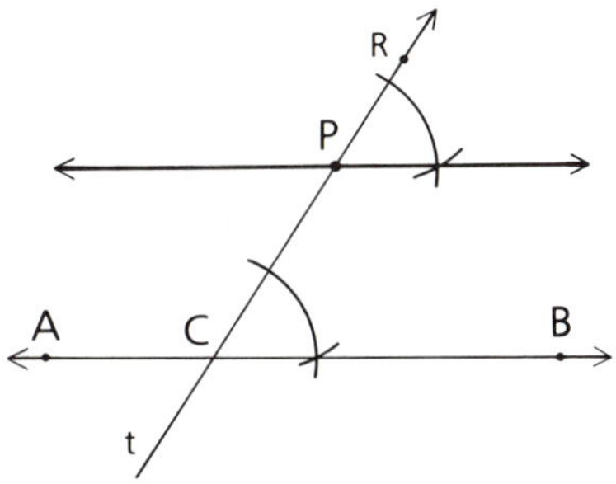

Construction 8 SEGMENT DIVISION

Dividing a segment into a given number of congruent segments.

Given: $\overline{AB}$

Construct: Points that divide $\overline{AB}$ into any number of congruent segments. (The figure shows division into *three* congruent segments.)

Analysis: If parallel lines cut off $\cong$ segments on some transversal, they cut off $\cong$ segments on any other. Work backwards, transversals first. $\overleftrightarrow{AB}$ is the "any other" transversal. On "some" transversal through A, mark off three $\cong$ segments of any length. Call their sum AR. Then $\overleftrightarrow{RB}$ determines the direction of the parallel lines.

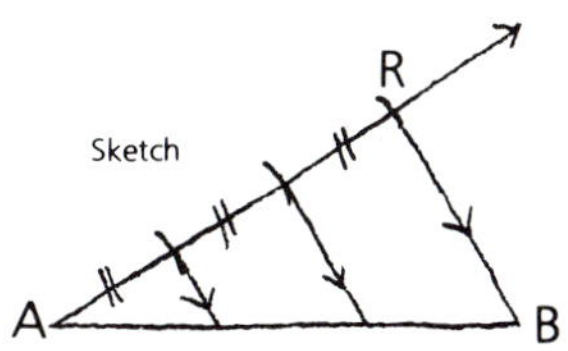

Procedure:

1. Draw any line ℓ through point A.
2. With a radius r, construct arc (A,r) intersecting line ℓ at a point P.
3. Construct arc (P, r), intersecting ℓ at Q.
4. Construct arc (Q,r), intersecting ℓ at R.

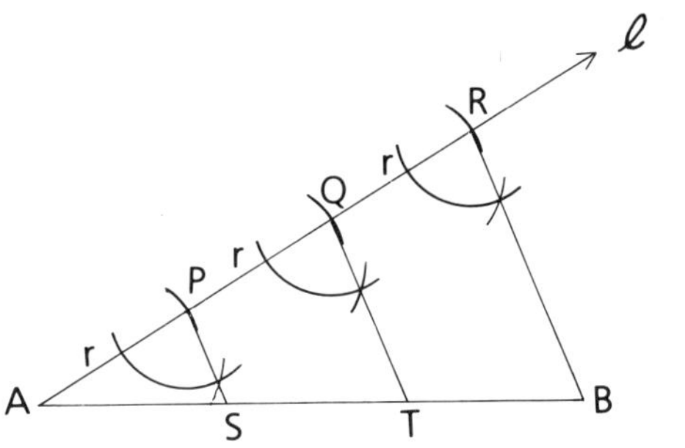

5. Draw $\overline{RB}$.
6. Using Construction 7, construct lines through P and Q parallel to $\overleftrightarrow{RB}$. Call the intersections of these lines with $\overline{AB}$, points S and T.
7. $\therefore \overline{AS} \cong \overline{ST} \cong \overline{TB}$. (Do you know the reason why?)

Construction 9 MEAN PROPORTIONAL

Construction of a segment whose length is the mean proportional between the lengths of two given segments.

Given: $\overline{AB}$ and $\overline{CD}$

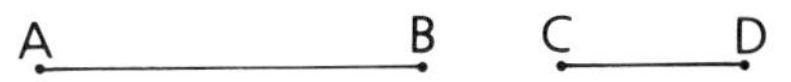

Construct: $\overline{VR}$, such that $(VR)^2 = (AB)\,(CD)$

Analysis: Mean proportional suggests an altitude on hypotenuse. We can find the length of h if we recall that an angle inscribed in a semicircle is a right angle.

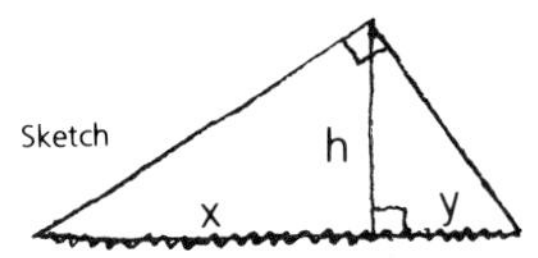

Procedure:

1. On a working line w, use **SEGMENT COPY** to construct a segment of length AB + CD. (Make TV = AB and VZ = CD)

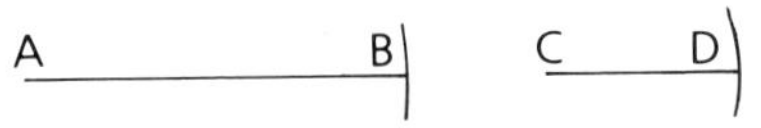

2. Use **PERPENDICULAR BISECTOR** to find the midpoint M of $\overline{TZ}$.
3. Construct semicircle (M,MT).
4. At V, **ERECT THE PERPENDICULAR** to $\overleftrightarrow{TZ}$ The perpendicular will intersect ⊙ M at R, and $\angle TRZ$ will be a right angle.
5. $\therefore h^2 = xy$, so $(VR)^2 = (AB)\,(CD)$

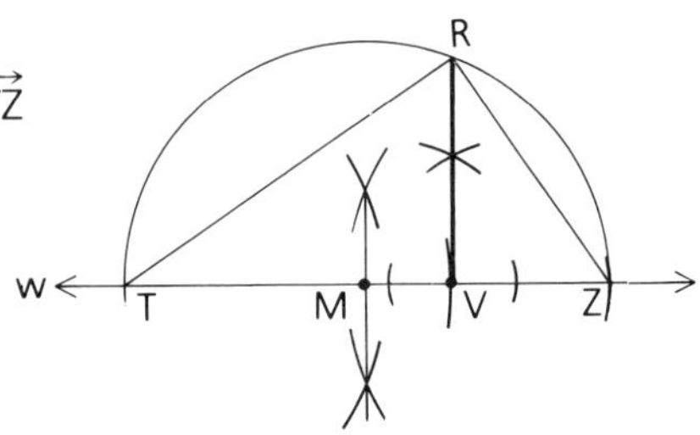

Construction 10 FOURTH PROPORTIONAL

Construction of a segment whose length is the fourth proportional to the lengths of three given segments.

Given: $\overline{AB}$ a $\overline{CD}$ b $\overline{EF}$ c

Construct: $\overline{TV}$, such that $\frac{a}{b} = \frac{c}{TV}$

Procedure:

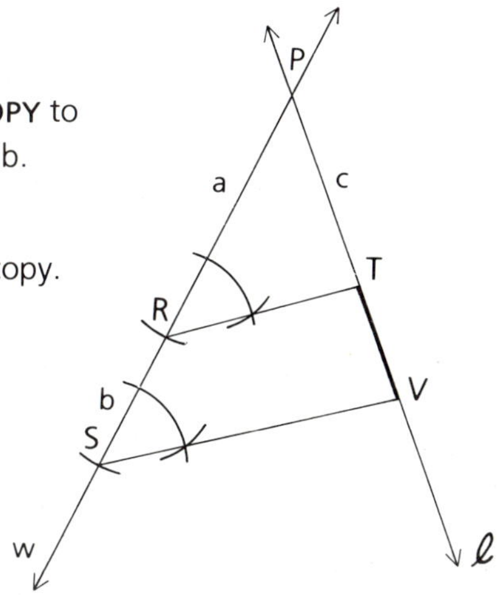

1 On a working line w use **SEGMENT COPY** to construct a segment $\overline{PS}$ of length a + b.

2 Draw any other line ℓ through P.

3 On ℓ, construct PT = c by segment copy.

4 Draw $\overline{TR}$.

5 Through S, construct a line parallel to $\overleftrightarrow{RT}$, intersecting ℓ at V.

6 ∴ $\overline{TV}$ is the required segment, since $\frac{a}{b} = \frac{c}{TV}$.

Section 14.5 Sample Problems

1 Inscribe a circle in given △ABC.

Analysis: The center of an inscribed circle is equidistant from the sides, so it is the point of concurrency of the angle bisectors.

Procedure:

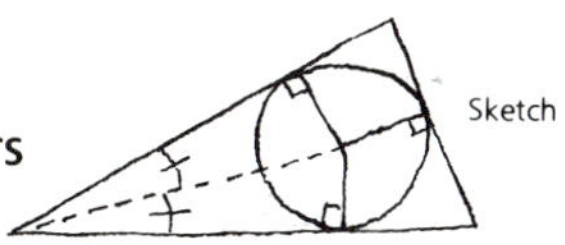

1 Construct the angle bisectors of ∠A and ∠C.

2 Their intersection will be T, the incenter of △ABC.

3 Construct a perpendicular from T to $\overline{BC}$. Call the foot F.

4 Construct ⊙ (T,TF).

5 ∴ ⊙ T is inscribed in △ABC.

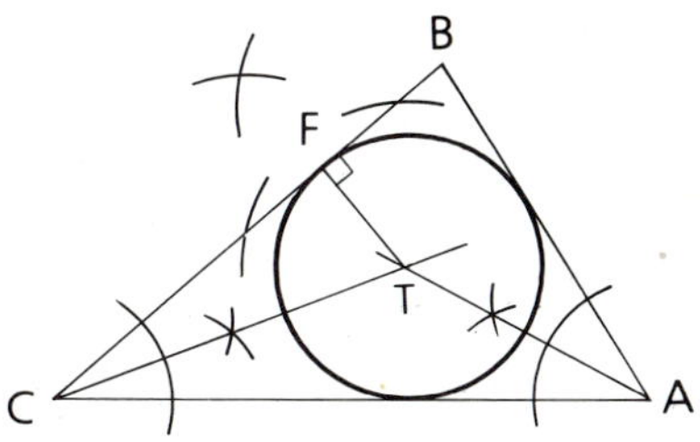

2 Given: 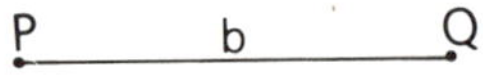

Construct: A segment whose length is $\sqrt{b}$.

Analysis: Since $x = \sqrt{b}$ is equivalent to $x^2 = b$, or $\frac{b}{x} = \frac{x}{1}$, use the MEAN PROPORTIONAL. To represent 1, choose any segment as a UNIT SEGMENT. Then b is whatever number of such units are in the given segment $\overline{PQ}$.

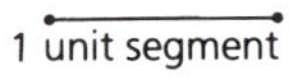

Procedure:

Using $\overline{PQ}$ and the UNIT SEGMENT, construct the mean proportional, x, between b and 1.

Thus, $x^2 = b \cdot 1$ and $x = \sqrt{b}$.

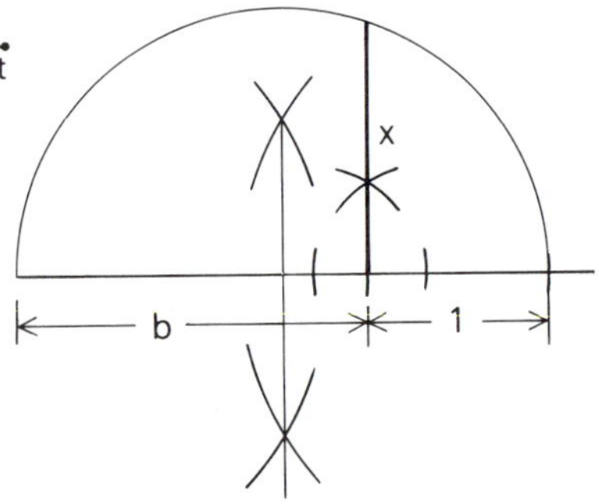

Section 14.5 Problem Set A

1 Given $\triangle ABC$, construct a line parallel to $\overleftrightarrow{AB}$ and passing through C.

2 Given $\triangle PQR$, trisect $\overline{QR}$.

3 Given $\overline{AB}$, with point C between A and B, construct a segment whose length is the mean proportional between AC and BC.

4 Given acute angle $\angle DEF$, with H between E and F, find by construction a point J between E and D such that $\frac{EJ}{JD} = \frac{EH}{HF}$.

5 Construct an equilateral triangle and its inscribed circle.

6 Construct a parallelogram given two sides and an angle.

7 Construct an isosceles right triangle and its circumscribed circle.

8 Construct a rectangle given the base and a diagonal.

9 Construct the centroid of a given triangle.

10 Use an object with a circular surface to trace the outline of a circle. By construction, locate the center of the circle.

11 Given a point P anywhere on a line w, construct a circle of radius r that is tangent to w at P.

Section 14.5 Problem Set B

12 Given a segment of length b (make it about 14 cm long), solve $5x = b$ for x by a geometric method.

13 Construct a rhombus, given its diagonals.

14 Construct an isosceles trapezoid given the bases and the altitude.

15 Given three non-collinear points, construct a circle that passes through all three points.

16 Given ▱ABCD as shown, construct:

a a rectangle with the same area as ▱ABCD.

b a triangle with the same area as ▱ABCD.

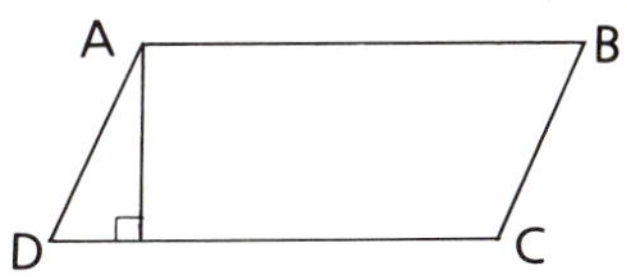

17 Where should a straight fence be located to divide a given triangular field into two fields whose areas are in the ratio 2:1.

18 Given a segment of length a, construct the geometric mean between 2a and 3a.

19 Given $\triangle ABC$, find by construction a point M on $\overline{AC}$ that divides $\overline{AC}$ in a ratio equal to $\frac{AB}{BC}$.

Section 14.5 Problem Set C

20 Given: a ―――――――――― 1 ――――――

Construct: a segment whose length is $\frac{1}{a}$.

21 Given a unit segment, construct a segment whose length is $\sqrt{3}$.

22 Find the centroid, circumcenter, and orthocenter of a large scalene triangle. What seems to be true about these 3 points?

23 Construct a square equal in area to a given parallelogram.

24 Construct a square that has an area twice as large as the area of a given square.

25 Circumscribe a regular hexagon about a given circle.

Section 14.5 Problem Set D

26 Construct a triangle equal in area to the given quadrilateral PQRS.

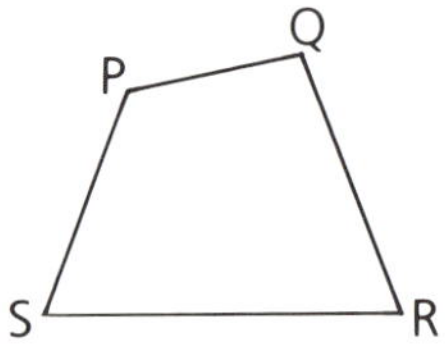

Procedure:

1 Draw diagonal $\overline{PR}$.

2 Construct a line parallel to $\overline{PR}$ through Q, intersecting $\overleftrightarrow{SR}$ at some point T.

3 Draw $\overline{PT}$.

4 $\therefore$ area ($\triangle$PST) = area (quad PQRS)

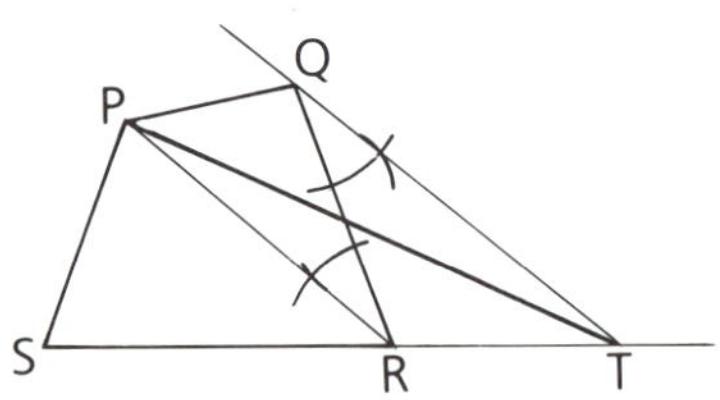

a Write a paragraph proof that shows the above procedure is valid.

b Construct a triangle that is equal in area to a given pentagon.

Section 14.6 Triangle Constructions

In this section you will construct triangles, given various combinations of parts and conditions. The notation for parts and associated segments, given below, will be helpful.

sides: a, b, c

angles: A, B, C

altitudes: h_a, h_b, h_c

medians: m_a, m_b, m_c

angle bisectors: t_a, t_b, t_c

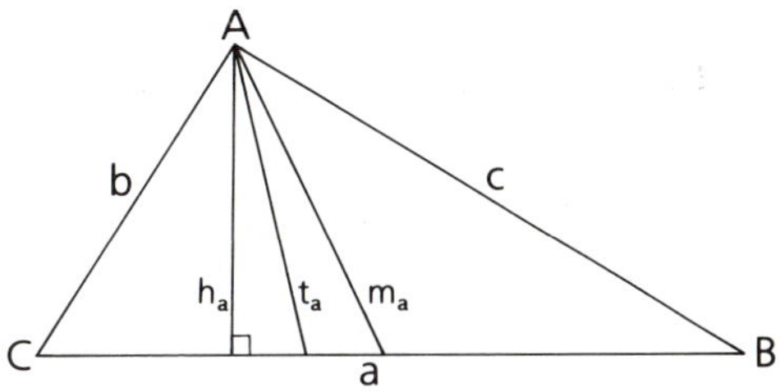

The side opposite vertex A is *a* units long.
The side opposite vertex B is *b* units long.
The side opposite vertex C is *c* units long.
h_a is the length of the altitude to side *a*.
h_b is the length of the altitude to side *b*.
Medians and angle bisectors have similar labeling.

The Sample Problems illustrate the importance of beginning with a sketch of the figure with its given parts and conditions.

Section 14.6 Sample Problems

1 Construct △ABC, given {a, C, b}. **(Note: This is SAS)**

Given: **a** ______________________

b ______________

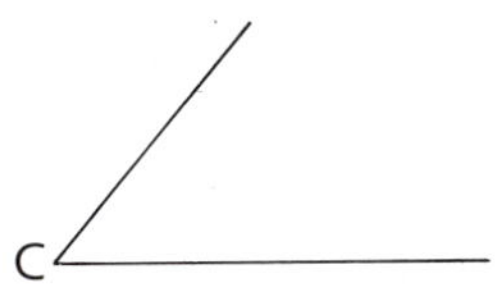

Construct: △ABC

Procedure:

1 Sketch the construction.

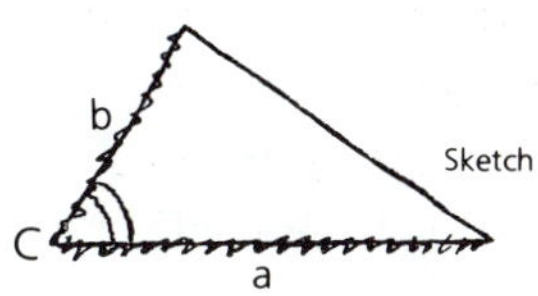

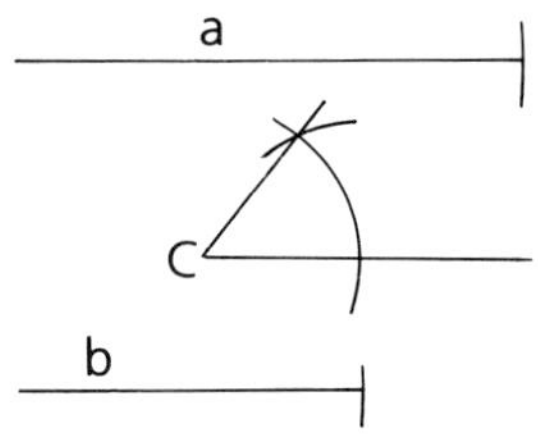

2 Copy side a on working line w.

3 Copy $\angle C$ at one end of side a. The other endpoint of a is B.

4 Copy side b on the second side of $\angle C$. The other endpoint of b is A.

5 Draw $\overline{AB}$

6 $\triangle ABC$ is the required triangle.

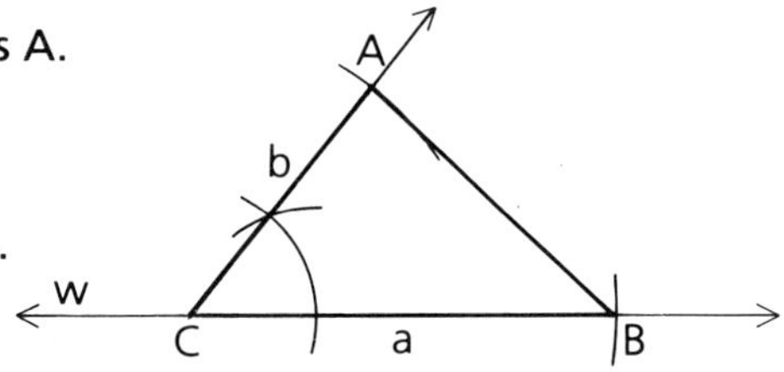

2 Construct $\triangle ABC$, given $\{a, h_a, \angle B\}$

Given: a

h_a

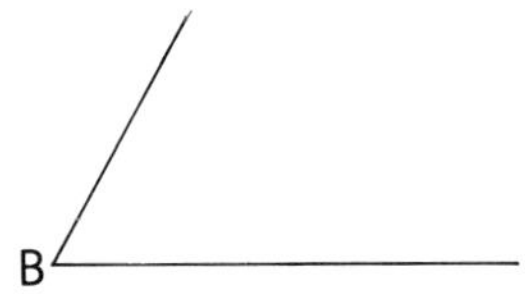

Construct: $\triangle ABC$

Procedure:

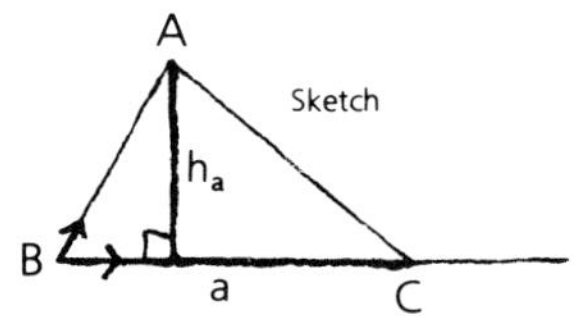

1 Sketch the required triangle. Use a Compound Locus to locate point A.
One locus of A is $\overrightarrow{BA}$, the side of $\angle B$ not containing C.
The other locus of A is the set of all points h_a units from $\overleftrightarrow{BC}$.
A is the intersection of the two loci.

2 Copy side *a* on the working line w, (next page).

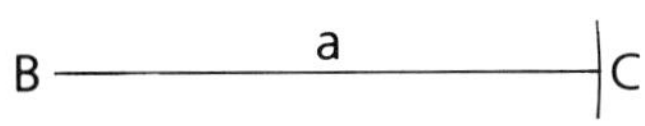

3 At some point P on w, construct the $\perp$ to w.

4 Use SEGMENT COPY to construct h_a on the $\perp$. Call it $\overline{PQ}$.

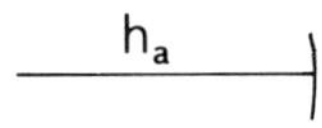

5 Construct the $\perp$ to $\overleftrightarrow{PQ}$ at Q.

6 Using ANGLE COPY, construct $\angle B$. The intersection of $\angle B$ and the parallel to w is A.

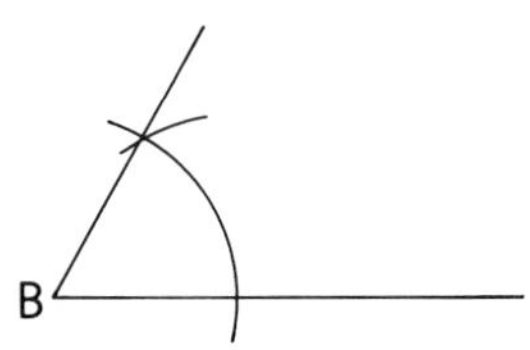

7 Draw $\overline{AC}$.

8 Name the required construction.

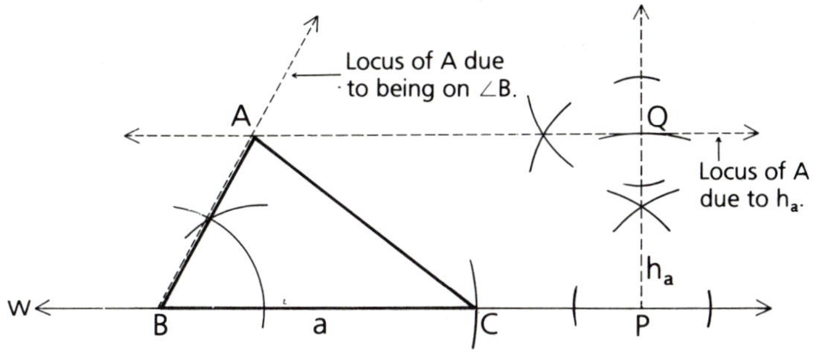

△ABC is the required △.

Section 14.6 Problem Set A

1 In Sample Problem 1 a triangle was constructed by SAS. In a similar manner, construct a triangle by each of these methods:

a ASA (Draw two different angles and a segment. Then construct a triangle in which the segment is the side included by the angles.)

b SSS

c HL

2 Construct an isosceles triangle, given:

a the vertex angle and a leg

b the base and the altitude to the base

3 Construct an isosceles right triangle given:

a a leg

b the hypotenuse

4 Construct a triangle equal in area to a given square.

5 Construct a 30°–60°–90° triangle.

6 Given $\triangle ABC$, construct a triangle whose area is twice as large as the area of $\triangle ABC$.

7 Given a triangle, construct a triangle that is similar, but not congruent, to the given triangle.

Section 14.6 Problem Set B

8 Construct a 30°–60°–90° triangle given the hypotenuse.

9 Construct an isosceles triangle given the base b and the radius R of its circumscribed circle, $b < 2R$.

10 Construct an isosceles triangle given the base b and the radius r of its inscribed circle, $b > 2r$.

11 Construct an isosceles right triangle given the median to the hypotenuse.

12 Construct a triangle, given AAS.
(Hint: Begin by constructing the third angle of the triangle.)

13 Construct an isosceles triangle given the vertex angle and the altitude to the base.

14 For each given set, construct a triangle:

a $\{a, c, m_c\}$
b $\{A, B, h_a\}$
c $\{h_b, t_b, a\}$, $h_b < t_b < a$
d $\{a, b, h_c\}$
e $\{A, B, h_c\}$
f $\{a, c, h_c\}$

15 Construct an isosceles triangle equal in area to a given triangle.

16 Construct an equilateral triangle given the altitude.

Section 14.6 Problem Set C

17 Construct an isosceles right triangle equal in area to a given triangle.

18 Construct a right triangle given the hypotenuse and the altitude to the hypotenuse.

19 For each given set, construct a triangle:
a $\{B, C, t_b\}$ b $\{a, m_b, m_c\}$ c $\{h_a, m_a, B\}$

20 Construct $\triangle ABC$, given a, b, and the point on b where t_b intersects b.

21 By construction, divide a given scalene triangle into a triangle and a trapezoid such that the ratio of the area of the triangle to the area of the trapezoid is 1:8.

Section 14.6 Problem Set D

22 Construct a triangle, given the three medians.

23 Construct a regular hexagon. Then construct an equilateral triangle whose area is equal to that of the hexagon.

Section 14.6 Problem Set E

24 Given an acute angle with a point P in the interior of the angle, construct a circle that is tangent to the sides of the angle and passes through P.

Chapter 14 Study Guide

As you study and review this chapter, be sure that you are able to:

1 Write the definition of a locus. (p. 631)

2 Describe the FOUR STEP PROCEDURE FOR LOCUS PROBLEMS. (p. 632)

3 Solve each of the following basic locus problems:

- **a** The locus of points equidistant from the sides of an angle. (p. 632)
- **b** The locus of points at a fixed distance from a given point. (p. 631)
- **c** The locus of points at a fixed distance from a given line. (p. 632)

4 Describe the two parts of the COMPOUND LOCUS PROCEDURE. (pp. 638–641)

5 State the four concurrence theorems. (pp. 642–644)

6 Define (for a triangle):

- **a** circumcenter (p. 642)
- **b** incenter (p. 643)
- **c** orthocenter (p. 644)
- **d** centroid (p. 644)

7
- **a** State the term that identifies the center of gravity of a triangle. (p. 645)
- **b** Describe how the centroid divides each median. (p. 644)
- **c** Find the coordinates of the centroid of a triangle, given the coordinates of the three vertices. (p. 647)

8 Describe the steps of the following:

- **a** the six basic constructions (pp. 650–653)
- **b** the four advanced constructions (pp. 657–659)

9 Construct a tangent to a circle from an outside point. (p. 654)

10 Identify for $\triangle ABC$: h_a, m_a, t_a. (p. 663)

11 Describe the steps of the TRIANGLE CONSTRUCTIONS. (pp. 663–667)

Chapter 14 Review Problems A

1 Given segment $\overline{AB}$, find the locus of points that are the vertices of isosceles triangles having $\overline{AB}$ as a base.

2 Find the locus of the centers of all circles that pass through two fixed points.

3 Find the locus of points 3 units from a given line and 5 units from a given point on the line.

4 In space, what is the name of the *surface* for which every point is a fixed distance from a given line?

5 What is the locus of points 2 inches from a circle with a radius of 2 inches?

6 A circle of given radius rolls around the perimeter of a given equilateral triangle. Sketch the locus of its center.

7 Write the equation of the locus of points 5 units from the origin in the coordinate plane.

8 Given scalene triangle $\triangle ABC$, construct each of the following:

a incenter **b** circumcenter **c** centroid **d** orthocenter

9 Given a 3 cm long line segment, draw the locus of points 1 cm from the segment. (Each point of the locus must be 1 cm from the point of the segment nearest to it.)

10 Construct a parallelogram given two sides and the angle they form.

11 Given a segment whose length is equal to the perimeter of an equilateral triangle, construct the triangle.

12 What is the locus of points that are a fixed distance from a fixed point and equidistant from two given points.

Chapter 14 Review Problems B

13 Write an equation of the locus of points for which the ordinate of each point is 5 more than 3 times the abscissa of that point.

14 What is the locus of points in space equidistant from all the points on a given circle?

15 Given segment $\overline{PQ}$. Find the locus of points each of which is the intersection of the diagonals of a rectangle that has $\overline{PQ}$ as a base.

16 In a circle with center P and radius 10 cm, all possible 12 cm chords are drawn. Describe the locus of the midpoints of all such chords.

17 Find the locus of the midpoints of all chords of a circle that have a fixed point of the circle as an endpoint.

18 A point outside a square 3 units on a side moves so that it is always 2 units from the point of the square nearest to it. Find the area enclosed by the locus of this moving point.

19 If the radius of a given circle is 10 cm, describe the locus of points 2 cm from the circle and equidistant from the endpoints of a given diameter of the circle.

20 Using coordinate geometry methods, find the locus of points 5 units from the origin and 4 units from the y-axis.

21 Inscribe a regular octagon in a given circle.

22 Construct a parallelogram given two sides and an altitude.

23 Explain how to construct an angle whose measure is:

a 30 b $18\frac{3}{4}$

24 Given three points, A, B, and C, find the locus of points that are equidistant from A and B and also equidistant from B and C.

Chapter 14 Review Problems C

25 Find the locus of the intersections of the diagonals of all possible rhombuses having a fixed segment $\overline{PQ}$ as a side.

26 Prove that the angle bisectors of a kite are concurrent.

27 Given a chord of a circle. Construct another chord parallel to the given chord and half its length.

28 Given scalene triangle $\triangle ABC$, construct a circle in the exterior of the triangle that will be tangent to one side and to the other two sides extended.

29 Construct a square whose area is equal to the sum of the areas of two given squares.

30 Given two parallel lines and a point P between them, construct a circle that is tangent to both lines and passes through point P.

31 Inscribe a square in a given rhombus.

Chapter 14 Review Problem D

32 Given two circles, construct a common external tangent.

21500A

Chapter 15

INEQUALITIES

Section 15.1 Number Properties

The statement $a < b$, *a is less than b,* is an inequality between the numbers a and b. $a < b$ is equivalent to $b > a$, *b is greater than a.*

We first review some of the properties of inequalities.

Law of Trichotomy

Given any two real numbers, x and y, exactly one of the following is true:
$x < y$, $x = y$, $x > y$

Transitive Property of Inequalities

If $a > b$ and $b > c$, then $a > c$. Similarly, if $x < y$ and $y < z$, then $x < z$.

■ Example: Given segments $\overline{AB}$, $\overline{PQ}$, and $\overline{XY}$.

If $AB < PQ$ and $PQ < XY$, then $AB < XY$.

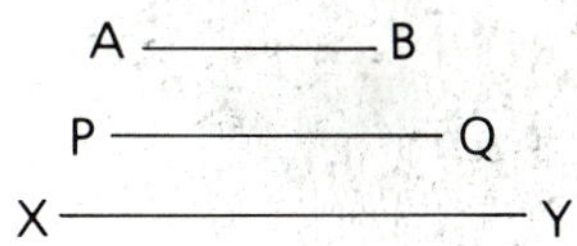

Addition Property of Inequality

If $a > b$ and $x > y$, then $a + x > b + y$.

■ Example 1: If $4 > -7$ and $10 > 5$, then $4 + 10 > -7 + 5$

$$14 > -2$$

■ Example 2: If $4 > -7$ then $4 + 9 > -7 + 9$

$13 > 2$

Positive Multiplication Property of Inequality

If $x < y$ and $a > O$, then $a \cdot x < a \cdot y$.

■ Example: $m\angle 1 < m\angle 2$

$3 \cdot m\angle 1 < 3 \cdot m\angle 2$

$\angle ABC < \angle DEF$

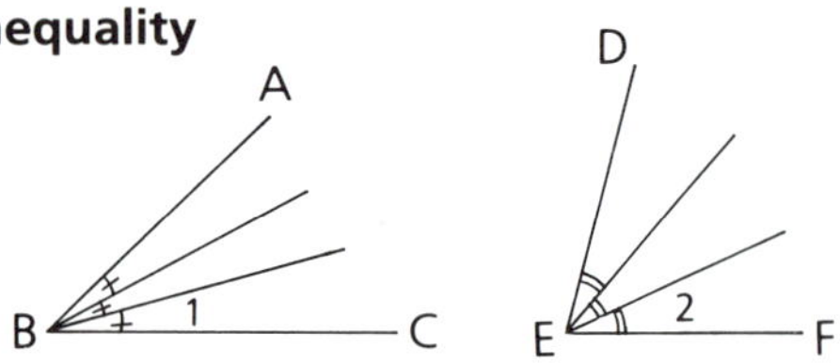

When we say that one angle is greater than (or less than) a second angle, we refer to their measures. Thus, $\angle ABC < \angle DEF$ means $m\angle ABC < m\angle DEF$.

Negative Multiplication Property of Inequality

If $x < y$ and $a < O$, then $a \cdot x > a \cdot y$.

Notice that the sense of the inequality REVERSES.

■ Example:

$$-5x < 15$$
$$(-\tfrac{1}{5})(-5x) > (-\tfrac{1}{5})(15)$$
$$x > -3$$

Section 15.1 Sample Problem

1 A given angle is greater than twice its supplement. Find the size of the given angle.

Solution: Let $x =$ measure of the given angle

$180 - x =$ measure of the supplement.

$$x > 2(180 - x)$$
$$> 360 - 2x$$
$$x + 2x > 360 - 2x + 2x \quad \text{(Addition Property)}$$
$$3x > 360$$
$$(\tfrac{1}{3})(3x) > (\tfrac{1}{3})(360) \quad \text{(Positive Multiplication Property)}$$
$$x > 120$$

Thus the given angle is greater than $120°$. Since it has a supplement, the given angle is also less than $180°$. Thus, $120 < x < 180$.

Section 15.1 Problem Set A

1 For each inequality below, solve for x.

a $\frac{3}{5}x > 15$

b $5x - 4 > 26$

c $-4x \leq 28$

d $10 - x < 8x - (2x - 3)$

2 a If $x + y < 30$ and $y = 12$, what is true about x?

b If $x + y = 30$ and $y < 12$, what is true about x?

3 If x exceeds y by 5 and y exceeds z by 3, how is x related to z?

4 If x is twice as big as y and y is three times z, how is x related to z?

5 If $\angle A = \angle 1 + \angle 2$, what is the relation between $\angle A$ and $\angle 2$?

6 a If X is between P and Q, then how is PX related to PQ?

b If X is the midpoint of $\overline{PQ}$, write the relation between PX and PQ as an inequality.

c Using the situation of part b, write the relation between PX and PQ as an equality.

Section 15.1 Problem Set B

7 The complement of an angle is less than the original angle. Find the restrictions on the size of the original angle.

8 If $\angle X < \angle Y$, what is the relation between their complements?

9 If $\frac{1}{x} > 5$, then x is between what two numbers?

10 An angle is greater than twice its complement. Find the restrictions on the angle and on the complement.

11 If $x \not< 3$ and $x \neq 3$, what can be concluded about x?

12 a What is the relation between an exterior angle of a triangle and the two remote interior angles?

b What, then, is the relation between an exterior angle and *one* of the remote interior angles?

13 Given: $\angle ABC > \angle ACB$
$\overrightarrow{BD}$ bisects $\angle ABC$
$\overrightarrow{CD}$ bisects $\angle ACB$

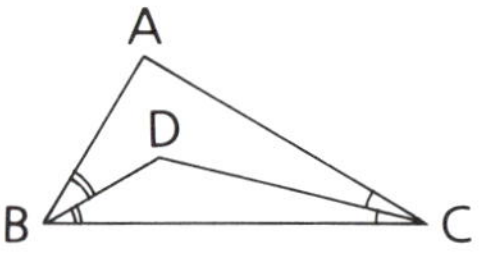

Find and justify the relation between $\angle DBC$ and $\angle DCB$.

Section 15.1 Problem Set C

14 Given: a, b, c representing real numbers with $a > b$.
Prove: $c - a < c - b$. (Special Subtraction Property)

15 Solve: $x^2 + x < 6$.

16 If $x > 3y + 7$ and $y > 6 - x$, find the restrictions on: a x b y

17 If x exceeds y by 20% and y exceeds z by 20%, by what percentage does x exceed z?

18 Solve: $18 - 3x > 3$ over the positive even integers.

19 Given: $\angle A >$ comp. $\angle A$, and comp. $\angle A > \angle B$.

a Compare the complement of $\angle A$ to the complement of $\angle B$.

b Compare the complement of $\angle B$ to $\angle A$.

c List $\angle A$, $\angle B$, comp. $\angle A$, comp. $\angle B$ in order of size, largest first.

Section 15.1 Problem Set D

20 Solve for x: $|2x - 7| > |x + 20| - 4$

Section 15.2 Inequalities in a Single Triangle

A THE TRIANGLE INEQUALITY POSTULATE
B THE EXTERIOR ANGLE INEQUALITY THEOREM
C ACUTE-RIGHT-OBTUSE TRIANGLES
D SIDE AND ANGLE RELATIONSHIPS

15.2 (A) The Triangle Inequality Postulate

P **The sum of any two sides of a triangle is always greater than the third side.**

In other words:
Traveling from A to B along segment $\overline{AB}$ is shorter than going first to X along $\overline{AX}$ and then to B along $\overline{XB}$; that is, $AX + XB > AB$

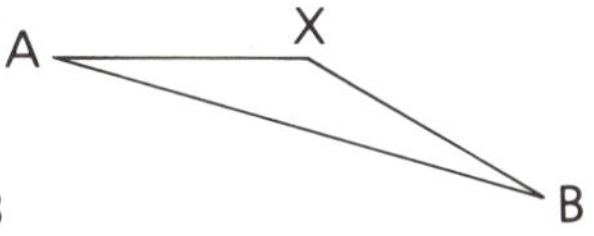

15.2 (B) The Exterior Angle Inequality Theorem

Theorem 125 Exterior Angle Inequality

T **An exterior angle of a triangle is greater than either remote interior angle.**

Given: $\triangle ABC$ with exterior $\angle 1$
Prove: $\angle 1 > \angle A$, $\angle 1 > \angle B$

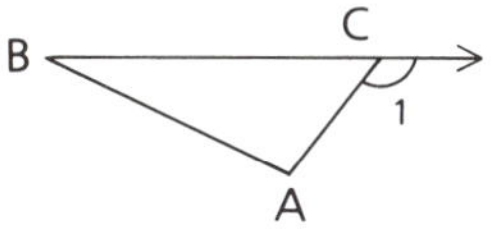

Proof: In Chapter 7 we learned that $\angle 1 = \angle A + \angle B$.
Clearly, $\angle A + \angle B > \angle A$, so $\angle 1 > \angle A$ by substitution. In a similar manner, $\angle A + \angle B > \angle B$, so $\angle 1 > \angle B$.

15.2 (C) Acute-Right-Obtuse Triangles

The converse of the Pythagorean Theorem (Section 9.3) determines whether a triangle is a right triangle. A test may determine if a triangle is acute or obtuse.

THE PYTHAGOREAN THEOREM TEST

To classify a triangle as acute, right, or obtuse, compute a^2, b^2, and c^2, where c is the longest of the three sides a, b, and c.
If $a^2 + b^2 = c^2$, then $\triangle ABC$ is right ($\angle C$ is right).
If $a^2 + b^2 > c^2$, then $\triangle ABC$ is acute.
If $a^2 + b^2 < c^2$, then $\triangle ABC$ is obtuse ($\angle C$ is obtuse).

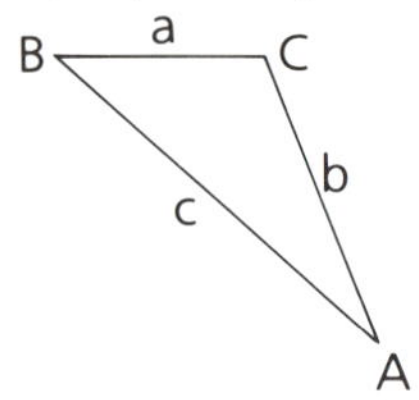

15.2 (D) Side and Angle Relationships

Theorem 126

T **If two sides of a triangle are not congruent, then the angles opposite them are not congruent, and the larger angle is opposite the longer side. (If  then )**

Given: $\triangle ABC$
$AC > AB$
Conclusion: $\angle B > \angle C$

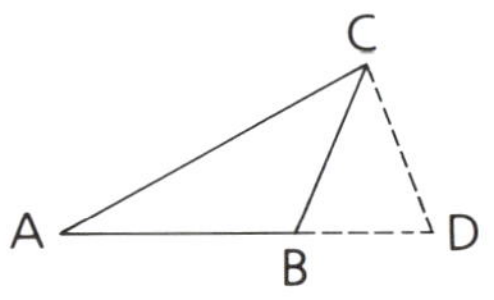

Proof: Since $AC > AB$, extend $\overline{AB}$ to D, so that $AD = AC$. Draw $\overline{DC}$.
$\angle ABC > \angle D$ by the Exterior Angle Inequality Theorem
$\angle D \cong \angle ACD$ by If  then  .
$\angle ABC > \angle ACD$ by substitution.
$\angle ACD > \angle ACB$ (see diagram)
$\therefore \angle ABC > \angle ACB$ by the Transitive Property of Inequality.

The converse of Theorem 126 is also true.

Theorem 127

T **If two angles of a triangle are not congruent, then the sides opposite them are not congruent, and the longer side is opposite the larger angle. (If  then )**

Given: $\triangle ABC$ with $\angle B > \angle C$

Conclusion: $AB < AC$

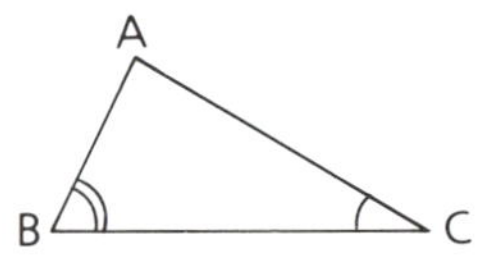

Proof: By the law of trichotomy there are exactly three possible conclusions:
$AB > AC$, $AB = AC$, or $AB < AC$.
We must test them:

Case 1: If AB $>$ AC, then by Theorem 126, $\angle C > \angle B$,
which *contradicts* the given information.
Thus, AB $>$ AC cannot be the correct conclusion.*

Case 2: If AB $=$ AC, then $\angle C \cong \angle B$ by If $\triangle$ then $\triangle$.
The given information is again *contradicted*
Thus, AB $=$ AC cannot be the correct solution.

All that is left is AB $<$ AC, which must be true by the Law of Trichotomy.

A simple extension of Theorem 127 enables us to say:
In a triangle the longest side is opposite the largest angle.

* This is an Indirect Proof method, presented formally in Section 16.2.

Section 15.2 Sample Problems

1 **Does a triangle with sides 2, 5, and 10 exist?**

Solution: The sum of *any* two sides must be greater than the third side, and $2 + 5 \not> 10$. Therefore the answer is *no.*

2 **Find the restrictions on $\angle A$.**

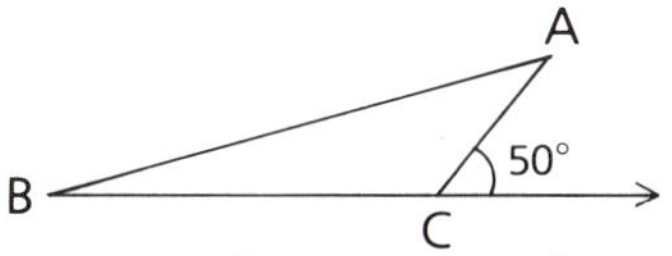

Solution: $50^\circ > \angle A$, because an exterior angle of a triangle exceeds either remote interior angle. To have a triangle, its angle must be greater than 0°, so $\angle A > 0^\circ$. Thus, $0^\circ < \angle A < 50^\circ$.

3 **In $\triangle ABC$, $\angle A = 40^\circ$, $\angle B = 65^\circ$.**
List the sides in order of their lengths, starting with the smallest.

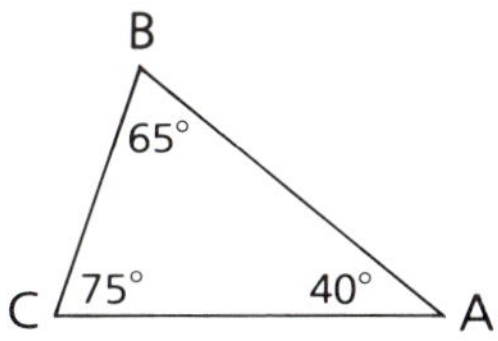

Solution: Draw a diagram listing all the angles. ($\angle C$ is easily found to be 75°.) The shortest side, $\overline{BC}$, is opposite the smallest angle, $\angle A$. The longest side, $\overline{BA}$, is opposite the largest angle, $\angle C$.

$\therefore$ $\overline{BC}$, $\overline{AC}$, $\overline{BA}$ is the correct order.

4 Given: $\triangle ABC$ with $AB < AC$
$\overrightarrow{BD}$ bisects $\angle ABC$
$\overrightarrow{CD}$ bisects $\angle ACB$
Prove: $BD < DC$

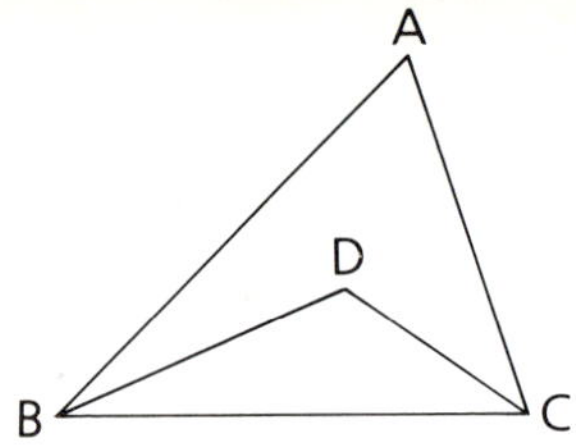

1 $AB < AC$	1 Given
2 $\angle ABC > \angle ACB$	2 If  then 
3 $\overrightarrow{BD}$ bisects $\angle ABC$	3 Given
4 $\overrightarrow{CD}$ bisects $\angle ACB$	4 Given
5 $\angle DBC > \angle DCB$	5 Positive Multiplication Property (by $\frac{1}{2}$)
6 $BD < DC$	6 If  then 

Section 15.2 Problem Set A

1 What are the restrictions on $\angle 1$?

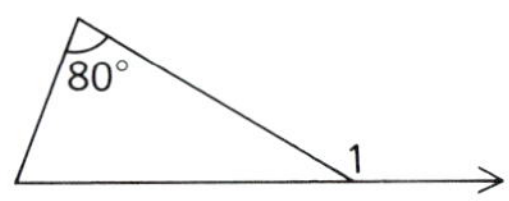

2 Which of these sets can be the lengths of sides of a triangle?
a 3, 6, 9 **b** 4, 5, 8 **c** 2, 3, 8 **d** $\sqrt{2}$, $\sqrt{3}$, $\sqrt{6}$

3 In $\triangle PQR$, $\angle P = 67^\circ$ and $\angle Q = 23^\circ$.
a Name the shortest and longest sides.
b What name is given to side $\overline{PQ}$?

4 Given: $AB > BC$, $BC > AC$
Prove: B is the smallest angle in $\triangle ABC$

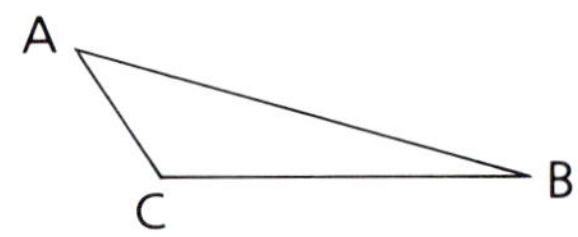

5 Name the longest segment in each diagram.

a
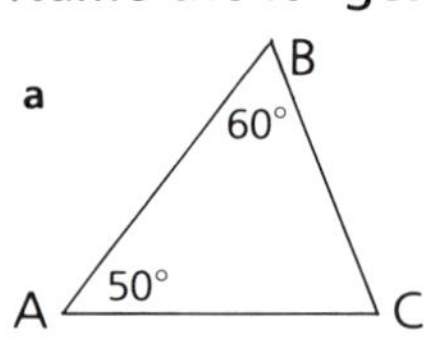

b
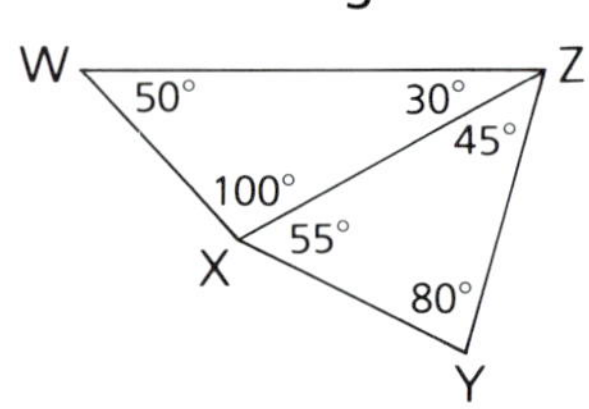

6 Given: $\angle 1$ is an exterior angle of $\triangle ACD$
$\angle 2$ is an exterior angle of $\triangle ABC$
Prove: $\angle 1 > \angle 3$

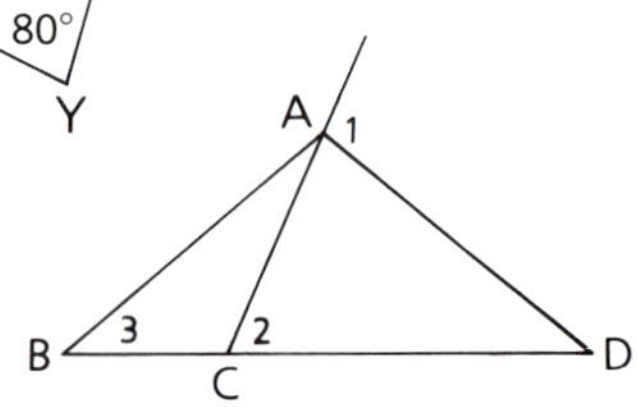

Section 15.2 Problem set B

7 A scalene triangle has a 60° angle. Is this angle located opposite the longest, shortest, or other side?

8 The sides of a triangle are 14, 6, and x. Find the set of possible values for x. __?__ $<$ x $<$ __?__

9 The vertex angle, A, of isosceles triangle ABC is between 40° and 88°. Find the possible values for $\angle$B.

10 **a** Name the longest segment. **b** Name the shortest segment.

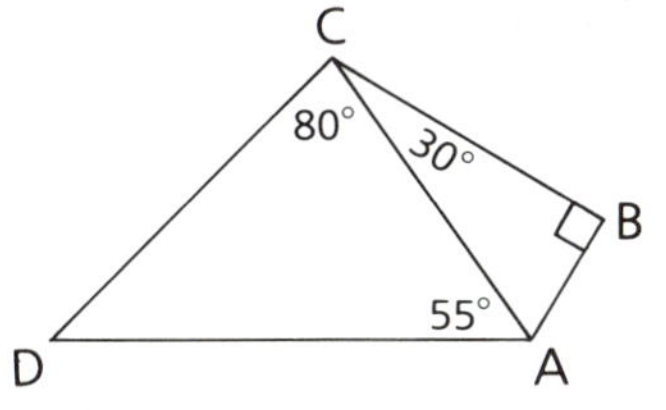

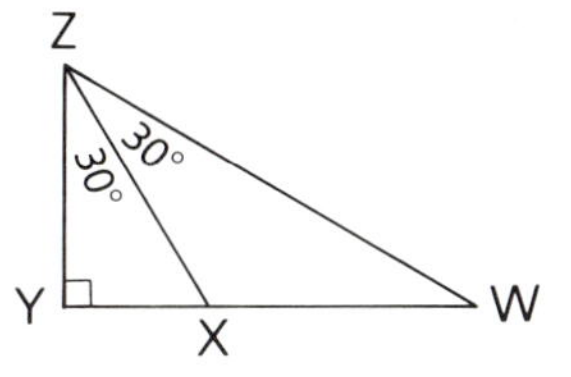

11 **a** List the angles in order of size, beginning with the smallest.

b At which vertex is the exterior angle the largest?

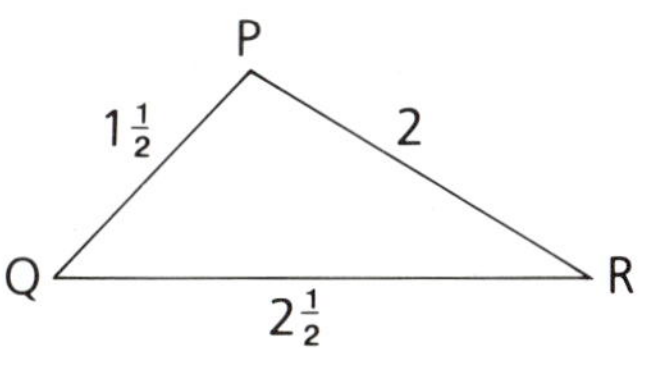

12 Find the restrictions on x:

a

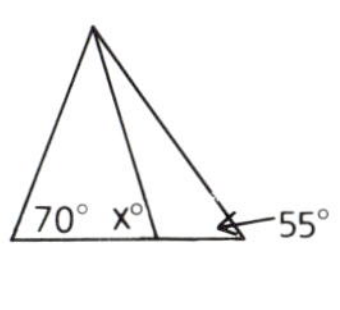

c

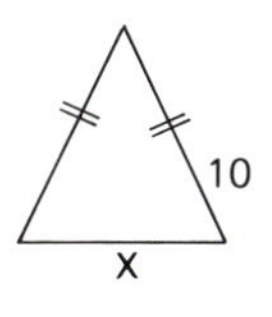

b

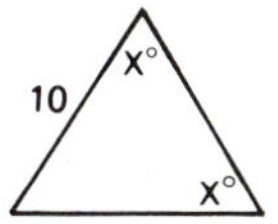

d

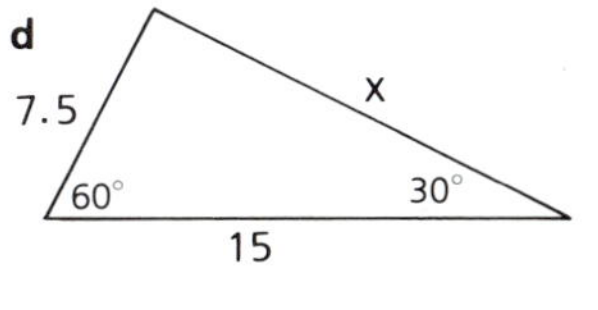

13 A stick 8 cm long is cut into three pieces of *integral* length to be assembled as a triangle. What is the length of the shortest piece?

14 For each triple of numbers, tell whether the numbers represent the lengths of the sides of an acute triangle, a right triangle, an obtuse triangle, or no triangle:

a 12, 13, 14 **b** 11, 5, 18 **c** 9, 15, 18 **d** $\frac{1}{2}, 1\frac{1}{5}, 1\frac{3}{10}$

15 Prove that an altitude of an acute triangle is shorter than either side that is not the base.

16 Prove that if ABCD is a quadrilateral, then $AB + BC + CD > AD$.

17 Given the diagram as shown.
Prove or Disprove that $\angle 2 > \angle 1$

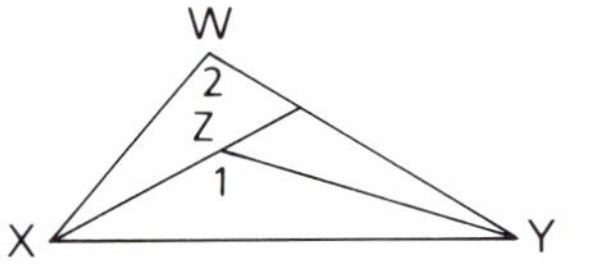

18 Given: $\overrightarrow{AC}$ bisects $\angle BAD$
Prove: $AD > CD$

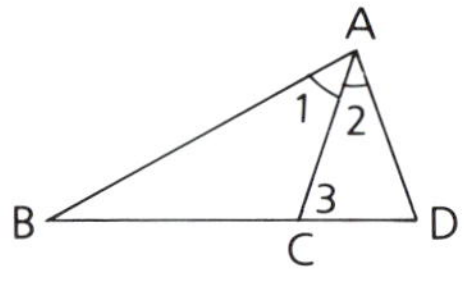

Section 15.2 Problem Set C

19 $\angle ACB$ and $\angle CDB$ are right $\angle$s, $\angle B = 20°$

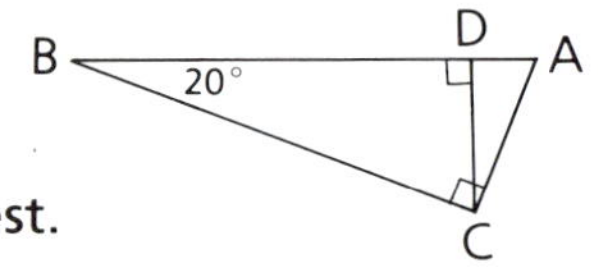

a List the segments $\overline{AC}$, $\overline{CB}$, $\overline{AB}$, $\overline{AD}$, $\overline{CD}$ in order of size, starting with the smallest.

b Where would $\overline{DB}$ fit into this list?

20 Given a point P in the interior of $\triangle XYZ$.
Prove that $PY + PZ < XY + XZ$.

21 If two sides of a triangle have lengths x and y, what is the range of possible values for the length of the third side?

22 Prove: The shortest segment between a point and a line is the perpendicular segment.

23 Given a point, P, in the interior of $\triangle XYZ$. Prove that $\angle XPZ > \angle Y$.

Section 15.3 The Hinge Theorems

Thus far we have discussed inequalities involving the sides and the angles of a *single triangle.* We now turn our attention to *two triangles.*

If an angle AHC is changed from its size in Figure 1 to that in Figure 2, what happens to the length of a spring connecting A and C?

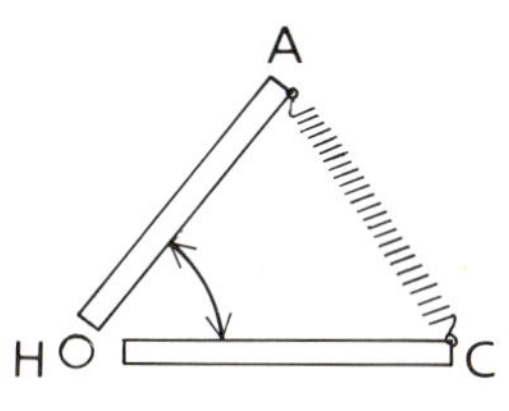

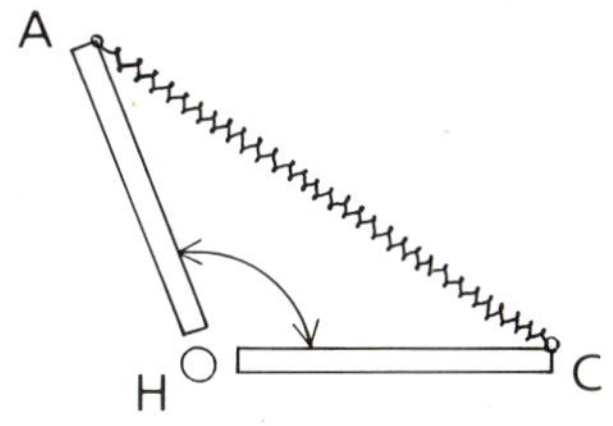

Theorem 128 The Hinge Theorem

T **If two sides of one triangle are congruent to two sides of another triangle and the included angle of the first triangle is greater than the included angle of the second triangle, then the remaining side of the first triangle is greater than the remaining side of the second triangle. (SAS $\neq$)**

The following set-up of Theorem 128 should be helpful in its application.

Given: $\overline{AB} \cong \overline{XY}$
$\overline{BC} \cong \overline{YZ}$
$\angle B > \angle Y$
Conclusion: $AC > XZ$

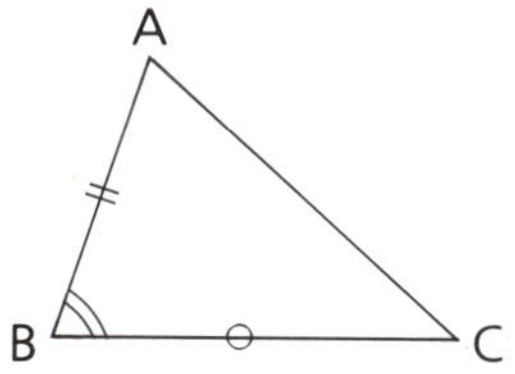

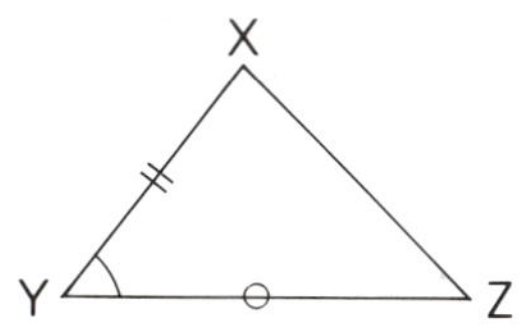

The reverse of The Hinge Theorem is also true.

Theorem 129 The Converse Hinge Theorem

T **If two sides of one triangle are congruent to two sides of a second triangle and the third side of the first triangle is greater than the third side of the second triangle, then the angle opposite the third side in the first triangle is greater than the angle opposite the third side in the second triangle. (SSS $\neq$)**

Given: $\overline{AB} \cong \overline{WX}$
$\overline{BC} \cong \overline{XY}$
$AC > WY$
Conclusion: $\angle B > \angle X$

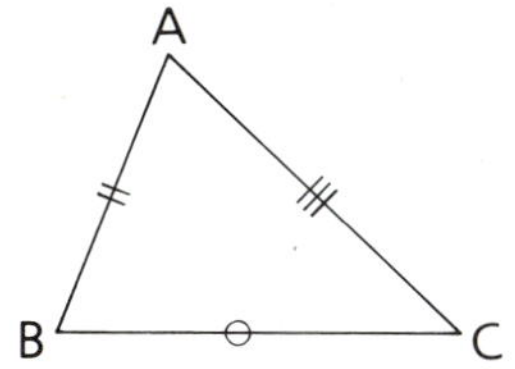

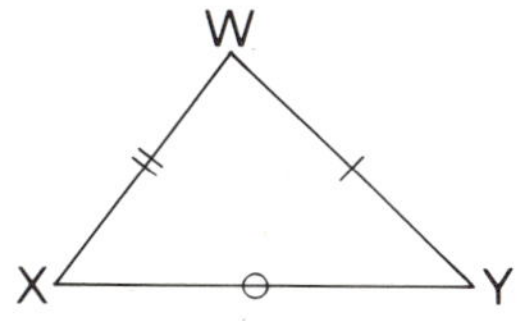

Section 15.3 Sample Problems

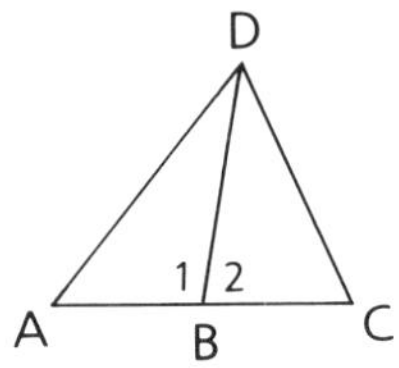

1 Given: $\overline{BD}$ is a median

$AD > CD$

Which is greater, $\angle 1$ or $\angle 2$?

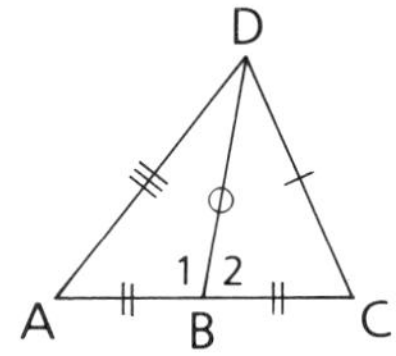

Solution:

Since $\overline{BD}$ is a median, $\overline{AB} \cong \overline{BC}$.

Also, $\overline{BD} \cong \overline{BD}$ and $\overline{AD} > \overline{CD}$.

Thus, by the Converse Hinge Theorem, $\angle 1 > \angle 2$.

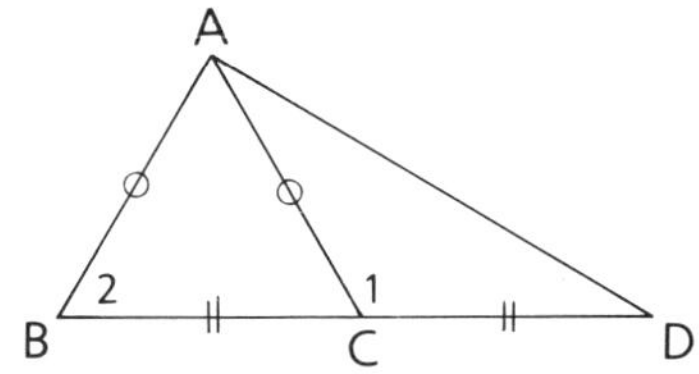

2 Given: $\triangle ABC$ is isosceles, base $\overline{BC}$

C is the midpoint of $\overline{BD}$

Prove: $AD > AB$

Statements	Reasons
1 C is midpoint of $\overline{BD}$	1 Given
2 $\overline{BC} \cong \overline{CD}$	2 A midpoint divides a segment into two congruent segments
3 $\triangle ABC$ is isosceles, base $\overline{BC}$	3 Given
4 $\overline{AB} \cong \overline{AC}$	4 The legs of an isosceles $\triangle$ are $\cong$
5 $\angle 1 > \angle 2$	5 Exterior Angle Theorem ($\angle 1$, $\triangle ABC$)
6 $AD > AC$	6 Hinge Theorem (SAS $\neq$)
7 $AD > AB$	7 Substitution, step 4 in step 6

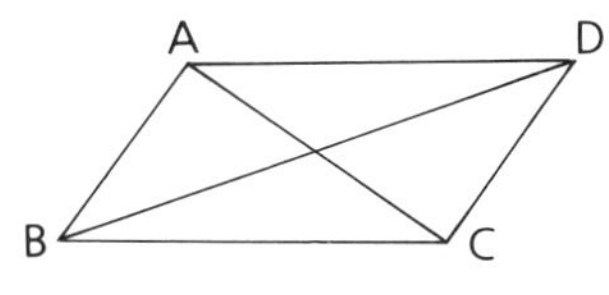

3 ABCD is a parallelogram.

$\angle BAD > \angle ADC$.

Which diagonal is longer, $\overline{AC}$ or $\overline{BD}$?

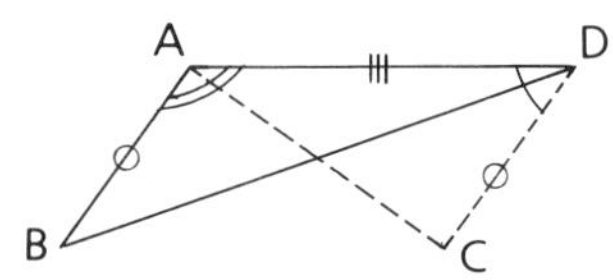

Solution: Consider the overlapping triangles as shown.

$\overline{AB} \cong \overline{DC}$ because the opposite sides of a parallelogram are $\cong$.

Also, $\overline{AD} \cong \overline{AD}$ and $\angle BAD > \angle ADC$.

So $BD > AC$ by the Hinge Theorem.

Section 15.3 Problem Set A

1 Which is longer, $\overline{AC}$ or $\overline{DF}$?

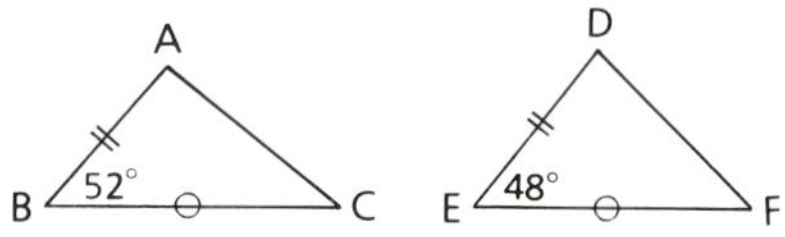

2 Which is larger, $\angle R$ or $\angle Y$?

3 Given: $\overline{AB} \parallel \overline{CD}$, $\overline{AB} \cong \overline{AD}$
$\angle DAC = 75°$, $\angle DCA = 45°$
Which is longer, $\overline{BC}$ or $\overline{DC}$?

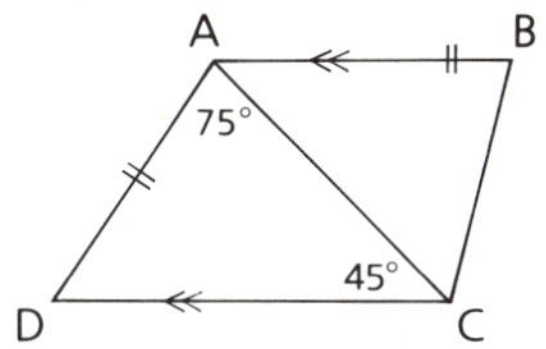

4 Compare AB in $\triangle ABC$ to XZ in $\triangle XYZ$,
where $BC = 7$ $AC = 9$ $\angle C = 75°$
$YZ = 7$ $XY = 9$ $\angle Y = 80°$

5 Given: $\overline{WX} \cong \overline{WZ}$
$\angle XWY > \angle ZWY$
Prove: $XY > ZY$

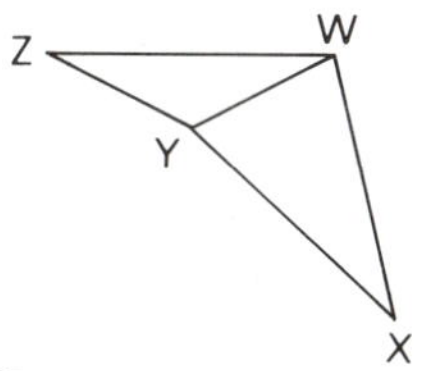

6 Given: $\odot$ O
$AB < CD$
Prove: $\angle 1 < \angle 2$

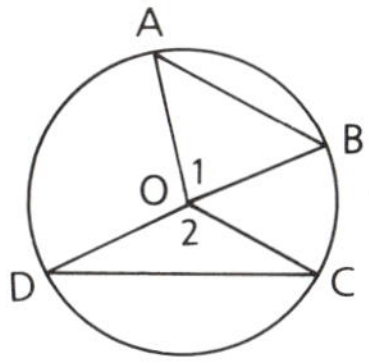

7 In $\triangle WXY$, $WX = 10$, $WY = 4$, $XY = 7$.
- **a** Name the largest and smallest angle.
- **b** Is the triangle acute, right, or obtuse?

Section 15.3 Problem Set B

8 In $\square$ WXYZ, $XZ > WY$.
Prove: **a** $\angle XWZ > \angle WZY$ (use a two-column proof).
b $\angle XWZ$ is obtuse (use a paragraph proof).

9 Given: $\overline{PQ} \cong \overline{PR} \cong \overline{RS}$
Prove: $QR < PS$

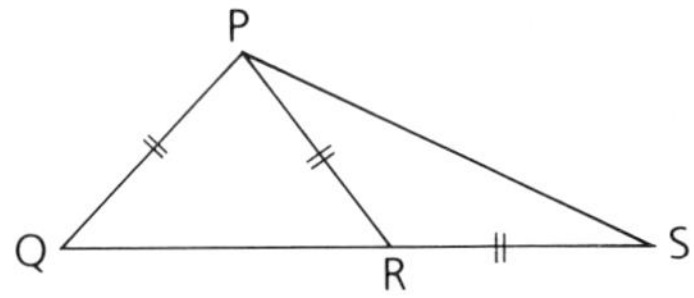

10 $\triangle WXY$ and $\triangle ABC$ are isosceles with bases $\overline{WY}$ and $\overline{AB}$, respectively. If $\angle X$ and $\angle B$ are each 50° and $\overline{WX} \cong \overline{BC}$, which triangle has

a the larger base? **b** a larger altitude to the base?

11 Given: $\overline{AB}$ and $\overline{BC}$ are tangent to $\odot Q$
$AD > DC$
Conclusion: $\angle ABD > \angle DBC$

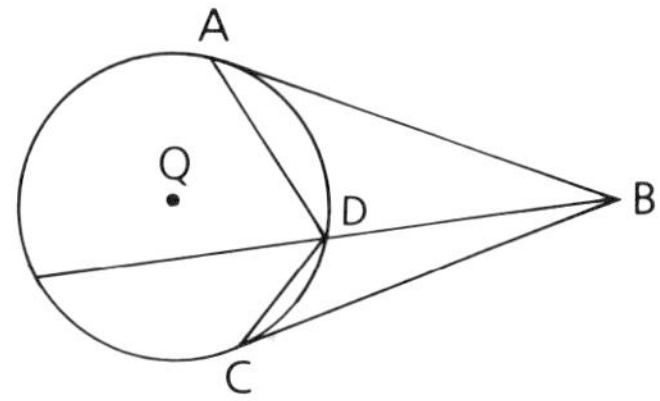

12 Given: $\angle 1 < \angle 3$
$\overline{BA} \parallel \overline{CD}$
$AC > AD$
Prove: $BC > AD$

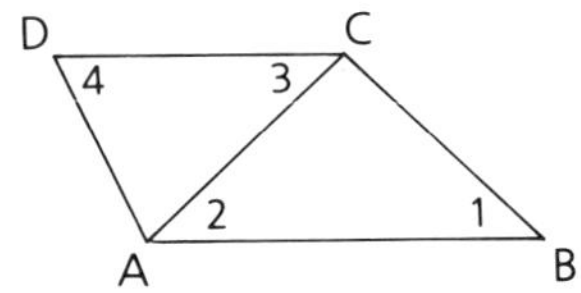

13 In $\triangle PQR$, $PQ = 1\frac{1}{2}$, $QR = 2\frac{1}{2}$, and $PR = 2$. Is $\triangle PQR$ acute, right, or obtuse?

14 In $\triangle ACE$, $AC < AE$ and D is the midpoint of $\overline{CE}$. B and F are points on $\overline{AC}$ and $\overline{AE}$ respectively, such that $\overline{CB} \cong \overline{FE}$. Prove $BD > FD$.

15 $\overline{AD}$ is a median of $\triangle ABC$, $\angle ADC = 2x + 35$, and $\angle ADB = 5x - 65$.

a Which side is longer, $\overline{AC}$ or $\overline{AB}$? **b** Which is larger, $\angle B$ or $\angle C$?

16 Given: $\angle C > \angle A$ and $\angle D > \angle B$
Prove: $AB > CD$

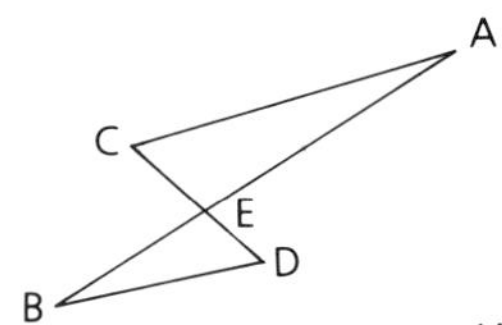

17 List the angles in order of size, starting with the largest. $\angle X$, $\angle Y$, $\angle XWY$, $\angle XWZ$, $\angle XZW$

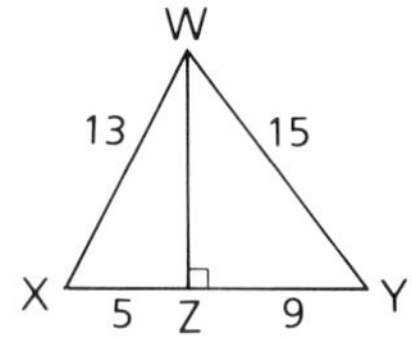

18 Given: $QT > TR$
$\overrightarrow{QS}, \overrightarrow{QT}$ trisect $\angle PQR$
$\overrightarrow{RS}, \overrightarrow{RT}$ trisect $\angle PRQ$
Prove: $PQ > PR$

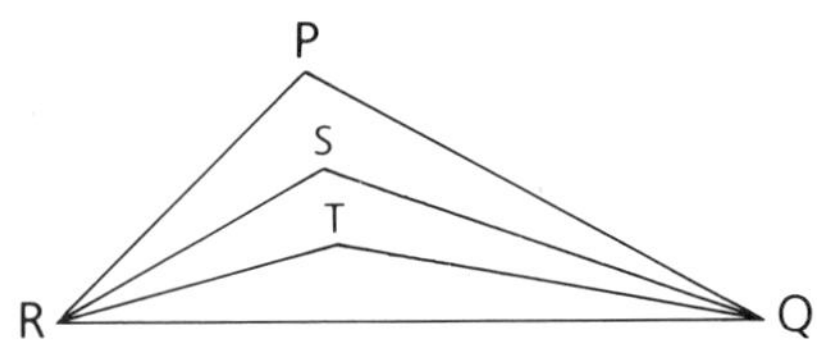

Section 15.3 Problem Set C

19 Prove that if chords $\overline{AB}$ and $\overline{BC}$ are in circle O and $AB > BC$, then $\overline{AB}$ is closer to the center than $\overline{BC}$. (Hint: Draw $\overline{MN}$.)

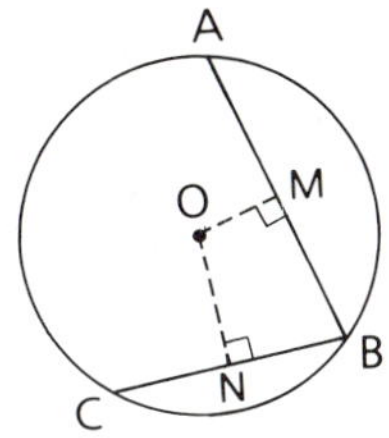

20 Given: ABCD is a square
$\overline{AF} \perp \overline{DE}$, $\overline{AE} \cong \overline{BF}$
Which of the following is correct?
a $DE < AF$.
b The figure is overdetermined.

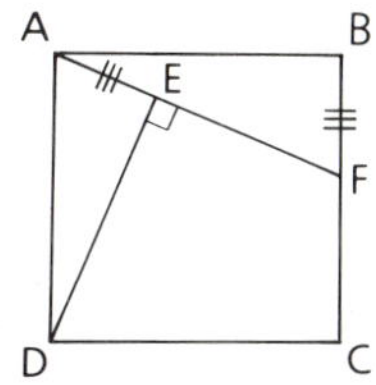

21 $\overline{WX}$ is a diameter of a circle with center P, and $\overline{YZ}$ is a diameter of a larger concentric circle. W, X, Y, and Z are non-collinear. Prove that $\angle YWZ > \angle XYW$.

22 Given: B, C, and D lie on plane *m*
$\triangle BCD$ is isosceles with base $\overline{CD}$
$\angle ABD > \angle ABC$
Conclusion: $\angle ACD > \angle ADC$

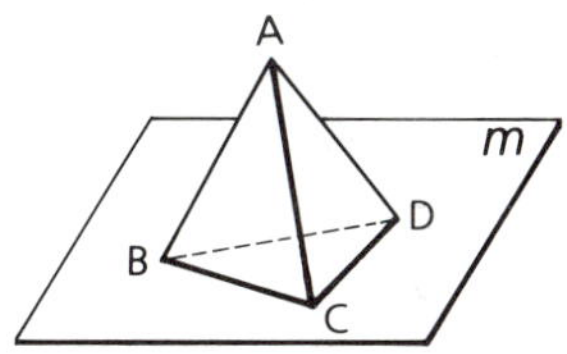

23 Given: $\overset{\frown}{AD} \cong \overset{\frown}{CD}$
$AE < EC$
Prove: $\overset{\frown}{AB} < \overset{\frown}{BC}$

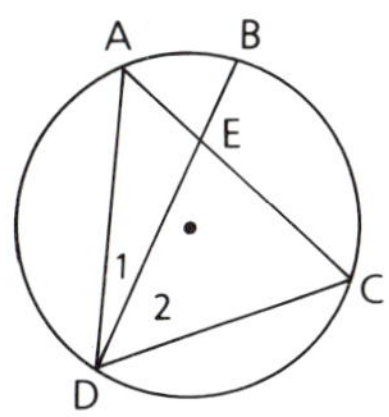

24 Given: $\overline{AD}$ is a median
$\angle ABD > \angle ACD$
Conclusion: $\angle 1 > \angle 4$

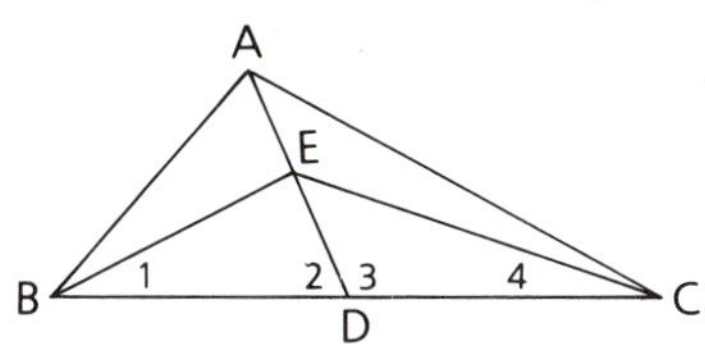

25 Given: $AC > CE$
$\overline{AB} \cong \overline{DE}$
Prove: $AD > BE$

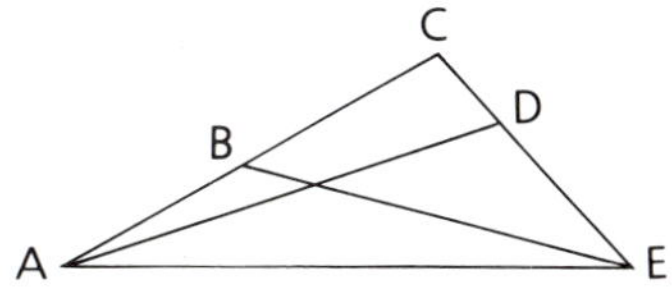

Chapter 15 Study Guide

As you study and review this chapter, be sure that you are able to:

1 State the Trichotomy Law and Transitive Property of Inequalities. (p. 673)

2 Make up and solve algebra problems using each property (p. 674):

a Addition Property **b** Multiplication Property

3 Describe how the supplements compare if $\angle A > \angle B$. (p. 676) What property does this problem illustrate?

4 Describe what is wrong with $\triangle$NOT and why. Describe what is true about m$\angle$EXT and why. State whether $\triangle$SHE is acute, right, or obtuse, and describe how you determined your answer. (pp. 677–678)

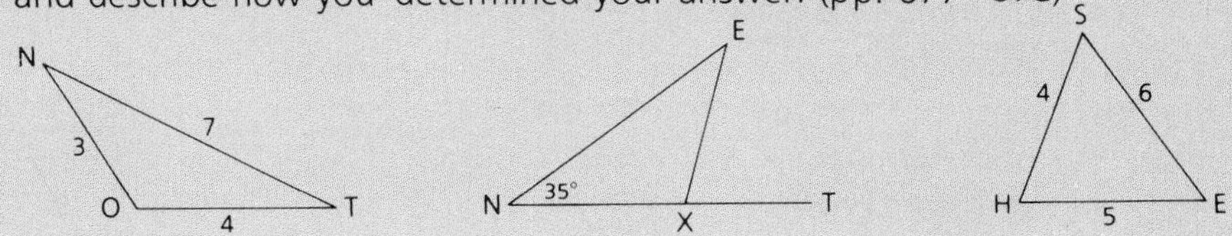

5 Find and solve a problem in the problem sets for each theorem:

a If △ then △. (p. 678) **c** Hinge Theorem. (p. 683)

b If △ then △. (p. 678) **d** Converse Hinge Theorem (p. 683)

The Royal Road: Algebra in Geometry

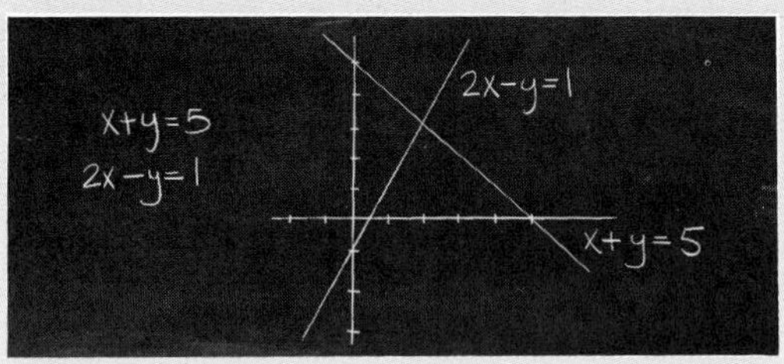

The solution to this **system of linear equations** is x is 2 and y is 3.

1 Solve the general system of equations for x and y:

$ax + by = c$

$dx + ey = f$

2 Write a computer program that asks the user to input the constants and prints the value(s) of x and y.

3 Under what conditions is there no solution for x and y? Under what conditions are there many solutions?

Chapter 15 Review Problems A

1 In $\triangle ABC$, $AB > AC > BC$. List the angles in order, smallest first.

2 In each case, decide which of the segments named is largest and state the reason for your decisions:

a

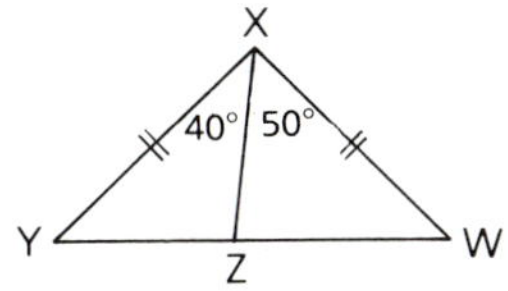

$\overline{WZ}$ or $\overline{ZY}$?

b

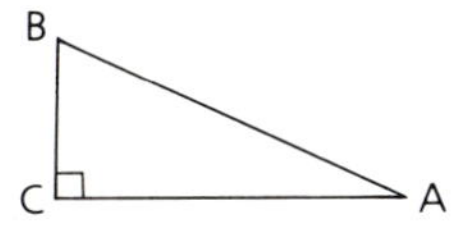

$\overline{AB}$ or $\overline{BC}$?

c

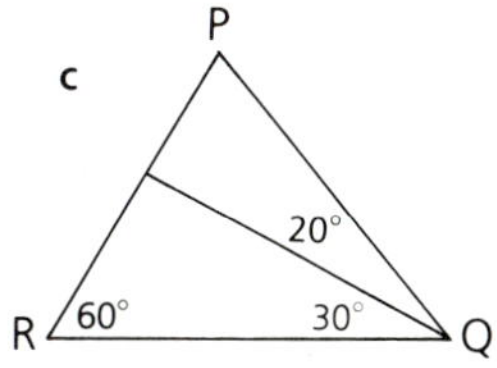

$\overline{PQ}$, $\overline{QR}$, or $\overline{PR}$?

3 In each case, tell which angle is largest and give the reason.

a

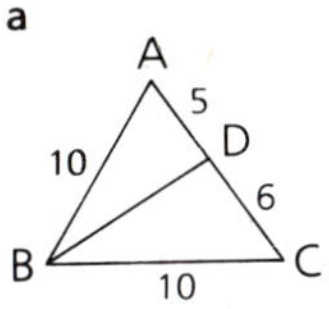

$\angle ABD$ or $\angle CBD$?

b

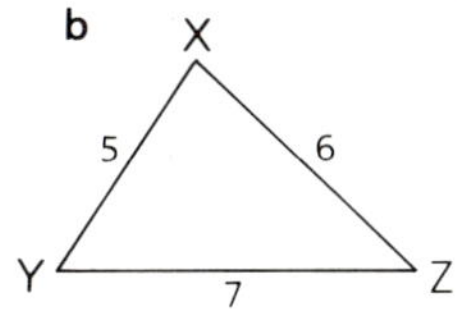

$\angle X$, $\angle Y$, or $\angle Z$?

c

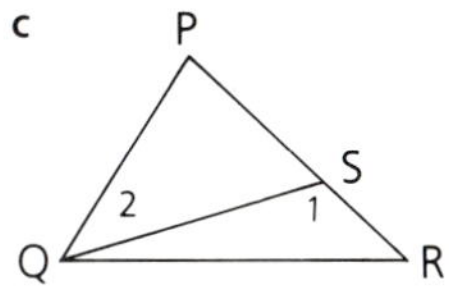

$\angle 1$, $\angle 2$, or $\angle P$?

4 If $x > 4$ and $x < y$, then y __?__ 4.

5 If $x \neq 6$ and $x \not< 6$, what can we conclude?

6 **a** Name all pairs of segments that we know to be congruent.
b Which is shorter, $\overline{BE}$ or $\overline{EC}$?
c What is the name of side $\overline{BC}$ in $\triangle BEC$?
d Which is longer, $\overline{AE}$ or $\overline{DE}$?
e What is the shortest segment in the figure?

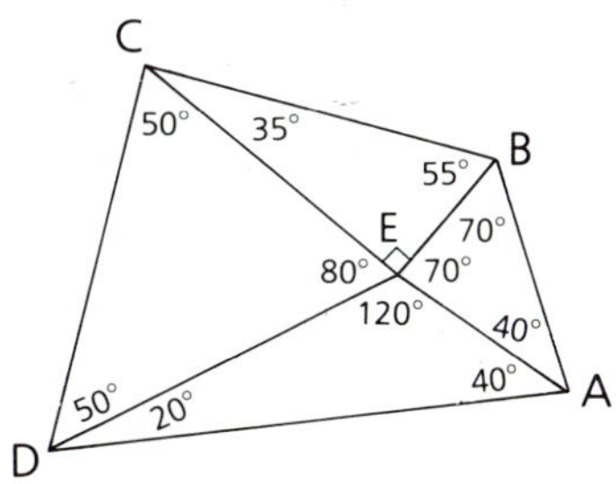

7 Which of these sets *cannot* represent the sides of a triangle?
a 20, 40, 20 **b** 30, 40, 20 **c** 20, 20, 20 **d** 30, 40, 50

8 Given: $\overline{PQ} \cong \overline{RS}$
Prove: $PS > QR$

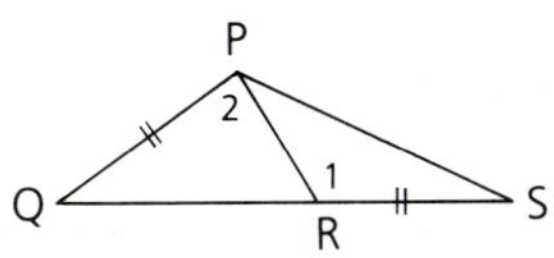

Chapter 15 Review Problems B

9 Given $\triangle ABC$ as shown. List the sides $\overline{AB}$, $\overline{AC}$, and $\overline{BC}$ in order of size, largest first.

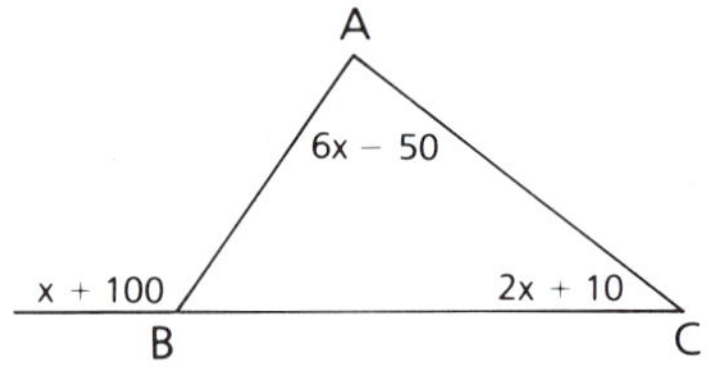

10 Give the limitations on x

$\underline{\quad ? \quad} < x < \underline{\quad ? \quad}$

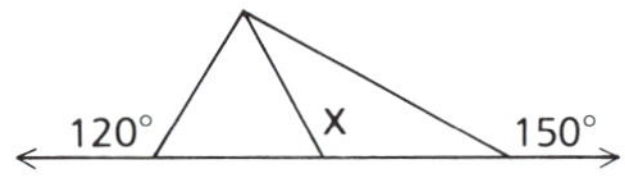

11 $\triangle ABC$ is isosceles with $\angle C$ obtuse. $AB = 6$.

a Which side is longest?

b The perimeter must be between what two numbers?

12 In $\triangle ABC$, $AB > BC$, $m\angle C = 4x - 4$, and $m\angle A = x + 9$. Find the minimum integral (integer) value of x.

13 A triangle has vertices $P(-1,-2)$, $Q(4,1)$, and $R(6,-2)$.

a Find PQ, QR, PR **b** Is $\triangle PQR$ acute, right, or obtuse?

c List the angles in order of size, smallest first

14 Given: $AB < AD$
$\overrightarrow{BE}$ bisects $\angle ABC$
$\overrightarrow{DE}$ bisects $\angle ADC$
Prove: $ED > EB$

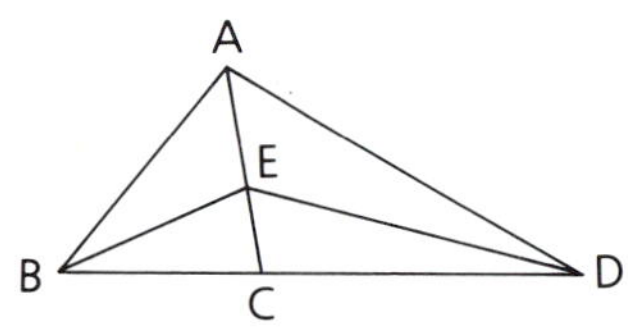

15 Given: $\overline{AB} \cong \overline{AD}$
Prove: $AC < AD$

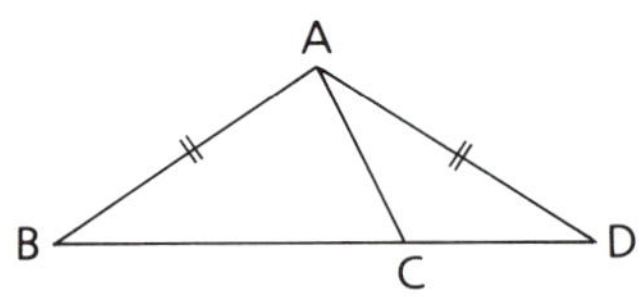

16 Given: $PR < QS$
$\overline{PQ} \cong \overline{SR}$
Prove: $\angle PQR < \angle SRQ$

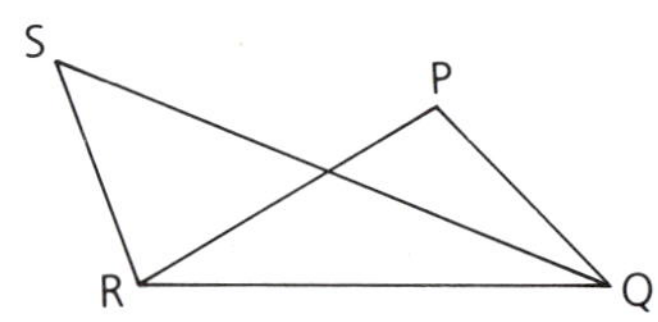

17 Given: $\overline{BC} \cong \overline{EC}$
$\angle A \cong \angle C$
Prove: $AE < EC$

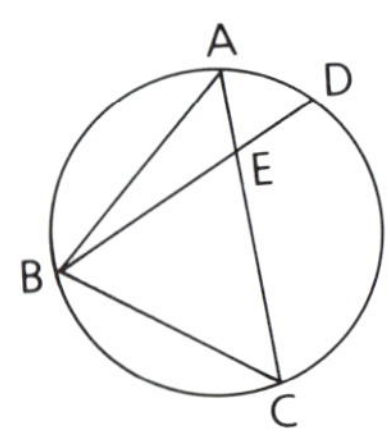

Chapter 15 Review Problems C

18 In an obtuse triangle, the side opposite the obtuse angle is 6. If all sides are integral, how many such triangles exist?

19 Given: $\triangle ABC$ as shown.
List the sides $\overline{AB}$, $\overline{AC}$, $\overline{AD}$, $\overline{BC}$, $\overline{BD}$, $\overline{CD}$ in order, largest first.

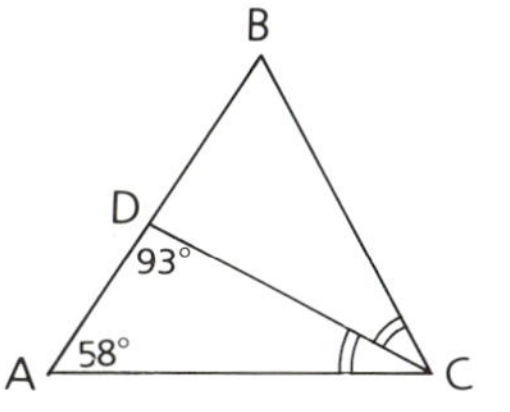

20 Prove that the median of a triangle is less than half of the sum of the two adjacent sides. That is, prove $AM < \frac{1}{2}(AB + AC)$.

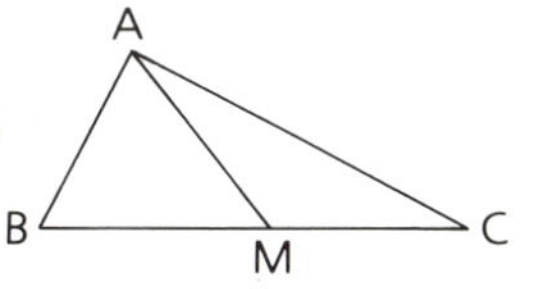

(Hints: (1) Draw an appropriate midline, or
(2) extend $\overline{AM}$ to point P so that AM = MP. Now form ▱ ABPC.)

21 Prove that in any quadrilateral, the perimeter is greater than the sum of the diagonals.

22 Given: $\overline{ZX} \cong \overline{ZY}$
WX < WY
Conclusion: XV > VY

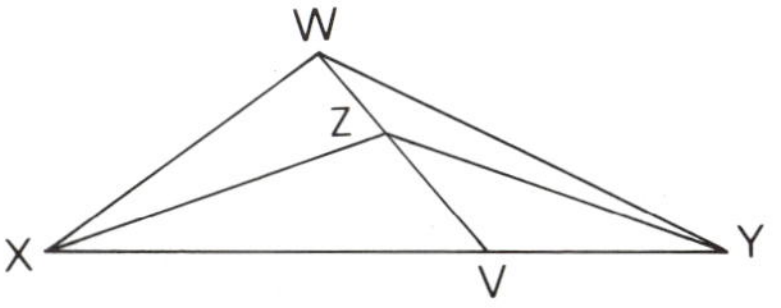

23 The sides of triangle ABC are integers. AB = 5, AC = 13. What is the probability that, if one of the possible values for $\overline{BC}$ is picked at random, the resulting triangle will be obtuse?

24 Given: Quadrilateral PQRS
PQ > PS
$\angle Q \cong \angle S$
Prove: RS > RQ

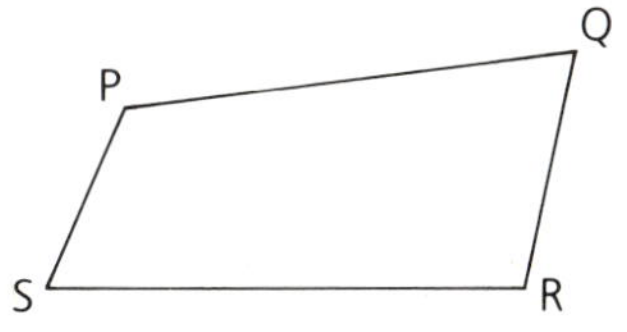

25 P is any point inside quadrilateral WXYZ. Prove that the sum of the distances from P to the four vertices (PW + PX + PY + PZ) is greater than or equal to the sum of the diagonals. (Consider all three cases for the position of point P within the quadrilateral.)

Chapter 16

ALTERNATE FORMS OF PROOF

Section 16.1 Statements of Logic

In this section we discuss:

A CONDITIONAL STATEMENTS

B NEGATION

C CONVERSE, INVERSE, AND CONTRAPOSITIVE

D LOGICAL SYLLOGISM

16.1 (A) Conditional Statements

This section continues the discussion of conditional statements begun in Section 4.2.

Recall that a *conditional statement* is a sentence that is in the form: "If . . . , then . . . ." Many simple sentences can be put into conditional form.

■ Example

Simple Sentence Form:		Conditional Form:
The base angles of an isosceles triangle are congruent.	$\Longleftrightarrow$	**If a triangle is isosceles, then its base angles are congruent.**

The clause following the word ''If'' is called the **HYPOTHESIS**, and the clause following the word ''then'' is called the **CONCLUSION**.

The conditional statement ''If p, then q'' can be written in symbols as $p \Rightarrow q$, which is read ''p implies q.''

16.1 (B) Negation

The **NEGATION** of the statement ''The quadrilateral is a parallelogram'' is ''The quadrilateral is not a parallelogram.'' The negation of a statement ''p'' is the statement ''not p.'' The symbol for ''not p'' is ''$\sim p$''. Notice also that the negation of ''It is not raining'' is ''It is raining.'' In general, not (not p) = p, or $\sim\sim p = p$.

16.1 (C) Converse, Inverse, and Contrapositive

Every conditional statement of the form ''If p, then q'' has three additional statements associated with it:

1 Converse (If q, then p)

2 Inverse (If $\sim p$, then $\sim q$)

3 Contrapositive (If $\sim q$, then $\sim p$)

■ Example

CONDITIONAL STATEMENT: If *you live in Atlanta,* then *you live in Georgia.*
(If p, then q)

CONVERSE: If *you live in Georgia,* then *you live in Atlanta.*
(If q, then p)

INVERSE: If *you don't live in Atlanta,* then *you don't live in Georgia.*
(If $\sim p$, then $\sim q$)

CONTRAPOSITIVE: If *you don't live in Georgia,* then *you don't live in Atlanta.*
(If $\sim q$, then $\sim p$)

You may have noticed that some of the statements are not necessarily true, although the original statement is true.

A useful tool for determining whether or not a conditional statement is true or false is a **VENN DIAGRAM**. Assume that the following statement is true: ''If Jenny lives in Atlanta, then Jenny must live in Georgia.''

All the people who live in Georgia are represented by points on the large circle and its interior. All the people who live in Atlanta are represented by points on the smaller circle and its interior. Notice that *every* person in set A, including Jenny, is also in set G.

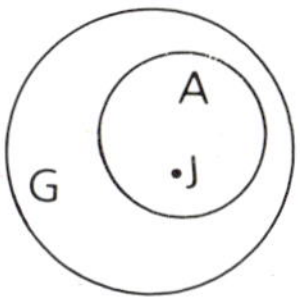

The Venn Diagram for the original conditional statement may be used to test whether the converse, the inverse, and the contrapositive are true or false.

The Converse: ''If Jenny lives in Georgia, then she must live in Atlanta.''

This converse is not necessarily true, as shown in the diagram. Notice that an element J may lie in G but *not* in A. This means that Jenny could live in Georgia, and yet not live in Atlanta.

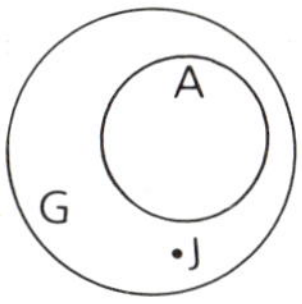

In general, the converse of a conditional statement is not necessarily true. Try a similar argument with the same Venn Diagram to convince yourself that the inverse of a conditional statement is also *not necessarily true.*

The Contrapositive: ''If Jenny does not live in Georgia, then she does not live in Atlanta.''

This time the element J lies outside of G, so J *cannot* lie in A. Thus, if J is not in G, then J is not in A, and the contrapositive is true.

This analysis suggests the following important result:

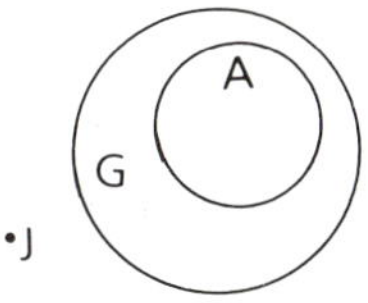

Theorem 130

T **If a conditional statement is true, then the contrapositive of the statement is also true. If p, then q $\Longleftrightarrow$ If $\sim$ q, then $\sim$ p.**

In summary, we may say that a statement and its contrapositive are logically equivalent.

16.1 (D) Logical Syllogism

Each proof that you do is a series of steps in a logical sequence.

Knowing that $p \Rightarrow q$ and $q \Rightarrow r$, we may conclude $p \Rightarrow r$.

This is a rule of logic called a LOGICAL SYLLOGISM or the CHAIN RULE.

■ Example: If we accept the two statements:
If you study hard, then you will earn a good grade. ($p \Rightarrow q$)
If you earn a good grade, then your family will be happy. ($q \Rightarrow r$)
We can conclude:
If you study hard, then your family will be happy. ($p \Rightarrow r$)

Section 16.1 Sample Problems

1 Find the converse, inverse, and contrapositive of the following true statement: "If a triangle has sides 3, 5, and 6, then the triangle is not a right triangle."

Solution:

Converse: If a triangle is not a right triangle, then it has sides 3, 5, and 6. (Statement false. There are many possible sets of sides.)

Inverse: If a triangle does *not* have sides 3, 5, and 6, then the triangle is a right triangle. (Statement also false.)

Contrapositive: If a triangle is a right triangle, then it does *not* have sides 3, 5, and 6. (Statement true—logically equivalent.)

2 Complete the following syllogism:

a If gremlins grow grapes, then elves eat earthworms.
b If trolls don't tell tales, then wizards weave willows.
c If trolls tell tales, then elves don't eat earthworms.

Solution: First, rewrite the statements in symbolic form.
(a) $g \Rightarrow e$
(b) $\sim t \Rightarrow w$
(c) $t \Rightarrow \sim e$
To make a chain of reasoning, rearrange the statements and use contrapositives as needed to match symbols. Thus, we have:

(a) $g \Rightarrow e$
(c) $e \Rightarrow \sim t$, since $t \Rightarrow \sim e$ is equivalent to $e \Rightarrow \sim t$.
(b) $\sim t \Rightarrow w$
$\therefore g \Rightarrow w$
Hence, If Gremlins grow grapes, then wizards weave willows.

Section 16.1 Problem Set A

1 Write each sentence in conditional form. (If . . . , then . . . .)
- **a** 18-year-olds may vote in federal elections.
- **b** Opposite angles of a parallelogram are congruent.

2 For each statement find the converse, inverse, and contrapositive. Determine the truth of each form.
- **a** If a quadrilateral is a rhombus, then the diagonals of the quadrilateral are perpendicular.
- **b** If two angles of a triangle are not congruent, then the sides opposite those angles are not congruent.
- **c** If a triangle is not scalene, then at least two sides of the triangle are congruent.

3 If a conditional statement *and* its converse are both true, then the conditional is BICONDITIONAL. Which of these are biconditional?
- **a** If two triangles are congruent, then their areas are equal.
- **b** If a triangle is equilateral, then it is equiangular.
- **c** If two triangles are congruent, then the triangles are similar.

4 Draw a Venn Diagram for the true conditional statement: "If a person lives in Chicago, then the person lives in Illinois."
Assume that each "given" statement below is true and determine the truth of the statement in parentheses.
- **a** Given: Penny lives in Chicago. (Penny lives in Illinois.)
- **b** Given: Benny lives in Illinois. (Benny lives in Chicago.)
- **c** Given: Kenny does not live in Chicago. (Kenny must live in Illinois.)
- **d** Given: Denny does not live in Illinois. (Denny lives in Chicago.)

5 Write a concluding statement for each of the syllogisms:

a
$a \Rightarrow b$
$d \Rightarrow \sim c$
$\sim c \Rightarrow a$
$b \Rightarrow f$

b
$p \Rightarrow \sim q$
$r \Rightarrow q$
$s \Rightarrow r$

c If weasels walk wisely, then cougars call their cubs.
If goats go to graze, then horses head for home.
If cougars call their cubs, then goats go to graze.
If bobcats begin to browse, then weasels walk wisely.

Section 16.1 Problem Set B

6 Write the converse, inverse, and contrapositive of: If  $\Rightarrow$ . Which statements are true?

7 Rewrite the sentence in conditional form and find the converse, inverse, and contrapositive: "A square is a quadrilateral with four congruent sides."

8 Find the converse, inverse, and contrapositive of each statement.

a If two circles are concentric, then they have no common tangent.

b If a segment is a diagonal of a parallelogram, then the segment bisects an angle of the parallelogram.

9 Complete this syllogism:

$\sim c \Rightarrow \sim f$ $\quad g \Rightarrow b$ $\quad p \Rightarrow f$ $\quad c \Rightarrow \sim b$

Section 16.1 Problem Set C

10 What conclusion can be drawn from the following?
If the line is long, then Quincy will go home.
If it is morning, then Quincy will not go home.
If the line is long, then it is morning.

Section 16.2 Indirect Proof

An INDIRECT PROOF may be useful in a problem where a direct proof would be difficult to apply.

■ Study the following example of an indirect proof:

Given: $\angle A \cong \angle D$, $\overline{AB} \cong \overline{DE}$, $\overline{AC} \ncong \overline{DF}$

Prove: $\angle B \ncong \angle E$

Proof: Either $\angle B \cong \angle E$ or $\angle B \ncong \angle E$.

Assume $\angle B \cong \angle E$.

From the given information, $\angle A \cong \angle D$ and $\overline{AB} \cong \overline{DE}$.

Thus $\triangle ABC \cong \triangle DEF$ by ASA.

$\therefore \overline{AC} \cong \overline{DF}$.

But this is impossible, since $\overline{AC} \ncong \overline{DF}$ is a given fact.

Thus, our assumption was false and $\angle B \ncong \angle E$, because this is the only remaining possiblity.

The following procedure will help you to write indirect proofs.

INDIRECT PROOF PROCEDURE

1 List the possibilities for the CONCLUSION.
2 ASSUME that the NEGATION of the desired CONCLUSION is correct.
3 Write a chain of reasons until you reach an IMPOSSIBILITY.

This will be a CONTRADICTION of either

a given information or

b a theorem, definition, or other known fact.

4 State the remaining possibility as the desired conclusion.

Section 16.2 Sample Problems

1 Given: $\overline{RS} \perp \overline{PQ}$

$\overline{PR} \not\cong \overline{QR}$

Prove: $\overrightarrow{RS}$ does not bisect $\angle PRQ$

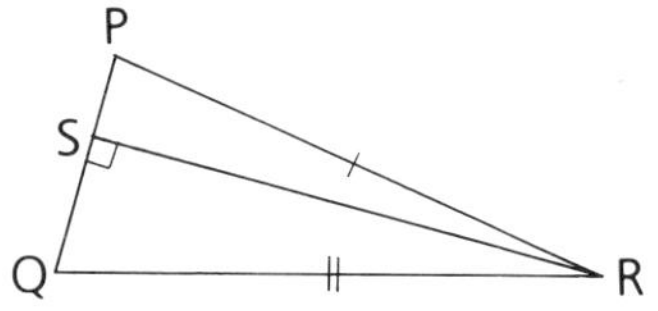

Proof: Either $\overrightarrow{RS}$ bisects $\angle PRQ$ or $\overrightarrow{RS}$ does not bisect $\angle PRQ$.
Assume $\overrightarrow{RS}$ bisects $\angle PRQ$.
Then we can say that $\angle PRS \cong \angle QRS$.

Since $\overline{RS} \perp \overline{PQ}$, we have $\angle PSR \cong \angle QSR$.

Thus, $\triangle PSR \cong \triangle QSR$ by ASA ($\overline{SR} \cong \overline{SR}$).

Hence, $\overline{PR} \cong \overline{QR}$ by CPCTC.

But, this is impossible because it contradicts the given fact that $\overline{PR} \not\cong \overline{QR}$. Consequently, the assumption must be false.
$\therefore$ $\overrightarrow{RS}$ does not bisect $\angle PRQ$, the only remaining possibility.

2 Prove: AIP (If two lines are cut by a transversal such that the alternate interior angles are congruent, then the lines are parallel.)

Given: $\angle 1 \cong \angle 2$

$\overleftrightarrow{AB}$ intersects ℓ at A and intersects m at B.

Prove: $\ell \parallel m$

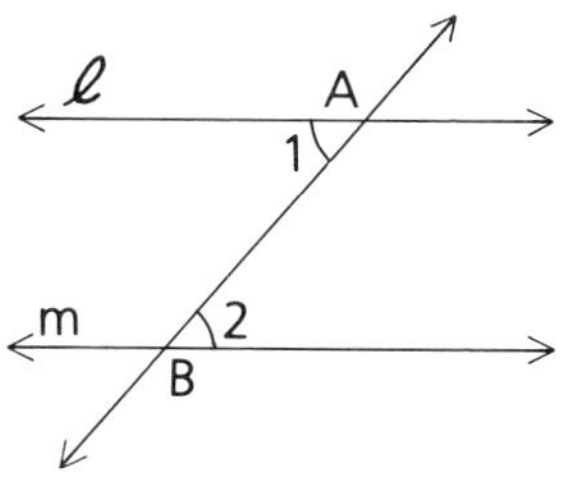

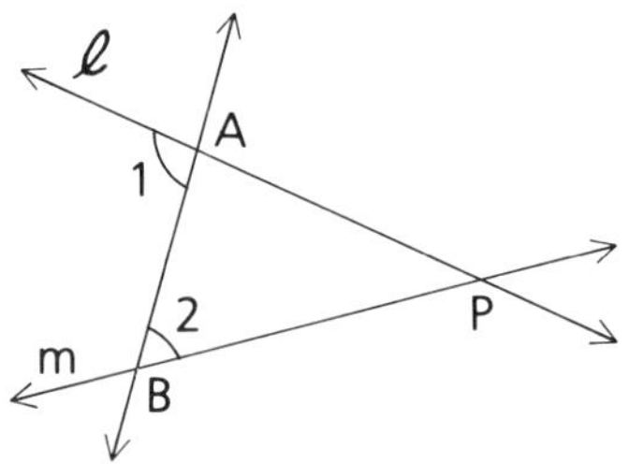

Assume $\ell \nparallel m$ (see diagram).
Then ℓ intersects m at some point P, and a $\triangle ABP$ is formed.
Thus, $\angle 1 > \angle 2$ by the Exterior Angle Inequality Theorem.
But, this is impossible since it contradicts $\angle 1 \cong \angle 2$, a given fact.

Hence, our assumption was false and $\ell \parallel m$ since this is the only remaining possibility.

Section 16.2 Problem Set A

1 Given: P is not the midpoint of $\overline{HK}$
$\overline{HJ} \cong \overline{JK}$
Prove: $\overrightarrow{JP}$ does not bisect $\angle HJK$

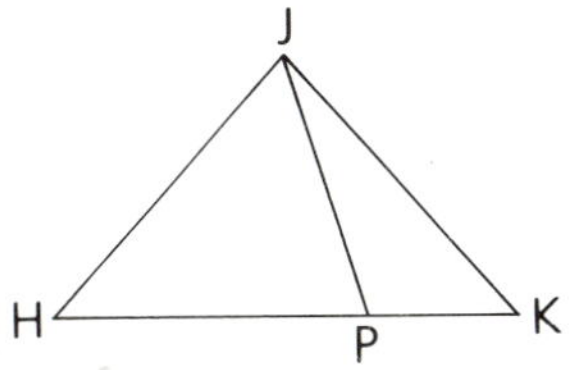

2 Given: $\overline{AC} \perp \overline{BD}$
$\overline{BC} \cong \overline{EC}$
$\overline{AB} \not\cong \overline{ED}$
Prove: $\angle B \not\cong \angle CED$

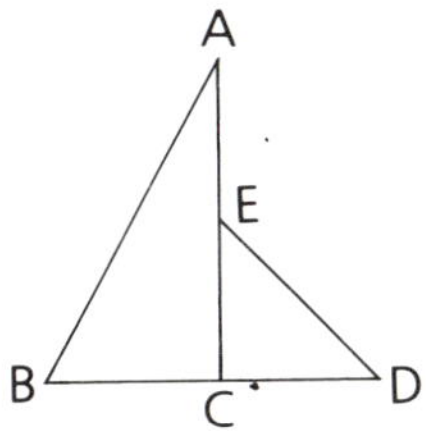

3 Given: $\angle B > \angle D$
Prove: ABCD cannot be a ▱

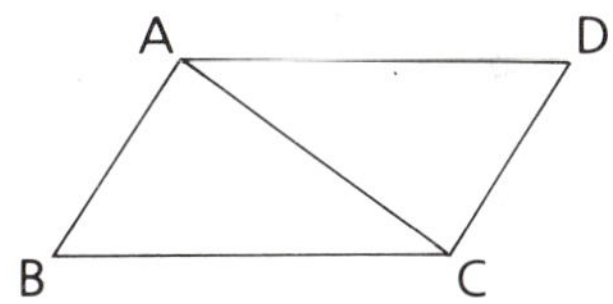

4 Given: $\overline{PR}$ is the median to $\overline{QS}$
$\triangle PQS$ is scalene
Prove: $\overline{PR}$ is not an altitude to $\overline{QS}$

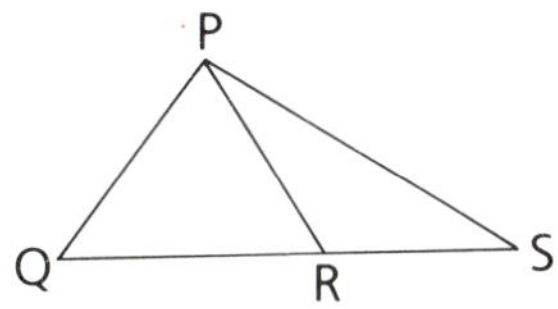

5 Given: $\odot$ O
$\overline{XW} \not\cong \overline{WY}$
Conclusion: $\overset{\frown}{XZ} \not\cong \overset{\frown}{ZY}$

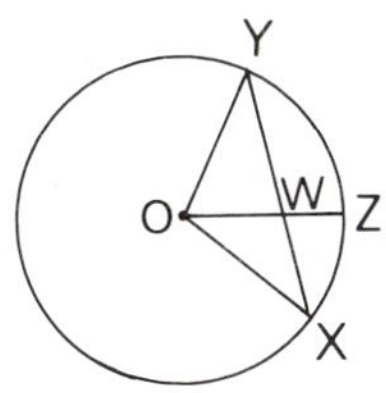

6 Prove that a triangle cannot have more than one right angle.

Section 16.2 Problem Set B

7 Given: $\overrightarrow{BD}$ bisects $\angle ABC$
$\angle ADB$ is acute
Prove: $\overline{AB} \not\cong \overline{BC}$

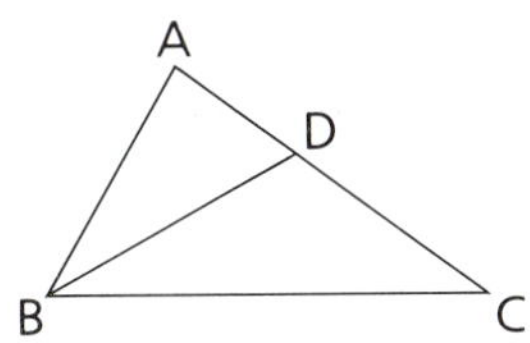

8 Given: $\odot$ O with $\overset{\frown}{PQ} \not\cong \overset{\frown}{PR}$
Prove: $\overline{PS} \not\perp \overline{QR}$

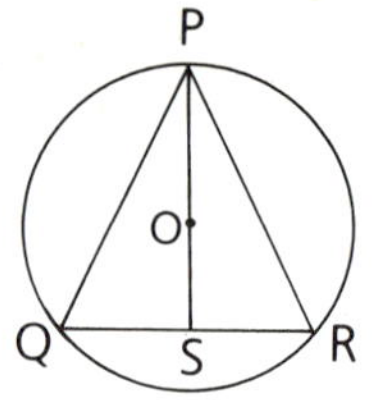

9 Given: $\overline{AC} \perp m$
$\angle B \not\cong \angle D$
Prove: $\angle BAC \not\cong \angle DAC$

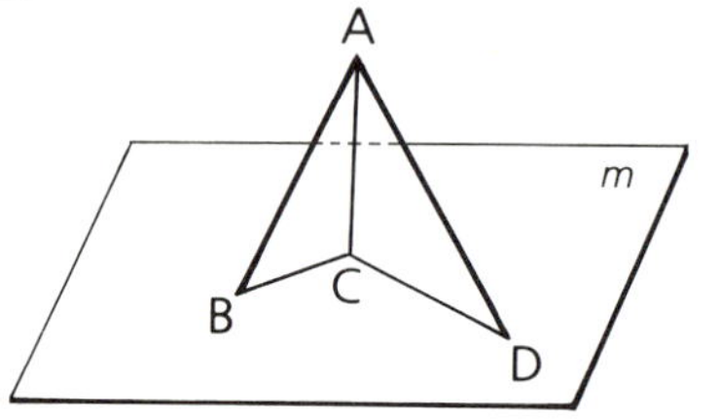

10 Given: $\triangle BEC$ is not isosceles
Prove: $\triangle ABC \not\cong \triangle DCB$

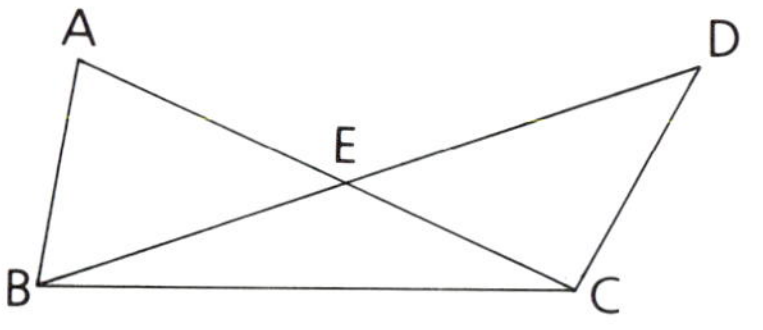

11 Prove that two distinct lines perpendicular to the same plane are parallel.

12 Prove that if $\triangle ABC$ is isosceles with base $\overline{BC}$, and P is a point on $\overline{BC}$ that is *not* the midpoint, then $\overrightarrow{AP}$ does not bisect $\angle BAC$.

Section 16.2 Problem Set C

13 If no two medians of a triangle are congruent, then the triangle is scalene.

14 Prove that two distinct lines cannot intersect at more than one point.

15 a Prove that if a line is perpendicular to one side of an angle, then it is not perpendicular to the other side.
b Give at least one example which disproves the statement in part a. (Proof by counterexample.)

Section 16.3 Analytic Proofs

Many theorems of geometry can be proved algebraically by use of coordinate geometry. Such proofs are called ANALYTIC PROOFS.

■ Example: Prove that the diagonals of a rectangle are congruent.
Let ABCD be any rectangle.
Place the rectangle on a coordinate system in a "convenient" place.

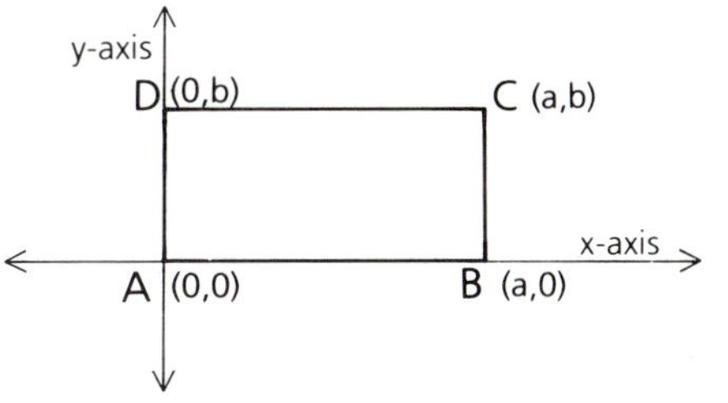

To prove that the statement is true for *any* rectangle, we must not restrict the lengths of the sides, so we call them a and b.

The lengths of the diagonals are:

$$AC = \sqrt{(a-0)^2 + (b-0)^2} = \sqrt{a^2 + b^2}$$

$$DB = \sqrt{(0-a)^2 + (b-0)^2} = \sqrt{(-a)^2 + b^2}$$

but $(-a)^2 = a^2$, so $DB = \sqrt{a^2 + b^2}$

Thus, AC = DB.

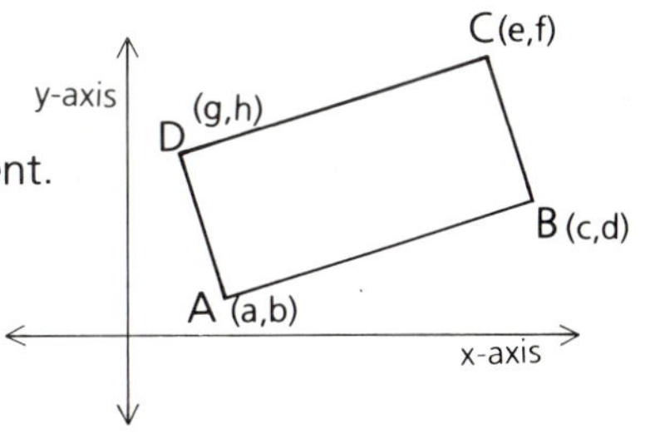

This diagram illustrates an inconvenient placement. Finding the lengths of diagonals will be harder, since all the coordinates are different.

Section 16.3 Sample Problem

1 **Prove: The medians to the legs of an isosceles triangle are congruent.**

Again we wish to place the figure on the coordinate system conveniently.

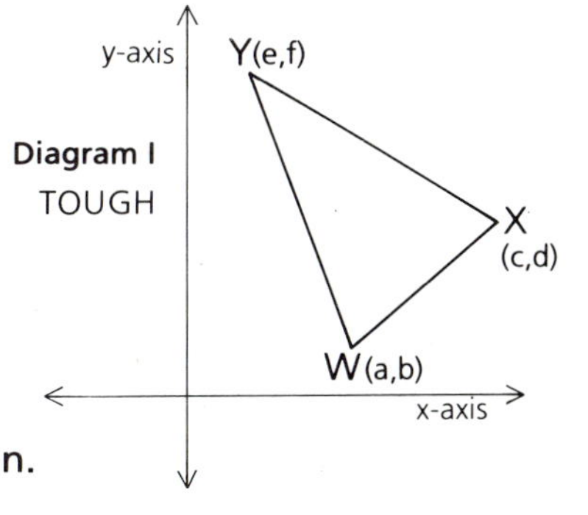

In Diagram I it is easily seen that the triangle is *not* in a convenient position.

In Diagram II the triangle is in a convenient position, but the coordinates of Y are *not* convenient. You will find that the coordinates for N, the midpoint of $\overline{XY}$, are even worse.

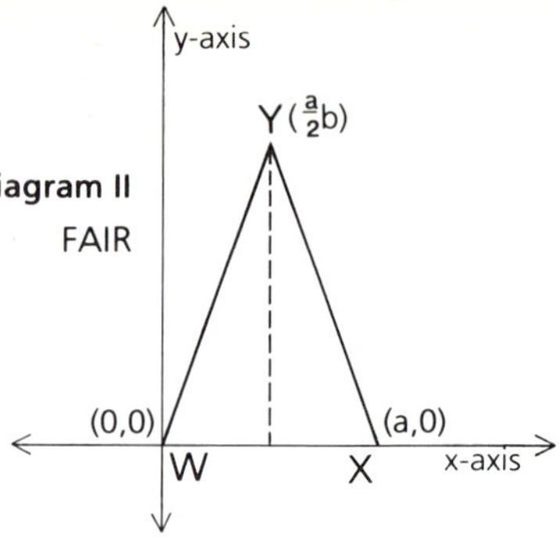

Ingenuity and experience will suggest that Diagram III is in a convenient position and has convenient coordinates. Midpoint M of $\overline{WY}$ is (a,b), and midpoint N of $\overline{XY}$ is (3a,b). The lengths of medians, $\overline{MX}$ and $\overline{NW}$, can be found by the distance formula:

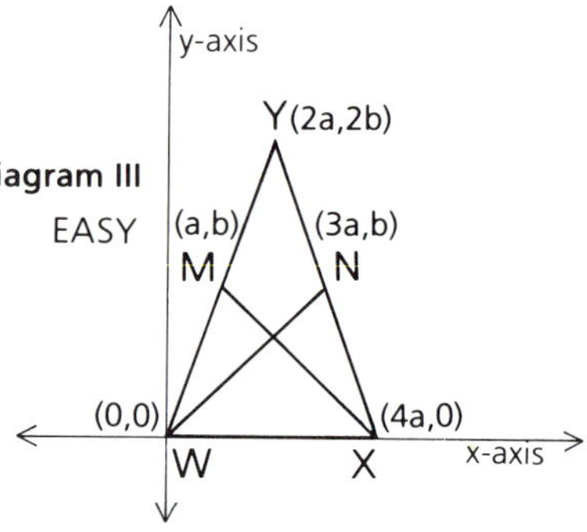

$$MX = \sqrt{(4a - a)^2 + (0 - b)^2} = \sqrt{9a^2 + b^2}$$

$$NW = \sqrt{(3a - 0)^2 + (b - 0)^2} = \sqrt{9a^2 + b^2}$$

Thus, $\overline{MX} \cong \overline{NW}$

Section 16.3 Problem Set A

1 Using the diagram as marked,

a show that $\overline{OQ} \cong \overline{QP}$.

b find the coordinates of the L, M, and N midpoints.

c find the slopes of $\overline{LM}$ and $\overline{MN}$.

d find the slopes of $\overline{OQ}$ and $\overline{QP}$.

e Prove that $\overline{MN} \parallel \overline{QO}$.

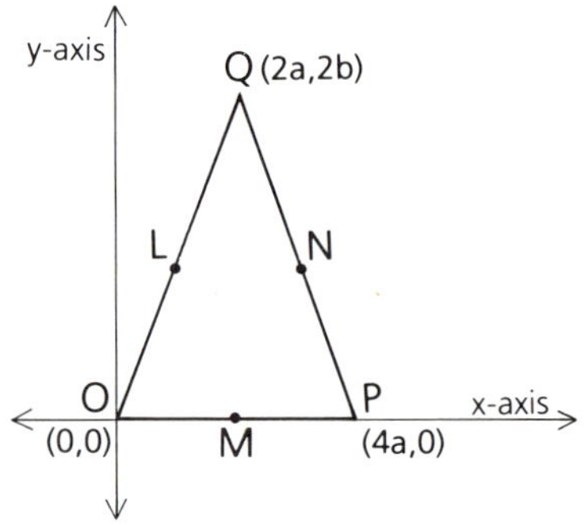

2 Given: Rectangle ABCO

Find: a the coordinates of A, B, C, O.

b M, N, P, and Q, the midpoints of the sides.

c the slopes of $\overline{MN}$, $\overline{QP}$, $\overline{MQ}$, and $\overline{NP}$. What can we conclude about MNPQ?

d the lengths of $\overline{MN}$, $\overline{QP}$, $\overline{MQ}$, and $\overline{NP}$. What can we now conclude about MNPQ?

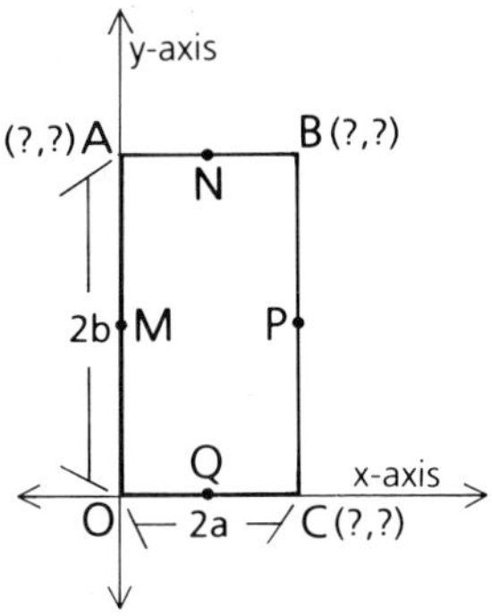

3 Given: Trapezoid PQRS

a Find PQ and SR and verify that PQRS is an isosceles trapezoid

b Prove that the diagonals $\overline{PR}$ and $\overline{QS}$ are congruent.

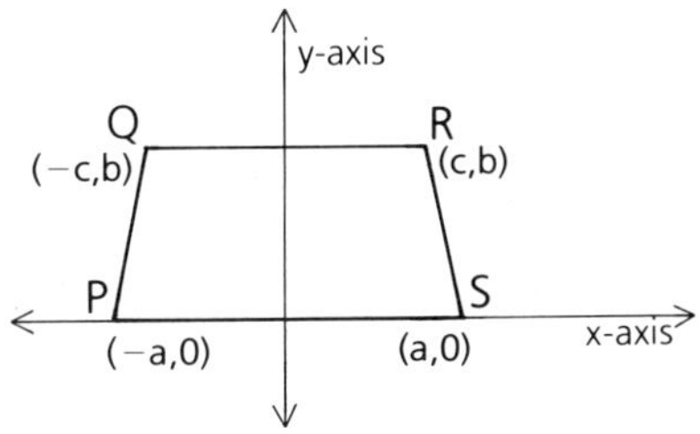

4 Given: Isosceles trapezoid ABCD

a Find the midpoints M and N of $\overline{AB}$ and $\overline{CD}$.

b Prove that the length of the median of trapezoid ABCD equals the average of the bases.

c Why did we use different coordinates from those used in problem 3?

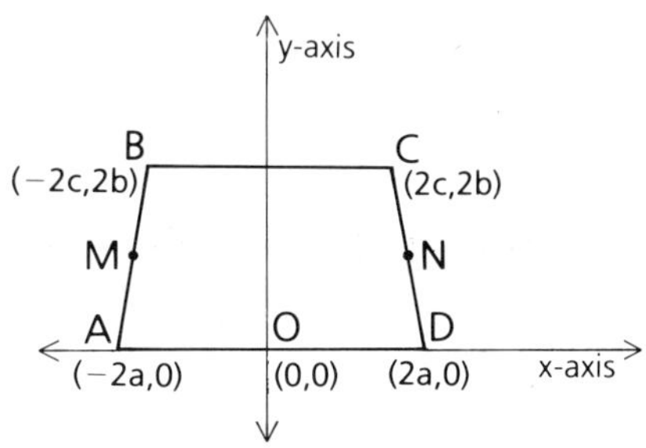

5 Without using any additional letters, label the remaining points of the parallelogram.

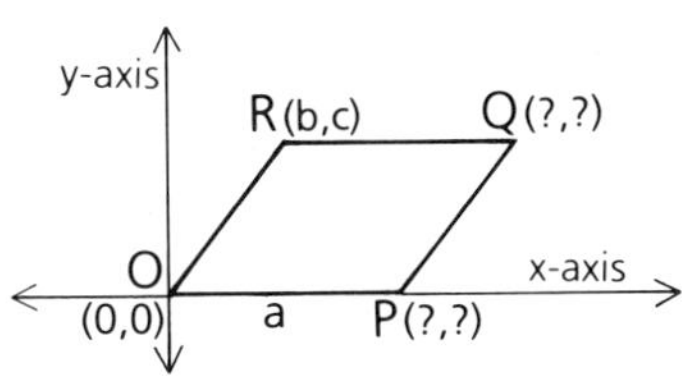

6 Prove that the diagonals of a parallelogram bisect each other.

7 Prove that the diagonals of a square are congruent and perpendicular.

Section 16.3 Problem Set B

8 Let the coordinates of $\triangle ABC$ be A(0, 0), B(2a, 0), and C(2b, 2c). Prove analytically that the line segment joining the midpoints of any two sides is parallel to the third side and half the length of the third side.

9 Prove that the segment joining the midpoints of the diagonals of a trapezoid is half the difference of the lengths of the bases.

10 Prove that the segments joining the midpoints of consecutive sides of an isosceles trapezoid form a rhombus.

11 Prove that the midpoint of the hypotenuse of a right triangle is equidistant from the three vertices.

12 In the diagram as marked, prove:

$(AC)^2 + (CB)^2 = \frac{1}{2}(AB)^2 + 2(CM)^2$

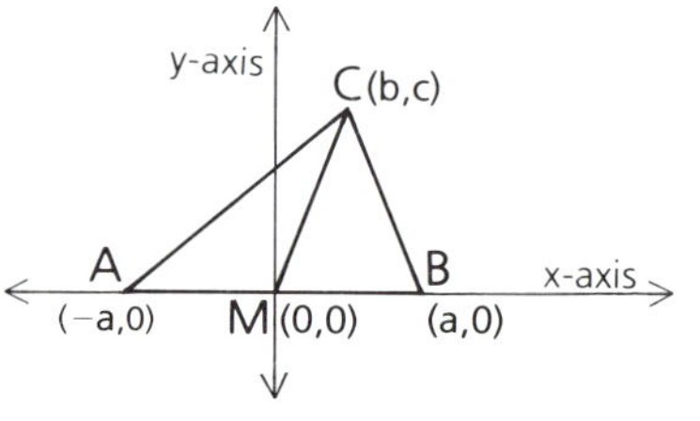

13 Prove that the sum of the squares of the sides of a parallelogram is equal to the sum of the squares of the diagonals.

Section 16.3 Problem Set C

14 Prove analytically that the diagonals of a rhombus are perpendicular.

15 Prove analytically that if the midpoints of any quadrilateral are joined in order, then the resulting figure is a parallelogram.

16 Prove that in any quadrilateral the sum of the squares of the sides is equal to the sum of the squares of the diagonals plus four times the square of the segment joining the midpoints of the diagonals.

Chapter 16 Study Guide

As you study and review this chapter, be sure that you are able to:

1. Change a simple sentence to conditional form. (p. 693)

2. Write the negation of a statement. (p. 694)

3. Write each of the following statements given a conditional statement (p. 694):

   a converse **b** inverse **c** contrapositive

4. Write each of the statements in problem 3 in symbolic form. (p. 694)

5. Tell which other statement in problem 3 must be true if a given conditional statement is true. (p. 695)

6. Describe what must be true if a statement is called biconditional. (p. 697)

7. Form a chain of reasoning from a set of statements in order to draw a conclusion. (p. 696)

8. Describe the temporary assumption made in the second step of an indirect proof, if the first step in an indirect proof is listing possible conclusions. (p. 699) Can you write a complete indirect proof?

9. Set up a geometric figure with convenient coordinates on a coordinate system to prove a theorem by an Analytic Proof. (p. 703)

10. Describe the formulas for finding, on a coordinate system:

    a the midpoint of a segment (p. 584)
    b the slope of a line (p. 587)
    c the length of a segment (p. 603)

Chapter 16 Review Problems A

1 Find the converse, inverse, and contrapositive of:
  a If two pentagons are regular, then they are similar.
  b If a parallelogram is inscribed in a circle, then it is not a "plain old parallelogram."

2 Given: $\overline{DE} \cong \overline{EB}$
ABCD is not a ▱
Prove: $\overline{AE} \not\cong \overline{EC}$

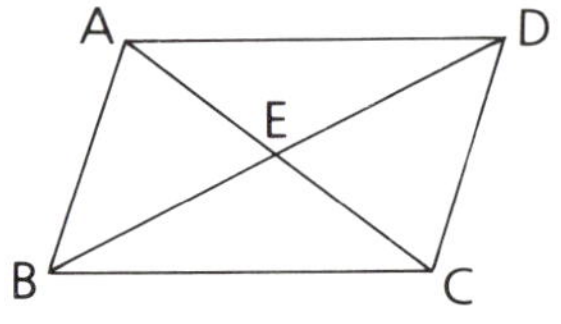

3 Given: $\angle X \not\cong \angle Z$
W is the midpoint of $\overline{XZ}$
Prove: $\overline{WY}$ is not an altitude to $\overline{XZ}$

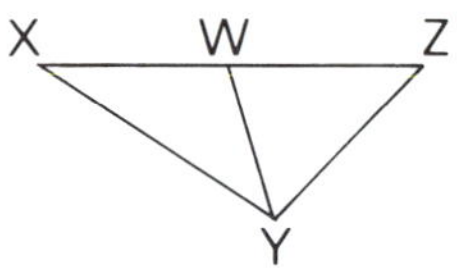

4 Let the vertices of a square be (0, 0), (0, 2a), (2a, 2a), and (2a, 0). Prove that the figure formed by joining consecutive midpoints of the four sides is also a square.

5 Prove analytically that the diagonals of a rectangle are congruent. (Let the lengths of the sides of the rectangle be *a* and *b*.)

Chapter 16 Review Problems B

6 In $\triangle ABC$, $\overline{AC} \cong \overline{BC}$, $\overline{AB} \not\cong \overline{AC}$, and $\overline{BD}$ is the median to $\overline{AC}$. Prove that $\overline{AD} \not\perp \overline{BD}$.

7 Prove the bisector of an angle of a scalene triangle does not bisect the opposite side.

8 Given: $\odot R$ is in plane *m*
$\overline{PQ} \not\cong \overline{PS}$
Prove: $\overline{PR} \not\perp m$

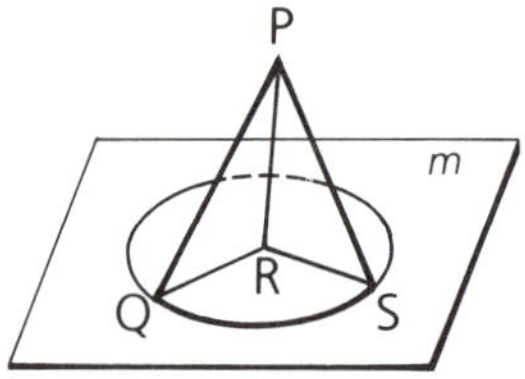

9 Prove that a base angle of an isosceles triangle cannot be obtuse.

10 Given: $\overline{WX} \not\cong \overline{WZ}$

$\overleftrightarrow{WY}$ bisects $\overline{XZ}$

Prove: $\overline{XY} \not\cong \overline{YZ}$

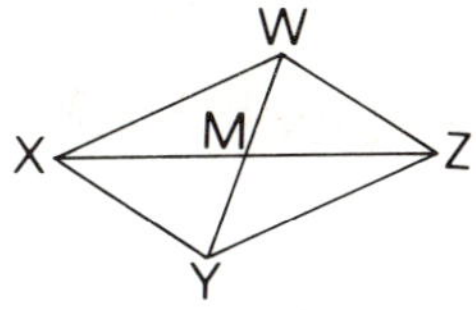

11 a Prove that ABCD is a kite.

b Prove analytically that the figure formed by joining consecutive midpoints is a rectangle.

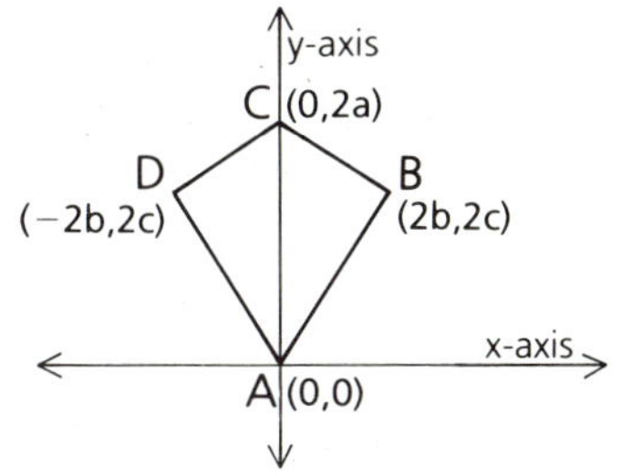

12 What can we conclude from the following statements?

If r is red, then b is blue.
If q is not green, then y is yellow.
If r is not red, then y is not yellow.
b is not blue.

13 Complete the following syllogisms.

a $f \Rightarrow \sim d$

$a \Rightarrow \sim b$

$\sim d \Rightarrow \sim e$

$\sim c \Rightarrow b$

$c \Rightarrow e$

b If Ergles are tardy, then Bergles are late.
If Gergles are not home, then Bergles are not late.
If Ergles are not tardy, then Mergles are prompt.
I know Mergles are not prompt.
What may I conclude?

Chapter 16 Review Problems C

14 Prove that the shortest segment from a point P outside a circle to the circle is the segment along the line from P to O.

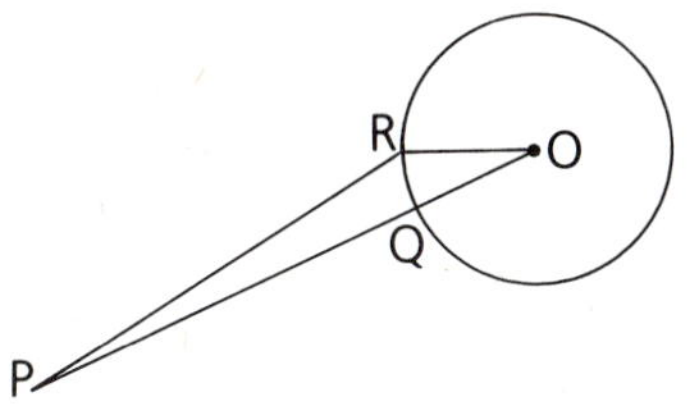

15 Given a square with vertices A(0, 0), B(6a, 0), C(6a, 6a), D(0, 6a). Line segments are drawn from A to the midpoints of $\overline{BC}$ and $\overline{CD}$.

a Find the coordinates of the points of intersection of these two segments with diagonal $\overline{BD}$.

b In part a, a pentagon is formed. Find the ratio of the area of the pentagon to the area of the original square.

Chapters 1–16 Cumulative Review Problems

A

1 Find the volume and total area of a circular cone with height 4 and base radius 3.

2 Given the line $2x + 3y = 12$. Find:

a the x-intercept b the slope

c if $(37, -21)$ lies on the line

3 Given: Concentric circles

$\overset{\frown}{CD} = 70°$, $\overset{\frown}{QR} = 54°$

Find: $\overset{\frown}{AB}$

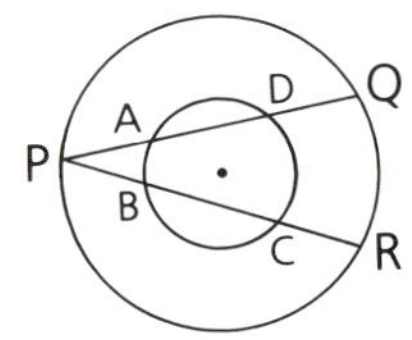

4 How far from the center of a circle with diameter 26 is a chord of 24?

5 If the length of tangent segment $\overline{JM}$ is 10, and JP = 4, find PQ.

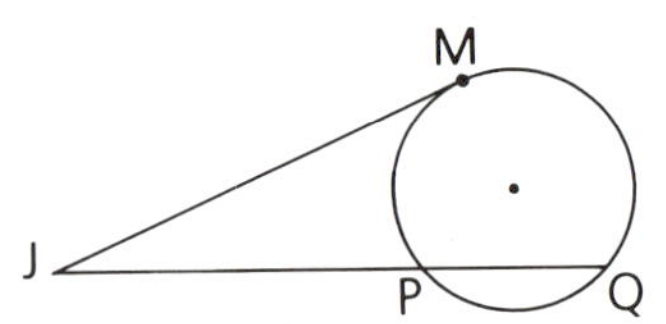

6 Solve for x.

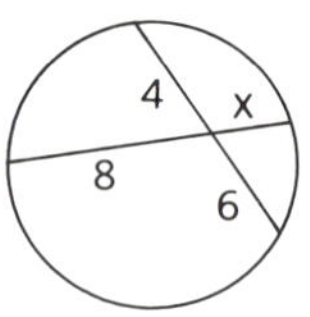

7 In △ADC, $\overline{BE} \parallel \overline{CD}$, AB = 8, BC = 4, AE = 6, BE = 9.

a Find DE b Find CD

c Is △ABE a right triangle?

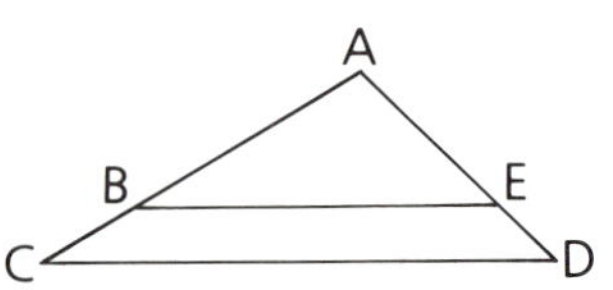

8 Given: ABCD is a trapezoid with $\overline{AD} \parallel \overline{BC}$

Prove: $AE \cdot BE = DE \cdot EC$

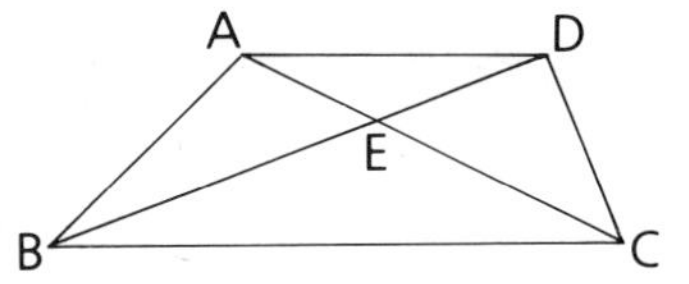

9 B and E are midpoints of $\overline{AC}$ and $\overline{AD}$ respectively. Find CD.

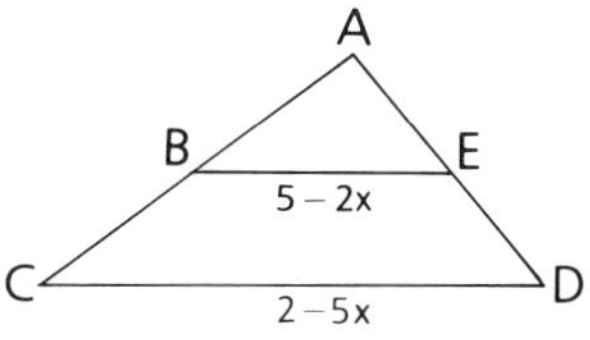

10 If a base angle of an isosceles triangle is twice the vertex angle, then find the measure of the vertex angle.

11 Given points A(3, 8), B(8, −4), and C(−6, −4), draw a graph of $\triangle ABC$ and solve each of the following:

a Find the lengths and slopes of $\overline{AB}$, $\overline{BC}$, and $\overline{AC}$.
b Is $\triangle ABC$ acute, right, or obtuse?
c Find the equation of $\overleftrightarrow{AC}$ and its x- and y-intercepts.
d Find the equation of $\overleftrightarrow{BC}$.
e Where does the altitude to $\overline{BC}$ cross $\overline{BC}$?
f What is the equation of the altitude to $\overline{BC}$?
g Find the length of the altitude to $\overline{BC}$.
h Find the midpoint of $\overline{CB}$ and the slope of the median to $\overline{CB}$.
i Find the area of $\triangle ABC$.

12 Two regular pentagons have areas 8 and 18. What is the ratio of their perimeters?

13 Each interior angle of a regular polygon is 160°. Find the number of diagonals.

14 Find the area of the sector formed by the hands of a clock at 2 o'clock, if the diameter of the clock is 12 inches.

15 Find the area of an equilateral triangle whose height is 6.

Chapters 1–16 Cumulative Review Problems B

16 a Find the area of the shaded region (Half washer).

b Find the perimeter. (There are two semicircles and two segments.)

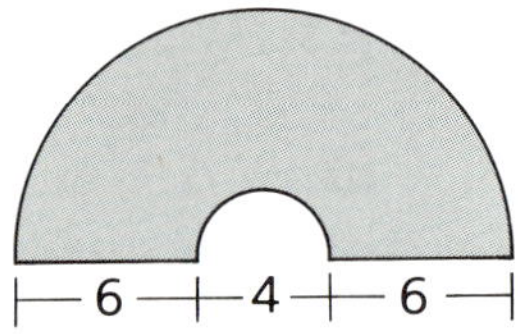

17 $\widehat{BD} = 80^\circ = \widehat{CE}$
$\angle CAB = 75^\circ$
Find $\widehat{BE}$

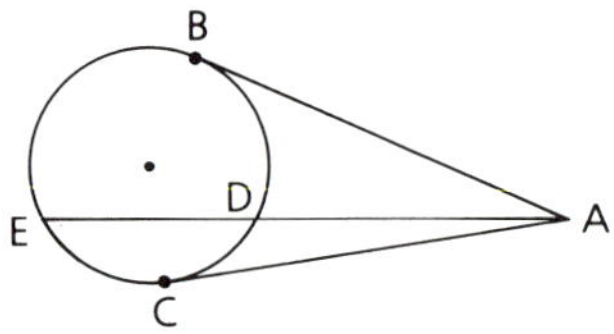

18 $\overline{BD}$ is a diameter
$\widehat{AB} = 10^\circ$, $\angle C = 40^\circ$
$\angle GFC = 80^\circ$
Find: a $\widehat{CD}$ b $\widehat{ED}$

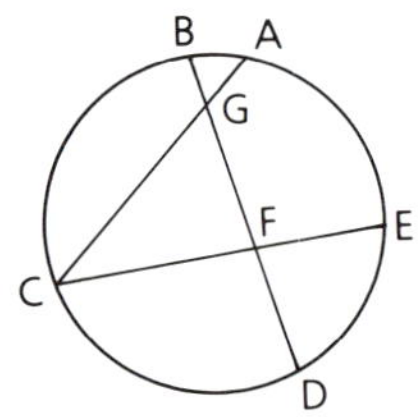

19 a Find $m\overline{RS}$.
b Find $\widehat{QTS}$.

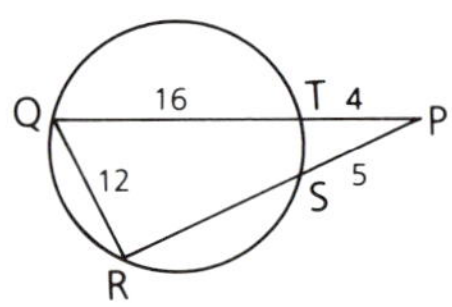

20 Find x.

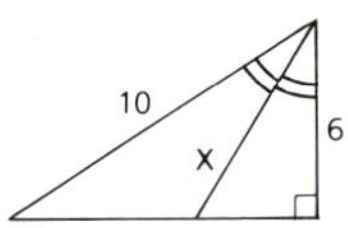

21 Find the area of ABCD.

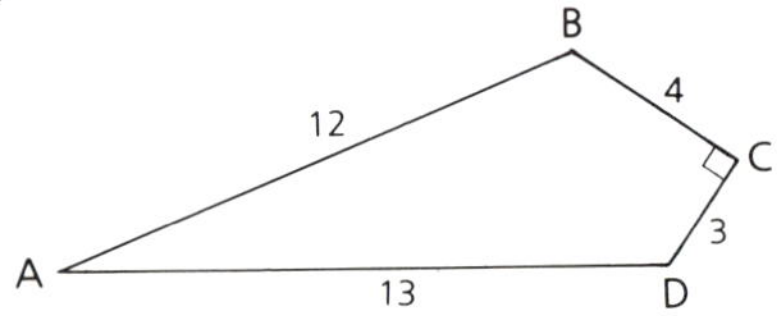

22 ABCD is a square with a side of 12. The midpoints of the sides of the square are centers of arcs drawn tangent to the diagonals. Find the shaded area.

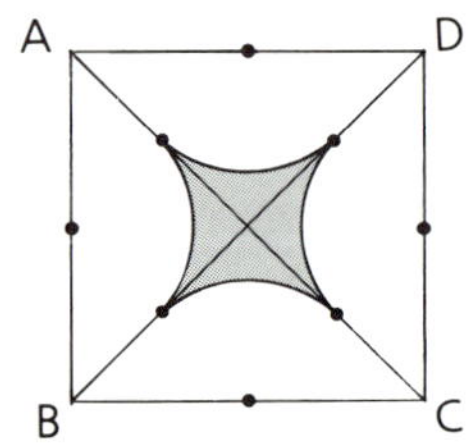

23 The vertices of $\triangle ABC$ are A(5, 4), B(11, 6), and C(9, 10). Find:

a the length of the median to $\overline{AB}$.

b the equation of the median to $\overline{AB}$.

c the equation of the altitude to $\overline{AB}$.

d the equation of the perpendicular bisector of $\overline{AB}$.

24 Given a kite with diagonals 6 and 14. Find the length of the segment joining the midpoints of two opposite sides.

25 Roger is 2 meters tall. When he stands atop a tower, the total length of his shadow and the tower's shadow is 14 meters. On the ground his shadow is 1 meter long. How high is the tower?

26 The diagonals of a rhombus are 8 and 12. Find its altitude.

27 Quadrilateral PQRS is inscribed in $\odot$ O. The measures of arcs PQ, QR, RS and SP are in the ratio 7:12:6:5 respectively. Find the acute angle formed by the diagonals of the quadrilateral.

28 Find the equation of the circle with center (2, 4) that passes through (1,7).

29 Is $\triangle ARO$ acute, right, or obtuse?

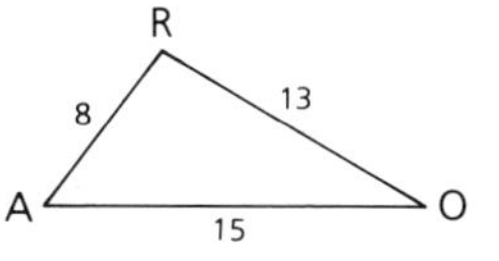

30 $\overline{CD}$ is an altitude to the hypotenuse of $\triangle ABC$. The coordinates of A, B, and D are given. Find the coordinates of point C.

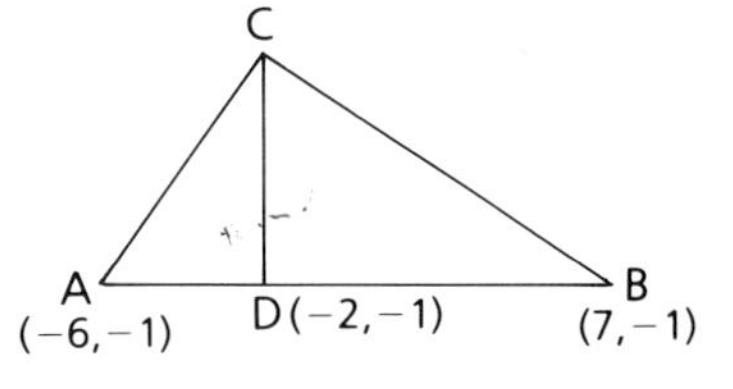

31 In $\odot$ O, find the ratio of the *length* of arc ARC to the length of diameter $\overline{AC}$.

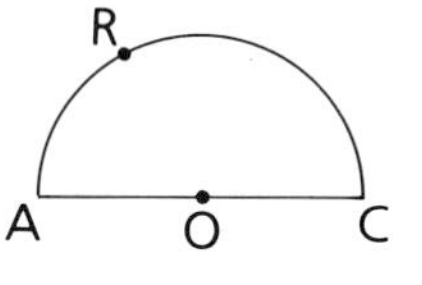

32 How far up the wall does the small ball touch the wall if the balls have radii of 4 cm and 9 cm?

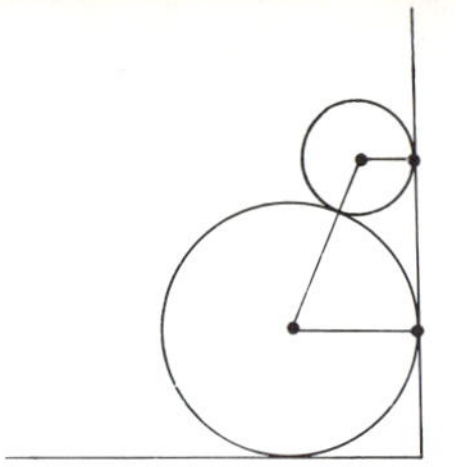

33 Given: Diagram as shown
Prove: ∠1 > ∠4

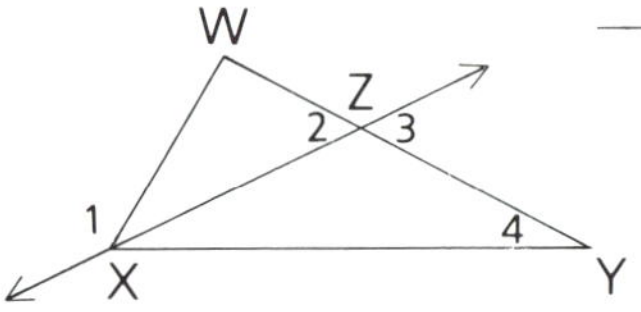

34 Given: $\overline{AD} \cong \overline{DC}$
∠ADB < ∠BDC
Prove: ∠A > ∠C

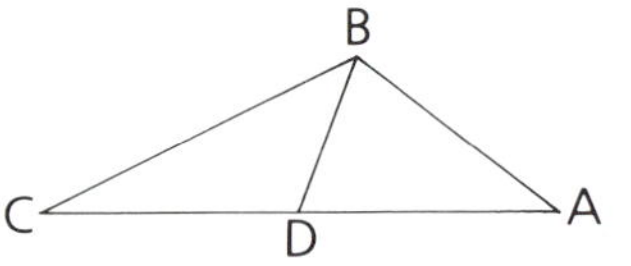

35 Find the locus of points a fixed distance from a given point and equidistant from the sides of an angle.

Chapters 1–16 Cumulative Review Problems C

36 Find the locus of points that are centers of congruent circles that are tangent to a given line if the centers lie on a given angle.

37 Each pair of ⊙s O, P, Q, and R are tangent. If the radius of ⊙ R is 11, find the difference between the areas of the shaded regions above and below $\overleftrightarrow{XY}$.

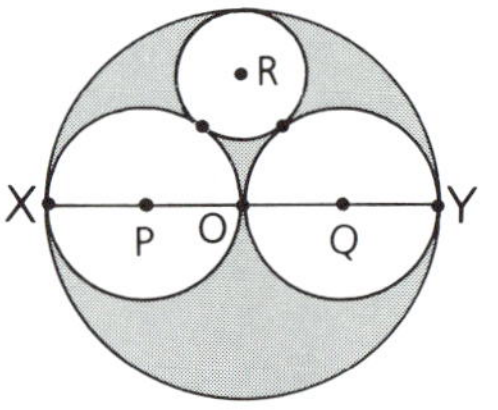

38 Prove: If two tangents are drawn to a circle from an external point, the triangle formed by these two tangents and any tangent to the minor arc included by them has a perimeter equal to the sum of the two original tangent segments.

39 In square ABCD, $\overline{HF} \perp \overline{AC}$. If the perimeter of the square is 32, and FC = BC, find the perimeter of quadrilateral HFEB.

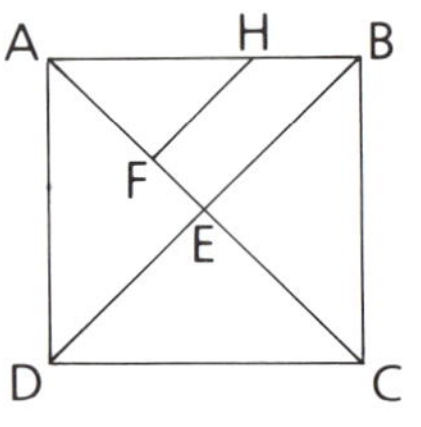

40 The diagonals of a parallelogram are 8 and 10 and intersect at a 60° angle. Find the area of the parallelogram.

41 Sixty-four one by one by one cubes are stacked together to form a four by four by four cube. The large cube is painted, and *then* broken up into the original sixty-four cubes. If two of the small cubes are selected at random, what is the probability that:

a exactly ten of the twelve faces will be unpainted?

b at least ten of the twelve faces will be unpainted?

42 **a** Describe the locus of points in space generated by obtuse $\triangle ABC$ if it is rotated about the altitude from A to $\overleftrightarrow{BC}$.

b Find the volume of the locus.

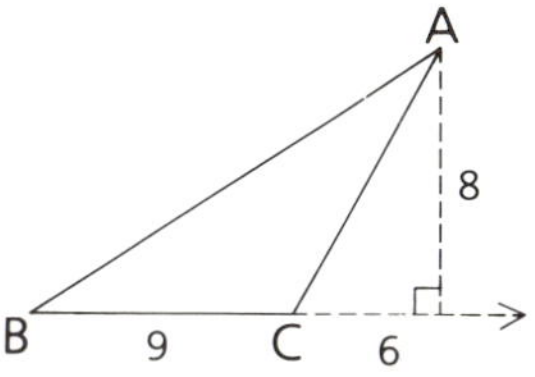

The Royal Road: All Triangles Are Isosceles!

Start with any $\triangle ABC$ (figure 1). Find point P, the intersection of the bisector of $\angle BAC$ and the $\perp$ bisector of $\overline{BC}$. From point P, drop $\perp$s $\overline{PR}$ and $\overline{PS}$. Draw $\overline{PB}$ and $\overline{PC}$.

First, show $\triangle TBP \cong \triangle TCP$ by SAS, so $\overline{BP} \cong \overline{CP}$. Then show $\triangle ARP \cong \triangle ASP$ by AAS, so $\overline{AR} \cong \overline{AS}$ and $\overline{RP} \cong \overline{SP}$. $\triangle RBP \cong \triangle SCP$ by HL, so $\overline{RB} \cong \overline{SC}$. Thus $AR + RB = AS + SC$ so the triangle is isosceles! But of course it isn't!

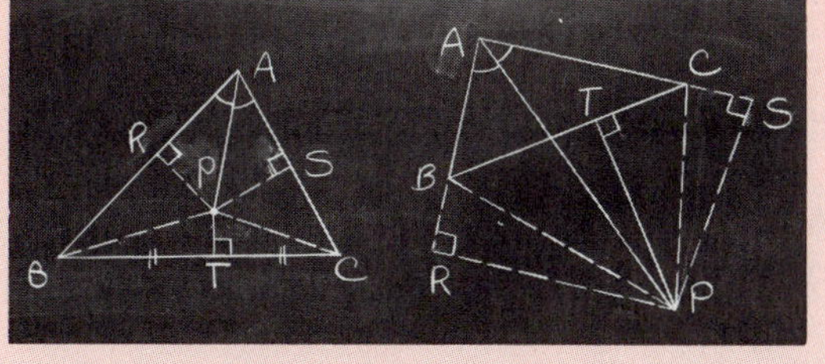

Is the error in the diagram? If $\overline{AP}$ and $\overline{TP}$ meet "outside" the triangle (Figure 2), repeat all the steps in the proof, and end with $AR - RB = AS - SC$. So where is the error?

The error really is in the diagram. Accurately drawn, exactly one of the segments $\overline{PR}$ or $\overline{PS}$ will be inside the triangle and the other will be outside. However, to prove this special result, and problems that depend on special diagrams, explicit definitions of *inside, outside,* and *between* must be added to the course.

Glossary

Abscissa The x-coordinate or first coordinate of an ordered pair of numbers. **580**

Acute angle An angle greater than 0° and less than 90.° **11**

Acute triangle A triangle in which all angles are acute. **137**

Adjacent angles Two angles (not necessarily congruent) formed by a ray dividing an angle. **177**

Alternate exterior angles Two angles formed by a transversal intersecting two lines. The angles must lie in the exterior region, must lie on opposite sides of the transversal, and must have different vertices. **201**

Alternate interior angles Two angles formed by a transversal intersecting two lines. The angles must lie in the interior region, must lie on opposite sides of the transversal, and must have different vertices. **201**

∠1 and ∠2 are alternate interior angles

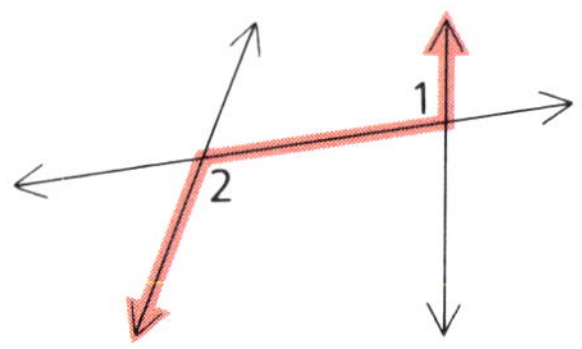

Altitude of a triangle A line segment drawn from any vertex of the triangle perpendicular to the opposite side, extended if necessary. **125**

Angle Two rays with a common endpoint. The endpoint is called the **vertex** of the angle. The rays are called the **sides** of the angle. **4**

Angle of depression If an observer at point A sights downward at an object at B, ∠A is the angle of depression. **392**

∠A is an angle of depression
∠B is an angle or elevation

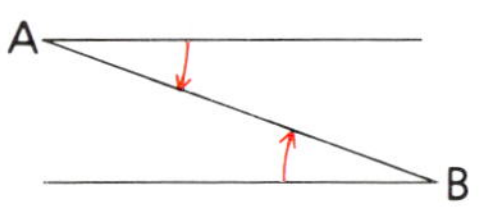

Angle of elevation If an observer at point B sights upward at an object at A, ∠B is the angle of elevation. **392**

Annulus The figure bounded by two concentric circles. **511**

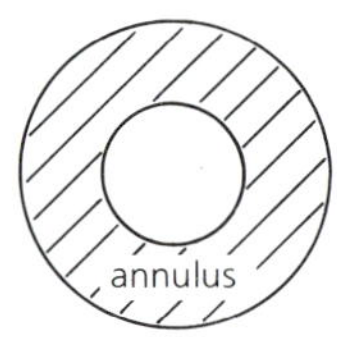

Apothem of a regular polygon A segment joining the center of the polygon to the midpoint of any side. **502**

Arc of a circle Two points on a circle and all the points on the circle between them. **420**

Area The number of square units contained within the boundary of a closed region. **482**

Arithmetic mean For any two numbers a and b, the arithmetic mean is $\frac{1}{2}(a+b)$. **358**

Auxiliary line A line introduced in a proof that does not appear in the original figure. **126**

Base of an isosceles triangle The side that is not one of the two congruent sides. **136**

Base of a trapezoid Either of the two parallel sides. **222**

Base angle The angle formed by the base and an adjacent side. **136**

Bisector of an angle A ray that divides an angle into two congruent angles. **29**

Bisector of a segment A point, segment, ray, or line that divides a segment into two congruent segments. **28**

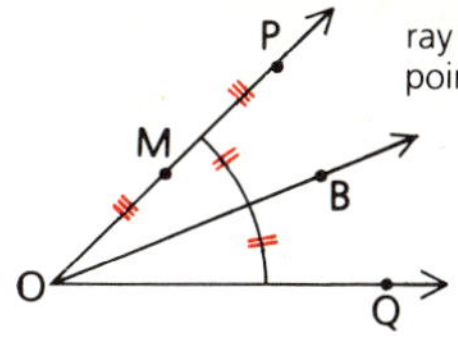

ray $\overrightarrow{OB}$ bisects $\angle POQ$
point M bisects segment

Center of an arc The center of the circle of which the arc is a part. **420**

point O is the center of arc $\widehat{RT}$
point M is the midpoint of arc $\widehat{RT}$

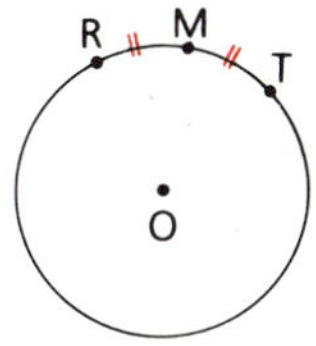

Center of a circle See **Circle**. **409**

Center of gravity of a triangle The centroid of a triangle. **645**

Central angle of a circle An angle whose vertex is at the center of a circle. **420**

Centroid of a triangle The point of concurrency of the medians of a triangle. **602**

Chord of a circle A segment joining any two points of the circle. **410**

Chord-chord angle An angle formed by two chords that intersect inside a circle, but not at the center. **440**

Circle The set of all points in a plane that are a given distance from a given point in the plane. That point is called the **center**. **409**

Circumcenter of a polygon The center of a circumscribed circle **457**

Circumference The perimeter of a circle. **469**

Circumscribed circle A circle is circumscribed about a polygon when each vertex of the polygon is a point of the circle. **457**

Circumscribed polygon A polygon each of whose sides is tangent to a circle. **457**

Collinear points Points that lie on the same line. **18**

Common tangent One line tangent to two circles (not necessarily at the same point). A tangent is a common **internal** tangent if it lies between the circles. Otherwise it is a common **external** tangent **431**

line $\overleftrightarrow{AB}$ is a common internal tangent
line $\overleftrightarrow{CD}$ is a common external tangent

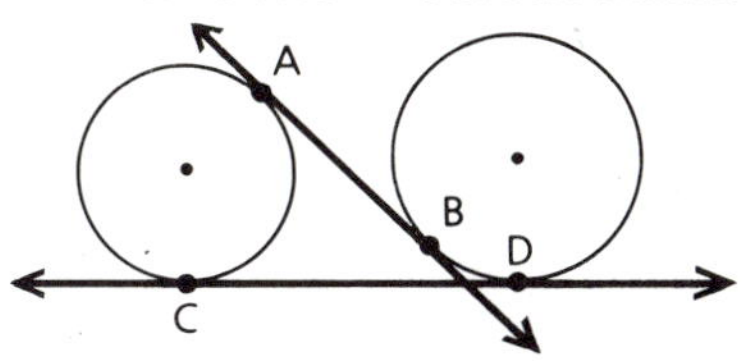

Complementary angles Two angles whose sum is 90° (or a right angle). **63**

Compound locus Two or more loci combined into one locus problem. **638**

Concentric circles Two coplanar circles with the same center. **410**

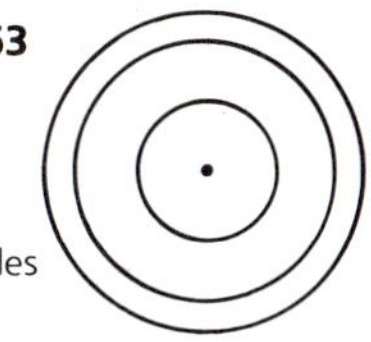

concentric circles

Concurrent lines Lines that intersect in a single point. **642**

lines k, l, m, and n are concurrent at point R

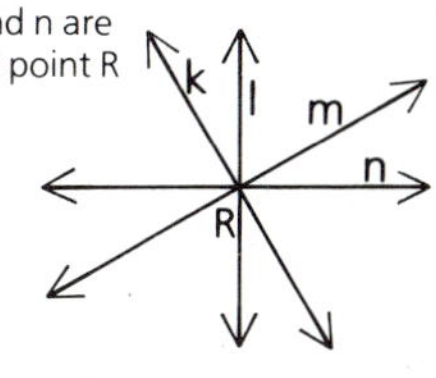

Concyclic points Points that all lie on the same circle. **463**

Conditional statement A statement that is in "If . . . , then . . . ." form. **173**

Congruent angles Angles that have the same measure. **12**

Congruent arcs Arcs that have the same measure and are in the same or congruent circles. **422**

Congruent segments Segments that have the same length. **13**

Congruent triangles Triangles that have all pairs of corresponding angles and corresponding sides congruent. **109**

Construction A drawing done with a compass and a straightedge, based on a classical Greek method. **649**

Contrapositive of a conditional statement The contrapositive of the statement "If p, then q" is the statement "If not-q, then not-p." **694**

Convex polygon Each interior angle is less than 180°. **221**

Coplanar Points, lines, segments, etc. in the same plane. **200**

Corresponding angles (formed by a transversal intersecting two lines). Two angles, one in the exterior region and the other in the interior region, on the same side of the transversal, and having different vertices. **201**

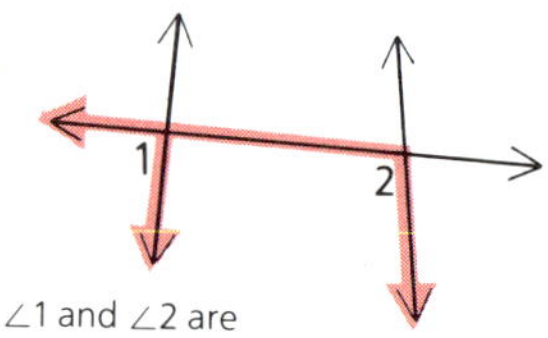

∠1 and ∠2 are corresponding angles

Diagonal of a polygon Any segment that connects two non-consecutive (non-adjacent) vertices of the polygon. **221**

Diagonal of a rectangular box A segment whose endpoints are vertices not in the same face of the box. **382**

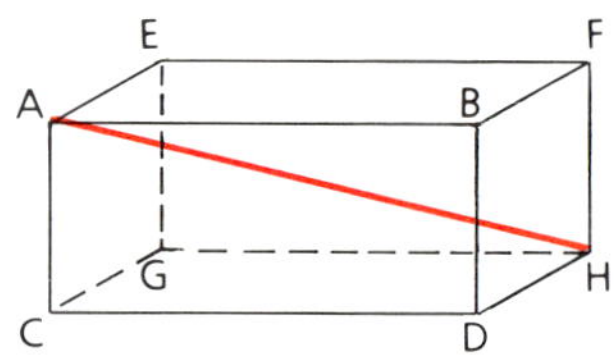

segment $\overline{AH}$ is a diagonal of the rectangular box

Diameter of a circle A chord that passes through the center of the circle. **410**

Distance from a point to a line The length of the segment from the point perpendicular to the line. **176**

Equiangular All angles are congruent. **137**

Equilateral All sides are congruent. **134**

Exterior angle of a polygon An angle that is adjacent and supplementary to an interior angle of the polygon. **280**

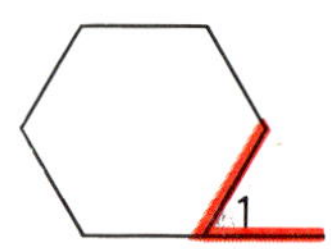

∠1 is an exterior angle

Extremes of a proportion The first and fourth terms of a proportion. In the proportion a:b = c:d (or $\frac{a}{b}=\frac{c}{d}$), a and d are the extremes. **310**

Face of a polyhedron BDHF is one of six rectangular faces of the box shown. The faces that intersect the base CDHG are called **Lateral faces. 382**

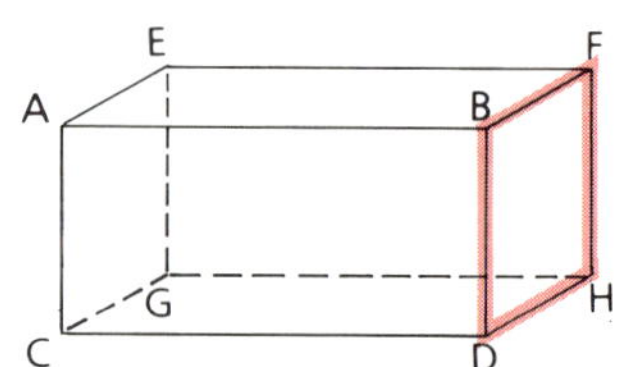

BDHF is a face of the polygon

Foot of a line The point of intersection of a line and a plane. **250**

Fourth proportional The fourth term of a proportion. In the proportion a:b=c:x (or $\frac{a}{b}=\frac{c}{x}$), x is the fourth proportional. **312**

Frustum That portion of a cone or pyramid between the base and a plane parallel to the base. **535**

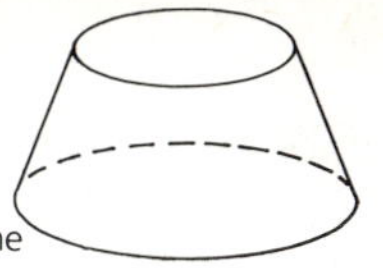

frustum of a cone

Geometric mean If the two means in a proportion are equal, either is called the geometric mean. Also called the **mean proportional.** **310**

Hypotenuse The side opposite the right angle in a right triangle. **137**

Hypothesis The clause following the word "If" in a conditional statement. **694**

Incenter of a polygon The center of an inscribed circle. **457**

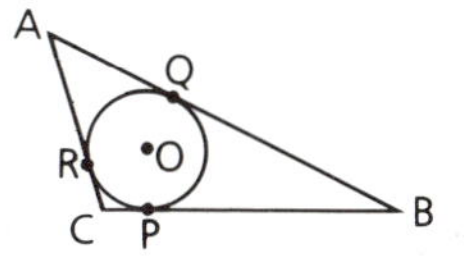

point O is the incenter of $\triangle ABC$

Inscribed angle An angle whose vertex is on a circle and whose sides are determined by two chords. **439**

Inscribed polygon A polygon is inscribed in a circle if all its vertices lie on the circle. **192**

Isosceles trapezoid A trapezoid in which the two non-parallel sides (legs) are congruent. **222**

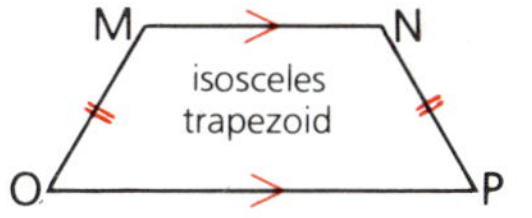

Isosceles triangle A triangle in which at least two sides are congruent. **136**

Kite A quadrilateral with two distinct pairs of congruent consecutive sides. **222**

Lateral surface area of a prism The sum of the areas of the lateral faces. **521**

Lattice point A point whose coordinates are integers. **624**

Legs of an isosceles right triangle The sides that form the right angle. **137**

Legs of an isosceles trapezoid The non-parallel, congruent sides of the trapezoid. **222**

Locus A set of points that contains all the points, and only those points, that satisfy specific conditions. **631**

Major arc An arc whose points are on or outside a central angle. **420**

Mean proportional see **Geometric mean.** **310**

Median of a trapezoid The line segment joining the midpoints of the non-parallel sides of a trapezoid. **494**

$\overline{MN}$ is the median in trapezoid ABCD

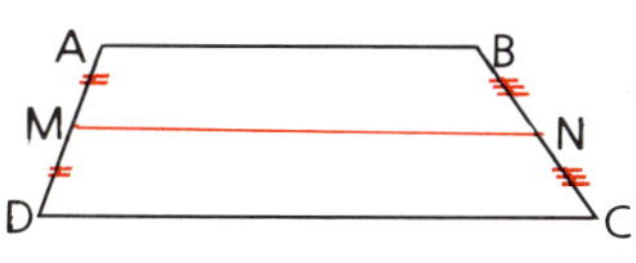

$\overline{GP}$ is a median in $\triangle EFG$

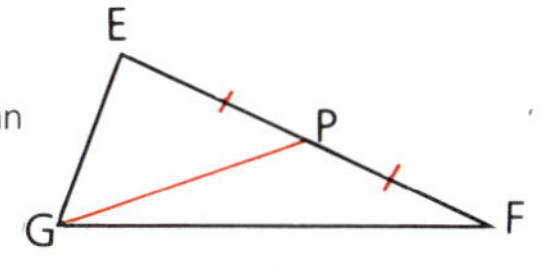

Median of a triangle A line segment drawn from any vertex of the triangle to the midpoint of the opposite side. **125**

Midpoint of a segment A point on the segment that divides it into two congruent segments. **28**

Minor arc An arc whose points are on or between the sides of a central angle. **420**

n-gon A polygon of n sides. **291**

Non-convex polygon A polygon that has at least one interior angle whose measure is greater than 180. **297**

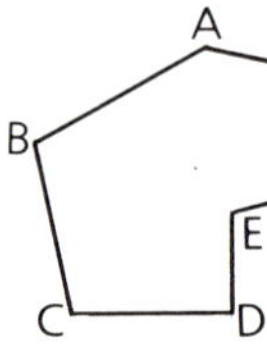

ABCDEFG is a non-convex polygon

Oblique lines Two intersecting lines that are not perpendicular. **62**

Obtuse angle An angle greater than 90° but less than 180°. **11**

Opposite rays Two collinear rays that have a common endpoint and go in different directions. **97**

rays $\overrightarrow{XY}$ and $\overrightarrow{XZ}$ are opposite rays

Ordinate The y-coordinate or the second coordinate of an ordered pair of numbers. **580**

Orthocenter The point of concurrency of the altitudes of a triangle. **644**

Parallel lines Coplanar lines that do not intersect. **202**

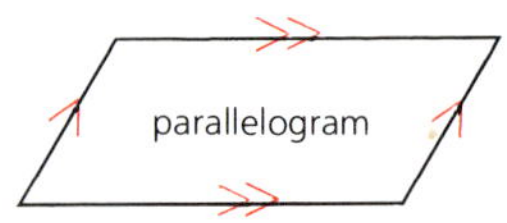

Parallelogram A quadrilateral in which both pairs of opposite sides are parallel. **210**

Perimeter of a polygon The sum of the lengths of the sides of the polygon. **8**

Perpendicular Lines, rays, or segments that intersect at right angles. **58**

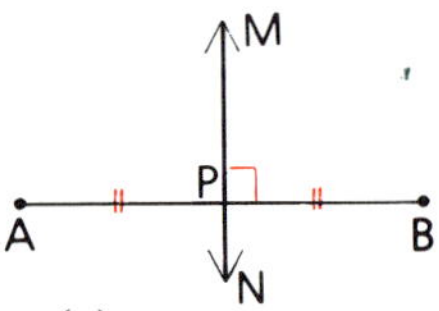

line $\overleftrightarrow{MN}$ is the perpendicular bisector of segment $\overline{AB}$

Perpendicular bisector A line that bisects a segment and is perpendicular to it. **181**

Plane A surface such that if any two points on the surface are connected by a line, all points of the line will lie on the surface. **199**

Postulate An assumption; a statement accepted as a starting point. **39**

Prism A solid figure that has two congruent parallel faces and a set of parallel edges that connect corresponding vertices of the two faces. **521**

prism

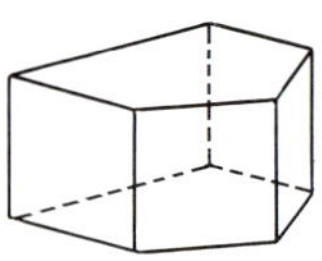

Proportion An equation in which two or more ratios are equal. **310**

Protractor An instrument, marked in degrees, used to measure angles. **9**

pyramid

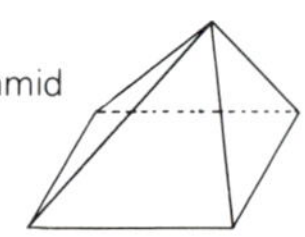

Pyramid A solid figure that has one base and a set of non-parallel edges that meet at a single point not in the base. **382**

quadrants

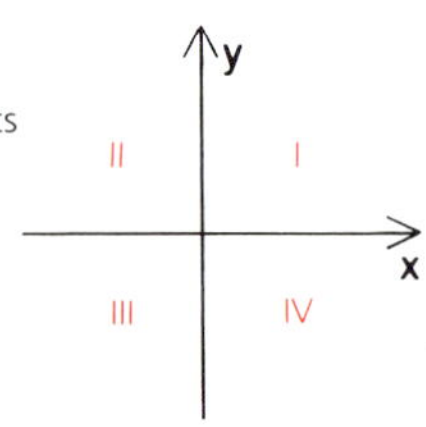

Quadrant Any one of the four regions into which a plane is divided by a pair of coordinate axes. **579**

Quadrilateral A four-sided polygon. **221**

Radius of a circle A segment joining the center of the circle to a point of the circle. **119**

Ray Begins at a point called its **endpoint** and extends infinitely in one direction. A ray is considered to be straight. **4**

Rectangle A parallelogram that has a right angle. **222**

Rectangular solid A prism with six rectangular faces. **382**

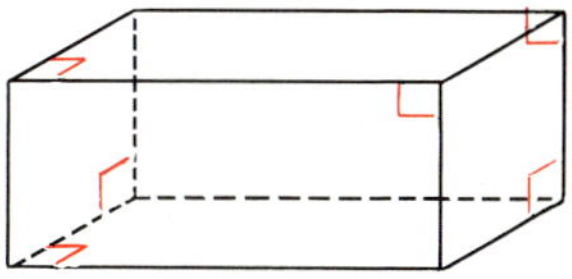

rectangular solid

Regular polygon A polygon that is both equilateral and equiangular. **298**

Rhombus A parallelogram with two consecutive congruent sides. **222**

Right angle An angle of exactly 90°. **11**

Right triangle A triangle in which one of the angles is a right angle. **137**

Scalene triangle A triangle in which no two sides are congruent. **136**

Secant A line that intersects a circle at exactly two points. **429**

Secant-secant angle An angle whose vertex is outside a circle and whose sides are determined by two secants. **441**

Secant-tangent angle An angle whose vertex is outside a circle and whose sides are determined by a secant and a tangent. **441**

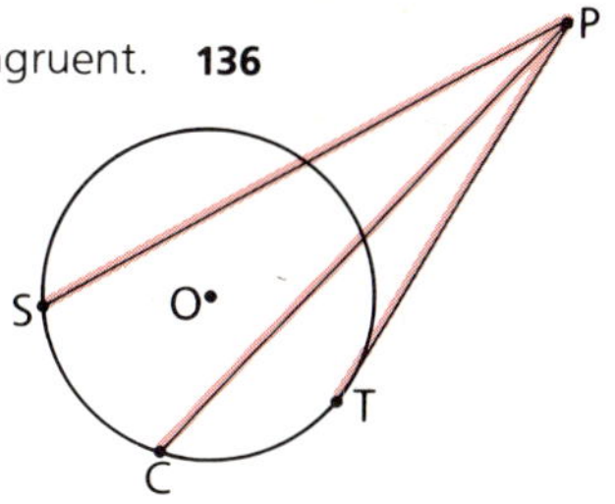

∠SPC is a secant-secant angle
∠SPT is a secant-tangent angle

Sector of a circle A region bounded by two radii and an arc of the circle. **508**

Segment of a circle A region bounded by a chord of the circle and its arc. **508.**

ACBO is a sector
ACBG is a segment

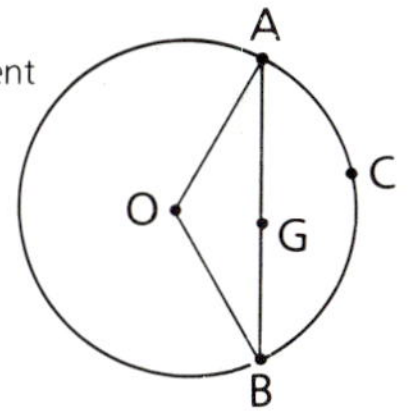

Semicircle An arc of a circle whose endpoints are the endpoints of a diameter. **420**

Similar polygons Polygons in which corresponding angles are congruent and ratios of the measures of corresponding sides are equal. **316**

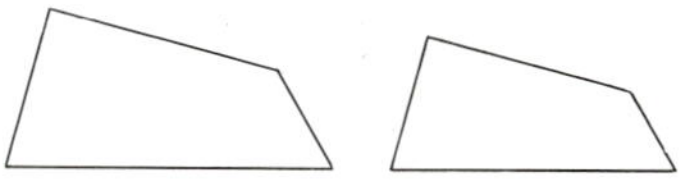

similar quadrilaterals

Skew lines Two lines that are not coplanar. **263**

Square A parallelogram that is both a rhombus and a rectangle. **222**

Straight angle An angle of exactly 180°. **11**

Supplementary angles Two angles whose sum is 180° (or a straight angle). **63**

Tangent A line that intersects a circle at exactly one point. **429**

Tangent-chord angle An angle whose vertex is on a circle and whose sides are determined by a tangent and a chord that intersect at that vertex. **439**

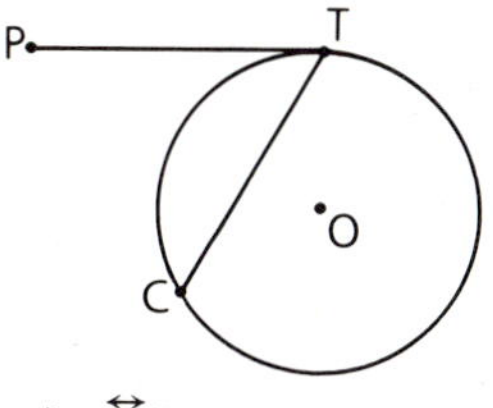

line $\overleftrightarrow{PT}$ is tangent to circle O
∠PTC is a tangent-chord angle

Tangent circles Two circles that intersect each other at exactly one point. **430**

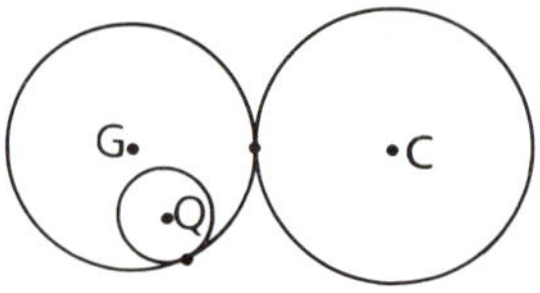

circle G and circle C are tangent
circle G and circle Q are tangent

Tangent segment The part of a tangent line between the point of contact and a point outside the circle. **429**

Tangent-tangent angle An angle whose vertex is outside a circle and whose sides are determined by two tangents. **441**

Tetrahedron A pyramid with triangles for all four faces. **530**

tetrahedron

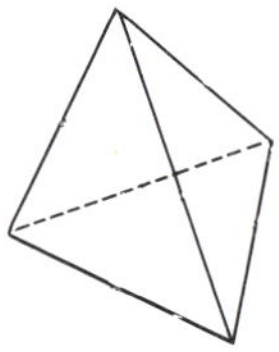

Theorem A mathematical statement that can be proved. **40**

Transversal A line that intersects two other coplanar lines in two distinct points. **200**

line t is a transversal

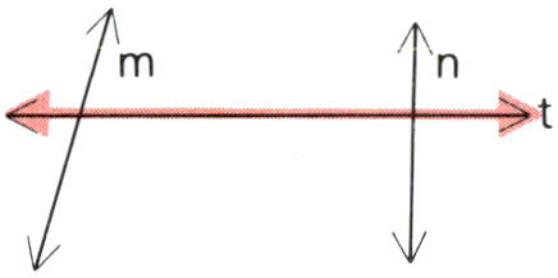

Trapezoid A quadrilateral with exactly one pair of parallel sides. **222**

Triangle A three-sided polygon. The sides are determined by joining three noncollinear points. **5**

Trisectors of an angle Two rays that divide an angle into three congruent angles. **30**

Rays $\overrightarrow{OX}$ and $\overrightarrow{OY}$ trisect $\angle AOB$

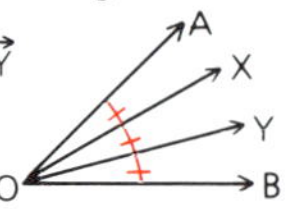

points P and Q trisect segment $\overline{CD}$

Trisectors of a segment Two points, segments, rays, or lines that divide a segment into three congruent segments. **29**

Vertex of an angle The common endpoint of the two rays that form the angle. **4**

Vertex angle of an isosceles triangle The angle opposite the base. **136**

Vertical angles Two angles in which the sides of one are opposite rays to the sides of the other. **97**

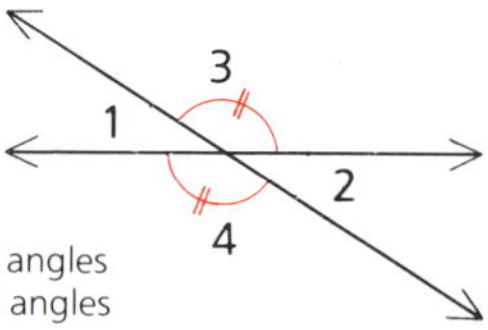

$\angle 1$ and $\angle 2$ are vertical angles
$\angle 3$ and $\angle 4$ are vertical angles

Volume of a solid The number of cubic units contained within the faces of a solid figure. **547**

23	If a point is on the perpendicular bisector of a segment, then it is equidistant from the endpoints of that segment.	182
24	If a point is equidistant from the endpoints of a segment, then it *lies on* the perpendicular bisector of that segment. (LIES ON THEOREM)	187
25	If two parallel lines are cut by a transversal, each pair of alternate exterior angles are congruent. (PAE)	205
26	If two parallel lines are cut by a transversal, each pair of corresponding angles are congruent. (PCA)	205
27	If two parallel lines are cut by a transversal, each pair of interior angles on the same side of the transversal are supplementary.	206
28	If two parallel lines are cut by a transversal, each pair of exterior angles on the same side of the transversal are supplementary.	206
29	In a plane, if a line is perpendicular to one of two parallel lines, it is perpendicular to the other.	206
30	If two lines are cut by a transversal such that a pair of alternate interior angles are congruent, the lines are parallel. (AIP)	212
31	If two lines are cut by a transversal such that a pair of alternate exterior angles are congruent, the lines are parallel. (AEP)	212
32	If two lines are cut by a transversal such that a pair of corresponding angles are congruent, the lines are parallel. (CAP)	212
33	If two lines are cut by a transversal such that a pair of interior angles on the same side of the transversal are supplementary, the lines are parallel.	213
34	If two lines are cut by a transversal such that a pair of exterior angles on the same side of the transversal are supplementary, the lines are parallel.	213
35	In a plane if two lines are perpendicular to a third line, then they are parallel.	213
36	If two lines are each parallel to a third line, they are parallel to each other. (TRANSITIVE PROPERTY OF PARALLEL LINES)	213
37	A line and a point not on the line determine a plane.	251
38	Two intersecting lines determine a plane.	252
39	Two parallel lines determine a plane.	252
40	If a line is perpendicular to TWO distinct lines that lie in a plane and that pass through its foot, then it is perpendicular to the plane.	258
41	If a plane intersects two parallel planes, the lines of intersection are parallel.	264
42	The sum of the measures of the three angles of a triangle is 180.	279
43	The measure of an exterior angle of a *triangle* is equal to the sum of the measures of the remote interior angles.	280
44	A segment joining the midpoints of two sides of a triangle is parallel to the third side, and its length is one-half the length of the third side. (MIDLINE THEOREM)	281

45 If two angles of one triangle are congruent to two angles of a second triangle, then the third angles are congruent. (We refer to this theorem as the NO CHOICE THEOREM, since the third angles have no choice except to be congruent.) **286**

46 If there exists a correspondence between the vertices of two triangles such that two angles and a non-included side are congruent respectively to the corresponding parts of a second triangle, the triangles are congruent. (AAS) **286**

47 The sum of the measures of the angles of a polygon with n sides is given by the formula, $S_i = (n-2)180$ **292**

48 Taking one exterior angle at each vertex, the sum of the measures of the exterior angles of a polygon is given by the formula, $S_e = 360$. **293**

49 The number d of diagonals that can be drawn in a polygon of n sides is given by the formula, $d = \frac{n(n-3)}{2}$. **293**

50 The measure of each exterior angle, E, of an equiangular polygon of n sides is given by the formula, $E = \frac{360}{n}$. **299**

51 In a proportion, the product of the means is equal to the product of the extremes. (MEANS-EXTREMES PRODUCTS THEOREM) **310**

52 If the product of a pair of non-zero numbers is equal to the product of a pair of non-zero numbers, then either pair of numbers may be made the extremes, and the other pair, the means. (MEANS-EXTREMES RATIO THEOREM) **311**

53 The ratio of the perimeters of two similar figures equals the ratio of any pair of corresponding sides. **317**

54 If there exists a correspondence between the vertices of two triangles such that two angles of one triangle are congruent to the corresponding angles of the other, then the triangles are similar. (AA) **322**

55 If there exists a correspondence between the vertices of two triangles such that the ratios of the measures of corresponding sides are equal, then the triangles are similar. (SSS ~) **322**

56 If there exists a correspondence between the vertices of two triangles such that the ratios of the measures of two pairs of corresponding sides are equal and the included angles are congruent, then the triangles are similar. (SAS ~) **323**

57 If a line is parallel to one side of a triangle and intersects the other two sides, it divides them proportionally. **333**

58 If three or more parallel lines intersect two transversals, they divide the transversals proportionally. **333**

59 If a ray bisects an angle of a triangle, it divides the opposite side into segments that are proportional to the other two sides. **334**

60 If the altitude is drawn to the hypotenuse of a right triangle, then: **353**

a the two triangles formed are similar to the given right triangle and to each other.

b the altitude to the hypotenuse is the mean proportional between the segments of the hypotenuse.

c either leg of the given right triangle is the mean proportional between the hypotenuse of the given right triangle and the segment of the hypotenuse adjacent to that leg (i.e., the projection of that leg on the hypotenuse).

61	The square of the measure of the hypotenuse of a right triangle is equal to the sum of the square of the measures of the legs. (THE PYTHAGOREAN THEOREM)	**359**
62	If the square of the measure of one side of a triangle equals the sum of the squares of the measures of the other two sides, then the angle opposite the longest side is a right angle.	**360**
63	In a triangle whose angles have the measures 30, 60, and 90, the lengths of the sides opposite these angles are respectively proportional to x, $x\sqrt{3}$, and 2x. (30°–60°–90° TRIANGLE THEOREM)	**374**
64	In a triangle whose angles have the respective measures of 45, 45, and 90, the lengths of the sides opposite these angles are respectively proportional to x, x, and $x\sqrt{2}$. (45°–45°–90° TRIANGLE THEOREM)	**374**
65	If a radius is perpendicular to a chord, then it bisects the chord.	**411**
66	If a radius of a circle bisects a chord *that is not a diameter,* then it is perpendicular to that chord.	**411**
67	The perpendicular bisector of a chord passes through the center of the circle.	**411**
68	If two chords of a circle are equidistant from the center, then they are congruent.	**416**
69	If two chords of a circle are congruent, then they are equidistant from the center of the circle.	**416**
70	If two central angles of a circle (or of congruent circles) are congruent, then their intercepted arcs are congruent.	**422**
71	If two arcs of a circle (or of congruent circles) are congruent, then their central angles are congruent.	**422**
72	If two central angles of a circle (or of congruent circles) are congruent, then their chords are congruent.	**423**
73	If two chords of a circle (or of congruent circles) are congruent, then their central angles are congruent.	**423**
74	If two arcs of a circle (or of congruent circles) are congruent, then their corresponding chords are congruent.	**423**
75	If two chords of a circle (or of congruent circles) are congruent, then their corresponding arcs are congruent.	**423**
76	If two tangent segments are drawn to a circle from an external point, then those segments are congruent. (TWO TANGENT THEOREM)	**430**
77	The measure of an angle whose VERTEX is ON a circle (an INSCRIBED ANGLE or a TANGENT-CHORD ANGLE) is *half* of the measure of its intercepted arc.	**439**
78	The measure of a CHORD-CHORD ANGLE equals *half* of the *sum* of the measures of the arcs cut off by the chord-chord angle and its vertical angle.	**440**

79	The measure of an angle whose *vertex* is OUTSIDE a circle (SECANT-SECANT ANGLE, SECANT-TANGENT, or TANGENT-TANGENT ANGLE) is *half* of the DIFFERENCE of the two intercepted arcs.	**441**
80	If two inscribed or tangent-chord angles intercept the *same* arc, then they are congruent.	**450**
81	If two inscribed or tangent-chord angles intercept CONGRUENT ARCS, then they are congruent.	**450**
82	An angle inscribed in a semicircle is a right angle.	**451**
83	The sum of the measures of a TANGENT-TANGENT angle and its minor arc is 180.	**451**
84	If a quadrilateral is inscribed in a circle, the opposite angles are supplementary.	**458**
85	If a parallelogram is inscribed in a circle, it must be a rectangle.	**459**
86	If two chords of a circle intersect inside the circle, then the product of the measures of the segments of one chord is equal to the product of the measures of the segments of the other chord. (CHORD-CHORD POWER THEOREM)	**464**
87	If a tangent and secant segment are drawn from an external point to a circle, then the square of the measure of the tangent segment is equal to the product of the measures of the entire secant segment and its external part. (TANGENT-SECANT POWER THEOREM)	**464**
88	If two secant segments are drawn from a point outside a circle, then the product of the measures of one secant segment and its external part is equal to the product of the measures of the other secant segment and its external part. (SECANT-SECANT POWER THEOREM)	**465**
89	The LENGTH OF AN ARC is equal to the *circumference* times the *fractional part* of the circle determined by the arc.	**470**
90	The AREA OF A SQUARE is equal to the square of a side.	**482**
91	The AREA OF A PARALLELOGRAM is equal to the product of the *base* times the *height.*	**487**
92	The AREA OF A TRIANGLE equals one-half the product of *base times* the *height* (or altitude) for that base.	**488**
93	The AREA OF A TRAPEZOID equals one half of the product of the *height* times the *sum of the bases.*	**494**
94	The measure of the MEDIAN OF A TRAPEZOID equals the *average* of the *bases.*	**495**
95	The AREA OF A KITE equals half the product of its diagonals.	**499**
96	The AREA OF AN EQUILATERAL TRIANGLE is the product of one-fourth of the *square of a side* times the square root of three.	**502**
97	The AREA OF A REGULAR POLYGON equals one-half the product of the *apothem* times the *perimeter.*	**503**
98	The AREA OF A SECTOR of a circle is the area of the circle times the fractional part of the circle determined by the arc.	**508**

99 If two figures are similar, then the *ratio of their areas* equals the *square of the ratio* of corresponding segments. (SIMILAR FIGURES THEOREM) 515

100 A median of a triangle divides the triangle into two triangles with equal areas. 516

101 The LATERAL AREA OF A CYLINDER is the product of the height by the circumference of the base. 531

102 The LATERAL AREA OF A RIGHT CONE is one-half the product of the slant height by the circumference of the base. 531

103 The VOLUME OF A RIGHT RECTANGULAR PRISM equals the *height* times the *area* of the base. 548

104 The VOLUME OF ANY PRISM equals the *height* times the *area* of the base. 549

105 The VOLUME OF ANY CIRCULAR CYLINDER equals the product of its *base area* by its *height*. 549

106 The VOLUME OF A PRISM OR CYLINDER equals the product of the *cross-section area* by the *height* of the solid. 549

107 The VOLUME OF A PYRAMID is one-third of the product of the *height* times the *area* of the base. 555

108 The VOLUME OF A CIRCULAR CONE is one-third of the product of the *height* times the *area* of its base. 555

109 In a PYRAMID OR CONE, the *ratio* of the area of a cross section to the area of the base equals the *square of the ratio of the distances* from the vertex, respectively. 556

110 The VOLUME OF A SPHERE is four-thirds of π times the *cube* of the radius. 561

111 If $A = (x_1, y_1)$ and $B = (x_2, y_2)$, then the midpoint $M(x_m, y_m)$ of $\overline{AB}$ can be found by using the averaging process: 584

$$M = (x_m, y_m) = \left(\frac{x_1 + x_2}{2}, \frac{y_1 + y_2}{2}\right)$$

112 If two non-vertical lines are parallel, then their slopes are equal. 589

113 If the slopes of two non-vertical lines are equal, then the lines are parallel. 589

114 If two lines are perpendicular, their slopes are NEGATIVE RECIPROCALS (with the exception of a horizontal and a vertical line). 590

115 If the slopes of two lines are NEGATIVE RECIPROCALS, the lines are perpendicular. 590

116 Y-FORM, or SLOPE-INTERCEPT FORM, of the equation of a nonvertical line is 594

$y = mx + b$ where b is the y-intercept of the line
m is the slope of the line

117 The formula for the equation of a HORIZONTAL LINE is 595

$y = b$ where b is the y-coordinate of every point on the line.

118 The formula for the equation of a VERTICAL LINE is 596

$x = a$ where a is the x-coordinate of every point on the line.

119 (The DISTANCE FORMULA) If $P = (x_1,y_1)$ and $Q = (x_2,y_2)$ are any two points, then the distance between them is **603**

$$PQ = \sqrt{(x_2-x_1)^2 + (y_2-y_1)^2}$$

or $PQ = \sqrt{(\triangle x)^2 + (\triangle y)^2}$

120 The EQUATION OF A CIRCLE whose center is (h, k) and whose radius is R is **615**

$$(x-h)^2 + (y-k)^2 = R^2$$

121 The PERPENDICULAR BISECTORS OF THE SIDES OF A TRIANGLE ARE CONCURRENT at a point that is equidistant from the vertices of the triangle. (The point of concurrency of the perpendicular bisectors is the CIRCUMCENTER of the triangle.) **642**

122 THE BISECTORS OF THE ANGLES OF A TRIANGLE ARE CONCURRENT at a point that is equidistant from the sides of the triangle. (The point of concurrency of the angle bisectors is the INCENTER of the triangle.) **643**

123 The lines containing the ALTITUDES OF ANY TRIANGLE ARE CONCURRENT. (The point of concurrency of the altitudes is called the ORTHOCENTER of the triangle.) **644**

124 The MEDIANS OF ANY TRIANGLE ARE CONCURRENT at a point that is two-thirds of the way from any vertex of the triangle to the midpoint of the opposite side. (The point of concurrency of the medians of a triangle is called the CENTROID of the triangle.) **644**

125 An exterior angle of a triangle is greater than either remote interior angle. **677**

126 If two sides of a triangle are not congruent, then the angles opposite them are congruent, and the larger angle is opposite the longer side. (If  then ) **678**

127 If two angles of a triangle are not congruent, then the sides opposite them are not congruent, and the longer side is opposite the larger angle. (If  then ) **678**

128 If two sides of one triangle are congruent to two sides of another triangle and the included angle of the first triangle is greater than the included angle of the second triangle, then the remaining side of the first triangle is greater than the remaining side of the second triangle. (SAS $\neq$) **683**

129 If two sides of one triangle are congruent to two sides of a second triangle and the third side of the first triangle is greater than the third side of the second triangle, then the angle opposite the third side in the first triangle is greater than the angle opposite the third side in the second triangle. (SSS $\neq$) **683**

130 If a conditional statement is true, then the contrapositive of the statement is also true. If p then q $\Longleftrightarrow$ If $\sim$q then $\sim$p. **695**

Answers to Short–Answer Problems

Pages 7–8 (Section 1.1)

1) $\overleftrightarrow{AB}$, $\overleftrightarrow{BA}$, ℓ **2)** $\angle CED$, $\angle DEC$, $\angle E$, $\angle 7$ **3)** No **4)** $\overline{RS}$, $\overline{ST}$, $\overline{RT}$ **5) a** B **b** $\overleftrightarrow{AC}$ or $\angle CEA$
5) c E **d** $\varnothing$ **e** $\overrightarrow{EC}$ **f** $\angle ABC$ **g** $\triangle BEC$ **6) a** $\angle RPO$, $\angle RPS$, $\angle SPR$ **b** 0 **c** 3 **d** $\angle TSO$, $\angle PST$, $\angle OST$ **e** 8
7) a Points **b** Rays, Endpoint **9) a** $37\frac{1}{2}$ **b** 20 **c** 28 **12) a** $37\frac{1}{2}$ **b** 25

Pages 15–17 (Section 1.2)

1) $61°40'$ **2)** $71°42'$ **3)** $84\frac{5}{6}°$ **4)** $\angle 1$ and $\angle 2$ **5)** $\angle XRS$ **6) a** $30°$ **b** $20°$ **c** $130°$
6) d Obtuse **7) a** $\angle 5$ **b** Same size **c** $\angle 4$ **8)** 55 **9) a** 90 **b** 45 **c** 100 **d** $142\frac{1}{2}$ **10) a** 8
10) b 2 **c** 10 **d** 4 **e** $\angle AEC$, $\angle BED$ **11)** 22 **12)** $r = 2$, $m = \frac{5}{3}$ **13)** $y = x + 17$ **14)** $67\frac{1}{2}$
15) Measurement is approximate. Either could be considered correct. **16)** $\approx$ 1 inch or $\approx$ 2.5 cm **17)** No
18) $x = -10$, $y = 40$ **19)** $10°$ **20)** $13°20'40''$ **21)** $21\frac{181}{360}°$

Pages 20–22 (Section 1.3)

1) 134 **3) a** B,D **b** No, Yes **c** G **d** $\overline{AB}$ and $\overline{BC}$ **e** Yes **f** Not necessarily
3) g $\angle C$, $\angle DCB$, $\angle DCA$, $\angle BCD$, $\angle ACD$ **h** B **i** G **j** $\overrightarrow{AF}$ **k** $\overrightarrow{EB}$, $\overrightarrow{ED}$ **l** E, B **4) a** No **b** Yes
5) 135 **8)** B **9)** $m\angle 1 = 80$, $m\angle 2 = 100$, $m\angle 3 = 80$ **10)** 11:10 A.M. **11)** 1:05′27″

Pages 32–35 (Section 1.5)

1) a $\overline{CO} \cong \overline{DO}$ **b** $\overline{WX} \cong \overline{WV}$ **2) a** $\angle NRO \cong \angle PRO$ **b** $\angle SXT \cong \angle TXV \cong \angle WXV$ **3) a** A **b** G
4) a $\overrightarrow{JG}$ **b** $\overrightarrow{OK}$ **5) a** $30°$ **b** $24°25'$ **c** $18\frac{1}{4}°$ **d** $14\frac{5}{8}°$ **6) a** 4 **b** 8 **c** 1 **d** 9 **e** $\overline{BD}$ **f** No
7) 43 **8)** Yes **18)** 80 **19)** 16 **20)** Yes **21)** 10 **22)** 60
23) $42\frac{6}{7}$ and $17\frac{1}{7}$ or $23\frac{1}{13}$ and $9\frac{3}{13}$

Pages 37–38 (Section 1.6)

5) Cannot be proved **6)** Can be proved false

Pages 41–43 (Section 1.7)

1) a Yes **b** No **c** No **2) a** Yes **b** No **3) a** T **b** D **12) b** If the two purrs "pil" and "til" are the same (not distinct), then more than one lilt could contain them. **13)** Wendy—purple monkey, Katie—red crocodile, Jody—green lizard

Pages 46–47 (Section 1.8)

1) $\frac{3}{5}$ **2)** $\frac{1}{5}$ **3)** $\frac{1}{5}$ **4)** 0 **5)** $\frac{3}{10}$ **6)** $\frac{2}{5}$ **7)** $\frac{1}{10}$ **8)** $\frac{9}{25}$ **9)** $\frac{1}{10}$ **10)** 18
11) a 1 **b** $\frac{1}{10}$ **c** 0

Pages 49–55 (Chapter 1, Review Problems)

1) a $\overleftrightarrow{AE}$, $\overleftrightarrow{AF}$, $\overleftrightarrow{AC}$, $\overleftrightarrow{EA}$, $\overleftrightarrow{EF}$, $\overleftrightarrow{EC}$, $\overleftrightarrow{FA}$, $\overleftrightarrow{FE}$, $\overleftrightarrow{FC}$, $\overleftrightarrow{CA}$, $\overleftrightarrow{CE}$, $\overleftrightarrow{CF}$ **b** $\overrightarrow{BA}$, $\overrightarrow{BC}$ **c** A and D **d** $\overrightarrow{DF}$ **e** $\overrightarrow{CB}$ **f** $60°$, $52°$, $120°$
1) g $\angle 4$ **h** No **i** We can't refer to $\angle B$ since it could be any of three angles. **j** E **k** $\overrightarrow{EF}$ **l** $\angle 1$ **m** A **n** $\overline{FE}$
2) a Right **b** Obtuse **c** Acute **d** Straight **e** Right **f** $\angle DEF$ is straight. **3)** $69°4'35''$ **4)** $50°59'43''$
5) a $\overline{BC} \cong \overline{RT}$ **b** $\angle A \cong \angle S$ **6) a** No **b** Yes **7)** 10 **8)** $m\angle 1 = 30$, $m\angle 2 = 90$, $m\angle 3 = 60$
9) No **17)** Can be proved false **18)** $13°24'$ **19) a** $30°$ **b** $140°$ **20)** 15 **21)** 30 **22)** 14
23) a $y = -\frac{3}{4}x + 22\frac{1}{2}$ **b** $x = 18$, $y = 9$ **26)** $48°50'44''$ **27)** 2 **30) a** $\frac{3}{10}$ **b** $\frac{2}{15}$ **c** $\frac{2}{15}$ **d** $\frac{4}{25}$ **e** No
31) $\approx$ 32 minutes, 44 seconds past 3:00 **32) a** 15 **b** 7 **33)** 69, 45, 66, 114 or 85, 45, 50, 130
34) $\approx$ 38 minutes, 11 seconds past 2:00

Pages 60–62 (Section 2.1)

1) **a** ∠s A, B, C, D **b** ∠s EHF, GHF, EFG **2)** **a** ∠MJK **b** ∠NOR, ∠POR **c** None **3)** **a** 20 **b** 45
3) **c** 21°42′26″ **4)** 150 **8)** ∠ROP **9)** 15, 30, 45, **10)** 5° **14)** 22 **15)** $152\frac{1}{2}$

Pages 65–68 (Section 2.2)

1) ∠A and ∠C **2)** 110° **3)** (90 − y)° **4)** 28°38′47″ **5)** 30 and 60 **10)** 60 **11)** 125
12) 110 and 70 **13)** 158 **15)** 27 **16)** Impossible **17)** 30 **18)** $163\frac{1}{3}$ **19)** 12
20) **a** $\frac{1}{5}$ **b** $\frac{1}{10}$ **21)** 70 **22)** **a** 0 **b** 6 **c** 3 **d** 12

Pages 76–78 (Section 2.4)

1) **a** 49 **b** 131 **c** 49 **d** 41 **e** 139 **f** 41 **g** 139 **4)** 144° **5)** 35 and 55 **12)** 40 **14)** 165
16) 98 **18)** $23\frac{2}{23}$ or 37 **19)** 3:2

Pages 83–85 (Section 2.5)

1) **a** $\overline{AD} \cong \overline{AC}$ **b** ∠JFG ≅ ∠JHG **c** $\overline{KO} \cong \overline{MP}$ **2)** **a** $\overline{RS} \cong \overline{TV}$ **b** ∠YXZ ≅ ∠CBD **c** $\overline{EK} \cong \overline{KF}$ **6)** 32
9) QS = 12, QT = 21 **14)** 57 **15)** **a** yes **b** ∠ABC is a straight ∠; $\overrightarrow{BF} \perp \overline{AC}$

Pages 88–91 (Section 2.6)

1) **a** 20 **b** 16 **13)** 165 **15)** Yes

Pages 94–96 (Section 2.7)

8) 68 **9)** 70 **14)** **a** 180 − x − y **b** 180 − y **c** 180 − x **15)** 50 **17)** Can be proved false.
18) $\{x \mid 60 < x < 90\}$

Pages 99–100 (Section 2.8)

1) **a** $\overrightarrow{FE}$ and $\overrightarrow{FC}$, $\overrightarrow{FD}$ and $\overrightarrow{FA}$, $\overrightarrow{BA}$ and $\overrightarrow{BC}$ **b** ∠EFA and ∠CFD, ∠EFD and ∠CFA **2)** 119°28′, 60°32′, 119°28′
3) 43 **7)** No **12)** 36, 90, 54 **13)** They are right angles. **15)** $132\frac{3}{4}°$ or 140°

Pages 102–105 (Chapter 2, Review Problems)

11) 6 cm **12)** 62 **13)** $22\frac{1}{2}$ **14)** oblique **15)** 4 **16)** **a** 97°, 7° **b** 137°44′22″, 47°44′22″
16) **c** 83°, None **23)** $x = 25\frac{5}{7}$, $y = 51\frac{3}{7}$ **24)** 50 **25)** 35 **26)** y = −2x + 10 **27)** $14

Pages 114–118 (Section 3.1)

1) **a** $\overline{GH} \cong \overline{KO}$, ∠J ≅ ∠M **b** $\overline{PS} \cong \overline{TR}$, ∠PVS ≅ ∠TVR **c** $\overline{BZ} \cong \overline{AX}$, ∠BWZ ≅ ∠AYX **d** $\overline{HJ} \cong \overline{EJ}$, ∠JFH ≅ ∠JCE
2) **a** SAS **b** None **c** None **d** ASA

Pages 121–124 (Section 3.2)

9) 20, 100 **10)** 67, 25, $13\frac{1}{2}$ **22)** 75, 30, 15

Pages 129–131 (Section 3.3)

1) **a** Median **b** Altitude **c** Altitude **d** Both **e** Neither **7)** 11, 11, 14 **14)** 48

Pages 139–141 (Section 3.5)

1) **a** Scalene **b** Isosceles **c** Equilateral **d** Scalene **e** Scalene **f** Isosceles **2)** **a** Right **b** Obtuse **c** Right
2) **d** Acute **e** Right **f** Acute **5)** Scalene **6)** 7, 63 **10)** Yes **11)** $\overline{VY}$ **13)** 60

Pages 146–149 (Section 3.6)

9) No **10**) 45 **19**) 77 cm **24**) 22, 8, 60°

Pages 156–158 (Chapter 3, Review Problems)

1) **a** S **b** A **c** N **d** N **e** N **7**) **a** 28, 60 **b** 114 **11**) Cannot be proved **12**) 30, 30, 120
13) 60

Pages 159–162 (Cumulative Review Problems, Chapters 1–3)

1) **a** F, G, H, D, or E, G, J, B **b** Many answers **c** Many answers **d** 92 **e** Obtuse **f** $\overline{AB}$ **g** Oblique
h C **i** $\overline{GJ}$ **j** ∠FAB **k** D **l** 0 **m** $\overline{FH}$ **n** △GEH **2**) scalene **3**) 36 **4**) 46°42′9″
5) 140° **6**) 94 **15**) 72, 60, 48 **16**) 41 **18**) 25 **19**) 35m
24) 4, 3 **25**) $\frac{2}{15}$

Page 180 (Section 4.3)

11) Yes **13**) 45°, 60° **14**) 7:21′49″

Page 186 (Section 4.4)

18) **a** $\frac{1}{5}$ **b** $\frac{1}{10}$ **19**) 112°

Page 191 (Section 4.5)

11) Cannot be proved

Page 203 (Section 5.1)

1) **a** ∠3 and ∠7, ∠4 and ∠8 **b** ∠1 and ∠5, ∠2 and ∠6 **c** ∠1 and ∠7, ∠2 and ∠4, ∠3 and ∠5, ∠8 and ∠6
1) **d** ∠7 and ∠8, ∠3 and ∠4 **e** ∠1 and ∠6, ∠2 and ∠5 **2**) **a** $\overleftrightarrow{AD}$ **b** ∠GBC and ∠FCB, ∠HBC and ∠ECB
2) **c** ∠ABG and ∠DCF, ∠ABH and ∠ECD **d** ∠ABG and ∠BCE, ∠GBC and ∠ECD, ∠ABH and ∠BCF, ∠CBH and ∠DCF
2) **e** ∠GBC and ∠ECB, ∠HBC and ∠FCB **f** ∠ABG and ∠DCE, ∠ABH and ∠DCF
3) **a** $\overleftrightarrow{JK}$ and $\overleftrightarrow{OM}$ **b** $\overleftrightarrow{JO}$ and $\overleftrightarrow{KM}$ **c** 3

Pages 208–210 (Section 5.2)

5) 41 **6**) 120 **7**) No **8**) 70, 70 **14**) ∠2 ≅ ∠5 **15**) Yes, 60 **18**) 32 **20**) **a** 28 **b** $\frac{2}{7}$
20) **c** $\frac{4}{7}$ (usually)

Pages 215–219 (Section 5.3)

11) 100, 80, 100 **12**) 64°26′, 25°34′, 25°34′ **13**) 150 **20**) 82 **21**) 70 **22**) 80
25) 65 or $95\frac{5}{9}$

Pages 223–225 (Section 5.4)

1) **a** A **b** S **c** S **d** A **e** A **f** N **g** N **2**) **a** $\overline{ST}$ and $\overline{RV}$ **b** $\overline{SV}$ and $\overline{RT}$ **c** $\overline{RS}$ and $\overline{VT}$
2) **d** ∠SRV and ∠TVR **e** ∠RST and ∠VTS **f** ∠STR ≅ ∠VRT, ∠TSV ≅ ∠RVS **3**) **a** Yes **b** No **c** Yes
3) **d** ABCDEFG **4**) **a** Square **b** Parallelogram **c** Trapezoid **d** Circle **e** Rectangle **f** Rhombus **g** Kite
4) **h** Equilateral triangle **i** Isosceles trapezoid **7**) 10 **8**) **a** x = 5, y = 7 **b** 34 **c** No
9) **a** It's a square. **13**) $\frac{10}{21}$

Pages 230–232 (Section 5.5)

5) 44 **6)** 134, 46 **7)** 8, 5, 5 **16)** 28 **17)** 16 **19)** **a** $a = 180 - y + x$ **b** $y - x < 90$

Pages 237–241 (Section 5.6)

7) **a** 9 **b** 17 **c** 56 **8)** Yes, Yes, Yes **9)** 10 **10)** **a** 50, 130, 50, 130 **b** 16
c 120, 60, 120, 60, Yes **19)** 145 **23)** **a** 2/7 **b** No, you will win 24/49 and lose 25/49 of the time.
24) $\overline{AB}$, $\overline{CD}$, parallel, $P = 37$

Pages 244–247 (Chapter 5, Review Problems)

2) 36 **3)** 124 **4)** 3 **5)** 110, 70, 110 **6)** Consecutive, 60, 120, 60, 120 **7)** **a** Yes **b** No
7) **c** Yes **14)** **a** 6 **b** 19 **15)** 106 **16)** **a** S **b** A **c** S **d** S **e** N **21)** 70 **23)** $\frac{3}{5}$ **26)** $\frac{1}{10}$

Pages 254–256 (Section 6.1)

1) **a** $\overleftrightarrow{AB}$ **b** $\overleftrightarrow{AB}$ **c** A, D, and B or C, E, and T **d** A, D, E, and either C or T, for example **e** s **f** s **g** E **h** C
1) **i** Yes **j** $\angle CDB$ and $\angle CDA$ **2)** No, No **3)** No, Yes **11)** Yes, No **12)** No **13)** Not necessarily
15) 7; ABC, ABP, BCP, CDP, DAP, ACP, BDP

Pages 259–262 (Section 6.2)

1) 12 **7)** No **15)** **a** $\overline{BC} \perp$ plane ADE **b** 5

Pages 265–268. (Section 6.3)

1) **a** F **b** T **c** F **d** T **e** T **3)** **a** Yes **b** Yes **c** Yes **d** Not necessarily **e** Parallelogram

Pages 270–272 (Chapter 6, Review Problems)

1) **a** F **b** T **c** F **d** F **e** F **f** F **g** F **h** T **i** F **j** F **k** F **4)** 4 **5)** Not necessarily
6) **a** Infinite **b** 1 **7)** **c** Isosceles △ **8)** d **9)** a

Pages 273–275 (Cumulative Review Problems, Chapters 1–6)

1) **a** Rhombus **b** Parallelogram **c** Right triangle **d** Square **2)** 105 **3)** 131 **6)** 105 **8)** 110
10) **a** S **b** N **c** A **d** A **e** A **f** S **g** S **11)** **a** 140 **b** 58 **14)** 110 **17)** $64\frac{1}{2}$

Pages 283–285 (Section 7.1)

1) 70 **2)** 50, 60, 120, 70, 110 **3)** 40, 140 **4)** 115 **5)** 48, 60, 72 **6)** 90 **7)** 48 **8)** 9
9) 20, 90, 70 **10)** 40, 50 **11)** **a** A **b** A **c** N **d** A **e** N **14)** 110 **15)** 117.5 or 92.5
16) **a** 40 **b** Rhombus **17)** 105 **20)** $25\frac{5}{7}$ **21)** **a** 55 **b** $m\angle H = 90 - \frac{1}{2}m\angle E$

Pages 288–290 (Section 7.2)

2) 40, 50, 90 **8)** 84 **9)** 50 **14)** 72.5 **16)** **a** Rectangle **b** Rectangle **c** Square **d** Rhombus
16) **e** Parallelogram **f** Parallelogram **g** Rhombus **19)** $a = 2b - c$

Pages 294–297 (Section 7.3)

1) **a** 360 **b** 900 **c** 1080 **d** 1800 **e** 16, 380 **2)** 40 **3)** 84 **4)** **a** 70 **b** 45 **c** 65 **5)** **a** 5
5) **b** 9 **c** 2 **d** 0 **6)** **a** 360 **b** 360 **c** 360 **d** 360 **7)** 3 **8)** **a** 1800 **b** 360 **10)** **a** 7 **b** 10
10) **c** 18 **d** $x - 2$ **e** Impossible **f** 5 **11)** **a** Quadrilateral **b** Hexagon **12)** 5 **13)** **a** Heptagon
13) **b** Decagon **c** 22-gon **14)** **a** A **b** N **c** S **d** A **18)** 40, 50, 130
19) {all polygons with 362 or more sides}

Pages 300–302 (Section 7.4)

1) **a** 120 **b** 90 **c** 45 **d** 24 **e** $15\frac{15}{23}$ **2)** **a** 108 **b** 120 **c** 140 **d** 150 **e** $162\frac{6}{7}$ **3)** **a** 6 **b** 9
3) **c** 10 **d** 180 **e** 48 **4)** **a** 10 **b** 6 **c** 15 **d** 20 **e** 50 **6)** Isosceles **7)** Equiangular decagon
10) 168 **11)** Pentagon **12)** Equiangular nonagon **13)** **a** A **b** S **c** A **d** S **e** S **f** N
16) $\{36, 60, 77\frac{1}{7}, 90, 100, 108\}$ **17)** $x = 30, y = 12$

Pages 304–307 (Chapter 7, Review Problems)

6) 120 **7)** 45 **8)** 50, 50 **9)** 50 **10)** 6, 80, 40 **11)** 20 **12)** 24 **13)** **a** 5580 **b** 360
14) 11 **15)** 90 **16)** Equilateral triangle **18)** **a** 32.5 **b** 122.5 **c** 25 **19)** 72 or 75 **20)** 115
21) **a** A **b** S **c** S **d** N **22)** It is non-convex. **23)** No **24)** Equiangular duodecagon **28)** $\frac{13}{18}$

Pages 313–315 (Section 8.1)

1) 9 **2)** 10 and 14, 7 and 20 **3)** Yes **4)** **a** 4 **b** $\frac{54}{7}$ **c** $\frac{47}{3}$ **5)** **a** 6 **b** 24 **c** $\frac{5b}{a}$
6) **a** $\frac{3}{2}$ **b** $\frac{3}{1}$ **c** $\frac{1}{3}$ **7)** $\frac{5}{36}$ **8)** $\frac{1}{4}$ **9)** Rectangle **10)** **a** 1:1 **b** 1:2 **11)** **a** ± 10 **b** $\pm\sqrt{15}$
11) **c** $\pm\sqrt{ab}$ **12)** 28 m **13)** 8, 12, 20, 28 **14)** $6\sqrt{2}$ **15)** $\frac{8}{3}$ **16)** 150 **17)** Yes
18) $\frac{c+d}{a+b}$ **19)** $\frac{f+h}{e-g}$ **20)** $\frac{x-3}{x+4}$ **21)** 12 **25)** 75 **26)** $\frac{1}{3}$

Pages 319–321 (Section 8.2)

1) **a** No **b** Yes **c** Yes **d** Yes **2)** b and c **3)** 90, 30 **4)** 5 **5)** 6, 9 **6)** **a** ± 10
6) **b** $\pm\sqrt{10}$ **7)** $\frac{5}{3}$ **8)** 70, 57, 53 **9)** $\frac{28}{5}$ **10)** **a** $\frac{3}{2}$ **b** 8 **c** 32 **d** $\frac{3}{2}$ **11)** 205 **12)** $\frac{4}{5}$
13) $\frac{11}{34}$ **14)** 33

Pages 325–326 (Section 8.3)

7) Cannot be proved **13)** **a** S **b** N **c** A **d** S **16)** $\frac{4}{15}$

Pages 330–332 (Section 8.4)

8) 4, 14 **9)** 8, 4, $5\frac{1}{3}$ **10)** 25 m **11)** $\frac{3}{4}$ **17)** $6\frac{6}{7}$ **18)** $4\frac{4}{9}$ **19)** 12 **20)** 12 **22)** 6
23) 17

Pages 336–341 (Section 8.5)

1) **a** 8 **b** $\frac{35}{3}$ **2)** 6, $\frac{72}{5}$ **3)** 3, 9 **4)** $\frac{10}{3}$, 5, $\frac{20}{3}$ **5)** 1 **6)** 4 **7)** 9 **8)** 24 m **9)** $\frac{5}{2}$, 10
10) $\frac{45}{13}, \frac{12}{5}$ **11)** 6, 5, 16 **12)** $\frac{28}{5}$ **13)** $\frac{27}{11}, \frac{72}{11}$ **17)** $\frac{32x}{5}, \frac{25x}{4}$ **18)** **a** 24 cm **b** $4x + 8$ **19)** 40
20) $\frac{20}{3}$ **24)** 120 m **25)** $42\frac{1}{4}$ **27)** $8\frac{3}{4}$ m

Pages 343–347 (Chapter 8, Review Problems)

1) b, c; a, d **2)** 12 **3)** ± 10 **4)** $\pm 3\sqrt{2}$ **5)** $\frac{4}{9}$ **6)** 9, $\frac{15}{2}$ **7)** $\frac{15}{2}$ **8)** 32 **9)** $\frac{2br}{3a}$
10) 20 m **11)** 4, 15 **12)** $\frac{90}{13}, \frac{40}{13}, \frac{72}{13}$ **13)** $\frac{20}{3}$ **14)** 14 **15)** $\frac{27}{2}$ **16)** $\frac{6}{5}$, 9 **20)** **a** A **b** A **c** N
20) **d** S **e** N **f** A **g** A **h** S **24)** 52 **25)** $\frac{43}{5}$ **26)** 30 **27)** $\frac{104}{5}$ **28)** 100 m, 125 m, 175 m
29) 225 **30)** $\frac{5}{2}, \frac{15}{4}$ **32)** **a** $\frac{1}{6}$ **b** $\frac{1}{3}$ **c** 1 **33)** $\frac{4}{5}$ **34)** $\frac{6}{1}$

Pages 351–352 (Section 9.1)

1) **a** 2 **b** $3\sqrt{3}$ **c** $6\sqrt{2}$ **d** $4\sqrt{2}$ **e** $7\sqrt{2}$ **f** $10\sqrt{2}$ **g** $2\sqrt{5}$ **h** $2\sqrt{6}$ **2)** **a** $15\sqrt{2}$ **b** $\sqrt{13}$ **c** 5
2) **d** 13 **e** $\frac{2\sqrt{3}}{3}$ **f** $7\sqrt{3}$ **3)** **a** $\frac{\sqrt{2}}{2}$ **b** $\frac{\sqrt{5}}{5}$ **c** $2\sqrt{2}$ **d** $2\sqrt{3}$
4) **a** $11\sqrt{3}$ **b** $8\sqrt{2} + 7\sqrt{3}$ **c** $5\sqrt{3}$ **d** $6\sqrt{2} + \sqrt{3}$
5) **a** ± 5 **b** ± 12 **c** ± 13 **d** $\pm\frac{1}{2}$ **e** $\pm 2\sqrt{3}$ **f** $\pm 3\sqrt{2}$ **6)** **a** ± 3 **b** ± 8 **c** ± 5 **d** ± 3 **e** ± 4
6) **f** $\pm 4\sqrt{5}$ **7)** **a** $\{6, -1\}$ **b** $\{2, -6\}$ **c** $\{5, 3\}$ **d** $\{6, -3\}$ **e** $\{12, -3\}$ **f** $\{9, -4\}$ **8)** **a** $\{0, 4\}$
8) **b** $\{0, 10\}$ **c** $\{0, 13\}$ **d** $\{0, 8\}$ **9)** **a** $\{5, -\frac{7}{2}\}$ **b** $\{-\frac{5}{3}, \frac{3}{4}\}$ **c** $\{\frac{1}{6}, 2\}$ **10)** $\{\frac{5}{2}, 5\}$ **11)** **a** $-h$
11) **b** $3 - x$ **c** pq **d** $-xy\sqrt{x}$

Pages 355–358 (Section 9.2)

1) **a** $\sqrt{21}$ **b** $\sqrt{77}$ **c** 4 **2)** **a** 6 **b** 8 **c** $4\sqrt{3}$ **d** 4 **e** 9 **f** Impossible **3)** **a** $2\sqrt{15}$ **b** 5
3) **c** 6 **d** $\sqrt{55}$ **4)** **a** $4\sqrt{5}$ **b** 3 **c** $3\sqrt{5}+8$ **5)** **a** 9 **b** 54 **c** $15+6\sqrt{3}$ **8)** 12 meters
9) **a** 5 **b** 12 **11)** **a** $2\sqrt{7}$ **b** $16\frac{2}{3}$ **c** $4\sqrt{6}$ **d** $7\frac{11}{12}$ **12)** $\frac{1}{3}$ **13)** 1.8 **15)** $2\sqrt{5}$
18) **a** 5, 4; $7\frac{1}{2}$, 6; $14\frac{1}{2}$, 10

Pages 362–366 (Section 9.3)

1) **a** $\sqrt{41}$ **b** 8 **c** 12 **d** 5 **e** 10 **f** 2 **g** $3\sqrt{2}$ **2)** $3\sqrt{2}$ cm **3)** 40 km **4)** 46 **5)** 9
6) $2\sqrt{3}$ **7)** 10 **8)** 14 dm **9)** **a** $\sqrt{x^2+y^2}$ **b** $\sqrt{4+x^2}$ **c** 5a **d** 12c **10)** **a** $2\sqrt{7}$ **b** 10
10) **c** 32 **d** 20 **11)** 10 km **12)** 60 **13)** $6\sqrt{2}$ **14)** 4.8 **15)** $\sqrt{5}$ **16)** $4\sqrt{3}$
17) $5\frac{19}{34}$ **18)** $(6+6\sqrt{2})$ cm **19)** $6\sqrt{5}$ **20)** **a** Obtuse **b** Impossible **22)** $18+6\sqrt{5}$
23) **a** 8 **b** No **24)** 10 **25)** 7 **26)** 10 **27)** 50 **29)** $\frac{6\sqrt{5}}{5}$ **30)** $12\sqrt{2}$ **31)** $\frac{799}{64}$
32) $\frac{5}{12}$ **33)** **b** Yes

Pages 370–373 (Section 9.4)

1) **a** 25 **b** 36 **c** 21 **d** $\frac{5}{3}$ **e** 60 **2)** **a** 10 **b** 78 **c** 36 **d** 65 **e** $6\frac{1}{2}$ **3)** **a** 250 **b** 48 **c** 28
3) **d** 2.4 **e** 264 **4)** **a** 51 **b** 3.4 **c** 75 **d** $\frac{4}{3}$ **e** 80 **5)** **a** 12 **b** 10 **c** 34 **d** 72 **e** $2\sqrt{7}$ **f** .5
5) **g** $5\sqrt{7}$ **h** 45 **i** $12\sqrt{7}$ **6)** 52 **7)** 50 dm **8)** 17 **9)** **a** 24 **b** $300\sqrt{5}$ **10)** 17
11) $\frac{9}{2}$, 10 **12)** 100 **13)** $3\sqrt{89}$ **14)** **a** 144 **b** $\frac{3}{8}$ **c** $\sqrt{7}$ **15)** **a** 7 **b** x = 20, y = 20 **16)** 40
17) $20+\sqrt{521}$ **18)** 29 km **21)** 140 cm **22)** $\frac{7}{3}$ **23)** **a** $\frac{2}{5}$ **b** $\frac{4}{15}$

Pages 376–381 (Section 9.5)

1) **a** 7, $7\sqrt{3}$ **b** 20, $10\sqrt{3}$ **c** 5, 10 **d** 346, $173\sqrt{3}$ **e** 114, $114\sqrt{3}$ **2)** **a** $2\sqrt{3}$, $4\sqrt{3}$ **b** $\frac{15}{2}$, $\frac{15\sqrt{3}}{2}$
2) **c** $5\sqrt{3}$, $10\sqrt{3}$ **d** $\frac{7\sqrt{3}}{3}$, $\frac{14\sqrt{3}}{3}$ **3)** **a** $2\sqrt{3}$ **b** $14\sqrt{3}$ **c** $13\sqrt{3}$ **4)** **a** $8\sqrt{2}$ **b** $4\sqrt{2}$ **c** $17\sqrt{2}$
4) **d** $19\sqrt{2}$ **5)** $11\sqrt{2}$ **6)** $3\sqrt{3}$ mm **7)** 16 m **8)** $4\sqrt{3}$ **9)** **a** $5\sqrt{2}$ **b** 13 **10)** **a** $\frac{\sqrt{3}}{2}$
10) **b** $\frac{\sqrt{2}}{2}$ **13)** 38 **14)** $28+12\sqrt{2}$ **15)** **a** $3\sqrt{3}$ **b** 9 **c** $6\sqrt{3}$ **d** 1 : 2 **16)** 6, $6\sqrt{3}$
17) $5\sqrt{3}$ cm **18)** $44+8\sqrt{3}$ **19)** $12\sqrt{3}$ dm **20)** **a** 48 **b** $6+6\sqrt{2}$ **21)** $3\sqrt{3}$
22) $x=5\sqrt{3}, y=8\sqrt{3}, z=3$ **23)** $x=3\sqrt{3}, y=\frac{3}{2}\sqrt{3}$ **24)** **a** $2+2\sqrt{3}$ **b** $2\sqrt{6}$
25) $\frac{40(12-5\sqrt{3})}{23}$ **27)** **a** $\frac{5}{14}$ **b** $\frac{25}{28}$ **c** $\frac{5}{14}$ **28)** **a** $\frac{1}{7}$ **b** $\frac{3}{28}$ **c** $\frac{1}{4}$ **d** $\frac{1}{2}$

Pages 383–385 (Section 9.6)

1) **a** 5 **b** 13 **2)** $5\sqrt{3}$, 10 **3)** **a** 17 **b** 17 **4)** $5\sqrt{2}$ **5)** **a** 14 **b** 7 **c** 25 **d** 56 **e** $14\sqrt{2}$
6) 13 **7)** **a** $5\sqrt{13}$ **b** $9\sqrt{11}$ **8)** **a** 12 **b** 8 **c** 6 **d** $2\sqrt{34}$ **9)** $2\sqrt{3}$ **10)** $5\sqrt{3}$ **11)** 7
12) 4 **13)** **a** $x\sqrt{3}$ **b** $\frac{x\sqrt{6}}{2}$ **14)** $d=\sqrt{a^2+b^2+c^2}$ **15)** 120, 160, 200 **16)** Impossible
17) $3\sqrt{2}$, $3\sqrt{3}$

Pages 388–390 (Section 9.7)

1) **a** $\frac{8}{17}$ **b** $\frac{15}{17}$ **c** $\frac{8}{15}$ **d** $\frac{15}{17}$ **e** $\frac{8}{17}$ **f** $\frac{15}{8}$ **2)** **a** $\frac{1}{2}$ **b** $\frac{\sqrt{3}}{2}$ **c** $\frac{\sqrt{3}}{3}$ **d** $\frac{\sqrt{3}}{2}$ **e** $\frac{1}{2}$ **f** $\sqrt{3}$
3) **a** $\frac{\sqrt{2}}{2}$ **b** $\frac{\sqrt{2}}{2}$ **c** 1 **4)** **a** $\frac{39}{89}$ **b** $\frac{39}{80}$ **5)** $\frac{4}{5}$ **6)** **a** $\angle A$ **b** $\angle A$ **c** $\angle A$ **7)** **a** $2\sqrt{6}$ **b** $\frac{2\sqrt{6}}{7}$
7) **c** $\frac{5\sqrt{6}}{12}$ **8)** **a** $\frac{12}{13}$ **b** $\frac{5}{13}$ **9)** **a** $\frac{7}{25}$ **b** $\frac{8}{17}$ **c** $\frac{4}{5}$ **10)** 129 m **11)** **a** 45 **b** 30 **12)** $\frac{4}{5}$
13) **a** $\frac{2}{3}$ **b** $\frac{3\sqrt{13}}{13}$ **14)** **a** $\frac{3}{5}$ **b** $\frac{3}{4}$ **15)** $\frac{4}{5}$ **16)** **a** A **b** N **c** S **19)** $\frac{1}{2}$ or $\frac{\sqrt{3}}{2}$ **20)** $\frac{1}{6}$
21) $\frac{1}{3}$ **22)** $\frac{3\sqrt{62}}{31}$

Pages 393–395 (Section 9.8)

1) **a** .3584 **b** 1.2799 **c** .9962 **d** 1.0000 **e** .8660 **2)** **a** 24 **b** 74 **c** 45 **3)** **a** 45 **b** 30 **c** 60
4) **a** 18 **b** 44 **c** 62 **d** 16 **5)** ≈ 15 **6)** **a** 17 **b** 120 **c** 75 **7)** 24°
8) **a** $\angle A = 38°$, AC = 16, BC = 12 **b** $\angle A = 52°$, AC = 20, AB = 33 **c** AB = 25, $\angle A = 56°$, $\angle B = 34°$
9) 37°, 53°, 90° **10)** 301 m **11)** **a** 67° **b** $5\sqrt{11}$ **12)** 5.6 and 11.0 **13)** 10 **14)** 6.88
15) 104 dm **16)** 668 **17)** **a** 36.4 **b** 37.75 **c** 74° **d** 68° **18)** 3.29 km **19)** **a** 35° **b** 60°
21) **a** 52° **b** 11 **d** $\frac{a}{\sin \angle A} = \frac{b}{\sin \angle B}$

Pages 397–401 (Chapter 9, Review Problems)

1) **a** 9 **b** 8 **c** 5 **d** $\sqrt{13}$ **2)** **a** 30° – 60° – 90° **b** 3-4-5 **c** 5-12-13 **d** 8-15-17
e 45° – 45° – 90° **3)** **a** 30 **b** 5, $5\sqrt{3}$ **c** 7 **d** 15 **e** $4\sqrt{5}$ **f** 9 **g** $5\sqrt{3}$, $10\sqrt{3}$ **h** $\frac{25}{2}$ **i** 26
j $4\sqrt{2}$, $4\sqrt{2}$ **4)** 40 **5** $3\sqrt{3}$ **6)** 5 km **7)** 7 feet **8)** $\sqrt{55}$ **9)** 48 **10)** $\sqrt{29}$
11) $\sqrt{85}$ **12)** $\frac{4\sqrt{3}}{3}$, $\frac{2\sqrt{3}}{3}$ **13)** 13 **14)** 120 **15)** $\sqrt{209}$ **16)** **a** 8 **b** 7.5 **c** 8.5
d 8.5 **e** 7.5 **17)** 75 km **18)** **a** 4.5 **b** 3 **19)** Swim directly across and walk. **20)** 12
21) 7.5 **22)** 51 Paces **23)** **a** 12 **b** $10\sqrt{13}$ **24)** $2\sqrt{29}$ **25)** 1:2 **26)** $\frac{1}{3}$

Pages 401–404 (Cumulative Review Problems, Chapters 1–9)

1) $67\frac{1}{2}$ **2)** 95 **3)** **a** 1260 **b** 30 **c** 14 **4)** **a** 20 **b** Yes **5)** 14.4 m **6)** 130° **7)** 52
8) **a** $\frac{81}{4}$ **b** $\frac{27}{2}$ **c** Acute **9)** **a** $\pm\frac{7}{2}$ **b** $\frac{20}{3}$ **10)** No **13)** **a** 55° **b** $5\frac{1}{2}°$ **14)** 10° **15)** 15
16) Impossible **18)** **a** 140, 160 **b** Yes **20)** $4\sqrt{3}$ **21)** **a** 15 **b** $4\sqrt{10}$ **c** 4 **23)** $3\sqrt{6}$
24) 25 **25)** 180, No **26)** $\frac{2}{7}$

Pages 413–415 (Section 10.1)

5) 8 mm **6)** 16 cm **11)** 8 m **12)** 15 cm **14)** 20 cm **22)** 25 cm **23)** 2 **24)** 16.9 cm
25) 24

Pages 417–419 (Section 10.2)

1) Same **2)** 8 **6)** **a** 8 cm **b** Circle **10)** **a** 8 **b** 5 **14)** 8, $24 + 6\sqrt{7}$

Pages 425–428 (Section 10.3)

1) f, b, e, d, c, g, a **2)** **a** $\overset{\frown}{QRP}$ or $\overset{\frown}{QPR}$ **b** $\overset{\frown}{BC}$ or $\overset{\frown}{AB}$ **c** 180 **d** $m\overset{\frown}{PQ}$ **e** No **3)** **a** 90 **b** 130 **c** 230
3) **d** 180 **e** 220 **4)** 130 **9)** **a** $\frac{1}{45}$ **b** $\frac{1}{5}$ **c** $\frac{1}{3}$ **d** $\frac{2}{5}$ **e** $\frac{2}{3}$ **f** $\frac{3}{4}$ **g** $\frac{5}{6}$ **h** $\frac{1}{360}$ **10)** **a** 216 **b** 200
10) **c** 252 **d** 9 **12)** 132 **13)** 12 **18)** **a** 60 **b** 72 **c** 45 **22)** 120
23) **a** $\frac{n(n-1)}{2}$ **b** $n(n-1)$ **c** $\frac{360}{n}$ **24)** $\frac{10}{3}$ **26)** **a** Neither chord is a diameter.

Pages 433–437 (Section 10.4)

1) 17 cm **2)** 12 **5)** **a** 2 Internal, 2 External **b** 1 Internal, 2 External **c** 0 **d** 2 External
5) **e** 1 External **9)** $4\sqrt{10}$ **10)** **a** 12 cm **b** Intersecting **11)** 6 **13)** 3, 5, 8 **14)** **a** 9 **b** 17
16) **a** $\frac{1}{2}$ **b** $\frac{1}{2}$ **18)** 48 **19)** 60 mm **20)** 10
21) $\triangle ABD$ is Equilateral and $\triangle CAB$ is a 30° – 60° – 90° triangle. **22)** $\frac{8\sqrt{10}}{3}$ **24)** **a** $\frac{c - a + b}{2}$
24) **b** Yes, since $\frac{c - a + b}{2}$ always exists.

Pages 443–449 (Section 10.5)

1) 62 **2)** 65 **3) a** 35 **b** 75 **4) a** 42 **b** 50 **c** 40 **5)** $w = 104°$, $x = 68°$, $y = 90°$, $z = 33°$

6) $a = 140$, $b = 80$, $c = 20$, $d = 81.5$, $e = 100$ **7)** 85°, 23° **8)** $57\frac{1}{2}°$ **9)** 10° **10)** 76°

11) 81° **12)** 125° **13)** 40° **14)** 110° **15)** 75 **16)** 20 **17)** 75 **18) a** 57.5°

18) b 57.5° **c** 72° **d** 108° **e** 54° **f** 90° **g** 93.5° **h** 86.5° **i** 245° **j** 137°

19) $x = 132°$, $\angle Y = 10°$, $\angle V = 20°$, $\angle W = 30°$ **20)** 42, 48, 132 **21)** 64, 120, 116 **22)** 98°

23) 104°, 74° **24)** 45° **25)** 25° **26)** 52° **27)** $175\frac{1}{2}°$ **28)** 120 **29) a** 70° **b** 110°

32) 90 **33)** 20° **34)** 132 **35)** 90 **36)** 100

Pages 453–456 (Section 10.6)

3) 10 mm **4) a** 17° **b** $81\frac{1}{2}°$ **5)** 97° **6) a** Yes **b** 6 **c** No **7)** 137° **8) a** Must **b** Could

8) c Must **d** Must **e** Could **f** Cannot **9)** $AB = 4\sqrt{2}$, $CD = 16$, $m\angle D = 60$ **10)** 1.25 **11) a** 6

11) b $18 + 6\sqrt{3}$ **12)** 15 **13)** $10\sqrt{2}$ **14)** $10\sqrt{3}$ **15)** 86° **16) a** 29° **b** 81° **23)** 4

27) 47°

Pages 460–463 (Section 10.7)

1) 113°, 76° **2)** 105, 75 **3) a** 70° **b** 110° **c** 85° **d** 80° **5)** No **6) a** 72 **b** 144 **c** 60

6) d 120 **e** 180 **7)** Square **8)** Isosceles trapezoid **9) a** 65 **b** 85 **c** 85 **d** 95 **10)** 24

11) 46 **14)** 18 **15)** 8 **16)** $12 + 12\sqrt{3}$ **17)** Yes, No, Yes **19) a** 20 **b** $\frac{1}{2}$

20) $6 + 6\sqrt{3}$ **23) a** A **b** S **c** S **d** N **e** S **f** S **24)** 48 **25)** 60 **26)** $\frac{3}{5}$

27) If the equiangular polygon contains an odd number of sides, it will be equilateral.

28) a Midpoint of hypotenuse **b** Interior of △ **c** Exterior of △

Pages 466–468 (Section 10.8)

1) $x = 15$, $y = 8$, $z = 25$ **2) a** 50 **b** 58 **3) a** 20 **b** 21 **c** 48 **4) a** 8 **b** 19 **c** 12 **5)** 3.5

6) 18, 4 **7)** 6 **8) a** 9 **b** Acute **9)** 1 or 6 **10)** 1 **11)** 5 **12)** 2.5 **13)** 12.5m

14) 900 miles **15)** 26, 39 **16) a** $12 + 6\sqrt{3}$ **b** $1:2:2 + \sqrt{3}$ **17) a** 4 **b** $12 - 4\sqrt{5} < y < 8$ **18) a** 26

18) b $\sqrt{313} - 13$ **19) a** Because the triangle cannot exist. (The sum of any two sides must be greater than the third.) **b** $\{AB : AB > 16\}$

Pages 471–473 (Section 10.9)

1) a 21π mm or ≈ 66 mm **b** 12π mm or ≈ 37.68 mm **2) a** 28 **b** ≈ 50 **c** 8.5 **d** ≈ 14 **3) a** 4π

3) b 5π **c** $\frac{10\pi}{3}$ **d** 10π **4) a** 94.2 cm **b** 9.42 m **c** 0.942 km **5) a** $40 + 6\pi$ **b** $24 + 4\pi$

5) c $12 + 3\pi$ **d** $6 + 7\pi$ **6) a** 3π **b** $12 + 3\pi$ **7)** $100 + 12\pi$ meters or ≈ 137.68 **8) a** 20π mm

8) b 72 **9) a** $4\pi\sqrt{3}$ **b** 6π **c** 9π **10)** ≈ 1256 cm **11)** ≈ 214.2 meters **12)** 13π

13) ≈ 96.52 meters, ≈ 71.4 meters **14)** $12 + 9\pi$ **15) a** $(24\sqrt{2} - 24)$ cm **b** 24cm **16)** $(24\sqrt{3} + 22\pi)$ cm

Pages 475–479 (Chapter 10, Review Problems)

1) a 94° **b** 94° **c** 43° **2)** 66°, 32° **3) a** 16 **b** 8 **c** 4 **4) a** 13 **b** 8.5 **c** 7

5) 125 **6)** 40°, 20° **7) a** $\frac{3\pi}{2}$ **b** $12 + 1.5\pi$ **8) a** 4 **b** C **12)** 12 **13)** $4\frac{2}{3}$ **14)** 8

15) a 5 **b** 7, 9, and 5 **16)** 45°, 105°, 135°, 75° **17) a** 140° **b** 30° **c** 10° **18)** 20 **19)** 110, 50

21) b 60 **c** 60 **d** $15\sqrt{3}$ **22)** 6, 18 **24) a** $6\sqrt{3}, 6\sqrt{2}, 6$ **b** $\frac{\sqrt{3}}{2}, \frac{\sqrt{2}}{2}, \frac{1}{2}$

25) a $a:b:c$ **b** $(a + b):(b + c):(c + d):(d + a)$ **28)** 8:5 **29)** $\frac{3}{5}$ **30) a** 4:1

30) b $x^2:y^2$ **c** Yes **32)** $\frac{3}{4}$

Pages 484–486 (Section 11.1)

1) 75 sq cm **2) a** 207 **b** 78 **3) a** 120 **b** 84 **4) a** 8 mm **b** 10 mm **5) a** 144 **b** 50
5) c 7 **d** 36 **e** 81 **6) a** 104 **b** 102 **c** 116 **7) a** 10 **b** 7 **8)** 256 sq cm
9) a $w = 5, x = 10, y = 25, z = 12.5$ **b** 50, 40, 58, 41 **c** The square **d** The rectangle must be a square.
10) 184 sq cm **11)** 9 m by 15 m **12)** $A = 32, P = 16\sqrt{2}$ **13)** 1 **14)** 1 by 72, 2 by 36,
14) 3 by 24, 4 by 18, 6 by 12, 8 by 9; $\frac{2}{5}$ **15)** $144\sqrt{3}$ **16)** $\frac{27}{65}$ **17)** 2

Pages 490–493 (Section 11.2)

1) a 198 sq mm **b** 102 sq cm **b** 35 **2)** 120 **3) a** 35 **b** 144 **4)** 6 **5)** 48 **6)** 187
7) 14 **8) a** 28 **b** 60 **9) a** $18\sqrt{3}$ **b** 72 **c** $36\sqrt{3}$ **10) a** $100\sqrt{3}$ **b** $60\sqrt{2}$
11) a 84 **b** 128 **12)** 12 **13)** Each has an area of 80. **14)** 85 **15)** 33 **16) a** 90 **b** 18
17) 12, 8 **18) a** 168 **b** 180 **c** 81 **19)** $\frac{225\sqrt{3}}{4}$ sq m **20) a** $42\sqrt{3}$ **b** $85\sqrt{2}$ **c** 120 **21) a** $39\sqrt{3}$
21) b 33 **c** $128 + 32\sqrt{3}$ **22)** 120 **23) a** 336 **b** $13\frac{11}{25}$ **24) a** 30 **b** 12 **25)** $A = 120, P = 52$
26) a 6 **b** $A = \frac{s^2\sqrt{3}}{4}$ **27)** $24 - 8\sqrt{3}$ **28)** 50 **29)** Square **30)** 120 **31)** 420
32) b Yes **c** No

Pages 496–498 (Section 11.3)

1) a 104 **b** 13 **2) a** 75 **b** 72 **c** 78 **d** 312 **3) a** 10.5 **b** 73.5 **4)** 9 **5)** 11 **6)** 22 feet
7) a 396 **b** 78 **8) a** $30\sqrt{3}$ **b** 115 **9)** 24 **10) a** 10 **b** 5 **c** 5 **d** 70 **11)** 89 **12)** 50
13) 72 **14)** 480 **15)** 16 **16)** 14, 70 **17) a** $42\sqrt{3}$ **b** $27\sqrt{3}$ **18)** $PQ = \frac{b_1 - b_2}{2}$

Pages 500–501 (Section 11.4)

1) a 60 **2) a** 56 **b** 85 **c** 160 **3)** 5 **4)** 168 **5) a** 336 **b** 500 **6)** 85 **7)** 78
8) $50\sqrt{3}$ **9)** $\frac{72}{5}$ **11)** Yes **12) a** $\frac{1}{2}$

Pages 504–507 (Section 11.5)

1) 36 **2) a** $9\sqrt{3}$ **b** $\frac{49\sqrt{3}}{4}$ **c** $16\sqrt{3}$ **d** $3\sqrt{3}$ **3) a** $108\sqrt{3}$ **b** $48\sqrt{3}$ **c** $27\sqrt{3}$ **d** $36\sqrt{3}$
4) a $54\sqrt{3}$ **b** $96\sqrt{3}$ **c** $72\sqrt{3}$ **d** $128\sqrt{3}$ **5) a** 12 **b** $6\sqrt{3}$ **c** $216\sqrt{3}$ **6) a** 100 **b** 576
6) c 49 **d** 50 **e** 72 **f** 9 **7)** 3 mm **8)** 6 km **9)** 324 **10)** $27\sqrt{3}$ **11)** $288\sqrt{3}$
12) a $\frac{81\sqrt{3}}{4}$ **b** 225 **c** $\frac{147\sqrt{3}}{2}$ **13) a** 11, 5.5 **b** 12 m, $2\sqrt{3}$ m **c** 4 cm, $2\sqrt{3}$ cm **14)** 32
15) a $108\sqrt{3}$ **b** 288 **c** $216\sqrt{3}$ **16)** $\frac{3}{4}$ **17) a** $36 - 9\sqrt{3}$ **b** $27\sqrt{3}$ **c** $36\sqrt{3}$ **18)** $\frac{2}{5}$
19) a $450\sqrt{3}$ **b** 8 **c** $A = \frac{s^2\sqrt{3}}{2}$ **20) a** $5 + 5\sqrt{2}$ **b** $200 + 200\sqrt{2}$
21) a $150 + 100\sqrt{2}$ **b** $50 + 100\sqrt{2}$ **22) a** $\sqrt{2} : 1$ **b** $4\sqrt{3} : 9$ **23)** $12\sqrt{3}$ **24)** $1764\sqrt{3} - 3024$

Pages 510–513 (Section 11.6)

1) a $\pi, 2\pi$ **b** $64\pi, 16\pi$ **c** $225\pi, 30\pi$ **2) a** 4 **b** 13 **3)** 20π cm **4)** 81π dm **5) a** π **b** 32π
5) c 2π **d** 27π **e** 75π **6) a** 18π sq m **b** 72 sq m **c** $(72 - 18\pi)$ sq m **7)** $150 + 25\pi$
8) a $2\sqrt{6}$ **b** $\frac{6}{\sqrt{\pi}}$ **9) a** 24π **b** 12π **c** 8π **10)** 144° **11) a** $16\pi - 32$ **b** $\frac{32\pi}{3} - 16\sqrt{3}$
11) c $12\pi - 9\sqrt{3}$ **12) a** 16π **b** $(R^2 - r^2)\pi$ **13) a** 18π **b** 50π
13) c $24\frac{1}{2}\pi$, shaded area is half the entire area. **14)** 12π **15) a** $100 - 25\pi$ **b** $25\sqrt{3} - \frac{25\pi}{3}$
15) c $100 - 25\pi$ **d** $25\sqrt{3} - \frac{25\pi}{2}$ **e** $50\pi - 100$ **f** $150\sqrt{3} - 75\pi$ **16) a** $\frac{375\pi}{4}$ **b** $\frac{125\pi}{2}$, $\frac{1}{25}$
17) 32 sq cm **18) a** $48\sqrt{3} - 16\pi$ **b** $18\sqrt{3} + 24\pi$ **c** $\frac{625\pi}{16} - 48$ **19) a** $50\pi - 50\sqrt{3}$ **b** 10π
20) a $60\pi + 36\sqrt{3}$ **b** 27π **c** $18\pi + 36\sqrt{3}$ **21) a** 120 **b** 120 **22)** $4\pi - 6\sqrt{3}$ **23)** $600 - 144\pi$

Pages 517–520 (Section 11.7)

1) a 1:1 b 1:2 c 15:8 d 5:6 **2)** a 25:4 b 4:9 c 4:1 d 1:9 **3)** a 1:1 b 1:2 c 2:1 **4)** 16:81 **5)** 3:4 **6)** No, she has a garden four times as big. **7)** 1:16 **8)** 49:16 **9)** a 64:225 b 1:2 c 1:1 d 4:9 **10)** a 4:81 b 2:9 c 1:3 d 4:9 **11)** 5:4 **12)** a 2:3 **12)** b 2:3 **13)** 250 **14)** 5:8 **15)** 16:81 **16)** 9:16 **17)** 4:9 **18)** a 1:1 b 1:1 c 1:1 **18)** d 9:16 e 3:4 **19)** a 4:1 b 8:1 c 1:1 d 1:2 e 30 **20)** 2:1 **21)** 5:6 **22)** a 5, 10, 20, 25 b $\frac{y^2}{2x^2 + 6xy + 4y^2}$

Pages 523–525 (Section 11.8 A)

1) a 550 sq cm b 282 sq cm c 810 **2)** a 150 b 126 **3)** a 550 b 120 c 790 **4)** a $90 + 18\sqrt{3}$ b $360 + 72\sqrt{3}$ **5)** a 150 b 294 **6)** a LA = 480, TA = 552 **6)** b LA = 120, TA = 132 c LA = 2500, TA = 2620 d LA = 360, TA = $360 + 108\sqrt{3}$ **7)** a 236 b $144 + 66\sqrt{3}$ **8)** 226 **9)** LA = 616, TA = 712 **10)** 170 **11)** a 0, 0, 0, 8, 12, 6, 1 b $\frac{20}{27}$ c 432

Pages 527–530 (Section 11.8 B)

1) a 60 b 240 c 340 **2)** a 120 b $64\sqrt{3}$ c $360 + 64\sqrt{3}$ **3)** a Because the base is not regular. **3)** b 936 c 1356 **4)** a No b 9 c 740 **5)** a 192 b 48 c 36 d 144 e 1:4 f 36 **6)** 105 **7)** a 72 b 432 c 324 d 756 **8)** a LA = 700, TA = 896 b LA = 544, TA = 800 **9)** a PR = 17, PS = 10, LA = 504, TA = 864 b 16 **10)** a $4\sqrt{2}$ b $\sqrt{5}$ c 16 d $8\sqrt{5}$ **11)** a $36\sqrt{3}$ b $2\sqrt{6}$ **12)** a $72\sqrt{3}$ sq mm b $6\sqrt{2}$ mm c $6\sqrt{2}$ mm d Square **13)** Cube

Pages 533–535 (Section 11.8 C)

1) a 196π b 36π c 36π d 25π **2)** a LA = 24π, TA = 33π b LA = 140π, TA = 238π **2)** c LA = 40π, TA = 72π d LA = 3π, TA = 4π **3)** 6 **4)** a 34π b 108π **5)** $35\pi + 20$ **6)** a 90π b 66π c 480π **7)** 24 cans **8)** 8π cm by 14 cm **9)** 93π **10)** $448 + 42\pi$ **11)** 140π **12)** a Frustum of a cone, 99π b Cylindrical shell, 60π c Same solid as problem 9, 424π.

Pages 537–538 (Section 11.9)

1) a 6 b $2\sqrt{5}$ c $10\sqrt{2}$ d $6\sqrt{3}$ e 60 f 84 **2)** $16\sqrt{3}$ **3)** a 36 b $4\sqrt{15}$ c 24 d $16\sqrt{3}$ **4)** a 0 (This is a collapsed triangle or line segment.) b $\sqrt{-385}$ (The two shorter sides have a sum that is less than the third side so no such triangle exists.) $\sqrt{-385}$ is not a real number. **5)** $4\sqrt{66}$ **6)** a $2\sqrt{14}$ **6)** b $30\sqrt{5}$ **8)** A = 84, altitudes are $\frac{168}{13}$, 12, and $\frac{56}{5}$ **9)** a $2\sqrt{6}$ **10)** a Closer and closer to the shape of a triangle. b Becomes Hero's Formula **11)** $\frac{75\sqrt{3}}{4} + 24$

Pages 540–545 (Chapter 11, Review Problems)

1) a 84 b 42 c 75 d 52 e 20 f 8 **2)** a 63 b 33 **3)** a 70 b 24 c $16\sqrt{3}$ **4)** $1200.00 **5)** $48\sqrt{3}$ **6)** 288 **7)** 18 **8)** $67\frac{1}{2}$ cents **9)** 196 **10)** 64π **11)** 81 sq m **12)** $\frac{49}{2}\pi$ sq mm **13)** a 9π b 16π c $100 - 25\pi$ **14)** a 24π b 16π c 4π **15)** a 5:8 b 9:16 **16)** a LA = 48, TA = 84 b LA = 56π, TA = 88π **17)** 360 **18)** $21\sqrt{2}$ **19)** 120 **20)** $\frac{49}{4}\sqrt{3}$ **21)** A = $77\sqrt{3}$, P = 50 **22)** a $4\pi\sqrt{6}$ b $14\sqrt{\pi}$ **23)** a 338 b 6 **24)** $648\sqrt{3}$ **25)** a $\frac{168}{5}$ **25)** b $27\sqrt{3} - 9\pi$ c $6 - \pi$ **26)** a 2:1 b 7:5 c 4:1 **27)** a $9\pi - 18$ b $6\pi - 9\sqrt{3}$ **28)** a 16:81 b 4:9 c 1:1 **29)** a LA = 180, TA = $180 + 25\sqrt{3}$ b LA = 60π, TA = 96π **29)** c LA = 195π, TA = $279\frac{1}{2}\pi$ **30)** a 75π b $45\pi + 60$ **31)** a 5:4 b 25:39 **32)** $143\sqrt{3}$ **33)** $120 + 48\pi$ **34)** $18\pi\sqrt{2} - 9\pi$ **35)** a $100\sqrt{3} - 50\pi$ b $6\pi + 9\sqrt{3}$ **36)** 432 **37)** a $\frac{25}{96}$ b $\frac{47}{96}$ **38)** $72\sqrt{3} - 24\pi$ **39)** 113π sq m **40)** $40\pi - 80$ **41)** $11\frac{1}{4}\pi$ **42)** 390 **43)** 36%

Pages 551–554 (Section 12.1)

1) **a** 300π **b** 720 **c** 288 **d** $45\sqrt{3}$ **2)** $(6+5\pi)$ m^3 **3)** 357 **4)** V = 300, TA = 280
5) **a** 343 **b** e^3 **c** 5 **6)** 22 **7)** **a** V = 600, TA = 620
8) **a** V = 243π, TA = 148.5π **b** V = 360π, TA = $156\pi + 240$ **9)** V = $540\sqrt{3}$, TA = $360 + 108\sqrt{3}$
10) 6 **11)** $250\sqrt{2}$ **12)** 6.75 cm **13)** 189 **14)** Yes, by $\approx$ 3.6 cu cm
15) **a** $480 + 45\pi$ **b** $316 + 39\pi$ **16)** 245π cu ft **17)** V = $\frac{2000}{3}\pi$, TA = $600 + \frac{1400}{9}\pi$
18) **a** 51.44 cu cm **b** 457.816 cu cm **c** 114.84 sq cm **d** His claim is not true (114.84 sq cm *vs.* 96 sq cm).
19) 90π **20)** 90π

Pages 558–560 (Section 12.2)

1) $490\sqrt{3}$ **2)** 100π **3)** **a** 400 **b** 360 **4)** 9 **5)** **a** 1080π **b** 369π **c** 450π **6)** $\frac{1600}{3}$ cu m
7) 462π **8)** $72\pi\sqrt{3}$ **9)** 200 **10)** Same **11)** 72π **12)** 1408π **13)** 720 cu m
14) **a** 15 **b** 9 and 15 **c** 324π **14)** **d** 1500π **e** 1176π **16)** **a** $18\sqrt{2}$ **b** $\frac{s^3\sqrt{2}}{12}$ **17)** $72\sqrt{2}$
18) 48 **19)** 228π

Pages 561–564 (Section 12.3)

1) **a** 36π **b** 972π **c** $\frac{500}{3}\pi$ **2)** V = 288π, TA = 144π **3)** 153π **4)** **a** 392π **b** 144π **c** 248π
5) 523 cu ft **6)** 6084π **7)** **a** 240π **b** 132π **8)** **a** $18{,}000\pi$ cu m **b** 900π sq m
8) **c** Twice as much **d** $30\sqrt{2}$ m **9)** 22.5π cu mm **10)** **a** 138π **b** 105π **11)** **a** $\frac{8}{125}$ **b** $\frac{4}{25}$
12) No, 12π cu cm for cone and $10\frac{2}{3}\pi$ cu cm for the ice cream. **13)** **a** $\frac{500}{3}\pi$ cu m **b** $500\pi\sqrt{3}$ cu m
14) 2:3 **15)** 2:1

Pages 566–568 (Chapter 12, Review Problems)

1) **a** 512 **b** 108 **c** 98π **d** 20 **e** 60 **f** $\frac{32}{3}\pi$ **2)** **a** V = 360π, TA = 192π **b** V = 540, TA = 468
2) **c** V = 100π, TA = 90π **3)** 90 **4)** **a** 5 **b** 6 **5)** 240π **6)** 562, 500 cu cm **7)** 512
8) 36π **9)** 48π **10)** 35 **11)** $4320\sqrt{3}$ cu cm **12)** 5040 **13)** 215 **14)** $70\sqrt{2}$
15) $150{,}459 + 927\pi + \frac{459\sqrt{3}}{4}$ **16)** V = $500\pi - 750\sqrt{3}$, TA = $300 + \frac{400\pi}{3} - 50\sqrt{3}$ **17)** $333\pi\sqrt{3}$
18) **a** $\frac{64\pi\sqrt{3}}{3}$ **b** 35π

Pages 569–574 (Cumulative Review Problems, Chapters 1–12)

1) 81 **2)** Isosceles **4)** **a** 5 **b** $\frac{32}{3}$ **5)** **a** 8 **b** $3\frac{1}{5}$ **c** 44 **d** 12 **6)** **a** $\frac{3}{5}$ **b** $\frac{3}{5}$ **7)** **a** $\frac{50}{3}$
b $\frac{32}{3}$ **c** 8 **8)** **a** 168 **b** $25\sqrt{3}$ **c** 169π **9)** 89° **10)** **a** Parallelogram **b** Kite **c** Rectangle
d Rhombus **11)** **a** 60 **b** 83 **c** 50 **d** 78 **12)** **a** 18 **b** ±12 **13)** **a** 25 **b** 16 **15)** 64
16) **a** 80 **b** 84 **c** $28\sqrt{5}$ **17)** 36 **18)** 16 or 10 **19)** **a** 60 **b** 135 **c** 120 **d** 45, No
20) 65 **23)** 2π **24)** **a** 165° **b** $32\frac{1}{2}$° **c** 42° **25)** 7 **27)** 100 m **28)** **a** 120 **b** $9\sqrt{3}$
c 9 **29)** **a** 3 **b** 1.2 **30)** **a** 98° **b** 82 **31)** 35 **32)** 25 **33)** 3 **34)** **a** 8 **b** 9:1
35) 32° **36)** $6\sqrt{5+2\sqrt{2}}$ **37)** 20 **38)** Yes **41)** $6\frac{1}{5}$ **42)** 50

Pages 582–583 (Section 13.1)

2) **a** E **b** None **c** A and B **d** Yes **e** 7 **f** 5 **g** $\sqrt{74}$ **h** $\frac{35}{2}$ **3)** IV **5)** **a** (3, 2) **b** (7, 10)
5) **c** (7, −3) **6)** (8, 5) **7)** (0, 0), (8, 0), (8, 5), (0, 5) **8)** (0,0), (a, 0), (a, b), (0, b) **9)** 9
10) **a** 3, 4, 5 **b** $\frac{12}{5}$ **11)** **a** (2, 5) **b** 14, 4 **c** 9 **d** 63 **14)** $\frac{2}{5}$ **15)** (−2, 6) or (16, 6)
16) **a** (−5, −6) **b** (5, 6) **c** (5, −6) **17)** **a** $-2\sqrt{3}$, 6 **b** 10 and −2; −20

Pages 585–586 (Section 13.2)

1) **a** (5, 0) **b** (1, −1) **c** (−6.5, 3) **d** $(4\sqrt{2}, 2)$ **e** $(\frac{3}{4}, 1)$ **f** (−.45, −3.95) **2)** (1, 4), (6, 2), (1, 1)
3) (a, 0), (2a + b, c), (a + 2b, 2c), (b, c) **4)** No **5)** (2, 14) **6)** (4, 1) **7)** (5, −9) **8)** (1, 8)
9) J = (9, −2), K = (17, 4) **10)** $(2, 2\sqrt{3}), (2, -2\sqrt{3})$ **11)** 4 **12)** (7, 4) **13)** (−3, 3)
14) $(7, -\frac{14}{3}), (1, -\frac{1}{3})$

Pages 591–593 (Section 13.3)

1) **a** $\frac{8}{9}$ **b** $\frac{1}{7}$ **c** $\frac{11}{6}$ **d** 0 **e** No Slope **f** $\frac{5c}{a}$ **2)** $-\frac{3}{5}$ **3)** −4 **4)** **a** $\frac{1}{4}$ **b** −4 **c** $-\frac{7}{2}$ **d** $\frac{1}{4}$
6) **a** .5 **b** .5 **c** Collinear **7)** 32 **8)** $\overline{AC}$ **9)** $\frac{y_3 - y_1}{x_3 - x_1}$ **10)** $-\frac{2}{7}$ **11)** $-\frac{2}{3}$ **12)** Yes
13) $\frac{2}{3}$ **14)** **a** 1 **b** −1 **15)** **a** No **b** Yes **16)** One line is horizontal, and the other line is vertical.
17) **a** (12, 6) **b** 72 **19)** $\sqrt{3}$ **22)** (10, 1), (−10, −5), (2, 15) **23)** $\frac{3}{5}$

Pages 599–602 (Section 13.4)

1) **a** 3; 7 **b** 4; 0 **c** $\frac{1}{2}; -\sqrt{3}$ **d** −6; 13 **e** −5; −6 **f** 0; 7 **2)** **a** y = 3x + 1; 3; 1 **b** y = −5x + 2; −5; 2
2) **c** $y = -\frac{2}{3}x + 2; -\frac{2}{3}; 2$ **d** $y = \frac{1}{2}x + \frac{1}{4}; \frac{1}{2}; \frac{1}{4}$ **3)** y = −6 **4)** x = 8 **5)** a and c **6)** **a** y = 4x + 2
6) **b** y = 5x − 2 **c** y = 10x + 1 **d** y = −2x − 5 **e** y = −x + 2 **7)** **a** 1 **b** y = x + 5 **c** $-\frac{1}{2}$
7) **d** $y = -\frac{1}{2}x - \frac{5}{2}$ **8)** **a** y = 3x − 5 **b** $y = -\frac{1}{2}x$ **c** y = 5 **d** y = 7x − 14 **e** y = −4x + 12 **f** x = −3
8) **g** $y = \frac{3}{2}x - 5$ **9)** $\frac{3}{4}$ **10)** $y = \frac{3}{2}x + \frac{5}{2}$ **13)** $y = \frac{1}{7}x + \frac{80}{7}$ **14)** y = −7x + 65 **15)** y = −7x + 40
16) y = −2x + 20 **17)** $\frac{1}{7}$ **18)** **a** y = 2 **b** x = −3 **19)** $y = -\frac{1}{2}x + \frac{13}{2}$ **20)** y = −3
21) No **22)** $y = x\sqrt{3} + 2$ or $y = -x\sqrt{3} + 2$ **23)** y = −.4x + 4
24) $y = \frac{d-b}{c-a}x + q - \frac{d-b}{c-a}p$ Except for: **a** If a = c and p ≠ a, then the equation is x = p.
24) **b** If a = c = p or if the 3 points are otherwise collinear, no parallel exists. **26)** (3, 13) **27)** (7, −9)
28) **a** $y = -\frac{3}{4}x + 1$ **b** $y = -\frac{3}{4}x - 1$ **c** $y = \frac{4}{3}x + \frac{4}{3}$

Pages 604–606 (Section 13.5)

1) **a** 2 **b** 4 **c** 5 **d** 10 **e** $\sqrt{29}$ **f** $2\sqrt{5}$ **2)** $P = 5 + \sqrt{34} + \sqrt{53}$ **5)** 225π **6)** **a** $\sqrt{85}$
6) **b** $3\sqrt{5}$ **9)** $2\sqrt{29} + 10\sqrt{2}$; kite **12)** 20 **14)** $-\frac{1}{7}$ **15)** $y = \frac{1}{5}x$ **16)** **a** 14 **b** $y = -\frac{4}{3}x + 18$
16) **c** $5\sqrt{5}$ **d** $5\sqrt{5} - 5$ **17)** **a** C = (2p, 2h), D = (−2p, 2h) **b** AB = 4a **c** 2a + 2p **d** 2a − 2p
18) $(4 + 2\sqrt{3}, 3 - 2\sqrt{3}), (4 - 2\sqrt{3}, 3 + 2\sqrt{3})$

Pages 609–610 (Section 13.6)

1) **a** (6, 4) **b** (2, 5) **c** (−3, −7) **d** (3, 2) **2)** **a** (4, ±3) **b** (1, 3) **3)** **a** (12, 0)
3) **b** {(x, y) | y = 3x − 7} **4)** **a** ∅ **b** (4, 17) **5)** (−2, 3) **6)** (a, 6 − 1.5a) **7)** (4, −2)
8) (4, −3), (4, 3), (−4, 3), (−4, −3) **9)** y = 5x − 9 **10)** $y = -\frac{2}{3}x + \frac{10}{3}$ **11)** (−1, 2)
12) (−8, −23) **13)** $\left(\frac{ce - bf}{ae - bd}, \frac{af - cd}{ae - bd}\right)$ **14)** $\sqrt{10}$ **15)** $\frac{4\sqrt{5}}{5}$ **16)** $(\frac{2}{3}, \frac{7}{3})$
17) 11 **18)** (−2, −3)

Pages 613–614 (Section 13.7)

2) **a** y < −3 **b** $y \geq \frac{1}{2}x + 2$

Pages 616–619 (Section 13.8)

1) **a** $x^2 + y^2 = 16$ **b** $(x + 2)^2 + (y - 1)^2 = 25$ **c** $x^2 + (y + 2)^2 = 12$ **d** $(x + 6)^2 + y^2 = \frac{1}{4}$
3) **a** (0, 0); 6; 12; 12π; 36π **b** (−5, 0); $\frac{3}{2}$; 3; 3π; $\frac{9\pi}{4}$ **c** (3, −6); 10; 20; 20π; 100π
d (−5, 2); 9; 18; 18π; 81π **4)** **a** $x^2 + y^2 = 25$ **b** $(x + 4)^2 + y^2 = 16$ **c** $(x - 5)^2 + (y - 6)^2 = 36$
5) **a** Yes **b** No **6)** **a** Point circle **b** Imaginary circle

7) A = (−6, 3), B = (8, 3), C = (1, 10), D = (1, −4)   **8) a** $x^2 + y^2 = 25$
8) b $(x - 3)^2 + (y - 13)^2 = 169$ **c** $(x + 1)^2 + (y - 7)^2 = 50$ **d** $(x - 2)^2 + (y + 3)^2 = 10$
9) a On   **b** Interior   **c** Exterior   **d** Exterior   **11) a** (0, 4); 5   **b** (−7, −3); 6   **c** (−5, 6); $\sqrt{51}$
11) d (4, −7); 10   **12) a** {(3, 4), (3, −4)}   **b** {(4, 3), (4, −3), (−4, 3), (−4, −3)}   **c** {(5, 3), (3, 5)}
12) d {(8, 6), (8, −6), (−8, 6), (−8, −6)}   **13)** $5\sqrt{2}$ **14) a** $y = -\frac{3}{4}x + \frac{25}{2}$ **b** 10π
14) c $\sqrt{53}$ **d** $\sqrt{53} - 5$ **e** $\frac{25\pi}{6}$ **15)** $y = \frac{6}{5}x - \frac{88}{5}$ **16)** $C = (-2, \frac{5}{6}), r = \frac{\sqrt{193}}{6}$ **17)** 17π
18) $(x + 5)^2 + y^2 = 16$ **19) a** $\frac{16}{3}\pi$ **b** $\frac{16\pi}{3} - 8\sqrt{3}$ **20) a** (9, ±12)   **b** $8\sqrt{3}$
20) c $\left(\frac{15}{7}, \frac{60\sqrt{3}}{7}\right), \left(\frac{111}{7}, \frac{52\sqrt{3}}{7}\right)$

Pages 620–624 (Section 13.9)

1) a $x^2 + y^2 = 25$ **b** 25π **c** 10π **2)** 3π **3) a** $64 - 16\pi$ **b** $25\pi - 48$ **4)** 25   **5) a** 5   **b** 5
5) c 5   **6)** $(3, 3\sqrt{3})$ **7)** 10   **8)** 12   **9)** −.5   **10)** 5   **11)** 12   **12) a** (13, 9)   **b** 52
13) a $(x + 3)^2 + (y - 3)^2 = 9$ **b** $\frac{36 - 9\pi}{4}$ **14)** 18   **15)** 35
16) a $(x - 7)^2 + (y - 1)^2 = 100$ **b** 100π **c** 20π **d** (1, −7)   **e** $y = -\frac{3}{4}x + \frac{75}{4}$ **f** 13   **g** 3
17) 50   **18)** (−6, 1) and (0, 1)   **19)** $\frac{24}{5}$ **20) a** 5   **b** 45   **21)** $(x + 4)^2 + y^2 = 36$
22) a (11, 9)   **b** 1:9   **23)** $\frac{8\sqrt{5}}{5}$ **24) a** 128π **b** 96π **c** $\frac{384\pi}{5}, \frac{336\pi}{5}$ **25) a** $5\sqrt{13}$ **b** $\sqrt{429}$
27) 20   **28)** $(0, \frac{35}{11})$ **29) a** (7, 0)   **b** $(\frac{20}{3}, 0)$ **c** $\left(\frac{28 - 2x_1}{3}, 0\right)$

Pages 626–629 (Chapter 13, Review Problems)

1) No   **2)** III   **3)** Above   **4) a** (10, 0)   **b** (2, 6)   **c** (−4, 4)   **5) a** 10   **b** (7, 2)   **c** $\frac{4}{3}$ **6) a** $\frac{4}{3}$
6) b (1, 6)   **c** $-\frac{3}{5}$ **d** $\sqrt{34}$ **e** $-\frac{3}{4}$ **f** $\frac{1}{8}$ **g** $-\frac{3}{4}$ **7) a** 50   **b** 27   **c** 18π
8) a $(x - 2)^2 + (y + 3)^2 = 16$ **b** $x^2 + y^2 = 100$ **c** $(x - 5)^2 + y^2 = 25$ **9) a** $y = 2x + 1$ **b** $x = 2$
9) c $x = -5$ **d** $y = 3x - 2$ **e** $y = \frac{1}{2}x - 2$ **f** $y = 3x - 7$ **g** $y = \frac{1}{2}x - 3$ **10) a** $m = -\frac{1}{2}$
10) b m = 2, the lines are perpendicular.   **11)** yes   **12)** $-\frac{2}{3}$ **13)** $-\frac{29}{4}$ **14)** ±2
15) (10, 5)   **16)** (−2, 2)   **17) a** $2\sqrt{5}$ **b** $y = 2x - 10$ **c** $y = -5x + 39$ **d** $y = -5x + 53$
17) e $y = \frac{1}{5}x + \frac{31}{5}$ **18) a** $(x - 3)^2 + (y - 2)^2 = 40$ **b** 40π **c** (−3, 0)   **d** $y = -3x + 31$ **e** 20
18) f $20 - 2\sqrt{10}$ **g** 60   **20) a** $\frac{3}{5}$ **b** $\frac{3}{10}$ **21) a** {(2, 7)}   **b** {(7, −1)}   **c** {(−3, 4), (5, 4)}
22) C = (−3, 2), r = 5   **23)** $\sqrt{85}$ **24)** Point circle   **25)** 48   **26) a** 60   **b** They are perpendicular.
27) III, IV, $(-\frac{10}{3}, -\frac{10}{3}), (\frac{10}{7}, -\frac{10}{7})$ **29)** $2\sqrt{10}$ **30)** $\sqrt{10}$ **31)** 108π **32)** $(1, \frac{1}{5})$
33) $(-\frac{10}{3}, \frac{7}{3})$ **34)** 66   **35)** $(\frac{44}{7}, 0)$ **36) a** (−5, −10)   **b** (−1, −8)   **c** (11, 2)
37) (2, 14), (−26, 42)

Pages 635–637 (Section 14.1)

1) The locus is two lines parallel to $\overleftrightarrow{AB}$ and 3 cm on each side of $\overleftrightarrow{AB}$.   **2)** The locus is a circle with the same center as the given circle and with a radius half as long as the given circle.   **3)** The locus is the perpendicular bisector of the segment joining the two given points.   **4)** The locus is a circle with the same center as the quarter and with a radius equal to the sum of the radii of the two coins.   **5)** The locus is two circles—each with the same center as the given circle and with radii of 2 inches and 22 inches respectively.   **6)** The locus is a line halfway between the two given parallel lines.   **7)** The locus is a circle concentric with the given circles and with a radius equal to the average of the two given circles. If the radii of the given circles are 3 and 8, the radius at the locus is $5\frac{1}{2}$.   **8)** $x^2 + y^2 = 16$
9) The locus is the union of a circle and its interior where the circle has the fixed point P as center and a radius of 14.
10) a The locus is a line halfway between a and b.   **b** The locus is the empty set.   **11)** The locus is a circle (minus the given point) tangent internally to the given circle at the given point and passing through the center of the given circle.   **12)** The locus is a circle with the same center as the given circle and with a radius equal to the distance from the center to one of the chords.   **13)** The locus is the perpendicular bisector of the segment joining the two points.
14) The locus is a segment known as the midline in the MIDLINE THEOREM.   **15) a** The locus is a sphere with the given point as center and a radius of 5.   **b** The locus is a cylinder with a radius of 5.   **16)** x = 2.5

Pages 635–637 (Section 14.1)

17) $(x+1)^2+(y-3)^2=36$ **18)** The locus is the 4 points: (5, 5), (5, −5), (−5, 5), (−5, −5). **19)** The locus is the UNION of a circle and a point. The circle has center Q and radius 18, and the point is Q. **20) b** $60+25\pi$ **20) d** $\frac{950\pi}{3}$ **21)** The locus is a circle with center at the foot of the perpendicular from P to M and with a radius of 3. **22)** $y=-\frac{1}{7}x-\frac{12}{7}$ **23)** The locus is a circle (minus T and V) with center at the midpoint of $\overline{TV}$ and diameter $\overline{TV}$. If $m\angle P=90$ then $m(\widehat{TV})=180$, and $\overline{TV}$ is a diam. **24)** $\{(x,y) \mid x^2+y^2=1 \text{ and } x\neq\pm1\}$ **25) a** Perpendicular bisectors of the sides **b** Angle bisectors **26)** $y=\pm\frac{4}{3}$ **27) b** $4\le(x+3)^2+(y-4)^2\le 25$ **c** 21π **28)** The locus is a quarter circle with center at the foot of the wall and radius of 3 m. **29)** $(x-2)^2+y^2=4$

Pages 640–641 (Section 14.2)

(Sketches are not shown.) **1) a** ∅, 1 point, or 2 points **b** ∅, 1 point, or 2 points **c** ∅, 1 point, or 2 points **1) d** ∅, 1 point, or a line **e** 4 points **f** ∅, 1 point, or a ray **2)** 4 points **3)** 1 point **4)** ∅ or 6 points **5) a** ∅, 1 point, or a segment **b** ∅ or a segment without its endpoints **6)** 2 points **7)** ∅, 1 point, 2 points, 3 points, or 4 points **8)** 4 points **9)** 2 points **10)** ∅, 5 points, or 10 points **11)** ∅ or 1 point **12) a** 2 circles **b** 32π **13)** ∅, 1 point, 2 points, 3 points, or 4 points **14)** ∅, a line, 2 lines, a circle, or an ellipse **15)** ∅, 1 point, 2 points, 3 points, . . . , 8 points **16)** ∅, 1 point, a ray, or a segment **17)** A sphere and its interior

Pages 645–648 (Section 14.3)

3) Find the INCENTER **4) b** Midpoint of the hypotenuse **5)** CENTROID and INCENTER **6)** CENTROID **7) a** 6 **b** 15 **c** 4 **d** $\sqrt{2}$ **8)** Right **9)** Equilateral **10)** None **11)** The locus is the circumcenter of the triangle formed by joining the three points. **12)** TN = 18 **13) a** $9\sqrt{3}$ **b** $162\sqrt{3}$ **13) c** $81\sqrt{3}$ **d** 30 **14)** (5, 4) **15)** $(\frac{28}{5}, \frac{8}{5})$ **16)** (4, 5) **17) a** $(\frac{4}{3}, 4)$ **b** $(\frac{4}{3}, 4)$ **18) b** $\frac{1}{3}$ **20) a** $\frac{1}{4}$ **b** Congruent

Pages 669–671 (Chapter 14, Review Problems)

1) The locus is the perpendicular bisector of $\overline{AB}$ (minus the midpoint of $\overline{AB}$). **2)** The locus is the perpendicular bisector of the segment joining the two points. **3)** 4 points **4)** A cylinder **5)** The locus is a point and a circle. **7)** $x^2+y^2=25$ **12)** ∅, 1 point, or 2 points **13)** $y=3x+5$ **14)** The locus is the line perpendicular to the plane of the circle at the center. **15)** The locus is the perpendicular bisector of $\overline{PQ}$ (minus the midpoint of $\overline{PQ}$). **16)** The locus is a circle with center at P and radius of 8 cm. **17)** The locus is a circle (minus the fixed point) that is tangent internally to the given circle at the fixed point and whose radius is half the original circle. **18)** $33+4\pi$ **19)** 4 points **20)** {(4, 3), (4, −3), (−4, 3), (−4, −3)} **24)** ∅ or 1 point **25)** The locus is a circle (minus P and Q) with $\overline{PQ}$ as diameter.

Pages 675–676 (Section 15.1)

1) a $\{x \mid x>25\}$ **b** $\{x \mid x>6\}$ **c** $\{x \mid x\geq -7\}$ **d** $\{x \mid x>1\}$ **2) a** $x<18$ **b** $x>18$ **3)** x exceeds z by 8 **4)** x is 6 times z **5)** $\angle A>\angle 2$ **6) a** PX < PQ **b** PX < PQ **c** 2PX = PQ **7)** $45<x<90$ **8)** Complement of ∠X > complement of ∠Y **9)** $0<x<\frac{1}{5}$ **10)** The angle is between 60 and 90; The complement is between 0 and 30. **11)** $x>3$ **12) a** = **b** > **13)** ∠DBC > ∠DCB **15)** $\{x \mid -3<x<2\}$ **16) a** $x>\frac{25}{4}$ **b** y is any real number **17)** 44 **18)** {2, 4} **19) a** Complement of B > Complement of A **b** Complement of B > A **19) c** Complement of B, A, Complement A, B **20)** $\{x \mid x<-3 \text{ or } x>23\}$

Pages 680–682 (Section 15.2)

1) $80<\angle 1<180$ **2)** b, d **3) a** $\overline{PR}$, $\overline{PQ}$ **b** Hypotenuse **5) a** $\overline{AB}$ **b** $\overline{WZ}$ **7)** Other **8)** 8, 20 **9)** $46<B<70$ **10) a** $\overline{AD}$ **b** $\overline{XY}$ **11) a** R, Q, P **b** R **12) a** $55<x<110$ **12) b** $0<x<90$ **c** $0<x<20$ **d** $7.5\sqrt{3}$ **13)** 2cm **14) a** Acute **b** None **c** Obtuse **d** Right **19) a** $\overline{AD}<\overline{CD}<\overline{AC}<\overline{BC}<\overline{AB}$ **b** Either right before or right after $\overline{AC}$ with current knowledge; actually, AC < BD < BC. **21)** $|x-y|<$ third side $<x+y$

Pages 685–687 (Section 15.3)

1) $\overline{AC}$ **2)** $\angle R$ **3)** $\overline{DC}$ **4)** $XZ > AB$ **7) a** $\angle Y$, $\angle X$ **b** Obtuse **10) a** $\triangle ABC$ **b** $\triangle WXY$
13) Right **15) a** $\overline{AC}$ **b** $\angle B$ **17)** $\angle XZW > \angle X > \angle XWY > \angle Y > \angle XWZ$ **20)** b

Pages 689–691 (Chapter 15, Review Problems)

1) $\angle A < \angle B < \angle C$ **2) a** $\overline{WZ}$, Hinge Theorem **b** $\overline{AB}$, Hypotenuse is longest side. **2) c** $\overline{RQ}$, from size of angles in $\triangle PQR$. **3) a** $\angle CBD$, Converse Hinge Theorem **b** $\angle X$, If  Then  **c** $\angle 1$ by Exterior Angle Inequality Theorem **4)** $>$ **5)** $x > 6$ **6) a** $\overline{AE} \cong \overline{AB}$, $\overline{ED} \cong \overline{EC}$ **b** $\overline{BE}$ **c** Hypotenuse **d** $\overline{DE}$ **e** $\overline{BE}$
7) a **9)** $\overline{BC}$, $\overline{AC}$, $\overline{AB}$ **10)** 60, 150 **11) a** $\overline{AB}$ **b** $12 < p < 6 + 6\sqrt{2}$ **12)** 5 **13) a** $\sqrt{34}$, $\sqrt{13}$, 7 **13) b** Obtuse **c** $\angle P < \angle R < \angle Q$ **18)** 4 **19)** $\overline{AC}$, $\overline{AB} \cong \overline{BC}$, $\overline{CD}$, $\overline{AD}$, $\overline{BD}$ **23)** 7/9

Pages 697–698 (Section 16.1)

1) a If a person is 18 years old, then (s)he may vote in federal elections. **b** If two angles are opposite angles of a parallelogram, then the two angles are congruent.

2) a Converse: If the diagonals of a quadrilateral are perpendicular, the quadrilateral is a rhombus.
Inverse: If a quadrilateral is not a rhombus, then the diagonals of the quadrilateral are not perpendicular.
Contrapositive: If the diagonals of a quadrilateral are not perpendicular, then the quadrilateral is not a rhombus.

2) b Converse: If two sides of a triangle are not congruent, the angles opposite those sides are not congruent.
Inverse: If two angles of a triangle are congruent, the sides opposite those angles are congruent.
Contrapositive: If two sides of a triangle are congruent, the angles opposite those sides are congruent.

2) c Converse: If at least two sides of a triangle are congruent, then the triangle is not scalene.
Inverse: If a triangle is scalene, then no two sides of the triangle are congruent.
Contrapositive: If no two sides of a triangle are congruent, then the triangle is scalene.

3) b **4) a** True **b** False **c** False **d** False **5) a** $d \longrightarrow f$ **b** $s \longrightarrow \sim p$ **c** If bobcats begin to browse, the horses head for home.

6) Converse: If two angles of a triangle are congruent, then the sides opposite those angles are congruent.
Inverse: If two sides of a triangle are not congruent, the angles opposite those sides are not congruent.
Contrapositive: If two angles of a triangle are not congruent, the sides opposite those angles are not congruent.
The converse, inverse, and contrapositive are all true.

7) Original Statement: If a polygon is a square, then it is a quadrilateral with four congruent sides.
Converse: If a quadrilateral has four congruent sides, then it is a square.
Inverse: If a polygon is not a square, then it is not a quadrilateral with four congruent sides.
Contrapositive: If a quadrilateral does not have four congruent sides, then it is not a square.

8) a Converse: If two circles have no common tangent, the two circles are concentric.
Inverse: If two circles are not concentric, then they have a common tangent.
Contrapositive: If two circles have a common tangent, then the two circles are not concentric.

8) b Converse: If a segment bisects an angle of a parallelogram, then the segment is a diagonal of the parallelogram.
Inverse: If a segment is not a diagonal of a parallelogram, then the segment does not bisect an angle of the parallelogram.
Contrapositive: If a segment does not bisect an angle of the parallelogram, then the segment is not a diagonal of the parallelogram.

9) $p \longrightarrow \sim g$ **10)** At least one of the given statements is false.

Pages 704–706 (Section 16.3)

1) b $L = (a, b)$, $M = (3a, b)$, $N = (2a, 0)$ **c** $-\frac{b}{a}, \frac{b}{a}$ **d** $\frac{b}{a}, -\frac{b}{a}$ **2) a** $A = (0, 2b)$, $B = (2a, 2b)$, $C = (2a, 0)$, $O = (0, 0)$
2) b $M = (0, b)$, $N = (a, 2b)$, $P = (2a, b)$, $Q = (a, 0)$ **c** $\frac{b}{a}, \frac{b}{a}, -\frac{b}{a}, -\frac{b}{a}$, MNPQ is a parallelogram.
2) d All have length equal to $\sqrt{b^2 + a^2}$, and thus MNPQ is a rhombus. **5)** $P = (a, 0)$, $Q = (a + b, c)$

Pages 708–710 (Chapter 16, Review Problems)

1) **a** Converse: If two pentagons are similar, then the two pentagons are regular.
Inverse: If two pentagons are not regular, then the two pentagons are not similar.
Contrapositive: If two pentagons are not similar, then the two pentagons are not regular.

1) **b** Converse: If a parallelogram is not a "Plain Old Parallelogram," then the parallelogram is inscribed in a circle.
Inverse: If a parallelogram is not inscribed in a circle, then it is a "Plain Old Parallelogram."
Contrapositive: If a parallelogram is a "Plain Old Parallelogram," then it is not inscribed in a circle.

12) q is green **13)** **a** $a \rightarrow \sim f$ **b** I may conclude Gergles are home. **15)** **a** (2a, 4a), (4a, 2a) **b** $\frac{1}{3}$

Pages 710–715 (Cumulative Review Problems, Chapter 1–16)

1) $V = 12\pi$, $TA = 24\pi$ **2)** **a** 6 **b** $-\frac{2}{3}$ **c** No **3)** 16° **4)** 5 **5)** 21 **6)** 3 **7)** **a** 3
7) **b** $13\frac{1}{2}$ **c** No **9)** 42 **10)** 36 **11)** **a** 13, 14, 15; $-\frac{12}{5}, 0, \frac{4}{3}$ **b** Acute **c** $y = \frac{4}{3}x + 4$; $-3, 4$
11) **d** $y = -4$ **e** $(3, -4)$ **f** $x = 3$ **g** 12 **h** $(1, -4)$; 6 **i** 84 **12)** 2:3 **13)** 135 **14)** 6π sq in
15) $12\sqrt{3}$ **16)** **a** 30π **b** $10\pi + 12$ **17)** 175° **18)** **a** 110° **b** 90° **19)** **a** 11 **b** 180°
20) $3\sqrt{5}$ **21)** 36 **22)** $72 - 18\pi$ **23)** **a** $\sqrt{26}$ **b** $y = 5x - 35$ **c** $y = -3x + 37$ **d** $y = -3x + 29$ **24)** $\sqrt{58}$
25) 26m **26)** $\frac{24\sqrt{13}}{13}$ **27)** 78° **28)** $(x-2)^2 + (y-4)^2 = 10$ **29)** **a** Acute **30)** $(-2,5)$ **31)** π:2
32) 21cm **35)** $\varnothing$, 1 point, or 2 points **36)** $\varnothing$, 1 point, 2 points, 3 points, 4 points, a ray, a ray and a point, a line
37) 121π **39)** $16\sqrt{2} - 8$ **40)** $20\sqrt{3}$ **41)** **a** $\frac{13}{56}$ **b** $\frac{43}{126}$ **42)** **b** 504π

Index

Credits

Senior Editor	Michael Green
Managing Editor	Kathleen Laya
Assistant Managing Editor	Geraldine Macsai
Assistant Editor	David Terebessy
Director of Design	Allen C[illegible]
Cover Design	Randi S. Brill
Cover Photograph	Earl Glass
Photographs	James L. Ballard, Frontispiece, chapter 1-8, 10, 12, 13, 15, 16; Lewis Baltz, chapter 11 detail; Treat Davidson, page 545; Historical Pictures Service, Chicago, page 276; Kipton Kumler, chapter 9 detail; Dan Morrill, chapter 14.

Symbols Used in Geometry

Symbol	Meaning
$\overline{AB}$	segment AB
$\overleftrightarrow{AB}$	line AB
$\overrightarrow{AB}$	ray AB
$\overset{\frown}{AB}$	arc AB
AB	length of $\overline{AB}$
m$\overline{AB}$	measure of $\overline{AB}$
$\angle$s	angles
$\angle$A	angle A
m$\angle$A	measure of $\angle$ A
$\triangle$	triangle
$\triangle$ with s	triangles
$\odot$	circle
ⓢ	circles
¢	cross-section area
$\square$	parallelogram
$\cong$	congruent
$\ncong$	not congruent
•—‖—• •—‖—•	congruent segments
$\measuredangle\ \measuredangle$	congruent angles
$\perp$	perpendicular
$\not\perp$	not perpendicular
⊾	right angle
$\parallel$	parallel
$\nparallel$	not parallel
⫽ ⫽	parallel lines
$\sim$	similar

Symbol	Meaning
$=$	is equal to
$\neq$	is not equal to
$>$	is greater than
$<$	is less than
$\ngtr$	is not greater than
$\geq$	is greater than or equal to
$\leq$	is less than or equal to
$\approx$	is approximately equal to
$\Longleftrightarrow$	is equivalent to
$\longleftrightarrow$	corresponds to
$\Longrightarrow$	implies
$\sim$p	not p, or p is false
$\therefore$	therefore
$\{\ \}$	set
$\varnothing$	null set
$\cup$	union
$\cap$	intersection
$\sqrt{x}$	square root of x
$\lvert a \rvert$	absolute value of a
$\lvert\ \rvert$	greatest integer in
$\triangle$x	change in x
π	Pi
$^\circ$	degrees
$'$	minutes
$''$	seconds
$\frac{a}{b}$	a $\div$ b, a : b, ratio of a to b